W0235030

Development and Reproduction in
Humans and Animal Model Species

Werner A. Mueller • Monika Hassel •
Maura Grealy

Development and Reproduction in Humans and Animal Model Species

 Springer

Werner A. Mueller
Centre of Organismal Studies
University of Heidelberg
Germany

Monika Hassel
Spezielle Zoologie
Universität Marburg FB Biologie
Marburg
Germany

Maura Grealy
Pharmacology and Therapeutics
National University of Ireland Galway
Galway
Ireland

Translation from the 5th German language edition 'Entwicklungsbiologie und Reproduktionsbiologie des Menschen und bedeutender Modellorganismen' Springer Spektrum, © Springer-Verlag Berlin Heidelberg 2012; ISBN: 978-3-642-28382-6

ISBN 978-3-662-43783-4   ISBN 978-3-662-43784-1 (eBook)
DOI 10.1007/978-3-662-43784-1
Springer Heidelberg New York Dordrecht London

Library of Congress Control Number: 2014958295

Cover: Human fetus, © Sebastian Kaulitzki / shutterstock

Printed on acid-free paper

Springer is part of Springer Science+Business Media (www.springer.com)

Preface

This book is written and designed for students of the biological sciences, undergraduates and graduates and for all those geneticists, biochemists, physicians, teachers and science journalists who are looking for a coherent and readily understandable description of human embryonic development and of the development of important animal model organisms. This book is not a collection of review articles composed for professional researches, although we consulted many current bibliographic databases and recent original publications. After an introduction into basic terms we choose an outline different from the organization chosen in other textbooks.

Long experience as teachers and researchers who are interested also in the findings obtained in other organisms tells us that nobody reads a textbook from the first to the last page without omitting this or that chapter. To comply with the actual demands of readers we first give a comprehensive description of the most important model species including experimental findings, and of the human embryo. This relieves the readers from having to reconstruct the complete development of, let's say, *Drosophila* or *Xenopus* by gathering together sections and figures scattered over several chapters. Thereafter comparative accounts are given to general topics such as, for example, sexual development, the construction of transgenic animals, the development of the nervous system, stem cell research and the evolution of development.

This concept, however, made it necessary to repeat this or that statement. The reader will profit from this concept in that (almost) every chapter can be read separately without first consulting other chapters.

Separate boxes are devoted to special areas in the periphery such as

- From the soul to information: The history of the developmental biology
- Prenatal diagnostics
- Disturbances of human sexual development
- Genetic and molecular methods used in the analysis of development
- Unsolved riddles: Primordial mouth and the direction of the body axes
- Only three germ layers?

For didactic reasons all figures were drawn not by professional graphic artists but by one of us (WM). These, the general outline, special areas and emphasis on human development are unique characteristics of this textbook. The literature list at the end of the book does not only contain references but also further reading. This 6th edition is a translated and updated version of the 5th German edition.

We are grateful to all those colleagues who gave us helpful comments and to the Springer editorial staff, in particular to Miss Anette Lindqvist and Anette Schneider.

Heidelberg, Germany   Werner A. Mueller
Marburg, Germany   Monika Hassel
Galway, Ireland   Maura Grealy

Contents

Development and Reproduction: An Introduction

1

1.1 Development as Self-Construction

1.1.1 Living Beings Construct and Organize Themselves on the Basis of Inherited Information

Development and reproduction are basic features of living beings. In the context of this book development means **ontogeny**, the development of an individual life, typically beginning with the fertilization of an egg, and ending with the death of the individual. Development in the sense of **phylogeny** or **evolution**, which becomes apparent as a gradual change of form and behaviour of the organisms in the course of long sequences of generations, is dealt with only insofar as we shall discuss some evolutionary variations of ontogenetic patterns in Chap. 6 and more so in Chap. 22.

The essential principles of development are conveyed by terms such as **self-construction** and **self-organization**. The development of a multicellular organism starts, as a rule, with a single, seemingly unstructured cell – the fertilized egg cell which hardly approaches the complexity of single-celled organisms. It was reserved to the eyes of former naturalists endowed with vivid imagination to see in the tiny sperm cell a complete **homunculus**, a little human being (Fig. 1.1). The present day scientist (cell biologist, biochemist, molecular biologist) finds in the fertilized egg cell constituents not much more diverse than those of any other cell. The size of an egg cell can be huge but its internal structures appear not to be more intricate and elaborated, even in the electron microscope, than those seen in many other cell types. Yet, without the intervention of an external designer, from this modest cell with its simple spherical shape a being arises with a highly complex form and many thousands of diverse cells. These arrange themselves to form the new organism in a highly ordered cooperation and interplay of forces.

Similarly, when a creature such as the freshwater polyp reproduces itself by a process of asexual reproduction through buds (Fig. 4.12), it is an association of only a few, poorly differentiated cells that constitutes the starting material for the development of the new organism. Whether egg or bud: the fully developed creature will have an internal complexity that surpasses our imagination. We are all familiar with the avian egg from our breakfasts – it is the amorphous yellow ball that constitutes the egg cell proper. But who is able to visualize the 100 billion nerve cells of our brain and their 10^{14}–10^{15} synaptic connections in their three-dimensional architecture?

All of the many and diverse **somatic** cells that originate from **generative** starting cells do not, however, attain autonomy as do the cells which arise from dividing single-celled protists. Somatic cells are viable only in the community of the cell

W.A. Mueller et al., *Development and Reproduction in Humans and Animal Model Species*,
DOI 10.1007/978-3-662-43784-1_1, © Springer-Verlag Berlin Heidelberg 2015

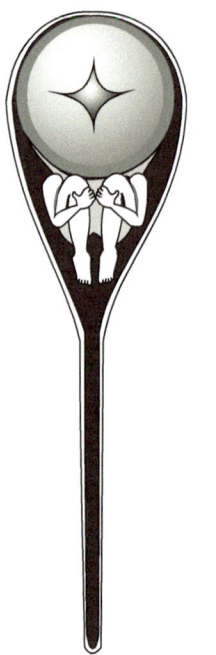

Fig. 1.1 Homunculus in a sperm cell, seen with the eyes of Nicolas Hartsoeker (1694)

association. For the sake of the supraindividual society – the organism or individual – the cells assume different responsibilities and tasks. They **differentiate** morphologically and functionally, together construct multicellular structures, tissues and organs. In this process, cells form the same structures and patterns in the same temporal and spatial order, from generation to generation.

The increase in complexity and diversity during embryogenesis and the autonomous shaping of form (**morphogenesis**) from simple, seemingly amorphous starting matter stimulated curiosity about the causal, organizing principles since thousands of years. Aristotle saw the ultimate form-determining principle in the soul (Box 1.1); nowadays the leading role is assigned to genetic information.

The ability to reproduce themselves is conferred upon organisms through genetic information (DNA in the nucleus and mitochondria of animal cells) and through further information-carrying structures in the egg cell (**maternal information, cytoplasmic determinants**); but

it is not the finished organization that this information encodes directly.

It is commonly stated that the genome incorporates a **Bauplan (construction plan, body plan)**: an architectural plan or blueprint of the body. Actually, this is not the case in the strict sense of the term: the genome is not a sketch or design of the finished body. The informational capacity of DNA is simply too low to store blueprints of the very complex final pattern of an organism. For example, a detailed design of the one hundred trillion to one quadrillion synaptic contacts in our brain alone would greatly exceed the capacity of the genomic memory.

What then does the **genome** (sum of genes) really encode?

The double-helix structure of the DNA provides information how to make two exactly identical replicas of the DNA itself which subsequently can be allocated from the mother cell to its two daughter cells in cell division.

The genome is subdivided in sections, called genes. Part of these genes, recently defined as '**coding' genes**, contains information on the sequential order in which amino acids have to be linked to form distinct chains, so that all the many proteins a cell needs get their correct primary and secondary structure. For this in the process of **transcription** the information of the gene is copied in form of **mRNA (messenger RNA)** and the copy exported into the cytoplasm. Mediated by this copy a specific protein can be produced (**translation**), instantly or at any appropriate time point later. Transcription and translation together are subsumed under the term **gene expression**.

Other types of genes contain information enabling the production of essential tools for protein synthesis, in particular tRNAs and rRNAs. Further sections of the DNA code for short-chain microRNA molecules which fulfil important regulatory functions that will be dealt with in Chap. 12.

- The genome embodies some hierarchical organization: master genes (Chap. 12) dominate, via their products, whole sets of subordinate genes.

- In the genome a spatio-temporal organization is realized. For instance, some classes of regulatory genes are arranged in an order that corresponds to the spatial and temporal pattern of their expression. Such correspondences (colinearity) apply to the *Hom/Hox* gene clusters (Chaps. 4 and 12).

- Also the 'non-coding' sections of the genome can contain information in a concealed form, for instance, when they offer binding sites for regulatory transcription factors and thus become controlling regions for coding genes. In this function these sections are called **promoters** or **enhancers**. Further potential roles of non-coding DNA sections will be discussed in Chap. 12.

How a developing organism arises by self organization on the basis of such minimal information, or how many organisms are able to regenerate lost structures surpasses our imagination.

1.1.2   Development Implies Increase in Complexity; This Gain of Complexity Results from Network of Gene Activities and Co-operation of Cells

How is it possible that a living system increases its complexity autonomously? Two partial explanations for this gain of complexity will be elaborated in the following chapters:

- **Combinatorics at the level of genes**. In the various cell types, organs and body regions, different combinations of genes become effective in time and space. With some thousand genes as all multicellular organisms possess the number of combinations of gene activities is practically unlimited. These activities are interconnected by many cross connections to form complex and robust networks.

- **Cell sociology**. Cells interact; they are parts of a society, in which the members influence each other mutually. The cells distribute tasks and social roles; they **differentiate**. The pattern of their behaviour is not only

defined by internal determinants such as genes, but also by the flow of energy and information arriving from neighbouring cells and, in later phases of the development, from the external environment of the living being.

1.1.3   In Natural Populations of Each Gene Several Variants, Called Alleles, Exist; the Unique Combination of Alleles Constitutes the Genotype of an Individual; as a Rule Different Genotypes Generate Different Phenotypes

In the course of thousands of years genes sustain mutations, the germ cells not excluded. A particular gene may mutate in this individual and, some hundred years later, in that individual, and it may mutate not only in somatic cells but also in germ cells or their precursors. Some of these mutations are not eliminated by natural selection. Therefore, of any given gene different editions (variants) exist in a population. These different editions, which once arose from one initial founder gene by mutations, are called **alleles**. Multicellular animals, and so also we humans, are, as a rule, **diploid**, meaning that our cells contain two alleles of each gene, one allele inherited from our mother through a certain chromosome of the egg cell, the other allele from our father through the corresponding (homologous) chromosome of his sperm cell. Of each gene the maternal and the paternal allele can be identical and the new individual is **homozygous** with respect to this particular gene. Typically, the two alleles are not identical and the new individual is **heterozygous** with respect to this gene.

Persisting different alleles constitute the **gene pool** of a species; as a rule it contains several variants of each gene. Alleles of one and the same gene differ from each other in one or several base pairs and possibly fulfil their function not equally well or in some other way differently. Each individual who owes his existence to an act of sexual reproduction comprises a unique combination of such alleles. This individual combination is known as **genotype**. The individual genotype can find its expression in an individual appearance, the **phenotype**. Certainly not each mutation becomes noticeable in the outward appearance of the organism or affects the function of the gene-coded protein, but the multitude of potential combinations of alleles enables an inexhaustible variety of individual traits within the framework of the species-specific variability.

1.1.4 The Genetic Inheritance of an Individual may Contain Enough Information to Develop More Than One Phenotype, That Is, More Than One Shape or Morph

During embryonic development the visible shape (**morphology**) of the germ and the structure of cell associations in epithelia, tissues and organs change continuously. Form and cellular composition change towards higher complexity. The cells stay together, collectively modelling the basic architecture of the new organism, and provide it with all essential organs to begin an autonomous life when the young animal hatches out of the envelope or the mother gives birth. One and the same genotype codes for dramatically differing phenotypes starting from the spherical fertilized egg cell and ending with a multitude of different cell types arranged in a species-specific pattern. However, besides the changes that constitute embryogenesis, most invertebrates develop two or more different, rather stable and autonomously living phenotypes. Embryogenesis results in a first phenotype which is able to begin an autonomous life. This first phenotype is called a **larva** when it does not reproduce itself. After a certain period of larval life it becomes transformed in a process called **metamorphosis**, into another phenotype, the **adult** (term preferred in vertebrates) or **imago** (term frequently used for invertebrates). This second and final phenotype then reproduces

sexually. Typically, the two phenotypes, the larval and the adult, occupy different ecological niches and exploit different resources in their environment. In several lower invertebrates such as the Cnidarians, besides larvae two different adult phenotypes, sometimes referred to as two different **morphs**, the polyp and the medusa, are capable of reproducing, though they do this in different modes. If the life cycle encompasses both forms, only the medusa can reproduce sexually (Fig. 1.3).

In spite of the multitude and growing number of completely sequenced genomes, and in spite of sophisticated and expanding algorithms of bioinformatics, it is at present not possible, and perhaps it will never be possible, on the sole basis of physical laws and logical rules, to deduce from the genome, – that is from the complete sequence of the base pairs in the DNA, – in which various phenotypes a given organism can exist and which shape and cellular composition it will have at any time during its life.

1.2 Reproduction: Sex Versus Natural Cloning

Among the essential traits of living beings is, besides the ability to self organize in the course of development, also the ability to reproduce; that is to give rise to new living beings of the same kind. We distinguish two fundamentally different kinds of reproduction:

- **Asexual, uniparental** (i.e. from one single parent originating) reproduction, and
- **Sexual, biparental** (from two parents originating) generation of offspring, and two rare forms of reproduction, derived from sexual reproduction,
- **Endogamy**, the uniparental generation of offspring in hermaphrodites, and
- **Parthenogenesis**, a kind of reproduction in which only female individuals contribute genes for the offspring.

1.2.1 Asexual, Also Called Vegetative, Reproduction Results in Offspring with a Genetic Inheritance Identical to That of the Parental Organism and That of Its Siblings; the Parental Organism and Its Descendants Constitute a Natural Clone

In asexual reproduction the offspring have only one parent; the reproduction is **uniparental**. Unlike in sexual reproduction there are not two parents who together contribute to the genetic makeup of the offspring. Asexual reproduction is based on conventional mitotic cell division (Fig. 1.2). In mitosis the entire genetic information is transmitted without change to each of the two new cells, which often are referred to as daughter cells. In preparation for mitotic cell division, in the S-phase of the cell cycle, the DNA is replicated and the chromosomes are duplicated with high fidelity; – in the traditional nomenclature the two strands of the duplicated chromosomes are called **chromatids**. In the M-phase each of the two daughter cells is provided with a complete set of chromatids and thus, of alleles. All descendants of a founder cell are equipped with the same genome. In unicellular organisms the founder cell and its mitotically derived offspring have been termed **clone**. The term was then transferred to multicellular organisms.

Clone: A clone is a collection of individuals who all are derived from one and the same individual by mitotic cell division and endowed with the exact same genetic information. However, if the mitotically generated cells stay together forming a multicellular organism, as it happens in embryogenesis, the resulting community is usually not called a clone but an individual. The term clone is justified when such an individual reproduces by mitotically produced (as a rule multicellular) buds. The founder individual and all its asexually generated offspring constitute a natural clone. Therefore, **asexual reproduction can also be defined as natural cloning**.

As asexual reproduction is particularly frequent in plants, it is also called **vegetative reproduction**. Plants frequently release descendants in the form of mitotically produced multicellular buds (Begonia), stolons or shoots (strawberries, blackberries), bulbs (potatoes) or tuberous roots. Also in many lower invertebrate animals, in particular in Cnidarians such as corals or the freshwater polyp *Hydra* (Fig. 4.12) multicellular buds are formed to produce naturally cloned offspring. In many laboratories all over the world *Hydra magnipapillata*, wild type 105, is grown and all the many thousands of polyps are descendants of one single individual (once found in a Japanese lake) and, therefore, represent one and the same clone.

> The verb "to clone" is not only used in the context of reproduction of living beings (reproductive cloning of organisms or cells) but also in the everyday language of molecular biologists. In his/her terminology cloning refers to processes used to create copies of DNA fragments (molecular cloning, recombinant DNA technology). In the laboratory the term "to clone" comprises the entire process of ligating a selected sequence of DNA into a vector and the (hopefully faithful) multiplication of the construct in bacteria or eukaryotic cells. Such cloning is intended to produce many identical copies of the selected DNA sequence.

1.2.2 Sexual Reproduction Results in Offspring with Newly Arranged Genetic Inheritance; Typically, the Two Parents and Their Offspring Have Differing Genotypes and Phenotypes

In the history of life, sexual reproduction got the upper hand over asexual reproduction. In contrast to asexual reproduction, sexual reproduction involves two parents, it is **biparental**, and it

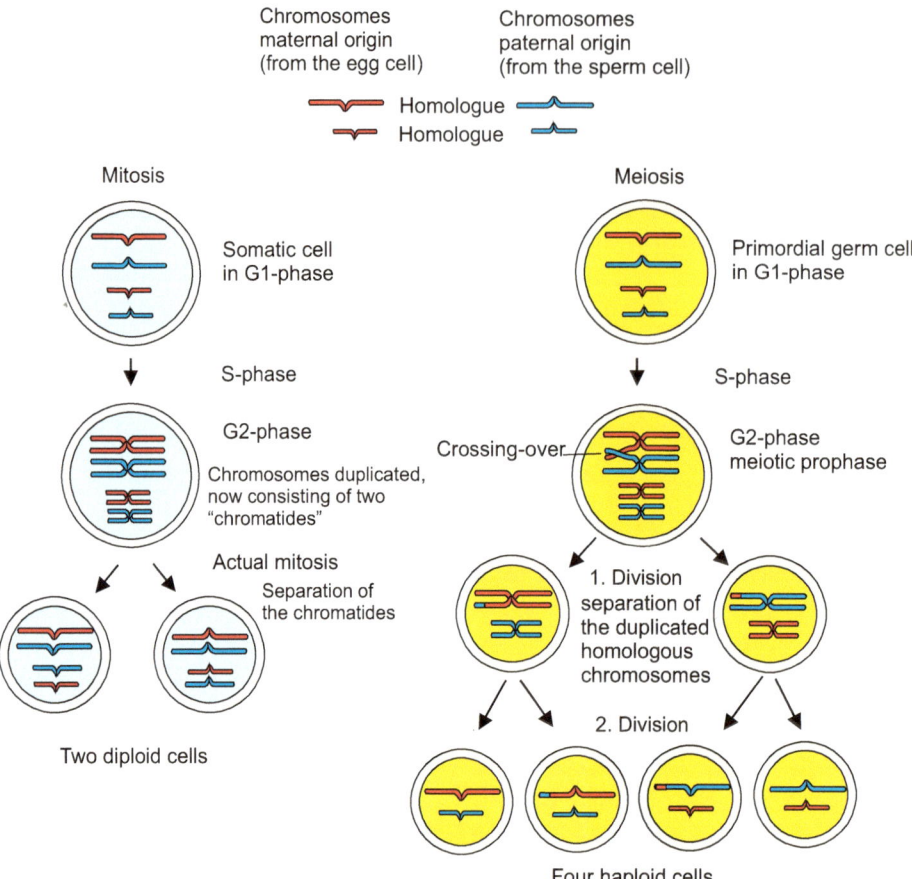

Fig. 1.2 Mitosis and Meiosis in comparison. From the two meiotic divisions four progenitors of germ cells (oocytes or spermatocytes) arise; these are haploid, possessing only one set of chromosomes with their genes. The genes may exist in different editions (alleles) because they are of different parental origin. From mitosis two cells with exactly the same genetic heritage arise

results in new genotypes similar to the parental genotypes but different enough so that natural selection can favour the momentarily fittest genotypes above less opportune combinations of genes.

The essential feature of sexual reproduction is that the new individual receives its genetic endowment from **two germ cells**, called **gametes**, with **two sets of chromosomes**:

- One set derived from its father and carried in a **male gamete**, the **sperm cell** or **spermatozoon** (plural: spermatozoa),
- The other set derived from its mother and carried in a **female gamete**, the **egg cell** or **ovum** (plural: ova).

These gametes come together and fuse at **fertilization** to form a **zygote**, the **fertilized egg cell**. It contains two set of chromosomes and is said to be **diploid**, whereas the gametes with their single set of chromosomes are said to be **haploid**. In a haploid set each chromosome is unique of its kind and embodies its specific alleles. A diploid set of chromosomes implies two sets of alleles which, however, are typically not identical.

In embryonic development the zygote undergoes many rounds of mitotic divisions, giving rise to thousands and thousands of diploid cells. These cells are **genetically equivalent**, meaning that they have an identical genetic endowment, at least initially. The cells divide up in two main categories:

Somatic cells together build the new organism, the offspring. In a process of division of labour they take on different tasks; they **differentiate**. Usually, somatic cells lose the ability to generate new organisms, and are determined to die.

The second group of cells are **generative cells**; they are not committed to participate in the construction of the body but are stem cells, in sexual reproduction known as **primordial germ cells**, destined to be used for the production of the next generation of gametes.

Two fundamental events characterize sexual development and make the outcome, the genotype of the new individual, to a great extent a matter of chance:

First: **Meiosis**, a sequence of two successive, particular cell divisions which reduce the diploid set of chromosomes present in the primordial germ cells to a single set, the haploid set, in the germ cells, and simultaneously sorts the maternal and paternal alleles into new combinations that are greatly subject to chance.

Second: **Fertilization**, the fusion of the two gametes into the single-celled zygote. This process is not based on targeted selection of the two partners but is subject to chance.

Meiosis: When the offspring gets ready to produce its own new germ cells, a type of cell division must be intercalated that reduces the diploid state of the **primordial germ cells** to the haploid state of the final gametes. Otherwise, the fusion of two gametes at fertilization would result in 2 + 2 sets of chromosomes, in the next generation 4 + 4 would be added up, then 8 + 8, and in a few generations the number of chromosomal sets would grow boundlessly.

This type of cell division is known as **meiosis** and consists of two sequential divisions. In the first division, previously duplicated chromosomes are separated and transferred into two daughter cells, in the next division chromatids are separated (Fig. 1.2). Thus, from a primordial germ cell typically four haploid gametes arise, though in female animals and

female humans three of the four haploid cells perish in favour of the fourth (Chap. 8).

Of paramount importance is that **in meiosis the chromosomes with their alleles are not separated and distributed according to their parental origin but become recombined**. We exemplify this by looking at the human chromosomes. Our primordial germ cells are still diploid, as are the somatic cells of our body, and contain 2×23 chromosomes, 23 inherited from our mother, 23 from our father. Each of the 23 chromosomes in the haploid set is unique and bears thousands of genes. In the diploid state each particular chromosome has its particular genetic content but is present twice, one inherited from the mother, the other from the father. Such two corresponding chromosomes are called homologous chromosomes or **homologues**. Both homologues harbour the same set of genes but frequently in different allelic editions.

In the new gametes, any particular chromosome can come from *either* parent; for instance, chromosome No 1, in the newly formed egg cell may be of paternal origin, once carried by the sperm, another chromosome, let's say chromosome No 2, was also of paternal origin, while chromosome No 3 was once received from the mother of the individual who produces the new gametes. Likewise, although in the testes of the sexually mature man the newly produced sperm cells are all haploid, each sperm cell contains chromosomes partly derived from his mother, partly from his father, and the particular mixture is subject to chance.

Considering all **23 chromosomes** a haploid germ cell embodies, the number of possible new combinations is huge. **It is 2^{23}, or 8,388,608 different** possible **combinations**. This means, a woman could produce approximately 8.4 million egg cells which differ from all others in at least one chromosome. Likewise, a man could produce 8.4 million different sperms.

In addition to recombination based on the redistribution of the chromosomes, physical exchange of chromosomal segments by crossing-over greatly increases the combinatorial possibilities. Let us say that the gametes of a particular species contain 10^4 genes. The diploid individual (e.g. a human) producing these gametes could possess 2×10^4 different alleles. Given that all these alleles can freely be recombined, the number of different gametes which potentially could be generated would be $2^{10,000}$, about **equal to $10^{3,000}$**. The number of elementary particles in the universe is estimated to be lower than this value.

Chance comes into play a second time at fertilization: Which sperm will meet which egg cell? Therefore, an essential feature of sexual reproduction is that the **outcome is unforeseeable** in detail, though the offspring take after their parents.

1.2.3 Most Organisms Retain the Options of Asexual Reproduction, Even Where Sexual Reproduction Occurs; Sexual Reproduction Is Preferred in Times of Distress As It Allows Genetic Adaptation

Asexual reproduction has many advantages for rapid and efficient propagation of species. Transient resources can quickly be exploited, as all individuals of a population are able to reproduce, and not only 50 % as in sexual reproduction. Why does sexual reproduction occur at all? There are severe costs for sexual reproduction. Males exploit resources without producing energy rich egg cells or viable offspring directly. Also the time and energy taken up with seeking a sexual partner, courting and mating are costly.

On the other hand, sexual reproduction may be advantageous at times of stress when maximum genetic flexibility is required for the population to survive. Sometimes this, sometimes that genotype is better adapted to cope with a changing environment. **In times of need and difficulty other genotypes might be called for than in times of abundance and glut. In addition, sexuality potentially enables producing descendants which compete more successfully with members of the same species than previously prevailing genotypes did.**

While most organisms retain a capability of asexual reproduction, even where sexual reproduction occurs, mammals have lost that option during evolution. One of the great puzzles of mammalian biology is to determine why.

1.2.4 By Modifying Sexual Reproduction, Several Organisms Have Developed Other Forms of Uniparental Reproduction: Uniparental Hermaphroditism or Diploid Parthenogenesis

The benefits of uniparental reproduction may be the reason why several organisms have altered sexual reproduction in such a way that particular males are superfluous.

Uniparental Hermaphroditism. One possibility is that females produce sperm themselves and use them to fertilize their own eggs. Species in which the individuals are capable of producing both eggs and sperm are called **hermaphrodites** (being male like the Greek demigod Hermes and female like Aphrodite). As a rule, hermaphrodites cannot use their sperm to fertilize their own eggs, either because the anatomical conditions do not allow meeting or fusion of sperm and eggs within the parental body, or because the two types of germ cells are produced at different times. However, in rare cases both types of germ cells are generated simultaneously and within the same compartment, and the sperm cells can fuse with their sister egg cells. This is the case in the roundworm *Caenorhabditis elegans* which became a favourite model organism in developmental biology (Fig. 4.19, Chap. 4.19). If this extreme mode of inbreeding (**endogamy**) occurs repeatedly, eventually the allelic composition of all descendants is almost identical. Genetically, they are very similar, almost like cloned individuals. In effect, hermaphrodites exhibiting endogamy have regained the option of asexual reproduction. (*Caenorhabditis elegans* in addition has the option of biparental sexual reproduction; Sect. 4.4)

Diploid Parthenogenesis. In several animals, for instance in plant-lice of the genus *Aphis* (below Sect. 1.2.5) and in the grasshopper *Pycnoscelus surinamensis* the germ cells dispense with meiosis altogether, forming diploid ova. Embryonic development starts and is finished in the absence of any sperm. If derived from diploid germ cells the offspring are genetically copies of their mother; all the descendants are naturally cloned individuals as are asexually generated individuals.

Diploid parthenogenetic development is also known from a few vertebrates, namely some turtles and the lizard *Cnemidophorus uniparens*. (This lizard became known because of its remarkable behaviour. Although the species consists of females only for many generations, the animals have not lost fun with sex. They consummate sham-marriage whereby one individual plays the role of a male the other of the female. From time to time the roles are exchanged.)

Can parthenogenesis also occur in human beings, and if it were to occur, what would be the outcome? These questions are discussed in Box 6.2.

1.2.5   Most Animal Organisms Develops Two or More Stable, Autonomously Living Phenotypes Which Reproduce Differently; One Speaks of an "Alternation of Generations"

The developmental pathways become more complex when a species exists in multiple reproductive morphological entities (phenotypes or morphs). In all such cases one phenotype reproduces by natural cloning (asexual reproduction or parthenogenesis), and the final phenotype reproduces sexually. One speaks of **alteration of generations**. This multitude of reproductive means is frequent in plants and, as a rule, associated with a regular alteration of haploid and diploid generations. This is not so in animals; if alternation of generations occurs both generations are diploid, only the gametes are haploid. The alteration between the two diploid

forms (diplont 1, diplont 2) occurs in two different ways, termed metagenesis or heterogony.

Metagenesis. For instance, in the cnidarian groups Hydrozoa and Scyphozoa the embryonic development of the fertilized egg produces larvae which metamorphose into a first reproductive phenotype or morph, the polyp. This first morph reproduces asexually by producing buds (in Hydrozoa) or constricting off body parts with larval traits (called ephyrae in Scyphozoa). They detach and develop into free-living medusae, the second morph. Medusae live in the free water. In the Scyphozoa the initially tiny ephyrae can achieve considerable sizes and are known as jellyfishes. Once grown up, the medusae reproduce sexually by producing gametes (Fig. 1.3). Differently looking, reproducing phenotypes are also known as different **morphs** of the respective species.

Heterogony. A seasonal alternation between parthenogenetically and sexually reproducing individuals is found among several species of plant-lice such as *Aphis fabae* (Fig. 1.4). In the spring the population consists entirely of females which clone themselves through parthenogenesis. Their egg cells do not undergo meiosis and therefore remain diploid; they develop without the genetic contribution of a sperm within the virginal mother (fundatrix) into small aphids. The viviparously born animals are again female and wingless and faithful copies of their mother because they are uniparentally generated by diploid parthenogenesis and, therefore, cloned. This type of reproduction allows a rapid increase of the population and, thus, a fast and extensive consumption of the young, soft and nutritious plants. In summertime and early autumn, when the leaves and shoots become harder and preparation must be made for hibernation, male as well as female, winged individuals appear, and the mode of reproduction is now sexual. The fertilized eggs are laid. They are covered with a hard shell, endowed with anti-freeze substances, and able to survive the winter. Likewise, the water flea *Daphnia magna* (a crustacean) reproduces uniparentally under

Fig. 1.3 Life cycle of the scyphopolyp or jellyfish, respectively, *Aurelia aurita*

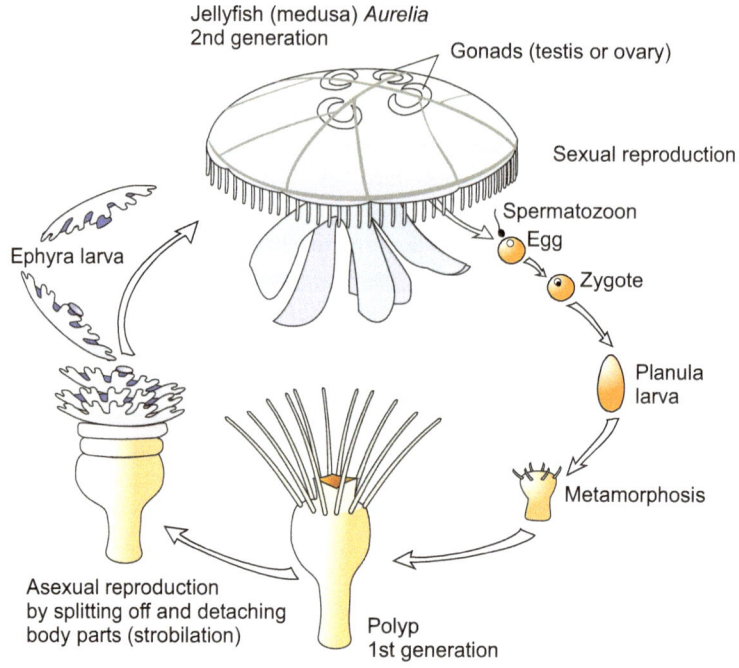

Jellyfish (medusa) *Aurelia* 2nd generation

Gonads (testis or ovary)

Sexual reproduction

Spermatozoon
Egg

Zygote

Ephyra larva

Planula larva

Metamorphosis

Asexual reproduction by splitting off and detaching body parts (strobilation)

Polyp 1st generation

Fig. 1.4 Modes of reproduction in the greenfly *Aphis fabae*

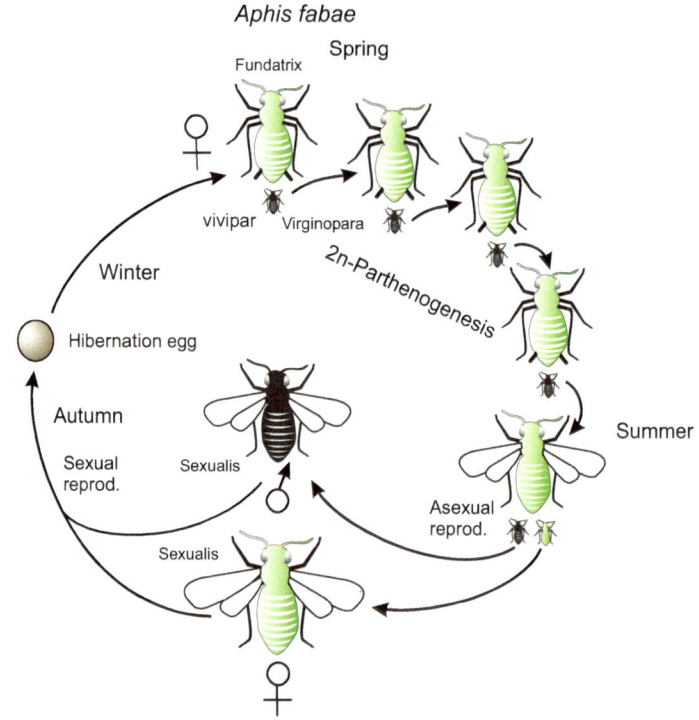

Aphis fabae

Spring

Fundatrix

vivipar   Virginopara

2n-Parthenogenesis

Winter

Hibernation egg

Summer

Autumn

Sexual reprod.

Sexualis

Asexual reprod.

Sexualis

favourable environmental conditions, bisexually under stress. An alternation of generations between uniparental parthenogenesis and biparental sexual reproduction is called heterogony. In the following chapters we shall focus on embryonic development starting from familiar, fertilized eggs.

Aphids exemplify an often realized principle of reproduction. **To make use of favourable but transitory environmental conditions and exploit seasonally restricted resources many animal organisms prefer uniparental natural cloning, be it in form of asexual reproduction or parthenogenesis. Sexual reproduction is on the agenda when harsh and dramatically changing conditions favour the availability of various genotypes to natural selection.** 'Hopefully' there are better fitted ones among them for survival.

1.2.6   Progenesis and Neoteny: Sexually Mature Larvae Goldmine for Evolution?

Progenesis (Paedogenesis). In the history of life it occasionally occurred that already larvae acquired sexual maturity and metamorphosis to the adult form was omitted. In these cases it might have been advisable not to leave the larval biotope or it became increasingly more difficult to leave it. Several marine small members of the phyla Annelida and Crustacea (Bathynellacea) look like the larvae of other species of the same taxonomic group.

Present day zoologist prefer to designate this precocious scheduling of sexual maturity **progenesis** or **paedogenesis** (in former times not distinguished from neoteny). Particularly many larvae capable of reproduction are known from flukes, i.e. parasitic flatworms (phylum Plathelminthes, group trematodes), for instance from the blood fluke *Schistosoma mansoni*, that causes schistosomiasis (bilharzia) in humans.

From the bisexually produced egg a larva (Miracidium) develops that lives freely in water, produces asexually huge amounts of secondary larvae and these tertiary larvae (Cercaria). Gradually the larvae adopt the appearance the adult form. Eventually, this reproduces sexually.

Progenesis is closely related, but not identical, to **neoteny** in the sense of present day zoology. In neoteny, larval characteristics persist in adults even when sexual maturity commences. Neoteny is known from two cavity occupying newts, the Mexican axolotl (*Ambystoma mexicanum*) and the cave salamander (*Proteus anguinus*). Both these amphibians do not leave the water and preserve gills life long.

Some scientists take the view that progenesis (paedogenesis) and neoteny provide chances to evolve new body plans in the history of life. Imagine a caterpillar would not transform into a butterfly. Since the adult construction is omitted the larva had to acquire the capability of generating offspring. The larval phenotype would serve as raw material based on which evolution could try out several new variants and extensions of the basic architecture. Soon caterpillars were to be expected that would greatly differ from conventional ones. However, to reconstruct a well-tried body architecture is not an easy task because deviation from the norm can easily result in reduced rates of reproduction or even chaos. To explain the evolution of new developmental patterns is a great challenge to biologists devoted to reveal the mechanisms of evolution. This topic will be dealt with in the last chapter of this book.

Summary

Development and reproduction are constitutive features of all living beings. In their **ontogenetic** development, their individual life, living beings construct themselves in a process of self-organization. For this self-construction internal, inherited sources of information are available. The **genome**, the complete inventory of genetic information laid down in the genes along the chromosomes (and in the DNA of the mitochondria), contains particular variants, termed **alleles**, of the species-specific genes. The individual allelic composition of a given genome constitutes the **genotype**, the individual genetic endowment. The individual genotype may be expressed in an individual **phenotype** (appearance). In many organisms the genome may code for several different phenotypes (larval and adult, for instance) or phenotypes, termed morphs, which not only display different shapes but also reproduce differently (e.g. asexually reproducing polyps and sexually reproducing medusae in Cnidaria).

In reproduction, genetic information is transmitted to the descendants. Two fundamental modes of transmission may occur: (1) **Asexual reproduction** is based on conventional mitotic cell divisions; therefore the genotype of all descendants is the same as that of their parents. Each descendant has only one single parent (**uniparental reproduction**), and each parent and all its descendants are members of a natural **clone**, that is a group of individuals endowed with identical genetic make-up. (2) In **biparental, sexual reproduction** two parents, typically occurring as male and female individuals, contribute to the genetic outfit of the offspring. Each parent contributes genetic information through its **gametes,** known as **egg cell**s and **sperm cell**s. Gametes are **haploid**, containing one set of chromosomes and thus one complete set of alleles. In **fertilization** one egg cell and one sperm cell fuse to form a **zygote** (fertilized egg cell). It is **diploid** as it contains two complete sets of chromosomes, one set of paternal and one of maternal origin, but these two sets contain alleles which may differ in most gene loci. All the **somatic** cells which arise from the zygote by mitotic divisions and together build the new individual are (initially) **genetically equivalent** and diploid, as are also the **primordial germ** cells, the founder cells from which new gametes derive.

Sexual reproduction involves two processes by which the genotype of the offspring becomes subject to chance. In the development of the gametes from the primordial germ cells the **meiotic cell divisions (meiosis)** not only reduce the diploid set of chromosomes to a haploid set but in addition recombine the allelic composition in such a way that a new germ cell partly contains originally maternal alleles and partly paternal. This recombination involves chance. Fertilization is another event that is subject to chance because it is not predetermined which sperm will meet which egg cell. In sexual reproduction parents and their descendants differ in their genotype and, as a rule, also in their phenotype. Sexual reproduction leads to diversity in a population but makes the individual outcome of any reproductive act to a great extent unforeseeable. Besides these main types of reproduction two more, rare types of uniparental reproduction, derived from biparental sexual reproduction, are known: **parthenogenesis** (birth from a virgin) and a particular type of **hermaphroditism** (endogamy) in which the individual's own sperm cells fuse with their sister egg cells. Uniparental modes of reproduction result in genetically homogenous offspring. In the alternation of generations as exhibited by many Cnidarians and insects such as Aphids, uniparental cloning and biparental sexual reproduction alternate.

As a rule, uniparental reproduction facilitates a quick increase of the population and extensive use of seasonal environmental resources, while biparental sexual reproduction allows long-term adaptation to changing conditions and more rapid evolution. In the evolution of the mammals the option of uniparental reproduction was lost.

Two types of abbreviated life cycles, **progenesis (paedogenesis)** and **neoteny**, are of interest in evolutionary biology. In progenesis, known from some marine invertebrates, already larvae acquire sexual maturity while in neoteny mature adult forms retain some larval characteristics such as gills in axolotls. It is thought that such

altered life cycles provide material for reconstruction of the body architecture.

Box 1.1 History: From the Soul to Information

The Dawn of Developmental Biology in Ancient Greece

Although embryos were described in ancient Sanskrit and Egyptian documents, the Macedonian **Aristotle** (384–322 BC), son of a physician, was the first to perform developmental studies in a systematic way, to interpret his observations in written words and to coin lasting terms. An acknowledged philosopher and academic teacher (e.g. of the prince that was to become King Alexander the Great) as well as an enthusiastic naturalist, he wrote the first textbooks of zoology and treatises on reproduction and development.

The multivolume "History of Animals" (*Peri ta zoa historai*, in Latin: *Historia animalium*) contains several chapters on development. Most treatises are compiled in five volumes entitled "On the Generation and Development of the Animals" (*Peri zoon geneseos*; Lat.: *De generatione animalium*). Some more expositions on development are found in the books "On the Soul (Mind)" (*Peri psyche*; Lat.: *De anima*) and in the Metaphysics (*Metaphysika*).

Aristotle distinguished four possibilities as to how organisms might arise: (1) spontaneous generation from rotting substrate, where flies and worms were thought to originate; (2) budding; (3) hermaphroditism; and (4) bisexual reproduction. In his view, the egg was the instrument of reproduction in **oviparous** species. Mammals, human beings and some other **viviparous** species lacked eggs. Females contributed to offspring by supplying unstructured material, males by supplying semen, the purveyor or causative principle of form.

Aristotle described the development of the chick in the egg. Eggs were incubated for varying periods of time and opened. According to his observations there is an initially unstructured material, which in the course of **morphogenesis** or **epigenesis** acquires form. In the midst of the emerging figure he observed a "jumping point", the beating heart.

He considered the aim of development to be the *ergon*, the finished work as it is the aim of the artisan. The modelling principle he thought of as *energeia* (energy), also called *entelecheia*, the principle bearing its aim, goal and end in itself. Energy is both the efficient and final cause. To reach a particular species-specific end, the forming principle must have a "pre-existing idea" of the final outcome. Hence the ultimate cause, the ultimate energy would be the soul (mind, spirit, *psyche*).

To quote Aristotle, "*It (the soul) causes the production . . . of another individual like it. Its essential nature already exists; . . . it only maintains its existence. . . The primary soul is that which is capable of reproducing the species*" (*De anima*, 416b1–b27).

Following Plato, Aristotle discriminated between the **vegetative soul**, which brings about life as such, the **animal sensitive soul**, which enables sensations, and the **spiritual soul**, which enables thinking.

The vegetative soul endows plants with the ability to regenerate. The vegetative soul also includes the formative power of animal development. In animals the mother (*mater*) supplies the matter (Latin: *materia*); in mammals the matter is supplied in the ▶

Box 1.1 (continued)

form of the menses (as the menses are cancelled if pregnancy occurs). The semen was thought to coagulate the female material and to trigger and govern its development. Aristotle's exposition on the residence and inheritance of the vegetative soul are not unequivocal: is it in the female matter or in the semen? In contrast, the second degree of soul, the animal soul, is inherent only in the semen and is transferred from the father to the future child in begetting. The animal soul then governs sensitivity and movement. The **spiritual soul is eternal**, immortal, painless, sheer energy, and enters human beings from outside "through a door".

Aristotle's imprint on the western educated world has endured for centuries. With all due respect for his great stature, what he said about begetting, fertilization and determination of the female sex may be courteously passed over in silence.

The Renaissance of Developmental Biology

Embryology revived in the sixteenth century. In the school of Padua (Vesalius, Fallopio, Fabricius de Aquapendente) the anatomy of ovaries and testes were studied. The idea of an egg arising in the ovary and the embryo arising in the egg was conceived by the Dutch physician **Volcher Coiter** (1514–1576) upon completing his detailed study of chick embryo development, a study for which he has finally been recognized as the father of embryology.

The English anatomist and physician **William Harvey** (1578–1657), best known as the discoverer of the (greater) blood circulation in the vertebrate body, resumed the embryological studies of Aristotle, extending his research to insects and mammals (sheep and deer). Though he was an admirer of "The Philosopher" (i.e. Aristotle), he maintained that spontaneous generation was restricted to lower organisms. But in insects, development

implies "*metamorphosin*" or **metamorphosis** – the transformation of already existing forms into other forms. Harvey considered the pupa as an egg, as did Aristotle before and several other investigators after him.

In higher animals, however, Harvey regarded development to be not merely transformation but "*epigenesin*" or **epigenesis** – creative synthesis, incremental formation of a new entity out of non-structured matter. And Harvey wrote, "***We, however, maintain … that all animals whatsoever, even the viviparous, and man himself not excepted, are produced from ova; that the first conception, from which the foetus proceeds in all, is an ovum of one description or another, as well as the seeds of all kinds of plants***".

Later literature shortened Harvey's phrase to "***omne vivum ex ovo***" ("all life from an egg"), probably inspired from the frontispiece of Harvey's embryological treatise "*Exercitationes de Generatione Animalium*", where an egg bears the inscription "*ex ovo omnia*" ("everything out of the egg"). However, Harvey's mammalian egg was not the same entity we have in mind today but the blastocyst (young embryo) within the "shell" of the uterus. It was Carl Ernst von Baer (below) who discovered the real mammalian egg.

Preformation and Mechanicism

When the Swiss naturalist and physician **Conrad Gessner** (1516–1565) author of the famous works *Historia animalium* and *Bibliotheca universalis*, following the Roman naturalist Pliny, reported that the female bear gives birth to a lump of meat, thereafter licking it into shape, he probably was not yet influenced by the philosophy of mechanicism. In contrast, his later compatriot scholar **Albrecht von Haller** (1708–1777) categorically maintained: "*nulla est epigenesin*" ("there is no epigenesis"). ▶

Box 1.1 (continued)

With this notion he followed the founders of microscopic anatomy. In 1683, **Anton van Leeuwenhook** (1632–1723) wrote *"that the human foetus, though not bigger than a pea, yet is furnished with all its parts"*. Leeuwenhook discovered *"**animalcules**"* or *"**zoa**"* within semen as did others before him. (Later, von Baer renamed the zoa **spermatozoa**). Several microscopists of those days saw or conjectured that they would see *"**homunculi**"* – minute preformed human beings within the animalcules (e.g. Hartsoecker, Fig. 1.1). Embryos were thought to result from enlargement of homunculi.

Likewise, in the 'eggs' (pupae) of insects, adult ants and butterflies were seen to be already prefigured in miniature, as are leaves and blossoms in the buds of trees. The pre-existing beings only needed to be 'evolved': unrolled and unwrapped. Such views advanced to meet the views of the **mechanicists**, who held that life merely obeys the laws of mechanics. Living beings were regarded as ingenious clockworks comparable to the marvellous astronomic clocks built by contemporary artisans. Whether these machines were viewed as soulless or animated entities depended on the religious and ideological position of the respective author.

In 1672 **Marcello Malpighi** published the first compelling account of chick development with detailed figures. For the first time he depicted the neural folds, the muscle- and vertebrae-forming somites and the blood vessels leading to and coming from the yolk. Nevertheless, Malpighi questioned epigenesis and adhered to the preformationist view. This view appeared to be better in accord with the narrations of the bible. God created living beings only once and then settled down, and because all beings and their organs were prefigured when created in the paradise, no further intervention of divine creativity is needed in development because prefigured structures need merely be unrolled.

The doctrine of preformation quickly led to some awkward problems:

- If ontogenetic development is only mechanical unwrapping of prefigured forms, must not all generations have been in existence from the very beginning of the world? *"**Emboitment**"* (encapsulation) was the answer: One generation lies within another generation like one Russian doll lies within another. According to computations of **Vallisneri** (1661–1730), the ovary of the primordial mother Eve contained 200,000,000 human burgeons, packed into one another. This stock should suffice until the end of all days. The French-Genevan **Charles Bonnet** (1720–1793), who accurately described **parthenogenesis** in aphids (Fig. 1.4) wrote in 1764: *"Nature works as small as it wishes"*.

- The microscope showed cells and with cells a lower limit to the size of the preformed organisms. Microscopists showed not only egg cells but also spermatozoa. Now the prefigured "homunculus" was claimed to be prefigured and visible in the egg (**ovists**) or in the sperm (**animalculists**, **homunculists**). Among the ovists were the renowned anatomists **Marcello Malpighi** (1628–1694) and **Jan Swamerdam** (1637–1680).

- How can regeneration of body parts be explained, if lost parts can only be made from preformed parts?

Lazzaro Spallanzani (1729–1799) was the first to perform **artificial insemination**. He reported that frog eggs degenerated in the absence of sperm. Working with dogs, Spallanzani finally laid preformation arguments to rest by proving that both the egg and the male semen are necessary to produce a new individual (although he erroneously believed that the animalcules swimming in semen were mere parasites). In its extreme form preformation soon was mere history. ▶

Box 1.1 (continued)

Epigenesis and Vitalism

Caspar Friedrich Wolff (1738–1794) who worked and taught in St. Petersburg and Berlin resumed the study of the chick embryo and again saw new formation – morphogenesis out of structureless yolk material. Wolff planted the seed, so to speak, of the germ layer theory by describing "*Keimblaetter*" (germ leaves) which later transform into adult structures. Like Aristotle before him, and all further vitalists after him, Wolff concluded that there are 'immaterial' (non-corpuscular) virtues, a "*vis essentialis*" or "*vis vitalis*" – a force specific for life. The academic colleague of Immanuel Kant, **Friedrich Blumenbach** (1742–1840) postulated a particular, physically acting "*Bildungstrieb*" (propensity, formative compulsion) that is inherited via the germ cells. Many important biologists were vitalists, among them **Carl Ernst von Baer** (1792–1876), who discovered eggs in several mammalian species, and performed extensive comparative studies. Von Baer concluded that all vertebrates develop in a fundamentally similar way from germ layers, and he established a rule, now known as von Baer's law, which states that all vertebrates go through a very similar embryonic stage and only thereafter do the developmental pathways diverge. Based on this rule, **Ernst Haeckel** (1834–1919) formulated his much disputed "*ontogenetic*" or "*biogenetic law*" (Chaps. 6 and 22). The theorem maintains that ontogeny is an abbreviated recapitulation of phylogeny.

Interest in human embryology was stimulated in 1880 by the Swiss anatomist **Wilhelm His** with the publication of *The Anatomy of Human Embryos*.

Experimental embryology began in France in the tradition of morphology. The former theologian and later zoologist **Etienne Geoffry Saint-Hilaire** (1772–1844), one of the founders of the theory of evolution, and opponent of the very influential **Georges Cuvier**, sought to elucidate the causes of developmental anomalies ("*terata*") and disturbed, with crude methods, the development of the chick. In 1886, his compatriot **Laurent Chabry** began studying teratogenesis in the more readily accessible tunicate egg (*Ascidia aspersa*). Although the tunicate egg is tiny (0.16 mm) Chabry succeeded in performing localized defects with self-made surgical instruments. Henceforth, invertebrates became preferred sources of eggs for studying very early animal development.

Impetus and Progress at the Turn of the Century

From 1860 on, numerous important discoveries were made, and the era of experimental embryology, cell biology and genetics began.

Experimental embryology. At the turn of the century 1899/1900 **Wilhelm Roux**, a German anatomist and embryologist attracted most attention with the first surgical interventions in a frog embryo. Using a hot needle Roux killed one of the two cells which result from the first division of the fertilized egg. As a consequence he saw embryos whose one half was normal but the second half almost lacking (if he had removed the killed cell he obtained tadpoles of half of the normal size but structurally quite normal). However, Roux was more theoretician than a practitioner. He drew up theoretical concepts and founded the first journal that was devoted to developmental biology (Archiv für Entwicklungsmechanik, Archives for Developmental Mechanics; today: Development, Genes and Evolution). Of equal rank as Wilhelm Roux was the scholar **Hans Driesch** who at the same time made similar experiments with sea urchin embryos (see below) and also wrote influential theoretical treatises and books. ▶

Box 1.1 (continued)

Cell biology, developmental genetics. The first pioneer on the way towards developmental genetics, however, was an almost blind theoretician. With a presentiment of the role of the genes, **August Weismann** (1834–1914) in his *"germ plasm"* theory (1892) ascribes his hypothetical, self-reproducing determinants to the chromosomes that had just been detected by investigators such as **Eduard Strasburger** and **Walter Flemming**. However, Weismann thought that the determinants become differentially distributed among the cells of the embryo, thus causing and directing cell differentiation. Weismann coined several basic terms of biology such as germline, primordial germ cells and stem cells.

Much of the credit for drawing attention to the **nucleus** as the seat of heredity goes to the **Hertwig** brothers, **Oscar** (1849–1922) and **Richard Hertwig** (1850–1937), who often worked together at the marine station in Roscoff (France). The Hertwigs supplemented the observations of **Otto Bütschli** on fertilization and recognized that the essential event is the fusion of the male and female gamete nuclei. Oscar Hertwig, working with sea urchin eggs also identified the polar bodies and saw the nucleus within these small sister cells of the ovum. He mixed sperm and egg suspensions together and repeatedly saw a sperm entering an egg and saw the two nuclei unite. He also noted that all the nuclei of the embryo were derived from the fused nucleus created at fertilization. The sea urchin embryo came to be the most important subject of research in embryology in that period (O & R Hertwig, T Boveri, H Driesch, TH Morgan).

Chromosomes versus cytoplasmic constituents. With his careful observations and sagaciously interpreted experiments on eggs of the maw-worm *Ascaris*, **Theodor Boveri** (1862–1915) advanced the **chromosomal theory**. He proposed that chromosomes were complex structures differing from each other even within the same nucleus and capable of producing qualitatively different effects in cells. Boveri for the first time experimentally demonstrated the significance of the chromosomes for development, and he also realized that cytoplasm and nucleus interact, a finding later verified for the sea urchin embryo by TH Morgan, H Driesch and himself. Boveri was the first to formulate the **gradient hypothesis** (Chap. 10; Box 10.2).

The importance of **cytoplasmic determinants** was also verified by careful observations and subtle surgical experiments done by **Edmund Beecher Wilson** (1856–1939) and his students on the eggs and embryos of marine invertebrates, in particular those with spiral cleavage such as the mollusc *Dentalium* (Sect. 3.5). **EG Conklin** performed similar studies on embryos of tunicates (Sect. 3.7). Wilson also discovered the sex chromosome in insects and wrote a very influential textbook, *The Cell in Development and Inheritance* (1896).

Regeneration studies: Most of the pioneers of modern developmental biology, including Boveri, Wilson, Driesch and Morgan, often met each other at the Stazione Zoologica di Napoli, founded by Anton Dohrn and opened in 1874. Besides studying eggs of sea urchins and other marine invertebrates there, they performed regeneration experiments on hydrozoans (mostly with *Tubularia*), following the example of the Swiss scholar **Abraham Trembley** (1710–1784), whose elegant and well documented regeneration and grafting studies of *Hydra* were carried out as early as 1740. These regeneration studies had marked the dawn of modern experimental biology. A compendium, still worth reading, of the regeneration studies of those days, was written by **TH Morgan** (*Regeneration*, 1901). ▶

Box 1.1 (continued)

From the "Soul" to Today's Information: Experimental Embryology and Developmental Genetics

Of particular relevance here are experiments on the egg of the frog by **Wilhelm Roux** (1850–1924), and on the sea urchin by **Hans Driesch** (1876–1941), and the establishment of *Drosophila* as the leading model organism of genetics by **Thomas Hunt Morgan** (1866–1945; the first biologist to win a Nobel Prize, in 1933).

In his classic experiment Driesch separated the first two daughter cells (blastomeres) that arise by division from the fertilized egg. Each of the separated cells gave rise to a whole sea urchin larva. This experiment proved that living beings are not machines as envisioned by the mechanicists, since no divided machine can restore itself.

Driesch revived the Aristotelian term entelechy, but did not assign a physical virtue to it (though entelechy was said to be able to "suspend" physical forces), instead attributing to it entities such as "knowledge" and "message". He thus anticipated the term "**information**", which was introduced as a scientific concept by Norbert Wiener only in 1942. However, the entelechy of Driesch was transcendental and not dependent on a physical carrier, in contrast to the material, molecular genetic information of today.

Driesch also anticipated the idea of **positional information** by stating that "*the prospective significance (fate) of a cell is a function of its position in the whole*" and "*each single elementary process or development not only has its specification but also its specific and typical place in the whole – its locality*".

Fate-determining events and inductive interactions between the parts of an embryo were subsequently investigated with delicate and novel surgical methods, using amphibian embryos, by a student of Boveri, **Hans** Spemann (1869–1941, Nobel Prize in 1935) and his student Hilde Mangold. Their experiments culminated in the discovery of the "**organizer**" (Chaps. 5.1 and 10, Fig. 5.13).

Classical, organismically-oriented developmental biology was predominant until 1970 and had many important practitioners:

- **Gradient theory**, T Boveri, Sven Hörstadius (in sea urchin embryos); TH Morgan, CM Child (in regeneration), K Sander, C Nüsslein-Volhard & Eric Wieshaus and followers (insect embryos);
- **Embryonic induction**: Hans Spemann, L Saxen, PD Nieuwkoop (in amphibians) CH Waddington (chick),
- **Cell interactions and cell cultures**: J Holtfreter, V Hamburger, Paul Weiss,
- **Transdetermination and transdifferentiation**: Ernst Hadorn, Tuneo Yamada,
- **Transplantation of cells and nuclei, cloning**: Robert Briggs, Thomas King, John B. Gurdon (Chap. 13, Fig. 13.2); Beatrice Mintz (mouse, teratocarcinoma cells; Fig. 13.4).
- **Biochemical and molecular developmental biology**: Pioneers included Jean Brachet (RNA in the amphibian egg), Alfred Kühn (chains of enzyme and gene activities in insects); W Beermann and A Ashburner (giant chromosomes in dipterans), H Tiedemann (inducing factors).

The role of **DNA** as the carrier of genetic information, mechanistically interpreted for the first time in 1953 by James D. Watson and Francis Crick; the interaction of cytoplasmic determinants with DNA; and the exchange of signals between cells, are points of major effort in recent research. The goal is a complete understanding of the network of gene activities and protein-protein interactions.

- *Drosophila* has become the model of reference in genetic and molecular developmental biology (Sect. 4.6) thanks to the pioneering works of the geneticists ▶

Box 1.1 (continued)

Edward B Lewis, Christiane Nüsslein-Volhard, Eric Wieschaus (shared Nobel prize in 1995), David S Hogness, Walter Gehring, and many others. Sydney Brenner succeeded in establishing the nematode *Caenorhabditis elegans* as another model system (Sect. 4.4). As the genomes of these two model organisms and many other organisms are now completely sequenced including the human and murine genome, the efforts of the life sciences are now focussed on the organization of genomes and chromatins, on the analysis of the proteome, the spectrum of all proteins coded by the genome, and on the comparative study of their various functions.

- Many investigators are now resuming classical studies, extending them by molecular methods; among them are Eric Davidson (sea urchin), Marc Kirschner (*Xenopus*, cell cycle), John B. Gurdon (embryonic induction, cloning), Eddy De Robertis (signalling factors), Lewis Wolpert (positional information, pattern formation in the avian limb bud).

- A pioneer of a very different kind of developmental biology was the British mathematician Alan Turing. Based on mathematically formulated theoretical concepts, models for computer simulation of biological pattern formation are being developed (Box 10.2). Followers are Alfred Gierer, Hans Meinhardt, James D. Murray, Hans G. Othmer, for instance.

Many more scientists with comparable achievements would deserve mention if space allowed.

Recent research in developmental biology can be followed in the current literature, notably by reading reviews offered by journals such as those listed in the references.

Stages and Principles of Animal Development: Terms of Developmental Biology

2.1 Stages of Development in Overview

2.1.1 Most Animals Pass Through an Embryonic Phase, a Larval Stage, Metamorphosis and an Adult Phase to Reach Eventually Sexual Maturity

In humans and in animals where offspring are generated through sexual reproduction the development of a new individual must be prepared in the **gonads** of both parents. The gonads must produce the **gametes** in processes subsumed under the term **gametogenesis** which occurs as **oogenesis** in females and **spermatogenesis** in males. In the female gonad, the **ovary**, oogenesis provides the egg cells, and in the male gonad, the testis, spermatogenesis supplies sperms. Details of oogenesis and spermatogenesis are described in Chap. 8.

The development of the new individual begins with **fertilization**. Fertilization is the fusion of two generative cells or gametes, the sperm and the egg cell (ovum), to form a **zygote**, the fertilized egg (Fig. 2.1). Sperm and egg cells are haploid; each contributes a single set of chromosomes, which is termed "paternal" in the sperm, and "maternal" in the egg cell. Gamete fusion provides the embryo with two complete genomes (the diploid set of chromosomes), a source of information sufficient to construct the new organism. In addition, fertilization induces the **activation** of the resting egg cell (Chap. 3). Only upon activation can the egg undergo embryogenesis.

Embryogenesis follows fertilization: within a protective envelope, and in **viviparous** organisms such as humans within the maternal body, numerous cells are produced from the zygote by continuous mitotic divisions. The resulting cells stay together, collectively modelling the basic architecture of the new organism, and provide it with all the essential organs necessary to begin an autonomous life when the young animal hatches out of the envelope or the mother gives birth. In the majority of animals, development occurs in the so called **indirect fashion**: embryogenesis results in a first phenotype, the **larva**. An example is the sea urchin, which can be taken as prototype of animal developments (Fig. 4.2). The larva itself undergoes development, which usually brings about only minor, unobtrusive changes, followed by a dramatic **metamorphosis** to a new phenotype, which is termed **imago** or **adult**, and which typically settles in an ecological niche different from that in which the larva lived. Larva and adult exploit different nutritive resources in their environments. Some groups of animals, among them the vertebrates, circumvent a larval state; in **direct development** the adult phenotype arises from the embryo step by step.

After a **juvenile** phase, life culminates when the adult reaches **sexual maturity**. **Senescence** finishes the phase of sexual maturity, and death ends the life of the **somatic cells** of the individual. Before death, its **generative cells** should have passed life on to a new generation.

W.A. Mueller et al., *Development and Reproduction in Humans and Animal Model Species*, DOI 10.1007/978-3-662-43784-1_2, © Springer-Verlag Berlin Heidelberg 2015

a   EGG MATURATION including MEIOSIS

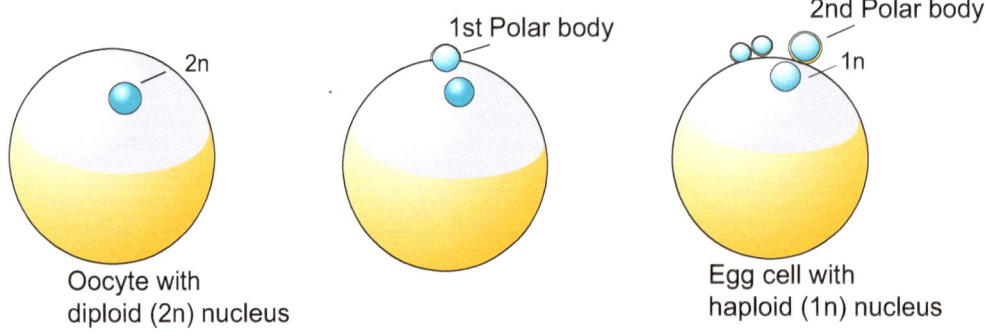

b   FERTILIZATION

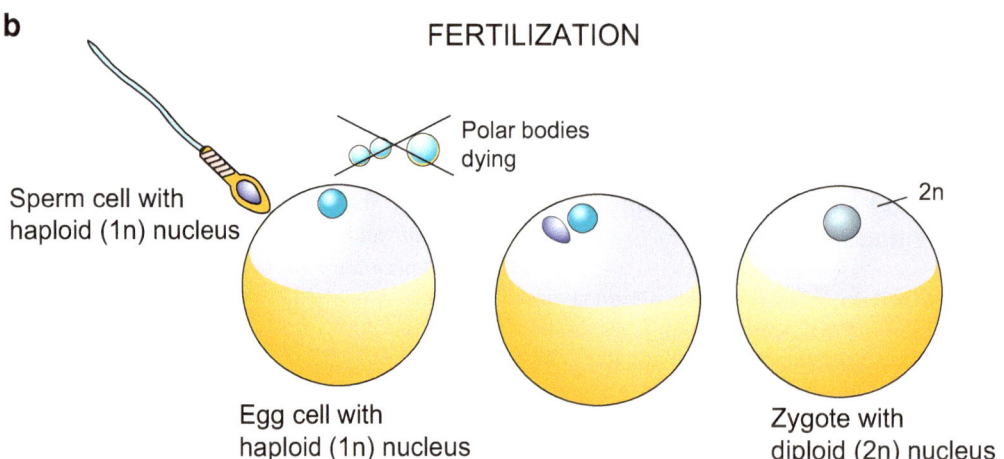

c   BILATERAL SYMMETRY
result of cytoplasmic segregation

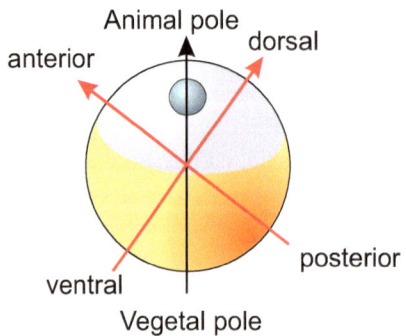

Fig. 2.1 General diagram showing egg maturation and meiosis, fertilization and establishment of the body axes, exemplified by the amphibian germ. Diagonally opposite to the point of the sperm entry into the egg the future back line will extend

2.1.2   The Egg Cell Contains, Besides the DNA of the Nucleus, Also Cytoplasmic Sources of Inherited Information; Structurally the Egg Becomes Polar, That is Asymmetrically Organized

Eggs contain yolk, a rich source of energy and molecular building materials. This is well known. Proposals to clone species which are about to become extinct, or even to clone prehistoric animals by removing nuclei from somatic cells of relics and implanting them into enucleated eggs of other, present day species show that knowledge about the various sources of inherited information is not very common. Although the DNA in the nucleus is the main source of information, it is not the only source, and not the only of significance. Ever since embryologists made surgical operations with eggs or early embryonic stages (e.g. T Boveri, EB Wilson, and EG Conklin; Chap. 1, Box 1.1), they became aware that eggs contain regulative determinants in their cytoplasm. In general terms these regulatory constituents are called **cytoplasmic determinants**. Pioneering work has been done with eggs of *Drosophila* (Sect. 4.6) and *Xenopus* (Sect. 5.1). Using refined techniques and the methodology of molecular biology these determinants are now being identified.

Essential constituents are mRNA's that code for proteins with regulatory function, collectively referred to as transcription factors. Such factors control gene activities. Since these mRNA's are produced by cells of the maternal organism and, therefore, are derived from genes of the mother they became known as **maternal RNA's**. The producing cells are the oocytes themselves or nursing cells in their neighbourhood: Their products (mRNA or proteins coded from them) are deposited in the egg in the course of oogenesis. Insofar as these exert regulatory functions one speaks also of **maternal information** or **maternal determinants**.

Effects of such stored cytoplasmic determinants become apparent when the Mendelian rule of reciprocity is invalidated and the genetic constitution of the mother is of

particular significance (e.g. hinny compared to mule, lefthand or righthand coiling in snails, Fig. 4.21).

One of the first identified cytoplasmic determinants is the mRNA of the gene *bicoid* in the egg of the fruitfly *Drosophila* (this maternal mRNA is indispensable but specific for *Drosophila melanogaster* and closely related flies, Sect. 4. 6). Several of such maternal factors have also been found in the eggs of the African clawed frog *Xenopus* (Sect. 5.1). According to present knowledge each animal egg harbours a collection of such maternal constituents, arranged in a three-dimensional pattern. As a rule, cytoplasmic determinants are not uniformly distributed in the egg but concentrated at defined sites. Experimentally working embryologists have to take the internal structure of the egg, the location of its cytoplasmic constituents and their possible species-specificity into consideration. In this introductory section we take a first look at the morphology of the animal egg without having an eye to ultramicroscopic details and without referring to biochemical data.

Animal-Vegetal Polarity. Even when the egg is spherical rather than elongated and elliptically shaped (the eggs of insects are elliptic, for example), in its internal structure an egg cell is always **anisotropic** or asymmetrical. In the terminology used in biology, an egg has a **polar** structure. At the very least, this polarity (anisotropy) is expressed in the location of the nucleus. Even in the immature **oocyte** (the diploid precursor cell) the nucleus is usually not centrally located but in the periphery near the surface of the cell. Because of its large size the nucleus of the oocyte is traditionally referred to as "**germinal vesicle**". In the course of the meiotic divisions which give rise to the mature egg, the germinal vesicle apparently disappears as it undergoes the two meiotic divisions. As a consequence of these divisions the **polar bodies** are formed at this peripheral location (Fig. 2.1). Polar bodies are miniature sister cells of the egg cell.

The location where the polar bodies are pinched off is usually shown as the 'North Pole' of the egg and is called the **animal pole**. The opposite 'South Pole' is named the **vegetative pole** or **vegetal pole**. Material is generally deposited at the vegetative pole which later in development is used in the formation of the primordial gut (**archenteron**), or is incorporated into the lumen of the gut.

In this context, the adjective '**animal**' refers to typical animal organs such as eyes or the central nervous system, which often are formed in the vicinity of the egg's animal pole. The adjective '**vegetal**' refers to the future 'vegetative' organs which derive from the primordial gut and serve 'lower' functions of life such as processing of food. The axis which extends from the 'North Pole' to the 'South Pole', and passes through the centre of the globe is termed the **animal-vegetal egg axis**.

However, when the site where the archenteron will be formed coincides with the location where the polar bodies are pinched off rather than being opposed to it, as for example in Cnidarians (Fig. 22.7), the traditional terminology often caused confusion and gave rise to erroneously oriented and labelled illustrations of eggs and embryos.

Bilateral Symmetry. Following fertilization the interior of the egg starts moving. Many constituents are rearranged and in the course of this **ooplasmatic segregation** the vast majority of animal eggs becomes organized in a bilaterally symmetrical way (Fig. 2.1c). In the amphibian egg, the prototype of a vertebrate egg, two polarity axes can be made out at this stage and one can distinguish between **anterior** or **cranial**, and **posterior** or **caudal**, and between the site of the future back (**dorsal**) and future belly (**ventral**). The egg shows **bilateral symmetry**. In insects the egg is already bilaterally symmetrical before fertilization.

Covers and Distribution of Yolk. Animal egg cells are surrounded by stabilizing and protective acellular envelopes. There are only a few exceptions; for instance the eggs of Cnidarians are covered only by an unobtrusive halo of glycoproteins. The innermost acellular sheet, which directly covers the surface of the egg cell, is made of glycoproteins and termed the **vitelline membrane** (this is not a phospholipid bilayer as is the cell membrane but a thin sheet composed of hydrophilic glycoproteins). In mammals the envelope of the egg is called the **zona pellucida**, in amphibians and fishes the **jelly covering**, and in insects the **chorion**. (The term **chorion** can refer to different, non-homologous structures. While in insects 'chorion' designates the acellular envelope of the egg; in reptiles, birds and mammals 'chorion' means a cellular 'extraembryonic' epithelial structure which is made by the embryo itself.)

Zoological textbooks like to list classifying terms, which inform readers familiar with Ancient Greek about the amount and distribution of yolk (end syllable *-lecithal*) in the egg. Prefixes are: *oligo* = few or little, *poly* = much or many, *iso* = uniform, *centro* = in the middle, *telo* = at the end, concentrated at one of the poles. Amount and distribution of the yolk affect the type and pattern of cleavage.

2.1.3 Cleavage Is a Series of Rapid Cell Divisions

Fertilization and activation of the egg are followed by **cleavage** (Figs. 2.2, 2.3, and 2.4). After fusion with the sperm, the fertilized egg, the diploid zygote, is still unicellular. Its task is to give rise to a multicellular organism that may comprise many millions of cells. There now follows a series of rapid cell divisions. At high speed the zygote is divided, without increase in volume and mass, into more and more cells which therefore are smaller and smaller. This stage of development is called **cleavage**. It is indicated by the appearance of furrows on the surface of the egg, for example on the surface of the yellow egg sphere of the chicken, indicative of proceeding cell divisions.

CLEAVAGE = Series of Cell Divisions

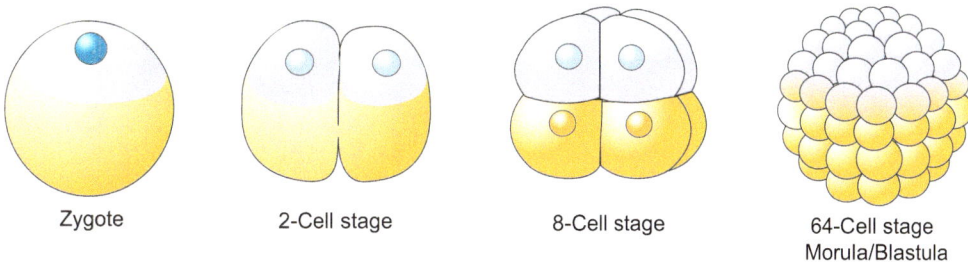

| Zygote | 2-Cell stage | 8-Cell stage | 64-Cell stage Morula/Blastula |

GASTRULATION: Formation of the Archenteron and "Germ Layers"

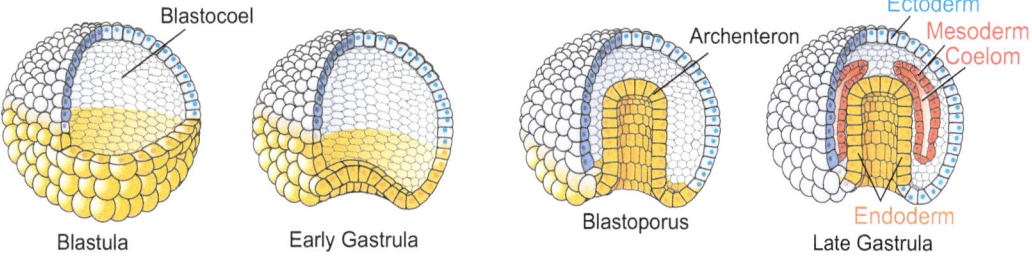

Blastocoel

Archenteron

Ectoderm
Mesoderm
Coelom

Blastoporus

Endoderm

| Blastula | Early Gastrula | | Late Gastrula |

Gastrulation: Formation of Mesoderm

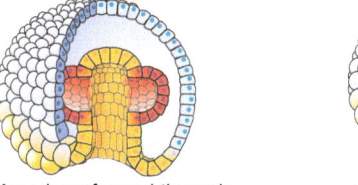

Mesoderm formed through evagination of vesicles (enterocoelia) from the archenteron e.g. in sea urchins, *Branchiostoma* and further Deuterostomia

Mesoderm arising from pairs of single, multiplying cells (primary mesoblasts) in annelids (e.g. *Platynereis*) and other spiraleans

Fig. 2.2 Schematic overview of early embryonic stages. The illustration basically follows the development of the Deuterostomia (sea urchin, amphibians) showing the primary mouth (blastopore) at the vegetal pole. The formation of the mesoderm can occur in different ways. The illustration shows two basic types

Cleavage can be

- **holoblastic** (=total); (Greek: *holos* = whole, *blastos* = germ-bud, seedling, embryo): the egg is completely subdivided into individual cells (Figs. 2.2 and 2.3a–c); or it can be
- **meroblastic** (=partial); (Greek: *meros* = part of), because of the huge amount of bulky yolk the egg is not divided in complete membrane-enclosed cells (Fig. 2.3d–f), at least not at the beginning of its development.

Whether or not holoblastic cleavage with regular cell divisions is possible depends on its spatial dimensions and its content of yolk. Where holoblastic cleavage occurs the first daughter cells are called **blastomeres**. Depending on whether the first daughter cells are equal or unequal in size, we refer to **equal** or **unequal cleavage**. Where unequal cleavage occurs a blastomere may give rise to large **macromeres** and small **micromeres**. Examples of meroblastic

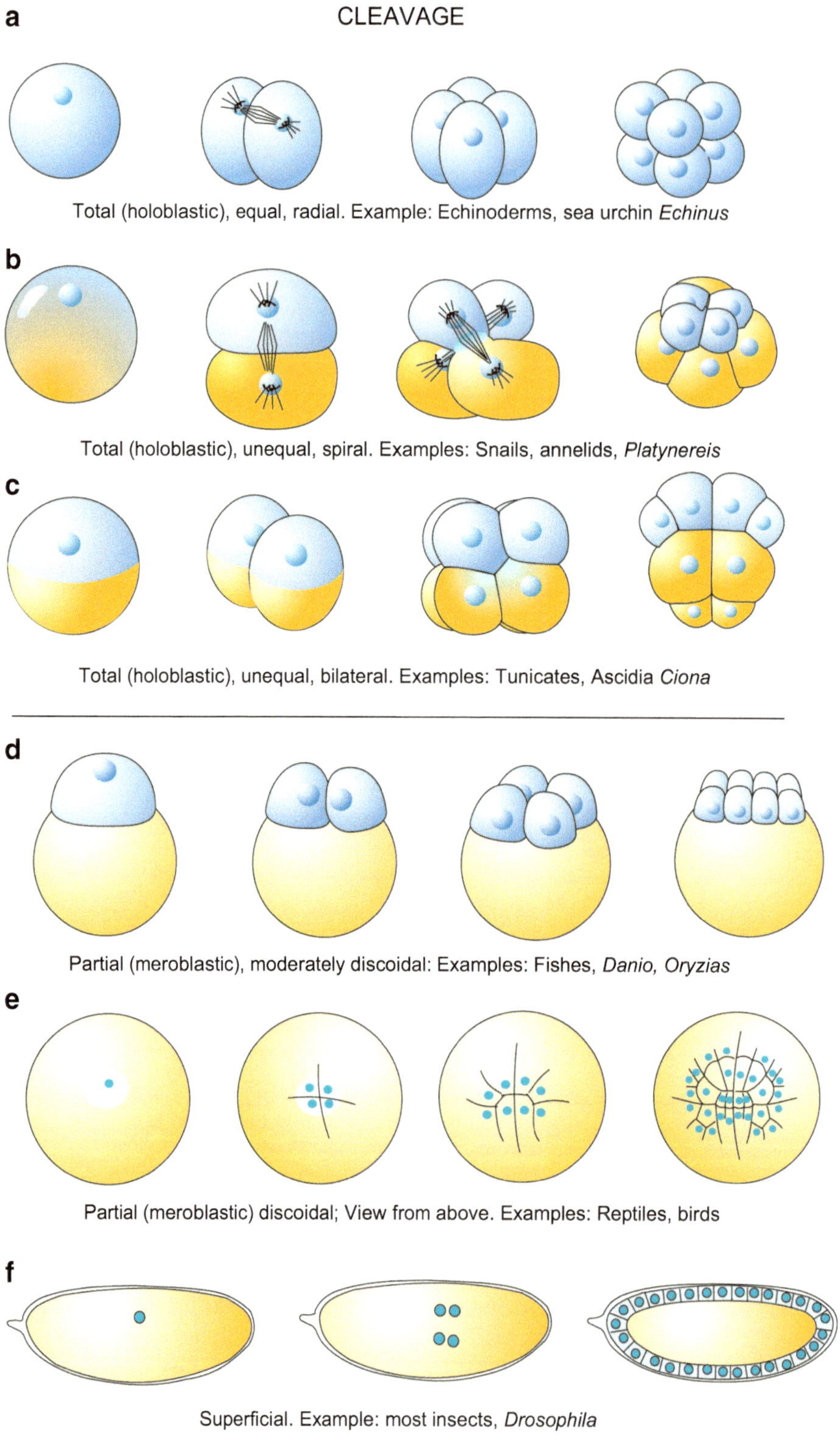

a   CLEAVAGE

Total (holoblastic), equal, radial. Example: Echinoderms, sea urchin *Echinus*

b

Total (holoblastic), unequal, spiral. Examples: Snails, annelids, *Platynereis*

c

Total (holoblastic), unequal, bilateral. Examples: Tunicates, Ascidia *Ciona*

d

Partial (meroblastic), moderately discoidal: Examples: Fishes, *Danio, Oryzias*

e

Partial (meroblastic) discoidal; View from above. Examples: Reptiles, birds

f

Superficial. Example: most insects, *Drosophila*

Fig. 2.3 Overview showing the most frequent modes of cleavage. (**a**) In radial cleavage the blastomeres come to lie in columns, one in top of the other. (**b**) In spiral cleavage (e.g. in annelids) in each division the daughter cells become displaced to the right or to the left at angles to the vertical axes caused by the oblique orientation of

Fig. 2.4 Types of cleavage according to the spatial pattern in which the first blastomeres become arranged. Holoblastic eggs usually contain little yolk and are entirely cleaved during cell division (cytokinesis). Radial and spiral cleavage patterns are most conveniently recognized looking down upon the top (animal pole) of the developing embryo. The bilateral and rotational cleavages are shown from a side view

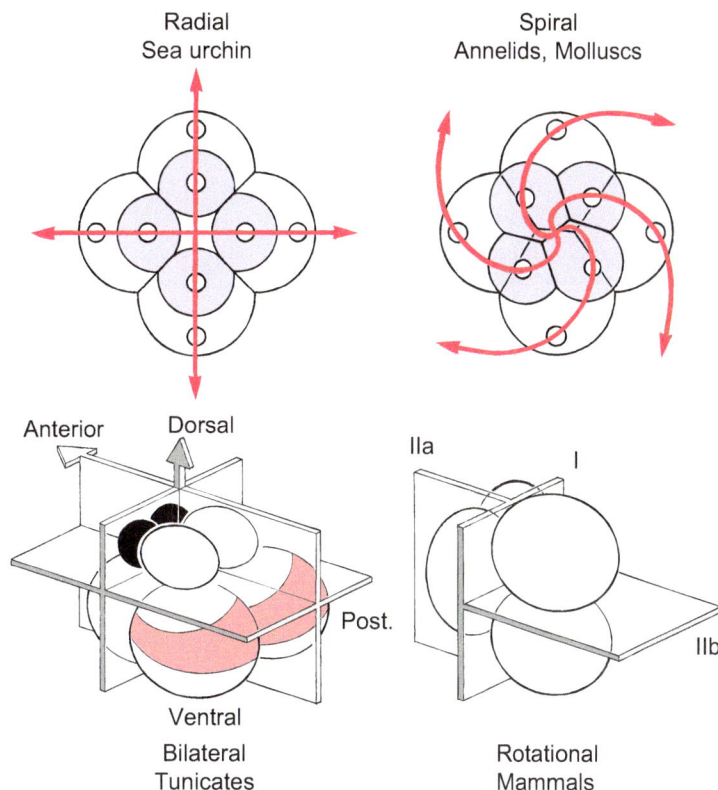

cleavage are the **superficial** cleavage in insects (Figs. 2.3f and 4.26), and the **discoidal** cleavage in fishes (Figs. 2.3d and 5.22), reptiles and birds (Figs. 2.3 and 5.22).

The time course of holoblastic cleavage and the spatial arrangement of the daughter cells often follow strict rules: directed by the centrosomes of the mitotic spindle, in each embryo of those species the blastomeres come to lie in a distinctive geometrical configuration. As a rule, the first two **cleavage planes** cut the egg along the animal-vegetal axis, starting at the animal and ending at the vegetal pole, the second cleavage plane being perpendicular to the

first (Figs. 2.2 and 2.3a). **Contractile rings**, consisting of actin and myosin filaments like those present in muscle cells, bisect the egg along the meridians. Accordingly, the first two cleavages are said to proceed **meridionally**. By this time the four-cell stage has been reached. The third cleavage, which gives rise to the 8-cell stage, proceeds in the equatorial plane or parallel to it and, therefore, is called **equatorial**.

Further development is not uniform among the various taxonomic groups of animals, but there are frequently observed basic patterns. The three dominant patterns are the **radial type**, **spiral type** and **bilaterial type** of cleavage (Fig. 2.4).

Fig. 2.3 (Continued) the mitotic spindle. (**c**) The bilateral cleavage (displayed for instance by the urochordates) produces a bilaterally symmetrical germ starting with the first division. (**d, e**) In eggs rich in dense yolk, as for example in those of birds, cleavage membranes initially separate the yolk mass incompletely. In the central area cells become entirely surrounded with a cell membrane first and thus complete cells occupy only the central area while in the periphery cells are still incomplete. The flat blastodisc lies on top of the yolk mass. (**f**) The superficial cleavage in *Drosophila* consists initially of division of nuclei only. Division of nuclei continues until 254 nuclei are present; these begin to migrate into the periphery to form the cellular blastoderm. Conventional cells are formed by wrapping cell membranes around the nucleus and a portion of cytoplasm

Termination of cleavage. While continuing cleavage, many animal germs arrive at a stage called **blastula**, a ciliated hollow sphere, the interior space of which is filled with fluid or liquefied yolk. The cellular, epithelial wall of the blastula is termed **blastoderm**, the interior space **blastocoel** or primary body cavity.

Although development into a blastula is typical, there are a number of different routes embryogenesis can take. The term blastula must not be confused with the similar term **blastocyst**. The mammalian blastocyst resembles the blastula of the sea urchin, but must not be equated with it, because the fate of the cellular wall of the cyst is completely different in sea urchins and mammals. Unlike the wall of a blastula the wall of the blastocyst does not participate in the formation of the embryo but gives rise to the placenta. Details are given in Sects. 5.4 ('Mouse') and 5.6 ('Humans'). Even though cell division will continue, cleavage is considered complete when the **blastula/blastocyst** stage is reached.

During cleavage the egg cell relies on maternal material including proteins stored in the egg. As time goes by this inherited material must be supplemented and eventually replaced by new material produced at the expense of precursor molecules of the yolk or, in mammals, produced by use of precursors supplied by the mother. Hence, genes of the embryo must be switched on. In amphibians this switch occurs at a time point before gastrulation starts and is known as **mid-blastula transition**, in general terms as **maternal-zygotic transition**. In most animal embryos there is no such fixed time point but a gradual transition that in mammals starts already with fertilization.

2.1.4 Gastrulation Prepares for the Building of Internal Organs

An animal needs internal tissues and organs. In the simplest animals, this may be a tubular space in which food can be enzymatically broken down and dissolved. Consequently, cells must be placed into the cavity of the blastula cyst, either singly or as a coherent sheet. The displacement into the interior of the blastula is termed **gastrulation**, and an embryo undergoing gastrulation is called a **gastrula**. The syllable *gastr* (Greek: *gaster* = belly, stomach) alludes to the future fate of the shifted cells: they give rise to the **primordial intestine**, the **archenteron** (**endoderm**), which will be largely responsible for forming the inner parts of the digestive tract and associated organs (such as the liver in vertebrates).

Gastrulation is a phase of dramatic events. Rapid RNA transcription retrieves large amounts of genetic information, marking the dawn of individuality at the molecular level. Gastrulation commonly is the point of no return for embryonic determination. Henceforth, cells move irreversibly down their developmental pathways. Actual cell movements make the dynamics visible to the eye of the inquisitive researcher.

The means by which cells are placed in the interior is quite different among various animals. Commonly, five basic modes of active movement or passive displacement are distinguished (Fig. 2.5):

- **Invagination**: infolding, inward buckling of a cell sheet into the embryo through active twisting (shared generation of bending moments by the cell community);
- **Involution**: inturning, creeping of a coherent cell sheet around an edge, typically the lips of a primordial mouth (blastopore);
- **Epiboly**: overgrowing, spreading of a sheet of cells over other, bulky cells or an uncleaved yolk mass that, thus, come to be internalized;
- **Delamination**: detachment of a sheet of cells from the outer blastoderm following a phase of asymmetric cell divisions;
- **Immigration or ingression**: ameboid movements of cells; cells immigrate singly and not as a coherent sheet as in involution;
- **Polar proliferation**: cell divisions taking place at one of the poles; daughter cells are released into the cavity of the embryo.

In most animal embryos, gastrulation occurs as a combination of several of these basic modes.

Germ-layer formation is another, traditional expression for this and the following embryonic stage. Gastrulation can be defined as a process by

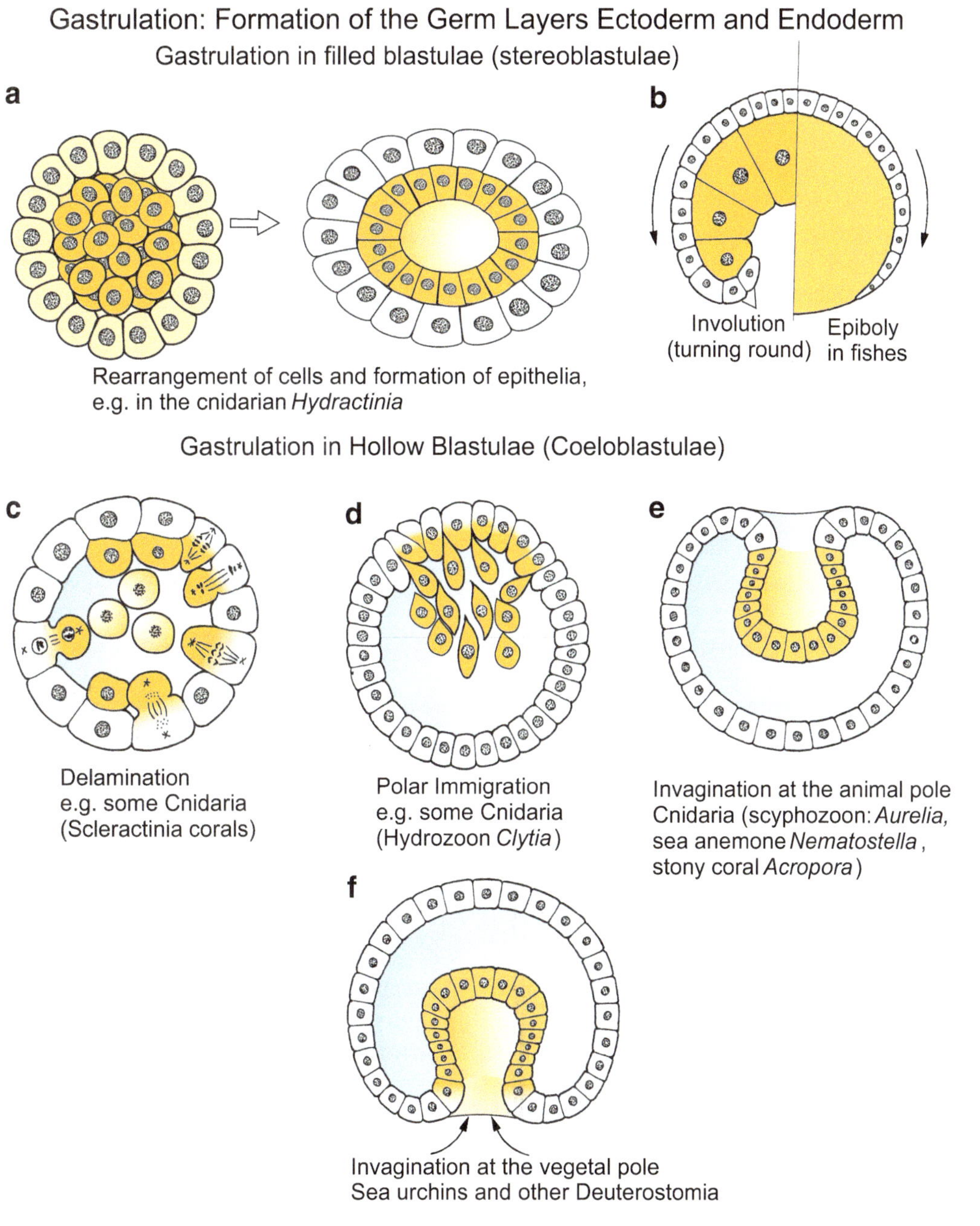

Gastrulation: Formation of the Germ Layers Ectoderm and Endoderm
Gastrulation in filled blastulae (stereoblastulae)

a

Rearrangement of cells and formation of epithelia,
e.g. in the cnidarian *Hydractinia*

b

Involution   Epiboly
(turning round) in fishes

Gastrulation in Hollow Blastulae (Coeloblastulae)

c

Delamination
e.g. some Cnidaria
(Scleractinia corals)

d

Polar Immigration
e.g. some Cnidaria
(Hydrozoon *Clytia*)

e

Invagination at the animal pole
Cnidaria (scyphozoon: *Aurelia,*
sea anemone *Nematostella* ,
stony coral *Acropora*)

f

Invagination at the vegetal pole
Sea urchins and other Deuterostomia

Fig. 2.5 Patterns of gastrulation, a process that brings cells into an interior cavity where they give rise to the archenteron and other internal parts of the embryo. (**a**) In embryos without a primary cavity a blastocoel can be formed through rearrangement of the cells. (**b**) In epiboly an epithelial superficial layer expands and spreads over other cells or the still uncleaved remainder of the egg cell, forming an envelope. In involution an outer layer of cells bend as a coherent sheet to cover yolky cells that come to lie interiorly. (**c**) In delamination cells are displaced into the interior cavity directed by the orientation of the mitotic spindle. (**d**) In immigration or ingression, cells move actively like amoebae (a case of polar immigration is illustrated). (**e**, **f**) In invagination, a sheet of cells buckles inward by actively generating bending moments at curves and edges. Expansion of the outer layer may be brought about by crawling movements supported by cell division and changes in cell adhesivities

which the embryo acquires two or more germ layers, thus becoming prepared to form internal organs. The expression "germ layers" comes from historic descriptions of the embryogenesis of the chick (e.g. by CF Wolff and CE von Baer, Chap. 1, Box 1.1). In the avian embryo, all three "germ leaves", **ectoderm** (in the chick and mammalian embryo called epiblast), **mesoderm** (mesoblast) and **endoderm** (hypoblast), are temporarily arranged in stratified layers one upon another (Fig. 5.25).

Animal phyla that early in evolution diverged from the main route of animal phylogeny, namely the Cnidaria and Ctenophora (formerly united as Coelenterates), manage with only two cell layers: an ectoderm forming the outer **epidermis**, and an endoderm forming the inner **gastrodermis**. They are known as **diploblastic**. In all animal organisms above the organizational level of the Cnidaria, gastrulation not only prepares the space for digestion and coats it with endoderm, but a further sheet is inserted between the gut rudiment (endoderm) and the outer wall (ectoderm) of the gastrula: the **mesoderm**. All these phyla are said to be **triploblastic** (but see Chap. 22, Box 22.1).

The **mesoderm** is an extremely versatile layer. As a rule, it will form muscles; connective tissue; blood vessels; the epithelial linings of the interior cavities, nephric, nephridial and other organs engaged in excretion and osmoregulation; and skeletal elements, provided the genetic program of the respective species includes such faculties.

The mesoderm is prepared and formed in many different ways (details in Chaps. 4, 4 and 6). In the Spiraleans, for example, it as a single, early separated cell (4d-cell) that takes the task to form the mesoderm; it gives rise to a pair of primordial mesoblasts (mesoteloblasts, Fig. 2.2) that proliferate to generate the segmental pairs of mesodermal packages (see Fig. 4.22). In the case of a multiphasic gastrulation, such as takes place in sea urchins (Fig. 2.2), cellular material is brought into the blastula cavity in several batches. In another mode of gastrulation, seen in amphibians, first the blastula wall buckles and displaces an archenteron consisting of mes-

endoderm into the interior cavity. This archenteron subsequently divides into the endoderm of the stomach and gut tube, and a mesoderm.

The remaining epithelial outer wall of the gastrula, the **ectoderm**, is used to form the epidermis, the outer layer of the skin, but may also give rise to some sensory organs and parts of the nervous system. Consequently, the dorsal part of the ectoderm in amphibians or the ventral part in insects is often termed **neuroectoderm**, because it gives rise to the nervous system and sensory organs (such as inner ear in vertebrates) (see Figs. 5.8, 16.2, and 16.16), before it becomes the epidermis of the skin.

It should be emphasized that germ layers are topographically defined structures. They do not necessarily have exactly the same fate or form the same tissue types in different species.

The commendable attempt of linguistically educated authors of developmental treatises to replace the often improper term "derm" (skin) by the term "blast" (building material) has not yet been consistently adopted. Nonetheless the term mesoblast is better than mesoderm.

2.1.5   The Formation of Organs and the Differentiation of Tissues Enable Autonomous Life

With gastrulation, the development of organs (**organogenesis**) is initiated. In vertebrates the first phase of organogenesis, which eventually leads to the formation of the central nervous system, is termed **neurulation**. Further stages cannot be subsumed under common, comprehensive terms because the results of organogenesis are species-specific. To achieve the final organization of different species, varied pathways must be taken.

Frequently, embryonic development results in a first 'edition' or phenotype of the animal, the larva. Larvae of marine invertebrates are, as a rule, only sparsely equipped with organs. The lack of organs can be attributed to the small size of these larvae. A small volume enables diffusion of oxygen into the body and leak of waste products such as ammonia out of it without the need of highly evolved organs for respiration, circulation and excretion. Evidence indicates that taking the long way through larval stages may also reflect ancient evolutionary paths (Chap. 22).

Thus, frequently embryonic development is completed with the hatching of the larva out of the egg envelope. The larva not only displays a form different than the adult, but has also acquired particular structural and physiological specialties that allow the larva to settle in another ecological niche - for instance, to make use of particular food supplies not accessible to the adult. Development that includes a larval stage is designated "**indirect**". Larvae are successful entities per se. But metamorphosis and the specific complex organogenesis associated with it enable it to acquire new biotopes that demand new organs (such as wings).

Some animal groups, including the true land vertebrates, omit a larval stage; in the course of **direct development** the adult form arises from the embryo gradually.

2.2   General Principles in Short

2.2.1 Though Developmental Pathways Are Diverse; They Display Common Basic and Recurring Processes

In spite of the great diversity of animal development, there are some basic and recurring events. The development of a multicellular animal organism consists of:

1. **Cell proliferation** (recurring cell divisions)
2. **Cell differentiation**, that is the development of functionally different cell types from uncommitted precursors. This becoming different must occur in a defined spatial order and therefore proceeds along with
3. **Pattern formation**: the various cell types are not chaotically arranged but occur in species-specific spatial patterns.

In animal development, unlike that of plants, cell differentiation often gives rise to

4. **Cell movements that lead to new shapes, and to cell migration**.

A close look will reveal that animal development also implies

5. **Programmed cell death (apoptosis,** Fig. 14.9). It is of significance, for example, in the breakdown of the hand paddles into fingers (Fig. 14.8).

All these events together lead to

6. **Morphogenesis**, the modelling of form.

Cell divisions must be precisely controlled as the organism's requirement for various cell types is different. Control of proliferation, therefore, is closely connected with control of differentiation.

Cell differentiation has a twofold meaning (Fig. 2.6). The term may mean (1) cells becoming different from each other, that is the divergence of developmental pathways taken by the various cell types; or (2) cells becoming different in time: each cell takes a cell-type-specific pathway until it reaches maturity and becomes **terminally differentiated**. The development of each cell begins with a **pluripotent founder cell** at the outset, the fertilized egg - which is capable of dividing and gives rise to several cell types. An act of decision, called **determination** or **commitment**, directs the developmental path of the descendants to different goals.

A cell that has become committed (determined, programmed) to follow a distinct developmental route is often designated by the suffix -**blast** (for example neuroblast, erythroblast, lymphoblast). Such a committed cell may retain the ability to divide. If so, it will be the founder cell of a cell clone whose members are all destined to form one and the same cell type. After some rounds of cell division, the derivatives of the founder cell proceed to undergo terminal differentiation. They acquire their specific molecular equipment, specific form, and function. In so doing, the cells will, as a rule, lose their ability to divide. The suffix -*cyte* indicates that this has happened.

Differentiation as Divergence of Developmental Routes

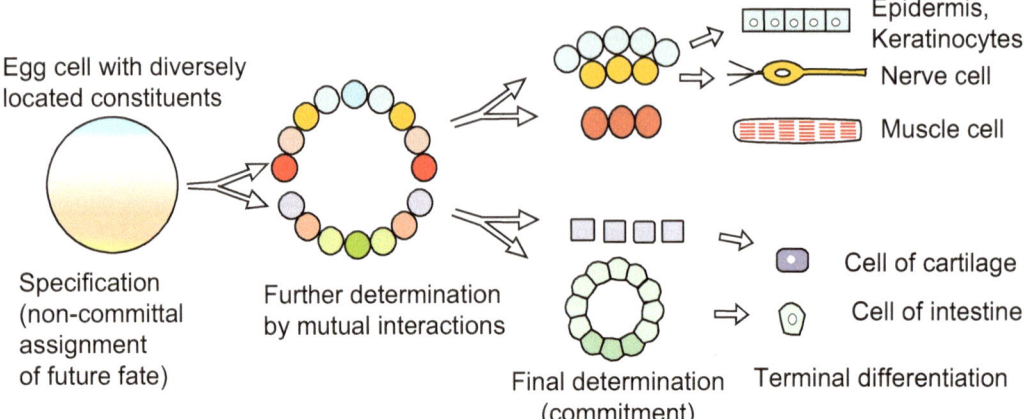

Differentiation as Individual Cell Development

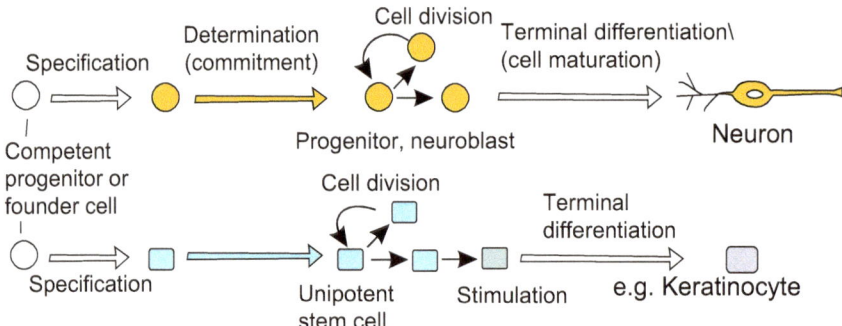

Fig. 2.6 Meaning and significance of the term differentiation for cell communities and for single cells. Divergence of the developmental paths in neighbouring cells is a prerequisite for cell differentiation in the sense of cell maturation. The coordination is accomplished through the mutual exchange of signals or on the basis of self-sufficient molecular clocks set going by cytoplasmic constituents

Determination proceeds along with **pattern formation**. Pattern is the ordered spatial configuration in which the various cell types and supra-cellular structures (tissues, organs) come to appear. Pattern formation in associations of cells is often based on **position-dependent assignment of specific duties**, i.e. on position-dependent **determination**. However, patterns can also result from active migration of already committed and differentiated cells to defined locations.

In the experimental analysis of developmental processes one starts from working hypotheses. Such hypotheses have to take into account the established fact that cell differentiation is based on **differential gene expression** ("differential gene activity", Chap. 12). Determination is viewed as a kind of programming, by which it is decided which genetic information can be recalled in the future and which set of genes will be blocked. However, not only genes of the embryo, referred to as **zygotic genes**, are of significance.

Fig. 2.7 Asymmetric mitotic cell division that separates motile mesenchymal cells from stationary epithelial cells. After Woolner & Papalopulu 2012 and Papalopulu & Woolner, oral presentation at a meeting

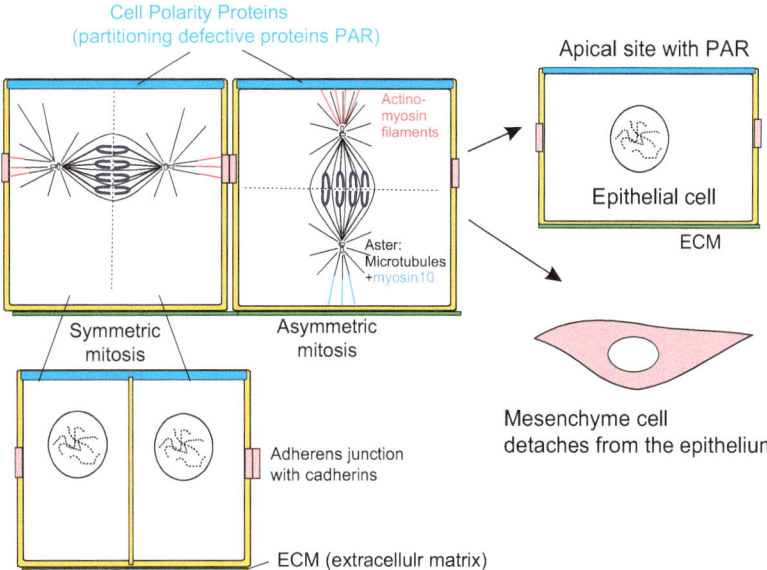

As stated above, programming or allotment of tasks may arise early in development from **cytoplasmic determinants**, which are deposited in the egg in a defined spatial pattern. **Maternal effect genes** provide maternal information, which in the course of **oogenesis** (development of the ovum in the ovary) becomes deposited in the egg cell as cytoplasmic determinants (details are given in Chap. 9). Maternal effects make oogenesis an important factor in subsequent development.

Autonomous or Mosaic Development and Asymmetric Cell Division. Traditional hypotheses proposed that a mosaic of determinants in the egg would enable cells that inherited a particular determinant to proceed in its developmental pathway independently of other cells. This is shown in Chap. 9. **Asymmetric cell division** can allocate such determinants differently to the two daughter cells and direct the fate of the daughter cells and their descendants in different roads. An example is the decision that separates mobile, amoeboid mesenchyme cells (precursors of connective tissues) from stationary epithelial cells (Fig. 2.7). A further example is the increase

of nerve cell precursors in the growing brain of vertebrates including man (Chap. 16, Fig. 16.7).

The programming of the cells can also be based on mutual consent. Such interactions are the topic of Chap. 10. **Cells communicate** with each other via signals and make agreements. Such interactions make the development of various cell types dependent on their neighbourhood, but also allow corrective regulation in case of disturbance. This aspect is expressed by the terms "**dependent**" and "**regulative development**".

A diagram of the events that start with maternal factors in the egg and proceed through mutual interaction in the embryo to an increasingly complex subdivision of the egg into different territories with different fates is shown in several Figs. of Chap. 9. Generally, these courses of events are summarized under the terms **positional information** and **pattern formation** and will be summarized in Chaps. 9 and 10.

The spectrum of meanings of the terms determination, specification and commitment is explained in Box 2.1.

Fig. 2.8 Sequence of determination in a generalized animal embryo. First, position-dependent specification of the future fate occurs through cytoplasmic determinants which in the course of an egg-internal pattern formation (ooplasmatic segregation) become rearranged by the cytoskeleton and molecular engines and differentially allocated among the various blastomeres. Later pattern formation and determination are based on the mutual exchange of signals, partly via cell adhesion and recognition molecules exposed on the surface, partly via gap junctions, partly via diffusion in the intersticies

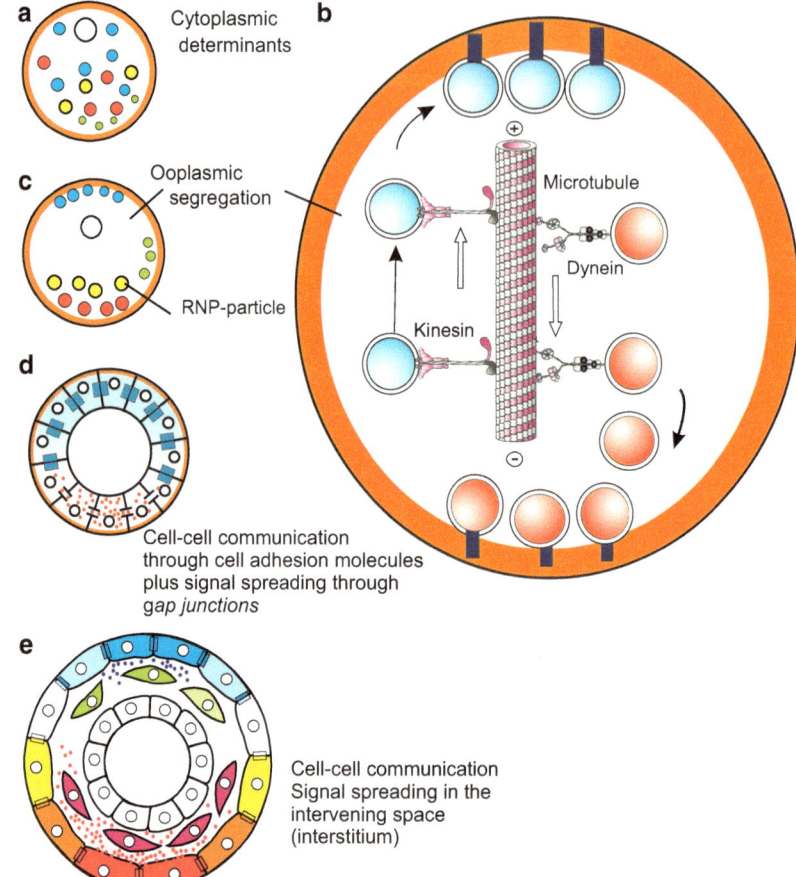

Cytoplasmic determinants

Ooplasmic segregation

RNP-particle

Microtubule

Dynein

Kinesin

Cell-cell communication through cell adhesion molecules plus signal spreading through *gap junctions*

Cell-cell communication Signal spreading in the intervening space (interstitium)

Summary

The chapter introduces fundamental terms and points to fundamental principles. Animal development proceeds by stages:

1. **Gametogenesis**, the generation of egg cells (**oogenesis**) or of sperm (**spermatogenesis**);
2. **Fertilization**, the entry of the sperm cell into the egg cell, followed by the union of the two cells' nuclei;
3. **Embryogenesis** (embryonic development) with
   - **Cleavage**, a series of quick, successive cell divisions,
   - **Gastrulation**, a process during which cells are displaced into the interior of the embryo to form inner organs, and the "germ layers" (ectoderm, mesoderm, endoderm) are segregated from each other;
   - **Organogenesis** and terminal **cellular differentiation**, the elaboration of the organs and the final development of the diverse cell types.

   Frequently embryogenesis ends in a larva which undergoes
4. **Metamorphosis** to an imago or adult.
5. **Sexual development** and **maturation** finishes the period in life in which new organs are formed and brought to operational readiness.

Further important terms of morphological embryology are:
   - Egg cell: animal and vegetal pole, "germinal vesicle"= the large diploid nucleus of the oocyte, polar bodies =

miniature, abortive sister cells of the haploid egg cell
- Embryo:
  - 2, 4, 8, 16, 32-... cell stage
  - Blastula with blastoderm and blastocoel
  - Gastrula with the "germ layers" ectoderm, mesoderm, endoderm
- Modes of cleavage:
  - Holoblastic (total): radial, spiral, bilateral cleavage
  - Meroblastic (partial): superficial, discoidal cleavage

- Modes of gastrulation: invagination, involution, epiboly, delamination, immigration, polar proliferation

General principles of animal development are:
1. Cell division
2. Pattern formation and determination (commitment)
3. Cell differentiation
4. Cell movements
5. Programmed cell death (apoptosis)
   All these events together lead to
6. Morphogenesis, the modelling of the body's shape

Box 2.1: On the Terms Determination, Specification, Commitment

In terms of classic experimental developmental biology, the processes through which cells become assigned different tasks and programmed to enter a distinct developmental pathway have been subsumed under the comprehensive term **determination**. About two to three decades ago the term **commitment** came into fashion and began to replace the classical term, determination. **Specification** (a term initially used by Hans Driesch – see Chap.1, Box 1.1 – and made popular by Lewis Wolpert) is another term meaning assignment of a particular task without, however, implying a definitive and lasting programming.

Many scientific terms come into use without having ever been defined in a generally committing form. This applies also to the three terms quoted above, although attempts were made to formulate definitions afterwards and to propose rules for their use. Unlike the names of species, other biological terms are not laid down with strict rules for their usage, as no committee exists endowed with the authority to prescribe imperative binding rules. In reading articles one has to sense and find out the actual meaning of a particular term from the context.

Specification and "**to specify**" can often be read in the context of pattern formation in areas of cells. In this context the term

specification points the way to the future which, however, is not yet irreversibly fixed.

Commitment means settling, laying down of a fate, as a rule in an (almost) irreversible manner, just like the classical term determination, but the term commitment is mostly used when speaking of a distinct type of cell.

Determination has been defined as a decision leading the way to a particular developmental path and remains a useful comprehensive term if specification is understood as initiation of a developmental program and commitment as definitive sealing of the fate. Moreover, in terms of classical embryology determination means programming (specification plus commitment) not only of certain cell types but also of tissues, organs or body parts, e.g. back versus belly, head versus trunk.

Mode and State of Determination Must Be Checked Experimentally. The observation that in cleavage some blastomeres get certain cytoplasmic constituents while others do not, does not per se provide reliable information about the mode of determination. The eggs of frogs of the genus *Rana* harbour black pigment granules in the peripheral condensed cytoplasm below the cell membrane known as cortex. These black granules are confined to the animal half of the germ and, ▶

Box 2.1 (continued)

therefore, get allocated only to animal blastomeres. But these granules are not of significance for processes of determination as albino eggs develop normally.

Whether or not components with fate-determining properties are present must be checked experimentally. This can be done before cleavage by shifting components of the egg to other sites, using pressure or centrifugation. Alternatively plasma from the region in question is sucked out with a micropipette and transferred to another region (Figs. 4.30 and 4.32).

The state of determination is tested with the following procedures:

Isolation Test. Embryonic cells or cell associations can be cultivated in media of suitable composition. Factors that permit or block differentiation are added to the medium. If the cells always give rise to the same terminally differentiated cell type, with or without the addition of the permissive or blocking factors, they are considered to be determined. One can breed myoblasts (C2/C12-cells) over years in amounts of kilograms in foetal calf serum which contains various growth factors. The syllable **myo-**points to muscle cells, the terminal syllable-**blast** to committed progenitor cells that still are able to proliferate. They always differentiate to muscle cells when the calf serum is removed. This example also gives evidence that in proliferating cell cultures the state of determination can faithfully be transmitted from cell generation to cell generation. One speaks of **cell heredity**.

Terminological Reference. Following Jonathan Slack (2001; p. 64) "a cell or tissue explant" should be "said to be **specified** to become a particular structure if it will develop autonomously into that structure after isolation from the embryo" and "The specification of a region need not to be the same as its **fate**

in normal development" because after isolation directing influences from the normal environment are cut off.

Transplantation Test. With the microsurgical techniques developed by Hans Spemann, cells or cell communities or larger parts of the embryo are moved to any other region of the same or a different embryo. **Donor** cells can be grafted into **host** embryos (**receivers**) of the same species (**autologous** or **homoplastic** or **orthotopic transplantation**). Alternatively, transplants can develop also in embryos of other species (**heterologous** or **heteroplastic** or **heterotopic** transplantations), as a rule, without causing immunological complications. Thus, successful transplantations have been performed between frogs and newts, between chick and quail.

The operational deduction of the state of determination is as follows: If transplants are **irreversibly determined** they develop in **accordance with their old position** autonomously, without being deviated by local influences exerted by their new environment. If a transplant **is not yet (irreversibly) determined** it develops **according to its new position** though it may previously have experienced specification in the sense of the above definition. Thus, an ectodermal piece taken from a frog gastrula that would have formed skin of the belly in its place of origin will form teeth if transplanted into the mouth region of a newt. Of course, these are horny teeth according to the genetic blueprint of the frog cells (Fig. 5.12) and not bony teeth characteristic of newts. The fitting into the new environment is also taken as evidence for the work of **positional information** (Chap. 10 and Box 10.2).

The results of isolation and transplantation tests are not always in agreement because a still unstable determination (specification) can be retuned in a new environment. ▶

Box 2.1 (continued)

According to the definition proposed by Jonathan Slack determination means stable programming and progression along a developmental path irrespective of any deviating influences from the neighbourhood.

Test of Developmental Potencies by Transfer of Nuclei. As a rule, with determination to a particular fate the ability to develop other cell types gets lost. **The potencies become restricted.** Whether or not this restriction is associated with an irreversible change of the genetic program can be tested by **transplantation of nuclei** (Chap. 13). Famous experiments first performed by R. Briggs, T. J. King, and J.B. Gurdon in the clawed frog *Xenopus*: Nuclei taken from differentiated somatic cells, for instance from cells of the intestine of tadpoles or from lung cells of adult frogs and transferred into enucleated eggs, enabled complete development in a considerable number of cases. There arose clawed frogs (Fig. 13.2). The nuclei of the successful donor cells were still **totipotent** (omnipotent). Determination did not lead to an irreversible loss of genetic potencies. Provided all the nuclei are taken from one and the same donor the thus generated descendants are genetically identical with the donor and with their siblings; together they constitute a **clone**.

Determination is similar to the choice of a particular education and profession after school. **With the decision for a particular job, with the adoption of a specific task, alternative possibilities become more and more restricted and eventually even entirely excluded.** Rare exceptions are found under the key words **transdetermination** and **transdifferentiation** (Chap. 18, Regeneration).

Appendix: Some Rules for Correct Notation

The scientific name of an organism is written in scientific biology with *italics*, for instance *Drosophila melanogaster*, the fruit fly. Also the designation of genes, and of the mRNA derived from them, is written in *italics*. The proteins derived from these genes are written in standard letters, increasingly frequently in capital letters (e.g. BICOID), as it is done in this textbook. More about rules for correct nomenclature can be read in Box 12.1.

The Start: Fertilization, Activation of the Egg and a First Series of Cell Divisions (Cleavage)

3

3.1 Fertilization

3.1.1 When Does Life Begin?

Beginnings are often difficult to pinpoint, particularly in life cycles, for life is continuous. Life does not tolerate a break (at most a transitory standstill in a stage of quiescence). Life is a continuing process that started some billions of years ago below the level of multicellularity and will end with the death of the last living being. Nevertheless, each individual life that is based on sexual reproduction has two discrete boundaries: fertilization and death.

Sperm cells and egg cells are not capable of starting an independent life on their own. Once released from the testes or ovary, the lifetime of sperm cells and egg cells is restricted to a few minutes, hours or days (in humans: egg cell 12 h, sperm up to a maximum of 3 days). Only the fusion of sperm and egg results in a cell with the potential to survive and to give rise to an independent individual as well as to a new generation. In biological terms a human is *Homo sapiens* from fertilization until death. No stage of development can be omitted, no non-human stage defined, at least not in biological terms.

3.1.2 Are Insemination and Fertilization the Same?

Maternal and paternal genomes are brought together in the process of fertilization.

Fertilization is preceded by insemination, but these two events are often not so clearly distinguished as some terminological purists want. According to their definition, **insemination** leads to the encounter of sperm cell and egg cell and culminates in the fusion of the sperm cell membrane with the egg cell membrane; **fertilization** is the fusion of the two haploid genomes to form the diploid genome of the zygote. A distinction between insemination and fertilization is meaningful and necessary in those organisms where the sperm enters an egg cell before the egg nucleus has completed meiosis. This is the case in mammals as well as in the nematode *Ascaris*. In such cases the haploid sperm nucleus has to wait until the egg nucleus is also haploid, before fertilization proper can take place. Usually, however, all processes starting with the first contact of sperm cell and egg cell and ending with the fusion of the two nuclei, are grouped under the heading 'fertilization'.

3.1.3 The Egg Cell Attracts Sperm; Attractants Are Manifold and Their Reception Is Based on Mechanisms Known from Photoreceptors or from Smell and Taste Receptors

The best investigated process of fertilization is that of the sea urchin. Fertilization takes place in the free water column (external fertilization). The freshly spawned egg emits an attracting

W.A. Mueller et al., *Development and Reproduction in Humans and Animal Model Species*, DOI 10.1007/978-3-662-43784-1_3, © Springer-Verlag Berlin Heidelberg 2015

pheromone in order to guide the sperm to its goal. Pheromones are inter-specific scents which are, as a rule, emitted to prepare mating or to distribute warning signals. In their function as attractants for sperm pheromones are also known as **gamones**. In the case of sea urchins the gamones are peptides. The flagellum of the sperm is densely equipped with molecular receptors for peptides. Activation of the receptor by bound scents causes a hyper-polarization of the electrical membrane potential, that is an increase in the voltage, as it is known from the photoreceptors (cones and rods) in the retina of our eyes. As mediator between the molecular receptors and the ion channels in the membrane of the flagellum the cGMP-dependent signal cascade is interposed.

Mammalian Sperm. Even more startling are recent findings about the chemotaxis of the mammalian sperm. It is not so surprising that the chemotactic signals are complex and highly efficient. In experiments sperm cells can be attracted by minute traces (picomol $= 10^{-12}$ M) of the steroid hormone **progesterone**. Progesterone is released from the cumulus cells which derive from the follicle and envelope the egg after it has been ejected from the ovary. Progesterone opens ion channels and this stimulates sperm to increase swimming activity. Not only that. Other emitted chemical signals appear to contribute to the attractive features of the egg, for instance prostaglandins and the gaseous nitrogen oxide NO. Most amazing of all, mammalian sperm are equipped with receptors that bind and respond to a substance called **bourgeonal**, an aromatic aldehyde known as scent emitted by the lily of valley. The substance is effective in extremely low concentrations. However, so far bourgeonal could not be detected in the female genital tract. What is the original odorant adapted to this receptor?

Since sperm swim in a watery fluid, not only typical odorants must be taken in consideration but also typical taste substances. Surprisingly receptors for the **taste quality umami** have been recently identified in the membrane of the sperm flagellum.

The swimming behaviour of the sperm is mediated by the sperm-specific **CatSper** channel that controls the intracellular Ca^{2+} concentration. In humans, CatSper is directly activated by progesterone and prostaglandins – female factors that stimulate Ca^{2+} influx. Other factors including neurotransmitters, chemokines, and odorants also affect sperm function by changing $[Ca^{2+}]i$. It appears that the CatSper channel complex serves as a polymodal sensor for multiple chemical cues that assist sperm during their voyage across the female genital tract.

Besides gradients of chemical signals a gradient of temperature along the oviduct is also said to provide guidance for the sperm.

3.1.4 The Sperm Acquires Qualification for Fertilization Only When It Has Undergone a Process of Priming, Called Capacitation

Irrespectively of how sperm is guided up the oviduct it has to swim along this route to become capable of fertilization. It is primed on its way through the uterus and oviduct by substances secreted by the female. This effect is called **capacitation** of the sperm.

In most mammalian species capacitation serves an additional purpose. For some time the sperm cells get caught by the viscous mucus secreted by the mucosa of the oviduct, and are stored there. Only when ovulation takes place modified products of the oviduct allow swimming to re-commence. Sperm now become hyperactive and escape the obstructive mucus, and they do this just when an egg capable of being fertilized has arrived in the oviduct. As stated above in Sect. 3.1.3, chemo-attractants emitted by the egg itself trigger speeding up of the swimming activity of the sperm.

Capacitation is also required in invertebrates and is, for example, in sea urchins accomplished by the slightly alkaline (pH 8.2) sea water.

3.1.5   The Acrosome, a Chemical Drill, Enables Sperm to Penetrate the Egg Envelopes

The contact of the head of the sperm with components (polysaccharides) of the egg envelope induces the **acrosomal reaction**. The acrosomal vesicle in the head of the sperm cell (Figs. 3.1, 3.2, and 3.3) opens and releases a collection of hydrolytic enzymes, among them many proteases, bundled in a large protein complex known as the proteasome. In mammals the collection of hydrolases includes glycosidases and hyaluronidases. The sperm possesses a chemical drill head and enzymatically lyses a path through the jelly layer and vitelline membrane, driven by its flagellum. In the sea urchin sperm the drill head elongates to a long boring-rod or finger: from the bottom of the opened acrosomal vesicle a finger-like structure, the **acrosomal filament**, projects through the enzymatically bored channel penetrating the jelly envelope. The stretching out is accomplished through a rapid polymerization of globular G actin molecules to F actin filaments in the evaginating acrosomal finger (Fig. 3.2). In mammals the zona pellucida is locally dissolved (Fig. 3.3).

3.1.6   Species-Specific Receptors of the Egg Envelope Evaluate the Captured Sperm

Sea urchins. In the free seawater of natural marine environments sperm of several species of sea urchin may contact an egg. It can happen, and is proofed in the laboratory, that sperm of foreign species may attach to the jelly coat and drill their way through the outer envelope of the egg. The jelly coat contains components which trigger the acrosome reaction in sperm from various species. As a final control, sperm and egg check each other out by displaying their identity card: on the outer surface of the acrosomal filament **bindin** molecules serve as distinguishing signs. Bindins from different species differ in

structural details. If the correct bindin is present on the surface of the sperm it will fit onto corresponding **bindin receptors** in the egg membrane (the extracellular domain of this receptor probably is integrated into the vitelline membrane which encloses the egg cell). This sperm or bindin receptor exists as a disulfide-bonded homotetramer of a transmembrane glycoprotein. Bindin and bindin receptor allow a mutual, species-specific recognition: only sperm of the same species can establish intimate contact with the receptor and thus with the cell membrane of the egg.

The species-specificity of the bindin protein is based on the polymorphism of its gene. Different species of sea urchins possess different variants of the gene which once derived from a common original gene by mutations (point mutations, duplications of certain sequences, deletions). The receptor protein on the egg cell must also have been adjusted to match with the species-specific edition of the bindin signal in a process of co-evolution.

Mammals. In mammals too the sperm cell must be prepared for its task to penetrate the egg envelopes, and it must be activated on its voyage across the female genital tract at the right moment. On the other side of the sought-after partnership, the egg must be enabled to catch and hold back the sperm. For that purpose glycoproteins are exposed in the zona pellucida of the egg which surrounds the egg cell proper. In the mouse these glycoproteins are known as ZP1, (ZP = zona pellucida), ZP2 and ZP3 (in humans an additional ZP4 has been found). The ZP3 protein has been considered to be the primary sperm-catching molecule but definitive proof is still lacking. Recent evidence indicates that the three-dimensional ZP complex as such, and not a single constituent, mediates sperm adhesion.

All ZP-proteins possess cystein-rich extracellular domains similar to those known from many other extracellular domains. Together the ZP proteins form a fibrillar complex that exists in a form allowing the head of the sperm to couple with it only in the zona pellucida of **non**-fertilized eggs. After fertilization some components of the

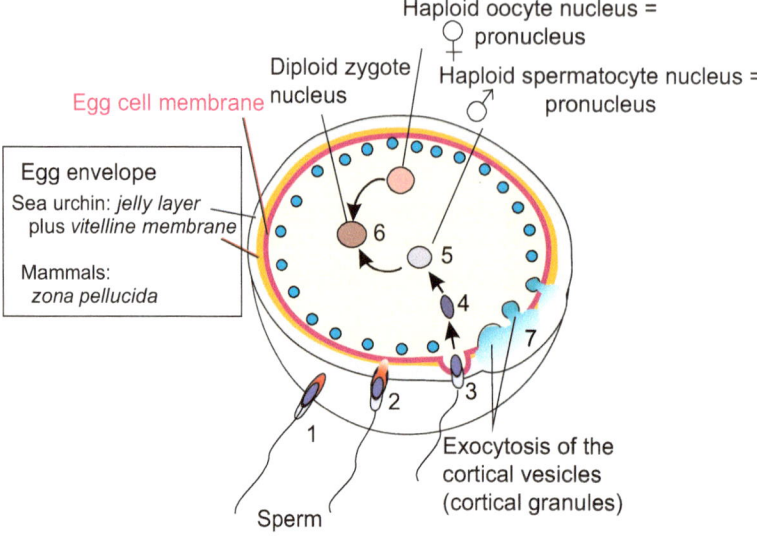

Fertilization

Fig. 3.1 Fertilization in overview. The numbers 1–7 relate to the temporal order of events. (1) Establishment of contact of the sperm with the external egg envelope. (2) Opening of the acrosomal vesicle and drilling through the jelly layer (*white*) and vitelline membrane (*yellow*) and contact with the cell membrane (*red*). (3) Activation of the egg, exocytosis of the cortical granules (*blue*), lifting off the vitelline membrane (7), uptake of the sperm head into the egg cell (4), release of the sperm nucleus (5) and fusion of the haploid pronuclei to form the diploid nucleus of the zygote

complex are enzymatically cleaved and the sperm-binding capacity of the zona pellucida decays.

The counterpart of the ZP-receptor-complex in the envelope of the egg is its ligand in the head of the sperm. In mammals the sperm comes in contact with the egg envelope not with its tip but with its equatorial zone. A particular binding zone extends along the surface of the equator that contains several factors indispensable for fertility. One of these proteins has been designated **ESP** (*Equatorial Segment Protein*) which in cooperation with a series of other proteins such as SED1 mediates the binding to the ZP complex of the egg. Binding triggers the acrosomal reaction and further proteins become exposed on the surface of the sperm.

However, the sperm catching proteins of the zona pellucida are not integral proteins of the egg cell membrane, the oolemma, but components of the extracellular envelope. Therefore, a further process must follow that allows the fusion of the sperm cell with the egg cell. Fusion is preceded by vesiculation of the outer acrosomal membrane

of the sperm. The ultimate terminal binding partners that mediate fusion of the vesicles with the egg cell membrane, the oolemma, are still unknown at present.

Hypotheses. Several molecules have been proposed as binding partners on the side of the egg cell, among them a member of the integrin family and fertilin-β, a protease. But subsequent analyses have defeated initial promising reports. A recent hypothesis assigns mediation of sperm-oolemma adhesion to a transmembrane protein of the immuno-superglobulin family (IZUMO) on the side of the sperm head and four-transmembrane proteins, collectively known as tetraspanins (actually CD9 and CD81) on the oolemma. A general property of tetraspanins is their ability to interact with one another and many other surface proteins and, thus, to mediate cell adhesion and fusion. Actual egg-sperm fusion is thought to share similarities with synaptic vesicle fusion. Integrins may play a supporting role via sustenance of the tetraspanin web in the oocyte cortex.

Fertilization in sea urchin

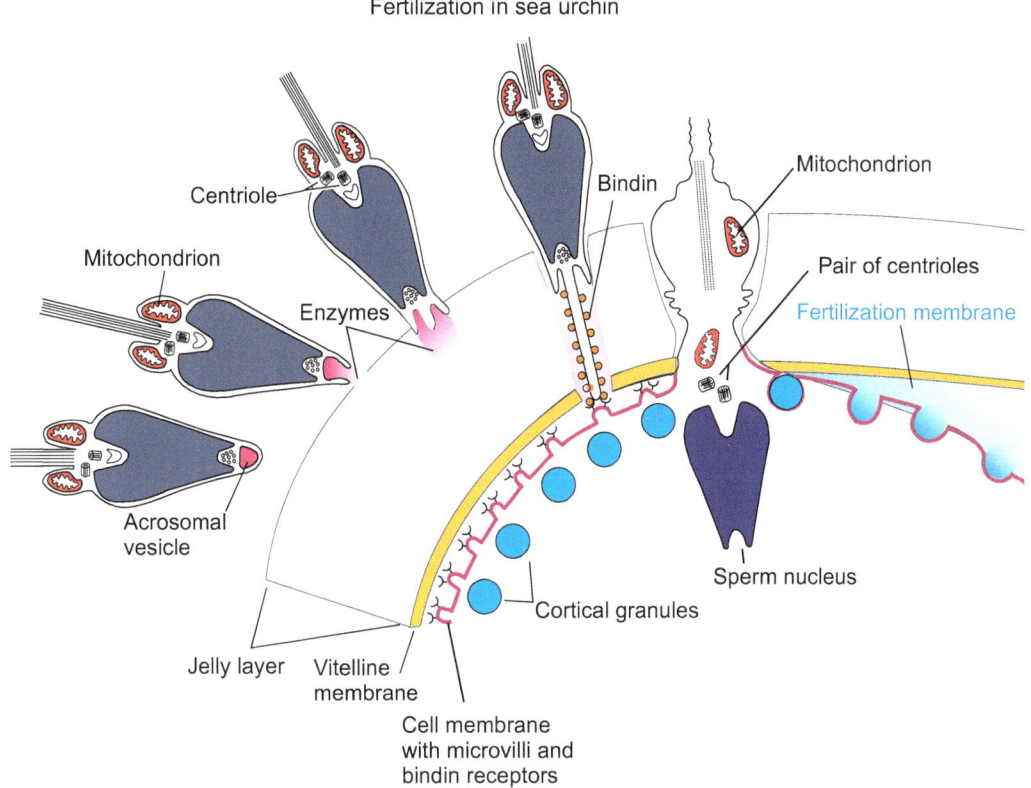

Fig. 3.2 Fertilization in the sea urchin. Contact of the sperm with the jelly coat of the egg induces acrosomal reaction: The acrosomal vesicle opens and release hydrolytic enzymes that begin to digest the envelope at the site of contact. An acrosomal filament is put out and perforates the jelly coat and vitelline envelope; bindin molecules on the surface of the filament establish contact with bindin receptors of the egg cell membrane. Fusion of the cell membranes of the sperm and the egg cell opens a passage through which nucleus of the sperm, together with the centrioles, is drawn into the egg cell. As the cortical vesicles are discharged (cortical exocytosis), the swelling content of the opened vesicles displace and inflate the vitelline envelope. It is now known as the fertilization envelope

A Preliminary Summary. In parallel to the events known from sea urchins (and from other 'fertilization models' such as ascidians) and supported by laboratory experiments with mouse gametes, the following sequence of events has been proposed: The mammalian sperm first communicates with the ZP-complex of the egg envelope. The ZP-complex attaches onto the sperm head and stimulates exocytosis of the acrosomal vesicle. A collection of hydrolytic enzymes is released, among them proteases bundled together to form a proteasome. The machinery of the proteasome attacks ubiquinated proteins of the egg envelope, the zona pellucida, and a channel leading to the egg cell is enzymatically cleared. Now the acrosomal membrane can establish immediate contact to the oolemma, the egg cell membrane proper. The acrosome exposes a protein, functionally comparable to the bindin in sea urchins, and presumably represented by a member of the highly polymorphic immunoglobulin family. The particular species-specific immunoglobulin serves as an identity card and becomes the ligand that couples the sperm cell to the ultimate receptor in the egg cell membrane. This receptor is represented by, or integrated in, a collection of tetraspan proteins. The ligand-receptor binding enables fusion of the membranes of the sperm cell and the egg cell.

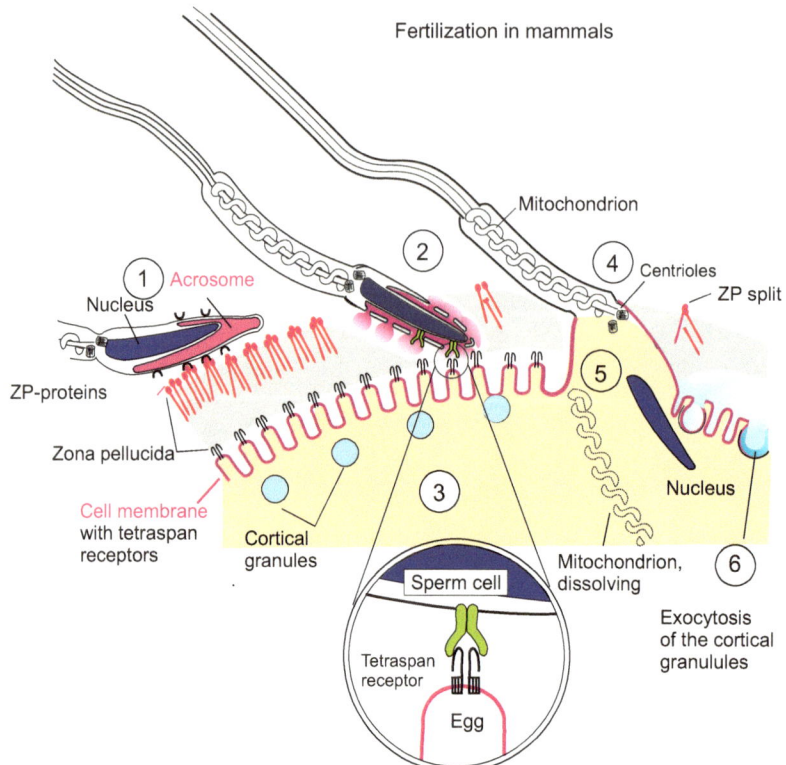

Fig. 3.3 Fertilization in a generalized mammal. (1) The sperm head docks with its equatorial zone onto the ZP-protein complex of egg envelope (zona pellucida). (2) Upon contact the acrosomal vesicle is fenestrated and releases a collection of hydrolytic enzymes. These digest a slit into which the rotating spermatozoan moves, propelled by the flagellum. (3) The opening of the acrosomal vesicle exposes a still unidentified ligand (*green*) that is bound by a complex of tetraspan protein complex in the egg cell membrane. (4) The mutual binding triggers vesiculation of the membranes of the spermatozoon and way is given to the interior compenents of the sperm. The egg cell engulfs (by endocytosis) the nucleus, mitochondrion and centrioles of the sperm. (5) Mitochondria supplied by the sperm are destroyed and degraded in the egg cell while the centrioles are left. (6) By means of exocytosis the cortical granules release enzymes (*blue*) that degrade the sperm-binding ZP-complex. A cortical reaction similar to that seen in the sea urchin egg and enzymatic linking of zona pellucida proteins provide a hardened fertilization envelope that hinders further sperm from entering the egg cell

Mutual adhesion triggers a signal transduction cascade similar to that shown in Fig. 3.5 for the sea urchin egg and described in Sect. 3.2. Following membrane fusion, the internal contents of the spermatozoon – its nucleus, mitochondrion and centrioles – are drawn into the egg. The egg's surface bulges outward into a **fertilzation cone**, and long microvillar processes erupt from the surface towards the sperm. The processes stretch out and grasp the head of the spermatozoon. Its nucleus, mitochondria and centrioles become engulfed.

3.1.7 As a Rule, Entrance Is Refused for Further Sperm Cells

When the cell membranes of sperm and egg fuse, entry of further sperm should be prevented. The biology of fertilization distinguishes between two mechanisms which both serve to block **polyspermy** (the entrance of multiple sperm). Both mechanisms are switched on during activation of the egg (Sect. 3.2):

1. The **primary, fast** but only **transient polyspermy block** is thought to be mediated

by a conformational change of the three-dimensional structure of the sperm-binding receptors on the egg, which now prevents additional sperm from attaching. This change is induced by a depolarization of the electrical cell membrane potential (decrease of the voltage), triggered by the sperm-egg adhesion.

2. The **secondary and permanent block** is realized by the rapid inflation of the so called **fertilization membrane** which suddenly sets apart from the egg surface and lifts off (more details in Sect. 3.2). This inflation can be seen clearly in the eggs of sea urchins and amphibians. In mammals the process of permanent blockage is less conspicuous and brought about by enzymatic degradation of the ZP-complex; thus, no further sperm can dock.

Often, inactivation of the binding molecules is vital. In most animals polyspermy is disastrous and causes early death of the embryo. Amphibian and avian eggs, however, appear to tolerate polyspermy; supernumerary sperm are destroyed within the egg.

3.2   Activation of the Egg

3.2.1   Sleeping Beauty Is, with Awakened With a Kiss

The unfertilized egg sleeps: transcription, protein synthesis, and cell respiration are at or near zero levels. Contact of the acrosomal filament with the vitelline membrane triggers a cascade of dramatic events; they are best studied in sea urchins but occur in the mammalian egg in similar way (Figs. 3.4 and 3.5).

- The membranes of the sperm cell and egg cell fuse at the site of contact. A passageway is established through which an activating factor is injected into the egg by the sperm. Subsequently, the **nucleus**, pair of **centrioles** and mitochondria of sperm are directed into the egg cell through this passageway.

- The egg membrane becomes electrically depolarized. A **wave of depolarization** starts at the point of sperm entry and spreads over the egg surface in the form of an action potential.

In sea urchins, voltage-dependent alterations of the binding receptors are thought to provide the early, fast polyspermy block: sperm that arrive late cannot establish contact with the voltage-deformed sperm receptors.

- At the sperm attachment site the **PI-PKC-signal transduction pathway** (Fig. 3.5) is initiated. Within seconds the second messengers **IP$_3$** and **diacylglycerol (DAG)** are produced, triggered by activation factors supplied by the sperm. One of these factors has been identified as phospholipase C, that enzyme that catalyses the generation IP$_3$ and DAG from phospholipids of the cell membrane (see also Chap. 11). Additional second messengers such as cGMP and cyclic ADP-ribose appear transiently. Since fertilization is a vital prerequisite for development, there are presumably redundant mechanisms to ensure signal transduction and egg activation.

- The second messengers IP$_3$ and cyclic ADP-ribose, or some further activating factors supplied by sperm, cause the release of **calcium ions** from the endoplasmic reticulum (ER) into the cytosol. A positive feedback loop results in an explosion of free calcium: the released Ca^{2+} ions evoke the release of further Ca^{2+} in the neighbouring parts of the ER. The calcium reserves in the ER behave as an assemblage of distinct compartments. Calcium released from the first compartment stimulates calcium release in the second compartment, and so forth. The process of calcium release spreads explosively across the egg (Figs. 3.4 and 3.5). Since the liberated calcium ions are very quickly pumped back into the ER, a circular propagating Ca^{2+} wave begins at the sperm entry point and travels across the egg towards the opposite pole. In the hamster and mouse eggs, the first calcium wave is followed by a burst of secondary calcium oscillations; repetitive waves occur at 1–10 min intervals.

- In the eggs of sea urchins and frogs the calcium waves induce an explosive exocytosis of the **cortical vesicles** (also called cortical granules). Beneath the egg membrane are thousands of cortical vesicles. Upon

Fig. 3.4 Activation of the egg following fertilization. In the right column the temporal events are shown. PIP_2 is the starting substance of the signal transduction system (Fig. 11.5). Positive feed back loops (*arrows*) indicate that calcium ions support the release of further calcium ions. While the depolarization of the cell membrane presumably is directly triggered by the sperm-egg contact the changes in the concentration of cytosolic Ca^{2+} and the rise in the pH are consequences of processes induced by the signal transduction cascade

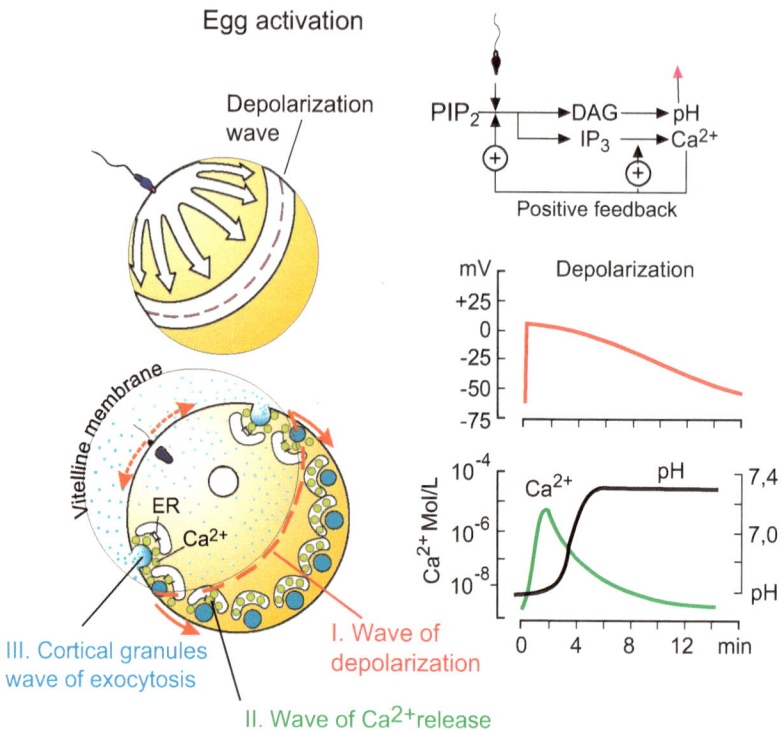

activation by calcium ions the vesicles release their contents, proteoglycans (hyalin) capable of swelling, into the space between the egg cell membrane and the shielding vitelline membrane. By taking up water hyalin suddenly swells up rapidly to form a spacious gelatinous layer known as jelly. The covering vitalline membrane, on the other hand, is highly elastic and has an extreme capacity to expand, like a balloon or air bag. The pumped-up balloon is known as **fertilization membrane or envelope**.

- In mammals, by contrast, no air bag is pumped-up. Instead, the cortical granules release enzymes which destroy the sperm-binding proteins of the zona pellucida. In addition, the coat that exists of tyrosine-rich glycoproteins is hardened through di-tyrosine coupling catalyzed by peroxidases. In any event, the egg envelope has lost its sperm-binding capacity (permanent block to polyspermy).

- The Ca^{2+} signal stimulates the **metabolic activation** of the egg. This metabolic activation is also mediated by IP_3 and diacylglycerol (DAG), although DAG remains in the egg membrane. A protein kinase C (PKC) is translocated from the cytosol to the membrane, associated with DAG and activated (Fig. 3.5, more details in Fig. 11.5). Subsequently, PKC stimulates a Na^+/H^+ antiport in the egg membrane by transferring phosphate onto serine and threonine residues of the antiporter. By the activity of this ion exchanger H^+ is extruded from the cell and Na^+ taken up instead. The pH in the cytoplasm of the egg rises. Present hypotheses assume that ion exchange and increased pH result in conditions under which the mRNA stored in RNP (Ribonucleoprotein) particles can be liberated and used for translation. Among the newly produced proteins are histones which are needed for reduplication of chromosomes in the course of cleavage.

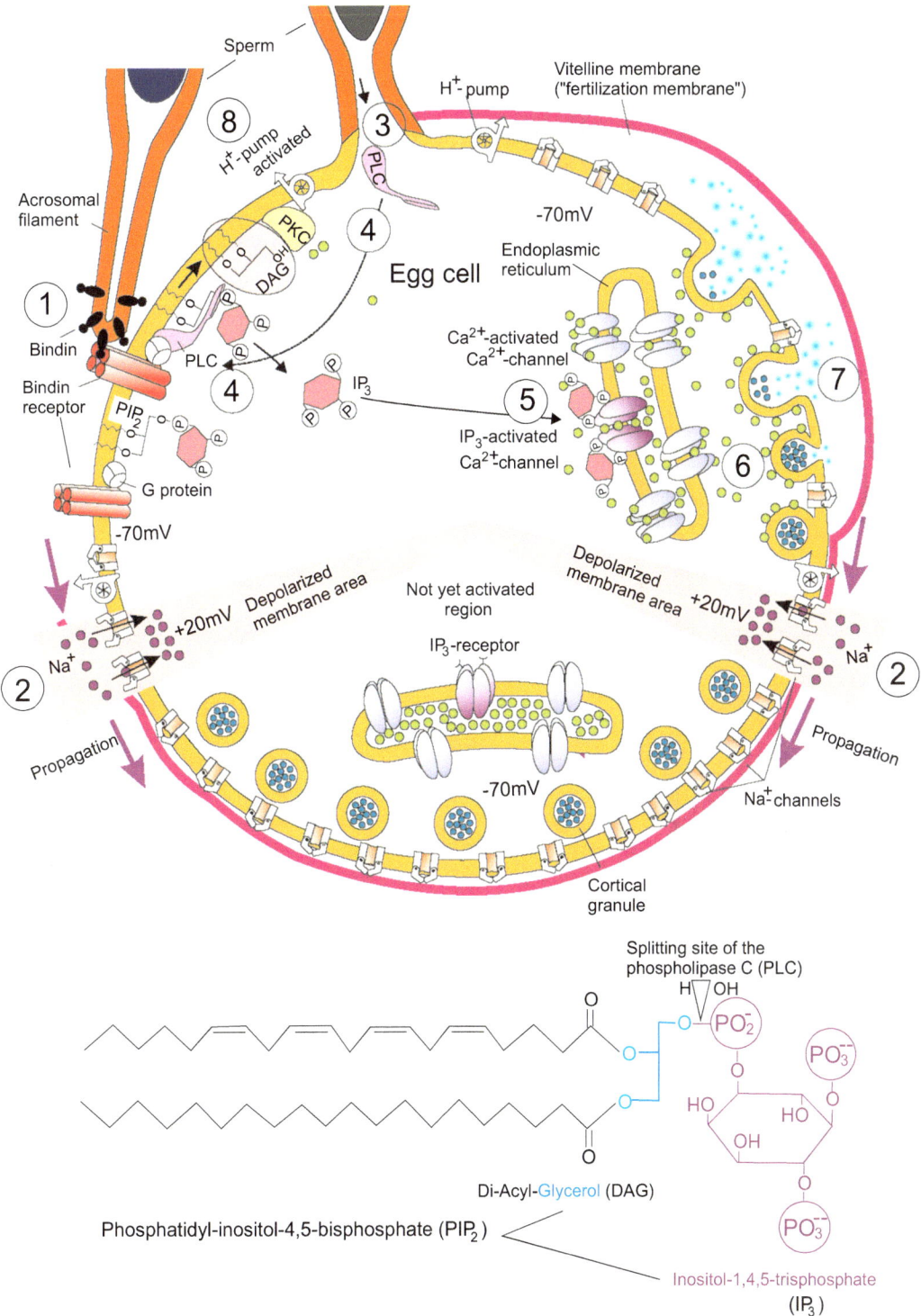

Fig. 3.5 PI-PKC-signal transduction inducing the activation of the egg. The encircled numbers indicate the temporal sequence. (1) The sperm docks onto the egg cell membrane by means of bindin molecules exposed by the acrosomal filament of the sperm; bindin interacts with bindin receptors in the egg cell membrane.

- In the mammalian egg the previously blocked second meiotic division is completed. The completion is manifested by the extrusion of the second polar body. Now the female nucleus remaining in the egg cell is haploid as is the male nucleus supplied by the sperm.
- Another cascade of events starting from activated PKC eventually results in the **start of the DNA replication**. DNA replication is initiated in the haploid nucleus of the sperm and the egg, and completed before the two nuclei meet each other to fuse. Directed into the egg, the sperm nucleus decondenses and is called the **male pronucleus**. It migrates to the **female pronucleus** guided by microtubules which emanate from the centrioles. **With the encounter and fusion of the male and female pronuclei, fertilization** in the sense of developmental biology **is concluded**.

3.3 Non-Chromosomal Sources of Information

3.3.1 In Mammals Only Maternal Mitochondria Survive

Sperm contain mitochondria which are introduced into the egg cell along with the sperm nucleus. However, genetic analyses indicate that in mammals only the maternal mitochondria survive. Therefore, paternal mitochondrial DNA (paternal mtDNA) is lost and mitochondrial genes are transmitted only in the female lineage. **With respect to mitochondrial genes we are children only of our mother**. The exclusivity of maternal mitochondrial inheritance has prompted population geneticists to construct matriarchal phylogenies. By analyzing mitochondrial genes which underwent frequent changes (so called hyper-variable sequences) geneticists deduce in which temporal order the succession of mothers should be arranged and from which geographical region did a population once originate. From such studies it was deduced – independently from archaeological and fossil records – that *Homo sapiens* has populated Eurasia by African ancestors in several waves, and likewise America by Asian ancestors. (In analogous means genes of the Y-chromosome can be compared to trace back the genealogy of a population according to the male lineage). Evidence has been found that the entire human-kind has descended from a few individuals who survived a narrow bottleneck in the history of evolution and migration.

Certainly, the paternal mtDNA does not get lost in all sexually reproducing organisms, for example not in *Drosophila* und not in diverse species of mussels. Also in offspring of mammals surviving paternal mtDNA-sequences could be detected, in a few instances even in human beings. But the egg cell brings along many mitochondria,

(continued)

Fig. 3.5 (continued) (2) Triggered by this interaction sodium channels in the egg cell membrane open one after the other. The Na$^+$-ions streaming into the cytosol depolarize the electric membrane potential (that is they decrease the voltage between the exterior and interior of the cell). The depolarization runs across the egg in form of a wave. (3) The sperm supplies a factor that renders signal relay possible. According to the proposal offered by some researchers this factor is a phospholipase C (PLC). (4) Mediated by a molecular switch, known as a trimeric G-protein, the PLC comes in contact with the bindin receptor and is activated. PLC splits membrane-anchored PIP$_2$ into two second messengers, DAG and IP$_3$ (5) The highly hydrophilic IP$_3$ diffuses in to the cytosol and binds to gated ion channels located along the endoplasmic reticulum ER. Ca^{2+}-ions pour out from the ER into the cytosol.

(6) Released calcium ions mediate a large variety of subsequent reactions, among them the fusion of the cortical vesicles with the egg cell membrane. The vesicles open and (7) release hyalin (blue) into the slit between the egg and vitelline membrane. By taking up water molecules hyalin swells up and the 'fertilization membrane' is blown up. (8) The apolar DAG remains integrated into the membrane and serves as a link onto which a specific enzyme attaches. The enzyme, known as PKC (protein kinase C) is translocated from the cytosol onto the membrane where it becomes activated in cooperation with Ca^{2+}-ions. PKC phosphorylates and activates proton pumps in the cell membrane. Protons (H$^+$-ions) are extruded from the cytosol and the internal pH rises. This rise is thought to render possible the start of many metabolic activities such as protein synthesis based on stored mRNA

Fig. 3.6 Structure of a centrosome. The central complex (*blue*) is in mammals usually derived from the sperm and thus supplied by the father, the other components are present in the egg and thus supplied by the mother

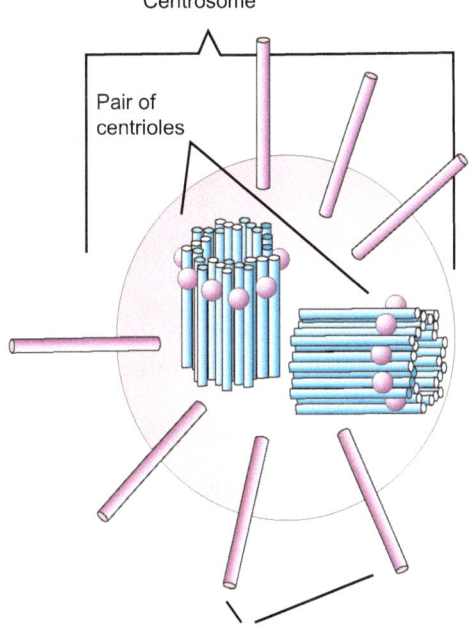

Centrosome

Pair of centrioles

Microtubules of the spindle apparatus (sperm aster) in construction

the sperm only one or a few. In the reproduction of mitochondria the initial quantity of maternal mtDNA is considerably larger than the paternal one from the outset and in the course of cell divisions the paternal mtDNA becomes diluted more and more.

3.3.2   The Paternal Centrosome and the Maternal Nucleolus Must Cooperate

In mammals the sperm cell brings along, besides its nucleus and mitochondrion, also a centrosome (Fig. 3.6). Unlike the paternal mitochondrion the centrosome is not destroyed. It contains a pair of **centrioles**. The paternal pair of centrioles assembles proteins from the oocyte and organizes the formation of the **sperm aster**. The term 'aster' points to the particular array of spindle-type microtubuli, filaments and motor proteins which brings together the two haploid nuclei and thus enables their fusion.

Already in gametogenesis a remarkable division of labour is prepared. The sperm cell retains its pair of centrioles but loses tubulin and other proteins that are required for the construction of a functional centrosome and spindle elements. And it loses the nucleolus required for any protein synthesis. On the other hand, the mammalian oocyte must settle for an almost degenerated pair of centrioles but stores a set of molecular components required for the assembly of a complete centrosome and spindle, and it is equipped with a nucleolus and thus endowed with the capacity to produce new and more proteins. Thus, in mammals components of the egg cell and sperm cell must cooperate to complete fertilization.

Without delay the first cleavage is initiated. To organize microtubules for construction of the spindle apparatus, the centrosome must repeatedly be duplicated. In some animal eggs the centrosome provided by the sperm is duplicated, in others the maternal centrosome, and in still others both the maternal and paternal centrosomes are used to direct and accomplish the distribution of the duplicated chromosomes.

3.3.3 Mammals: For Normal Development a Sperm and, Thus, a Father Is Indispensable as Long as Cloning Does not Replace Natural Reproduction

In mammals, and thus also in humans, three reasons can be adduced why under natural conditions a child can only be born if a father has previously made his contribution.

1. The sperm has to contribute its haploid genome to supplement the haploid genome of the egg cell for completion of the whole diploid genome of the zygote. A haploid genome is insufficient to give rise to a living being that comes out of embryogenesis without harm and remains capable of surviving in the long term.

2. The sperm introduces a protein into the egg required for its activation. Present hypotheses assign this role to a paternal phospholipase C, at least in the sea urchin egg, as indicated in Fig. 3.5. In addition, at least in mice the sperm introduces essential miRNA's and causes influx of external Ca^{2+} needed for full egg activation.

3. As revealed above in Sect. 3.3.2, the paternal centriole is responsible for the guiding together and merging of the two haploid nuclei. But with this work its task is not yet finished. Immediately following activation of the egg the chromosomes are duplicated and the first cleavage is initiated. To ensure an orderly distribution of the chromosomes, a perfect double spindle apparatus must be formed with two spindle poles. To organize their construction the zygote requires two centrosomes as does every eucaryotic cell, and in animal cells each of the two centrosomes should contain a pair of centrioles. In the majority of mammals including human beings the egg cell contributes the one, the sperm cell the other pair of centrioles which usually both become multiplied again and again. Two pairs of centrioles at opposite poles of the mother cell ensure segregation of the replicated chromosomes into the daughter cell with

perfect fidelity and, thus, ensure correct divisions in the course of cleavage. However division of labour between the paternal and maternal pair of centrioles is not a strict condition in all animals for development to start. In some animals a duplicated maternal pair can compensate for the lack of a paternal pair; this is a prerequisite for parthenogenesis to occur.

3.3.4 Is Parthenogenesis in Mammals Possible?

As just stated, in mammals a father is indispensable as a rule. In rare cases, however, even in mammals a parthenogenetic jumpstart is possible. The following possibilities are known:

- Being still diploid, germ cell precursors in the ovary, but also in the testes, can start developing. In these cases, however, no normal embryos arise but **teratomas** (deformed embryos) displaying chaotically disorganized tissues and organs, or even tumour-like **teratocarcinomas** (ovary or testis cancer).
- Oocytes are pushed out from the ovary, start development without a sperm having made its contribution, and haploid or diploid embryos may arise. They may be
  Haploid, when meiosis has been completed and the second polar body expelled before the oocytes start cleaving;
  Diploid, when the first polar body finds its way back into the oocyte cell.

In the mouse (and other mammals) parthenogenesis can artificially be triggered, for instance, by collecting oocytes that have been released from the ovary and activating them with electric stimuli, 7% ethanol or some other substances. When the germs have reached the blastocyst stage (Fig. 5.28), they are transferred into the uterus of a surrogate mother.

Result up to now: whether haploid or diploid, development of mouse embryos comes to a standstill after 10 days (in the 10th to the 13th somite stage). In general, parthenogenetic mammalian embryos are invariably lost in midgestation, possibly due to the lack of the paternal genome and the consequent induction of aberrant gene expression.

In 2012 viable mice were born derived from egg cells that were fertilized with a haploid embryonic stem cell (haESC) instead with a sperm cell. This haESC was of parthenogenetic paternal origin and thus paternally imprinted. Goal of such experiments is to introduce artificially modified genes (transgenes) previously generated in the haESC, in order to produce transgenic animals (Leeb et al. 2012).

The cause of the early death of parthenogenetic embryos is their inability to develop a complete and functional placenta without the assistance of a paternal genome. Present interpretations attribute this failure to **genomic imprinting** that is dealt with in Chap. 8. In the mouse a paternally imprinted genome is required in the trophoblast (Fig. 5.28), the precursor of the placenta; an exclusively maternally imprinted genome is insufficient. If two genomes are united that both have been imprinted in the female germ line the embryo only forms a stunted placenta that is unable to provide the embryo with all it needs.

Genomic imprinting causes fewer problems if cloning using somatic nuclei succeeds (Chap. 13). Somatic nuclei contain both a "paternal" and a "maternal" genome that once experienced different genomic imprinting in spermatogenesis and oogenesis, respectively, and transmit their particular state of imprinting to the cells of the growing child. Provided **cloning** were successful also in man as it is in other mammals a father would no longer be required (except if one wished to clone men, for which purpose the Y-chromosome of a male cell line would suffice). By repeated cloning a father- and men-less society could be generated.

Let us, for the time being, concede the right for natural reproduction to both sexes. With a naturally activated egg, two complete centrosomes and two complete and properly imprinted genomes, we can now start the first rounds of cell division, known as cleavage.

3.4   Cleavage and the MPF Oscillator

3.4.1   Cell Cycles in the Early Embryo Only Consist of S- and M-Phases

The fertilized egg is a single cell but must give rise to a multicellular organism; and this should happen as fast as possible because the embryo cannot evade predators (if it is not carried inside a viviparous mother). The egg utilizes stored molecules to jumpstart development. As a rule, until gastrulation the early embryo can abstain from transcribing genetic information, because all needed mRNA is already present in the maternal RNP particles. From these particles mRNA is liberated and used for translation. As a rule, in this early phase of animal development chromosomes are not needed for directing protein synthesis and the embryo can focus on their multiplication.

A normal cell cycle, that is the time span from one cell division to the next, is subdivided in four phases (time sections): **G1, S, G2 und M-phase** (Fig. 3.7). In the S-phase (synthesis phase) the DNA is replicated and the chromosomes become duplicated, in the M-phase (mitosis) the cells are divided. In G1 and G2 (gap phases) transcription of genes occurs if needed. During cleavage the cell cycle is shortened. In rapid succession, S-phase and M-phase alternate This applies to the eggs of sea urchins, the clam *Spisula*, the zebrafish and the clawed frog *Xenopus* that are popular for research because of their rapid and synchronous cleavage.

Since mRNA for histones and other chromosomal proteins is present in the cytoplasm in many copies and these proteins can therefore be supplied rapidly, chromosome replication is finished in 20-30 min. Immediately thereafter the chromosomes are condensed to a form suitable for movement during mitosis. The early embryonic cell cycle actually consists only of S-phase and M-phase. In *Xenopus*, the G1 and G2 phases are inserted only in the blastula stage, when new transcripts of zygotic genes (genes of the embryo itself) are needed. This stage is known as the

Fig. 3.7 The embryonic cell cycle (*left*) and the MPF oscillator. Not shown is the periodic coupling and decoupling of further kinases and phophatases. There are actually multiple sites where the amino acids threonine, serine and tyrosine are loaded with, and later deloaded from, phosphate. Activation of the MPF complex demands both phosphorylation and dephosphorylation of several distinct amino acids in the cyclin moiety. *CDK* Cyclin-Dependent Kinase, *G1* G1 phase, *G2* G2 phase, *S* S-Phase, *M* Mitosis; *MPF* Maturation- (or Mitosis)-promoting factor

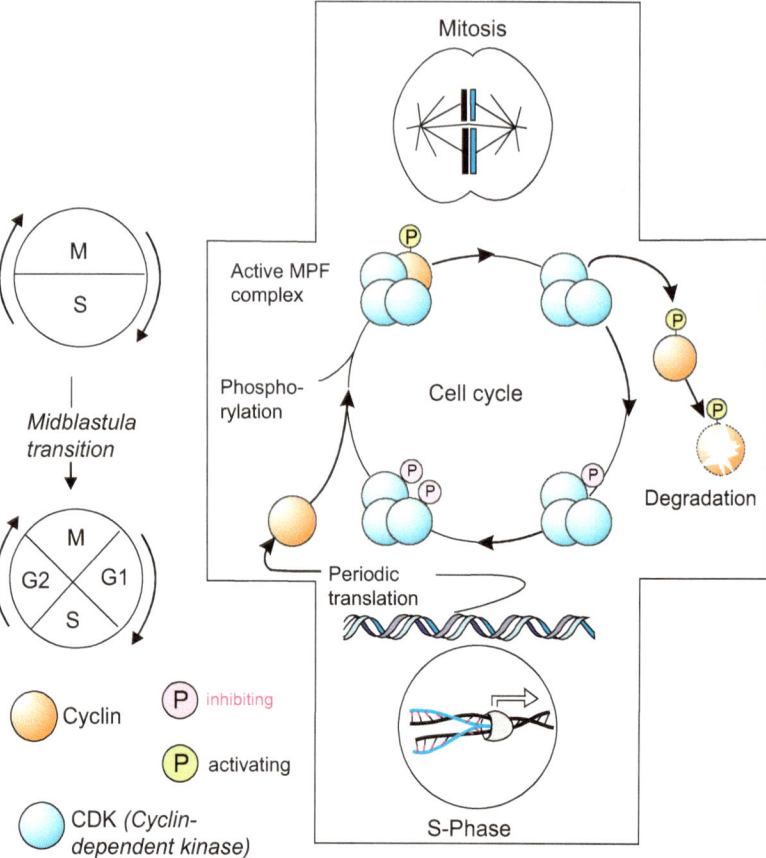

midblastula transition. (Equivalent expressions used in the literature are **zygotic gene activation (ZGA)** or **embryonic genome activation**). Now the cell cycles become longer and the divisions of the various cells asynchronous. Increasingly, the **start of a new cell cycle becomes dependent on external cues**.

Mammalian eggs are exceptions to this rule, transcription of zygotic genes commences early and the cell cycle in cleaving mammalian eggs is unusually long.

3.4.2　Cleavage Divisions Are Driven by a Clock, the MPF Oscillator

The internal oscillator that drives the cell cycle in the early embryo uses the same molecular components that underlie the controlling circuitry of any cell cycle (Fig. 3.7). Component

of the oscillator is a multitude of proteins named **cyclin**; in *Xenopus* the main component is called cyclin B. As their designation indicates the cyclins are periodically (in cycles) synthesized, whereas other components are continuously present, as are several protein kinases of the type **CDK (Cyclin-Dependent Kinases)**. Thanks to their kinase function these enzymes transfer phosphate residues from ATP onto amino acids of proteins. The phosphates are received by the amino acids serine and threonine in located near to proline at defined positions of the protein. The CDK's in turn are controlled by other enzymes. Phosphatases which remove phosphate, such as CDC7 (Cell Division Cycle 7) and CDC25 also play roles in the cell cycle. Extensive studies have shown that each organism possesses its own spectrum of such components. In no case has the construction of the clock been completely disclosed. Behind all diversities a common basic principle can be recognized; it is illustrated in Fig. 3.7.

With the activation of the egg cyclin B synthesis is initiated. The newly synthesized cyclin B couples to CDC to form **MPF** meaning **mitosis-promoting factor** (synonyms: **meiosis-promoting factor** or **maturation-promoting factor** because MPF controls also meiosis). However, the complex is still inactive; it becomes activated only after DNA replication has finished. Activation of MPF is mediated by an alteration in the pattern of its phosphorylation. Shortly before the cells enter the M-phase to undergo mitosis the amount of the periodically produced cyclin (in *Xenopus* cyclin B) is highest. For MPF to be activated, threonine-161 of cyclin B must be loaded with phosphate (positive activation) and threonine-14 and tyrosine-15 must have their phosphates removed (release of inhibition). In this state the complex has come to be called active MPF After mitosis has finished, cyclin B is destroyed by proteases and the cycle closed.

Mitosis is only a short phase of the entire cell cycle. In the fully developed cell cycle, DNA replication is prepared in G1 phase and actually accomplished in the following S-phase DNA. In G2 the following actual division (cytokinesis) is prepared. A cell must proceed through the whole cycle in an orderly sequence. For example, cytokinesis must not start until DNA replication is finished and the chromosomes are condensed into a transportable form. Therefore, several key events in the cell cycle are regulated to ensure an orderly progression. At decisive checkpoints diverse sets of cyclins, CDKs and phosphatases appear and play key roles. All these exist in multiple variants (isoforms). Particular assembles of such enzymes together with co-factors are thought to control the commitment of the cell to DNA replication, an event called START in yeast and **restriction point R** in animal cells.

The cycles of cell division are associated by waves of calcium-ions released shortly before each cytokinesis from the store of the endoplasmic reticulum ER. Each swarm of calcium ions travels across the cytoplasm, and then is pumped back into the ER.

3.4.3   Tightly Patterned Sequences of Cell Divisions Allow Cell Genealogies to Be Traced in Some Species

In the small embryos of small organisms careful observation has revealed fixed patterns in the temporal progression of the timing of cell divisions and the spatial arrangement of daughter cells. The diverse lineages always lead to the same types of cells, tissues or organs. In nematodes such as *Caenorhabditis elegans* (*C. elegans*), for example, the P-lineage always terminates in the primordial germ cell P4 (Fig. 4. 20). In the spiralian embryo the D-line leads to the primordial mesoblast d4 which in turn gives rise to all mesodermal tissues (Fig. 4.21). In the ascidian embryo a certain cell line leads to the musculature of the tail, another line to the notochord (Fig. 4.43). However, even in *C. elegans*, where a particularly precise and rigid order governs the sequence of cell divisions and subsequent events of cell differentiation, not all muscle cells are descendants of one and the same founder cell and not all nerve cell descendants of another founder cell. Instead, several different founder cells contribute to the final inventory of muscle cells and nerve cells. Most tissues are thus said to have a **polyclonal provenance**.

A cell lineage faithfully repeated from generation to generation does not from the outset permit reliable statements about the time course of determination. When is the fate of the cells programmed, and when is their destiny irreversibly fixed? However, experimental evidence suggests that exact genealogy is correlated with early determination.

3.4.4 In Other Embryos, for Instance in Vertebrates, Much Variability Is Allowed

When one follows the course of cleavage in the embryo of amphibians, let's say in the mountain newt *Ichthyosaura alpestris*, in time lapse videos motion and compares different embryos much variability can be noticed. Phyla that diverged from the main line of animal phyla early in evolution, that is in sponges and Cnidaria the variability is particularly high. Anything that does not cause major problems appears to be permitted, and much deviation can be seen even in sibling embryos (e.g. in *Hydractinia echinata*, Fig. 4.17) – up to what is said to be a 'catastrophic cleavage'. High flexibility in the execution of cleavage is, as a rule, associated with high powers of regulation. Such high regulative ability is not restricted to organisms considered to be 'ancient' in terms of evolution, the mammalian germs also possess high powers of regulation. Well known proof is the development of complete **identical twins** when early embryos fall apart in two halves or after experimental separation of the blastomeres up to the 8-cell stage (identical octuplets). In spite of lacking precision at the start eventually a highly complex organism in miraculous order will arise. In the following chapters we try to find out how such powers of self organization are possible.

3.4.5 In the Course of Cleavage Maternal Developmental Determinants Are Distributed

Already during the first cell divisions factors are allocated to the daughter cells that initiate a first differentiation of the cells – differentiation in terms of divergence of developmental paths. Often these factors are so-called transcription factors which bind onto the controlling region (enhancer, promoter) of various genes and determine their activities in the cells to which they have been allocated. The well-ordered allocation of these factors is associated with the phylotypic pattern of cell division (superimposed on constraints of the yolk as shown in Figs. 2.3 and 2.4) and associated with the specification of the future body axes (polarity axes), where the head pole and where the tail pole will come to lie.

A particular feature of these early acting factors is that they become deposited and stored in the egg already before fertilization in the course of oogenesis. Therefore, for their production the genome of the sperm is of no relevance, instead these components are generated under the governance of the maternal genome. They are said to be **maternal determinants**. Often nursing cells in the environment of the oocyte produce these materials and place them at the disposal of the growing oocyte, sometimes it is the diploid oocyte itself which generates these factors while it still harbours the complete maternal genome in its nucleus. That varies from case to case and, therefore, these maternal determinants will be specified in the following Chaps. 4 and 5 which introduce model organisms; a comparative summary will be given in Chap. 9 (Positional Information and Pattern Formation by Maternal Determinants). After these maternal factors have accomplished their work the further subdivision and differentiation in the growing embryo are based on concerted cell-cell interactions. These will be dealt with in Chap. 10 (Positional Information and Pattern Formation by Cell–Cell Interactions).

Summary

New life starts with **fertilization**, the fusion of the sperm cell with the egg cell. The sperm is guided to the egg by attracting pheromones called **gamones**, and caught by the egg envelope. In mammals the female hormone **progesterone** is component of the attracting bouquet. The egg catches the arriving sperm by means of specific molecular receptors; in mammals this is the multi-meric **ZP-protein** complex that is located on the surface of the egg envelope (zona pellucida). The attachment of the sperm triggers the **acrosomal reaction**. The acrosomal vesicle at the tip of the sperm head releases enzymes that clear the way for the sperm through the envelopes including the elastic vitelline membrane that covers the egg cell membrane. In order to prepare the fusion of the cell membranes of the sperm and egg cell the sperm head exposes ligands that interact with receptors in the membrane of the egg. The fusion of the cell membranes opens a passage through which the nucleus of the sperm is actively engulfed by the egg together with the sperm's centrosome and mitochondrion. The fusion of the haploid 'paternal' pronucleus provided by the sperm with the haploid 'maternal' pronucleus of the egg cell yields the diploid nucleus of the zygote, and fertilization is completed.

In mammals the paternal mitochondrion brought along by the sperm gets decomposed and the mitochondrial **mtDNA** degraded. As a consequence, with respect to the mitochondrial genes we are children only of our mother only, and by analysing mitochondrial DNA sequences geneticists can trace back the female genealogy (using genes of the Y-chromosome also the paternal lineage). On the other hand, as a rule in mammals the centrosome of the sperm is required to guide the two pronuclei together and to ensure correct subsequent cell divisions.

The contact of the sperm with the egg cell triggers an **electric depolarization** of the egg cell membrane that travels as a wave across the egg surface comparable to an action potential in excitable cells, and activates a signal transduction cascade of the PI-PKC-type. Transduced signals initiate metabolic activity, lead to a wave-like release of calcium ions from the ER, to the exocytosis of materials stored in the cortical egg granules, to the subsequent inflation the vitelline-mambrane to a 'fertilization membrane' (in sea urchins and amphibians) or, in mammals, to the enzymatic degradation of the sperm binding receptors. Fertilization membrane or degradation of binding sites ensures a permanent block against the entry of further sperm (polyspermy).

The subsequent cleavage is a rapid series of cell divisions. Since the egg can make use of stored maternal mRNA initially no transcription is needed and the cell cycle is abbreviated to S- and M-phase. Shortly before gastrulation, in mammals earlier, G1 and G2 are inserted as the germ starts to make use of its own, 'zygotic' genes. The cell cycles are driven by an internal clock. A key role is played by a complex of proteins consisting of cyclins, cyclin-dependent protein kinases (CDK) and phophatases. The complex has become known as **MPF (Mitosis-Promoting Factor, M-phase-Promoting Factor** or **Maturation-Promoting Factor)** and is periodically activated and degraded in the course of cell divisions. The spatial and temporal pattern of the first cell divisions often is very precise and strictly determined, for instance in spiral cleavage. Maternal components of the egg specify the pattern.

Early in development it must be specified which side of the embryo should become anterior and which posterior, which ventral and which dorsal. The specification of the coordinates, called **body- or polarity axes**, is accomplished in many animal organisms already in the course of oogenesis by components of the egg that are produced under the guidance of the maternal genome and therefore called **maternal determinants** or factors. Examples are given in Chaps. 4, 5, and 9.

In the separate Box 3.1 cases of self fertilization and parthenogenesis , i.e. development without the participation of a sperm, are presented. In mammals (mice) parthenogentic development comes to a standstill in embryonic stages; no viable parthenogenetic mammals have seen the light of the day up to now (human beings merely in legends).

Box 3.1: Self-Fertilization and Parthenogenesis

We resume a theme that was outlined already in Chap. 1. In many animals, particularly in invertebrates, embryonic development can start and come to an end without a paternal parent having had the opportunity to contribute to the genetic makeup of offspring. Supplementing what had been said in Chap. 1 we here list possibilities of uniparental reproduction. Three basic means can be distinguished:

- Self-fertilization (endogamy) in hermaphrodites (for example in round worms)
- Haploid parthenogenesis (for example in male bees)
- Diploid parthenogenesis (for example in aphids)

In endogamy, haploid parthenogenesis and certain forms of diploid parthenogenesis a rigorous elimination of defect genomes will take place.

In Hermaphrodites Capable of Self-Fertilization Repeated Inbreeding Causes Descendants to Become More and More Similar and Eventually Identical

By defintion, hermaphrodites (named so after the ancient Greek deities Hermes and Aphrodite), can produce both eggs and sperm. In most cases incestuous self-fertilization is prevented by anatomic conditions, time differences in maturation of the gonads, or incompatibility of sperm and eggs rendering fusion impossible. In rare cases, though, self-fertilization (endogamy) can occur. Well-known examples are endoparasites such as tapeworms (Cestodes) and one of our model organisms, the round worm *Caenorhabditis elegans* (Fig. 4.19). In its gonads a mature egg captures a sperm when it passes the sperm chamber and immediately undergoes development up to the larva already in the uterus of the mother.

In repeated inbreeding, say after 6–7 cycles of reproduction, almost 100 % of the alleles are present in the homozygous state. In most instances homozygous defect mutations are lethal. This results in elimination of the individuals bearing them and, thus, also of the defective alleles. This 'wanted' effect, however, is paid for with the risk that the entire population could die out. Through occasional insertion of biparental reproduction this risk is mitigated.

Haploid Parthenogenesis: Haploid Male Insects Are Subjected to Fitness Tests to a Particularly High Degree

In bees and other hymenopterans (wasps, ants) it is often within the discretion of the queen to decide which sex a descendant should have. In insects, like in terrestrial vertebrates, an internal insemination takes place. The egg is fertilized before it leaves the maternal body. For that the queen uses sperm stored after mating in a chamber called receptaculum seminis. A queen bee or ant can lay unfertilized eggs, too. In spite of being haploid the egg develops to a vital individual that is always male (a drone) and committed to once undertake nuptial flights.

In the nuptial flight drones must prove their fitness. In haploid organisms each defective gene will reduce the fitness of the individual. Drones with defective alleles scarcely get a chance for mating. The defective allele becomes eliminated from the population.

Diploid Parthenogenesis: The Result Can Be a Kind of Natural Cloning

There are three means by which females can give birth to diploid descendants without admitting sperm:

- Mode 1: Oocytes develop before having undergone meiosis. Offspring is a clone of the maternal organism. Example is the grasshopper *Pycnoscelus surinamensis*.
- Mode 2: After completion of the first meiotic division the polar body returns and fuses with the egg cell thereby bringing back ▶

Box 3.1 (continued)

its set of chromosomes and pair of centrioles. This fusion can, in addition, trigger egg activation. Also in this case the descendants of an individual are members of a clone.

- Mode 3: Egg cells undergo both meiotic divisions and become haploid, but thereafter later an up-regulation takes place, be it that the second polar body fuses with the egg cell, be it that, after the first reduplication of the chromosomes, division of the nucleus and cell does not take place. (In this case all alleles would be homozygous including defective ones and the individual subject to rigorous fitness tests as in case of haploid parthenogenesis.)

Diploid parthenogenesis (after mode 1 or 2) is known from many invertebrates, in particular from greenflies: In spring a completely female population rapidly brings about a large number of offspring (Fig. 1.4). But even some vertebrates are familiar with parthenogenetic reproduction. In particular some species of fish can abstain from a male partner and even some terrestrial vertebrates. Recently (2012) two viviparous American snakes (*Agkistrodon contortrix* and *A. piscivorus*) were reported to have given birth to parthenogenetically generated young. Another terrestrial reptile that attracted attention is the American lizard *Cnemidophorus uniparens*. (This lizard caused even more of a stir because of its sexual behaviour. Even though the entire population consists of only females they do not abstain from sex. One partner takes the part of a male, the other behaves as a female; after some time the roles are exchanged). In a population reproducing in such a way soon this, then that allele gets lost, because not all individuals have offspring in the same number. The genetic variability decreases.

As parthenogenesis occurs in vertebrates one may ask whether it can happen in mammals as well, perhaps even in humans? We shall discuss this in Box 6.2.

Model Organisms in Developmental Biology I: Invertebrates

4

The ability to understand developmental processes requires appropriate organisms. The field of genetics established the precedent of focussing research on a few reference or "model" organisms such as *Drosophila* or maize. Recent developmental biology has depended on a small number of organisms for much of its spectacular progress. This concentration of effort facilitates attempts to advance in the analysis of the basic processes down to the molecular level. On the other hand, there is no single organism which could be selected to study all fundamental events and aspects of development; for each embryonic development leads to a single particular species, and not to an animal in general. From the egg of *Drosophila* a fruit fly arises, not an insect in general, not a fish and not a human being. General principles are only recognized when the events of development are studied in several diverse organisms, which develop differently yet display some common features. In addition, laboratory work quickly reveals that even the best model organism exhibits, in addition to its particular advantages, specific disadvantages. Thus the zebrafish has transparent embryos but large-scale genetic studies require hundreds of aquariums and a staff of workers, and the mouse provides a wealth of developmental mutants but cannot be conveniently frozen for long-term storage.

The main invertebrate model organisms that are currently used are shown in Fig. 4.1.

4.1 Sea urchin: Basic Model of Animal Development in General and Subject of Historically Important Experiments

4.1.1 Sea urchin Gametes and Embryos Are Models of Reference for Fertilization and Early Embryogenesis

The sea urchin embryo occupies an important place in the history of developmental biology. The translucent embryos of sea urchins not only captivate us because of their beauty; in addition, they triggered many seminal ideas, and were found to be favourable material to study very early development.

As early as 1890 pioneering work on fertilization was performed with eggs and sperm of sea urchins. Even today, sea urchin eggs remain the best investigated system for studies on fertilization, egg activation and embryonic cell cycle. However, sea urchins are not good model organisms in all respects. Mutational analysis is not practical because of the long generation times and the difficulties in rearing larvae in the laboratory through metamorphosis. Mature sea urchins are not available throughout the year and must be removed from their natural marine habitat. This may pose problems with respect to species protection.

W.A. Mueller et al., *Development and Reproduction in Humans and Animal Model Species*,
DOI 10.1007/978-3-662-43784-1_4, © Springer-Verlag Berlin Heidelberg 2015

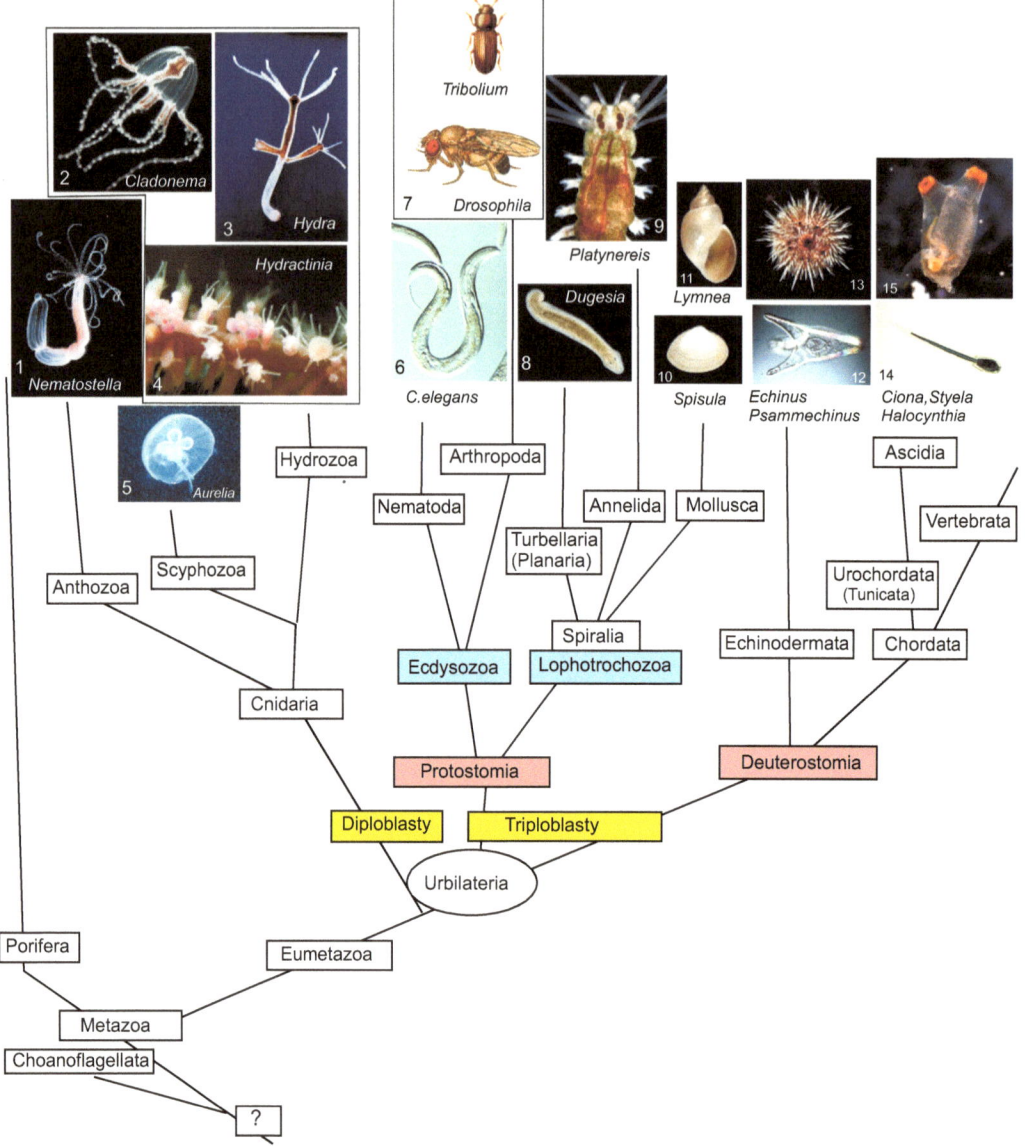

Fig. 4.1 Model organisms—invertebrates, that are dealt with or mentioned in this book, categorized according to pedigrees that are presently preferred on the basis of molecular-genetic data. Teaching picture collection of WM

Because in the various regions of the oceans different species are found there is no single species that would represent "the sea urchin". In littoral states of the North Sea much work is done *Paracentrotus lividus* and *Echinus esculentus*. In rim nations, Japanese and American researchers prefer to work with the Pacific species *Strongylocentrotus purpuratus*. In this introductory book we do not discriminate between species since species-specific differences are small in early embryogenesis. Sea urchins can make use of about 12,000 genes in controlling development, consequently the

(continued)

selection presented here is very restricted and determined by the current literature which focuses on homologies to developmental genes in vertebrates.

These disadvantages are balanced by several particular advantages: eggs and sperm can be harvested separately and in large amounts. Eggs are small (0.1 mm in diameter), transparent and surrounded by a translucent and easily removed envelope (jelly coat). The eggs develop in water, even under the microscope, and after artificial insemination, in perfect synchrony. Development up to the hatching larva takes 1–2 days.

Shedding of the Gametes, Fertilization. Eggs and sperm are released through genital pores on the dorsal side of mature sea urchins. In the laboratory, sea urchins are placed upside down onto beakers, and stimulated to shed gametes by intracoelomic injection of 0.5 M KCl. Eggs are released into sea water, while sperm is shed 'dry' into dishes and can be used to inseminate eggs at a chosen time by mixing the milky sperm suspension into the egg suspension. The transparency of the envelope and the embryo facilitates in vivo observation of the events of fertilization, cleavage and gastrulation under the microscope.

The egg of sea urchins and the pattern of its cleavage have become prototypes in textbooks for animal eggs and development in general (Fig. 4.2). The egg is covered by an elastic, acellular coat, called the **vitelline membrane**, and further wrapped up in a **jelly coat**. In its internal structure the egg is polarized along its **animal-vegetal axis**. By definition, the **animal pole** is that pole where the **polar bodies** are extruded during the course of the meiotic divisions, when the egg still is in the ovary. Polar bodies are miniature sister cells of the egg cells; they will decay. The haploid nucleus of the egg is found near the animal pole. Another marker of polarity is the subequatorial band of orange pigment found in some batches of the Mediterranean species *Paracentrotus lividus*.

In sea urchin eggs, unlike mammalian eggs, all of the polar bodies are formed before fertilization. Insemination, fertilization and activation of the sea urchin egg have been described in Chap. 3.

Start of Development, Cleavage. The sperm nucleus is gated and transported into the egg. Triggered by the contact of the sperm with the egg membrane, a **fertilization envelope**, serving as a barrier against the penetration of further sperm, is raised from the surface of the egg, and the egg becomes **activated**. Following activation, cleavage divisions dissect the egg into successively smaller cells, without appreciable changes in its volume. Sea urchins undergo synchronous, radial, holoblastic cleavages until the blastula stage (Fig. 4.2a). The first daughter cells which can be individually recognized in the microscope are called **blastomeres**. Doubling of their chromosomes takes place rapidly, and the cells divide every 20–30 min, driven by an internal molecular clock. The embryo thus passes through the 2-, 4-, 8, -16, 32-, 64- and 128-cell stages. As development proceeds, the cell cycle lengthens and cleavage is no longer synchronous in the various regions of the embryo. A fluid-filled cavity now develops. Cells rearrange themselves to construct an outer epithelial wall enclosing the central cavity: the embryo has arrived at the **blastula** stage. The blastula's wall of cells is termed the **blastoderm**, and its central cavity the **blastocoel**. On the exterior surface of the blastoderm cilia are formed; the coordinated beat of the cilia causes rotational movements of the blastula within its envelope.

At the animal pole the first larval sensory organ is formed, recognizable as a bundle of long, non-motile cilia: the **apical tuft**. At the vegetal pole, the blastoderm flattens and thickens to form the **vegetal plate**. The vegetal plate includes a group of cells, micromeres, which are filled with ribosomes. The micromeres begin to move and thereby open the phase of gastrulation.

Gastrulation begins shortly after the blastula, now comprising about 1,000 cells, hatches from its fertilization envelope. Gastrulation takes place in several phases (Fig. 4.2b).

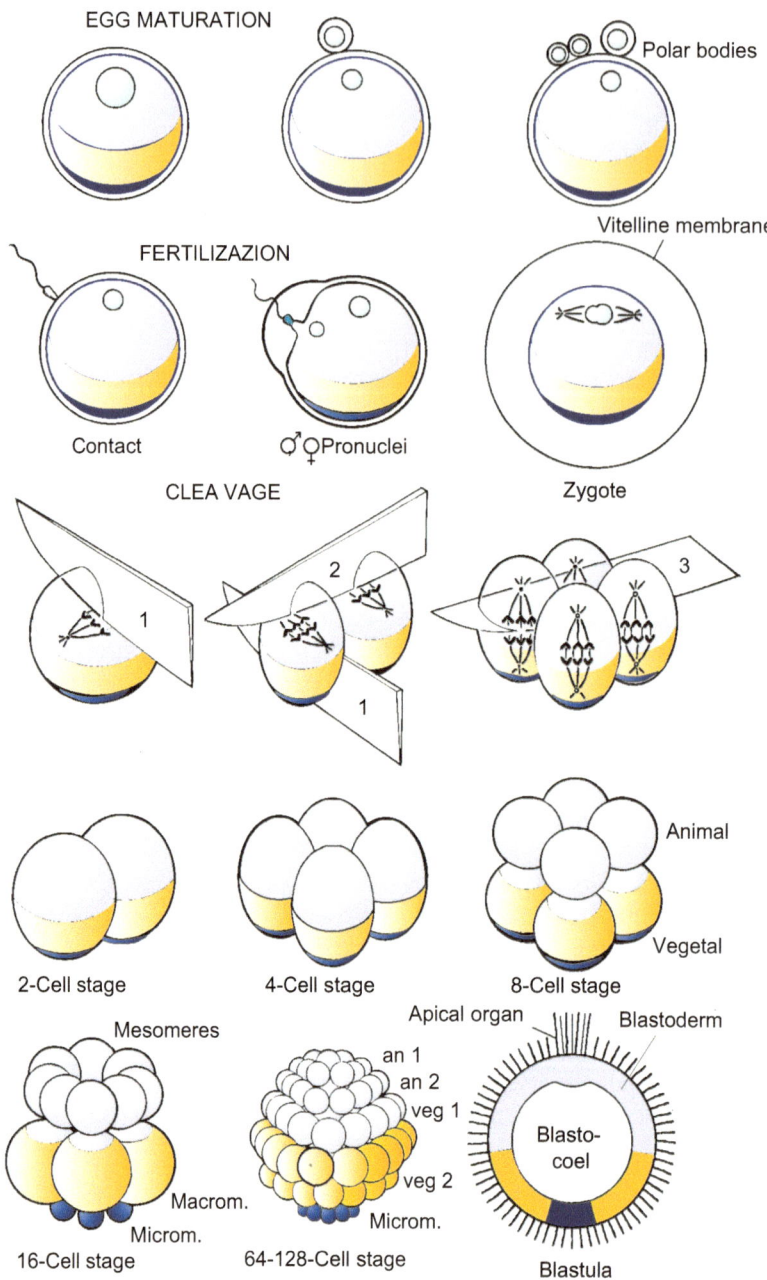

Fig. 4.2 (continued on p. 63)

1. The onset of gastrulation is marked by the immigration of the descendants of the micromeres into the central cavity. The cells become bottle-shaped, release their attachment to the external hyalin layer and to their neighbouring cells, and eventually detach to migrate individually into the blastocoel. They move in the fashion of amoebae. The immigrated cells represent the **primary mesenchyme**. The descendants of the smallest micromeres, which were previously located around the vegetal pole, are thought to

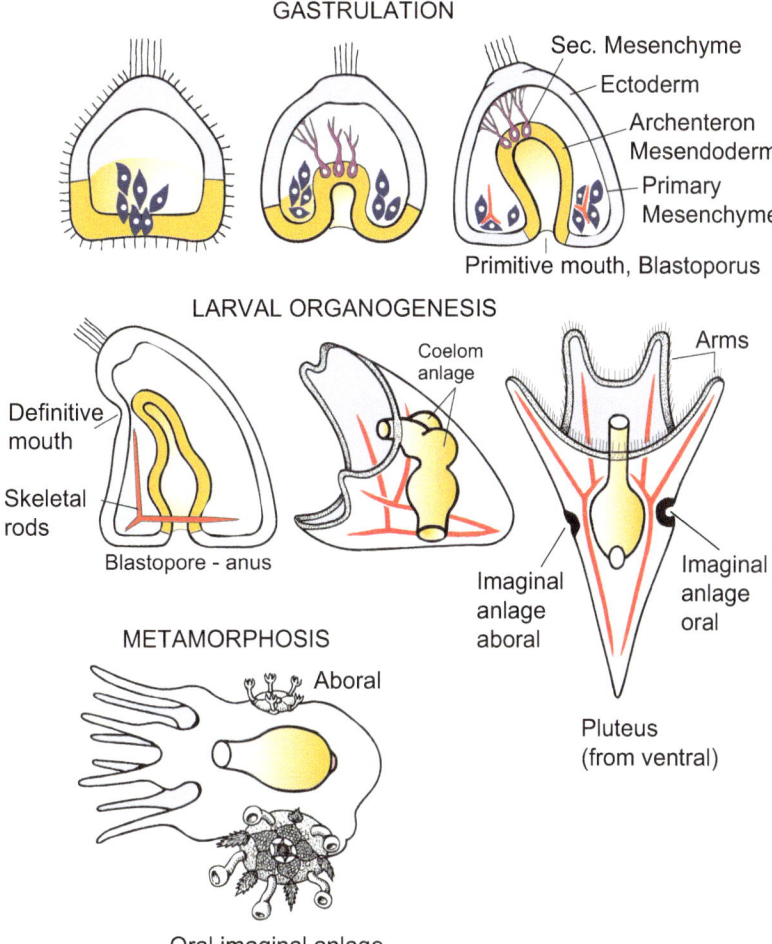

Fig. 4.2 Development of the sea urchin. (**a**) Cleavage: The first two cleavages pass through the animal-vegetal axis and are called meridional because they divide the egg along meridians. The third cleavage is equatorial and perpendicular to the animal-vegetal axis. Further cleavages feature oblique spindle orientation, become asynchronous and unequal, and give rise to several tiers of cells. Finally, the cells adhere more strongly to each other, form an epithelial layer, the blastoderm, and give rise to a central cavity (the blastocoel). (**b**) From gastrulation to the onset of metamorphosis. Gastrulation takes place in several phases and provides the interior cavity with cells from which the interior organs arise. The resulting larva is called the pluteus. From the metamorphosis of the bilaterally symmetrical larva to the pentameric adult sea urchin, only an initial stage is shown

undergo terminal differentiation only after metamorphosis. But most of the cells of the primary mesenchyme soon become skeletogenic: clusters of reaggregated mesenchyme cells fuse into syncytial cables. These cables form the larval skeleton by secreting insoluble calcium carbonate that crystallizes into glittering spicules.

2. **Micromeres** are emitters of inducing signals: Transplanted to another site of the blastoderm, they induce their neighbours to invaginate (Sect. 4.1.9). **Invagination** (Fig. 4.2b) is the mode by which the **archenteron** is formed in this second phase of gastrulation. The vegetal plate, consisting of descendants of the **tier veg1**, buckles inward and elongates within the

cavity of the blastocoel to form the tube-like larval intestine. The invagination is radially symmetric around the animal-vegetal axis. When isolated surgically from the rest of the larva, the vegetal plate undergoes autonomous bending and involution.

The archenteron stretches along the length of the blastocoel. Lobe-like pseudopods on the cells at the tip of the archenteron are transformed into filamentous pseudopods or **filopods**. Reaching out and moving around the inner surface of the blastoderm, the filopods explore the underside of the blastoderm cells. The filopods direct the tip of the elongating archenteron to a site where the definitive mouth will later break through. The primary mouth, the blastopore, will function as the larval anus. Sea urchins and all other members of the echinoderms belong to the **deuterostomes**, as does our own chordate phylum.

3. Having fulfilled their function as pathfinders, the cells at the tip of the archenteron reduce their filopodia and detach. They represent the secondary mesenchyme, which gives rise to muscle cells surrounding the gut and some other cell types.

With the ingression and involution of cells into the blastocoel, germ layer formation still is incomplete. In terms of comparative embryology the formation of the coelomic cavity is of particular interest. As the larva develops, **mesoderm** is formed by **coelomic pouches** which balloon out from the archenteron as evaginations (Fig. 4.2b). Following the segregation of the mesodermal vesicles, the remainder of the archenteron constitutes the **endoderm**.

Mesodermal vesicles evaginating from the archenteron called enterocoelia (see Fig. 2.2) occur in similar way also in *Branchiostoma* and 'lower' chordates, especially in tunicates. These taxa are grouped together under the term **Archicoelomata**. Epithelial coelomes are always considered to be derivatives of the mesoderm.

In the tradition of classic embryology the sea urchin is said to be **triploblastic**, meaning that it is formed by three tissue layers although three stratified sheets can never be observed. The outer layer of tissue is said to be an **ectodermal** epithelium, the archenteron is said to represent the **endoderm** and cells occupying the space between to represent the **mesoderm**.

4.1.2 During Metamorphosis Imaginal Discs Give Rise to the Adult Animal; the Bilateral Symmetric Larva Becomes Transformed into the Pentameric Sea urchin

Embryogenesis terminates with the hatching of the beautiful larva, which is transparent and fairy-like. It is called **pluteus**. The larva floats through the water and uses its cilia to swirl microscopic food into its mouth.

The transformation of the free swimming (planktonic), bilaterally symmetrical larva into the pentameric sea urchin requires a fundamental reconstruction. The new construction starts from groups of cells that were set aside from participation in embryogenesis itself. The process resembles metamorphosis in "holometabolous" insects such as butterflies, and the groups of cells that build up the new body are termed **imaginal discs** in sea urchins as well as in insects. The complicated, 'catastrophic' metamorphosis will not be described here. Experiments with embryos of sea urchins end with the observation of the pluteus larva.

4.1.3 Landmark Experiments 1. Embryos Are Not Machines and Are Capable of Regulation

The following experiment of historical significance was performed by Hans Driesch at the Stazione Zoologica in Naples. If the first two blastomeres are separated from each other at the two cell stage, each blastomere gives rise to a complete larva. Although reduced in size to half of the volume of a normal larva, both larvae are harmoniously diminished (Fig. 4.3). They are identical, monozygotic twins. If the blastomeres are separated at the 4-cell stage, quadruplets are formed. Even blastulae can be bisected, resulting in identical twin plutei, provided the embryo is

Fig. 4.3 Experimental bisection of an early sea urchin embryo at the 8-cell stage. (**a**) Bisection perpendicular to the animal-vegetal egg axis in the equatorial plane results in two unequal embryos. (**b**) Bisection along the animal-vegetal axis results in two normal larvae of half the usual size. Experiments made by Driesch (1892) and Hörstadius (1935)

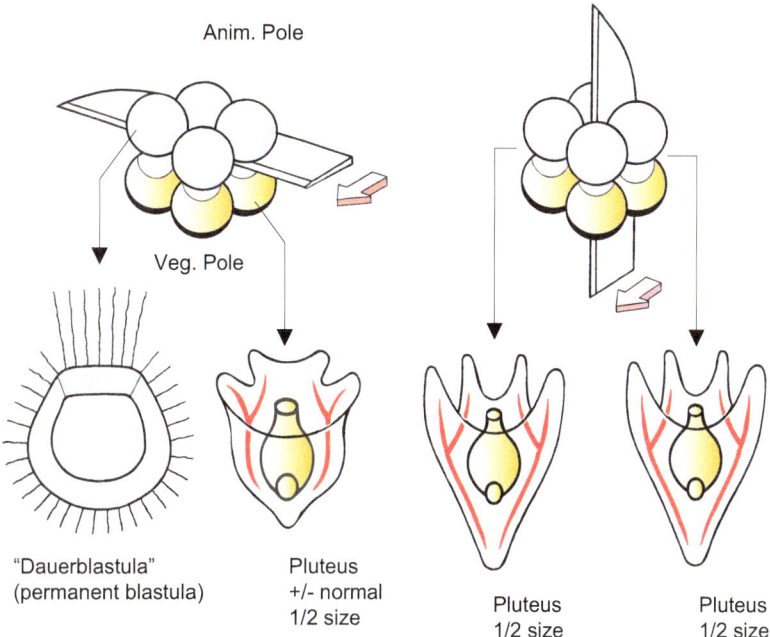

Anim. Pole

Veg. Pole

"Dauerblastula" (permanent blastula)

Pluteus +/- normal 1/2 size

Pluteus 1/2 size

Pluteus 1/2 size

bisected before the beginning of gastrulation and the cut is performed along the animal-vegetal axis. Driesch concluded that living organisms are not mere machines, because pieces of machines cannot supplement themselves autonomously to restore the complete machine. This interpretation was in striking contrast to the prevailing view of the mechanicists (Chap. 1, Box 1.1). Biologists now understand that regulation is possible because all cells are provided with the whole set of genetic information and because both halves receive animal and vegetal cytoplasmic components when the cut is made along the egg axis.

From the 8-cell stage on, it is technically possible to bisect the embryo in the equatorial plane at a right angle to the egg axis. Now the result is different: The animal half develops into a blastula-like hollow sphere, which, however, is incapable of forming an archenteron. Conversely, the vegetal half is able to form an archenteron, but the emerging pluteus displays considerable deficiencies. For instance, it is mouthless and has short arms (Fig. 4.3). By displacing cytoplasm through pressure or centrifugation, Theodor Boveri, Hans Driesch and Thomas Hunt Morgan accumulated experimental

evidence suggesting that **cytoplasmic components** are responsible for the different developmental potentials. Driesch furthermore concluded that the future fate of a cell, its "*perspective significance*", is "*a function of its position in the whole*". Boveri launched the idea of graded potentials along the animal-vegetal axis. The graded differences of the cells along the animal-vegetal axis were in the focal point of the following experiments.

In today's terms in the embryo of sea urchin an early regionalization takes place that subdivides the sphere of the blastula into zones with different developmental potentials. As predicted by the founders of the experimental developmental biology maternal factors deposited in the cytoplasm at different locations prepare a first regional subdivision of the blastula. Thereafter exchange of signals between the cells make behaviour according to the local position possible. Today one speaks of **positional information** (Chaps. 9 and 10). For the transmission and spreading of signals gradients in the concentration of diffusible signal molecules are thought to be of particular significance. This has been deduced from the following classical experiments.

4.1.4 Landmark Experiments 2: Experimentally Deduced Interactions of Embryonic Parts Provided Reasons to Propose the Gradient Theory

Among the most remarkable experiments in the history of embryology were those performed by Sven Hoerstadius investigating the developmental potential of cells along the animal-vegetal axis of the early embryo. He separated tiers of cells from cleavage stages by transverse cuts, and juxtaposed tiers of cells taken from different locations along the animal-vegetal axis. To explain his results, he advanced the gradient theory (first formulated by Boveri). The experiments are described here in some detail because they document the scientific approach of the classic developmental biology and prompted similar experiments in many animal systems (Fig. 4.4).

If the animal cap *an1* is isolated by micromanipulation at the 64-cell stage, the cap gives rise to a hollow sphere unable to gastrulate. A close look at the blastula-like sphere reveals a strange anomaly: the long cilia which mark the first larval sensory organ (the apical organ) are not concentrated in a small area forming the apical tuft. Instead, the whole blastula bears these long cilia. Even the tier *an2*, whose descendants are normally equipped only with short cilia, develops, following isolation, into a blastula that is largely covered with long cilia. The ability to form an apical organ has spatially expanded. Such an exaggeration of structures characteristic of the animal sphere is called **animalization**. The tendency to form exaggerated animal structures diminishes with distance from the animal pole.

In the intact embryo the strong animal potency must be repelled or compensated by influences originating from the vegetal part of the embryo. These influences have been systematically investigated by Runnstroem, Hoerstadius and other Scandinavian researchers, who performed delicate transplantation studies. By juxtaposing micromeres onto isolated tiers of cells derived from the animal hemisphere, the exaggerated animal tendency can be compensated. Transplanted micromeres were by far the best **vegetalizers**. By determining the number of micromeres needed for compensation, one can titrate the strength of the animal potency (Fig. 4.4).

The tier *an1* is normalized by adding 4 micromeres; the conglomeration develops into a largely normal dwarf pluteus. To normalize *an2*, two micromeres suffice. In *veg1* one micromere suffices, but the embryo has such a weak animal component that the resulting ovoid larvae have short arms and show further defects. Apparently, an animal component is required for normal development. This requirement is even more apparent when the experiment is extended to *veg2*. The tier *veg2*, which ordinarily gives rise to a rudimentary, pluteus-like creature, is "**vegetalized**" by the addition of micromeres. It loses the ability to form arms and instead forms an intestine which is oversized relative to the rest of the body. As a result, the intestine cannot be invaginated into the interior cavity (**exogastrula**).

These results gave rise to the so-called **double-gradient model**: two physiological activities along the animal-vegetal axis occur as mirror-image gradients. The activities were assigned to **morphogenic substance**, nowadays called **morphogens** following a proposal by A Turing (Chap. 1, Box 1.1). The molecular identities of these hypothesized substances are now partly known (Sect. 4.1.6).

An "animalizing" morphogen that pushes determination toward animal properties has been thought to peak at the animal pole, while a vegetalizing morphogen pushes vegetal properties and peaks at the vegetal pole. Because the opposite morphogens neutralize one other, the strength of each factor decreases with distance from its origin. The local developmental pattern is therefore determined by the ratio of two morphogens.

A popular hypothesis maintains that the morphogens can spread from cell to cell, freely crossing cell membranes. Alternatively, the

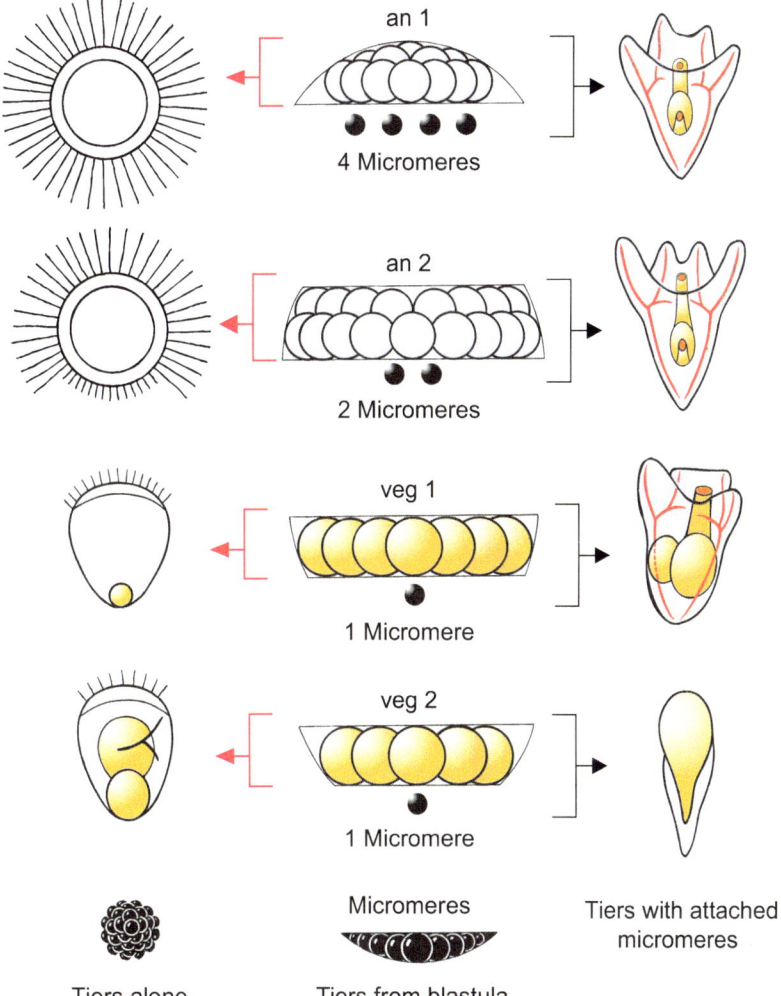

an 1

4 Micromeres

an 2

2 Micromeres

veg 1

1 Micromere

veg 2

1 Micromere

Micromeres

Tiers with attached
micromeres

Tiers alone   Tiers from blastula

Fig. 4.4 Sea urchin. Classic experiment supporting the gradient hypothesis [conducted by Sven Hörstadius (1935, 1939)]. From middle to the left column: Tiers of cells from a 60-cell stage blastula were isolated and their developmental potentials observed. For example, tier an1 gives rise to an "animalized" blastula with an expanded apical tuft of long cilia covering the entire surface. Tier veg2 gives rise to incomplete larvae lacking the oral arms and the mouth but with an enlarged hindgut; they show features of "vegetalization". Isolated micromeres are unable to continue development. From middle column to right column: Development of the same tiers following addition of micromeres. If 4 micromeres are added to an1, the enhanced animal potential is compensated and a normal pluteus results. To normalize an2, it takes only 2 micromeres. Addition of one micromere to veg2 leads to a highly "vegetalized" development: the 'larva' lacks all oral structures and the enlarged archenteron cannot be incorporated into the interior. Such a malformation is called an "exogastrula"

relevant signals mediating interaction might spread only in the interstitial spaces between the cells or must be exposed on the cell surface and directly presented to neighbouring cells. In either case, the signal molecules have to be picked up by receptors. Such a view connects the classic gradient hypothesis with the new data.

The molecular identities of these hypothesized substances are now partly known as outlined in the following sections.

Fig. 4.5 Sea urchin: Fate map of a blastula and the corresponding structures in the pluteus larva. Various regions of the egg cell and the early blastula are equipped with different sets of maternal gene products. The different areas give rise to different parts of the pluteus larva. After investigations by Davidson

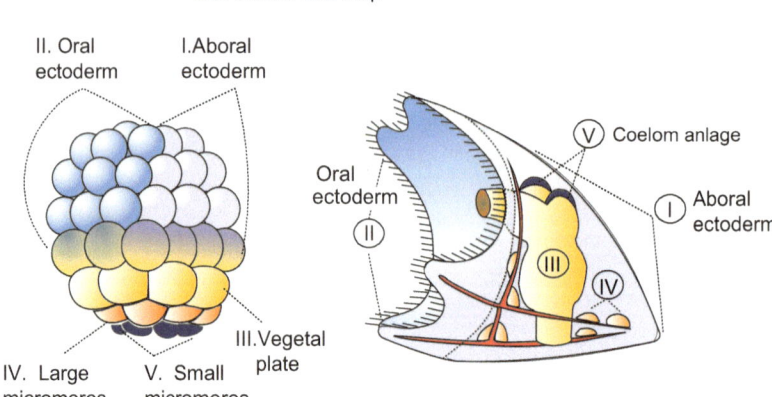

Sea urchin: fate map

II. Oral ectoderm

I. Aboral ectoderm

V Coelom anlage

Oral ectoderm

II

Aboral ectoderm

I

III

IV

IV. Large micromeres

V. Small micromeres

III. Vegetal plate

4.1.5 Cytoplasmic Determinants Stored in the Egg Are of Significance for the Orientation of the Body Axes; mRNAs of Transcription Regulators are Already in the Egg Present in Form of Gradients and Specify the Fate of the Region

Aiming at detecting factors controlling development that are stored in the egg, proteins have been sought possessing DNA-binding domains, and mRNA has been extracted encoding such proteins. Such proteins are candidates for factors controlling further gene activities and became generally known as transcription factors. The following picture emerged (Figs. 4.5 and 4.6):

As in *Drosophila* (Sect. 4.6) and *Xenopus* (Sect. 5.1) the unfertilized egg contains maternal mRNA transcribed and stored during oogenesis which codes for a large array of different transcription factors. These maternal components are deposited within the oocyte at different locations and differentially distributed among the blastomeres, but in the unfertilized oocyte wrapped in ribonucleo-protein-(RNP)-complexes. They are unwrapped after fertilization and several are found in opposite gradients along the animal-vegetal axis.

- In the upper, animal half of the egg diverse ATF's (*Animalizing Transcription Factors*) are found.

- In lower, vegetal halve the transcription-cofactor β-CATENIN is found. In the blastula β-CATENIN has entered the nuclei of the lower hemisphere, the highest density is seen in the nuclei around the vegetal pole where the micromeres are positioned and the blastopore begins to be formed.

Beta-CATENIN is known as essential element of a signal cascade that normally starts with an externally arriving WNT signal (Chap. 11; Fig. 11.6). When the cascade is activated by WNT, β-CATENIN is taken up into the nucleus and controls gene activity in cooperation with another factor, TCF. It may be that in the sea urchin embryo maternal β-CATENIN is taken up before a WNT signal arrives as also appears to be the case in the amphibian embryo. Up to the 16-cell stage β-CATENIN is still in the cytoplasm, in highest concentration in the micromeres, and ingresses the nuclei between the 16- and 60-cell stage.

Among all the determinants identified so far β-CATENIN is the most effective one. Installed into the micromeres β-CATENIN enables them to differentiate to skeletogenic cells idependently of their environment (autonomous development) and to initiate gastrulation (Sect. 4.1.9).

β-CATENIN is also of fundamental significance for the spatial subdivision of the embryo of the amphibians and other organisms (summary in Chaps. 11 and 22; Fig. 22.6). **We see a remarkable congruency. In the sea urchin as well in the vertebrate embryo and even in the cnidarian**

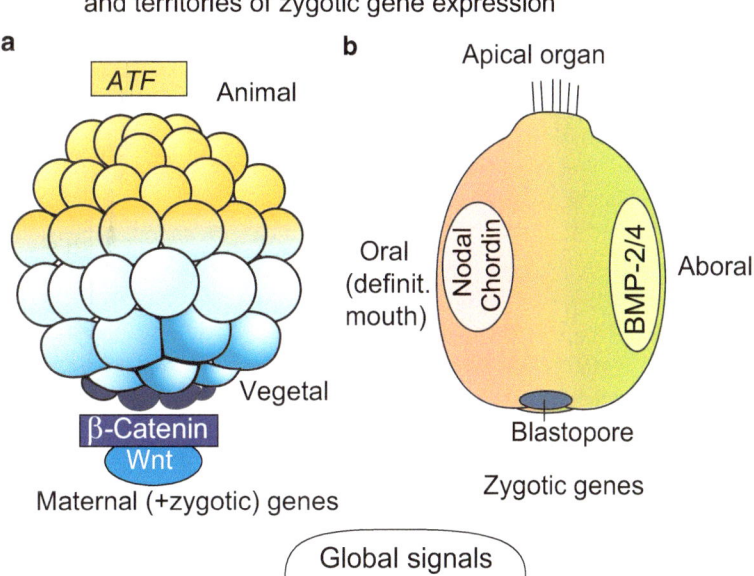

Fig. 4.6 Sea urchin, (**a**) Pattern of distribution of region-specifically deposited maternal gene products in the blastula. (**b**) Expression of zygotic genes that encode signal substances; these are also expressed in the vertebrate embryo and fulfil the criteria of antagonistically acting morphogens

polyps the blastopore is formed in the area where stored mRNA for β-CATENIN is located (**Fig. 22.7**). In addition, the cells harbouring these factors are enabled to produce signal molecules known as inducers. These are emitted and specify the fate of neighbouring cells.

4.1.6   Long Sought Morphogens Are Being Tracked: Of Particular Significance Are the WNT-Signalling System and Signals of the TGF-β Class (BMP, Nodal) Which Are of Central Significance in Vertebrates as Well

The existence of morphogens has long been a matter of controversial dispute. Being biological signal molecules they are effective in very minute quantities and exert the expected effects only if present in the form of gradients or other spatial patterns. Gradients of morphogens have been

chemically identified only a few decades ago (not until 1980) in *Drosophila* (Sect. 4.6) and *Xenopus* (Sect. 5.1).

Potential morphogens along the animal-vegetal axis:

1. **The WNT/β-CATENIN system**. As just mentioned, in the blastula maternal β-CATENIN enters the nuclei and a gradient becomes established with its high point at the vegetal pole. But β-CATENIN is not an extracellular signal molecule; it cannot disperse over the boundaries of cells and consequently not act as a morphogen in terms of a signalling molecule. Yet, in the nuclei β-CATENIN switches on the gene for the WNT3a signal and this can be released into the interstitial spaces or transmitted to neighbours via transcytosis (Chaps. 10 and 11).

In this context age-old experiments have gained new interest and found a new interpretation. Sea urchin embryos that develop in sea water enriched with **lithium ions** become vegetalized. The expression "vegetalized" points to increased expression of vegetal morphological traits: the oral arms become too small, the intestine too large. Animal halves of blastulae are normalized not only by adding micromeres but also by bathing them in solutions of lithium-ions, and in high concentrations form even exogastrulae the large intestine of which extrudes from the gastrula instead of being invaginated.

Treatment with lithium is known to simulate the arrival of a WNT signal by interfering with signal transduction. It inhibits the enzyme GSK-3, a member of the cascade, and thus facilitates the uptake of β-CATENIN into the nucleus: Eventually WNT-dependent target genes are activated.

2. **Calcium-ions**. Parallel to the gradient of nucleus-resident β-CATENIN a gradient is established in the concentration of cytoplasmic calcium-ions. Also this gradient has its high maximum in the micromeres and contributes to the specification of their fate.

3. **NOTCH system**. Before the blastopore is formed the micromeres occupy the vegetal pole. They have to leave and migrate into the interior. Before they segregate and leave the micromeres talk with their neighbours by means of the NOTCH signalling system. This system has been detected in *Drosophila* but turned out to be a universal system of mutual arrangement if the aim is to segregate from each other. As an example the segregation of nerve cell precursors from future epidermal cell will be presented in Chap. 10 (Fig. 10.1).

4.1.7 Being a Bilaterally Symmetric Organism the Sea urchin Embryo Requires a Second Polarity Axis, an Oral-Aboral Axis. For Its Establishment the Morphogen NODAL Plays a Decisive Role

The adult sea urchin displays a pentameric radial organization but the pluteus does not; it is bilaterally symmetrical. In the structure of the egg only the animal-vegetal axis is anticipated. At the lower end of this primary body axis the blastopore is formed. In the gastrula the tube-like archenteron

bends with its tip towards the outer body wall. At the position where the tip of the archenteron touches the wall the definitive mouth is formed. Now a second axis can be recognized: an oral-aboral axis which extends in about the perpendicular direction to the primary axis.

Already in the early blastula the future oral field is marked when the expression domain of the gene *nodal* is made visible with the technique of in-situ-hybridisation (Chap. 12). When the expression is suppressed no mouth will appear. The protein NODAL coded by the gene *nodal* is a member of the TGF-β family of signal molecules known from vertebrates as is BMP which becomes expressed at the opposite site of the blastula. We will meet NODAL and BMP again in the vertebrate embryo where they play indispensable roles in the establishment of the dorso-ventral body axis.

The *nodal* mRNA is not yet present in the egg as a maternal transcript. Which factors determine that the *nodal* gene is switched on in the oral field? The determining elements are factors of quite different quality and nature. In the region of the future oral pole more mitochondria have accumulated than at the opposite pole. The high oxidative activity, it is assumed, causes the switch of the gene nodal (and of several other genes) to be shifted to the "on" position. NODAL in turn switches on the gene for the protein CHORDIN, and we arrive at the

BMP/CHORDIN System. Transcripts (mRNA) have been found for signal proteins of the CHORDIN-class at the oral pole and for a protein with high similarity to BMP2/4 at the opposite aboral pole (Fig. 4.6). Here we see a further remarkable parallel to vertebrates. In the amphibian embryo we will get to know two antagonistically acting morphogens which radiate from the ventral or dorsal site, respectively, and exert antagonistic functions as they might do in the sea urchin. In the amphibian embryo these are the morphogen BMP-4, a member of the TGF-β family, that becomes secreted in the ventral region of the blastula/gastrula, and CHORDIN that appears in the dorsal region. Emanating from opposite sites these two morphogens form concentration gradients running and decreasing in opposite directions

(Figs. 5.19 and 10.4). CHORDIN binds and inactivates BMP. In the sea urchin, however, CHORDIN appears not to be of significance for the specification of the oral-aboral axis, and the definitive mouth is formed in the NODAL domain, in the amphibian it is formed in the BMP domain. In both groups of animals CHORDIN is responsible for an initial specification of the **nervous system**.

4.1.8   Eventually the Embryo Is Subdivided into Territories in Which Different Gene Activities Give Rise to Different Tissues and Organs

The primary gradients cease to have effects on the global body pattern after the sixth cleavage when the embryo has reached the 64-cell stage. The blastula is now subdivided in five territories that are differently supplied with transcription factors (Figs. 4.5 and 4.6).

1. A first territory is directed by local determinants to form the ectodermal epithelium of the mouth region and the ciliated band that will extend along the arms.
2. A second territory will form the aboral, cilia-free ectoderm.
3. The third territory, consisting of the subequatorial vegetal plate, gives rise to the archenteron and its derivatives (e.g. the epithelium of the coelom and the secondary mesenchyme). Surprisingly, this region expresses a gene homologous to the gene *brachyury* that previously had been found to be expressed first in the entire mesoderm and eventually in the precursor cells of the notochord of chordates. The Echinodermata and Chordata once arose from common ancestors but echinoderms don't have a notochord.
4. The fourth territory defines those cells that later construct the larval skeleton.
5. The fifth territory comprises the small micromeres which initiate the formation of the archenteron as specified in the following section.

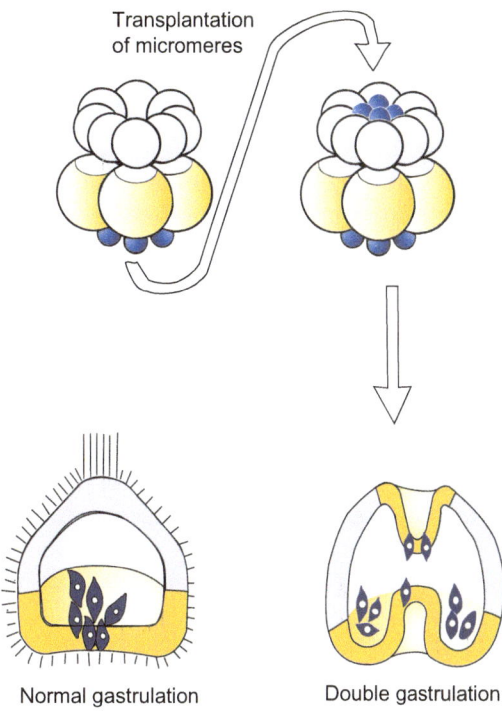

Transplantation of micromeres

Normal gastrulation   Double gastrulation

Fig. 4.7 Sea urchin: Induction of a second archenteron after transplantation of additional micromeres onto the animal pole of a blastula

4.1.9   Micromeres Give Origin to Short-Range Inductive Signals Which Trigger the Formation of the Archenteron

The micromeres are not only the source of morphogens that extend over long distances. Shortly before or while they migrate into the interior of the blastula they give the order for the neighbouring vegetal plate to bulge inwards to form the archenteron. When micromeres are transplanted to a different (ectopic) site they induce the formation of a second archenteron (Fig. 4.7). We will find comparable phenomena in the amphibian embryo. One speaks there of **organizer** effects, and similar terms are used here if the equivalence of the two events is emphasized. In vertebrates as well as in sea urchins the establishment of the organizer is associated with the uptake of the transcription factor β-CATENIN into the nuclei of the organizer cells. In the case of the sea urchin, however, the molecular identity of the signals emitted by the organizer is still unknown.

4.1.10 The Ratios in Which Different Cell Types Are Represented Is Subject to Control; Lateral Inhibition Is Involved

When in the late blastula the micromeres that later would secrete the skeletal spicules are removed, the cells of the tier veg2 which normally participate in the formation of the archenteron step into the breach and take over the task of the lacking micromeres. Viewed the other way round, if enough skeletogenic micromeres are present they prevent the neighbouring future intestinal cells from fulfilling their potential for producing spicules.

However, the maternal transcription factors that have been allocated to the various territories do not enable autonomous development (except in the 5th territory). Rather the cells of the various territories exchange signals, leading to a sequence of **inductive interactions**. For example, by presenting signal molecules the micromeres induce, in the neighbouring tier veg2, enhanced expression of some genes and repression of others, leading to the suppression of the most vegetal developmental potency in veg2. In more general terms, by the mutual exchange of signals, the state of determination experiences fine tuning and stabilization. The ratio in which the various cell types are formed is controlled. These principles are realized in all animal organisms.

4.2   An Outsider: *Dictyostelium discoideum*

4.2.1   Cooperation Helps Survival: From Amoeba-Like Unicellular Organisms to a Multicellular Community

Dictyostelium is a simple eukaryotic soil-living amoeba belonging to the animal group ('phylum')

that usually is called "cellular slime moulds". Slime moulds (Mycotozoa) are characterized by spore-bearing fruiting bodies. However, the phylogenetic relationships are still a matter of debate. The "cellular slime mould" *Dictyostelium* is unrelated to multinucleated slime moulds such as *Physarum*, nor is it a member of yeasts or filamentous fungi. Some zoological textbooks speak of *Dictyostelium* as a member of the "social amoebae" (Acrasiales); recent classifications prefer to speak of Amoebozoa, subgroup Mycetozoa. Genomic studies and proteome-based phylogeny place the Amoebozoa after the plant–animal split and classify them as a sister out-group to both the animals and the fungi. The haploid genome comprises 34MB and encodes 12,000–13,000 proteins.

The organism is unusual in several aspects. Under standard laboratory conditions, the life cycle does not start from fertilized eggs, but from single haploid amoebae liberated from an envelope of **spores**. The life cycle of *Dictyostelium* is shown in Fig. 4.8. Note that it is normally entirely asexual. (A curious type of sexual reproduction is known but does not occur under standard conditions: Two cells fuse and enlarge to a giant cell by cannibalistically engulfing neighbouring amoebae. The giant cell encysts and undergoes later meiotic and mitotic divisions which give rise to new haploid amoebae.)

During the vegetative life cycle, an unusual transition from the unicellular to the multicellular state takes place: Numerous individual amoebae assemble to form a social community capable of making provisions for unfavourable environmental conditions. *Dictyostelium* has been said to be a 'part-time multicellular organism' formed from previously independent unicellular amoebae.

Dictyostelium lives in rich organic soils and decaying leaves. When moistened, spores seeded by fruiting bodies release haploid cells that exhibit the appearance and mode of living of amobae. They live in films of water, eating bacteria and multiplying by binary fission (**vegetative phase**). Only when their food supply is exhausted, or the substrate is exposed to the risk

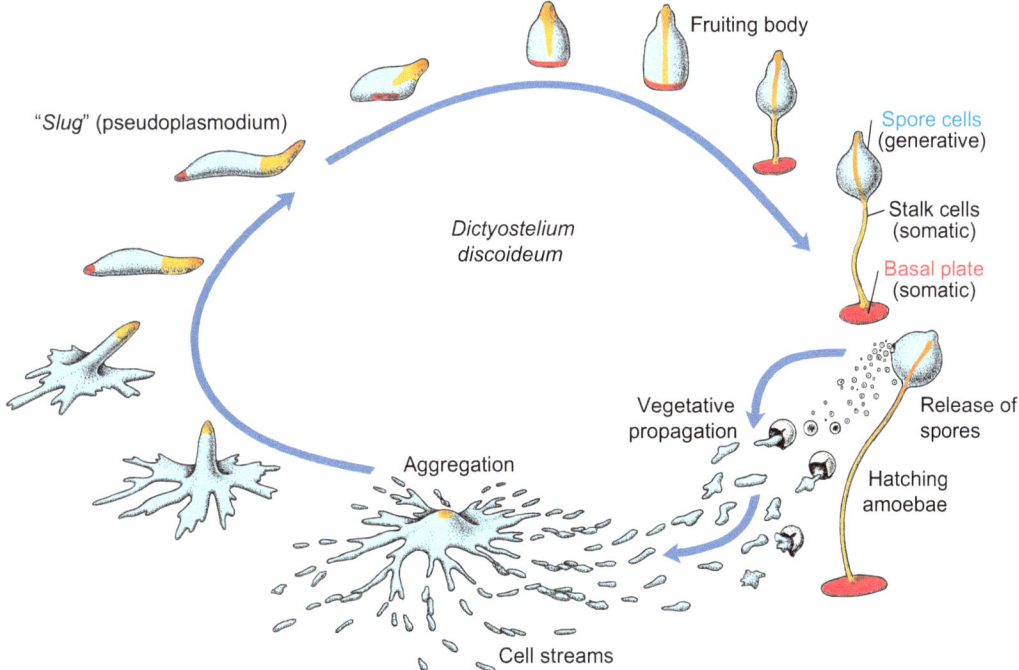

Fig. 4.8 Life cycle of *Dictyostelium discoideum*. Haploid cells hatch from the spore envelope, live as soil amoebae and reproduce asexually by mitotic cell division. As the food supply is exhausted, the amoebae present within a certain area assemble at a point into an aggregate. The size of the area is determined by the range of chemotactic signals emitted by the cells in the centre of the aggregate. The aggregate takes the form of a slug, migrates to an appropriate place and forms a fruiting body that releases new spore cells. (After Gilbert, Developmental Biology, redrawn and modified)

of drying out, do hundreds to thousands of amoebae assemble, migrating singly or collectively like caravans, towards a common meeting place. The aggregate absorbs all of the converging streams of cells. Eventually the aggregate forms a multicellular association that eventually acquires the form of a slug. Correspondingly, the association is colloquially called **slug**. In scientific terms, the association is called grex or **pseudoplasmodium**. The slug is surrounded by a slimy acellular sheet and is able to move much like a true slug. It migrates to a bright location where it transforms into a **fruiting body**, consisting of a basal plate and a stalk that bears a spherical accumulation of new spores beneath its tip. The basal plate and stalk consist of **somatic cells**. The somatic cells form walls made of cellulose, and eventually die. By contrast, the spore cells survive; they are **generative cells** whose formation and release perform the functions of asexual reproduction. Thus, the fruiting body fulfils the **criteria defining a true multicellular organism: Generative, potentially immortal cells segregate from mortal somatic cells, and there occurs division of labour among the somatic cells.**

Although the life cycle of *Dictyostelium* is not at all typical for multicellular animals (in contrast to the life cycle of sea urchins) several features of "*Dicty*"(lab slang) development can be taken as paradigms for similar events in higher eukaryotes. These include: aggregation, cell differentiation, and pattern formation, the spatial order in which the various cell types occur.

4.2.2 In Aggregation Pacemaker Cells in the Meeting Place Rhythmically Emit Chemotactic Signals (cAMP); These Are Strengthened by Amoebae in Their Environment and Relayed in the Periphery

Approximately five hours after removal of the food supply, hungry cells emit an attractant chemical serving to guide nearby starving cells to a central location. In *Dictyostelium discoideum* the chemical signal is cyclic adenosine monophosphate (**cAMP**). cAMP is emitted by the starving cells in synchronous pulses every 5–10 min, and spreads by radial diffusion in the water film. The signal is detected by a surface receptor protein having seven transmembrane domains like many receptor proteins in animal cells. The receptor is coupled to a signal transduction system that uses elements known from animal cells and human leukocytes, in particular elements of the adenylate cyclase-PKA system (Fig. 11.5) and phosphatidylinositol lipids. Two of them, namely phosphatidylinositol (1,4,5)-trisphosphate IP_3 and phosphatidylinositol (3,4,5)-trisphosphate PIP_3 are liberated from the cell membrane and used as intracellular messengers (Figs. 4.9 and 4.10) (More about signal-transducing systems can be read in Chap. 11). The responses of the amoebae include the production and release of cAMP molecules by the receiving cell. Neighbouring amoebae that have received a signal at their surface receptors respond to it by releasing cAMP of their own.

Upon receiving a signal neighbouring cells synchronize their transmitting frequency. About 100 cells effect a hundredfold increase in the signal strength. Such a contingent of synchronized emitters becomes the **pacemaker.** Synchronization serves to increase the speed and range of diffusion of the released extracellular cAMP.

Nevertheless the signalling range would be restricted had *D. discoideum* not invented a second, highly efficient trick. To further speed up the propagation of the signal and to increase the range of signalling, *D. discoideum* has evolved a courier

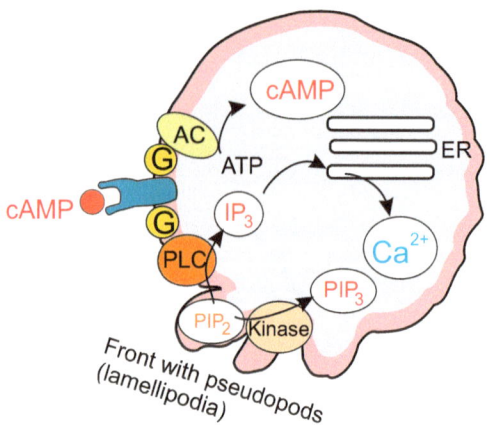

PIP$_2$ = Phosphatidylinositol bisphosphate

IP$_3$ = Phosphatidylinositol (1,4,5)-triphosphate

PIP$_3$ = Phosphatidylinositol (3,4,5)-triphosphate

Fig. 4.9 *Dictyostelium*. Extracellular and intracellular signal molecules that are of significance for chemotactic orientation and migration towards the aggregation centre

system for relaying the signal. After having received a signal each cell becomes an emitter itself. Thus, the local cAMP concentration is increased, the signal is amplified and diffusion accelerated. To ensure that the signal is transmitted from the pulsing central pacemaker to the periphery of the aggregation field, but not backward, each amoeba is unresponsive (refractory) to a further cAMP pulse for about 3 min. The transitory "deafness" prevents the cells from being disturbed by their own signal, or that emitted by their neighbours located farther from the meeting place. Only when the signal has reached inaudible remoteness are the amoebae sensitive to a new signal arriving from the centre.

Each pulse of cAMP induces a jerky movement toward the centre. As each wave arrives, the amoebae take another step towards the central meeting point. Although each cell of the population can approach the meeting place independently, they usually migrate with a caravan. Cells join with each other to form streams, the streams converge into larger rivers of cells, and eventually they all merge at the centre. The final aggregate may comprise up to 100,000 cells.

Fig. 4.10 *Dictyostelium*: Aggregation. (**a**) Amoebae in the pace maker centre (exemplified by 3 out of about 100 amoebae) synchronously emit cAMP in intervals of 2.5–5 min. cAMP signals caught by amoeba in the surroundings stimulate the receivers (1) to move in the direction of the signalling centre and (2) to emit cAMP signals themselves. While the signals spread from the centre into the periphery the amoeba move in opposite direction from the periphery towards the centre. (**b**) In dense aggregates spiral-shaped figures form. In the light of a phase contrast microscope momentarily quiescent cells contrast as dense clusters with stretched cells being in motion at this moment

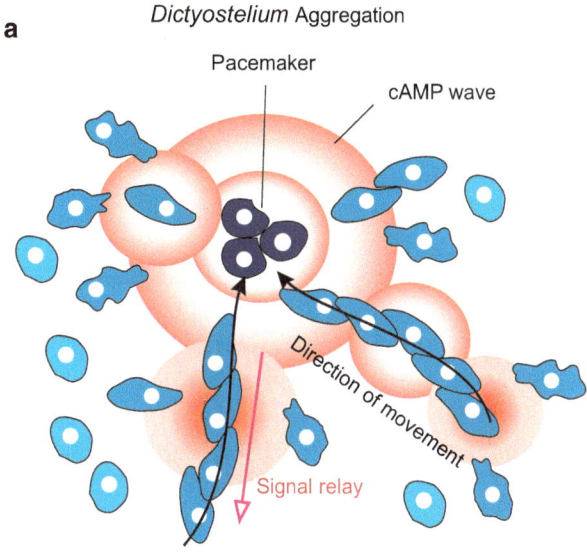

Dictyostelium Aggregation

Pacemaker

cAMP wave

Direction of movement

Signal relay

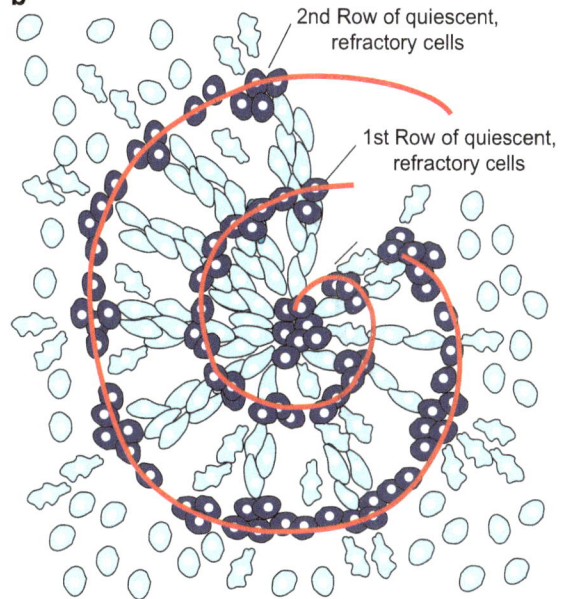

2nd Row of quiescent, refractory cells

1st Row of quiescent, refractory cells

In dense aggregations often charming spiral patterns can occur. Cells being at a given moment refractory, immobile and spherical scatter light differently compared to long-stretched, moving cells. The refractory cells can arrange themselves in spiral tracks (Fig. 4.10). By developing partial differential equations to describe the spreading of the cAMP waves and the responses of the cells, mathematically talented scientists succeeded in simulating these patterns on the screen of their computer. Computer simulations are a special field of modern developmental biology (Chap. 10, Box 10.2)

4.2.3   In the Course of Aggregation Cell Differentiation Takes Place That Requires Communication and Agreement

During the aggregation phase which ends in the formation of the "slug", the genetically uniform cell population segregates into several sub-populations: **prestalk cells**, representing about 20% of the population and positioned at the tip of the migrating slug; **prespore cells**; and the cells of the future **basal plate**, cells which resemble the stalk cells but form the rear of the slug.

Two hypotheses have emerged regarding the division of the amoeboid cell population into prestalk, prespore, and basal plate cells:

(a) The **positional information** hypothesis: the location of the cell within the slug determines its fate.

(b) The **sorting out** hypothesis: cells differentiate before or during aggregation and seek out their place within the slug according to their future role.

Both hypotheses must explain how a characteristic ratio is achieved in the numbers of the three cell types. The ratio is restored after experimental manipulation such as bisecting the slug. Surgical removal of either anterior or posterior cells causes the remaining cells to revise their developmental commitment. Whichever cells find themselves at the anterior end will become stalk cells, whichever are in the posterior region will become spore cells or basal plate cells. Eventually prespore cells and prestalk cells are present in normal ratio (Fig. 4.11). Numerical regulation is possible until cell differentiation has become irreversible in the developing fruiting body. Much evidence favours hybrid hypotheses that combine features of both positional information and sorting out.

Initiation of aggregation appears to require, besides starvation, the accumulation of a **secreted protein** in the cell's environment. The concentration of protein may be an indirect measure of cell density (**quorum sensing**), so that development only starts if sufficient cells are present to form a fruiting body. Experimental

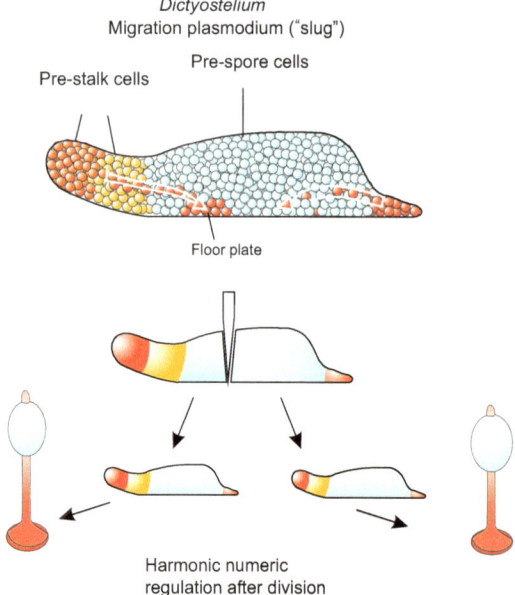

Dictyostelium
Migration plasmodium ("slug")

Pre-spore cells
Pre-stalk cells

Floor plate

Harmonic numeric
regulation after division

Fig. 4.11 *Dictyostelium*. Position of the future stalk and spore cells in the migration plasmodium, the "slug". Upon early division of the plasmodium numerical regulation of the ratio of spore cells to stalk cells is possible

evidence also emphasizes a pivotal role for low molecular weight signal substances in the numerical control of cell differentiation and in positioning of the various cell types. Thus, cells are committed to become stalk cells in the presence of high concentrations of **cAMP** and **differentiation inducing factor (DIF)**, but low concentrations of **ammonium (NH₃)**, conditions that are thought to exist at the tip of the slug.

The choice between prestalk and prespore differentiation is critically dependent on the concentration of DIF. DIF is a chlorated aromate, the nucleus of which is a phenol moiety [1-(3,5-dichloro-2,6-dihydroxy-4-methoxy-phenyl)-hexan-1]. Like cAMP, DIF is liberated from starving and aggregating amoebae. NH₃ is formed when surplus protein is catabolized, for instance in future stalk cells, which synthesize the N-free cellulose of their rigid and enduring cell wall before they die. The conversion of amino acids to cellulose releases ammonium.

There is evidence that additional low molecular weight substances are released to convey information in the transactions that are negotiated among the cells of the community. For instance, the nucleoside **adenosine** is produced from cAMP at the tip of the migrating slug and inhibits the formation of spore cells at the place of its origin. Evaporation of ammonium at the tip of the slug, on the other hand, favours the formation of stalk cells there. Evaporation is facilitated at the elevated tip. Thus, both chemical and physical conditions determine the mode of cell differentiation and the location where a particular cell type will arise.

In sum, *Dictyostelium* first has become a model to study periodic emission of signals, signal relay, chemotaxis and the establishment of cell contacts through cell adhesion molecules. In recent studies aspects of cell differentiation and pattern formation prevail.

4.3   The Immortal Hydra and Further Cnidaria, and the Dawn of Modern Experimental Biology and Stem Cell Research

4.3.1   Systematically Planned Regeneration Experiments With Hydra in 1740 Heralded Experimental Developmental Biology

Regeneration and grafting studies on the small freshwater polyp *Hydra*, performed under a magnifying glass with forceps and scalpels by the Swiss scholar Abraham Trembley 20 years before Mozart was born, rang in the era of modern experimental biology. Trembley had a green hydra fished from the Lake of Geneva. To test whether it was a plant or animal he cut the unknown creature with a scalpel in pieces. In those days European scholars were familiar with ancient Greek literature, and following Aristotle thought that only plants were capable of regeneration. Although he observed an incredible power of regeneration he eventually concluded that the creature was an animal. (It was Linneus who later named it *Hydra* with respect to the ancient water monster who quickly replaced each cut off head with two). Trembley's observations are well documented in detailed descriptions and masterful engravings. Two and a half centuries later, developmental biologists still make use of the enormous capacity of reconstitution and regeneration displayed by *Hydra* and its marine relatives. *Hydra* readily replaces lost body parts – head, foot, or any other part of its tube-like body.

Regeneration studies are supplemented with transplantation experiments (grafts) like those shown in Figs. 4.15 and 10.15. Tissue grafts performed by the American Ethel Brown led in 1909 to the discovery of the phenomenon of **induction**; today also known as **organizer** activity (Chap. 10).

However, *Hydra* is not always co-operative. It is not a useful model to study early embryogenesis from an egg, and a classical genetic approach is very tedious if not impossible. As a rule, gametes are produced only under adverse and stressful environmental conditions, and only in small numbers. The eggs are enclosed in a rigid, non-transparent envelope. The investigator interested in embryogenesis (from the egg to the planula larva) or metamorphosis (from the planula to the polyp) should consider several marine relatives (Sect. 4.3.7).

4.3.2   *Hydra* Represents an Ancient Metazoan 'Bauplan'; Yet, the Cellular Composition of Its Body Is in a Permanent Steady State and the Body Is, Thanks to Stem Cells, as a Whole Potentially Immortal

The tube-like body (Fig. 4.12) displays a unipolar, radial organization with a hypostome encircled by tentacles (colloquially called the "head") at the upper end of the body column and an adhesive basal disk ("foot") at the lower end. The opening at the upper pole serves as mouth through which prey (mainly small crustaceans caught with the surrounding tentacles) is engulfed (or as an anus through which remnants are expelled). The body wall consists of two epithelial sheets, **ectoderm** (epidermis) and **endoderm** (gastrodermis); *Hydra* is said to be diploblastic in contrast to the triploblastic bilaterians which all develop from embryos

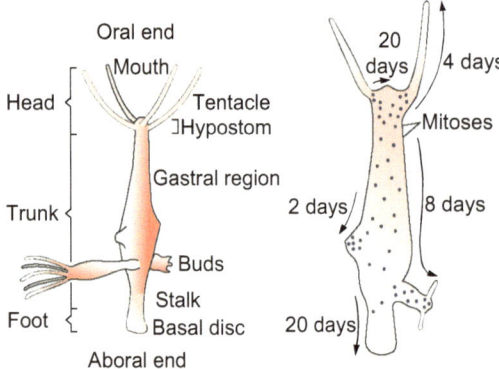

Fig. 4.12 *Hydra* with buds. Right: Dots along the animal indicate sites of cell division; the arrows indicate the direction in which the epithelial cells are displaced. The shift of epithelial cells results from two events: (1) In the terminal body regions used up or aged cells undergo cell death and are phagocytosed or sloughed off, while (2) new cells are born in the body regions between the terminal regions. A surplus of cells is exported in the form of buds. Once a bud has formed a head and a foot, it detaches to live as a new, independent individual. It is genetically identical with its parent (it is cloned)

with three germ layers. The genome of *Hydra* comprises about 20,000 protein-encoding genes.

Hydra is member of the phylum Cnidaria: the simplest metazoans (multicellular organisms) possessing **typical animal cells such as sensory, nerve- and epithelio-muscle cells**.

The inventory of cell types is compiled in Fig. 4.13. Traditionally the cells of a hydra are subdivided into two main categories:

1. **Epithelial cells** determine the basic architecture of the body. Two adjoined epithelial layers, the ectoderm and the endoderm, form the walls of the tube-like body. As these cells also serve as muscle cells mediating the movements of the body column they are also called epithelial-muscle-cells.

2. **Interstitial cells** occur in spaces (interstitia) between the epithelial cells. Interstitial cells include sensory nerve cells, ganglionic nerve cells, four types of stinging cells, a type of gland cell, gametes, and the stem cells of all these cell types.

Despite being equipped with mortal cells such as nerve cells *Hydra* as a whole is **potentially immortal**. Immortality is achieved by an

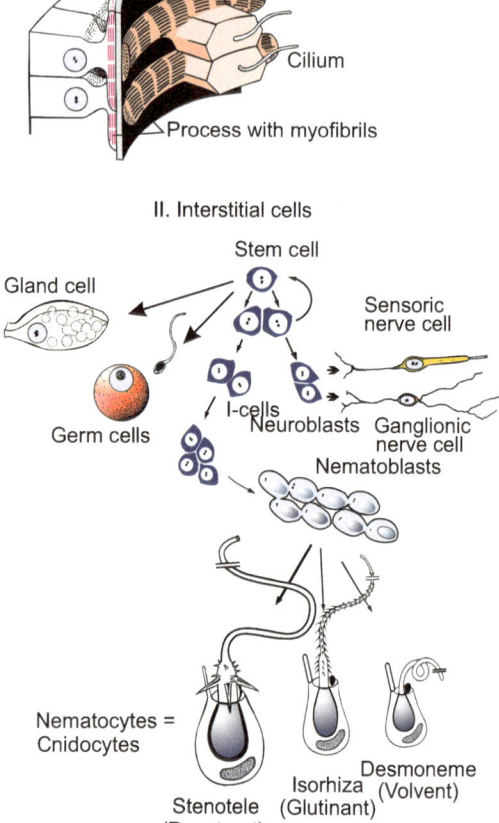

Fig. 4.13 *Hydra*. Basic cell types of a hydra. Diagrammatic presentation; sizes are not to a consistent scale. In the ectoderm, muscle fibres run longitudinally and mediate the contraction of the body column; in the endoderm muscle fibres run circumferentially and mediate the expansion of the body column. Derivatives of the interstitial stem cells reside in the spaces between the epithelial cells (except the gland cells that become integrated into the endodermal epithelium, and the mature nematocytes that are mounted within ectodermal 'battery cells' of the tentacles)

unlimited capacity for self-renewal: the polyp can replace any aging cell, or cells having fulfilled their tasks, by substitutes generated from immortal **stem cells**. In this process of perpetual renewal all cells are replaced even the nerve cells. Cell proliferation, cell differentiation and cell migration never come to a standstill. The hydra is a **perpetual embryo**, and although its terminally differentiated cells die, the cell community survives.

According to the cellular inventory, three categories of stem cells are distinguished in the genus *Hydra*. 1st Ectodermal epithelial stem cells, 2nd endodemal epithelial stem cells, 3rd pluripotent interstitial stem cells (abbreviated i-cells). The interstitial stem cells give rise to the interstitial cell lineages: nematocytes, sensory and nerve cells, gland cells and gametes.

The epithelial cells, even those of the head and the foot, arise from epithelial stem cells in the middle of the body column. These cells are not stem cells in the conventional sense because they are differentiated; but they retain the capability of dividing and self-renewal.

The cells born in the middle of the body column are displaced towards the mouth; they become integrated into the growing tentacles or into the mouth cone. Other cells are shifted towards the foot. Having arrived in the terminal body regions (in the mouth cone, the tentacles or the foot), they undergo terminal differentiation and eventually die. They are sloughed off or phagocytosed by neighbours. As a hydra has no defined lifetime this steady state of production and loss of tissue goes on continuously when the animal is fed regularly.

In well-fed polyps a surplus of cells generated in the gastric region is used to produce buds and thus the surplus is exported. Buds detach as self-reliant animals. By budding, *Hydra* clones itself.

4.3.3   Hydras That Lose Their Nerves Survive

Hydras can be deprived of their interstitial stem cells by treatment with chemicals (hydroxyurea) or, in a particular sensitive strain, by a transient increase of the temperature. As a consequence, the animals can no longer replace used up nematocytes and aged nerve cells. Astonishingly, if artificially fed on a regular basis the animals survive in spite of becoming, in the long term, deprived of their entire nervous system. Nerveless hydras can regenerate their head and foot, and even produce nerve-free offspring by

budding, and a clone of nerve-less hydras can be established. When new stem cells are introduced through transplants new nerve cells and nematocytes occur and the polyps regain a normal appearance and behaviour.

4.3.4   Hydras Can Be Dissociated Into Single Cells; a High Capacity for Self-Organization Enables Aggregates of Such Cells to Give Rise to Whole Animals

Astonishingly: The hydra polyp can be dissociated experimentally into single cells, which sink to the bottom of the dish, creep around like amoebae, re-establish contact with one another and form aggregates of cells, called **re-aggregates**. Initially, re-aggregates are disorderly arranged clumps of cells and inconspicuous. But within days or weeks, the aggregates organize themselves to form new, viable polyps. Reorganization takes place in steps.

Sorting Out. First ectodermal and endodermal cells segregate. Amoeboid movements of cells and selective cell adhesion enable and cause endodermal cells to form an inner ball-like mass that becomes enveloped by an extending sheet of ectodermal cells. Eventually a blastula-like hollow sphere is formed with an endodermal inner and an ectodermal outer layer.

Pattern Formation. Following the reconstitution of an epithelial organization a process of pattern formation takes place, beginning chaotically and ending with order (Fig. 4.14). Bunches of tentacles emerge in initially irregular arrangements. Mouth cones push their way between them; the more that do so the larger an aggregate is. The positions of the tentacles are shifted in circles around the mouth cones, superfluous tentacles are reduced, new ones fill the gaps. Each "head" so formed becomes the organizer of a body axis; this emerges from the aggregate and lifts up the head at its end. Eventually feet are formed and these have the faculty to segregate. From the multi-headed monster

Fig. 4.14 *Hydra.* Reorganization in an aggregate of dissociated cells. At the end of this process that takes several days complete polyps of different sizes detach by autonomous splitting of the feet. After a preparation by the author WM. (Since the emerging polyps were fed buds also appeared which are omitted for clarity.) *Wnt* expression after Hobmayer et al. (2000)

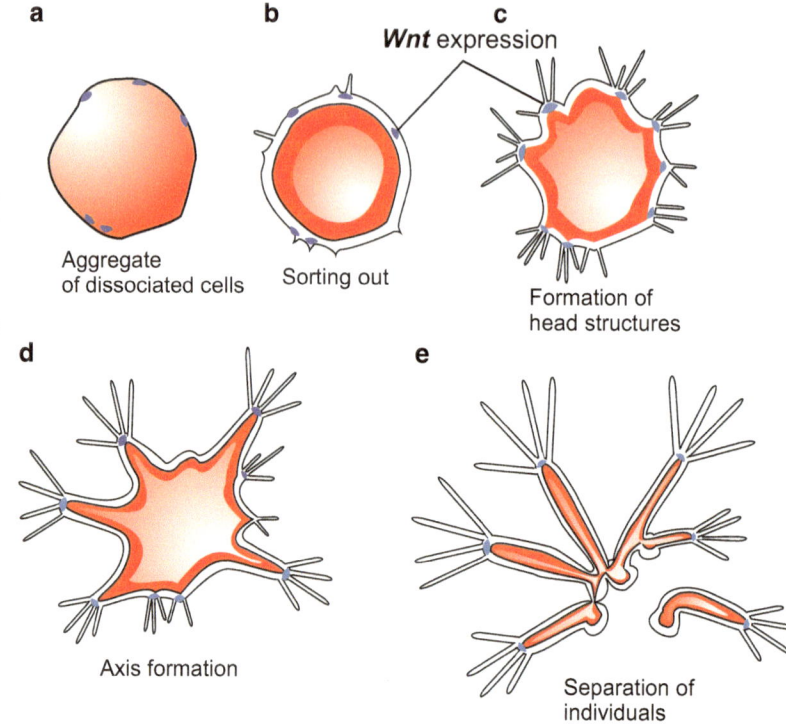

a

Aggregate
of dissociated cells

b

Sorting out

c

Wnt expression

Formation of
head structures

d

Axis formation

e

Separation of
individuals

single, normally-organized polyps detach one after the other. This process of reorganization is called **reconstitution**.

4.3.5 Permanent Cell Renewal Presupposes Positional Information; This Is Analyzed in Classic Regeneration and Transplantation Experiments

Recent *Hydra* research has emphasized, besides self-renewal through stem cells, patterning and the regulation of cell differentiation. **Pattern formation** and **positional information** have been investigated in the context of regeneration and reorganization. Depending on where a group of cells is, they can organise themselves to form a hypostome with tentacles ("head"), gastric region, stalk, or basal disk (the "foot") (Fig. 4.15a). But how do the cells recognize where they are? How positional information is supplied and how the body pattern is corrected

after experimental interference will be discussed in Chap. 10. Here some findings are summarized.

- The head functions as an **organizer**, comparable to the Spemann organizer in the amphibian embryo (Sect. 5.1 and Chap. 10). Small pieces taken from the head region and transplanted into the trunk influence the trunk cells in their environment in such a way that they participate in formation of a complete additional head (Figs. 4.15b, 10.19, and 10.20).

- The capability of forming and organizing a head decreases along the body column in a graded fashion while the capability of differentiating a foot increases. These reciprocal capabilities are rather stably encoded in the tissue along the body column but can be changed in regeneration. In terms of L Wolpert"s **positional information theory** (Chap. 10, Box 10.2) these reciprocal, locally changing capabilities are united and called **positional value**. A high positional value means a high competence to organizing a head and trunk coupled with an inverse, low competence to form a foot. When a hydra is

a

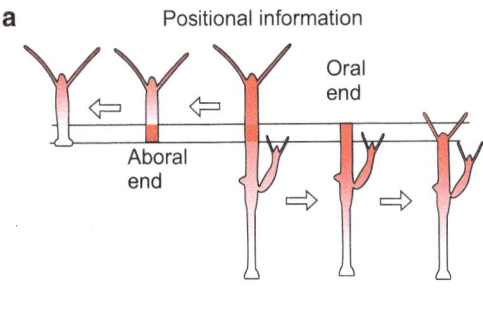

Positional information

Oral end

Aboral end

b

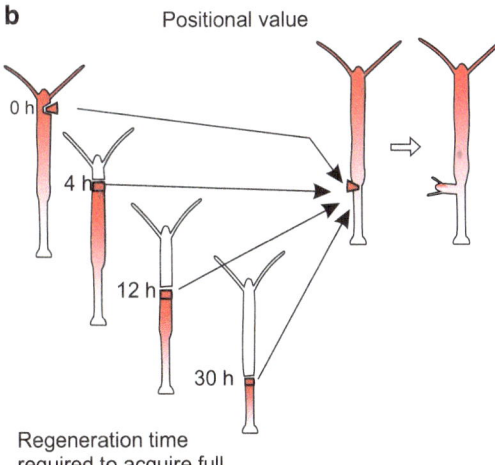

Positional value

0 h

4 h

12 h

30 h

Regeneration time
required to acquire full
potency for head induction

Fig. 4.15 *Hydra*. Positional information and positional values. (**a**) Simple regeneration experiment providing evidence for the existence of positional information. One and the same group of cells forms gastric region, or head, or foot according to its position along the body column. (**b**) The positional values become evident in the position-dependent capability of forming head structures and inducing head-bearing secondary axes when transplanted into a lower (standard) position of other polyps. The inducing capacity is highest in the head region (hypostome and close beneath the tentacles). The lower down the body column of a polyp is cut and its apical part removed, the longer takes it until the apical tissue of the regenerating body column gains the full power to induce an axis

cut and the trunk caused to regenerate the lost head the positional value at the regenerating tip increases, as measured by the speed and frequency it induces a head in the grafting assay (Fig. 4.15b).

- There are long range interactions between the body parts. An existing head suppresses the formation of another head along the body while it aids in forming a foot. The head-inhibiting activity decreases with distance from the head in a gradient-like fashion.

4.3.6   In the Establishment of the Head and the Body Axes Signalling Systems Known from Vertebrates Are of Significance, in Particular the WNT/Beta-CATENIN Cascade

Tools of molecular biology are used to search for molecules putatively involved in the control of regeneration and budding. Using methods of inverse genetics (Chap. 12) elements of the 'canonical' WNT signalling system were found and attracted high attention. "Canonical' means literally "according to normal standards", practically it means "controlling gene activities through β-catenin". As will be summarized in Chap. 22 (evo-devo, Figs. 22.6 and 22.7) the canonical WNT system is used in a wide variety of animal groups to direct and pattern one of the body axes; in Cnidaria it is the main oral-aboral body axis of developing embryos and of emerging buds. As in other organisms, in cnidarians the signal pathway comprises (among other molecules) the **signal molecule WNT3a**, the intracellular enzyme GSK-3, the transcription factor TCF and the **transcription cofactor β-catenin**.

In the cnidarian *Hydractinia* maternally transcribed mRNA for WNT3a, TCF and β-catenin is deposited at the upper egg pole that gives rise to the tail pole of the larva and in metamorphosis to the head of the polyp (see Fig. 9.5). Increased activation of the pathway results in multi-headed polyps. Likewise, in adults still growing polyps of *Hydractinia* as well as of *Hydra*, stimulation of the WNT/beta-CATENIN cascade results in the formation of **supernumerary heads** (Fig. 4.16b). In the growing buds of regenerating animals, and in aggregates of dissociated *Hydra* cells the appearance of organizing power is heralded by and associated with the occurrence of mRNA for beta-CATENIN and TCF, followed by a strictly local spot of WNT3a transcripts at the tips of the emerging heads. Upon stimulation of the cascade, colonies of *Hydractinia* (Fig. 4.17) develop giant buds which segregate into many multi-headed polyps. The significance of the WNT/beta-CATENIN cascade for patterning will be further discussed in Chap. 10.

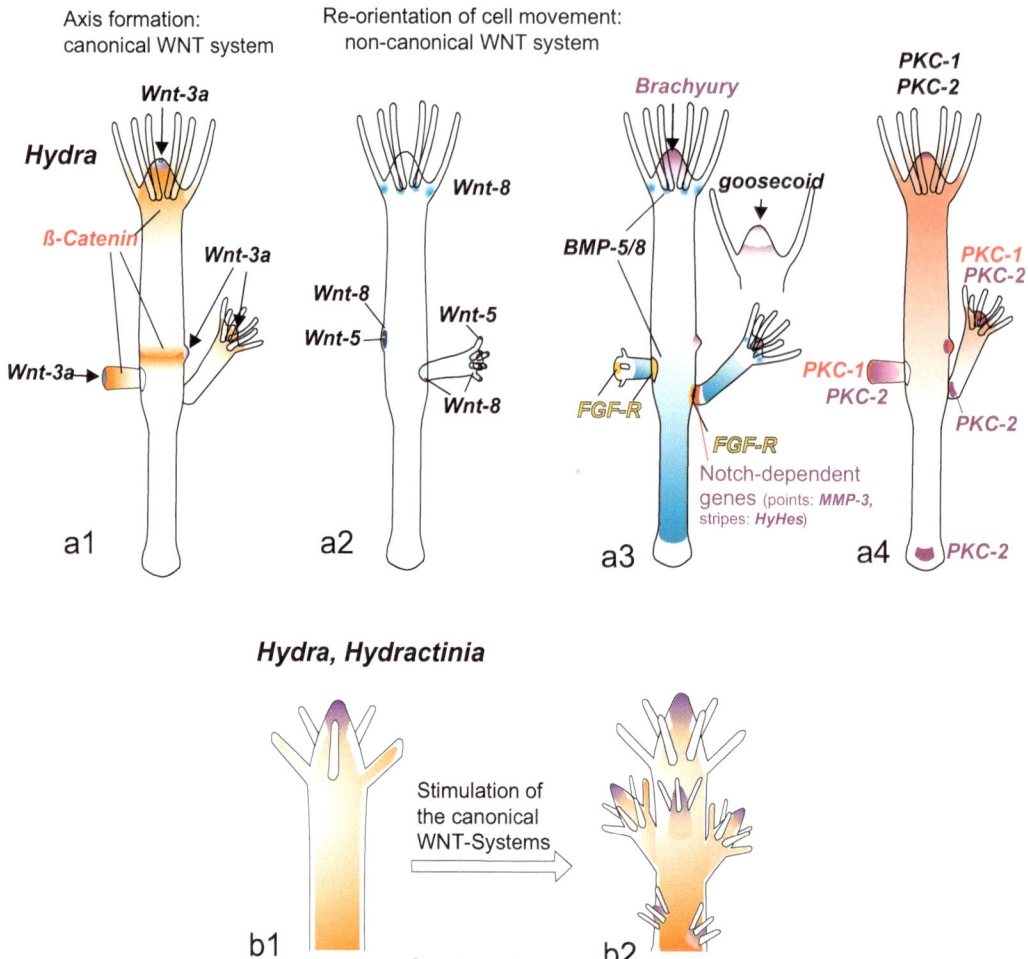

Fig. 4.16 *Hydra* and *Hydractinia*: Expression of genes playing significant roles in the construction of the body during budding and in the reconstruction during regeneration. Only gene activities are shown that are present in homologous form also in vertebrates. (**a1**) Sites of expression (expression domains) of elements of the canonical WNT-system responsible for the formation and patterning of the body axis; (**a2**) Non-canonical WNT system that correlates with redirection of cell movements (In vertebrates Wnt8 is classified among the canonical WNT's but in *Hydra* its amino sequence corresponds more with the non-canonical WNT11 of the vertebrates – Hobmayer, pers. information). (**a3**) and (**a4**): Expression sites for further genes conserved in evolution, actually for secreted factors (BMP, FGF), elements of signal transduction pathways (PKC) and for transcription factors (BRACHYURY, GOOSECOID). More about signal systems in general can be read in Chap. 11. (**b**) Multiple head structures induced by stimulating the canonical WNT system

As a rule basic body patterns are not controlled by one signalling system alone but by networks of mutually interacting controlling systems. Thus, in *Hydra* as well as in *Hydractinia* multi-head formation can also be induced by stimulating the PI-PKC-signalling system (Figs. 3.5 and 11.5) using, for instance, the second messenger diacylglycerol DAG.

In reorienting movements of cells as it occurs in budding also the **non-canonical WNT signalling system** is involved. This system (exposed in Chap. 10, Sect. 10.5.5) directs polarity in sheets and guides movements in many systems. For **detachment of the buds** another well known signalling system is of significance, the **NOTCH**-system, in cooperation with

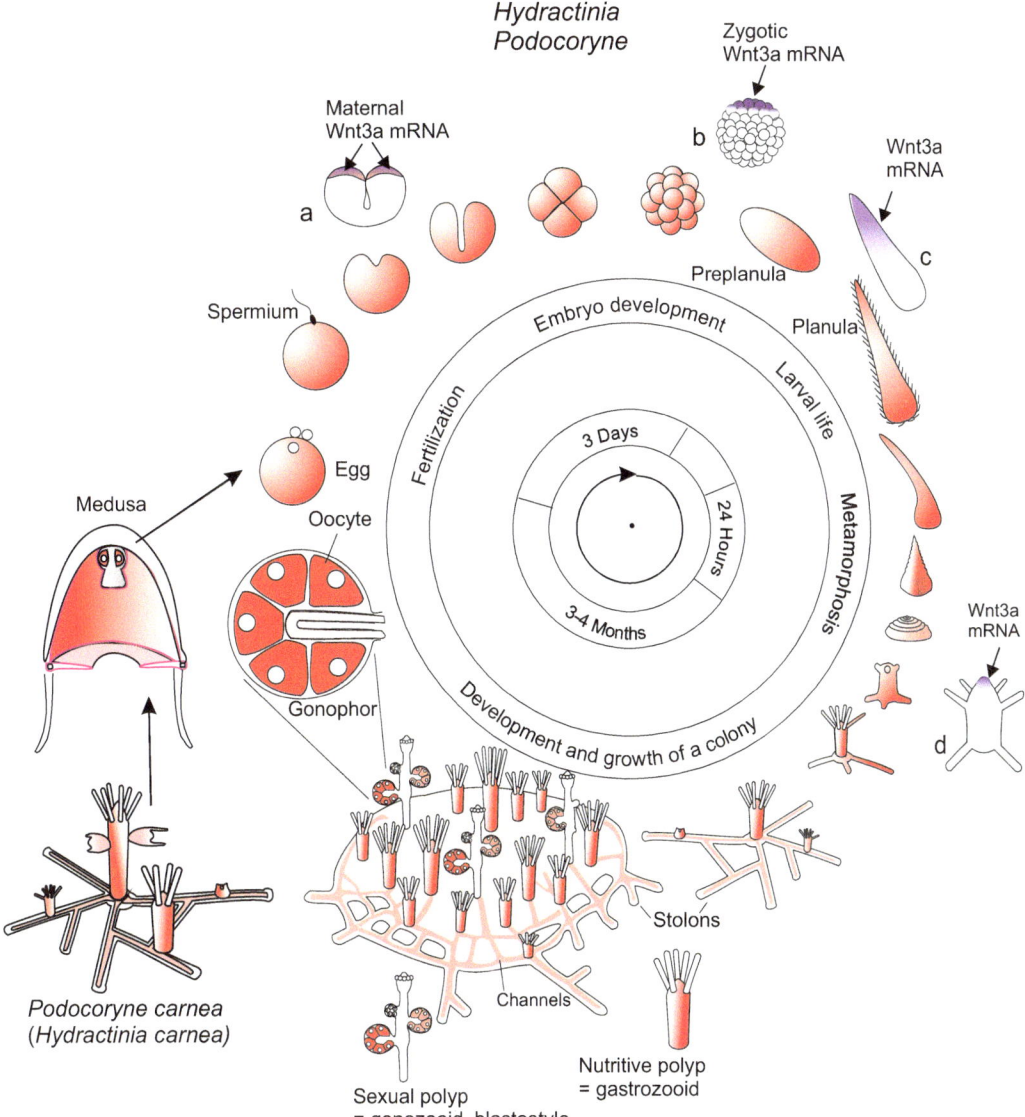

Fig. 4.17 *Hydractinia echinata*, a colonial hydrozoan frequently found along the coast of Northern Europe on shells inhabited by hermit crabs. A closely related species with identical life cycle, *Hydractinia symbiolongicarpus*, is found along the Atlantic coast of North America. *Podocoryne (Hydractinia) carnea* is another sibling species of warmer seas, that exhibits a metagenetic life cycle including a medusa (jellyfish). The colonies of all these species grow preferentially on shells inhabited by hermit crabs but can easily be cultured in the laboratory on any substrate. The additional figures (**a–d**) show the expression pattern of WNT-3a

receptors for signalling molecules of the **FGF** class. Genomic data indicate that many more signalling systems known from vertebrates are at work also in the cnidarians. Examples of some gene activities are shown in Fig. 4.16.

The site of expression of ***brachyury*** raises an interesting and long debated question: **Can the**

head of a hydra be set equal (homologized) with of the head of the vertebrates? Studies with the Anthozoon *Nematostella* (Sect. 4.3.8) led to the assumption that the mouth of the Cnidaria corresponds to the blastoporus. In the Deuterostomia, including sea urchins and vertebrates, the region of *brachyury* expression

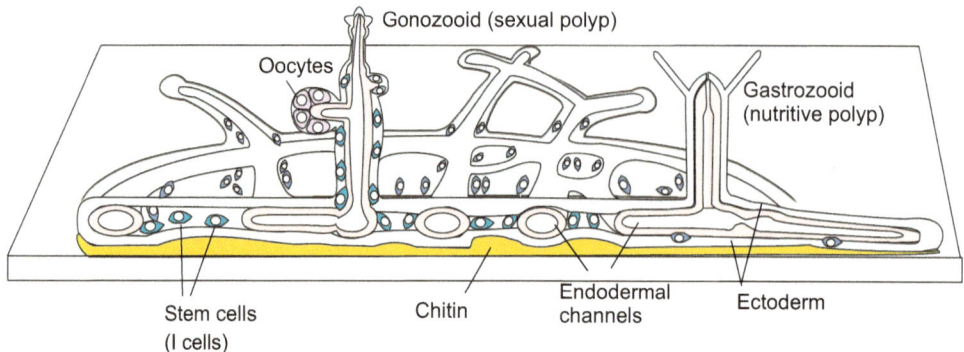

Fig. 4.18 *Hydractinia*: Structure of a colony and locations where stem cells multiply and migrate

marks the site where the blastoporus is formed. Yet, in the Deuterostomia this porus becomes the anus and a second, final mouth is perforated later. Cnidaria do not belong to the Deuterostomia; the primary mouth stays to function as the definitive mouth. The discussion about homology or non-homology of heads and of the mouth will be resumed in the final Chap. 22.

4.3.7 *Hydractinia*: In This and Related Hydrozoan Species Stem Cells Were First Detected; Their Observation Led to the Proposal That a Germ Line and Primordial Germ Cells Exist

Pioneering Role in Stem Cell Research. In the history of developmental biology marine species of the Hydrozoa once played important roles. Many well known researchers such as Thomas Hunt Morgan performed regeneration studies using, for example, polyps of *Tubularia*. In an extensive study the zoologists August Weismann, known as founder of neo-Darwinism (Darwinism without inheritance of acquired characteristics), investigated the occurrence of germ cell precursors in a large array of marine species, among them *Hydractinia echinata* (Fig. 4.17). This representative marine colonial hydroid was the first organism in the history of biology in which migratory precursors of germ cells were described and termed "*Stammzellen*"

("stem cells"). They were interpreted as being **primordial germ cells** and as such members of **the germ line** (Chap. 8).

Recent Stem Cell Research. The stem cells, now known as interstitial cells (i-cells, Figs. 4.13, 4.18, and 18.12) are not only germ cell precursors that express general germ line markers such as *Vasa, Piwi* and *Nanos* but can differentiate into any cell type of adult animals including epithelial cells (in contrast to *Hydra's* i-cells). Evidence has been deduced experimentally. Colonies were deprived of their own stem cells by treatment with alkylating cytostatics. The colonies, prone to die in the long term, were recovered by transplanting i-cells from the same clone, or from (histocompatible) siblings having the opposite sex, or from mutant donors that differed from the host in morphological traits (forming, for instance, multi-headed polyps). With time the receiver not only recovered but adopted all the qualities of the donor including its sex. Forced expression of the Oct4-like transcription factor in epithelial cells transformed them back into stem cells that develop neoplasms, that are tumour-like aggregates.

Life Cycle *Hydractinia* is a colonial form consisting of hundreds of cloned polyps linked together by vascular channels called stolons (Fig. 4.17). Mature colonies of are preferentially found on shells of hermit crabs. In the European North Sea the genus is represented by *Hydractina echinata*, along the Atlantic coast of North America by *H. symbiolongicarpus*.

Gametes are shed into the free water where embryonic development takes place. The larva settles on substrates covered with certain substrate-bound bacteria and undergoes metamorphosis into a primary polyp, the founder of a new colony. In the lab, bacteria are replaced by agents that depolarize and thus stimulate sensory cells of the larva. By releasing internal neuropeptides these sensory cells trigger and synchronize the internal processes of metamorphosis.

Allorecognition. Colonies of *Hydractinia* are endowed with the ability to recognize genetically non-identical (allogeneic) colonies to which they come in contact when both colonies grow on the same substrate. Recognition is based on surface-exposed molecules of the immunoglobulin class, an astonishing finding. Colonies identified to be allogeneic are attacked and destroyed by discharging toxins from a special type of nematocyte. The immunoglobulins are highly polymorphic - each individual can have its own edition - serve as identity cards and mediate, in addition, histo-compatibility or -incompatibility. In rare encounters of compatible sibling colonies migrating interstitial stem cells can invade the neighbour, displace its germ cell precursors and take over germ cell production (germ cell parasitism). It has been proposed that the potential of stem cells to invade neighbouring colonies occupying the same substrate may have provided selection pressure for the evolution of allorecognition and histo-incompatibility. Hence, *Hydractinia* have now attained the position of a powerful model in stem cell research, axis formation and allorecognition.

Other Hydrozoan Species. Related species used in various labs are *Podocoryne* (*Hydractinia*) *carnea* and *Clytia johnstoni*. Their life cycle includes a medusa (jellyfish) (Fig. 4.17). As in other animals, the state of differentiation is not always irreversibly fixed. Rather cells can change their state and a muscle cell may transform into other cell types. Such processes of transdifferentiation have been studied using isolated cross-striated muscle cells prepared from medusae, for instance of the species *Podocoryne carnea* (Fig. 18.13).

4.3.8   *Nematostella*: This Sea Anemone Represents the Anthozoa, (Corals) Considered to Be a Basal Group Among the Phylum Cnidaria

Besides the hydrozoan species introduced above a further cnidarian species has gained the position of a model organism. It is the small, about 2 cm long sea anemone *Nematostella vectensis* (see Fig. 4.1). The solitary animal inhabits brackish estuaries of England or America digging itself into the sediment. It can easily be raised in the laboratory to sexual maturity, and releases gametes into the surrounding water. The Anthozoa generally lack a medusa generation (jellyfish) and therefore are considered by some biologists who construct phylogenetic trees as the most basal member of the phylum Cnidaria. Embryonic development follows standards and gastrulation occurs in the archaic mode of invagination (Fig. 2.4). However, gastrulation takes place not at that pole that in the traditional terminology of developmental biology is called the "vegetal" pole but at the "animal" pole where the polar bodies are given off. In this respect anthozoans disregard the traditional standard textbook model, as displayed, for example, by the sea urchin embryo. This inverse relationship is shown in Fig. 22.7 and will further be discussed in Chap. 22. It has not always been taken into account in treatises aiming at reconstructing phylogenetic lines and branchings.

Irrespective of the fact that gastrulation in *Nematostella* and the sea urchin take place at opposite poles, in both organisms gastrulation starts, and the blastopore forms, in that region of the embryo where maternal β-CATENIN is taken up by the nuclei and expression of *Wnt* commences. Embryonic development in *Nematostella* leads to a classic planula larva which undergoes metamorphosis into a polyp similar to the metamorphosis of *Hydractinia* shown in Fig. 4.17. The blastopore of *Nematostella's* planula is identical with the definitive mouth.

Particular stem cells were not yet identified in *Nematostella* or other anthozoan species. The task of stem cells to refill the pool of nematocytes and to replace aged nerve cells is assigned to apparently

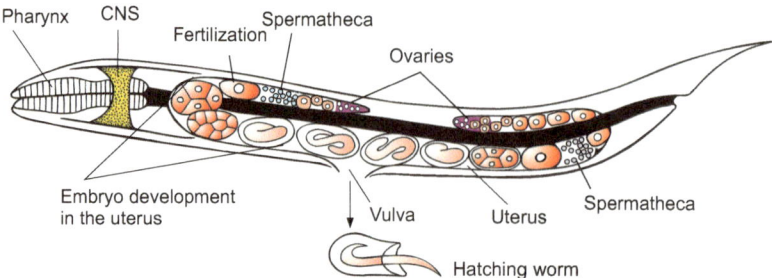

Fig. 4.19 *Caenorhabditis elegans (C. elegans)*, a nematode (roundworm, colloquially called "the worm"). Morphology of the hermaphroditic, viviparous nematode. The gonad has two arms; each arm contains not only oogonia but also sperm stored in the spermatheca. Embryogenesis takes place in the last part of the gonadal tubes, the uterus

differentiated cells that have retained high plasticity and high powers of transdifferentiation.

Nematostella's genome is sequenced. A survey of the inventory of genes in the Cnidaria has brought an unexpected finding. **Cnidaria possess a large array of genes that display homology, that is high sequence similarity, to mammalian, including human, genes**. This applies in particular to genes encoding signal molecules, members of signal transduction cascades (Chap. 11) and transcription factors. With respect to certain classes of genes Cnidaria resemble vertebrates more than the invertebrate model organisms *Caenorhabditis elegans* and *Drosophila melanogaster* do.

4.4 *Caenorhabditis elegans,* Example of Invariant Cell Lineages

4.4.1 A Small Roundworm (Nematode) With Extremely Precise Embryonic Development Got Ahead in the Laboratory

Forty years ago the molecular biologist Sydney Brenner proposed intensive investigation of the small transparent roundworm *Caenorhabditis elegans* (Fig. 4.19). The worm, whose name is usually abbreviated *C. elegans*, is now among the acknowledged reference models in developmental biology, and has proven to be a superb subject for studying eukaryotic developmental genetics, cell biology, neuroscience, and genome structure.

In nematodes, embryogenesis proceeds in a precise, species-specific pattern that is faithfully repeated in every generation. Through careful observation the ontogenetic tree for every somatic cell of the body could be reconstructed. The most complete reconstruction of the cellular pedigree has been achieved in *C. elegans* (Fig. 4.20). This success rests on the worm's anatomical simplicity, its transparency, its genetic tractability and especially on the high precision with which **cell lineages** are reproduced in each individual and each generation (apart from few occasional minimal deviations). Development terminates in each adult with an identical number of cells (**cell number constancy**).

C. elegans is easy to keep in large numbers. However, eggs are fertilized within the uterus of the mother: embryos in cleaving stages need to be dissected from the mother and from the tough outer coat. But they survive and develop perfectly outside the egg shell.

4.4.2 *C. elegans* Permits Sophisticated Genetics; Its Short Life Cycle Starts in Hermaphrodites or With Two Sexual Partners; Mutants Are Easily Produced

The natural environment of *C. elegans* is the soil. Like *Dictyostelium, C. elegans* feeds on bacteria,

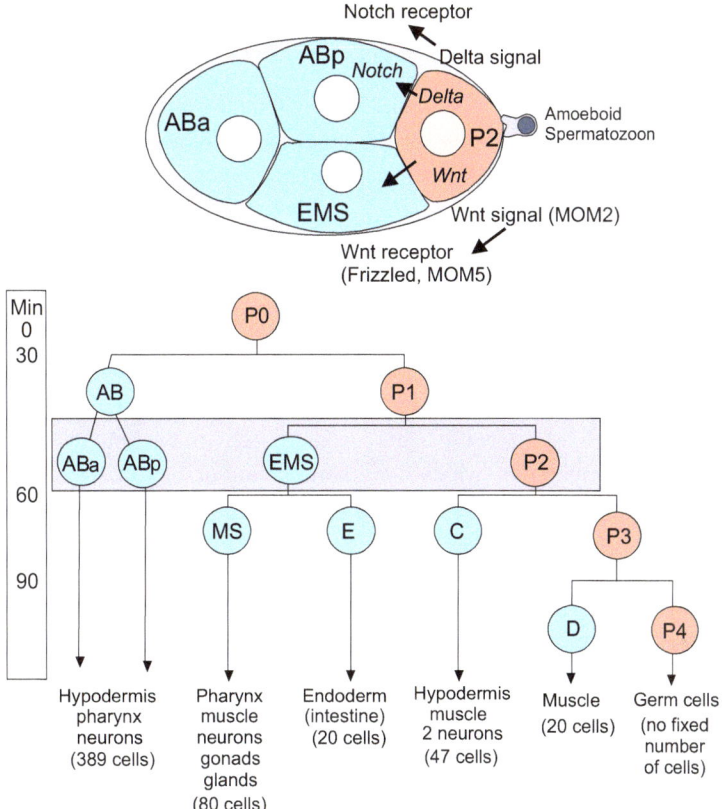

Fig. 4.20 Cell genealogy (*cell lineage*) of *C. elegans*. The P line (*red* cells) represents the germ line; the blue cells represent somatic founder cells. The lineage and exact number of cells is nearly invariant. In addition signalling systems that are effective in the 4-cell stage are indicated. The P2 cell acts upon its ABp neighbour by means of membrane-bound DELTA signals, that are received by the ABp-cell by means of NOTCH receptors. Cell P2 directs a WNT signal onto the EMS cell; the signal catching receptor for WNT is generally known as FRIZZLED. However, traditionally the literature on *C. elegans* uses other names. DELTA is usually called APX-1, NOTCH is called GLP-1; WNT is called MOM-2, and FRIZZLED MOM-5. The signals specify the further fate of the addressed cells. The general anterior-posterior polarity has been determined by the spermatozoon which in *C. elegans* moves like an amoeba and penetrates the egg envelope at one of the egg poles; this pole is thus selected to become the anterior pole of the animal

and in the laboratory a similar technique is used to culture both organisms: bacterial lawns are grown on agar plates, and then the plates are inoculated with the organisms of interest: spores of *Dictyostelium* and in *C. elegans*, embryos enclosed in their transparent egg shell or hatched worms. The life cycle is short (3.5 days); embryogenesis lasts about 12 h at 25°C and 18 h at 16°C.

C. elegans is usually hermaphroditic. When self-inseminated, the wild-type worm will lay approximately 300 eggs in its life from which

L1 larvae hatch. Separated by moults three more larval stages (*L2*, *L3* und *L4*) follow. When inseminated by a male, the number of progeny can exceed 1,000. How do males come into play?

The hermaphrodites have XX sex chromosomes and are female in appearance and anatomy, but they are capable of producing not only eggs but also sperm in their tubular gonads. Self-fertilization results in endogamy, and as a convenient consequence of repeated inbreeding, mutated genes (new alleles) are homozygous as early as in the F2 generation. The occasional loss of an X chromosome by non-disjunction results in XO males at a frequency of 0.2% (the corresponding XXX embryos are not viable). The X0 males mate with hermaphrodites, which now play the part of true females. Thus, in *C. elegans* cross-fertilization as well as self-fertilization is possible. By cross-fertilization new alleles can be imported.

Mutants can conveniently be produced using transposable elements (transposon-like DNA sequences) which are injected into the syncytial gonad and, if inserted into a gene or into the promoter region of a gene, can destroy its function (Chaps. 12 and 13). Following self-fertilization a recessive mutation which in hermaphrodites is present in the heterozygous state will be automatically homozygous in a quarter of offspring in the F2 generation and thus become phenotypically apparent. Samples from mutant strains can conveniently be stored in the freezer for long-term storage.

4.4.3 In Embryogenesis Exactly Determined Cell Lineages Result in Individuals Consisting of a Constant Number of Cells; Programmed Cell Death Contributes to Cell Constancy

Embryonic development takes place in the proximal half of the gonadal tube, called the uterus. In addition to the transparency of the mother and embryo, the embryo can be removed from the uterus and its envelope without killing it. A favourite method facilitating the mapping of cell fates is to inject permanent markers such as a fluorescent dye, labelled antibody or reporter genes into the embryo. Following injection of reporter genes such as *GFP* or *LacZ* (Chap. 12) not only will the injected blastomeres be labelled, but their descendants as well. Some cell lines have natural differentiation markers: in the E-cell lineage leading to the gut, for example, rhabditin granules (of unknown function) can be detected with polarization optics. The original descriptive analyses of cellular ontogeny in *C. elegans* have been supplemented by the

study of numerous mutants displaying disturbed cell lineages. For eliminating defined cells, a laser beam can be used. This powerful combination of tools from genetics and cell biology is unique to *C. elegans* and has justified Brenner's prescient enthusiasm on its behalf.

These patterns of cell lineage are largely invariant between individuals. The worm terminates its embryonic development, and is born, with 556 somatic cells and 2 primordial germ cells. It undergoes 4 larval stages separated by moults. The larval period lasts 3 days. At the end of its development, the mature worm has 959 somatic cells and about 2,000 germ cells if it is a hermaphrodite, or 1,031 somatic cells and about 1,000 germ cells if it is a male. The nervous system consists of 302 nerve cells; these arise from 407 precursors, of which 105 undergo a **programmed cell death** (apoptosis).

4.4.4 The Cell Lineage Leading to the Primordial Germ Cells Is (Worthy) of Particular Note; the Germ Line Is Marked by Specific Molecular Components

Cell divisions frequently are **asymmetric** (Fig. 4.20; see also Fig. 9.7). That is, the two daughter cells resulting from mitotic division of a progenitor cell are equivalent with respect to their inherited genetic information but not with respect to their cytoplasmic contents, and they have different destinies. The **germ line** is particularly impressive in nematodes. By definition, a germ line leads from the fertilized egg to the primordial germ cells and thus comprises the cells capable of forming gametes (Fig. 4.20). In *C. elegans* the cells lying on this line are termed P0, P1, P2, P3, P4. The last P cell, P4, is the primordial germ cell. The cells P0 to P4 are characterized by a peculiar legacy: asymmetrical cell divisions allocate cytoplasmic **P-granules** only to these generative cells but not to their sister cells destined to become somatic cells.

The P-granules are not enclosed with a typical bi-layered lipid membrane; they contain proteins, various species of RNA and mitochondria. The granules resemble and correspond to the germ-line specific granules found in many other organisms, for example in the annelid *Platynereis* (Sect. 4.5.3). Some components are known to be essential for the germ line to give rise to germ cells. One of the components that occur in the entire animal kingdom and marks the germ line are products (mRNA or protein) of the gene *vasa*. The protein VASA, in *C. elegans* called GLH-1, has been identified as an RNA helicase which binds and disentangles RNA.

In the roundworm *Ascaris* (synonym *Parascaris*) the chromosomes remain complete only in the germ line, whereas the somatic cells generated by asymmetric divisions lose some chromosomal material (**chromatin diminution**). (Theodor Boveri who described and analyzed this phenomenon in 1910 suggested the abandoned chromatin might be of significance in the development of the germ line. The possible significance of this strange phenomenon will be discussed in Sect. 12.6.2)

We return to *C. elegans* where chromatin diminution is not known. Analysis of the cell pedigree in *C. elegans* revealed that a few founder cells produce only one tissue: blastomere E gives rise to all of (and only) the gut. But this specialization is not the rule. Normally, equivalent lineages give rise to cells of more than one type along the length of the embryo. Conversely, most tissues are derived from several founder cells, which in turn give rise to other tissues. Such tissues generated by several embryonic founder cells are of **polyclonal origin**. For instance, nerve cells originate from cells whose descendants can also participate in the generation of muscles. Muscle cells, on the other hand, originate from three pluripotent founder cells. Thus blastomere C produces, besides muscles, neuronal and hypodermal cells.

In tracing cell lineages the term "stem cell" is avoided in the recent literature on *C. elegans* because no daughter cells with the capacity for self-renewal is left in the hatched worm; instead the term "**founder cell**" is used. In the adult animal there seems to exist only one cell type that corresponds to the traditional definition of stem cells, giving rise to differentiated cells and further stem cells as well, this is the primordial germ cell.

4.4.5 Cell Type Specification Is Not Only Based on Cytoplasmic Determinants Transmitted in the Genealogy; Initially Also Cell Interactions Between Neighbouring Cells Contribute to Cell Type Commitment

The high precision with which pedigrees of cell lines can be reconstructed for long supported the traditional view that *C. elegans* represents a standard mode of development termed "autonomous" or "mosaic development". It was supposed that visible components (such as the P-granules) or invisible determinants (such as mRNA) were present already in the unfertilized egg in a precise spatial pattern. In cleavage these would be allocated to the founder cells and would determine the fate of their descendants. This view was seemingly corroborated by first experimental interferences: Killing of certain blastomeres resulted in irreparable defects.

Yet, a careful analysis of numerous mutants as well as surgical deletions of founder cells have led to the view that the fate of each cell is determined not only by cytoplasmic constituents allocated to it in early embryogenesis but to a great extent also by early and precise interactions between adjacent cells. These are based (mainly or exclusively) on surface-exposed signalling molecules presented to receptors in the cell membrane of neighbouring cells. Two such signalling systems, well known from many animal organisms including vertebrates, are installed already in the 4-cell stage; these are the NOTCH/DELTA system and a WNT system. DELTA (in *C. elegans* known as APX-1) and WNT (in *C. elegans* called MOM2) designate proteins serving as signalling molecules, while NOTCH (GLP-1) and FRIZZLED (MOM5) designate the corresponding receptors. (More about these often employed signalling systems can be learned from Chaps. 11 and Fig. 11.6).

4.4.6 After Programming of Cell Fates Cell Migration Over Distances Brings Together Cells With Shared Identities; This Is a Particular Mode of Pattern Formation

When the position of a blastomere is shifted to another location it can adopt a different life course dependent on its new environment. Yet, already in the 32-cell stage the fate of the cells is fixed; a killed cell cannot be replaced. Amazingly, before and even after the 28-cell stage single cells move and shift their position. Cells with the same destiny aggregate in groups. This is a rather unusual mode of pattern formation. In *C. elegans* often cells get an identical or similar fate allocated far apart from their final destination; then they move and join other cells committed to the same fate.

4.4.7 *C. elegans* Was the First Animal to Have its Genome Completely Sequenced

C. elegans is small and so is its genome. The total amount of DNA is merely 1/38th of the human genome. Therefore, in a pioneering project sequencing of its genome was started early and finished in 1998. Thus *C. elegans* became the first animal to have its genome completely sequenced. The search for potentially protein-coding sequences found about 19,000 candidate sequences, not much less than found in *Drosophila* and even in humans. The genome sequence of *C. elegans* is an established high-quality reference genome, available in the data base WormBase.

A useful feature of *C. elegans* is that a relatively straightforward method can be applied to disrupt the function of specific genes by RNA interference (RNAi). The method has been developed with this worm and is still most successfully applied in *C. elegans*. By introducing selected mutant or alien genes of interest also various transgenic strains have been produced and made available to researchers world wide.

4.5 Spiralians: A Recurring Cleavage Pattern

4.5.1 Spiral Cleavage Characterizes Several Animal Phyla

In the animal kingdom, spiral or oblique cleavage (Figs. 4.21 and 2.4) occurs in several phyla: in the Acoelomorpha (former acoelous turbellarians), plathelminths, nemerteans, annelids, and molluscs (excepting cuttle fish). These phyla are grouped under the supra-phylum **Spiralia**. As a rule, the embryonic development of spiralians passes through a classic blastula and gastrula, and terminates in a **trochophore larva** (Fig. 4.22) or a trochophore-like larva such as the veliger of snails.

4.5.2 The Founder Cell 4d Gives Rise to the Mesoderm

As in *C. elegans*, an almost invariant pattern of cleavage and cell genealogy makes it possible to trace the origin of various parts of the body back to distinct founder cells. Particular attention has long been paid to blastomere **D** of the four cell stage and to its descendant **4d**. The cell 4d is the *primordial mesoblast*, the founder cell of the mesodermal inner organs. In a number of molluscs a conspicuous part of the egg's vegetal cytoplasm is sequestered in the form of a **polar lobe**. If the vegetal cytoplasm is removed with a micropipette prior to its translocation into the lobe, the embryo fails to develop a dorso-ventral body axis.

In many mussels and snails, in particular in the scaphopode *Dentalium,* (tusk shell), a conspicuous lobe is extruded periodically in the process of cleavage. During cell division, the lobe remains attached to one of the two daughter cells, into which it is absorbed after cytokinesis. The lobe extrudes again prior to the next cleavage. The cells marked by the pulsating bulb are the cells of the D line. The lobe contains components of unknown identity that are essential for future mesoderm formation. When the lobe is cut off and fused with, e.g., cell A, the cell D loses the

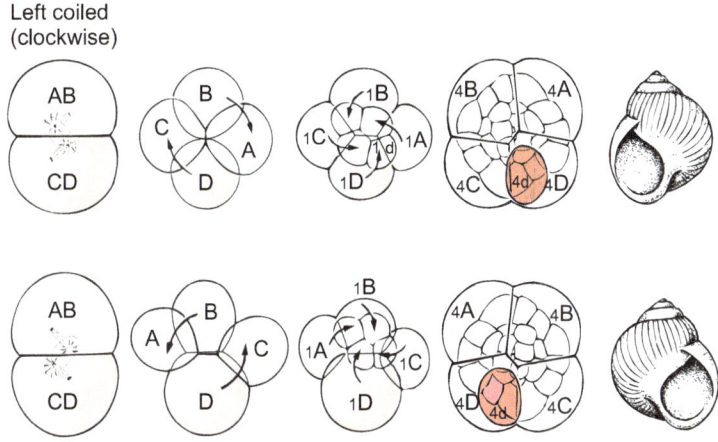

Fig. 4.21 Spiral cleavage in right- and left-handed coiling snails. We are looking down upon the animal pole. The left-handed and right-handed snails display mirror-image patterns of cleavage (after Morgan 1927). The red-labelled cell 4d represents the founder cell of the mesoderm

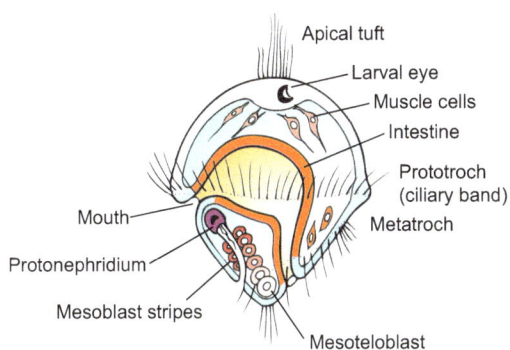

Fig. 4.22 Trochophora larva of an annelid, schematic illustration. The mesoteloblast is a descendant of the 4d cell and gives rise to mesodermal tissues

interest. Among the annelids, the marine polychaete worms *Chaetopterus* and or the tropical leeches *Helobdella* and *Haementeria*, have often been brought into the laboratory. In the leech embryo microinjecting tracers enabled reconstruction of the pedigree of each neuron back to its founder cell. Recently the situation changed and the annelid *Platynereis* got the position of an established model organism.

ability to give rise to mesoderm. Instead, cell A can now generate the mesoblast.

For long, among the spiralians no single species has gained the position of a dominant reference model. Of the molluscs, development has been examined in the pond snail *Lymnea stagnalis*, the land slug *Bithynia*, the marine snails *Littorina* and *Ilyanassa*, the scaphopode *Dentalium* (tusk shell), and the marine mussel *Spisula* (trough shell). In *Spisula* activation of the egg has been of particular

4.5.3 *Platynereis Dumerilii*: An Annelid Representative for a Large Group of Invertebrates, the Lophotrochozoa, Exhibits Unexpected Parallels to Vertebrates

Among the spiralians the annelid *Platynereis dumerilii* (Fig. 4.23) is being about to gain the position of a dominant model organism representative for a large group of invertebrate phyla, now collectively known as Lophotrochozoa (Fig. 4.1). *Platynereis* has its home in marine habitats but, as a good model organism, can be reared in the laboratory. Here it discloses its

Fig. 4.23 Life cycle of *Platynereis dumerilii*, a polychaete worm representing the phylum Annelida and the group of phyla called Lophotrochozoa

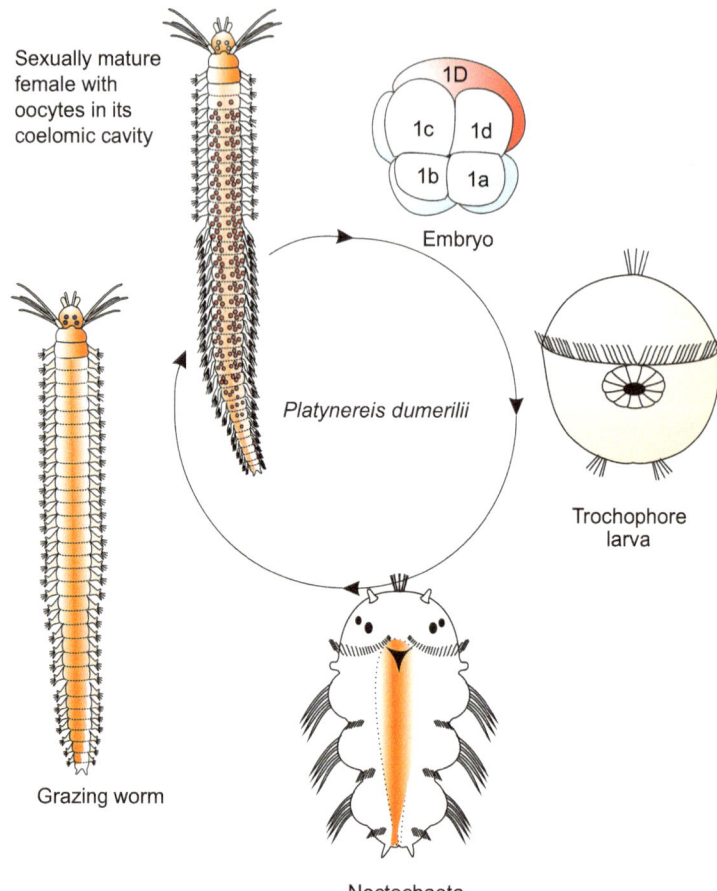

Sexually mature female with oocytes in its coelomic cavity

1D

1c   1d

1b   1a

Embryo

Platynereis dumerilii

Trochophore larva

Grazing worm

Nectochaeta

complete life cycle provided the light regime under which it is reared takes the lunar periodicity of its sexual maturation into account.

The transparent worms, in the adult state 5–6 cm long, are equipped with many "feet" or paddles (parapodia). In the posterior body regions of mature animals these parapodia are transformed to swimming paddles when the animals prepare to swim to the water surface to liberate egg or sperm synchronously in a certain new moon night. This, however, takes place only once in their life. After spawning, the animals die.

Before dying the females release about 600 eggs. These cleave according to the standard spiralian mode. The stretched blastopore gives rise to both the definitive mouth and the anus, as has been proposed by theoreticians to apply to the ancient Urbilateria (primitve bilaterians, see

Fig. 22.14). The gastrula gets a pair of primordial mesoblasts that derive from the cell 4d in the early embryo and start to proliferate in the gastrula thus giving rise to pairs of mesodermal vesicles. The gastrula proceeds developing to form a trochophore-like larva.

As in other model organisms extensive studies are performed to detect genes relevant for controlling development or being of particular interest in comparative studies and theories on evolutionary relationships. In particular, *Platynereis* has been considered a model suited to reconstruct the hypothetical primordial bilaterians (Chap. 22, Box 22.3). Much attention was given to its eyes (Fig. 22.15A). Surprising correspondences and even homologies were found in ultrastructural details and molecular components of the photoreceptors with vertebrate photoreceptors, and in the inventory of neurosecretory cells in the brains of both the annelids and vertebrates.

4.5.4   Uniformity Versus Diversity: Apparently Uniform Early Embryos Can Give Rise to a Large Variety of Adult Animals

The spiralians in general and annelids in particular are superbly suited models for investigating how animal diversity arose by developmental changes during evolution. From embryos that in their appearance look quite similar such diverse adult animals as worms, snails and mussels arise.

When arthropods are taken in view as well, particular controversial questions arise. For instance, the segmental body plans of arthropods and annelids were in former times presumed to have arisen from common segmented ancestors, yet the type of segmentation can differ extensively among the various members of these groups and some scientists refuse to take such traits in account. It is still a matter of controversial debate whether the mechanisms of segmentation in the animal kingdom follow common principles and to what extent homologization is justified.

To describe and explain the evolution of diversity in terms of molecular biology and in terms of ecology is a huge task for the future biological sciences.

4.6   *Drosophila*: Still the Model of Reference in Genetic and Molecular Developmental Biology

4.6.1   A Small Fly Makes History

One may ask: How can a researcher work with a small, medically insignificant fly life long and even get a Nobel Prize for medicine (TH Morgan in 1933, EB Lewis, C Nüsslein-Volhard and E

Wieschaus in 1995)? Public opinion demands that publicly supported research should be of benefit to humans. Even geneticists and embryologists were stunned when they became aware that flies and human beings share so many genes being of relevance for development and health.

More than 100 years have passed since Thomas Hunt Morgan made *Drosophila* the acknowledged leading model of classic genetics. The rapid life cycle, ease of breeding, and giant polytene chromosomes which allow gene localization, made the fruit fly superbly suited to analyze heredity. However, *Drosophila* came to be the most important model subject in developmental biology only in the past decades, even though EB Lewis introduced the complex of homeotic genes (see below) as early as 1978. Two complementary lines of research were particularly successful: the search, by mutagenesis, for genes that specifically control development, carried out by Christiane Nüsslein-Volhard and Eric Wieschaus, and the use of new tools of molecular technology by many researchers world wide. Initially, genes are defined as physically unknown entities responsible for certain phenotypic traits. Using newly developed molecular approaches several groups of researcher were successful in identifying the physical genes, cloning and sequencing them, in visualizing the pattern of their expression and in unravelling the function of the proteins encoded by them.

Soon after the genome of *C. elegans* the genome of *Drosophila* was sequenced. According to computer-based analyses of the sequences the entire genome comprises 13,600 protein-coding genes of which about 5,000 are indispensable for normal development. When both alleles of these genes are defective no vital fly develops; the embryo or pupa dies. A fraction of about 100 genes explicitly controls development. Before more is said about genes controlling development we first should have a look at the normal life cycle of the fruit fly.

4.6.2 A Short Curriculum Vitae: In 24 Hours Embryonic Development Is Finished

The life cycle of the fruit fly from hatching to hatching takes 2–3 weeks (Fig. 4.24). Embryogenesis takes only 1 day; the larva passes the three larval stages, separated by moults, in 4 days. The larva then encloses itself in a new cuticle, called the puparium, in which it undergoes metamorphosis in 5 days. The adult fly lives for about 9 days. Many peculiarities of development in *Drosophila* reflect paths taken during the evolution of its unusually rapid development. *Drosophila* has also developed a particularly efficient means of producing mature eggs at high speed.

4.6.3 In Oogenesis Nurse Cells Feed the Oocyte and Detailed Care Is Taken for the Future

Production of the egg cell in *Drosophila* already anticipates rapid embryonic development. Many features of fly developmental genetics make eminent sense when seen from this perspective.

The eggs are formed within tubes called **ovarioles**, which are divided into chambers by transverse walls (Fig. 4.25). The cells constituting the wall of an ovariole and surrounding the future eggs are termed follicle cells by analogy to the follicle cells in the ovary of vertebrates. In each chamber there is one female primordial germ cell, the **oogonium**. The oogonium undergoes four rounds of mitotic divisions, resulting in 16 cells. These remain interconnected by cytoplasmic bridges, called **fusomes**. In the centre of the cluster, two of the 16 cells are connected with 4 sister cells; one of these two cells will become the **oocyte**, the future egg cell. The remaining 15 sister cells are fated to become **nurse cells**. While the oocyte remains diploid, and later will become haploid in the course of two meiotic divisions, the nurse cells become polyploid by replicating their DNA repeatedly; this amplification of the genome enables high transcriptional activity. The nurse cell will provide the oocyte with huge amounts of

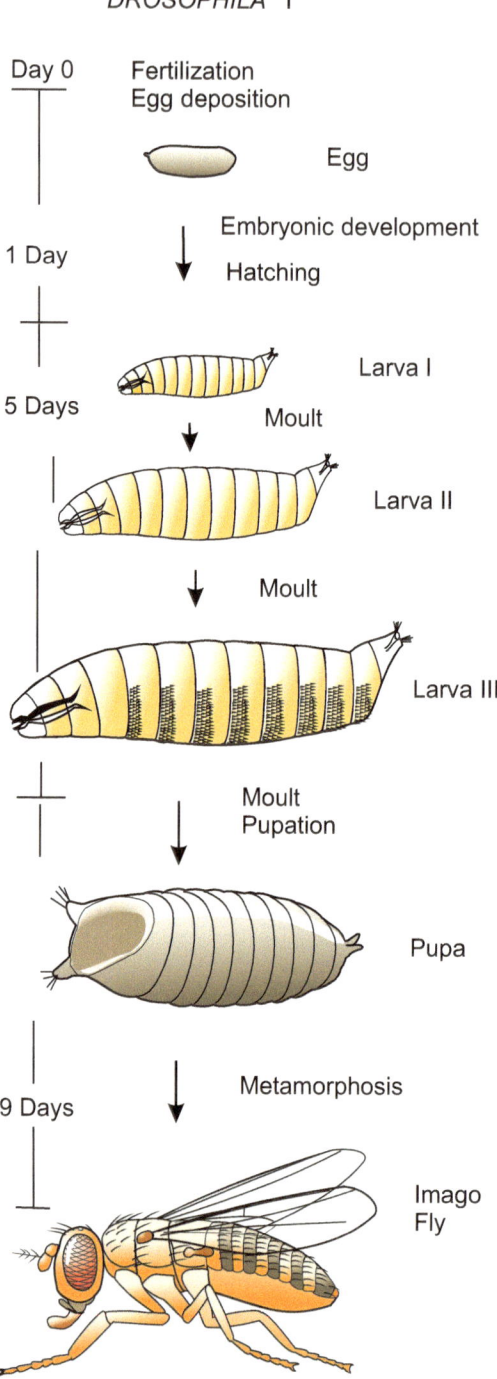

DROSOPHILA I

Fig. 4.24 *Drosophila melanogaster*. A short curriculum vitae

ribosomes and mRNA enclosed in **ribonucleoprotein particles** (**RNP**). Ribosomes and RNP are exported and shipped by the nurse cells

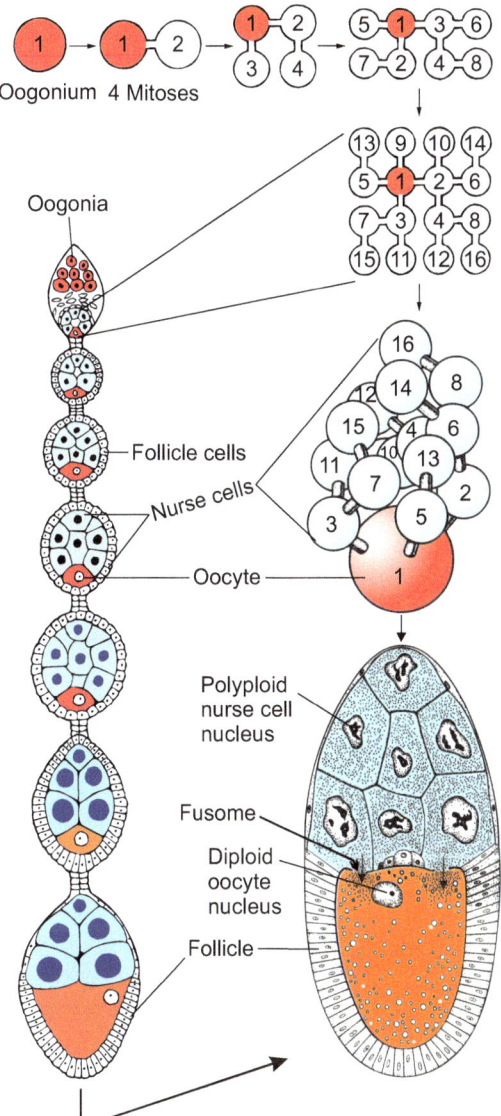

In the late stages of oogenesis, the follicle cells help nourish the oocyte. They mediate the supply of **yolk** – the major constituent of the large egg. Yolk is synthesized outside the egg cell in the so-called fat body, in the form of **vitellogenins** and phosphovitins. These lipoproteins and phosphoproteins are liberated into the fluid of the central body cavity (the hemolymph), gathered by the follicle cells and passed over to the oocyte. The oocyte takes over the materials, storing them in yolk platelets and granules to build up a stock of amino acids, phosphate and energy.

The egg cell is merely a consumer; its own nucleus is transcriptionally inactive. The maternal nurse cells, follicle cells and the cells of the fat body use their own genes and cellular resources to manufacture all of the materials that are exported into the oocyte. The gene products, therefore, are maternal and carry **maternal information**. When such genes affect the embryonic development of the egg, they are termed **maternal effect genes**.

Finally, the follicle cells secrete the multilayered **chorion**, a rigid envelope encasing the egg. At the anterior end of the egg shell, a canal called the **micropyle** is left open, allowing the entry of a sperm before the egg leaves the mother fly. Insemination occurs, as in terrestrial vertebrates, in the last section of the ovary which serves as an oviduct. Sperm, stored during copulation in the receptaculum seminis, is used throughout the life of a female fly.

Fig. 4.25 *Drosophila*. Oogenesis. Schematic drawing of one tubular ovariole, containing oogonia close to its distal tip. Along the tube oocytes of progressive size and maturity are seen. The oocytes are accompanied by nurse cells and surrounded by follicle cells. Enlarged details are shown outside the tube of the ovariole

4.6.4   Embryogenesis: In 'Superficial Cleavage' First Only Nuclei Are Generated at High Speed, Then the Nuclei Become Encased by Membranes to Form Cells, the First Complete Cells Being the Primordial Germ Cells

through the fusomes into the oocyte and are thereby placed at the disposal of the developing egg, enabling it to pass the dangerous phase of embryogenesis, when escape from predators is impossible, rapidly.

Embryonic development in *Drosophila* (Fig. 4.26) starts immediately following egg deposition and leads within only one day to a larva able to hatch: a remarkably rapid process. Cleavage, called superficial, is unusual: nuclei are duplicated at a high frequency with intervals of only 9 min, until after 13 rounds of replication about 6,000 nuclei are present. In this phase the egg represents a syncytium. When the stage of 256 nuclei is surpassed, nuclei begin to move to the periphery of the egg and settle in the cortical layer. The first nuclei to reach the cortex are those moving to the posterior end of the egg. Previously, under the organizing influence of the maternal gene *oskar*, **pole granules** containing several species of RNA accumulated in this region, among them RNA of mitochondrial origin, mRNA of *vasa* which will have a decisive function in specification of the germ line, and *nanos* which in addition aids in specification of the posterior body region.

At the posterior pole of the normal egg, the immigrant nuclei and pole granules are encased by cell membranes and segregated as **pole cells** by a process of budding (Fig. 4.26b, c). These first embryonic cells are committed to become the primordial germ cells from which the next generation will arise. During gastrulation, the pole cells migrate through the midgut epithelium to arrive in the embryonic gonad (Fig. 4.27).

Specification of the pole cells is accomplished by local cytoplasmic factors. Transplanted posterior-pole cytoplasm from oocytes can instruct early embryonic nuclei at any position within the syncytium to adopt the germ cell fate (Fig. 4.30), even at the anterior egg pole However, at this position the primordial germ cells do not find the way into the ovarial tubes and cannot proceed developing up to germ cell. But if they are transferred with a pipette to the posterior pole they give rise to fertile germ cells.

The other nuclei are arrayed in the cortex of the egg as a monolayer, and the egg arrives at the stage of the **syncytial blastoderm**. After some more divisions the final 6,000 nuclei are produced. Next, the plasma membrane is pulled down between the nuclei. Membrane invaginations continue to deepen past the nuclei and close off a compartment of cytoplasm around each nucleus, thus creating cells. This is the stage of the **cellular blastoderm**.

Some nuclei are left in the central yolk mass; they represent the **vitellophages** (a misnomer, as nuclei do not eat yolk). Later in development, these nuclei are enclosed in cells, too. In several groups of insects, these central vitellophages are thought to participate in the formation of the midgut, but recent literature does not assign such a role to these cells in *Drosophila*.

The ventral part of the cellular blastoderm constitutes the **germ band**, which gives rise to the embryo proper. Thus, the first, the ventral part, of the embryo becomes constructed.

4.6.5 In Gastrulation the Digestive Tract is Formed by Anterior and Posterior Invaginations While the Mesoderm and Primary Neuroblasts Are Folded into the Interior by a Ventral Furrow

The formation of the body begins at the ventral midline of the germ band, and proceeds to the dorsal side of the egg. Gastrulation (Fig. 4.26 f–h) comprises two main, independent events.

Formation of the digestive tract. The digestive tract is formed by anterior and posterior midgut invaginations, classified as endoderm. Both invaginations deepen into the interior, approaching each other. Eventually they fuse, forming the midgut. As the indentations deepen, the surrounding blastoderm is drawn in, forming the ('ectodermal') stomodeum, which will give rise to the foregut, and the ('ectodermal') proctodeum, which will give rise to the hindgut.

Formation of the mesoderm and the ventral nerve cord (Figs. 4.20, 4.21 and 4.36): Along the ventral side a wide band of cells undergoes coherent morphogenetic movements. The cells change shape to form an indentation in the epithelium, creating the **ventral furrow** (also called **primitive groove**), which deepens and invaginates. The invaginated groove is incorporated into the interior of the egg and forms the band-shaped **mesoderm**. Soon thereafter, the band becomes dispersed into groups of cells which will form the larval muscles. Two further bands of cells, which formerly accompanied the mesoderm

Fig. 4.26 *Drosophila.* Summary of embryo and postembryo development. In the "superficial" cleavage nuclei multiply to establish a preliminary syncytial blastoderm, and only later are complete cells formed constituting the cellular blastoderm. Before and during gastrulation a longitudinal "germ band" is formed that expands longitudinally so that the tail comes to lie close to the head (i). Later, the germ band is retracted (k). Gastrulation (f to h) takes place as invagination at several sites; most cells are incorporated into the interior of the embryo through a ventral furrow that appears along the ventral midline. The furrow is known as the primitive groove and is considered to be homologous to the blastopore in other animals. The cells giving rise to the mesoderm (*red*) and the central nervous system (*yellow*) invaginate in this primitive groove

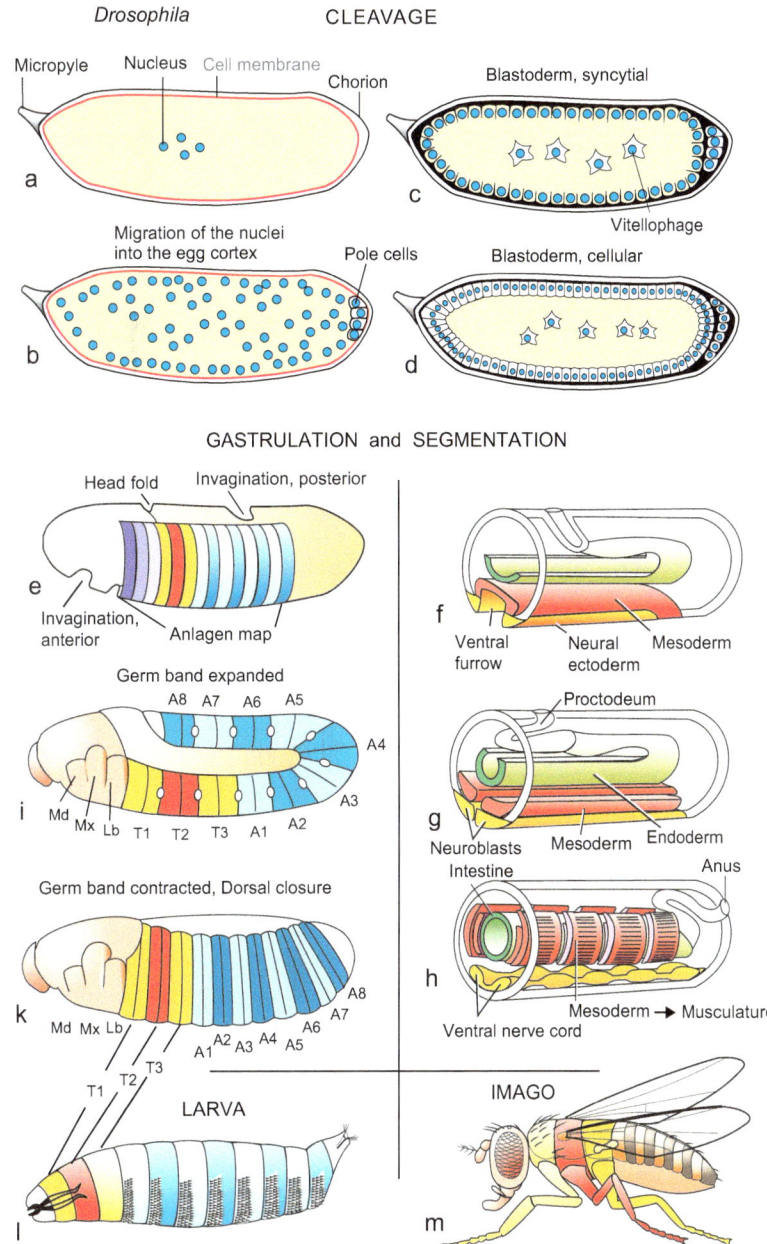

band of the ventral midline on each side, contain the **neurogenic cells**. Neurogenic cells are singled out (Figs. 10.1 and 16.10), segregate from the remaining epidermoblast cells, delaminate into the interior and come to lie between the mesoderm and the ectoderm. Incorporated into the interior, the neurogenic cells separate and give rise to

groups of **neuroblasts** from which the **ventral nerve cord** derives.

The Brain. The primordium of the optic lobe appears as a placode (local thickening of the blastoderm) and is invaginated. Delaminating neuroblasts provide further building material to construct the brain. An aggregate of neuroblasts

Fig. 4.27 *Drosophila.* Gastrulation and early steps of organ formation in an idealized embryo. Schematic longitudinal section. Gastrulation takes place as invagination at three sites. (1, 2) Ahead and behind the primitive groove, local invaginations give rise to the foregut and hindgut. Both extend into the interior and approach each other. (3) Along the ventral midline a long furrow appears. The cells giving rise to the mesoderm (red) and the central nervous system (orange) invaginate in the primitive groove. The invaginated mesodermal band segregates in discrete segmental (metameric) packets. These are soon subdivided into smaller units that give rise to the various muscles. Transitory changes associated with germ band extension and retraction are omitted

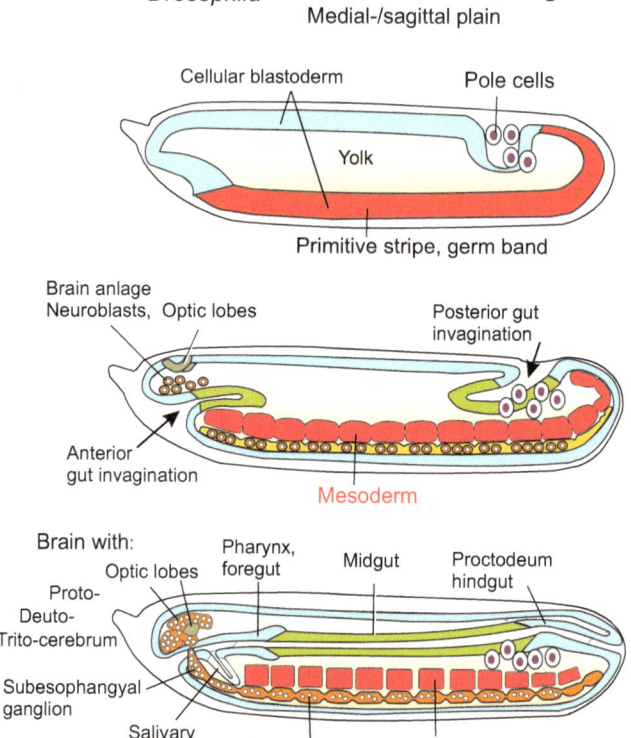

forms the cerebral ganglion, comprising three distinct regions: the protocerebrum, the deuterocebrum, and the triticerebrum. The subesophageal ganglion is composed of fused ganglia; they are connected to the two strands of the ventral nerve cord which bears pairs of segmental ganglia along its length.

4.6.6 Only Briefly Mentioned: Dorsal Closure, Segmentation, Movement of the Germ Band

The dorsal edges of the ectoderm as well as the edges of the internal organs grow over the central yolk dorsally until they meet each other and fuse along the dorsal midline.

Events of minor importance, such as the transitory expansion of the germ band shown in Fig. 4.26i, and the formation of an internal envelope called the amnionserosa will not be described in detail in this book.

4.6.7 The Body Is Subdivided into Segments Which Initially Are Similar But Are Very Different in the Adult Fly

In the meantime **segmentation** has begun: the body comes to be divided into periodically repeated units. Segmentation begins at the level of gene expression at the stage of the syncytial blastoderm but is morphologically visible only when the mesoderm is divided into packets and furrows appear in the outer ectodermal epithelium. In the outward body pattern 14 segments can be distinguished; these are initially quite uniform (homonomous) but become dissimilar (heteronomous).

The Basic Body Pattern. The germ band subdivides itself into three main groups of segments called **thagmata**. In examining the fly larva the various segments and groups of segments can be recognized only by identifying the numerous visible specializations of the larval cuticle, including dorsal hairs, ventral denticles, tracheal spiracles and sense organs. After metamorphosis, the differences are conspicuous in the imaginal fly. The fusion of the terminal **acron** (bow) with presumably **7 cephalic segments** results in the head. The number of 7 fused segments has been derived from the expression pattern of genes such as *engrailed* and *wingless*, which normally are expressed along the borders of segments (see below, segment polarity genes). All 7 cephalic segments (three pregnathal and four gnathal) supply neuroblasts for forming the central nervous system. The CNS consists of the supra-esophageal ganglion (brain) and the sub-esophageal ganglion. The three posterior (gnathal) segments, termed **mandibular, maxillar and labial, segments (Mb, Mx, Lb)**, manufacture the tools for eating around the mouth. In the larva, however, the head is retracted into the interior of the body. In its outward appearance, the body of the larva begins with the three thoracic segments: **T1 = prothorax, T2 = mesothorax, and T3 = metathorax**. In the finished fly, each thoracic segment bears a pair of legs, the mesothorax a pair of wings, and the metathorax a second pair of structures which once were wings but now are evolutionarily reduced to oscillating bodies called **halteres**, structures which are equipped with sensory organs to control wind-induced torsions during the flight.

The **abdomen** of the fly or its larva consists of eight segments (**A1–A8**); the terminal **telson** (after-deck) at the rear is not classed with true or complete segments, nor is the terminal acron at the anterior end.

4.6.8 Genes Controlling the Body Pattern: Master Genes Control Subordinate Genes

Based on extensive mutagenesis and crossing experiments geneticists, in particular C. Nüsslein-Volhard and E. Wieschaus, succeeded in collecting and identifying an almost complete compendium of development controlling genes. In traditional genetics the existence of a gene is indicated, and tracking of its inheritance facilitated, if its mutation causes a characteristic, phenotypically recognizable defect. While in former times preferentially mutagenic agents were used to induce mutations, today introduction of transposable P-elements is preferred. These can insert into any genes, disrupt the reading frame, and thus destroy it.

We restrict our review to embryonic development. The early embryonic development of *Drosophila* is governed by a rather small set of **key genes (master genes = selector genes)** which affect, via their protein products, the state of activity of other, subordinate genes. These regulatory gene products contain DNA-binding domains and act by controlling transcription, hence they are called **transcription factors**.

At present, three main classes of genes controlling development are distinguished (Fig. 4.28):

I. **Maternal polarity genes affecting the establishment of the body's coordinates (axis determination)**. These genes are classified into subgroups

   1. *Genes affecting anterior-posterior polarity:*
     • Anterior group with *bicoid,*
     • Posterior group with *nanos, oskar*
     • Terminal group with *torso* and *caudal*
   2. *Genes affecting dorso-ventral polarity*

II. **Zygotic genes affecting segmentation (segmentation genes)**

III. **Homeotic genes affecting segment identities**

     • with *dorsal (dl)* and *toll.*

Maternal (effect-) genes means: The genes do not affect the body of the mother herself but of hers progeny: Actually; the product of these

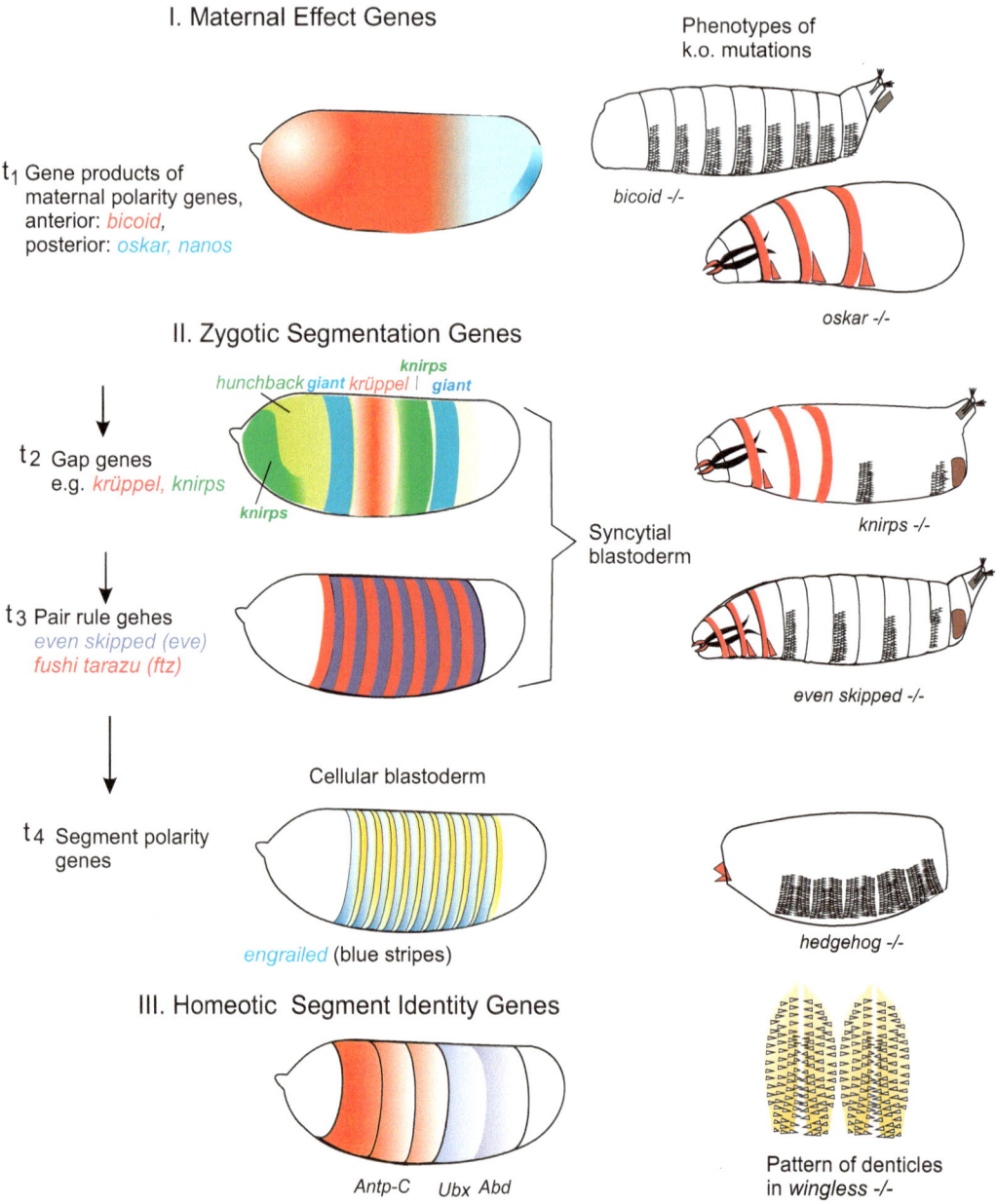

I. Maternal Effect Genes

t_1 Gene products of maternal polarity genes, anterior: *bicoid*, posterior: *oskar, nanos*

Phenotypes of k.o. mutations

bicoid -/-

oskar -/-

II. Zygotic Segmentation Genes

hunchback giant krüppel | knirps giant

knirps

t_2 Gap genes e.g. *krüppel, knirps*

Syncytial blastoderm

knirps -/-

t_3 Pair rule gehes *even skipped (eve) fushi tarazu (ftz)*

even skipped -/-

Cellular blastoderm

t_4 Segment polarity genes

engrailed (blue stripes)

hedgehog -/-

III. Homeotic Segment Identity Genes

Antp-C   Ubx Abd

Pattern of denticles in *wingless -/-*

Fig. 4.28 *Drosophila*. Classes of genes involved in embryonic pattern formation. Temporal sequence in the expression of master genes (selector genes) controlling embryonic pattern. The stripes show the distribution of the proteins encoded by those genes, The body pattern is specified (1) by the maternal genes (maternal effect genes), that establish the body axes and induce the expression of (2) zygotic gap genes; these define broad territories and turn on (3) pair-rule genes that are expressed in alternating stripes and forecast the location of future segments. (4) The segment polarity genes initiate the actual segmentation and their subdivision into smaller units. (5) The homeotic genes ultimately define the individual identities of the segments

genes are used in oogenesis and affect the early embryonic development of the child.

Zygotic genes means: Segmentation genes and homeotic genes are classified among the zygotic genes, meaning that genes of the embryo itself come into play and become decisive.

More information on all these genes and their products are given in the following sections.

4.6.9 Initially the Mother Has the Say: Maternal Genes Are Competent for Establishing the Coordinates of the Future Body (Axis Determination)

In its basic architecture the fly displays **bilateral symmetry** as does the human body. To reach such an organization, two axes must be set up: a polar (asymmetrical) anterior-posterior axis and, perpendicular to it, a polar (asymmetrical) **dorso-ventral axis;** the **left and right** halves of the body along the anterior-axis will display **mirror image symmetry**.

Seemingly, the basic bilaterally symmetrical architecture is already anticipated in the morphology of the egg. One can easily distinguish an anterior pole from the posterior pole, and the belly side from the back side. Yet, mutants inform us that this external shape of the egg and the inner organization of the embryo can completely disharmonize.

In *Drosophila* the coordinates of the future embryo are established under the influence of particular genes the products of which (mRNA, or Protein) are deposited in the growing egg cell in a distinct three-dimensional pattern. In *Drosophila*, unlike in several other animal eggs, it is not information of genes in the nucleus of the oocyte itself that is used to produce the internal constituents but genes in the surrounding cells of the mother's body (nurse cells and follicle cells, Fig. 4.25). The constituents include developmental determinants. The significance of such **maternal genes** (also called **maternal-effect genes**) is demonstrated when eggs, produced by mutant females, give rise to embryos that display wrongly oriented axes or fail to form distinct body regions, or to form them in correct locations, although the eggs may appear normal in their shape. Genetic analysis reveals that only the mutant genotype of the mother is responsible for this failure, and not that of the embryo (or its father).

For instance, instead of an abdomen a second, posterior head comes to appear in a mirror image orientation to the anterior head (in females bearing the mutation *dicephalic*), or, inversely, instead of

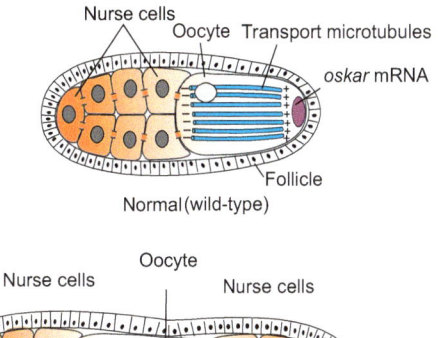

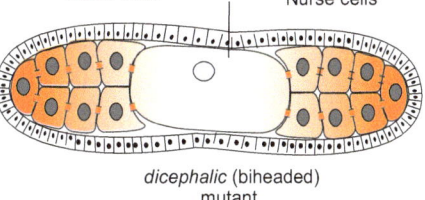

Fig. 4.29 *Drosophila.* Chamber of an ovariole in wild-type animals (left) and in the maternal mutant *dicephalic.* After Alberts et al. redrawn and modified

a head a second abdomen will be formed (in females bearing the mutation *bicaudal*). When the ovary (egg chamber) of a *dicephalic* fly is opened, revealing observations can be made. The oocyte is fed by nurse cells not only at its anterior but also at its posterior pole (Fig. 4.29).

How factors controlling development are unearthed (tracked down) and their significance revealed. We ask: which materials deposited in the oocyte are candidates to become fate determinants, which genes or gene products are decisive in controlling development, where is the mRNA transcribed, in which position of the egg is it guided, when is a protein produced and which function has it to fulfil?

Evidence for the presence of fate determining constituents can in some instances be gained without genetics. When at the posterior pole of a freshly laid egg cytoplasmic material is sucked into a pipette and injected at the anterior pole of another egg pole cells, that is primordial germ cells, appear at this position (Fig. 4.30). On the other hand, if cytoplasm is withdrawn from the anterior pole (or the anterior body region is irradiated with a beam of hard UV) larvae arise without a head. However, to figure out which components were decisive for the observed effects, and how such components came to lie at

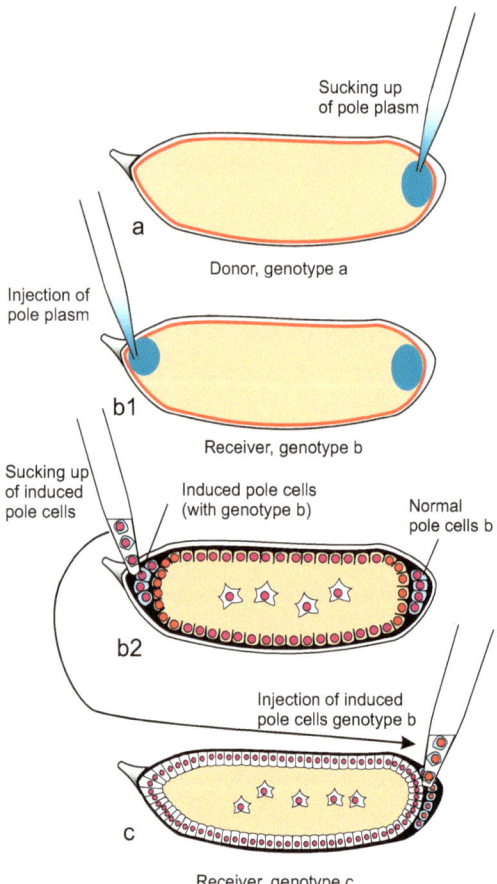

Sucking up
of pole plasm

a

Donor, genotype a

Injection of
pole plasm

b1

Receiver, genotype b

Sucking up
of induced
pole cells

Induced pole cells
(with genotype b)

Normal
pole cells b

b2

Injection of induced
pole cells genotype b

c

Receiver, genotype c

Fig. 4.30 *Drosophila*. Transplantation of pole plasm from the posterior pole of a donor egg into the anterior pole of a receiver egg. Pole cells arise at this location which become functional germ cells if they are transplanted into the posterior region of another host

the posterior or anterior pole, respectively, such experiments are not elaborated and sophisticated enough. In the cytoplasm at the posterior pole, for example, hundreds if not thousands of diverse substances may be present but which constituents are of decisive significance in specifying the fate of cells? In this case, studies with mutants pointed to a particular function of the gene *oskar*.

The *oskar* **mRNA**, copied from the maternal *oskar* gene in the nurse cells, comprises an *Utr* sequence that is not translated into an amino acid sequence (*utr = untranslated region*) but serves as a **zip code** and mediates the transport of this mRNA to the posterior pole. The transport takes place along microtubules and is accomplished by

means of motor proteins (see Fig. 9.2). Once arrived at the posterior pole, the *oskar* gene product organizes recruitment and accumulation of further RNA species, for example the maternal mRNA of the genes *vasa* and *nanos*. When the gene product of *vasa* is defective, no germ cells are produced. When the gene *nanos* is defective or not concentrated at the posterior pole a larva without an abdomen arises and, thus, also without germ cells (Figs. 4.31 and 4.32). Under the organizing activity of *oskar* not only germ cell determinants accumulate at the posterior pole but also substances required for the programming of an abdomen.

In examining thousands of mutants one can also find the opposite phenotype: Larvae without a head and thorax. Caused by several alleles of the responsible gene an abdomen like structure is found instead of a head; hence this gene was designated *bicoid* (=b*icaudal* like), abbreviated *bcd*. If a larva is descendant of a *bcd-/-* mother it misses the anterior body part with the head (Fig. 4.31).

With *bicoid* and *nanos* two genes were identified that apparently participate in the **organization of the basic architecture of the body**. Initially, however, such genes are merely names; with a classical genetic analysis genes are not yet in our hand as material entities. But the interest and ambition of so many working groups skilled in classical genetics as well as in the methodology of molecular biology were awakened. A tedious search for the physical gene was eventually rewarded with success. A gene must be mapped on the chromosomes, cloned (amplified) sequenced and expressed (translated in a protein). By means of in-situ-hybridization using anti-sense RNA the location where a particular mRNA is deposited in the egg can be examined; using antibodies to the expressed protein in the technique of microscopic immunofluorescence, the location of the corresponding proteins in the embryo and to which sites they are transported can be seen (Chap. 12, Figs. 12.5 and 12.11).

The result looked promising: In the first minutes of the embryonic development maternal mRNA is translated into protein; and gradients in the distribution of the proteins form. The BICOID protein gradient has its maximum at

Fig. 4.31 *Drosophila*. Mutants of body architecture (*bottom left* and *right*) in comparison to wild type (*top, middle*). Illustrated are the distribution of the (defective) maternal mRNA, the distribution of the (defective) development-controlling proteins encoded by this RNA, and the resulting (mutant) phenotype

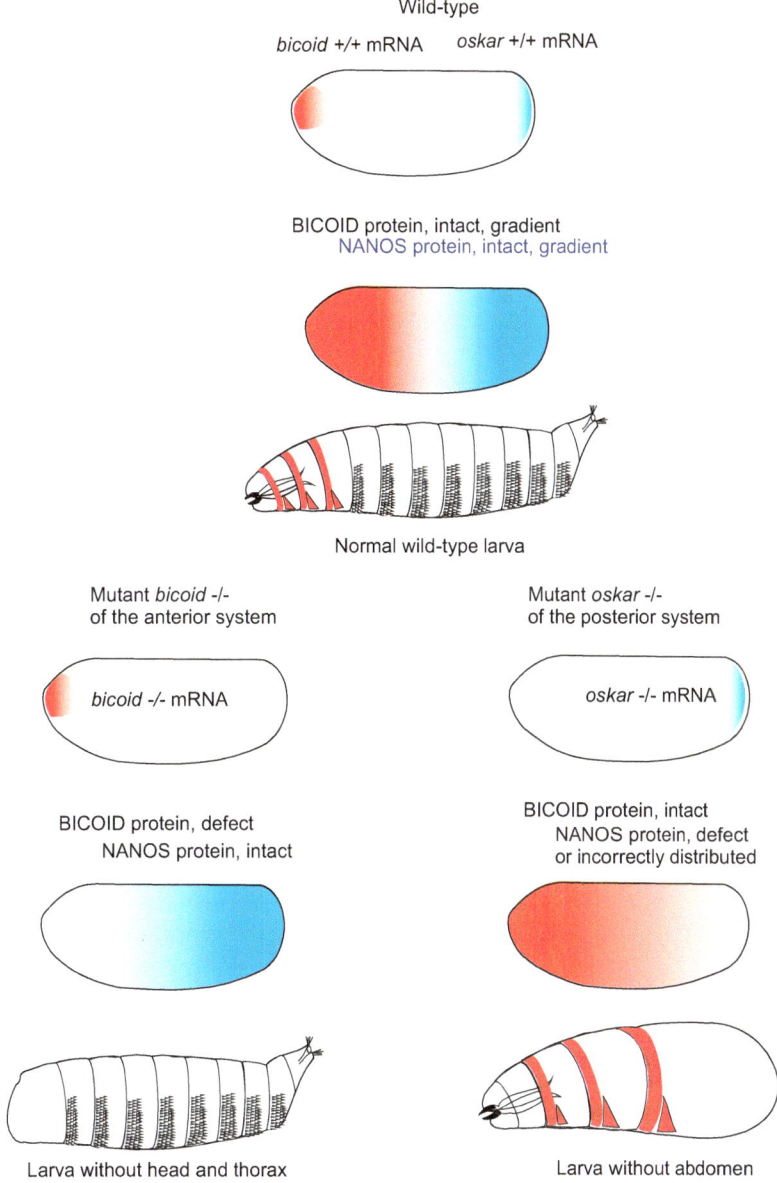

the anterior pole, the NANOS and CAUDAL gradients at the posterior pole (Fig. 4.31).

But had one really found the gene responsible for the genetic effects and did the microscopic preparations really show the right mRNA and the right protein? Take *bicoid* for example. Definitive proof that one was successful in obtaining a gene governing head formation is attained if defective eggs of a *bicoid*-/- mother are rescued when *bicoid* + mRNA derived from the wild

type (*bicoid* +/+) genome has been injected. Or when, following the injection of *bicoid* + mRNA at the posterior pole, a second head arises there (Fig. 4.32).

A skilful strategy using tools and tricks of molecular biology can cause nature itself to guide mRNA to another place. Genes may contain a particular mRNA localization sequence that serves as a zip code to guide the transport of the transcribed mRNA to certain locations in

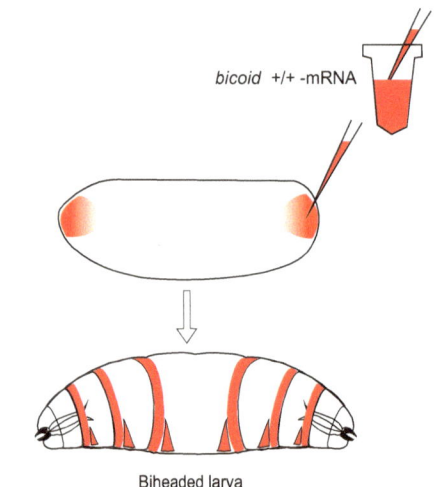

Biheaded larva

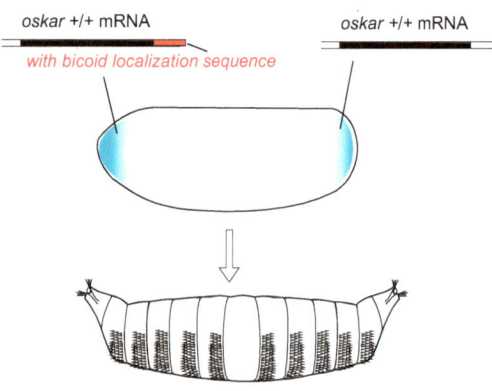

Larva with two abdomina

Fig. 4.32 *Drosophila*. Mirror image duplication of the anterior body caused by ectopic (ectopic = outside the normal location) expression of the *bicoid* gene and of the abdomen, respectively, caused by ectopic expression of *oskar*. The ectopic expression of *bicoid* was achieved by injection of wild type *bicoid* mRNA at the posterior pole of the freshly laid egg, an ectopic expression of *oskar* was achieved by providing the fly mother with an *oskar* gene supplemented with a *bicoid* localization sequence

the cytoplasm. When, for example, in the laboratory the localization sequence of the *oskar* DNA is removed and replaced by the localization sequence of *bicoid*, and subsequently measures are taken that this construct is properly integrated into the genome of female flies, such transgenic flies can lay eggs that deposit *oskar* at the anterior pole, thanks the attached localization

sequence of *bicoid*, as well as at the posterior pole guided by its own normal localization sequence. As a result, larvae with two, bipolar abdomens (Fig.4.32) arise. Such larvae are known also to derive from eggs laid by *bicaudal-/-* females. Thus, a phenocopy (imitation) of the *bicaudal* mutant has been produced.

Gene, responsible for the realization of the basal architecture of the body. Here we summarize relevant findings of gene screens.

Axis determination genes are active in the ovary tubes of the maternal organisms – in the nurse cells or in the follicle cells. The products encoded by these genes are channelled into the oocyte, as a rule, in form of mRNA enclosed in RNP particles, and translated into protein after the egg is fertilized and laid. The proteins produced with maternal information do not directly participate in the construction of the embryo but are transcription factors controlling the activity of subordinate ("downstream") genes, or encode elements of signal cascades that are established in the embryo and render cell responses possible according to their position.

Maternal polarity genes affecting the anterior posterior polarity comprise:

- An anterior group with *bicoid (bcd)*. Embryos derived from mutant *bcd -/-* mothers miss the entire anterior body with head and thorax;

- A posterior group with *nanos (nos)*. Embryos derived from mothers with disturbed posterior system miss an abdomen. The posterior organizing centre is dominated by the products of the maternal effect genes *nanos* and *oskar*. The message of the genes is localized at the posterior pole. Transplantation of posterior pole plasm to the anterior pole can result in larvae with a second mirror-image abdomen instead of a head. Such a phenotype is also known from the mutation *bicaudal* in which a complete second mirror-symmetrical abdomen is formed instead of an anterior body.

- A terminal group with *torso* and *caudal*. Mothers who bear a defect in the gene *torso*, the key gene of the terminal group, have progeny which lack the terminal body parts acron and telson. The *torso* gene codes for a trans-membrane protein serving as the receptor for an extracellular signal molecule. The signal molecule or its precursor has been deposited by follicle cells at either end of the egg into the perivitelline space between the cell membrane of the egg and the vitelline membrane covering the egg. Occupied by the ligand, the TORSO receptor (a tyrosine kinase) mediates the formation of the terminal body structures, the acron and the telson.

In addition to the few genes listed here more than 30 different gene products have been identified that bring about the subdivision of the egg space and determine the basic architecture of the embryo.

4.6.10 The *Bicoid* Makes History: The BICOID Protein Is Considered to be the First Identified Morphogen – and Is, in Addition, an Example of a Transcription Factor With a Homeodomain

The product of the *bicoid* gene has attracted particular attention. Its manifold features and functions are the follows:

- **Localization**. As stated above, *bicoid* mRNA is directed into the egg by means of its untranslated 3'-region, and anchored at the egg's anterior pole. The *bicoid* message enables the anterior end to become an organizing centre. In the first minutes of development, the message is translated into protein. The BICOID protein has some restricted capacity to diffuse; yet, it can diffuse because the embryo is in the state of the syncytial blastoderm and not yet subdivided into cells. Its half-life is short - 30 min, thus, a **gradient in BICOID protein distribution is established** having its highest concentration at the anterior pole and extending over half the length of the egg (Fig. 4.33). In the *bicoid* mutant the product is defective. If both alleles are defective (a homozygous null mutation)

the larva lacks head and thorax, and the acron is replaced by an inverted telson.

- **Transcription factor**: The BICOID protein is incorporated into the embryonic nuclei in the anterior region of the egg. The gene contains a particular DNA sequence known as a homeobox; this partial sequence encodes a part of the protein called a **helix-turn-helix domain** or a HELIX-TURN-HELIX motif HTH (Fig. 12. 14) With this domain the BICOID protein attaches onto the enhancer/promoters of certain other genes (e.g. of *hunchback*), thus bringing them under its control (Fig. 4.33). The activated subordinate genes are **zygotic genes**, that is genes of the embryo itself.
- **RNA-binding factor**: The BICOID-Protein binds the mRNA of *caudal* and blocks its translation. By this binding caudal is taken out of business and an impressive subdivision of the embryo along longitudinal axis in stripes of zygotic gene expression commences (following Sects. 4.6.1 and 4.6.14).
- **Morphogen:** The gradient of the BICOID protein is said to provide **positional information**. High concentrations of BICOID protein turns on different genes than lower concentrations. BICOID fulfils the function of a **morphogen**. By definition a morphogen is a substance that, thanks to its concentration profile, enables here this and there that developmental program to start (see Chap. 10, Box 10.2). In *Drosophila* the concentration profile of BICOID determines the position and spatial dimension of the future head and thorax region. By means of genetic manipulation mothers have been generated who contained more or less copies of *bicoid* genes in their genome and who supplied their eggs with more or less bicoid mRNA as a dowry. If, with increased total amount of BICOID protein, a certain critical threshold is shifted in a posterior direction the head-thorax region is shifted posteriorly as well (Fig. 4.34). However, BICOID does not alone specify subsequent position-dependent development; it interacts with several other factors, for example CAUDAL and CIC (Fig. 4.33).

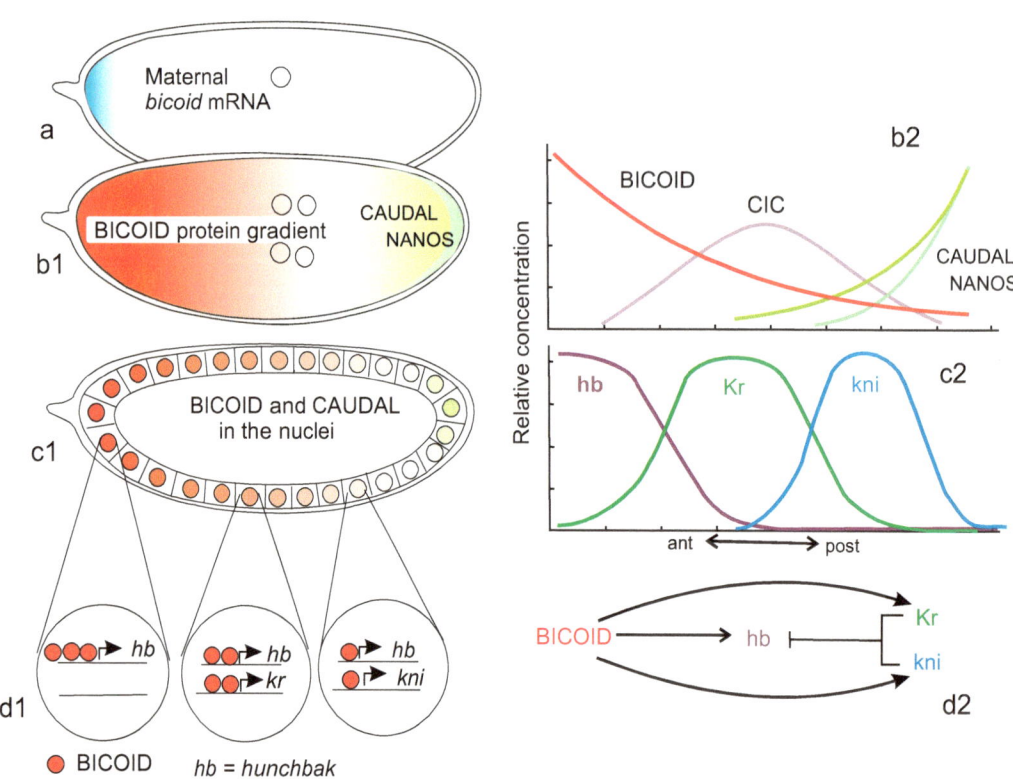

Fig. 4.33 *Drosophila*. Role of the product of the maternal *bicoid* genes in the specification of the anterior body region. The protein migrates into the nuclei and acts there, in a concentration-dependent manner, as activator or repressor of other, zygotic genes, for example, as activator of the gap gene *hunchback*. In the cellular blastoderm the gradient of the BICOID protein is counteracted by opposite gradients of the proteins CAUDAL and NANOS. The range of their distribution is not only a function of their diffusion coefficient but also restricted by complicated mutual interactions of the gene products. In contrast to the *bicoid* mRNA the maternal mRNA's of *caudal* und *hunchback* are initially ubiquitously distributed. In the cellular blastoderm the maternal mRNA of *hunchback* is replaced by zygotic transcripts. The zygotic transcription of the *hunchback* gene gets started by nuclear BICOID protein; on the other hand *hunchback* transcription becomes restricted to the anterior body region by the combined influence of KRÜPPEL and KNIRPS. The activating action of BICOID on its target genes is limited by further factors, for instance by the maternal factor CIC (Capicua) that participates in the fixation of the posterior border of target gene expression. Combined after diverse authors, e.g. Driever and Nüsslein-Volhard 1988; Jaeger and Martinez-Arias 2009; Grimm et al. 2010; Loehr et al. 2010

4.6.11 As Well As controlling Transcription; Inhibition of Translation Also Contributes to the Patterned Distribution of Proteins. Example: NANOS

Unlike the BICOID protein, the NANOS protein does not enter the cell nuclei but influences gene activities in an indirect way. Among the zygotic (embryonic) genes turned on by the BICOID transcription factor the ***hunchback*** gene is one of the first to be expressed. Initially the *hunchback* message is widely distributed in the syncytial egg chamber, but its spatial expression domain in the body becomes restricted to the anterior two-thirds of the egg by the suppressing influence of the NANOS protein. Actually NANOS prevents *hunchback* expression by

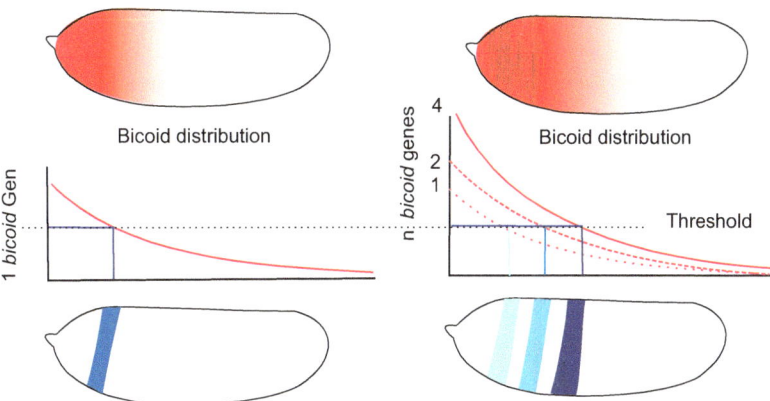

Expression stripe of the gene *empty spiracles* (*ems*)

Fig. 4.34 *Drosophila*. Gradient in the distribution of the BICOID protein. The gradient specifies the position and dimension of the head-thorax region. If more *bicoid* genes have been active in oogenesis due to the presence of up to four copies of the *bicoid* gene by genetic manipulation of the parental flies, the height and range of the gradient are enlarged. As a consequence, the dimension of the head-thorax region also becomes enlarged, indicated by the expression stripe of the homeobox gene *ems* (*empty spiracles*) that marks the future head region in the blastoderm stage. After Driever and Nüsslein-Volhard 1988, Walldorf and Gehring 1992

binding and thus blocking its mRNA. HUNCHBACK protein would repress genes required to construct the abdomen and, therefore, its translation must be prevented in the posterior body. On the other hand HUNCHBACK is permitted to exert its repressing function in the anterior region of body.

Positively the development of the abdomen is rendered possible by gene products derived from the maternal effect genes *oskar*, *pumilio* and *caudal.* The BICOID protein binds *caudal* mRNA and thereby prevents the translation of the caudal mRNA. Since, however, the concentration of the inhibiting BICOID decreases from anterior to posterior, an inverse gradient of CAUDAL protein is established (Fig. 4.33). **This is a recurrent principle in development: A gradient decreasing in the one direction gives rise to the formation of a second gradient running in the opposite direction**. We refer to the embryos of sea urchins (Fig. 4.6) and amphibians (Fig. 10.5) as further examples. More factors such as CIC (CAPICUA)

participate in restricting the effective range of BICOID and in sharpening of the expression domain of target genes. How gradients might be transposed in body segments with sharp boundaries cannot be discussed here; at present this is a topic of theoreticians who construct computer models.

4.6.12 The Back-Belly Axis Is Predetermined by an External Signal That Directs the Transcription Factor DORSAL Into Ventral Nuclei

For the correct construction of **ventral** structures the maternal gene ***dorsal*** is indispensable. For those unfamiliar with the terminology of genetics the designation *dorsal* may be misleading, as the protein defines the ventral side in the normal embryo. Bear in mind: As a rule, genes are named so in allusion to the phenotype of (defective) mutations. With a non-functional protein encoded by a defective *dorsal* gene the embryo is unable to make ventral structures (such as a ventral nerve cord). It will have a dorsal appearance all around; the embryo is said to be **dorsalized** – hence the name of the gene and its product.

As in case of BICOID, we eventually see a gradient of DORSAL accumulated in the nuclei, but for now the highest concentration is in the nuclei on the ventral side of the embryo. Yet, in contrast to BICOID the newly translated, native DORSAL protein is initially found throughout the egg and exists there in a state that does not allow its uptake into nuclei. Immediately following its translation the DORSAL protein is bound and locked up as a DORSAL/CACTUS heterodimer. DORSAL must be set free from the complex but only on the ventral side, and for that a signal is needed that tells where the ventral side is.

The external cue specifying the ventral side has been laid down and anchored in the perivitelline space surrounding the egg cell. The signal itself has got the endearing name SPAETZLE (a German word that can be translated, as one wishes, as Suebian pasta or as darling) and is liberated from the anchoring complex by means of a protease. The liberated molecule becomes ligand of a receptor called TOLL (German for fantastic). This receptor is a protein tyrosine kinase like the insulin receptor; it has been translated from a maternal *toll* mRNA in the first minutes of development, and inserted in the cell membrane of the egg.

Since the SPAETZLE precursor is present on the ventral side but not on the dorsal side of the egg, only the receptors on the ventral side will find a ligand. Receptors occupied by ligands then will organize and direct internal mechanisms effecting the redistribution of the DORSAL protein. This redistribution is enabled by receptor-mediated phosphorylation and subsequent degradation of the CACTUUS protein in the ventral region of the egg. The DORSAL protein is liberated and accumulates in the nuclei of the syncytial blastoderm.

Antibodies to DORSAL, coupled to dyes, find and mark nuclear DORSAL in form of a concentration gradient (Figs. 4.35 and 4.36). In the nuclei DORSAL takes on the function of a transcription factor; activating zygotic downstream genes in a concentration dependent manner, comparable to BICOID. DORSAL is homologous to the vertebrate factor NF-kappa-B.

4.6.13 Subdivision of the Embryo Along the Dorso-Ventral Axis into Ectodermal, Neurogenic and Mesodermal Bands Is Accomplished by Means of Two Morphogens Which Occur in Similar Form Also in Vertebrates

The DORSAL gradient is used to subdivide the, now cellular, blastoderm in several zonal territories, beginning from the ventral midline over the flanks up to the dorsal side. In these zones different sets of embryonic genes channel the further type/kind of differentiation. These zones announce four future fates (Fig. 4.36).

- The ventral stripe in which the genes *twist* and *snail* are switched on is fated to become the mesoderm; this will be involuted through the primitive groove into the interior of the embryo.

- Adjoining dorso-laterally on the left and right flank are the bands of the future nerve cells (neuroblasts) and future epidermal cells (epidermoblasts). These two cell types will later segregate from each other. (Information about the mechanisms that involves the NOTCH/DELTA signalling system is given in Sect. 10.2 and Fig. 10.1.) The neuroblasts will ingress into the embryo to form the ventral nerve cord; the adjacent two stripes will stay as outer layer and form the epidermis. For genes associated with neurogenesis and epithelial differentiation, respectively, see Figs. 16.10 and 16.12.

- The last stripe along the dorsal midline forms a thin transitory membrane (amnion serosa) that will disappear upon the dorsal closure of the embryo.

In this subdivision proteins play a role that are produced by cells of the dorsal and ventral halves

Fig. 4.35 *Drosophila.*
Establishment of dorso-
ventral polarity. Cross
section. The process is
started with the liberation
of the SPÄTZLE factor by
a protease in the
extracellular perivitelline
space. The factor had been
deposited during oogenesis
in the chorion along the
future ventral side of the
egg cell. The released
factor is bound by TOLL
receptors in the egg cell
membrane and triggers a
signal-transduction cascade
causing the segregation of
the DORSAL protein from
its CACTUS companion
and the uptake of the
liberated DORSAL protein
into the nuclei. In the nuclei
wild-type DORSAL acts as
transcription factor
activating, in a
concentration-dependent
manner, zygotic genes that
are required for the
construction of ventral
structures

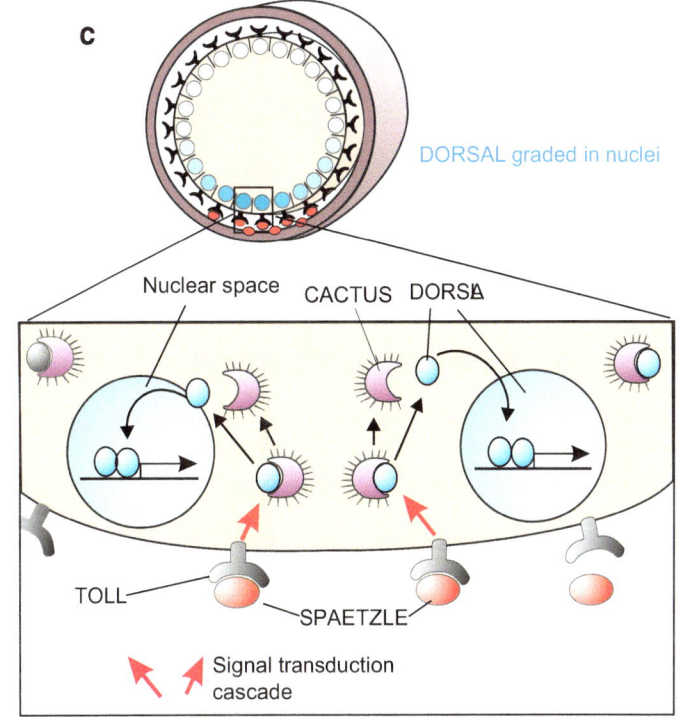

Dorso-ventral polarity

a   Egg cell
     Follicle
     Chorion
     Spätzle factor

b   Egg with many nuclei
     (syncytial blastoderm)
     Chorion
     DORSAL homogenous in cytoplasm

c   DORSAL graded in nuclei

Nuclear space   CACTUS   DORSAL

TOLL
SPAETZLE
Signal transduction
cascade

of the embryo and secreted as **morphogens** into
the extracellular spaces.

- Cells in the blastoderm of normal embryos not
  occupied by the dorsal protein liberate a morphogen into the dorsal perivitelline space. It is
  the **DECAPENTAPLEGIC DPP** protein
  encoded by the ***ddp*** gene, its counterpart in
  vertebrates is the morphogen **BMP-4**. (Both
  DPP and BMP-4 belong to a family of
  proteins named after a third member of this

group, the Transforming Growth Factor-beta,
TGF-β family, Chap. 11).

- Ventral cells produce the protein **SOG**
  (encoded by the gene *short gastrulation,
  sog*). Its counterpart in vertebrates is the protein **CHORDIN** (CHD).

In *Drosophila* DPP diffusing ventrally and
SOG diffusing dorsally meet each other and
unite forming DPP/SOG dimers (see Fig. 10.3).
We will find corresponding BMP-4/CHD dimers

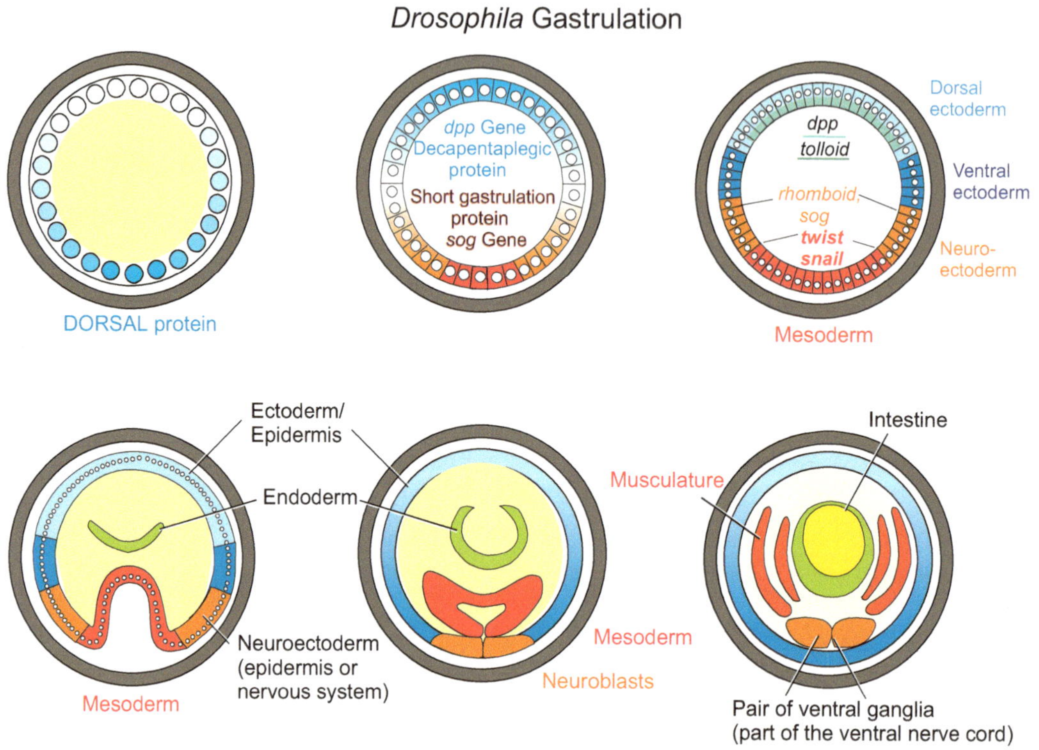

Drosophila Gastrulation

Fig. 4.36 *Drosophila*. Dorso-ventral patterning and gastrulation. Cross sections showing (upper row, left to right) the distribution of the DORSAL protein, the expression domain and secretion of the DECAPENTAPLEGIC factor, and the fate map along the dorso-ventral body axis.

The lower row shows gastrulation: formation of the primitive groove, invagination of the mesoderm (red), and of the future nervous system (orange). Blue: ectoderm = epidermis; green: endoderm = intestine

in the amphibian embryo, however in inverse orientation. The possible significance of this system of double gradients will be discussed in the context of the specification of the vertebrate body axes. Actual mechanisms of BICOID and DPP gradient formation are shown in Chap. 10, Box 10.1.

4.6.14 Along the Longitudinal Body Axes Cascades of Gene Activities Prepare the Segmentation of the Body

Along its main axis, the body of arthropods, including insects, is composed of **repetitive modules (metamery)**. These are organizational and structural units, which eventually become visible as **segments**. The zygotic (embryonic) genes turned on by the transcription factors encoded by maternal effect genes, such as *bicoid*, are involved in **pattern formation**, eventually resulting in the establishment of those segments.

For long, however, bands of gene expression do not exactly coincide with the future visible segments but are shifted anteriorly against the visible segments by half a segment. The early embryonic segments of a fly are called **parasegments**. A definitive segment is formed at the end of the embryonic development from the posterior half

(continued)

of a given parasegment and the anterior half of the posteriorly adjoining parasegment. In other words, the periodicity of the whole series of segments is shifted by half a wave length (phase). Such a phase shift is not observed in other Arthropoda.

Segmentation occurs in steps. In the syncytial blastoderm the embryonic cells begin to produce mRNA and protein of their own. Many of the new proteins are again gene regulatory factors. These are not uniformly expressed along the body but in spatially restricted **expression zones**. First these zones are broad; later expressed products appear in smaller but more numerous bands (Fig. 4.28). *Drosophila* presents wonderful, exactly sectionalized patterns to the eyes of the researcher, patterns that soon disappear and become replaced by others. About 25 genes have been identified which participate in the elaboration of future segments. In the morphologically uniform blastoderm the sequence of expression is as follows:

1. **Gap gene** activities emerge in broad, overlapping zones. These include the expression zones of the genes *hunchback (hb)*, *Krüppel (Kr*; German for cripple), and *knirps (kni*; German for doll or dwarf). Defective products made by the eponymous mutants cause groups of consecutive (para-)segments to be deleted in the larva. In *hb-/-* null mutants head- and thorax segments are missing and, in addition, the segments A7 and A8. In *Kr-/-* und *kni-/-* mutants a zone comprising several (para-)segments is missing in the middle of the body, and another zone in the posterior third. The expression of the gap genes is followed by the expression of the

2. **Pair-rule genes**. The embryo is not yet cellularized; nevertheless the future segmentation is heralded by repetitive bands of gene expression in the embryonic nuclei. The transcription pattern is unexpected and striking: pair-rule genes are expressed in patterns of 7 stripes in alternating segments: one vertical stripe of nuclei expresses a gene, the next stripe does not express it, the next adjoining segment expresses it, and so forth.

The pair-rule genes comprise remarkable smart genes such as *even skipped* and *fushi tarazu (ftz*; Japanese for 'too few segments'). *Fushi tarazu* is expressed in the odd-numbered parasegments, *even skipped* in the even-numbered. As the embryo is going to make 14 visible segments, it expresses 7 stripes of *fushi tarazu* and 7 of *even skipped*. In fly embryos mutant for *even skipped*, 7 stripes are of no use and will be skipped, and the larva will be left with the remaining 7 *fushi tarazu* segments. Conversely, embryos with mutant *fushi tarazu* genes give rise to larvae with 7 stripes that correspond to the expression bands of the intact *even skipped* gene.

3. **Segment polarity genes**. Once the stage of the cellular blastoderm is reached, segment polarity genes subdivide the various segments into smaller stripes. Several genes are of particular significance in the demarcation of the final, visible segment boundaries in the middle of the former parasegments. These genes are **engrailed (en), wingless (wg),** and **hedgehog (hh)**, supplemented by the gene *patched (ptc)* (Figs. 4.28, 4.37, and 4.38). The genes are found in other arthropods as well, apparently fulfilling similar tasks, but genes related to *engrailed, wingless* and *hedgehog* are also found in vertebrates (Chap. 12).

The ENGRAILED protein appears in 14 narrow stripes, only a few cells wide (Fig. 4.37). In the absence of proper ENGRAILED protein, a stripe in the posterior part of each segment is replaced by a duplicated and inversely-oriented anterior stripe. Hence the designation segment **polarity** genes. Homozygous *engrailed* mutants display segments with mirror-image duplicated anterior stripes.

In normal embryos, the stripes expressing *engrailed* are adjacent to stripes expressing the WINGLESS protein. The border between these adjacent stripes marks the future boundary between two visible segments. WINGLESS is

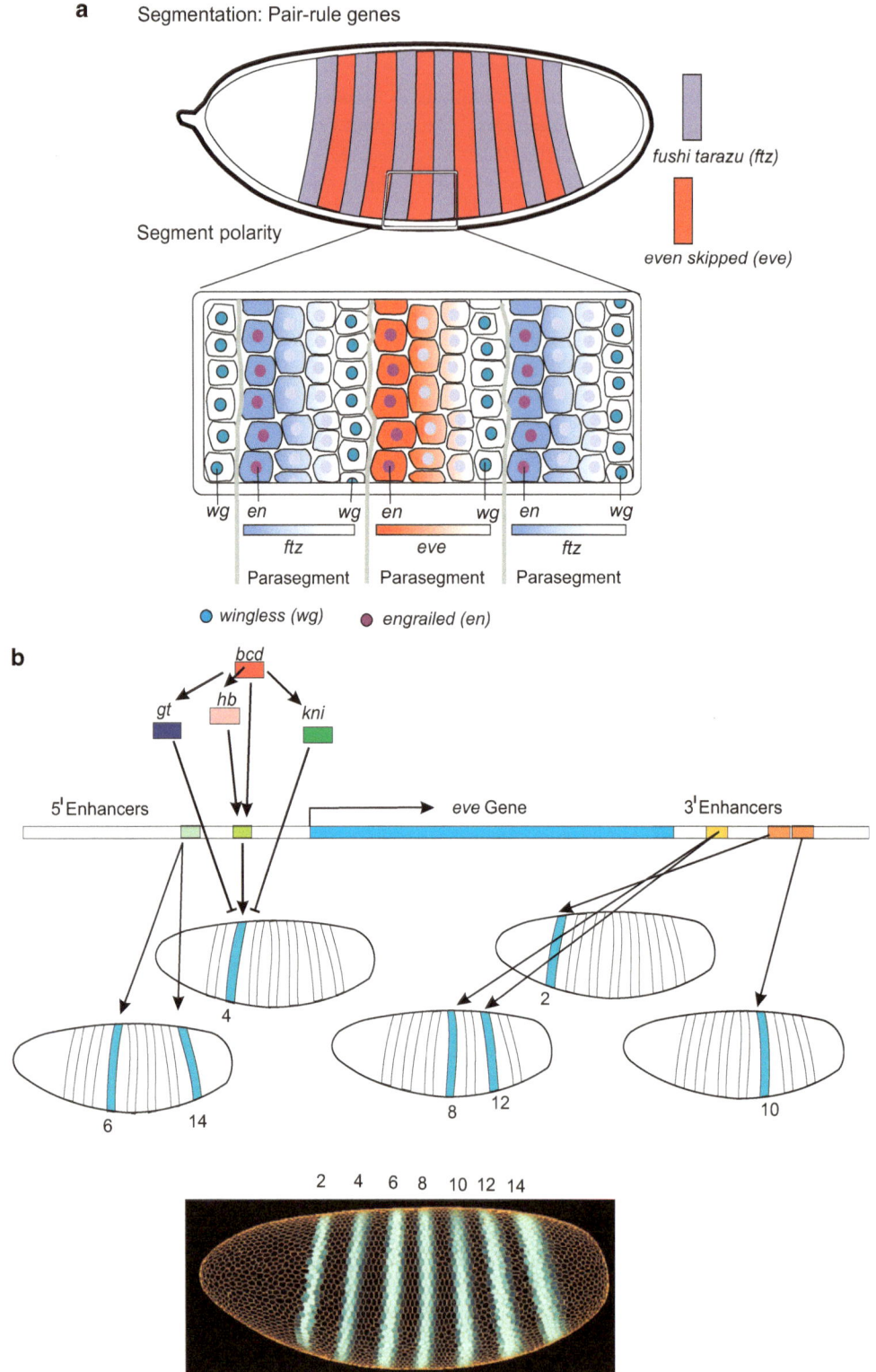

Fig. 4.37 *Drosophila.* Subdivision of the embryo into repeated territories along the anterior-posterior axis and segmentation. (**A**) Expression patterns of the zygotic pair-rule genes The diagram refers to a detail in the larva. The transcription factor ENGRAILED (en) is only present in a single row per (para-)segment; the proteins EVEN

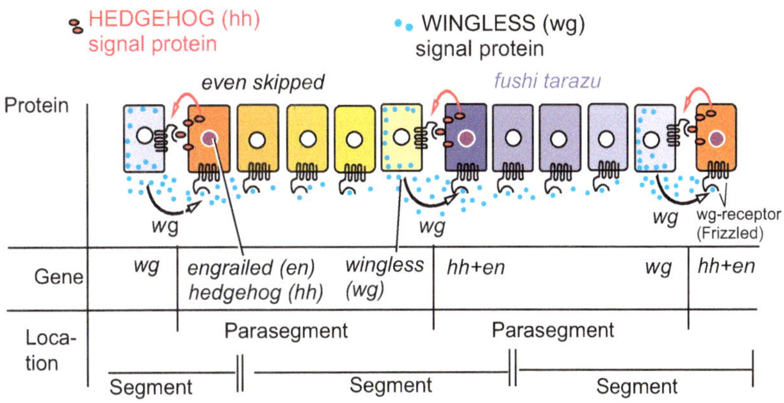

Fig. 4.38 *Drosophila*. Detail showing the expression of various genes in (para-)segments and at their borders. Note that the definitive segments are phase shifted against the early embryonic parasegments. A definitive segment arises from the posterior half of the parasegment X and the anterior half of the adjoining parasegment Y.

HEDGEHOG is involved in segment formation as a membrane-associated signalling molecule, and WINGLESS as a secreted, diffusible molecule. The transcription factor ENGRAILED remains stationary in the nuclei of the cell row defining the anterior border of a parasegment

not incorporated into nuclei. Instead, it is spit out from the producing cells as a signal molecule (Fig. 4.38). The genes corresponding to ***wingless*** in other organisms are called ***wnt*** (spoken went); they code for a signal molecule WNT that is often addressed in this book (summarized in Chap. 11).

However, signals are not emitted unidirectionally. There is a back and forth pattern. For example, the cells producing ENGRAILED also produce HEDGEHOG. The HEDGEHOG protein is exposed on the cell surface and presented to the anterior neighbour expressing WINGLESS. Stimulated by the presented signal, the

anterior cells in turn continue to emit WINGLESS. Thus, the expression zones of *wingless* and *hedgehog* stabilize and hold each other in check.

In other parts of the body, such as the **wing imaginal discs**, another variant of HEDGEHOG is made. This second version of HEDGEHOG is cut off at the cell surface and spreads by diffusion in the interstitial spaces between the epithelial cells. For example, it spreads in imaginal discs which later in metamorphosis are used to

(continued)

Fig. 4.37 (Continued) SKIPPED (eve) und FUSHI TARAZU (ftz) form steep gradients extending over a few (about 3) cell stripes. The border between *wg* and *en* marks the boundary between parasegments. The boundaries of the final segments are localized posterior to the stripes in which *engrailed* is transcribed. Besides showing the expression patterns of genes, the figure also suggests interactions: The *engrailed* gene is expressed when the cells contain high amounts of either EVEN-SKIPPED or FUSHI TARAZU proteins. The *wingless* gene is transcribed when neither of these two genes is active. (**B**) Expression stripe of the gene *eve* (*even skipped*) in the blastoderm of *Drosophila*. The gene is not controlled by a single controlling sequence (enhancer) but by several enhancer sequences partly localized

upstream, and partly downstream of the gene proper. Several transcription factors bind to the same controlling element; for example the activating BICOID protein binds to element 4, while the inhibiting GIANT and KRÜPPEL proteins bind to the neighbouring element thereby restricting the expression of *eve* to a small stripe in this body region. Note that the normal wild-type gene *even skipped* is required for constructing normal, even-numbered segments; "*even skipped*" in the literal sense applies to the mutant in which the 7 even -numbered segments are skipped and the 7 odd-numbered *fushi-tarazu* stripes remain. After Gaul and Jaeckle 1989; Carroll 1989; Small et al. 1993, 1996; Arnosti et al. 1996; Papatsenko and Levine 2008

construct the adult fly. In some imaginal discs the row of cells adjacent to the row of cells producing HEDGEHOG respond to the HEDGEHOG signal by expressing WINGLESS, as it is the case in the body segments. In other discs cells adjoining HEDGEHOG cells respond by secreting DECAPENTAPLEGIC protein. In *Drosophila* mutual interactions between *engrailed* and *wingless* expression is also implicated in the segregation of mesoderm from ectoderm. Mesoderm, newly formed in gastrulation, emits WINGLESS; under its influence, the ectoderm continues to express *engrailed*.

Like the BICOID protein most of the proteins encoded by the segmentation genes (and homeotic selector genes, Sect. 4.6.15) include a DNA-binding domain, for instance a **zinc finger domain** (HUNCHBACK), or a **helix-loop-helix domain** derived from the **homeobox** sequence of the respective gene (*fushi tarazu, even skipped, engrailed*).

Thanks to their DNA-binding domains these proteins in turn function as **transcription factors** controlling other, subordinate genes. This is a remarkable result of recent developmental genetics: **in early development a hierarchical cascade of gene activation is initiated, whereby early expressed genes turn on, or off, batteries of subordinate genes to be expressed later**.

4.6.15 Homeotic Genes of the *Antennapedia* and *Bithorax* Class Confer Unique Identities to the Segments

Homeotic genes ultimately determine which particular type of body segment will appear. For instance, will a given segment become a wingless prothorax or a winged mesothorax, a metathorax with halteres, or an abdominal segment? The genes responsible are the **homeotic selector genes** or **homeotic master genes**; they define the individual identity of each segment. In *Drosophila* most of the homeotic genes are on the third chromosome, arranged in two clusters. One cluster is called the

* ***Antennapedia complex (Antp-C)*** and another one
* ***bithorax complex (BX-C)***

Both together constitute the **HOX complex** (Fig. 4.39).

The genes of *Antp-C* and *BX-C* are arranged on the third chromosome one after another. The order of their alignment in the chromosome roughly reflects the temporal and spatial order of their expression (**colinearity**). Moving along the *Drosophila* body from the head to the rear, one first will see ANT-C proteins, and these continue to be present. In the posterior thorax, the first BX-C proteins are met. In the last abdominal segments A8 proteins encoded by several genes of the HOX-complex are present, most abundantly those encoded by *Abd-B*, the last of the *BX-C* genes.

The five genes ahead of *Antp proper* are expressed in segments destined to form the head and the thorax; the three genes of *BX-C* are expressed in the thorax and the abdomen. Defects in these genes may lead to spectacular **homeotic transformations: a morphologically correct structure is made at the wrong place.** For example: the wild-type (normal) *Antennapedia* gene is needed to specify the wing-bearing mesothorax. In a dominant *Antennapedia* mutant, the gene is expressed in the head as well as in the thorax. Accordingly, parts of the head are transformed into thorax segments, but thorax segments are committed to bear legs and not antennae: Therefore, a pair of legs rather than a pair of antennae arises from the head (Fig. 4.40b; for a more detailed interpretation of this effect see Sect. 12.2.2). On the other hand, a *loss-of-function* mutation of the *Antp* gene cluster causes the conversion of the mesothorax in a prothorax and the hatched fly misses wings.

Most spectacular is the occurrence of four-winged flies. In mutant flies in which several homeotic mutations were collected by inbreeding, the metathorax becomes transformed into a mesothorax and, consequently, the halteres are transformed into wings (Fig. 4.40a). Thus, **the evolutionarily original state of four-winged insects is re-established** (a phenomenon generally known as **atavism**).

Fig. 4.39 *Drosophila.* Effective range of the homeotic genes of the *Antennapedia-complex* and the *bithorax-complex*. The expression strength of each gene starts with highest intensity at the anterior border of its domain, and declines in posterior direction. For comparison with vertebrates see Fig. 12.16

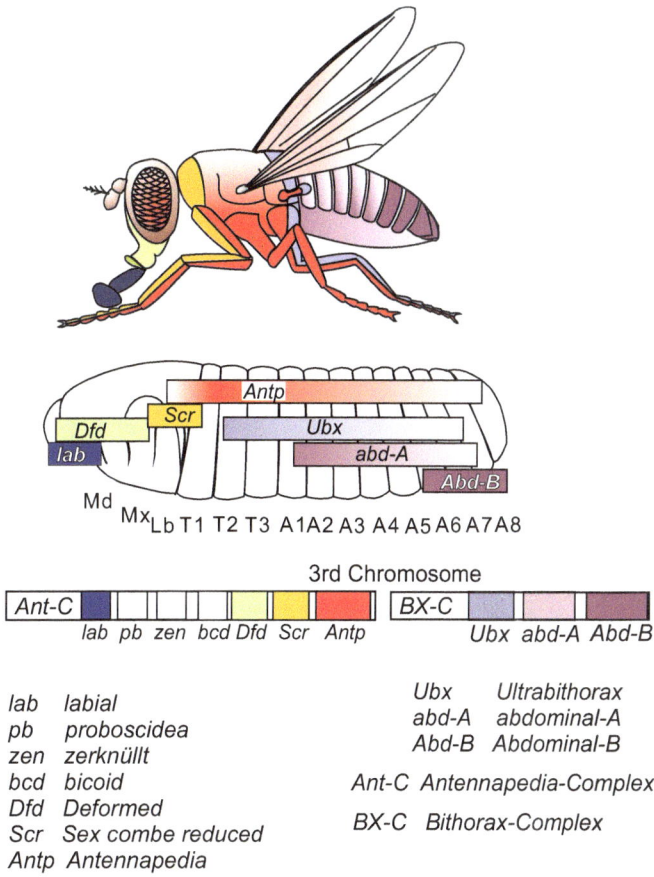

lab	labial
pb	proboscidea
zen	zerknüllt
bcd	bicoid
Dfd	Deformed
Scr	Sex combe reduced
Antp	Antennapedia

Ubx	Ultrabithorax
abd-A	abdominal-A
Abd-B	Abdominal-B

Ant-C Antennapedia-Complex

BX-C Bithorax-Complex

These examples also show that the effect of these classic homeotic genes (discovered by Edward B. Lewis) often become obvious only after the metamorphosis of the larva to the fly. A model aimed at explaining the causes of homeotic transformations in accordance with the **rule of posterior dominance** is presented in Chap. 12, Fig. 12.17.

In *Drosophila* a daring experiment succeeded. For the first time in the history of biology, artificial induction of organ formation was achieved not by transplantation of tissue but by directed ectopic expression of a gene. The experiment was based on the detection of an eyeless mutant, and the identification of the corresponding gene, *eyeless*. When the functional wild-type allele is expressed in the imaginal discs committed to develop antennae, legs or wings, flies arise with extra eyes on their antennae, legs or wings (Fig. 12.19). More about this spectacular experiment can be read in Sect. 12.3. The term "imaginal disc" is explained in more detail in the following section.

4.6.16 Metamorphosis Can Be Considered a Dramatic, Second Embryogenesis

As early as the first larval stage a fraction of cells remain diploid instead of becoming polyploid or

a

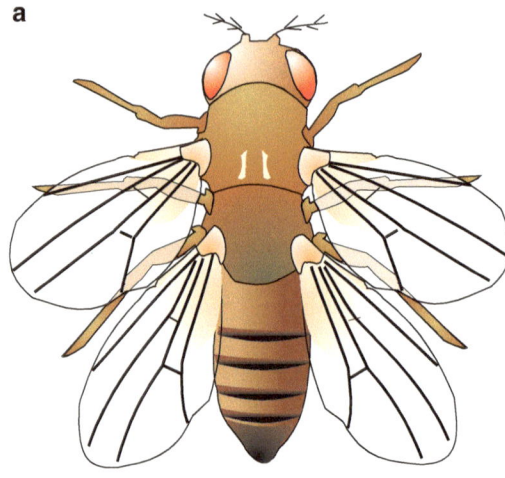

b

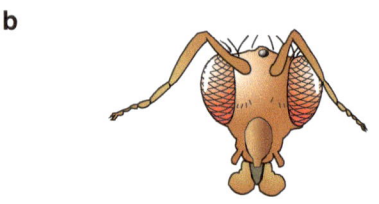

Fig. 4.40 *Drosophila.* Homeotic transformations caused by inbreeding mutated homeotic genes of the *Antennapedia complex* (*bottom*, after a photograph by Lawrence) or of the *bithorax complex* (*top*, after a photograph by Lewis). In the four-winged fly the halteres are transformed back into wings

polytene, as do most of the larval somatic cells. A particular task is assigned to groups of such cells: to construct, beneath the pupal envelope, the **imago**: the adult fly. In the interior cavity of the larva flat, rounded bodies can be found wrapped in thin epithelial bags, the **imaginal discs** (Fig. 4.41). In addition, single **imaginal cells** are dispersed within the inner organs such as intestine or Malpighian tubules. The imaginal discs are unwrapped in the course of metamorphosis; they evaginate and expand, and piece together the exterior imago. The fly is a mosaic composed of expanded imaginal discs.

The numerous events of metamorphosis are triggered and synchronized by **hormones**, whose release is controlled by the brain. Hormonal control is outlined in Chap. 20.

4.6.17 With Transplanted Imaginal Discs the Phenomenon of Transdetermination and the Heredity of the State of Determination Were Discovered

To understand the experiments described in the following, it suffices to know that in the adult fly the production of **juvenile hormone** is resumed. It now functions as a gonad-controlling (**gonadotropic**) hormone, but exerts the same effect on imaginal discs as it did in the larva: the hormone permits the growth of the discs but prevents their premature metamorphosis into adult structures. The experiments in question make use of this effect to multiply discs (Fig. 4.42). Imaginal discs are surgically removed from larvae, cut in pieces, and implanted into the body cavity of adult flies. Under the influence of juvenile hormone, the pieces regenerate and grow to normal size. Thus from one single leg disc a clone of hundreds of leg discs can be raised.

If one wants to know whether a leg disc is indeed still a leg disc, that is, whether the state of determination has been maintained and transferred to the disc's offspring, the cloned discs can be placed back into larvae ready to metamorphose. Here they undergo metamorphosis together with their host, and one can find a leg, a wing, or some other fragment of an imago within the body cavity of the hatched fly. By such experiments it has been shown that normally the **state of determination is inherited and maintained over many cell generations**. One speaks of **cell heredity**.

Occasionally, however, the discs forget their imprinted tasks, and one finds a wing instead of the expected leg, or a leg instead of the expected wing. The phenomenon became known as **homeotic transdetermination**. Apparently, homeotic selector genes are involved, but how?

- **Heredity:** As a rule, the state of determination remains unchanged following their multiplication (through division and regeneration) and is transferred to the daughter discs; one gets clones of identically determined discs.

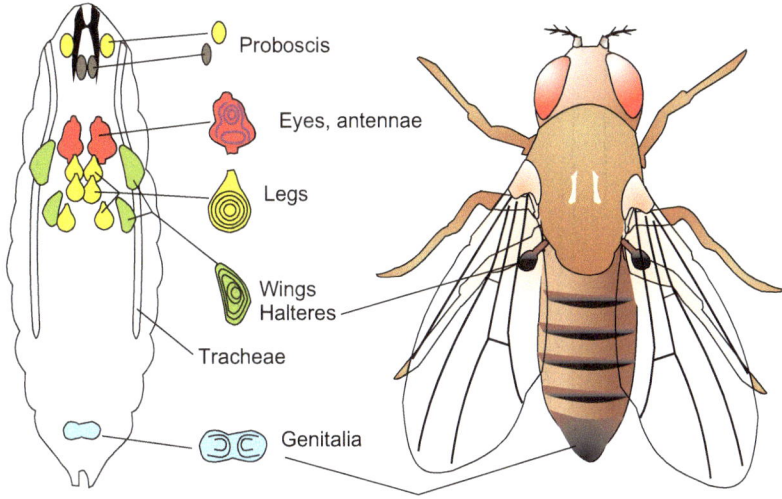

Fig. 4.41 *Drosophila*. Imaginal discs in the larva and structures in the adult fly originating from them

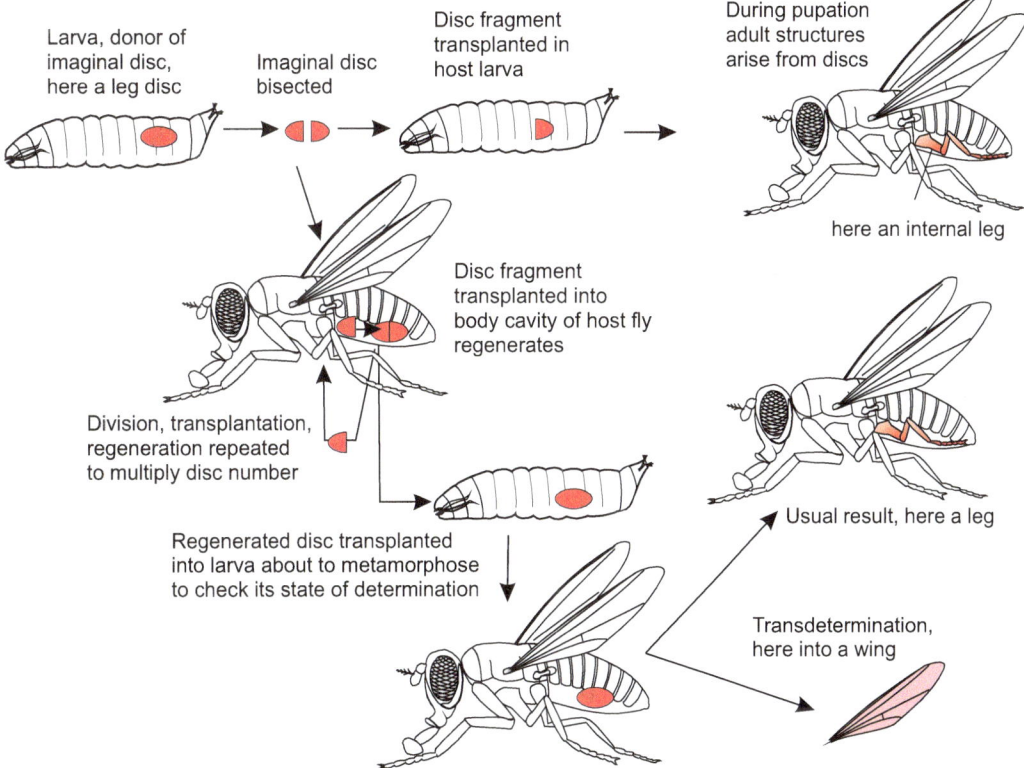

Fig. 4.42 *Drosophila*. Transplantation and multiplication of imaginal discs, and assay of the state of determination. Experiments by Hadorn 1966 (explained in the main text)

- **Transdetermination**: In rare cases one picks a wing instead of an expected leg, an antenna instead of an expected wing, a wing instead of the expected antenna. According to present knowledge transdetermination occurs only after repeated regeneration, and, thus, cell divisions are evoked again and again. Though transdetermination takes place suddenly like a mutation it is unlikely that it is caused by a mutation since they occur more frequently than expected by known mutation rates, and also reversion of the effect is observed at higher rates. Strong evidence suggests that mechanisms of gene silencing are disturbed – a matter of epigenetics (Sect. 12.5). At any rate, the program of determination in which homeotic genes are involved has changed. Transformations occur similar to those observed after inbreeding homeotic mutations.

How is the state of determination programmed and how is the program replicated in cell division? *Drosophila* is still a leading model in research aimed at answering these questions (Chap. 12) and will continue to provide answers that might be more difficult to obtain in other systems.

4.7   Tunicates: Often Quoted as an Example of 'Mosaic Development' in the Phylum of Chordates

4.7.1   Tunicates Are Marine Organisms That: Though Being Invertebrates – Belong to the Phylum of Chordates

Tunicates (Urochordates) can come into the range of vision when one observes a sea life aquarium. One may see ascidians which are barely recognized as animals, not to speak as members of a 'higher' animal class. The lumpy bodies attached on a substratum resemble sponges both in their appearance and life style (suspension feeders). But their tiny free-swimming tadpole larvae (Fig. 4.43) possess a notochord, a dorsal hollow nerve cord, a light-sensitive ocellus, and a pharyngeal basket with gill slits (used for filtering small food particles only after metamorphosis). Essentially these features are characteristic for the phylum of chordates. The ascidian tadpole represents the prototype of a chordate.

To observe embryonic development adult specimens, often of the species *Styela picta* and *Halocynthia roretzi,* are brought into the laboratory of marine biological stations. These solitary ascidians are hermaphroditic but, being proterandric, desist self-fertilization. Weakly yolked eggs are produced in high numbers and shed to the sea. Fertilization occurs externally. In most species cleavage is radial, holoblastic, and slightly unequal. Cleavage leads to a hollow coeloblastula which undergoes gastrulation by invagination. The blastopore forms at the presumptive posterior end (but closes as development proceeds). The complex process of metamorphosis is described in textbooks of zoology.

4.7.2   An Internal Pattern of Maternal Determinants Anticipates an Early Determination; But Also Cell Interactions Are of Significance

The expression 'mosaic development' refers to the traditional theory that the egg is subdivided into regions of different qualities (Box 1.1). In the course of cleavage, these qualities would be differentially allocated to the cells. In this way, the blastomeres would be enabled to continue development autonomously, independently of their neighbourhood.

This view rested on old experiments, done with fine needles (today with laser beams) and aimed at deleting certain cells or separating them from others. In the 19th century French researchers were interested in **teratology** – the causes of embryonic malformations. Seeking an experimentally accessible system, in 1886 Laurent Chabry set out to produce malformations by puncturing blastomeres in the tunicate

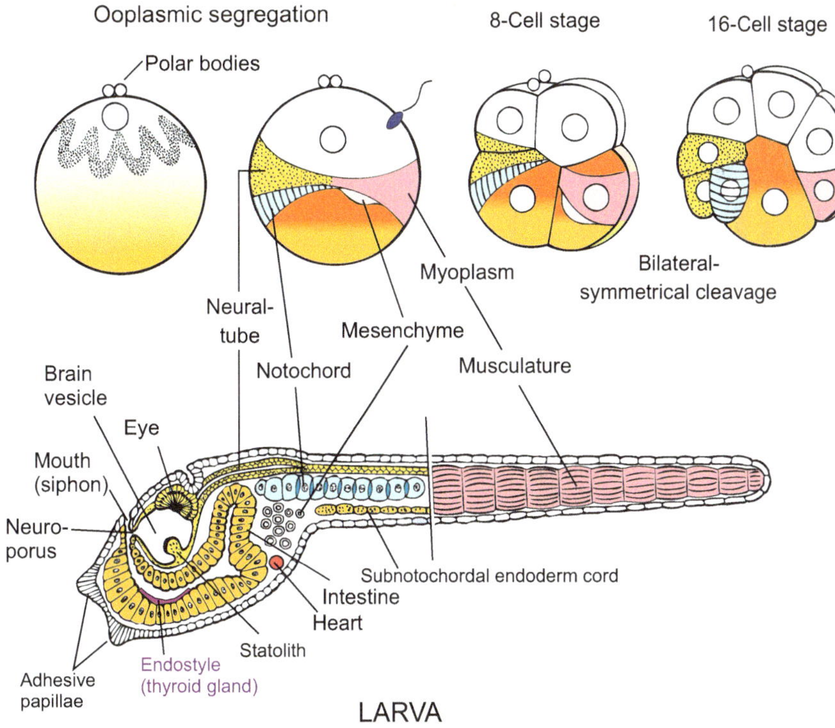

Fig. 4.43 Tunicate: an ascidian (phylum Chordata, subphylum Urochordata), showing development through the larval stage. The larva can be considered a prototype displaying the basic characteristics of the chordates

embryo. Because the defects were permanent and not corrected by the remaining blastomeres, he concluded that each blastomere is responsible for generating a particular part of the body. In 1905, E.G. Conklin (USA) described how coloured plasma is allocated into various blastomeres. By following the fate of these blastomeres Conklin came to the conclusion that each of the coloured regions of cytoplasm contains specific "organ-forming substances".

The ascidian egg indeed does contain **cytoplasmic morphogenetic determinants** responsible for programming cell fates: These determinants are almost homogenously distributed in the oocyte but become arranged in a distinct spatial pattern after fertilization but before cleavage begins, through a process of sorting out termed **ooplasmic segregation** (Figs. 4.43 and 9.6). In particular, the yellow plasm contains a component initiating muscle-specific development. The component is a member of the **MyoD1/myogenin** family of transcription factors (Sect. 12.3) and it shares

with these factors the ability to release the program of muscle cell differentiation. The ascidian factor comes to be segregated into 8 cells in the ventral-posterior part of the 64-cell embryo. These cells early acquire molecular components indicative of muscle cells (e.g. acetylcholinesterase, f-actin, myosin) and they give rise to the cross-striated muscle tissue in the tail of the larva. Upon isolation from other cells, cells containing myoplasm autonomously continue to develop into muscle cells. However, one also finds secondary muscle cells that only arise if the founder cells have contact with neighbouring cells.

The situation is similar to that in sea urchins, in *Caenorhabditis elegans* and in the spiralian embryo: the work of maternally inherited cytoplasmic determinants, even if it is as elaborate as in tunicates, must be supplemented by cell-cell interactions between the cells of the embryo. Such interactions are needed for a more detailed allocation of individual cell fates, and are dominant in the vertebrate embryo.

Summary

In Chaps. 4 and 5 the development of so called model- or reference organisms is described and important experiments are introduced that have been performed with the respective organisms. Chapter 4 introduces important invertebrate model organisms.

- The sea urchin is considered to represent the prototype of animal development in general; in addition it became the classic case in point for fertilization and metabolic activation of the quiescent egg. On the basis of transplantation studies with the sea urchin embryos the gradient theory, and associated with it, the morphogen hypothesis was developed. The equipment of its genome with genes for regulatory proteins (e.g. BMP, CHORDIN, NODAL and WNT), as well as the existence of many other homologue genes substantiate a close relationship of the echinoderms to the chordates/vertebrates. In terms of the developmental history both phyla constitute the main groups of the deuterostomes.

- *Dictyostelium*, often designated as cellular slime mould or social amoeba, is an amoebozoon, that exists as single-celled amoebae or as multi-cellular structures. Research has been focused on the chemo-tactically guided aggregation of single amoebae to a multicellular community, called plasmodium or "slug", and on the differentiation of different cell types (spores, various stalk cells) within this community while it gives rise to a fruiting body. Up to 1,000,000 cells signal each other by releasing chemoattractants such as cyclic AMP (cAMP), and coalesce by chemotaxis to form an aggregate that may move collectively (hence "slug") before establishing the fruiting body. The ratio of the various cell types is regulated.

- *Hydra*, the fresh water polyp, is the only hitherto known animal endowed with potential immortality, because it replaces all used up and aged cells by new ones that derive from stem cells. Studies focus on the derivation of differentiated cell types, in particular of nerve cells, from

multipotent stem cells. Regeneration and transplantation experiments are designed to analyse positional information and patterning. Pattern formation comprises the subdivision of the body column into head (hypostome), trunk and foot (basal disk) as an expression of self organization which involves long-range interactions of the body parts. Down the body column decreasing positional values confer positional memory to the tissue, thus render position-dependent differentiation and regeneration possible. The unmatched capability of regeneration and self organization enable chaotic aggregates of single cells derived from dissociated hydra to reorganize themselves to whole, viable animals. In the phylum Cnidaria, besides *Hydra*, also the hydrozoan *Hydractinia* and the sea anemone *Nematostella* have gained the position of a model organism. In the Cnidaria, the elaboration of the primary body axis in embryogenesis (*Hydractinia*) as well as the processes of budding of new polyps and head formation are controlled by the canonical WNT-β-CATENIN signalling system. In the Hydrozoa among the Cnidaria the developmental potencies of the multipotent (*Hydra*) or totipotent (*Hydractinia*) stem cells have been studied for a long time. Also transdifferentiation has been studied, for example the transformation of isolated cross-striated muscle cells into several other cell types (in *Podocoryne*). In the Cnidaria an unexpectedly high number of genes are found that in homologous form are also present in vertebrates including humans (conserved genes).

- *Caenorhabditis elegans* is a small nematode amenable to classic genetics (and RNAi interference) whose genome was the first to be completely sequenced. It contains 19,100 protein-coding genes. The precise, faithfully reproduced embryonic development terminates with a constant number of 556 somatic cells; this precision enabled the reconstruction of all cell pedigrees; the germ line segregating from the somatic lineages already in the two-cell stage. Early cell interactions effect early determination after which cell differentiation proceeds autonomously. But even after final fate commitment cells change positions, a particular mode of pattern formation.

- **Spiralia** (snails, clams, annelids and other groups), whose eponymous characteristic is spiral cleavage, are also known for predictable cell genealogies. The mesoderm originates always from the founder cell 4d. The position of a leading model organism representing the phylum Annelida and the more comprehensive group of Lophotrochozoa has recently been assigned to the polychaete worm *Platynereis dumerilii*. Also in this animal numerous genes have been identified that previously were considered vertebrate-specific.

- ***Drosophila melanogaster***, the fruit fly, is the leading model of reference in the genetic and molecular developmental biology. Its genome comprises 13,600 protein-coding genes. The embryonic development proceeds within 1 day passing a superficial cleavage and an atypical gastrula to a larva that undergoes three moults and eventually metamorphosis within a pupa envelope, terminating in the adult fly.

In *Drosophila* for the first time specific development-controlling genes have been identified. An example is ***bicoid***. This gene is not activated in the egg cell or embryo itself but in nurse cells of the ovary and as such the gene belongs to the maternal genome. The maternal mRNA is channelled into the egg cell and fixed at the anterior pole. The translated BICOID protein is distributed in the form of a concentration gradient the profile of which specifies the dimension of the head-thorax region. Therefore BICOID is considered to be a morphogen. It is also a transcription factor since it enters nuclei and binds with its homeodomain which derives from the homeobox of the *bicoid* gene to promoters of zygotic (embryonic) target genes controlling their expression.

Genes controlling embryonic development are classified in three main categories:

1. **Maternal effect genes that establish the body axes**
   - with *bicoid, oskar, nanos, caudal,* and *torso* establishing the anterior-posterior polarity,
   - and with *spätzle, toll* and *dorsal* establishing the dorso-ventral polarity;
2. **Segmentation genes,**
   - with **gap genes** such as *hunchback, Krüppel, knirps,*
   - and **segment polarity genes** with *engrailed* and *wingless* for example,
3. **Homeotic genes** (in the original sense of the word) that determine the particular quality of the body segments. Mutations of such genes (such as *Antennapedia*) effect correct structures appearing at wrong positions, for instance, legs sprouting from the head instead of antennae. The homeotic genes are arranged on the chromosome in two clusters, the ***Antennapedia complex*** (*Antp*-C) and the ***bithorax complex*** (*BX*-C). Astonishingly homologues of these genes are found in all bilaterian animals including humans.

During metamorphosis the external parts of the adult fly originate from imaginal discs that were stored within the larval body in a predetermined state. Occasionally such discs can experience transdetermination, i. e. a change of their fate, and, for instance, a leg may be formed instead of a wing.

- **Ascidia** (group Urochordata, phylum Chordata, subphylum Tunicata). Unlike the adult, sessile organism the tiny tadpole larva displays characteristics typical for chordates. In the cytoplasm of the egg cell fate determining factors are present that are redistributed after fertilization in a three dimensional pattern (ooplasmatic segregation). Allocated to distinct blastomeres these factors specify the program of differentiation, for example, a factor known as myoplasm commits cells to become muscle cells.

Development of Important Model Species II: Vertebrates

5

5.1 *Xenopus*: Standard of Reference for Vertebrate Development

5.1.1 Amphibian Embryogenesis Is Textbook Prototype for Understanding Vertebrate Development

Amphibians represent the archetype of vertebrate development. Figure 5.1 quotes the phylogenetic positions of vertebrates that are often investigated in the laboratory. From the Ancient World (Aristotle, Box1.1) up to the eighteenth century incubated chick eggs were the gold standard for animal development; since 1900 amphibians took over this position. The modified development of reptiles, birds and mammals can be deduced readily from the amphibian model exemplified by frogs and newts. The amphibian egg is rich in yolk and large (often 1–2 mm in diameter), yet the egg is able to undergo holoblastic cleavage, and cleavage converts a typical blastula to a gastrula displaying textbook features. In addition, amphibian embryos develop outside the mother and are therefore accessible to experimentation at all stages. The transparent jelly coat can easily be removed, surgically or chemically (after fertilization the envelope can be removed with 2.5 % cysteine-HCl, pH7.4). Pieces cut out from an embryo are able to continue development in sterile salt solutions without added nutrients, thanks to a supply of yolk in each cell. The eggs are suitable for microsurgical operations, which can be performed by hand with glass needles prepared in the laboratory; no complicated and expensive apparatus is needed.

At the beginning of experimental embryology eggs and embryos of frog specimens of the genus *Rana* (*R. temporaria* in Europe, *R. pipiens* in America) were the preferred subjects that had to tolerate still rough surgical interventions. Under **Hans Spemann**, who was to win the Nobel Prize for his experimental studies with amphibians, embryos of newts (genus *Triturus*) became the principal actors in classic developmental biology. Between 1950 and 1970 the leading role was assigned to a new protagonist, the African clawed toad *Xenopus laevis* (Figs. 5.2 and 5.3). Although surgery may be more difficult with this species, it has many other favourable features. Xenopus always stays in water, is easily maintained, and because Xenopus is bred in the laboratory, there is no need to take endangered animals from natural populations (species conservation). In particular, Xenopus can be induced to spawn at any time by injection of chorionic gonadotropic hormones (Fig. 5.2). (Even human gonadotropin can be used; and injection of female urine to induce spawning in Xenopus was the first assay for pregnancy in human medicine.)

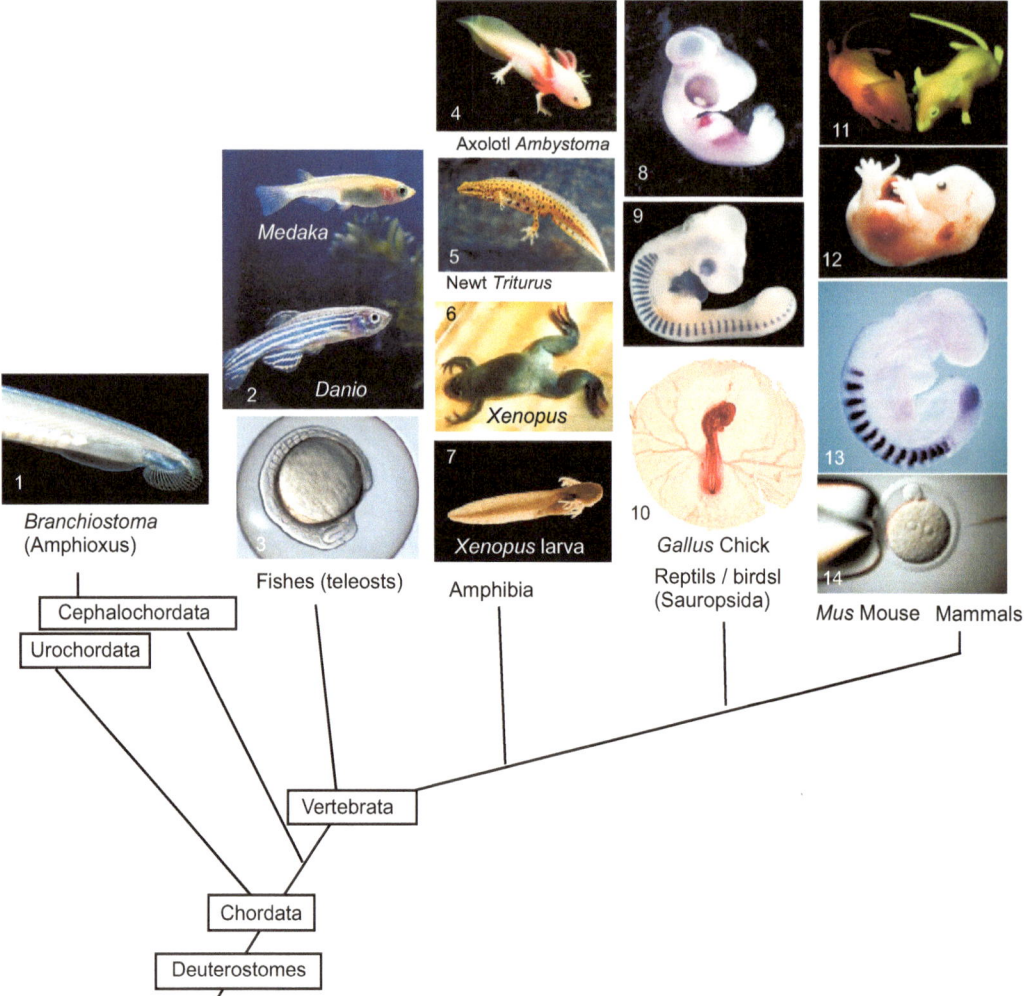

Fig. 5.1 Model organisms II. Vertebrates, that are dealt with or mentioned in this book, placed in the phylogenetic tree that is currently preferred on account of molecular genetic data. (The Cephalochordate *Branchiostoma* represents a side line of the Chordata that does not belong to the Vertebrata). Image copyrights belong to 1 Jonathan Veracaripe, 5 Julius Kramer, 11 ethlife, ETH-Zürich, 12 Michael F McElwaine, 13 Dr Alexander Aulehla, EMBL; Images reproduced with kind permission of the authors. Further pictures are part of the study collection of WM

Xenopus laevis displays, however, an unwanted feature, too. The species is (pseudo) tetraploid. If one wished to generate homozygous mutations, not only two but four identical alleles had to be brought together in extensive and time-consuming cross-breeding programs. Practically, this is almost impossible. Therefore, recent genetic work is being focussed on the diploid sibling species **Xenopus tropicalis**. For this species methods have been developed to introduce transgenes, but these methods are amenable to specialists only.

Generally, experimental results are scored in embryos up to the tadpole stage. The larva of *Xenopus* (Fig 5.3) is transparent. It feeds by sweeping in planktonic algae and other microorganisms. As in other amphibians, the fundamental reorganization into the adult clawed toad is controlled by the hormones prolactin and thyroxine (Chap. 20).

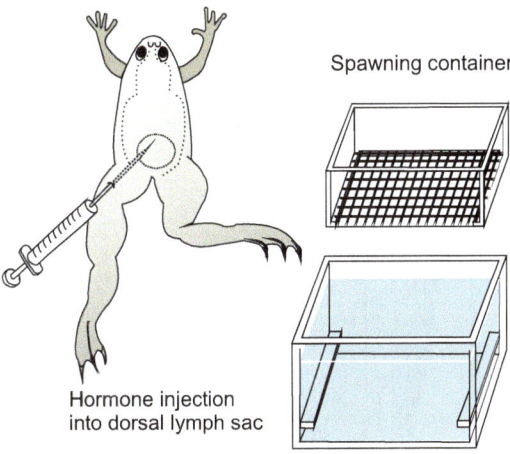

Spawning container

Hormone injection
into dorsal lymph sac

Fig. 5.2 *Xenopus.* Spawning container and injection of gonadotropic hormones to induce spawning

5.1.2 The Study of Oogenesis in *Xenopus* Yielded Basic Knowledge, Applicable to Oogenesis Also in Other Vertebrates

Amphibian embryogenesis is the textbook prototype for understanding vertebrate development. One scarcely would understand oogenesis in humans if oogenesis in *Xenopus* had not been studied thoroughly already (Fig. 6.1). Embryos of *Xenopus* make extensive use of maternal gene transcripts stored in the oocyte.

In 1958 researchers at the University of Oxford discovered a mutant whose cells, although diploid, contained only one nucleolus in their nuclei instead of two nucleoli present in the wild type. Nucleoli are the factories where ribosomes are manufactured. These factories are constructed at two homologous chromosomes that contain 450 copies of the ribosomal genes (18S-, 5,8S-, 28S-rDNA) clustered in the nucleolus organizer region of the chromosomes. The heterozygous mutant, called 1-nu, was viable, because

one set of ribosomal genes suffices to make enough ribosomes in oogenesis.

- Crossing 1-nu × 1-nu resulted in homozygous 2-nu specimens (containing two nucleoli)
- Heterozygous 1-nu and 0-nu specimens (one nucleolus) and
- Homozygous 0-nu/0-nu individuals (no nucleolus)

in the Mendelian ratios 1:2:1. The 0-nu/0-nu offspring who lacked both nucleoli were unable to produce ribosomes and, therefore, were not viable in the long term. However, surprisingly, embryonic development proceeded rather normally and the embryos even reached the stage of free-living tadpoles. Then they died. This surprising finding prompted the working hypothesis that oocytes might be provided with ribosomal RNA and the message for ribosomal protein even before they undergo meiosis. This would enable embryos arising from *0-nu* eggs and *0-nu* sperm to manufacture ribosomes without making use of their own genome, which lacks the corresponding genes. Perhaps, in the oocytes from which the *0-nu/0-nu* embryos arose, message for the production of ribosomes was transcribed before meiosis led to the haploid stage. Before completion of meiosis in the oocytes of the *0-nu/ 1-nu* mother, a normal chromosome was still present besides the mutant chromosome that was left after meiosis in the oocyte (while the normal chromosome was allocated to a polar body cell). The analysis of the phenomenon has led to a deeper understanding of oogenesis.

In fact, amphibian oocytes have a supply enabling the embryo to pass through embryogenesis even without any contribution of its own genome. However, normal zygotic transcriptional activity commences in the stage called the '**midblastula transition**' shortly before gastrulation begins (see as follows).

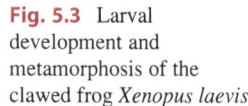

Fig. 5.3 Larval development and metamorphosis of the clawed frog *Xenopus laevis*

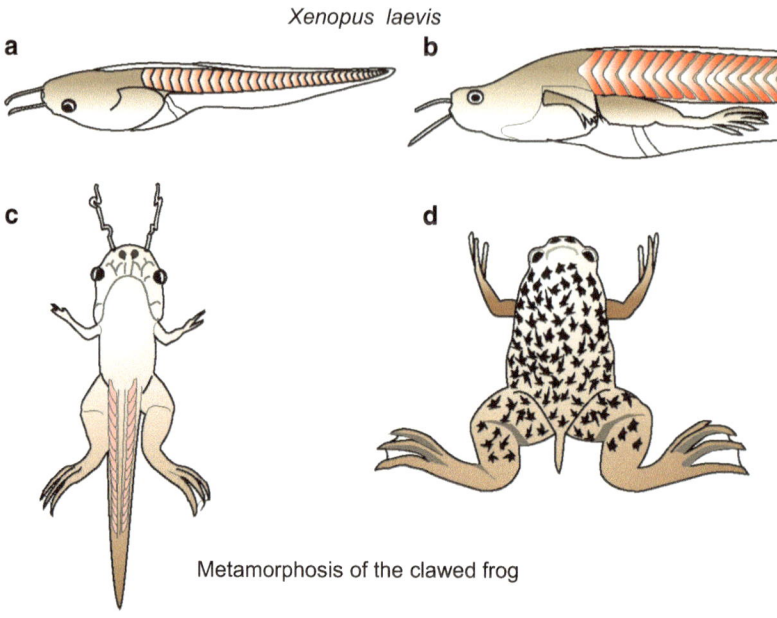

Xenopus laevis

Metamorphosis of the clawed frog

In amphibian oogenesis meiosis starts in the ovary of a young female soon after metamorphosis and takes several months. The meiotic prophase proceeds up to the diplotene stage. Then the prophase is interrupted for a long time during which the oocyte undergoes extensive growth. In the nuclei of the oocytes extensive transcriptional activity is initiated. As an expression of this high activity, **lampbrush chromosomes, rDNA amplification**, and **multiple nucleoli** occur in the nucleus of the oocyte (Figs. 8.4, 8.5 and 8.6). Furthermore, the liver of the female supplies maternal **vitellogenins** (protein) that are carried to the ovary by the blood stream, taken up by the oocyte and stored in the form of yolk granules. Oogenesis in vertebrates is described in more detail in the Chaps. 6 (human) and 8 (in general).

5.1.3 External Cues, Gravity and Sperm Provide Instructions for the Specification of the Body Axes and Bilateral Symmetry

Even before cleavage starts, processes are initiated which lead to the establishment of the egg's spatial coordinates – that is to the establishment of **bilateral symmetry**. Unlike the egg of *Drosophila*, in the amphibian egg the maternal stores do not completely predetermine the future bilateral body organization. In its outward appearance the *Xenopus* egg, and more so frog eggs, presents a dark (black) animal hemisphere and a white vegetal hemisphere. Besides this rotationally symmetrical pigmentation pattern, no other polarity axis is visible. Yet, being a bilaterally symmetrical organism with right and left sides, a belly and a back, the vertebrate animal needs two axes of asymmetry: an **antero-posterior axis** and a **dorso-ventral axis**. The dorso-ventral median plane separates **mirror-image symmetrical body halves** the axes of which point from the midline to the outside.

How are these axes established? What is the relationship between the **animal–vegetal axis** and the two **final coordinate systems** of the body?

If we view the egg as a globe of the earth, and project the future embryo onto this globe, the head of the embryo will come to lie near the animal 'North Pole' (in other amphibians it will lie on the Tropic of Cancer) and the head–tail

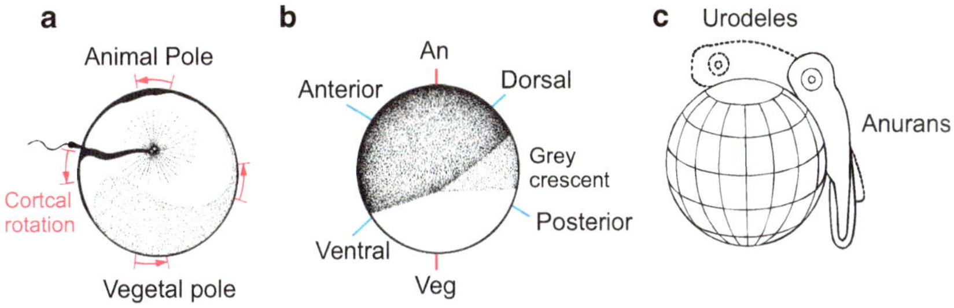

Fig. 5.4 Amphibian. Determination of bilateral symmetry. The future dorsal-posterior side (tail position) is diagonally opposite the entry point of the sperm. (**a**) After the sperm has entered, a sperm aster (*black*) appears in the centre of the egg, and cortical rotation commences. (**b**) Side view of the grey crescent which appears during cortical rotation. (**c**) The projection of the embryos onto the egg sphere does not take into account cell movements during gastrulation and, therefore, does not represent a true fate map. The projections merely superimpose the initial egg sphere and the finished tail bud larva

line will extend along a longitudinal line over the Northern Hemisphere, cross the equator, and terminate in the Southern Hemisphere near the Tropic of Capricorn (Fig. 5.4). The general directive that specifies that the head should be situated in the Northern and the tail in the Southern Hemisphere is laid down in the internally visible organization of the egg (animal–vegetal asymmetry). However, the embryo needs a further instruction to determine along which longitude the line should extend. The specification of this head-tail line is called **dorsalization**, and is accomplished by the interaction of several events. Besides the given animal–vegetal egg axis, there is also:

1. the **point where the sperm enters the egg**, and
2. **the gravitational force** of the (real) earth

   The sperm can only attach on the animal hemisphere, but the exact point is determined by chance. After its entry, the 'fertilization **membrane**' is elevated, just as in the sea urchin egg. This elevation enables the egg to rotate so that the yolk-rich, heavy vegetal hemisphere points downward in response to gravity.

For dorsalization to occur, a uniform rotation of the entire, unaltered egg would not be much help. However, not all components of the egg move to the same extent and in the same direction. Induced by the sperm and influenced by gravity, active movements within the egg lead to an asymmetrical rearrangement of egg components. While the inner cytoplasm, heavy with dense yolk, sinks down and remains stabilized by gravity an active **cortical rotation** commences. **Cortex** in this context designates the mass-dense peripheral layer of cytoplasm associated with the cell membrane. The layer contains dense assembles of actin filaments and bundles of microtubules. Microtubules arranged parallel to the cell membrane are displaced against each other driven by **kinesin motor proteins**. The coordinated movement of the microtubules causes the egg membrane and outer cortical layer of the cytoplasm to shift relative to the inner cytoplasmic mass about 30° towards the point of sperm entry (Fig. 5.4). This rotation causes an **asymmetrical rearrangement of egg components including maternal RNA. Plasma that in the unfertilized egg occupied the area around the vegetal pole is displaced to a site diagonally opposite the sperm entry point where the blastopore will form. Bilateral symmetry is now established; anterior and posterior, dorsal and ventral are defined**.

The rotation of the cortical layer relative to the inner mass also causes a redistribution of pigment granules in the animal hemisphere. In eggs of the frog *Rana* the redistribution sometimes leads to a partial depigmentation on the future dorsal side opposite the point of sperm entry. The region of diminished pigment is referred to as the **grey crescent**. The grey crescent or its spatial equivalent (in *Xenopus*, a grey crescent is not visible) marks the region where gastrulation will be initiated and the blastopore will form. In *Rana*, the egg now has not only an invisible internal but also a visible bilaterally symmetrical organization: A line drawn from the grey crescent over the animal pole to the sperm entry point coincides with the line extending from the tail bud over the back to the head.

Such projections, however, must not be misunderstood as a definitive fate map. The material located between the animal pole and the grey crescent will largely be displaced into the interior during gastrulation. Cellularized egg material located around the grey crescent will shift into the head region, and material near the poles will come to lie in the tail region.

The cellular and molecular means by which sperm and gravity affect the molecular organization of the egg are partially known. The cytoplasmic rotation that occurs during fertilization leads to a segregation and repatterning of maternal cytoplasmic determinants, in particular an enrichment of components specifying the location of the Spemann organizer (see Sect. 5.1.10).

5.1.4 Cleavage and Gastrulation Come into the Picture as Textbook Prototypes; During Gastrulation Cell Material Is Incorporated into the Interior Giving Origin First to the Archenteron and Subsequently to the Mesoderm

Radial, holoblastic cleavage leads to a blastula (Fig. 5.5 illustrates real cleavage, Fig. 5.6 a didactically idealized cleavage). The subsequent gastrulation combines processes of **invagination**, **involution**, and **epiboly**. It starts in the centre of the area corresponding to the grey crescent in frogs. A group of cells sinks into the embryo, forming a groove and pulling along adjoining cells: the **blastopore** (primitive mouth, first mouth) is formed. The crescent-shaped groove of the blastopore becomes elongated and eventually forms a circle (Fig. 5.7). The dorsal sector of the circle is referred to as the **upper blastopore lip** which in the world of developmental biologist earned some fame (Sect. 5.1.10). **The cells constituting the lip are constantly changing, because they are travellers merely passing through**. Another picture that illustrates the dynamic structure of the lip is a waterfall: apparently constant when viewed from distance, constantly changing at a closer look.

In traditional terminology one speaks of **invagination** when the crescent area buckles inward to form the primitive mouth and its edges (lips). Attracted by signals emanating from the region around the blastopore, cells of the blastoderm stream together towards the blastopore (**convergence movement**); arriving at the blastopore, they roll over the lip and squeeze into the interior. The turning around the lip is called **involution**.

Once inside the embryo, the cells move in an anterior direction as a sheet and spread along the entire inside surface of the blastoderm, thereby forming the vesicular **archenteron** (primary intestine). Involution and gliding are partly the result of changes in cell adhesiveness and are supported by the formation of lobopodia by the cells located at the leading edge. These leading and pioneering cells are known as **bottleneck cells** due to their shape (Figs. 5.7a2 and 14.1). Guided by these cells the invaded cell cohorts move as coherent sheet along the inner surface of the blastoderm forming the wall of the archenteron.

This active involution is supported by **epiboly** of the outer blastoderm: Although more than half of the blastula disappears in the interior, the circumference of the embryo does not decrease; because the cells of the outer blastoderm spread out by division and flattening, and expand over the yolk-rich archenteron encompassing it, and thus compensate for the loss in terms of surface area.

By such active migration, cells of the animal hemisphere and of a belt of cells around the equator, termed the marginal zone, invaginate. The marginal cells passing over the lip pull along yolk-rich cells of the ventral blastoderm. All of these cells participate in forming the roof and floor of the archenteron. The archenteron becomes the principal cavity at the expense of the blastocoel.

I. Cleavage

II. Gastrulation

Blastopore

Grey crescent

a   b   c   d

III. Neurulation

Neural plate

Neural folds

Neuroporus

IV. Tail Bud Stage
Organogenesis

Eye anlage

Gill branches

Eye

Suction cup

External gills

Musculature

e   f

g

V. Larval period
Tadpole

Breath hole (spiraculum)

h

Inner gills

h

Fig. 5.5 Embryo development of a frog *Rana* (after Ch Houillon). (**c**) The grey crescent lies above the slit-shaped blastopore

Germ layer formation: The archenteron sometimes is termed mesendoderm because it gives rise to both the endoderm and mesoderm. The roof of the archenteron detaches from the floor; the roof comes to be the **mesoderm** (Fig. 5.7). We now arrive at the stage that displays the three conventional germ layers:

- The **endoderm**: After the detachment of the roof of the archenteron, the walls of the remaining archenteron expand and converge to replace the lost roof. The archenteron has become the endoderm, it is identical with the primary gut where yolk released by the vegetal cells is dissolved.
- The **mesoderm** (in early gastrulation also called the **chordamesoderm**) is the detached roof of the archenteron, an initially flat sheet of cells on the dorsal side lying between the closed archenteron and the outer blastoderm. Without delay, the median stripe forms the **notochord**, the adjoining lateral stripes form the metameric series of **somites** (Sect. 5.1.6).
- The **ectoderm**: The outer wall of the gastrula is designated ectoderm; in the animal region of the gastrula it is also known as **neuroectoderm**, because the animal region will subsequently give rise to the nervous system.

Now the classical "germ layers" of a triploblastic animals are established.

Xenopus Early development

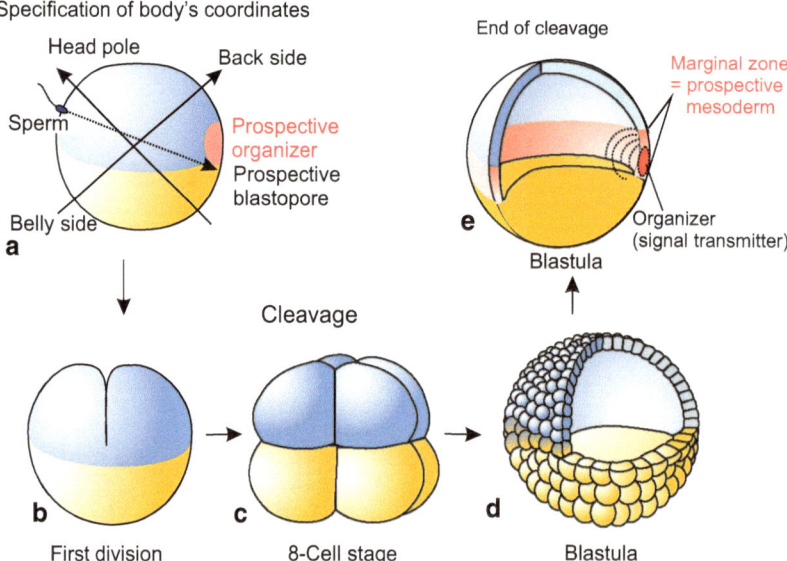

Specification of body's coordinates

Head pole

Back side

Sperm

Prospective organizer

Prospective blastopore

Belly side

a

End of cleavage

Marginal zone = prospective mesoderm

e

Organizer (signal transmitter)

Blastula

Cleavage

b First division   **c** 8-Cell stage   **d** Blastula

Fig. 5.6 (**a–e**) Amphibian. (**a**) Polarity axes in the fertilized egg (zygote); (**b–e**) Cleavage; (**e**) Blastula at the beginning of gastrulation

Xenopus Gastrulation

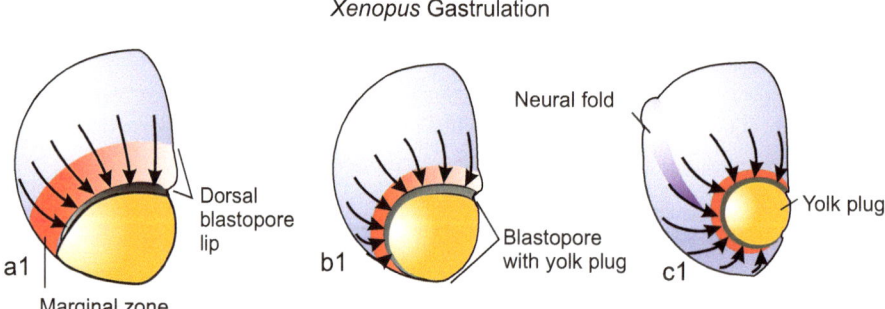

a1   Dorsal blastopore lip

Marginal zone

b1   Neural fold   Blastopore with yolk plug

c1   Yolk plug

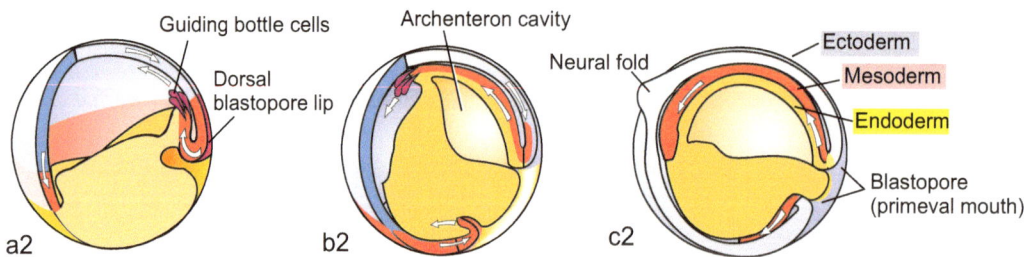

a2   Guiding bottle cells   Dorsal blastopore lip

b2   Archenteron cavity   Neural fold

c2   Ectoderm   Mesoderm   Endoderm   Blastopore (primeval mouth)

Fig. 5.7 (**a–c**) Amphibian. Gastrulation. (**a1–c1**) View from outside onto the blastopore; (**a2–c2**) Embryo dissected, lateral view

Xenopus Neurulation - Phylotypic Stage

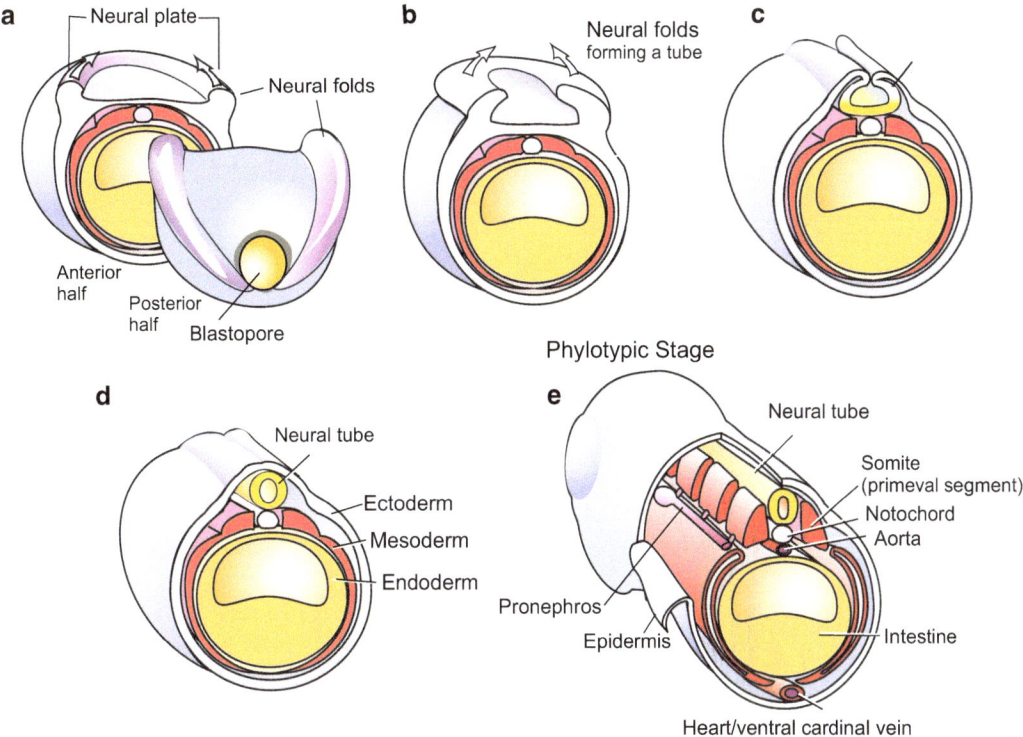

Fig. 5.8 (**a–e**) Neurulation in *Xenopus*. View onto the posterior half of the embryo. In (**e**) the phylotypic stage is reached

5.1.5   Neurulation Prepares the Formation of the Central Nervous System

Neurulation is a process by which cellular material is sequestered to form the brain and spinal cord. The keyhole-shaped **neural plate** (Fig. 5.8) is formed above the detaching roof of the archenteron. The plate is delimited by **neural folds** that rise like surging waves and converge along the dorsal midline of the embryo. Here the folds adhere to each other and fuse, forming the hollow **neural tube**. An embryo undergoing these processes is called a **neurula**. While the tube is being formed, it sinks into the interior of the embryo and detaches from the surface. For some time closure of the neural tube is incomplete: an anterior and a posterior **neuropore** are left open.

Above the neural tube, the ectoderm re-closes the outer wall. The anterior, expanding half of the neural tube will form the **brain**, while the posterior half will elongate to form the **spinal cord**. Residual cell groups along the neural tube, termed **neural crest cells**, will supplement the central nervous system by forming the spinal ganglia and the **autonomic nervous system** (Figs. 15.3 and 16.16). Further developmental potentials of the fascinating neural crest cells will be described in the Sects. 15.2 and 16.5.

5.1.6   The Mesoderm Gives Rise to Many Internal Organs: Notochord, Skeleton, Musculature, Heart, Blood, Kidneys and Some More

The detached chordamesoderm subdivides itself in a process of self-organization into several parts (Figs. 5.8, 5.9, and 5.10).

Fig. 5.9 Amphibians/
vertebrates. Generalized
basic body plan or
phylotypic stage. By this
stage, the basic traits of a
chordate are elaborated:
dorsal neural tube,
notochord, foregut with gill
slits, ventral heart (not
shown). In addition, eye
vesicles and somites are
formed in all vertebrates

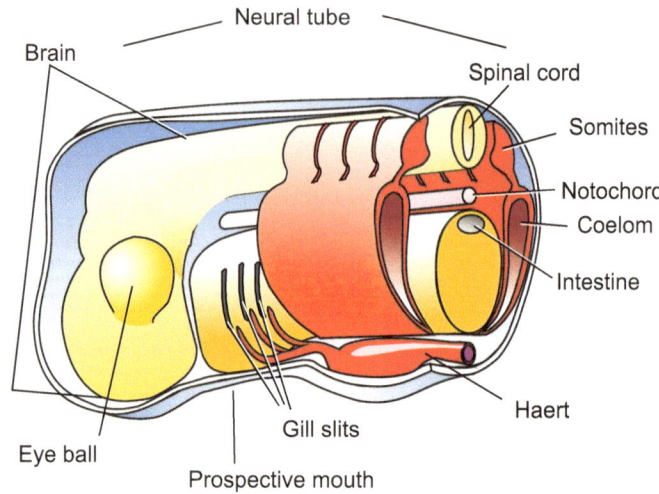

Fig. 5.10 Amphibian,
generalized vertebrate.
Development following
neurulation. Fate of the
'germ layers' (after
Portmann, top, and
Huettner, *bottom*)

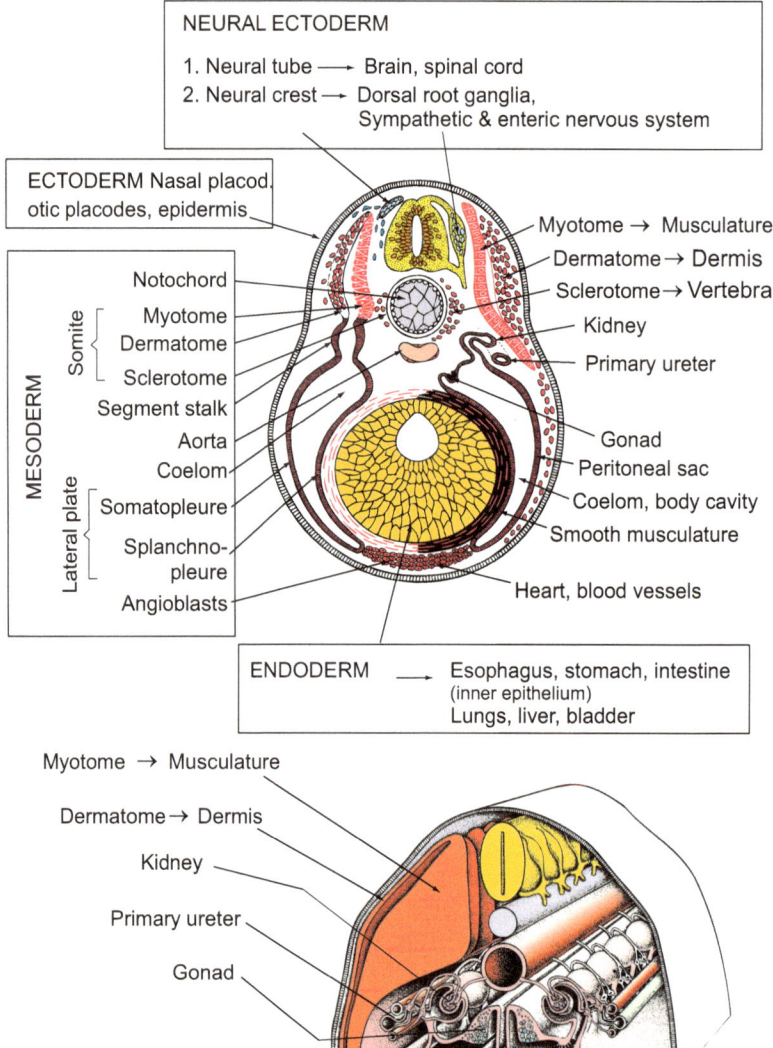

1. The anterior part of the mesoderm which was the first mesoderm to involute, gives rise to the **prechordal head mesoderm** that forms the muscles of the head region.

2. The head mesoderm is followed by the more posterior **chordamesoderm**. Along its midline, the round rod of the **notochord** detaches; the notochord is a transient structure needed to organize the development of the central nervous system CNS and the vertebral column.

3. and 4. The dorsal axial mesoderm that is involved in notochord formation is flanked by two lateral bands, the **paraxial** or **somitic mesoderm**. These bands separate into block of cells called **somites**. Somites are also transient structures, but they are extremely important in organizing the basic segmental body pattern, and they give rise to many tissues: vertebrae, muscles, and the inner layer of the skin, called the dermis. Notochord and the two series of metameric somites are often subsumed under the term **axial organs**.

5. and 6. The adjoining residual cells derived from the roof of the archenteron become flat, expanding sheets called **lateral plates**. Both plates spread laterally and ventrally into the space between the ectoderm and the endoderm. While doing so, a cleft within the plates transforms them into flattened bags. The bags enclose a new cavity, called the **coelom** or secondary body cavity. The coelom will subdivide into the pericardial cavity enclosing the heart, and the large body cavity. In mammals, this cavity is subdivided into the pleural cavity enclosing the lungs, and the peritoneal or visceral spaces around the intestine.

The wall of the lateral plate snuggling up and clinging to the ectoderm (**somatopleure**, or **parietal mesoderm**) will form the peritoneum and pleural lining, whereas the wall clinging to the endoderm (**splanchnopleure**) will provide the circular musculature of the stomach and the intestine, as well as the mesenteries needed to suspend the digestive tract in the coelom.

Somites and lateral plates are connected by intermediate mesoderm that forms a longitudinal series of thickenings, the **pronephrotic rudiments**. The solid rudiments hollow out and form funnels opening into the coelom. The pronephrotic rudiments will later be transformed and supplemented by further mesodermal derivatives to yield the **urogenital system**.

In the anterior body beneath the archenteron, where the two lateral plates approach each other, migratory mesodermal cells accumulate. They are myogenic and organize themselves to form the **heart** and the adjoining major **blood vessels** (Fig. 17.11).

The most extensive developmental potential belongs to the **somites**. Somites are transient but the cells constituting them do not vanish. They separate into two main groups: **sclerotome** and **myodermatome**. The myodermatome further subdivides into the **myotome** and **dermatome** (Fig. 5.10).

- **Sclerotome**. The cells of the sclerotome lose contact with each other, emigrate, and accumulate at the notochord, thereby surrounding it. They become **chondrocytes** committed to construct the **vertebral bodies**. Most of the enclosed notochordal cells die, but some are left to form the nucleus of the **intervertebral discs**. These are the discs that slip in painful back injuries.

- **Myotomes**. The packets of the myotomes expand and give rise to the cross-striated musculature of the body that occupies the largest spaces in the trunk and tail region.

- **Dermatome**. The (dividing) cells derived from the dermatome migrate and spread extensively, clinging to the inner surface of the ectoderm. They will give rise to some additional muscle tissues but will mainly form the **dermis**, while the epidermis will be formed by the outer ectoderm covering the dermis.

The complete derivation of all connective tissue masses and **skeletal elements** is complex.

Most precursors of the chondrocytes and osteocytes that construct the cartilage and bones of the body derive from the somites, but those osteocytes which form dermal bones appear to arise in the dermis. The skeletal elements in the ventral head and pharynx region, summed up as **visceral skeleton**, do not derive from mesoderm but from **neural crest cells** (Fig. 15.3).

5.1.7   From the Endoderm Arise: Gastrointestinal Tract, Lung, Liver, Pancreas and Bladder

The archenteron forms the **pharynx** with **gills** (of the tadpole) and **lungs** (of the adult), and the **digestive tube** with adhering organs such as **pancreas**, **liver** with **gallbladder** and the **urinary bladder**. The roof of the pharynx participates in forming the pituitary gland by contributing the anterior lobe, the **adenohypophysis** (Fig. 6.20). Yet, it is only the interior, epithelial lining of the gastrointestinal tract that derives from the endoderm; the musculature derives from the mesoderm (splanchnopleure), the nerve net from immigrated neural crest cells.

5.1.8   The Most Famous Embryological Experiments Were Done on Amphibians; Here: Cloning by Transplanted Nuclei, Identical Twins and Chimeras

Nuclear Transplantations. Experiments that led to the production of cloned frogs are justly famous. Originally, the experiments were planned (by JB Gurdon (Nobel Prize 2012), R Briggs, TJ King, and others, see Box 1.1) not to clone animals but to examine whether somatic nuclei remain totipotent in the course of development, or if genetic information is irreversibly lost or inactivated (Fig. 13.2).

Identical Twins and Chimaeras. One of the first experiments performed in developmental biology was carried out in 1888 by the German anatomist Wilhelm Roux. Under the magnifying glass were frog germs in the 2- (to 4-)cell stage. Using a hot needle, Roux killed one of the first

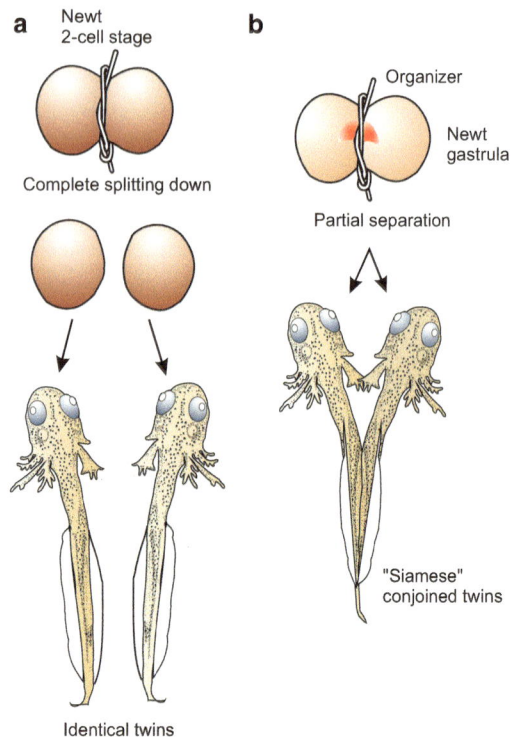

Fig. 5.11 Amphibian. Constrictions of newt embryos by a ligature performed by H. Spemann. Complete division results in monozygotic, identical twins, partial division leads to partial duplication of the body ("Siamese" twins in common parlance). Division of a blastula or early gastrula leads to a (entire or partial) duplication only when both halves get a piece of the Spemann organizer

two blastomeres. Because he did not remove the killed cell, he got crippled embryos with only one healthy half. Roux interpreted his observation as evidence that the fate each blastomere was already fixed and continued its development independently of other parts by "self differentiation". His notion became known as "mosaic theory" of development. However, he also saw indications of some regulatory powers like those Hans Spemann could observe after him.

At about 1920 Hans Spemann re-examined Roux's findings using a different technique. He separated early amphibian embryos (from the 2-cell stage up to beginning gastrulation) into halves using lassos made from baby hairs to make a ligature (Fig. 5.11). If the embryos were divided along the animal–vegetal axis so that each half was provided with material of the grey crescent, **monozygotic (one-egg) identical** twins arose. Partial constriction resulted in "Siamese" **conjoined twins**. By pressing together two cleaving embryos from the newts *Triturus cristatus* and *Triturus vulgaris*, **chimaeras** composed of a mosaic of genetically different cells were obtained.

Experiments with tissues from different origin, including tissues taken from different species, were continued in extensive transplantation studies (as follows).

5.1.9   By Means of Transplantations the Progress of Determination can be Analysed: Do Transplants Behave According to Their Location or Their Origin?

Among the most exciting experiments in the history of developmental biology were those carried out by Hans Spemann and his students between 1920 and 1940, using amphibians. The experiments yielded the first information on the significance of signals sent out by parts of the embryo to instruct other parts (**embryonic induction**). By microsurgery pieces were taken from donor embryos and inserted into host embryos at various places (Figs. 5.12, 5.13, and 5.14). The experiments were aimed at finding out whether tissue pieces would behave **according to their (new) location**, meaning that they still could be reprogrammed by positional cues, or whether they would behave **according to their descent**, meaning that they already were irreversibly committed.

In an experiment which addressed the question of **positional information** (Spemann: "**development according to location**") the non-determined, prospective epidermis of the belly region was transplanted to the area fated to become the mouth region of a newt. The transplant formed mouth and teeth according to its new location (but horny teeth according to the genetic potency of the donor; tadpoles of frogs have horn teeth and not teeth of dentin as do salamanders! Fig. 5.12). In the course of such studies the phenomenon of induction was observed.

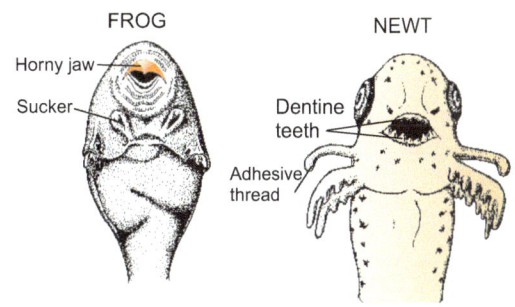

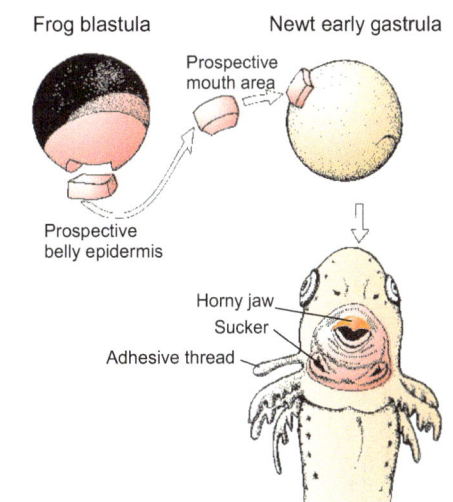

Fig. 5.12 Amphibian. Demonstration of positional information. Classic transplantation experiment of Spemann. A future (presumptive or prospective) piece of the epidermis of the belly, taken from a frog blastula, is transplanted into the future mouth region of a newt. The frog piece becomes committed in the mouth region of the host to form mouth tissue. However, it is a frog mouth that is formed, according to the genetic potential of the frog cells. Positional information is not species-specific, but the positional information received by a cell can only be interpreted according to the cell's own genetic program

5.1.10   Systematic Transplantations Led to the Discovery of Embryonic Induction and of the Most Effective Signalling Centre, Now Known as the Spemann Organizer

Embryonic induction means: one part gives orders, the other responds. Systematically planned transplantation studies which began in 1918 lead to the discovery of extensive

Spemann-Mangold Organizer Experiment

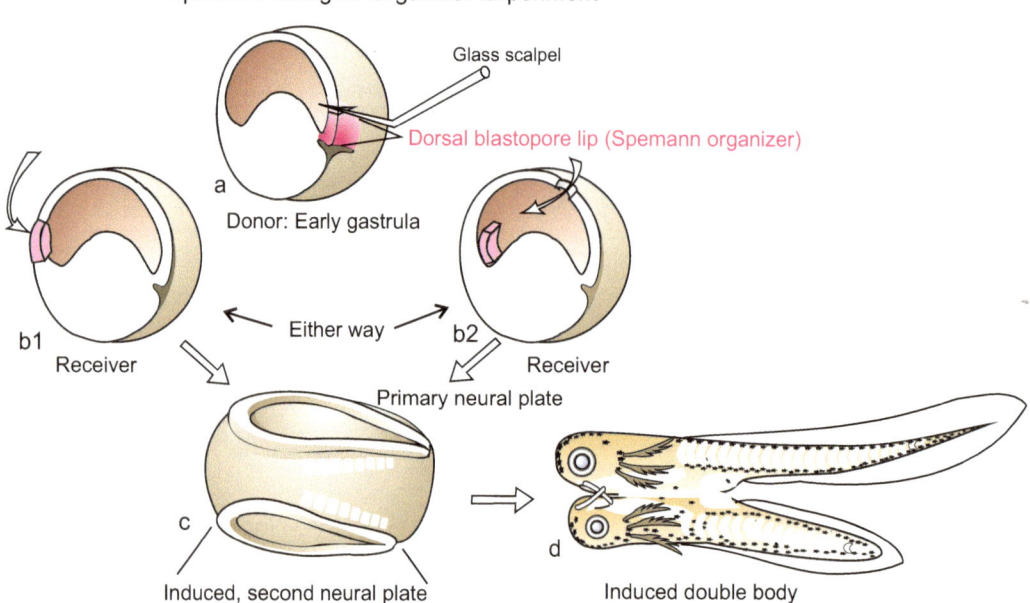

Fig. 5.13 Amphibian. Classic experimental induction of a secondary embryo by transplantation of an upper blastopore lip ('organizer'). The transplanted tissue is marked in red, so it was shown that the secondary embryo is mainly formed by the host. Experiment proposed and interpreted by Hans Spemann and carried out by Hilde Mangold in 1924

embryonic induction (published in 1924; for this discovery Spemann won a Nobel Prize for medicine in 1935, the second zoologist after Thomas Morgan to win it). In a classic example, the roof of the archenteron induces the overlying animal ectoderm to form the neural plate and, hence, the central nervous system (this is so-called "primary" or neural induction). But induction begins earlier and has a spectacular culmination when the early gastrula forms the blastopore. At this time and place the **organizer** becomes active.

Disregarding the actual sequence of inductive events in the early embryo for a moment, we tell the story of discovery in historical order. Prompted by Spemann, his student Hilde Mangold transplanted the **upper blastopore lip** of the darkly pigmented *Triturus taeniatus* into the ventral ectoderm or (accidentally) into the blastocoel of the white blastulae of *Triturus cristatus* (Fig. 5.13). The result was exciting: When properly inserted into the ventral ectoderm, the donor tissue did not adapt to the new surrounding. Instead, it started invaginating; soon axial organs (notochord, somites) and a head emerged, and eventually a complete second embryo formed, joined to the other: a Siamese twin possessing head, axial organs (notochord, neural tube), somites, etc. But even when the inserted lips dropped into the interior cavity, they initiated gastrulation in the surrounding host tissue. When the host embryo developed further, it was found that an additional system of axial organs appeared on the side where the graft came to lie. Partial or complete supernumerary embryos arose. Since black donor and white host embryos had been used, the supernumerary embryos were readily determined to be composed of host as well as donor tissue. In fact, the largest part of the supernumerary embryo was made by the host; host cells were induced to participate in formation of the second body.

Spemann referred to the upper blastopore lip as the **organizer** because it induces host cells to change their fate and is able to release an entire, coordinated developmental program leading to a well organized, complete embryo. A general term designating the source of an inducing signal is **inductor** or **inducer**.

It may be explicitly emphasized that organizer activity is not carried out by the same cell population all the time, because the cells that at this instant constitute the dorsal lip will in the next moment disappear into the interior and will be replaced by newly arriving cells. **The term organizer designates a zone travellers pass through thereby being informed (induced) which path to take subsequently.** In another paradigm the organizer is viewed as a classroom which students (cells) enter, and leave with new learned lessons.

A number of authors take the view that an intermediate signalling centre is inserted between the cortical rotation and the work of the Spemann organizer. First a preparatory emitter would be established in the area of the grey crescent (or its equivalent area in *Xenopus*), called Nieuwkoop centre. In the 32-cell blastula signals radiate from this centre committing the dorsally adjacent cells to act as the Spemann organizer. Other authors consider the **Nieuwkoop centre** and Spemann organizer to be two consecutive stages of the same signalling system. In both centres identical signalling molecules were detected, for instance WNT signalling molecules whose production is triggered by maternal β-catenin protein shifted into the posterior region of the egg during the cortical rotation. Read more about the important WNT-β-catenin (that often will be addressed in this book) in the following Sect. 5.1.12; a summary can be read in Chap. 11

5.1.11 The Organizer Separates into Head- and Trunk Organizer

Already in the young gastrula the Spemann organizer begins to display regional specificity. It splits into two to three initially overlapping fields that contribute differently to the induction of a second embryo.

- A first field with inductive power is the region that encompasses the lip of the blastopore at the moment it is just being formed. The cells of this field are the first to flock together and to move into the blastocoel; they move anteriorly getting into the future head region and forming there the head mesoderm. This vanguard cohort of cells is responsible for the induction of the **head with forebrain and midbrain**.

- The second field is initially located just above (in animal direction) the first field. Following the first cohort these cells invade the interior, too, and come to rest in the roof of the mid-archenteron; while they themselves are fated to become notochord they are responsible to **induce the trunk**.

- Some authors distinguish a third discrete field, separated from the trunk field, and call it **tail organizer**.

In the course of the gastrulation the fields expand and separate from each other. If successive regions of the archenteron roof of late-gastrulae are transplanted into the blastocoel of early-gastrulae (Fig. 5.14), they no longer induce complete secondary embryos but merely heads (induced by anterior roof parts) or trunks (induced by posterior parts).

5.1.12 Processes of Induction Are Turned on by Maternal Factors; the Organizer Is Installed by Means of Components of the WNT Signalling System

Organizing is a complex event. A whole bundle of signal molecules appears to be involved, either simultaneously or consecutively effective. A **cascade of inductive events** already begins with fertilization. In fact, recent knowledge dates the starting point of the story back to fertilization or even oogenesis. During fertilization, dorsoventral specification is accomplished by rotation of the egg cortex (Sect. 5.1.3) and displacement of

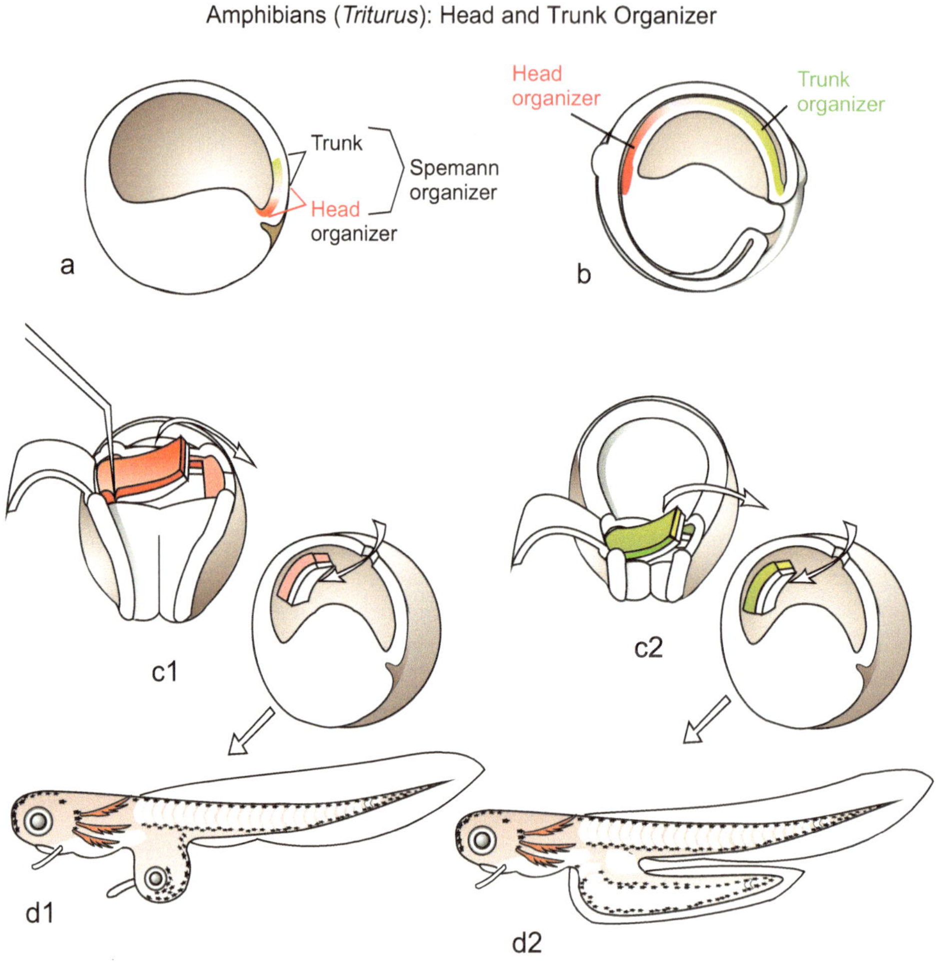

Amphibians (*Triturus*): Head and Trunk Organizer

Fig. 5.14 Amphibian. Segregation of Spemann's organizer into head organizer and trunk organizer. Upper part: After experiments by C. Niehrs 1998; lower part: Transplantation of parts of the roof of the archenteron, after O. Mangold 1933

inner cytoplasmic constituents. Maternal mRNA coding for signal molecules and receptors (among many other proteins) becomes localized in a distinct pattern. Incorporated into the cells of the blastula, the differentially disseminated mRNA species enable the cells in the various regions of the embryo to generate and to receive different sets of signal molecules. Important maternal information is concentrated in and around the area exposed as grey crescent or its equivalent.

Inductive interactions are studied at several levels with methods reflecting the history of experimental developmental biology (following

Sect. 5.1.13). From a large variety of experiments at the level of biology (surgery, transplantations) and biochemistry, a coherent picture must be reconstructed. The picture must be dynamic like a film. Different researchers tell stories differing in detail, but in general the play is divided into three to four acts (Fig. 5.15):

1. **The mesodermalizing induction**. Signals spread from the vegetal cells of the blastula and cause a broad, ring-like zone, called **marginal zone**, around the equator to become the mesoderm in the future. This signalling function is currently assigned to the secreted proteins **Vg-1** and **NODAL** released from

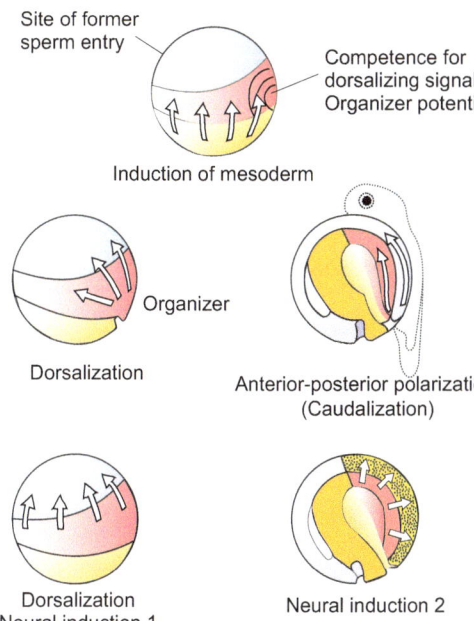

Site of former
sperm entry

Competence for
dorsalizing signals
Organizer potential

Induction of mesoderm

Organizer

Dorsalization

Anterior-posterior polarization
(Caudalization)

Dorsalization
Neural induction 1

Neural induction 2

Fig. 5.15 Amphibian. Sequence of inductive events in the blastula and gastrula, suggested by the outcome of experiments in which different parts of early embryos were combined. Arrows indicate the assumed spread of inductive signals

cells around the vegetal pole and radiating in the animal direction. These proteins are members of the large **TGF-β protein family**, termed after a factor supporting tumour growth and called TGF-β (*Transforming Growth Factor beta*). The marginal zone responds by expressing, for example, the gene ***Brachyury***, a marker of mesodermal progenitor cells. During gastrulation this marginal zone is shifted into the interior, forming the roof of the archenteron which will give rise to the notochord, the somites, and the lateral plates. The gene *Brachyury* remains active in the notochord for a prolonged time.

2. **The "dorsalizing induction" and the establishment of the Spemann organizer**. Among the cytoplasmic constituents shifted during egg activation into the posterior-dorsal position of the grey crescent is mRNA of components of the canonical WNT signalling system; they are considered to be the decisive components: mRNA for the signalling

molecule **WNT-1**, Dishevelled-Protein (Dsh) wrapped in vesicles, and mRNA for the transcription-cofactor **beta-CATENIN (β-CATENIN)**. As exposed in Chap. 11, β-CATENIN taken up in cells activates its own gene and, beyond that, genes of the entire signalling path to which it belongs. And see: Microinjection of *β-catenin* **mRNA at an antero-ventral place causes twinned axes and hence a more or less perfect supernumerary embryo,** just as if an early blastopore lip had been implanted (see Fig. 5.17).

In the traditional terminology of research on amphibian embryos one speaks of "**dorsalizing induction**", because the organizer is located at the dorsal rim of the blastopore and, thus, marks the future back. Since the blastopore will become the anus, the dorsalizing induction also specifies the tail pole. Therefore, one may read: the **antero-posterior body axis becomes fixed**.

Among the genes activated by β-CATENIN, besides genes of the WNT signalling system, are further genes whose products synergistically aid in the establishment of the Spemann organizer, for example the genes for the transcription factors GOOSECOID and SIAMOIS. Yet, being proteins with DNA-binding homeodomain (see Fig. 12.14) they are not exported and, therefore, cannot convey information to neighbouring cells without the intervention of signal molecules that are secreted or exposed on the cell surface. GOOSECOID and SIAMOIS find their way into the nuclei of the cells in the organizer region and turn on genes for factors to be exported. If mRNA of *goosecoid* or *siamois* is injected into a blastomere located at an anterior-ventral site of a cleaving embryo, later, in the neurula stage, dorsal structures such as notochord and neural tube will appear there, as if a blastopore lip had been implanted there. Among the secreted factors are NOGGIN and CHORDIN. The expression of the *noggin* and *chordin* genes starts in Spemann's organizer subsequent to that of *goosecoid*. Both NOGGIN and CHORDIN are expressed

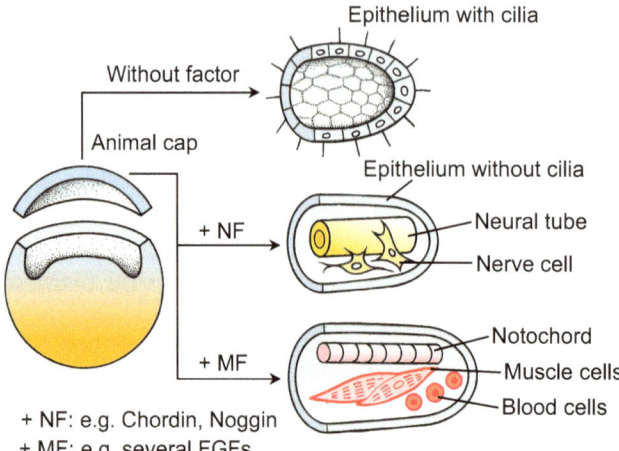

Fig. 5.16 Amphibian. Assay to test the inducing potential of extracts or purified factors. Uncommitted animal caps are removed from blastulae and exposed to solutions containing putative factors. The caps form hollow spheres with ciliated walls in the absence of inducing factors (*top*). If inducing factors are present in the solutions, the exterior layer forms an epidermal epithelium enclosing neuronal or mesodermal cells. Neuralizing factors NF or mesodermalizing factors MF have been extracted from heterologous sources (i.e. from other species) or from autologous sources (i.e. from the same species from which the animal cap was taken)

Amphibians (*Xenopus*): Inducing factors

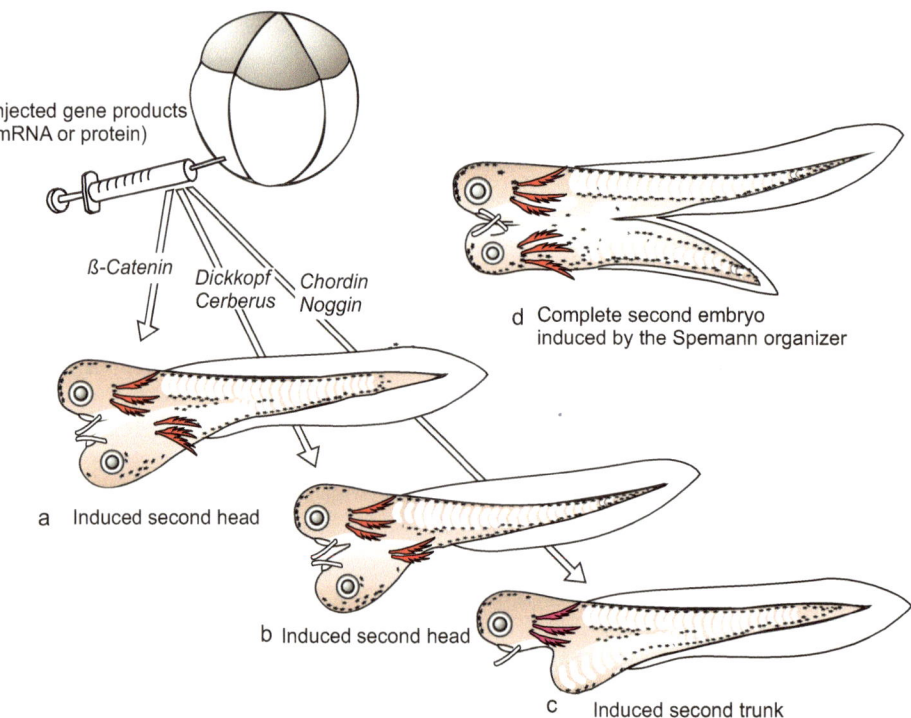

Fig. 5.17 (**a–c**) Amphibian. Induction of secondary heads (**a, b**) or secondary trunks (**c**) by factors obtained from the Spemann organizer; (**d**) Whole secondary embryo induced by living organizer tissue for comparison

at the right place and at the right time to regulate the onset of gastrulation.

3. Planar signals and convergence movements. From the blastopore lip, signals cause the cells destined to become the mesoderm to stream along the future back meridian in a way called convergence movements towards the blastopore in order to submerge there into the cavity of the blastocoel. One speaks of **planar signals** (Compare Fig. 10.7). While the archenteron invaginates, and further on after completion of the involution, the upper blastopore emits a different kind of signals that spread in the plane of the archenteron's roof and thus, in the young mesoderm. The function assigned to these signals is to establish an anteroposterior polarity in the mosoderm, resulting structures in the anterior body region (such as skeletal elements of the skull and heart) different from posterior structures (such as trunk and tail musculature). On the other hand signalling molecules are also produced that leave the plane of the roof of the archenteron at the dorsal side and spread through the interstitial cleft to reach the overlying neuroectoderm. We arrive at the neuralizing induction.

4. **Neuralizing induction**. The induction of the central nervous system with brain and spinal cord is also a multistep process.

   • **Acquisition of competence**. The signals radiating into the animal hemisphere enable the animal cap to deviate from the pathway towards the epidermis. Instead, the animal cap acquires the competence to form neuronal tissue as the default state.

   • **Regionalization**. While moving along the ectoderm, the mesoderm confers gradual regional differences upon the neuroectoderm.

Since these events are main topics in developmental biology, a separate Sect. 5.1.15 devoted to *Xenopus*, and, considering also comparative aspects, Chap. 16 will take up the topic in more detail.

5.1.13 By Means of Biological Assays Many Factors Exhibiting Inductive Potency Have Been Identified; Here a Brief List

Inductive interactions between the cells begin in the early blastula. They are studied at several levels with methods that reflect the history of experimental developmental biology.

1. **Surgical methods** such as explantations: Different parts in the mid- and late- blastula stages are isolated, rejoined in various combinations and cultured, and the result is analyzed microscopically. Antibodies recognizing defined differentiated cells or cells about to differentiate facilitate the evaluation of the outcome. A historically important example of this type of experiment was the combination of isolated, uninduced animal caps with pieces comprising yolk-laden vegetal cells. While neither animal caps nor vegetal pieces by themselves gave rise to mesodermal structures, the combination did. The vegetal partner caused descendants of the animal cap to develop into mesodermal cells such as notochord cells, muscle cells and blood cells. By combining different parts of the embryos at different times consecutive inductive interactions were disclosed. The observations were brought into a coherent picture (Fig. 5.15).

   Experimentally, mesodermal or neural gene expression can be released not only by living cells. We already mentioned the possibility to inject mRNA or protein of putative inducing factors into blastomeres (Fig. 5.17). But how can one get on the track of a putative factor?

   It is a **bioassay**, called the **animal cap assay**, that has been used for years, and still is used, for identifying inducing factors.

2. **Bioassays of possible inducers using animal caps**. In the hunt for inducers appropriate bioassays are crucial. The "**animal cap**

assay" was both the source of many pitfalls and of the final success. Caps are excised from the animal hemisphere of blastulae and maintained in cell culture. For decades the caps were taken from newt blastulae. Pieces of gastrulae, that were excised from other regions of donor embryos and supposed to be source of inductive substances, were attached onto the caps. Alternatively, the animal caps were bathed in buffered solutions that contained the putative inducers. After some time of contact or incubation, the caps were microscopically examined for the presence of differentiated cells and complex structures. Even eyes or brain structures could be identified in the isolated and treated caps. Thus the beginning was promising.

However, the hunt for inducers, in particular the hunt for neural inducers, was soon confounded by the fact that just about any culture condition triggered neural development, an effect that became known as "autoneuralization". Fortunately, *Xenopus* ectoderm or "animal cap" tissue does not readily "autoneuralize" or develop mesodermal cells such as muscle cells, blood cells or cells typical for the notochord. Uninduced animal caps are exposed to solutions of various "growth factors", and the result is again examined microscopically (Figs. 5.16 and 16.1B). Originally, growth factors were either supernatants from *Xenopus* cell cultures (autologous source) or from alien donors (heterologous sources such as chick embryos, foetal calf serum, or supernatant of mammalian cell cultures), or fractions thereof (more about the history of discovery can be read in Chap. 10.4). The result of such experiments was the differentiation of distinct cell types.

3. **Injection of putative inducers, or of mRNA coding for such inducers**. Putative inducers are injected into a site that still contains uncommitted cells and normally does not develop the expected structures (Fig. 5.17). If the putative inducer is a polypeptide, mRNA can be injected instead of the protein. Injection of anti-sense RNA, on the other

hand, can be used to weaken or eliminate an inducing interaction.

The large, ever growing spectrum of methods supplied by modern molecular biology rendered it possible to trace inducing factors. Here a (still incomplete) list:

- **FGF-family** (*Fibroblast Growth Factor family*); several members of this family induce, to different degrees, formation of structures and cells derived from ventral mesoderm, such as blood cells, mesenchyme (prospective connective tissue), cartilage, bone, and musculature.

- **TGF-β family**; besides the already mentioned maternal factors Vg-1 and NODAL that reside in the ventral region of the early blastula, factors that are manufactured in the late blastula and gastrula also belong to this family, coded by zygotic genes, Of particular significance are the **BMP's** (*Bone Morphogenetic Proteins*). They had already been known as factors supporting bone formation in mammals, hence their name. Among them **BMP-4** is rated as a powerful morphogen that in the late blastula and neurula is present in the form of a gradient extending from anterior-ventral to posterior-dorsal (Figs. 5.18 and 5.19). In its function as a morphogen it evokes various local responses according to its local concentration (for a definition of the term morphogen s. Chap. 10, or Glossary). **In the ventral ectoderm BMP's suppress the inherent propensity to form nervous tissue**. The highest concentration says: "here epidermis has to be made", the lowest concentration permits to initiate the formation of dorsal structures including the central nervous system.

- **WNT family** (in *Xenopus* designated Xwnt), the members of which display sequence similarity (homology) to the *Drosophila* gene product WINGLESS. *Xenopus* possesses an assortment of at least 15 diverse WNTs with partially different ranges of effects, and it too possesses a multitude of WNT receptors,

Amphibians: Controlling Factors in Early Development

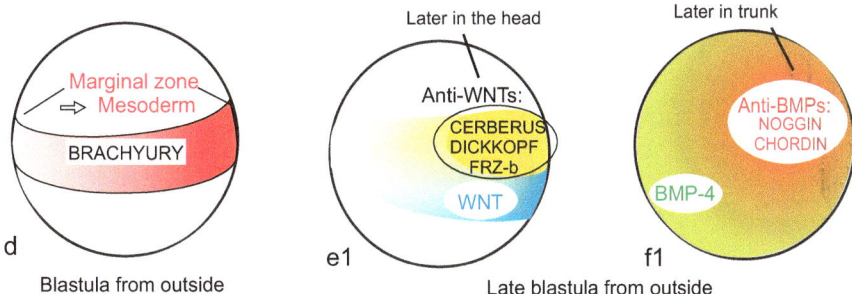

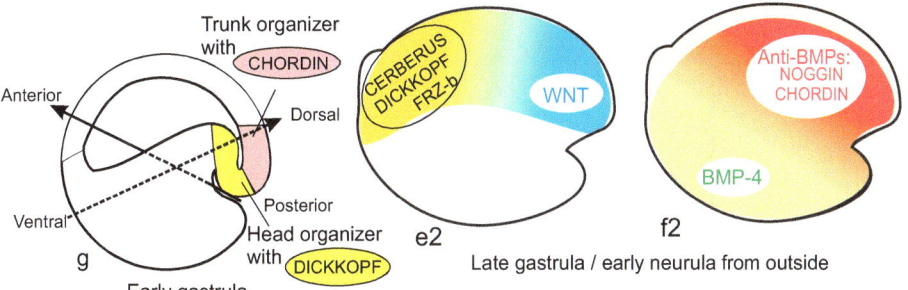

Secreted protein   Transcription factor

Fig. 5.18 (**a–g**) Amphibian. Early events specifying the bilateral symmetrical organization. (**a**) Location of maternal factors in the unfertilized egg. (**b**) Cortical rotation after entry of the sperm; β-catenin and a β-catenin-stabilizing agent are shifted into the region of the future blastopore. (**c**) Sphere of action of maternal factors, establishment of the Spemann organizer above the prospective blastopore, and expression of *Wnt/β-Catenin*. (**d**) Marginal zone = expression domain of mesodermal genes, e.g. of *Brachyury*. (**e1**) Expression domain of WNT morphogens (WNT-3 and WNT-8), and of their antagonists CERBERUS, DICKKOPF, and FRIZBEE-B in the blastula. (**e2**) Relocation of the expressions domains of the WNTs and anti-WNTs in the gastrula. (**f1**) Expression domain of the morphogen BMP-4 (Bone Morphogenetic Protein 4) and of its antagonists CHORDIN, NOGGIN and FOLLISTATIN in the blastula. (**f2**) Translocation of the expression domains of BMP-4 and of anti-BMP's in the gastrula. (**g**) Expression domains of typical representatives of the anti-WNTs (Dickkopf) and the anti-BMPs (Chordin) at the beginning of gastrulation. Secreted proteins are surrounded by an ellipse, transcription factors by a rectangle

Fig. 5.19 (**a–c**) *Xenopus.* Gradients of the morphogens BMP and WNT, and of their antagonists in the neurula. After a figure from Niehrs 2004, redrawn and upgraded

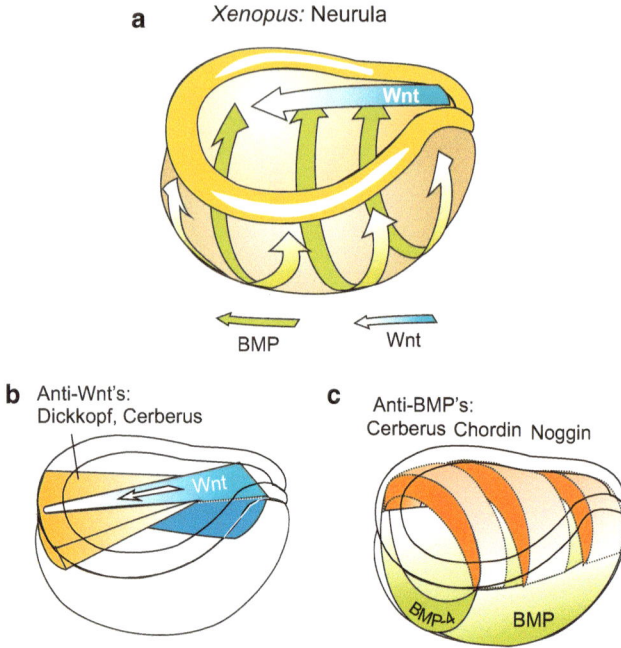

generally called FRIZZLED. In the organizer and around the blastopore initially WNT-1 and WNT-11 are found, in the neurula Wnt-3a and **WNT-8**. In its function as a morphogen the highest concentration of WNT-8 defines the posterior end (tail pole) (Figs. 5.18 and 5.19). Knowledge gained from other systems suggests that WNT-11 confers planar signals to guide convergence movements of gastrulating cells to the blastopore. Transcripts for WNT-11 are prepared as maternal constituents in the *Xenopus* egg and ready to act.

- SONIC-HEDGEHOG. Its function in the upper blastopore lip is unknown, but something is known about its function in the notochord and its surroundings (Fig. 10.6). According to transplantation studies described above and shown in Fig. 5.14, the organizer separates into head- and trunk organizer. Accordingly, factors have been found corresponding to this subdivision.
- **Head formation**: The mRNA's of the genes *cerberus*, *dickkopf* and *frizbee-b*, or

corresponding CERBERUS-, DICKKOPF- or FRIZBEE-b protein, evoke, when injected into an anterior-ventral quadrant of the blastula, (Fig. 5.17) the formation of a second head. In the normal gastrula, these proteins are produced in the tissue where, according to transplantation assays, the head-activator resides. These cystine-rich proteins class among the anti-WNT's (following section).

- **Trunk formation**. mRNA's of the genes *chordin* and *noggin* induce trunks (Fig. 5.17c). The proteins CHORDIN and NOGGIN count among the **anti-BMPs** (following section).

Because the induced secondary embryos (heads and/or trunks) possess a central nervous system, a notochord and somites, the collection of the above listed factors must have the potency to induce the formation of neural plate, notochord and somites, either directly or in cooperation with local factors. Animal caps bathed in solutions of CHORDIN or NOGGIN differentiate nerve cells (and occasionally also notochord structures); hence, these factors are also classified as **neuralizing factors**.

5.1.14 Several Inducing Factors Generated at Opposing Sites Bind and Neutralize Each Other in the Extracellular Space, Thus Forming Reverse Gradients

In the previous sections it was occasionally mentioned that several inducers may act antagonistically. Some factors bind each other in the extracellular space forming heterodimers thus neutralizing each other mutually. Such antagonistically acting factors are emitted from regions positioned at opposite locations. In this way gradients are established of non-neutralized and therefore still effective factors (see for example Fig.10.3c). We summarize:

- **Wnt-8 and Wnt-3** are emitted in the region of the blastopore. On their diffusion-driven way towards the head they become more and more neutralized by anti-Wnt's (Figs. 5.18 and 5.19), especially by **CERBERUS, DICKKOPF, FRIZBEE-b** and some more factors. Thus, head formation is rendered possible in the anterior region of the embryo, while the formation of trunk and tail is enabled in the posterior region of the embryo.
- **BMP factors** (BMP-4, BMP-2) support the formation of the belly region. Factors enabling the formation of the nervous system are antagonists to the BMP's. These neuralizing factors are initially emitted by the Spemann organizer, later from the roof of the archenteron (Figs. 5.18 and 5.19). **BMP-4** in particular is neutralized by the **anti-BMP's: NOGGIN, CHORDIN**, FOLLISTATIN, NODAL-related 3, and CERBERUS.

Bond to their antagonists BMP-4 and WNT-8 cannot couple to their receptors and thus are unable to activate them. In areas into which the organizer-born factors spread BMP-4 is no longer in charge. In the sphere of action of the organizer the concentration of free BMP-2/4 molecules decreases and a gradient of active BMP factors appears outside the organizer domain. In the head region, on the other hand, WNT-8 is trapped and a gradient decreasing from posterior to anterior appears (Fig. 5.19).

5.1.15 Neural Induction and Regionalization: The Ectoderm of the Dorsal Blastula Hemisphere Has an Autonomous Tendency to Form Nervous Tissue; This Tissue Needs to be Regionally Subdivided

Maternal determinants, for example the mRNA for the transcriptionfactor SOX3, confer the cells of the animal hemisphere a bias for becoming nerve cells. This latent proclivity, however, must be activated in due time.

From the mesodermal ring (marginal zone) of the blastula signals radiate in direction of the animal hemisphere which in the area of the prospective neural plate turn on transcription of **proneural genes** (Chap. 16) and also the gene for the **cell adhesion molecule N-CAM** (see Fig. 14.4). With that a first step is made toward neural differentiation. Neural differentiation is furthermore supported by signals emanating from the Spemann organizer. When in the course of gastrulation the future mesoderm rolls over the upper blastopore lip, enters the interior of the embryo and glides, while forming the roof of the archenteron, along the overlying ectoderm the future mesoderm continues to emit neuralizing signals (Fig. 5.20). Eventually, these signals lead to definitive determination and initiate neurulation.

Between 1930 and 1990 much effort was spent on extraction and isolation of neuralizing factors from tons of chick embryos and other sources. Yet, the specificity of the factors were questionable; because animal caps of newts, and to a lesser extent also of *Xenopus*, disclose their dormant predisposition for differentiating nervous tissue upon treatment with any stimuli and to produce autonomously nerve cells and glia cells without being instructed by specific external cues. One speaks of the *default option*. In the early blastula the animal blastoderm must be restrained from doing this. According to present understanding this suppressive function is fulfilled by BMP-4 (in cooperation with members of the WNT family). CHORDIN and NOGGIN, which are expressed not only in the organizer but

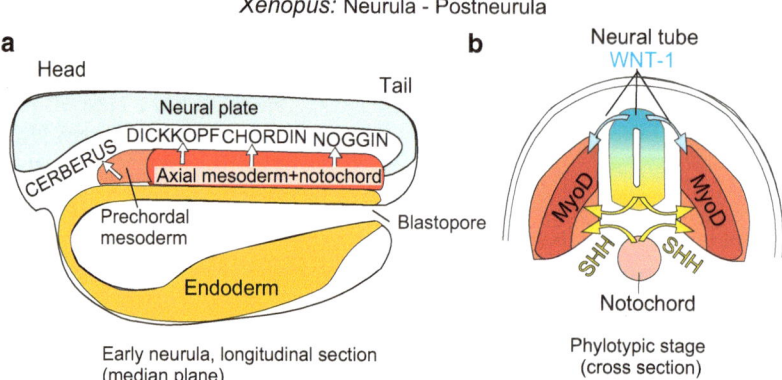

Xenopus: Neurula - Postneurula

Fig. 5.20 (**a, b**) Amphibian. (**a**) Neurula in lateral view; inducing factors emanating from the mesoderm; (**b**) Induction of the musculature by SONIC HEDGEHOG (SHH) and WNT-1; Expression of the muscle-determining transcription factor MyoD

further on in the axial mesoderm, that is in the notochord primordium, (Fig. 5.20), release neurulation by neutralizing BMP-4. In order to convert the anterior neural tube into brain, and the posterior neural tube into spinal cord further signals are needed that specify regional differences along the longitudinal axis. Members of the WNT and FGF families play roles, as well as a non-peptide factor **retinoic acid (RA)**. Overdoses of this vitamin A derivative cause headless embryos in fish, frog and chicken.

Besides these permissive organizer-born factors, instructive neuralizing factors are also involved, for instance several FGFs, which become active in the absence of BMP-4 and switch on neural-specific transcription factors such as Zic.

5.1.16 There Are Cascades of Primary, Secondary and Tertiary Induction Processes: A View to the Eye

When Spemann coined the term "primary induction" he was not yet aware that the cascade of inductive interactions leading to the formation of the central nervous system starts in the uncleaved egg. However, he was aware that further inductive events, called secondary and tertiary, follow primary induction because Spemann detected the

classic example of such a secondary induction: the optic vesicle induces the ectoderm to form the **lens** to help complete the eye (Figs. 5.21 and 16.21). After the lens placode has been formed the story continues with a tertiary induction: the lens induces the overlying epidermis to supplement the optical apparatus of the eye by forming the transparent cornea.

However, in *Xenopus*, unlike other amphibians, poor lens formation would start at the correct place even in the absence of the optic vesicle. In this case inducing signals are transmitted by the authorized cells when they still are included in the neural plate and before they are recognized as optic vesicle. Removal of the optic vesicle merely shortens the time period of induction

Twisted eyes. Among many experiments done with later stages of amphibian embryos, we have space to mention only one: By removing, rotating and retransplanting optic vesicles we gain insights on how the eye is neurally connected with the brain (**retino-tectal projection**, Fig. 16.25). Recent interest in neuroscience prompted similar investigations in fish and in the avian embryo.

Induction of the Eye Lens

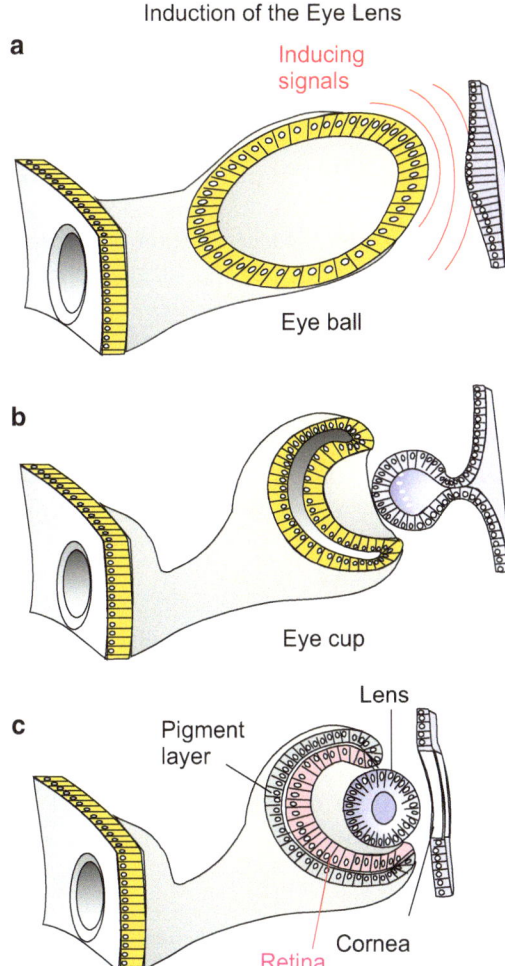

a

Inducing
signals

Eye ball

b

Eye cup

c

Lens

Pigment
layer

Retina

Cornea

Fig. 5.21 Amphibian. Induction of an eye lens. When the optic vesicle, which is formed by the posterior prosencephalon (future diencephalon), comes in contact with the ectoderm, it induces the formation of a lens. The lens in turn induces the formation of the translucent cornea in the overlying epidermis

5.1.17 Homeo Box Genes and Further Genes Coding for Transcription Factors Participate in Induction Processes

The actual realization of induced structures requires the turning on of many genes and their control through transcription factors. Many possess homeodomains as DNA-binding sequences. We mentioned, for example, GOOSECOID and SIAMOIS. However, since transcription factors are acting, as a rule, within the nuclei of cells and are not transmitted to neighbours it is not easy to discriminate in any given cases whether they qualify cells to produce signals or to respond to signals. Qualification to respond, termed **competence**, may be based on the expression of receptors to receive signals.

In the late gastrula and neurula the complete range of homeotic genes, in *Xenopus* called *Xhox*, is expressed between the head and the tail. The function of these genes is dealt with in Chap. 12.

5.2 Favorite Asian Fishes: The Zebrafish *Danio rerio* and Medaka *Orycias latipes*

5.2.1 Fishes Can Provide Many Advantages: Embryos Are Transparent, Mutations Can Easily be Induced and Genetic Studies Conducted

The zebrafish *Danio* is a cyprinid found in rivers of India and Pakistan. The adult fish is 4–5 cm long, displays a pleasant striped pattern, and lives socially in schools. In industrial laboratories all over the world, zebrafishes are widely used for standard toxicological assays. However, increasingly the zebrafish is used for embryological and genetic studies and as a disease model. In parallel to *Danio* the medaka *Orycias latipes*, a fish found in brackish lagoons and rice fields, and also known as Japanese killifish, is a favoured, because robust, aquarium fish in many laboratories around the world. These fishes have been bred in the laboratory for years and natural populations remain unmolested. Adult fishes, unlike female mice, do not suffer harm when embryos are required.

Those who promote fishes for embryological studies list the following advantages, exemplified by *Danio*.

- The fish is easy to breed in the laboratory. Mature animals may spawn every morning and a female releases up to 200 eggs. Typically, 1 female and 2–3 male fishes are put in a prepared container in the evening. The bottom of the container is replaced by a grid, through which the eggs can drop but which prevents the voracious males from eating the eggs. After spawning, an exhausted female needs 1 week for recovery.

• The eggs are transparent and have a diameter of 0.6 mm; embryo development is fast and completed in 2–4 days, depending on the water temperature. Transparency is one of the features that turns out to *Danio*'s advantage in comparison with *Xenopus*. Another feature is its suitability for extensive genetic analyses.

• For genetic studies recessive mutations can be produced using mutagenic agents: males are allowed to swim in a solution of the point mutation-inducing mutagen ethyl-nitroso-urea. For large-scale studies, mutagenized males are crossed with normal females; the F1 and F2 generations are intercrossed and homozygous mutants appear in the F3 (Fig. 12.2).

• Knockdown of genes can be accomplished using morpholinos (modified antisense oligonucleotides), and more recently genes can be knocked out using zinc finger nucleases, TALENs and CRISPER-CAS systems See Chap. 12)

For small-scale studies, homozygous off-spring can be obtained from mutagenized females by parthenogenetically activating their eggs with UV-irradiated sperm. Methods have been developed to double the haploid set of 25 chromosomes to the diploid set of 50 by applying heat shock or mechanical pressure to suppress the first mitotic division. Haploid eggs would develop until the hatching stage.

The zebrafish genome is fully sequenced (26,247 coding genes; 70 % of them being homologous to human genes). It is now used extensively to study gene expression and wonderful pictures have been made showing the products of genes containing homeoboxes, zinc fingers, POU boxes, and so on. Besides *Danio* the medaka (*Orycias latipes*) also came to the fore of developmental geneticists. Its small genome (about 700MB, half the size of Danio's genome but encoding ca. 20.000 genes) was completely sequenced early on. Medaka was the first vertebrate species to produce offspring in space, aboard the space station (in 1994). Whether *Danio* or *Orycias*, in both species the embryos are transparent and the course of development is very similar. The route and fate of GFP-labelled proteins or cells can be traced in living specimens under the fluorescence microscope (Fig. 12.11). It is particularly useful in the study of the development of the nervous system, the eye, the cardiovascular system, stem cells in general and the hematopoetic system in particular.

Our introductory treatment will provide a brief description of the embryo development in zebrafish.

5.2.2 In the Course of Cleavage a Blastodisc Is Formed on Top of the Egg Sphere

Teleosts (bony fishes) have taken their own evolutionary pathway. In comparison with amphibians and land vertebrates they are an out-group, as reflected by peculiarities in their ontogeny.

Determination of the body axes. After fertilization, the cytoplasm streams to the animal pole of the egg and forms a cap of clear material in which the zygotic nucleus is incorporated. Beneath, yolk material accumulates in the vegetal part of the egg. The egg rotates and usually the animal–vegetal axis of the egg becomes horizontally oriented (Fig. 9.4b). The head will be positioned in the region of the animal pole; the midline of the animal's back will extend along the uppermost meridian.

Early embryonic development (Figs. 5.22 and 5.23). Cleavage starts 40 min after fertilization; and occurs only in a thin region of yolk-free cytoplasm called the blastodisc. Early cleavage

Fig. 5.22 The zebrafish *Danio rerio* . Adult fish and gametes (*top*); embryonic developmental stages

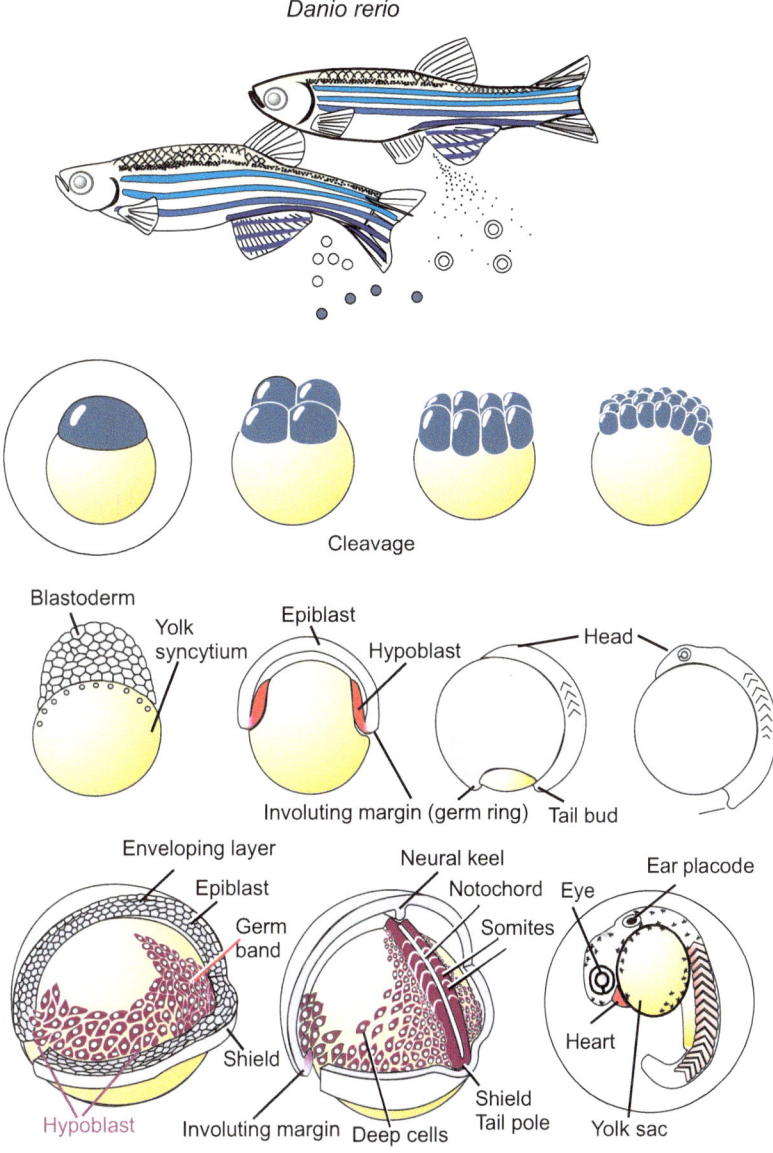

Danio rerio

Cleavage

Blastoderm
Yolk syncytium
Epiblast
Hypoblast
Head
Involuting margin (germ ring)   Tail bud

Enveloping layer
Epiblast
Germ band
Hypoblast
Shield
Involuting margin   Deep cells
Neural keel
Notochord   Eye
Somites
Ear placode
Heart
Shield
Tail pole   Yolk sac

divisions are rapid (about 15 min each), synchronous and partial (meroblastic). Because they occur only in the blastodisc they are called discoidal, resulting in the formation of a cap-shaped **blastoderm**, containing about 1,000 cells, called the sphere stage, which lies on top of the yolk. Midblastula transition occurs at the 10th cell cycle, and cell divisions become asynchronous. At around this stage too, the vegetal cells of the blastoderm merge with the cytoplasmic cortex of the underlying yolk ball to form a yolk syncytial layer (YSL).

As cell divisions continue the blastoderm and the YSL expand and spread over the yolk ball during **epiboly**. Embryos may be staged according to the portion of the yolk covered by the blastoderm; one speaks of 30 %, 50 %, 70 % or 90 % **epiboly**.

The early blastoderm is already composed of several cell layers.

1. **EVL = enveloping layer** or covering layer (some authors use the term periblast which, however, is used by others to designate the yolk syncytial layer). This superficial, thin

cell layer does not participate much in the construction of the embryo.

2. **Deep cells**. Below the covering layer are cells which are summarized as deep cells. Their number increases in gastrulation by cells immigrating at the margin of the disc. Deep cells segregate into two layers:

- **Epiblast**: The top layer of deep cells adjacent to the EVL is called epiblast; the epiblast can largely be homologized with the blastoderm of the amphibian blastula and becomes the ectoderm after gastrulation is terminated.

- **Hypoblast**: Once the covering envelope and epiblast have expanded over half the yolk hemisphere (50 % epiboly) **gastrulation** commences, during which the hypoblast is formed (as described in the following). The hypoblast gives rise to meso- and endoderm.

5.2.3 Gastrulation and the Formation of the Axial Organs Appear Initially Unfamiliar but Eventually Lead to a Similar Phylotypic Organization

Gastrulation. The process of gastrulation only remotely resembles that seen in amphibians. Cells of the blastoderm stream towards the margin and **involute** along the margin. This involution is gastrulation. Under the transparent blastoderm, a hem appears; it broadens to form the hypoblast while the overlying blastoderm remains epiblast as long as it is not yet involuted. More epiblast cells stream to, and converge at, the margin, turn under, join the hypoblast and move towards the animal pole. The epiblast and the hem of the hypoblast together form the **germ ring** around the circumference of the blastoderm (Fig. 5.23).

Having in view gastrulation in amphibians the gastrulation events in fishes appear unfamiliar at a first glance. Yet, one can recognize similarities. If one looks at mid- and late gastrulation stages in *Xenopus*, one sees a ring-shaped blastopore

surrounding the yolk plug (Fig. 5.7). Like the germ ring in fish, the ring-shaped lip of the blastopore is an edge over which cells roll into the interior. The whole germ ring can be equalized with the blastopore, and the yolk ball of the fish with the yolk plug of the frog. The upper blastopore lip of the amphibian gastrula is congruent with the site where in *Danio* the involution starts. At this site genes are expressed such as *siamois* and *goosecoid* that characterize the upper blastopore lip with its Spemann organizer. In the fish embryo this location becomes thickened and is known as **embryonic shield**. It is endowed with similar organizing powers as is the Spemann organizer in amphibians.

Convergence Movement and Formation of the Primitive Streak. With the appearance of the embryonic shield epibolic spreading of the blastoderm and gastrulation are not yet finished. While the margin of the blastoderm and with it the involuted hem move in direction of the vegetative South pole, in the interior the immigrated cells converge at the dorsal side in front of the shield and move in the direction of the animal North pole. The hypoblast in fish gastrulae is only a loose and rather transient association of cells. Deep cells are able to disperse themselves as amoeboid cells by migration over the surface of the yolk. The cells can also reassemble. They converge and aggregate along a meridian that marks the future head-back-tail line, incorporating more and more arriving cells. The elongating aggregation together with the overlying band of epiblast cells is called the **primitive streak**, and it is here that the primary axial organs will be formed. As epiboly continues and the blastoderm spreads over the entire yolk mass to enclose it, the primitive streak lengthens and extends over the animal and vegetal pole. Together with the margin of the blastoderm (the germ ring), the shield shifts to the caudal end of the embryo and eventually becomes the **tailbud**.

Formation of the Axial Organs and the Basic Body Construction. The embryo arises from the primitive streak. Its developmental potential

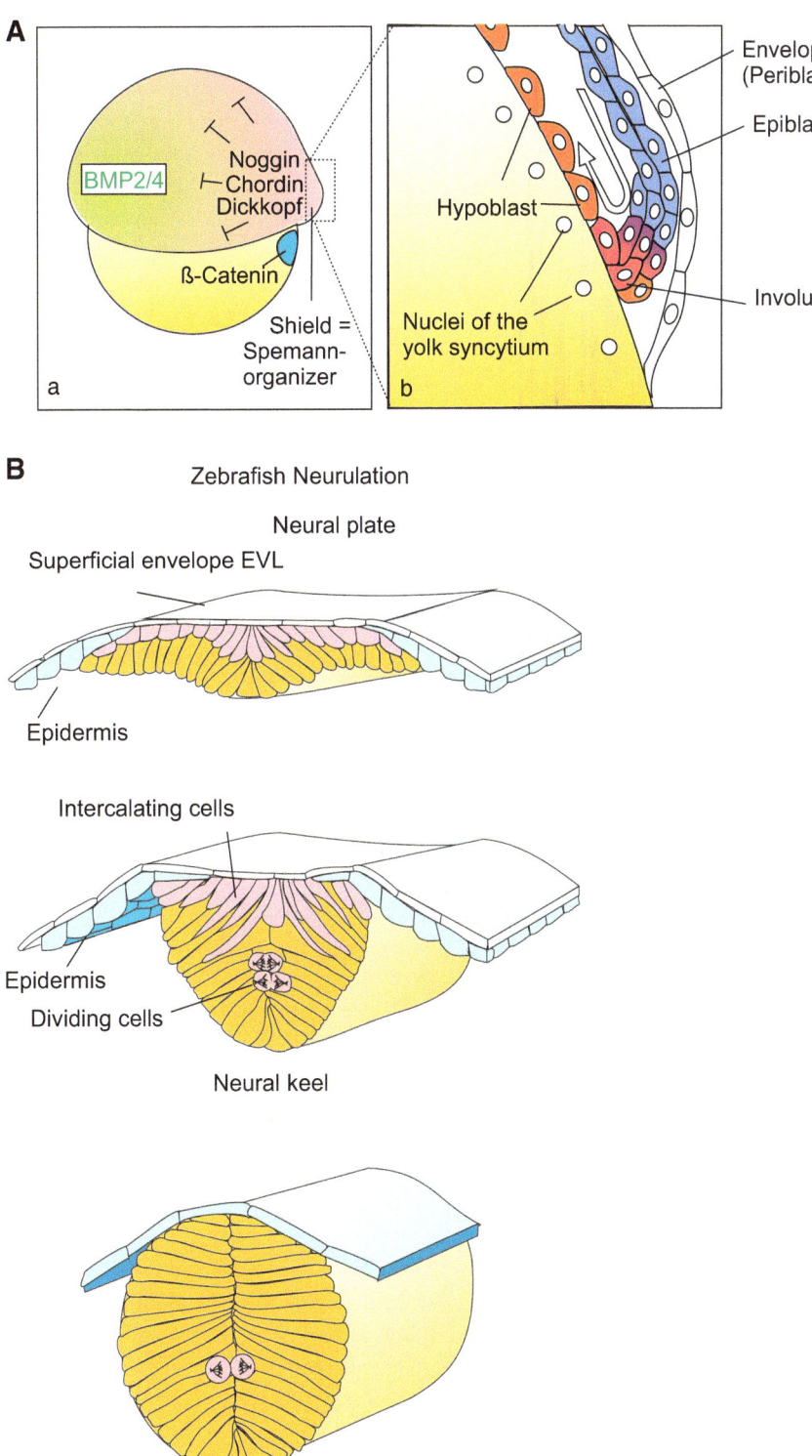

Fig. 5.23 **A** (**a, b**) *Danio rerio*. (**a**) Sphere of action of developmental-controlling proteins known also from *Xenopus* embryos. (**b**) Structure of the embryo at the margin of the blastoderm where involution takes place **B** Neurulation

includes neuroectoderm (in the epiblast) and mesoderm (the aggregated deep cells), but neural folds do not appear. Instead, a solid **neural keel** detaches from the epiblast. Beneath the keel, aggregated deep cells organize themselves into notochord and somites. Later arriving deep cells supplement the mesoderm by forming the lateral plates. Endodermal epithelial layers form underneath the mesoderm.

For some time, the detached neural keel persists as neural rod. Later, a central canal transforms the rod into a neural tube. In its anterior part the tube widens and gives rise to the brain. The transparency of the fish embryo makes it possible to observe the development of the eyes in living fish (see Fig. 16.20) and of the inner ear from ear placodes and even the outgrowth of nerves such as the spinal nerves.

5.2.4 In the Expression of Genes Controlling Development Many Congruencies Between Fish and Amphibians Are Evident

Numerous genes that are important in *Xenopus* embryonic development are active in *Danio* and *Oryzias* as well: There expression pattern can be visualized by in-situ-hybridisation. For example:

- *siamois* and *goosecoid*: They code for transcription factors that are expressed in the shield in fish and in the Spemann organizer in *Xenopus*;
- *Brachyury*: a mesoderm-indicating gene,
- Several *wnts*, *Bmp-4*, *chordin* and *noggin*, genes which in *Xenopus* code for far-reaching morphogens;
- *Fgf-8* as another example of a gene that gives rise to a signal protein, a member of the fibroblast growth factor family. The variant FGF-8 plays a role in the subdivision of the brain into sections.
- Genes of the *Hox A* and *Hox D*-families; they encode homeotic transcription factors which mark positions along the axial organs conferring them particular identities.

Perspectives. Due to their transparency and the possibility to generate targeted mutations and to

produce transgenic animals with targeted gene exchange fishes have become the best-investigated models in tracing the fate of the neural crest cells including the emergence of coloured patterns of the skin (described in Fig. 15.10) and in the development of the eye (described in Fig. 16.20). The functional analysis of excitable cells and other cells fitted with ion channels can be analysed with the novel technologies of optogenetics (Fig. 12.10). Moreover, the development of the nervous and cardiovascular system can conveniently be observed in the translucent living specimens, whether normal wild-type, or mutant, or transgenic.

5.3 Amniotes: Chick, Quail and Chimaeras of Both

5.3.1 We Not Only See Huge Egg Cells; for the First Time We Deal With Land Vertebrates, That Is With Amniotes

The large size of the avian egg and its availability made it possible to observe the development of the chick even in ancient times (Egyptians, Aristotle; see Box 1.1), though in the first two days of incubation not much can be seen without a magnifying glass. In comparison with amphibians, reptiles and birds have evolved some features that permit them to live completely on land and to omit a larval stage. The egg is huge – the egg cell proper (Fig. 5.24) is the yellow ball enveloped by the fortifying acellular, elastic and translucent vitelline membrane. The egg cell is surrounded by albumen (egg white) and the whole is contained in further envelopes including the outer calcareous shell. Moreover, within the shell the developing embryo becomes wrapped in a fluid-filled cyst, the **amnion** (see Fig. 5.27). Reptiles, birds and mammals are collectively referred to as **amniotes**.

Nowadays, fertilized eggs must be purchased from specialized farms or breeding stations, because eggs sold by grocery stores are generally not fertilized. The cock has to do his job and contribute sperm timely, shortly after an egg is liberated from the ovary and before it is enveloped in the oviduct. Figure 5.24 shows how to prepare

Fig. 5.24 Egg of the chick and a simple method for preparation the blastodisc. The incubated egg is cautiously broken into a cup with preheated mild salt solution; the blastodisc is separated from the yellow yolk with forceps and scissors, and drawn onto a curved watch glass. For the method of fenestration see special literature on practical methods

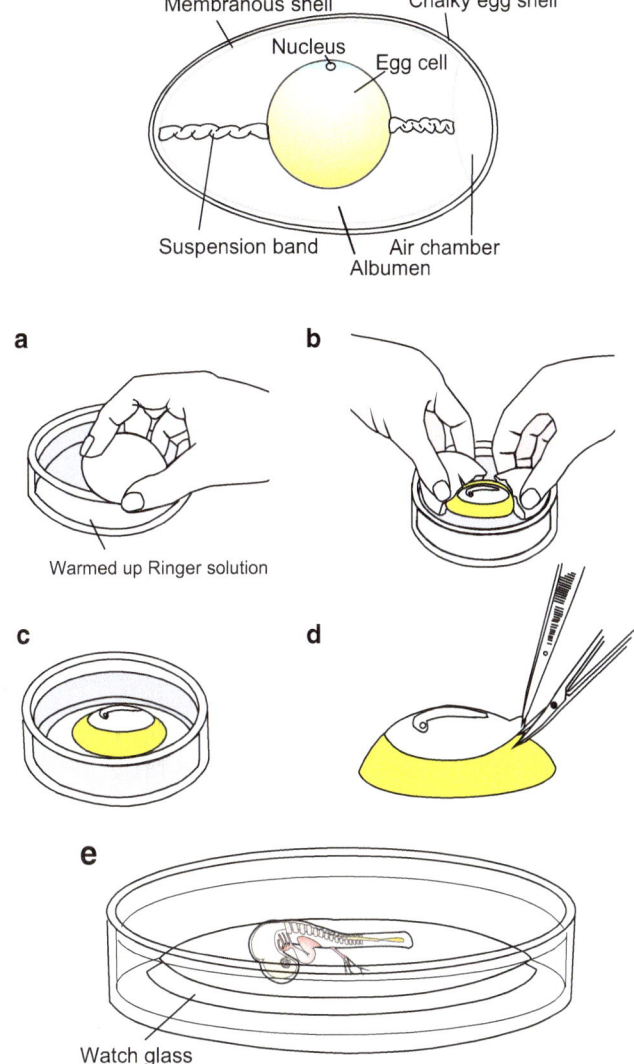

an egg for observations without intending demanding experiments . The chickens hatch, if a constant breeding temperature of 37 °C is maintained, 19–21 days after beginning incubation.

5.3.2   On the Blastodisc First the Backside of the Embryo Is Formed

Cleavage begins immediately after fertilization in the oviduct of the hen. This is a disadvantage of working with avian embryos. Very early stages are not accessible to experimental

manipulation. The morphological events of cleavage have been studied in eggs removed from the oviduct.

As in the zebrafish, cleavage is meroblastic and discoidal. Cytoplasmic division is at first by furrows produced by downward extensions of the cell membrane, leaving the blastomeres open to the yolk for some time. The progressive cellularization of the area around the animal pole results in the development of a blastodisc. As in the fish embryo, at the edge of the disc is a circular syncytial zone where the cells are still open. By continuing cell divisions the blastodisc (Fig. 5.25) expands peripherically, while below the centre a cavity – the subgerminal cavity or

Fig. 5.25 (**a–d**) Avian egg. Discoidal cleavage, blastodisc, primitive groove. (**a–d**) Cross section through the forming blastodisc while it is lifting off the yolk, cleaving and forming the early embryonic stages (**a, b**) Progressing cleavage, expanding the blastodisc; below the blastodisc a subgerminal cavity representing the blastocoel is formed. (**c, d**) Gastrulation through the primitive groove. Prospective mesodermal and endodermal cells immigrate through the groove into the subgerminal cavity, migrate in an anterior direction and eventually give rise to the mesodermal rudiments and the endoderm. At its anterior margin the groove forms Hensen's node, an organizer corresponding to the amphibian upper blastopore lip. By the growing embryo, the node is relatively shifted to a more and more posterior position. Neurulation commences at the anterior pole of the embryo, while at the posterior end the primitive groove still persists

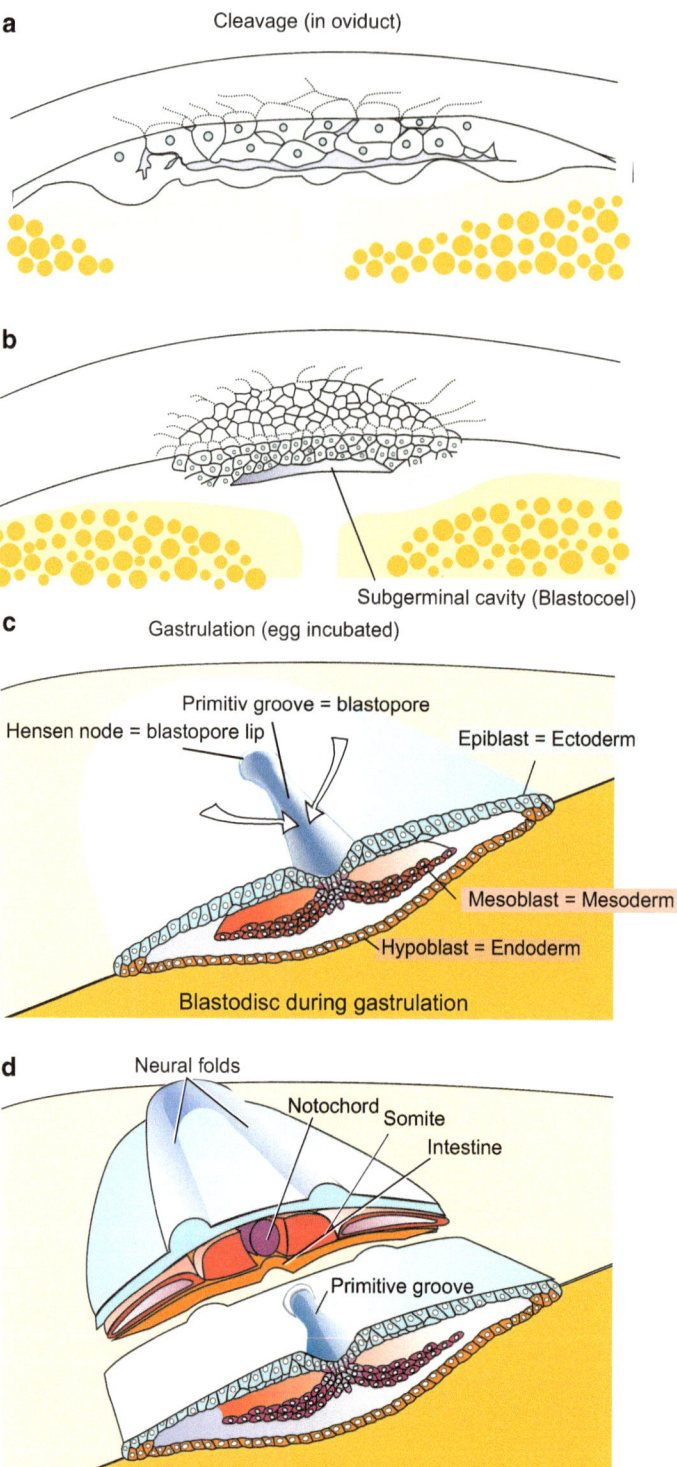

blastocoel – is formed. (The central, translucent area above the cavity is called the area pellucida, and the peripheral opaque zone the area opaca.)

The cellular roof of the cavity is called the **epiblast**. At its posterior margin cells detach and settle on the floor of the subgerminal cavity, forming the **hypoblast** and covering the uncleaved mass of the yolk-rich residual egg cell. The hypoblast cells are thought to play a directing role when the endo- and mesodermal cells immigrate; but they themselves are thought not to participate in the construction of the embryo.

In Gastrulation, Endodermal and Mesodermal Cells Migrate in a Cleft Between Epi- and Hypoblast. When the egg is laid, the blastodisc has a diameter of 2 mm. Gastrulation (Fig. 5.25) begins when the egg is incubated and warmed up. During gastrulation endodermal and mesodermal cells colonize the space between epi- and hypoblast. The location where immigration happens is not an open blastopore but a **primitive groove** that remains closed, although it is homologous to the amphibian blastopore. How can cells pass an apparently closed door? Cells of the epiblast stream toward the groove, dip down in close contact to their neighbours (therefore, the groove seems to be closed), but once inside they migrate individually like amoebae or the deep cells in the fish embryo, to colonize the available space. The largest area free for colonization is located in front of the primitive groove. At the anterior end of the groove is a small region, known as **Hensen's node**, where ingression is particularly active. This node corresponds to the amphibian upper blastopore lip and is endowed with similar inductive capacity.

The germ layers are separated and the embryo is formed in front of the primitive groove. The stream of immigrating cells bifurcates into two branches; one branch advances anteriorly on the floor of the cavity pushing aside the cells of the hypoblast. These deep-moving cells form the definitive **endoderm**, which will give rise to the lining of the gut. The other stream flows into the space between epi- and hypoblast, broadens

and forms the sheet of **mesoderm**. As the mesoderm continues to spread anteriorly away from the primitive groove, it incorporates more and more ingressing cells and becomes densely packed. In front of the primitive groove and along the midline of the mesodermal layer, cells coalesce to form the mesodermal axial organs: the stage of the **primitive streak** is reached.

The elaboration of the large-scale organ rudiments is accomplished in a similar manner as in amphibians. A view from outside shows neural folds emerging in front of the primitive groove and above the inducing mesoderm. The neural folds merge along the midline of the embryo, forming the neural tube (Figs. 5.25 and 5.26). In the meantime the mesoderm subdivides itself into notochord, somites and lateral plates. The endoderm covering the yolk sphere begins to form a longitudinal fold which approaches the notochord. Later, the endodermal fold is closed into a tube: the primitive gut is formed.

The elaboration of the body parts in the avian embryo does not take place simultaneously; instead it starts at the anterior pole and proceeds posteriorly (Fig. 5.26). In the external appearance the progress in elaborating the body shape is rendered visible in the posteriorly advancing emergence of neural folds and the successive addition of new somites. The successive generation of new somites – a *highlight* of contemporary research – **is driven by an inner clock** (Fig. 10.20).

5.3.3   Two Novelties Are Yolk Sac and Allantois

The embryo has now elaborated its basic shape, but it is still open on its ventral side. Early closure on this side is prevented by the rest of the original, uncleaved egg. Endoderm, mesoderm and ectoderm (epiblast) gradually grow over this residual ball. Enclosed, the yellow ball is known as the **yolk sac**. It is connected with the primitive gut, sharing with it a common endodermal lining. The body of the embryo raises itself away from the surface of the yolk sac. Thus a

Fig. 5.26 Avian embryo: Neurulation. Looking down upon the blastodisc. The time-displaced, successive development of the embryo from anterior to posterior is obvious in the cross sections

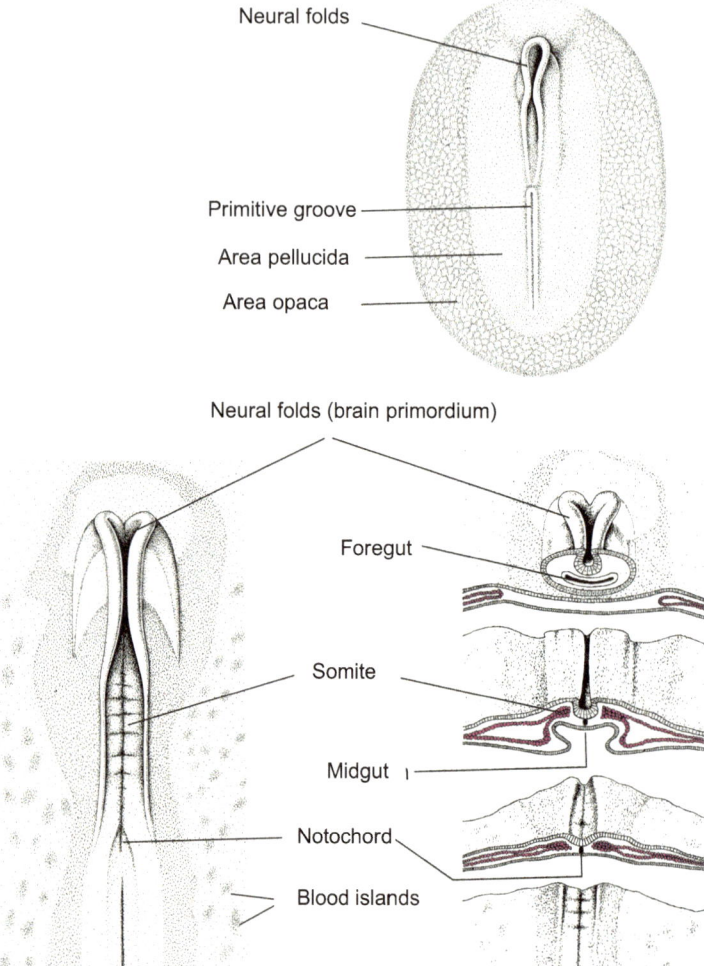

partial separation is introduced between the yellow sac and the body of the embryo proper. The broad connection becomes restricted later; the embryo remains connected with the yolk sac by a stalk enclosing an endodermal canal and later, blood vessels as well. The yolk sac gradually diminishes in size as liquified yolk is taken up by the gut and used by the embryo.

A second canal, the urinary duct, connects the posterior intestine with an evaginated cyst, known as **allantois** or embryonic urinary bladder (Fig. 5.27). An egg enclosed in a shell has no means of disposing of the waste products of protein breakdown and water formed in oxidative metabolism by removing them from the egg. A solution was found to the problem by producing, instead of urea, insoluble crystals of uric acid and storing them in the allantois with a little water until the time of hatching. However, the allantois not only stores waste products but also serves as an embryonic organ of respiration. The growing allantois is accompanied by blood vessels and establishes contact with the air chamber at the blunt end of the egg.

Fig. 5.27 (a–c) Avian
embryo. Formation of the
amnion, the yolk sac and
the allantois (**a1–d1**)
Sagittal sections (lateral
view), (**a2–c2**) cross
sections

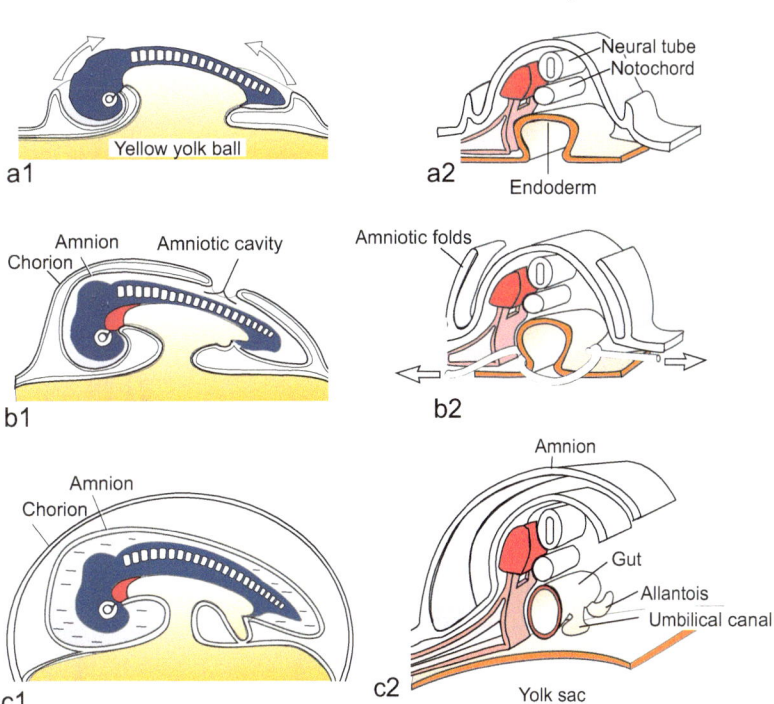

Formation of Amnion and Yolk Sac in the Avian Embryo

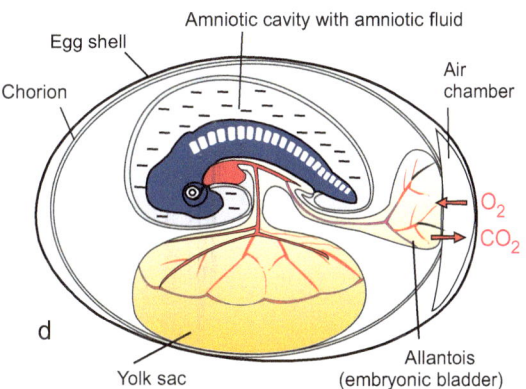

5.3.4   A Further Novelty, Persistent in all Land Vertebrates, Is the Amnion

While in the emerging embryo the neural tube is formed, dermal folds arise from the surrounding extraembryonic epiblast. They are pulled over the embryo, the free edges of the folds are welded together, and eventually the embryo is completely enclosed in a double-walled cavity – the **amniotic cavity**. In order to protect the tender embryo, the cavity is filled with fluid. The water is provided by the inner wall of the cavity; this inner wall represents the **amnion** proper. The **amniotic cavity has come to be a private pond**, so that embryonic development of land animals still proceeds in water. The fluids of the amniotic cavity and allantois are familiar as the fluid pouring out when the chicken hatches and the thin walls of these extraembryonic containers disrupt.

5.3.5   Experiments With the Avian Embryos I: Fate of Neural Crest Cells Analysed by Means of Chimeras

Taking advantage of features unique to avian development, the differentiation potentials of the **neural crest cells**, a group of highly versatile cells present in all vertebrates, have been most successfully analyzed in the chick. Their migration routes and roles in constructing the sympathetic nervous system and producing chromatophores are understood (Figs. 15.3 and 15.4). In order to trace the fate of the roaming neural crest cells as well as other mobile embryonic cells, small pieces taken from quail embryos were implanted into embryos of the chick. Quail cells are tolerated without any problems and participate in the construction of a **chimeric bird**, but can readily be identified and distinguished from the host cells, for the nuclei of quail cells contain considerable dense heterochromatin. Today, however, own species-specific (autologous) transplants bearing fluorescing transgene proteins (e.g. GFP fluorescence, Chap. 12) replace the heterologous quail cells.

5.3.6   Experiments With the Avian Embryos II: Pattern Formation Decoded

We focus on two types of experiments:
1. Like all vertebrates, the chick displays a high degree of flexibility and regulation in its embryonic development. If a pre-primitive streak is cut into several fragments, most of the fragments are able to produce complete, albeit somewhat smaller, embryos. As a primitive groove has formed, the capability of complete regulation is restricted to that portion containing Hensen's node (see Fig. 10.9). The result will be discussed in terms of **morphogenetic fields** (Chap. 10).
2. Pattern formation can be studied effectively on buds of the extremities, in particular the readily accessible **limb buds**. The

morphogenetic action of an array of signalling molecules including retinoic acid and secreted signalling proteins (SONIC HEDGEHOG, FGF), and the significance of homeobox genes are being investigated using limb buds as model systems in vivo (Figs. 10.11, 10.12, and 10,13). A further question addressed concerns the nervous connection of the eye with the brain, the so called **retinotectal projection** (see Fig. 16.22).

5.4   The Mouse: A Proxy for Humans

5.4.1   Medical Interest, the Availability of Mutants and Fast Development Constitute the Mouse the Model for Mammals; As Such it Establishes Close Relation to Its Mother

The development of mammals is greatly modified compared to that of other vertebrates, due to adaptations to **vivipary**. In viviparous animals the embryo receives an adequate supply of nutrition from the mother while it is retained in the uterus; therefore, the embryo **can increase in mass and grow during development**.

Supplied with nourishments by its mother, the embryo can do without yolk. As yolk supply of the egg is superfluous and therefore reduced, total (holoblastic) cleavage is rendered possible. On the other hand the embryo is confronted with the need to establish intimate contact with the nourishing mother. Long before forming the embryo proper, the germ first develops a **trophoblast** (Greek: *trophein* = to nourish) as an extraembryonic organ for food intake. From a partial territory of the trophoblast the **placenta** of the growing child will arise. The late embryo of mammals is called the **fetus**; in man the term is used from the 8th week onward.

Mammalians are not well suited for developmental studies. Embryo development takes place hidden in the maternal body. For investigations and experimental manipulations surgery must be performed, and such an intervention is subject to authorization in many countries. The embryos are inconvenient to handle and difficult to maintain in the laboratory for long periods of time. Only early embryos can be rinsed out from the oviduct; after implantation, embryos must be removed surgically and will scarcely be uninjured. In spite of all these inconveniences, the mouse is now the developmental model to which the greatest attention is being paid. Above all, it was medical interest that made the development of the mouse the paradigm for experimental mammalian embryology. Compared to other mammals mice do not need much space, nourishment and care. Nonetheless, the logistics of breeding and regulations about space and standards in facilities mean that technician time and equipment for significant mouse operation are rather costly. On the other hand, rapid, season-independent development, availability of a great number of mapped mutants, and the relative ease with which transgenic animals can be generated have made the mouse the leading model of mammalian development. However, it must be pointed out that mouse development differs in several ways from that in most mammals. With regard to oogenesis and pre-implantation stages, schematic sketches of the development of mouse and man appear to be fundamentally similar. Of course, the time-scales are different.

(For a better understanding of mammalian development in a more generally valid mode, you may first read the section on development of man (see following Chap. 6), as the human embryo is not so different from a generalized mammalian archetype as is the mouse embryo.)

5.4.2   Mice Can Reproduce Soon After Birth and Over the Entire Year; Generation Time Is Merely Nine Weeks

Five days after the birth of a female mouse, her oocytes have already duplicated their chromosomes in preparation for meiosis (DNA

amount $= 4C = 4\times$ the haploid mass). The oocytes enter prophase of meiosis I. However, prophase is interrupted in the diplotene stage; lampbrush chromosomes and multiple nucleoli become visible (see Figs. 8.4 and 8.5). Of the oocytes initially present, at least 50 % perish, but some 10,000 remain. At 6 weeks the mouse reaches maturity. The ovulation cycle is very short: every 4 days 8–12 oocytes finish the first meiotic division, extruding the polar body. The stimulus required by the egg to proceed to the second meiotic division is provided by the fertilizing sperm.

After 19–20 days of development mice are born. Theoretically, per year up to 40 generations can see the light of day. The (minimum) generation time is 9 weeks.

5.4.3   The Embryonic Development of a Mouse Is Strange and Not Easy to Comprehend

Early Embryonic Development up to Nidation (Fig. 5.28). Mice mate only when the female is in oestrus. Some 50 millions of spermatozoa are released in coitus and seek an egg. Successful fertilization is indicated by the explosive elevation of the vitelline membrane, which is blown up by the swelling of the discharged contents of the cortical granules (Chap. 3). Once the egg is fertilized and activated, the second polar body is extruded, indicating the completion of the second meiotic division. Cleavage in mammals is extremely slow and takes days. It starts about 18 h after fertilization (compared to 1 h in the sea urchin) and is accompanied by **early transcription of zygotic genes**. There is no *midblastula transition* known in mammals. In mice zygotic genome activation is observed as early as the S/G2 phase in the male pronucleus of the 1-cell zygote and becomes robust at the 2-cell stage.

The Blastomeres Are Still Totipotent Through the 8-Cell Stage. If separated, they can give rise to 8 genetically identical mice. At the transition

Fig. 5.28 Mouse I. From cleavage to the blastodisc stage which is about to implant into the uterine wall of the mother

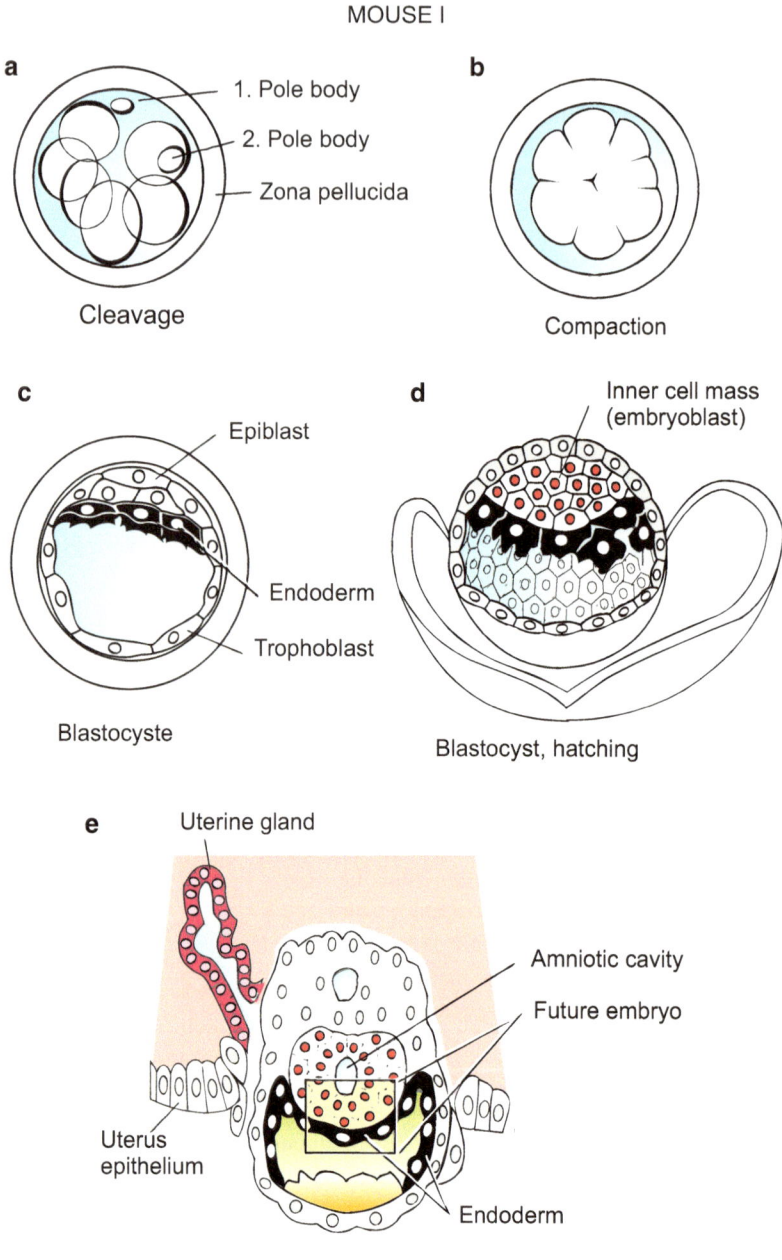

MOUSE I

a

1. Pole body
2. Pole body
Zona pellucida

Cleavage

b

Compaction

c

Epiblast

Endoderm

Trophoblast

Blastocyste

d

Inner cell mass (embryoblast)

Blastocyst, hatching

e

Uterine gland

Amniotic cavity

Future embryo

Uterus epithelium

Endoderm

Nidation (implantation)

to the 16-cell stage **compaction** occurs (Fig. 5.28c), in which the blastomeres are tightly cemented together by cell adhesion molecules (uvomorulin = cadherin E). The embryo becomes a **morula**. A cavity appears inside the morula and we arrive at the **blastocyst**. In the blastocyst the first irreversible event of differentiation occurs: the cells of the outer epithelial layer, called the **trophectoderm**, undergo 'endoreduplication': they amplify their entire genome and become polyploid. The trophectodermal layer surrounds a cavity, the blastocoel. In its interior, the blastocyst contains an eccentrically located **inner cell mass (embryoblast)**;

these inner cells remain diploid. The blastocyst liberates itself from its envelope (zona pellucida). Hatched, it is ready for implantation.

Postimplantation Development. After **implantation (nidation)**, the trophectoderm forms **giant cells**; it is now called the **trophoblast**, and will eventually give rise to the **placenta**. In the terminology of embryologists the embryo plus the extraembryonal structures are referred to as the **conceptus.**

When buried in a uterus crypt the blastocyst gives rise to three different cell lines:

1. Polyploid cells of the trophoblast
2. The diploid cells of the inner cell mass (= embryoblast). The inner cell mass gives rise to the embryo proper; now pluripotent embryonic stem cells can be taken out (see Fig. 18.1)
3. Cells of the prospective endoderm.

The cell lines of the trophoblast and to a large part also of the primitive endoderm do not participate in the construction of the embryo; they provide extrambryonic structures, above all eventually the placenta. How the segregation of these cell lines is accomplished is still a matter of debate and not definitively clarified. Besides influences of the microenvironment, that is of the surrounding maternal tissues, also accidental processes are taken into consideration. Elucidation is impeded not only by the inaccessibility of the embryos; moreover, cells of the 16-cell stage, and perhaps also in later stages, change their positions relative to each other (as in the round worm *C. elegans*).

In comparison with other mammals, the development of the mouse (Fig. 5.29, 5.30, 5.31, and 5.32) displays some peculiarities which may be misleading if they are taken to be typical. For instance, the embryo takes the form of an "**egg cylinder**". This cylinder comprises an 'ectoplacental cone', destined to become the trophoblastic part of the placenta, and several extra-embryonic cavities. The head-back-tail line is extremely curved, giving the embryo an abnormally concave back for some time. The belly side remains open for a prolonged period after the trophectoderm in the region opposite to

the ectoplacental cone has dissolved, as has its thick basement membrane, called 'Reichert's membrane' (this structure is derived from the trophectoderm, Fig. 5.29). For some time the exterior cell layer is composed of endodermal cells; this is referred to as seeming **germ layer inversion** (Figs. 5.29 and 5.30).

The turning movements of the embryo are very complicated. It becomes rotated around its longitudinal axis leading to its ventral closure. The concave hollow back is converted into a convex back (hyperkyphosis) and the exit of the umbilical cord displaced by 180° (Fig. 5.31).

5.4.4 Mice Seem to Need a Father, (Unless They Become Cloned)

Parthenogenesis? Or Do Mice Need a Father? Occasionally, development can start without fertilization. This can happen in the oviduct or even in the ovary. If precocious development starts in the ovary before the onset of the first meiotic division, the embryo will be diploid. Nonetheless, it develops abnormally, yielding a tumour-like creature, called **teratocarcinoma**.

Even when an egg has completed meiosis regularly and has been released into the oviduct in time, various kinds of abnormal development are possible. A parthenogenetic embryo is not necessarily haploid. Perhaps the second meiotic division was omitted, or a polar body instead of a sperm cell fused with the egg thereby bringing back its haploid genome to its previous diploid form. Fusion of a polar body with the egg can result in pseudo-fertilization; the egg becomes activated and starts cleaving. Such diploid but **bimaternal** egg cells lacking a paternal genome may undergo complete cleavage, implant into the uterus and even develop up to the stage where the buds of the extremities appear. At this stage or earlier, the embryo dies. No living mouse has been born which definitely underwent parthogenetic development. Thus, despite the parthenogenetic potential they exhibit, evidence has accumulated indicating that mice need to have a father and a mother. However, the

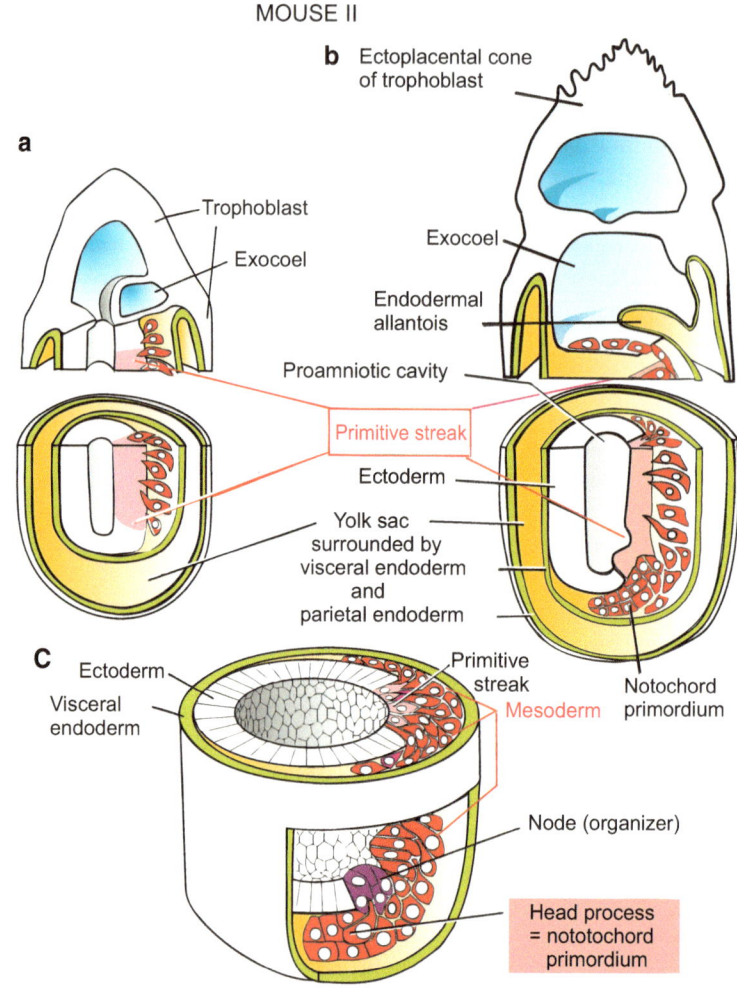

MOUSE II

mouse is not a paradigm for mammalian development in all aspects and experimental results with mice must not prematurely be extrapolated to other mammals or man.

Related to the question whether mice (and other mammals) need to have a father is the question of whether the genomes contributed by both parents are equivalent. The answer is that they are not, to be more precise: not in all aspects. In the following section we refer to evidence on the biological level:

Two Fathers or Two Mothers? Experimentally working researchers were faced with failures when they went on to fuse two haploid nuclei in the egg, but both were of maternal origin (supplied by two oocytes) or of paternal origin (supplied by two sperm). All such experiments to generate **bimaternal** or **bipaternal** mice failed; the eggs may have started developing, but the embryos did not survive. It is not surprising that two sperm nuclei are insufficient if both bring along a Y chromosome but no one contains an X chromosome, because YY embryos lacking an X would never survive. Yet, experimental evidence disclosed that even the XX constitution is only unproblematic if one X chromosomes stems from an egg cell, the other one from a sperm cell. In Chap. 3 (fertilization) a reason was given for that. The sperm must deliver a centrosome with centrioles which subsequently duplicate to establish a complete

Fig. 5.30 Mouse III. Neurulation and early organogenesis. Neural folds (*yellow*) close to form the neural tube. Underneath the neural tube (that is still open at its anterior and posterior end) mesodermal cells aggregate in groups forming the notochord, somites and heart. In the mouse, the ventral parts of the trophoblast dissolve (remnants are found as inconspicuous Reichert's membrane) so that actually the endoderm comes to lie at the outer side, while the ectoderm forms the innermost epithelium facing the amniotic cavity (apparent germ layer inversion)

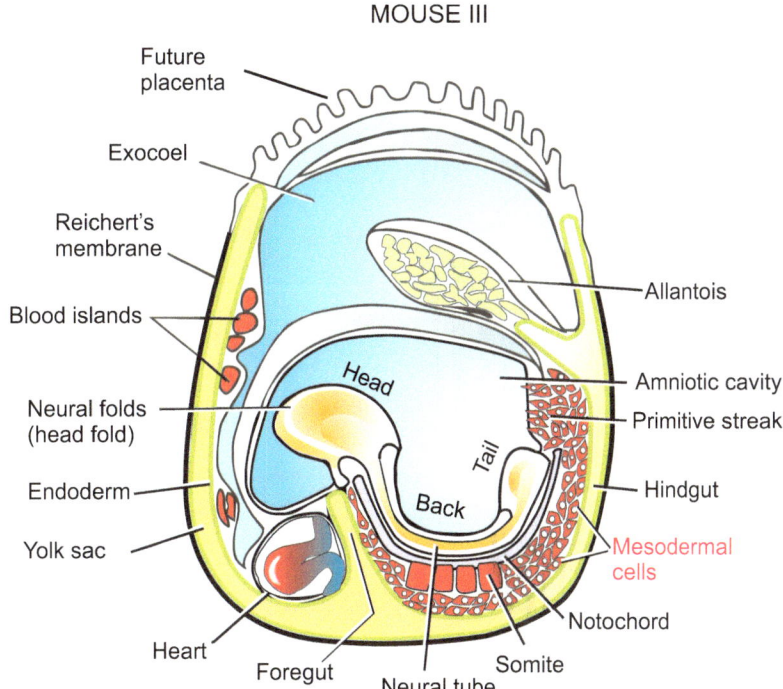

MOUSE III

Future placenta
Exocoel
Reichert's membrane
Blood islands
Neural folds (head fold)
Endoderm
Yolk sac
Allantois
Amniotic cavity
Primitive streak
Hindgut
Mesodermal cells
Notochord
Head
Tail
Back
Heart
Foregut
Neural tube
Somite

spindle apparatus for the first cleavage. In addition, the problems with bimaternal embryos must be considered in the context of **maternal**, or **paternal imprinting** which the DNA had experienced because of different methylation patterns in the testis of the father and the ovary of the mother. This will be discussed in Chap. 8 and Sect. 12.5 which describe the phenomenon of genomic imprinting that confers nonequivalence on the maternal and paternal genomes.

When, in fertilization, both a maternally and a paternally imprinted genome are brought together the condition for normal development is met. If these differently imprinted genomes are inherited by the daughter cells essentially unchanged, cloning by transfer of nuclei into enucleated eggs is possible (Chap. 13).

5.4.5 Chimeric Mice Provided Means to Genetically Manipulate Mice

Chimeric mice with heterogenic tissues (mosaic tissues of different genetic origin) can be produced by mechanically mixing early embryonic cells taken from different blastocysts (see Fig. 13.4). The method has been developed by Beatrice Mintz. Using a micropipette, foreign cells are injected into the cavity of a host blastocyst. The foreign cells are derived from the inner cell mass of a donor blastocyst or from a teratocarcinoma, and are called **ES cells** (ES = embryonic stem). A teratocarcinoma results from a miscarried embryo (but not all miscarried embryos develop into teratocarcinomas). ES cells injected into a host blastocyst may be integrated into the inner cell mass of the host, proliferate and contribute to a viable **chimerical mouse**. Such a mouse will be composed of heterogenic tissues; even primordial germ cells may originate not only from host cells but also from foreign cells. If previously a cloned gene of, for example, human origin had been introduced into the donor ES cells and inserted into its genome (by homologous recombination, Chaps. 12 and 13), the host embryo may include areas containing and replicating this artificially introduced human gene. In case those transgenic regions give rise to primordial germ cells, whole **transgenic mice**, carrying the human gene in all

Fig. 5.31 Mouse IV.
Torsion of the embryo
during and after
neurulation. The concave
bending of the back line is
converted into a convex
curvature, and the placenta
is formed. For reasons of
clarity, the illustration
separates the torsion of the
longitudinal axis (3–7)
from the torsion of the
dorso-ventral axis (8–10).
The figures are drawn
partly after sketches found
in the literature (Theiler
1989), partly after photos
from original preparations

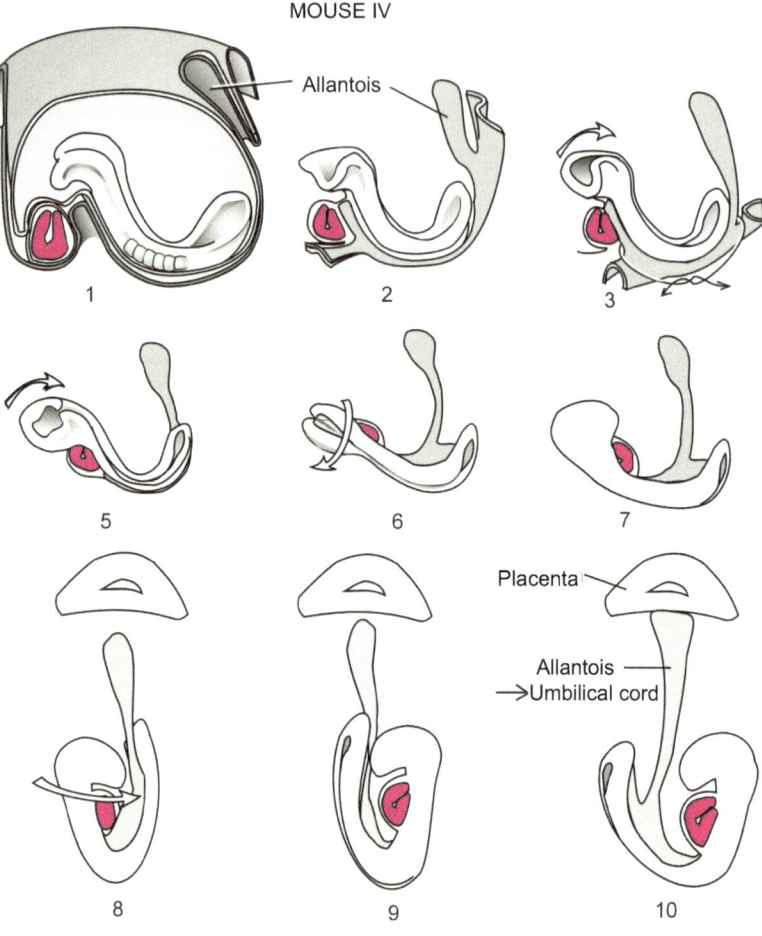

MOUSE IV

their cells, can appear among the offspring. If the human gene was randomly integrated into the genome of the donor ES cell, it will probably be present in only one of two homologous chromosomes. By inbreeding, homozygous animals can be obtained. With transgenic mice the role of a single gene in the development of a phenotype (e.g. sex) or the contribution of a gene to a hereditary disease can be studied.

Summary

In this chapter the embryonic development of those vertebrates is described whose development is extensively studied and in the laboratory experimentally analysed. The developmental stages as such cannot be recapitulated here. In this respect we refer to the illustrations in the figures.

Xenopus, the clawed frog, and amphibians in general are archetypes of vertebrate development. At any rate, Sect. 5.1 should be read if one wants to understand development of a vertebrate including human. Embryonic development proceeds through a classic textbook blastula and a gastrula to reach a **phylotypic stage** which in similar form also occurs in other vertebrates and displays the prototype of the vertebrate organization. This stage is characterized by the following structures (global primordial organs; No. 1–4 = **axial organs**):

Fig. 5.32 Mouse V. Embryo in the phylotypic stage

MOUSE V

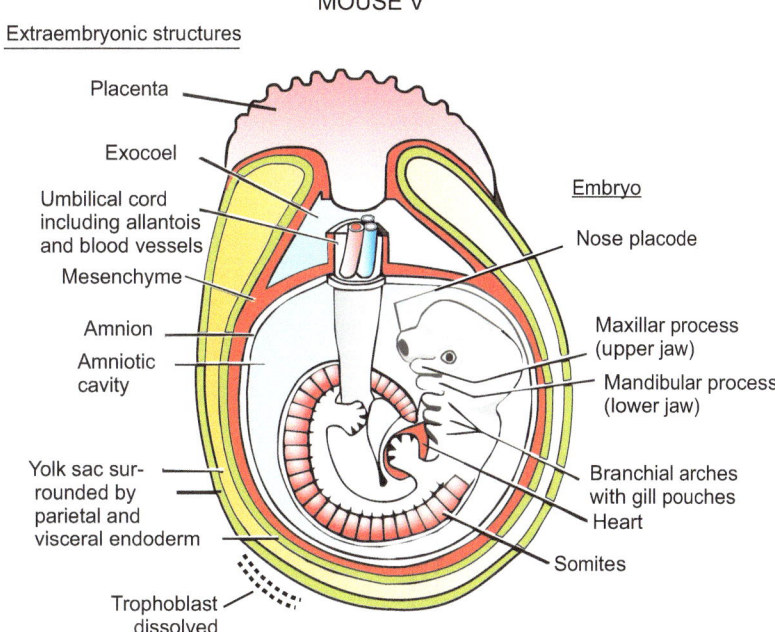

Extraembryonic structures

Placenta

Exocoel

Umbilical cord including allantois and blood vessels

Mesenchyme

Amnion

Amniotic cavity

Yolk sac surrounded by parietal and visceral endoderm

Trophoblast dissolved

Embryo

Nose placode

Maxillar process (upper jaw)

Mandibular process (lower jaw)

Branchial arches with gill pouches

Heart

Somites

1. **Neural tube**, which gives rise to the CNS including brain and spinal cord
2. **Neural crests**, a cohort of migratory cells which gives rise to the peripheral nervous system, adrenal medulla, the chromatophores of the skin and the skeletal elements of the branchial region of the body
3. **Notochord**, the embryonic placeholder for the vertebral column
4. **Somites**, which give origin to the vertebral bodies, the cross-striated musculature of the body, and the subcutaneous tissue (dermis)
5. **Lateral plates**, from which the epithelia lining the body cavities (coelomic cavities) and parts of the blood vessel system arise, including the
6. **Ventral heart**
7. **Foregut with gill pouches** (branchial pouches)

The structures 3rd–6th are mesodermal, 7th is endodermal.

With embryos of amphibians famous experiments were conducted. It was shown by Spemann how identical and Siamese twins arise. **Transplantation** experiments with embryos of newts and frogs by Spemann and his co-workers lead to the discovery of **positional information** and of **embryonic induction**: A distinct area of the embryo releases signals that stimulate neighbouring areas to develop structures appropriate to that location. A particular powerful transmitter is the upper blastopore lip of the early gastrula, known as **Spemann organizer**. When transplanted into a donor at a site opposite to its normal location it induces the development of a **complete secondary embryo**. While invaginating into the interior forming the roof of the archenteron the organizer becomes subdivided into **head- and trunk organizer**.

By transplanting small pieces of the organizer areas or injecting genetically engineered inducing factors one can induce the development of

- **Complete embryos** (with beta-CATENIN, a transcription factor present in the early organizer), of an

- **Additional head** (e.g. with DICKKOPF or CERBERUS, factors secreted from the anterior roof of the archenteron), or of an
- **Additional trunk** (with NOGGIN and CHORDIN, factors secreted from the posterior roof).

These factors are antagonized by neutralizing factors. In the posterior-dorsal region of the gastrula/neurula **WNT-8** signals suspend the action of the head inducers, while in the anterior-ventral region the ventralizing morphogen **BMP-4** suppresses the action of NOGGIN and CHORDIN which in their own domain induce dorsal structures including the spinal cord. Agonists and antagonists are produced at opposite sides of the embryo, diffuse toward their opponents, bind and thus neutralize each other mutually. Thus inversely oriented gradients of free, effective factors are established.

All known inducing factors are also **morphogens** as they launch different programs of subsequent differentiation dependent on their local concentration. In the neurula, for instance, the head inductors DICKKOPF and CERBERUS, and the trunk inductors CHORDIN and NOGGIN aid in cooperation with other factors such as retinoic acid and various FGFs, to subdivide the neural tube in forebrain, hind brain, and spinal cord.

Numerous further factors have been identified, that are active not only in amphibians but in other vertebrates as well, for example NODAL that is active already in early stages, and various FGFs, which help to segregated the germ layers and to specify cell types.

In *Xenopus* **cloning of a vertebrate was successful for the first time** (e.g. by Gurdon).

The Zebrafish (*Danio rerio*) and **medaka fish** (*Orycias latipes*) advanced to the position of the most recent model organisms among the vertebrates; since these animals spawn in the free water, have transparent embryos, develop rapidly and are amenable to genetic studies. The translucent egg envelop permit observation of a development typical for fish but different from the familiar amphibian type in several aspects. At the animal pole a multi-cellular and multilayered **blastodisc** (blastoderm) forms atop the yolk sphere. Expanding more and more the blastoderm progressively covers the yolk sphere. Beginning at a point on the future dorsal site, called the **shield**, cells of the blastoderm involute along the blastoderm margin, The blastoderm becomes double layered (epiblast, and hypoblast = deep cells). The margin is homologized with the amphibian blastopore, the shield with the upper blastopore lip, and it shares with the lip the inductive powers of the Spemann organizer, emitting homologous signal proteins and expressing homologous genes. Beneath the epiblast migratory **deep cells** form an elongated aggregate in front of the shield, from the outside visible as **primitive streak**. The streak gives rise to the mesodermal axial organs (notochord and somites) and to the gut. In the translucent fish the development of the nervous system, of the eye and of the cardiovascular system can conveniently be observed in living specimens.

The Chick. The egg cell proper is huge and identical with the yellow, yolk rich sphere. The animal cap is the **blastodisc** that, in the course of cleavage, fractionalizes itself into cells, Below the blastodisc a subgerminal cavity forms. Indicating the onset of gastrulation, **a primitive groove** appears on the blastodisc. This groove is homologous to the amphibian blastopore, the anterior margin of the groove, known as **Hensen's node**, is homologous to the amphibian upper blastopore lip and, hence, to the Spemann organizer. Cells of the upper sheet of the disc (epiblast) move towards this groove and submerge through the groove into the depth, forming there a loose association, collectively called hypoblast, and spreading on top of the yolk into the subgerminal cavity. Most hypoblast cells move away from the Hensen's node in anterior direction, aggregate and eventually form the axial organs and endodermal organs like they arise in fish. When the phylotypic stage is reached a protective **amniotic** cavity filled with fluid is formed in which the embryo continues to grow. A **yolk sac** and an **allantois** (embryonic urinary bladder) are further structures characteristic for all **amniotes** (reptiles, birds, mammals). Avian embryos have for long been favoured to study the development of the nervous system and the fate of the migratory neural crest cells. Furthermore, pattern formation in limb buds, and recently, the periodic, clock-driven formation of the somites are studied extensively on the organismic and genetic level.

The Mouse. The development of the viviparous, **placental mammals** takes place in the oviduct and the uterus of the mother. The egg is almost devoid of yolk, nevertheless the embryo can grow to a large size due to nutrients supplied by the mother. The egg matures in the oviduct after fertilization by extruding the second polar body, undergoes an extremely slow, holoblastic cleavage while already switching on zygotic gene activities in the 2-cell stage. Cleavage results in a spherical **blastocyst** (not be confused with a blastula). Its outer wall becomes the polyploid, nutritive **trophoblast** that encloses an **inner cell mass**, also called **embryoblast** as it will give rise to the true embryo. The inner cell mass begins to build the embryo not until the blastocyst has lodged itself in the maternal uterus, has developed trophoblast villi on its surface as precursors of the placenta, and the inner cell mass has scooped out an **amniotic cavity**. Now a **blastodisc with primitive groove** is formed on the floor of the amniotic cavity, similar to the avian and reptilian embryos. In the mouse, however, peculiarities in details occur that are unfamiliar in other species, for instance, the initially convex, then concave curvature of the embryo and its complex turning. As in other placental mammals, the **placenta** has its origin in the trophoblast of the germ. Via its placenta, the embryo, in later stages called **fetus**, receives nutrients and oxygen from the mother and disposes waste metabolites the mother has to remove.

In the mouse many strains with defined and mapped mutations are available. Genes can be exchanged by a technique known as homologous recombination. Often genes are exchanged for genes possessing a targeted mutation. Transgenes can be generated by introducing foreign genes. In addition, mice can be cloned.

The Human

6

6.1 Human Embryos and Model Organisms

6.1.1 Does Knowledge of the Development of a Fly Help to Understand Human Development?

Naturally, our main interest is to learn how we ourselves have once developed. How could information of the development of a mouse, or even of a frog or fly, help to understand human development?

The zoologist knows that human beings are members of the mammalian family of primates and, therefore, is not surprised to learn that early human embryogenesis exhibits considerable differences to that of mice, but shows only little differences to that of other primates. Self-evidently, disparities between the development of humans, frogs and flies are huge – in the final outcome. It is more astonishing that comparing embryonic development of forms such diverse as humans, frogs and flies discloses so many and fundamental commonalities. Modern molecular genetics has disclosed uncanny likeness and unexpected homologies. The common genetic basis will be dealt with in Chap. 12.

Experiments with model organisms introduced in the Chaps. 4 and 5 facilitated, or even enabled in the first place, the disclosure of many fundamental control mechanisms and genes involved also in human development. A pioneering finding was that many developmental malformations known to physicians since ancient times can be evoked in animal model organism experimentally. A classical historic example was the generation of "Siamese twins" in amphibians (see Fig. 13.1). Further examples are the many known hereditary symptoms of diseases that in similar form can be evoked "in animal disease models" by targeted mutagenesis and genetic crosses.

Although laws, ethical and religious reservations forbid experimental interference, the development of man is the subject of morphological, histochemical, molecular and genetic investigations. For example, the pattern of gene expression can be studied with in-situ-hybridization (see Fig. 12.5) in embryos lost from pregnant women. In addition to aborted embryos, living cells isolated from human teratocarcinomas are available for studies. The following description provides information on peculiarities of mammalian development in general and human development in particular, compared to traditional generalizing descriptions of animal development.

For a better understanding, the description of amphibian and avian development should be read first. On the morphological as well the molecular biological level many congruities have been found between mammals and amphibians, whose ancestors also gave rise to land vertebrates. And since a reptilian ancestor was inserted in the phylogenetic lineage leading from amphibian to mammalian ancestors, avian development can

W.A. Mueller et al., *Development and Reproduction in Humans and Animal Model Species*,
DOI 10.1007/978-3-662-43784-1_6, © Springer-Verlag Berlin Heidelberg 2015

also confer information about our evolutionary heritage (zoologists conjoin reptiles and birds to a comprehensive taxon Sauropsida).

6.2   From Primordial Germ Cells up to Fertilization

6.2.1   Oogenesis: Long Before Their Own Birth, Females Already Possess a Large Store of Egg Cells; but Their Growth and Maturation Still Take Much Time

While still babies or embryos, girls store thousands of eggs which pause in meiotic prophase I for many years. **Primordial germ cells (oogonia)** destined to give rise to the next generation are generated and laid aside early in embryo development; they can be identified in the pre-natal female embryo in the 3rd week of development. By this time, the primordial germ cells immigrate from the yolk sac by amoeboid movements into the gonadal ridges (see Figs. 8. 3f and 15.2). In the 5th month of pregnancy the female embryo harbours some seven million oogonia in her ovaries. By the 7th month of development the majority have perished. The surviving 0.7–2 million oogonia stop mitotic multiplication and become **primary oocytes**. These **enter the prophase of meiosis I** (Fig. 6.1) but **pause at diplotene** (Figs. 6.1 and 8.4). There follows an interval lasting 12–40 years, in which lampbrush chromosomes and multiple nucleoli (Fig. 8.5) may be present in oocyte nuclei (not firmly established for human oocytes). Much transcription takes place, many ribosomes are produced, and the oocyte is fed with yolk material by the surrounding follicle cells. Even the relatively small human oocyte undergoes a 500-fold increase in volume.

During childhood, many more oocytes perish. At puberty some 40,000 are left; they are enveloped by nourishing **primordial follicles**. Stimulated by the pituitary gonadotropin **follicle-stimulating hormone (FSH)**, 5–12 oocytes begin to mature in each ovarian cycle. Later, the follicles are exposed to another gonadotropin, the **luteinizing hormone LH** (Fig. 6.2). These hormones signal the egg to resume meiosis and to initiate the events culminating in the release of the mature egg from the ovary. As a rule, in one ovarian cycle, which lasts some 28–30 days, only one (or up to three) of the follicles attains the state of a mature **Graafian follicle**. In the middle of the menstrual cycle, about 14 days after the last menses, one of the oocytes finishes the first meiotic division by extruding the first polar body, and the egg is released from the ovary (**ovulation**). In mammals including man the second polar body is not constricted off until the egg is fertilized.

Before fertilization takes place, the egg is caught at the mouth of the oviduct. If not, a fertilized egg that falls into the abdominal cavity dies or may give rise to a malformation exhibiting tumour-like features, called a **teratocarcinoma**.

The time point of ovulation, the short window of 6–12 h in which the egg can be fertilized, and the short, at the maximum 3 days, life span of the sperm, restrict the conceptive period in the cycle of a women to about 4 days in the middle of the menstrual cycle (Fig. 6.2b). Physicians assume the optimal time point for conception to be the day before ovulation.

An Annotation to Present Medical Reproduction Biology

In industrial countries about 12 % of the women suffering breast cancer are still in the reproductive age of 20–34 years, and even young girls can develop breast cancer, boys may develop other kinds of cancer. Yet, the unavoidable chemotherapy destroys the ovaries or testes, respectively. Therefore, present reproduction biology develops methods (in animal experiments) to remove tissue from the ovary or testis before cytostatic agents are given, to store the removed tissue deep-frozen (cryopreservation) and to re-implant it upon successful therapy.

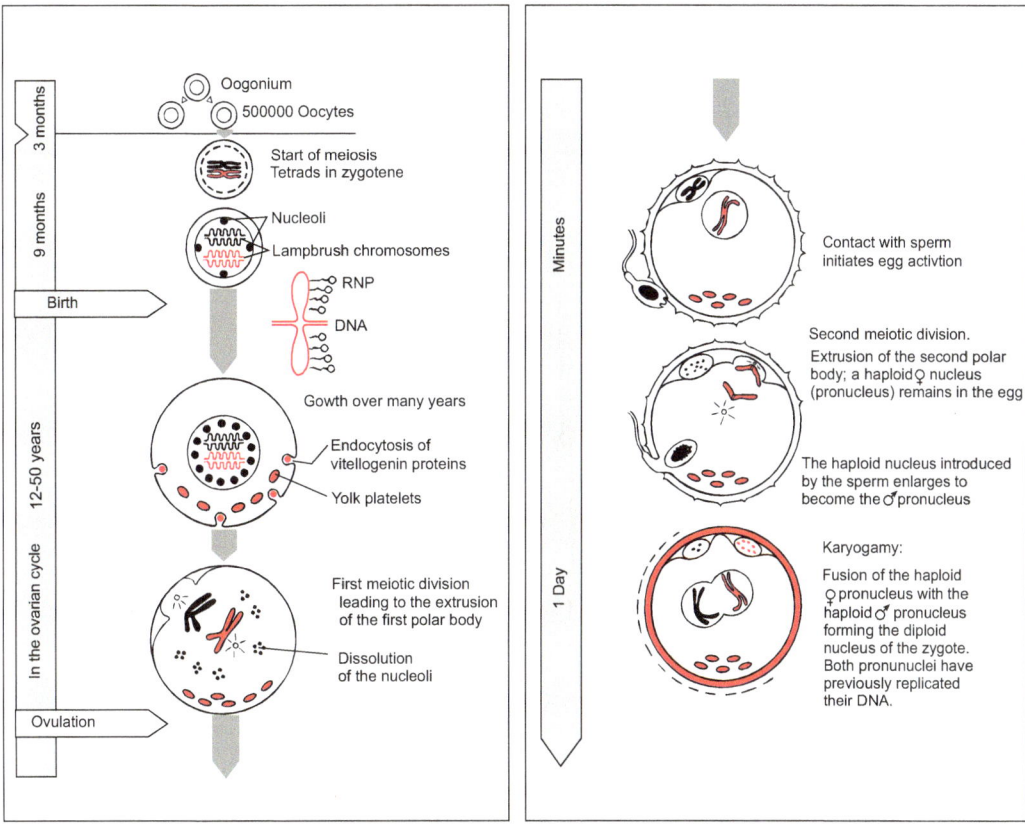

Fig. 6.1 Human. Oogenesis in review. Occurrence of lampbrush chromosomes is typical for vertebrates but not yet definitively established for humans

6.2.2   Only One of Millions of Sperms Gets a Chance

Once caught by the mouth of the oviduct, the egg awaits the sperm with reservation: it is surrounded by a tough envelope, called the **zona pellucida**, and by a corona of mucous material, called the **corona radiata**. Both these envelopes have been produced by the ovarian follicle, and must be penetrated by the sperm.

From 200 to 300 million sperm released into the vagina in each ejaculation about 1 % advance to reach the egg. On their long journey through the vagina, uterus and oviduct the sperm cells undergo **capacitation**, that is, they acquire the competence or capacity to fertilize through the influence of female secretory products. Only one sperm succeeds in entering the egg cell. The other

sperm may help to dissolve the egg's envelopes by releasing acrosomal enzymes (Chap. 3).

Whether in humans pregnancy can occur naturally without the contribution of a male individual (parthenogenesis) is discussed in Box 6.2.

6.3   From Fertilization up to the Phylotypic Stage

6.3.1   The Egg First Develops "Extraembryonic" Organs to Tap the Mother, Therefore the Embryo Can Grow Enormously

Pre-implantation development (Fig. 6.3). Because the embryo will be provided with nutrients by the mother early, the egg can be

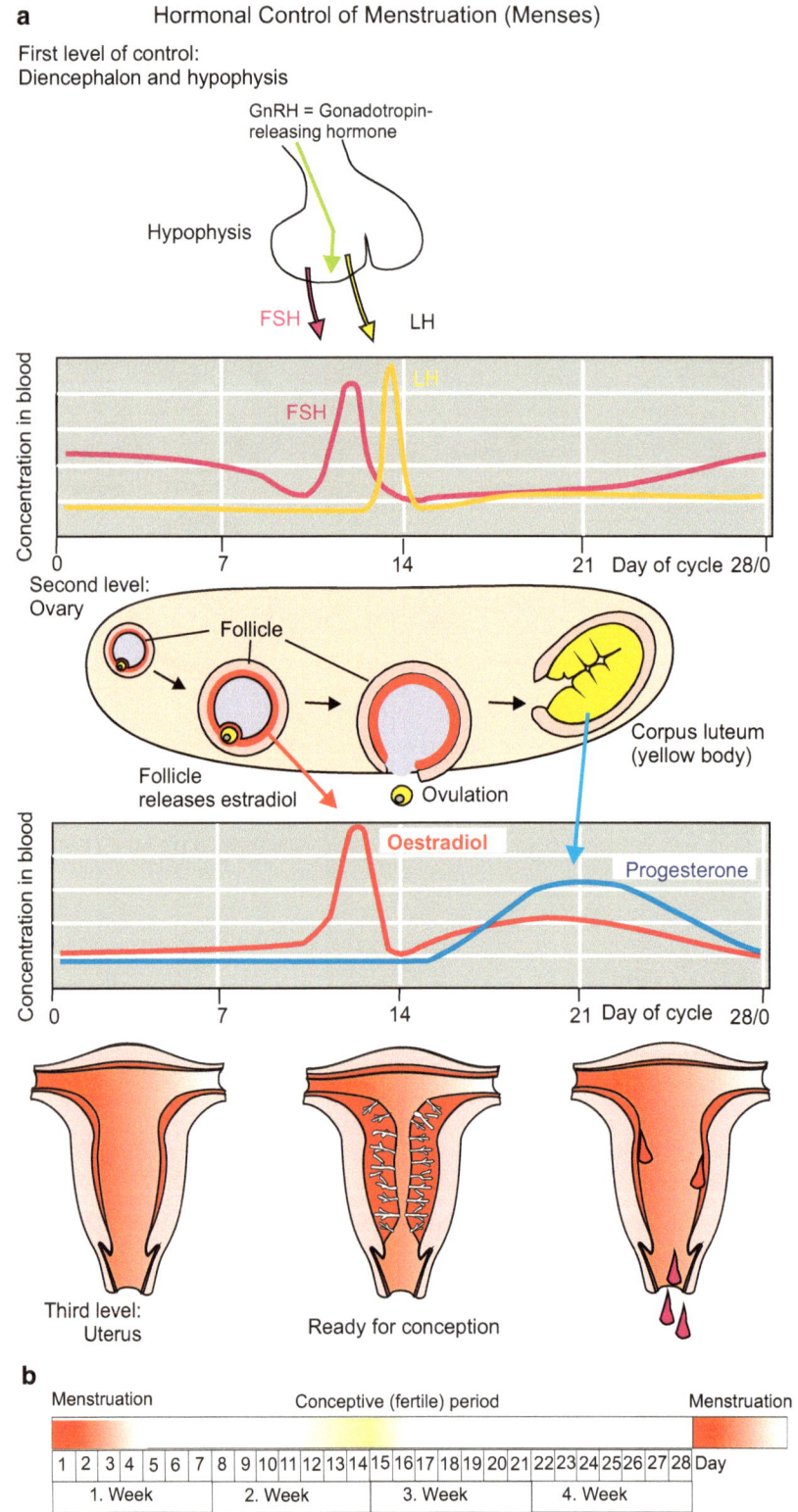

a   Hormonal Control of Menstruation (Menses)

First level of control:
Diencephalon and hypophysis

GnRH = Gonadotropin-
releasing hormone

Hypophysis

FSH   LH

Concentration in blood

LH

FSH

0   7   14   21   Day of cycle 28/0

Second level:
Ovary

Follicle

Corpus luteum
(yellow body)

Follicle
releases estradiol   Ovulation

Concentration in blood

Oestradiol

Progesterone

0   7   14   21   Day of cycle   28/0

Third level:
Uterus   Ready for conception

b
Menstruation   Conceptive (fertile) period   Menstruation

| 1 | 2 | 3 | 4 | 5 | 6 | 7 | 8 | 9 | 10 | 11 | 12 | 13 | 14 | 15 | 16 | 17 | 18 | 19 | 20 | 21 | 22 | 23 | 24 | 25 | 26 | 27 | 28 | Day |

| 1. Week | 2. Week | 3. Week | 4. Week |

Fig. 6.2 (**a**) Human. Hormonal control of the ovarian cycle in preparation of pregnancy. Provisions are made at three levels: Level I: Pituitary gland, the hormones FSH and LH trigger the final maturation of the egg cell and its

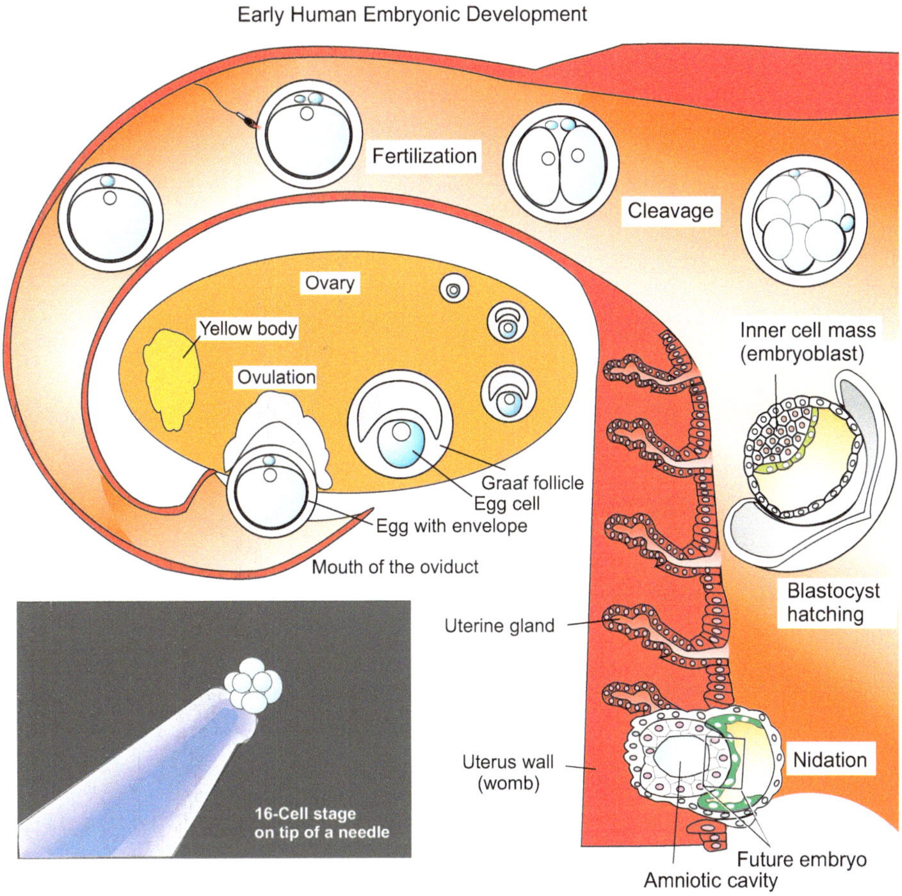

Early Human Embryonic Development

Fig. 6.3 Human: From ovulation to implantation. A mature egg contained in a Graafian follicle of the ovary is liberated and picked up by the tube of the oviduct. Fertilization, formation of the second polar body and cleavage take place in the oviduct, which leads into the uterus. Once arrived in the uterus, the embryo has reached the blastocyst stage. The blastocyst hatches out from its envelope (zona pellucida) and attaches to the wall of the uterus. Now the blastocyst begins to invade the uterine wall

void of yolk, and it is in fact almost free of yolk. The human egg is amazingly small, just 0.1 mm in diameter and thus not larger than the egg of sea urchins, and considerably smaller than the egg of the small fruitfly. The egg and early embryo can sit enthroned on the tip of a needle (Fig. 6.3).

Although the mammalian egg lacks significant quantities of yolk and has reverted to **holoblastic cleavage**, cleavage in mammals is **extremely slow**. This apparently retarded development is correlated with an early start of transcription of zygotic genes; no midblastula transition as it occurs in amphibians (and other animals) is known from mammals. Not until the third day is the 12–16 cell stage of the **morula** reached. By the 6–7th day the **blastocyst** stage is attained. The blastocyst looks just like a blastula, but it isn't a blastula that would proceed to form

Fig. 6.2 (continued) release from the ovary (ovulation). Level II: In the ovary the female sexual hormones are produced, oestradiol in the follicle, progesterone from the *yellow body* (corpus luteum). The hormones are released into the circulation and reach the uterus. Level III: The sexual hormones induce the growth of the nutritive mucosa which is shed if no germ lodges itself in it. (**b**) Conceptive period

Fig. 6.4 Human: Implantation (nidation), amnion, yolk sac. The invasion of the uterine wall is accomplished by an egg-derived 'extraembryonic' tissue or organ, the trophoblast. It shows features of a terminally differentiated tissue, including polyploidy. The peripheral invasive trophoblast forms a syncytium by dissolving cell membranes. The amniotic cavity is not formed by amniotic folds (as in the avian and insectivore embryo) but by the separation of cells of the inner cell mass. Hypodermal (endodermal) cells surround the yolk sac cavity, forming the (empty) yolk sac. An embryo proper is not yet formed

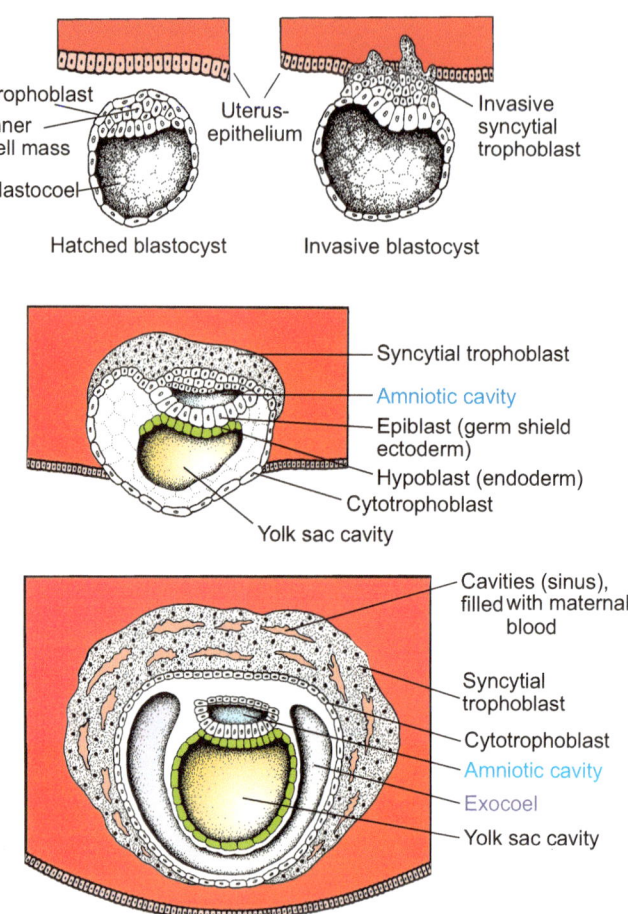

an archenteron through invagination as do the true blastulae of sea urchins and amphibians. Rather, the cellular wall of the blastocysts undergoes differentiation to become the **trophoblast (trophectoderm)**. This is an 'extra-embryonic' organ destined to invade the wall of the maternal uterus, absorb maternal nutrients, and dispose of waste. After implantation the trophectoderm undergoes several rounds of endoreplication and develops **polyploid giant cells**. Part of the trophoblast will eventually give rise to the **placenta**, a foetal organ formed to optimize all of these functions (Sect. 6.4). The trophoblast encloses a cavity and the **inner cell mass**, also called the **embryoblast** because those cells, which remain diploid, eventually will generate the embryo proper. The inner cell mass is a

group of cells eccentrically concentrated at a location beneath the 'polar' trophoblast. The blastocyst hatches, attaches to the uterine wall and sinks into it, a process known as **implantation** or **nidation**.

If the mouth of the oviduct fails to catch the egg, and the blastocyst settles down in the abdominal cavity (**ectopic abdominal pregnancy**) the embryo dies or develops to a tumour-like **teratoma**, a risk a woman has to bear; a teratoma eventually must be surgically removed.

After **extracorporeal fertilization** the fertilized egg can, appropriate culture media provided, develop *in vitro* up to the blastocyst stage. Then, according to the current state of reproductive medicine, further development must take place in the uterus.

Fig. 6.5 (a–c) Human. Embryonic development up to the somite stage (phylotypic stage). The embryo is formed by a blastodisc on the floor of the amnion. The blastodisc is composed of epiblast and hypoblast. Gastrulation occurs by a process highly reminiscent of gastrulation (primitive streak formation) in birds (see Fig. 5.25c, d). Extraembryonic organs are yolk sac, allantois (partially), trophoblast-derived villi and eventually the placenta

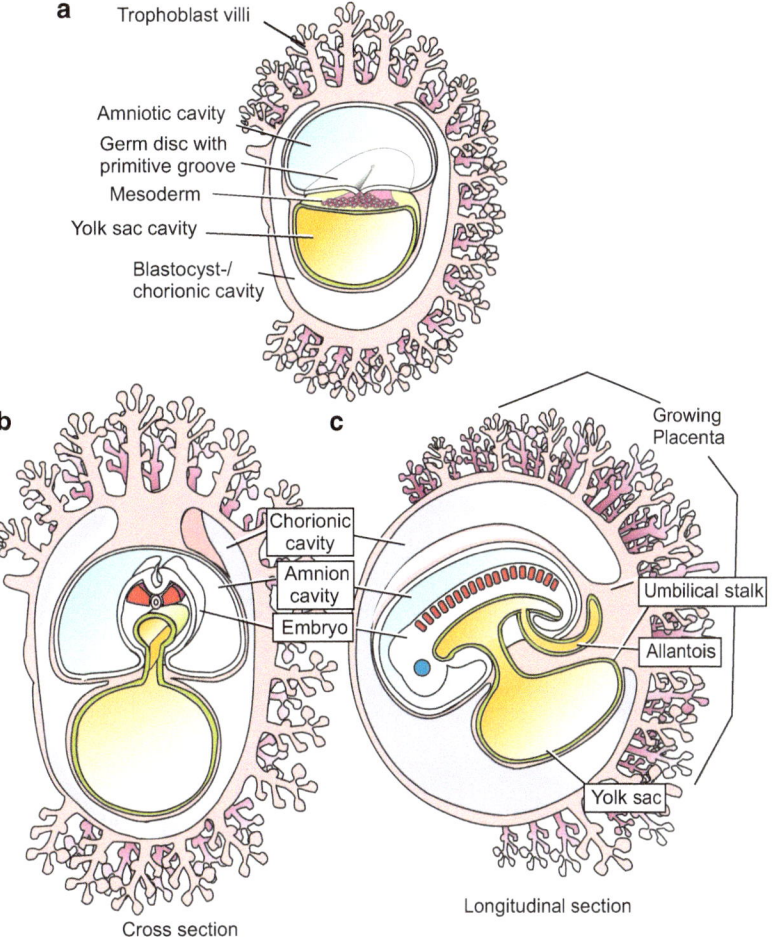

a
Trophoblast villi
Amniotic cavity
Germ disc with primitive groove
Mesoderm
Yolk sac cavity
Blastocyst-/ chorionic cavity

b
c
Chorionic cavity
Amnion cavity
Embryo
Growing Placenta
Umbilical stalk
Allantois
Yolk sac

Cross section
Longitudinal section

Post-implantation development (Figs. 6.4, 6.5, 6.6, 6.7, 6.8, 6.9, 6.10, 6.11, and 6.12). The hatched blastocyst attaches with its polar trophoblast to the wall of the uterus. Entry is facilitated by lytic enzymes released from the trophoblast. The outer layer of the trophoblast proliferates into the uterine tissue. While doing so it transforms into a **multinucleated syncytial trophoblast**; the inner layer of the trophoblast, called the **cytotrophoblast** (Figs. 6.4 and 6.5), remains cellular and encloses the blastocoel. Later the wall of the blastocyst is supplemented by an inner layer, called the **parietal mesoderm**. The cavity of the blastocyst will not be enclosed in the embryo and is therefore called the **exocoel** or **chorionic cavity**.

The invasive syncytial trophoblast prepares the space into which the embryo can grow. The solid epithelial cytotrophoblast becomes reinforced and supplemented by mesodermal cells that immigrate from the embryo: the whole organ is known as the **chorion**. From its outer surface **primary chorionic villi** (Fig. 6.5) project and extend through the loose syncytiotrophoblast to establish contact with the surrounding maternal uterine tissue. With these chorionic villi the embryo takes up nutrients and O_2 from maternal blood lacunae of the decidua (layer of the uterus surrounding the germ). In human embryological illustrations the term trophoblast usually is now replaced by the term **chorion**. The nutritive organ of the chorionic

Fig. 6.6 Human: Neurulation, organogenesis. The surrounding envelopes (trophoblast and chorion) are removed and the amniotic cavity opened to show the blastodisc. The neural folds form in front of the primitive groove, just as in the avian embryo (compare Fig. 5.25c, d). The notochord is formed by the roof of the archenteron; in an intermediate stage of its formation the tubular notochord encloses the notochordal canal. The rudimentary allantois and the yolk sac become integrated into the umbilical cord. During the entire development shown the embryo increases in length from about 1 to 5 mm

HUMAN

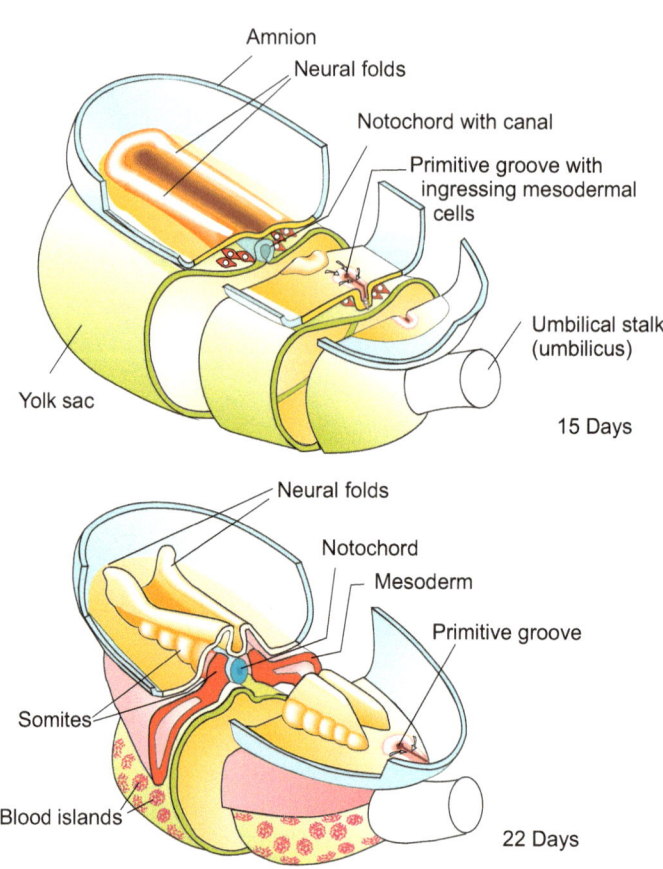

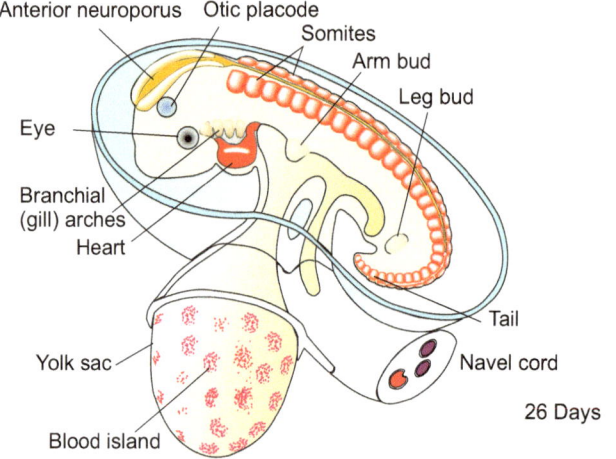

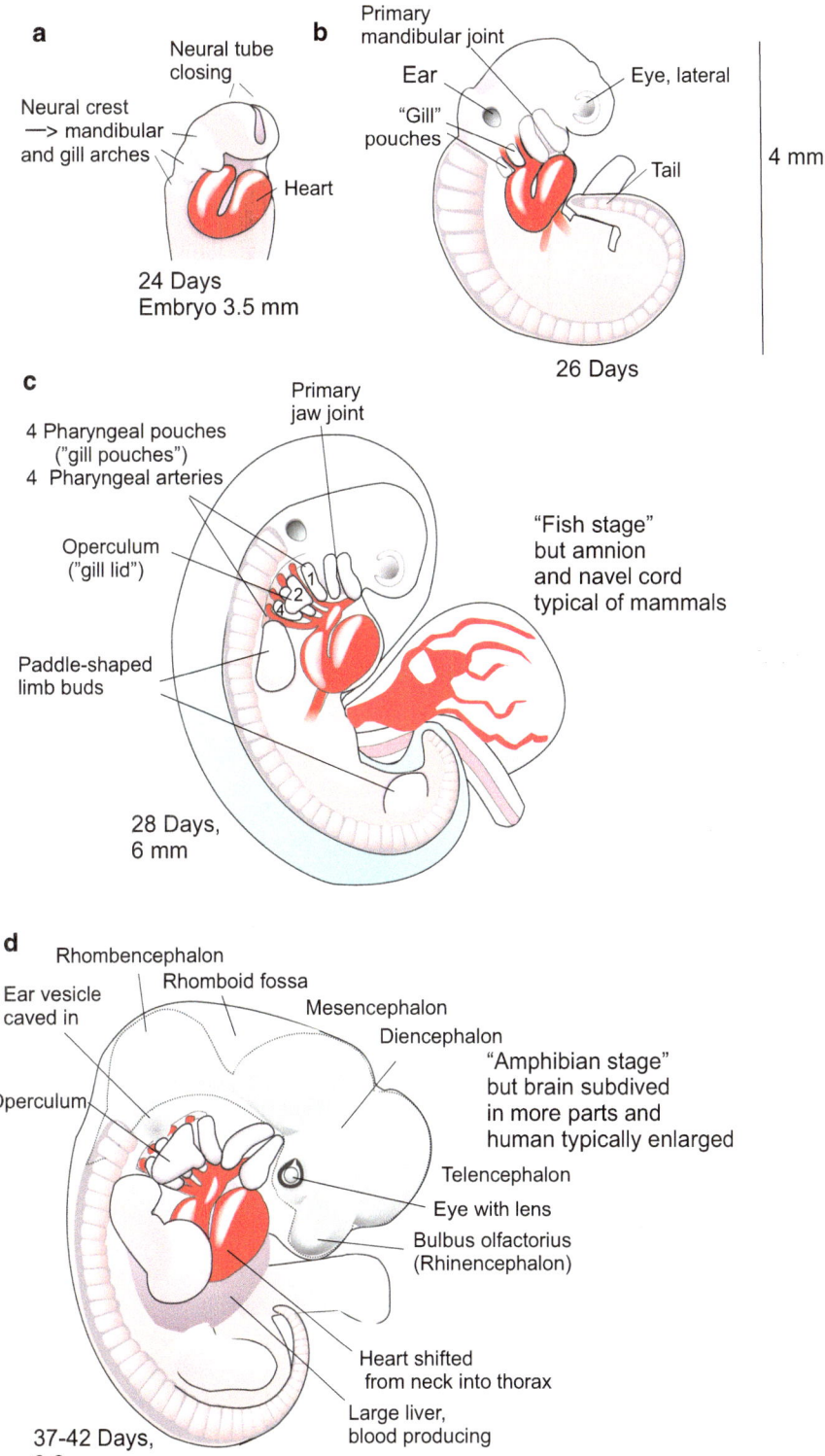

a

Neural tube
closing

Neural crest
—> mandibular
and gill arches

Heart

24 Days
Embryo 3.5 mm

b

Primary
mandibular joint

Ear

Eye, lateral

"Gill"
pouches

Tail

4 mm

26 Days

c

4 Pharyngeal pouches
("gill pouches")
4 Pharyngeal arteries

Primary
jaw joint

Operculum
("gill lid")

"Fish stage"
but amnion
and navel cord
typical of mammals

Paddle-shaped
limb buds

28 Days,
6 mm

d

Rhombencephalon

Rhomboid fossa

Ear vesicle
caved in

Mesencephalon

Diencephalon

"Amphibian stage"
but brain subdived
in more parts and
human typically enlarged

Operculum

Telencephalon

Eye with lens

Bulbus olfactorius
(Rhinencephalon)

Heart shifted
from neck into thorax

Large liver,
blood producing

37-42 Days,
9,6 mm

Fig. 6.7 (**a–d**) Human. Development of the embryo in the first 6 weeks post fertilization

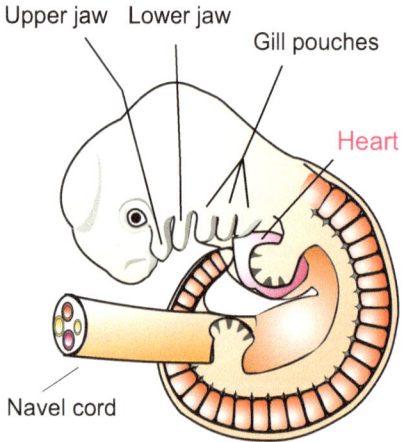

Upper jaw   Lower jaw

Gill pouches

Heart

Navel cord

Phylotypic stage
Day 26-28, 4-6mm

Fig. 6.8 Human. Phylotypic stage, length 4–6 mm. The global organ anlagen are present; but later subdivided, elaborated and refined through different growth rates and different paths of terminal differentiation

epithelial layer called the **hypoblast** separates. It expands, lining the central cavity and forming the **parietal endoderm** or **yolk sac** (a misnomer since it does not contain yolk!).

The volume of the amniotic cavity increases with the increasing size of the embryo. Its fluid is known as **waters** or **forewaters**. Unlike in the mouse, in humans the evolutionary old primary yolk sac bursts open and becomes replaced by a secondary sac. Eventually, the yolk sac is enclosed by extraembryonic endoderm and mesoderm, integrated into the umbilical stalk and accompanied by blood vessels; all these structures together are twisted to yield the **umbilical cord**. Into the umbilical cord also the **rudimentary allantois** is integrated. The allantois, in reptiles and birds serving as embryonic urinary bladder, can in mammals be rudimentary because the embryo can hand over oxidation water and its metabolic end products to its mother for disposal.

villi is designated **extra-embryonic**, meaning that the villi are not organs of the embryo proper; this does not yet exist at this stage. The chorionic villi are the functional precursors of the much larger villi that are later formed by the placenta.

6.3.2 In Human Embryonic Development Evolutionary Ancient Structures Emerge: Primitive Groove, Yolk Sac, Amnion and Allantois

Amnion, Yolk Sac, Umbilical Cord. Before the embryo develops, the human blastocyst, like most mammalian blastocysts, makes some preparations for the benefit of the coming embryo. The embryos of reptiles and birds have enclosed themselves with a self-made, water-filled amniotic cavity, and they wore a yolk sac. The mammalian embryos conserved these innovations. However, the germs of most mammals establish an amniotic cavity and a yolk sac before an embryo proper exists at all, and so it is in humans, too. By the 9th day the **amniotic cavity** appears as a cleft in the midst of the inner cell mass. From this inner mass an

Embryogenesis. What happens inside the (former) blastocyst? Where is the embryo formed? The **embryo** proper is formed from two minute epithelial areas: from an area in the floor of the amniotic cavity, now called the **epiblast**, and the roof of the yolk sac, known as the **hypoblast**. Together these two adjoining areas, one on top of the other, represent the **blastodisc**.

From this point on, generation of the actual embryo follows the path paved by our non-mammalian ancestors.

Gastrulation, germ layer formation, and **neurulation** are similar in many ways to avian embryo development (Figs. 6.5 and 6.6; compare Fig. 5.25). In gastrulation, epiblast cells move to a **primitive groove** and pass through it, colonizing the cleft between the epiblast and the hypoblast. In front of the groove, a **primitive streak** appears. Underneath, the invading cells give rise to the **mesoderm**, and perhaps contribute also to the endoderm which essentially derives from the hypoblast. The mesoderm subdivides itself in the traditional manner giving rise to notochord, somites, and lateral plates. Atop the notochord the epiblast forms **neural folds** that close to form the neural tube (Fig. 6.6). **Thus, gastrulation (through a primitive groove), neurulation,**

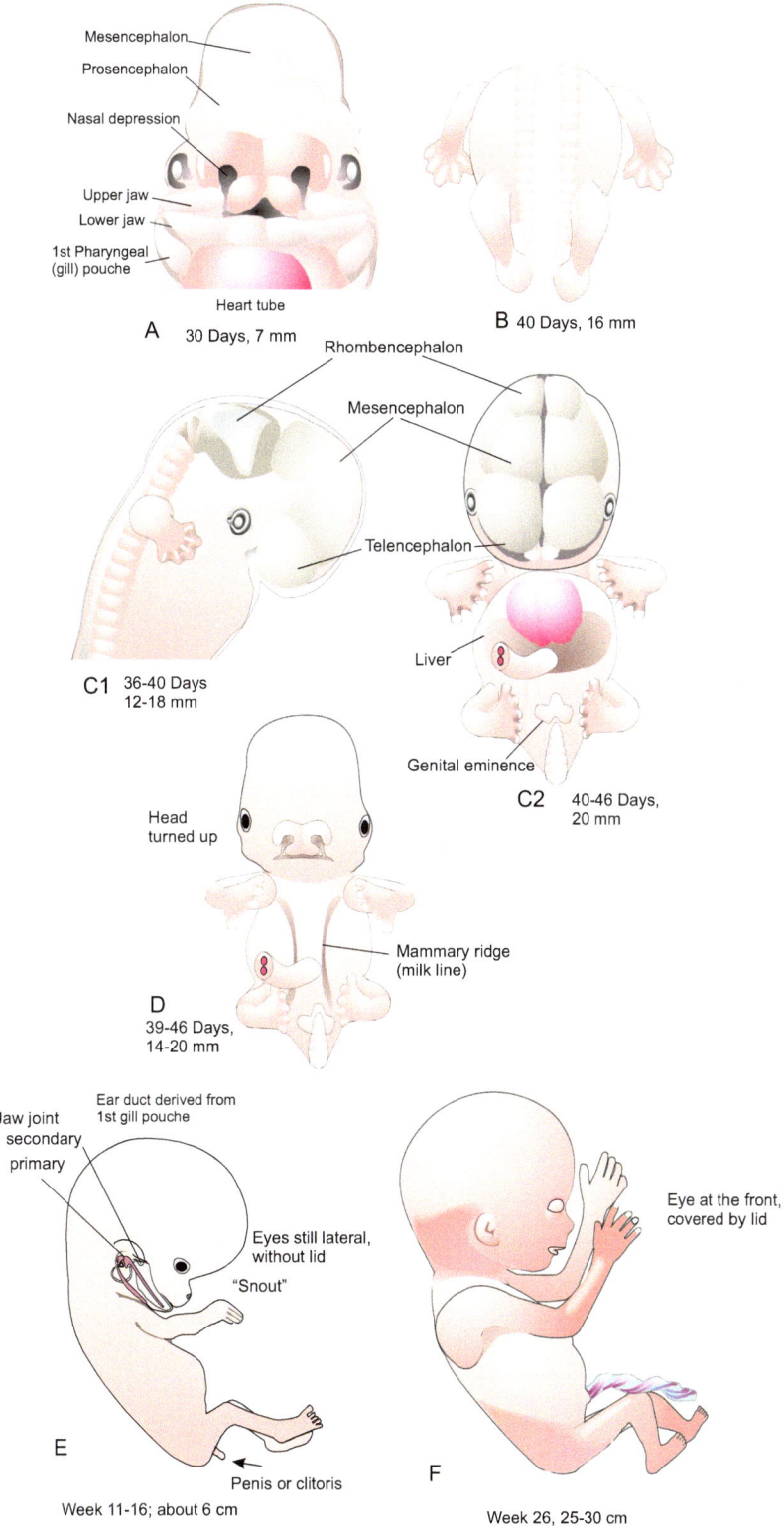

Fig. 6.9 (**a–d**) Human. Development from day 30 to day 46. (**e–f**) Human. Development from week 11 to 26

Fig. 6.10 (**a**, **b**) Human: Placentation 1. The chorionic villi (tree-like structures) are reduced on one side but enlarged on the side that faces the maternal uterine wall. They concentrate in a disc-shaped area called placenta. The amniotic cavity ousts the chorionic cavity

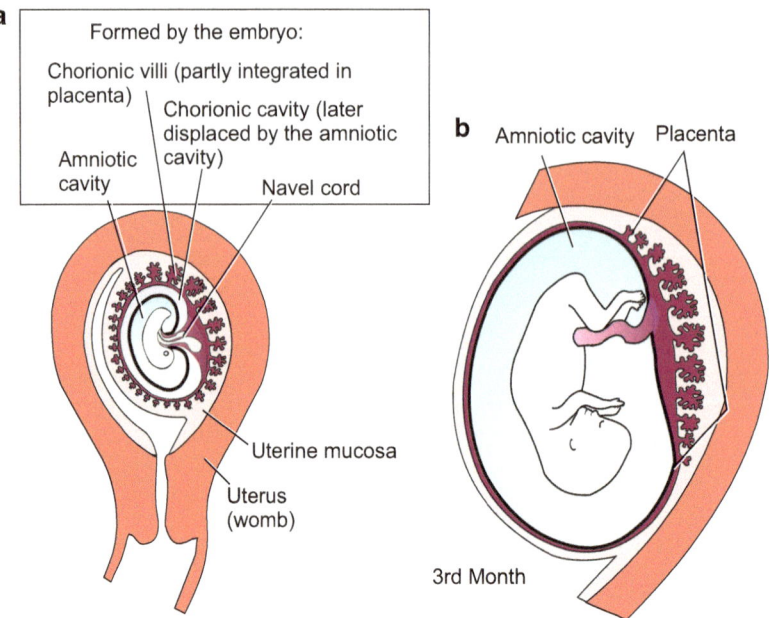

a

Formed by the embryo:

Chorionic villi (partly integrated in placenta)

Chorionic cavity (later displaced by the amniotic cavity)

Amniotic cavity

Navel cord

Uterine mucosa

Uterus (womb)

b   Amniotic cavity   Placenta

3rd Month

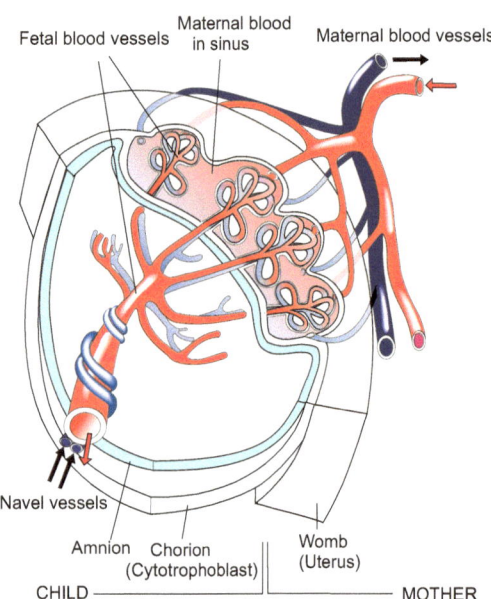

Fetal blood vessels

Maternal blood in sinus

Maternal blood vessels

Navel vessels

Amnion   Chorion (Cytotrophoblast)

Womb (Uterus)

CHILD ———————————————— MOTHER

Fig. 6.11 Human: Placentation 2. The villi are provided with blood vessels (umbilical vessels) and together constitute the major part of the placenta. The maternal endometrium contributes to the placenta in a negative way as it partially dissolves. The resulting spaces (lacunae) are filled with maternal blood. The chorionic villi of the foetal placenta dip into the maternal blood

and formation of notochord and mesodermal primordia, follow traditional basic patterns of vertebrate development.

There are, however, mammalian-specific peculiarities. The anterior part of the notochord, known as the "head process" pushes anteriorly and contacts a prechordal plate which will give rise to head muscles and to the pharyngeal membrane. Initially and temporarily, the notochord is hollow. The **chordal canal** begins with an open pore in the primitive groove close to the node and ends with another open pore in the yolk sac. A similar chordal canal (**canalis neurentericus**) is found in reptiles and some birds.

When the neural folds merge above the notochord to form the **neural tube**, this tube also remains open at its anterior and posterior ends. Both **the anterior and posterior neuropores** connect the neural canal with the amniotic cavity. The pores are closed when the majority of the **somites**, the **optic vesicles**, the **ear placodes** and the **pharyngeal pouches** ("**gill arches**", Figs. 6.7, 6.8, 6.10, 6.11, 6.12, 6.13, 6.14, 6.15, 6.16, 6.17, 6.18, 6.19, 6.20, and 6.21) are visible, and the **heart beats**.

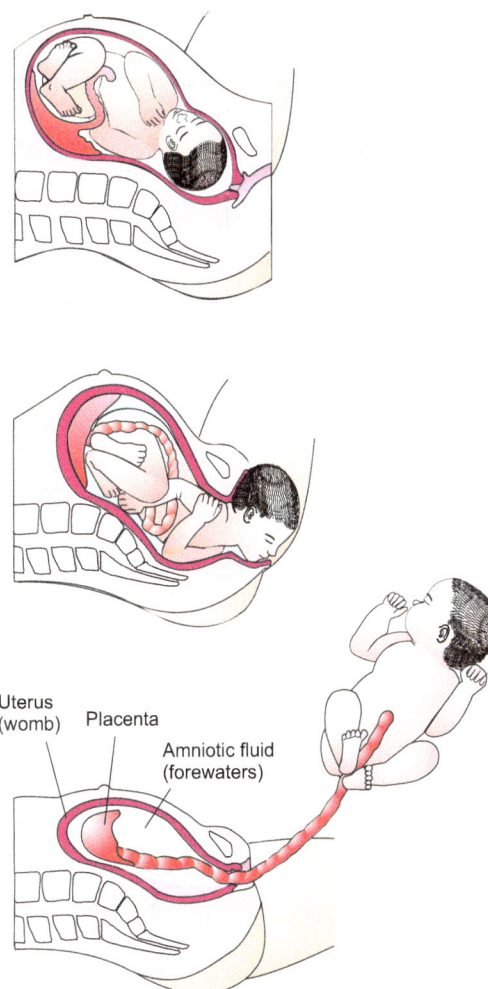

Fig. 6.12 Birth. Note: the "waters" flowing out consist of amniotic fluid

Uterus (womb)

Placenta

Amniotic fluid (forewaters)

extremely slow cleavage. Compared to other vertebrates cleavage is slow in all mammalian species. The germ commences transcription of its (zygotic) genes shortly upon fertilization, a first wave starting at the two-cell stage. Transcription takes time. In subsequent development the mammalian human passes stages that can be understood only owing to its evolutionary past. Also in the outward appearance (Fig. 6.7) such traits become apparent, for instance **gill pouches** (Figs. 6.6, 6.7, 6.8, and 6.21), **eyes positioned at lateral sides** and for long time **uncovered by a lid**, **arms and legs laterally protruding** like in amphibians and reptiles, **tail** and **milk lines** (Fig. 6.9d).

A further, significant feature is the enormous growth of the **cerebral hemisphere**s (Fig. 6.9), which for their part expand the still malleable head enormously. The diameter of the head, which principally increases in the frontal region, sets limits to length of time the child can stay in the womb of the mother. The birth canal cannot be expanded to any width, and it is indeed rather often too narrow, compelling surgical intervention (Caesarian delivery).

Another peculiarity occurs on the side of the mother. Especially in human mothers, in the uterus wall-less lacunae are formed, filled with maternal blood, that facilitate the exchange of substances between the mother and the child (see following section).

By the 28th day of development the blood vessels have proliferated and spread inside and outside the embryo; embryonic circulation is connected to the placenta through the umbilical cord (Figs. 6.5, 6.6, 6.10, and 6.11).

6.3.3   A Peculiar Feature of the Human Development Is the Enormous Prenatal Growth of the Brain

In human development embryologists find unique features compared to other mammals. One peculiarity of the human embryo is the

6.4   Interface Mother/Child: The Placenta

6.4.1   The Placenta Is the Organ by Which the Embryo Is Anchored and Through Which the Embryo Exchanges Substances with Its Mother

The developing child lies within a double-layered cyst. Both layers are of embryonic origin: the inner layer is the wall of the amniotic cavity and the outer layer is the chorion, largely a derivative of the trophoblast (Fig. 6.5 and 6.10).

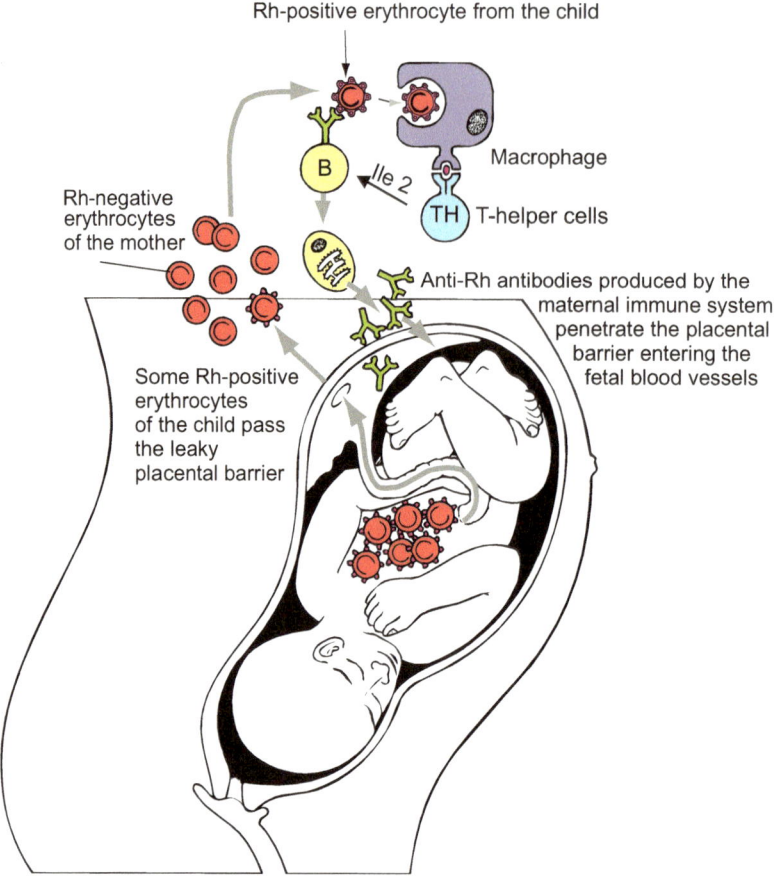

Rh-positive erythrocyte from the child

Macrophage

Ile 2

T-helper cells

Rh-negative
erythrocytes
of the mother

Anti-Rh antibodies produced by the
maternal immune system
penetrate the placental
barrier entering the
fetal blood vessels

Some Rh-positive
erythrocytes
of the child pass
the leaky
placental barrier

Fig. 6.13 Danger for the child by the immune system of the mother exemplified by the immune response elicited by the Rhesus factor. The Rhesus factor is a complex of glycoproteins on the surface of the red blood cells consisting of 50 defined blood-group antigens, among which the D antigen is the most important. The commonly used terms Rh factor, Rh positive and Rh negative refer to the D antigen. Rh positive erythrocytes of the child bearing the D antigen are dealt by the immune system of a Rh negative mother as foreign. When Rh positive erythrocytes get into the blood of a Rh negative mother via a leaky placental barrier (or on occasion of the birth of a former Rh positive sibling) the maternal immune system generates anti-RH antibodies (= anti-D antibodies). These can infiltrate the child causing haemolytic disease of the fetus and newborn

In a region of the chorion, now called the **placenta**, the tiny primary chorionic villi are replaced by larger secondary villi, and eventually by tertiary villi. The branching, tree-shaped villi grow into cavities inside the uterine wall. The villi become vascularized: Two arteries (umbilical arteries) leave the body of the foetal child at the point of the future navel, traverse the amniotic cavity inside the umbilical cord, enter the villi, ramify and re-join to form the umbilical vein, which returns to the embryo to supply it with nutrients and oxygen. In exchange for maternal nutrients and oxygen, waste products such as CO_2 and urea are transferred to the mother across the villi.

All this is of benefit to the child, indeed the mother has to pay a blood price because in birth the lacunae are not immediately closed. When a mother gives birth (Fig. 6.12) maternal blood gets lost. In other mammalian species, especially in non-primates, birth is not accompanied with much blood loss, and appears to be less painful.

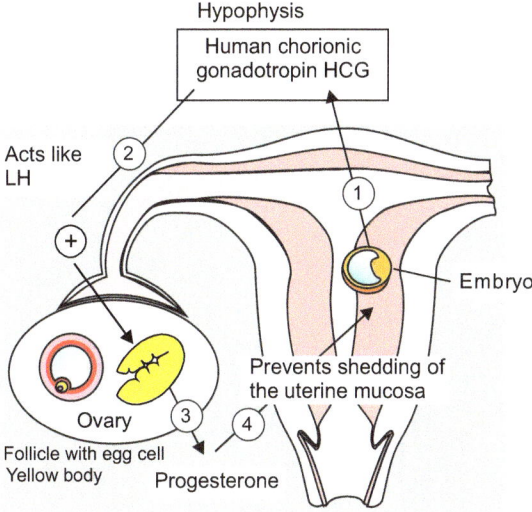

Fig. 6.14 Hormones preventing further menstruation after a germ has implanted in the uterus

6.4.2   There Are Further Dangers: The Immune System of the Mother Must Successfully Be Suppressed

From a pure biological view, and disregarding all joyful anticipation, the human embryo like any viviparous embryo might be considered a parasite that has to overcome the host's defence. In fact, the removal of tissue barriers between mother and child is not entirely beneficial. The fetus comes into dangerous contact with the immune system of the mother; therefore, reasonable compromises must be found.

Although a child of its mother the fetus is not molecularly identical because half of the maternal genes have been singled out in the meiosis of the oocyte, and were replaced by paternal genes in fertilization. This applies also to the genes of the **MHC** (*Major Histocompatibility Complex I*, (in humans also known as *human leukocyte antigen HLA*). The complex codes for multimeric proteins that are exposed on the surface of somatic cells as an individual identity card. The barcode characterizing each individual encompasses a set of 14 different MHC proteins. Checking this card the immune system identifies foreign cells and tissues, for instance transplants,

as being foreign. Foreign tissue is destroyed by cytotoxic T cells at the boundaries of transplants taken from a genetically different donor (heterologous transplants). The residual transplant is rejected. Why not the embryo?

One reason is that the cells of the trophoblast do not express the set of classical MHC molecules. On the other hand, cells lacking the classic MHC, as those of parasites, would evoke aggressive behaviour of the natural killer cells. To dampen the aggressive behaviour of the maternal natural killer cells the embryo expresses a set of peculiar embryonic HLA-G molecules. The mechanism by which these molecules locally shut down the immune system of the mother is currently under investigation.

But even when protection of the 'transplant child' against rejection by the maternal immune system is successful, the immunological coexistence of mother and child remains endangered; because a reliable barrier against antibodies, which are much smaller than immune cells, does not exist.

We refer to the **Rhesus factor** (Fig. 6.13). In the case of incompatibility it can lead to haemolytic disease of the fetus and newborn.

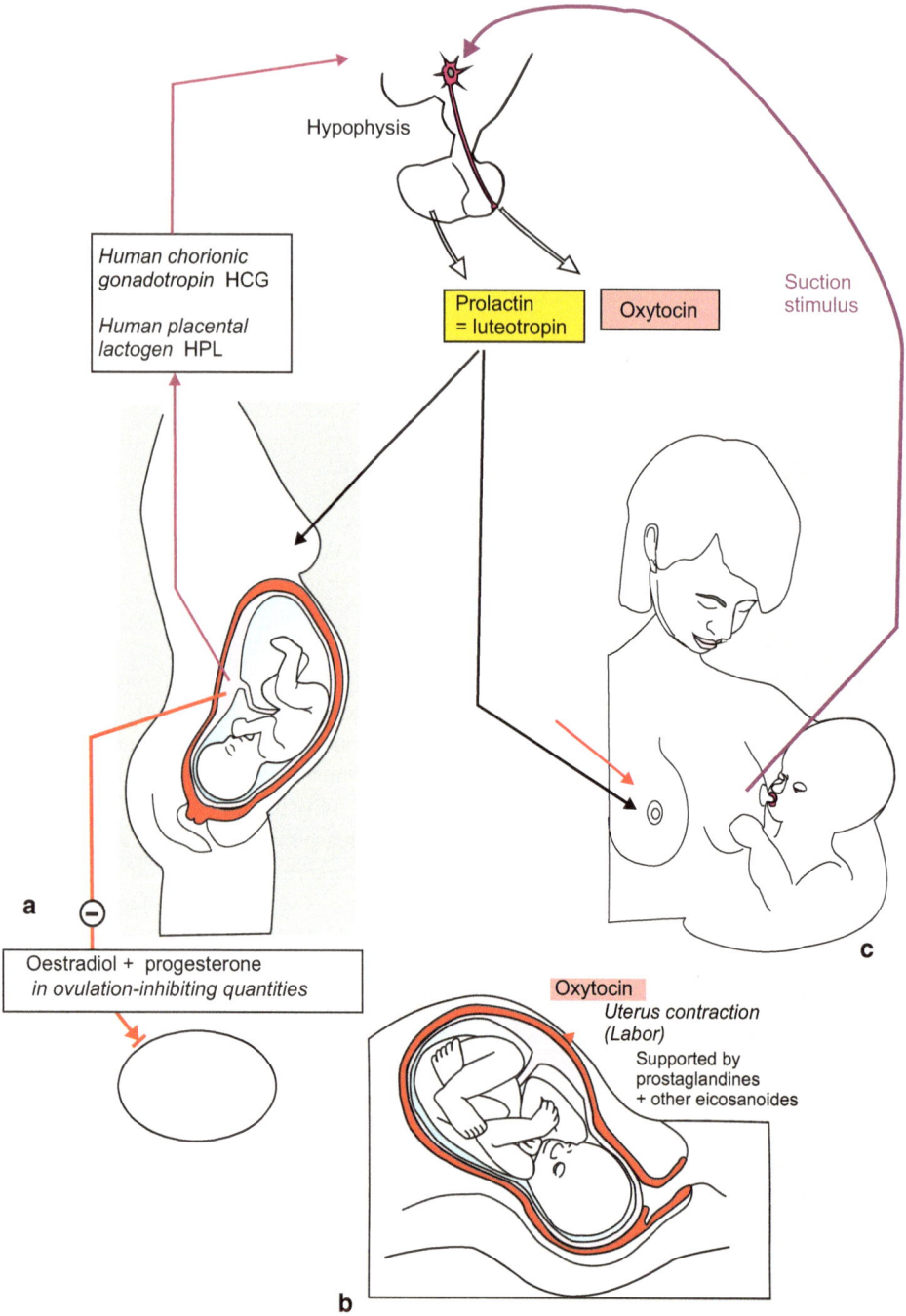

Fig. 6.15 Hormonally controlled events before, during and after birth. Hormones of the child and the mother interplay

This occurs when foetal Rh-positive erythrocytes find their way into the blood system of a Rh-negative mother; their immune system produces anti-Rh antibodies and these enter the foetal blood, passing the placental barrier. The antibodies release an immune response, the attacked red blood cells release haemoglobin and this causes a yellow colour of the skin (jaundice). Problems raised with chromosomal disorders in the child, such as trisomy 21 (Fig. 6.16) are discussed in Box 6.3.

Mother-Child-Chimeras

In this context a bewildering new discovery may be mentioned. In examining cells from diverse tissue samples of woman occasionally cells were found bearing a Y chromosome. All these women had once given birth to a boy, in some cases years ago. Apparently, foetal cells able to divide (putative stem cells) found their way through the placental barrier into the body of the mother, had been tolerated by her immune system on unknown grounds, and survived for years by self-renewal, as it is known from stem cells.

On the other hand in male offspring sporadically cells with XX chromosomes derived from the mother were found, testifying in retrospect to an imperfect placental barrier. Another source of XY cells in women and XX cells in men might have been blood stem cells once introduced by transfusion from donors belonging to the opposite sex.

6.4.3 Human Beings Live 9 Months Under Water, and During This Period the Unborn Child Has a Circulatory System Similar to That of a Fish

Teaching experience, gained in elementary and high schools as well, tells us that most children (and adults) are not aware that before birth they lived completely under water without any access to air, for 9 months. (As far as children in the womb hear voices or music, the voices and sounds are acoustically heavily distorted; therefore, some popular allegations by psychologists and the media about prenatal communication should be heard with due reservation).

The fluid in which we lived was the **water** of the amniotic cavity. Life under water and without atmospheric oxygen was possible because we got oxygen and disposed of carbon dioxide through the umbilical cord. With the increasing size of the fetus also the placenta grows so that efficient exchange is always possible.

The child is connected with the placenta by three main vessels. Two umbilical arteries leave the child at the site of the future navel, cross the amniotic cavity within the umbilical cord, and ramify in the placental villi into capillaries to facilitate exchange of substances. Enriched with oxygen the foetal blood returns to the embryo through the umbilical vein.

Thus, oxygen-loaded blood does not arrive from the lungs. Blood enriched with oxygen flows through the venous system into the undivided heart, is distributed in the body, collected and pumped back through the umbilical arteries to the villi of the placenta. The fetus has a circulatory system in the form of a single circle, like a fish, with the placental villi serving as gills. Because the lungs of the fetus do not yet function, but have to take over their vital task immediately after birth, the circulatory system must be prepared for a rapid conversion into a double-circle (body and lung) system. The solution to this serious technical problem is described in Chap. 17.

6.4.4 Pregnancy and Birth Are Attended with Great Dangers

In humans, the human blastocyst leaves the oviduct and settles itself into the uterine wall about one week after fertilization. The time point of nidation defines the start of pregnancy (gestation). The embryo is usually called **fetus** with beginning of the 10th week of pregnancy, that is 56 days after fertilization (about 10 days after Fig. 6.25d). At this age the heart beats already since a month. The embryo has attained a size of about 45 mm, is recognizable with ultrasound and begins to move its arms. However, prior to this stage 40–50 % of the implanted embryos perish as consequence of faulty development.

Most disappear undetected with some blood during menses. Presumably a considerable part of mal-development is caused by the immune system of the mother.

The child is ready to be born after 38 weeks (266 days) of pregnancy. Birth represents another serious risk.

It is a risk to the mother: when the placental villi are pulled out from the uterus, blood flows out from the lacunae and is lost. The mother might be in danger of bleeding to death. Will the uterus succeed in sealing the large openings in time?

It is also a risk to the baby: in an instant, the lungs must be blown up and blood pumped through the lungs. At this time point a short circuit in the circulation (ductus botalli, see Fig. 17.10) must be closed, so that the pulmonary circulation is completely separated from the great body circulation. If it fails to close a congenital heart condition results, called patent ductus arteriosus. In spite of all these risks, the benefits of being harboured within a protecting and nourishing mother is worth these risks. The evolutionary success of the placental mammals, and of man in particular, is a testament to the value of this developmental innovation.

6.5   Hormonal Relations Between Child and Mother

The intimate relationships between mother and child are biologically and mentally prepared and strengthened. The biological components that establish and stabilize these relationships include hormones.

6.5.1   Hormones of the Embryo Help to Prevent Its Physical Rejection by the Menses

The hormone-mediated establishment of a particular relationship between mother and child commences early with the implantation of the germ into the wall of the uterus – and this with

good reason. The embryo or fetus must insure that pregnancy is not prematurely terminated by the cyclic shedding of the endometrium, the nourishing mucosa of the uterus. The tiny embryo is heavily endangered to be washed out from the mother's body. In the immediate surroundings of the embryo the endometrium must locally persist, and the embryo achieves this through a hormonal signal.

In times without pregnancy the shedding of the endometrium happens when at the end of an ovarian cycle the yellow body (corpus luteum) in the ovary no longer releases the hormone **progesterone**. The yellow body terminates production of progesterone when it no longer urged to continue production by the pituitary hormone **Luteinizing Hormone LH**. Consequently, the shedding of the endometrium could be prevented when LH continues to be present in the blood of the mother. The embryo "knows" this and provides a LH substitute itself (Figs. 6.14 and 6.15). This is the **Human Chorionic Gonadotropin HCG**, that belongs to the LH protein family, having high sequence similarity with LH.

Amazingly, the tiny germ, the diameter of which initially amounts to merely 0.1 mm, can produce such amounts of HCG that it becomes effective in spite of the enormous dilution in the maternal blood. Moreover, traces of HCG occur in the urine. Apparently, the pores in the ultrafiltration membranes in the kidney are not narrow enough to retain all HCG molecules in the maternal blood. The traces of HCG in the urine are measured in a common **pregnancy test**.

In the further course of pregnancy the placenta of the fetus takes over the production of LH-equivalent hormones such as **Human Placental Lactogen HPL** (Fig. 6.15).

6.5.2   Further Ovulation Is Prevented by Foetal Steroid Hormones; This Is Simulated by the "Pill"

The placenta of the growing child secretes, besides HPL, also oestrogens (oestradiol) and gestagens (progesterone), that is steroid hormones that are generated in the ovary of women, and released into the blood, in rhythms with 28 day periods. If both hormones are present in the blood in high doses, the endometrium in the uterus persists and a further ovulation is prevented. It is exactly these two purposes that are accomplished in times of pregnancy for 9 months because the placenta of the child itself provides for a high production of oestrogens and gestagens (Fig. 6.15).

The "**contraceptive pill**" contains oestrogens and gestagens in chemically modified forms and various mixture ratios selected by the producing company. The pill mimics pregnancy therefore prevents ovulation.

> The pill contains artificial hormones such as ethinyloestradiol which have manifold side effects and are persistent in sewage plants. Residuals getting into drinking water are held accountable for the increasing infertility of men in industrial countries.

6.5.3   Maternal and Paternal Care Are Reinforced by Hormones

Let us first view parental behaviour excluding humans. In birds and mammals which give birth to dependent and even naked cubs (nestlings) such as bears and rodents, a cave must be looked for, or a nest prepared, in time. Surgical interventions in the hypophysis to disconnect potential sources of hormones, and postoperative injection of putative hormones in order to rescue observed defects, have drawn attention to prolactin. In female mammals **prolactin** controls maturation of the mammary glands. In addition, it reinforces disposition for nest-building activities and supports postnatal maternal behaviour. Mammals – it is mainly hamsters and rats which are subjected to experimentation – still showed maternal behaviour after removal of the pituitary and ovary, however not immediately

after birth but rather after being in contact with their cubs for several days. Mixtures of oestradiol, progesterone and prolactin induced instant acceptance of the cubs and appropriate maternal care.

Recently in hamsters strong indications were found that also the intensity of paternal care depends on the level of prolactin in the blood of the fathers. If this finding is verified in further studies, the long debated question of what function prolactin in males might have, would no longer remain unanswered. To what extent findings in mammals which give birth to nestlings can be applied to humans is open.

6.5.4   Pheromones Stabilize Relationships Within Families

Mother and newborn catch sight of each other frequently, and keep contact with sounds. Visual and acoustic contact in the early phases of life can take place also between other members of the family and supports socialization within the family. In addition the senses of smell play important roles. In mammals equipped with a well-developed vomeronasal organ, the second olfactory system besides the main olfactory bulb, pheromones build and reinforce family connections. Pheromone-based family smell acts *via* olfactory pathways that begin in the vomeronasal organ and leads to adjacent old olfactory centres in the brain. Both a child-specific and a mother-specific bouquet of odours are thought to aid in reinforcing mother-child relationships, in distinguishing one's own mother from foreign mothers, and one's own child from other children. Family-specific bouquets of odours are thought to later help the children to avoid family members with own nestmate-odour in partner choice. Perhaps similar events are of significance for building social connections in humans as well, though subconsciously and presumably not mediated by the vomeronasal organ because this is degenerated in humans, but mediated by the classic olfactory epithelium.

Box 6.1: Assisted Reproduction

If a couple is unable to conceive this may have several causes. It may be because the oviducts of the woman are irreparably sealed. Or the oviducts had to be removed in consequence of a dangerous tubal pregnancy, and a natural conception is no longer possible. On the other hand, often sperm of the man are produced in insufficient quantities (oligospermy or aspermy) or the sperm cells are almost immobile. In rare cases also a strong immune response of the women against sperm may cause infertility. In clinics three medical methods are employed to help such couples to have a child.

Artificial Insemination   This is the placing of sperm in the vagina or (more usually) in the uterus (intrauterine insemination, IUI) of the woman *via* a long narrow tube. The sperm may be fresh or frozen. This method is useful if the cervix is scarred, if the man has low sperm counts, or sperm with low motility, or if he has physical problems preventing normal ejaculation. Sperm from the woman's partner or from a donor may be used.

In-Vitro Fertilization (IVF)   Hormonal treatments are used to stimulate the maturation of many more eggs than normal. The gynaecologist surgically removes the mature egg cells directly from the ovary. The egg cells are then inseminated extracorporally, i. e. outside the body of the woman (in vitro). The embryos are cultured in vitro for 2–5 days before being transferred into the woman's uterus with a blunt cannula.

The sperm may come from the woman's partner, or a donor. However, in cases where the sperms of the partner are unable to fertilize the egg normally, for example if the sperms are immotile, a single sperm can be injected into the egg cell with a micropipette (**intracytoplasmic sperm injection, ICSI**)

under the dissecting microscope. In suboptimal cases the gynaecologist takes sperm from **anonymous donors**. In this case the husband can only be stepfather of the child.

In 1977 in vitro fertilization was reported (Louise Brown) for the first time – in those days a sensation. Today many thousands of human beings exist that were conceived in an assisted way.

(In animal breeding artificial insemination is the rule. In many countries only approved high performance cows and bulls are admitted for reproduction. A bull from a breeding line could have 50,000 offspring; the semen is sold to farmers who do not have to incur the costs of purchasing a high-quality bull.)

Surrogate Mothers   Some mothers are physically unable to bear the fetus up to term. In such cases a surrogate mother, who is prepared for pregnancy by hormonal pretreatment, can be of help. The extracorporally fertilized embryo of the natural mother is then transferred into the uterus of the surrogate mother. (Like the extracorporal insemination, also surrogate parenting had first been established in stockbreeding. Reproduction techniques developed for stockbreeding plays a pioneering role for the development of new techniques in human medicine.) Conflicts arise when the surrogate mother develops maternal emotions towards the child and wishes to retain it as hers own.

Selection of the Sex?   In stockbreeding techniques have been developed, and are being improved, to separate male-determining from female-determining sperm. A technique patented in the U.S.A. ("Microsort", The Genetics & IVF Institute, GIVF) is based on the DNA content that is minimally larger in sperm bearing an X chromosome than in sperm bearing the smaller Y chromosome. ▶

Box 6.1 (continued)

The sperm suspension is labelled by DNA-bound fluorescing molecules and subsequently segregated in the flow cytometer. Sperm cells bearing an X chromosome exhibit a minimally brighter fluorescence than sperm bearing the smaller Y. Such techniques are employed for human sperm in some clinics in the U.S.A and some other countries (but are forbidden in other countries such as Germany). Using such pre-selected sperm in artificial insemination the probability is increased that the child will be born with the sex wanted by the parents. The method, however, does not yield 100 % separation. The probability to give birth to a girl is 88 %, to give birth to a boy merely modest 73 %.

A high accuracy is provided by a method based on prenatal diagnostics (PID). Subsequent to artificial *in vitro* fertilization, from the eight-cell embryo one cell can be removed and examined whether it contains an X or a Y chromosome. Unwanted germs are discarded, a wanted one gets implanted. However, surgical invasion is not without risk, even in early embryos.

The initial incentive to develop methods for sex selection has been said to be the wish to help couples having sex-coupled hereditary diseases to produce a healthy child. For example, in women being conductors of haemophilia, sperm selection might ensure that only sperm with an X chromosome get a chance of fertilization. Even if in the arising embryo the maternal X would bear a mutated allele, the paternal 'healthy' allele could compensate the defect and the daughter would not manifest the disease. Soon the demand for sex selection in some countries led to the extension of the methods to healthy couples.

Whether parents will take responsibility for their children's sex is a moral issue and question of conscience that hardly will be answered and judged uniformly worldwide. Moreover, discarding an embryo merely because its sex is momentarily unwanted will meet strong reservation in many countries and cultures.

Breeding and Cloning of Supermen?
Journalists, philosophers or anxious laymen often discuss whether assisted breeding or cloning of human beings exhibiting extraordinary abilities might be possible or is about to be planned using targeted genetic manipulation. Some points for consideration: In the development of any organ, for example of the brain, thousands of genes are involved. The activities of these genes change in countless spatial and temporal patterns. How many musical pieces can be composed with only 88 piano keys? How many activity patterns could be generated with $2 \times 21,000$ protein-coding genes/alleles of the human genome (not to speak of the much more non-protein-coding DNA sequences)? Which combinations would give rise to an Einstein, which to a Mozart?

In contrast to forward looking planning of a targeted genetic makeup, cloning of an extraordinary personality would have a realistic perspective. However, to what personality would a Wolfgang Amadeus Mozart have grown if he were not born into the family of Leopold Mozart, which position in the history of science had Albert Einstein acquired if his interest were not guided to physics?

Box 6.2: Born from a Virgin?

In many cultures and tribes legends, religious narrations and holy scriptures tell of demigods, heroic tribal founders, kings, Wise Men, founders of a religion and **other worshipped personalities who were born to a virgin**. In many cultures ancient legends and holy scripts report of a human who arose from the womb of a virgin without participation of a man. These women are said to have conceived immediately from a deity (Buddha according to a legend from a white elephant). Of divine descent and born from a virgin were, for example, Gilgamesh, the mythic hero and king of the ancient Babylonian kingdom Uruk. Also some ancient Greek deities didn't have a father. In the Persian mythology the eschatological saviour Saoshyant is announced as a son born from a virginal maiden named Eredat-fedhri. She will enter a lake and sitting in the water, the girl, who has not associated with men, will receive "victorious knowledge." Her son, when born, will not know nourishment from his mother, his body will be sun-like, and the "royal glory" will be with him. The late Pharaohs were accorded the honour of being descendant of the sun god Amun-Re. Also Alexander the Great accredited himself to be descendant from Amun-Re (or from Zeus) and when the Roman emperor Augustus declared himself Pharaoh of Egypt it was soon rumoured that he was the son of the sun god and born from a virgin. The name of the emperor got the official label *DIVI filius* (Son of God). At the same time (40 BC) the Roman poet Virgil announced the early virginal delivery of a son with whom the last age would commence. The story best known among Christianity is offered by the Gospels according to Matthew and Luke. The virgin mother of Jesus is said to have conceived her son by the Holy Spirit.

In a few special cases a parthenogenetic development, known from some lower vertebrates such as certain reptiles, can commence even in mammals, when, for instance, an immature egg cell starts developing before meiosis, or when after the first meiotic division the first polar body re-fuses with the egg cell rendering it diploid and activating it. Such parthenogenetic embryos often give rise to teratomas; that is chaotically disorganized embryos exhibiting tumour-like behaviour (even in testes teratomas can arise from primordial germ cells).

Occasionally, parthenogenetic egg cells are released into the oviduct and start a seemingly normal development. This has been observed in mice where parthenogenetic development can be triggered artificially. However, in mice **parthenogenetic development**, whether it started from a haploid or from a diploid egg cell, **always came to a standstill** by day 10–11, halfway through the mouse gestation. A beating heart, somites and forelimbs were sometimes present but subsequently the embryos deteriorated and none survived to full term. Apparently, mouse development cannot be completed solely with mother-derived chromosomes. Mice need to have a father and this need is at present attributed to the phenomenon of parental genomic imprinting: Some few but essential genes are expressed only if inherited by the father, because the maternal allele is permanently silenced by heritable imprinting.

In 2004 sensational news was distributed among the international press that apparently contradict the above statement. In a Japanese laboratory for the first time parthenogenetic mice were born alive and one even became sexually mature (Kono T et al., NATURE 428: 860–864). These embryos, however, did not arise from single egg cells but from two fused haploid egg cells. By fusion biparental diploid egg cells arose, composed of differently imprinted cells. One of these two egg cells contained a mutated gene called H19 which normally becomes paternally imprinted. The authors conclude that paternal ▶

Box 6.2 (continued)

imprinting of this gene normally prevents parthenogenetic development up to term. Besides, the only surviving mouse was female because both egg cells contained an X chromosome.

But human beings are not mice, and the relevant experiments are not carried out, nor would they be ethically justified. Medical evidence suggests that even in the testes of male humans sometimes diploid primordial germ cells start embryogenesis; however, this results in tumour-like teratomas (as happens in women in cases of ectopic embryonic development). Only in the uterus of a female can embryonic development proceed to beings capable of surviving. Could it happen that a diploid oocyte dispenses with meiosis (or regains a diploid state by fusion with a polar body), and then is released from the ovary, transported through the oviduct into the uterus, activated by unspecified environmental stimuli, and undergoes complete embryonic development in the absence of any sperm? However, what would be the outcome? Non-fertilized mammalian eggs **could never contain a Y-chromosome**, since only a sperm cell can contribute a Y-chromosome to the child's chromosomal complement. **A parthenogenetically grown human being would be female and genetically a cloned copy of her mother**. This contradicts all the legends and scripts which tell of a virginal birth of a **man**. Therefore, if one believes in virginal birth of a human male, one also has to believe in a divine miracle. (In line with this proposition, the bible describes virginal conception of Jesus as a divine providence).

Box 6.3: Dangers to the Developing Embryo

With the decision to have a child the mother accepts great responsibility that she can only partially share with the father and society. Here we point out dangers for the unborn child.

Teratogenic Substances. The thalidomide tragedy conveyed the urgent problem of substances of hidden, long-term hazardousness to public attention in frightening way. In the years 1958/1959 more than 10,000 children in 46 countries were born with deformed, shortened or absent extremities. Their mothers had taken thalidomide-containing pills during the first 6 months of pregnancy for the prevention of nausea or 'morning sickness'. The sedative drug and effective painkiller was also used to aid sleep, and for coughs, colds, and headaches. The disaster once more showed, what was long-established knowledge in embryological studies, that embryos are affected by drugs orders of magnitude more sensitively than adults.

Teratogenic influences (substances, radiation exposure) evoke strong defects especially in distinct early phases of development (sensitive or perceptible phases) and in certain organs. The central nervous system is particularly endangered because its development extends over the entire prenatal period of time. Yet, the thalidomide disaster also shows the limited validity of animal testing. In rodents such effects were not observed.

Yearly, the chemical industry develops thousands of new substances. In most countries laws do not mandatorily prescribe tests with pregnant animals or animal embryos maintained in culture vessels. But even when such animal tests are performed and no malformations or physiological defects become obvious, harmful effects in humans, in particular in human brains, cannot be excluded because a human being is not a mouse.

Hormones and Hormone-Like Substances. Among the substances being under suspicion ▶

Box 6.3 (continued)

of causing defects are sex hormones and surrogates of such hormones, often used by sportswomen. Such substances may have harmful side effects in the woman herself (see Chap. 7) but may even more profoundly influence the somatic and psychological traits of the unborn. This applies especially to the male sexual hormone testosterone used to bulk up muscle-masses and increase sporting prowess. In men, on the other hand, environmental estrogens (derived from the pill) and xenoestrogens (present in many industrial products) can cause infertility (see also Box 7.3). Even vegetable "bioproducts" can contain substances being under suspicion of disturbing hormonal control, for example flavones from soy and cabbage.

Unfortunately even cautious behaviour cannot always prevent harm. Several years ago in women of about 20-years old genital carcinoma was diagnosed. The careful exploration of the life history of their families revealed that 20 years earlier their mothers had received the estrogen surrogate diethylstilbestrol (DES) to prevent threatened abortion. In the long term the child had to bear the consequences.

Possible harmful side effects of many years of taking the **contraceptive pill** are widely ignored in wide sections of the population, and do not meet much response in the media. The pill usually contains artificial hormones such as ethinyloestradiol which are not readily decomposed and can cause infertility in the long term ("post-pill-infertility"). Accompanying package leaflets specify as possible side effects: tiredness, increased frequency of migraine attacks and phases of depression: Some instruction leaflets point to the increased risk to suffer pulmonary embolism, stroke or heart attacks. In addition, some leaflets advise against intake in breast cancer because hormone-dependent tumours are positively "fed" by the pill (see above the case with ethinyloestradiol). Lastly, upon intake

during – a perhaps not yet recognized – pregnancy miscarriages can happen.

High risks of miscarriage refer to several possible causes: (1) Even before ovulation oocytes may become impaired. (2) The artificial hormones of the pill may penetrate the placental barrier and cause harm to the embryo in early phases of its development. (3) The hormones cause permanent alterations of the uterine epithelium.

Viruses, Infections, and Radiation. Among the teratogenic agents are a number of viruses, for example those causing rubella. Another danger derives from toxoplasmosis; this is a parasitic disease caused by the protozoan *Toxoplasma gondii*. The parasite is mainly hosted by cats but can also infect humans. Toxoplasmosis is risky during pregnancy because the parasite may infect the placenta and the unborn baby. For that reasons young women wishing to become pregnant are advised to undergo medical screenings targeted for rubella and toxoplasmosis (and in many countries the rubella vaccine is offered to all girls at 11–13 years).

Among well-known teratogenic agents are x-rays and ionizing radiation. A doctor will not expose the womb area of a young, let alone pregnant, women to x-rays except in urgent emergencies.

Alcohol and Other Drugs. Environmental chemicals, viruses and radiation can be avoided only to a limited extent. By contrast, the mother bears full responsibility when she smokes during pregnancy, drinks alcohol or takes harder drugs. When a child is born with an underdeveloped brain, reduced mental performance, reduced motor functions, heart defects, underweight, or psychological impairment, it is not always genes or insufficient nutrition that have caused these interferences.   ▶

Box 6.3 (continued)

Age of the Parents. The more often in our society women include children in their life planning not until 30, or even far beyond this age, or women do not successfully conceive earlier for natural reasons, the larger is the risk to give birth to a disabled child. Recent studies that compared children from young mothers and old fathers have shown that also a higher age of the father (beyond 55 years) is a risk factor. Plausible reasons for these risks can be proposed.

The Woman. An oocyte underwent a last replication of its DNA in preparation of the first meiotic division. Replication is followed by a comprehensive repair of faults. This happened when the respective woman still was a tiny embryo. In the long meantime of 35–40 years chromatin defects have accumulated; in addition, proteins of the cytoplasm and the cytoskeleton may have experienced (partial) denaturation. Many defects do not emerge because heavily aged oocytes fail to undergo maturation, can no longer be fertilized, or embyos arising from them die early and are flushed out in the menses without being noticed. Permanent defects remaining until term are certain aneuploidies: faults in the number of chromosomes (see below) that occurred during the meiotic divisions of the egg cell.

The Man. Occasional faulty distribution of chromosomes can occur in spermatogenesis as well, but most mitotic and all the final meiotic divisions do not take place until puberty, meaning that they start 15 years later than in females, and they proceed for long years to come. Beyond that natural selection may help. A sperm bearing the burden of a supernumerary chromosome has a reduced chance in winning the swimming contest compared to normal sperm. On the other hand, if a sperm is short of a chromosome the embryo will die early (with the exception of the 45/X0 karyotype; see Box 7.1).

There are some more reasons why humans in advanced age should be wondering if they

are elected to become parents. The child will soon have very old parents, and it may soon be orphaned.

Cytogenetic studies show that 50–60 % of all spontaneous abortions are chromosomally conspicuous whereby numerical anomalies prevail. Most often aneuploidies such as trisomy (three homologous chromosomes instead of two) or monosomy (one of two homologous chromosomes ▶

Age of the mother	Risk for Trisomy 21 (Down's-Syndrome)	
	Risk at birth	Risk in the 12th week of pregnancy
20	1 of 1,526	1 of 1,018
25	1 of 1,351	1 of 901
30	1 of 894	1 of 596
32	1 of 658	1 of 439
34	1 of 445	1 of 297
36	1 of 280	1 of 187
38	1 of 167	1 of 112
40	1 of 96	1 of 64
42	1 of 55	1 of 36
44	1 of 30	1 of 20

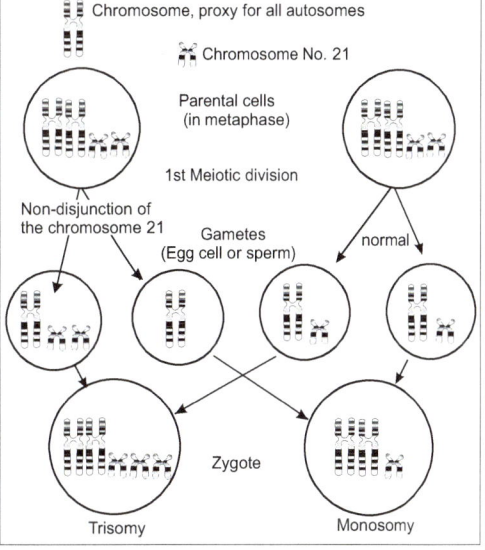

Fig. 6.16 Trisomy 21 caused by meiotic nondisjunction. The small chromosome symbolizes chromosome No 21

Box 6.3 (continued)

lacking) are diagnosed (Fig. 6.16). Trisomy has been observed for all chromosomes except No 1. The most cases of trisomy detected in abortive embryos concern chromosome 16 (in 32 % of cases) followed by trisomy 22 (14 %), trisomy 13, and trisomy 21.

Only trisomy 13, 18 and 21 as well as aneuploidy of sex chromosomes are tolerable up to term. But even in the most frequently identified trisomy 21 the number of live births is reduced by abortive births (see table). In all other numerical anomalies natural selection reduces live births to zero. Due to early negative selection during the period of pregnancy the statistical probability decreases that a child with a chromosomal anomaly will be born. Abortions due to chromosomal aberrations can be considered to be a regulative means of nature.

Trisomy 21 causes **Down's syndrome (DS)**. It became the best known trisomy because the majority of the affected children come into the world alive. To prevent misunderstandings it may be emphasized: **Down's syndrome is not a hereditary disease** in the sense that defective genes existed that could be transmitted to offspring. Rather, intact but supernumerary genes are present; the gene dose is disturbed. There are only few genes which make their presence felt when being active in overdose.

Down's syndrome encompasses serious disadvantages (delay in cognitive ability, particular set of facial characteristics, defective vision, often heart defects, premature aging associated with dementia), but can live normal productive lives if provided with support. Social problems may arise when the parents die before the child. In most countries laws permit assessing the risk of genetic defects or chromosomal abnormalities by prenatal diagnostics and, possibly, terminating pregnancy.

Box 6.4: Prenatal Diagnosis: Practical and Ethical Issues

According to common understanding responsibility for her growing child includes the willingness of the expectant mother to undergo prenatal tests at regular intervals and to have this confirmed in a maternity log (pregnancy record). The term pre-natal diagnostics PD encompasses all types of diagnostic screenings beyond routine gynaecologic care, performed to spot, or exclude, any morphological, biochemical, chromosomal, and genetic disorders before birth. PD aims at getting information whether in the case of elevated risks (for instance advanced age of the mother) a healthy child is to be expected or a child with disabilities.

Diagnostic prenatal testing can be by invasive or non-invasive techniques. Non-invasive investigations offered by gynaecologists use ultrasonic scanning and determination of relevant biochemical markers in the serum of the pregnant woman. Through repetition of the investigations at regular intervals it can be assessed, for example, whether the child grows well. It may be emphasized from the outset that in the vast majority of cases women, who are sure enough of their pregnancy because of repeated amenorrhoea, will give birth to a healthy child.

Certainly, 10–15 % of all pregnancies come to an end before term; they mostly terminate as early abortion before the end of the 12th gestation week. These numbers are valid only for pregnancies that were clinically recognized using ultrasound scanning or determination of hCG (human chorionic gonadotropin). Some studies arrive at the ▶

Box 6.4 (continued)

conclusion that up to 70 % of all fertilized eggs and early embryos respectively die soon after conception and are flushed out unrecognized in the next menses. Among the numerous known causes for abortion (anatomical, immunological, humoral causes or disturbed blood supply) those attributable to chromosomal aberrations are the most frequent ones.

Non-Invasive Methods

Ultrasound Scanning (Sonography). Ultrasonic waves are reflected by media to different degrees dependent on the density of the respective medium. From the pattern of reflected waves an image of the subject is reconstructed by computer-based image processing. Ultrasound scanning is considered to be entirely harmless for mother and child.

Some physicians and hospitals recommend a first scanning between the 11th and 13th week of gestation. As a rule, the child is depicted through the abdominal wall. Occasionally, however, the sight is not clear and a re-examination through the vagina is required.

As a rule, ultrasound scanning confirms foetal wellbeing. In addition, the gestational age is determined by measuring the vertex-rump length. Repeated scans are scheduled between the 19th to 22nd, and 29th to 32nd gestation weeks. Major physical defects can be excluded, and the position of the placenta, and with it the posture of the child, can be determined. With Doppler sonography movements of the child can be displayed. Occasionally also **fMRI** (*Functional Magnetic Resonance Imaging*) is done.

In reasonable doubts measurement of **nuchal translucency** with high resolution ultrasound scanning or fMRI can be added. Ultrasound transparency results from (normal) accumulation of fluid in the foetal back

of the neck the extension of which steadily increases between 10 and 14 gestation weeks. If the area of nuchal translucency is larger than normal elevated risk of Down's syndrome or some other failure (for example heart deficiency) is indicated.

Tests in the Blood Serum of the Pregnant Woman: The Triple (Quadruple) Test. Some tests carried out in the first trimester are based on enrichment of foetal cells or foetal DNA which circulate in maternal blood. Since foetal cells contain all the genetic information of the developing fetus they can be used to perform prenatal diagnosis. However, the tests are restricted to well-defined genetic risks. Testing can potentially identify foetal aneuploidy.

Measurement of foetal proteins in maternal serum is a part of standard prenatal screening for foetal aneuploidy and neural tube defects. A widely used screening paradigm is the triple or quadruple test carried out in the pregnancy week 15–16. Three to four parameters are measured in the maternal blood (1) the level of non-conjugated (= free) estradiol, (2) the concentration of PAPP-A (pregnancy-associated plasma protein A), (3) of an isoform of free human chorionic gonadotropin β-hCG, and (4) the content in alpha-fetoprotein AFP). High levels of alpha-fetoprotein most likely signify neural tube defects in the fetus; a low maternal serum alpha-fetoprotein, a low estradiol and high hCG levels are associated with increased risk of Down's syndrome. Yet, the tests are based on statistical correlations and do not permit reliable predictions. Combining these values with transparency measurement are said to increase the rate of correct prediction to up to 90 %.

In addition, checking distinct markers in the maternal blood can identify risks resulting from the state of health of the pregnant ▶

Box 6.4 (continued)

woman herself (for example: Diabetes mellitus, autoimmune diseases).

Physicians recommend:

- First trimester (trimenon, first three months): Ultrasound scanning in gestational week 11–13, plus PAPP-A and β-hCG test.
- Triple test in week 15–16.
- Second trimester: Ultrasound scanning in gestational week 21–23.

Several new tests are being developed. Many advanced clinics provide a useful guide to most of the currently available screening paradigms.

If non-invasive methods give cause for concern and the expectant mother wishes more certainty, medicine offers invasive methods of analysis.

Invasive Methods

Amniocentesis. Invasive methods involve probes or needles being trans-abdominally inserted into the amniotic cavity of the fetus, a procedure known as amniocentesis, in order to sample amniotic fluid and foetal cells. Amniocentesis can be performed from about 14 weeks gestation up to about 20 weeks. The samples are subjected to cytological, biochemical and molecular genetic screenings. Biochemical markers in the amniotic fluid can point to inborn errors of metabolism.

More cells can be removed with the

Chorion Villi Biopsy/Placenta Biopsy. Not only the embryo proper but also the chorionic villi and subsequent placental villi derive from the fertilized egg and contain the same genome (though multiplied) and, if any, the same chromosomal anomalies. Sampling of chorionic cells can be done earlier (between 9.5 and 12.5 weeks gestation) but is slightly more risky to the fetus compared to amniocentesis. The fetus itself is not impaired but the samples may be contaminated by maternal cells.

Foetoscopy and Chordocentesis. An endoscope and needle traverse the abdominal wall of the pregnant woman and the amniotic cavity and allow access to the fetus, the umbilical cord, and the placenta. Foetoscopy allows biopsy or medical interventions such as a laser occlusion of abnormal blood vessels. In chordocentesis foetal blood is sampled form the umbilical chord with a cannula. Compared to amniocentesis and chorioncentesis this method is associated with the highest risk for foetal loss.

Cytogenetic Examinations of Foetal Cells

At present investigations on numerical anomalies (aneuploidy diagnostics in the interphase of the cell cycles) have priority. A limited range of anomalies can rapidly be identified, or excluded, with high reliability in amniotic or chorionic interphase cells using **fluorescence-*in-situ*-hybridization (FISH)**. The FISH technique uses fluorescent probes that bind to only those chromosomes or parts of them with which they show a high degree of sequence complementarity. Fluorescence microscopy can be used to find out where the fluorescent probe is bound to one, two or three of the homologous chromosomes under investigation. Routinely probes for the chromosomes 13, 18, 21, X and Y are used. Any other chromosomes can be investigated as well if corresponding questions are posed.

Besides trisomies also other chromosomal anomalies can be identified, for example aneuploidies of the sex chromosomes (karyotypes 47/XXY or 47/XYY) and structural aberrations of the Y chromosome. Which consequences arise from such anomalies for the future child will be discussed in Box 7.1. Whether such anomalies justify premature termination of pregnancy appeals the conscience of the pregnant woman and her partner.   ▶

Box 6.4 (continued)

Hereditary Diseases, Unfavourable Prognosis for Life Expectancy and Life Quality

In some genetic conditions, for instance cystic fibrosis, an abnormality can only be detected if DNA is obtained from the fetus. Usually an invasive method is needed to do this.

The methods to analyse minute differences in the base-pair sequence of the DNA, differences scattered throughout the entire genome and comprising also non-protein-coding sequences, are becoming more and more sensitive (thanks to PCR-based techniques, see Box 12.1) and more and more comprehensive (thanks to new chip technologies). Thus it would in principle be possible to make predictions on the probability that the growing child sometimes will suffer from any one of the 30 known hereditary diseases, or will have a disposition for genetically fostered cancer diseases, for inferior aptitude and intelligence, for Alzheimer's or merely conspicuousness in behaviour. Whoever would expect herself/himself to bear the use of such a flood of information to take an irreversible decision of terminating pregnancy? Any such a decision based on a genome-wide analysis of DNA-sequences with insufficiently known functions and uncertain consequences would exceed the responsibility of the parents and the counselling physician as well.

We must not delude ourselves – **we all are bearers of hidden alleles which in adverse conditions favour outbreak of a disease, or adversely influence progression of an acquired disease, cause early onset of complaints of old age, or are responsible for underperformance in this or that circumstance of life.** Perhaps parents merely want a certain gender for their planned child. Responsible human geneticists and physicians recommend undergoing invasive investigations only, when non-invasive pre-investigations cause anxiety, when familial medical history report of frequent miscarriages, stillbirths or congenital diseases, or when the pregnant woman already has given birth to a child with a hereditary disease.

Pre-Natal Diagnosis Performed in Artificially Fertilized Pre-Implantation Embryos

Artificial in vitro-fertilization provides the possibility to remove a cell from the four- to eight-cell embryo und subject it to investigation. The early embryo can compensate for the loss and survives, as a rule, unscathed. The removed cell can be examined for chromosomal anomalies or for any grave and well diagnosable genetic defect. If such a defect was recognized, laws in several countries would allow discarding the embryo; while other countries (for instance Germany) forbid killing of any human embryo (whereas abortion is permitted under certain conditions; see Box Ch6E).

Evaluation of Pre-Natal Diagnostics PD

Reasons for PD. The possibilities of pre-natal investigations offered by modern techniques are far from being positively assessed in all societies and by all individuals. Providers of investigations give the following reasons for prenatal diagnostics:

- In most cases anxiety of the parents can bealleviated.
- PD enables timely medical or surgical treatment of a condition before or after birth (There are, however, only few and rare diseases which can already be treated in the womb.) Or arrangements can be made in time so that delivery can be scheduled in a hospital where the baby can receive appropriate care. This applies, for example, when Rhesus incompatibility requires blood ▶

Box 6.4 (continued)

exchange immediately upon birth. Down's Syndrome is associated with cardiac defects that may need intervention immediately upon birth as well.

- PD can give parents the chance to "prepare" psychologically, socially, financially, and medically for a baby with a health problem or disability. Perhaps they may join a self-help group and receive advice from experienced parents.

Reasons Against PD. Arguments against PD, often adduced by companies struggling for right to life for disabled persons, are the following:

- Invasive methods may induce abortion. The risk is estimated to be 0.1% to 5% dependent on the particular method used.
- Sometimes the diagnosis is wrong.
- An expected elevated risk of having a disabled child often results in abortion. Promises that pre-natal diagnosis (PND) guarantees a healthy planned child

can often be realized only by "taking prophylactic measures".

- In particular disabled people feel treated as rejects unworthy to live. One speaks of "selection" and "antedated euthanasia".
- Parents will be blamed if a disabled child is born into the world.
- In general, the multitudinous and advancing methods of prenatal diagnosis did not increase confidence and cheerful mood in expectant and potential parents, but led to uncertainty, helplessness, and anxiety. The burden of knowledge causes hitherto hardly known conflicts.

In conclusion it may be emphasized again that in most cases the unborn who survived the 12th week of gestation and makes its presence felt in the womb will come into the world as a healthy baby. It is exactly those responsible-minded mothers-to-be who avoid smoking, alcohol and similar avoidable impairments, who can be confidently EXPECTING.

6.6 Comparative Review: The Phylotypic Stage of Vertebrates, Common Versus Distinct Features, Aspects of Evolution

6.6.1 In Spite of All Differences in Their Early and Late Phase of Development Vertebrates Pass a Phylotypic Stage Common to the Whole Phylum

Observation of a Sharp Eye: Von Baer's Law. Carl Ernst von Baer (1791–1876, see Box 1.1) is considered the pioneer and father of comparative embryology. Von Baer discovered the notochord and the mammalian and human egg. He collected embryos of vertebrates, observed them with magnifying glasses and compared them in the stage when tetrapod limbs were

imminent to appear (Fig. 6.17). Von Baer observed remarkable resemblances; he observed that early vertebrate embryos exhibit common features: He observed:

"Die Embryonen der Säugethiere, Vögel, Eidechsen und Schlangen, und wahrscheinlich auch der Schildkröten, sind in früheren Zuständen einander ungemein ähnlich im Ganzen, ... Je weiter wir also in der Entwicklungsgeschichte der Wirbelthiere zurückgehen, desto ähnlicher finden wir die Embryonen im Ganzen und in den einzelnen Theilen."

"The embryos of mammals, birds, lizards and snakes, and probably also of tortoises, are, in early stages, tremendously similar on the whole...The more we go back in the history of development (meaning ontogeny, translated by WM) *the more similar we find the embryos on the whole and in their different parts."*

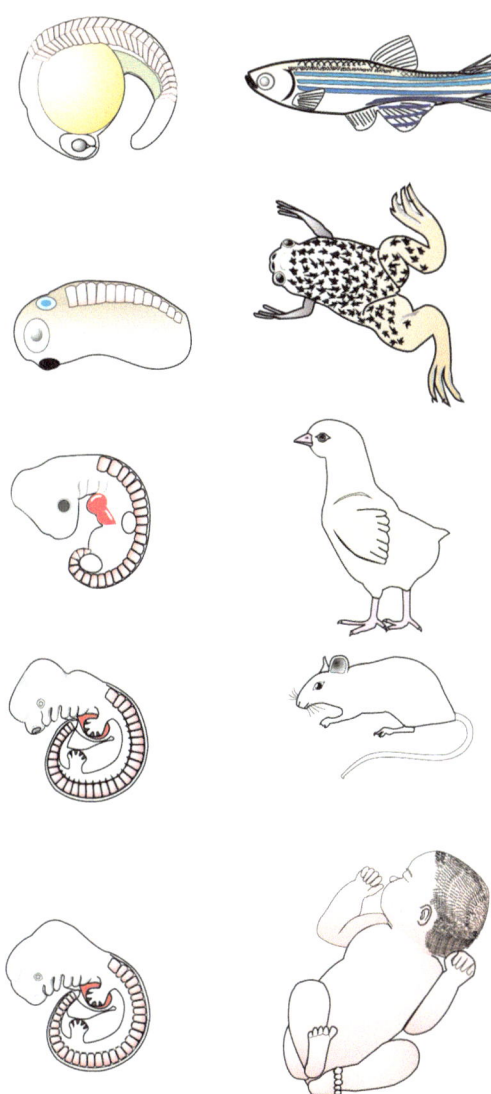

Fig. 6.17 Embryos and juvenile stages of various vertebrates illustrating an observation of Carl Ernst von Baer that a certain early embryonic stage (now termed the phylotypic stage) is very similar among the various vertebrates, while divergence occurs during later development. The phenomenon prompted Haeckel to formulate his 'biogenetic law'

New structures emerge in a temporal order which correlates with the position of the animals in the zoological kingdom: Features that characterize all vertebrates such as dorsal brain and spinal cord, notochord, somites and aortic arches are seen earlier in development than features distinguishing the various classes (limbs in tetrapods, feathers in birds, hair in mammals). Such observations caused Karl Ernst von Baer to point out, "*daß das Gemeinsame einer größeren Thiergruppe sich früher im Embryo bildet als das Besondere ..., bis endlich das Speziellste auftritt.*" "*that features common to a larger group of animals develop earlier than peculiarities ..., until eventually the most special feature occurs.*"

This statement is known as "**von Baer's law**". Thus, at an early stage following gastrulation and neurulation, embryos of fish, amphibians, reptiles, birds and mammals all look alike. As development progresses, their developmental paths diverge and embryos become recognizable as members of their class, their order, their family and finally their species (Fig. 6.17). However, as shown below Von Baer's law is valid only if the **phylotypic stage** is considered the starting point. Apparently, this phylotypic stage is a bottleneck stage which all vertebrates must pass through.

Haeckel's "Biogenetic Law". In 1880 Ernst Haeckel, a German zoologist, artist and enthusiastic admirer of Charles Darwin drafted his much disputed "**biogenetic law**". In its succinct and catchy form, the law states that "ontogeny recapitulates phylogeny" in a condensed and abbreviated way. Haeckel proposed that ontogeny – the individual development of an organism from the egg to the adult – is a shortened version of the specie's evolution. Multicellular organisms start their development from a unicellular egg and pass through a blastula and gastrula stage because their common ancestry first existed as an unicellular protist, followed by a *Volvox*-like colonial hollow sphere ('blastula') and by a (putative) beaker-shaped "gastraea".

In his argument Haeckel made several serious mistakes (that are still recapitulated worldwide in popular treatises on evolution). Not well versed in embryology he erroneously maintained that the human embryo would undergo gastrulation through invagination, thereby creating an archenteron.

Divergent Early Development. A **blastula** is known from amphibians. We got to know a hollow sphere designated blastula also in sea urchin development (Chap. 4). It is considered to be typical for animal development in general, and is observed also in non-vertebrate Chordata such as Ascidia and *Branchiostoma* (Amphioxus). Yet, such a **blastula is does not occur in mammals, although it seems so at a first glance**!

In fact, the blastocyst of the placental mammals does not undergo a classic gastrulation through invagination. Instead, the entire outer wall of the blastocyst is used to form a first and voluminous 'extraembryonic' organ, the **trophoblast**, by which the embryo corrodes its way into the wall of the maternal uterus. Next, an epithelial **amnion** enclosing an amnionic cavity is segregated from the residual "inner cell mass", and beneath it a **yolk sac** (enclosing a non-existent yolk) is formed.

The amnion and yolk sac are also found in reptiles and birds, but the amnion of the sauropsides is not formed until the embryo has established its basic body architecture (Fig. 6.18). In mammals, by contrast, an embryo proper is not yet present at the time the amniotic cavity appears.

Remarkably and confusingly, the procedure by which the amnion is formed is very different even among the various mammals (Fig. 6.19). In the shrew, considered by taxonomists to be a relatively archaic mammal, the amniotic cavity is formed by amniotic folds in much the same way as in reptiles and birds. But even in the shrew an embryo proper does not yet exist at the time the amnion is formed. The embryo will arise from a small area in the bottom of the amniotic cavity (epiblast) supplemented by a corresponding area in the underlying roof of the yolk sac (hypoblast). This double-layered area develops into a blastodisc. The following processes take place in much the same way as they do in and below the blastodisc of reptiles and birds: as an expression of the beginning of gastrulation, a furrow-like depression in the epiblast indicates the downward streaming of those cells which will give rise to the endoderm and mesoderm. The furrow is known as the **primitive**

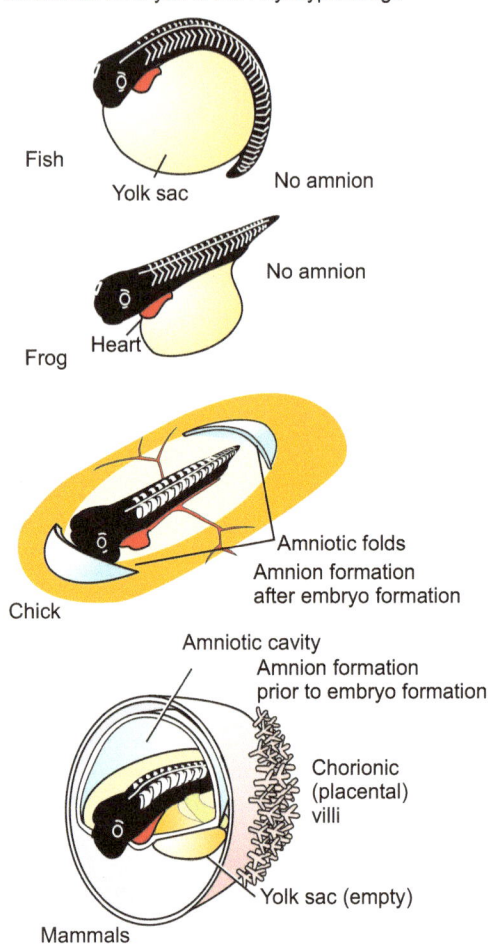

Vertebrate Embryos in the Phylotypic Stage

Fish — Yolk sac — No amnion

Frog — Heart — No amnion

Chick — Amniotic folds — Amnion formation after embryo formation

Mammals — Amniotic cavity — Amnion formation prior to embryo formation — Chorionic (placental) villi — Yolk sac (empty)

Fig. 6.18 Embryos of non-amniotes (fish and frog) and of amniotes (chick and mammals). Note that in chick amniotic folds do not emerge until the embryo has reached the phylotypic stage, while in mammals embryonic development commences within a previously formed amniotic cavity

groove and interpreted as the homologue to the blastopore in the amphibian gastrula. A Hensen's node develops at the anterior end of the groove and in front of the node a **primitive streak** emerges.

To Summarize: In mammals it was very early development that underwent a considerable alteration through evolution, adapting to the protective and nutritive environment of the mother. In mammals, including man, there are structures such as amnion, yolk sac, allantois,

Fig. 6.19 Amniotes: Differing formation of the amniotic cavities and other extraembryonic organs in different land vertebrates

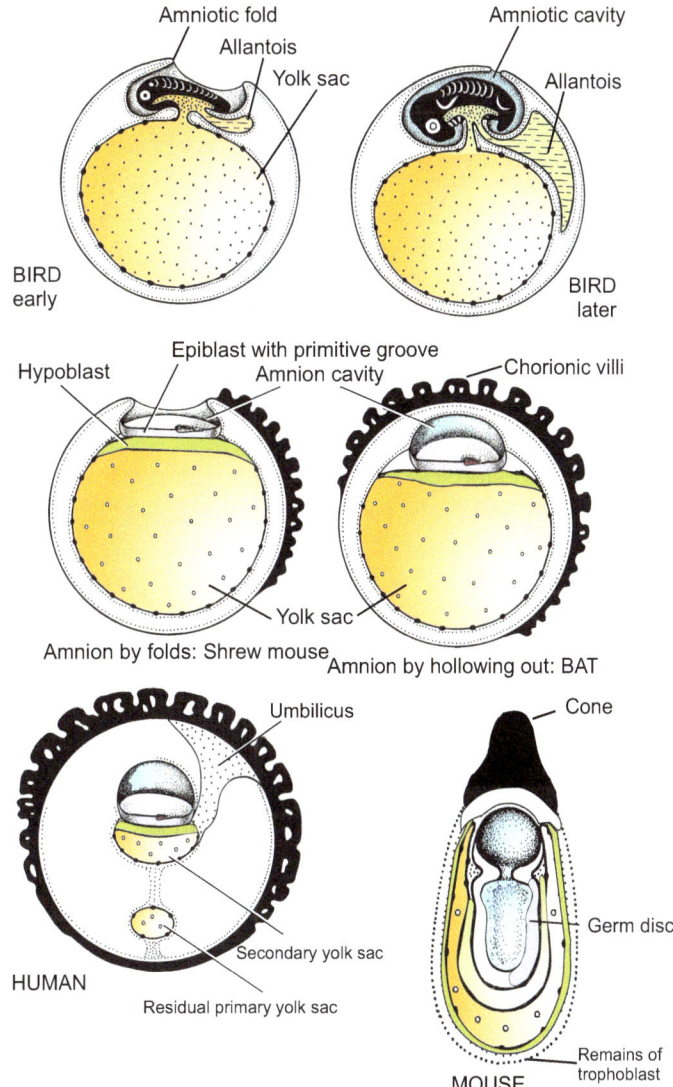

blastodisc, or primitive streak, which are novel compared to amphibians but ancient if one looks at reptiles. The formation of an empty yolk sac and a non-functional allantois in the mammalian embryo can only be understood, if the development of reptiles is inserted between amphibian and mammalian embryo development.

If these observations are taken into account, Haeckel's biogenetic law merits acknowledgement as it points to the evolutionary

context of developmental biology, but it must be corrected: each organism's ontogeny does not repeat phylogeny of a species but rather previous ontogenies. In each generation all species recapitulate their own ontogeny which, compared to the ontogeny of related species, is more or less modified.

The Phylotypic Stage. All vertebrates pass through a highly conserved common stage that displays a uniform basic body architecture characteristic of all vertebrates. Therefore, the biogenetic law is valid if it is modified by stating that **all vertebrates including humans recapitulate certain embryonic traits of their ancestors – in particular, a common phylotypic stage**. All developmental pathways, although initially so different in amphibians, reptiles, birds and mammals, converge in a basic stage, exhibiting a **neural tube, notochord, somites, lateral plates, foregut with gill pouches and arches**, and **a ventral heart** (Figs. 5.6, 6.8, 6.17, and 6.18). In vertebrates the vertebral column will be built up subsequently around the notochord.

However, a close look at the embryos will always detect differences. For instances, in the tail bud stages the lengths of the embryos differ up to tenfold. Careful examinations will find subtle differences in almost all anatomical structures. On the genetic and molecular level one will find individual differences at all stages, because even homologue genes display individual differences in parts of their base pair sequence (with the exception of monozygotic, identical twins).

Thus, the phylotypic stage is a conceptual model that emphasizes common over disjunctive traits. The researcher applying molecular biological methods identifies more and more position- and time-specifically expressed genes that emphasize the common basic organization of all vertebrates.

6.6.2 In the Phylotypic Stage Many Processes of Induction Take Place, and It Reflects Constraints in the Mode by Which a Complex Feature Can Be Organized

Why a uniform phylotypic stage in spite of all the divergence in the earlier and later stages of development? In particular in humans most phylotypic anatomical structures, and the preceding patterns of gene expression, are transient. For instance, the somites disintegrate in groups of cells the future development of which is differently canalized. The notochord gives way to the vertebral column.

Recent research finds plausible explanations for this conservatism: **Transitional structures such as the notochord are needed for the organization of further development**. For instance, in the midline mesoderm being about to form the notochord the gene *Brachyury* is expressed. When both alleles of this gene are defective no notochord and no useful vertebral column are formed; more defects concern development in other organs. The notochord or its precursor, the mesodermal midline cells, release signals which initiate the segregation and detachment of the sclerotome from the somite, and to attract the migrating sclerotome cells towards their goal (Fig. 10.6). The signals emanating from the notochord, represented by SONIC HEDGEHOG, instruct the sclerotome cells to build the vertebrae around the notochord, and they induce the differentiation of motor neurons in the spinal cord (see Figs. 5.20 and 16.11).

Without much probability of error one can assert that evolutionarily old pathways are not chosen by the embryo for love of tradition but on serious physiological grounds. But even when an organizing function can be assigned to transitional structures, there remains the question: Why this way and not another way? Apparently, a difficult undertaking such as the organization of development cannot easily and arbitrarily be changed and abbreviated. Solutions found in millions of years of evolution cannot easily be altered or replaced. One speaks of "*constraints*"– restricted possibilities. An illuminating example: birds and bats could develop wings only at the expense of forelegs. Three pairs of extremities remained reserved to the phenotypically similar mythic dragons.

6.6.3 In the Phylotypic Period Many Organ Rudiments Are Formed or Prepared

In the phylotypic stage, the following structures are formed or prepared:

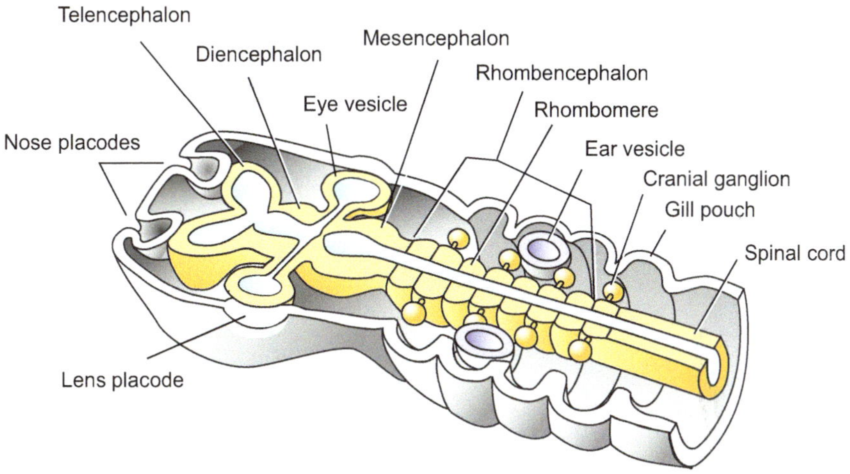

Derivatives of Ectoderm and Neural Plate

Fig. 6.20 Ectodermal and neural plate derivatives in the anterior body showing the segmental structure of the hindbrain and pharyngeal region. In the hindbrain (rhombencephalon, medulla oblongata) seven segmental constrictions mark off 7–8 repeating units called rhombomeres or neuromeres. Branches originating from the first even-numbered rhombomeres (r2, r4, r6) and the associated ganglia give rise to cranial (trigeminal, facial, vestibuloacoustic) nerves. The subsequent posterior branches and ganglia are known as the glossopharyngeal nerve and vagus; the subsequent nerves are called spinal nerves (dorsal and ventral roots, dorsal root ganglia) in that part of the central nervous system known as the spinal cord. The segmental organization of the medulla oblongata corresponds to the segmental organization of the pharyngeal pouches and organs

- The **nervous system** is in the form of a closed tube, broadened anteriorly where the brain will develop. The parts of the brain are beginning to be differentiated by thickenings and constrictions of the neural tube into three parts: **prosencephalon, mesencephalon, metencephalon (= rhombencephalon)**. The metencephalon is compartmentalized into **rhombomeres** and accompanied by the metameric sequence of **cranial- and spinal ganglia**. Posteriorly the spinal cord is accompanied by the sympathetic chain with its segmentally organized ganglia (Fig. 6.20).
- **Main sensory organs associated with the brain. Eye vesicles** bulge laterally from the forebrain/midbrain boundary. The epidermal epithelium has formed pairs of plate-shaped thickenings, called **placodes**. These become invaginated to form the **eye lens** and sensory organs: the **nasal placodes** develop into the olfactory sacs, the **auditory or otic placodes** into the otic vesicles from which the inner ear with the labyrinth and cochlea will arise.

- **Axial organs**. The **notochord** stretches underneath the neural tube from the midbrain level to the posterior end of the body. To each side of the notochord the mesoderm is subdivided into the modules of the **somites**.

  Most conspicuous, however, is the structure of the pharyngeal region: In acknowledgement of its evolutionary past even the human embryo displays in its foregut structures known as
- **Pharyngeal (branchial) 'gill' pouches** (Figs. 6.20, 6.21, and 6.22). These pouches open in fishes and tadpoles to become gill clefts; the bridges between the clefts form gills. In land vertebrates the pouches do not open, or they are secondarily closed (the first cleft by the ear drum).
- **Cartilaginous 'gill' arches** (visceral arches). The cartilaginous skeletal elements (gill braces) including the **primary jaw** apparatus (derived from the first cartilaginous branchial arch) are referred to as **visceral skeleton**.
- **Circulatory system**. Beneath the foregut, the cardiac tubes fuse to form the ventrally

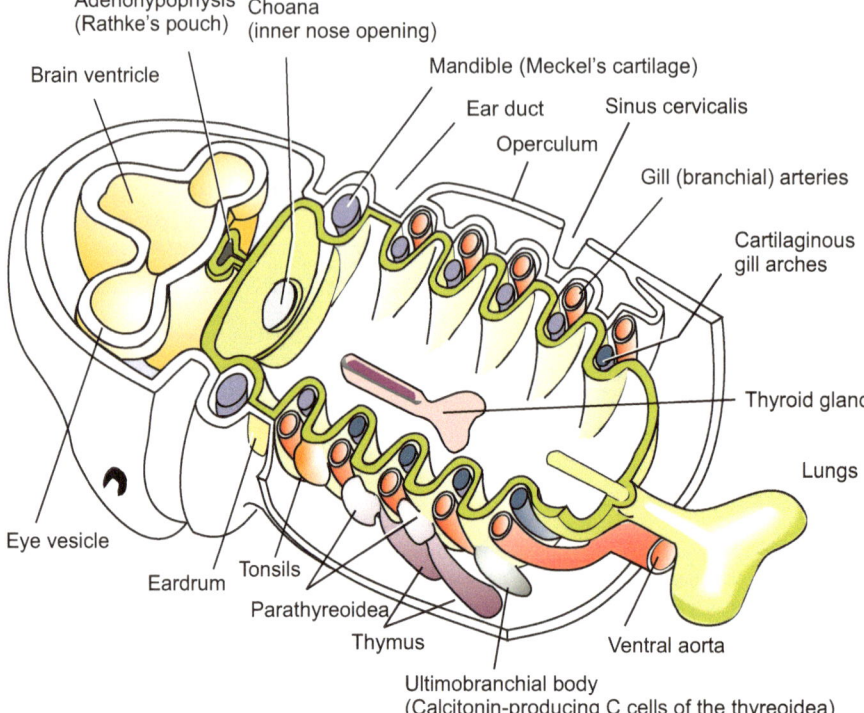

Fig. 6.21 Phylotypic stage of vertebrates, showing the derivation of the branchiogenic organs in the anterior body (head and pharyngeal region). The lymphatic organs (tonsils and thymus) and the hormone glands (parathyroids and ultimobranchial bodies or C cells) derive from the endodermal walls of the pharyngeal pouches. The thyroid arises from the floor of the pharynx and is considered homologous to the hypobranchial groove of the tunicates, Branchiostoma and cyclostomes. Also, the trachea and lungs are evaginations of the foregut and are thought to have evolved from the last pair of pharyngeal pouches. The operculum that in fishes and amphibians covers the gills is only a transient structure in mammals

located **heart. 'Gill' arteries**, mostly four pairs, originally presumably six, cross the gill pouches.

• **Hormonal system**: In fish and in all higher vertebrates the **adenohypophysis** (anterior pituitary gland) develops from an area in the roof of the pharynx (known as Rathke's pouch), while the **thyroid gland** develops from a pocket in the floor of the pharynx.

of *Gallus domesticus*, or an individual of the house mouse *Musculus musculus*. In order to attain differing traits, subsequent to the phylotypic stage the developmental paths of the various classes, orders, families, species, and individuals diverge more and more. Structures characteristic of the phylotypic stage experience considerable transformations.

6.7   Conservative Paths Versus Novelties

The vertebrate embryo does not develop to an abstract scheme but to a unique zebrafish of the species *Danio rerio*, to an individual clawed frog of *Xenopus laevis*, to an individual domestic fowl

6.7.1   Did the Skull, as Goethe Assumed, Derive from Vertebrae? Embryonic Evidence Says: "Partially Yes"

With lively concern and his own contributions to the discussion J W von Goethe participated in a scholarly dispute, that flared up among the

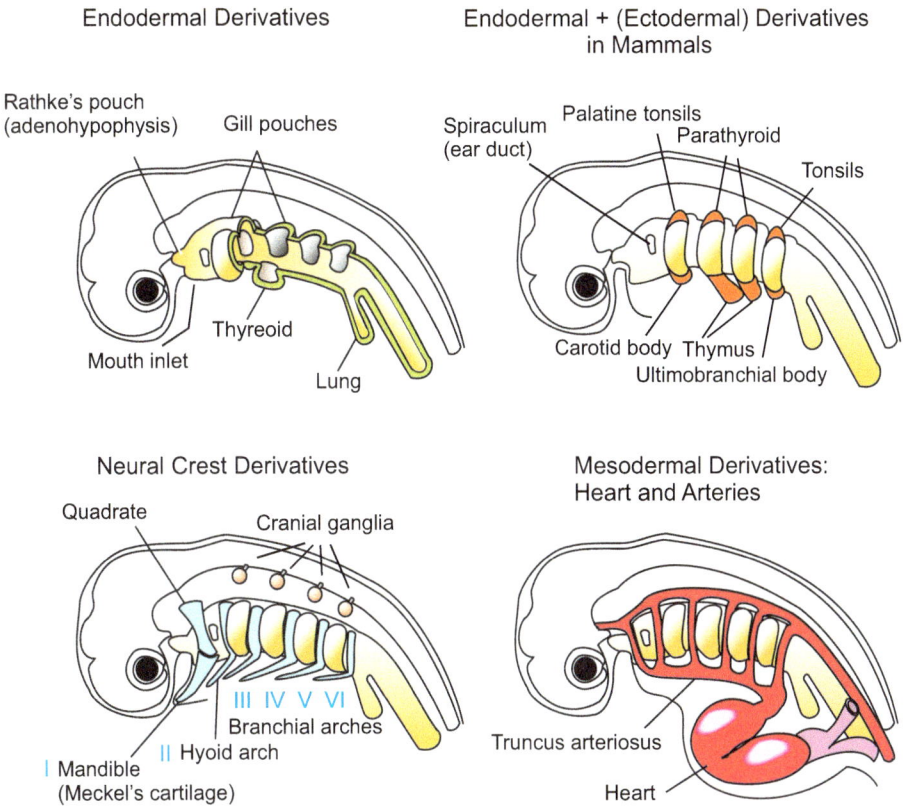

Endodermal Derivatives

Rathke's pouch
(adenohypophysis)   Gill pouches

Thyreoid

Mouth inlet

Lung

**Endodermal + (Ectodermal) Derivatives
in Mammals**

Spiraculum   Palatine tonsils
(ear duct)   Parathyroid

Tonsils

Carotid body Thymus
Ultimobranchial body

Neural Crest Derivatives

Quadrate

Cranial ganglia

III IV V VI
Branchial arches
II Hyoid arch
I Mandible
(Meckel's cartilage)

**Mesodermal Derivatives:
Heart and Arteries**

Truncus arteriosus

Heart

Fig. 6.22 Derivation of organs in the head and pharyn-geal region. The endodermal walls of the pharyngeal pouches give rise to lymphatic organs and hormone glands. The cranial and spinal ganglia as well as the cartilaginous elements of the primary jaw and of the gill arches are derivatives of the neural crest. Heart and aortic organs are of mesodermal origin

anatomists of the eighteenth and nineteenth century: Is the mammalian skull composed of segments (vertebrae)? Present-day studies of the embryonic development give answers that would delight so many scholars of those days. The occipital region at least has a segmental origin.

We see this, and recognize more unexpected evidence, viewing two transient structures and patterns of gene expression:

- **Rhombomeres and *Hox* genes**. In the region of the embryonic rhombencephalon we can catch clear sight of segmented packets, rhombomeres, just as in the adjoining spinal region. In all these genes of the *Hox* and *Pax* groups are expressed (though in somewhat strange combinations, see Fig. 16.3).

- **Vertebral bodies and skull**. In the anterior and mid-region of the head no convincing evidence for a former segmental organization exists. Neither the boundaries between different parts of the brain, nor the pattern of gene expression (see Figs. 16.13 and 16.14), nor the exit sites of the cerebral nerves reflect an originally segmental organization. Attempts to interpret the paired cartilaginous braces, called trabeculae and parachordalia, as former vertebral bodies would not remain without contradiction. By contrast the occipital region of the skull derives from somites even today.

The cellular material for the formation of the vertebral bodies is supplied by the sclerotome compartments of the somites (Fig. 6.23). The

Differentiation of Vertebrae in Amniotes

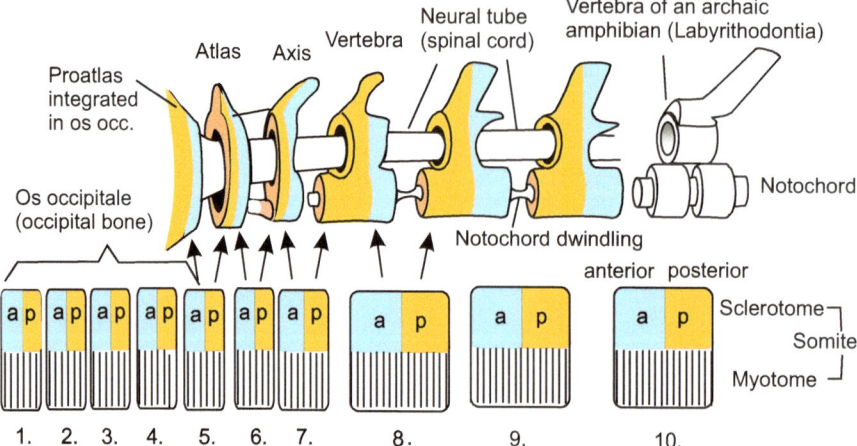

Fig. 6.23 Derivation of the occipital region of the skull, of atlas, axis and the following vertebral bodies from the sclerotome parts of the somites. The sclerotome is divided into two parts. Each definitive structure gets composed of a posterior part of a distinct sclerotome and an anterior part of the next sclerotome

sclerotomes for their part are subdivided into two compartments; an anterior (cranial) and a posterior (caudal) compartment. Each vertebral body arises from the posterior compartment of one distinct sclerotome and the anterior compartment of the adjacent posterior sclerotome.

Yet, the first four somites/sclerotomes do not form distinct vertebral bodies but fuse giving rise to the hind bone (os occipitale) of the skull. The following three vertebral bodies undergo a reshaping enabling increased movability of the head. The anterior part of the **proatlas** (5th/6th somite) forms the paired hunched joint (condylus) at the posterior skull, against which the **atlas** (6th/7th somite) can be flexed; thus nodding movements of the head are rendered possible. The fish must tilt its whole body axis, if it wishes to pick up something from the floor (what is not difficult to do in water but arduous on land). In mammals the posterior part of the proatlas joins the **axis** (7th/8th somite) and forms the dental process (dens axis). In mammals the atlas can rotate around the cone of the dental process: We can turn our heads to and fro, answering in the negative.

6.7.2   The Ear Ossicles Are Often Presented as an Instance of Evolutionary Transformations That Can Be Reconstructed in Embryonic Development

Even the human embryo is said to have a **primary jaw joint** (Fig. 6.24) as do not only the embryos but also the adults of fish, amphibians and the non-mammal-like reptiles. Yet, mammals are not satisfied with the primary jaw. They develop a second one, and use the first for a very different task: to optimize the transmission of sound from the ear drum to the inner ear. When the dental and squamosal bones connect to one another in the mammalian embryo and form the secondary jaw joint, the elements of the primary jaw are freed to acquire a new function, becoming the **auditory ossicles** (Fig. 6.24).

Let's have a closer look at the ontogenetic as well as phylogenetic events:

Neural crest cells of the head region colonize spaces ahead of and between the gill clefts, aggregate and form cartilaginous two-(or more-) part braces (Fig. 6.22) to support the gills in fish and

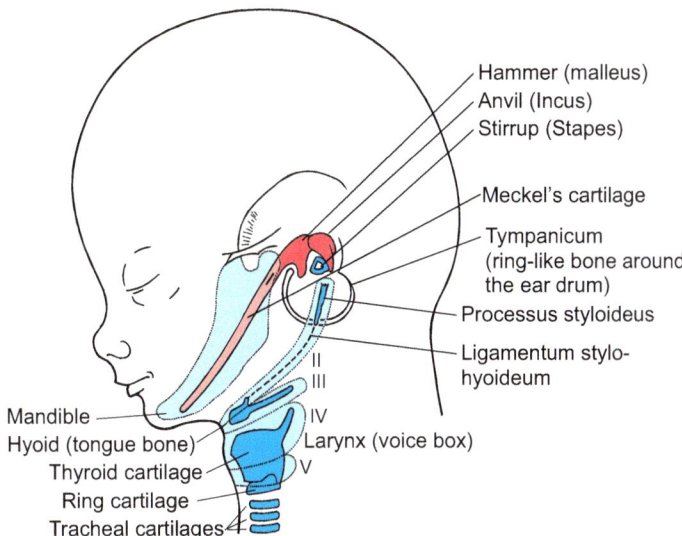

Fig. 6.24 Transformation of the visceral skeletal elements in the pharyngeal region. When the dentale bone contacts the squamosal bones of the skull and, thus, the secondary jaw joints are formed, the first arch – called mandibular arch (containing the Meckel's cartilage) – is modified and ossified, and gives rise to the auditory ossicles hammer (malleus) and anvil (incus). In lower vertebrates (fishes, amphibians, reptiles) these two elements bent around the mouth, forming the primary lower and upper jaws and jaw joint. The third auditory ossicle, the stirrup (stapes) is a remnant of the second or hyoid arch. The dorsal element, known as hyomandibular cartilage or bone in fishes, once anchored the jaws to the braincase, and braced against the otic capsule. In amphibians (or even in ancestral fishes) the hyomandibular bone was used to transmit sound waves to the ear region (and is known as the columella in amphibians, reptiles and birds). In mammals, the hyomandibular element or columella, respectively, is used as the stapes. The malleus will attach to the eardrum, transmit the sound to the incus, which in turn transmits the sound to the stapes and the stapes to the oval window of the inner ear. The ventral elements of the hyoid arch and of the subsequent posterior arches are used to form the hyoid bones (2 + 3), and the thyroid (4 + 5) and cricoid cartilage (6) of the larynx

tadpoles. The most anterior 'gill' braces undergo dramatic transformations. The first four pairs deserve particular interest:

- Further neural crest cells from the region of the rhombomeres r1 and r2, form the first cartilaginous arch, the **mandibular arch** consisting of the **quadrate** (upper jaw) and **Meckels cartilage** (lower jaw). These constitute the **primary jaw joint** that persists in fish and amphians. Yet, in mammals, the elements of the primary jaw are transformed into two of the **auditory ossicles** (Fig. 6.24), into **hammer** (malleus) and **anvil** (incus).

- Another cohort of neural crest cells arriving from the region of the rhombomer r1 encircles the opening of the first gill canal, the one lying behind the mandibular arch, and forms the cartilaginous, later ossified **tympanal ring** in which a closing membrane, known as **ear drum**, is inserted and stretched. The closed canal is known as the **auditory cavity** (a widened part of the canal) with linked Eustachian tube.

- Neural crest cells arriving from the r4 region form the second branchial arch, the **hyoid arch**. The dorsal element of this arch, known as **hyomandibulare**, serves in fish to anchor the primary jaw onto the skull. In amphibians the hyomandibulare is transformed into the **columella**, that is a rod-shaped bone which connects the outer eardrum (which encloses the first gill cleft) with the inner eardrum and transmits sound to the hearing organ. In mammals the hyomandibulare/columella becomes the **stirrup** (stapes).

- The following braces of the branchial arches become, in humans, **hyoid bone** (tongue bone) and **larynx** (**voice box**) (Fig. 6.24).

In mammals/humans, **hammer, anvil and stirrup** enter the tympanic cavity, a derivative of the first gill cleft, to transmit sound from the outer eardrum (that is the just mentioned membrane sealing the first gill cleft) to the **fenestra ovale** (oval window, inner eardrum), and eventually to the cochlea. In contrast to the stiff columella, the three auditory ossicles form flexible joints enabling reducing the amplitude of sound waves while increasing the force so that airborne sound is more effectively transmitted to the incompressible fluid of the inner ear.

Transformations of such kinds were in land vertebrates possible because a gill apparatus no longer had to be constructed that required a supporting bracket.

6.7.3 Extensive Transformations of the Branchial Epithelia Give Rise to Further 'Branchiogenic Organs', to Hormonal Glands and Lymphatic Organs

The pharyngeal region offers some more fascinating recapitulations of old ontogenetic pathways. This applies to the endodermal derivatives (Figs. 6.20 and 6.22) and to the system of blood circulation in the embryo (Fig. 6.22). Of course, in other regions of the body interesting transformations can also be seen. As an example, we outline the development of some hormonal glands:

The phylogenetic derivation of the hormonal glands is very diverse and sometimes curious. This is also apparent in ontogeny. In land vertebrates, several hormonal glands arise from gill pouches or other parts of the pharyngeal foregut, tissues which are no longer needed to construct functional gills and can adopt new functions or optimize old secondary functions because requirements of respiration no longer must be taken into account and compromises are no longer necessary.

In fish and in all higher vertebrates the following structures arise from the branchial region of the body:

- **The hypophysis**: The **adenohypophysis (anterior pituitary gland)** develops from an area in the roof of the pharynx known as **Rathke's pouch**, while the **neurohypophysis** derives from a process of the diencephalon called infundibulum (see Fig. 16.5).
- **The thyroid gland** develops from a pocket in the floor of the pharynx (Fig. 6.22). The developmental progeny of the pocket, and with it of the thyroid gland, is the **hypobranchial groove** (endostyle) in the branchial pharynx of the acranial chordates (ascidians and Amphioxus) and in the larvae of the cyclostomes. Already these precursors of the thyroid gland produce the hormone **thyroxin**.
- **The parathyroid gland and the C-cells** within the thyroid gland (hormone: calcitonin), as well as their homologous counterparts in the non-mammalian vertebrates, the **ultimobranchial corpuscles**, develop from cells that detach from the walls of the pharyngeal pouches; in humans they associate with the thyroid and eventually become enclosed by it.
- **The thymus and tonsils**. In addition to hormonal glands, the pharyngeal pouches give rise to organs of the **lymphatic system**, such as **tonsils** and **thymus**.

The thyroid, parathyroid and thymus are shifted downward and backward until they come to lie in the neck or thorax region.

Very different developmental pathways create the **adrenal gland**: the **central medulla** is formed by aggregating sympathetic cells that once emigrated from the neural crest together with the precursors of the sympathetic nervous system. It will produce the hormone adrenaline (epinephrine). By contrast, the **adrenal cortex**, that will produce steroid hormones such as cortisol and corticosterone, is formed by descendants of the mesodermal lateral plates.

6.7.4 The Blood Vessels of the Branchial Region Also Experience Extensive Transformations

These transformations occur in the context of the switch from gill breathing to lung breathing. In times of urgent need, when the body of stagnant water was about to dry up, in some fishes a ventral sac on the oesophagus came into use enabling the fish to engulf air. Lung fishes developed. Since this air sac initially was merely an additional breath organ, it sufficed to make a vascular shunt and fuel it with blood diverted from the 6th gill artery. For mammals such a shunt was not enough; they have to drive all their blood through the lungs to ensure complete unloading of carbon dioxide and loading with oxygen. On the other hand, land vertebrates no longer need vessels to provide for gills. Therefore, some vessels can be skipped, others are reconstructed (see Figs. 17.8–17.10).

6.7.5 After the Phylotypic Stage the Temporal Sequence of Developmental Events Largely Reflects the Sequence of Evolutionary Changes

Errors in Haeckel's treatises (and aversions to his personality) caused a wide-spread refusal to acknowledge the credits of his 'biogenetic law'. Actually, ontogeny reflects phylogeny in many aspects. After all, the finished body is generally considered the result of evolution.

Some data concerning human development may point out the apparent parallelism between ontogeny and phylogeny (Figs. 6.7, 6.8, 6.9, and 6.25):

- By day 20 the neural folds begin to close; the first three somites of the neck region emerge, ear placodes are formed, the first gill slits appear. Between the gill clefts neural crest cells form the mandibular and hyoid arch, but jaws are not yet formed. Beneath the pharyngeal intestine two parallel running tubes

fuse forming the unpaired heart tube. The embryo approaches the phylotypic stage.

- Between day 23 and 26 an embryonic circulation is established with a construction like a fish would develop. The brain is composed of several lobes separated by constrictions, the largest and most posterior of which is the **rhombencephalon**, the next the **mesencephalon** and the most anterior the **prosencephalon**. The heart lies within the cup-shaped posterior end of the branchial basket and consists of **one atrium** and **one ventricle**. A hypophyseal pouch approaches the midbrain; the **thyroid gland** sinks down. The intestinal tract has **gill pouches**, and continues to the **cloaca** into which also the ducts of the **pronephric kidney** lead. The notochord is present but no other skeletal elements: **jaws are absent**. The putative fish embryo apparently will become a lamprey or slime eel (order Agnatha – jawless fishes).

- By day 27 the 'fish' apparently prepares to immigrate into shallow fresh water areas with often low oxygen levels and seasonal droughts. A lung develops. The kidney transforms into a mesonephros. In addition, jaws are forming and a tongue. Apparently, a lung-fish or an amphibian species is about to develop.

- By day 34 we see a **pectoral** and a **pelvic limb** developing a **palm**. The pelvic fin is soon going to become a tri-part leg. The intestine is associated by a pancreas which is needed to digest 'dry' terrestrial food. By day 38 the **heart is divided into two auricles** but still has **one ventricle**. Its position is shifted posteriorly. Apparently, an amphibian is about to conquer land. Its travel on land is facilitated by some other features: The nasal sacs are shifted from a lateral to a frontal position, the eyes from a dorsal to a lateral position (ultimately, in the human fetus the eyes will be shifted anteriorly).

- By day 42 the **lung** is fully developed; the **trachea is separated from the oesophagus**. The gill slits disappear with the exception of the **first slit** which will become the **middle**

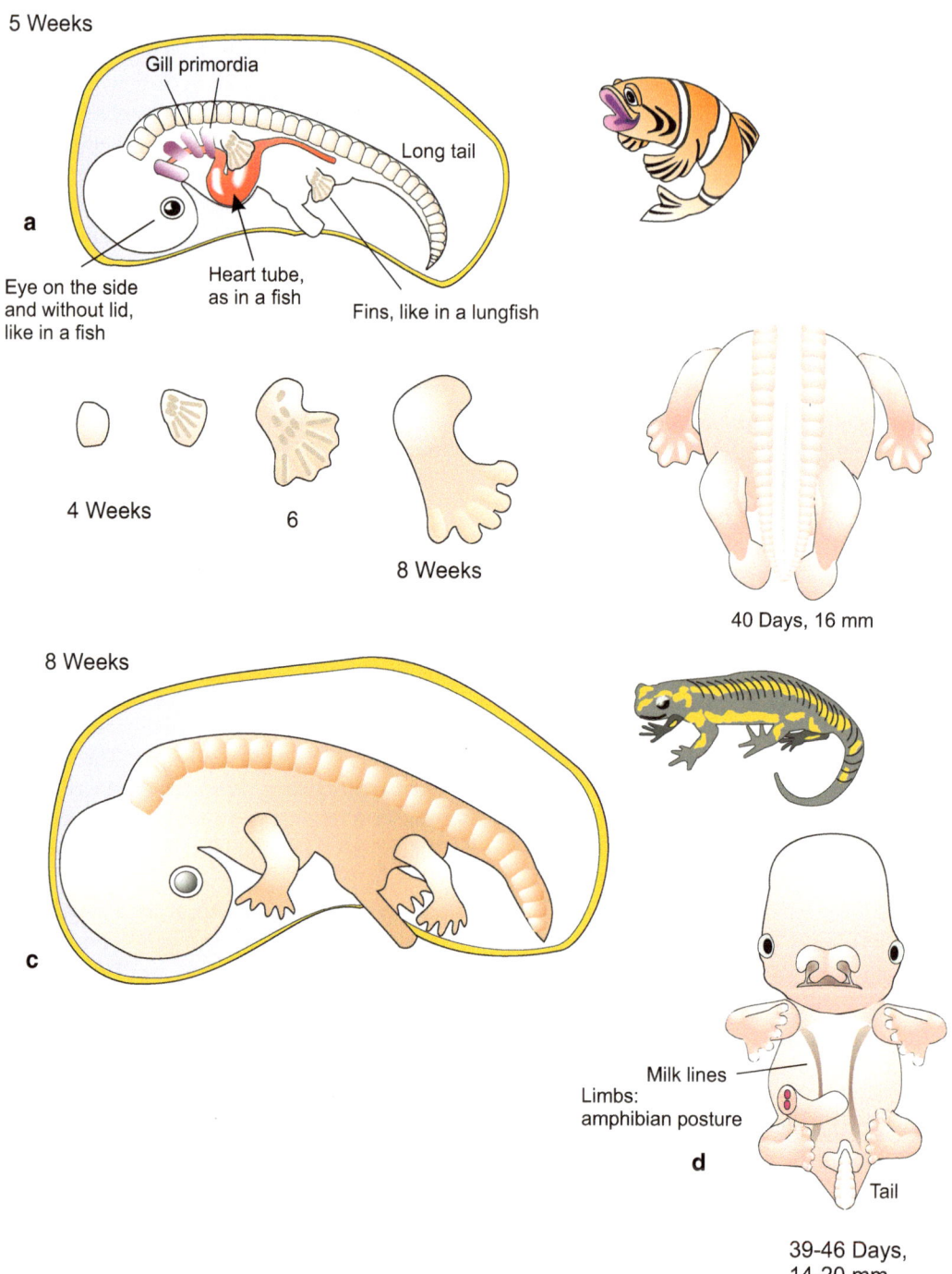

5 Weeks

Gill primordia

Long tail

a

Eye on the side
and without lid,
like in a fish

Heart tube,
as in a fish

Fins, like in a lungfish

4 Weeks

6

8 Weeks

40 Days, 16 mm

8 Weeks

c

Milk lines

Limbs:
amphibian posture

d

Tail

39-46 Days,
14-20 mm

Fig. 6.25 (**a–d**). Aspects of human evolution as documented by stages of embryonic development. Schematic illustrations emphasizing common traits. (**a**) "Fish stage" embryo; (**b**) development of an extremity from a paddle-like fin to a tetrapod limb; (**c**) "amphibian stage"; (**d**) example of a heterochrony, here the forward displacement of a typical mammalian trait, the milk line, in the "amphibian stage"

ear cavity (tympanal cavity) and Eustachian tube. The human fetus has a long tail and the tail has now reached its longest length. **Hands and feet are directed inwards while the elbow and the knees are directed outwards, as it is the case in newts, salamanders and reptiles**.

- By day 45 the nasal sacs are connected with the pharyngeal cavity through **choanae (internal nares)**, an achievement first made by the amphibians. The secondary bony palate which separates the nasal from the mouth cavity is finished by day 57. Such a secondary palate is characteristic of crocodiles, mammal-like reptiles and mammals. A **four-chambered heart** is characteristic of crocodiles and mammals as well, and is now gradually developed.
- The stage of a putative reptilian is further indicated by the appearance of finger nails ('claws') by day 60. Features of mammal-like reptiles are, for example, **longer, rotated extremities with elbows pointing posteriorly and knees pointing anteriorly**; the development of different types of teeth; of the **ear ossicles hammer and anvil**, and of an external ear canal.
- By day 45 the genitalia begin to develop, but a penis or vagina with a clitoris are recognizable only at the end of the 4th month. Amphibians and reptiles do not have such devices. **Sexually differing outer genitalia have been developed only by mammals**. Concomitantly, the gonoducts are now separated from the hindgut (only the egg-laying mammals preserve a cloaca).
- **Other mammalian features also develop late**. From about day 100 onward the fetus sucks its thumb. By day 110 it develops a – transient – **dense hairy coat**, called the **lanugo**. The disappearance of this coat marks the beginning of the final human phase of the foetal development. It includes legs growing longer than arms.

When we spoke of "fish-, amphibian- or reptile-stage" these abbreviating statements must not be misunderstood. **Thanks to his/her genetic complement the human is human from the very beginning up to the death**. (The same applies to any other species). And no one of these stages would survive in free nature as a fish, amphibian or reptile. **The human embryo in these stages is very tiny**. Many organs are not yet fully formed and most cells will terminate differentiation much later. In these stages **no functional nerve cells are present and no functional sensory organs**. We saw merely embryos resembling other embryos, but not resembling adults.

6.7.6   Not All Features Develop in Temporal Patterns Reflecting Faithfully Evolutionary Changes; One Finds Also Heterochronies

It must be pointed out that the time-course of a few ontogenetic events deviates from that of the evolutionary course. Such deviations are known as **heterochronies**. For example, already by day 42, when many other traits resemble those of reptiles, auditory ossicles begin to form, and while the first hairs occur in the 3rd month, milk lines can be recognized as early as in the 1st month. A striking example is the reduction of the tail in the human fetus. It occurs from day 42 to 70 – that is, in the 'reptilian phase' of the foetal development, whereas in evolution the tail was reduced only when the higher primates emerged from the level of monkeys.

Yet, the most comprehensive and dramatic temporal shift concerns the extremely early development of the extraembryonic organs, trophoblast and amnion. Their development is antedated even before embryogenesis proper commences.

Conservative paths versus innovations in development will be discussed in more global terms of evolution of developmental processes in the final Chap. 22 ("Evo-Devo"= Evolution of Development).

Summary

While growing in the ovary, the egg cell is prepared for its particular tasks. Starting already in the oocytes of the still unborn girl, the meiotic prophase is interrupted for years until in the ovarian cycle month per month 1 oocyte resumes meiosis and is released into the oviduct. The last meiotic division is completed not until the egg is fertilized.

Fertilization and early embryonic development take place in the oviduct. Cleavage of the yolk-less, only 0.1 mm in diameter measuring egg is holoblastic, dividing the egg completely in into daughter cells and leading to a hollow sphere called **blastocyst**. The blastocyst implants into the mucosal layer of the maternal uterus. The outer epithelial layer of the blastocyst becomes the polyploid **trophoblast** which encloses an **inner cell mass** the cells of which, also known as **embryonic stem cells**, remain diploid and pluripotent. The trophoblast gives rise to extraembryonic structures designed to establish intimate contact with the mother. An outer multinucleated layer of the trophoblast, called **syncytiotrophoblast**, paves the way into the uterine tissue with lytic enzymes; the inner layer of the trophoblast, called **cytotrophoblast**, forms **villi** to absorb nutrients supplied by uterine glands. The villi increase giving rise to branched **chorionic villi**. Later, the embryo in the interior is connected with the villi through the **umbilical cord**, and the villi in the disc-shaped area contacting the uterus grow to huge **placental villi** which dip into lacunae filled with maternal blood.

Before forming the embryo proper, the inner cell mass of the blastocyst leaves blank an **amniotic cavity**. Its floor, called **epiblast**, forms a **blastodisc** similar to that known from birds and reptiles. Underneath the epiblast an epithelial layer, termed **hypoblast** and classified as extraembryonic endoderm, extends giving rise to the lining of the (empty) yolk sac. **Gastrulation** of mesodermal and endodermal cells occurs by migration of epiblast cells through a **primitive groove**. The anterior end of the groove, called the **node**, is considered to be homologous to Hensen's node in the chick and to Spemann's organizer in amphibians. Above the axial mesoderm neural folds emerge forming a neural tube. At the end of gastrulation and neurulation the human embryo reaches the phylotypic stage with **neural tube, notochord, somites, and gill pouches** (which later are transformed).

Development of mammals in general and humans in particular takes place exceptionally slowly. Already cleavage takes much time presumably due to an early onset of transcription. During the entire period of gestation the embryo/fetus must **resist rejection by the maternal immune system**. It achieves this by presenting on the surface of the trophoblast a particular MHC molecule called HLA-G, and emitting hormone-like substances such as human **chorionic gonadotropic hormone HCG**. On the other hand the embryo, unlike that of non-placental animals, can grow enormously. In humans, it is **especially the cerebrum (telencephalon) that grows extraordinarily**, and with it the head.

In preparation, and stabilizing of an intimate mother-child relationship, the hormone **prolactin** is involved, in establishing close familiar ties also pheromones are thought to be involved.

In further sections of Chap. 6 human development is compared with that of other vertebrates and under aspects of evolution. Vertebrates pass, irrespective of their very different early embryonic phases and the subsequent divergent foetal phase, a **common phylotypic stage** that displays a common basic construction (Bauplan) of the body. **Conserved characteristics are: dorsal neural tube, notochord, somites, pharyngeal region with gill pouches, cartilaginous gill braces, 4–6 gill arteries, and a ventral heart**.

These similarities, first assessed by Carl Ernst von Baer in the early nineteenth century, has caused Ernst Haeckel to formulate a **"biogenetic basic law"** according to which ontogeny recapitulates phylogeny in an abbreviated way. Considering the example of human development the validity and limits of this "recapitulation" are demonstrated and discussed. It is pointed out that **in the phylotypic period the basic development of many organs is initiated and many processes of induction take place** albeit **heterochronies** may now and then hamper comparability. For instance, in primates the amniotic cavity is formed earlier (and in a different way) than in reptiles, birds and shrews.

Subsequent to the phylotypic stage fundamental reconstructions take place, in mammals especially in the region of the head and pharynx. From the first seven vertebral primordia the **occipital region** of the skull, and the **atlas and axis** of the vertebral column arise. From the cartilaginous braces of the **first branchial arch**, known as **mandibular arch** with its **primary jaw**, and further braces of the branchial foregut **the auditory ossicles** and the **voice box** develop. The first gill cleft becomes the auditory canal. From Rathke's pocket in the roof of the foregut the **adenohypophysis** arises, and from a pocket in the floor of the pharynx the **thyroid gland**. From parts of the gill pockets the **parathyroid gland** and **tonsils** develop.

More about the human embryonic development can be read in Chap. 3 (fertilization), Chap. 7 (sexual development), Chap. 16 (nervous system), and Chap. 17 (heart and circulation). Medical and legal matters are discussed in the accompanying Boxes 6.1, 6.2, 6.3, and 6.4, medical issues also in Chap. 18 (stem cells and regenerative medicine).

In Preparation for New Life I: Sex Determination and Sexual Development

7

Beginning new life must be prepared in the parents. The sex organs must be fully developed and grown, and the gametes maturated. In humans the development of the sex organs commences in the embryo but the organs acquire full maturity as late as puberty. As an introduction we first consider the meaning of sexual reproduction in general.

7.1 The Essence of Sexuality

7.1.1 Sex Is Devoted to the Exchange and Recombination of Genetic Information in Offspring; As a Rule Mother and Father Contribute Genes, the Selection of Which Is Subject to Chance

Sex in the sense of genetics primarily is a process by which new combinations of alleles are assembled and assigned to descendants. Most eukaryotic organisms exist in two complementary forms, females and males, who invest genetic material in a mutual game of chance. Unlike asexual reproduction that is exclusively based on mitotic division, sexual reproduction incorporates two characteristic events enabling novel genetic combinations: (1) fusion of two genetically different cells (gametes) in fertilization, and (2) meiosis, a process that usually takes place in preparation for fertilization (in several "lower" eukaryotes meiosis occurs at some other point of the life cycle) and enables the rearrangement of individual chromosomes and genes.

Most multicellular animals are diploid (exception: bee drone). The nuclei of their cells harbour two sets of chromosomes. Each distinct chromosome is represented twice, known as a pair of homologous chromosomes, the one homologue being designated maternal, the other paternal, with respect to their provenance in the previous act of fertilization. As a rule, one female and one male each transmit one set of chromosomes, that is one haploid genome, with their **gametes (germ cells, sex cells)** to the next generation. Each sexual partner may only pass one homologous chromosome of each pair to the future descendant and it is subject to chance whether the originally maternal or the paternal chromosome is allocated to a distinct germ cell.

We repeat what had been said in Chap. 1. In the new gametes, any particular chromosome can come from *either* parent; for instance, chromosome No. 1, in the newly formed egg cell may be the originally paternal chromosome, once carried by the sperm, another chromosome, let's say chromosome No. 2, was also of paternal origin, while chromosome No. 3 was once received from the mother of the individual who produces the new gametes. Likewise, although in the testes of the sexually mature man the newly produced sperm cells are all haploid, each sperm cell contains chromosomes partly derived from his mother, partly from his father, and the particular mixture is subject to chance.

Considering all **23 chromosomes** a haploid germ cell embodies, the number of possible new combinations is huge. There are 2^{23}, **or 8,388,608 different** possible **combinations**. This means, a woman could produce approximately 8.4 million egg cells each of which differs from all the others in at least one chromosome. Likewise, a man could produce 8.4 million different sperms.

Additionally, chance is involved when homologous chromosomes exchange genes by cross-over since there are no strictly fixed positions along the chromosomes where crossover occurs. Furthermore, chance is at play when two germ cells fuse in fertilization, since it is not predetermined which gametes will meet each other.

The random rearrangement of genetic material is associated with production of offspring and, therefore, is called **sexual reproduction**. The offspring literally embody the new combination of genes and subject it to the test of natural selection.

Gametes occur as two types of highly specialized cells, the egg and the sperm. In a number of plants and "lower" animals, each individual produces both types of gametes. This condition is called **hermaphroditism**, combining the mythological figures Hermes and Aphrodite; it is exhibited by **monoecious species** (Greek: *mono* = one, *oikos* = house). In the vast majority of animals, however, the role of producing eggs and sperm is allotted to two distinct types of individuals, and the species are **dioecious** (Greek: *di* = two, twofold, *oikos* = house) or **heterosexual**, with females and males.

As a rule, the differences between the sexes are not restricted to the reproductive organs but extend to other traits such as size, ornaments and weaponry, and always include sex-specific, reproduction-associated behaviour. Thus, the typical animal species displays **sexual dimorphism**.

7.1.2 Most Organisms Have a Bisexual Potential

Two understand aberrant but also normal sexual development in humans, it is important to appreciate that each individual is basically endowed with **bisexual potential** and possesses almost all of the genes required to develop both female and male characteristics. Bisexual potential is testified in hermaphroditism but exists in heterosexual organisms including humans as well. Otherwise, a mother could not transmit "male" genes from her father to her sons, and a father could not transmit "female" genes from his mother to his daughters.

7.2 Sex-Determining Systems and Genes

Given the fact of bisexual potential, what is the ultimate decisive event? In all organisms, beyond the sets of genes required to develop both types of gonads and sex organs, additional **key genes**, (**selector genes**, **master genes**) determine which of the two sets of genes is actually being used in developing an individual's sex. Typically, one decisive key gene is present but allotted to only half of the offspring. This is referred to as

- **Genotypic sex determination**. Alternatively, master genes are differentially activated by environmental influences. Such cases are subsumed under the heading.
- **Environmental or phenotypic sex determination**.

7.2.1 In Phenotypic Sex Determination: Environmental Variables Take the Decision

In environmental sex determination the ultimate decision is made only after fertilization, and **external cues** such as temperature or presence of a sexual partner make the choice, presumably by influencing the activity of master genes. For instance, in the polychaete annelid *Ophryotrocha puerilis* two individuals that encounter each other by chance influence each other by means of chemical signals (pheromones) and eventually one individual plays the role of the male and one the role of the female. Environmental sex determination has been discovered in several invertebrate species but also is known to occur in fish and reptiles (all crocodilians, many turtles and some lizards). Environmental sex determination is indicated, but not proven, by sex ratios deviating significantly from 1:1.

The Mississippi-alligator make micro-organisms to produce breeding heath. The female animal establishes a compost heap to nourish the micro-organisms. At 30 °C all the eggs give rise to females, at 34 °C all give rise to males. At temperatures between 32 and 33 °C females and males arise in varying ratios. How an egg-laying crocodile manages to arrange the clutch of eggs in the pile in such a way that offspring hatch in a balanced ratio of female and male individuals, is not known. Biologists interested in evolutionary theories ask: is a ratio of 50:50 always appropriate? It is supposed, and supported by findings in the Australian lizard *Amphibolurus muricatus*, that fitness of a population might increase if sometimes more, sometimes less females make use of limited nutritive resources for growing large eggs. Perhaps breeding temperature serves as an environmental counterbalance. Zoologists with sense of humour like to refer to further examples of phenotypic sex determination. Adult, full grown anglerfish of the deep sea (genus *Edryolichnus*) are always females. To get an abiding partner without fail they catch a larva of their own species. The larva attaches onto the body of the female and both specimens grow together. The larva becomes a mini male which supplies sperm at call. There is no need for the male to look after food himself; on the other hand he is captured life long. If the larva escapes this fate it will become female and an angler. A similar destiny is granted the larvae of the annelid *Bonellia viridis*. The dwarf male lives in the nephridia of the green female. A change of sex, as it occurs in *Ophryotrocha*, is also known from the slipper snail *Crepidula*. Several animals form a pyramid, each younger one sitting on top of the older animals. The animal at the bottom is the oldest and largest one and female, the animal on the top is the youngest and male. In between one finds (infertile) intersexual specimens. Sex determination is thought to occur via pheromones.

7.2.2 In Genotypic Sex Determination the Decision Is Left to Chance in Fertilization; Chance Allocates a "Sex-Chromosome" Bearing Key Genes to Only Half of the Offspring

Traditional belief, offered by ancient writers such as Aristotle, ascribes sex determination even in humans to non-genetic determinants such as nutrition or the heat of passion during intercourse. How such factors ensure a sex ratio of 1:1 is usually not the subject of much attention in superstitious explanations.

But how can chance yield a ratio consistently close to 1:1 (in humans, 105 male births per 100 female births)?

In genotypic sex determination certain particular key genes (selector genes, master genes) bring about the decision whether the set of genes required for male, or the set of genes required for female development will be expressed. Such a key gene is, as a rule, found on one chromosome of a pair of homologous chromosomes while its partner lacks this gene.

Such a pair of unequal chromosomes is called

- **Heterosomes,** also called **allosomes** or **gonosomes**, while all the other chromosomes of the species are collectively called
- **Autosomes**; from which sex-specific differences are not known.

In haploid germ cells only one heterosome of the pair can be present, it is either the chromosome containing the key gene or the chromosome lacking this gene. As a consequence, in conception only one half of the descendants will get this key gene.

The principle may be explained more specifically using humans as the example. All humans carry $2 \times 23 = 46$ chromosomes in their diploid cells. Of these, 2×22 chromosomes are indistinguishable in males and females and, therefore, categorized as autosomes. In females, the 44 autosomes are complemented by a two X. In men, besides 44 autosomes two morphologically distinct chromosomes are found: one is an X, like those found in women, while the other is a modified X, termed the Y chromosome.

When meiosis takes place, of the complete diploid set of 44 autosomes +2 heterosomes one haploid set of 22 autosomes + 1 heterosome is parcelled out to each future egg or sperm cell. In the oocytes this set of 22 autosomes is always supplemented by one X. Spermatocytes also get a total of 23 chromosomes, but the set includes, besides 22 autosomes, either the X or the Y. On the average, 50 % of the sperm cells obtain the X chromosome by chance while the other 50 % obtain the Y chromosome. As a consequence, while all diploid offspring will have 44 autosomes, with respect to the heterosomes, 50 % of the diploid offspring will be marked by XX, the other 50 % by XY. It then will turn out that XY babies will become boys, while XX babies will become girls. With respect to the two heterosomes, in humans and other mammals, females are said to be **homogametic** (XX) while males are **heterogametic** (XY). The suffix 'gametic' suggests consequences in gametogenesis. With respect to decisive key genes one speaks of **homozygou**s or **hemizygous** constitution.

X and Y chromosomes are often called "**sex chromosomes**" but this designation is highly misleading. Actually in mammals the majority of genes required for the development of sexual traits are located on the 44 autosomes. A better, but still misleading designation for the mammalian heterosomes is "**sex-determining chromosomes**". As discussed in Sect. 7.3.3, it is not the entire X or Y chromosome that is of significance in sex determination but a dominant male-determining key gene **Sry**, located on the

Y. Female development is the default program switched on by female-specific key genes such as **R-spondin1** when the male-determining gene is absent or defective.

Unexpectedly, intensive research into the nature of genotypic sex determination has shown that the role of chance in distributing chromosomes or genes is the only common denominator among the various types of genotypic sex determination identified in the animal kingdom. Despite nature's tendency to retain established "solutions" to biological problems, and despite the universality of sexual dichotomy, there is no universal, molecular mechanism of genotypic sex determination. In most animal species the mechanism is complex and includes cascades of mutually dependent events. Frequently, the sex-determining genes occur on two morphologically different chromosomes, usually called, as in mammals, the X and Y, or they are designated W and Z indicating that these two systems are different in basic features. In both cases one of the two sexes is homogametic and the other heterogametic – but in the XY and the WZ system the correlation is inverse. Despite the consistent XY or WZ formalism, the actual roles of these chromosomes vary in different organisms, and they bear different genes. By definition in systems designated XY the males are heterogametic, in systems designated WZ males are homogametic. A chromosome designated Y in mammals and a chromosome designated Y in *Drosophila* bear the same name signature but different genes; they merely obey a superficially similar formalism in sex-determination in that males are heterogametic. In birds, incidentally, females are (usually) heterogametic and the males homogametic. While in mammals the male-determining key gene *Sry* is located on the Y and males are XY, in birds the male-determining key gene (presumably) *DMRT1* is located on the Z and therefore males are ZZ. Teleost fishes display an amazing diversity of genetic as well as environmental sex-determining systems, and combinations of such systems. Information about the occurrence of heterosomes is given in Table 7.1.

Table 7.1 Heterosomal Constitution

Animal group	The homo- gametic sex	becomes	The hetero-gametic sex	becomes
Mammals, humans	XX	♀	XY	♂ by Y-*Sry*
Birds	ZZ	♂ by Z-*DMRT1*	WZ	♀
Fish	XX or ZZ	♂	XY or WZ	♀
Xiphophorus variatus	YY	♂	XY	♀
Drosophila	XX	♀	X (Y insignificant)	♂
Butterflies	ZZ	♂	WZ	♀

To exemplify variability only two model organisms will be presented:

In the nematode *Caenorhabditis elegans* XX individuals are hermaphrodites; pure female individuals are not known.

X0 individuals are male (arisen by loss of one X in gametogenesis).

In *Drosophila*, the ratios between the number of X chromosomes and the two sets of autosomes (AA) in each individual diploid cell is decisive in determining the sex of each individual cell.

XX/AA (a ratio of 1:1) determines female

XY/AA (a ratio of 1:2) determines male

X0/AA (a ratio of 1:2) determines male, but the males are infertile.

The Y chromosome in *Drosophila* is of significance in spermatogenesis in the adult male but not in early sex development. Sexual determination is cell autonomous and not mediated by hormones. If one X is lost due to a faulty mitosis (nondisjunction) in an XX embryo, the resulting X0 cell and all its descendants become male in female surroundings (while the descendants of the corresponding XXX cells are "superfemale"). In Drosophila and other insects, individuals composed of a mosaic of female and male cells are called **gynanders** (Greek: "wifeman") or gynandromorphs (Fig. 7.1). In insects unlike vertebrates, there are no sex hormones that could moderate differences in X chromosome numbers and ensure a uniform phenotype. An individual with an abnormal ratio of XX/AAA develops an intersexual phenotype with genetically uniform cells but an intersexual organismal appearance. In each somatic cell X-linked genes code for factors aptly called **numerators**, while genes located in the autosomes code for **denominators**. In the presence of two X chromosomes more numerator proteins are produced (by genes that apparently escape dosage compensation, or before dosage compensation is accomplished). When numerator proteins predominate, the master gene *sxl* (*sex lethal*) is switched on. The SEX LETHAL protein SXL directs a cascade of biochemical events. These include the correct splicing of its own pre-mRNA by which SXL makes it possible to maintain its own synthesis. The cascade of events set going by SEX LETHAL terminates in the activation of female-specific genes. In contrast, male development is the default program, resulting from insufficient levels of SEX LETHAL protein to trigger female development.

Papilio dardanus

Gynander

Fig. 7.1 Gynander in a sexually dimorphic butterfly

7.3 Early Sexual Development of Mammals and Humans

7.3.1 Primarily the Gonads are Sexually Indistinct; They Develop in Cooperation with the Embryonic Kidneys Resulting in the Urogenital System

In mammalian development the gonads and embryonic kidneys develop in close proximity and partially from common primordia; the primordial kidneys contribute their ducts thus giving rise to the common **urogenital system** (Fig. 7.2). The primordial material of the system derives from the intermediate mesoderm (segment stalk) between the somites and the lateral plates (see Fig. 5.10). Both the gonads and the embryonic kidneys arise from the tube of the holonephros with its ciliated funnel, and the adjoining nephrogenic mesenchyme plus the gonadal primordium. From the holonephros three parts segregate in temporal order:

1. The **pronephros** which uses the holonephric duct with its ciliated funnel, now called

Müller's duct; in females this duct is the precursor of the **oviduct.**

2. The **mesonephros** which contributes the **Wolffian duct**, in males this duct is the precursor of the **spermatic duct** (vas deferens).

3. The **metanephros**, the final kidney in both sexes which develops its own duct, the final urinary duct called the **ureter** (see Fig. 7.4).

It must be pointed out, that this sequence of events starts in female and male individuals from the same primordia including the precursors of the oviduct and the spermatic duct.

7.3.2 Sexual Development in Mammals Proceeds in Steps From an Indistinct Initial State Over Divergent Paths to Sex-Specific Behaviour; a Brief Introductory Overview

The basic bisexual potential that our genetic heritage harbours is morphologically expressed. Initially our somatic traits, the gonads included, are the same in both sexes. In the gonads this indistinguishable state is documented by the initial presence of both oogonia and spermatogonia. Also the primordia of the genitalia and other somatic characteristics that serve sexual propagation are prepared for both genders and initially indistinguishable (Figs. 7.2, 7.3, 7.4, and 7.5). Beginning with the activity of a male- or female-determining key gene in the gonads the further development is directed in alternative paths and eventually end up in the known male or female traits.

In mammals, particularly as studied extensively in mice and humans, sexual development is a multistep process. Several stages occur prior to birth

1. the genetic sex,
2. the gonadal sex,
3. the somatic sex,
4. the psychological sex.

Gonadal, somatic and psychological sex of humans develop in two distinct phases of life,

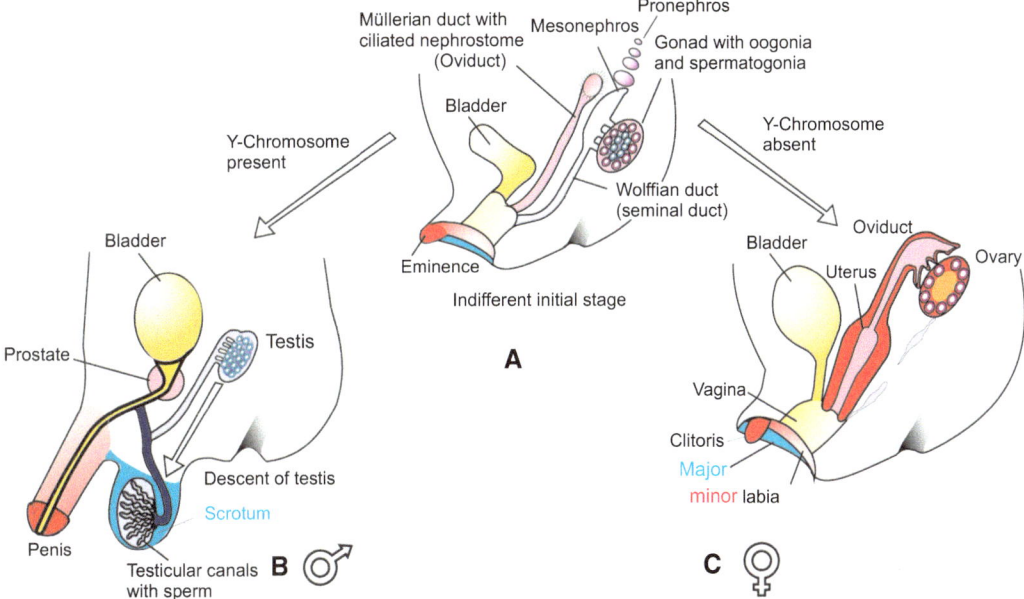

Fig. 7.2 (**a–c**) Sexual development in humans. (**a**) Divergent development starts from a common, indistinct stage (*middle*). (**b**) Development proceeding in male direction. In the male gonad, oocytes in the cortex begin to degenerate while the testis cords in the centre of the gonad continue to develop into seminiferous tubules. Under the influence of the testosterone-induced AMDF (Anti-Müllerian Duct Factor) the Müllerian duct regresses, the Wolffian duct comes to be the *vas deferens*. The gonad is converted into a testis. The former mesonephros is transformed into the epididymis. (**c**) In the female gonad, the central testis cords disappear, while the oocytes in the peripheral cortex survive and become surrounded by follicle cells; the gonad is developing into an ovary. The Müllerian duct is enlarged to become the oviduct and uterus, while the Wolffian tube is reduced. The external genitalia also arise from a common, indistinct stage

(a) an embryonic phase in which the **primary sexual characteristics** find their expression, and

(b) maturation in puberty, when the testes start producing sperm, ovaries bring eggs to maturity, and the secondary sexual characteristics appear.

7.3.3   The Genetic Sex: A Single, Dominating Gene Determines Whether One Becomes a Man Or a Woman

As all eggs have an X chromosome, the genetic basis of sex depends on the genotype of the sperm cell. If the sperm contributes another X, the embryo will be XX and eventually become female. If the sperm cell contributes a Y, the embryo gets the constitution XY and will become male. The decisive significance of the Y chromosome arises from the fact that it carries a master gene *Sry* (*Sex-determining region of the Y chromosome*).

Sry activity generates the **transcription factor SRY**, also known as **TDF (Testis-Determining Factor)**. TDF contains a DNA-binding domain of the HMG (High Mobility Group) class that enables the factor to direct the activity of subordinate genes (most of which remain unidentified). A 14 kb DNA fragment containing the *Sry* sequence and no other gene, was sufficient to cause XX mouse embryos to develop male traits. It appears **that a single gene ultimately determines whether an individual will be a man or a woman.** This does not mean that other genes are not involved in the realization of the sex; but one particular gene's activity tips the scale. However, with a *Sry* gene alone, male mice are infertile. One additional Y-resident gene, called the *spermatogonial proliferation factor Eif2s3y* suffices to provide the animals with the capabability to produce sperm.

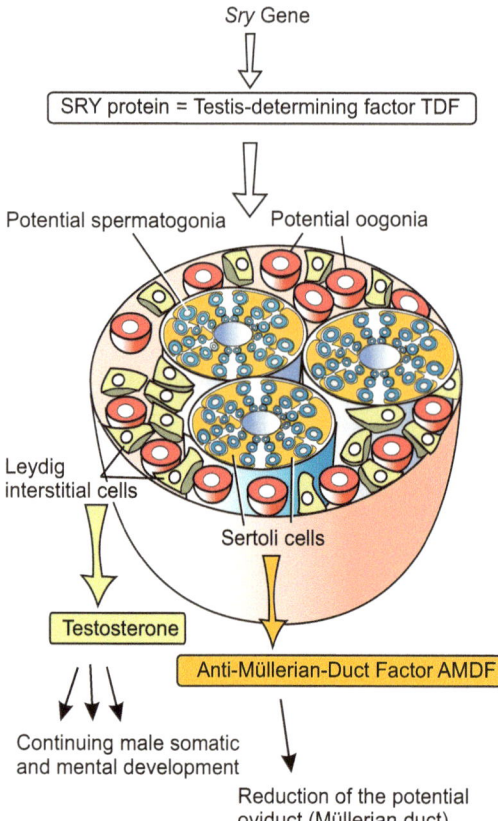

Fig. 7.3 Sexually indistinct gonad and effect of the SRY factor = TDF (Testis-Determining Factor) encoded by the *Sry* gene on the gonad. The Sertoli cells are induced to emit the hormonal factor AMDF, the Leydig interstitial cells are stimulated to produce and release the male sexual hormone testosterone

With respect to sex determination in mammals researchers were on the right scent by chasing genetically caused syndromes of maldevelopment.

The particular significance of the Y chromosome could be deduced from

– **XXY Klinefelter syndrome.** Phenotypically such individuals are male.
– **X0 Turner syndrome.** In case the Y chromosome is lacking (0 means lacking of a chromosome) the phenotype is female (though these individuals are infertile).

Klinefelter- and Turner-Syndromes are known not only from humans but also from other mammals, for instance cats.

The *Sry* gene especially could be identified by analyzing sexual aberrations causing disharmony between the physical and psychological sex.

XX but Notwithstanding Male   Putatively, aeons ago X and Y chromosome had been homologous chromosomes since they still are equivalent at the end of the short chromosomal segment (Fig. 7.6). Abnormal sexual development can result from events that have taken place in the father of the child. During meiosis in spermatogenesis, the *Sry* gene may have been **translocated** from the Y to the X chromosome. Fortunately, crossover between X and Y is infrequent. But if a sperm carrying an X-linked *Sry* wins the race and fertilizes an egg cell, the seemingly normal XX embryo will develop male traits in their outward appearance and their behaviour. In the few cases known to result from such a crossover, the person remained infertile, indicating that additional Y-linked genes are required to support male differentiation. These genes are required in spermatogenesis. However, not all genes needed for complete spermatogenesis are concentrated on the Y chromosome, others are linked to the X chromosome.

Translocations of this type were also observed in mice and this made searching for the decisive chromosomal locus, and eventually for the ultimate gene, possible. The putative locus was called *"Sex-determining Region of the Y chromosome"*, in short **Sry**. The ultimate gene was identified with the sophisticated method of positional cloning (see textbooks of genetics).

XY but Nevertheless Female: The *R-spondin1* Gene   The female sex is, in mammals and humans, the default ground condition. It ensues when no *Sry* gene is present (in XX constitution) but also when in XY individuals the *Sry* gene that normally blocks the start of the female cascade is defective. Being a Y-linked gene *Sry* does not have a homologous allele that could compensate for the deficiency (to a certain degree a lacking SRY can be compensated by SOX9 complex, Fig. 7.7b, details that cannot be dealt with in an introductory book). According to the

Fig. 7.4 Progressing development of the sexual organs starting from an indistinct common condition. From the paired organs only the left or the right organ is shown

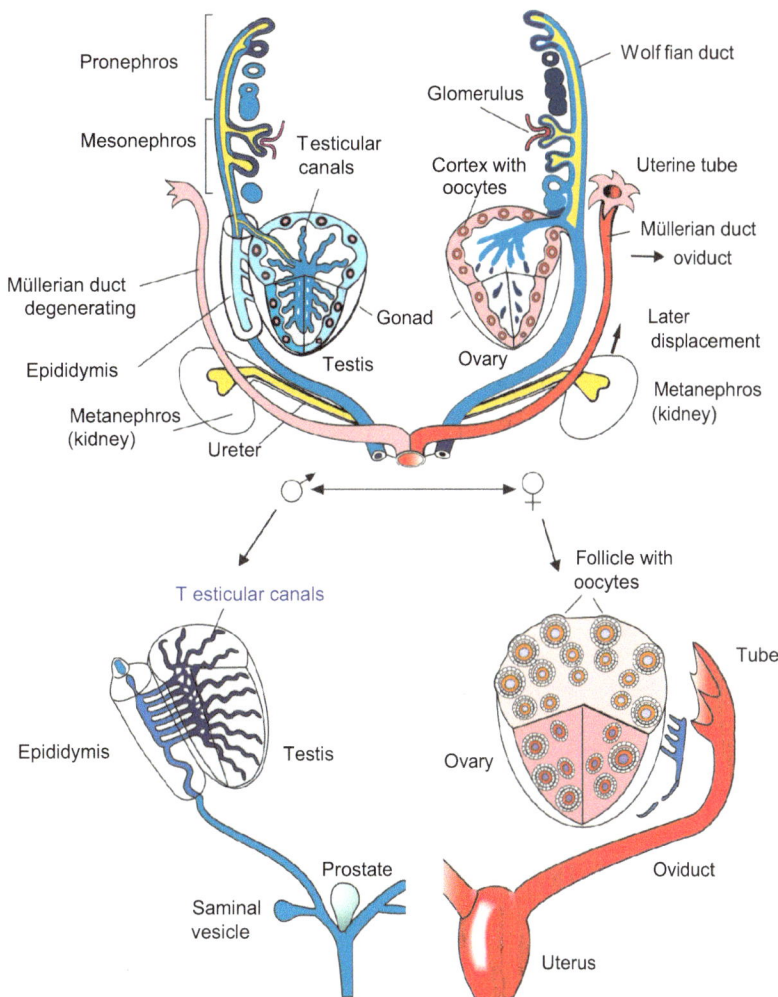

chromosomal figure an XY constitution is diagnosed, but failure of the *Sry* function suffices to direct the development into the female direction because female-determining key genes become effective by default. The somatic cells of such women lack the **Barr body** (the inactivated second X chromosome, Chap. 12, Fig. 12.24), characteristic of normal female cells.

Present knowledge assigns the function of a **female-determining key gene** to the autosomal gene *R-spondin1*, that in humans is found on chromosome 1 and codes for the peptide RSPO1. This peptide is among a group of peptides enabled to activate WNT signal cascades. Putatively the RSPO1 peptide acts synergistically with WNT4 in guiding the further development of the gonads in the direction of ovaries. In the presence of an active *Sry* gene the *R-spondin1* gene gets thwarted and with it female sex determination (Fig. 7.7).

7.3.4 The Gonadal Sex: From a Hermaphroditically Laid Out Gonad Alternatively a Testis or Ovary Arises

In early embryogenesis, the gonads of XY and XX individuals do not exhibit morphological differences. Moreover, the gonad contains both

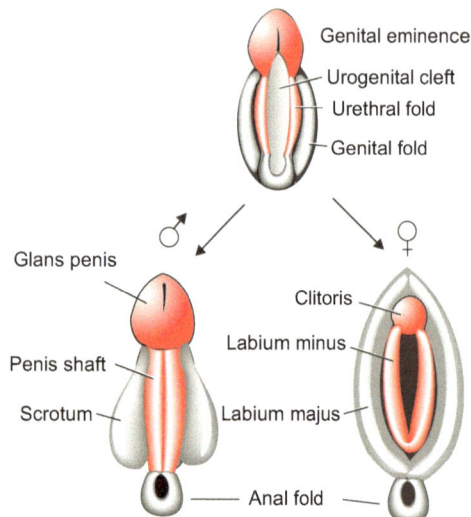

Fig. 7.5 Development of the external genitalia starting from an indistinct common condition

potentially male and potentially female germ cells. Primordial germ cells colonize the gonadal cortex where they can become oogonia, as well as the central medulla where they can become spermatogonia (Figs. 7.3 and 7.4).

Male Development In XY embryos the primordial germ cells in the cortex perish under the influence of the Testis-Determining Factor whose synthesis is *Sry*-dependent. The gonad becomes a testis.

* The central medulla develops testis cords; at puberty, these cords will hollow out and become the seminiferous tubules in which sperm are produced.
* The supporting cells differentiate into **Sertoli cells**; these produce the hormonal **Anti-Müllerian-Duct Factor** (**AMDF**, also known as **MIS** or Müllerian duct-Inhibiting Substance);
* The steroid precursor cells (interstitial cells) develop into **Leydig cells**, that begin to produce **the male sex hormone (androgen) testosterone**.

Female Development When the *Sry* gene is not present or is defective, the female key genes such as ***R-spondin1*** are automatically activated and the gonad becomes an ovary:

* The primordial germ cells in the cortex stop mitotic division, become oocytes and enter meiosis. In contrast to males where meiosis in the spermatocytes is delayed until the onset of puberty, in females meiosis begins in the ovaries as early as the 12th week of embryonic development.
* The supporting cells assume the function of **follicle cells** and surround the oocytes.
* The steroid precursor cells (interstitial cells) become **theca cells** and produce the female sex hormone **oestradiol** and related oestrogens (Fig. 7.8).

Further gonadal development depends on whether or not testosterone is present.

Testosterone is an essential **androgen**. Its presence triggers male development; its absence allows female development to take over.

7.3.5 The Alternative Decision Is Brought About by External Signals That Tip the Balance

Many a malformation in sexual development, also of humans, did not find an explanation until signal molecules were discovered which act onto the gonad from opposite topographical sides steering its development in the male or female direction dependent on the genetic XX or XY constitution.

If the *Sry* gene is in business the somatic cells in the periphery of the gonad facing the coelomic cavity send off the signal molecule **FGF9** (Fibroblast Growth Factor 9); subsequently somatic cells within the gonad itself take over and continue its production (Fig. 7.7). FGF9 suppresses the **WNT4/β-catenin system** known to favour female development (WNT-systems are introduced in Chap. 11). As a further

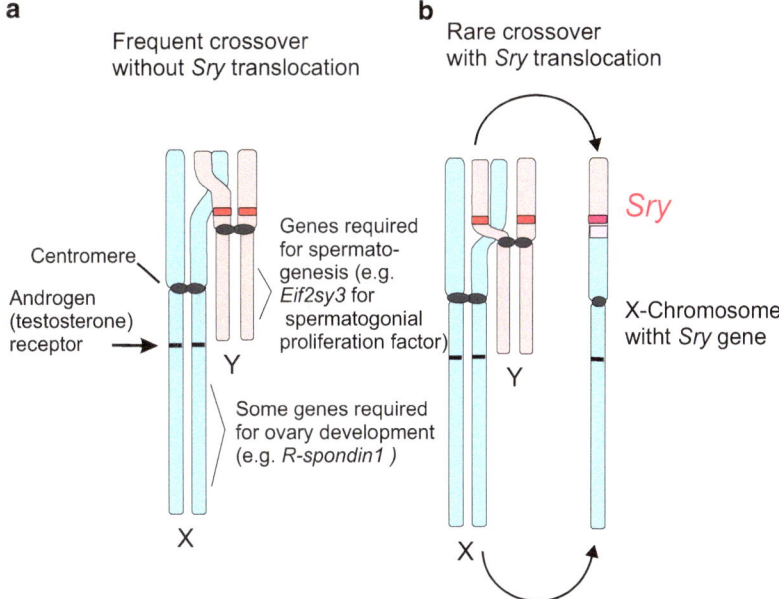

Fig. 7.6 (**a, b**) Crossover between X and Y chromosome. (**a**) Crossover between homologous regions without transfer of the *Sry* locus from Y to X; (**b**) Translocation of the male-determining *Sry* region from Y chromosome onto an X chromosome by a rare crossover

consequence of the FGF9-signalling downstream genes are switched on making the supporting cells differentiate as Sertoli and Leydig cells.

If, on the other hand, in the XX constitution the *Sry* gene is lacking a chain of reactions is set going which eventually tips the balance in direction of female sexual development. In somatic cells located at the opposite side of the gonad and facing the mesonephros (Fig. 7.4) the gene *R-spondin1* is activated. The translated peptide RSPO1 stimulates the WNT4/β-catenin signalling system which in turn disrupts the expression of FGF9. Therewith the development of a male gonad is prevented and an ovary is formed instead with follicle and theca cells. *R-spondin1* lies in humans on chromosome 1, meaning that it is autosomal and present in both sexes; yet, it can become active only in the absence of SRY. If both alleles of *R-spondin1* or *WNT4* are defective the FGF9 transmitter continues signalling and the gonad grows, as can be derived from Fig. 7.7, to a testis even in the XX fetus. Eventually the individual becomes masculinized.

7.3.6   The Somatic Sex: Hormones Direct Further Development; of Particular Significance Is Testosterone

The development of the inner and outer sex organs that are designed to support the function of the ovary or the testis also starts from an indistinct primordium (Figs. 7.4 and 7.5).

The embryo has initially two developmental options and its gonads are accompanied by two different kinds of tube-shaped channels. One of these tubes, called the **Müllerian duct**, can grow to become an oviduct, the other tube, called the **Wolffian duct**, has the potential to become a vas deferens and conduct sperm. Externally, the genitalia are initially indistinct in their morphology as well.

In XY individuals, the Müllerian duct atrophies, mediated by AMDF, while in the presence of testosterone the Wolffian duct forms the vas deferens and the pronephros forms the epididymis.

a

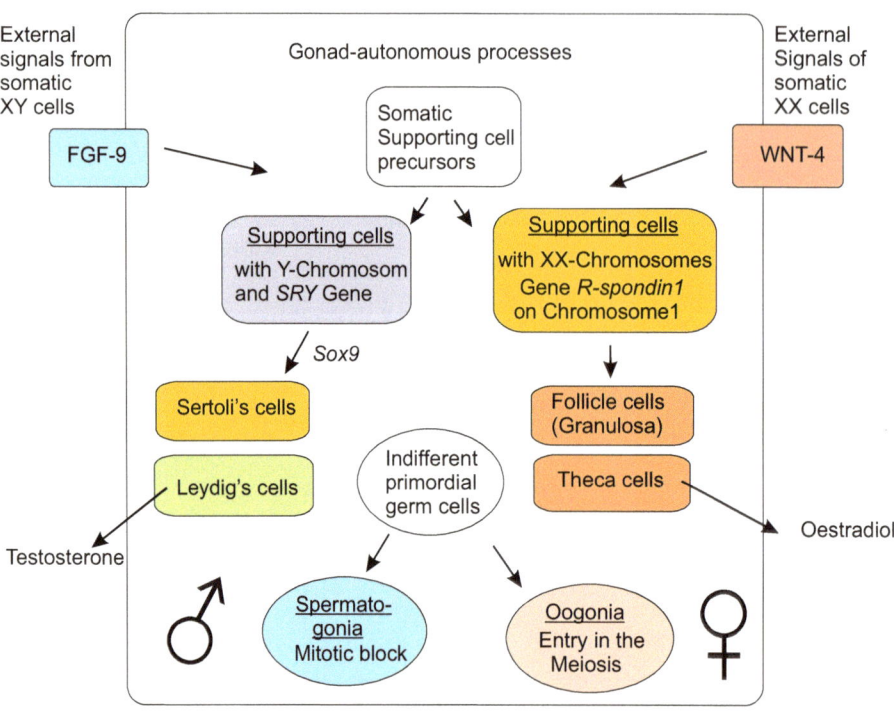

b

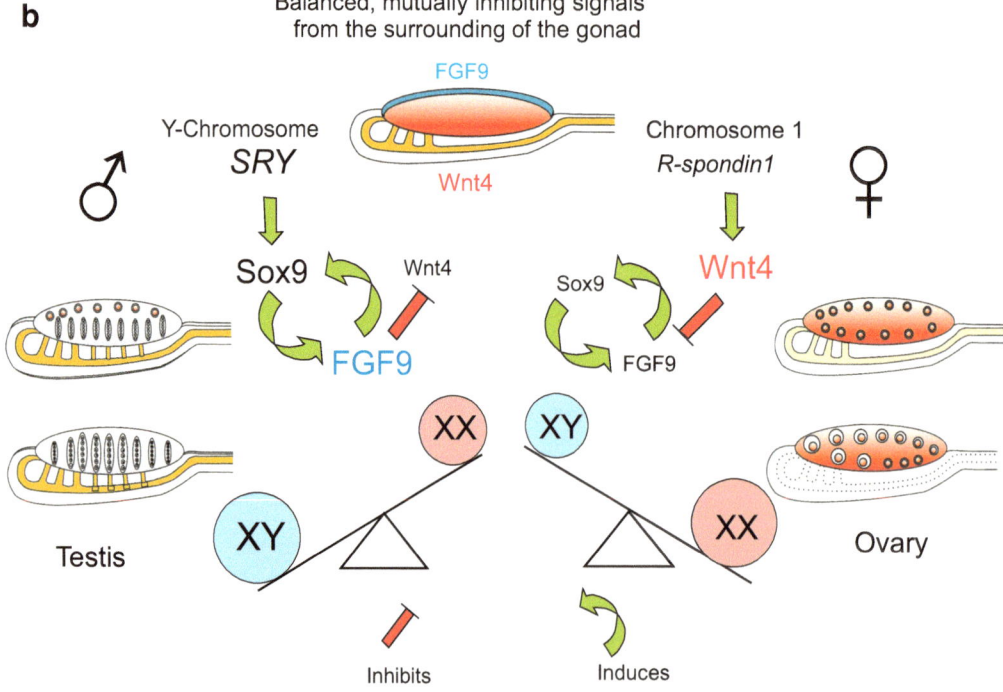

Fig. 7.7 (**a, b**) Decisive factors in sex determination in the foetal human. (**a**) Of primary significance are factors emitted by somatic cells of the environment of the gonad and by the gonad itself. The primordial germ cells are submitted to secondary processes of selection and either the spermatogonia or the oogonia perish. (**b**) Balance model of sex determination after Kim et al. (2006) and Lau and Li (2008), modified

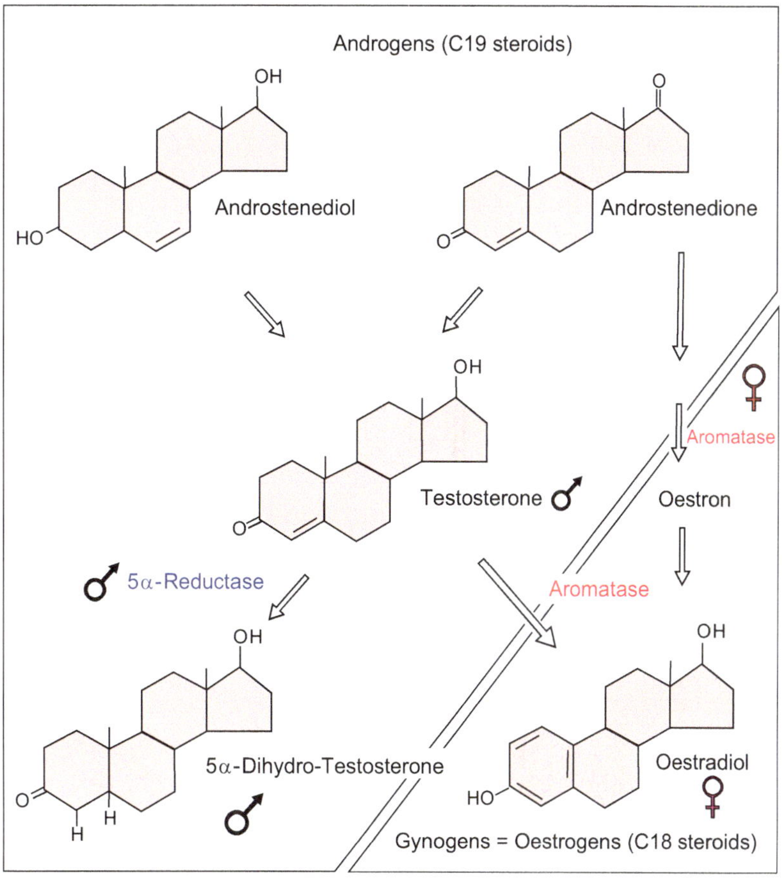

Fig. 7.8 Steroid sexual hormones

The external genitalia form the scrotum and penis under the influence of **dihydro-testo-sterone**, derived from testosterone (Fig. 7.8) by means of a 5-alpha-reductase in the early uro-genital primordium (genital ridge). Deficient testosterone levels due to lacking 5-alpha-reduc-tase activity, or defective testosterone receptors may lead to **phenotypic feminization (testi-cular feminization)** in spite of an XY genotype. Testicular feminization is the obverse of those cases in which in spite of a XX constitution a male phenotype arises, because one X bears a translocated *Sry* gene.

The statement that female development is the default program must not be misunderstood as meaning that female sexual hormones are not of significance. When in the absence of a functional *Sry* the gene *R-spondin1* on chromosome 1 finds expression, a cascade of events eventually leads to the production of oestrogens (e.g. oestradiol) and gestagens (progesterone).

Oestradiol is indeed important. When in female mice the production of oestradiol prematurely ceases, in the ovary male somatic cells such as Sertoli- and Leydig cells can be detected. On the other hand, several environmental substances that act like oestradiol (xenoestrogens) impair the fer-tility of male individuals (see Box 7.2).

7.3.7 Male and Female Hormones Are Present in Both Sexes

The gonads are the most important source of sex hormones. But also the cortex of the adrenal gland supplies such hormones, especially at the start of puberty, although in much lower amounts. With increasing sensitivity of analytic methods further sites of synthesis were detected scattered in the body, surprisingly even in the brain. Details of this topic can be called up from databases using the key word

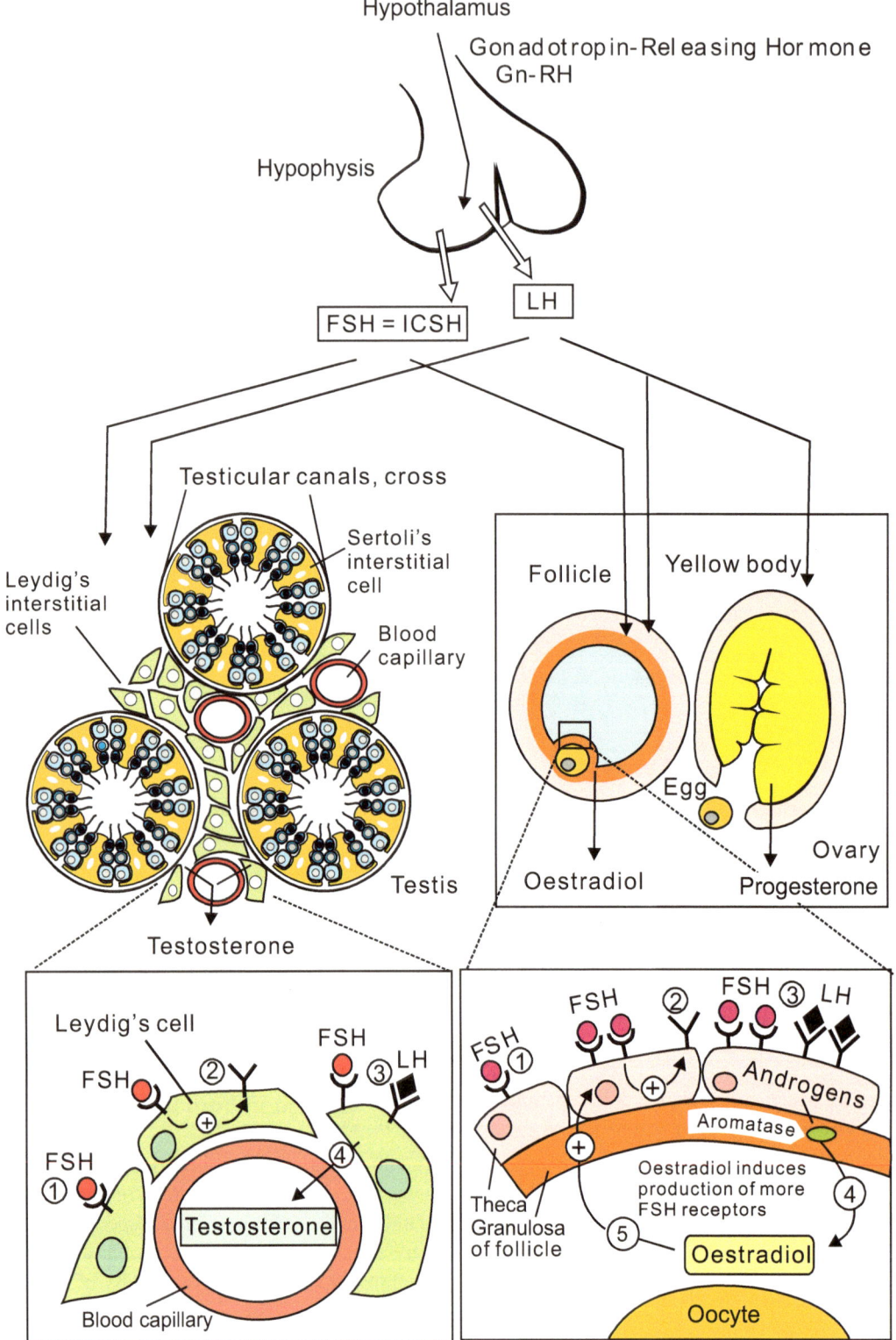

Fig. 7.9 Overview over hormones controlling sexual development of mammals/humans (from Müller and Frings: Tier- & Humanphysiologie, Springer 2009)

neurosteroids. Currently, the significance of such findings cannot be assessed.

Surprisingly, the most important **biochemical source of the female sex hormone oestradiol is the male sex hormone testosterone.** In the female organism testosterone is converted into oestradiol by means of aromatases (Fig. 7.8). These enzymes are found, besides in the ovary, at several other locations, for example in adipose tissue of the breast and in bones. As a rule, such conversions of testosterone into oestradiol are not one-hundred percent and, therefore, traces of testosterone are detected in the blood of women as well. Conversely, in the male sex aromatase is also expressed to a certain degree and consequently traces of oestradiol are found also in males.

> Hypotheses. Androgens have been said to foster the libido in women. In the "pill" progestin, a derivative of 19-nor-testosterone, is added to compensate for the libido-damping effect of oestradiol. On the other hand, oestradiol is said to promote fatherly behaviour in cooperation with the pituitary hormone prolactin. Prolactin became known as a female hormone since it promotes the growth of the milk glands in gestation. Now evidence is augmenting that this hormone fulfils biological functions also in men.

7.4   The Psychological Sex and Postnatal Sexual Development

7.4.1   The Brains of Women and Men Display Sex-Correlated Differences

When testosterone is administered to pregnant rats shortly before they give birth, both XY and XX offspring behave like males. To facilitate coitus, a female rodent normally displays a behaviour known as lordosis. She arches her back by elevating her head and rump, and moves her tail to one side. Following prenatal exposure to testosterone, females do not show this behaviour. Instead, they show the mounting behaviour typical of males. Similar observations have been made in other mammals. The phase of hormonal imprinting is generally shortly before birth but may extend in some species into postnatal life. In rats, hormonal influence affects a sexually dimorphic nucleus in the optic area. The brain develops differently in males and females and this different development may determine different behaviour in later life.

The phase of imprinting is generally ahead of birth albeit in some animals the subsequent behaviour can be modified with hormonal injections even after birth. When sexual maturity is reached (in humans in puberty) imprinting is refreshed and reinforced by augmented release of testosterone. Loss of production of testosterone in phases of primary imprinting and secondary reinforcement, or deficient testosterone receptors in the brain, might lead to female behaviour even in XY individuals.

Imprinting in terms of behavioural biology is likely reflecting synaptic connections in certain brain areas. For instance, sexually dimorphic ultrastructural differences in female and male brains have been reported in the olfactory brain and in a certain area of the optic centre in rats. In songbirds testosterone-dependent ultrastructural differences have been diagnosed in areas concerned with the programming of the male courting song. These areas commence proliferating in each spring.

Also in humans more and more differences in the structure and expression patterns of sex-related genes are being described. The whole brain of men is, in the statistical average, more voluminous and heavier than the brain of women. The male brains weigh 1,375 g on average and have a volume of 1,446 cm^3, the female brains tip the scales at 1,245 g on average and have a volume of 1,130 cm^3. However, larger overall volume and weight of the male brain do not mean a larger number of nerve cells and nervous connections in those areas of the brain which in evolution experienced expansion and elaboration in parallel to the mental abilities of men. The enlargement does not, or merely minimally, apply to the prefrontal cortex which has particular significance in cognitive processes and long-term memory. In several areas, however, correlations to focal points in aptitudes appear to exist. Here a few examples:

- Three dimensional functional magnetic resonance imaging- (fMRI-) based computation indicated that **gyri in the right prefrontal and right temporal cortex** of women are more complex than those of men. These gyri are assumed to reflect the number of nerve cells in the peripheral layer of the cortex and is, by the researchers referred to, correlated with cognitive, musical and lingual abilities. Also the volume of the **speech centres (Broca's and Wernicke areas),** related to the total volume of the brain, is larger in women than in men, and it is suggested to correlate this with the greater verbal agility of women.

- The **amygdala**, known to be of particular significance in emotions, is often larger in men than in women. This particular sex difference is also found in other mammals, whereas the neighbouring **hippocampus** to which the function of a switch box is assigned in the establishment of the long-term memory seems to be more prominent in the female brain.

- The psychologist and cognitive scientist finds some more notable differences for which, however, no material correlations have been found as yet. This applies, for instance, to the better ability of women to recognize faces and to retain social networks and relationships in memory, or the better visual thinking and spatial imagination of men.

- Here and there local groups of cells are found which express genes linked to the X or Y chromosome differently without the significance of these observations being obvious currently.

Significant and remarkable differences are found in the evolutionarily older regions of the brain known to be correlated with instinctive, sexuality-dependent behavioural patterns. Two examples are presented in the following:

- A small cluster of neurons in the anterior region of the **hypothalamus** (INAH3 = *Interstitial Nuclei of the Anterior Hypothalamus No. 3*) is in heterosexual (= non-homosexual) men twice as large as in women. The hypothalamus is considered the centre of sexual instincts. The specific function of this neuronal INAH cluster is currently not known. In the hypothalamus also 5α-reductase activity is found which locally converts testosterone into its active form. Erotic movie scenes are said to activate this area in men significantly stronger than in women.

- Another cluster of neurons that in women is smaller than in men was found in the *stria terminalis*. This cluster is designated **BSTc** (**B**ed nucleus of the *Stria Terminalis, part centralis*). This stria connects the emotions governing amygdala with the hypothalamus. The neurons of this cluster produce the neuropeptide somatostatin the significance of which at this location is not known. It would not be surprising if in the future neurons were identified correlated with feminine and maternal behaviour.

Literature sources about further sex-related differences in the brains of men and women are found in the reference list at the end of this book.

Transsexual humans have the irrefutable feeling to be born with the false physical sex (about potential causes see Box 7.1). Autopsy of transsexual individuals has spotted "feminine" characteristics in certain brain areas, such as the above mentioned BSTc, and in the *nucleus uncinatus* of the hypothalamus. For example, the number of neurons in the BSTc amounted to a value similar to that counted in women.

Yet, it is controversial whether the psychological sex, the **gender identity,** is exclusively hormonally imprinted. Studies on gene expression in the brain prompt hypothesizing that in some areas of the brain the XX or XY constitution controls sex-correlated gene activities without hormonal mediation. In particular it is unclear and **controversial whether and to what extent postnatal influences of the social environment co-determine the sexual ground disposition of humans.**

7.4.2   Puberty Is a Kind of Metamorphosis Through Which Sexual Development Is Finalized

During the final sexual maturation, in humans called puberty, numerous transformations occur. "Larval" cartilaginous elements become ossified, and many new structures and functions develop, collectively called **secondary sex characteristics.** These changes are also developmental processes. In humans the existence of a hormone has been postulated that, by analogy to prolactin in amphibians and juvenile hormone in insects (Chap. 20), prevents premature sexual maturation. When the production of this putative hormone ceases, the pituitary gland is hypothesized to emit increasing amounts of the gonadotropic hormones FSH (Follicle-Stimulating Hormone) and LH (Luteinizing Hormone). This takes place in boys as well in girls. Stimulated by the gonadotropic hormones, the gonads augment the production of testosterone or oestrogen, respectively. The main function of these steroid hormones is to trigger spermatogenesis or promote maturation of the oocytes.

Puberty proceeds particularly dramatically in boys/young men. Androgen hormones, especially the locally acting **dihydro-testosterone,** effect in boys, in functional interaction with growth hormones,

- Growth spurt at the age of 12.5–16 years; completion of length growth of the bones.
- Development of the secondary sex characteristics such as beard growth, voice break by enlargement of the voice box and reinforcement of the vocal cords.
- Increase of muscle growth and of protein synthesis in general (anabolic effects of testosterone).
- Enhanced production of sebum (perhaps acne).
- Aggressive behaviour.
- Production of sperm, capacity for libido and ejaculation.

Oestrogens effect in girls, in cooperation with growth hormones and prolactin,

- Growth spurt at the age of 10–14 years, that is beginning and ending about 2 years earlier than in boys; completion of bone growth.
- Development of secondary sex characteristics with breast, widened pelvis, fat depots at known sites (where they delusively enhance optic traits indicating fecundity).
- Enhanced production of sebum (perhaps acne).
- Start of the menstrual cycles.
- Compared to the effects of testosterone seemingly less striking influence of the oestrogens on the behaviour.

7.4.3   Periodic Hormonal Cycles Coordinate Cycles of Sexual Development

Cyclic periods of reproduction in invertebrates is a topic so rich in variants that special textbooks and monographs have been written to describe them. Cycles of reproduction are known to be annual (mayflies, many butterflies), lunar and semilunar (the marine midge *Clunio*). Enthusiastic naturalists can have the pleasure of seeing colourful mating displays in many fishes and amphibians. Sex-related, periodically repeating developmental processes occur in mammals as well, as is well known to those who collect horns and antlers. The monthly reproductive (menstrual) cycle of women is described in physiology textbooks.

Cycles of sexual reproduction are, as a rule, synchronized within a population and fitted into the seasons. In the geographical latitudes of Europe, North America and Asian regions of these latitudes correlations of behaviour and seasons and the biological meaning of synchronization are evident. Birds for instance must begin the duty of breeding in spring early enough so that offspring is capable of migrating before onset of winter, or have accumulated enough fat reserves to overwinter in northern latitudes. In this respect human beings are an exception. Although the ovarian cycle of the woman approximately corresponds to one lunar month, the individual cycles are not bound to a distinct lunar phase and they are not synchronized among different women, instead they are "free running" meaning that over the year menses shift more and more away from a distinct phase of the moon and the magnitude of this phase shift is different in different individuals.

Summary

Sexually propagating multicellular organisms generally possess bisexual potential. Though they appear as male and female individuals they are, as a rule, endowed with the whole inventory of genes required to develop both forms of sex. In **phenotypic sex determination** external cues decide which sets of genes, the ♂- or the ♀-specific sets, find expression. In **genotypic sex determination** an additional **key gene (selector gene, master gene)** takes the decision; it is transmitted to, or activated in, only one sex. As a rule, such key genes are located on "sex chromosomes" which differ from the other chromosomes (the autosomes) in bearing this gene and frequently occur as pair of morphologically and/or genetically differing chromosomes (**X versus Y**, W versus Z) or occur in different numbers (X versus XX). The chromosomes of a pair are transmitted by chance to the offspring, so that 50 % of the offspring get the X (or W), the other 50 % get the Y (or Z, or none = 0).

In *Drosophila* sex depends on the number of X chromosomes that complement the autosomes (XX → ♀; XY or X0 → ♂).

In humans and other mammals XX (and X0) determines ♀, while XY determines ♂. In spite of differing genetic constitution the bisexual constitution is expressed in that the primordia of the inner and outer genital organs initially are indistinguishable and competent to give rise to male or female organs. Subsequent divergent development takes place in several steps, these encompass:

(1) The genetic sex (XX or XY), (2) the gonadal sex (testis or ovary), (3) the somatic (corporal) sex with the inner sex-related structures such as oviduct or spermatic duct, the outer genitalia and the secondary sex characteristics, and **(4) finally the psychological sex (gender identity).**

The key gene that in humans takes the primary decision is the male-determining *Sry* gene. It is located on the Y chromosome and downregulates the antagonistically acting gene *R-spondin1* on chromosome1. In the absence of the inhibitory *Sry* the autosomal *R-spondin1* activates the WNT4/β-catenin signal cascade and thus promotes female development.

Sry codes for the transcription factor **SRY**, also known as **Testis-Determining Factor TDF.** Under the influence of SRY the indifferent gonad that contains both potential oogonia and spermatogonia proceeds to become a testis and this produces the **AMDF** (Anti-Müllerian Duct Factor) and the steroid hormone **testosterone**. While AMDF causes regression of the Müllerian duct, the potential precursor of the oviduct, testosterone directs the subsequent development of the somatic and psychological characteristics in a masculine direction.

The female development is the default option that is actuated if a Y chromosome and thus a *Sry* gene is absent, or the *Sry* gene is defective, and the *R-spondin1* gene can take over the leading role. Sexual development proceeds in a female direction also in later phases of foetal life when testosterone effects are lacking or too weak for whatever reasons. Normally in the XX constitution testosterone is converted by aromatase into the female sexual hormone oestradiol.

Sexually dimorphic development embraces also certain areas of the brain. These differences pertain to the ultrastructure, the number of certain neurons or the expression patterns of XX- or XY-related genes. Thus, sex-specific differences are found in clusters of neurons in the hypothalamus and the limbic system. They correlate with differing sexual behaviour. According to present knowledge divergences in the sexual development from the norm can causally be explained in many instances, though in humans often only hypothetically.

Potential genetic causes of divergent and disturbed sexual development are discussed in Box 7.1, disturbances caused by environmental substances in Box 7.2, and problems associated with the Y chromosome in Box 7.3.

Box 7.1: Disorders in Sexual Development and Possible Causes

Here terms are explained which are used in medical treatises, and some cases of atypical sexual development frequently encountered by physicians will be introduced.

A Somatic Deviations from the Norm Caused by Prenatal Disorders

Infertility in Women

Causes related to physical conditions include rudimentary ovaries associated with progressive loss of primordial germ cells (medical term: *gonadal dysgenesis*). The accompanying hormonal failure also prevents the development of secondary sex characteristics in either sex, resulting in a sexually infantile female appearance and infertility. The symptoms include lacking menses (*primary amenorrhoe*). Primary cause can be numerical or morphological anomalies in the set of chromosomes, for example

- **Triple X syndrome**. The complete set of chromosomes encompasses 44 autosomes plus 3 X chromosomes (caused by non-disjunction in the meiosis of the egg cell). **Karyotype** with the usual notation: 47/XXX, meaning that in the total number of 47 chromosomes three X are included. Relatively frequent (3 of 2000).
- **Turner syndrome** (synon. Ullrich-Turner syndrome). Growth-restricted women with 44 autosomes and a single X (karyotype 45/XO). A second X or a Y is lacking. Normal pubertal development and spontaneous menstrual periods do not occur in the majority of children. Individuals with Turner syndrome report an increased incidence of fractures in childhood and osteoporotic fractures in adulthood. Frequency: 1 Turner syndrome in 2000–2700 newborn girls.

- **XX Gonadal dysgenesis.** In spite of seemingly normal karyotype (46/XX) and normal feminine appearance the gonads are empty and do not induce puberty and menstrual cycles. In the individual case causes are unclear but certain X-linked genes required for normal ovary development are suspected to be non-functional; several other causes are discussed.
- **Swyer syndrome. Masculine karyotype (46/XY),** but feminine phenotype. People with this disorder have female external genitalia and mental gender identity but do not have functional gonads. According to present knowledge several causes must be taken into consideration:
  The *Sry* gene on the Y chromosome is defective or is lacking due to deletion.
  The Sox9-FGF9 system (Fig. 7.7b) was disturbed or insufficient in embryonic development.
  In the foetal period testosterone production was turned off or androgen receptor signalling did not work.

More genetically based causes of infertility are listed below under intersexuality.

Non genetic causes for female infertility can be manifold. For instance, it is suspected that long-term intake of the "pill" over several years causes infertility ("post-pill-sterility").

Infertility in Men

Infertility in men (formerly usually called sterility) can have numerous causes. The primary cause may be insufficient production of androgens (testosterone, dihydrotestosterone), or reduced responsiveness to androgens, for example because of low quantity or quality of androgen receptors. ▶

Box 7.1 (continued)

Consecutive implications may be: lacking or reduced sperm production (*oligozoospermy* or *azoospermy*) or deficient quality of the sperm, lacking or reduced production of accompanying substances normally supplied by the epididymis or prostate.

Possible genetic causes:

Klinefelter Syndrome Karyotype: Total of 47 chromosomes, among them 2X and 1Y. (47/XXY). The testis does not produce sperm (*azoospermy*); the stature displays a partially feminine appearance with enlarged breasts, and wide hips. Klinefelter syndrome occurs in about 1 out of 1,000 males.

More genetically based causes of infertility are listed below under intersexuality.

Non-genetic causes:

Xenoestrogens and Endocrine Disruptors.
In the last decades it was repeatedly reported that in industrial countries the number of sperm per ejaculate has decreased down to half. Many indications and animal assays direct responsibility to exogenous substances with diverse chemical structure but all being capable of disturbing the hormonal system if they enter our body by food, breathing air or skin. Among these substance collectively referred to as endocrine disruptors are many industrial products listed in Box 7.2 under "hormones and hormone-like substances". Among them are persistent derivatives of oestradiol and related oestrogens of the "pill" (see also Box 6.3). Many of these substances impair fertility also in domestic animals and even induce feminization of genetic male offspring in fishes, amphibians and birds. Supposedly some populations of fish and amphibians became extinct because no fertile males were in charge.

Intersexuality

Under this generic umbrella term all cases are subsumed in which the inner and/or external sexual organs have an intermediate appearance between the male and female norm. For instance, if the gonads are diagnosed as being ovaries whereas the external genitalia are more akin to the male phenotype. Conversely, if a testis (which may continue to stay in the body cavity) is associated with a feminine stature. Primary causes are, as a rule, hormonal disorders during the embryonic and foetal development. Frequently it is difficult to clearly assign the newborn a distinct sex and gender. Sometimes in puberty one or other sexual trait may gain the upper hand. Distressing cases became known in which the gender assignment of the parents in name-giving, in the choice of clothes and in sex-related education had been misguided and drove the adolescents during and after puberty in great psychological conflicts. In textbooks of gynaecology regularly the following forms of intersexuality are described:

- *"True hermaphroditism"* (**hermaphroditismus verus**). "True hermaphroditism" in medical terms is an intersex condition in which an individual is born with ovarian and testicular tissue, often in one and the same gonad called ovotestis. External genitalia are often ambiguous, although female traits prevail but the clitoris is unusually large. Biologists by training must reject the term "true" as used in the medical literature, because in biology true hermaphroditism means that an organism simultaneously or consecutively produces fertile sperm as well as viable egg cells and is capable of discharging both tasks, the male and the female, successfully. This is never the case in humans naturally (and only possible upon extensive corrective genital surgery) even when both ▶

Box 7.1 (continued)

spermatocytes and oocytes are present in an ovotestis or in different gonads. Spermatocytes do not give rise to fertile sperm. Any condition which prevents a complete tipping of the "balance" (Fig. 7.7b) in the one or the other direction could be the cause. Encountered karyotypes are 47XXY, 46XX/46XY, or 46XX/47XXY.

In rare cases intersexualism is based on chimerism. Two blastocysts one with the XX the other with the XY constitution have fused and given rise to a 46 XX/46XY chimera. Chimeras may exhibit various degrees of mosaicism.

- **Female intersexuality (*pseudohermaphroditismus femininus*) due to virilizing adrenal hyperplasia (adrenogenital syndrome).** In most cases reported such people are looked upon and educated as girls and are genetically indeed female (karyotype 46/XX). Yet, in the body cavity testes instead of ovaries are found. Further symptoms of masculinization (virilization) may be low voice, amenorrhea and muscle growth. As in 'true' hermaphrodtism the external sex organs look intermediate between the typical clitoris or penis. Such conditions arise when excessive secretion of adrenocortical androgens of the mother cause a somatic masculinization effect on the fetus or baby in the womb. Steroid hormones pass the placental barrier, enter the blood circulation of the foetus and influence its corporal and psychological condition. High androgen levels can result from tumours of the adrenal gland or from androgen-containing medications or contraceptive agents (or from doping in sports). A supporting cause may be insufficient aromatase activity in the mother or the girl herself causing insufficient conversion of testosterone into oestrogens. Incidence in Europe is said to be 1 to 7,000. Androgen-mediated virilization can also affect behavioural dispositions (see below Sect. 7.2)

The most frequent genetic disorder engendering accumulation of androgens in the female blood is lack of 21-hydroxylase, a basic enzyme of the steroid synthesis. Shortage of steroid hormones, especially shortage of cortisol, causes the pituitary to release increased amounts of ACTH (adrenocorticotropic hormone). As a consequence of the permanent stimulation by ACTH the adrenal gland experiences an enlargement. This hyperplasia in turn results in increased levels of androgens in the blood. The aromatases in the diverse tissues cannot completely convert this oversupply into oestrogens and androgens prevail. Long-term high stress which is associated with increased levels of cortisol in the blood is thought to have similar effects.

- **Male intersexuality (*pseudohermaphroditismus masculinus*), formerly designated testicular feminization syndrome.** The individuals are genetically normal male (karyotype 46/XY) but are born looking externally like girls. Feminine characteristics of various degrees develop due to insufficient or lacking synthesis of testosterone during the foetal phase, or insufficient conversion of testosterone into dihydrotesterone in the target areas, or, most frequently, insensitivity to androgens due to insufficient receptor-mediated responses. In this particular case the term **complete androgen insensitivity syndrome** has been proposed. The gene for the syndrome is on the X chromosome and codes for the dihydrotestosterone receptor. If the mutation is homozygous, target tissues cannot respond to testosterone. Internally, there is a short blind-pouch vagina and no uterus, fallopian tubes or ovaries. Testes do not ▶

Box 7.1 (continued)

migrate down into the scrotum but remain in the abdomen. The androgen insensitivity syndrome is usually detected at puberty when a girl should but does not begin to menstruate. The syndrome is not frequent (1: 20,000 up to 1: 60,000).

B Deviations of the Psychological Sex and Gender Identity from the Norm

Psychological Intersexuality, Transexualism

Transsexual humans have the strong feeling, often from childhood onwards, of having been born the wrong sex. Having a male body but feeling like a woman and desiring urgently to be a woman. Or one is a girl but wants to belong to the group of boys. A person's assigned sex at birth conflicts with their psychological gender. Potential causes may be:

- **The genotype is XY.** All sex-related events in embryonic and foetal development proceed as usual. A boy grows up. At the end of the foetal period production of testosterone is restrained and ceases up to puberty. Yet, shortly before and after birth, release of testosterone sinks down under a critical threshold, imprinting of the psychological gender should take place. But when in the proper brain areas too few androgen receptors are expressed, or when the production of testosterone ceases too early, the default program of female imprinting is turned on. The default program implies production of female sexual hormones such as oestradiol. In puberty, refreshment and reinforcement of the gender identity may also be under the regime of female hormones.
- **The genotype is XX.** As in case of the adrenogenital syndrome the XX fetus is exposed to undue high levels of testosterone but this excess' occurs not until the psychological gender is imprinted. The individual is physically a normal girl but prone to boyish behaviour. In some instances the cause has been found in a too high androgen production in the cancerous adrenal gland of the mother. (However, being a tomboy is a behaviour in the normal range of female behaviour as is a soft attitude in boys).

Transvestism (also called transvestitism) is the practice of wearing clothing traditionally associated with the opposite sex or gender. Transvestite refers to a person who cross-dresses. Since transvestites often display overt transgender feelings and behaviour typical of the opposite sex the word transvestitism often has additional connotations indicating psychological intersexuality.

Homosexuality

Homosexuality refers to "*a person who is sexually attracted to people of their own sex.*" (Oxford English Dictionary). If psychological intersexuality is not obvious and men identify themselves as men, women identify themselves as women, homosexuality is difficult to explain. Traditional explanations hold the brain of a gay being partially female structured or imprinted, the brain of lesbian being partially male structured or imprinted. Several studies on ultrastructural peculiarities of certain sex-related areas in homosexuals as well as studies based on magnetic resonance imaging appear to support such notions (see list of references at the end of this book). Yet, currently, there is no scientific consensus about the specific factors that cause an individual to become homosexual. Sexual orientation probably is not determined by any one factor but by a combination of genetic, hormonal, and environmental influences.

Box 7.2: Disorders of Sexual Development and Fertility by Hormone-like Environmental Chemicals

In Box 6.3 (Dangers to the developing embryo) the key words "hormones and hormone-like substances" pointed to the existence of environmental chemicals exerting hormonal effects or interfering with hormonal balance if they enter the body of animals or humans. Since such substances can disrupt endocrine control circuits they are designated **endocrine disruptors**. Any system in the body controlled by hormones can be derailed by hormone disruptors. Many of such substances exert influences similar to oestrogens and, therefore, are called **xenoestrogens** or, when of vegetable origin, **phytoestrogens.** Besides, there are also substances acting as anti-oestrogens, androgens or anti-androgens.

With arguments of varying strength endocrine disruptors have been made responsible for

- The decrease of sperm counts in industrial countries (among others Denmark, Germany, Norway, Japan, U.S.A) in the period of 1950 to 2000. Reported decreases were from 40 million/mL down to 20 million/mL ejaculate.
- Increase in breast cancer and osteoporosis (suspected, among others, DDT, artificial hormones of the contraceptive pill).
- Increase in incidence of testis and prostate carcinomas.
- Malformations of the genitalia and urogenital system in wild animals and also in humans; hypospadias (split of the urethra, a hypospadic urethra opens anywhere along the penis).
- Premature puberty (uncertain correlations).

Numerous compounds from the long list of endocrine disruptors (see below) diminish fertility in laboratory mice and domestic animals, and cause feminization of genetically male offspring in fishes, amphibians and birds.

An effect that in the laboratory is well reproduced and well quantified is the induction of vitellogenins in males by xenoestrogens. Vitellogenins are phosphoproteins which normally are produced by females (in land vertebrates in the liver), distributed by the blood circulation and eventually taken up by oocytes and stored in the yolk mass. It has been conjectured that some populations of fishes and amphibians became extinct because the number of fertile males sank below a critical value. (In some marine snails such as the waved whelk *Buccinum* and the common periwinkle *Littorina* these substances cause virilization by inhibiting aromatase activity and, thus, the conversion of testosterone into oestradiol).

Endocrine disrupting compounds encompass a variety of chemical classes, including medical drugs, pesticides, compounds used in the plastics industry and in consumer products, industrial by-products and pollutants, and even some naturally produced plant chemicals. Synthetic xenoestrogens include widely used industrial compounds, such as PCBs, BPA and phthalates, which have oestrogenic effects on a living organism.

Examples:

1. **Phytoestrogens** are found, for example, in form of flavones, in soy, hops, and cabbage. A chemical offered in catalogues of chemical companies is the tyrosine-kinase inhibitor genistein.

2. **Environmental chemicals,** identified as being, or suspected to be, endocrine disruptors are numerous, so for example

Plastic Components, Plasticizers

- **Bisphenol A (BPA)**, is precursor substance of epoxy resins and polycarbonates used to produce plastic bottles, glasses, casings for electronics, paints, enamels, and adhesives. Even paper is frequently coated with a BPA containing clay for printing purposes. Bisphenol A is ▶

Box 7.2 (continued)

omnipresent. In the bioassay it diminishes the fertility of rodents. Correlations point to elevated rates of diabetes, mammary and prostate cancers.

- **Polychlorinated biphenyls (PCBs)** are a class of chlorinated compounds used as industrial coolants and lubricants, rubber-like sealants, and cable sheathings. A worldwide ban on PCBs concerns their use in open systems, but they still are omnipresent in closed systems. They have neurotoxic effects and cause feminization of male animals.

Tensides, lubricants

- Alkylphenols (**nonylphenol**, octylphenol, pentylphenol). Long-chain alkylphenols are used extensively as detergents, additives for fuels and lubricants, polymers, and as components in phenolic resins. They are present in numerous everyday products such as cleansing agents, PVC foils, paints, textiles, leather, paper, and pesticides. They accumulate in the food chain and resist complete degradation in sewage treatment plants. They diminish fertility in fishes and exert teratogenic effects in embryos. A well known octylphenol in our laboratories is **triton-X**.

Further industrial chemicals

- **Polybrominated diphenyl ethers (PBDEs)** are a class of compounds found in flame retardants used in plastic cases of televisions and computers, electronics, carpets, lighting, bedding, clothing, car components, foam cushions and other textiles. PBDE's are structurally very similar to PCBs, have similar neurotoxic and feminizing effects
- **Tributyltin**. Uses include wood preservation, antifungal action in textiles and industrial water systems, such as cooling tower and refrigeration water systems. Once widely used as anti-fouling paint to protect marine vessels it has been banned by the International Maritime Organisation. Tributyltin exerts strong teratogenic and masculinizing effects in fish and marine snails (*Buccinum, Littorina*).

- **Pesticides. DDT** (Dichloro-diphenyl-trichloroethane) was first used as a pesticide against Colorado potato beetles on crops beginning in 1936. An increase in the incidence of malaria, epidemic typhus, dysentery, and typhoid fever led to its use against the mosquitoes, lice, and houseflies that carried these diseases. Use of DDT is no longer permitted in the U.S.A and European countries but DDT is still used as anti-malarial insecticide in Africa and parts of Southeast Asia. It kills the larvae of mosquitos of the genus *Anopheles* known to be transmitters of malaria. Side effects of DDT include neurotoxic and weak oestrogenic effects, while androgen effects are assigned to its degradation product DDE.

Medical drugs

- **Diethylstilbestrol DES, an informative example of bitter disappointments.** There are more than a few women who are unable to bear a fetus up to regular term. Early labour causes high risks of miscarriage. To prevent abortion the drug diethylstilbestrol (DES), a non-steroidal oestrogen, was developed. Prior to its ban in the early 1970s, physicians prescribed DES to as many as five million pregnant women to block spontaneous abortion. It was discovered after the children went through puberty that DES affected the development of the reproductive system and caused vaginal cancer.
- **"The pill"** with its mixture of chemically modified and persistent oestrogens, gestagens and androgens. These modified steroids, in particular 17α-ethinylestradiol and progesterone-derivatives, are not sufficiently decomposed by microorganisms ▶

Box 7.2 (continued)

in sewage treatment plants and is detected not only in effluent flows but is traceable even in drinking water. 17α-ethinylestradiol causes feminization in fishes with even higher effectiveness than physiological levels of the natural hormone 17β-estradiol. They potentially diminish fertility in men.

Biochemical and Physiological Effects. Endocrine disruptors block, or interfere with, hormonal control circuits at many possible levels. Details cannot be listed here. We refer to the list of references at the end of this book.

Box 7.3: Men: A Gender Becoming Extinct?

Evidence is strong that the Y chromosome is a shortened X chromosome. In gleefully presented articles in diverse magazines, but also in contributions to discussions found in scientific journals, the view is taken that the Y chromosome is a crippled, stubby X, men were supplied with fewer genes and, therefore, be the underprivileged gender.

Yet, this argumentation is not correct, at least not with respect of gene numbers. On the one hand, men possess an X as do women, though only one X while women have two. But quantitative differences in X-linked alleles become in women levelled by heterochromatization (silencing) of one of the two X chromosomes as early as in the cleaving stage of embryonic development (this X becomes visible as Barr body, Fig. 12.24). On the other hand the Y chromosome bears about 78 genes which are absent in the X chromosome, so for instance, the *Sry* gene and some genes involved in spermatogenesis. Besides, if findings in rodents apply to humans as well, a paternally imprinted set of chromosomes is indispensable for development of a functional placenta.

However, women are indeed genetically at an advantage over men in one respect. Even if in females one of the two X chromosomes in each cell is almost quiet, in case of need two editions from all X-resident alleles are available. And when in one cell the paternal X is inactivated, in other cells this paternal allele may be active and the maternal allele instead

inactivated. In one organ, for example the liver, it may suffice when a part of the cells is able to supply a fully functional protein. In the eye it may suffice when a part of the rods contain a particular functional opsin variant to ensure a basic level of colour discrimination. In the long-term, crossover can ensure preservation of functional alleles in a population from generation to generation.

Not so in men. In men an X-linked defective allele does not have a partner. The result can be, for example, red-green blindness, or the effect may even be lethal. It may also be worrisome that many X-resident genes are required for brain development. Here the growing embryo is left without alternatives. Possibly this is one of the reasons why male embryos die more often than female embryos. Certainly, more boys are born than girls (105 to 100), but in the embryonic phase of life XY embryos come in even more superior numbers compared to XX (120 to 100). According to the prevailing view this initial imbalance is due to faster swimming activity of Y-carrying sperm compared to the heavier X-carrying sperm, but subsequent lethality is not a function of the previous swimming speed.

Quite serious is the discussion whether in the long term Y-linked genes are subject to degeneration. Though the Y chromosome bears backups from many of its specific genes in palindromic tandem arrangement, an exchange between Y chromosomes of different provenances by crossover is not possible. ▶

Box 7.3 (continued)

Moreover, the Y chromosome is prone to deletions and the shortening might continue until the male sex dies out (as in certain parthenogenetically reproducing lizards). The female part of human kind could continue to exist with the help of clinics being skilled at cloning. Men would then exist as genetically separated clonal lineages.

According to our prognosis, the contemporary generation, and after it some more generations, may look forward the extinction of the male sex and gender with the same serenity as it may look forward the often predicted imminent end of the world.

In Preparation for New Life II: Gametogenesis – The Development of Egg Cells and Sperm and Their Provision with Heritable Reserves

8

8.1 Germ Line and Primordial Germ Cells

8.1.1 Primordial Germ Cells Are Set Aside Early in Development or Derive from Stem Cells Late in Development: Maternal "Germ Plasm Determinants" Versus Inductive Determination by the Somatic Environment

Frequently in embryo development cells which are not used to construct the **soma** – the body of the new individual – are put aside and remain in stock as **primordial germ cells**. They are spared from constructing the mortal body because their task is to transmit potentially immortal life in the germ line (Figs. 8.1 and 8.2) to the next generation before the soma, the individual's body, dies.

Early versus late commitment of primordial germ cells. Nowadays distinction is drawn between early predetermination dependent on **maternal factors**, collectively referred to as **germ plasm**, and late assignment to stem cells of germ cell fate by the somatic surroundings.

Extreme cases of early commitment are known from some nematodes and from *Drosophila*.

- **Nematodes**. When observing the chromosomes in the cleaving embryo of the roundworm *Ascaris* (now, *Parascaris equorum* or *Parascaris univalens*) at the turn of the last century, Theodor Boveri (1862–1915) observed unusual behaviour in those chromosomes. This nematode has only two pairs of large compound chromosomes (i.e. several chromosomes sticking together). The first divisions are **asymmetric** (compare *C. elegans*, Figs. 4.20 and 9.7). After the first cleavage, the compound chromosomes in one of the two blastomeres fragment into many pieces and many of the fragments are lost. This phenomenon of **chromatin diminution** occurs in the cell destined to participate in the construction of the body. The eliminated chromatin predominantly contains repetitive sequences and a few genes that are required in the development of the germ cells but not in soma cells. The task of giving rise to germ cells is assigned to the second blastomere and in this cell the chromosomes remain intact. This second blastomere, designated P1, is the founder cell of the **germ line**. During the next divisions the story is repeated: only the cells on the direct route to the primordial germ cells preserve the entire chromosomes, whereas all future somatic cells lose chromatin. **The P lineage leads to the primordial cell P4**. This P4 cell is set aside and later in the adult worm used for producing germ cells. By experimental interference, such as centrifugation, Boveri was able to demonstrate that localized "cytoplasmic determinants" protect chromosomes of the germ line cells from being fragmented. Moreover, these cytoplasmic factors were thought to commit their host cells to become germ cells.

Fig. 8.1 Germ line (*red or blue*), shown using the example of great crested newt (*Triturus cristatus*)

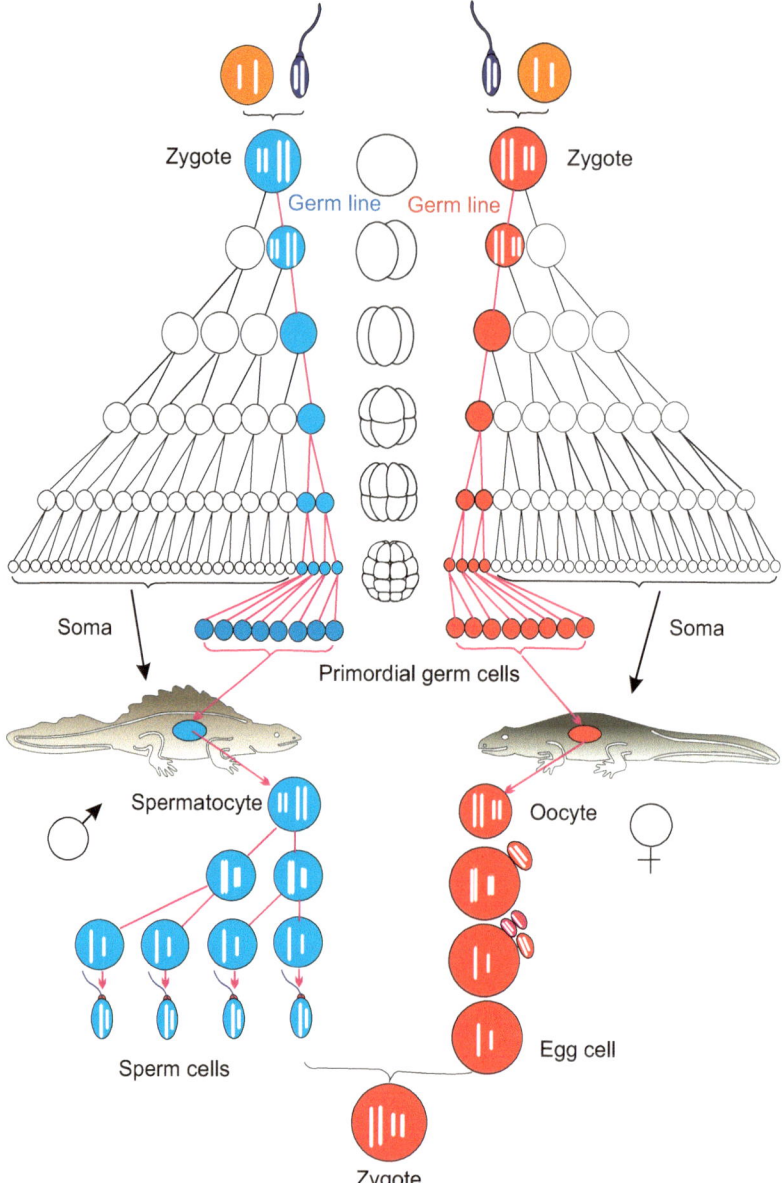

In the nematode *Caenorhabditis elegans* (Figs. 4.19 and 4.20) which displaced *Ascaris* as model organism, chromatin diminution is not known. Nevertheless also in the eggs of *C. elegans* such germ line-determining components exist embedded in granules. These are allocated to the P line (Figs. 8.3a and 4.20). These granules embody, among other components, mRNA of genes that are homologous to the *Drosophila* genes **vasa**, **piwi** and **nanos**. Their activity is distinctive of the germ line in animals in general as discussed below under Sect. 8.1.2.

- In **Drosophila** the primordial germ cells are the first cells to be formed in the cleaving embryo. They are called **pole cells** because they arise at the posterior pole of the egg (Fig. 8.3b, also Figs. 4.26 and 4.30). As the

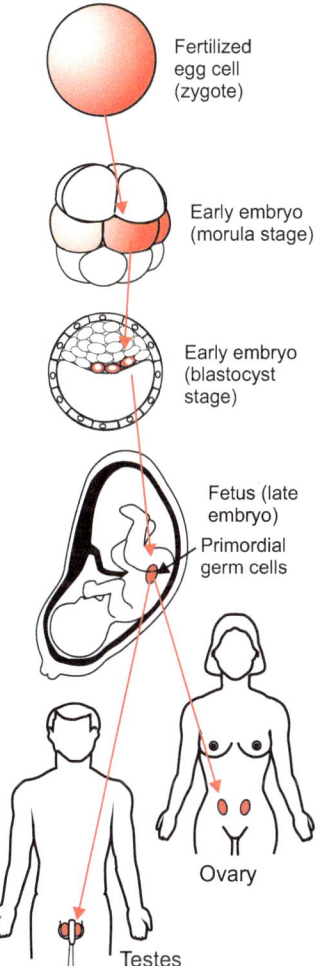

Germ line, general

Soma cells (body cells)

Primordial germ cell

Germ line, human

Fertilized egg cell (zygote)

Early embryo (morula stage)

Early embryo (blastocyst stage)

Fetus (late embryo)
Primordial germ cells

Ovary

Testes

Fig. 8.2 Germ line in humans

pole cells form, they enclose **pole granules** similar to those seen in nematodes. These are fibrous conglomerates containing considerable protein and RNA, and embody **germ cell determinants**. If pole plasm is removed with a micropipette from the posterior egg pole and transferred into the anterior pole of another egg, oogonia or spermatogonia arise at this ectopic site (Fig. 4.20) (ectopic from Greek *ek*, *ekto* = outside, *topos* = place; outside the normal place). Not all of the determining components are identified as yet but it is established that the products of the maternal genes *vasa*, *nanos* and *piwi* are of particular significance. An unusual component of these granules is RNA of mitochondrial origin which has been released by mitochondria. But this mitochondrial RNA probably is not a decisive germ cell determinant. The determinants are allocated to the posterior region of the egg under the organizing influence of *oskar*. When injected beneath the anterior pole of the egg, the *oskar* product causes ectopic development of pole cells at the anterior end of the embryo. These, however, will not find their way into the gonads. Only pole cells properly located at the posterior pole are eventually incorporated into the gonads.

- **Amphibians.** We discussed earlier the remarkable work that has made it possible to follow the developmental heritage of every cell in nematodes, but even in vertebrates the lineage of the primordial germ cells can be traced back to the uncleaved egg. In *Xenopus*, certain staining procedures demonstrate a particular "germ plasm" – granular cytoplasmic constituents close to the vegetal pole of the egg (Fig. 8.3c). Like the pole cytoplasm in *Drosophila* this region harbours RNA-loaded granules. Later these granules are found in the primordial germ cells that arise in the region of the embryonic hindgut and migrate along the mesenteries (devise to suspend the intestine) into the gonadal ridges (the rudiments of the ovaries or testes).

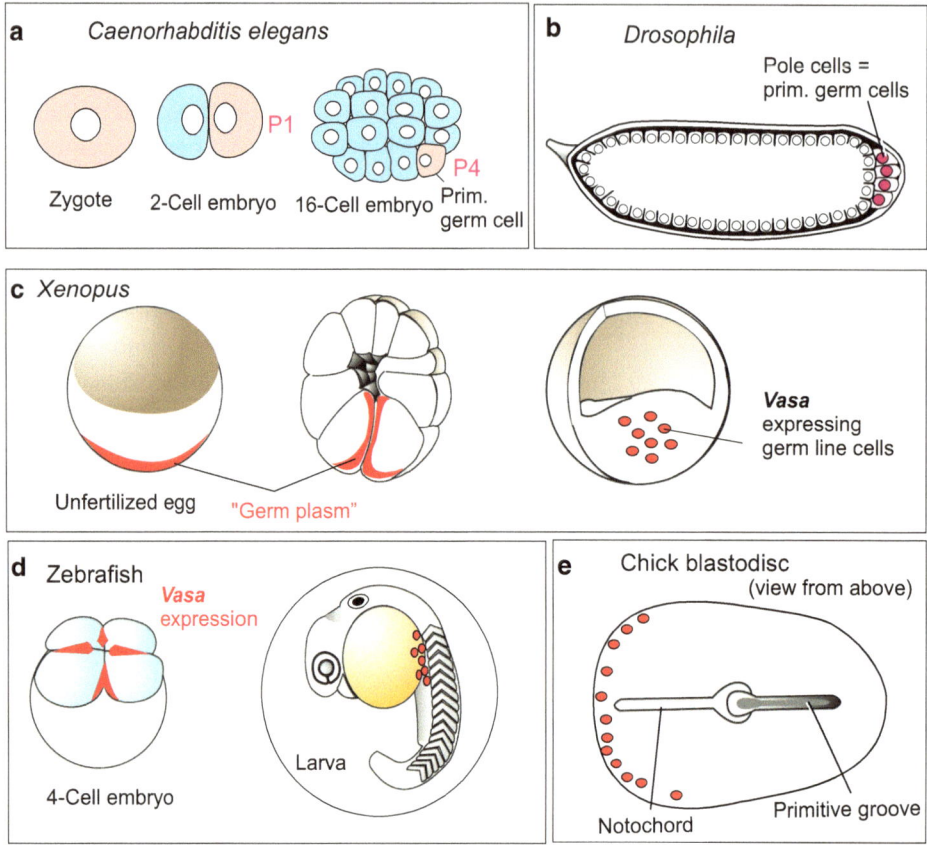

Fig. 8.3 (continued on p. 245)

The term **germ plasm** ("*Keimplasma*") was coined by the zoologist August Weismann at about 1893 but had a meaning very different from the meaning the term obtained in the subsequent literature, and has in the present literature. In Weismann's terminology germ plasm is the entirety of all – in those days still unknown – material carriers of hereditary information. His "germ plasm" can rather be equalized with the term "genome" of our days. Presumably stimulated by Theodor Boveri's observation on chromatin diminution in *Ascaris* Weismann assumed only the cells of the **germ line** ("*Keimbahn*", his term also) would be equipped with the entire hereditary material, while soma cells would obtain only those parts required for, and being the cause of, differentiation.

Late determination:

- **Hydrozoa.** Particularly late is the segregation of soma and germline in *Hydractinia echinata*. Here in the gonads a last asymmetric division of pluripotent precursors generate primordial germ cells and somatic envelope cells. Up to this point the germ line passes through the population of pluripotent interstitial stem cells which can give rise to somatic and generative cells in the mature animal at any time.
- **Vertebrates, birds.** In the avian (and likewise the mammalian) embryo the first cells that can

Fig. 8.3 (**a–f**) Origin of primordial germ cells in various animals. (**a–e**) The gene *vasa* is expressed in the germ line possibly of all animals. Sites where *vasa* products have been found are marked in red. (**f**) In mammals/humans traditional notions see the origin of the primordial germ cells in the allantois, recent reports find stem cells with the potency to give rise to germ cells in blood islands which are dispersed in the cover of the yolk sac

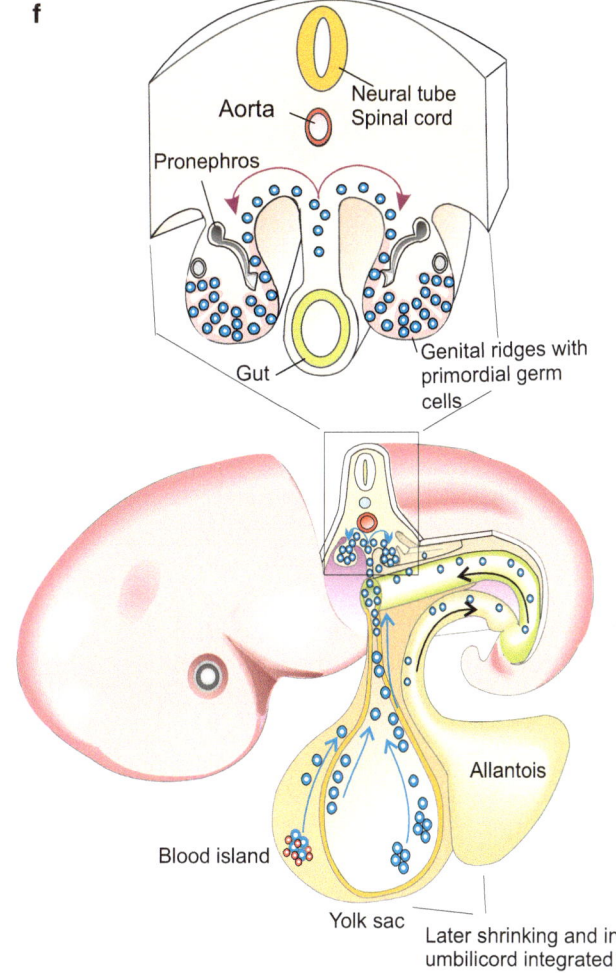

be identified as germ cell precursors are found in the extra-embryonic mesoderm at the site where the allantois, the embryonic precursor of the urinary bladder, begins to form. Subsequently germ cell precursors appear in the posterior region of the yolk sac (Fig. 8.3). Compared to amphibians this emergence of germ cell precursors appears to be a late event. Recent studies attribute commitment to inductive influences of the microenvironment, called a niche. Induced by these local influences some cells withdraw from companions in fate to enter the path to germ cells. Once committed, primordial germ cells set off to find the distant primordia of the gonads. Starting from their home place they take a similar route as in the amphibian embryo (see Fig. 15.2). In part they use, like the cells of the immune system, the blood stream to drift near to their destination. The destination sites send chemotactic signals (SDF-1-factor), to pinpoint the location of the destination.

- **Mammals** appear to take an intermediate position between early specification and late commitment. An early though weak expression of germ line specific genes is already seen in the blastocyst. The first morphologically identifiable primordial germ cells are found in the blood islands dispersed over the yolk sac

(Fig. 8.3f). They are descendants of pluripotent stem cells which also give rise to blood cells (Figs. 17.2 and 17.3). From here they emigrate, taking a similar route as in amphibians and birds, to reach the genital ridge where the gonads form. During migration expression of germ line specific genes is intensified.

8.1.2 Early Determination: Maternal Constituents, Referred to as "Germ Plasm" and a Complex Called "Nuage" Imprint Primordial Germ Cells

As stated above in *C. elegans* "P granules", in *Drosophila* a "pole plasm" and in amphibians a "germ plasm" have been identified. In all these cases the particular plasm marks future germ cells. Similar types of "germ plasm", also called "germinal granules", are found in germ cell precursors of increasing numbers of other animal species as well.

The pole or germ plasm is an accumulation of granules that are not enclosed by membranes and contain dense aggregates of RNA and protein (RNP particles). This applies also to the P granules in the egg cell of *C. elegans*.

In amphibians, fishes and also invertebrates such as the *Platynereis* one finds, in addition to, or instead of, these granules, an accumulation of fibrous structures and mitochondria. Such aggregates have been described in the oocytes of more than 80 animal species and are referred to as **nuage** (from the French word for cloud). They harbour a large collection of enzymes and RNA species, especially DEAD box helicases, Tudor domain proteins and PIWI/argonaute proteins which bind piRNAs (piRNA = piwi-interacting RNA). In the ovary of the zebrafish 42,856 different piRNAs were found; they belong to the category of non-protein-coding, small RNA's and in part derive from transposons.

Piwi proteins and piRNAs are also components of the **chromatoid bodies CB** in the spermatocytes and are considered to be generally germ line specific.

Germ plasm, nuage aggregates and CB particles contain, besides differing constituents, also identical gene products, for example **VASA** (a RNA helicase) and the transcription factor **OCT-4**. In mammals, no dense conspicuous granules attracted attention in the migrating germ cell precursors up to now, but cloudy and granular nuage-like material is described in oocytes and spermatocytes, and products of the genes VASA and Oct-4 are present.

Currently, the following functions are assigned to components of the germ plasm:

- A leading part in **committing cells to become germ cell precursors** is assigned to the genes *vasa* and *nanos*.
- **Control of transcription and translation in the germ line** cells is thought to be a main function of the *piwi* RNAs and proteins (*piwi* stands for *P-element induced wimpy testis;* in humans they are also called *hiwi* and in mice *miwi*). It is assumed that binding of small *pi*RNAs onto messenger RNAs interrupts protein synthesis.
- **This translational control encompasses control of transposons** in the germ line cells. In particular retrotransposons (see Chap. 12) need to be prevented from making mischief by jumping in the genome and disrupting genes.
- **Preservation of pluripotency in the germ line**.

Commitment to germ cell fate and preservation of pluripotency are considered in more detail in the following section.

8.1.3 Products of the Genes *Vasa, Nanos* and Pluripotency-Conferring Genes Such as *Oct-4* Make Cells of the Germ Line Primordial Germ Cells: Also in Humans

The pole plasm in the egg of *Drosophila* and the germ plasm in the embryo of *Xenopus* are known for decades. But not until 1999 a gene product promising to be a decisive component of the germ cell determinant was found. It is the RNA derived from the gene **vasa**. Homologues of this gene, initially identified in *Drosophila,* have been found in members of many animal phyla, in cnidarians, planarians, in the ecdysozoans *Caenorhabditis elegans* and *Drosophila*, in the annelid *Platynereis* and in chordates such as tunicates (Ascidia), the zebrafish *Danio*, the clawed frog *Xenopus*, the chicken, the mouse and humans. Defects in both alleles result in infertility.

Most remarkably, the products of the gene *vasa*, especially the VASA protein, often accompany the entire germ line, from the egg cell up to the mature sperm or egg cells produced to give rise to the next generation. Already the egg cells contain maternally generated *vasa* mRNA, meaning that the germ line is specified under the regime of maternal egg components. Antibodies to VASA protein detect VASA in the embryo of the zebrafish along the first cleavage furrows (Fig. 8.3a), in the cytoplasm of founder germ line cells (for example the in the P line of *C. elegans*), in migrating primordial germ cells (Fig. 8.3b, c) and eventually in maturing sperm and oocytes. VASA protein or *vasa*-mRNA are markers of the germ line also in organisms in which morphological tags such as germ line specific granules are not seen.

Besides VASA also the gene products of *nanos, Nanog* and *Oct-4* are indicators of germ line cells. We had met the gene *nanos* in *Drosophila* where its product plays a role in the development of the abdomen. Now two more functions can be added:

- NANOS enables primordial germ cells to migrate to and into the gonads.
- NANOS and VASA in cooperation with the **transcription factor OCT-4** aid the cells of the germ line in **preserving stem cell characteristics**. (NANOS and VASA themselves are not transcription factors but intervene in the utilization of RNA.) Knockout mutations of *nanos, Nanog, vasa* or *Oct-4* incapacitate mice to generate germ cells.
- **The transcription factor OCT-4** is already working in the embryonic stem cells and enables pluripotent stem cells in cooperation with other factors to preserve pluripotency

in the future life. Remarkably a gene homologous to *Oct-4* confers puripotency to embryonic cells and stem cells not only in vertebrates but already in Hydrozoa (*Hydractinia*). We will later in Chap. 18 (stem cells) see, that *Oct-4* is functionally supported by further pluripotency-conferring genes, in particular by **c-myc**, and members of the **Sox** and **kfl** families.

8.1.4   In Vertebrates the Decision Whether the Path of Differentiation Leads to Egg Cells or Sperm Is Finalized Only in the Gonads

In the early vertebrate gonad the immigrating primordial germ cells colonize the peripheral layer of the 'female' **cortex** as well as the internal 'male' **medulla** in the centre of the still indistinct gonad (Figs. 7.2 and 7.4). At this stage the embryo's sex has not yet definitively been determined. If the gonad develops to a testis, directed by the master gene **Sry** on the Y chromosome (Chap. 7), the primordial germ cells in the cortical layer perish, and those of the medulla differentiate into **spermatogonia**. In females there is no Y chromosome and, therefore, no *Sry* gene. In the absence of *Sry,* and thus of the SRY factor, the gonad develops into an ovary by default. Only the primordial germ cells of the cortex survive and become **oogonia**.

8.2   Oogenesis: Manufacturing and Provisioning of the Egg Cell

8.2.1   In Many Animals Including Vertebrates, the Nuclei of the Oocytes Exhibit Lampbrush Chromosomes, rDNA Amplification and Multiple Nucleoli

Oogonia undergo a phase of mitotic proliferation (Fig. 8.4; see also Fig. 6.1). After these rounds of

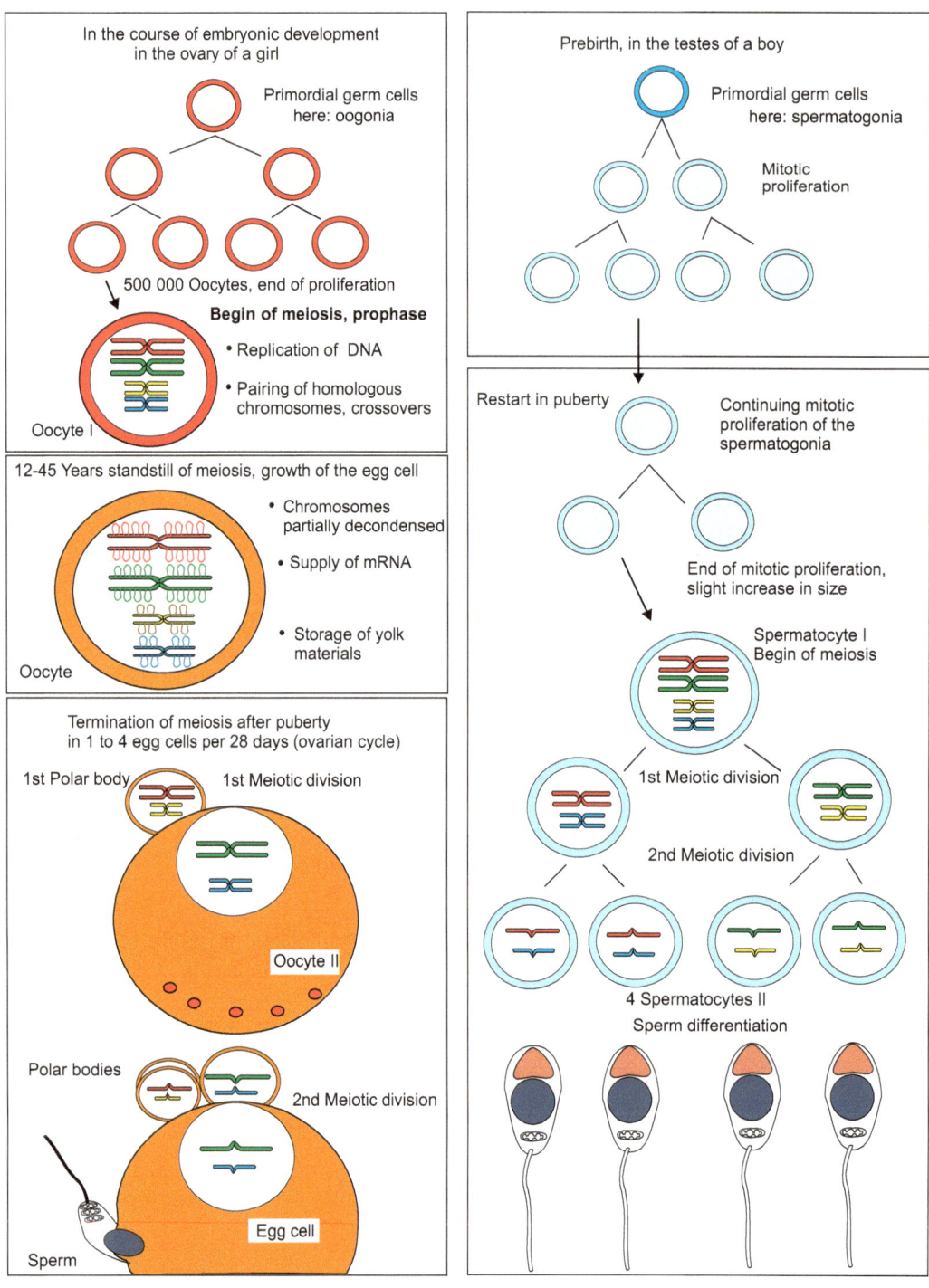

Formation of germ cells

Fig. 8.4 Gametogenesis (oogenesis and spermatogenesis) in humans in overview. Oogonia and spermatogonia still are diploid and proliferate mitotically. Upon termination of proliferation one speaks of oocytes I or spermatocytes I; these enter meiosis. In spermatogenesis meiosis does not take place until puberty. In oocytes meiosis starts very early in the embryonic phase of life but then is locked for 12–45 years before egg cells undergo terminal maturation. Meiosis results in four haploid cells. In oogenesis divisions are extremely asymmetric. Three of the four cells, called polar bodies, are small and eventually perish, and only the large oocyte II survives. Spermatogenesis gives rise to four equal spermatocytes II which undergo differentiation to four sperm cells

division have ceased, the germ cells are termed **oocytes**. The ovary of a female human fetus contains about 500,000 oocytes. After the birth of a girl the proliferation ceases, in contrast to spermatogenesis in a boy, where proliferation resumes in puberty. With the beginning of puberty about 90 % of the oocytes die.

As early as in the 3rd to 7th month of human female foetal development, oocytes enter prophase of meiosis. The chromosomes duplicated earlier during S-phase. In zygotene of prophase the homologous chromosomes, each consisting of two chromatids, pair along their long axes, forming **tetrads** (bivalents) and held in correct alignment by a **synaptonemal complex**. But then, remarkably, **prophase is interrupted for a prolonged time period lasting, in humans, 12–45 years**, The chromosomes remain paired but decondense and form lateral loops, taking on the appearance of 'lampbrushes' – brushes once used to clean oil lamps – , therefore their designation **lampbrush chromosomes** (the occurrence of lampbrush chromosomes in humans appears to be not yet definitively established). Simultaneously **multiple nucleoli** appear.

- **Lampbrush chromosomes** (Fig.8.5): The lateral loops extruding from the paired meiotic chromosomes represent stretches of largely uncoiled DNA and indicate busy engagement in RNA synthesis. The transcripts are packed in RNP particles and shipped into the cytoplasm. Eventually, the egg cell is filled with RNP particles, enclosing many species of mRNA, in preparation for the unfolding of the program of early embryo development.
- **rDNA amplification and multiple nucleoli**: A large quantity of **nucleoli** (multiple nucleoli) appear, indicating production of an enormous number of ribosomes. Their appearance is the manifestation of a previous selective amplification of ribosomal genes, called **rDNA amplification**.

In the genome of *Xenopus*, the best studied species in this respect, the genes for the ribosomal 18 S, 5.8 S and 28 S RNAs are clustered together in a single transcriptional unit (Fig. 8.6). This transcriptional unit has already been multiplied in evolution by repeated gene duplications,

resulting in tandem array of up to 450 copies of the rDNA genes, even before the rDNA becomes additionally amplified in oogenesis. The array is sometimes referred to as the **nucleolar organizer** because the factory to produce ribosomes is constructed along this region of the chromosome. Normally, one factory is built on each of the two homologous chromosomes in somatic cells. Thus two nucleoli can be seen in the nucleus. Since *Xenopus* is tetraploid, up to four nucleoli can be expected in somatic cells. But in oocytes many more are seen. This is because an additional multiplication, called **amplification**, takes place. Through selective replication by way of a rolling circle of DNA, from each of the preexisting $4 \times [450]$ copies about 250 copies are generated. Eventually, 1,000 additional pieces of rDNA are present. Each of these rDNA copies detaches from the coding DNA strand, is closed into a circle and supplemented by proteins. Thus 1,000 factories for producing ribosomes are generated. The 1,000 nucleoli each contain 450 copies of three (18 S, 5.8 S, 28 S) of the ribosomal genes. The missing 5 S rRNA is procured as transcripts of the 24,000 copies of the linearly multiplied 5 S gene present in the regular genome. These 24,000 copies are clustered on another chromosome and have been multiplied in evolution. In the additional 1,000 nucleoli about 10^{12} ribosomes are assembled. Without gene multiplication the frog would require 500 years instead of a few months to produce such an amount! The egg is prepared to undergo rapid development because it has an enormous potent machinery to produce proteins.

8.2.2   Somatic Cells Often Assume Additional Nursing Duties and Help to Raise Giant Egg Cells

The oocyte of vertebrates produces mRNA and ribosomes by itself. But yolk proteins and yolk lipids, which serve as energy stores or building materials, are not made by the oocyte. **Vitellogenins**, the protein precursors of the yolk materials, are produced by the liver, transferred via the blood to the ovary, taken up by follicle cells and handed over to the oocytes (Fig. 8.7).

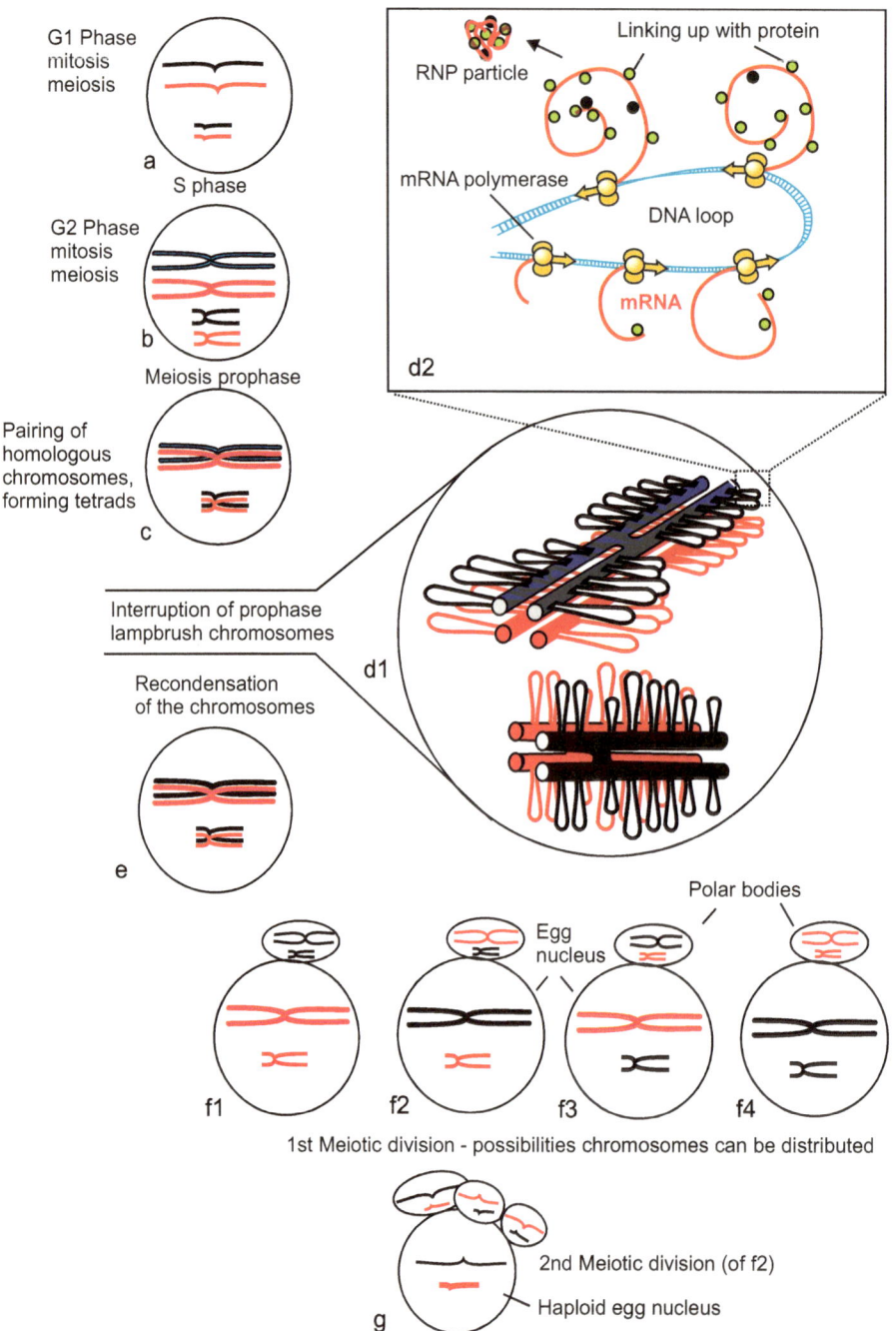

Fig. 8.5 (**a–g**) Oogenesis in vertebrates I. Meiosis is interrupted in prophase and the stage of lampbrush chromosomes is intercalated. Along the DNA loops protruding from the lampbrush chromosomes transcription takes place. Polymerases migrate along the DNA loops and transcribe mRNA. This mRNA is parcelled together with protein and transported into the cytoplasm in the form of RNP particles. The number of possible ways in which chromosomes become distributed in the subsequent meiotic divisions is a function of the number of chromosomes present. In any case two divisions are required to reduce the diploid 4C condition to the haploid 1C condition. C means amount of DNA; 1C = amount in the haploid condition; in the diploid condition DNA amounts to 2C before replication and to 4C after replication. 4C is the condition at the onset of meiosis because the DNA is already replicated = duplicated. Therefore two divisions are needed to reduce 4C down to 1C

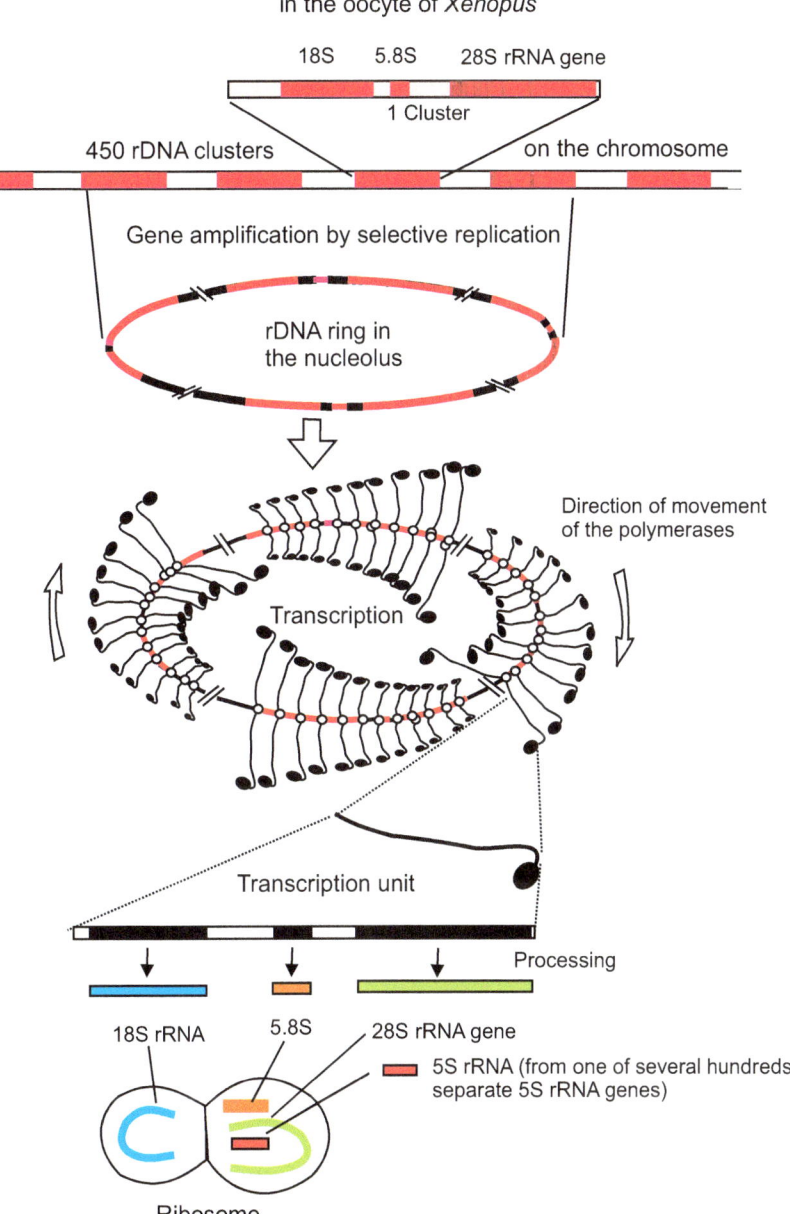

Fig. 8.6 Oogenesis II. Amplification of the rDNA for rapid production of numerous nucleoli and ribosomes. In the nucleoli the amplified and ring-shaped cluster of rDNA genes is utilized to produce, via transcription, large amounts of rRNA that is required to construct ribosomes. Because several polymerases migrate along the rDNA ring simultaneously one sees protruding rRNA pieces of increasing length – the length being a function of the transcribed section ("Christmas tree structure")

Vitellogenins are found only in female blood. Incorporated by endocytosis, the vitellogenins are split in the oocytes and converted into the heavily phosphorylated **phosvitin** protein and the lipoprotein **lipovitellin**. Encased in membranes, these two materials constitute the main content of the **yolk platelets**. In addition, the maturing egg deposits **glycogen** granules. The oocytes develop

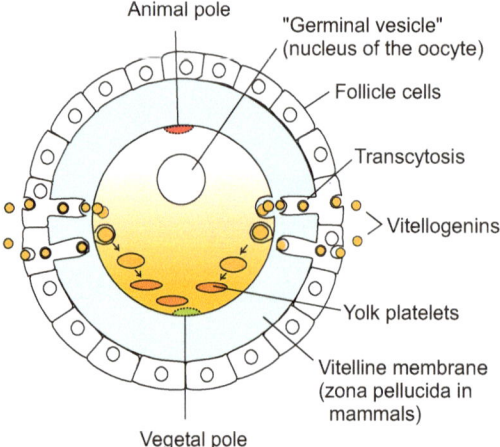

Animal pole

"Germinal vesicle"
(nucleus of the oocyte)

Follicle cells

Transcytosis

Vitellogenins

Yolk platelets

Vitelline membrane
(zona pellucida in
mammals)

Vegetal pole

Fig. 8.7 Provision of the oocyte with vitellogenins (yolk protein precursors) by the follicle cells. The vitellogenins are taken up by the follicle cells and transmitted to the oocyte via transcytosis. Finally, the follicle cells secrete the vitelline membrane around the egg cell

into the largest known cells. The egg cell of the chick (the yellow sphere in the centre of the egg) has a volume 9×10^9 times greater than that of a normal somatic cell! The egg cell in the ostrich is the largest animal cell known today.

8.2.3 In *Drosophila* All the Needs of Oocytes Are Provided by Nurse Cells

In insects such as crickets, which give themselves time to complete oogenesis, we observe lampbrush chromosomes and multiple nucleoli like those in vertebrates. Not so in *Drosophila*; this fly lives for only 14 days and produces an egg in 12 h. The oocyte is entirely supplied with all it needs by nurse cells.

The oogonia are packaged in tubes termed **ovarioles** (Fig. 4.25). Each oogonium divides mitotically four times to generate an association of 16 cells. The cells remain connected through thin tubes called **fusomes**. Two of the centrally located cells are connected with four neighbours; one of these two cells becomes the oocyte, whereas the other 15 sister cells are destined to become nurse cells. These amplify their genome, become polyploid and eventually contain

500–1,000 copies of the genome. This multiplication permits a high level of transcriptional activity.

The products of their synthetic activity, including RNP particles and protein, are directed into the swelling oocytes. Among the gene products are development-controlling molecules such as *bicoid* mRNA, deposited at the anterior pole of the egg, and *nanos* and *oskar* mRNAs, translocated to and accumulated at the posterior pole. These RNAs will then specify where head and abdomen are to be made. All of these products are of **maternal** origin, as are the products procured from the **follicle cells** in the wall of the ovarioles.

As in vertebrates, yolk proteins in *Drosophila* are manufactured outside the ovary in the form of vitellogenins. In insects the factory is the fat body; the vitellogenins are released into the haemolymph, and forwarded to the oocyte by the follicle cells.

8.2.4 Oocytes Become Polar (Asymmetrical) and Enclosed by Extracellular Membranes and Envelopes

The constituents of the yolk are asymmetrically deposited in the oocyte. In *Drosophila* the nucleus stays in the centre of the egg. However, antero-posterior and dorso-ventral polarity are manifested by the distribution of yolk materials and the elliptic form of the egg, which is stabilized by the hardy envelope of the **chorion** secreted by the follicle. But the apparent bilateral symmetry of the egg is not decisive for the bilateral organization of the future body. In certain mutants (*bicoid, dorsal*) or as a result of experimental interference (injection of *bicoid* mRNA into the posterior region of the egg), the architecture of the body can be fundamentally altered in an ostensibly normal egg (see Fig. 4.32).

Most animal egg cells, including those of vertebrates and sea urchins, appear spherical; but in their internal structure they are endowed with an **animal-vegetal polarity**: yolk platelets and glycogen granules are accumulated in the vegetal hemisphere. The **huge nucleus** of the oocyte, traditionally called the **germinal vesicle**,

is located near the animal pole. The extent to which later development is influenced by the non-uniform distribution of the oocyte constituents must be tested experimentally (Chap. 9).

Finally, the egg cell is surrounded by an acellular, strengthening **vitellin membrane** (Fig. 8.7), and by additional enveloping layers of various consistency. In mammals these layers are designated the **zona pellucida** and **corona radiata** which derive from remnants of the follicle. In reptiles and birds, albumen and egg shell are wrapped around the egg cell by the wall of the oviduct only after fertilization.

8.2.5   In Vertebrates, Hormonal Signals Initiate the Formation of the Polar Bodies and the Final Maturation of the Egg

The vertebrate oocyte undergoes a 500-fold increase in volume, a process that can take several years (12–45 years in human females). When the oocyte has reached its final size, meiosis is resumed. The chromosomes condense into a form suitable for transport. The **germinal vesicle breakdown** indicates the decomposition of the nuclear membrane as a prelude of commencing meiosis. Subsequently, the first meiotic miniature sister cell of the egg cell, the first **polar body**, is extruded. The mammalian egg segregates the second polar body only after fertilization.

Resumption and completion of the interrupted meiosis is controlled hormonally. Stimulated by an internal clock in the brain, or in a few mammals such as rabbits, by the act of intercourse, the **pituitary gland** first releases the **gonadotropic** (gonad controlling) hormone

- **FSH (follicle stimulating hormone)**. In response to this hormone, maturing follicles undergo the last phase of growth and granulosa cells express receptors for the second gonadotropic hormone
- **LH (luteinizing hormone)**. FSH and LH stimulate the production and release of a second set of hormones: the **steroids**

- **Oestradiol** in the theca cells of the follicle and
- **Progesterone** in the granulosa cells of the follicle (Fig. 6.2a).

These steroids prepare the uterus for implantation. A certain species-specific ratio of FSH, LH and steroid hormones induces the extrusion of the first polar body and the subsequent **ovulation**: the release of the egg from the ovary. In amphibians, progesterone plays the dominant role in breaking the dormancy of the egg. In human females, LH induces some (estimated up to 50) oocytes to resume meiosis in the first half of the **menstrual cycle** after years of arrest. But as a rule, only one mature egg is liberated. Ovulation occurs in the middle of two menstrual bleedings. In most other mammals ovulation takes place in the period of **oestrus** ('heat'), which must not be equated with bleeding but indicates the time of ovulation and, therefore, readiness for conception.

8.3   Spermatogenesis: The Sperm, a Genome with a Motor

8.3.1   In Mammals Sperm Are Produced Continuously; Meiosis Takes Place Rapidly in the Final Stage of Spermatogenesis

Primordial germ cells become spermatogonia in the seminiferous tubules of the testis (Fig. 8.8). **Spermatogonia are stem cells** that retain the ability to divide: one daughter cell remains a stem cell, while the other becomes a **spermatocyte**. The stem cells lie along the inner surface of the tubules; the segregated spermatocytes are displaced in the direction of the inner lumen of the channels. The spermatocytes derived from one stem cell remain interconnected and undergo their further development synchronously. Together, the spermatocytes cease mitotic division but grow only slightly, unlike oocytes. At puberty, the spermatocytes undergo meiosis. Each spermatocyte gives rise to four

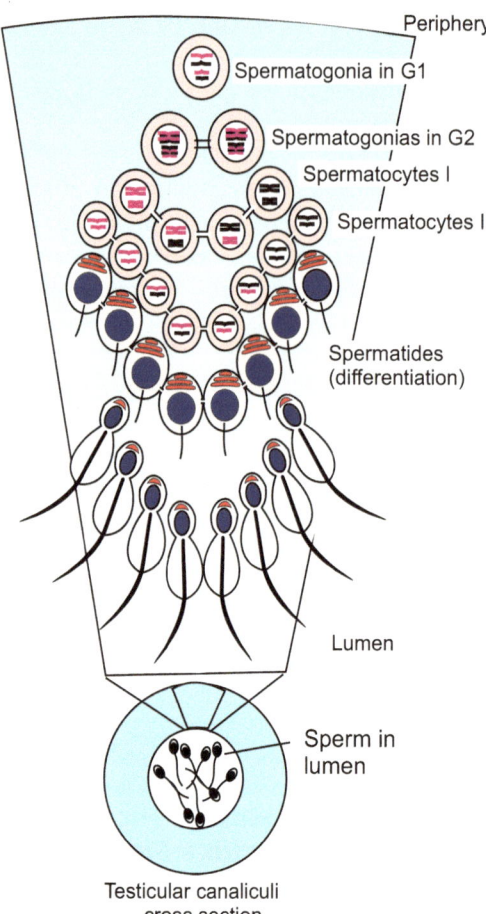

Fig. 8.8　Spermatogenesis. Detail from a testicular cana-liculus. Spermatogenesis takes place in the wall of the canals; the finished sperm are released into the lumen of the canals. Note: In spermatogenesis meiosis is not interrupted, no stage with lampbrush chromosomes occurs, and all cells arising from meiosis give rise to germ cells (sperm)

equal spermatids. In a process of terminal differentiation four spermatids give rise to four sperm cells. During their entire development the descendants of a primordial germ cell remain attached to each other by cytoplasmic bridges (just like the oocyte remains connected with the nurse cells in the ovarioles of *Drosophila*). Only the finished sperms separate from each other. While sperm cells accumulate in the centre of the tubules, the stem cells along the tubule walls

provide a fresh supply, continuously (in man) or during the mating period.

A typical animal sperm cell, such as that of the sea urchin (Fig. 8.9), possesses an acrosomal vesicle beneath its tip or **acrosome**; proceeding posteriorly are the highly **condensed haploid nucleus**, the neck with a pair of centrioles, the midpiece with the mitochondrial power station, and the propulsive **flagellum**. The sperm is ready for delivery and fertilization.

Certainly, not all animal species have sperm displaying the features of the textbook prototype shown in Fig.8.9. Sperm without a flagellum but being motile like an amoeba are found in nematodes and several arthropods. Very elongated are the sperm in *Drosophila* (Fig. 8.10). In spite of the large differences in morphology displayed by the sperm of the fly, fishes and the mouse, biochemical investigations have detected a commonality that possibly pertains in the entire animal kingdom. In the chromatin of the sperm nucleus histones are replaced, completely or partially, by **protamines**. Protamines are small, alkaline, i.e. electrically positively charged, proteins that attach densely onto the negative charged nucleic acids, enabling a particularly high package density.

8.4　An Invisible Inheritance: Imprinted Methylation Patterns from Father and Mother

8.4.1　Sperm and Egg Cells Are Genetically Not Entirely Equivalent; One Difference Can Be Based on Differences in the Methylation Pattern

The 'maternal' genome supplied by the egg and the 'paternal' genome supplied by the sperm do not always contribute equally to the programming of the characteristics of the new organism. In the mouse embryo, for instance, the **'paternal'**

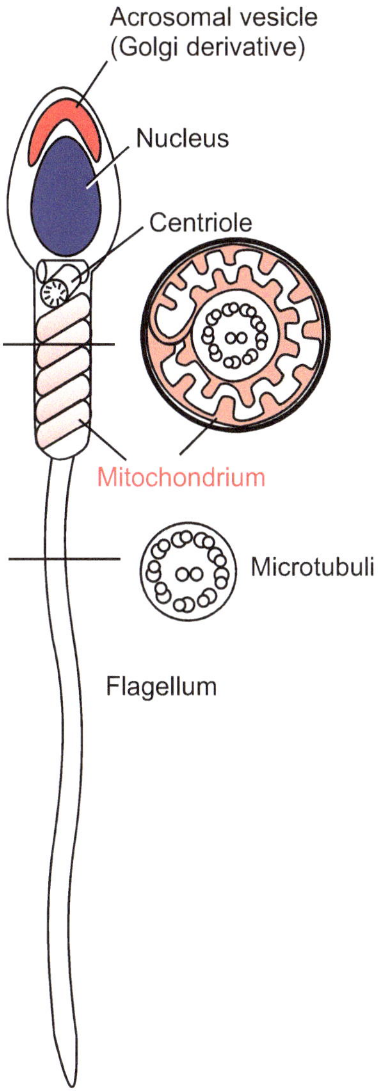

Fig. 8.9 Spermium. Prototype of a generalized sperm as represented by the sperm of sea urchins

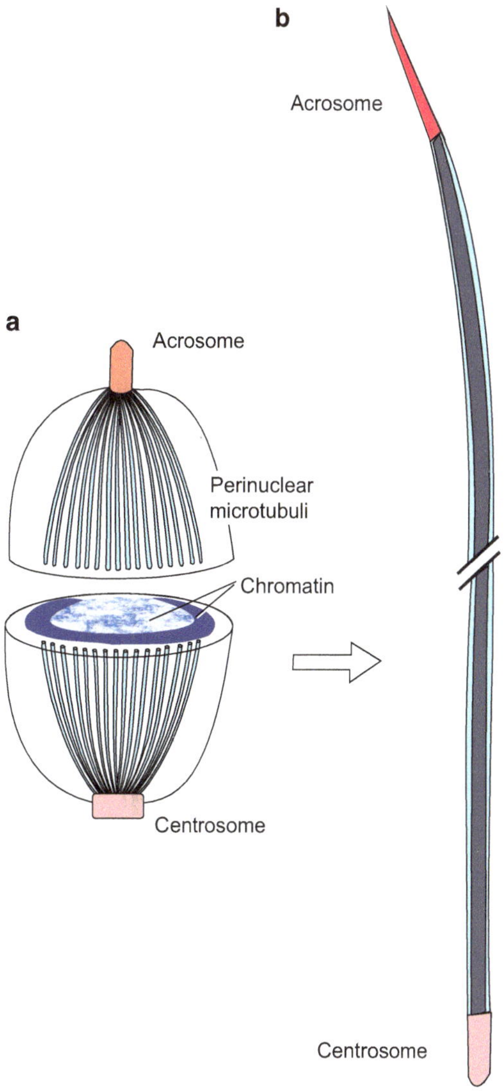

Fig. 8.10 (**a**, **b**) Sperm maturation in *Drosophila*. (**a**) Haploid spermatocyte II; microtubules growing in longitudinal direction begin to stretch the cell. In the chromatin histones are replaced by protamines. (**b**) Finished, extremely elongated sperm

genome predominates in the extraembryonic structures (trophoblast, placenta) while in the embryo proper the 'maternal' genome appears to have greater influence. On the other hand, in man the expression of the gene for **Huntington's disease** is stronger if the (autosomal dominant) congenital disease is inherited from the father. This phenomenon, known as **genomic imprinting**, is due to differing **methylation patterns** in the DNA of the egg and the sperm, supplemented by other, additional mechanisms of genomic imprinting (Fig. 12.22).

8.4.2 Mitochondria: Source of Additional Genetic Information. In Mammals Mitochondria Are Inherited Only via the Oocytes and, Thus, Only in the Maternal Genealogy

In the course of evolution mitochondria arose from endosymbionts. They possess remnants of their own genome. However, mitochondria are not completely autonomous, since several mito-chondrial enzymes are not (or no longer) encoded by mitochondrial genes but by genes of the nucleus (according to prevailing hypotheses in evolution these genes were trans-ferred to the nucleus). Egg cells are equipped with many mitochondria, sperm with only one or a few. And this sparse equipment gets lost in fertilization as the mitochondrion of the sperm is dissolved (or diluted out in the subsequent cell divisions). Therefore the mitochondrial genes are considered to be "purely maternal" and can be consulted in studies on the ancestry of populations in the human genealogy (Sect. 3.3).

8.5   Genetic Consequences of the Soma: Germ Line Segregation

8.5.1   Mutations and Artificially Inserted Genetic Constructs (Transgenes) Are Transmitted to the Next Generation Only If They Are Transmitted in the Germ Line

Generative cells in general, sexual **germ cells** in particular, convey genetic information accumulated over millions of years of evolution to the next generation. In the ontogenetic devel-opment of multicellular organisms a line can be drawn from the fertilized egg cell to the germ cells. This lineage is now known to the reader as **germ line** (Figs. 8.1 and 8.2). It is a matter of discussion whether a germ line exists in all organisms that definitively separates somatic cells from future primordial germ cells, or whether cases exist where cells first fulfil a somatic function and only later take over the function of germ cells (to our knowledge, clear evidence for this option is currently lacking). But even if somatic cells could be converted to pri-mordial germ cells, a linear series of DNA replications undoubtedly connects the egg cell with the germ cells. Only mutations, epigenetic modifications and genetic manipulations that happen in the germ line, and only transgenes (foreign genes) that are introduced into the

germ line, get transmitted to the next generation. This point is relevant to any informed discussion about the capacity to alter the genetic constitu-tion of organisms and introduce alien genes permanently.

Summary

Germ cells, **egg cells** or **spermatocytes (sperm)**, originate from **primordial germ cells,** a particular set of stem cells. In some species (*Drosophila, Caenorhabditis elegans, Xenopus*) primordial germ cells are set aside early in embryonic development and only migrate into the gonads later in life. In other organisms, such as the mouse, stem cells are specified and committed to become germ cells by inductive influences exerted by their envi-ronment (niche) late in development. Primor-dial germ cells do not give rise to somatic cells but to germ cells only. The cell line leading from the fertilized egg to the primordial germ cells, the cellular origin of the next generation, is called **germ line**. In *Drosophila* the germ line is extremely short, because primordial germ cells, known as pole cells, are the first complete cells in the embryo and are set aside immediately upon formation. Their commit-ment occurs under the regime of the maternal gene *oskar,* the mRNA of which is directed to the posterior pole of the egg in oogenesis. Here *oskar* attracts germ cell-determining RNA spe-cies, among them RNA derived from the genes *vasa* and *nanos*. At the posterior egg pole mRNA of all these genes is incorporated into **pole granules**. Similar granules, supplemented or replaced by cloudy material (nuage) and containing mRNA from *vasa, nanos, Oct-4* and further genes is deposited in the eggs of many, if not all, animal species and collectively referred to as **germ plasm**. The products (mRNA or protein) of the genes *vasa, nanos* and *Oct-4* accompany and mark the germ line also in mammals including humans. The **tran-scription factor OCT-4** confers (in coopera-tion with further factors) **totipotency** to the germ line cells. In mammals (and birds) the first morphologically recognizable primordial germ cells are found outside the embryo proper in the primordium of the allantois and in the yolk sac whence they migrate into the gonads. Only in the gonads is the decision made whether the initially sex-neutral primordial germ cells become stem cells for eggs (**oogonia**) or for sperm (**spermatogonia**).

When the gonad becomes an **ovary,** already in the unborn female proliferating oogonia give rise to numerous oocytes most of which, however, perish, and **already in the embryo the oocytes enter the meiotic prophase** but subsequently prophase is interrupted for years (in humans), the chromosome become decondensed and transcriptionally active, often passing a structure called **lampbrush stage**, along with **amplification of ribosomal genes** and appearance of **multiple nucleoli** in which numerous ribosomes are manufactured for storage in the cytoplasm of the egg cell. In *Drosophila* the oocyte does not display lampbrush chromosomes and multiple nucleoli but is provided with all storage material by adjacent nurse cells.

In the final growth phase proteins supplied by the maternal organism (the liver in vertebrates, the fat body in insects) in form of **vitellogenins** are taken up by the egg cell and stored in yolk platelets. In vertebrates **hormonal signals** prompt the full grown oocyte to continue **meiosis**. This consists of two extremely asymmetric divisions whereby in each division a tiny, abortive daughter cell is pinched off as **polar body** from the large oocyte.

In the testis the meiotic divisions and the subsequent terminal differentiation of the sperm cells do not take place until puberty. While in oogenesis only one haploid cell remains, each spermatocyte gives rise to four haploid sperm cells. A finished sperm is little more than a genome with an engine. To facilitate a particularly dense compaction of the DNA, in the chromatin of the sperm histones are replaced by **protamines**.

As a further legacy, egg cells have multiple mitochondria while the single mitochondrion supplied by the sperm is dissolved upon fertilization. Therefore, female and male germ cells are not entirely equivalent since **mitochondrial DNA is of maternal origin only**. In addition, differential chemical (epigenetic) modification of the DNA, such as different degrees and patterns of methylation, known as **genomic imprinting**, can effect differences in the penetrance of maternal and paternal alleles in the offspring. Independently of that, mutations, epigenetic modifications and artificially introduced transgenes are of significance for the offspring only if they occur, or are generated, in the paternal or maternal germ line.

Specification of Body Axes and Localized Fate Allotment by External and Maternal Cues

9

9.1 Start of Differentiation Programs in Accordance with Position

9.1.1 How Is Development in Accordance with Location (Normotopic Development) Accomplished in Spite of Genomic Equivalence?

In the course of embryonic development all cells are allocated the entire genetic information, as cell divisions are based on mitotic cell cycles. In the S phase of the cell cycle the DNA is faithfully replicated; the two daughter cells obtain the exact identical genetic information since mitochondria are also apportioned to both cells. Initially there is genomic equivalence, that is the cells of an embryo are initially genetically identical and totipotent (= capable of giving rise to any cell type; proof: nuclear transplantations, Chap. 13).

Yet, the tasks the descendants of a founder cell have to take over are not at all identical. Cells of the embryo must behave in accordance with their location. Here they construct the nervous system, there they form a muscle, at this place they must jointly produce an element of the skeleton giving it a distinct shape, and at that place they must commit suicide to create a cavity. How do cells know where they are? Do their genes somehow tell them what their position is at each moment? This is not possible, for the DNA

of the nucleus does not contain a street map, and the cells of the embryo cannot look into their nuclei to find out where they are. Somehow position-dependent cell differentiation must be initiated from the outside. By "from outside" we mean by signals having their origin outside the nucleus of the cell in question.

Different tasks require switching on or off of different genes, as each cell type has its own set of proteins. The initially totipotent cells must make decisions about which genetic subprograms to call into action.

9.1.2 Before Being Committed to a Distinct Fate, Cells Need Information About Their Location in the Embryo. Teratomas Show: Without Proper Positional Cues Chaos Will Arise

Teratomas are failed, chaotically organized embryos. They derive from unfertilized cells of the germ line in the testes or ovaries. Occasionally germ line cells start embryo development precociously. Another source of teratomas are fertilized eggs which fail to be taken up by the tube of the oviduct and instead implant anywhere in the abdominal cavity. Experimentally, in mice teratomas are produced by removing blastocysts from the uterus and implanting them into the abdominal cavity. (Teratomas may develop into

W.A. Mueller et al., *Development and Reproduction in Humans and Animal Model Species*,
DOI 10.1007/978-3-662-43784-1_9, © Springer-Verlag Berlin Heidelberg 2015

malignant teratocarcinomas, tumours which even can metastasize. Cells derived from teratocarcinomas, such as the widely used murine 3T3 or F9 cells, are frequently immortal and can easily be propagated as cell cultures.) Teratoma cells are, as a rule, genetically still intact. If implanted into normal blastocysts, they may integrate into their new surroundings unobtrusively and participate in the construction of the new animal. In spite of intact genomic information chaos arises if a correct sequence of appropriate cues does not guide locality-pertinent utilization of genetic subprograms.

In its protein-coding genes the genome contains information about the order in which amino acids have to be linked to get a distinct protein. Beyond this basic information the genome is organized in such a way that entire programs can be called up to manufacture, for example, a muscle cell or a nerve cell (Sect. 12.2). But since the genomic information initially is identical in all cells, a cell needs positional information enabling it to behave in accordance with its location. This information cannot directly be derived from the cell's own nucleus.

9.2   Defining the Body's Coordinates

9.2.1   First the Body Axes (Antero-Posterior, Dorsal-Ventral) Must be Established; This Can Take Place in Oogenesis or Subsequent to Fertilization

The first decision to be made is the future location of the head and the tail, the back and belly. Most animals are bilaterally symmetrical (Fig. 9.1). Perpendicular to an antero-posterior axis extends a dorso-ventral axis, while the left and right sides are mirror-inverted symmetrical. Bilateral symmetry means capable of being split into two equal parts so that one part is a mirror image of the other. In developmental biology axes of asymmetry or anisotropy are called polarity axes.

Egg cells are always organized in a polar manner. Frequently, oocytes in the ovaries are not completely surrounded by nourishing nurse cells. The oocyte is fed predominantly from one side (Fig. 4.25 as an example), which suggests a source of cues for asymmetric organization. Even gravity has an influence in certain cases. Thus in the oocyte various substances, whether produced by the egg itself or provided by nurse cells, are not uniformly deposited. Special transport systems manage their asymmetric distribution as is known from the egg of *Drosophila* (Figs. 9.2 and 4.12). But also in the oocytes of other animals the internal constituents are not uniformly distributed. Heavy yolk granules often accumulate near the vegetal pole while the nucleus of the oocyte, in traditional terminology called the germinal vesicle, comes to lie near the animal pole, where the polar bodies are constricted off later in the course of the meiotic divisions (actually, the location of the polar bodies defines the animal pole).

In most eggs it is not difficult to recognize the animal-vegetal axis under the microscope. But in the vast majority of egg cells only one unipolar axis is visible, not a complete bilaterally symmetrical architecture. The egg of *Drosophila* is exceptional, displaying a bilateral shape and internal organization (Fig. 9.2). Most animal eggs, however, display a recognizable animal-vegetal axis but neither a bilaterally symmetrical shape nor bilateral organization in their internal constituents.

The animal-vegetal axis may coincide with either the future antero-posterior or the dorso-ventral body axis, or with neither of them. In molluscs, ascidians and birds the animal-vegetal axis of the egg coincides with the dorso-ventral axis of the embryo. But this coincidence is merely rule of thumb and not a generally valid rule.

9.2.2   In *Drosophila*, the Mother Decides in Advance the Future Polarity Axes of Her Child; The Decisions Thus Become Dependent on Maternal Genes

In the invertebrate model *Drosophila* symmetry determination is markedly different than

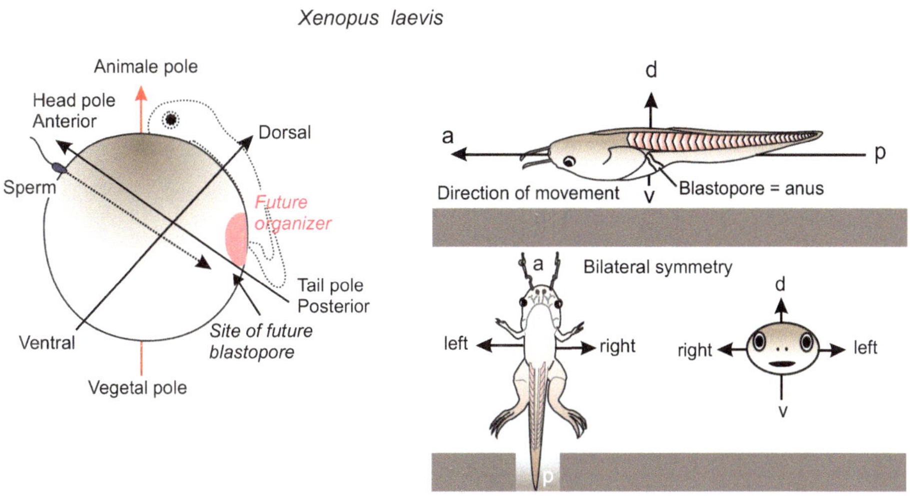

Fig. 9.1 Determination of bilateral symmetry in the amphibian embryo. Decisive for the specification of the antero-posterior and the dorso-ventral axis are (1) the site of sperm entry and migration path of the sperm centriole, and (2) gravity: When the contact with the sperm releases rotation of the egg cortex, gravity determines the direction of the rotation

described below for vertebrates. A significant difference is that the process of specifying bilateral symmetry is completely under the control of the maternal genome. As *Drosophila* has a dominant position in developmental biology, there is the risk that peculiarities found in *Drosophila* are taken as a paradigm for animal development in general. In particular such fundamental decisions as the location in which the head, the tail, the back and the belly are to be formed, must be, so we are inclined to think, put under the control of the genome and *Drosophila* appears to testify the veracity of our intuitive assumption. Is it really so, that these fundamental decisions are laid down in our genome? If we have a closer look on the events in oogenesis even in *Drosophila* the decisive factors originate outside the egg cells.

The antero-posterior polarity axis is specified by the spatial distribution of particles enclosing maternally generated mRNA (RNP- particles). *Bicoid* message is deposited at the front pole, the mRNAs of *nanos* and *oskar* at the tail pole (Fig. 9.2). Viewed from the egg cell, the origin of these gene products is external, but the determinants are now internalized. When *bicoid* mRNA is translated into protein, this protein acquires the function of a transcription factor controlling gene activities in the early embryo

(Chaps. 4 and 12). In addition, the embryo produces the TORSO receptors using maternally supplied *torso* mRNA; these receptors are exposed on the surface of the egg cell and recognize external cues. The cues are factors produced in the ovary by follicle cells and stored in the space between the egg cell and the egg envelope, in front of and behind the egg cell.

Likewise, specification of the dorso-ventral polarity is mediated by signal molecules that have been secreted by follicle cells at the future ventral side, stored in the 'perivitelline' space between egg cell and egg envelope, and picked up by receptors (coded by the maternal gene *toll*) of the egg cell membrane. The signals trigger a mechanism which redistributes a determinative cytoplasmic factor: The maternal factor DORSAL, initially homogenously distributed in the pre-cleavage egg cell, migrates into the ventral nuclei of the blastoderm (Fig. 9.2b; see also Fig. 4.35).

Since the messages for each of these factors (derived from *bicoid, nanos, oskar* and *dorsal*) and each of these receptors (derived from *torso* and *toll*) are products of genes, the *Drosophila* genome is directly involved in the establishment of the coordinates of the body. It is, however, the maternal genome that is involved, and not the embryo's own (zygotic) genome. From the

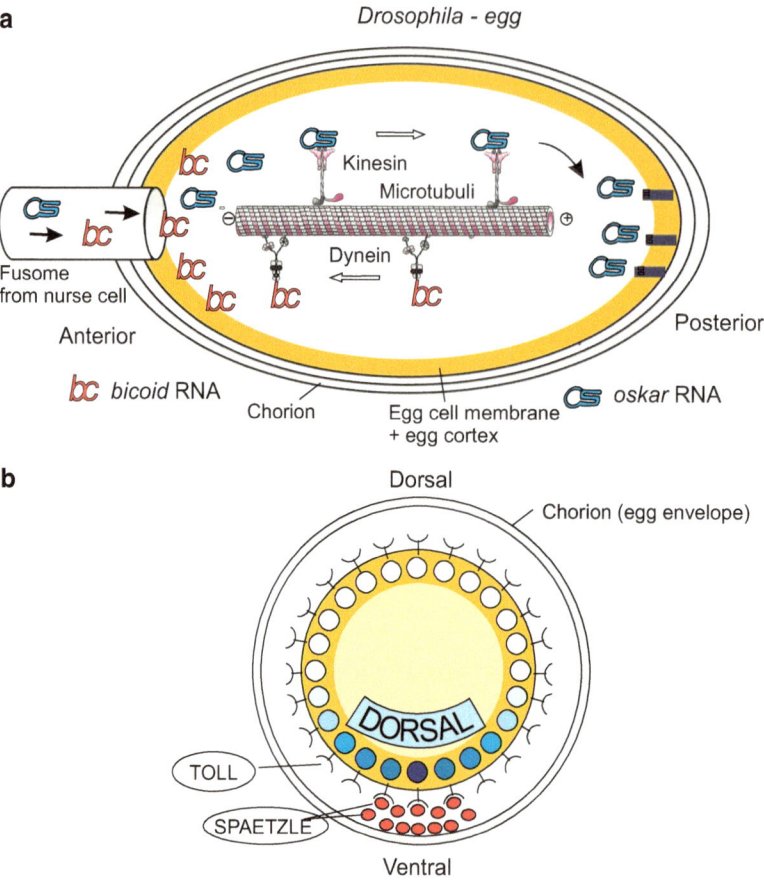

Fig. 9.2 (**a**, **b**) Maternal determinants of the body axes in *Drosophila*. (**a**) Antero-posterior axis: Of particular significance are transport systems which direct the diverse species of mRNA supplied by the nurse cells to different locations in the egg. The *UTR* (*UnTranslated Region*) of the transcripts serves as a postal code. Kinesin motor proteins transport *oskar* mRNA to the posterior egg pole where it is moored at the cytoskeleton. In turn, dynein motor proteins arrange accumulation of *bicoid* mRNA near the anterior pole. Many more components, not shown here, are involved in the specification and stabilization of the longitudinal body axis. (**b**) Dorso-ventral axis: localization of important maternal elements in the cross-sectioned egg

perspective of the oocyte, even in *Drosophila* the decisive cues for orientation come from outside, from the nurse cells and follicle cells of the maternal ovary.

It is genetics that testify the significance of maternal genes. Even if any one of the above mentioned genes is homozygously defective, the mother herself is not affected, because these genes are required only in oogenesis. The mother herself displays a normal phenotype, only the offspring manifests the mutation. One speaks of maternal-effect mutations or genes.

9.2.3 If the Coordinates Are Established Only After Fertilization, External Cues Can be Utilized: Amphibians as an Example

In the simplest case the sperm can play the role of an orientation guide, and in the round worm *Caenorhabditis elegans* the site where the sperm enters the egg marks the future posterior pole of the embryo (see Fig. 9.7). Also in vertebrates environmental cues are significant.

In amphibians (and in the zebrafish *Danio*) the animal-vegetal axis enables an observer to predict, approximately, where the head will be formed. The head will be located within a short radius of the animal 'north' pole (Figs. 9.1 and 9.3). The spine will extend over the animal hemisphere, cross the equator and end at about the Tropic of Capricorn. But initially it is not determined along which of the 360 possible longitudinal lines the spine will extend. The position of this 'zero meridian' is determined at the moment of fertilization. Two external parameters are involved: the point of sperm entry and gravity.

The sperm attaches randomly at a point anywhere in the animal hemisphere (it cannot attach to the vegetal hemisphere). The nucleus of the sperm pushes its way through the cytoplasm of the egg, guided by the centriole, to the nucleus of the egg. Subsequently egg constituents of the cytoplasm are displaced by an event called cortical rotation. These movements encompass the egg cortex and are driven by a kinesin microtubule motor system. Under the additional influence of gravity the grey crescent, or its functional equivalent, appears diagonally opposite to the sperm entry point. The grey crescent harbours several important determinants of development and marks the site where the blastopore will appear in gastrulation (see Fig. 5.4). By experimentally rotating an egg the location of the grey crescent and hence of the future blastopore can be displaced. The grey crescent (unfortunately not visible in *Xenopus*) marks the future location of the spine.

In the internal organization elements of the WNT-system (mRNA of *Wnt1* and *Wnt11*, and β-catenin) are displaced to the site of the grey crescent enabling this area to become the Spemann organizer (Fig. 9.3). Below this the blastopore is formed. Since vertebrates belong to the deuterostomes the blastopore eventually will become the anus; and dorsally to it the tail bud will form.

The zero meridian is now fixed. The head-back-tail line will start at the animal pole (urodeles such as newts and salamanders) or near the Tropic of Cancer (anurans such as frogs and toads) and extend along the zero meridian down to the grey crescent, or Spemann organizer, at the Tropic of Capricorn. In sum, an internal architecture, the animal-vegetal asymmetry, determines bilateral symmetry in cooperation with external cues (point of sperm entrance and gravity).

Fish (*Danio*) eggs usually lie on the ground horizontally. The head-tail axis will run along the uppermost meridian. The directing forces are not known but likely include gravity. Internal factors determining the site of the Spemann organizer are the same as in the amphibian embryo.

In the avian egg the dorso-ventral polarity is predetermined by the structure of the yolk-rich egg. The animal pole marks the centre of the upper surface of the blastodisc and defines the dorsal side. The location of the head-tail line is thought to be specified by the combined effect of gravity and rotational movements: as the egg descends along the oviduct and uterus, it rotates. But only the shell and some internal constituents rotate, not the yellow egg cell. Thus, the blastodisc remains at the top of the egg cell but its plane is tilted; the angle to the horizontal is 45° and therefore the bastodisc is subjected to shearing forces. The direction in which the egg is rotated is believed to specify the direction of the head-tail line (Fig. 9.4).

Mammalian and human embryos appear to leave to chance the site where the inner cell mass will segregate. The position of the inner cell mass defines the future dorsal side. How the head-tail polarity is specified is unknown. The egg of the mouse embryo is the only known animal egg from which material can be removed at the animal and vegetal pole without significantly disturbing development. Apparently little fate is contained in the cytoplasm. Although WNT signalling is also of significance in the mammalian embryo (see following section) it is currently unknown how the signal emitter is installed. According to time lapse recordings the shape of the elliptic egg envelope (zona pellucida) and external mechanical cues guide the orientation of the anterior-posterior axis in early mouse embryos.

Maternal Transcripts in the Egg of *Xenopus*

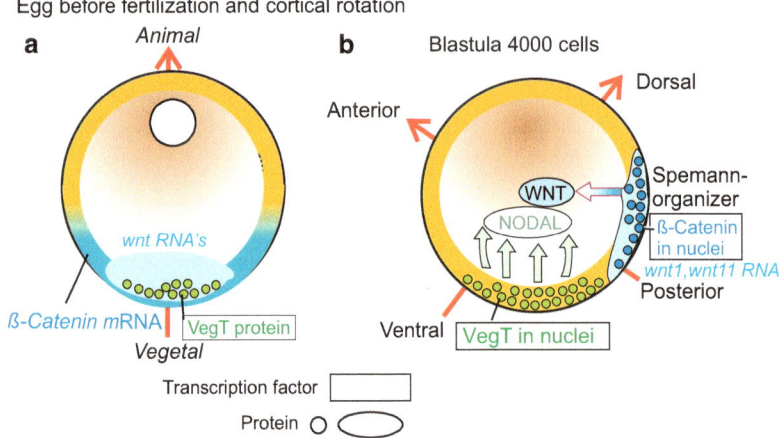

Fig. 9.3 (**a, b**) Significance of maternal transcripts for establishing the body coordinates (polarity axes) in the embryo of *Xenopus*. (**a**) Distribution before cortical rotation. (**b**) After redistribution of the maternal transcripts (mRNA) subsequent to fertilization, the future bilateral symmetry of the amphibian embryo is specified. In addition, the region in which β-catenin is taken up into the nuclei determines the future position of the blastopore and of the overlying Spemann organizer

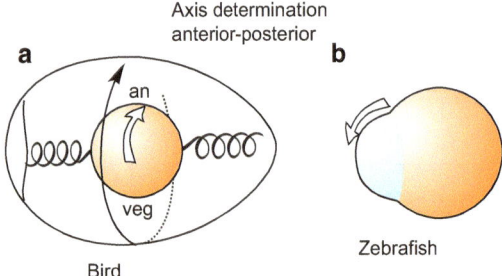

Fig. 9.4 (**a, b**) Determination of the head-tail axis in the avian egg and the zebrafish egg. In the avian egg the direction in which the egg is rotated when passing the oviduct is decisive. In the horizontal egg of the fish the uppermost meridian is decisive

9.2.4   Aamong the Determinants in the Cytoplasm of Eggs Often Components of the WNT/beta-catenin System Are of Particular Significance

A series of studies performed in members of several phyla, from the cnidarians up to the vertebrates, disclosed unexpected parallels. Often components of the so called canonical WNT (pronounce went) signalling system are stored in oocytes, for instance maternal mRNA

of a WNT signal molecule or the protein of the transcription factor β-(beta)-catenin (but also of other components such as DSH, GSK3, TCF). The WNT system will be introduced in detail in Chap. 11, Fig. 11.6, and its significance further discussed in Chap. 22. Here we introduce briefly its significance in the establishment of body axes.

Several members of the WNT protein family are encoded in the genome of various animals. From the phylum of Cnidaria that diverged from the main tree of the eumetazoan animals as early as 500 millions of years ago, 12 WNT members are known (actually in *Nematostella*). The spatio-temporal expression pattern (see Fig. 22. 8) suggests overlapping functions. For the establishment of the body axis mainly the isoform WNT-3a is responsible, with subsequent activation of the canonical WNT signalling pathway.

An experiment exemplifying the role of WNT in installing a body axis illustrated in Fig. 9.5. We speak of the oral-aboral axis, the only body axis displayed by the members of the phylum Cnidaria. The oocyte contains maternal mRNA for WNT3a as well as for other components of the pathway. Through a positive feedback loop (compare Fig. 12.20) these maternal transcripts enable the cells at the upper pole of the gastrula

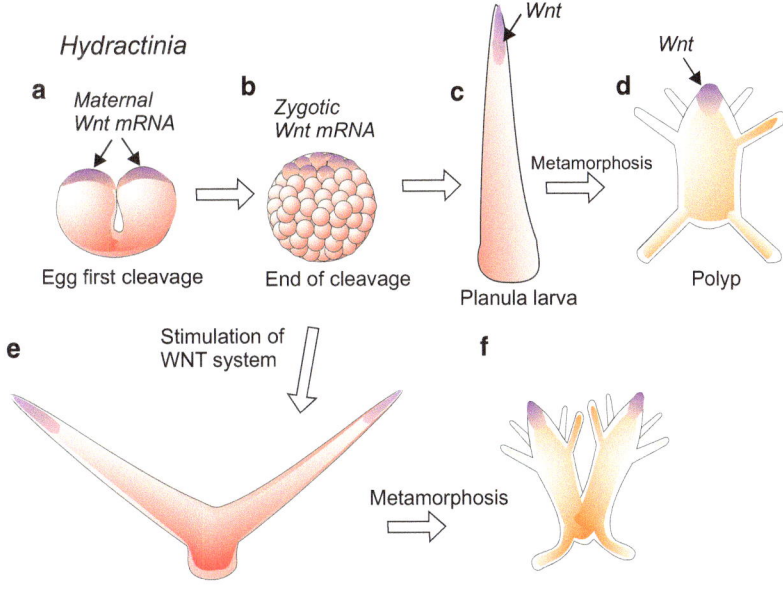

Fig. 9.5 Axis determination by the canonical WNT signalling system in the hydrozoan *Hydractinia echinata*. Initially maternal transcripts are decisive that had in oogenesis been deposited at the animal pole of the oocyte, followed by zygotic transcripts in the gastrula. When phosphorylation of β-catenin and its subsequent degradation is globally prohibited by inhibitors of the kinase GSK3, and thereby the amount of non-phosphorylated β-catenin is left at high levels, more nuclei, and nuclei in ectopic places, take up β-catenin. As a consequence two or more tails form and the polyp arising from metamorphosis will have two or multiple heads. After Plickert et al. (2006)

to continue WNT signalling using their own zygotic genes. Gradients of expression are seen running from the upper (oral) pole downwards. The WNT signals appeal to their neighbours to participate in the construction of the body axis. If WNT emitters are experimentally evoked at other than normal (ectopic) locations several body axes are formed and as a consequence polyps with multiple heads arise.

WNT signals specify one of the two body axes also in several bilaterally symmetrical organisms (one of the few exceptions is *Drosophila*). As a rule this is the axis which connects the head pole with the tail pole and it extends, as a rule, in the direction the animal moves forward – as a rule but not in all organisms.

- In the bilaterally symmetrical sea urchin embryo and larva the peak of WNT production is around the blastopore (which later will become the anus). Developmental biologists have long known that treatment of sea urchin embryos with lithium ions renders them "vegetalized", meaning that organs arising from the vegetal part of the embryo grow too large. Lithium is known to interfere with WNT signalling strengthening its activity.
- In amphibians the situation is not easily understood as at different times different isoforms of WNT appear. Initially maternal WNT-1 plays its part (and in addition the 'planar' signals WNT-5 and WNT-11). Later in the gastrula and neurula zygotic WNT-3a plus WNT-8 come into play. Whichever WNT isoform actually is of decisive significance in axis specification, it presumably forms a gradient running from the posterior pole in the anterior direction (similar to the WNT-8 gradient in the neurula shown in Figs. 10.3, 10.4 and 10.5). There are two arguments supporting this notion. (1) First (weak) argument: Immediately upon fertilization maternal mRNA of *Wnt-1*, *Wnt-11* and β-catenin protein are

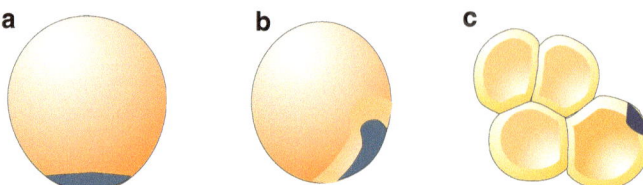

Displacement of maternal, musculature-determining mRNA **Macho-1** (blue)
in the Ascidian embryo by ooplasmic segregation

Fig. 9.6 Redistribution of maternal mRNA in the ascidian egg by ooplasmic segregation following fertilization. As an example the location of the *macho-1* mRNA in the embryo of *Halocynthia roretzi* is shown. Macho-1 protein commits, in cooperation with further factors, cells to which it is allocated to become muscle cells of the tail. After Nishida and Sawada (2001)

translocated at the posterior pole where the Spemann organizer becomes established and the blastopore is formed (Fig. 9.3, see also Fig. 22.7). In this area WNT signalling continues. (2) Second (strong) argument: Injection of β-catenin into an anterior blastomere of an early blastula induces the formation of a second blastopore, the establishment of a second, ectopic organizer and, as a further consequence, the formation of a second body axis with a head (Fig. 5.17). By positive feedback β-catenin activates *Wnt* genes.

- Also in fishes, birds and mammals a WNT system participates in the establishment of the Spemann organizer and, hence, in the specification of the antero-posterior body axis, though it is not known in each case where and when exactly the signal emitter is positioned.
- Further examples for the role of WNT systems in the establishment and patterning of a body axis in invertebrates (annelids, planarians) are introduced in Chap. 22 (Figs. 22.6 and 22.7).

As vertebrates and most invertebrates are bilaterally symmetrical organized, a second axis, dorso-ventral axis, must be defined perpendicular to the antero-posterior axis. Following the amphibian model it is suggested that this second body axis is specified by a BMP4 protein gradient. In the frog blastula, perpendicular to the WNT8 gradient a BMP4 gradient forms (see Fig. 10.5). Because there is no evidence that this BMP4 gradient arises from maternal components its significance will be discussed in the next Chap. 10 in the context of the question how the embryo is subdivided into regions of different destinations by cell-cell-communication.

9.3 Early Commitment of Cell Types by Maternal Factors

9.3.1 In Some Organisms Commitment of the First Somatic Cell Types Is Programmed by Maternal Factors

An egg cell may not only be provided with orientation aids specifying the global coordinates but may in addition get a maternal endowment enabling the embryo to make cell types in accordance with the location. The paradigm is the ascidian embryo. In ooplasmic segregation a distinct fate determinant, represented by the *macho-1* mRNA, is targeted to a distinct area in the egg (Fig. 9.6) where it is allocated to distinct cells in cleavage. Transcripts of *macho-1* commit the cells to which they are allocated to form tail musculature. Other examples are found in sea urchin embryos. Factors allotted to the micromeres, including β-catenin, cause these cells to become skeletogenic (see Fig. 4.6). Also the pool of maternal mRNA encoding induction factors (Fig. 9.3) can be classified among the cell-type specifying determinants in that they prepare the development of the nervous system.

9.3.2   In Several Organisms Also the Primordial Germ Cells Are Programmed by Maternal Factors

Of particular significance are determinants committing cells not to participate in somatic differentiation programs but to become primordial germ cells instead. We refer to the pole cell determinants in the embryo of *Drosophila* and the P granules in the egg cells of *Caenorhabditis elegans* and *Xenopus*. Decisive maternal constituents are maternally generated mRNA's of the genes **vasa, nanos** and **oct4**. More about this topic can be read in Chap. 8.

9.4   Autonomous Versus Dependent Development, Asymmetric Cell Division Versus Cell Interactions

9.4.1   Autonomous and Dependent Development are Closely Linked to Two Basic Mechanisms of Determination: Asymmetric Cell Division and Cell Interactions

Maternal factors can enable a cell to go its own way towards terminal differentiation autonomously, that is independently of its neighbourhood. To be sure, blastomeres of the early embryo are founder cells giving rise to different cell types in spite of their heritage in maternal determinants. Well documented examples are found in *Caenorhabditis elegans* (see Fig. 4.19). Careful examination of the cell lineages and cell movements in the early embryo, combined with cell ablations have disclosed two fundamental means by which divergence of fates and differentiation pathways can be accomplished.

Asymmetric cell division and autonomous development: In the founder cell the cytoplasmic determinants are asymmetrically localized and in cell division allocated to only one of the two daughter cells. As a consequence two differently programmed cells arise. A paradigm of an asymmetric cell division is the segregation of the germ line cells (P cells) from the soma cells (S cells) in nematodes (Fig. 9.7, also Fig. 4.20). Each division gives rise to one S and one P cell which retain germ line quality by inherited P granules. Asymmetric cell divisions are not confined to early embryos the maternal legacy of which is differentially bequeathed among the descendants. Asymmetric cell divisions occur also in later stages of development when maternal factors don't play a role any more. The reason for unequal treatment of the two daughter cells can be an asymmetrical structure of the starting cell itself. For instance, an asymmetrical structure, generally called polarity, is displayed by epithelial cells sitting on a basal lamina such as the layer of epithelial stem cells (*stratum germinativum*) which renew our skin (see Figs. 2.8 and 18.2) or the stem cells in the central nervous system (Fig. 16.7).

Once committed to their tasks descendants of asymmetric division may be enabled to undergo autonomous development independently of their environment. If such cells have retained the ability to divide and are grown in cell culture they may preserve their identity over uncountable cell generations. This type of cell heredity will be the subject of Sect. 12.5.

Cell interactions: The daughter cells are initially not only by their genetic heritage equivalent but also contain equal cytoplasmic components and, thus, are equivalent in their initial developmental potential. They become different upon mutual interactions and enter different pathways in mutual arrangement. One speaks of dependent development. This type of fate allotment is flexible and allows high diversity. For example, cells of the animal hemisphere of the amphibian embryo can give rise to epidermis, nerve tissue or musculature dependent on influences exerted upon them by their environment (see Fig. 5.16).

9.4.2   Embryos Displaying Mosaic or Regulative Behaviour Differ in the Temporal Progress of Determination

Supposedly Mosaic Embryos. Often, in the embryos of the ascidians, the nematodes and the

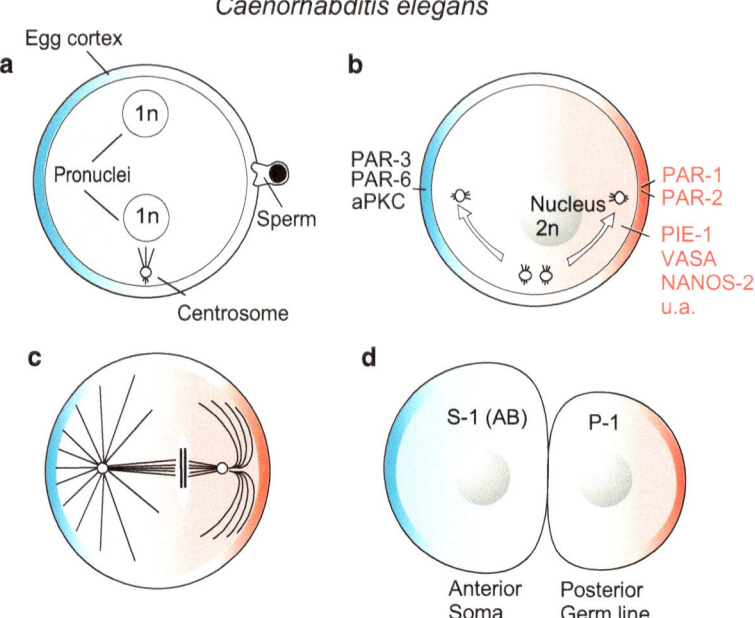

Fig. 9.7 Asymmetric cell division in the first cleavage of the egg of *C. elegans*. The first division determines the direction of the antero-posterior axis and separates the first soma cell S-1 (AB) from the germ line cell P-1. PAR (=PARtitioning-defective) proteins, styled so after defective mutants, are localized in the egg cortex and direct, in cooperation with other factors, the centrosomes to their correct positions. The centrosomes line-up the spindle for mitosis horizontally. Division segregates anterior from posterior cytoplasmic determinants. Among them are products of the genes *vasa* and *nanos-2*, which are allocated to P1 and will accompany the germ line

spiraleans the removal of a founder cell cannot be compensated for by reprogramming other cells. On the other hand isolated parts develop according to the fate map. Such observations prompted the idea that the eggs contained a mosaic of determinants located in a distinct pattern. We have seen that in fact such patterns exist or are created following fertilization in the event of ooplasmic segregation. A paradigm of this type of development is *Caenorhabditis elegans* and the egg of the ascidians. Yet, careful observations and subtle experiments such as displacing a distinct cell into another neighbourhood have shown that the fate of a cell is not only a function of its lineage but also of – though early and subtle – mutual interactions of the cells. During the first cleavage stages cells enter into agreements with their immediate neighbours using membrane-associated signalling systems (in the manner of the NOTCH/DELTA systems, see Figs. 10.1 and 11.7b). Lineage trees reflect the sequence of decisions made in the early stages. In *C. elegans*, as early as in the 28-cell embryo the fate of the cells is fixed and the embryo is a community of rather independent individuals. Removal of an individual cell cannot be tolerated. Astonishingly the cells still move around to seek and ally to other cells with the same destiny.

Regulative Embryos. In members of phyla as diverse as Cnidaria and Vertebrates, and in many other animal taxa, determination predominantly proceeds by cell interactions as they are exemplified in Chap. 10. Moreover, the phase of programming is long and is associated with high powers of regulative compensations. Stem cells, for instance of the blood-forming system (Chap. 18, Figs. 18.5 and 18.6) preserve pluripotency life long.

On the other hand, maternally deposited determinants are also found in these animals (see above Sect. 9.3) directing the cells which inherit them into different pathways. They may deviate or change direction upon influences by neighbours but in general they are predisposed for a certain profession and cut off from other options.

Differences between mosaic and regulative development are gradual.

Summary

Early in development decisions must be taken as to where anterior and posterior, dorsal and ventral shall be. The establishment of the body's coordinates, also called body or polarity axes, is in some organisms predetermined in oogenesis by components of the egg that are manufactured under the control of the maternal genome, therefore called maternal factors or determinants, and deposited in the egg in a distinct spatial pattern. The best known paradigm is the egg of *Drosophila*. The maternal organism specifies in advance the position of the coordinates in that nurse cells and follicle cells of the ovaries channel determinants into the egg. These factors are transported to distinct locations and moored on the cytoskeleton. Mutations in the genes required to produce these factors can cause deficiencies in the offspring and therefore are called maternal effect mutations.

The antero-posterior body axis in *Drosophila* is specified by mRNA of the transcription factor *bicoid* located at the anterior egg pole and by the mRNA of the genes *oskar* and *nanos* anchored at the posterior egg pole. The dorso-ventral body axis is predetermined by an external cue, the maternal SPÄTZLE factor located in the space between the egg envelope and the ventral side of the egg cell. Its presence is received by TOLL receptors of the egg cell membrane. Upon reception a factor DORSAL is redistributed within the egg and taken up by the nuclei on the ventral side.

In many animal species and taxa components of the WNT signalling system, in particular mRNA for WNT signals and the transcription factor beta-CATENIN, are deposited in the egg cell at distinct sites. On the basis of this maternal heritage distinct cells produce WNT signals. In cnidarians the WNT system establishes and patterns the oral-aboral body axis, in the sea urchin they mark the vegetal egg pole and thus the position of the blastopore, in the amphibian embryo the Spemann organizer at the blastopore. Often these maternal factors are redistributed upon fertilization by cytoplasmic flows, a process called ooplasmatic segregation, and allocated to distinct cells in the course of cleavage. After the sperm's entry into the amphibian egg various mRNAs coding for WNTs and β-catenin are translocated into the region where subsequently the blastopore forms and the Spemann organizer is established. Artificial translocation of β-catenin at other sites induces the formation of a second body axis at those sites.

With the establishment of the body axes diverging programs are turned on. This programming proceeds over an orienting specification to a finally stable determination (with respect to cell types also called commitment). Also the specification of certain cell types (in sea urchins and ascidians) and the determination of primordial germ cells (in *Caenorhabditis*, *Drosophila*, *Xenopus*) can be effected by maternal cytoplasmic determinants. In programming germ cell precursors homologues of the genes *vasa*, *nanos* and *Oct-4* are significant (Chap. 8). Asymmetric cell division can parcel out such determinants differently among the daughter cells. Asymmetric cell division is an important means of differential determination. Other means of differential fate allocation are based on cell-cell interactions. In embryos once classified as mosaic embryos and thought to be completely under the regime of maternal determinants (nematodes, spiralians, some insects, ascidians) determination occurs early and thereafter the cells proceed towards differentiation autonomously, no longer change their state of differentiation but still may change their position in the embryo. In regulative embryos, (Cnidaria, sea urchins, vertebrates) disturbances are easier to correct because the final determination occurs late and the development of cells, tissues and organs is dependent on their environment for a long period (Chap. 10).

Box 9.1: Where Is Front, Where Is Back, Where Is Up, Where Is Down? Sources of Primary Positional Information

A cell does not possess a street map or a compass in its nucleus to define its position autonomously. Correctly localized (normotopic) cell differentiation must be launched from outside. Substantiated working hypotheses help to trace the origin of positional cues.

1. **External cues**. Information comes from the external environment. For example, gravity, light, site of sperm entry, or site where an oocyte contacts the wall of the gonad or a blastocyst contacts the maternal uterus, could provide landmarks to guide orientation. Though external cues are more important in plant development, they are also used in early animal development in several instances.

2. **Patterns of maternal determinants laid down in the egg cell**. Maternal cytoplasmic determinants (represented by RNA or protein) are assigned differentially to the cells. They function like airline tickets, specifying route and destination. However, various tickets are needed and must be allocated to the cells in an orderly spatial pattern.

Two mechanisms for the creation of spatial order can be imagined and both have been identified. (a) The cytoplasmic determinants, whether synthesized by the oocyte itself or supplied by neighbouring nurse cells, are deposited in the egg at distinct places during oogenesis. Strictly confined localization in the egg leads to asymmetric distribution of determinants and enables specification of body axes. (b) A process of sorting out and internal patterning, called **ooplasmic segregation,** takes place within the egg after fertilization.

3. **Cell interactions and concerted behaviour**. Ballet dancers can create beautiful patterns guided by mutual arrangement and exchange of gestures. Cells may also 'look' at their neighbours who already have their position defined, and make agreements with them to create coordinated patterns. We will devote a separate chapter to the fascinating problem of how ordered patterns can arise from mutual influencing (Chap. 10).

Positional Information, Embryonic Induction and Pattern Formation by Cell-Cell-Communication

10

10.1 Positional Information and Generation of New Patterns

10.1.1 How Are Cells Enabled to Differentiate Well-Organised and in Accordance with Their Location? We Develop Working Hypotheses

We have discussed in the previous Chap. 9 the problem of how cells can behave in accordance with their location in the whole embryo. Here they have to construct nerve tissue, there to form muscle, and over there manufacture a skeletal element. But the DNA of the nucleus does not contain information about where a cell is at a given moment.

Two basic mechanisms ensure that at the end of development the various cell types can be found at their correct place:

1. The cells are assigned a distinct differentiation program in their location by local influences or by their provenance; they might have inherited cytoplasmic determinants from their mother cell. During or after termination of differentiation they actively search for the place to settle temporarily or permanently. Examples are the melanocytes of our skin, nerve cells of the peripheral nervous system, the cells of the immune system or the germ cells (Summary in Chap. 15).

2. More often cells choose their 'profession' in accordance with their residence and take a differentiation pathway correspondingly. It is the local environment that tells them what should be done at this particular place. Position-pertinent cell differentiation is always launched from outside, from **activating** or **permitting** influences originating in the cell's exterior.

In Chap. 9, we already became acquainted with two sources of guiding cues: (1) External cues: Information comes from the external physical conditions: gravity, light, for example. In animal development, unlike development in plants, the influence of external physical cues on patterning appears to be restricted to very early embryonic development. (2) Maternal cytoplasmic determinants stored in the oocyte, for instance mRNA coding for transcription factors. (3) Now in this Chapter we add **cell interactions and concerted behaviour** as main and sustained means to create new patterns. In the majority of embryos, most probably of all animals, there is no detailed system of regional plasm in the egg, yet neither is there complete cytoplasmic homogeneity. There is instead some system consisting of a few reference points to which developmental events throughout the embryo are related. Development proceeds through interactions between cells which may act variously as the source, destroyer, transmitter or receiver of developmental signals. The most direct sort of interaction is **embryonic induction**. Signals provided by neighbours which determine the fate of cells formatively, are known as **inductive signals**

W.A. Mueller et al., *Development and Reproduction in Humans and Animal Model Species*, DOI 10.1007/978-3-662-43784-1_10, © Springer-Verlag Berlin Heidelberg 2015

(inductors) or as **permitting signals**; these are dealt with in the following sections.

10.1.2 In the Establishment of the Body's Architecture New Patterns Are Created Which Are Not Yet Predefined in the Egg Cell

Pattern Formation as Result of Cell–Cell-Communication As stated above the pattern of cytoplasmic determinants can only be a first guide for orientation. Think of a *Hydra*: by budding, the parental animal gives rise to hundreds of offspring; hundreds of offspring give rise to thousands of offspring and so forth, until after thousands of years vast numbers of polyps have been produced without intervention of an egg cell and sexual reproduction. How could the detailed development and shape of all these hydras be specified by cytoplasmic determinants laid down in the one original egg cell many years ago? Or imagine apple-trees of the variety Golden Delicious. Trees all over the world have been grown by grafting and are the cloned offspring of one egg cell that existed decades ago. How could the pattern of cytoplasmic determinants in the founder egg cell determine in advance the exact pattern of all these trees with all their branches, leaves and blossoms?

Pattern formation is a reproducible "epigenetic" accomplishment. Epigenetic refers to processes that are determined by events 'above' (Greek *epi* = on, above) the level of genetic information. (Since in recent years the expression "epigenetic" has got a very restricted meaning, pointing to certain chemical modifications of the DNA and associated proteins (chromatin, Sect. 12.5) we henceforth will follow this usage and confine the term "epigenetic" to posttranslational modifications of the chromatin.) The processes decisive for pattern formation are interactions of cells with other cells or with extracellular substrates (extracellular matrices). (A general synopsis is given in Figs. 2.8 and 11.1–11.4.) Of course, genetic information is required to produce those molecules which mediate cell interactions. For example, genes specify the signal molecules that can be exposed on the cell's surface or secreted into the surroundings, and receptors that can be used to receive and recognize such signals. But all these interactions are significantly determined also by non-genetic principles such as the degree of hydrophilicity or lipophilicity of the interacting molecules and the physical laws of diffusion or the physical forces of adhesion.

10.1.3 Cell-Cell Interactions Bring About Physical Morphogenesis and Continue Maternally Arranged Steps of Determination

Among the first molecules made by the embryo, produced utilizing maternal mRNA or its own zygotic genes, are molecules enabling strong adhesion between the cells. In the mouse embryo this becomes visible under the microscope. In the 8-cell embryo the process of **compaction** takes place (see Fig. 5.29b): cells adhere strongly along their contact surface. Synthesis of special **cell adhesion molecules (CAM)** results in large-area adhesion. In the blastocyst of mammals, and likewise in the blastula of amphibians or sea urchins, cell adhesion molecules confer an epithelial organization to the outer cellular layer. Changes in the local structure of the cells with such molecules, together with secreted lubricants, enable **cell movements** in gastrulation. The mechanical forces generated with such molecules are dealt with in Chap. 14; they serve the physical creation of shapes during the entire embryonic development. An important class of physically acting cell adhesion molecules is known as **cadherins**, molecules anchored in the cell membrane and protruding into the extracellular space. Other membrane-associated molecules mediate signal transfer thus enabling locality-pertinent behaviour as explained below in Sect. 10.2.

10.1.4 In the Embryo Positional Information Is at Work. The Effect It Takes Is Demonstrated by a Classic Transplantation Experiment

The term positional information was coined by Lewis Wolpert (see Chap. 1, Box 1.1). In terms of content the idea was proposed by Hans Driesch (see Box 1.1 and Sect. 4.1). The experimental verification was presented by Hans Spemann. In a famous experiment which addressed the question of positional information (Spemann: "**development according to location**") non-determined, prospective epidermis of the belly region was transplanted to the area fated to become the mouth region of a newt. The transplant formed mouth and teeth according to its new location (but horny teeth according to the genetic potency of the donor; tadpoles of frogs have horn teeth and not teeth of dentin as do salamanders! (see Fig. 5.12).

How positional information is encoded is a matter of dispute, and no generally accepted theory exists, simply because there is no single system providing such information. Some speculative proposals are discussed in Box 10.2. Compared to such speculative models the actual examples discussed in the following sections show that in fact positional information is drawn from various sources.

10.2 Pattern Formation by Exchange of Signal Between Adjacent Cells: Lateral Inhibition and Lateral Help

With termination of the first cleavage, in *Drosophila* with the formation of the cellular blastoderm, further processes of pattern formation rely on signal exchange between the cells and mutual agreements between neighbours. Such agreements are found even in classic "mosaic embryos" such as *Caenorhabditis elegans* (Sect. 4.4) and the ascidians (Sect. 4.7). Here we refer to two examples in the development of *Drosophila*. Of general significance are neighbourly interactions termed **lateral inhibition** (a term borrowed from the neurophysiology of vision) or **lateral help**. Examples are given in the following sections.

10.2.1 Nerve Cell or Epidermis, That Is the Question. The NOTCH/DELTA System Effects Binary, Alternative Decisions Through the Principle of Lateral Inhibition

When the egg of *Drosophila* has reached the stage of cellular blastoderm and gastrulation commences through the ventral primitive groove (see Fig. 4.36) the cells of the blastoderm compartmentalize into three groups:

- Along the ventral midline cells which switch on the genes *twist* and *snail* immerse into the depth of the embryo to give rise to mesodermal tissue (e.g. muscles).
- The cell bands adjoining on the left and right side also have a tendency to immerse into the interior to become there **neuroblasts** and construct the ventral nerve cord.
- The residual cells of the blastoderm should stay on the surface, adopt the task of **epidermoblasts** and form the epidermis.

The segregation of the neuroblasts from the epidermoblasts requires a process of decision because in the border area the cells are still undecided and tend to oscillate. How is the decision made? Initially, all of the cells are nearly equivalent and have the potential to become either epidermis or nervous system, although in the ventral side there is an initial bias for becoming neuroblasts. But the cells are not yet irreversibly committed to one path or the other. If transplanted into a more dorsal position, ventral cells adopt the features of a dorsal cell and participate in forming the epidermis. Likewise, if displaced ventrally, prospective epidermal cells comply with the social rules and instructions valid at the new location, and become nerve cells.

Both groups segregate definitively from each other by the exchange of signals while they still are more or less clustered in intermingled groups. In this process of separation signal molecules exposed on the surface of the cells are of particular significance (Fig. 10.1). One of the surface molecules is encoded by the gene *Notch*. When the NOTCH protein is defective due to a mutation in the *Notch* gene, all cells develop into neuroblasts instead of a mixture of neuroblasts and epidermoblasts. NOTCH interacts with another membrane protein exposed by neighbouring cells and encoded by the gene *delta*. Both NOTCH and DELTA are transmembrane proteins containing in their extracellular moiety repetitive sequences, called EGF repeats because they had previously been found in the epidermal growth factor EGF. The extracellular domains of the NOTCH and DELTA proteins exposed by adjacent cells bind mutually and mediate a loose cohesion; but they have multiple additional functions. In the present example NOTCH plus DELTA serve the supression of the neurogenic bias in the epidermoblasts. In the recent literature DELTA is classified as an inhibitory signal, NOTCH as its receptor. This classification was chosen because upon mutual attachment and activation, an intracellular domain is cleaved from the NOTCH molecule, enters the nucleus of the cell and acts as gene controlling transcription activator. This cleaved off intracellular domain is abbreviated **NICD** (Notch Intracellular Domain).

Activating of NOTCH signalling is needed by the cells to become epidermis. In order to reach this end, the NOTCH receptor – once stimulated by the DELTA ligand – initiates via its NICD a cascade of events resulting in the switching on of a gene complex called *enhancer of split* (*ESPL-C*). This complex suppresses the inherited neurogenic bias and confers the ability to adopt instead epidermal features. When NOTCH is defective the intrinsic preference to take the neurogenic path is not properly suppressed. DELTA is used as a competition suppressing signal by cells being slightly in advance in taking the neurogenic path themselves. It enables them to suppress the neurogenic tendency in their neighbours by showing them their DELTA-decorated cell surface and thus to induce the

NOTCH pathway in the neighbour. To realize their own neurogenic potency, the DELTA-decorated cells must continue to express another complex of genes, called *achaeta scute-complex* (*AS-C*).

The two cell-type programming gene complexes (the epidermis-specific *ESPL-C* and the nerve-cell specific *AS-C*) cannot be simultaneously activated: their expression is mutually exclusive. Eventually, the differently programmed cells separate mechanically. The nerve cell precursors are shifted into the interior of the embryo (Fig. 10.1), whereas the epidermal cells remain outside and secrete the shielding cuticle.

We still shirked answering the question: How is the decision reached which cells are allowed to enter the interior of the embryo becoming definitive neuroblasts and which cells are destined to stay outside forming epidermis? The hypothesis presented as an explanation refers to a basic principle in biology: **Initially minor differences are enhanced by competition between the participating cells. Lateral inhibition brings about winners and losers**.

In the ventro-lateral border area where undecided cells are located all cells expose NOTCH as well as DELTA. The future neuroblasts possess slightly more inhibitory DELTA molecules, either due to their position, provenance or chance. Presenting a few more DELTA is a slight starting advantage in competition: the neighbour is slightly more suppressed than the cell presenting a higher amount of inhibitory signals. As a consequence of received inhibition, in the neighbour expression of DELTA is reduced while NOTCH molecules for reception of the inhibitory signal remain preserved. Thus a weak difference experiences progressing self-enhancement by positive feed back. The neighbours remain susceptible for inhibitory signals (because they retain NOTCH), but their ability to produce inhibitory signal declines (because DELTA expression is more and more curtailed). Eventually they will be the losers.

Reaching alternative decisions using the NOTCH/DELTA system is not restricted to this particular case. Genes coding for members of

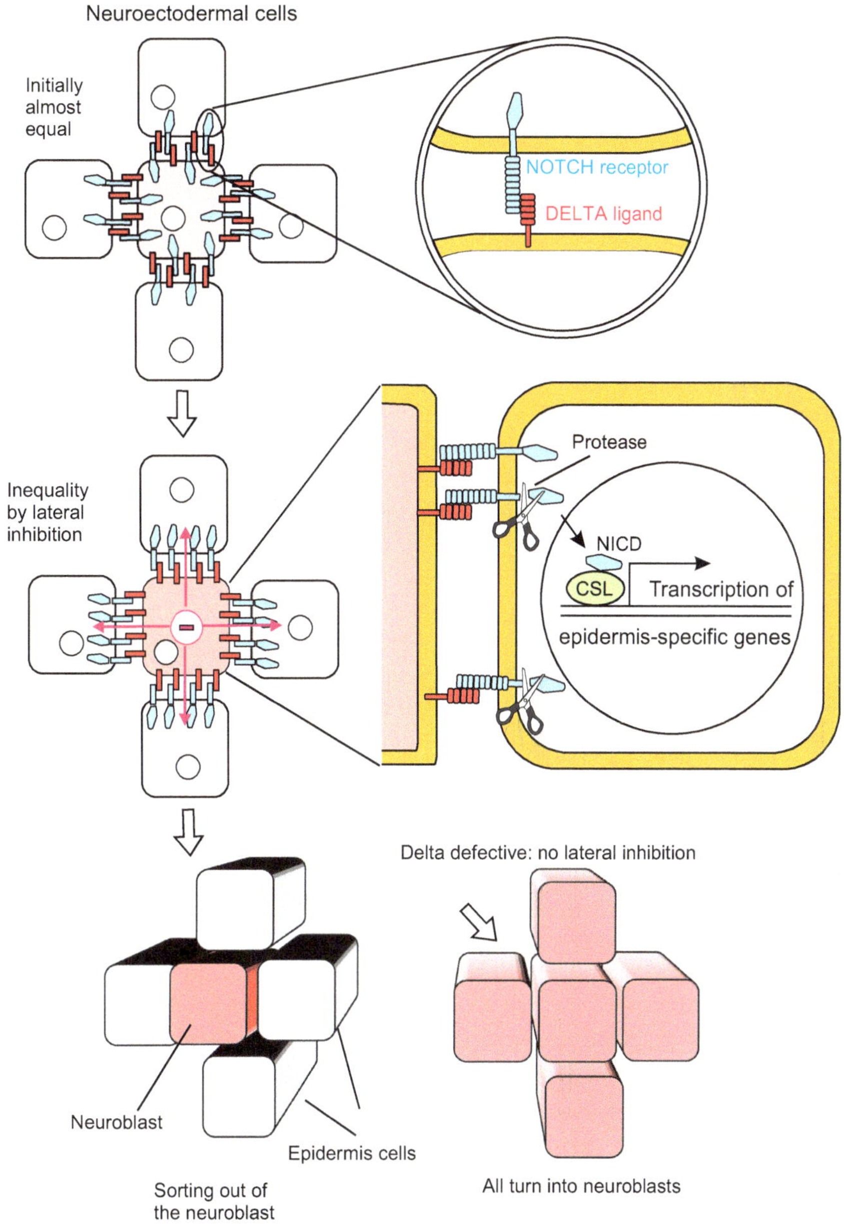

Fig. 10.1 NOTCH/DELTA signalling. Segregation of a neuroblast from epidermoblasts in the ventral blastoderm of a *Drosophila* embryo. Initially, in the ventro-lateral blastoderm all cells are (almost) equivalent, with a bias toward becoming neuroblasts. This bias is indicated by the presence of transcription factors that are encoded by the neurogenic *achaete–scute* gene complex. Local inhibitory cell interactions mediated by cell surface proteins restrict the expression of the neurogenic factors to one cell, the neuroblast. The surface proteins in question NOTCH and DELTA are encoded by the *Notch* and *delta* genes, whereby NOTCH is viewed as receptor and DELTA as its ligand. Upon activation of Notch signalling, the intracellular domain NICD of NOTCH is cleaved off, enters the nucleus and acts as gene regulatory transcription factor. The particular cell that is fated to become a definitive neuroblast is thought to be specified by random fluctuations in some signalling activity. Once a definitive neuroblast is specified, it suppresses the neurogenic bias in adjacent cells by DELTA/NOTCH-mediated signalling. Finally, the neuroblast is singled out and delaminated into the interior of the embryo. In *Notch* as well as in *delta* null mutants all cells of the ventral blastoderm continue to express the *achaete–scute* gene complex and adopt the neuroblast cell fate, which leads to lethal hypertrophy of the nervous system

the NOTCH and DELTA families were found in all animal organisms in which homologous genes have been searched for. Also in **vertebrates** segregation of future neuroblasts and epidermoblasts along the border of the neural plate is mediated by a NOTCH/DELTA system (see Fig. 16.10). Moreover, the system plays a role not only in the segregation of future nerve cells and epidermal cells, but in **many processes of alternative decisions when boundaries are drawn between territories of different destiny**. There exist several different isoforms of NOTCH as well as of DELTA enabling different fields of application; an example will be presented in Fig. 10.25.

10.2.2 Here Is the Question: Photoreceptor No. 7 or *Sevenless*? A Case of Lateral Help

The compound eye of *Drosophila* comprises about 800 small eyes called the ommatidia. Each **ommatidium** is composed of 20 cells. Of these, four cells are used to construct the dioptric apparatus consisting of lens and crystal body – in the following "lens" for short – while eight cells are induced to become photoreceptors. These cells are arranged in a precise, consistent pattern.

In cross sections of the eye always only 7 photoreceptors are seen, R6 to R7 forming a circle and R7 or R8 taking the centre position. R7 and R8 are arranged in tandem order along the longitudinal axis, R7 taking the top and R8 the bottom position (Fig. 10.2). All photoreceptors form with their hem of microvilli (rhabdomeres) the light absorbing mid zone of the ring.

The first photoreceptor to develop is the central cell, called R8; it is the BOSS cell. BOSS refers to the gene *bride of sevenless*. The protein encoded by this gene is exposed on the surface of R8 and shown to the adjacent cell R7 as an identity card. R7 in turn is equipped with a membrane-associated receptor to check the identity card of its neighbour. The receptor is a transmembrane tyrosine kinase derived from the gene *sevenless*. If photoreceptor R8 is unable to offer the correct signal due to a mutation, or R7 is unable to recognize the message, the cell R7 will not develop properly. Instead of developing into a photoreceptor, it makes a lens. Lens cell is the default option if the photoreceptor program is not turned on (Fig. 10.2).

These two examples of neighbourly inhibition or help could well also be classified as induction phenomena. This topic is addressed in the following sections.

10.3 Embryonic Induction and the Spemann Organizer

10.3.1 The Principle of Embryonic Induction: Cells Emit Signals into Neighbouring Areas

Induction is a matter of great significance. In induction, signal-emitting (inducing) cells instruct or permit signal-receiving (responding) cells to take a specific developmental path. Such instructive or permissive interactions are among the fundamental events leading to the cooperative formation of tissues and organs. The term 'induction' is used in biology in diverse contexts (e.g. induction of an enzyme refers to turning on transcription of a gene whose product is an enzyme). To distinguish between the various meanings, developmental biologists often speak of **embryonic induction** (although inductive influences are not restricted to embryogenesis). In terms of embryology induction embraces emission of a development controlling signal by (inducing) emitters and the response by the signal receiving cells

Transfer of inductive information can be accomplished by signal molecules exposed on the cell surface. In Sect. 10.2 we gave two examples: in the *Drosophila* ommatidium the cell R7 is instructed to develop into a photoreceptor by the adjacent photoreceptor R8, and cross talk between neuroblasts and epidermoblasts in *Drosophila* facilitates their segregation from each other. Both are examples of induction. However, the term is usually associated with signal molecules capable of

Fig. 10.2 (**a**)–(**g**) Pattern formation and commitment in the *Drosophila* ommatidium by close-range interactions. The photoreceptor R7 is formed only when the cell R8 expresses the gene *boss*⁺, and the cell number 7 the gene *sevenless*⁺. The proteins coded by these genes are exposed on the cell surface and presented to the neighbours. If either of the two gene products is absent or defective, photoreceptor R7 is replaced by an additional lens cell. Lens cell formation is the default option if the photoreceptor program is not turned on

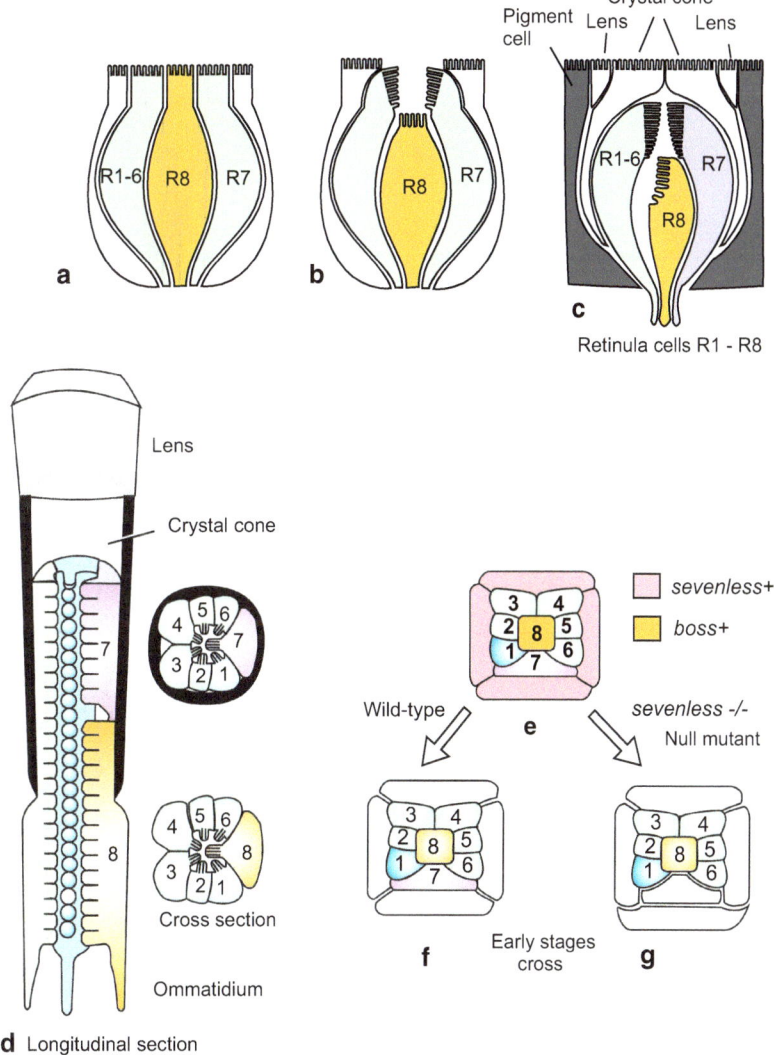

spreading into a larger area by diffusion, though spread by diffusion is not a strict criterion defining inductive molecules.

By transplanting cells or pieces of tissue, and by eliminating putative emitters of inductive signals, inductive interactions have been revealed in all animal organisms that have been investigated with adequate methods. Attempts to cause the development of structures in the wrong places have been successful in sponges, hydrozoans, sea urchins and vertebrates. Even in embryos previously considered to represent a genuine mosaic type of development,

such as *Caenorhabditis elegans* and ascidians, more and more inductive interactions between adjacent cells are being detected.

A structure artificially induced in the wrong location is called **ectopic**. The first discovery of ectopic induction was in 1909 in *Hydra*: a small piece taken from the mouth cone (hypostome) or the subtentacular region and inserted into the gastric region organizes the formation of a second head and body axis (see Fig. 10.19). For many years inductive phenomena have been most extensively studied in amphibian embryos

(Sects. 5.1 and here 10.3.3). These famous experiments were paradigmatic for similar experiments in other animal embryos.

10.3.2 The Receiver Must Be Competent

In radio broadcasting a message is heard only if the receiver is tuned to the transmitting station sending the message. Likewise, in embryonic induction the response to a signal requires competence of the recipient cells, where **competence** is defined as the ability to respond to an inductive signal. The acquisition of competence implies the production and display of **receptors** ('antennae') for the inducing signal. Competence often is restricted to a narrow window of a few minutes or hours.

10.3.3 The Spemann Organizer and the Induction of Heads and Trunks in the Vertebrate Embryo

Not only in the simple *Hydra*, also in the highly complex vertebrate embryo induction of an additional head, and even of an additional, almost entire embryo can be brought about by transplantation of only a small piece of tissue. As we have seen in the Sect. 5.1 in the late blastula and early gastrula there is a small area, known as the **upper blastopore lip or organizer**, which by itself will form the anterior segment of the notochord (plus prechordal plate), has been shown to be an outstanding signal emitter. In 1924 Hans Spemann and coworkers drew the attention of the scientific world to the astonishing inductive power of the upper blastopore lip of the amphibian gastrula: a blastopore lip inserted into a blastula at an ectopic site organizes the surrounding cells into a supernumerary gastrulation, resulting in a secondary body axis.

The equivalent signal emitter is called embryonic shield in fish, Hensen's node in birds, and simply node in mice. All these structures are homologous; they correspond to each other both in terms of evolution and function. **Even a small transplant taken from the organizer of any vertebrate can release the development of complex structures such as heads and trunks in any other vertebrate**. The designation "organizer" was chosen by Spemann because the upper blastopore lip prompted its surroundings to initialize a second site of gastrulation culminating in the development of a second embryo (Siamese twin) with head and trunk (see Fig. 5.13).

In the amphibian embryo the organizer is established at a site located diagonally to the sperm entrance. In the course of the cortical reaction components of the WNT signalling system had been relocated to this diagonally located point. These components include mRNA for WNT signalling molecules and components of the associated signal transduction cascade such as Dishevelled (Dsh) and β-**CATENIN** (Fig. 10.3). The action of these components turn this area into the organizer. When, for example, the mRNA for β-*catenin* in the organizer area, or beforehand at the vegetal egg pole, is inactivated by anti-sense RNA or UV irradiation organizer activity fails. On the other hand, injection of mRNA for β-*catenin* (or β-CATENIN protein) into blastulae rescues the emitter. If injected ectopically, β-*catenin* mRNA leads to the establishment of a second, ectopic organizer (see Fig. 5.17). Oocytes and young embryos are well suited to such experiments because the translation machinery is ready to work with high efficiency and even ectopic mRNA is translated into much protein.

In the conventional scheme of the canonical WNT signal transducing pathway β-CATENIN takes a downstream position (see Fig. 11.6). In spite of this seemingly subordinate position it can release the entire signalling system because β-CATENIN is a transcription factor and among its target genes are *Wnt* genes and the genes of the other components of the pathway, including its own β-*catenin* gene. The cart is put before the horse.

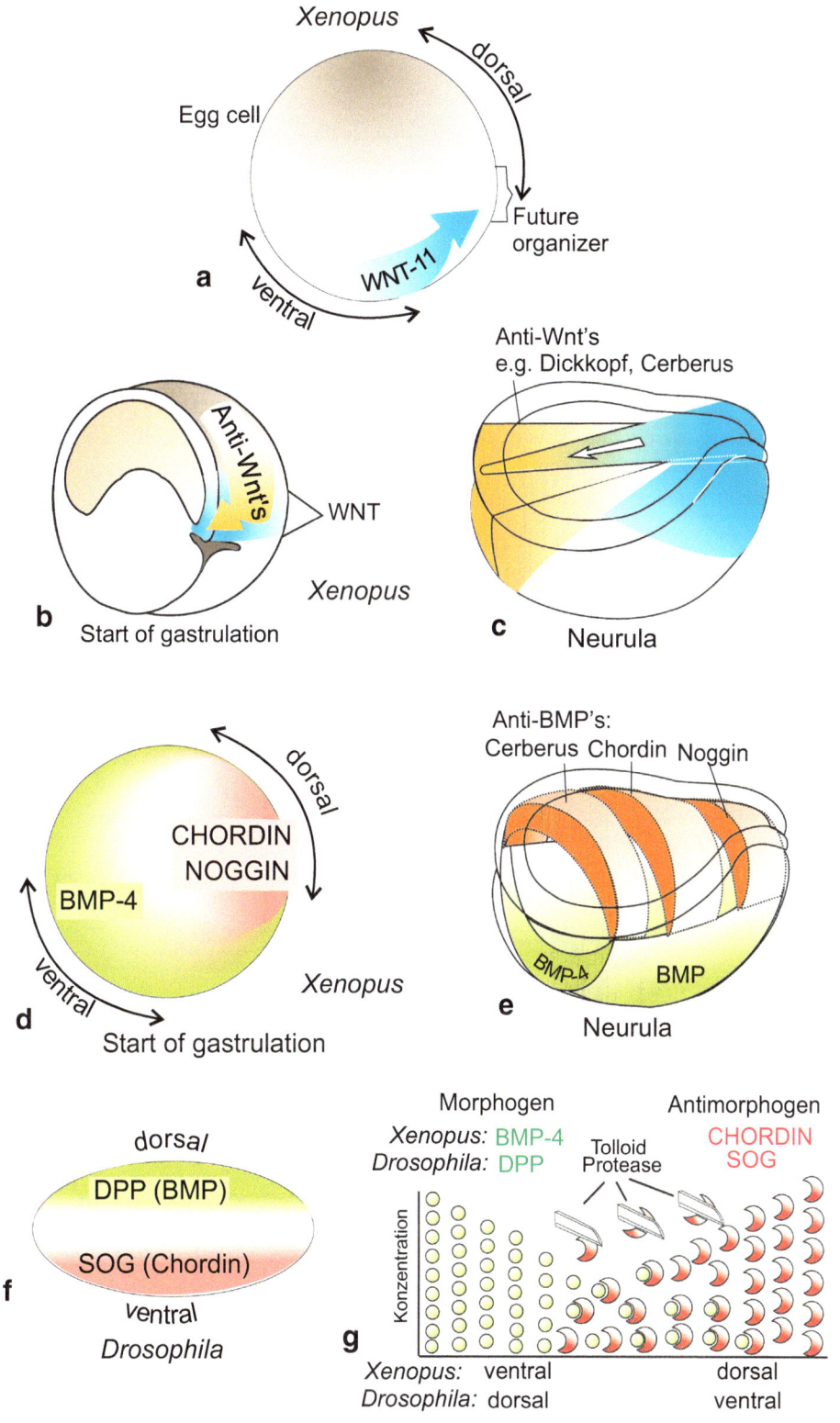

Fig. 10.3 (**a**)–(**g**) Morphogens in the early embryo (shortly before gastrulation starts) in *Xenopus* and *Drosophila* by comparison. (**a**) Induced by maternal mRNA stored in the oocyte, a signal generator known as

In birds and mammals, the part of the amphibian upper blastopore lip is played by **Hensen's node** (**node** in mammals) at the anterior end of the primitive groove. Hensen's node is homologous to the upper blastopore lip, is endowed with similar inductive power when transplanted into an amphibian blastula, and generates similar signal molecules. In addition, all these "organizers", the amphibian upper blastopore lip and the node in birds and mammals, express certain homeobox containing genes such as *goosecoid* and *siamois*. These genes confer the ability to produce inducers. Note, however, that the final structures induced by these organizers, such as a forehead or a second head-trunk structure, can only be initiated by the primary inducing signals. Many events must follow to realize such complex structures.

10.3.4 The Spemann Organizer Is Subdivided in Head and Trunk Organizer

The statement of this headline has been first shown to be valid in the amphibian embryo but is valid in other vertebrate embryos as well. As in Sect. 5.1 described in detail, in gastrulation the organizer area successively moves through the blastopore into the interior. Already while streaming towards the blastopore it is protracted and lengthened and subdivided in two sub-areas:
- a **head organizer**, which in the early gastrula is located near the blastopore lip, and
- a **trunk organizer**, which is located behind and above the head organizer (see Fig. 5.14).

When the cell material that rolls around the blastopore lip into the interior has become the middle stripe of the roof of the archenteron and thus the primordium of the notochord, the action ranges of the head and trunk organizers are well separated. The anterior notochord primordium induces head, the posterior primordium induces trunk. Several factors have been identified exerting corresponding inductive powers:

1. **Head formation**: The effect of the head organizer is in all vertebrates mediated by a cocktail of cysteine-rich proteins with CERBERUS, DICKKOPF-1, FRIZBEE-b and further secreted proteins as ingredients. When one injects the mRNA of, for instance, *dickkopf* (and simultaneously suppresses the inhibitory BMP-4), a complete head emerges, but no trunk (see Fig. 5.17). **In sum head-formation supporting factors are antagonists of the canonical WNT-8 signalling system** (Fig. 10.3, see also Fig. 5.18).

This statement may be confusing, since the organizer includes head organizing power and is established, as said above, by means of a WNT-system. Subsequently this topographical region is a source of WNT-8, now said to prevent head formation. Apparently, this region has the capacity not only to produce WNTs but also antagonists which bind and neutralize WNTs. To resolve this apparent contradiction and to understand anti-WNT activity we have to

(continued)

Fig. 10.3 (continued) Spemann organizer is established. Already before cleavage the process of cortical rotation, released by the sperm's entry, displaces maternal mRNA from the vegetal egg pole to the site of the future organizer. These translocated factors comprise, besides mRNA for WNT-11, some more components of the WNT pathway, in particular β-catenin. (**b**) Gastrulation movements shift cohorts of cells which produce anti-WNT morphogens using their zygotic genes, into the interior of the embryo. These cohorts of cells become the prechordal mesoderm and the anterior roof of the archenteron. (**c**) In the organizer a secondary WNT signalling system (*blue*) is set up producing mainly WNT-8. In the region of the prechordal mesoderm plus anterior intestinal roof anti-WNT's are produced which neutralize WNT-8 and thus stabilize the head-tail body axis. (**d**)–(**f**) The antagonistically acting morphogens BMP-4 and CHORDIN in *Xenopus* are sequence homologous to the corresponding antagonistic morphogens DPP and SOG in *Drosophila*. They are expressed in patterns inverted by 180°, corresponding to the apparently inverted anatomical body plan of vertebrates and arthropods. (**g**) In the subframe beneath it is schematically illustrated how the antagonistic morphogens diffuse towards each other and associate in the overlap area forming heterodimers. Proteases contribute to the sharpening of the distribution pattern. BMP (Bone Morphogenetic Protein) is homologous to DPP (DecaPentaPlegic). SOG (ShOrt Gastrulation) is homologous to CHORDIN

take into consideration the fact that the cells of the upper blastopore lip are not stationary but flow. The cells that pass the lip at the start of gastrulation occupy at the end of gastrulation the anterior roof of the archenteron forming thereafter the prechordal mesoderm. Here they appear to release the anti-WNTs, enabling head formation and aiding in establishing a gradient of WNT-8 extending from posterior to anterior.

2. **Trunk formation**. The effect of a trunk organizer is brought about by a collection of other cysteine-rich proteins, in particular of NOGGIN, CHORDIN, FOLLISTATIN and NODAL-related 3. When in the *Xenopus* egg stored mRNA for these inducing factors is destroyed by UV irradiation and subsequently mRNA of, for instance, *noggin* is injected into the prospective belly region of the blastula, the embryo develops back structures such as notochord and neural tube. But a head is lacking. NOGGIN induces structures of the trunk only. **Trunk-inducing factors are antagonists of the BMP factors** (BMP-4, BMP-2), which radiate from the opposite, ventral side of the embryo (Fig. 10.3, see also Fig. 5.18).

The interaction of WNTs with their antagonists generates double gradients with peak activity at opposite poles and running in directions reverse to each other (Figs. 10.3 and 10.4). The Wnt antagonists aid in programming head structures. By working in opposite direction the **WNT and anti-WNT gradients arrange**, in cooperation with other factors such as **retinoic acid** and diverse **FGF's**, that the **anterior part of the neural tube will form the brain and the posterior part the spinal cord** (see also Fig. 16.1).

Simultaneously and perpendicular to the WNT-/anti-WNT gradients double gradients of the factors BMP and CHORDIN are installed in reverse orientation (Fig. 10.3). This is accomplished in that the BMP and CHORDIN molecules diffuse towards the equator, and bind and neutralize each other in the region of overlap. Together the two gradient systems, the **WNT/**

anti-WNT gradients and the BMP/CORDIN gradients stabilize irrevocably the body axes, the antero-posterior and the dorso-ventral axis.

Under aspects of evolution it is remarkable that homologues of these factors which provide for regional specification in the amphibian gastrula, are called into action also in echinoderms and *Drosophila* (Fig. 10.3f). The factors BMP-4 and CHORDIN in amphibians correspond with, and are homologous to, DPP (DECAPENTAPLEGIC) and SOG (SHORTGASTRULATION) in *Drosophila*. Like in amphibians, in the fly embryo both factors are essential for the establishment of the dorso-ventral axis but with inverse algebraic sign. Homologous factors specifying the dorsal side in amphibians specify the ventral side in the fly and conversely, homologous factors specifying the ventral side in amphibians specify the dorsal side in the fly. How this inversion might have arisen in evolution will be discussed in Chap. 22 (Fig. 22.14).

In the neurula the gradients of WNT-8 on the one hand, and BMP on the other hand, form a system of **Cartesian coordinates** (Fig. 10.5). The possible significance of such a system will be discussed below in the Sect. 10.5.3.

10.4 Cascades of Induction and Identification of the Signal Molecules

10.4.1 Inducing Signal Molecules Are Difficult to Identify; Methods of Biochemistry and Molecular Biology Combined with Appropriate Bioassays Made the Breakthrough

Evidence for inductive events is usually deduced from transplantation studies: When host tissue touches transplanted tissue with inducing power, it responds by developing a structure which would not otherwise form. Although such experiments have been performed many times and in many organisms, identification of the inducing signals turned out to be very

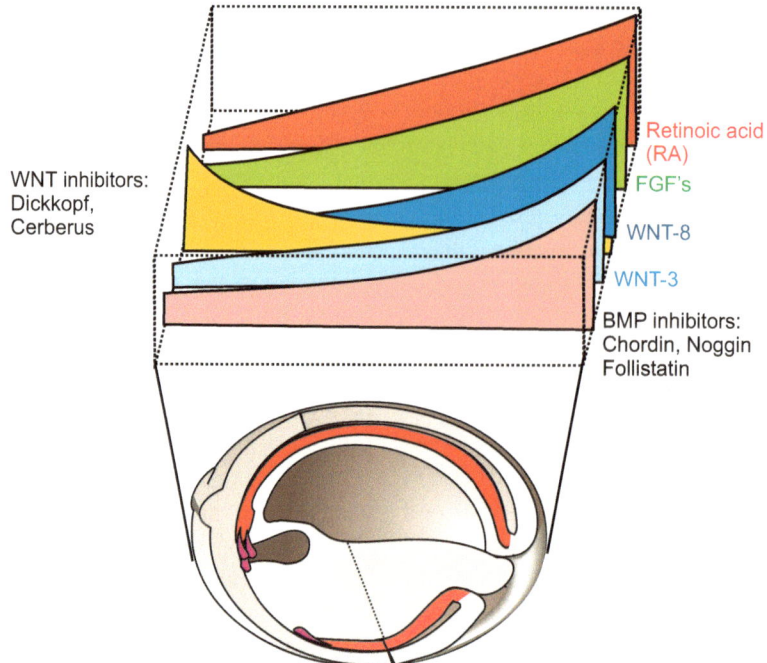

Fig. 10.4 Parallel and reverse gradients of signalling molecules and of neutralizing anti-factors in the dorsal area of a *Xenopus* neurula

Cartesian Coordinates

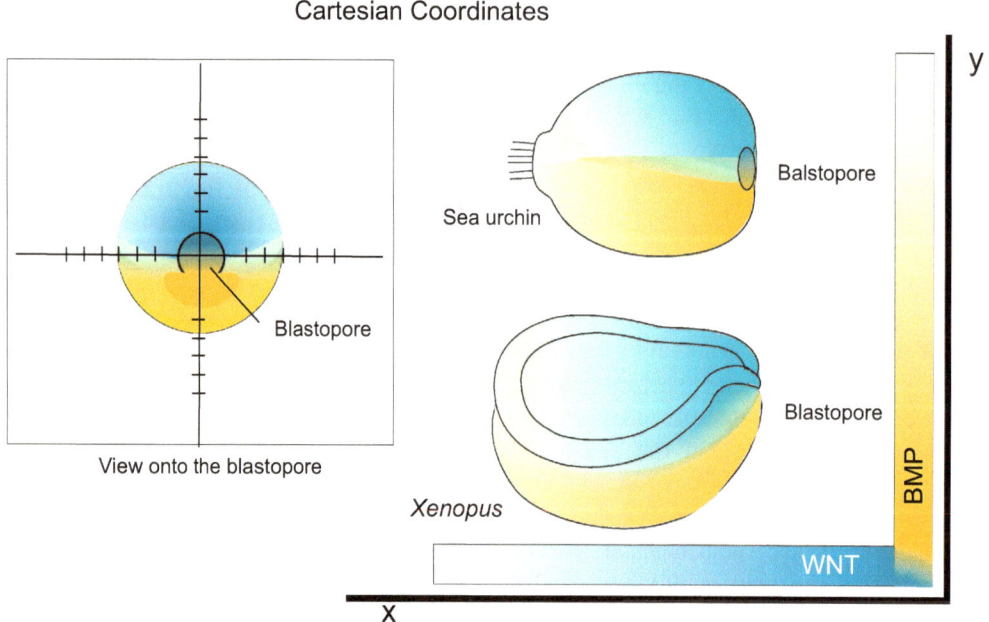

Fig. 10.5 Cartesian coordinates by perpendicular gradients of WNT and BMP, exemplified by the embryos of sea urchins and *Xenopus*. Both the animals belong to the deuterostomes, therefore the blastopore can serve as site of reference

difficult. Embryos are tiny, surgical interventions difficult, and signal molecules present in only minute quantities, so that tons of embryos would have to be extracted to obtain chemically analyzable traces of the inducing substance.

Despite these unfavourable circumstances, impressive progress has been made in *Drosophila* and *Xenopus*. In *Drosophila*, mutagenesis helped identify genes for signal molecules and their receptors. Using the methods of molecular biology (Chap. 12, Box 12.2) eventually the genes of several inducing factors and receptors were sequenced, and more signal molecules are still being identified. Inducing factors and receptors can be expressed in large amounts in appropriate experimental systems such as transfected bacteria or animal cells.

In vertebrates, progress also depended on methods from molecular biology, although classic biochemical methods have also been employed with some success. The discovery of induction stimulated an intense race to identify the putative inducer. In fact the amphibian inducers proved remarkably elusive and the first were isolated only in 1987.

In the hunt for inducers appropriate bioassays are crucial. The "**animal cap assay**" (see Figs. 5.16 and 16.1b) by which competent but still undifferentiated tissue is used to identify the presence of inductive substances, was both the source of many pitfalls and of final success. Caps are excised from the animal hemisphere of blastulae (for decades newt blastulae) and placed in small bowls containing neutral buffer solution. Without exposure to induction factors the animal cap forms a ciliated epithelium as it is characteristic of the blastoderm of a blastula. In classical experiments the Dutch scientist P Nieuwkoop placed explants taken from different regions of the blastula near the cap. After some time of contact with added tissue, or incubation in supernatants, the caps were examined microscopically for the presence of differentiated cells. Various types of differentiated cells were found and even complex structures such as cartilage, brain tissue or eyes. Thus, the beginning was promising.

However, the hunt for inducers, in particular, the hunt for neural inducers, was soon confounded by the fact that just about any culture condition triggered neural development, an effect that became known as 'autoneuralization'. Fortunately *Xenopus* caps do not autoneuralize readily or develop mesodermal cells, such as muscle cells, blood cells and cells typical of the notochord.

Such mesodermal cells were seen by Nieuwkoop when he placed tissue taken from the vegetal hemisphere onto animal caps. Hence from the added explant mesoderm-inducing factors had been released.

In the end, two lines of research converged:

The first line of research was based on hard work. From tons of incubated chick eggs and other biological sources a mesoderm-inducing substance was extracted, tested using the animal cap assay and purified. Animal caps formed mesodermal cells such as muscle cells when treated with traces of the substance. Identification was facilitated by a happy coincidence: animal caps excised from *Xenopus* blastulae were maintained in culture. When the medium was supplemented by certain growth factors used in the culture of mammalian cells, the caps differentiated into mesodermal cells. A similar substance was found in the supernatant of a cultured *Xenopus* cell line. The mesoderm-inducing substances from the mammalian and the amphibian source turned out to be related to each other. They shared a particular property, binding to heparin, with known growth factors isolated from supernatants of mammalian cells, especially with factors belonging to the FGF (fibroblast growth factor) and TGF-β (transforming growth factor beta) families. A partial sequencing of the amino acids pointed to a protein related to ACTIVIN; this protein is in fact a member of the known TGF-β family. The TGF-β prototype is produced by blood platelets and supports growth of transformed (cancerous) cells. Both FGF and TGF-β showed inductive properties. These findings suggested that other known 'growth factors' should be assayed for inductive properties, and that similar molecules should be sought in the amphibian embryo using molecular probes.

The second path of successful investigation began in about 1990 when the powerful methods of molecular biology became available also for *Xenopus laevis*, although in this tetraploid frog identification of corresponding genes by mutagenesis and positional cloning, as performed in *Drosophila*, is not possible. A breakthrough was achieved using the cDNA subtraction procedure. Messenger RNA extracted from organizer region was converted to cDNA and subtracted from cDNA derived from indifferent tissue (the principle of the method is explained in Box Ch12.2 and specialist books). Organizer-specific cDNA was translated and the resulting proteins obtained in quantities that allowed tests of inductive effectiveness in bioassays. In a modified and simplified bioassay the mRNA of putative induction factors is injected into uncleaved eggs of *Xenopus*, or into cleavage stages, at sites where presence of mRNA of this sort is not to be expected. When the mRNA is accepted by the embryo and properly translated it can become effective at the selected site. In addition, meanwhile a large array of methods is available to deactivate the endogenous (body's own) factors, e.g. by means of antibodies, anti-sense oligonucleotides, morpholinos or RNAi (see Box 12.2).

10.4.2 The Development of an Embryo Is Based on Cascades of Successive Induction Processes and Communication by Means of Numerous Signalling Substances

A remarkable, unexpected observation was the fact that not only transplanted tissue with its numerous components but a single isolated substance is capable of releasing the development of complex structures composed of many different cell types such as a head or a trunk, and that such complex structures arise in due sequential order. **When a single substance induces the development of a complex structure this is possible because the inducing signal releases a cascade of subsequent events that follow automatically like a domino effect**. Thus the action of the Spemann organizer is followed by many secondary and tertiary processes of induction (for instance lens and cornea induction, Sect. 10.4.4).

Yet, also before the Spemann organizer starts its work, induction processes take place.

Processes of induction before the establishment of the organizer: The embryonic development of *Xenopus* is governed and controlled by cascades of inductive processes. The cascade starts as early as fertilization, or at the latest in the early blastula. P. Nieuwkoop and thereafter Japanese researches have revealed the origin of the mesodermalizing induction (also called vegetalizing induction): a broad ring of equatorial cells in the blastula is specified to develop mesodermal tissues by signals emanating from the vegetal pole region of the blastula. Which molecules actually confer this effect is a matter of dispute yet. Candidates are molecules which later are produced also in the organizer, for instance NODAL and CHORDIN. In the animal cap assay these factors effect differentiation of mesodermal but also of neural fate. Whether factors such as CHORDIN evoke nervous tissue or notochord or both may be a function of concentration, or a sequence of events is launched: the factor assayed may induce axial (dorsal) mesoderm, the precursor of the notochord, and this induces nervous tissue on its part. We recall: signals emitted by the roof of the archenteron (axial mesoderm) cause the overlying animal ectoderm to develop the central nervous system.

10.4.3 Absence of Certain Factors Might Be a Signal as Well: According to a Widespread Hypothesis Ectodermal Cells Become Nerve Cells Spontaneously If Not Prevented to Follow Their Inherent Propensity

The development of the nervous system was, and still is, in the centre of interest for experimentally working embryologists. Spemann once classified the neural induction radiating from the upper blastopore lip and thereafter from the roof of the archenteron as "primary induction" when the result was a central nervous system embedded in a twinned embryo, or in a head or a trunk. He spoke of secondary induction when the eye vesicle induced a lens in the overlaying ectoderm as described in Sect. 10.4.4.

Factors that undoubtedly support neural differentiation are NOGGIN and CHORDIN. They bind and thus neutralize the ventralizing factors BMP-4 and BMP-2. BMP-4 on its part causes animal cap cells to form regular epidermis instead of ciliated blastoderm. The neuralizing NOGGIN and CHORDIN are thought to accomplish their task by binding BMP-4 in the extracellular space and overriding its anti-neuralizing effect, so that cap cells can follow their inherent propensity and differentiate into nerve cells. Thus, NOGGIN and CHORDIN would be **permissive factors**.

Unfortunately not all vertebrate embryos behave equally well. Already in classical arduous experiments aimed at identifying neural inducers using the assay with animal caps from newts one was confronted with the confusing experience that it was more difficult to prevent than to induce neural differentiation, quite contrary to *Xenopus* or the chick who do not develop neural structures spontaneously. It would therefore not be surprising when in future besides permissive factors also positively inducing factors will be found. Under discussion are various FGFs and retinoic acid.

10.4.4 The Classical Textbook Example of a Secondary Embryonic Inductive Event Is the Induction of the Eye Lens

Development of the eye (Figs. 5.21 and 16.21). In its inner 'kernel' the eye is a derivative of the brain and thus of the neural tube, whose development is initiated by the 'primary' induction emanating from the roof of the archenteron. Two bulges of the lateral wall of the midbrain enlarge, giving rise to the optic vesicles. By a process of invagination the vesicle transforms into the double-walled optic cup. The inner wall gives rise to the layers of neurons and photoreceptors, collectively called retina, whereas the outer wall will form the pigmented layer. Mesodermal cells enclose the optic cup

and supplement the eye ball with the layers of the vascular choroid coat and the sclera. Among the solid structures, only the lens and cornea are still missing.

Inductive events. The optic cup comes into contact with the overlying ectoderm. In a process once classified as secondary induction, the cup stimulates the formation of an ectodermal lens placode (placode = thickening) in the right time, at the right place and of the right size. As a consequence of the induction process in a circular area of the ectoderm a lens-specific transcription factor (L-MAF) is expressed, which on its part switches on genes for lens proteins (crystallins). A circular lens placode is formed that sinks down.

As the lens placode detaches from the ectoderm, it becomes covered by the surrounding ectodermal epithelium. A tertiary signal emanating from the lens causes the covering ectodermal epithelium to transform into a translucent cornea.

There was a long-lasting scientific dispute about lens induction. Surprisingly, it was discovered that in some vertebrates, and in fact in some amphibian species, lens formation does not depend on having received an inducing stimulus from the eye cup. Lens formation takes place even when the eye vesicle is removed. This applies also for *Xenopus*, though the lens remains meagre in the absence of the eye vesicle. Apparently, the cells of the future optic vesicle begin to emit inducers before the optic vesicle is detectable. The finished optic cup has a subsidiary, amplifying influence which in some species is still strong and required, whereas in other species it can be omitted.

Some of the signals controlling induction and formation of the eye structures are well-known: SHH (Sonic hedgehog) defines in cooperation with BMP-4 the dorso-ventral axis of the eye; the lens forms where dorsal and ventral signals meet. FGF/WNT control later steps of differentiation. We will deal with eye development and its neural connection with the brain in Sect. 16.7.

10.4.5 Signal Molecules Such as SONIC HEDGEHOG May Be Used in Different Organisms and Organs for Different Purposes

Nature does make it not easy for the researcher or the student. On the one hand a biological function can be controlled by several synergistically acting, even non-homologous factors; on the other hand one and the same factor can be used in different locations, for different functions and in the formation of different, non-homologous structures, dependent on where and when the factor is used to convey a signal. Informative examples are the factor encoded by the gene *sonic hedgehog* (*shh*) and the lipid retinoic acid.

Sonic hedgehog

Contemporary developmental biology benefits much from molecular techniques known collectively as 'reverse genetics'. In particular the use of heterologous gene probes (Chap. 12, Box 12.2) allows the rapid determination of whether a gene known from, say *Drosophila*, is also present in another organism by screening gene banks or using *Drosophila* sequences as heterologous probes to fish genes from genomic banks of other species. Such investigations yielded many surprising results, for example in the case of the potential signalling molecules coded by the *Drosophila* gene *hedgehog* and the signal molecules coded by corresponding genes in vertebrates, with *sonic hedgehog* as a prominent example.

The mutation *hedgehog* makes a *Drosophila* larva look like a hedgehog with a crew cut; the back is covered by chitinous spines. In the embryo the gene is expressed in various stripes and areas. First, *hedgehog* is expressed at the anterior border of the (para-)segments in those cells whose nuclei contain the transcription factor ENGRAILED (see Fig. 4.38). In other regions of the larval fly such as the imaginal discs, another variant of HEDGEHOG is produced. This variant detaches from the cell surface. HEDGEHOG is not only a signal molecule but also a proteolytic enzyme and cleaves itself into a domain remaining in the membrane and a component that floats away. The detached HEDGEHOG domain serves as a signal molecule, capable of reaching distant targets by diffusion. HEDGEHOG is used for short-distance as well as for long-distance communication in vertebrates (*Danio*, *Xenopus*; chick, mouse) also.

In vertebrates *sonic hedgehog* is expressed in the upper blastopore lip and subsequently in the derivative of the invaginated lip cells, the notochord. The notochord presents SONIC HEDGEHOG protein as an inducing cue to the adjacent neural tube in a contact-dependent manner. In response to this cue the ventral-most midline cells of the neural tube form the **floor plate** (Fig. 10.6). In the neural tube the induced floor plate takes over the production of HEDGEHOG. The signal stimulates the development of **motoneurons** on either side of the floor plate (see also Fig. 16.11).

The soluble diffusing component of SONIC HEDGEHOG reaches the somites. In the somites, the HEDGEHOG signal induces the sclerotome cells to detach. They migrate towards the emitter of the signal, enclose the notochord and construct the **vertebrae** (Fig. 10.6).

SONIC HEDGEHOG gets another chance to demonstrate its versatility when the limbs are formed. In the **limb bud** of the chick embryo HEDGEHOG is expressed at the posterior border, where a signal emitter known as **ZPA** (Zone of Polarizing Activity) is located. When cells expressing HEDGEHOG are transplanted to the anterior border of the limb bud, supernumerary

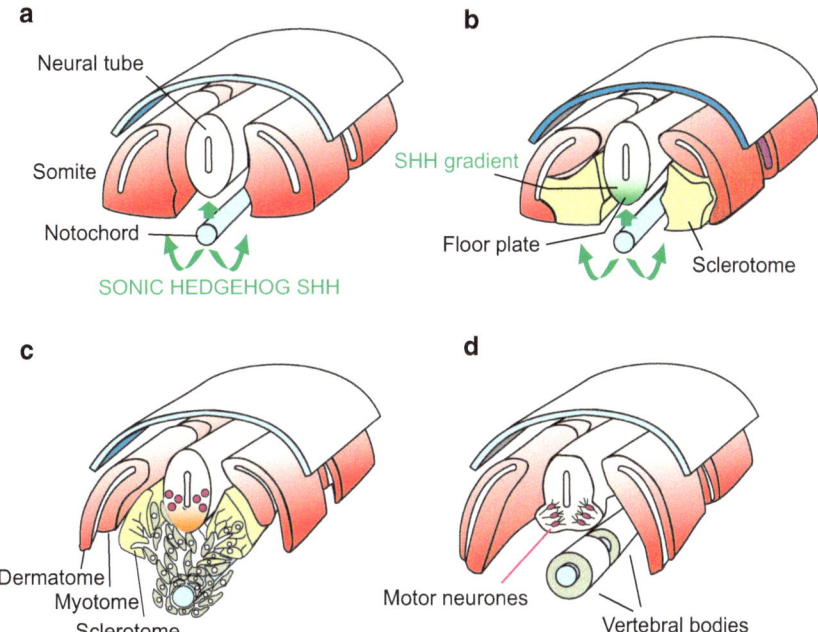

Fig. 10.6 (**a**)–(**d**) SONIC HEDGEHOG signalling in vertebrates. The protein encoded by the gene *sonic hedgehog* can appear in a membrane-associated form or in a form that detaches from the cell surface and spreads into the surrounding spaces where it reaches distant target cells. SONIC HEDGEHOG directly presented by the notochord to the adjacent neural tube induces (**a**) the formation of the floor plate in the neural tube. Subsequently the floor plate also produces HEDGEHOG, inducing (**b**) neuroblasts to differentiate into motoneurons (**c**), (**d**); see also Fig. 16.11c. SONIC HEDGEHOG released by the notochord into its surroundings induces (**c**) the sclerotome cells to detach from the somite, migrate around the notochord and form the vertebral bodies (**d**)

fingers are formed (see Fig. 10.13). This intriguing phenomenon is discussed further in Sect. 10.8.

Also in the establishment of the left-right asymmetry in our body SONIC HEDGEHOG is an essential factor (Sect. 10.6).

Some of the various functions of retinoic acid are addressed in Sect. 11.4.

10.4.6 Concentration-Dependence, Synergy, Antagonism, and Redundancy Are Frequent Principles of Induction and Pattern Formation

To realize region-specific effects, for example, to induce heart muscle cells (cardiomyocytes) or skeletal muscle cells, several basic principles are utilized:

- **The concentration profile** of a factor. When a locally produced factor diffuses into three dimensions of space **concentration gradients** will form which can be elongated or shortened by secondary conditions. Different concentrations may have different quantitative effects but may also have **different qualitative** effects. Read more about gradients as important determinants of pattern formation in Sect. 10.5.
- **Synergy**: Several factors may have similar functions and support each other in a common task. For example all the 23 different representatives of the FGF family in the mouse display a similar array of functions. Yet, synergistically acting factors do not always belong to the same protein family. The five factors which bind BMP-4 and thus enable dorsal structures to form, namely NOGGIN,   CHORDIN,   FOLLISTATIN,

NODAL-related 3 und CERBERUS (see Sect. 5.1.14) are members of different protein families.

- **Antagonism**: Factors which mutually deactivate or contain each other act antagonistically. The factors presented above which govern patterning along the dorso-ventral axis in vertebrates and insects are represented by **homologous antagonistic systems** (Figs. 10.3 and 10.4).

On the one side of the embryo (vertebrate anterior-ventral, insect dorsal) a factor of TGF-β-family is expressed, namely,

BMP-4 (Bone Morphogenetic Protein) in the vertebrate,

DPP (DECAPENTAPLEGIC) in the insect.

Towards these factors other factors diffuse binding and neutralizing the first-mentioned factors; these are:

CHORDIN in the vertebrate,

SOG (SHORT GASTRULATION) in *Drosophila*.

A further antagonistic system of mutually neutralizing factors is the system **WNT** (WNT-3 + WNT-8) contra anti-WNTs with DICKKOPF in *Xenopus* (Figs. 10.3, 10.4, and 10.5; see also Figs. 5.18 and 5.19).

In both systems enzymes are also involved, namely proteases. For example CHORDIN and SOG, respectively, are cleaved by the TOLLOID protease but not BMP-4 or DPP. Production rate of factors, diffusion velocities, binding kinetics of the antagonistic factors and enzyme-mediated decay, and the emergence of receptors create dynamics that hamper predictions about the distribution patterns. Also mathematical models for computer simulations are not easily developed since many parameters are not known and matters of mere speculation.

- **Redundancy**. The principle of synergism is often so well elaborated that many a researcher experiences disappointment. One has in a spectacular experiment succeeded in inducing an additional head, let's say with a factor X1. After targeted loss-of-function mutagenesis

one expects failing head formation or complete blockade of development. But in fact often no or only a minimal effect is observed, because a redundant factor X2 with similar efficacy compensates for the loss of X1. Such redundancy safeguards the embryo against the fatal consequences of mutations and can aid in correcting faults. However, redundancy is often not perfect, and many a mutated allele can cause malformations or is lethal if present in homozygous dose.

10.5 Pattern Formation, Morphogens and Gradient Theory

10.5.1 Many Inducers Can Be Classified as Morphogens Acting Differently Dependent on Their Local Concentration

By definition, a morphogen acts locally to organize the spatial pattern of cell differentiation by virtue of spatial concentration differences. A high concentration of a morphogen at location 1 specifies cell type or structure A, a lower concentration at location 2 specifies cell type or structure B (Box 10.2).

Several inducing factors – if not all of them – in the early embryo not only act upon the area adjacent to the emitter but also have a patterning function in the area where they are produced. In at least some instances the pathway of differentiation is a function of the concentration of the inducing factor. This has been shown, for example, using the animal cap assay. When exposed to very high concentrations of activin B, the uncommitted cap cells develop into endoderm-type cells. Intermediate concentrations of activin B cause cap cells to differentiate into dorsal mesoderm such as notochord and somite-derived muscle cells, and very low concentrations cause them to form epidermis. Thus activin B is also acting as a morphogen.

Best-known example of a morphogen is **BICOID** in the egg *Drosophila*. At the egg pole where concentrations are high, the head will be formed, at the location of lower concentration the thorax segments are formed. A shift in the concentration profile causes a shift in the boundaries (see Fig. 4.34). Another example of a morphogen is **DECAPENTAPLEGIC** in the dorsal region of *Drosophila* embryos and its homologous counterpart **BMP-4** in the blastula and gastrula of *Xenopus* (Fig. 10.3). Likewise, **WNT** and retinoic acid can be designated morphogens, as well (Figs. 10.4 and 10.5). There is hardly an inducing substance which cannot also be classified as a morphogen.

> When in articles and textbooks of developmental biology in sections dealing with amphibians the terms **inducer** or inductive substance is preferred and the term morphogen only hesitantly used, this usage is accounted for by historical reasons. What transplants conspicuously evoked in the classical experiments were responses in the tissue **adjacent** to the transplants. In recent literature the term morphogen is used with increasing frequency.

10.5.2 The Gradient Theory: A Long Disputed Theory of Developmental Biology Is Now Verified by Many Examples

The term morphogenetic field is closely associated with the gradient theory which once was derived from classical experiments in the sea urchin embryo (Sect. 4.1) and ranks among the basic theories in developmental biology. By definition, a gradient is given when a measurable magnitude (y value), for example the quantity of a factor, declines over a stretch (x axis). **The gradient theory implies that the quality of differentiation is specified by the local quantity of a graded factor**. Quality of differentiation means sort of cell types that are specified but can also mean type and dimension of entire organs or body regions (e.g. scope of the archenteron in the sea urchin larva, or dimension of the head region in *Drosophila*). Quantity of a factor means, in the **simplest case**, local **concentration of a**

morphogen being present in non-uniform distribution such as in the form of a gradient. For instance, the concept has been proposed that in the sea urchin blastula an animal morphogen would have its peak concentration at the animal pole, and a vegetal morphogen its highest concentration at the vegetal pole. Both substances would diffuse towards the equator and would there silence themselves mutually.

When the gradient theory was proposed biochemical or molecular tools to identify the putative morphogens or other graded properties of significance were not yet available. Even today in the sea urchin not all of the putative morphogens are known. However, with BMP2/4 in the dorsal territory of the gastrula, with CHORDIN and NOGGIN in the ventral territory, and with WNT-8 in the vegetal half some morphogens have been traced. Remarkably, these morphogens were already known from the amphibian embryo. When it is about morphogens in the first instance one has to refer to *Xenopus* and *Drosophila*. The current ideas about the antagonism of

- BMP-4 versus CHORDIN,
- DPP versus SOG
- WNT-8 versus CERBERUS and DICKKOPF

  correspond exactly with the vision of the traditional gradient theory (Figs. 10.3, 10.4, and 10.5). And in studies with factors of the TGF-β family (activins and BMPs) and the FGF family it was shown experimentally that the quality of differentiation can be a function of the morphogen concentration.

However, the gradient theory must not be reduced to the notion of concentration gradients of diffusible morphogens. "Quantity of a factor" may also mean the quantity of structurally bound molecules such as receptors, amount of energy carriers, quantity of mitochondria or other organelles, or number of a distinct cell type. The first identified morphogen, BICOID in *Drosophila* (1988 identified by Christiane Nüsslein-Volhard et al.), is an instructive example. The BICOID proteins the concentration profile of which specifies the head-thorax region in the larva (see Figs. 4.33 and 4.34), is present in diffusible form immediately after being translated, and a concentration gradient forms in the cytoplasm of the syncytial egg. But subsequently the protein is taken up by the nuclei of the blastoderm and bound to DNA thus losing its diffusibility. And only in its bound form does it fulfil its task as transcription factor. The amount of factor taken up by the nuclei determines which genes are switched on (see Fig. 4.34).

10.5.3 Two Gradients Along Cartesian Coordinates Enable Location Determination

In bilaterally symmetric organisms gradients are found before and during gastrulation which are arranged perpendicular to each other and therefore can be entered (inscribed) into a Cartesian system of coordinates with X and Y axes. In *Drosophila* for example the BICOID gradient extends along the x-axis (= antero-posterior axis) and the DORSAL gradient along the y-axis (= dorso-ventral axis). In embryos of many and quite different taxa perpendicular gradients of WNT and BMP have been visualized, for instance in the sea urchin embryo and in the amphibian embryo (Fig. 10.5). Associated with such a system is the (theoretical) possibility that cells can find out their position by measuring the local concentration values of WNT and BMP. This was once predicted by a then well-known protagonist of the gradient theory, the American embryologist C Child (1869–1954). To date the use of such a tracking system is not strictly proven but we will find good evidence for a two-dimensional tracking system when we discuss path finding of axons of nerve cells (Chap. 16, Fig. 16.22). Migrating cells, however, must not only measure local concentrations but also figure out the vectors of concentration changes.

How gradients can be read with high precision and reacting cells can sharply demarcate from neighbours to take different pathways of differentiation is the concern of several models. These models operate with partial differential equations and are too complex to be dealt with and critically evaluated in an introductory textbook. Anyone interested is referred to the publication list at the end of this book.

10.5.4 Inducers and Morphogens Spread over Large Areas Enabled by Unexpectedly Diverse Mechanisms

How can a signalling substance spread in an area in which cells are densely jammed together? If cells are only attached to each other at selective points and the meshwork of merely loosely packed cells leaves blank open spaces (interstitia), it is not difficult to imagine diffusion being the physical principle of spreading. Certainly, nature utilizes the diffusion potential of signalling molecules where possible, to let them spread in the surroundings without much expenditure of energy. Theorists who design mathematical models of spread preferentially work with terms of diffusion (see Box 10.2).

Ways in which diffusion gradients form:

- when molecules are emitted by a spatially restricted source and spread in a wide three-dimensional space;
- when molecules disintegrate spontaneously with a brief half life,
- when molecules are trapped, inactivated by other molecules which diffuse against them, or are enzymatically decomposed; in simplified form this case is illustrated in Fig. 10.3.
- when molecules during their spread are trapped by receptors of adjacent cells or by components of the extracellular matrix and therefore the concentration of free molecules gradually decreases (see Figs. 11.3c and 11.6). This happens, for instance, with FGFs and WNTs. If bound FGF molecules are enzymatically released from the matrix their range of spreading enlarges considerably, whereas the lipophilic WNTs may become more movable in the water-filled interstitium by binding to other molecules.

The conditions become complex when strong barriers against diffusion such as epithelia exist. In epithelia cells are connected and stick tightly together by means of adhesion belts. In such cell communities small molecules such as O_2, CO_2 and retinoic acid may cross cell membranes or diffuse from cell to cell via **gap junctions**;

Box 10.1: How Can Gradients Be Recognized and Made Visible?

In the history of biology gradients of morphogens were merely hypothetical for a long time and a matter of highly controversial disputes. Gradients had been proposed by several scientists (see Chap. 1, Box 1.1) because regulators of development exhibit properties of a gradient or double gradient as in the case of the sea urchin (Sect. 4.1) and account for size-invariant regulation within embryonic fields. The existence of gradients was deduced from the results of transplantation studies, for instance in sea urchins, showing that cells can influence their neighbours to different degrees dependent on their origin (Fig. 4.4). According to their position along the animal-vegetal axis the power of influence transplanted cells exerted upon their new neighbours was graded. Further historically important experiments were transplantation of tissue pieces with graded inductive potency in *Hydra* (as in Fig. 10.19) and in the avian limb bud (Fig. 10.13). Several efforts were made to visualize gradients but without convincing results for a long time. The gradients actually observed, for instance by measuring the susceptibility to toxic substances or the time required for oxidative reduction of vital dyes to colourless derivatives, could not be shown to be relevant to the problem. The acceptance of gradient models declined.

Morphogen gradients were first made directly visible in embryos of *Drosophila*. The pioneer paradigm was the BICOID gradient (see Fig. 4.33). Two methods are available to render a gradient visible provided the morphogen in question is a protein: (1) in-situ-hybridization for localization of the mRNA and (2) immunocytochemistry using antibodies to visualize the distribution pattern of the protein itself (for technical procedures see Chap. 12, Box 12.2). Yet, these procedures are performed with chemically fixed embryos. Temporal changes can only be reconstructed with specimens fixed at different time points. Rapid kinetics are scarcely apprehensible. One

wants real-time observations and exact quantitative measurements of local concentrations and their alterations in time. First, pioneering procedures are developed. For example, in 2009 in embryos of the zebrafish the spreading of the morphogen NODAL could be observed, however, in an indirect way in that its intracellular action was visualized with a fluorescing construct. In the same year a working group (in Dresden) succeeded in working out a direct proof. In preparation for the experiment, constructs of the gene for the signalling molecule FGF8 with attached EGFP (Enhanced Green Fluorescent Protein) tag were made and injected into a blastomere of the 32-cell stage zebrafish. Subsequently in the embryo locally FGF8-EGFP fusion protein was expressed, and the labelled protein was exported by the producing cells due to export labels attached to it. Coupling of FGF8 with EGFP did not compromise its physiological function. The connected EGFP rendered it possible to film the movements of single FGF8-EGFP molecules. They emanated from their source, they flowed in the extracellular space in the manner of Brownian molecular movements, were caught by neighbouring cells and internalized, and thus removed from the diffusion space. Hence, internalization did set limits to the range of spreading. In this case the establishment of the gradient took place according to a source-sink model (Box 10.2).

By contrast, in case of DPP, the BMP4 homologous signalling molecule in *Drosophila*, it was shown that this molecule was transmitted by transcytosis and correspondingly migrated much slower through the imaginal disc. DPP-GFP constructs had been injected into the discs at sites where DPP is produced naturally. Although the DPP-GFP molecules did not spread in the extracellular space, a gradient of internalized DPP emerged extending from the source along the rows of cells with decreasing intensity. This decline in intensity happened because not all molecules ▶

Box 10.1 (continued)

taken up by endocytosis were transmitted by exocytosis to the adjacent cell. The gradient spread over a wide area. Not so in the case of the signalling molecule WINGLESS (WG): WG-GFP-fusion molecules did spread quickly and presumably through the extracellular space though WG is thought to be insoluble in its chemically unaltered form. It may

be that the WG molecule was chemically modified. Irrespective of that, spread was facilitated by the removal of signalling molecules by the receptors, and thus by steepening of the slope of the concentration gradient. Thus it appears that different mechanisms are used to create and modify gradients (more about gradients in Chap. 11).

but for larger molecules special transport systems are required. Several, partly hypothetical, mechanisms of targeted signal spread (lipid rafts, transcytosis, nanotubes, filopodia) are discussed in Sect. 11.2.

10.5.5 In Two-Dimensional Cell Communities Often Planar Cell Polarity Develops, That Aligns, for Instance, Hairs, Spines and Sensilla, and Frequently Is Based on "Non-canonical" WNT Signalling Systems

Even without boosting and adjusting with a brush the hairs of a fur are oriented in groups in a distinct direction, likewise the feathers of a bird skin, and the bristles, scales and sensilla (small sensory organs) in the cuticle-covered epidermis of insects are oriented in parallel. Examples of planar polarity are shown in Fig. 10.7. Parallel to the surface mechanical forces and/or signalling systems are at work which orient the primordia of these structures or the entire epithelium in a common fashion. Frequently there are **signalling waves** which originate from the edge of the epithelium and assign the cells a desired vector, a distinct alignment. Currently a signalling molecule is brought into focus that is named after the *Drosophila* mutant *Wingless* and in other organisms, from *Hydra* to humans, is called **WNT**.

Once more WNT? This time it is the matter of a different, "non-canonical" WNT system. The signalling molecule in question is (often) WNT-

5a. It acts also through membrane-anchored receptors called Frizzled. With its intracellular domain Frizzled is connected to the cytoskeleton. But it is not the β-catenin-dependent pathway that is used upon ligand-receptor binding, instead a signal relay system is started which transmits the signal in the plane from cell to cell. More helpful than listing the various components of this relay system may be a look at Fig. 10.7 and its legend. More details may be learned from the original papers. The planar WNT signalling wave causes a wave of increased cytoplasmic calcium-ion concentration. Planar signals are likely also in play when – as approached in the next section – left-right asymmetries are generated.

10.6 To Have One's Heart in the Right Place: Left-Right-Asymmetries

10.6.1 Mirror Image of the Body Halves Is Broken by Several Local Left-Right Asymmetries in the Location of Inner Organs

Being bilaterally symmetric organisms we humans possess, as do the majority of all animals, a body plan with two asymmetrical (polar) axes oriented perpendicular to each other, the head-foot(tail) axis and back-belly axis. Along the antero-posterior axis left and right sides are mirror-inverted symmetrical – in the outward shape and in general. Yet, in the interior organization there are exceptions. The gastrointestinal tract is not symmetrically

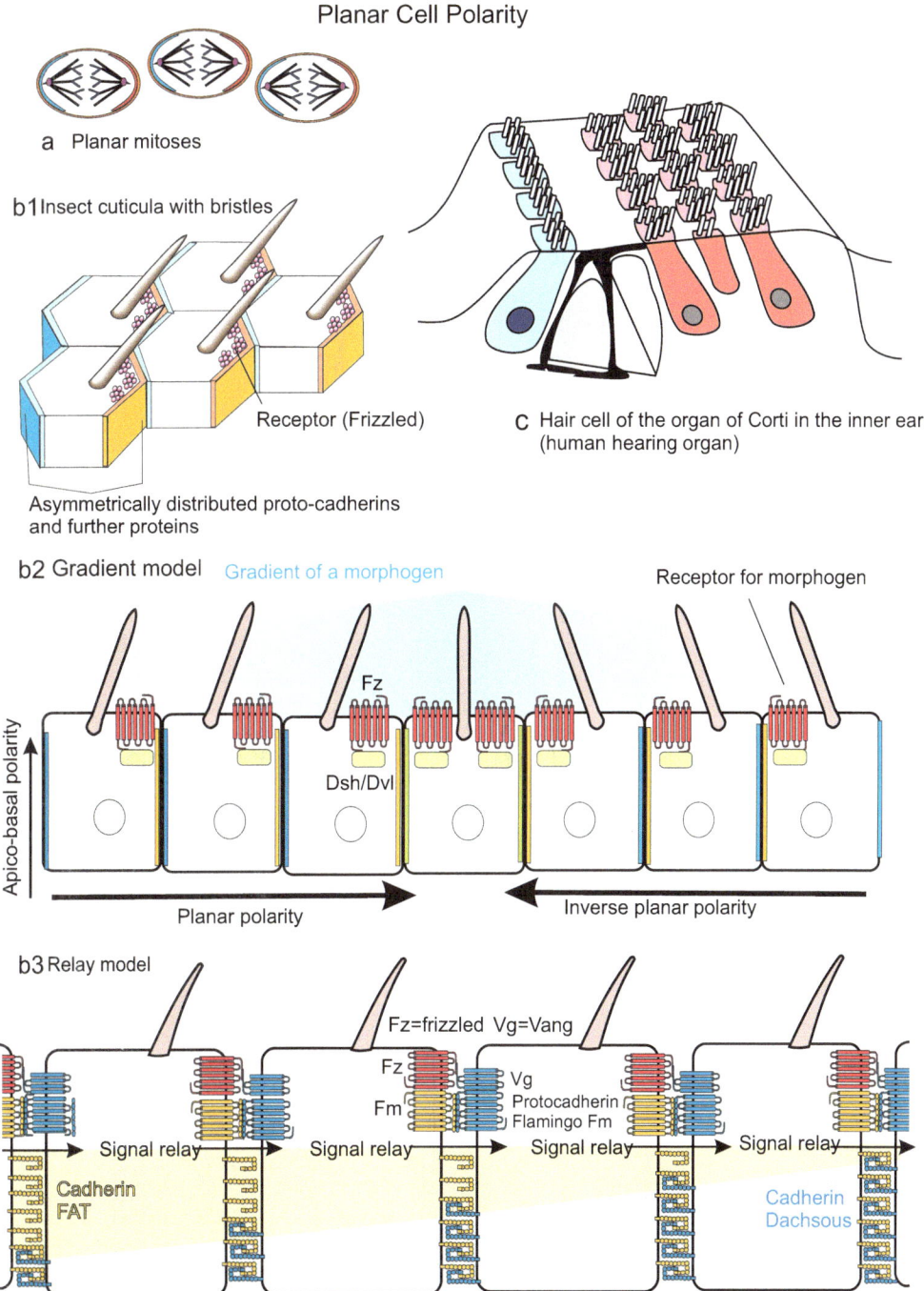

Planar Cell Polarity

a Planar mitoses

b1 Insect cuticula with bristles

Receptor (Frizzled)

Asymmetrically distributed proto-cadherins
and further proteins

c Hair cell of the organ of Corti in the inner ear
(human hearing organ)

b2 Gradient model   Gradient of a morphogen

Receptor for morphogen

Fz

Apico-basal polarity

Dsh/Dvl

Planar polarity

Inverse planar polarity

b3 Relay model

Fz=frizzled Vg=Vang

Fz
Fm

Vg
Protocadherin
Flamingo Fm

Signal relay

Signal relay

Signal relay

Signal relay

Cadherin
FAT

Cadherin
Dachsous

The reverse gradients of the cadherins FAT and DACHSOUS code positional values

Fig. 10.7 (**a**)–(**c**) Examples of planar cell polarity (PCP).
(**a**) Planar oriented mitoses generate a plane epithelium. (**b1**)
Polarized structure of an insect cuticle. (**b2, b3**) Two alternative models have been proposed to explain polarization of
the insect epidermis with its bristles and denticles (wing
imaginal disc of *Drosophila*). (**b2**) Here a morphogen gradient (e.g. of WINGLESS) directs the orientation. This model,
here shown in simplified form, also accounts for the

orientation of scales and hairs in vertebrates, whereby
morphogens represented by WNT-4, WNT-5a or WNT-11
are taken into consideration. (**b3**) According to proposals
illustrated here a signal relay mechanism is responsible for
the orientation. The signals are transmitted by variants of the
protocadherin FLAMINGO (Fm2) on the one hand, and
variants of the cadherin FAT. But FAT is an active transmitter only if not too firmly or too weakly bound by the cadherin

arranged, the liver is largely located at the right side, and the heart prefers the left position – the right position of the heart is the left position!

There are rare deviations from this rule.

- The normal organization is called the *situs solitus*.
- Seldom (1:9000) the physician finds a *situs inversus*: This condition is a mirror image of the normal case. The heart is tilted to the right, the liver located in the left side. Even more rare, and associated with discomfort is
- **heterotaxy**: asymmetries are randomly distributed.

Often deviations from the norm can be assigned to this or that mutated gene.

10.6.2 Already in the Gastrula Asymmetric Expression of Several Genes Predicts the Future Asymmetry of the Organs, and One Can Influence Their Location

The competence and intention of molecular biologists to identify more and more genes that control development and to demonstrate their expression patterns has visualized some unexpected patterns. In the gastrula of *Xenopus*, and later when the phylotypic stage is reached the expression domains of several genes, some of which are already familiar to us, are transitorily asymmetric.

- **Leftist expression**. The gene *nodal*, that codes for a member of the TGF-β family, and the gene *Sonic hedgehog* are transiently expressed only in the left half of the gastrula. In a genetic constitution predicting a *situs inversus*, a laterally reversed expression of

nodal and *Sonic hedgehog* was observed. Also on the blastodisc of the chick and along the primitive streak of the mouse *nodal* expression emerged on the left side, left of the node, the homologous equivalent of the Spemann organizer (Fig. 10.8a). The *nodal* expression is preceded by the disappearance of *Sonic hedgehog* expression on the opposite side. In the mesoderm ahead of the node *Sonic hedgehog* expression is initially symmetrical but it persists on the left side longer than on the right side.

- **Rightist expression**. The factor FGF8 is expressed right of Hensen's node for a longer time than left of the node (Fig. 10.8a).

Later in development further genes for signalling molecules are expressed asymmetrically, for example *lefty-1* in the left wall of the neural tube, and *lefty-2* in the left lateral plate. Also genes for some transcription factors experience a unilateral expression, for example *PitX2*. This transcription factor is first expressed in the left lateral plate of the mesoderm and subsequently in the left heart primordium and in left areas of the forearm. There is, however, even earlier evidence for later asymmetric position of organs, as shown in the following section.

10.6.3 Rotating Cilia, Left Hand Twist of Cell Movements and in the Blastula Unilaterally Produced Serotonin Are Considered to Be the Primary Cause

An intriguing though not generally valid explanation for the occurrence of left-right asymmetries has been suggested by careful observation of anatomical details of the *Xenopus*

Fig. 10.7 (continued) DACHSOUS (DS) of the neighbouring cell. (Merely weakly bound Dachsous variants are omitted for clarity's sake.) Because the strongly binding Dachsous is present in form of a gradient, FAT molecules free for signalling are present in form of an inverse gradient. Binding strength between FAT and DS is controlled by phosphorylation of the cadherins. After various authors (Cho et al. 2006; Wei-Sheng et al. 2008; Strutt 2009; Simon et al. 2010) combined and simplified. In the insect leg the two inverse gradients provide positional values which gradually vary from position to position (Lawrence 2008; Bando et al. 2011). The two models shown in (**b2**) and (**b3**) are not mutually exclusive but may supplement each other. The elements FZ (Frizzled), Dsh/Dvl (Dishevelled), Vang (Vg, vanGogh = stbm strabismus) and Flamingo Fm are found in homologous form also in polarized planar systems in vertebrates

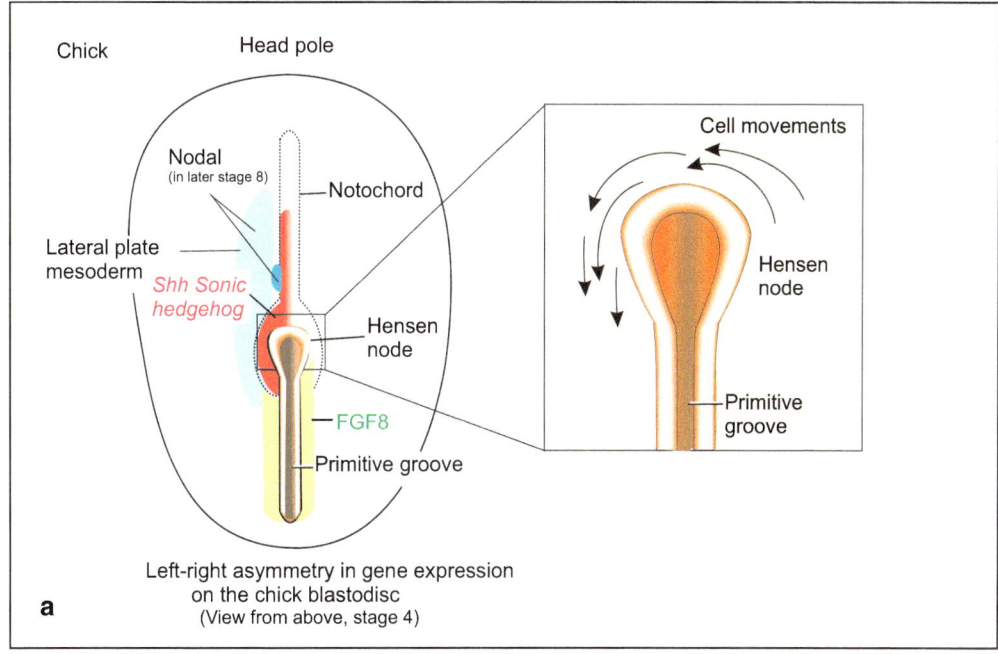

a   Left-right asymmetry in gene expression
on the chick blastodisc
(View from above, stage 4)

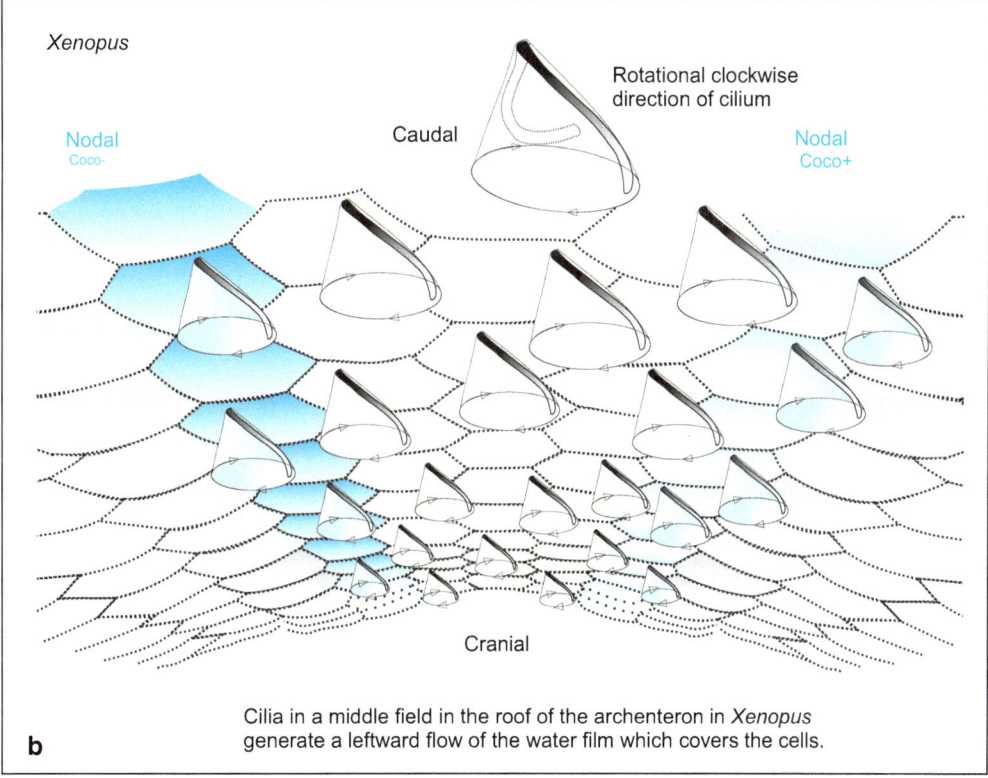

b   Cilia in a middle field in the roof of the archenteron in *Xenopus*
generate a leftward flow of the water film which covers the cells.

Fig. 10.8 Left-right asymmetries as putative origin of asymmetries in the position of heart, liver and other organs. (**a**) Asymmetric expression of SONIC HEDGEHOG, FGF8 (in stage 4) and NODAL (in stage 8) on the blastodisc of the chick. In stage 4 cells move around the node from right to left (after Levin et al. (1995) and Gros et al. (2009), combined). (**b**) In gastrulae of fishes, amphibians and mammals a field of cilia generates a stream of fluid directed from right to left. Shown is the field of cilia at the soffit (underside of the roof) of the archenteron in the gastrula. In the roof of the archenteron of the gastrula a small field has been detected the cells of which bear a particular type of cilia, called 9 + 0 monocilia. These cilia protrude down into the intestinal cavity (Fig. 10.8b) and are covered by a fluid film. The cilia rotate clockwise and generate a leftward directed flow of the fluid. The rotational direction is determined by the molecular motors within the cilia. A similar field with monocilia is also found at a corresponding site, that is close to the node, of the mammalian embryo. In rare cases, correlated with genetic mutants leading to *situs inversus*, the direction of rotation is counter-clockwise. This observation suggested the hypothesis that the fluid flow would float dissolved morphogens or ions to the – in the normal situation – left side.

However, there remains need for clarification. In the avian blastodisc no such field of cilia is found. Here fluid does not stream from right to left but cells do (Fig. 10.8a). Moreover, in the amphibian embryo asymmetries occur already in earlier stages. A first asymmetry pertains the cortical rotation released by the sperm entrance and is again based on molecular motors. However, cortical rotation generates an asymmetry with respect to the dorso-ventral body axis but with respect to left-right asymmetry in this starting stage no reliable observations have been reported as yet.

The earliest fate determining asymmetry found to date has been detected in the embryo of *Xenopus*. In the 32-cell stage in a cell located at a left-ventral position produces **serotonin** (= 5-hydroxytryptamine), a molecule known as transmitter in the nervous system. Silencing of the serotonin receptor prevented development of the cilia field in the archenteron and the future establishment of regular left-right asymmetries. **It is a common principle in developmental biology that small differences in early stages cause large remote effects**.

There are, however, still more needs for clarification. For instance, if the blastodisc of the chick is separated by a cut along the midline two embryos will form in mirror image disposal (Fig. 10.9). Also Siamese twins are oriented towards each other in mirror image symmetry. In the one organism left-right asymmetry dominates, in its partner the reverse right-left asymmetry dominates.

10.7 Morphogenetic Fields

10.7.1 There Is Nothing Supernatural with Morphogenetic Fields But They Have Remarkable Features

In developmental biology quite often the term **morphogenetic field** emerges. At times this term met with disapproval in science because it was misunderstood as preternatural field (especially by persons attached to esotericism). Yet, the concept can be defined in terms of natural science, and it is useful in the same way that the term field is useful in physics.

- **Definition 1**: A morphogenetic field is an area whose cells can cooperatively bring about a distinct structure, for example an extremity, or a set of structures through self organization.
- **Definition 2**: Another way to define a morphogenetic field is as an area in which signal substances – "inducer", "morphogen" or "factor" – become effective and contribute to the subdivision of the field into subregions.

Why two definitions? A few words do not suffice to comprehend the connotation of the term. A morphogenetic field has the following attributes:

- It gives rise to a complex structure such as an eye, a heart or a limb. The processes of subdivision, spatial restriction and developmental

---

Fig. 10.8 (continued) gastrula of *Xenopus*. The cilia generate a leftward flow of the thin water film covering the epithelium. When the flow is interrupted a clear arrangement of the asymmetrical organs gets lost (after Blum (2009a, b) and Schweickert et al. (2010) combined)

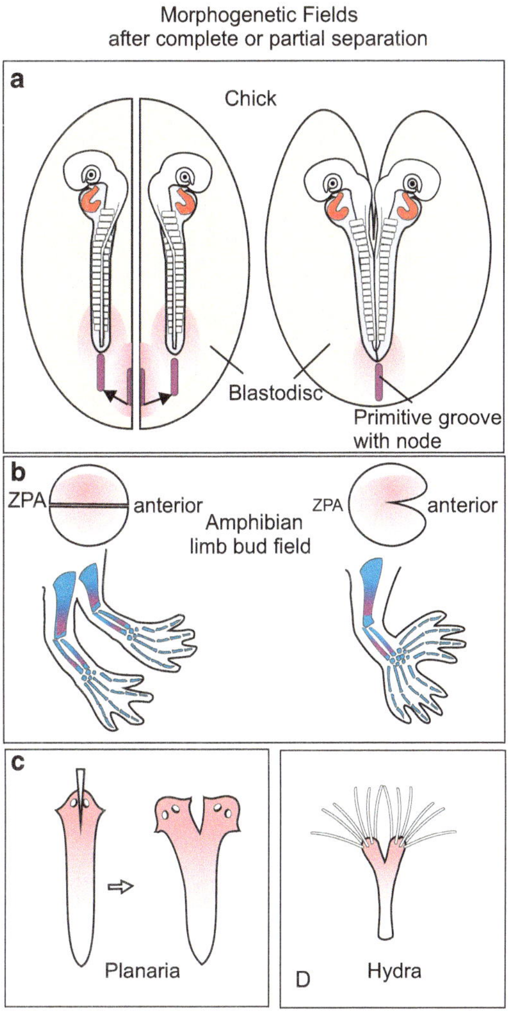

Morphogenetic Fields
after complete or partial separation

Fig. 10.9 Morphogenetic fields. Examples of partial or complete duplication of structures after partial or complete separation of the starting area or starting material. Often duplicates are mirror-inverted

restriction are phenomena of **self-organization** and **pattern formation**. The entire process of organ formation consists of sequences of such phenomena. The capacity of self organization is amazingly manifested when, for example, an aggregate of embryonic stem cells develops into an eye without any organizing influence from outside (see Fig. 18.9).

- Morphogenetic fields define developmental potencies; these are not necessarily exactly coincident with developmental fates. Usually fields are initially larger than the area fated to construct a particular organ.

- Fields exhibit the faculty for regulation. When cut into two, each half will give rise to a complete structure, albeit of half the size (Fig. 10.9). A morphogenetic field of large dimension is the blastodisc of the chick before gastrulation commences. When such a disc is divided in several separate areas each part may give rise to a complete though small embryo.

- With time it became evident that a morphogenetic field is congruent with the area in which morphogens act. The range of signal spread is often initially larger until it is restricted by lateral inhibition

In organogenesis often fields of secondary and tertiary order emerge. The classic example is the circular area in the trunk wall where the morphogen FGF10 is in charge and from which the forelimb originates (see Fig. 10.11a). The area is subdivided into concentric rings, an outer ring destined to become the shoulder girdle, and an inner area destined to become the free limb. A secondary morphogenetic field in the limb is the hand field, which subdivides itself into palm and fingers.

10.7.2 How to Create a Field, Subdivide It and Define a Point Within It

To pinpoint a location, such as a spot on the wall where we wish to drive a nail, we often draw vertical and horizontal lines that cross at the desired point. Nature can do the same. In the embryo of *Drosophila* the future development of a leg is prepared by the formation of an **imaginal disc**, a circular group of cells which is specified to form an appendage. The imaginal disc is subsequently invaginated and stored inside the larva until the larva undergoes metamorphosis. How is the location of a leg disc in the blastoderm of the embryo determined?

At the border of two parasegments two vertical stripes of cells producing signal molecules are juxtaposed: a stripe secreting WINGLESS (WG) and a stripe of cells expressing *engrailed*

Division of a Morphogenetic Field

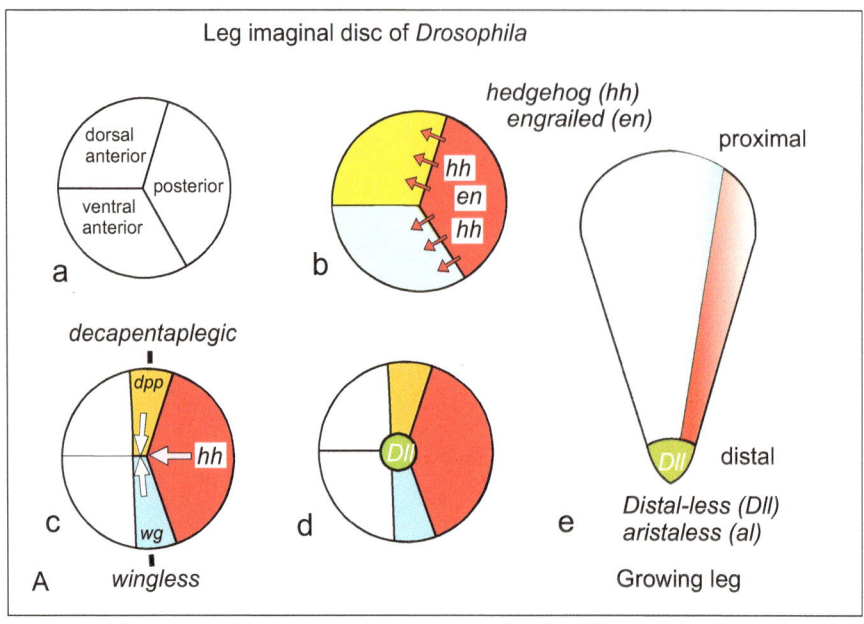

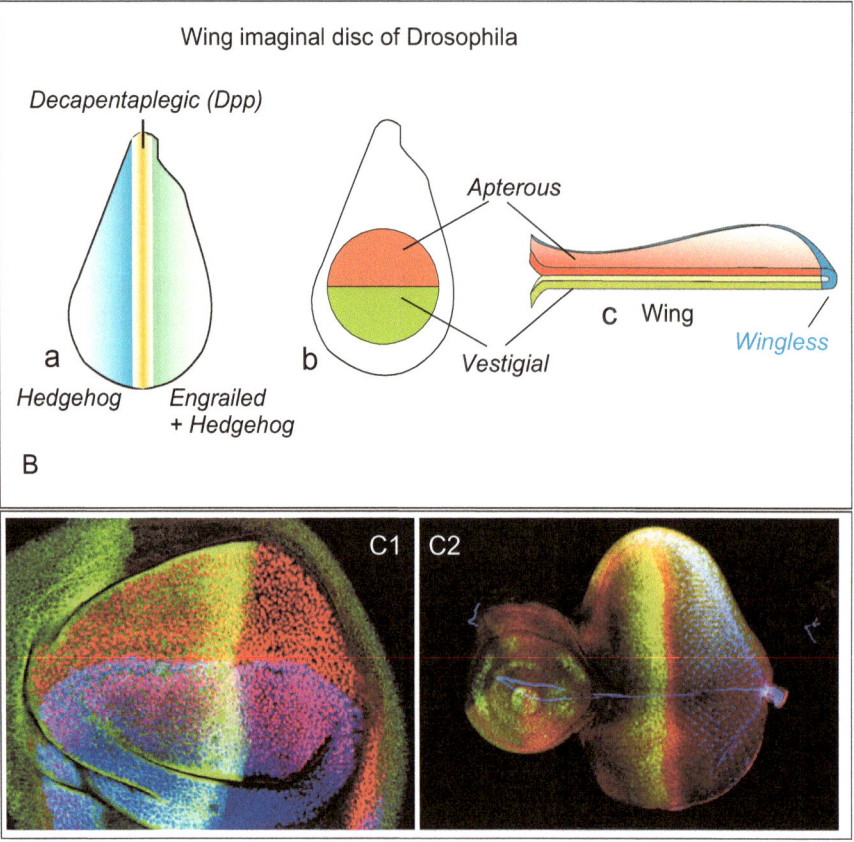

Fig. 10.10 (**A–c**) Specification of a developmental field in an circular area of the blastoderm fated to become an imaginal disc (leg or wing). The disc is specified where

compartments expressing different genes border on each other. In the centre is a small region where cells expressing *hh* (*hedgehog*) and *wg* (*wingless*) are in close association and exposing HEDGEHOG (HH) on its surface (Fig. 10.10). Due to another process of patterning a second, horizontal coordinate subdivides the anterior compartment into a dorsal and a ventral area (the molecular tools used to draw this line are unknown). The stripe of cells secreting HEDGEHOG induces the adjacent stripe of cells in the dorsal anterior compartment to secrete a third signalling molecule, DECAPENTAPLEGIC (DPP).

The site where the boundaries of the three compartments (anterior dorsal, anterior ventral, and posterior) intersect is the only site where cells expressing DPP contact those expressing WG, and this site defines the centre of the disc. In a small circle around this focal point the homeobox genes *Distal-less* (*DLL*) and *aristaless* are switched on; they denote the future distal tip of the extremity, the tarsus of the insect leg. The designation "*less*" refers to a defect mutant. In the wild type the intact gene is expressed at the distal tip but *DLL* is initially expressed also in the proximal leg while *aristaless* is only expressed at the tip. A working hypothesis proposes that the central point becomes the source of a morphogen (DLL is a stationary transcription factor and as such not a morphogen); a high concentration in the centre of the disc specifies the future 'fingers' (tarsalia).

There is evidence that also in vertebrates limb fields are specified at borders between compartments.

10.8   The Avian Wing as a Model Limb

The development of limbs is a highlight in developmental biology. The avian wing bud is particularly accessible to surgical manipulations by the experimenter; the outcome of disturbed pattern-forming processes can readily be read off from the pattern of the skeletal elements.

As a three-dimensional structure, the wing has three polarity axes:

1. The proximal-distal axis from the shoulder to the digits.
2. The antero-posterior axis from the first to the third finger (fingers 4 and 5 are missing in the avian wing).
3. The dorso-ventral axis from the upper side to the lower side (this axis is not considered here).

10.8.1 A Morphogen, FGF-10, Denotes a Field in the Flank of the Embryo from Which the Limb Bud Originates

A limb originates from a limb bud. To form a bud and subsequently construct an entire extremity material is required. First cells are recruited which are enticed away from the mesodermal somites (these cells give rise to the muscles) and lateral plates (these cells give rise to the skeletal elements) and piloted onto the right place to settle (Fig. 10.11a). Shielded by an ectodermal epithelium, the limb bud incorporates these immigrating mesodermal cells. Here they congregate, form a loose, non-epithelial tissue that in terms of histology is classified as mesenchyme. The collection of mesenchyme cells and the epithelial cells covering them are stimulated to proliferate by growth factors.

Both the chemotactic lure and the proliferation stimulating factors are members of the fibroblast growth factors **FGF** (Fibroblast Growth Factors). Fibroblasts are a type of

Fig. 10.10 (continued) with those expressing dpp (*decapentaplegic*). At this site, the homeobox genes *aristaless* and *Distal-less* are activated and a proximo-distal organizing centre is established that controls the growth of the leg during metamorphosis. (**B**) Subdivision of the domains and localization of the factors in charge in the wing imaginal disc. (**a**) Domains of *hedgehog*, *engrailed* and *dpp* define the antero-posterior polarity.

(**b**) and (**c**) The wing eventually arises from a circular area; upper and lower side are specified by different gene activities, actually *apterous* and *vestigial*. The activity of *wingless* (*wg*) is restricted to the rim. The *wg-/-* mutant is wingless as its name implies. (**C**) Fluorescence microscopic image of *Drosophila* imaginal discs. (**C1**) Wing disc, (**C2**) Eye disc. Images by courtesy of Stephen (Steve) Paddock, University of Wisconsin

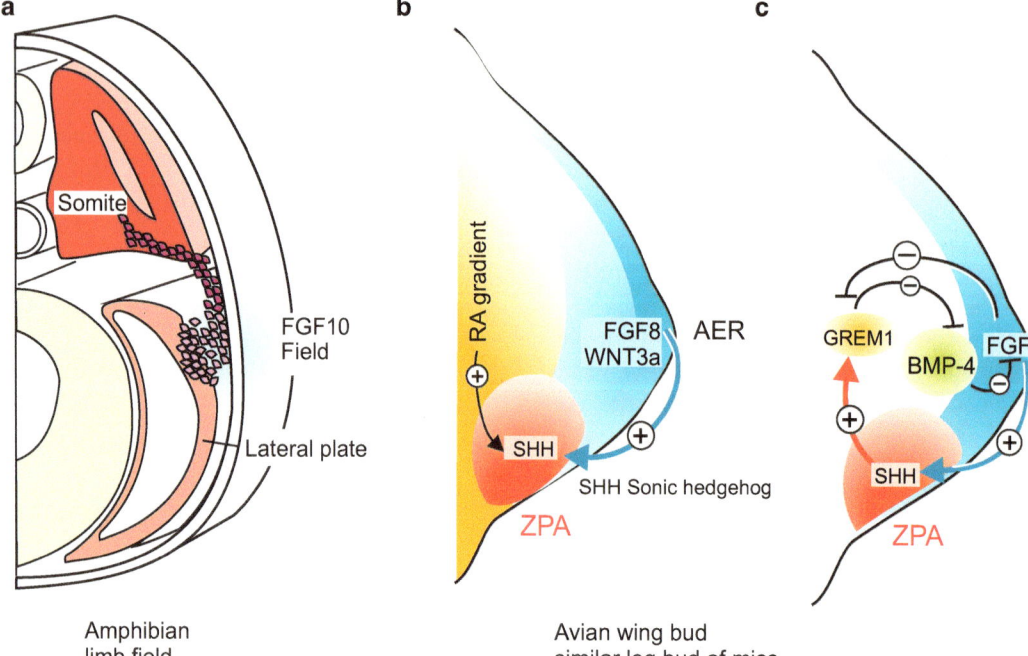

a Somite / FGF10 Field / Lateral plate / Amphibian limb field

b RA gradient / FGF8 WNT3a / AER / SHH / SHH Sonic hedgehog / ZPA / Avian wing bud similar leg bud of mice

c GREM1 / BMP-4 / FGF / SHH / ZPA

Fig. 10.11 (**a**)–(**c**) Extremity field and limb bud in the vertebrate embryo. (**a**) Primordial field in the flank of an amphibian embryo. In the FGF-10 expressing field (*blue*) cells cluster that emigrated from the somites and will give rise to the skeleton and musculature in the extremity. (**b**) Further subdivision of the field, as shown for the wing bud, is initiated by various signalling molecules which stimulate their production mutually. An important broadcasting centre is the ZPA, the Zone of Polarizing Activity. *AER* apical epidermal ridge, *FGF* fibroblast growth factor, *RA* retinoic acid. (**c**) Further factors come into play, for example Grem1 = Gremlin 1. The mutual, partly supporting, partly inhibiting interplay of all these factors can only be made predictable by computer simulations based on mathematical models

stem cells which can give rise to cartilage and bone. In the early limb bud several FGFs are in charge.

Before a bud emerges in the mesodermal lateral plate and the ectodermal cover circular areas can be spotted where the factor **FGF10** is expressed. These areas coincide with the morphogenetic field that can be detected by transplanting the areas to another location where they give rise to an ectopic limb. Instead of a transplant also **FGF10 locally administered onto the flank of a chick embryo induces there the development of an additional limb**. In response to FGF10 stimulation the overlying ectoderm produces another member of the FGF family, FGF8. With that a lively interplay of mutual stimulation starts. Both factors are true growth factors: the bud grows.

10.8.2 Accepted Opinion Assumes That Along the Proximal-Distal Axis a Temporal Program Is Translated into a Spatial Pattern, But Also Signalling Systems Are Considered

Shielded by an ectodermal epithelium, the limb bud incorporates mesodermal cells that reaggregate to form a 'mesenchyme'. This mesenchyme, initially unstructured, must be properly organized so that humerus, radius and ulna, the elements of the palm, and the elements of the phalanges are laid down in the correct sequence.

In birds and mammals the limb bud is covered by an ectodermal cup over which an **apical ectodermal ridge AER** extends (Fig. 10.11b, c). The ridge continues to produces FGFs and secretes them into the underlying mass of mesenchymal cells. Upon removal of the ridge the outgrowth of the bud stops.

The signalling range of the growth factors defines the limit of the **progress zone** in which cells proliferate. The resulting cells become gradually displaced out from the shelter of the apical cup and, thus, from the signalling range of the growth factors. As a consequence, proliferation ceases.

The first cells to leave the space of the sheltering cup and cease proliferation form the

- upper arm with the **humerus**;
- the next group forms the lower arm with **radius and ulna**,
- the third group forms the **carpals** and the
- last group forms the **phalanges**.

Various experiments indicated that a **temporal program** is run in the growing limb. When a very early bud with its young progress zone is transplanted onto the stump of a later-stage bud, the young bud begins with the humerus according to its own autonomous program, and it does so even when the stump already contains cells specified to form a humerus. A medium-stage progress zone continues its program by laying down radius and ulna; a late-stage progress zone grafted onto a young stump merely adds fingers without taking into account that the removed progress zone of the young bud did not yet have the chance to provide cells for ulna and radius; these are now missing (Fig. 10.12).

The nature of the temporal program is unknown, though several indications point to a gene-controlled mechanism: in the growing limb a series of homeotic genes (*Hox* genes) is activated sequentially (Chap. 12, Fig. 12.18). In the course of the proximal-distal pattern specification, continuously changing **positional values** are assigned to the cells. These values can be used in amphibians to supply patterning information to regrow an amputated limb (Sect. 18.7, Fig. 18.16).

The notion of a temporal program (clock) claimed interpretational sovereignty for long time but was recently (2011, 2014) challenged by two working groups. These prefer the notion that two signal emitters placed at opposite poles control outgrowth of the limb and its subdivision

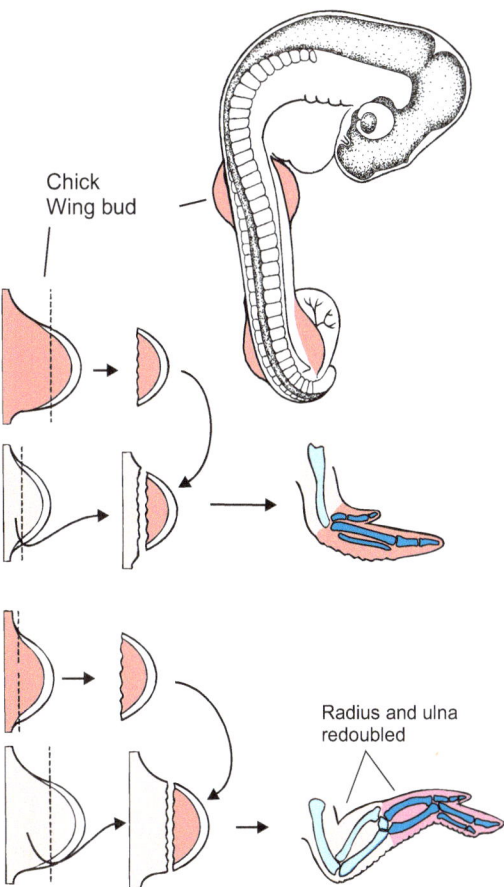

Chick
Wing bud

Radius and ulna
redoubled

Fig. 10.12 Chick wing bud I. Patterning along the proximo-distal axis during growth of the wing bud. A cap cut off from a bud and grafted onto the stump of another bud continues its developmental program irrespective of the pattern elements that are still lacking, or already laid down, in the stump (after Alberts et al., modified). On the basis of these experiments the time-program model was proposed (by L Wolpert)

in upper arm, lower arm and hand. As shown in Fig. 10.11b at the base of the bud retinoic acid RA is produced, and at the tip of the bud FGF8 and WNT-3a. These molecules are said to form opposing gradients. **The retinoic acid gradient has its maximum at the proximal position, the FGF8 and WNT gradients have their maximum concentration at the distal most position**. With the outgrowth of the limb, the distal cells escape from the influence of RA, the proximal cells escape from the influence of FGF8 and WNT3a. However, a satisfactory notion how the results of the transplantation experiments shown in Fig. 10.12 might be explained is still lacking. A hypothesis combining gradients and a clock mechanism like the segmentation clock shown in Fig. 10.25 might attain the best explanation.

10.8.3 The Order of Fingers 1–3 Is Specified by an Emitter of Signal Molecules

At the posterior margin of the limb (where in our hand the little finger would appear) an emitter of signals is located: the **ZPA (zone of polarizing activity)**. A small group of cells near the posterior junction of the limb bud and the body wall is the source of one or more secreted signal molecules. Two hypotheses have been advanced to explain finger development.

A traditional hypothesis sees the signal intensity from the ZPA decreasing from the posterior margin to the anterior margin. This gradient of signal strength specifies the sequence and characteristics of each finger. If an additional ZPA is implanted into the anterior margin where signal intensity would ordinarily be low, it elicits the appearance of an additional set of fingers. A mirror-image duplication of the finger pattern results in the sequence of digits 321123 or 32123 (Fig. 10.13).

Retinoic acid (RA, vitamin A acid) can substitute for a transplanted ZPA. A porous bead soaked with RA and implanted into the anterior margin induces a mirror-image duplication of the digits just like a ZPA does. Since retinoic acid, derivatives of retinoic acid, receptors for retinoic acid, and even a graded distribution of bound ^{3}H-RA have been found in the limb bud, retinoic acid has been assumed by some researchers to be the first identified true morphogen.

However, this position has been disputed by the suggestion that the bead soaked with RA causes cells in the anterior marginal area to adopt features of ZPA cells. These artificially induced ZPA cells would emit the natural signal which is not RA, and thus would trigger a cascade of secondary events. **SONIC HEDGEHOG**

SHH has been proposed as the natural signal. Indeed, SONIC HEDGEHOG is expressed in the ZPA, and cells secreting SONIC HEDGEHOG can replace a ZPA in its polarizing activity. A cascade of subsequent signalling processes could follow. SONIC HEDGEHOG might trigger the release of further signal substances (e.g. peptide growth factors such as FGF4 and BMP). According to current models there is a network of mutual stimulation and inhibition (lateral help and lateral inhibition) as shown in Fig. 10.11b, c. This mutual dependence in turn ensures that the expression of all these factors is restricted to the field of mutual interaction.

Whether or not retinoic acid is a primary signal in the specification of proximo-distal (and antero-posterior) polarity in limb buds, the numerous morphogenetic effects of RA are remarkable. RA influences proliferation, pattern formation and cell differentiation in many rudimentary organs. It does so presumably by stimulating gene expression in a way known from steroid hormones (see Fig. 11.11). RA contributes to the regional subdivision of the central nervous system (e.g. Fig. 10.4), while SONIC HEDGEHOG organizes, among many other functions, the formation of the vertebrae (Fig. 10.6).

10.8.4 In the Course of Pattern Specification Step by Step Genes of the *Hox* Class Are Activated

Starting at the posterior flank of a wing bud (ZPA), or the posterior flank of the extremity bud of a mouse, step by step genes coding for transcription factors of the *Hox* class are switched on. In the fly embryo *Hox* genes program the segment identities along the head-abdomen axis and fulfil similar functions in the vertebrate body (see Fig. 12.16). Now we learn of a similar function in the vertebrate extremity. The expression domains of the *Hox* genes spread from the posterior edge in the form of waves and eventually an antero-posterior and proximo-distal pattern is seen.

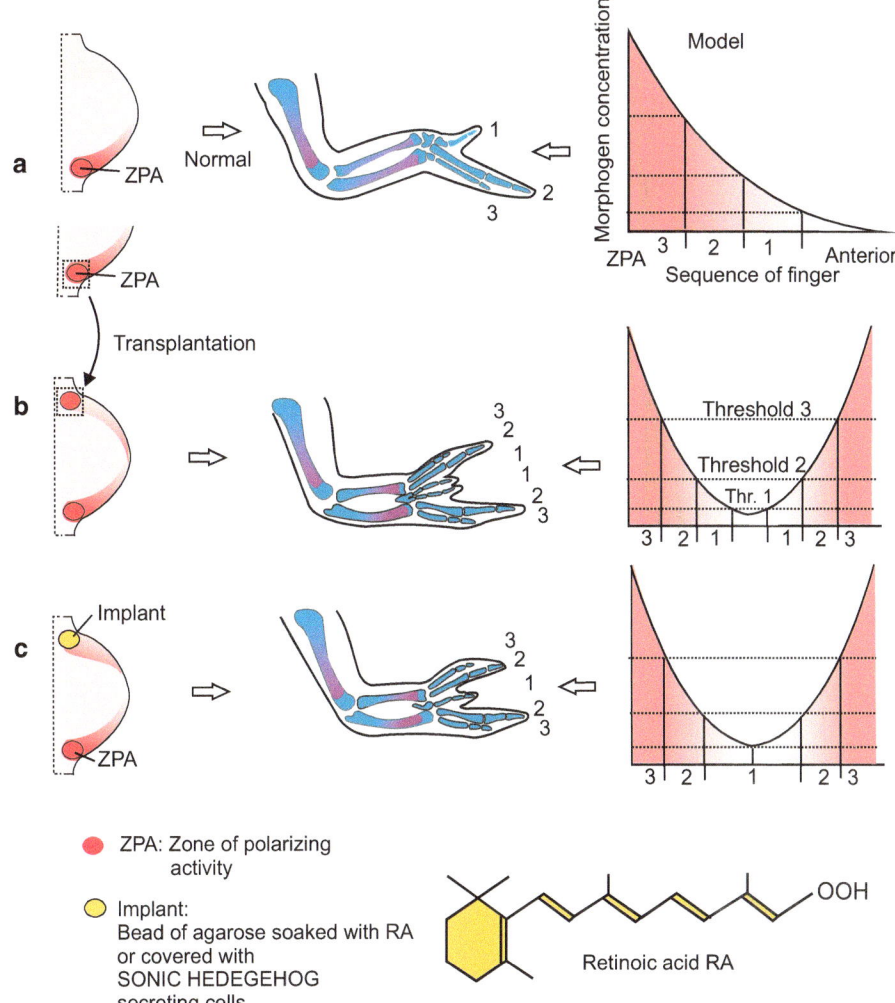

Fig. 10.13 Chick wing bud II. Pattern specification along the antero-posterior axis. (**a**) The type of fingers and their sequence are specified under the influence of a zone of polarizing activity (ZPA) that is located at the posterior margin of the bud. Note that the wing contains three fingers only also in the wild-type animal (formerly it was assumed that finger 1 and 4 are lacking but recent studies suggest that in evolution the fingers 1–3 were preserved). The coloration of the skeletal elements is based on the classic alizarin stain. (**b**) Implantation of posterior tissue into the anterior margin of a host bud results in the mirror-image duplication of the sets of digits. (**c**) A resin bead soaked with retinoic acid or a cluster of cells secreting soluble SONIC HEDGEHOG mimic the effect of an implanted ZPA. The results have been interpreted in terms of a concentration gradient of a morphogen, presumably SONIC HEDGEHOG, that is released by the ZPA (*right diagrams*). Above certain threshold concentrations (*parallel vertical lines*) the cells respond by forming the digits 1, 2, or 3 (after Wehner & Gehring: Zoologie, Stuttgart, 1990, modified and extended; sequence of fingers after Tamura et al. 2011)

- **From posterior to anterior**. In the establishment of the antero-posterior pattern the *Hox-D* group finds expression, one by one, starting with *Hox-D11* (see Fig. 12.18). All expression domains start at the posterior margin at the position of the future last digit but spread differing distances. *Hox-D11* spreads to the anterior margin of the bud, *-D12* stops in the middle, *-D13* remains restricted to the field of the ZPA.

- **From the tip of the bud to the base**. The proximo-distal expression pattern comprises the sequences:
  - *Hox-A9, -A10, -A11, -A12, -A13* and
  - *Hox-D9, -D10, -D11, -D12, -D13*.

The *Hox-D* group is again utilized and now completely. Again, expression starts in the ZPA with the gene with the lowest number. While the bud grows the expression zones spread and widen, and transcripts of genes with increasing higher numbers appear. Eventually, the expression zones of *Hox-A9* and *Hox-D9* extend from the shoulder to the finger tip. Activities of *Hox* genes with higher numbers spread less far (*Hox-A11* up to the end of the forearm, *Hox-A13* up to the end of the hand).

Which factors effect spread, and which factors set limit to the range of spreading is not known as yet. Being transcription factors bound to DNA HOX proteins most probably do not spread by themselves. Presumably the propagation of the *Hox* expression follows soluble signals, for instance SONIC HEDGEHOG, which can spread by diffusion.

In the course of *Hox* gene expression continuously changing **positional values** are assigned to the cells (Box 10.2). In case of disturbances these values allow for pattern correction thus facilitating, for instance, correct limb regeneration in urodeles.

10.9 Pattern Control and Positional Memory in Hydra

10.9.1 Cells of a Hydra Need Positional Information Permanently

The freshwater polyp *Hydra* needs positional information not only in embryogenesis but for later life as well. It needs information to guide the perpetual renewal of its body from stem cells and to be ready to replace body parts whenever regeneration is required. Accidents resulting in damage to the body can easily happen to these unshielded and soft animals.

New cells for replacing used up cells are not produced at their definitive destination but derive from stem cells being located far away. Thus, nerve cells and nematocytes (nettle cells) that are needed in the head with its tentacles derive from so called i-cells (= interstitial stem cells) in the trunk. For homing, migrating precursors need positional information.

Mitotic proliferation of epithelial cells also takes place preferentially in the middle of the body whereas loss of aged or used up cells takes place at the terminal regions of the body and at the tips of the tentacles. By these dynamics the cells experience a continuous shift along the body column from the middle of the body either towards the upper or the lower end of the body or into the tentacles. The body is in a **permanent steady state condition or flow equilibrium**. Position-pertinent transformation of the cells is partly based on substitute cells supplied by stem cells, partly on position-pertinent transdifferentiation. Also for that purpose positional information is needed.

When in excessive injury entire body parts are lost positional information is required to ensure regeneration in accordance with the local need. Wherever the *Hydra* body column is bisected by a transverse cut, a head is restored at the upper end of the lower fragment, and a foot at the lower end of the upper fragment. If we cut at various levels we will see that the entire body column between the existing head and foot has the potential to form both head and foot. Why are these potentials not expressed in the intermediate gastric region? Traditional views assign to the **head an organizing power** that includes a suppressing activity, comparable to the apical dominance phenomenon known from plants, and systems of gradients providing lateral inhibition and positional information (Box 10.2). The body column is checked for integrity and completeness by an unknown system of control: If necessary, corrections are made (**pattern control**).

Box 10.2: Models of Biological Pattern Formation

Hypotheses Computer Models. Pattern formation is a central topic of developmental biology. Pattern formation refers to spatially organised (non-random) cell differentiation that results in ordered arrays of structures. Complex patterns are the synergistic outcome of many interacting molecules and cells. Because the complexity of reality often surpasses the faculties of our intuition, simplified models have been constructed that allow computer simulations. As examples, two simple, historically significant models are outlined here.

Positional information model. according to Wolpert. In 1969 L Wolpert reintroduced (after H Driesch; see Chap. 1, Box 1.1) the idea that cells should behave in accordance with their location in the whole and coined the term **positional information** for his main conceptual innovation. A system which may be basic and universally distributed in animal embryos informs the cells of their position, upon which information the cells act in deciding which of the pathways of determination open to them they should enter. His actual model (Fig. 10.14) was as follows:

An emitter releases a signal in the form of a soluble substance, the hypothetical **morphogen S**. The concentration of S diminishes with increasing distance from the source. A continuous gradient may be set up from the source to a sink. The morphogen source may be located at one end of a string of cells, with the 'sink' destroying the morphogen at the other end. In this configuration, under equilibrium conditions the concentration decreases linearly along the string. Alternatively, all cells along the string remove S, in which case the concentration decreases exponentially. The resulting gradient of S provides positional information. Cells are capable of measuring the local concentration and, by doing so, of locating their position along the gradient. The positional information thus deduced is 'interpreted' by the cells in that

they behave in accordance with their position (and their previous history).

In addition, the S gradient is a guiding principle and used to adjust a second gradient of **positional value, P**. Positional value is a relatively stable tissue property that serves as positional memory in case of need – regeneration, for example. The concentration profile of S determines the shape of the P gradient; the local S value acts as inhibitor determining the upper limit of the local P value.

When the source of S is removed and S drops below a critical value, P automatically rises. Cells acquiring the maximal P value first begin to emit S, thus suppressing an increase of P elsewhere.

S and P are involved in pattern formation. If there are threshold values above which the cells respond by becoming, say, "red", and below which they respond by becoming "white", a sharp segregation into two populations, "red" and "white", is possible. ▶

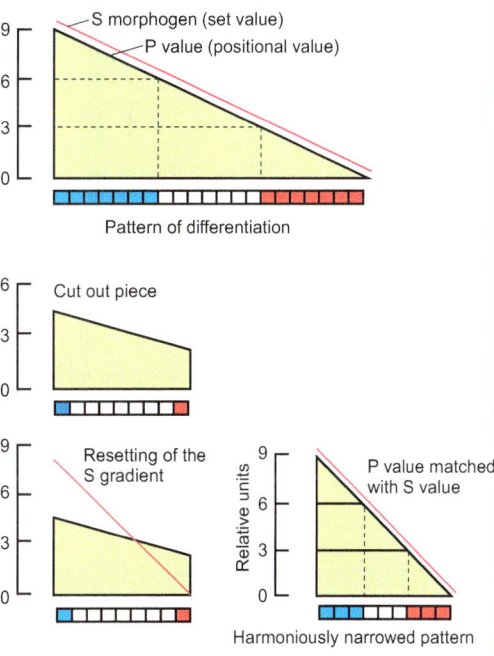

Fig. 10.14 Wolpert's positional information model

Box 10.2 (continued)

Accordingly, Wolpert's model became known as "*French Flag*" model. A feature of the model is that the same grid may be used in different situations, e.g. in the wing and the leg, or in different species, but the cells in each case would respond differently.

The hypothesis does not explain how a gradient is established for the first time, and does not propose a mechanism as to why and how P values increase when no S brake is applied.

Reaction-diffusion models according to Turing and followers. The primary aim of these models is to propose mechanisms by which patterns could arise from homogenous or chaotic initial conditions. Ideas as to how patterns of orderly yet different cells might arise in initially uniform populations of genetically identical cells have been developed mainly by mathematicians, starting with a landmark model by **Alan Turing**, or physicists. The basic concept of reaction-diffusion models is the **morphogenetic field**, in which several biochemical reactions create **pre-patterns** in the distribution of chemicals, generally called **morphogens**. Cell determination and differentiation would follow these chemical pre-patterns.

The simplest versions of **reaction-diffusion models create stable, non-uniform concentration patterns by combining the production of at least two interacting substances which feed back onto their own production as well as the production of the other substance.** In a basic Turing-type model developed by Gierer and Meinhardt (Fig. 10.15) the generation of an activator, **a**, is an **autocatalytic** event. Through a process of non-linear amplification, **a**'s presence generates the production of more **a**. The explosive rise in **a** is limited by the decay of **a**, its diffusion into neighbouring areas, and by the production of an inhibitor **i**, which is **heterocatalytically** initiated by **a**. The inhibitor **i** sets limits to

the production of **a**. Furthermore, it is assumed that the diffusibility of the two substances is different: the activator **a** diffuses slowly and only over short distances, while the inhibitor **i** diffuses rapidly and has a long range. Due to these properties, **i** allows **a** to increase in concentration up to a limit, but suppresses the start of an autocatalytic **a** increase outside the first peak. Hence, **i** prevents the occurrence of a second, competing peak (**lateral inhibition**).

The behaviour of the two substances in time and space is described by two partial differential equations on which computer simulations are based (Fig. 10.15). By selecting appropriate parameters (basic rates of production, rates of decay, diffusion constants, size of the field), and, if necessary, by introducing more interacting substances, a multitude of patterns can be created with a computer. For example, gradients are generated that remain stable and, after experimental disturbance, regenerate (Fig. 10.16). Periodic patterns such as fields of cuticular spines (Fig. 10.17), patterns of stripes, or ▶

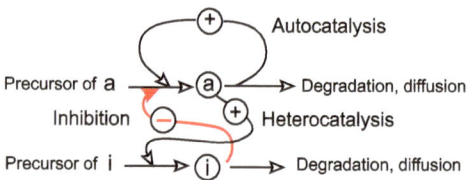

$$\underset{\substack{\text{Alteration of the}\\\text{concentration of a}\\\text{with time}}}{\frac{\delta a}{\delta t}} = \underset{\substack{\text{Source}\\\text{density}}}{\rho_a} + \underset{\substack{\text{Auto-}\\\text{catalysis}}}{\underset{\text{Inhibition}}{\frac{a^2}{i}}} \underset{\text{Degradation}}{- \mu a} + \underset{\text{Diffusion}}{Da \frac{\delta^2 a}{\delta x^2}}$$

$$\frac{\delta i}{\delta t} = \rho_i + a^2 - vi + Di \frac{\delta^2 i}{\delta x^2}$$

Fig. 10.15 Gierer-Meinhardt model (ref. in Further Reading)

Box 10.2 (continued)

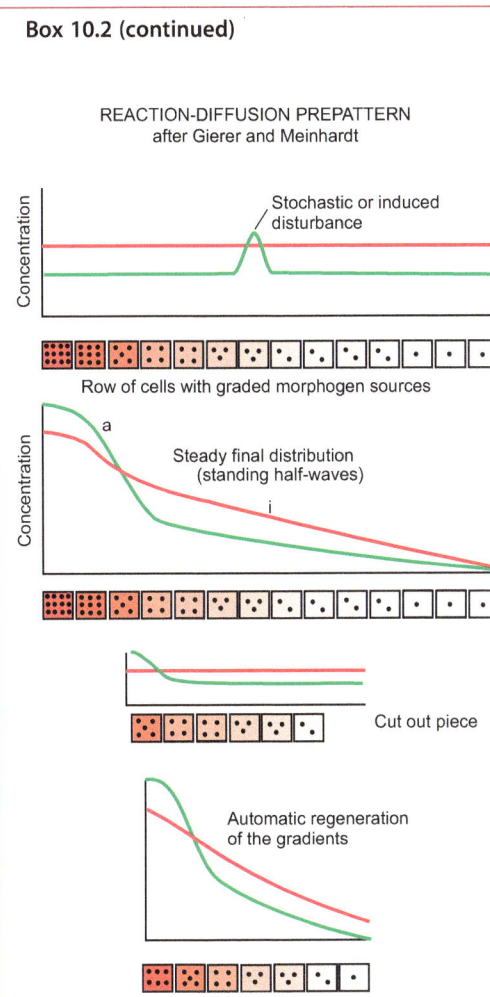

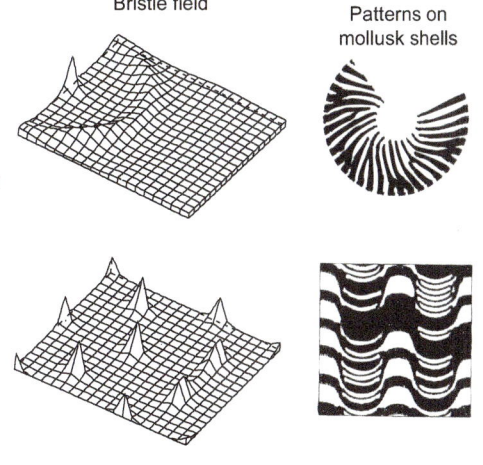

Fig. 10.17   Selection of patterns generated with the Gierer-Meinhardt model (ref. in Further Reading)

Fig. 10.16   Establishment and regeneration of a gradient simulated with the Gierer-Meinhardt model (ref. in Further Reading)

branching patterns have been generated with remarkable fidelity to those observed in nature. Periodic patterns arise in these models if the range of the inhibitor is shorter than the length of the field, and when a relatively high level of activator initiates a new, self-enhancing activator peak outside the inhibited area. Many patterns have been simulated successfully, such as patterns on mollusc shells (Fig. 10.17) or eye spots on wings of butterflies (Fig. 10.18).

However, chemical reactions have not yet been identified experimentally that would be as simple as recent models propose. It is likely that reaction-diffusion models describe some general formal principles, such as the principle of self-enhancement (autocatalysis) rather adequately, but not actual mechanisms.

Extended models. Pattern formation can be the result of a multitude of physical forces and chemical processes as well as resulting from the behaviour of motile cells. Accordingly, a multitude of explanatory models have been designed, including mechanical and mechano-chemical models (see References). Many models share formal basic properties: they embody

- a process of **self-enhancement** ('autocatalysis', positive feed back) and
- a process that **sets limits to the self-enhancement**. Restrictions to prevent unlimited increase could be achieved by production of an inhibitor, depletion of a substrate or cell type, or saturation.

Models of biological pattern formation are reminiscent of models used in ecology to ▶

Box 10.2 (continued)

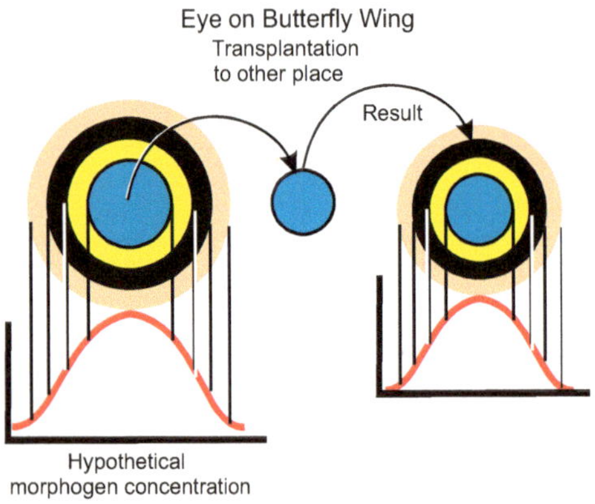

Eye on Butterfly Wing
Transplantation
to other place

Result

Hypothetical
morphogen concentration

Fig. 10.18 Formation of an eye spot on the butterfly wing. Computer simulation by Papatsenko (2009)

describe prey-predator relationships and the spread of epidemics, populations or genes.

Many arguments have been presented against too simplified models. One may doubt whether the concerted behaviour of thousands of cells of which each produces many hundred thousands of different substances can be derived from the reaction of a few substances. One single mutation could cause the collapse of the whole system. In nature robust networks rather than the simplest solutions are required. Furthermore, homogenous conditions for diffusion are found in the computer but not in living organisms. The first generation models (like those of Alan Turing and Gierer & Meinhardt) produce patterns of free substances and ignore stationary receptors, the number or density of which may vary in space and time. For instance HEDGEHOG induces up-regulation of its receptor PATCHED which in turn sets limits to the signalling range of HEDGEHOG by mediating its internalization by endocytosis. Generally local receptor densities influence the profile of the morphogen.

Many more objections may be raised. On the other hand it is good and approved practice in science to start with minimal conditions and minimal interactions and to upgrade the model

under the command of experimental results. A putative "activator" for instance may turn out to be a complex process of activation. Stepwise improved or alternative models are being developed world wide.

Of great benefit are new methods utilizing fluorescence dyes or molecular fluorescence tags attached to proteins (such as GFP, see Chap. 12, Box 12.2), to track the emergence and disappearance of gradients in video recordings in living and transparent embryos and to measure local concentrations. Such video recordings inspired new models which do not derive location-pertinent behaviour from a static morphogen profile but by its temporal progression, duration and decay.

Alternative Models. There are a number of difficulties in the way of accepting global positional information models, at any rate as a universal mechanism. With the separation of a special process of information signalling from a second process of interpretation and implementation, the burden of complexity is placed on the interpretive mechanism as the price for simplicity in the signalling system.

Alternative models place more emphasis on **short-range interactions** between cells, each of which has no information about its position relative to some distant and special group of cells but only about what its neighbours are doing, and its internal state. Suggested alternative solutions are known as **cell contact models**.

If we have a look at actual examples analysed with genetic, molecular and biochemical methods we see, on a case-by-case basis, arguments in favour of short-range interactions (for example Figs. 10.1, 10.2, and 10.7), of long-range interactions (for example Figs. 10.3, 10.4, 10.5, 10.11, 10.13, 10.22, and 10.24b, c), and of combinations of short- and long-range interactions (for example Figs. 10.19 and 10.25). This suggests that no single type of model can be used to explain all actual processes of biological pattern formation and pattern control.

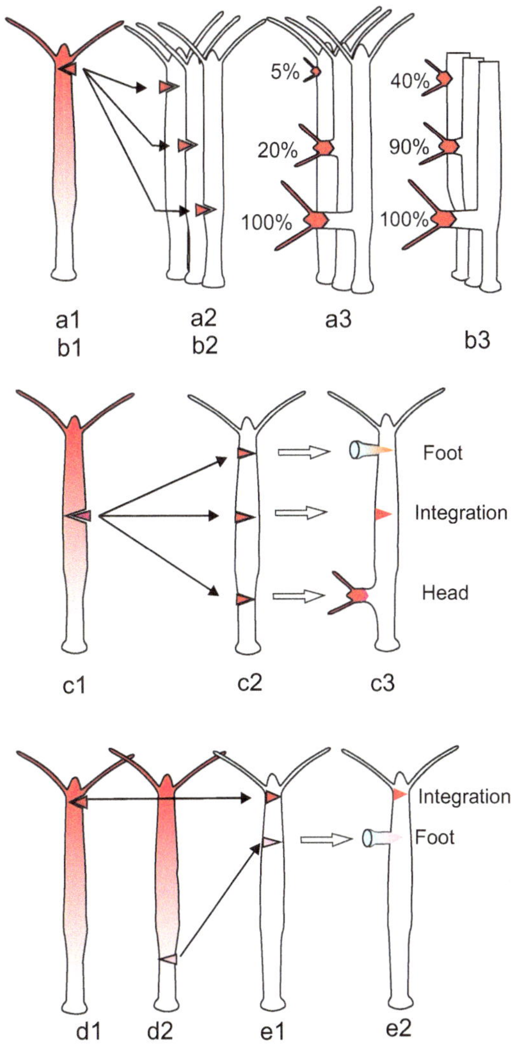

Fig. 10.19 Head and foot induction in the hydra (after Mueller 1990, 1995b) (**a**)–(**e**) Synopsis of transplantation studies in *Hydra*. (**a**) and (**b**) Pieces of tissue taken from donor animals are implanted into host animals at various positions along their body axis. If the positional value of an implant is higher than that of its new surroundings, it may form a head; if it is lower, it may form a foot. Implants whose positional value corresponds to that of the surrounding host tissue, integrate themselves into the host unobtrusively. (**a1**) and (**b1**) Donors of the transplants. (**a2**) and (**b2**) Receivers (with own head). The magnitude of the differences in the positional values of implant and their new environment is reflected by the frequency in which heads (**a3**) are formed in multiple experiments. In the series (**b**) the head of the receivers was removed after implantation of the donor tissue (**b2**). This promotes the formation of ectopic heads. (Experiments performed by Mueller 1995, 1996). Transplantation of head-proximal tissue into lower body regions repeats the oldest induction experiment known (performed by Ethel Browne in 1909)

10.9.2 Pattern Control Comprises Long Distance Interactions Between the Body Parts

Already in the fresh water polyp transplantation and regeneration studies revealed principles of developmental control similar to those found in other, "higher" evolved systems.

- **Principle of organizer agency**. A small fragment of tissue with organizing power, here a small piece taken from or close to the head (hypostome), can induce the development of a whole head and, beyond, of a second body axis (Fig. 10.19). The organizing activity is, at least in part, due to **WNT signals** emitted by this tissue.

- **Gradient of inducing potency and of positional value**. Pieces taken from the trunk exhibit a latent potency of head induction if inserted into the body wall below the place of their origin and distantly from it (Fig. 10.19). The decreasing potency is subsumed under the term positional value.

- **Principle of lateral inhibition**. The induction of an additional head is facilitated when the existing head of the host polyp is removed (Fig. 10.19b3). An extant head inhibits the development of an additional, competing head.

- **Principle of lateral help**. An existing head suppresses the formation of another, competing head, while it promotes the formation of a foot at the opposite end of the body column. Additional heads grafted onto the body column can even evoke the development of supernumerary feet, while additional feet do not evoke additional heads. In between two heads at opposite ends, the body region having the relatively lowest positional value is caused to further lower its value until it reaches the value zero which allows foot formation (see Fig. 10.22). This is also a mechanism for pattern correction by intercalation.

10.9.3 In the Establishment of the Head Organizer the Canonical WNT Systems Plays a Leading Part

The head of *Hydra* and presumably the head of any cnidarian polyp cannot be equated with the head of a vertebrate. Vertebrates are deuterostomians, the blastopore becoming the anus, whereas in the Cnidaria the head (hyopstome) is formed around the blastoporus (and therefore could be classified as protostomes if only this criterion is considered; yet in the Cnidaria the mouth has also the function of an anus). When we equate the mouth of cnidarians with a blastopore (see Fig. 22.7), it is not surprising that around the mouth of *Hydra* or *Nematostella* some genes are active that in the deuterostomes are turned on around the blastopore. Nevertheless many a finding was surprising. In the establishment of the head organizer of a polyp a canonical WNT system is of significance – exactly as in the installation of the Spemann organizer in the vertebrate embryo. According to accumulating evidence, the WNT cascade participates in many organisms in setting up a body axis (Chap. 22, Figs. 22.6 and 22.7). In case of *Hydra* the following reasons can be presented:

- Genes of the WNT/β-catenin cascade are expressed where a new axis emerges and a new mouth is formed: i.e. at the upper end of a trunk being about to regenerate a lost head, at the tip of a bud, in aggregates of dissociated cells at sites where head structures announce themselves in advance (see Figs. 4.14 and 4.16).

- There is a simple method to indirectly stimulate the WNT cascade by blocking a component of the system, the enzyme GSK-3, for instance with lithium ions or certain pharmaceuticals (alsterpaullone, azakenpaullone). By doing this with appropriate dose additional tentacles or whole heads arise along the body column (Fig. 10.20a–e). Such effects appear also in the hydrozoan *Hydractinia*. Segments cut out from the body column and subjected to stimulation of the WNT/β-catenin system form heads at

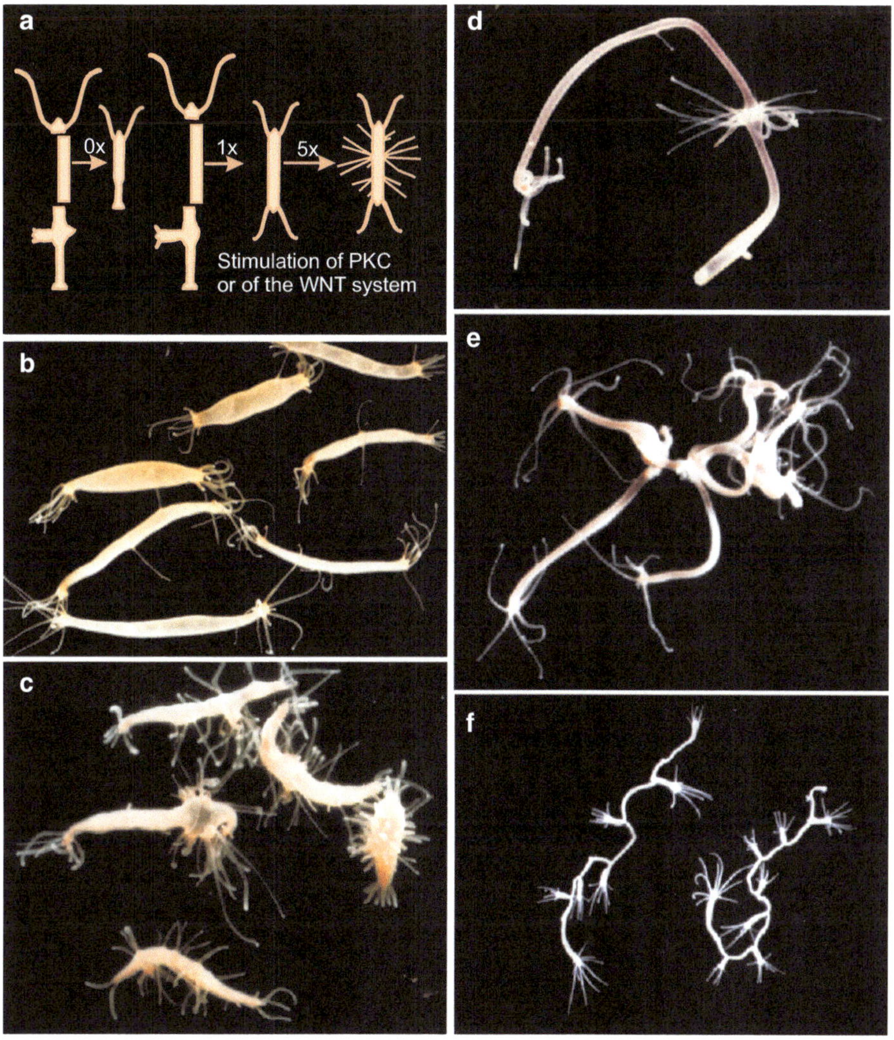

Fig. 10.20 Bipolar and multipolar forms of regenerating hydroid polyps of *Hydra* and *Hydractinia*. (**a**) Experimental procedure. (**a**)–(**c**) *Hydra magnipapillata*; before segments of the body column were cut out the polyps received multiple treatments with diacylglycerol a natural substance stimulating PKC (Protein Kinase C). (**d**) and (**e**) Since hydras are permanently growing and engaged in self renewal, ectopic head structures can be induced even in adult polyps. (**f**) Nine-headed hydra resulting from repeated treatment with diacylglycerol. Polyp of *Hydractinia* after multiple stimulation of WNT signalling using azakenpaullone. Photos by author WM. Details of the procedure in Mueller (1982), Mueller et al. (2004, 2007)

both ends or even transform into multi-headed creatures (Fig. 10.20f). Moreover, in colonies of *Hydractinia* the formation of giant buds can be evoked which eventually give rise to several multi-headed polyps. In embryogenesis multi-tailed larvae develop which in metamorphosis also give rise to multi-headed polyps (see Fig. 9.5).

With respect to theoretical notions and models of biological pattern formation introduced in Box 10.2 a recent (2011) finding is of great interest. Ahead of the gene concerned, WNT-3a, two controlling sequences (enhancers) are placed, one sequence through which WNT-3a can up-regulate its own production in an autoregulatory manner via β-catenin, and a sequence which enables down-regulation.

In the establishment of a high positional value that mediates the capacity of head formation also the PI-system of signal transduction is involved as activators of the protein kinase C such as diacylglycerol or tumour promoting phorbol esters also induce the development of ectopic heads. Such spectacular results should not mislead us to ignore that much remains to be clarified. For example the expression domain for WNT-3a is restricted to a small area around the mouth. On the other hand, according to theoretical considerations WNT-3a should have long-range effects. Wnt molecules are rated as not readily soluble; how can they act at long distances? Which system generates lateral inhibition?

Signalling systems involved in the control of development, in pattern formation and pattern control are (unfortunately) never simple even when allegedly simple organisms such as a hydra are investigated. Besides genes of the canonical and the non-canonical WNT systems more factors are in play which are familiar to the professionals but could also be known to those who have read previous sections and chapters. There exists a DICKKOPF-WNT antagonism, a BMP-CHORDIN antagonism, there are NOGGIN and FGF signalling proteins, among them a particular receptor for FGF is in *Hydra* responsible for the release of the bud. Besides β-CATENIN and TCF of the WNT system, further known transcription factors such as BRACHYURY are found but it still has to be clarified which genes they control. In addition, components of several signal transduction systems which transfer messages in the cell's interior are found, for example PKC (Chap. 11, Fig. 11.5) and stimulation of which can also evoke ectopic head formation as does stimulation of the canonical WNT system.

The genomes of *Hydra* and *Hydractinia* (and of further Cnidaria such as *Nematostella*) contain thousands of genes homologous to human genes. To enumerate all potentially development controlling genes would not make sense, and given the long list, hardly possible.

10.9.4 Relatively Stable Positional Values Mediate Positional Memory

Though in principle heads and feet can be made along the entire trunk column, the capacities are not uniformly distributed. They are graded: a position close to the head confers a high capacity to form and induce head structures but a low capacity to form a foot. In contrast, tissue located near the foot end has a low potential for head formation but a high potential for foot formation.

This gradient specifies polarity as well. In an excised section of the body column, whether from the upper, middle or lower part of the column, the head-forming potential is always highest at the upper end, while the foot-forming potential is highest at the lower end. These properties can be subsumed under the term **positional value**. A high value combines a high capacity for head formation with a low capacity of foot formation. A low positional value means a high foot forming potential combined with a low head forming potential. The received positional information, where ever it comes from, is in *Hydra* used to adjust positional values.

The relative positional values along the body column can be measured in two ways:

1. **Transplantation**. Pieces of the body wall are excised and transplanted into another body region. In a surrounding with the same positional value the pieces simply preserve their identity. In a surrounding with moderately different positional value the pieces fit into the new community by adopting the identity of the new neighbours (position-dependent behaviour). However, in a neighbourhood with significantly different positional value, they behave strangely and change their own

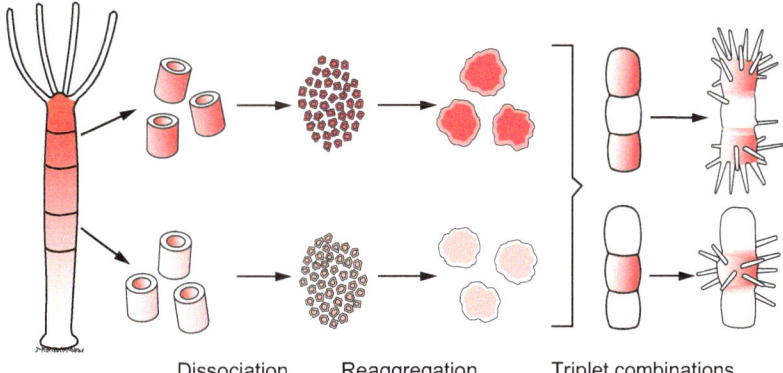

Dissociation   Reaggregation   Triplet combinations

Fig. 10.21 Experiments with dissociated cells in *Hydra*. Sections of the body column taken from two different levels are dissociated; from the resulting cell suspensions aggregates are prepared, and subsequently the aggregates are combined in triplet arrangements. Tentacles are formed preferentially by those cells that originally possessed the relatively highest positional value. Experiments of Gierer et al. (1972)

state completely. In an environment of significantly lower positional value, the pieces form ectopic heads, and in an environment of significantly higher value they form feet (Fig. 10.19). The frequency of ectopic head and foot formation reflects the sign and degree of the differences in positional value.

2. **Dissociation-reaggregation**. Pieces of tissue are excised from the trunk, dissociated into single cells, and allowed to reaggregate. These arrange themselves in new orders, as shown in Fig. 4.14. When experimentally placed in a distinct order those cells which had occupied the position nearest to the head in the animal will form head structures such as tentacles, whereas the cells from the position nearest to the foot will form feet (Fig. 10.21). Dissociated cells thus preserve **a memory of their previous position**. The newly-formed structures are a function of the relative position the cells had in the previous body. Polarity is determined by a scalar parameter: the slope of positional value.

The molecular bases of positional information and positional memory are in Cnidaria largely unknown, as are the mechanisms of pattern control. More information about positional values is available for insects (Fig. 10.7).

10.10   Pattern Correction by Intercalation

10.10.1   Positional Values Enable Controlling the Completeness of Body Parts. Missing Positional Values Can Be Inserted Through Intercalation

If a bipolarly biheaded Hydra is created as shown in Fig. 10.22, but the sequence of body regions is incomplete because the lower body parts are missing, the missing structures are intercalated. Eventually, pattern correction leads to two complete hydras which separate from each other. More clearly in the marine hydrozoan *Hydractinia*, a reaction can be observed as an expression of pattern control evocative of the spectacular experiments with the Spemann organizer. If tissue with significantly different positional values are brought in direct contact, in the extreme case head tissue with foot/stolon tissue, a trunk column is inserted (**intercalated**) which realizes the missing positional values (Fig. 10.23). The sharp contrast 10/1 is transformed into a continuous sequence 10, 9, 8, 7, 6, 5, 4, 3, 2, 1 (arbitrarily chosen symbolic numbers).

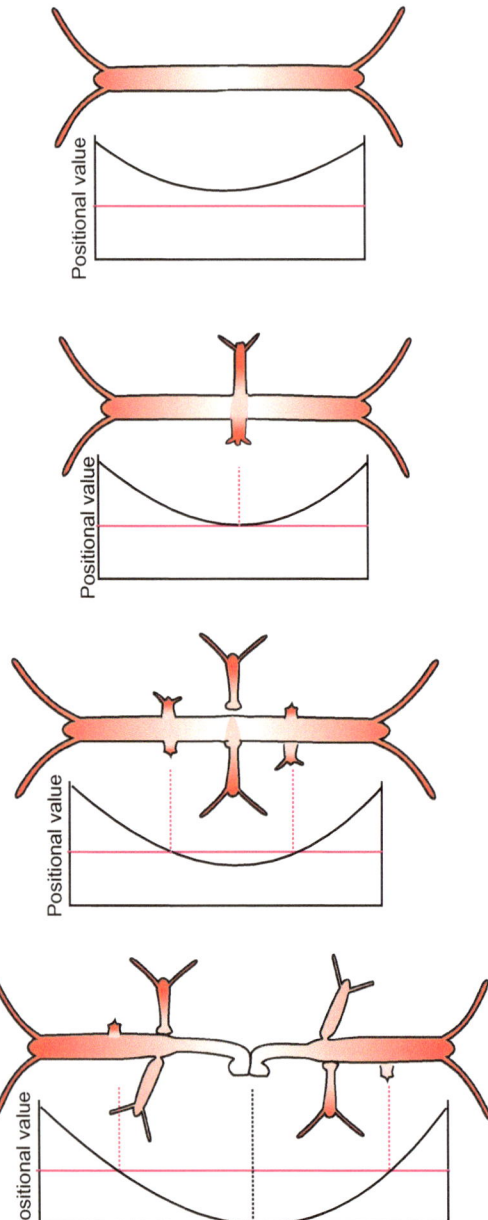

Fig. 10.22 Pattern correction in mirror-image duplicated upper body halves of *Hydra*. Double headed hydras can be generated by transplantation or by over-stimulation of WNT- or PKC-dependent signalling pathways in the cut out middle piece (Fig. 10.15a) of a polyp. Supported by the "helping" activity of the both heads positional value in the middle sinks more and more until the value zero releases foot formation which in turn permits segregation of the duplicated polyps. The capacity of budding is associated with a distinct mid-level positional value. Experiments by Mueller (1989)

10.10.2 In Insect Appendages Positional Values Are Considered to Be Encoded in the Density of Cadherins

Positional values as an expression of positional memory also play a role in the extremities of vertebrates and insects. When a fly hatches from the pupa after its holometabolous metamorphosis (definition in Chap. 20) its extremities are finished structures that cannot be corrected. Not so in hemimetabolous insects who have legs already as baby creatures and whose legs have to grow in each moulting. In such insects intercalation of leg parts can be evoked.

In the legs of insects, discontinuities in positional values are generated by grafting. For example, a leg may be bisected experimentally, a piece removed and the distal leg fragment grafted onto the remaining stump. Or, an additional piece of a leg taken from another specimen can be inserted in either the correct or wrong orientation. In any case, the confrontation between cells with different positional values results in the intercalation of leg structures. Dissonances stimulate cell proliferation, and the missing positional values are assigned to the additional leg structures until discontinuities are smoothed and the positional values are no longer interrupted by a gap. The experimenter's trickery may even force the insect to intercalate a piece with reverse polarity (Fig. 10.23b).

How is this possible? On the basis of the experiments referred to, and further experiments not described here, the hypothesis has been proposed that along the segments of the legs a molecular code denoting their normal position is assigned to the cells. This code would be represented by molecules on their surface, molecules which also mediate cell adhesion and for their part are controlled by *HOM/Hox* genes. Especially in case of insect extremities a concrete concept exists about the molecular nature of positional value. They are thought to be encoded in the density of cell adhesion molecules of the cadherin class, namely FAT and DACHSOUS; these are

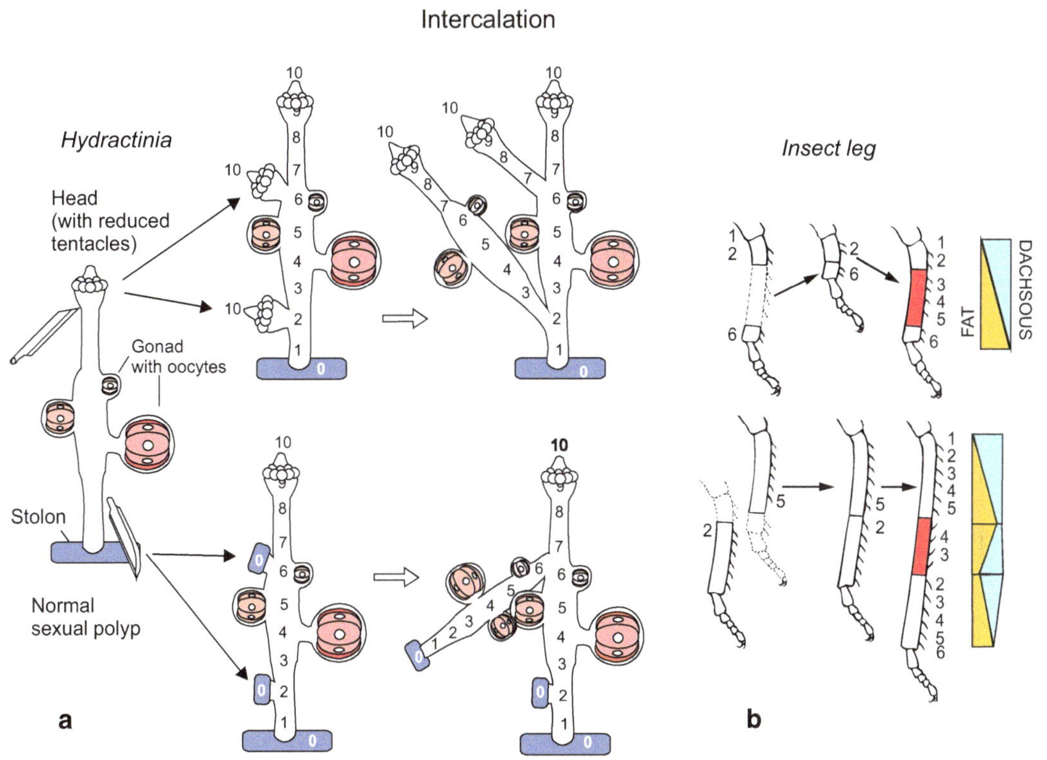

Intercalation

Fig. 10.23 (**a**) Intercalation in sexual polyps of the hydrozoan *Hydractinia*. If leaps in adjacent positional values are created by transplantation growth responses intercalating the missing values are induced until the sequence of positional values is complete and the leap smoothened. Experiments by Mueller (1964, 1982). (**b**) Intercalation of missing parts in the regenerating cockroach leg. Disparities in positional values are generated by removing a segment or replacing the distal part of the leg by a longer part taken from a donor animal. Disparities induce the insertion of parts during moults. After one or several moults, the structure of the intercalated segment is such that the sequence of adjacent positional values no longer shows a gap. *Note*: The intercalated segment displays a reversal of polarity. Experiments by Bohn (1971). In molecular terms positional values may be encoded by the inverse gradients of the cadherins FAT and DACHSOUS, that is by the molecular outfit of the cell surface with cell adhesion molecules. After proposals by Lawrence (2008), Bando et al. (2011)

known to form opposing gradients (Fig. 10.23b; see also Fig. 10.7 regarding planar polarity).

10.11 Periodic Patterns

10.11.1 In the Majority of Multi-cellular Organisms Components Are Found Which Are Repeatedly Manufactured as Constructive Modules

Patterns composed of repeating units are frequent in organisms.

- In colonies of hydrozoans and corals, polyps (hydranths) are arranged in regular and beautiful patterns.
- Periodic patterns are displayed in the segments of articulated animals, and
- by many segmental units in vertebrates such as somites, vertebrae and spinal ganglia.
- Repeating units include in large numbers manufactured structures such as spines, bristles, scales, feathers, hairs.

Several mechanisms of periodic patterning, which may also occur in combinations, can be imagined (Fig. 10.24).

1. **Transformation of temporal rhythmicity into a spatial periodic pattern**. In growing

GENERATION OF PERIODIC PATTERNS

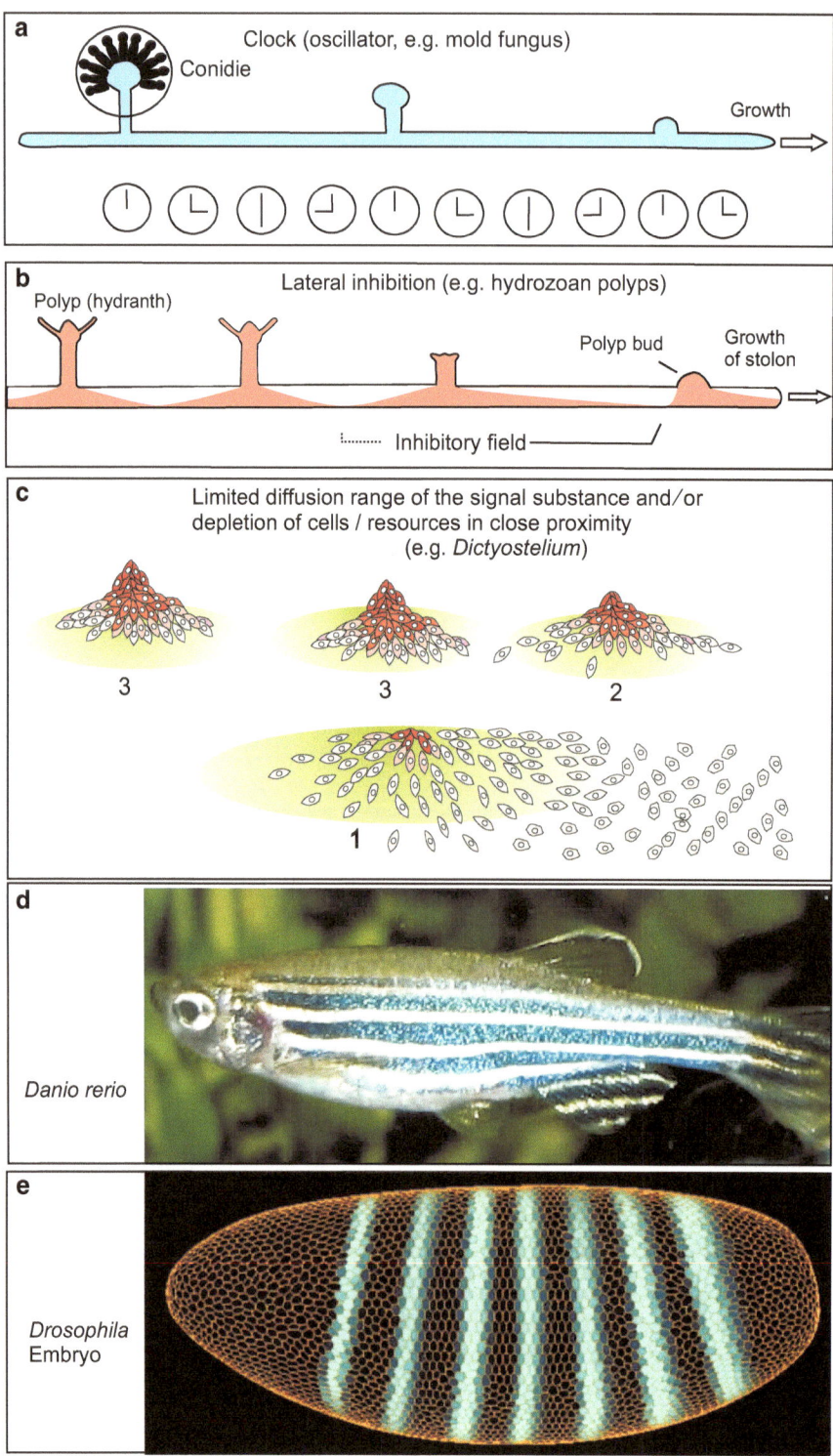

Fig. 10.24 Formation of periodic patterns. (**a**) Oscillator model: a new structure is formed at regular time intervals. An example is the periodic development of spore-forming conidiophores in the fungus *Neurospora*. New conidia are

systems internal oscillators (internal clocks) trigger the generation of new units at regular intervals. Such oscillators might be coupled to the cell cycle or to an internal clock. A particularly fascinating example is described below in Sect. 10.11.2.

2. **Lateral inhibition**. Existing structures are sources of inhibitory signals. As the signals spread into the neighbouring area, their intensity decreases with distance, due to enzymatic degradation, spontaneous decay or dilution. Outside the range of the inhibiting signals, new structures are made spontaneously or induced by stimuli acting from outside. Such models are proposed also by botanists who want to explain the spacing patterns of stomata on leaves, or of leaf primordia developing from a shoot apical meristem (phyllotaxis, see also Box 10.2). In the feather primordia of birds BMP-4 has been identified to be the conveyor of the inhibitory signal (whereas SONIC HEDGEHOG and FGF-4 are said to be activators of feather formation).

3. **Spacing by depletion**. A developing structure uses up essential substances or incorporates irreplaceable precursor cells. The area around the emerging structure becomes depleted. Farther away, such substances or precursor cells are still available and another structure can be made. This type of periodic patterning is found in competing aggregates of *Dictyostelium* amoebae and in similar multiple cell aggregations. Certainly, the spacing of aggregates primarily is a function of the signalling range of the chemoattractant cAMP (Figs. 10.24c and 4.10). But by these signals several aggregation centres also compete for cells. Soon in the area covered by the signal no more cells can be recruited; only outside this area new aggregates can form.

4. **Complex interactions**. The longitudinal stripes in adult zebrafish (Fig. 10.24d) are brought about by interactions of three chromatophores (see Fig. 15.10). Very different and strange is the emergence of stripes of gene expression in *Drosophila*. When one observes the precise stripe patterns of pair rule genes expressed in the blastoderm of *Drosophila* (Fig. 10.24e), it seems to be a safe bet that these are generated by a common mechanism, driven by an oscillator. Yet, molecular genetic analyses prompts us to draw a quite different picture. Each stripe is specified by a different set of transcription factors and particular mutual interactions. An example is shown in Fig. 4.37b. Primarily induced by the maternal BICOID gradient mechanisms of **lateral help** and **lateral inhibition** are activated and interconnected in a complex manner. Tentatively, some

Fig. 10.24 (continued) added in daily intervals (circadian rhythm). A further example of oscillatory pattern formation is shown in Fig. 10.20. (**b**) Inhibitory field model: the spacing of hydranths (polyps) in colonial hydroids is specified by inhibitory influences that originate in existing hydranths and decrease in intensity with distance. The inhibitory influence may be mediated by a substance emitted by the existing hydranth into the stolon; alternatively, inhibition may be based on the depletion of an essential factor that is taken up by the hydranths and removed from the stolon. After experiments by Mueller and Plickert (1982). (**c**) Depletion model: spacing of aggregates of *Dictyostelium* cells is (in part) due to depletion of cells in the vicinity of the aggregates. An aggregate emits attracting signal molecules. The distance between aggregates depends on the range the chemoattractant can spread with sufficient concentration. Beyond the signalling range of an aggregation centre, a new centre is established by the autonomous activity of pacemaker cells. The distance between neighbouring aggregates primarily is a function of the range the diffusing signal can overcome in sufficient strength but also a function of the number of cells that can be recruited. (**d**) and (**e**) Two stripe patterns with very different genesis. (**d**) Zebrafish: the stripes are formed by three types of chromatophores which are derived from the neural crest (see Chap. 15), migrate into the skin and settle in alternating stripes due to mutual interactions and bilateral arrangements. (**e**) Expression stripes of the pair-rule gene *even skipped* in the blastoderm embryo of *Drosophila*. The cells are stationary. Striped expression patterns result from network interactions of gene products as indicated in Fig. 4.37b. Images from study collection of WM

interactions are formulated in partial differential equation to simulate stripe formation on the monitor of the computer but the whole pattern cannot be simulated or explained with words up to date. Astonishingly, an endogenous oscillator operates in another, unexpected case, as described in the next section.

10.11.2 The Segmentation Clock, Based on Genes of the *Hairy/Her/Hes* Family, Assists in Generating the Series of Somites and Counting Their Number

On the blastodisc of the chick, and likewise on the blastodisc of mammals, a primitive groove (homologous to the amphibian blastopore) is seen. At the anterior end of this groove a **primitive node** (Hensen's node in the chick, homologous to the upper blastopore lip) is located (see Fig. 5.25). Ahead of the primitive groove and underneath the uppermost cell sheet (epiblast) mesodermal cells, that had immigrated through the primitive groove into the depth and crawled beneath the epiblast, form three stripes of cells. The middle stripe forms the notochord, the two stripes located at both sides of the notochord form the **paraxial mesoderm**, also called the **presomitic mesoderm** because their task is to form the series of somites in numbers appropriate for the given species.

Visible somite formation commences in the future thorax region far ahead and away from the node. While the node moves backward – in relation to the emerging embryo – at the anterior region of the presomitic mesoderm somites are sculptured one after another and segregated from the unstructured younger parts of the stripes behind them (Fig. 10.25). One by one the somites detach. As soon the number of somites characteristic of the species is formed, the counting mechanism stops. How is periodicity in somite formation achieved?

Actually, somite formation does not start where it is morphologically visible to our eyes. If we use methods to visualize gene expression such as in-situ hybridization (see Fig. 12.5) we see that somite formation starts with periodic gene expression at the node.

Here at the node a gene is periodically activated that is homologous to the *Drosophila* gene *hairy*, called *chairy* in chick, *her* in fish and *hes* in the mouse. Ahead of the node *chairy/her/hes*-mRNA appears, disappears and reappears at regular intervals. Activation of *chairy/her/hes* transcription does not stay at the node but travels wavelike over the stripes of presomitic mesoderm. During migration the initially broad band of expression shortens to a smaller zone while speed of movement decelerates. Activity of *chairy/her/hes* is accompanied by activity of *Notch* and *delta*. When the wave approaches the anterior end the zone of expression freezes in place and transcription of *chairy/her/hes* is supplemented by *Mesp2* expression. Cooperative interactions of HER/HES products, MESP2 and the NOTCH/DElTA system eventually cause the zone to detach as a new somite. After a due interval, 90 min in the chick, the next wave of *chairy/her/hes* transcription starts at the primitive node. When the wave has arrived at the anterior end of the presomitic mesoderm the next somite is formed and detaches. At the node, cell proliferation replaces presomitic cells lost at the anterior end until enough somites are made and the oscillator comes to a standstill. Each wave is accompanied and followed by expression of 40 to 100 more periodically activated genes. Over all gradients of FGF and WNT-3a and an inverse gradient of retinoic acid are thought to control locally changing gene expression.

The analysis of the molecular construction of the oscillator has been elaborated enough to enable developing models for computer simulation (Fig. 10.20b). In these models *her/hes* genes are core elements of the clock, generating their own oscillatory expression via feedback loops, while the NOTCH/DELTA system is assigned the function of coupling the rhythms of the individual cells. Experimental interference (for instance transplantations) showed in addition that the NOTCH/DELTA system assists in boundary formation. Transplanted cells

Molecular Oscillator in Somite Formation
(*Segmentation clock*)

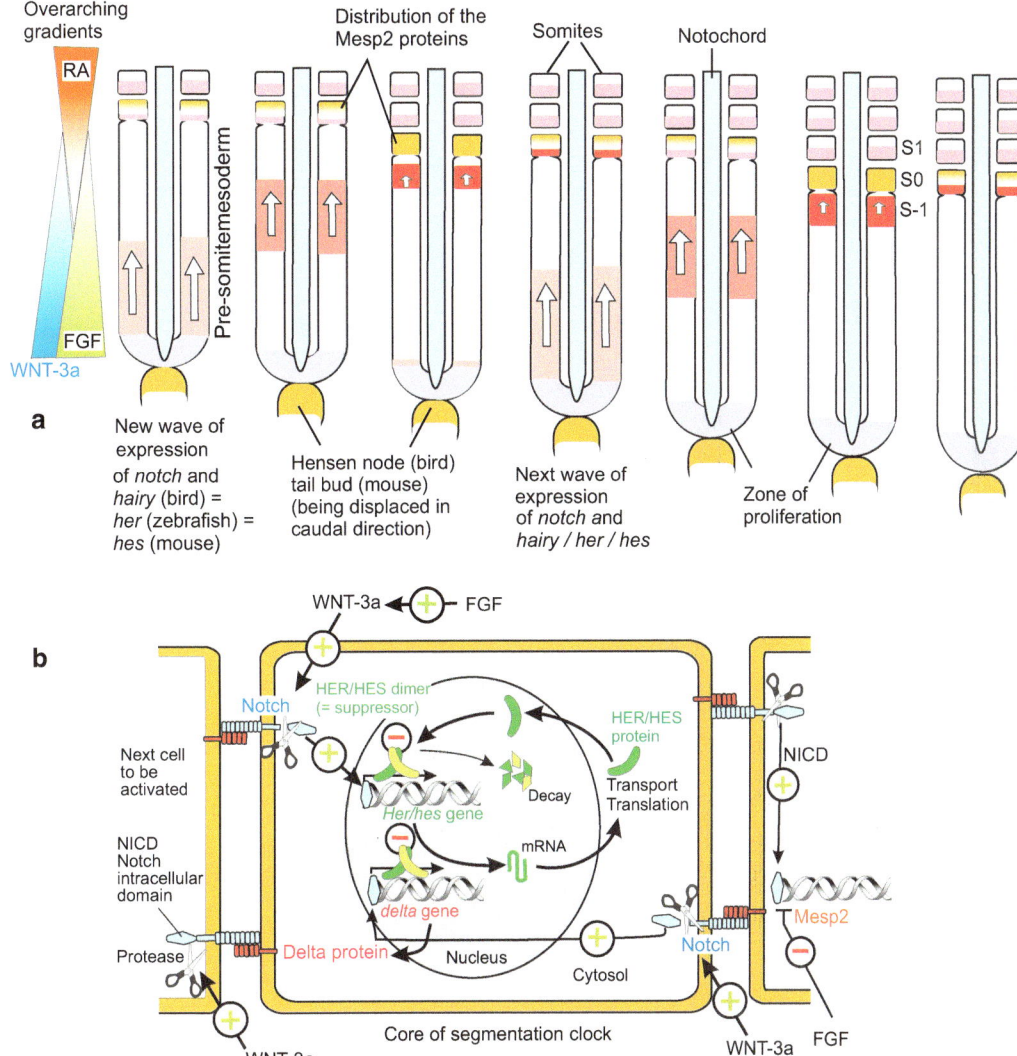

Fig. 10.25 (**a**) and (**b**) Segmentation clock. Function of the molecular oscillator in somite formation. (**a**) On the chick blastodisc expression waves of the gene *chairy* (*her* in zebrafish, *Hes* in mice) start near the node. Each wave spreads headwards in the still structureless paraxial mesoderm, also called presomitic mesoderm. While migrating the expression wave decelerates, contracts and eventually halts behind the most recently formed somite, forming

expressing *Notch* and *delta* eventually segregate from surroundings not expressing these genes.

Summary

In the course of development new systems of positional information are set up enabling position-pertinent behaviour of the cells. New patterns are generated through cell-cell interactions. As soon as in the embryo several cells are present these cells exchange signals to denote those locations where origin has to be given to certain organs and cell types and to bring about ordered patterns of differentially committed cells. We learn about the principle of **lateral inhibition** through competition and **lateral help** from the examples of nerve cell versus epidermis decision and photoreceptor versus lens commitment in an ommatidium in *Drosophila*, from the example of left-right asymmetry in the arrangement of some inner organs in vertebrates, and from the example of head-foot asymmetry in *Hydra*. Lateral assistance is related to **embryonic induction** where a small group of cells emits signals to neighbours in order to call on them to participate in constructing a body part. A particular mighty inducing action originates in the **Spemann organizer** in the upper blastopore lip of the amphibian gastrula or the corresponding homologous structures in other vertebrates, the embryonic shield in fishes, the Hensen's node in the chick and the node or tail bud in mammals. The organizer is established at a site of blastula whereto previously maternal mRNA of elements of the canonical signalling pathway (among others WNT-11, beta-Catenin) have been displaced. The Spemann organizer induces +/− complete second embryo (like Siamese twins) if transplanted to the opposite site of a host gastrula. While lengthening and approaching the blastopore the organizer is subdivided into head and trunk organizer. Summarizing accounts given in Chap. 5 antagonistically acting factors subdivide the embryo in regions of different fate. These are: (1) head inducers (DICKKOPF, CERBERUS) versus tail inducing WNT signals (WNT-8, WNT-3), (2) dorsal back organizing inducers (CHORDIN, NOGGIN) versus BMP-4, the ventralizing counterpart in the anterior-ventral region of the gastrula/neurula.

Fig. 10.25 (continued) a standing pre-somite band. Synchronously with the activity wave of the gene *chairy*, an expression zone of *Notch* + *Delta* travels. In the domain of the standing presomite band, the protein Mesp2 appears; it stops the expression of the Notch-Delta systems and aids in separating the band from the presomite mesoderm. Eventually the presomite segregates as new somite from the presomite mesoderm behind it. After an interval of about 90 min, and again starting at Hensen's node, the next *chairy* expression wave travels along the presomite mesoderm leading to the segregation of the next somite. The gene homologous to *chairy* is called *her* in zebrafish and *HES* (*HES = hairy/enhancer of split*) in the mouse. Also in these animals waves of activity travel from the caudal origin headwards, accompanied or followed by 40–100 further periodic gene activities. (**b**) Model of molecular interactions which essentially contribute to oscillations of the system. Like in many technical oscillators (clocks) periodic activity of *chairy/HER/HES* is achieved by interleaving of positive start-up processes with negative feedback loops. Model of a single-cell oscillator in which gene products repress gene activity. This negative-feedback loop involves a sequence of steps. The *Her/Hes* gene is transcribed with a basal rate and mRNA is transported from the nucleus and translated. Proteins are transported back into the nucleus where they dimerize to form a transcriptional repressor. This repressor accumulates and binds the gene promoter to inhibit gene expression. Finally, when gene products (mRNA, protein, dimer) decay, the gene repression is released and the cycle can start over again. Elaborate mathematical models are required to explain the emergence of oscillations and their transmission from cell to cell. In these models the Notch/Delta system is assigned the function of coupling the rhythms of the individual cells. *Her/Hes* genes induce oscillatory expression of the NOTCH ligand DALTA-C in zebrafish and the Notch modulator LUNATIC FRINGE in mice. In the mouse HES7 induces coupled oscillations of NOTCH and FGF signalling, while NOTCH and FGF signalling cooperatively regulate HES7 oscillation, indicating that HES7 and NOTCH and FGF signalling form the oscillator networks. Notch signalling activates, but FGF signalling represses, expression of the master regulator for somitogenesis MESP2. These results together suggest that Notch signalling defines the prospective somite region, while FGF signalling regulates the pace of segmentation. After Kageyama et al. (2012), and Oates et al. (2012)

Comparative investigations show that the agonist-antagonist system CHORDIN versus BMP-4 in the vertebrate corresponds to a homologous system in *Drosophila*. The vertebrate factor CHORDIN corresponds to the homologous *Drosophila* factor SOG, the vertebrate factor BMP-4 to the homologous *Drosophila* factor DPP. However, the expression domains of these factors in the *Drosophila* embryo are rotated through 180° against the vertebrate embryo.

In general inducing factors act synergistically or antagonistically and release cascades of secondary processes. The induction of the eye lens is classical example of a subsequent event in a chain of successive processes of induction.

Using the example of SONIC HEDEGEHOG we learn that one and the same signalling molecule can be used for several different purposes, for example to organize the formation of vertebral bodies around the notochord, to subdivide the neural tube in a dorso-ventral pattern, and to specify the sequence of fingers. A further example of a versatile signalling molecule is retinoic acid RA which, among other factors, is responsible for the subdivision of the central nervous system into brain and spinal cord, and for the proximo-distal as well antero-posterior polarity of the limb.

When secreted molecules not only act in adjacent areas but also within the area of production, such as BMP-4 (DPP), WNT or retinoic acid RA, and their various effects are dependent on the local concentration, the factors are also designated **morphogens**. The area in which morphogens bring about their effects is called **morphogenetic field**. The **gradient theory** assumes that morphogens (or some other determinants such as mitochondria) are present in the form of concentration gradients and the type of launched differentiation is a function of the local concentration. Many of once proposed but hypothetical gradients are now verified. Examples are the BICOID gradient in the early embryo of *Drosophila* and the gradients of WNT,

CHORDIN and BMP-4 in the embryos of sea urchins and vertebrates.

In planar organs such as the skin often **planar signals** are propagated that imprint a common physical orientation (a vector) onto repetitive structures such as hairs, feathers, scales or bristles. This task is often assigned to a "non-canonical" WNT system which evokes waves of increased intracellular calcium concentration in the plane. A summary of the theme planar polarity is given in Fig. 10.7.

An organizing centre equivalent to the vertebrate Spemann organizer is found at the oral end ("head") of the Cnidaria (*Hydra*, *Hydractinia*, *Nematostella*). In embryonic development, in the budding of new polyps, in head regeneration and in the emergence of future head structures in disorganized aggregates of dissociated cells the set up of an organizing centre is enabled and brought about by a "**canonical" WNT signalling system**. Overstimulation of this system results in the formation of multiple and ectopic heads along the body column. Using these examples the existence of continuously decreasing **positional values** which mediate **positional memory** is pointed out and the phenomenon of **intercalation** introduced. When discontinuities in positional values are generated by grafting, the confrontation between cells with different positional values results in the intercalation of missing structures. Dissonances stimulate cell proliferation, and the missing positional values are assigned to the additional structures until discontinuities are smoothed and the positional values are no longer interrupted by a gap.

Furthermore it is discussed which mechanisms create **periodically repeating structures** and multiple developmental modules. We refer to temporal cycles (oscillators, clock) which lay down repeated structures in space, we refer to inhibitory fields (e.g. BMP-4 surrounding feather primordia), to mutual activation and inhibition of neighbouring areas, and to competition for limited resources as mechanisms of spacing. An impressive example of a temporal program is the segmentation clock that generates the series of somites in the vertebrate embryo. Near to the posterior margin of the avian or mammalian blastodisc a clock is located close to and in front of the (Hensen's) node. Cycles of activity of the genes *chairy/her/hes* emerge periodically in intervals of 90 min (chick) at this point and travel along the paired stripes of unstructured presomitic mesoderm anteriorly until they reach the last formed somite. Here the expression domain stops forming a standing band in which other genes (Mesp.2) are activated. The cooperative action of CHAIRY/HER/HES, MSP2 and of NOTCH/DELTA brings about segregation and detachment of the band which forms a somite. Wave after wave starts at the posterior border, travels to the anterior end of the presomitic stripes and gives rise to the next somite. This continues until the number characteristic of the species is complete. A model of the oscillator is shown in Fig. 10.25.

The complexity of the themes of positional information and pattern generation is supplemented by Box 10.2 which introduces influential pioneering models of biological pattern formation.

Controlling Signals, Signal Propagation and Signal Transduction

When thousands and billions of cells have to fulfil the task to collectively construct an organism with its complex tissues and organs a manifold of mutual agreements are necessary. Controlling cells, for example organizing centres in early embryos, must emit **coordinating signals**. In order to respond to the signals cells must have the correct equipment including receptors and downstream intracellular signalling systems, which are cell-type specific.

In addition, control instances are required to monitor:

- where and when certain types of cells should differentiate,
- how many copies of each cell type should be generated and at which time point proliferation of the progenitors should be restricted or stopped,
- when and where programmed cell death has to clear spaces, and
- which routes migrating cells, outgrowing nerve fibres and sprouting blood capillaries have to take to reach their targets.

A general view of various means of cell communications is given in Figs. 11.1, 11.2, 11.3 and 11.4. The most important signalling molecules and released signalling cascades are summarized in Figs. 11.5, 11.6, 11.7, 11.8, 11.9, 11.10, and 11.11 under various aspects.

11.1 Signalling Systems and Mechanisms of Signal Propagation: An Overview

11.1.1 Membrane-Bound Signalling Molecules Convey Messages to Adjacent Neighbours, While Soluble Signal Molecules Spreading in Space Mediate Long-Distance Effects. Signal Transduction into the Cell's Interior Is Always Needed

Signalling molecules are released, transmitted and received by many different means. Figures 11.1, 11.2, 11.3 and 11.4 illustrate basic mechanisms. In each instance it is necessary that the target cell forwards the message into the cell's interior.

Independent of the respective biological field of application, intercellular signalling systems follow, as a rule, a chain of events:

1. The signal is passed on to the neighbours by means of an extracellular molecule or molecule moiety; this molecule becomes the ligand of a receptor at the target cell: in the case of hydrophilic signal molecules the receptor is provided on its surface, in the case of lipophilic molecules it is present either within

W.A. Mueller et al., *Development and Reproduction in Humans and Animal Model Species*, DOI 10.1007/978-3-662-43784-1_11, © Springer-Verlag Berlin Heidelberg 2015

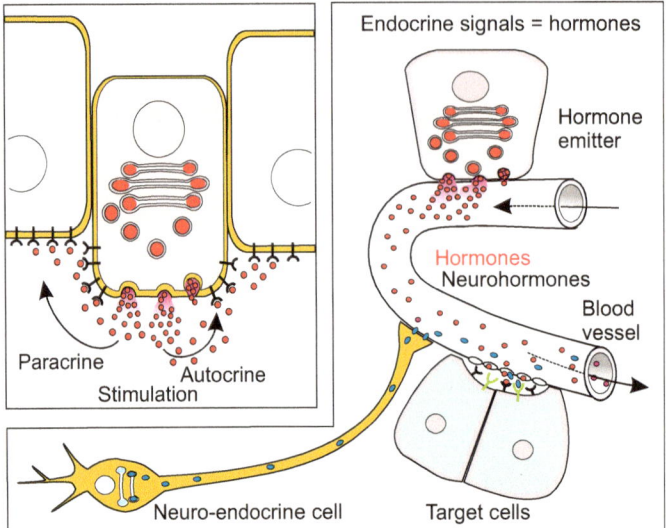

Fig. 11.1 Illustration and terminology regarding propagation of signalling molecules. In autocrine stimulation a cell releases signalling molecules, receives and responds to them itself with enhanced release of the signalling molecule (positive feedback) or by stopping release (negative feedback). The term paracrine stimulation designates dispersal of a factor by an emitting cell and its action in the neighbourhood. The term endocrine secretion is used for the secretion of (neuro-) hormones which are released into the blood or haemolymph and transported to distant targets

an interior compartment or also as an integral membrane protein.

2. In the case of membrane-bound receptors a system is needed that forwards the message away from the membrane into the cytosol of the receiving cell (Figs. 11.5, 11.6, 11.7, 11.8, 11.9, 11.10 and 11.11). In addition, as a rule, these systems amplify the signal strength through downstream enzyme-catalysed processes.

3. Eventually in many cases gene activities are turned on, others turned off, by means of transcription factors. In parallel, other functionally important proteins, for instance ion channels or metabolic enzymes, may be modified: ion channels opened or closed, enzymes activated or inactivated. In the implementation (realization) of such changes often phosphorylation of the proteins by kinases plays an important role.

Before we take a closer look at actual signalling molecules and signalling cascades we will summarize the general nature of signalling molecules and become acquainted with the mechanisms that lead the signals from the emitter to the receptors of the receivers.

11.1.2 A Large Array of Morphogens, Inducers and Other "Factors" Control Development and Growth

A look back to the processes of induction in the amphibian embryo and to pattern formation in the avian wing bud, and a look forward on the development of the nervous and blood system and the regulation of the supplies of blood cells reveals that a multitude of molecules can adopt the function of a messenger.

As biological signalling molecules are effective in minute quantities their identification is difficult, even though their number is growing incessantly. Initially, they reveal their presence only by the effect exerted by a piece of tissue, an extract or the supernatant of a cell culture. Since at the beginning of the search the chemical nature of the effective agent is unknown, the noncommittal designation "**factor**" gained currency, and when once a substance is labelled "factor" this label remains attached even when the molecule is chemically identified. As a rule, one can assume that a "factor" is a polypeptide (peptide, protein).

Fig. 11.2 (**a–c**) Communication between cells mediated by direct contact. (**a**) Communication by means of homotypic (of the same kind) or heterotypic (different) cell adhesion molecules CAM, or (**b**) by means of receptor-ligand pairing such as Notch with Delta. (**c**) Stationary signalling systems enable migrating cells to orient themselves by scanning components of the extracellular matrix or characters on signpost cells. Soluble molecules may enhance attraction

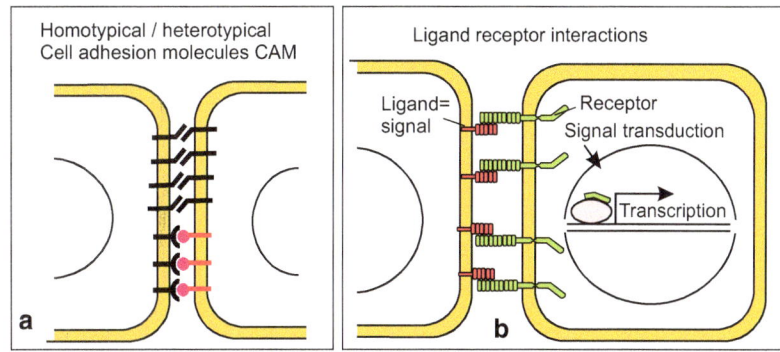

Communication by Means of Membrane-bound Molecules
Stationary Molecules Facilitating Orientation

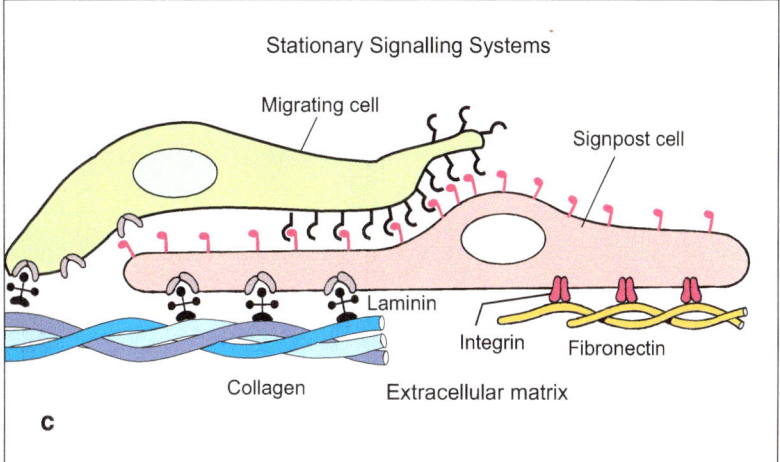

Therefore it is unable to directly enter the cell but is bound to a receptor on the cell's surface.

Rich sources of soluble signalling molecules are cells that are propagated in culture and release growth factors into the medium (**conditioned media**). By means of such factors the cells stimulate themselves mutually to proliferate and therefore the factors were designated **growth factors**. Other factors induce differentiation or chemotactic movements. The majority of factors identified so far are polypeptides with 500 to 1,000 amino acids. When their sequences were compared it often turned out that one and the same factor was detected with different methods by different research groups and named differently. All factors described so far are multifunctional and used in the organism for very diverse biological purposes.

Examples

- The **Bone Morphogenetic Proteins BMP** stimulate growth of chondroblasts (cartilage precursors) and osteoblasts (bone precursors) and they have functions in bone formation. In addition, members of the BMP family are already found in the early embryo (blastula) well before skeletal elements are being formed. In the early embryo BMPs function as morphogens aiding in realizing the basic body plan of a bilateral symmetric organism (see Figs. 5.18, 5.19, Fig. 10.3, Chap. 10).

- **Insulin**, a well-known hormone regulating blood sugar levels, is found in the embryo long before the Langerhans' islets are formed in the pancreas, quoted as the place of insulin production in textbooks of physiology.

Diffusion-driven Propagation

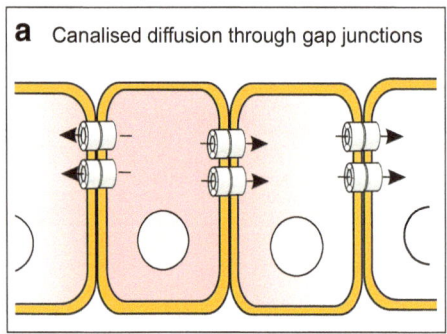

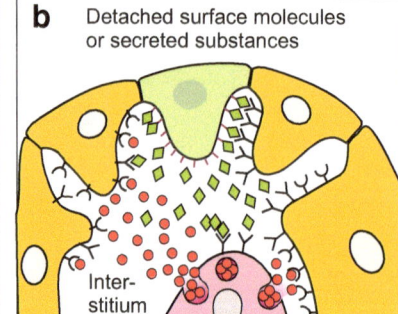

c Gradient Formation by Successive Internalization of Signalling Molecules

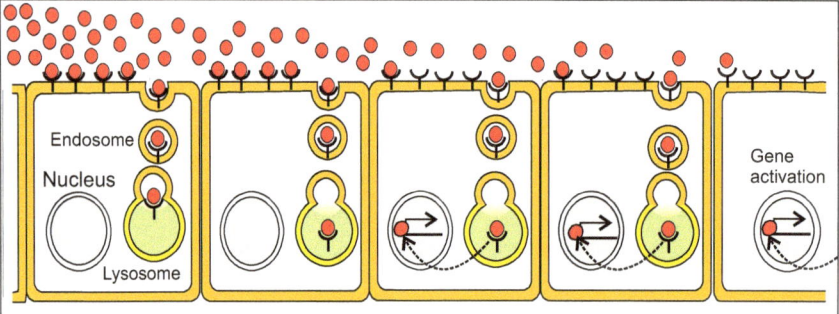

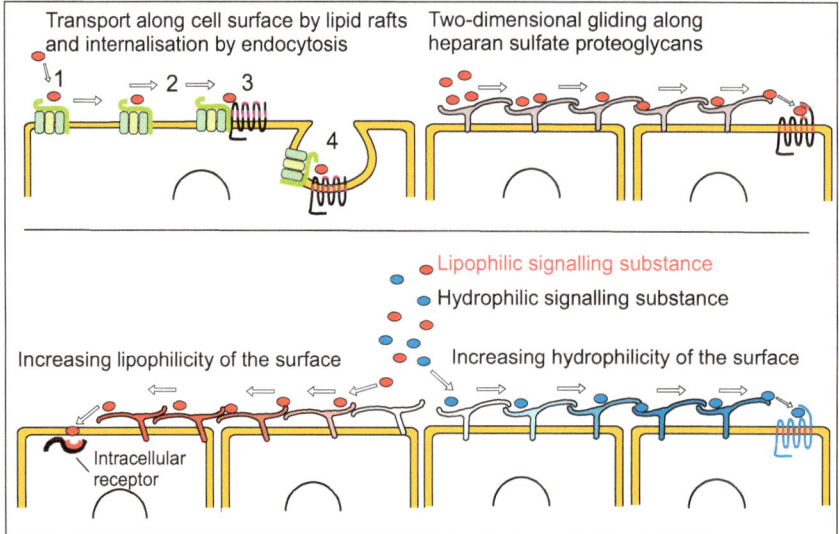

Fig. 11.3 (**a–d**) Mechanisms of signal transmission. (**a**) Diffusion-driven intracellular spreading through gap junctions or (**b**) extracellular spreading in the interstitial space. (**c**) Gradients of molecules arise, for example, through binding by antagonists, through enzymatic degradation, or through endocytosis of signal-receptor complexes along a diffusion route, and alternatively by binding of the signalling molecules to the extracellular matrix ECM (for example FGF binds to heparan sulfate proteoglycans). (**d**) Guided diffusion of signalling molecules is achieved, for example, by receptor binding and transport along cell surfaces, or through attraction or repulsion by hydrophilic or lipophilic cell surfaces

Fig. 11.4 Targeted and directed transfer of signalling molecules. Transcellular transfer is enabled by, e.g., transport vesicles (called by some authors argosomes). These vesicles transport signalling molecules directionally along the cytoskeleton to the neighbouring cell. Transport to distant cells can also happen through tenuous bridges called cytonemes or TNT (Tunneling Nano Tubes) which extend in the manner of filipodia, make contact to adjacent cells and fuse with the membrane of the targeted cell

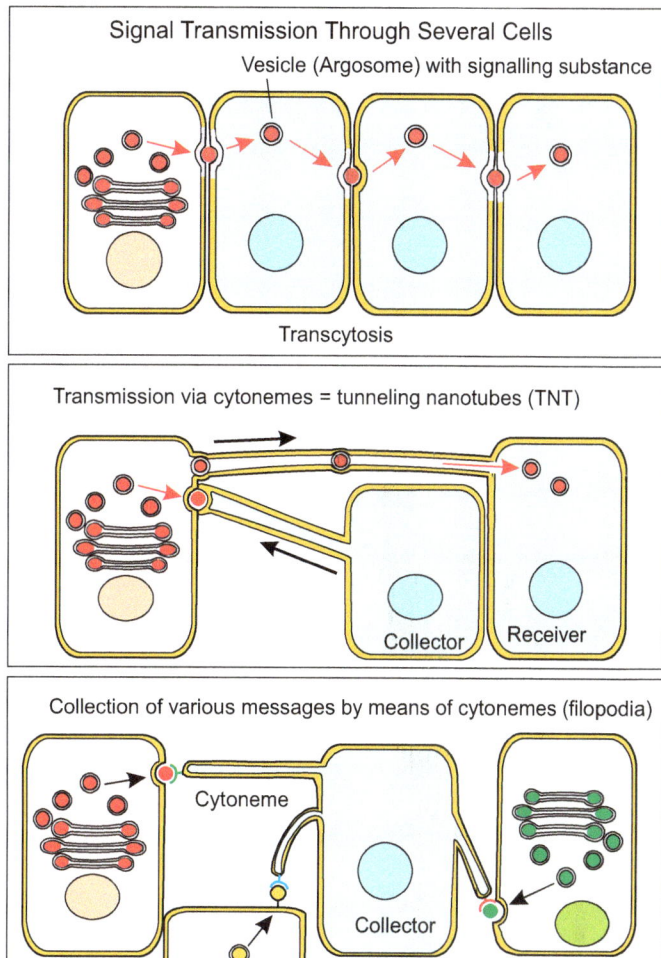

Multi-functional factors, therefore, often cannot be assigned to a single category but have to be classified among different categories dependent on developmental stage and local function.

11.1.3 Determination Factors, Morphogens, Inducers, Differentiation Factors and Chemokines Are Distinguished According to the Biological Functional Range

One possibility of classification is to categorize development-controlling molecules according to their function in the early embryo.

- **Determination factors** specify the future fate of a cell or group of cells. Morphogens and inducing factors are subsumed within this category.
- **Morphogens** act within a morphogenetic field from which complex structures such as extremities or eyes arise, and they contribute to pattern formation; their locally different effects being a function of their locally different concentration.
- **Inducers** are produced by local emitters and act on neighbouring cells.
- **Growth factors and cytokines** are involved in the control of proliferation (control of cell division).
- **Differentiation factors** initiate terminal differentiation of pre-determined (comitted) cells.

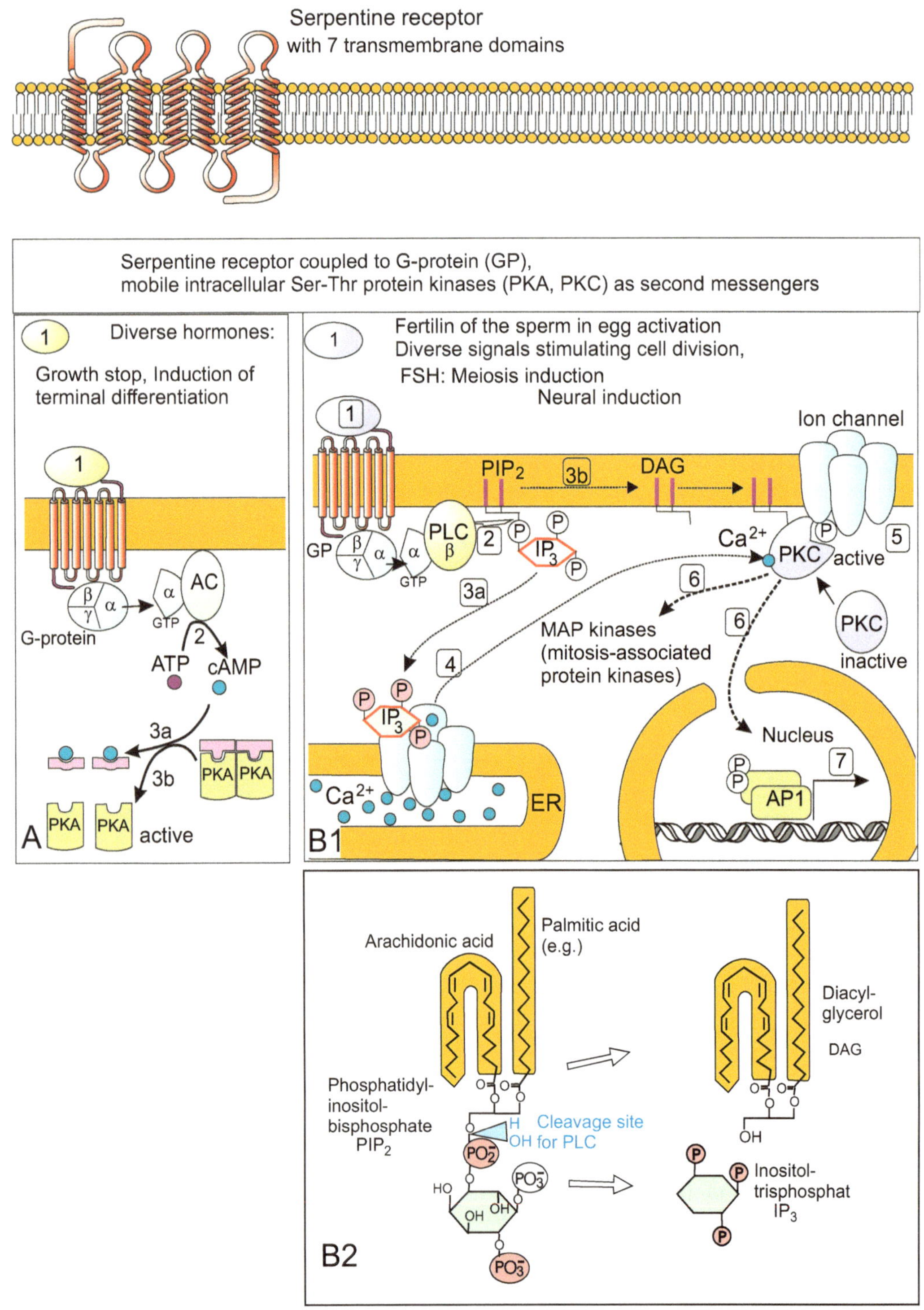

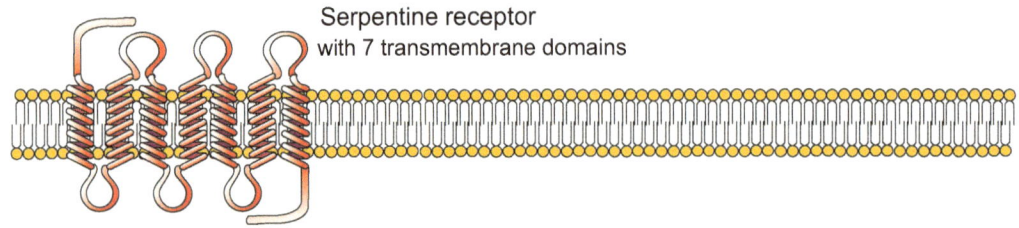

Serpentine receptor
with 7 transmembrane domains

Serpentine receptor coupled to G-protein (GP),
mobile intracellular Ser-Thr protein kinases (PKA, PKC) as second messengers

1   Diverse hormones:

Growth stop, Induction of
terminal differentiation

1   Fertilin of the sperm in egg activation
Diverse signals stimulating cell division,
    FSH: Meiosis induction
               Neural induction

Ion channel

1

1

PIP_2 **3b**   DAG

GP   β/γ   α   α   PLC β   **2**   P IP_3 P

GTP   **3a**

Ca^{2+} P   PKC active   **5**

G-protein   β/γ   α   AC   GTP

2

ATP   cAMP

4

6

MAP kinases
(mitosis-associated
protein kinases)

6

PKC
inactive

3a

3b   PKA PKA

Nucleus

IP_3 P P P

P P   AP1   **7**

A   PKA   PKA   active

B1 Ca^{2+} ER

B2

Arachidonic acid   Palmitic acid
(e.g.)

Diacyl-
glycerol

DAG

Phosphatidyl-
inositol-
bisphosphate
PIP_2

H   Cleavage site
OH   for PLC

PO_2^-

PO_3^-

HO   OH   OH

PO_3^-

OH

P

P   Inositol-
trisphosphat
IP_3

P

Fig. 11.5 (**a**, **b**) Signal transduction 1: Systems with serpentine receptors (seven pass transmembrane receptors) are often involved when peptides act as signalling molecules. The name serpentine receptors points to seven transmembrane helices. Upon reception of a signal a heterotrimeric G-protein binds to the receptor, dissociates in Gα(−GDP) and Gβ/γ subunit; these associate with further proteins, in particular enzymes. (In the

- **Chemokines** are signalling molecules emitted by cells of the target areas, or cells along the way to the target, to guide migrating cells. We refer to the chemokines which guide primordial germ cells (see Fig. 15.2), neural crest cells (Chap. 15, Fig. 15.3), neuroblasts (Chap. 16) or cells of the blood and the immune system (Chap. 17) to their target places.

In practise this subdivision is often not applicable without overlap. Depending on dose and operating place as well as competence of the target cell, but also dependent on tradition and school of thought that had imprinted a researcher, only one of the multiple effects is placed in the foreground or one of many designations is preferred. The "inducers" in the *Xenopus* embryo are factors belonging to multiple categories all at once.

11.1.4 Autocrine, Paracrine and Endocrine Factors (Hormones) Are Distinguished According to Operating Range and Targets

Another classification accounts for the range and target of the signalling molecules (Fig. 11.1):

1. **Autocrine factors** retroact on the producers themselves. The emitter cell possesses receptors for its own signals. These may retroact on the disposition to divide, or cause the emitter to continue or stop signalling. For example, via autocrine stimulation members of the TGF (Transforming Growth Factor) family promote tumour formation.

2. **Paracrine factors** diffuse in the interstitial spaces towards neighbouring cells. Here again, the morphogens and inducers in the early embryo could be quoted. In postnatal mammals a protein called PDGF (Platelet-Derived Growth Factor) which plays a role in the regeneration of injured blood vessels and injured tissue is among the paracrine factors.

3. **Endocrine factors = hormones. To get signals to distant targets use is made of circulating fluids in blood vessels** (convection, perfusion) Fig. 11.1). Examples are **erythropoietin** which regulates the amount of red blood cells (Sect. 18.2.7) and the growth-promoting **IGFs** (Insulin-like Growth Factors, also called **somatomedins**) in the fetus. When the liver has taken over their production these factors are designated hormones. **Regular classic hormones** that are distributed with the blood or other body fluids (lymph) intervene, as a rule, only late in development because the distribution systems must be created first.

Circulation hormones act to initiate and synchronize large-scale processes of transformation, for example control of pregnancy (Chap. 6), sexual development and maturation (Chap. 7), and metamorphosis (Chap. 20); one speaks of **endocrine control**. Hormones can reach many and distant receivers but cannot

---

Fig. 11.5 (Continued) illustration GDP and GDP-GTP interconversions are omitted for reasons of simplification.) (**a**) The $G\alpha$(-GTP) subunit activates adenylyl cyclase which catalyzes the conversion of ATP to the second messenger cAMP. cAMP on its part activates protein kinase A by binding and removing the inhibitory component of PKA complexes. PKA (Protein Kinase A) phosphorylates various substrates (actual substrates depending on cell type), in particular enzymes of the carbohydrate metabolism. (**b**) Activation of the PI-PKC system. Upon binding of G-protein to the ligand-activated receptor the $G\alpha$(-GTP) subunit of the G-protein activates phospholipase C (PLC). This enzyme hydrolyses the membrane associated phospholipid phosphatidylinositol-bisphosphate (PIP_2) resulting in two second messenger molecules. (1) In the membrane residual diacylglycerol (DAG) recruits a protein kinase C (PKC) to the membrane, where PKC activates diverse proteins (e.g. opens ion channels) through phosphorylation of hydroxyl groups of serine and threonine amino acid residues. In addition, eventually PKC can enter the nucleus to phosphorylate nuclear proteins. (2) Inositoltrisphosphate, IP_3, the other second messenger, remains in the cytosol, binds to an IP_3 receptor in the endoplasmic reticulum and thereby induces sudden release of calcium ions. These serve as tertiary messengers and co-activate certain subtypes of PKC. The Ca^{2+}-ions induce in different cell types different responses and are pumped back into the ER with lightning speed (compare egg activation in Fig. 3.5). **B2** Structure of PIP_2 and IP_3 in simplified depiction and indication of cleavage (hydrolysis) site of IP_3 and DAG from PIP_2

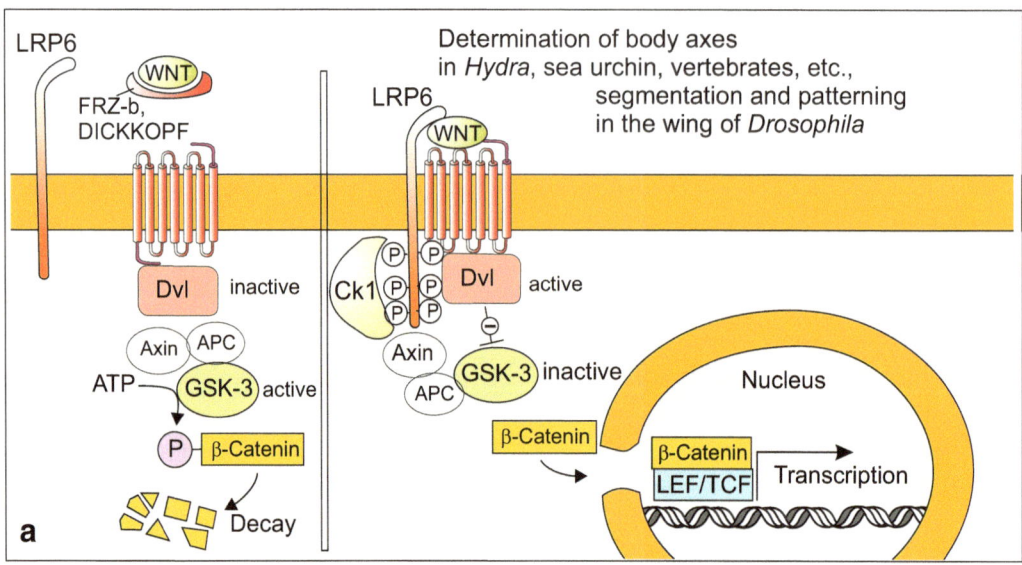

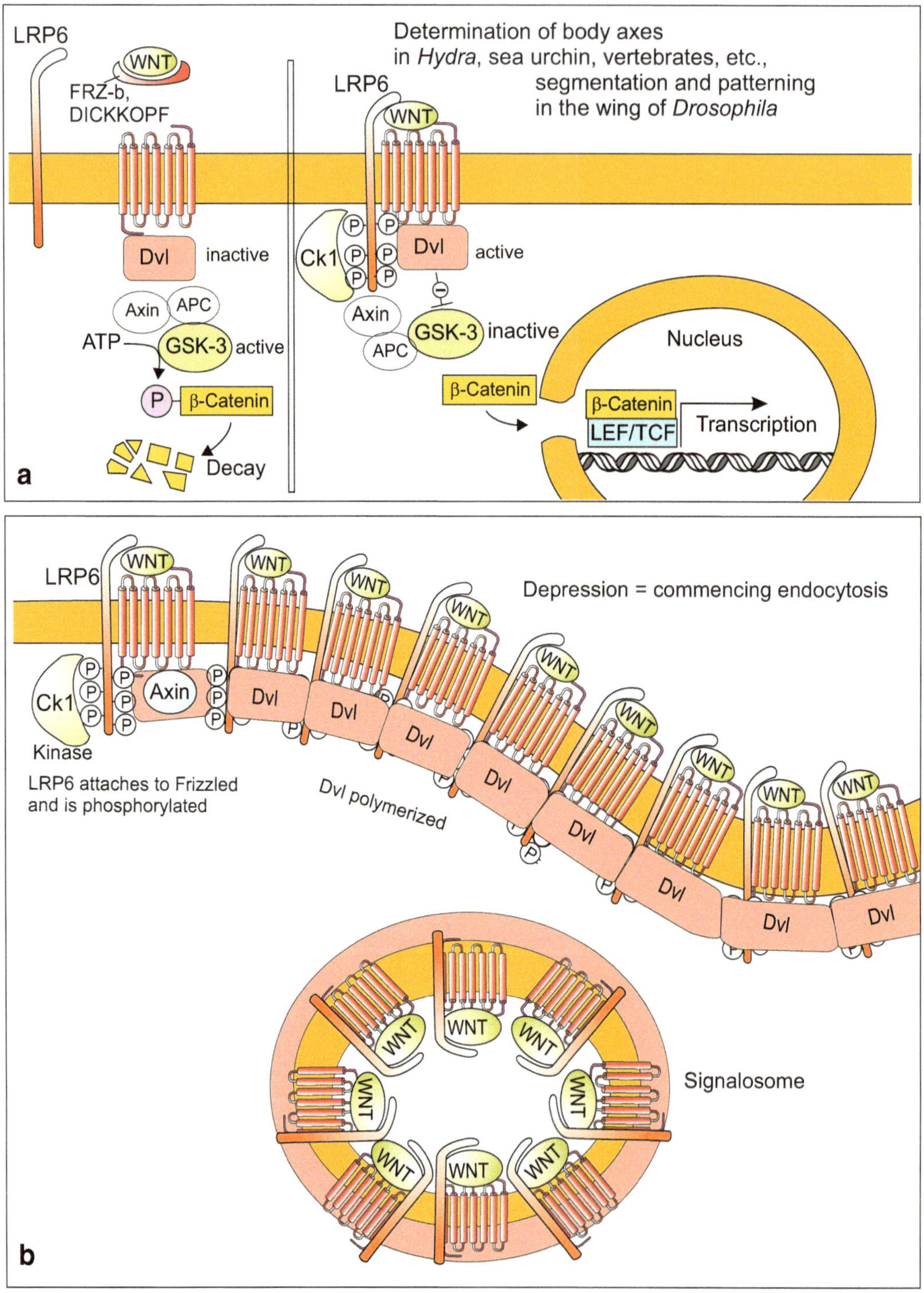

Fig. 11.6 (**a, b**) Signal transduction 2: The canonical WNT system. (**a**) Simplified depiction as it is introduced in many publications. If no WNT binds to the receptor FRIZZLED, the kinase GSK3 can unrestrictedly phosphorylate cytosolic β-catenin which thus is subject to degradation by the proteasome. If, however, a WNT signal arrives at the receptor (or GSK3 is inhibited by pharmaceutics), a chain of reactions inactivates GSK3 (chain shortened here). As a consequence β-catenin is no longer degraded, accumulates, enters the nucleus and

provide positional information. A general introduction to the topic of hormones is given in Chap. 20. With regard to its effects on gene activities the paracrine factor retinoic acid, RA, takes a mid-position between "factors" and "hormones".

11.2   How Cells Communicate: Signal Transmission and Propagation

Well-ordered cell differentiation, directed cell locomotion and the coordinated formation of shape, are true morphogenetic processes in which many cells participate. These processes require communication and interaction between the cells. The transfer of information from cell to cell is called **signal transmission**. If the message is distributed over large distances one speaks preferentially of **signal propagation**. The passing on of the message from the cell surface into the interior is designated **signal transduction**.

In every organism a multitude of signalling molecules is used for a multitude of purposes. Accordingly, there are many means of signal transfer from the emitter to the receiver too, dependent on the physical properties of the signalling molecules and the physical and chemical properties of the tissue which they have to pass and to affect. For a first overview we choose a classification based on the mechanism and the range of signal transmission.

11.2.1   Contact Molecules Mediate Direct Communication With Neighbours

Cells interact with adjacent neighbours by means of signalling molecules anchored in their cell membranes. All cell adhesion molecules CAMs (Fig. 11.2; and Chap. 14) have signalling characteristics. Closely related to CAMs are signal-receptor systems like DELTA-NOTCH (Figs. 11.2b, and 11.7b). Both the receptor NOTCH, and its ligands DELTA and Jagged are anchored in the cell membrane. The system serves to enhance slight differences between neighbours, to drive both cells in varying differentiation pathways, and to demarcate the boundaries between groups of cells so that they can physically separate from each other. In pathfinding of migrating cells further membrane-associated molecules play roles, for example cadherins, ephrins and ephrin receptors, but also components of the extracellular matrix, for example fibronectin which interacts with cell associated proteins such as integrins (Fig. 11.2c). (In addition, soluble molecules present in form of gradients can facilitate orientation).

11.2.2   Stationary Information Carrier and Signposts in the Extracellular Matrix Fulfil Leading Roles

In the clefts and spaces between cells macromolecules such as proteins, glycoproteins and proteoglycans are released which not only constitute an "extra-cellular matrix" ECM and serve as stuffing and stabilising material but also serve as trail signs for migrating cells, outgrowing nerve fibres, and sprouting blood capillaries (Fig. 11.2; Chaps. 16 and 17). Collagen IV, fibronectin, laminin and **heparan sulphate** proteoglycans are well-known molecules of the ECM which assume a signalling function on demand. Frequently specific signalling molecules released by emitters are bound to the ECM, e.g. netrin and other molecules that

Fig. 11.6 (Continued) activates, in cooperation with the transcription factor TCF/LEF, a set of WNT-dependent downstream genes. (**b**) Model how WNT might be internalized derived from investigations in *Xenopus* embryos. The transmembrane protein LRP6 attaches on the FRIZZLED receptor the intracellular domain of which is phosphorylated by a kinase. The thus activated FRIZZLED + LRP6 complex recruits DISHEVELLED (Dvl) at its intracellular domain. Adjacent Dvl-molecules polymerize and hold together the complex in form of a vesicle. The vesicle is internalized by endocytosis and called signalosome

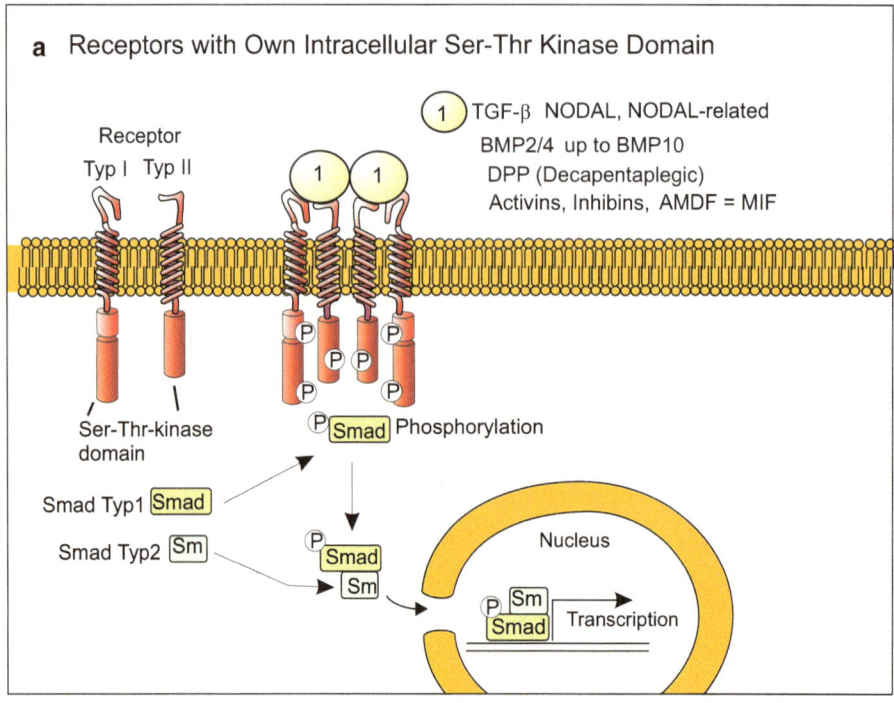

a   Receptors with Own Intracellular Ser-Thr Kinase Domain

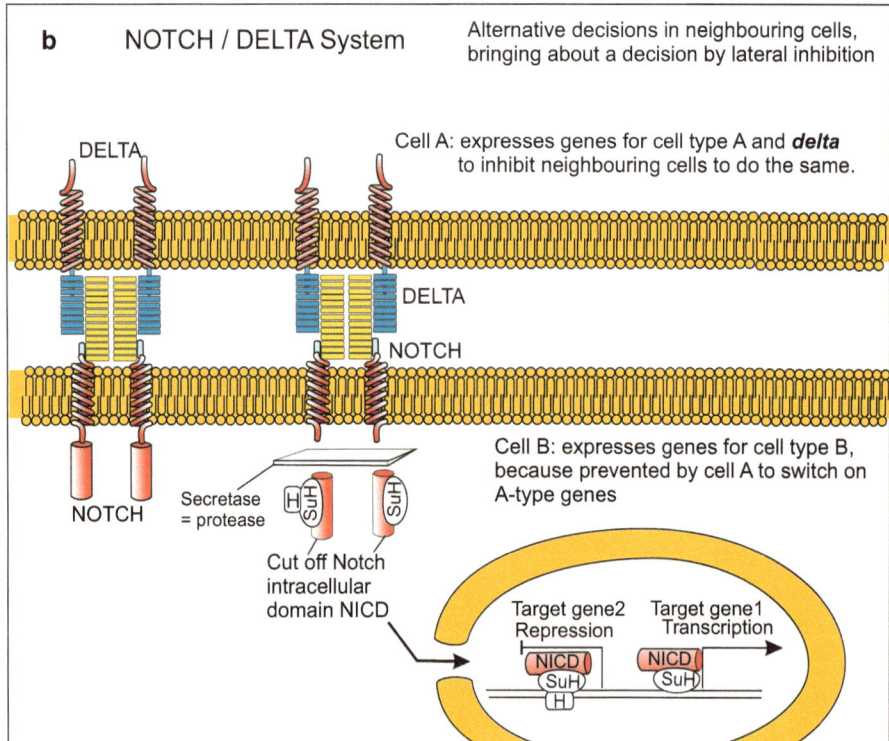

b   NOTCH / DELTA System

Fig. 11.7 (**a**, **b**) Signal transduction 3: (**a**) Receptors with intracellular Ser-Thr-kinase domain. Such receptors bind, for example, TGF-β, BMP or NODAL arriving from the exterior. They mediate, following multimerization of receptors and SMAD phosphorylation, transduction of the message into the nucleus where transcription of target genes is switched on. (**b**) The Notch/Delta system, well-known by its function in lateral inhibition and in alternative decisions,

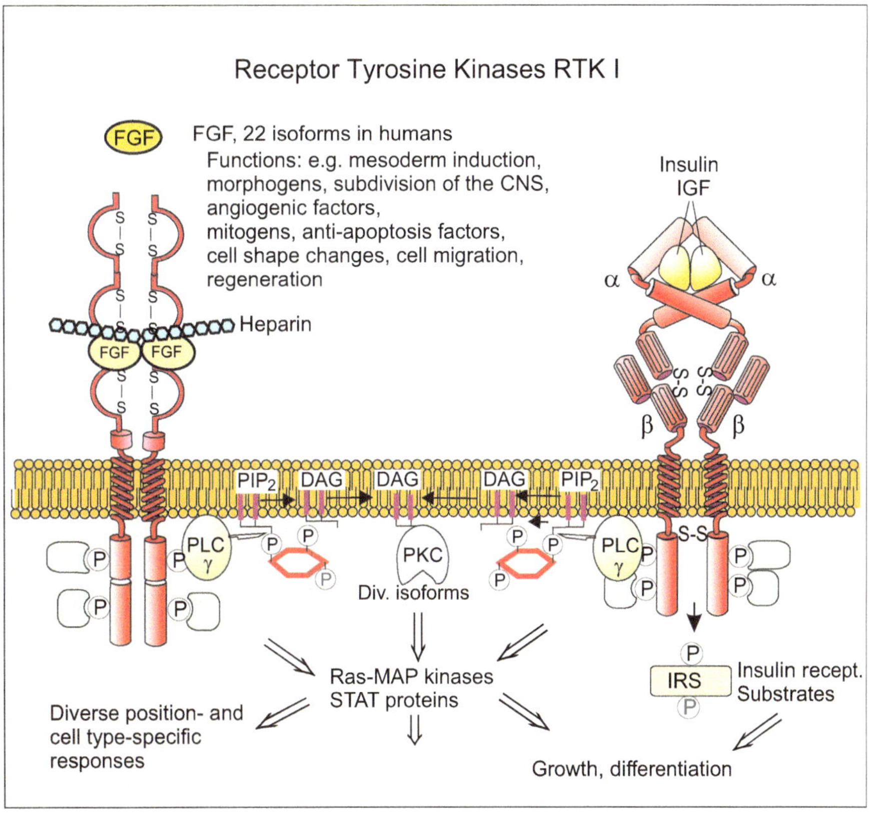

Fig. 11.8 Signal transduction 4: Receptor-tyrosine kinases RTK I dimerize upon ligand binding and activate either a PI/PKC system, the RAS/MAPK pathway, or the PI3-kinase pathway. Whereas the latter induces apoptosis or launches physiological responses, the RAS/MAPK pathway frequently causes changes in the cytoskeleton and thereby in the cell shape, e.g during migration

label places for growing nerve fibres (Sect. 16.5). They confer specific information, comparable to words on signposts. Some specific factors such as FGFs are bound to the ECM and stored until they are enzymatically released.

11.2.3 Communication Via Liberated, Diffusing Signalling Molecules

The above mentioned DELTA also belongs to those molecules that are partly released from the surface of the cell by means of proteases and thus enabled to reach distant targets via diffusion (Fig. 11.3b). FGF, EGF and SONIC HEDGEHOG are further examples for diffusible extracellular factors. However, free diffusion is often limited and requires certain preconditions:

Small lipophilic molecules can directly pass cell membranes and move from cell to cell if driven by a diffusion gradient. A signalling function is allotted, for example, to derivatives of **retinol (vitamin A)** as well as to many low molecular weight lipophilic molecules such as

Fig. 11.7 (Continued) is based on the membrane-bound ligand DELTA und its membrane-bound receptor Notch. Upon Delta-Notch binding, the intracellular domain of NOTCH (NICD) is proteolytically cleaved off. NICD enters the nucleus acting there as suppressing or activating transcription co-factor. As a rule, NOTCH and DELTA are exposed on opposite sides of adjacent cells so that the signal is transmitted directionally

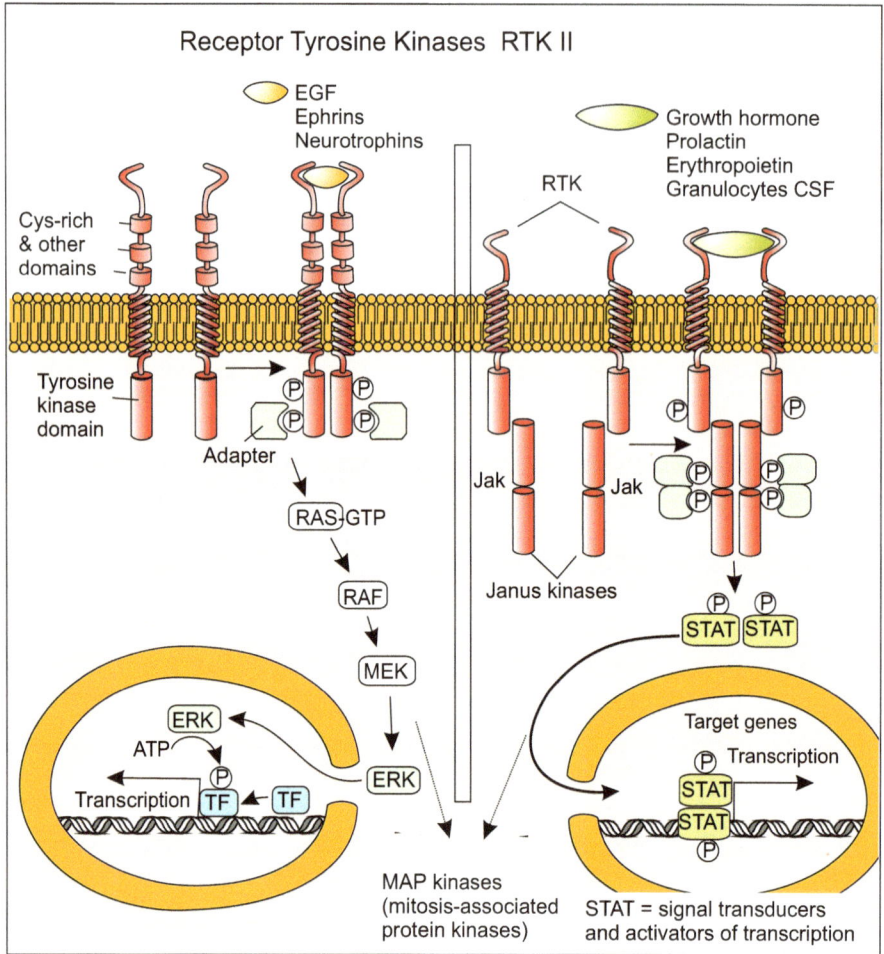

Fig. 11.9 Signal transduction 5: Receptor-tyrosine kinases RTK II activate genes via the RAS-MAPK-ERK-pathway. Alternatively they activate other sets of genes via Janus-kinases and STAT-transcription factors. Result may be, for example, an alteration in the structure of the cytoskeleton or transition of a stem cell into a differentiation path

arachidonic acid and its many derivatives known as leukotrienes, prostaglandins and hydroxy fatty acids, together with arachidonic acid subsumed under the collective term **eicosanoids**. Arachidonic acid derivatives influence, among other physiological events, cell motility. The significance of low molecular weight substances for embryonic development is currently underexplored because it is difficult to identify them, to quantify them in situ, and to trace their spreading.

Larger, +/− apolar (=lipophilic) molecules like thyroxin and steroid hormones are traditionally said to penetrate cell membranes and to be bound to cytosolic receptors (see Fig. 11.11). However, in many cases, shuttle proteins for lipophilic molecules are required: for many functions, especially for gene activation, it is not sufficient that such molecules enter the lipid bilayer of the cell membrane, the signal must be forwarded into the watery cytosol. How this might be accomplished is discussed below in Sect. 11.4.1. On the other hand, for many well-known lipophilic molecules receptors in the cell membranes also have been identified which mediate (fast) responses by starting signal transduction pathways.

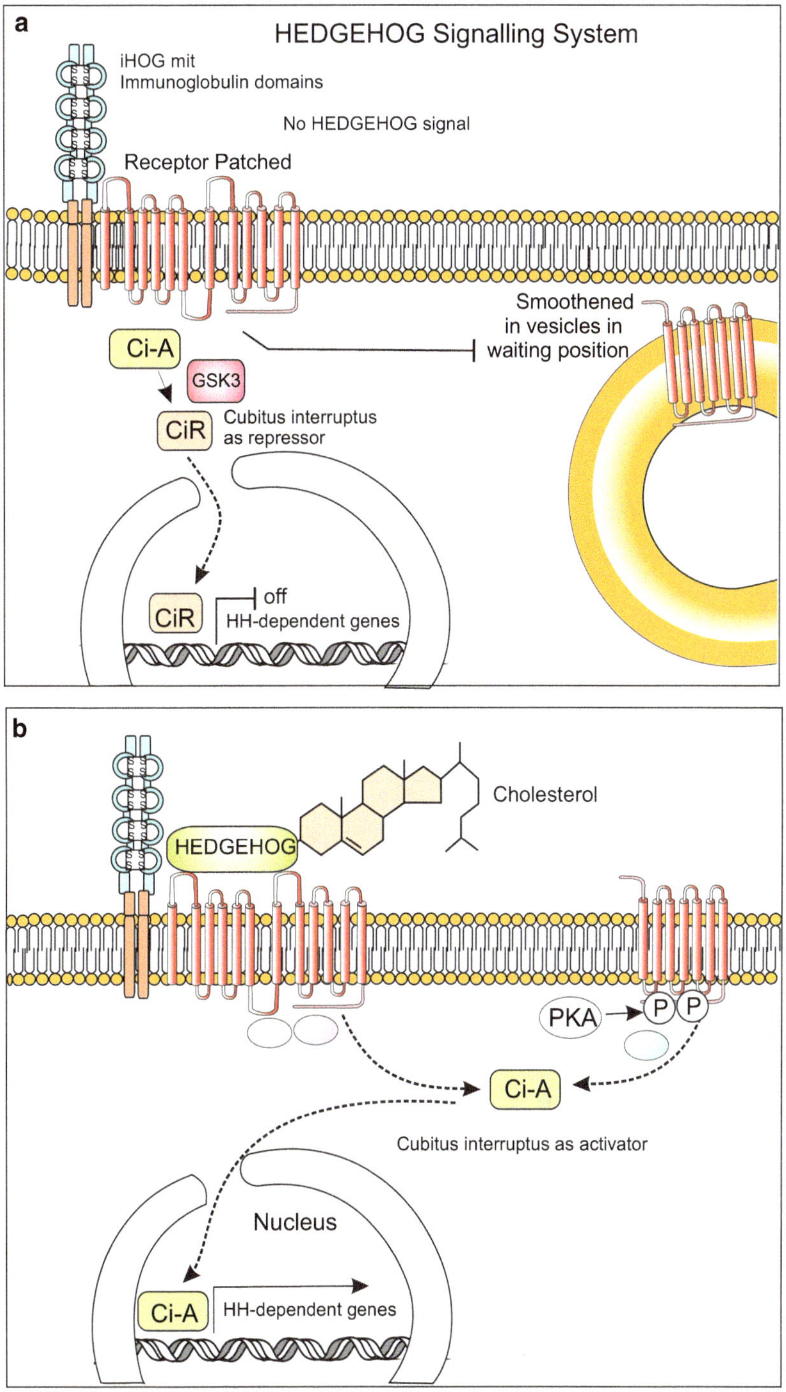

Fig. 11.10 (**a, b**) Signal transduction 6: The HEDGE-HOG system. Ligand is hedgehog, in vertebrates also its homologue SONIC HEDGEHOG. The receptor PATCHED has 12 transmembrane helices and is connected to a further membrane-resident complex, interference Hedgehog (iHOG) the extracellular domains of which exhibit immunoglobulin type repeats. (**a**) In the absence of a HEDGEHOG signal the transcription factor Ci (Cubitus interruptus) acts as suppressor of hedgehog-dependent genes and the 7-transmembrane protein SMOOTHENED is held in standby condition enclosed in vesicles. (**b**) Upon binding of the hedgehog ligand, which is covalently linked with cholesterol, to the PATCHED receptor, SMOOTHENED is integrated into the cell membrane. By processes unknown in detail or under debate, the transcription factor Ci is transferred in the active form and taken up by the nucleus. Here it activates genes including the gene *patched* for the PATCHED receptor (positive feedback)

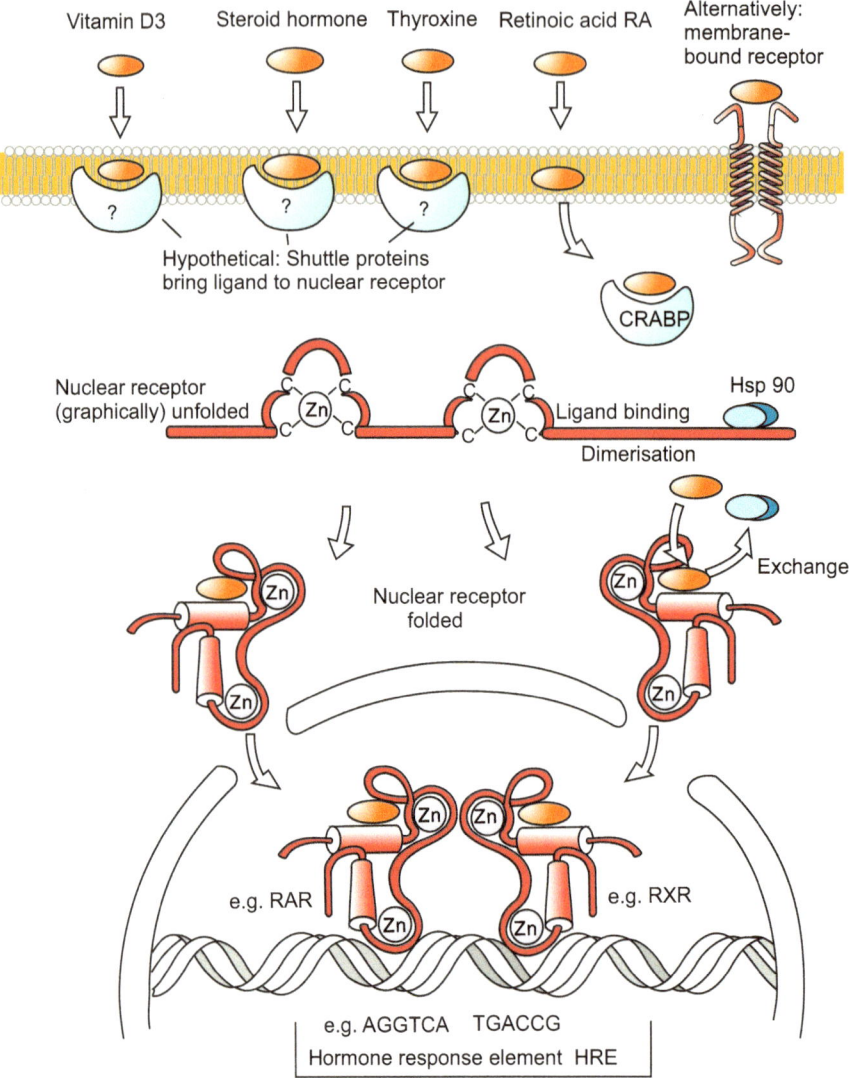

Fig. 11.11 Signal transduction 7: Lipophilic signalling molecules and nuclear receptors. A series of biological signalling molecules, in particular those with (predominantly) lipophilic properties such as steroids, vitamin-A derivatives, thyroxin or corticoids penetrate the cell membrane. In unknown manner (perhaps by means of shuttle proteins with a lipophilic binding domain) they arrive in the cytosol where they bind to intracellular receptors. The signalling molecules displace an inhibitory ligand (Hsp90) thus enabling their receptor to be taken up by the nucleus and binding to control sequences of genes. The receptors for the depicted signalling molecules all belong to the same protein family and possess very similar structures. At the controlling sequences that are marked by HRE sequences, the receptors form dimers, frequently heterodimers whereby the two receptors are loaded with different hormones. In this way a synergistic control of gene activities by two different hormones is possible

Large hydrophilic molecules, in particular polypeptides are able to reach surprisingly distant target cells. The majority of the known signalling molecules are peptides/proteins, for example those listed below, the WNTs, FGFs and the members of the TGF-β family including the BMPs. In the traditional concept that underlies most computer simulations of biological pattern formation (Chap. 19, Box 10.2) propagation happens by diffusion.

Precondition for diffusion-driven propagation of such large molecules is the existence of a fluid filled interstitium that offers the necessary free space. If this precondition is met, the range of spread can be between a few nanometres and millimetres. Some protein factors can also serve as chemotactic attractants (chemokines), for instance the Nerve Growth Factor (NGF) which guides the sprouting of the fibres of sympathetic neurons.

However, some proteins have unexpected properties making it unlikely that their propagation is based on simple diffusion. Thus, WNT molecules are post-translationally linked with palmitic acid, therefore being lipophilic and in watery environment not readily soluble. Possibilities to propagate such molecules are pointed out below in Sect. 11.2.4. Recent work shows that also the extracellular matrix plays an important role by binding or repelling signalling molecules thus restricting or increasing the range of diffusion. For example the ECM binds FGFs and thus restricts their range of diffusion in the developing brain (from 40 cell width to only 5). By degrading ECM components proteases release the factors in due time and at due places.

Channeled Restricted and Facilitated Diffusion. Free spaces for unrestricted diffusion are rare or lacking in tissues where cells are positioned side by side. Cells in epithelia adhere to each other by means of dense **adhesion belts**. Therefore, free diffusion is only possible in the baso-lateral clefts but not across the epithelia.

Channeled Diffusion Via Gap Junctions. Injection of fluorescing, low molecular weight dyes or of isotopes in single cells revealed possibilities for diffusion along certain epithelia. Electrolyte-ions (Na^+, K^+, Ca^{2+}, Cl^-) and small polar (=hydrophilic) molecules, such as inositol-trisphosphate (IP_3), cAMP, cGMP and ATP, can diffuse from cell to cell when the cell membrane of adjacent cells are equipped with conducting gap junctions (tube-shaped transmembrane channels, Fig. 11.3a).

11.2.4 Guided Diffusion and Forwarding by Means of Physical Effects

Extracellular Matrix (ECM). Diffusion is channeled not only via gap junctions. Also in the extracellular space, where little restriction is caused by the size of molecules, spatially restricted and directed spead of molecules is possible (Fig. 11.3c,d). Epithelia are always mounted on an extracellular basement membrane, a barrier for signal transmission. In addition, in most tissues spaces are filled with cross-linked macromolecules of the ECM which set limits to diffusion and influence the direction of spreading.

Proteoglycans as Hiking Guides and Companions. There are components of the extracellular matrix which bind to surfaces and restrict spreading of growth factors to two dimensions. Molecules known to act in this way are heparan-sulfate proteoglycans and gangliosides. Proteoglycans are thought to act as catwalks (Fig. 11.3d) bridging long rows of cells and mediating target-oriented gliding of signalling molecules. Sometimes they act as their companions and agonists. To what extent the flow of such signal-and-companion complexes is driven by diffusion or requires other sources of energy, such as ATP, is unknown to date. At their target, proteoglycans can act as co-factors, synergistically activating signal transduction pathways and being translocated into the cell's interior by endocytosis. Propagation of members of the **TGF**-β family including the **BMP**s, of the **FGF**s and putatively of the **WNT** factors is directed in such a way.

Guidance by Means of Hydrophilic or Lipophilic Interaction. In the cell membrane macromolecules are often anchored, and this alters the physical properties; they may, for example, make the cell surface more hydrophilic or confer enhanced lipophilicity to it. Such features may increase or decrease along cell rows. Since lipophilic molecules are attracted by lipophilic surfaces (but repelled by hydrophilic environments), and *vice versa*, since hydrophilic molecules seek contact to hydrophilic surfaces but avoid lipophilic surfaces, physically driven migration along cell rows is possible. Thereby microphysical forces are in play that are insufficiently studied to date but can be predicted to be of great significance. Energies acting on and along surfaces are in physics summarized under the terms surface tension or wetting tension. Adsorption and cohesion have not been sufficiently considered in computer simulations of signal propagation and biological pattern formation up to now.

Effects of Chromatography and Electrophoresis. Key word such as hydrophilicity and hydrophobicity remind biochemists of known mechanisms of chromatography and suggest new hypotheses. A mixture of more or less hydrophilic and lipophilic substances can be separated on +/− hydrophilic matrices, similar to hydrophobic interaction chromatography and the related reversed-phase chromatography. The various components of the mixture are separated according to their degree of hydrophilicity or lipophilicity. Similarly, in tissues molecules might move in different directions and to targets at different distances (Fig. 11.3d).

Similar to a gradual varying hydrophobicity, a sequential change of binding sites with decreasing or increasing affinity to a given signalling molecule would act. Such a mechanism would have a counterpart in chromatography, actually in ion-exchange chromatography or affinity chromatography. Such effects can be brought about by proteoglycans and gangliosides that are present in the ECM or anchored in the cell membrane.

Finally there are indications and theories about electrophoretic effects. A positively charged cell surface would attract and bind proteins negatively charged through, e.g., phosphorylation.

Lipid rafts are molecular complexes consisting of sphingolipids, cholesterol and

proteins, which float and drift like rafts in the fluid cell membrane and bring along components such as coupling proteins required for signal transduction and internalization (Fig. 11.3d). They can weakly bind signalling molecules, carry them along as cargo and dock on receptors of the seven-pass membrane proteins. The lipid rafts together with their connected receptors are translocated into the interior through endocytosis. In this way, the signalling molecule is also internalized, degraded and thus removed from the diffusion equilibrium.

11.2.5 Endocytosis of Ligand-Laden Receptors Contributes to the Schaping of Gradients and Aids in Opening the Way to the Nucleus

Proteins including morphogens can be labelled with reporter molecules such as GFP (Green-Fluorescent Protein). As a rule, the compact structure of the GFP does not disturb spreading and function of a labelled reporter protein so that its propagation can be observed live under the fluorescence microscope (*live cell imaging*). Among the morphogens that spread (in the transparent zebrafish) via diffusion is FGF8. This factor is picked up by receptors of the target cells which, upon binding, turn on a rapid signal transduction system. Subsequently the receptor-ligand complex is internalized and thus removed from the diffusion space. Often signalling molecules enter the interior linked to heparan-sulfate proteoglycans or to lipids. In this way WNT (see Fig. 11.6b) or TGF-β/BMP are organized in complexes, guided to their receptors and internalized. As a rule endosomes are marked with certain molecules (Rab proteins) and thus posted into certain compartments of the cell. If the endosomes are transported to the enzyme-filled lysosomes (Rab 7), the internalized complexes are degraded. But if the endosomes are labelled for the *recycling compartment* (Rab 11), the signalling molecule-receptor-complexes are transported back to the cell surface.

Complexes are not always recycled or degraded. Cleaved-off moieties of the receptor (NOTCH, see Fig. 11.7b), or the entire receptor (EGFR) can enter the nuclear space through pores, become bound to the chromatin and act as a transcription controlling complex or alter the three-dimensional structure of the chromatin. Retinoic acid and steroid hormones are among the signalling molecules that are translocated into the nucleus bound to their receptors and function as transcriptional switch there (see Fig. 11.11).

11.2.6 Targeted Transport to Neighbours by Means of Transcytosis, Nanotubes and Filopodia

Transcytosis. Some working groups assume that several well-known signalling molecules consisting of proteins are guided from cell to cell via transcytosis, so for example **WNT**, **BMP** and **SHH**.

In favourable conditions the just mentioned recycling and other transport processes can be visualized in living organisms or tissues surgically removed from them. The imaginal discs of *Drosophila* produce the signalling molecule DPP, a homologue of the vertebrate BMP-2/4, in a distinct area (shown in Fig. 10.10). In Box 10.1 experiments with imaginal discs taken from *Drosophila* strains bearing a *dpp-gfp* transgene are described which show the expression of *dpp-gfp* at the correct location, the export of the translated, green fluorescent DPP-GFP fusion protein and its propagation in the neighbouring cell areas by means of transcytosis. The video recordings showed fluorescing endosome vesicles moving from cell to cell. Since not all of the received DPP-GFP is forwarded, eventually a gradient of internal DPP-GFP was established.

Another research group demonstrated that GFP was anchored in the cell membrane of imaginal discs with a GPI (glycosylphosphatidylinositol) anchor. While the cells secreted WINGLESS small GFP-labelled, fluorescing vesicles emerged first in the adjacent cells, then in cells farther away. The vesicles were interpreted as transport vesicles for WINGLESS (the research group, apparently familiar with the ancient Greek mythology, named them **argosomes**, Fig. 11.4). Now they are named exosomes (see below Sect. 11.3.2).

Mailing Through Nanotubes and Filopodia Experiments using GFP-labelled molecules brought a further, unexpected mode of signal transmission into the focus of cell biologists (Fig. 11.4). In imaginal discs of *Drosophila* cells sprouting fine filopodia-style processes and extending them through cell free spaces towards other cells were observed. The authors named the filopodia-like processes **cytonemes** and assumed they would serve long-distance transport of vesicles. Using colour-labelled receptors the authors concluded that cytonemes are formed not only by emitter cells but also by target cells to collect a specific message. In mixed cell cultures, cells taken from wing imaginal discs formed filopodia in response to the morphogen DPP (i.e. had receptors for DPP), cells taken from eye discs formed filipodia that responded to (and had receptors for) an eye-relevant signalling molecule ("Spitz"; *Drosophila* EGF), and tracheal cells had filopodia with various receptors on their tip. The study showed that filopodia of the target cells can have multiple receptors for different signalling molecules enabling the cells to respond to various specific messages.

A similar phenomenon has been investigated extensively in mammalian cells. Various loosely distributed cells grown in cell culture sent out fine tube-like filopodia – just like antennae. By elongating, these antennae span unusually long distances through the culture medium and contacted distant target cells. A subpopolation of the cells was engineered to express a fluorescing synaptophysin-GFP product (synaptophysin is a protein characteristic for vesicles). Small amounts of this product and also other fluorescent membrane proteins, found their way from the producers into those non-producers to which they were connected via these fine tubes. The very thin tubes (50–200 nm diameter) were designated **tunnelling nanotubes** (**TNT**); putatively they correspond to the cytonemes or filopodia, respectively, of the *Drosophila* cells. Contrary to the expectation, low molecular weight substances did not pass through these nanotubes. On the other hand, evidence indicates that even macromolecules including DNA can pass through such tubes. Are they, as these researchers assume, a tube post system for special packaging vesicles? Or do these filopodia serve as fishing rods with their receptors as bait?

Surprisingly, in the zebrafish neural plate moving cells extend long **filopodia** that contact other cells of the swarm. The filopodia expose membrane-associated **WNT** signalling molecules at their tip; the contacted cell in turn exposes the WNT receptor Frizzled at the contact site. Likewise in the developing limb bud of the chick embryo filopodia-like thin extensions emanating from mesenchymal cells span several cell diameters to reach distant targets and bring membrane-associated **Sonic Hedgehog** (**SHH**) signalling molecules in contact with the target cells. Notably, particles containing SHH travel along these extensions with a net anterograde movement.

Immunologists have observed a similar phenomenon. B-lymphocytes stretch out fine filopodia and establish contact to antigen-presenting dendritic cells. The filopodia bear at their tips antigen-recognizing receptors known as BCR (B-Cell Receptors). (BCRs are prototypes of future antibodies anchored in the cell membrane of B lymphocytes). One side of the dendritic cell MHC-bound antigen is presented to the BCR. (MHC, the Major Histocompatibility Complex type 2, is a molecular complex present in the cell membrane of the antigen-presenting cells of the immune system, such as dendritic cells and macrophages. The complex consists of several transmembrane molecules which bind antigens at their exterior part). The contact structure

between B-cell and dendritic cell is called **immunological synapse**. At this synapse the B-cell extracts antigen from the membrane of the dendritic cell, internalizes it, and presents it, bound to its own MHC, to a T-cell. If the antigen is considered (to be) a signal a messenger relay service is created starting from the dendritic cell, passing the B-cell and terminating at the T-cell. Considering the antigen as signal is fully justified since the antigen stimulates the B-cell to proliferate. The daughter cells migrate into the lymph nodes and spleen to manufacture and secrete antibodies in large numbers, or to stay there as memory cells. Likewise, antigen-stimulated T-cells proliferate and set up a subpopulation of T-memory cells.

Given the capability of cells to sprout and extend long processes, one may ask how these processes find and reach their targets. The growth cone at the tip of outgrowing nerve fibres (Chap. 16) and the terminal filopodia of blood capillaries (Chap. 17) stretch out and recognize their targets by sensing diffusible signalling molecules and scanning directing signals exposed by stationary signposts. With regard to the recently discovered transport systems in B cells, be they called cytonemes, TNT or filopodia, it is yet unknown how they function.

11.3 Signalling Molecules and Associated Transduction Systems

11.3.1 Signal Transduction Forwards Messages in the Cell's Interior and Enhances Signal Strength

The term **signal** *transduction* is often misunderstood. **It does not mean** forwarding a message from cell to cell – here one would speak, as stated above, of signal *transmission* or signal *propagation* – rather it is about **forwarding the signal from the cell's exterior across the cell membrane into the interior of the cell**. As a rule, the external signal arrives at a membrane-bound receptor. The binding of the signalling molecule to a receptor (or to a receptor dimer) causes a conformational change of the receiver's intracellular domain, and thus its activation. As a consequence of activation, **second messengers** are generated for the purpose of directing the message into the depths and distributing it into various compartments of the receiver cell. While the receptor with its outer domain binds a single signalling molecule, a large number of second messengers are generated intracellularly thus **amplifying the signal strength** and enabling **diversification of the message into several inner compartments**.

The repertoire of responses that a cell can show depends on which receptors it possesses, how these are coupled to signal transduction pathways, and how these pathways are coupled to gene regulation.

In any case, in all the pathways essential functions are assigned to **protein kinases**; these are enzymes that **decouple phosphate from ATP and transfer it to other proteins (phosphorylation)**. Three types of amino acids can potentially be coupled with phosphate: **serine (Ser, S)**, **threonine (Thr, T)**, and **tyrosine (Tyr, Y)** (common denominator is the presence of a free OH-group in the amino acid residue). Kinases that transfer phosphate to either Ser- or Thr-residues are correspondingly called **Ser/Thr kinases**; whereas those that phosphorylate tyrosine belong to the **Tyr kinases**. Not every Ser, Thr or Tyr can be phosphorylated, it depends on a particular arrangement of surrounding amino acids, which together with the Ser, Thr or Tyr present a so called consensus environment consisting of (e.g.) a certain constellation of basic amino acids (lysines or arginines).

Phosphorylation may act as a switch, turning a protein "on" or "off". It does this by conferring a double negative electric charge to the protein with each added phosphate thus causing a conformational change in the protein. When, for instance, the protein is component of an ion channel, the channel may be opened or closed; when the protein is an enzyme or a transcription factor phosphorylation may turn it in the active state or refold it back in the inactive state. Moreover, autophosphorylation of receptor tyrosine kinases

like FGFR or EGFR generates docking sites for intracellular proteins following ligand binding.

With all the Diversity of Possible Transducing Systems Some Systems Exhibiting Key Functions Dominate. Cell-surface receptors catch and bind **polypeptides** or other polar (water-soluble) signalling molecules (e.g. neurotransmitters, adrenaline). In addition, more and more non-canonical examples are found of lipophilic molecules binding to membrane-exposed receptors, for instance various steroid hormones. The following classes of receptors and connected transducing systems dominate:

- **Seven-pass transmembrane receptors**, also called serpentine receptors, or G-protein coupled receptors are composed of a long, single polypeptide chain crossing the membrane seven times and without possessing their own kinase function (Fig. 11.5). To this class of receptors belong receptors long well-known to physiologists, namely the receptors for the blood sugar mobilizing hormones glucagon und adrenaline (epinephrine), and also the molecular receptors of the sensory cells in our olfactory organ where they capture odorants. In development seven-pass transmembrane receptors are competent to bind various proliferation stimulating growth factors. As a rule these receptors are linked to intracellular switching devices known as **G-proteins**. G-proteins are trimeric proteins composed of α, β and γ subunits. When a ligand is bound, the activated receptor activates the G protein by causing exchange of guanosine diphosphate (GDP) bound to the α subunit for guanosine triphosphate (GTP), the activated α subunit is released and can interact with other membrane components.

  - **The classic cAMP-PKA** system (Fig. 11.5a) shows how signals are amplified. When a single hormonal molecule is caught by the receptor, via G-protein (α subunit + GTP) several molecules of **adenylyl cyclase** are activated which convert thousands of adenosine triphosphate **ATP** to thousands of cyclic adenosine monophosphate **cAMP** molecules.

– **The inositol phospholipid pathway, in short the PI-PKC system** (Fig. 11.5b), plays an important role in activation of the sleeping egg in the course of fertilization (see Fig. 3.5).

Receptors with seven-pass transmembrane domains are also the receptors known as FRIZZLED for collecting signalling molecules of the WNT class (Fig. 11.6); they release intracellular cascades via docking proteins other than G-proteins, actually via Dishevelled. However, the type of cascade and response released is cell type-specific.

- **Transmembrane receptors with intracellular Ser/Thr kinase function** (Fig. 11.7a). All of these proteins possess a single transmembrane domain, and the enzymatically active site on the cytoplasmic domain. Ligands for the exterior domain of such receptors are, for example, members of TGF-β family. Functional receptors associate with captured ligands to form heterotetramers.

- **Transmembrane receptors with intracellular Tyr-kinase function** (Figs. 11.8 and 11.9). The receptors are present as dimers (insulin-R) or associate to dimers after having captured ligands. They then undergo **autophosphorylation** whereby each receptor molecule phosphorylates tyrosine residues of its partner, thus activating it. On the phosphorylated dimer intracellular enzymes can couple, for instance, phospholipase C-γ, which then generate second messengers. In other cases docking proteins mediate entrance to further signalling pathways.

- **Receptors of which the intracellular domain is cut off and translocated into the nucleus.** Such receptors accomplish signal transduction without many mediators. The proteolytically cleaved off domain itself enters the nucleus and controls target genes, in cooperation with other transcription

factors. We refer to the NOTCH-DELTA system (Fig. 11.7b).

There are many network interactions between these systems. One and the same cell can be equipped with several different receptors and possess several intracellular forwarding systems. For example phosphorylated transmembrane tyrosine kinases of the FGF receptor class (FGFR) can recruit phospholipase C, thus activating the PI-PKC system. Also the Frizzled-dependent wingless cascade, responsible for responses to signalling molecules of the WNT class, is connected with the PI-PKC system (Fig. 11.6b). The effect of one pathway upon others is often called "**cross talk**". Most signalling pathways terminate in controlling gene activities.

11.3.2 Important Families of Signalling Molecules

In the course of evolution repeated duplication of genes that code for 'growth factors' and the corresponding receptors yielded large gene families. Since most known signalling molecules are polypeptides, comparing the amino acid sequence can reveal such families including their genealogy. Often the sequence identity of ligands over the entire sequence is not high, for example the 22 vertebrate FGFs. However the family members can be identified by comparing certain *core* sequences, for example transmembrane domains. There is a multitude of abundance of protein families and each family is likely to have members that fulfil important functions in embryonic development. Here we present a list of the most quoted signalling and receptor protein families and of the associated signal transduction pathways.

The WNT Family (Fig. 11.6)

Concealed behind the term WNT is a large superfamily of isoforms. In the human genome no less than **19 *wnt*-genes** have been identified, even in the Cnidaria 11 isoforms have been found (see Fig. 22.8). Accordingly there is hardly any process in development and adult body in which WNT is not involved. Even in cancer research WNT is among the most frequently mentioned molecules.

Wnt was first discovered through the wingless mutant of *Drosophila*. Due to defective *wg* genes/alleles the mutant is incapable of producing a correct WINGLESS signalling molecule. The *wg* gene is expressed along the edge of the growing wing (see Fig. 10.10) in the pupa stage, and before this phase, it is expressed in the embryo at the borders of parasegments (see Fig. 4.38). This expression pattern is a peculiarity of insects and WINGLESS thus did not appear to be significant for humans. Separately the *int1* gene was found in mouse tumours, and eventually it turned out that *int* was the human homologue of WINGLESS. In vertebrates and other animals the proteins are called **WNT** (a mix of int and wg). They are used in many, if not all, animals including vertebrate embryos, and they are exported in the extracellular space to serve as signalling molecules. The corresponding membrane-associated receptors are called FRIZZLED (Fig. 11.6). Before being secreted, WNT proteins acquire lipophilicity by coupling with palmitic acid. The means by which these partly hydrophilic, partly lipophilic molecules are released by signalling cells and reach distant targets are still a matter of debate. Strong evidence points to different modes of signal exposure and propagation realized in different animals, tissues and locations. The following mechanisms have been proposed or uncovered that transform the insoluble WNT molecules into long-range signalling molecules that can spread several cell diameters away from their source of production:

1. Once released from the emitting cell WNT molecules are bound to extracellular cargo proteins such as secreted Frizzled-related proteins. However, how these large complexes spread is unknown.

2. As suggested by video recording under the fluorescence microscope, in imaginal discs of *Drosophila* GFP-labelled WNT is forwarded in a series of consecutive steps to adjacent and farther away neighbours via **transcytosis** (Fig. 11.4). Transcytosis appears to be related to, or identical with, the following mechanism.

3. A mode verified by electronmicroscopic pictures are vesicles called **exosomes** (formerly argosomes). Exosomes are small 40-100 nm microvesicles. Cultured cells of *Drosophila* as well as cultured mammalian cells (embryonic kidney, microglia of the brain) bud off vesicles into the extracellular space. These exosomes contain WNT molecules that are associated with a multipass transmembrane cargo protein known as **Evi** or **WNTless**. In epithelia the vesicles are released at the basal side where they may move along the basal membrane. The physical forces that effect their propagation are unknown. When WNT containing vesicles are presented to target cells in culture they become internalized together with Evi. Moreover, target cells are stimulated to produce and release WNT-carrying vesicles themselves. Thus a signal relay mechanisms similar to that known from *Dictyostelium* (Fig. 4.10) can be envisaged.

4. As mentioned above in the previous section in zebrafish embryos cells can be seen which extend long specialised filopodia to distant target cells. The filopodia carry WNT molecules at their tips, the responding cell exposes the WNT receptor Frizzled.

With respect to **WNT gradient formation** several mechanisms must be envisaged including the observation that in several tissues cells along a row express WNT, or elements of WNT signal transducing systems such as β-catenin, in a graded manner presumably due to some other directing gradient. For example, gradients of β-catenin expression were seen in intestinal crypts (see Fig. 18.3).

Independent of the mode of release and propagation with respect to their biological function and signal transducing pathways there exist several WNT-based signalling systems. Two terms refer to the two main types of WNT-based signalling systems; these are the **canonical** and the **non-canonical** pathways:

1. **The canonical WNT system**. The canonical (canonical = according to the norm or dogma), refers to a defined signal transducing system (as shown in Fig. 11.6a). The system operates in vertebrates predominantly with the WNT-3a and WNT-8 isoforms. At its intracellular domain Frizzled recruits a multifunctional protein called **dishevelled (Dsh)**. More factors such as **axin** and adenomatous polyposis coli (**APC**) join Dsh. The complex inactivates the **kinase enzyme glycogen synthase kinase 3 (GSK3)**. Its function in the inactive state of the cell is to phosphorylate β-**catenin** which thus is branded for degradation by a proteasome, without having fulfilled an apparent function. This seemingly wasteful mechanism is nonetheless advantageous for the cell. The non-stimulated system is in a **stand-by mode**. As soon as a WNT ligand docks to FRIZZLED, GSK3 is inactivated by the Dsh-Axin-APC complex; the non-phosphorylated β-catenin translocates instantly into the nucleus where it binds to, and activates, the transcription factor TCF/LEF. There is no need to first produce β-catenin via transcription and translation. WNT-dependent transcription of target genes starts immediately once a WNT signal arrives.

The canonical system acts in many processes. We refer to only a few characteristic examples.

- **Establishment of body axes and pattern formation along these axes.** The canonical WNT system is of decisive significance in the establishment, orientation and regional subdivision of one body axis. It defines the **posterior pole**. This applies to very different and in evolutionary terms remote animal phyla. Examples mentioned in this book are *Hydra*, *Hydractinia* and *Nematostella* from the phylum Cnidaria (Figs. 4.16, 9.5, 22.7 and 22.8), flatworms (Planaria, Fig. 22.6), annelids (*Platynereis*, Fig. 22.7), sea urchins (Figs. 4.6 and 22.7) and *Xenopus* embryos (Figs. 5.18, 5.19, 10.32, and 22.7). Mammalian embryos including human make use of the system to adjust the antero-posterior body axis. A remarkable exception is *Drosophila* where the gene first was identified but does not fulfil an axis-specifying function.

We recall pioneering experiments with embryos of *Xenopus*: Injection of mRNA for WNT or β-catenin at the belly side of the *Xenopus* blastula causes the establishment of an additional Spemann organizer at the injection site and therewith induction of a second body axis with head and trunk (Fig. 5.17). Apparently, **WNT has a decisive function in establishing the body axis which leads through the blastopore** (see Fig. 22.7). This function is conserved throughout evolution. Already in *Hydra* and, in evolutionary terms less ancient, in the sea urchin and *Amphioxus* (like the vertebrates deuterostomes) Wnt-producing and secreting cells are situated around the mouth or blastopore, respectively (see Fig. 22.7), (but not so in *Drosophila*).

In the amphibian neurula WNT-3 and WNT-8 are expressed in the postero-dorsal region at the blastopore and spread in form of gradients anteriorly (towards the head). In cooperation with other factors this WNT gradient ensures that the posterior part of the neural plate will become the spinal cord (see Figs. 5.19, 10.4 and 16.1).

- A second operational field is the **orientation of branches** of tubular structures such as blood capillaries and growing nephric tubuli (Fig. 14.7).

- In the play "**Development and Function of the Nervous System**" WNT takes over a bewildering array of major and minor roles. It participates in the first act when it comes to subdivide the neural tube into brain and spinal cord, and in the last act when new synapses form to fix the long-term memory in the adult brain.

- The **control of stem cells** is a further, important operational range which will be dealt with in Chap. 18.

2. **Non-canonical WNT signalling systems**. The term non-canonical refers to several alternatives for inter- and intra-cellular signal relay (see Fig. 10.7).

   The best known field of application for such systems is the **planar polarization of epithelia**. This was reviewed in Sect. 10.5.5 and depicted with examples in Fig. 10.7. In sum, it is about the following: In epithelia frequently a preferred and coordinated orientation of skin structures such as bristles, scales and hairs can be observed. For orientation a WNT signalling system is responsible. This system operates in the plane and is designated as non-canonical, Ca^{2+}-dependent system. In many cases **WNT-5a** is said to carry and convey the signal (in other cases WNT-4, WNT-11). Directionally spreading WNT signals provide those epithelial structures a common orientation. Receptors are again members of the FRIZZLED family and also other components of the canonical system such as Dishevelled may take part. Unfortunately, it is not possible to show a common diagram of schedule, cause and effect as a diversity of variants in the pathways exist. But there is one common feature: the chain of events does not terminate with translocation of β-catenin into the nucleus, but in the recruitment of factors which reorganize the cytoskeleton. Associated with that, WNT signals of this category frequently trigger a pulse-type increase of cytosolic Ca^{2+}-concentration (Ca^{2+}-transients). Such calcium waves can be elicited through activation of the PI-PKC system (Fig. 11.5). **Wnt-5a signals and associated cytosolic Ca^{2+}-waves run across the epithelia**.

   Further operational fields of non-canonical systems are the processes that induce and control the reorganization of cell positions during **convergent extension** (see Fig. 14.2), and the polar orientation and guidance of migrating cells (Fig. 15.8).

The TGF-β Family (Transforming Growth Factor-Beta, Fig. 11.7a)

Like the WNTs, the TGF-β factors also form an extended family in vertebrates with more than 20 members. Many of the essential factors introduced as inducers or morphogens in the Chaps. 5 and 10 belong to this family, for example:

- **Veg-1**, the maternally transcribed mRNA of which is located at the vegetal pole of the egg cell of *Xenopus* (not to be confused with Veg1 multivitamin supplement supplied by the Vegan Society).

- **NODAL**, a factor produced in blastula cells near the vegetal pole which diffuses towards the equator and programs, in cooperation with other factors, an equatorial band of cells (marginal zone, Fig. 5.6) to form future mesodermal tissue. In this function Nodal is supported by other members of the family known as **activins**. The newly specified marginal zone switches on *brachyury* as one of the early genes (see Fig. 5.18).

  NODAL is of significance again in the organization of left-right axial structures (see Fig. 10.8); the asymmetry manifests in the location of, for example, the heart.

  In the sea urchin embryo NODAL essentially participates in the establishment of dorso-ventral polarity. Its presence marks the position of the future, definitive mouth (Fig. 4. 6).

- **BMP-4** is a ventralizing morphogen in the early vertebrate embryo (and in the sea urchin embryo predominantly expressed at the aboral (dorsal) side).

- **DPP (DECAPENTAPLEGIC)** in *Drosophila*, the homologous (orthologous) counterpart to BMP-4 in vertebrates, defines the dorsal side in insects and the ventral side in vertebrates (e.g. Fig. 10.3).

- **AMDF.** The **Anti-Müllerian Duct Factor** is another representative of the family and acts as part of a **negative growth factor**, which in male mammals causes regression of the potential oviduct (Chap. 6).

Not all members fulfil their task for our benefit. After all, the prototype of the family, Transforming Growth Factor TGF-β1, induced certain cultured cells to behave like tumour cells, to proliferate and to overgrow their neighbours forming foci (small three-dimensional agglomerates).

TGFs are proteolytically excised (processed) from large precursor molecules. After their processing most isoforms form homodimers and as such bind to Ser/Thr-kinase transmembrane receptors (Fig. 11.7A).

The NOTCH-DELTA System (Fig. 11.7B)

This system differs from the others listed in this section in that both the ligand **DELTA** and the receptor **NOTCH** are integral membrane proteins. Their interaction therefore only takes place if the cells exposing them are in immediate contact. Notch and Delta are distant relatives of the EGF family. In *Drosophila* they mediate "lateral inhibition" in the process that separates future neuroblasts from epidermal cells (see Fig. 10.1, also Fig. 16.10). The NOTCH-DELTA system mediates mutual agreements between neighbours and alternative decisions ("I take this developmental pathway, you the alternative one"). According to models developed for computer simulations the system coordinates cell-internal oscillators of groups of cells in periodic somite formation and aids in segregation of somites (Fig. 10.25). Both molecules, Notch and Delta, are structurally quite similar and can also be classified as cell adhesion molecules. Upon mutual binding the intracellular domain of NOTCH (**Notch Intra Cellular Domain NICD**) is cleaved off by a protease. NICD serves as an intracellular signal and eventually enters the nucleus to control gene activities. In recent literature NOTCH is classified as receptor and DELTA as its ligand.

Occasionally, the NOTCH-DELTA system can be used also for long-distance signalling. In these cases the extracellular domain of DELTA is cleaved off and reaches distant targets per diffusion or transcytosis. The same applies for some representatives of the SHH-(Sonic Hedgehog) family.

The FGFs (Fibroblast Growth Factors), VEGF, PDGF (Fig. 11.8)

The first representatives of this superfamily of developmental control factors were detected by their ability to stimulate growth of fibroblasts in cell culture. Fibroblasts are precursor cells that give rise to connective tissue, cartilage or smooth muscle fibres. Similar to the WNTs, there is hardly an embryonic process during which FGFs are not used to convey a message. Their role lies in the control of cell migration, morphogenesis of branching, establishing of borders (together with NOTCH-DELTA), reorganisation of the cytoskeleton, or induction of apoptosis. Recently, even physiological functions in glucose and fluid homeostasis have been reported for one FGF subfamily. We refer to a few developmental examples:

Participation in the **induction and subdivision of the central nervous system CNS**, establishment of the midbrain-hindbrain border (see Fig. 16.14).

Specification of mesodermal patterns (**mesodermal induction**) in the vertebrate embryo.

Stimulation of migration of mesodermal cells and recruitment of myoblasts from the somites in the **budding and outgrowth of extremities** as well as **pattern formation** within the extremities.

Stimulation of cell proliferation in fibroblasts.

Sprouting of blood vessels: diverse FGFs, in particular FGF-2, act as **angiogenic factors.**

Formation of further tubular structures such as nephric tubuli.

Formation of the **tracheal system** and of the dorsal vessel (heart) in *Drosophila*.

Control of bud release in *Hydra*, presumably in cooperation with Notch.

In the human genome there are 22 different FGF genes. Almost all are also **heparin-binding** factors. Their receptors, the FGFR, are counted among the receptor tyrosine kinases (Fig. 11.8) precursor proteins of which are already found in the sister group of Metazoa, the unicellular choanoflagellates. Having bound a ligand, the receptor tyrosine kinases dimerize, phosphorylate and mutually activate each other. Subsequently the phosphorylated amino acids are used either to couple PLC and thus to obtain access to the PI-PKC system, or they turn on, mediated by several docking proteins, the RAS-MAPK or the PI$_3$-kinase system. The first causes reorganization of the cytoskeleton, the second cell migration and/or cytoskeletal reorganisation, the last activates apoptosis.

In terms of their evolutionary origin, the FGFRs are near relatives to the VEGFRs which bind VEGFs (Vascular Endothelial Growth Factors); these are essential signals involved in the formation of branched blood vessels and branched bronchioles in the lungs. In distant kin receptors of PDGF (Platelet-Derived Growth Factor) and EGF are included in the receptor tyrosine kinase family.

Strikingly, though starting at different receptors, the signalling pathways frequently converge on the same components. Nonetheless there is no traffic jam since not all pathways are simultaneously open. While one cell makes use of the RAS/MAPK kinase pathway following activation of an FGFR, the other uses it following stimulation of the insulin receptor by IGF. Moreover, not every cell keeps ready all types of receptors, they rather differentially express the receptors required for their function.

Insulin Family with IGF (Insulin-like Growth Factor I, IGF II, Fig. 11.8)

Insulin-like growth factors are produced at many locations in the embryo; they induce local growth processes. Like the FGFs, EGFs and ephrins the IGFs bind to receptor tyrosine kinases and activate similar intracellular signalling pathways. The IGFs function as growth factors as well as classic hormones.

The EGF Family (Epidermal Growth Factor Family, Fig. 11.9)

Further members of this superfamily of growth factors are **KGF** (Keratinocyte Growth Factor) and **TGF**-α (Transforming Growth Factor-alpha, biochemically not related to TGF-β). EGF

stimulates proliferation and thereby expansion of epithelia. The EGF molecule is a peptide comprising 53 amino acids (aa) that is cleaved off from a precursor molecule of 1200 aa. This precursor protein is bound to the **surface** of the producing cell and contains the characteristic EGF motif in many repetitions. EGF motifs are cleaved off by extracellular proteases piece by piece like slicing a salami. The signal transduction system starts with binding of the ligand to an EGF type receptor tyrosine kinase similar to FGF signals.

Ephrins and Ephrin Receptors (Fig. 11.9)

Ephrins are of vital significance as signpost molecules for growing nerve fibres and blood vessels, for the fusion of arterioles with venoles in the network of capillaries, or in marking targets. Ephrins (Eph) are exposed on the surface of signpost cells or target cells and are anchored in the cell membrane of these cells. Molecules of the Eph-A class anchor by means of an attached lipid (glycosylphosphatidylinositol, GPI, anchor), the molecules of the Eph-B class anchor the membrane through their own protein transmembrane domain. As sensors for recognition of guidepost molecules the homing cells expose ephrin receptors (EphR) on their surface. Densely arranged sensors are located on the growth cone of nerve fibres and on the terminal cell of blood capillaries (Chaps. 16 and 17). Ephrin receptors are again family members of the transmembrane tyrosine kinases. In certain instances the functional roles of the ephrins (Eph) and the ephrin receptors (EphR) are exchanged: The Eph is positioned on the searching cell and the corresponding EphR on the guidepost or vice versa.

The HEDGEHOG(HH) Family (Fig. 11.10)

The HEDGEHOGs were first identified because mutations of the corresponding gene in *Drosophila* disrupted the segmentation pattern and made the larvae look like hedgehogs. The family comprises several isoforms with SONIC HEDGEHOG, SHH, of vertebrates the best-known member of the family. In terms of evolutionary history HH-related molecules belong, like the WNTs and FGFs, to an ancestral protein family. They are found in cnidarians and planarians, and even in choanoflagellates, the sister group of Metazoa. Members of the HH family participate in segmentation processes in *Drosophila*, in arthropods in general and in annelids. They contribute to the dorsal-ventral subdivision of the neural tube (see Fig. 16.11) and are involved in patterning the vertebrate extremities (see Fig. 10.11) as well as the permanent renewal of the intestinal crypts (see Fig. 18.3).

A remarkable feature of the HH protein is its ability to function as an **auto-protease** cleaving itself in two, an N-terminal and a C-terminal part. Cholesterol is coupled to the N-terminal fragment HH-N – a most unusual process. This happens already in the HH-producing cell, and HH is exported linked to cholesterol. In the extracellular space several HH-cholesterol molecules form aggregates that are bound to a proteoglycan ("Dally") on the cell surface. How they spread and form gradients is currently unknown. Both gliding on extracellular proteoglycan slipways and transcytosis are discussed. As mentioned above, one means to present SHH to distant cells is via long, filopodia-like extensions. The intracellular signalling cascade is highly complex and in no case sufficiently elucidated that a complete chain of events, starting with HH-cholesterol bound to the extracellular IHOG-PATCHED receptor complex down to gene regulation, could be shown. Therefore, Fig. 11.10 is still fragmentary. A close connection between SHH-signalling and arachidonic acid derivatives, recently discovered in the context of cell migration, does not clarify and simplify the picture.

Strange stories are told about HEDGE-HOG. Based on unexpected evidence, it was suggested the factor would be sensed with chemosensory primary cilia that appear early in development. As discussed in Sect. 10.6, in amphibians and mammals such cilia participate in processes which generate left-right asymmetries and bring the heart to its correct, left place. In fact, in vertebrates the primary cilia are equipped with HEDGEHOG receptors. The cilia are considered as homologous to the chemosensory cilia in the olfactory epithelia of the nose. According to this view HEDGEHOG is an "odorant" and is "smelled" by the target cells. Hence, a close connection would be established between signal reception in embryology and signal reception in the sensory physiology of adults.

11.3.3 The Large Number of Non-Peptidic Potential Signalling Substances Is Still Barely Explored With Respect to Function in Developmental Physiology

Up to now, among the non-polypeptides the steroid hormones and, the here often quoted, retinoic acid attracted interest. Yet, an uncountable number of lipid or sugar molecules constitutes a completely blank area, ready to be explored by future researchers. This applies especially to low molecular weight substances. We refer to the example of arachidonic acid and its derivatives and to **serotonin** (5-hydroxytryptamine), that appears in a ventral-left blastomere of the 32-cell stage of the *Xenopus* embryo and predicts the left displacement of the heart and other left-right asymmetries (Sect. 10.6.3).

11.4 Lipophilic Signalling Substances and Control of Gene Activities

11.4.1 Retinoids, Steroid Hormones and Thyroxin Control Gene Activities Via Nuclear Receptors

Apolar (lipophilic) carrier of messages and signalling molecules of low polarity such as retinoic acid, steroid hormones or thyroxin penetrate the cell membrane, are able to **activate membrane associated enzymes** and **regulate the opening state of ion channels**. But eventually they are collected by **cytosolic receptors**. How lipophilic substances can enter and cross the watery cytosol is currently unknown. Speculation may consider entrance in a manner of pinocytosis (=micro-endocytosis), collection and transport by shuttle proteins (as known for retinoic acid), or collection at the membrane directly by those intracellular receptors to which the lipophilic molecules eventually are bound. These receptors have, in addition to a hormone-binding domain, a DNA-binding domain and a transcription activating domain. Having entered the nucleus, the receptors are called **nuclear receptors**, form dimers and act as transcription factors.

Remarkably, in spite of their quite different chemical structure and their different operational range, all signalling molecules mentioned and shown in Fig. 11.11, are received by the target cells in a similar way. The intracellular receptors belong to the same protein family and turn on physiological responses by similar basic mechanisms. The signalling system will be discussed using retinoids as an example.

Derived from carotene (also carotin), retinoids are present in three oxidation states, as retinol (vitamin A), as retinal (vitamin A aldehyde) and **retinoic acid** (synonym **vitamin-A acid**). Retinoic acid exists in three conformations: as *all-trans*-retinoic acid, or its metabolic derivatives 9-cis-retinoic acid or 3,4-didehydro-retinoic acid. All three variants fulfil a manifold of functions as signalling molecules in embryonic development (Chap. 10).

- Step 1: Being a lipophilic substance retinoic acid invades the cell membrane, is collected by the shuttle protein Cellular RA-Binding Protein **CRABP** and carried into the cytoplasm. In the cytoplasmic compartment it can function in metabolic processes, for example as cofactor in glycosylation of proteins within the endoplasmic reticulum.
- Step 2: RA is bound to cytoplasmic receptors which also act as potential transcription factors. They exist in two similar forms (1)

RAR (Retinoic Acid Receptor) und (2) **RXR** (Retinoid X Receptor). Both forms have a ligand-binding domain, two zinc finger domains which stabilise the three-dimensional structure, and a DNA-binding domain.

When the RAR receptors are loaded with *all-trans*-retinoic acid and the RXR receptors with, e.g., 9-cis-retinoic acid, both are taken up by the nucleus and attach with their DNA-binding domains to distinct control sequences **RARE** (RA-Response Elements) in the promotor region of target genes. As a rule two different receptors join forming a heterodimer RAR/RXR. Further factors join the complex and supplement or modify the gene regulatory function.

In a similar way **steroid hormones** and **thyroxin** fulfil their tasks. These molecules also penetrate the cell membrane and eventually are found in the nucleus bound to dimeric receptors. Surprisingly, the **receptors for retinoic acid, steroid and thyroxine are members of the same family of zinc finger proteins**. There even exist heterodimers of zinc finger proteins, one of which is loaded with RA, the other with a non-retinoid. For example, a retinoid laden RAR/RA may associate with a thyroxin-laden RXE/TR forming a heterodimer RAR/RA + RXR/TR and together control a distinct gene.

Yet the effects at the organismal level are quite different, since the biological context in which these signalling substances are used is quite different.

- **Retinoic acid** influences pattern formation and differentiation in the extremities (see Figs. 10.11 and 10.13), in the central nervous system, in the eye, and in the differentiation of skin organs (scales, hairs, feathers). In the avian leg, for example, upon treatment of embryos with overdoses of RA feathers are formed instead of scales. In addition RA can have strong teratogenic (deformity causing) effects.
- **Thyroxin** is released from the thyroid gland to boost general energy metabolism. In

amphibians, the hormone also controls metamorphosis (Chap. 20), and in birds the seasonal moult.

- **Steroids** have a leading part in the sexual development of vertebrates (Chap. 7) and in the metamorphosis of insects (Chap. 20).

11.4.2 Not All the Effects of the Lipophilic Retinoic Acid and the Steroid Hormones Are Due to Gene Regulation

Besides gene activation, steroid hormones may also act via membrane-associated receptors and fast signal-transducing systems. Some effects take place so rapidly that no time is left for prior transcription and translation which take several minutes to hours. Today (in 2014) hardly a steroid hormone can be quoted from which, besides slow gene regulatory functions, rapid responses are not known.

Summary

In this chapter the various modes are reviewed by which cells communicate and interact. Focus is laid on soluble protein signalling molecules which are produced by emitter cells and exported, examples for membrane-associated and lipophilic molecules are also presented. According to their function soluble factors can be classified as

- determination factors including morphogens and inducing factors,
- differentiation factors, and
- growth factors.

  According to signalling range and addressed targets one distinguishes

- autocrine,
- paracrine and
- endocrine factors.

  Protein families comprising particularly important signalling molecules are

- The **WNT** (pronounced wint) superfamily; these are proteins homologous to the *Drosophila* WINGLESS proteins with many members (in humans 19). So called canonical WNT systems are in many animal groups (cnidarians, planarians, annelids, sea urchins, and vertebrates) of significance in the establishment of the body axis that runs through the blastopore. WNTs have many further functions including the decision whether stem cells remain stem cells or proceed to differentiation. Other, non-canonical WNT signalling molecules act in the plane of epithelia conferring a common orientation on structures such as scales, hairs, or feathers (**planar polarity**).
- The **TGF-β** (Transforming Growth Factor beta) family which includes important morphogens (BMP, DECAPENTA-PLEGIC, NODAL) which pattern the basic body plan, and also the Anti-Müllerian Duct Factor (AMDF).
- The **FGF** (Fibroblast Growth Factor) family, with 22 members in humans, which perform a large array of functions in embryonic development, including patterning of the CNS and organogenesis
- The **EGF** (Epidermal Growth Factor) family which promotes formation and expansion of epithelia.
- The **Ephrin** family, all of them membrane-associated proteins involved in attraction and repulsion which aid in guiding growing nerve fibres and blood vessels to their targets.
- The **insulin** family with the IGFs (Insulin-like Growth Factors) which promote local growth processes.
- The **NOTCH/DELTA** family, integral membrane proteins serving cell adhesion and mutual agreements between neighbouring cells, for example in the separation of neuroblasts from the epidermis, or in other alternative decisions.
- The **HEDGEHOG**-(HH) family which is involved in many processes of pattern formation and cell migration.

The functions as growth factor or classic hormone are fluid as discussed using the examples of the **IGFs** and **retinoic acid RA**. Classic hormones come into play when the hormonal glands and the blood circulation are developed. Neurosecretory cells mediate between the hormone and the nervous system.

In the embryonic development many of the above listed factors become effective in the form of gradients, the **quality (type) of local differentiation being a function of the local concentration** of the factor. **Gradients** can form in various ways, for example as a result of local release of signalling factors into the surrounding extracellular space and their diffusion-driven spread, frequently supported by stepwise binding of the factors to extracellular proteins, or their attachment to the extracellular matrix **ECM** (e.g. FGF, WNT, HH), or by binding to receptors of the target cells and their subsequent **internalization**. Frequently signalling molecules, together with their receptors, are internalized via endocytosis (e.g. WNT, FGF) and can arrive in the nucleus (EGFR). The molecular outfit and physical property (hydrophilicity versus lipophilicity) of the surface of cells can direct the spread of extracellular signalling molecules in preferred directions. Particular significance belongs to **proteoglycans of the ECM** which guide the flow of signalling molecules or store growth factors until they are proteolytically released. There are mechanisms to **guide signalling molecules directly to target cells**. With regard to low molecular weight substances this can happen through **gap junctions**, with proteins through **transcytosis** (e.g. with BMP proteins). Several cells export signalling proteins such as WNT and SHH by means of microvesicles called **exosomes**, or transmit them through filopodia-like extensions that initially have been called **tunneling nanotubes TNT**, or cytonemes. By means of specialised **filopodia** cells can **actively transmit membrane-associated signalling molecules to distant target cell or collect different messages**.

Bound to the extracellular domain of integral membrane receptors signalling molecules activate **signal-transducing systems** which forward the message into interior compartments of the target cells. Irrespective of the large diversity of signalling molecules and fields of application, some basic mechanisms of signal transduction prevail.

Protein factors and all polar (hydrophilic) signal substances are collected by membran-anchored receptors.

Seven-pass membrane proteins. A class of such receptors possess seven transmembrane domains. Frequently the intracellular part of these receptors is linked to **trimeric G-proteins**; these mediate the enzymatic production of a large number of **second messengers** (cAMP in the cAMP-PKA system, diacylglycerol and IP$_3$ in the PI-PKC system).

Single transmembrane receptors with kinase function in their intracellular domain are the **Ser/Thr-kinase receptors**, responsible for signalling using for example TGF- β, BMP, DPP and Nodal. Another class with kinase function are the **receptor tyrosine kinases**, responsive to FGF and IGF. Upon binding their ligand these receptors aggregate, forming dimers or (in the case of ephrins) multimers and activate pathways which eventually end in the recruitment and activation of transcription factors in the nucleus. Thus, **FGF**, **EGF**, **ephrins** and growth hormones use their intracellular domain to switch on a tyrosine kinase function, which may initiate activation of transcription factors to control downstream genes.

In the **Notch/Delta** system the intracellular part of the Notch receptor is cleaved off and directly becomes a (co)transcription factor.

Apolar (lipophilic) signalling molecules such as **retinoic acid**, **steroid hormones**, and **thyroxin** are received by intracellular receptors. Loaded with a ligand, the receptors enter the nucleus becoming **nuclear receptors** with gene regulatory function. All these receptors belong to the same protein family and possess a hormone- binding domain, a zinc finger domain and a DNA-binding domain. The latter binds to the control region of target genes forming dimers, frequently heterodimers so that, for example, retinoic acid and thyroxine synergistically control the function of certain genes.

Development and Genes

12

Signals controlling development eventually turn on genes – or they turn genes off. In this chapter we are dealing with genes directing global events, and with genes paving the path to differentiation and executing the differentiation. This is an issue so extensive and diverse that any textbook can only provide a basic introduction.

Box 12.1: On the Nomenclature Regarding Genes and Their Products

We first briefly recapitulate some rules for correct notation outlined in the appendix of Chap. 2.

The **designation of genes**, and of the mRNA derived from them, is written in *italics*. In the history of *Drosophila* genetics capital letters were used to denominate dominant genes (e.g. *Antennapedia*), while recessive genes were written with small letters (e.g. *bicoid*) provided statements about the relative expressivity of genes were possible. In addition, it has become common usage, but not imperative rule, to contract the designation of genes to a few – most frequently three – letters (*bicoid* to *bcd*).

Guidelines for human gene nomenclature were proposed by Wain et al. (2002).

The proteins derived from these genes are written in standard letters, increasingly frequently in capital letters (e.g. BICOID), as it is done in this textbook. This allows, for instance, to distinguish right away the protein WINGLESS, as coded from the wingless allele, from a wingless fly.

Shorthand symbols indicating the genetic constitution of an organism used in genetics are **plus** (+) for functional wild type alleles, **minus** (−) for defect alleles (*loss-of-function* **alleles**). Using these symbols in diploid organisms the following constitutions are possible: +/+ homozygous for the wild-type allele; −/− homozygous for the defect allele, and +/− heterozygous combining one wild-type with one defect allele. For instance the designation *bc* −/+ or *bcd*/+ means, that a defect *bcd* allele and a functional wild-type *bcd* allele are present. The specimens with the constitution −/− are frequently called **k.o. (knock out) mutants**. Dominant mutations may generate *gain-of-function* **alleles**.

Please note: **The name of genes is usually derived from the phenotype (appearance) of mutants**. Therefore, the designation can be very misleading if one would deduce the function of the wild type allele from the mutant-derived name. The wild-type gene *dorsal* of *Drosophila*, for instance, and the wild-type **DORSAL** protein coded by the *dorsal* gene are not required for the development of dorsal but of **ventral** structures. In the (*dorsal*-/-) mutant no functional protein is produced and **dorsal instead of ventral structures** are made, for whatever reasons. ▶

W.A. Mueller et al., *Development and Reproduction in Humans and Animal Model Species*, DOI 10.1007/978-3-662-43784-1_12, © Springer-Verlag Berlin Heidelberg 2015

Box 12.1 (continued)

Many names of genes refer to the phenotype without giving hints to the particular function of the protein. For example mutations in the *Antennapedia* complex may result in flies bearing legs instead of antennae.

The function of the ANTENNAPEDIA protein, however, was mysterious for a long time, and eventually only revealed by arduous research work.

Box 12.2: Genetic and Molecular Methods in Developmental Biology

Detecting Genes Relevant for Development

We speak of relevant genes if their products essentially contribute to the control of development but also of genes the expression of which denotes a developmental path and the products of which are useful markers of differentiation paths leading to distinct cell types. In case of muscle tissue, for example, the members of the *myoD/myogenin* family are regulatory genes, while the appearance of muscle-specific actin and myosin filaments indicates terminal differentiation in progress.

Genes relevant to development can be identified by two broad approaches: forward genetics and reverse genetics. Forward genetics traces from the phenotype to the genotype. The mutant phenotype may be produced as a result of natural variation or breeding, or as a result of mutagenesis. The mutants are then analysed by positional cloning to find the gene responsible for the phenotype. Following this the wild-type gene is searched for, including its control region (enhancer, promoter).

Reverse genetics proceeds in the opposite direction to forward genetics; we start with a known sequence and trace the phenotype. Due to genome sequencing, we now have sequences for many genes whose function we do not know. In reverse genetics the function of a specific known gene is manipulated, for example by gene silencing, knockout or overexpression, and the resulting phenotype examined to find the function of the gene.

How Mutants Are Generated and Analysed

Profound analysis of many mutants is only realisable with reasonable expenditure if the animals propagate sexually, have a short life-cycle duration (preferentially less than 4 months), can be kept in large numbers without demanding much space and physical and human resources. Also the ability to make "knockouts" of selected genes has become a critically important technique. The procedures have been most frequently and successfully applied in the fly **Drosophila** and the small nematode **Caenorhabditis elegans** which provides the convenient possibility to store defined genetic stocks in the freezer. The zebrafish **Danio rerio** and the **medaka fish Orycias latipes** are easily crossed, but need expensive aquarium facilities for large-scale genetics. With considerably higher expenditures genetics is done with **mice**, proxy of humans. The technician time and space required for profound genetic analyses, removal of embryos from the maternal animal and thorough search for altered traits indicating a developmentally relevant mutation are very high.

Chemical Mutagenesis. This is the classical procedure. As a rule, mutagenic agents are administered to male individuals only. The agents are mixed with the food or given into the drinking water; fishes have to swim in solutions of such agents. Progeny is carefully examined for abnormal phenotypes, and not only the F1-generation. Since in ▶

Box 12.2 (continued)

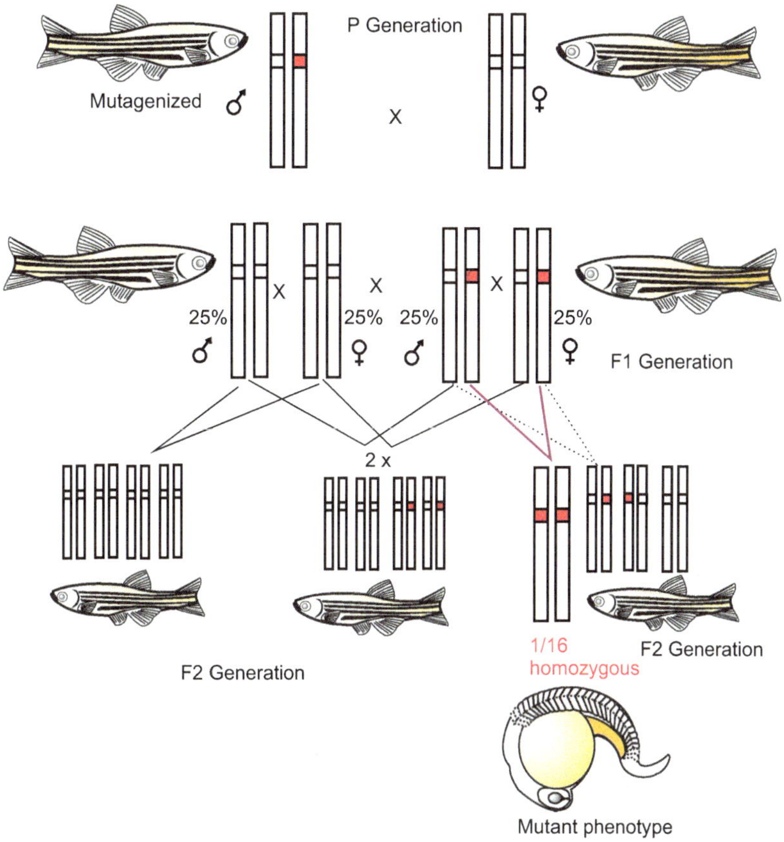

Fig. 12.1 Cross-breeding to arrive at homozygosity of an induced, recessive mutation. Recessive mutations become phenotypically obvious in the homozygous condition only. Statistically this condition is found in 1/16 of the individuals of the F2 generation

all likelihood a mutation affects only one allele of a pair of homologous alleles, as a rule mutations come to light only in the F2 when the mutated allele is present in some individuals in homozygous condition (Fig. 12.1). Today, targeted genetic finger-printing can detect mutations already in het-erozygous F1, and facilitates selection of individuals for continuing cross-breeding.

The particular skill a researcher needs to develop is a sense for judging which mutation is of specific significance. A defect-causing mutation in a general household gene will evoke considerable, often lethal disturbances, but these are not development-specific. Indic-ative for relevance are, for example, localized defects or cessation of development at a cer-tain stage. A strong indication can be deduced from the genetics. If the Mendelian rule of reciprocity is infringed and the genetic consti-tution of the mother decisive (but not of the father or the child itself) strong interest is justified. Then, classic cross-breeding must aid in narrowing down the chromosomal locus affected by the mutation. ▶

Box 12.2 (continued)

From Mutations to Genes. Searching for the actually affected gene is tedious, time-consuming, and its definite identification difficult. With regard to successful strategies for searching and the large array of available genetic tools we must refer to textbooks of molecular genetics. Suggested key words are: linkage group, chromosomal walk, positional cloning; correlation of a trait with **AFLP** (Amplified Fragment Length Polymorphism).

The gene searched for is definitively identified only if a functional analysis succeeds.

- The gene is expressed at the right time and the correct place.
- The defect is curable through injection of wild-type mRNA or injection of protein produced (in the lab) using the wild-type gene (**rescue experiment**).
- Anti-sense RNA directed against the wild-type gene or RNA interference (RNAi, see Fig. 12.6) simulates the mutant phenotype (phenocopy) in wild-type embryos.
- Gene products (mRNA or the protein derived from it) generate additional structures if injected at wrong (ectopic) locations, or they alter developmental fates of body regions dramatically, for example, by inducing head formation at the posterior body pole (see *Drosophila*, Fig. 4.32).

P-element Mutagenesis and Transgenesis. In *Drosophila melanogaster* (and similarly in *Caenorhabditis elegans*) a different kind of mutagenesis facilitates search for the mutated gene. This procedure is known as **P-element transformation**. A P-element is a virus-like **transposon** that is naturally present in wild-type *Drosophila melanogaster*. The P transposon is used to create genetically modified flies. P-elements occur in two forms: (1) in form of ring-like plasmids or (2) linearly inserted in the host's genome. The transposon

is flanked by terminal inverted repeats which are important for its insertion and excision. Insertion takes place by a "cut and paste" mechanism, and can occur at many sites in the genome causing disruption of genes. For movement of the element a **transposase** is required and a *transposase* gene is naturally part of the P-element. Complete transposons with integrated *transposase* are life-threatening for a *Drosophila* population, because in the permanent presence of *transposase* the P-elements can move beyond the fly's control and destroy genes (Many natural populations died out, but some survived because they developed some kind of inhibition to uncontrolled movements or had transposons lacking the *transposase* gene).

For mutagenesis in the lab two types of plasmids were isolated from *Drosophila* strains and engineered according to the researchers desire: plasmid type 1 lacking the transposase gene, and plasmid type 2 comprising *transposase* but lacking the transposon-type flanking sequences and therefore unsuitable to be inserted into chromosomes. Based on plasmid type 1 constructs were made containing, between the flanking sequences, a *Drosophila* reporter gene, often a red-eye marker (the product of the *white* gene) and, optionally, any other transgene chosen by the researcher, for example a selectable marker gene, often some kind of antibiotic resistance. Both types of plasmids, the plasmid type 1 lacking *transposase* and the plasmid type 2 containing *transposase*, are injected at the posterior pole of freshly laid eggs before pole cells form (Fig. 12.2). A pole cell (primordial germ cell) that accidentally takes up a transposon type 2 is qualified to produce transposase enzyme. If the cells in addition has taken up a transposon type 1 this transposon will be inserted into the genome, mediated by the transposase, and often disrupt a protein-coding gene. Such fly strains can be used ▶

Box 12.2 (continued)

Fig. 12.2 P-element-induced mutagenesis in *Drosophila*. (**a**) When injected into the cytoplasm of a still uncleaved egg, P-element-type plasmids can be taken up by the forming pole cells, can integrate into chromosomes aided by transposase, insert into genes and cause mutations by disrupting the message. Offspring that inherited a plasmid can be recognized if the plasmid brings along a marker gene, for example the bacterial *LacZ* gene which codes for β-galactosidase, an enzyme that can produce a blue dye in an established lab protocol

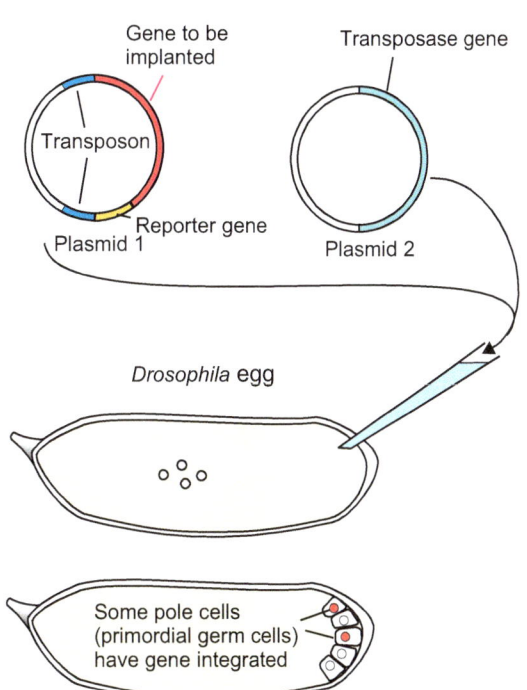

to generate mutated offspring only for a limited time; they will not survive in the long term if the transposase is not outcrossed. (Transposase-bearing individuals can cause genetic instability of a strain). Once the gene of interest, for instance a transgene, has been inserted and the *transposase* gene removed, the transgene-containing plasmid is no longer mobile and the exchanged mutated gene and/or the introduced transgene is a permanent part of the genetic heritage of the respective strain.

Following mutagenesis and selection of mutants of interest from progeny DNA is extracted and cut into pieces by means of restriction enzymes to establish a genomic library (see following section). This is screened for sequences containing sequences that are partially indicative of the fly, partially indicative for the inserted plasmid, containing a fly-specific sequence linked with a part of the transposon-type flanking sequence or a part of the reporter gene. Such sequences, containing both host-specific and plasmid-specific sequences, could only derive from those genes into which the transposon was inserted. These partial sequences of the affected host gene can, after their cloning and labelling, be used as probes to search for overlapping sequences in the genomic library. With some diligence and luck a whole set of overlapping sequences can be collected until the whole gene including its promoter and enhancers is complete. Available databases of the complete genome facilitates the search considerably. ▶

Box 12.2 (continued)

How Genomic or EST Databanks Help to Quickly Identify and Isolate a Gene or its Relatives

Genomic Banks. To establish a genomic bank the entire DNA of an organism is extracted, fragmented in pieces by a collection of restriction enzymes, or randomly by applying shearing forces, splitting long DNA helices into small workable portions. The fragments are ligated into a vector, typically a bacteriophage. Vectors have multiple restriction enzyme consensus sites to either side of the insert so that the inserted DNA fragment can later be cut out. Using these phage vectors the DNA fragments are transferred into bacteria and 'cloned' (=multiplied, amplified) by breeding the bacteria. Samples of the bacteria are stored frozen: a '**genomic library**' or '**genomic bank**' is established.

Sequencing and Databases. For sequencing, the fragments are cut out from the bacterial clones by restriction enzymes, and subjected to any of the available (automatized) sequencing procedure.

To establish a complete library and database, many thousands of fragments must be sequenced. Bioinformatics provide tools to reconstruct from this chaotic collection of sequences the entire genome by assembling overlapping sequences.

More and more methods are available or in development to quickly establish unknown DNA sequences, that is to determine the precise order of nucleotides within a stretch of DNA, with high speed using fluorescence-based sequencing methods and automatic machines. Selected key words for the reader's own enquiries : *Second and third generation sequencing or high-throughput sequencing, Single-molecule real-time sequencing, Sequencing by ligation (SOLiD), Cyclodextrin-hemolysin nanopore technology, Nanoball sequencing, Sequencing with mass spectrometry, RNA polymerase-based sequencing (RNAP), RNA sequencing (RNA seq); Microscopy-based techniques.*

Using appropriate methods the search for putative genes is simplified and speeded up, in particular the search for genes with high sequence similarity such as members of gene families or other homologous genes, but also detecting novel genes is facilitated.

However since large genomes including the human genome contain giant sections of non-protein-coding sequences, effort and expenditure are large, but the yield is low, if one is looking for a distinct protein-coding gene (without the help of an attached marker as in case of P element transformation). By contrast, under aspects of gene regulation non-protein-coding sequences are of particular interest.

The feasibility of large-scale genome sequencing has led to an increasing number sequenced entire genomes. **Yet, genome sequencing as such does not reveal much about the function of a gene**. To unravel the function of a particular gene one has to mutate it, or to manipulate it in some other way, as discussed below. Important hints as to the function can be deduced from EST banks.

EST Projects and cDNA Libraries. EST is the abbreviated term of *Expressed Sequence Tag*. In EST projects the sum of all expressed genes, the **transcriptome**, of an entire organism, of a developmental stage, or of an organ or a tissue, is analysed. The starting material is extracted mRNA; the mRNA is transcribed into cDNA using an enzyme called reverse transcriptase, and often subjected to a first amplification step by PCR (Polymerase Chain Reaction). The resulting cDNA molecules are engineered so that they have restriction enzyme recognition sites at each end of the molecule, which allows them to be inserted into a vector and to be excised after cloning in bacteria. By means of ▶

Box 12.2 (continued)

vectors the cDNA specimens are transferred into bacteria and cloned (amplified) to establish a cDNA bank (library). In such libraries only sequences are present derived from mRNA and, thus, mRNA that was with high probability transcribed from genes relevant for the developmental stage and biological source under investigation. Supplemented by parallel *in-situ* hybridisation studies EST projects enable the survey of gene expression patterns in the course of development and cell differentiation.

How Reverse Genetics Allow Tracing Genes

There are, as mentioned above, only a few organisms which allow the identification of a large array of developmentally-relevant genes by mutagenesis and subsequent crossings, and from which many genes have been cloned and sequenced.

Of course, researchers working with organisms less or not suited for genetics, for example *Xenopus laevis* or *Hydra*, seek possibilities to trace such genes and to study their expression also in their organism of choice. Such possibilities are provided by **reverse genetics**. If in the genome of the organism of choice, let's say in *Xenopus*, a gene is present displaying sequence similarity to a known gene of, for example, *Drosophila* (100–70 % sequence identity), reverse genetics helps. One chooses a cloned piece of the *Drosophila* gene, for example the conserved homeobox of the *Antennapedia* gene, flags it by coupling an easily identifiable marker, and uses it as a heterologous (derived from another species) **probe** to screen a genomic or cDNA bank (library) made from DNA or mRNA extracted from *Xenopus* in a search for similar, autologous sequences. Today, this is done on a DNA chip (see below). If such a sequence is found it is amplified by molecular cloning or by means of Polymerase Chain Reaction PCR-based amplification, and sequenced.

However, reverse genetics cannot hunt out completely unknown sequences such as those endemic to the species in question. Unknown genes of relevance can be tracked down with the following procedure.

From mRNA Via cDNA to Unknown, Novel Genes and Study of Their Expression

Search for Differentially Expressed Genes by Means of cDNA Subtraction

A standard technique to identify and isolate differentially expressed genes is **cDNA subtraction** that enables enrichment of cell- or stage-specifically expressed mRNA. First mRNA is extracted from two different cell populations (e.g. early versus late gastrula). During synthesis of cDNA using reverse transcriptase the arising cDNA derived from a reference source (e.g. late gastrula) can be labelled with biotin. If one now adds a small amount of cDNA to be studied (e.g. derived from the early gastrula) and conditions are chosen (heating) to transiently separate DNA into single strands, all cDNA molecules present in both sources will find a partner and hybridise, when conditions (cooling) allow, forming double strands. Since they are marked with biotin they bind to streptavidin-coated beads (streptavidin binds biotin) and can be removed. As only unlabelled cDNA, specific for the source of interest (here early gastrula), is left unbound and survives, one speaks of cDNA subtraction. The remaining, specific cDNA molecules are cloned into a vector and eventually characterized using the tools of molecular biology.

Using the subtraction technique, however, all gene expression that takes place in overlapping stages or simultaneously in the two compared sources remains hidden; in our example mRNA present in early as well as in late gastrula, gets lost. Several configuration levels have been proposed ▶

Box 12.2 (continued)

promising higher efficiency, for example differential display (DDRT-PCR. However, these methods have largely been replaced by chip-based methods including DNA-microarrays and more recently by RNA-Seq combined with software to identify differentially expressed genes (for example DESeq).

DNA-Microarrays (DNA-Chips)

DNA microarrays, also known as DNA chips, are applied to comparative genome analysis, diagnosis of genetically-caused diseases and studies of differential gene expression. Further, DNA microarrays can be used, for example, to measure changes in gene expression levels, to detect single nucleotide polymorphisms (SNPs), or to genotype organisms in biodiversity studies.

In the sense of molecular biology a microarray is a collection of microscopic DNA spots firmly attached to a solid surface in a well-defined pattern on a grid (*spotted microarrays*, Figs. 12.3 and 12.4). Each DNA spot contains picomoles (10^{-12} moles) of a specific DNA sequence, known as a **probe** or reporter or 'oligo'. The spotted and immobilized sequences are characteristic single-stranded sections of genes or other DNA elements complementary to 'target' sequences to be detected and identified. In standard microarrays, the probes are synthesized and then attached via surface engineering to a solid surface by a covalent bond to a chemical matrix which covers a glass or a silicon chip. Alternatively, microarrays can be constructed by the direct synthesis of oligonucleotide probes on solid surfaces. Each DNA probe is plotted twice.

The **target sequences** to be analysed attach to the probes on the chip by hybridisation, and if they match perfectly they remain attached even under high-stringency conditions. Since an array can contain tens of thousands of probes, a microarray experiment can accomplish many genetic tests in parallel. The target

nucleic acids are usually linked to fluorescent dyes. Normally there are two samples which are compared, one tagged with red fluorescent dye and the other with green. The probe-target hybridisation pairs are detected and quantified by measuring the wavelength ratios, typically the red-to-green fluorescence ratio, and strength of fluorescence emitted by the spots.

Which probes are spotted onto the chips, and how many, depends on the question to be answered, on exploratory work previously done, and on previous achievements. Probes could be, for example

- original DNA sequences amplified by polymerase chain reaction PCR,
- synthetic DNA sequences (*oligonucleotide microarrays*) or
- cDNA samples of differentially expressed genes obtained by cDNA subtraction.

If one wants, for example, to know whether in a suspicious-looking tissue mRNA of an oncogene (cancer gene) is present, then the following procedure can be used: Probes of synthetic oligonucleotides are prepared, each probe being characteristic for a distinct oncogene, and immobilized on pre-defined positions of the grid. Then mRNA is extracted from the suspicious tissue, transcribed into cDNA and linked with a fluorescence dye. The thus marked collection of cDNA is hybridized with the oligo spots on the array. Because you know which spot on the grid represents which oncogene, you can deduce from the pattern of fluorescing spots which oncogenes had been active in the tissue sample. Furthermore, because each probe was plotted twice on the array, digitalized comparisons assessing the signal strengths allows evaluating which oncogene was expressed particularly strongly.

Possible applications of microarrays are manifold. Any conceivable application can be realized, whereby always sense-antisense partners must be chosen for hybridisation and, as a rule, mRNA is used in the form of its cDNA: ▶

Box 12.2 (continued)

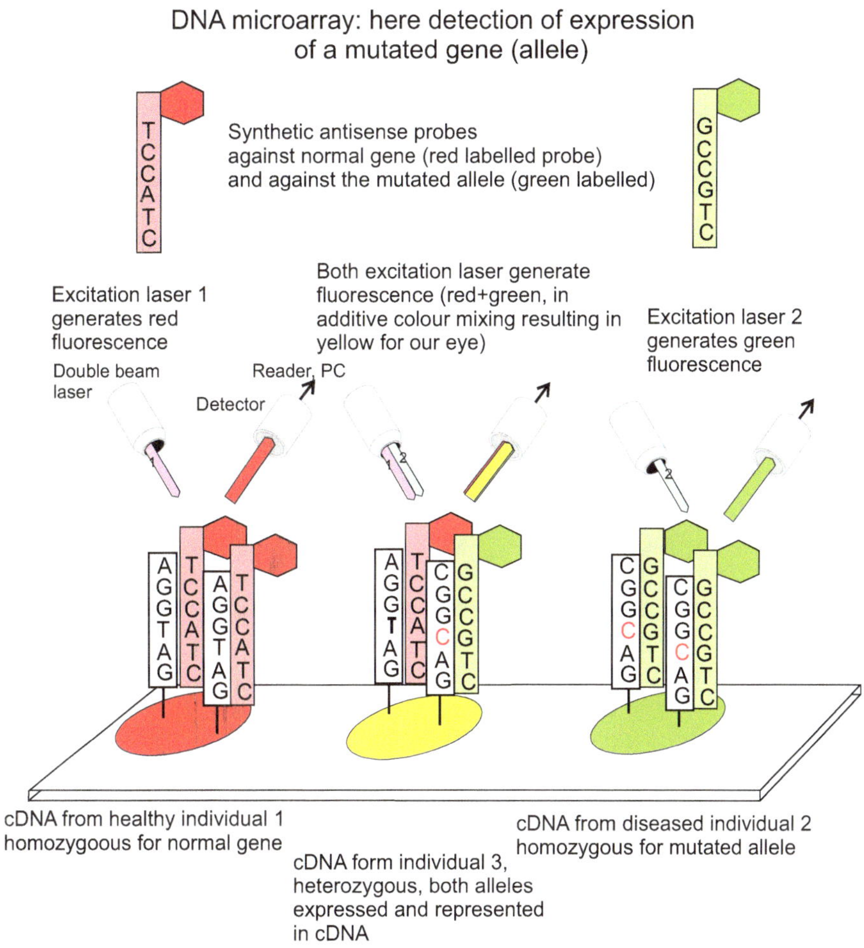

DNA microarray: here detection of expression of a mutated gene (allele)

Synthetic antisense probes
against normal gene (red labelled probe)
and against the mutated allele (green labelled)

Excitation laser 1
generates red
fluorescence

Both excitation laser generate
fluorescence (red+green, in
additive colour mixing resulting in
yellow for our eye)

Excitation laser 2
generates green
fluorescence

Double beam
laser

Reader, PC

Detector

cDNA from healthy individual 1
homozygoous for normal gene

cDNA form individual 3,
heterozygous, both alleles
expressed and represented
in cDNA

cDNA from diseased individual 2
homozygous for mutated allele

Fig. 12.3 DNA microarray (DNA chip) I. On a support, typically on a small silicon plate (chip), small amounts (spots) of DNA are loaded as reference probes, in the present case single-strand cDNA corresponding to selected genes and/or mutated alleles of these genes. The reference cDNA probes are fixed on the spots. In the second step single-stranded cDNA samples from sources to be assayed are labelled with fluorescing dyes and hybridized with the reference cDNA. In the example shown it is checked whether the examined sources contain mutated alleles or sequences identical to the reference probes. From the colour of the reflected laser beam it can be deduced whether a mutated allele is present, and whether it is present in the assayed source in homozygous or heterozygous condition

- DNA versus DNA from a sample to be diagnosed versus reference DNA; by choosing appropriate reference DNAs it can be explored whether a mutated allele is present in the sample and, if so, whether it is present in homozygous or heterozygous condition.

- DNA versus DNA from two different individuals or species to see whether they share genes with high sequence similarity (putative homologue genes).
- DNA versus mRNA (transcribed into cDNA) from different sources. The DNA probes represent a collection of genes ▶

Box 12.2 (continued)

the expression of which is investigated, for example, as outlined above, a collection of oncogenes; mRNA is extracted from putative cancer tissue and normal tissue, transcribed into cDNA and labelled. By analysing the pattern of signals it can be inferred which gene had been active in the samples. (This case is illustrated in Fig. 12.3.)

- DNA versus mRNA (transcribed into cDNA) from different developmental stages. The DNA probes represent developmentally relevant genes the expression of which is investigated; mRNA is extracted from different developmental stages, let's say from a gastrula and from a blastula. By comparing the strength of the hybridisation signals hints can be deduced as to which gene activities are down-regulated or up-regulated in the course of a developmental process (This case is illustrated in Fig. 12.4).

- mRNA versus mRNA from different sources, for instance as exploratory work for cDNA subtraction but also to identify genes up- or down-regulated in diseased vs normal states, in drug-treated vs untreated cells or animals.

However, microarrays do not provide information on where exactly in the embryo a gene is expressed. For that mRNA would have to be extracted from single, exactly located cells. A very useful, supplementing technique to localize expression sites is the *in-situ* hybridisation technique described in the following section.

Screening by Whole Mount In-Situ Hybridisation Visualizes Cell Type and Stage-Specific Gene Expression in Whole Animals

A strategy aiming at a complete expression analysis has been called *in situ* **screening** or **expression screening**, and proved successful in *Xenopus*. From the tissue of interest, let's say

the upper blastopore lip of the early gastrula, mRNA is isolated and used to construct a plasmid-cDNA bank in which the orientation of the inserted and cloned genes within the plasmids is identical and known. The used plasmid vector possesses an inserted promoter at its $3'$-terminus that binds a T3 RNA polymerase, and a further promoter at its $5'$-terminus where a T7 RNA polymerase can dock. These promoters enable the *in-vitro* synthesis of labelled (for example digoxygenin- or biotin-labelled) *sense-* and *antisense*-RNA. (*Antisense*-RNA runs antiparallel to mRNA and by hybridisation both together can form a double strand). The labelled *antisense*-RNA is used as a probe in the procedure known as ***in-situ* hybridisation**, to demonstrate endogenous mRNAs that are expressed at various locations in the embryo, and at different times (Fig. 12.5).

Whenever possible *in-situ* hybridisation is performed in intact embryos or animals (**whole mount preparations**), so that the three-dimensional arrangement (pattern of expression) of the labelled cells is preserved. If the specimen is too large or not transparent it is dissected to prepare microscopic slices. From a series of two-dimensional pictures the three-dimensional arrangement of the structures must subsequently be reconstructed with the aid of a computer using an image processing program.

Messenger RNAs, that are differentially expressed (region-specific, stage-specific or cell type-specific), can be investigated further. Compared to the method of cDNA subtraction this strategy has the advantage that also mRNA from genes which are simultaneously expressed in several cell populations, territories and developmental stages is rendered visible and does not remain hidden. The targeted search for groups of genes that are expressed in identical or similar spatial and temporal patterns (**synexpression groups**) and therefore are likely functionally linked has provided instructive results. ▶

Box 12.2 (continued)

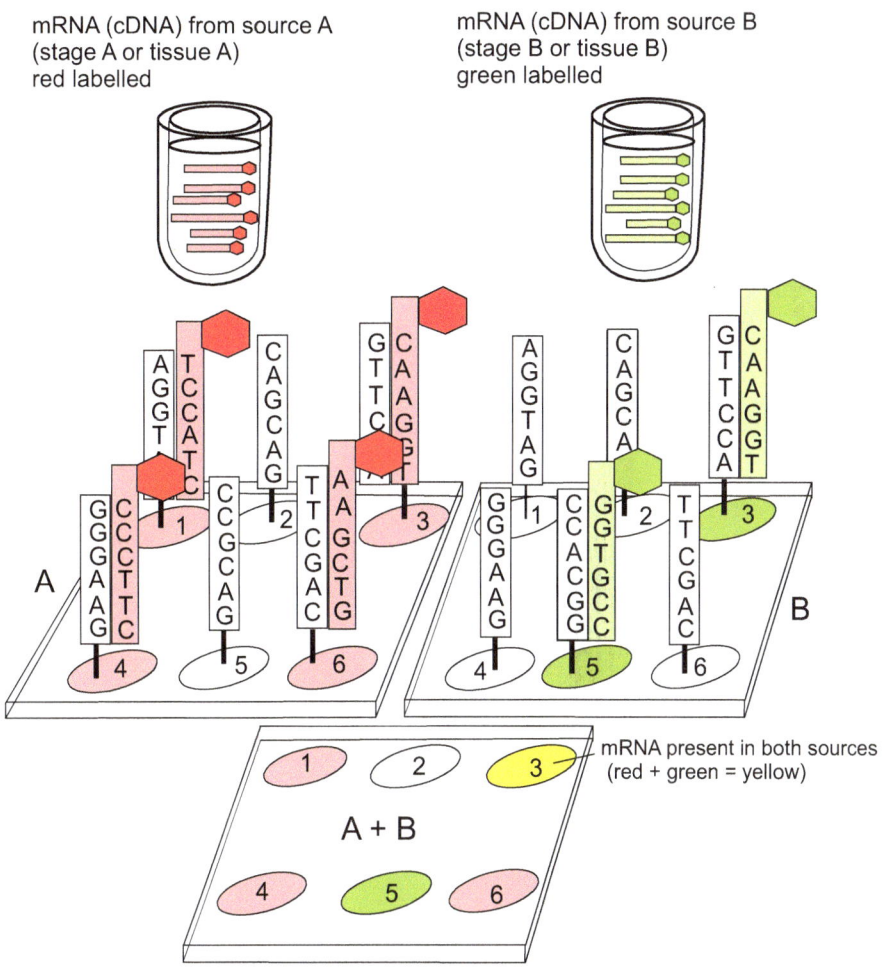

Spots with chromatic fluorescence: synthetic *antisense* oligonucleotides
generated against the gene the expression of which is examined.

White: synthetic *antisense*-oligonucleotides against other genes

Fig. 12.4 DNA microarray (DNA chip) II. Here sequences of DNA are fixed onto the support spots that are unique sequences characteristic for the selected genes. One single chip may be loaded with thousands of spots representing thousands of different genes. Messenger RNA extracted from various sources to be compared is transcribed in cDNA and thereby labelled with different fluorescing dyes. In the example shown cDNA from the first source is marked red, cDNA from the second source labelled green. Through hybridisation it can be examined whether and to what extent the two sources contain identical or different mRNAs

There are more means and procedures to study gene expression, among them a "shining example" in which particular reporter genes are involved. An example will be outlined below in the section "Demonstration of proteins without using genetic engineering". ▶

Box 12.2 (continued)

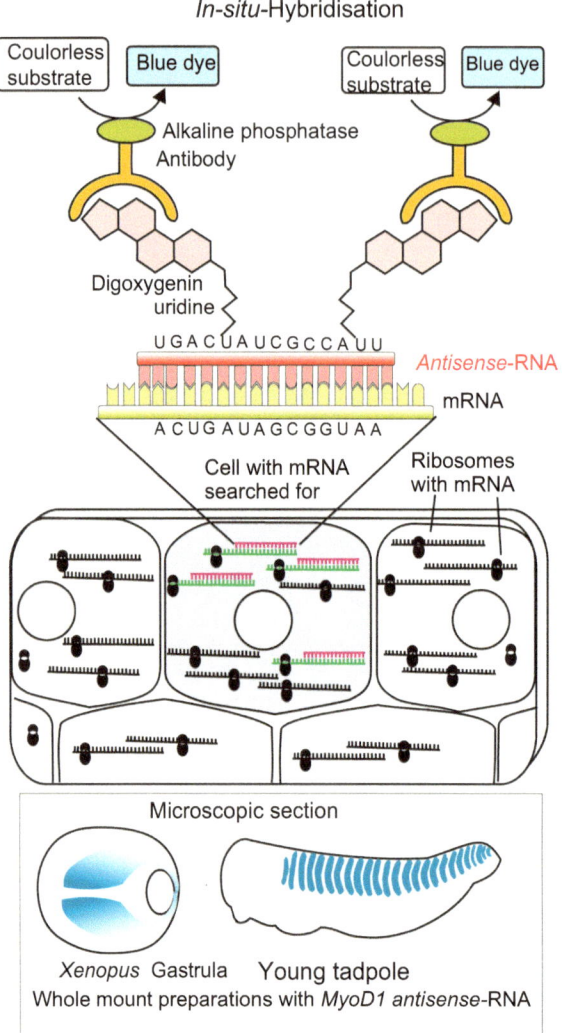

Fig. 12.5 *In-situ*-hybridisation. The technique is used to visualize the presence of a certain endogenous mRNA in (chemically) fixed samples of cells or whole objects (for example whole embryos or microscopic tissue slices). In the lab synthetic antisense-RNA is generated linked with—in the present case—digoxigenin-UTP. The antisense-RNA binds (hybridizes) to the sense-RNA (=mRNA), if present in the object. The hybrids are detected by means of anti-digoxigenin antibodies. These are coupled to an enzyme that catalyses the formation of a blue or fluorescing dye

Protein Microarrays and Immuno-Fluorescence

Microarrays and *in-situ* hybridisation refer to the level of DNA and RNA. Yet, since transcription and translation can considerably diverge in space and time, and proteins often do not remain at the place of their synthesis, the detection and localization of proteins is also required. For that purpose also several types of micro-arrays can be made. Partners for mutual binding can be antigen-antibody, enzyme-substrate, receptor-ligand, or protein-protein in general. ▶

Box 12.2 (continued)

Proteins extracted from a biological source are immobilized on spots of the chip and solutions of defined partner proteins such as a fluorescently labelled antibody, is added. Or conversely, a collection of defined test proteins, for example a collection of various selected antibodies, is fixed on the solid surface in an ordered arrangement, and protein sample extracted from the biological source in question added. One of the two binding partners, the antibodies for example, are coupled with fluorescent dyes, and the pattern of fluorescent spots, indicating binding, is analysed by means of a fluorescence reader.

The counterpart of detecting mRNA in microscopic slices by *in-situ* hybridisation is detecting the location of defined proteins in microscopic preparations by immunocytochemistry using antibodies (see Fig. 12.11).

Studies of Gene Function, Applications

How the Function of Genes Can Be Tested Without Conducting Genetics

Inhibition Through Antisense-Oligonucleotides. Translation of selected mRNA species in living cells can be inhibited through hybridisation (annealing) with short, complementary single-stranded DNA pieces (antisense-oligonucleotides) if they pair with sequences at, or close to, translation codons (ATG). The protein is no longer synthesized. Lack of the respective protein can cause a **phenocopy** of knock-out mutations, that is, evoke symptoms including traits in the appearance identical with those seen in mutants. Antisense RNAs can be of natural origin. Natural antisense transcripts (NATs) are a group of RNAs encoded within a cell that have transcript complementarity to other RNA transcripts. They are often found among the category of the small miRNAs or siRNAs. Antisense species complementary

to mRNA can suppress translation of the corresponding mRNAs or induce their degradation (Fig. 12.6); they are known to be transcribed for that purpose in development and cell differentiation.

For introduction in cells, synthetic, single-stranded antisense-oligonucleotides are enclosed in lipid micelles suited to merge with cell membranes and to release their content into the cells (lipofection = liposome transfection). Alternatively, the oligos are transferred from concentrated solutions into the host cells by making them transiently permeable by a pulse-type application of an externally applied electrical field (electroporation or electropermeabilisation). This procedure is highly efficient also for the introduction of foreign genes in tissue culture cells, especially mammalian cells. In studies on gene function with antisense oligonucleotides usually not natural oligos are used because these are quickly degraded; instead artificial, chemically modified nucleotides (phosphorothioate, **morpholinos**) are linked together yielding a class of rather stable antisense probes. These are more resistant to degradation by nucleases than are the natural transcripts. Morpholinos are widely used in developmental biology to knock down genes. They can be targeted against the ATG start site to block translation, or against a splice site to block mRNA splicing. They are injected into the embryo at the one-cell stage and persist for about 3–4 days. This allows time to see the effects of transient gene knockdown in models with rapid development, for example zebrafish.

Inhibition by RNA Interference (RNAi) Using siRNA. RNA interference (RNAi) is a method to induce destruction of specific mRNA molecules. RNAi enables the knockdown of gene function without targeted mutagenesis (Fig. 12.6). Historically, it turned out that the method, developed in the laboratory, imitates natural phenomena, especially the ▶

Box 12.2 (continued)

RNAi METHOD

I. Generation of dsRNA in the lab

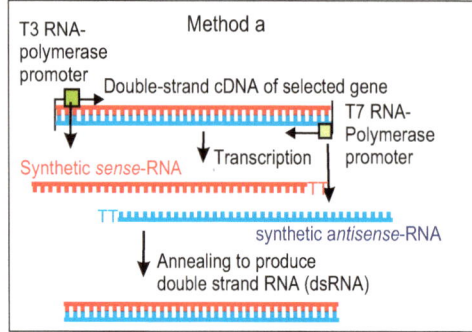

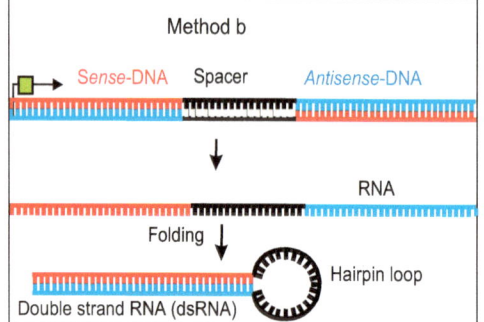

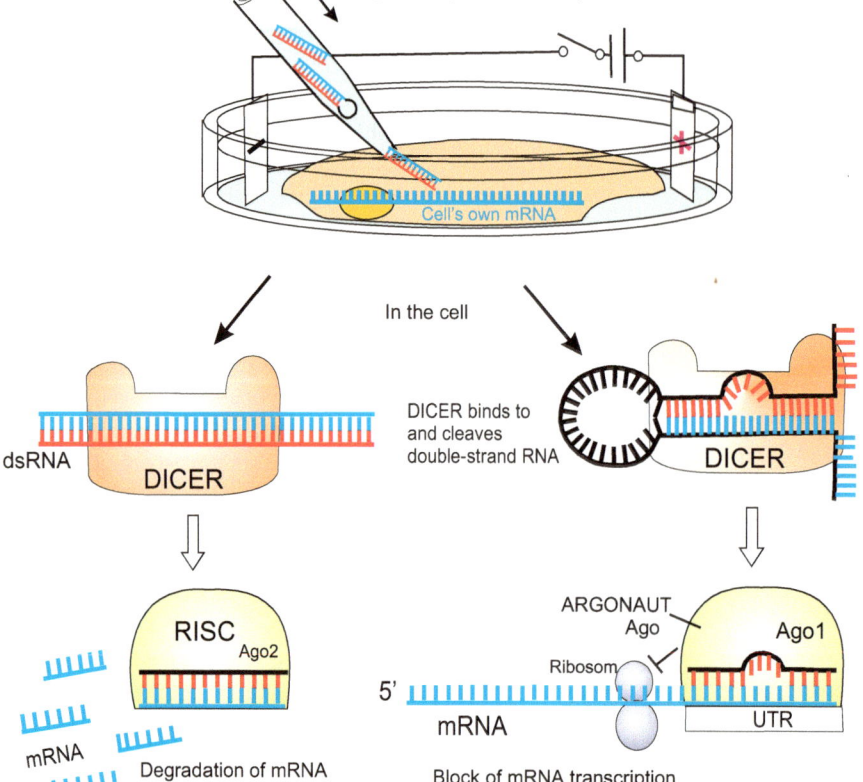

Fig. 12.6 RNAi method to deactivate a gene function on the level of its mRNA. In the laboratory a corresponding double-stranded RNA (dsRNA, with or without a hairpin loop) is synthesized and introduced into the target cell – as a rule an egg cell or a stem cell. In the cell the DICER complex cuts out small RNA pieces from the dsRNA. Transferred to an ARGONAUT protein Ago (see Fig. 12.21) these ▶

Box 12.2 (continued)

function of natural siRNA (see Fig. 12.21). The method gained currency when in 2006, A. Fire and C. C. Mello shared the Nobel Prize for their work on RNA interference in the nematode *C. elegans*, and protocols were developed to apply the method in other organisms as well.

In practise, a sequence derived from a gene (or a cDNA) is used as template to synthesize corresponding, complementary sense and antisense RNA strands, both folded together to form double-stranded dsRNA molecules. Introduced into a cell or organism by micro-injection or lipofection, dsRNA is recognized as exogenous genetic material and activates the RNAi pathway. An internal complex of enzymes known as **DICER** fragments the dsRNA into short pieces, each comprising 21–28 nucleotides, generally called siRNA (defined as *short inhibiting RNAs* or *small interfering RNAs*). Subsequently, these pieces are transferred to proteins of the ARGONAUTE (Ago) family by an endogenous mechanism. One of the two strands of the siRNA attaches to the UTR (*UnTranslated Region*) in the start region of an existing mRNA. If the siRNA and mRNA strands match, the Ago protein is complemented by further proteins and a complex called **RISC** (RNA-Induced Silencing Complex) is established. The RISC complex functions as a nuclease and the mRNA is cleaved resulting 21- to 25-mers. These are materials from which new and more siRNA pieces are made. The process proceeds avalanche-like, or in scientific terms, autocatalytically. The primary reaction gives rise to further siRNAs which attack further parts of the mRNA yielding the next set of siRNAs, and so forth, until all endogenous mRNA is

destroyed. No more protein can be made. The phenotype corresponds to the phenotype of a **knock out** mutation, like the phenotype evoked by antisense nucleotides, except that a DNA mutation is heritable. Modifications of the method allow, instead of down-regulation of gene functions, activation of translation beyond the normal level, a kind of gene **overexpression**.

RNAi effects last up to 5 days following introduction of dsRNA. In organisms that have a short life-cycle duration such as *Caenorhabditis elegans*, the effect can be "inherited" over 1–2 generations. Good results can also be achieved when, instead of a long dsRNA that reflects the length of the original mRNA, straightforwardly short siRNAs are introduced from the outset. Problematic with the method are difficulties to recognize and control non-specific inhibitory effects.

RNAi Technology Enables Thousands of Parallel and Simultaneous Experiments on a Chip and Genome-Wide Screening of Gene Functions. The just outlined method of RNA interference was soon modified yielding several variants, and is still being further developed. For example, thousands of tiny stem cell cultures are placed on a chip and in each single culture one distinct gene is knocked out: Subsequently, all cultures are exposed to a stimulus, for instance to a differentiation factor, that opens the developmental path to, e.g. nerve cells. It is examined whether a given gene is indispensable for the researched process, here for nerve cell development. Automated microscopy facilitates quick evaluation and identification of all those genes needed for the production of ▶

Fig. 12.6 (continued) pieces can bind to complementary sequences in the untranslated starting region UTR of the targeted mRNA and activate there a RISC complex. In cooperation with a set of additional proteins (not shown) the mRNA is cleaved into small

pieces which are not translated. A dsRNA with a hairpin loop which does not fully fit to the UTR of the mRNA leads to a block of its translation. The procedures make use of natural ways to stop translation. For details of these procedures see Fig. 12.21

Box 12.2 (continued)

a desired cell type. If all gene functions are eliminated, step by step, one speaks of a **genome-wide screening**, and genome-wide RNAi libraries.

Overexpression or Ectopic Expression After Micro-Injection of Synthetic, Capped mRNA. We ask, whether a gene, the expression of which is subjected to spatial and temporal variation, is of significance for the control of development, or whether its expression is a mere consequence and sign of progressing differentiation. Selected species of cDNAs derived from embryos, or synthetic mRNA, are injected into cells of a host embryo, for example into a cell (blastomere) of the 4- or 8-cell stage, to study its effect. For injection a site is selected where this mRNA normally does not occur (**ectopic expression**). Or by injection of additional mRNA at the normal site an **overexpression** is evoked. In embryos of the clawed frog RNA injection has been routinely applied to study gene function.

It even proved successful to begin with mixtures of mRNA that were produced synthetically from pooled cDNA clones. When an interesting effect was observed through repeated subdivision of the mixture (**sib selection**) eventually the effective mRNA could be isolated. In analogous manner, effects of mixtures of proteins, manufactured in expression systems using mRNA, were checked. Using such strategies the induction factors NOGGIN and DICKKOPF were detected.

Targeted Introduction and Excision of a Gene: The Cre/loxP System

Cre-Lox recombination is a site-specific recombinase technology widely used to carry out deletions, insertions, or translocations of genes. It allows insertion of a (trans)gene to be targeted to a specific organ, tissue or cell type; some modifications of the procedure in addition allow recombination to be triggered by a specific external stimulus. And eventually the tool allows the excision of a previously inserted gene, often the *neomycin resistance* gene *neo*, if temporal restriction of its function is intended (in the case of *neo* to restrict its function to the selection procedure performed to isolate successfully transformed cells or individuals).

A selected foreign gene, for example the gene for the Green-Fluorescing Protein GFP (Fig. 12.7) is in the lab flanked by bacteriophage-derived ***loxP* sequences** whereby both *lox* sequences are oriented in the same direction (only with this orientation the gene can later be excised). Then the construct is introduced into cells, typically into stem cells (the procedure to introduce genes into mice is intricate and will be outlined in Chap. 13). By inserting flanking target sequences corresponding to the gene to be exchanged, the construct is suited to be integrated into the genome by homologous recombination (see Fig. 12.8). However, this happens seldom. To pick out from a cell culture those few cells which have successfully integrated constructs, the whole culture is treated with the antibiotic neomycin. Only cells having taken up and expressed the *neo* gene survive the treatment. Subsequently a **Cre recombinase** is introduced into these transformed and selected cells. Introduction is performed either in the form of Cre recombinase protein enclosed in liposomes, or in form of its gene enclosed in a viral vector. Now the *loxP* site-specific recombinase can remove the neomycin gene out of the genome by a complex recombination event occurring between the *loxP* sites (Fig. 12.7) (In a first approach one may imagine that the double stranded DNA is cut at both *loxP* sites by the Cre protein, and the strands are then rejoined with DNA ligase). In Fig. 12.7 a case is shown where, after the excision of the *neo* selection gene, a reporter gene, be it *Lac-Z* or a *GFP* gene, is transcribed. The reporter protein bears witness ▶

Box 12.2 (continued)

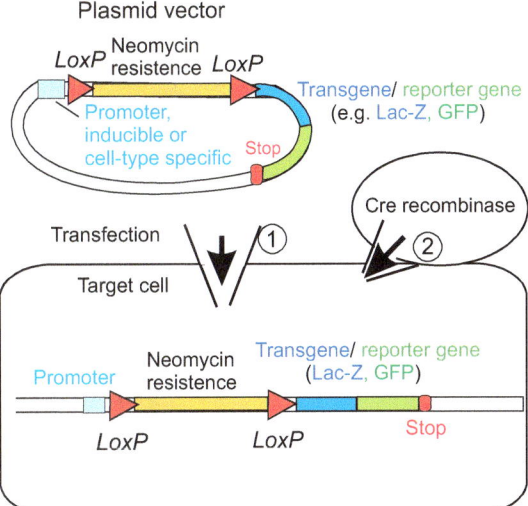

(1) Integration of the construct via somatic recombination Selection of cells with successful integration by neomycin treatment
(2) Thereafter introduction of Cre recombinase

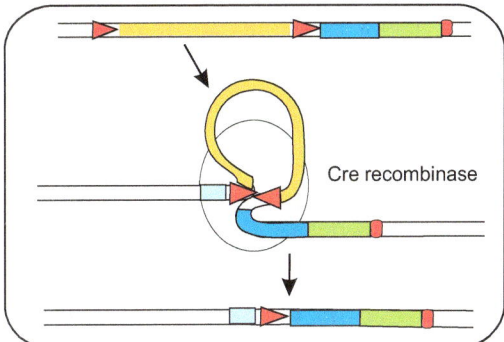

Removal of the neomycin resistence gene

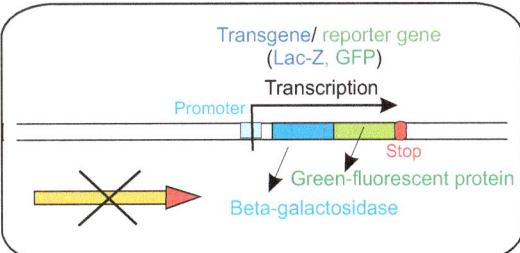

Transcription of the transgene/ reporter gene

Fig. 12.7 Cre/Lox-System. In the laboratory a construct (vector) is generated which contains the target gene framed by *LoxP* sequences oriented in the same direction. In addition, the construct can harbour a transgene fused with a *GFP* reporter gene. Using such vectors transgene *LoxP*-mice can be prepared (method see Chap. 13). Another group of mice gets a *Cre*-recombinase gene placed under the control of a tissue-specific promoter. These *two* groups of transgenic mice are crossed. In the offspring it can happen that the *LoxP*-framed gene is removed by the activity of the recombinase whereas the transgene-GFP fusion gene remains, is activated through the tissue-specific promoter and causes that tissue to fluoresce green ▶

Box 12.2 (continued)

Targeted Mutagenesis or Introduction of a Transgene via Homologous Recombination

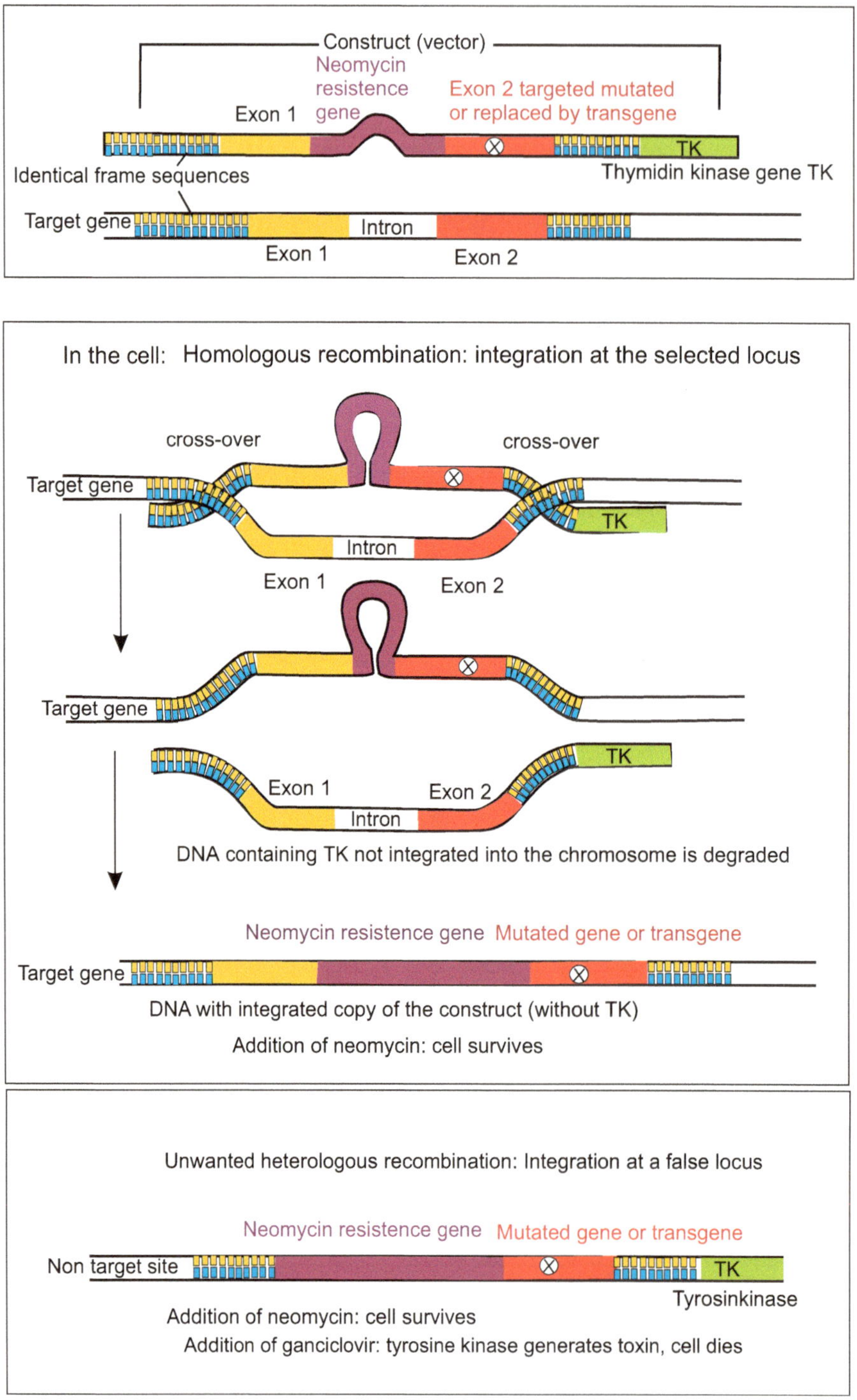

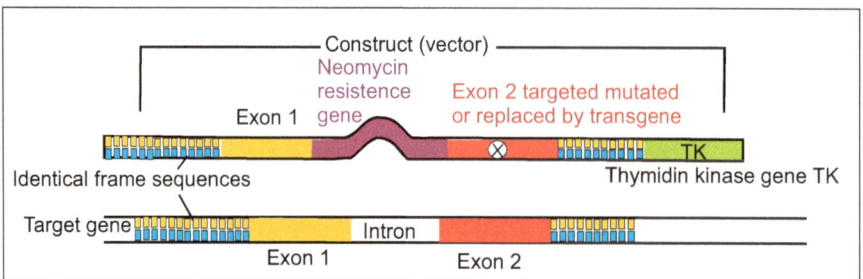

Fig 12.8 Homologous recombination is used in gene targeting, a technique for introducing genetic changes into target organisms. When an investigator wants to replace one allele with an engineered construct but ▶

Box 12.2 (continued)

to the success of the endeavour. One can easily conceive various variants of the procedure to achieve one's own goals. Placing *Lox* sequences appropriately allows genes to be activated, repressed, or exchanged for other genes. The activity of the Cre enzyme can be controlled so that it is expressed in a particular cell type or triggered by an external stimulus like a hormone or a heat shock. In elegant variants of the system a cell type-specific promoter is positioned upstream of the recombinase gene. Only in this cell type recombinase will be expressed, and only in those tissues composed of this cell type the *neo* gene is excised enabling transcription of the downstream reporter gene and any further transgene. In Fig. 13.7 (Chap. 13) a mouse is depicted with a heart-specific promoter placed upstream of a GFP reporter gene, causing the heart of the mouse to fluoresce green. Targeted DNA changes are useful in cell lineage tracing and can provide a first survey of whether a selected gene is required or not for a certain differentiation program or functional task.

How Genes Can Be Exchanged for Mutated Variants, and Foreign Genes Introduced

The aim of such experiments is to introduce gene variants, be they natural alleles, foreign genes or artificially designed constructs, permanently into the genome to study their effects in all developmental stages and in following generations (today, preferentially inducible genes or constructs are introduced, see below). How does one proceed?

We already mentioned several technical means to introduce genetic materials into cells, be these materials mutated versions of a gene, foreign genes (transgenes), or artificial constructs, be they naked or enclosed in carriers, the basic techniques are the same: direct injection with micropipettes, electroporation, liposome-mediated uptake, virus-mediated transfer. One common problem remains: Without particular arrangements and designs introduced DNA inserts, if at all, anywhere in the genome and the effects are hard to predict and to control.   ▶

Fig 12.8 (continued) not affect any other locus in the genome, then the method of choice is homologous recombination. To perform homologous recombination, the DNA sequence of the gene to be replaced must be known. With this information, it is possible to replace any gene with a DNA construct of choice. In the technique frequently used in developmental biology an allele with a targeted mutation or a transgene (foreign gene) is first introduced into stem cells through a plasmid-type construct. This construct may contain, besides the target DNA sequence, additional genes (e.g. for antibiotic resistance, or green fluorescent protein GFP). In any case the construct must include some flanking DNA that is identical in sequence to the targeted locus. The construct is introduced in stem cells through injection or electroporation. By mechanisms that are poorly understood but are similar to what occurs during meiosis and mitosis when homologous chromosomes align along the metaphase plane to repair strand breaks, the engineered construct finds the targeted gene and recombination takes place within the homologous (meaning identical in this case) sequences. The recombination may take place anywhere within the

flanking DNA sequences. In the example shown here, the construct contains besides the gene to be exchanged two genes for selection of those few cells in which homologous recombination took place: (a) a neomycin resistance gene enabling survival of only those cells that have integrated the construct and express resistance when the cell culture is treated with neomycin, and (b) a thymidine kinase (TK)-gene, that converts the base-analogous ganciclovir into a toxin. Provided recombination took place at the correct locus, the TK sequence is left over and degraded. If, however, the targeting vector aligns in a non-homologous region of the genome, then recombination is random and the negative selection marker may become incorporated into the genome. Those cells die after administration of ganciclovir. Stem cells that responded according to desire can be introduced into mouse blastocysts where they are expected to give rise to germ cells. These can carry the genetic modification to descendants. Yet, as a rule only one of two alleles will be exchanged by the introduced construct. By repeated inbreeding in the long term also homozygous offspring are to be expected

Box 12.2 (continued)

Non-Random Integration by Transposons.
We used transposons to evoke mutations or as
vectors to introduce transgenes into primordial germ cells of *Drosophila* (Fig. 12.2). The
integration of the introduced DNA into the
host genome was mediated by transposase
introduced by another carrier. But in those
days large-scale mutations were carried out
for large-scale genetic mapping of developmentally significant genes, it was not possible
to determine the locus of integration; the loci
were random – almost but not completely.
Experience taught that transposons preferentially haunt loci on chromosomes with active
genes, perhaps because the chromatin is
relaxed there. If by chance a transposon is
integrated downstream of the control region
of a gene but upstream of the gene proper, an
introduced foreign gene can be expressed in a
tissue- and stage-specific manner. This
applied to the experiment in which supernumerary eyes in flies were successfully induced
(see Fig. 12.19).

**Targeted Exchange by Homologous
Recombination.** In general terms, homologous recombination is a type of genetic recombination in which nucleotide sequences are
exchanged between two adjacent similar or
identical molecules of DNA. It is used by
cells during mitosis to accurately repair harmful breaks that frequently occur on both strands
of DNA, known as double-strand breaks. We
focus on practical aspects and the use of the
principle in developmental biology.

In the pioneer model mouse it is possible to
render a gene useless at a researcher's or
customer's option (**targeted mutagenesis**),
or to exchange a gene for another one. The
method is based on the rare event of homologous recombination between the introduced
and the endogenous gene. The gene vector is
linearised, the introduced gene to be
exchanged is flanked by sequences which are
exactly identical with a large part of the initial
and the terminal sequence of the endogenous

gene. The procedure for targeted manipulation of mice will be described in Chap. 13
(Fig. 13.6). Here we focus on the procedure
in general as it can be carried out in cell
cultures (Fig. 12.8). Cells are stimulated to
proliferate in the presence of the introduced
construct. Occasionally, proliferating cells
accomplish exchange following DNA replication but before mitosis using the introduced
DNA stand as template for repair of breaks. In
order to facilitate selection of those few cells
in which recombination took place, a selection marker is included in the construct. When
neomycin is present in the culture medium
also those cell survive which have integrated
the construct and express the resistance gene
properly.

In combination with the **Cre/loxP** system
homologous recombination enables excision
of a gene only in a particular tissue or organ.
After integration of the construct and selection, the neomycin resistance gene (or any
other gene) can be removed if this was flanked
by *loxP* sequences. For that a *Cre*
recombinase must be introduced with any
vector (see Fig. 13.7).

Methods such as ZFN, TALENs, and
CRISPR which are not specifically, or not
primarily, developed for analysing developmental processes but to alter the genetic
makeup of an organism and to produce **transgenic animals** are described in Chap. 13.

Permanent Labelling of Living Cells and Their Descendants Using Reporter Genes

There are many reasons to permanently label
living cells, for example to track cell
genealogies, to follow migrating cells and
outgrowing nerve fibres, and identifying
them even when they enter and infiltrate a
dense network of fibres.

Labelling with dyes is transient even if the
dye itself is stable, because with each cell
division the concentration of the dye is
halved. Not so with reporter genes that ▶

Box 12.2 (continued)

are introduced (by injection, electroporation or lipofection). Well-designed constructs are replicated and therefore not diluted out, and are expressed. Skillfully designed constructs have a promoter taken from a cell type-, tissue-, or stage-specific gene and placed upstream of the reporter gene. Often used reporter genes are

- **β-Galactosidase gene** *lacZ*, of bacterial origin but linked to a eukaryotic promoter (see Fig. 12.11, photo 2), and the
- **luciferase** gene.

After following expression of these genes and addition of appropriate ingredients the presence of these enzymes is detected by the emergence of a blue dye or by emission of light.

Even more impressive and used in innumerable applications is the

- **GFP (Green-Fluorescent Protein)**. The gene of this 238 amino acid protein and related genes were isolated from bioluminescent Cnidaria, the prototype from the pacific medusa *Aequorea victoria*. The protein encoded by this gene spontaneously forms a chromophoric group by rearrangement of three amino acids. The chromophore absorbs excitation light in the blue to ultraviolet range and emits green fluorescent light. The particular advantage is that no separate chromophore or other ingredients must be introduced into the living cells. Once introduced into an organism it is maintained in their genome through breeding.

The particular pleasure is that the glow of the GFP can be seen in the living animal, provided it is transparent (or the light emitting molecules are near the surface, see Fig. 13.5). Under the microscope migrating cells and growing nerve fibres can be observed *live* and their behaviour documented with a video-recorder.

Research all over the world derives most benefit from the accidental circumstance that the **GFP gene, or part of it, can be coupled to a cell's own gene and expressed with it in the form of a fusion protein,** also called a **tagged protein** (see Fig. 12.11, picture 3, 4, 6; and Fig. 13.5). Astonishingly often, the cell's own protein retains functionality. Another possibility is to place the GFP gene under the control of a selected cell-type-specific promoter. Expression of the *GFP* gene is controlled **in parallel** to the cell's own (unattached) gene, they are switched on and off simultaneously. In addition, by chemical mutation and modification several GFP variants were made available fluorescing in various nuances and intensities of blue, green, yellow and red, so that under changing excitation waves varying patterns of coloured structures can be observed, each colour representing a distinct gene product.

Inducible Promoters: How Genes Can Be Switched On and Off at the Will of Researchers

Experimental interventions in processes of gene expression are possible when the introduced gene is placed under the control of an inducible promoter. As their name suggests, the activity of these promoters is induced by the presence or absence of an externally added chemical or physical factors. Inducible promoters are very powerful tools in genetic engineering because the expression of genes controlled by them can be turned on or off at certain stages of development of an organism or in a particular tissue.

We refer, for example, to a promoter that can be turned on by the well-known WNT signalling system (Fig. 12.9a). Or to the promoter of the heat shock gene *hsp-70*, or to the promoter of a steroid hormone receptor gene (Fig. 12.9b). When installed in a cell, the gene controlled via such an inducible promoter can be switched on by a WNT signal, a heat shock, or a steroid hormone. ▶

Box 12.2 (continued)

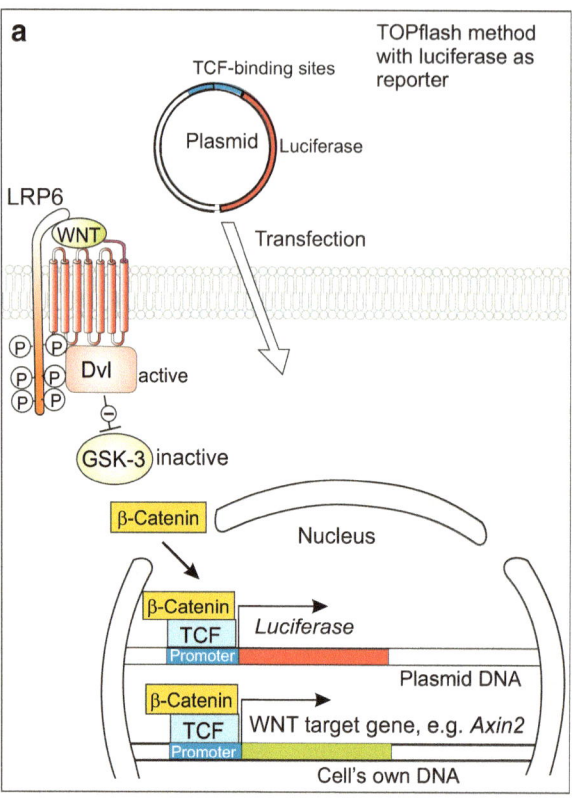

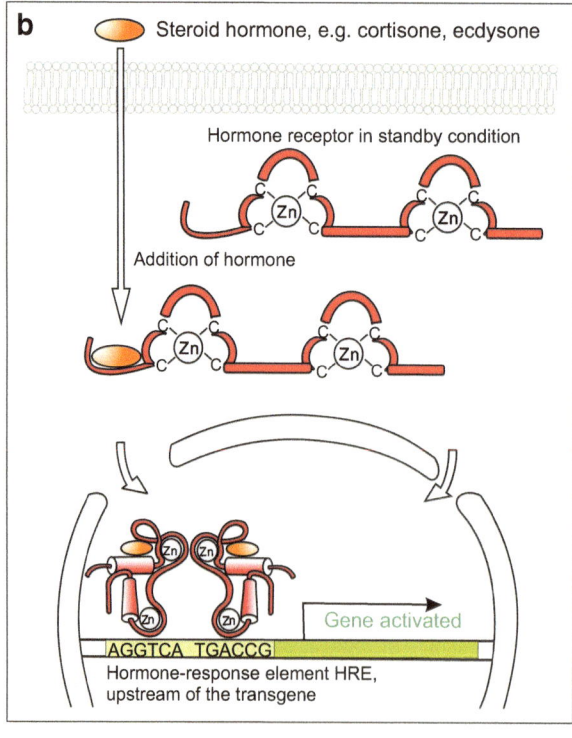

Box 12.2 (continued)

Since, however, heat shock or a steroid hormone may switch on also other genes of the cell, more specific solutions are required.

The promoter of a mammalian steroid-dependent gene can be substituted by the promoter of the insect-specific steroid hormone ecdysone. Now this gene, and only this gene, can be activated by administering ecdysone).

Many laboratories strive to design more types of artificial promoters that can be handled at the will of the researcher. Example of successful constructs are sophisticated chimeric constructs composed of prokaryotic (bacterial and/or viral) and eukaryotic sequences such as the Tetracycline Response Element (TRE) used in **tetracycline-controlled transcriptional activation,** also known as **tetracycline resistance system,** in short **Tet system**. This is a method of **inducible** gene expression where transcription is reversibly turned on or off in the presence of the antibiotic tetracycline or one of its more stable derivatives, actually **doxycycline Dox**. Two variants of the system, named **Tet-Off** and **Tet-On** are currently the most commonly used inducible expression systems for research of eukaryote cell biology. **Tet-Off** activates expression of downstream genes in the absence of Dox, whereas **Tet-On** and Tet-On Advanced activate expression in the presence of Dox. The benefits of the system are (a) an **inducible on or off promoter**. The investigators can choose to have the promoter always activated until Tet is added or always inactivated until Tet is added. (b) The second benefit is the ability to regulate the strength of the promoter. The more Tet added, the stronger the effect is. Thus the Tet Technology provides efficient, precise and reversible control over both timing and level of gene expression in eukaryotic cells.

Currently more and more inducible systems are being developed. For instance the following:

Optogenetics: Switching Functions On and Off with Light Impulses in Living Cells and Animals

The term optogenetics refers to newly-introduced technologies to localize and regulate cellular components and to switch functions on and off in milliseconds in living cells or animals that have previously been prepared with genetic technologies. Optogenetic tools allow high-resolution spatial and temporal manipulation of individual cells or subcellular parts such as single synapses. Pioneering experiments were designed to localize synapses equipped with distinct ion channels (Fig. **12.10**a,b) and to switch the electric signals emanating at synapses on and off on demand. Through genetic technologies the cells or animals were provided with channel rhodopsins of microbial origin. These are light-gated ion channels that contain retinal and combine sensitivity for light pulses of a certain wavelength with selectivity for either cations or anions (Fig.**12.10**b). At present the technology is mainly used in neurosciences to analyse and control neuronal activities in cell cultures, brain slices and in living animals. For controlling the behaviour of freely-moving rodents extremely fine optical fibres are introduced in selected parts of the brain.

Optogenic technologies are being extended by fusing vertebrate opsins to G-protein coupled hormonal receptors and to create novel photo-switchable enzymes. The emerging repertoire of optogenetic probes will allow ▶

Fig. 12.9 (**a**) Luciferase-based reporter gene. In parallel to the natural target genes of the WNT signal the introduced luciferase gene is activated as well, since it is under the control of the same promoter. After addition of luciferin substrate the reaction of the substrate with the expressed luciferase produces light signals. (**b**) Inducible expression systems placed under the control of steroid hormones. A promoter (hormone-responsive element HRE) is set ahead the transgene to be activated. The HRE binds intracellular receptors laden with steroids, and thus the transgene is activated

Box 12.2 (continued)

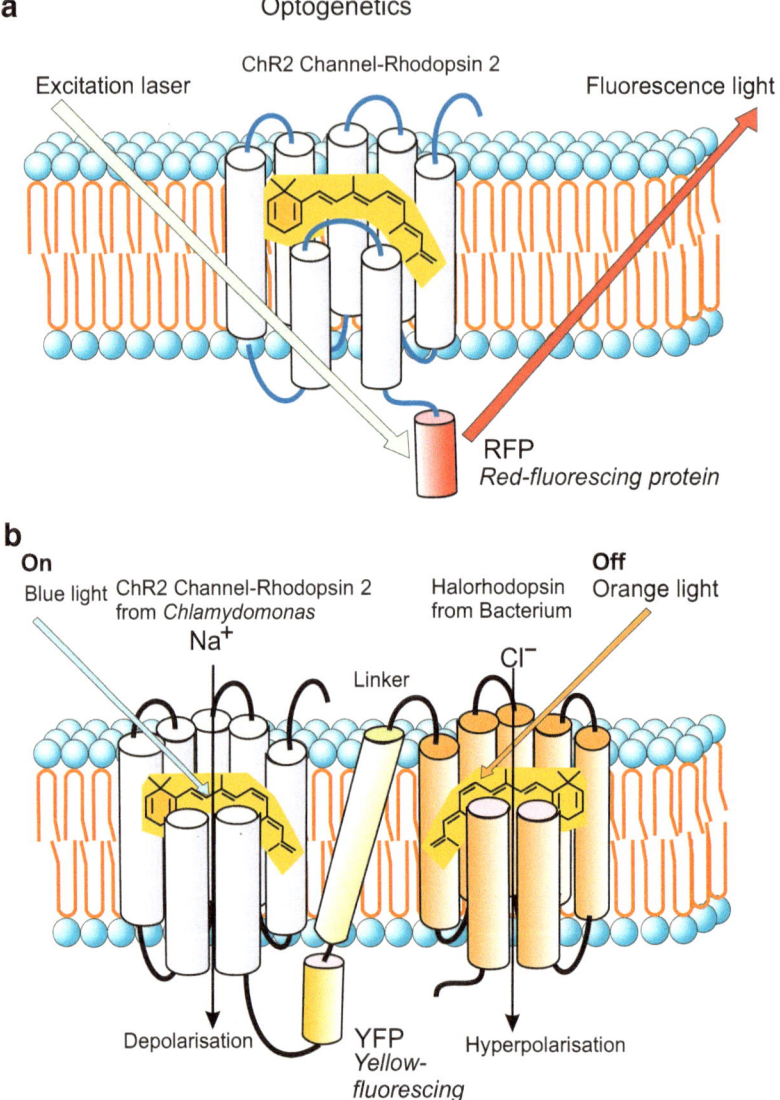

Fig. 12.10 Light-based technologies and optogenetics. (**a**) Light pulses localize a distinct ion channel to which a derivative of the GFP has been added. (**b**) On-off switch in the synapse (postsynaptic membrane) of a neuron. The switch consists of a complex of two different photosensitive ion channels of the rhodopsin-channel family coupled with a linker. The ion channel derived from the unicellular alga *Chlamydomonas* is permeable for cations; blue light induced opening of the channel causes depolarization of the electric membrane potential and thus stimulation of the neuron to fire action potentials. The channel derived from a halobacterium is selective for chloride anions. Its opening through orange light causes hyperpolarization and thus silencing of the neuron. The channels were introduced in human neuronal cell cultures or embryos of rodents by inserting the microbial genes into the genome of the host and placing it under the promoter of specific neurons. After Kleinlogel et al. (2011) as regards content. Own composition by WM ▶

Box 12.2 (continued)

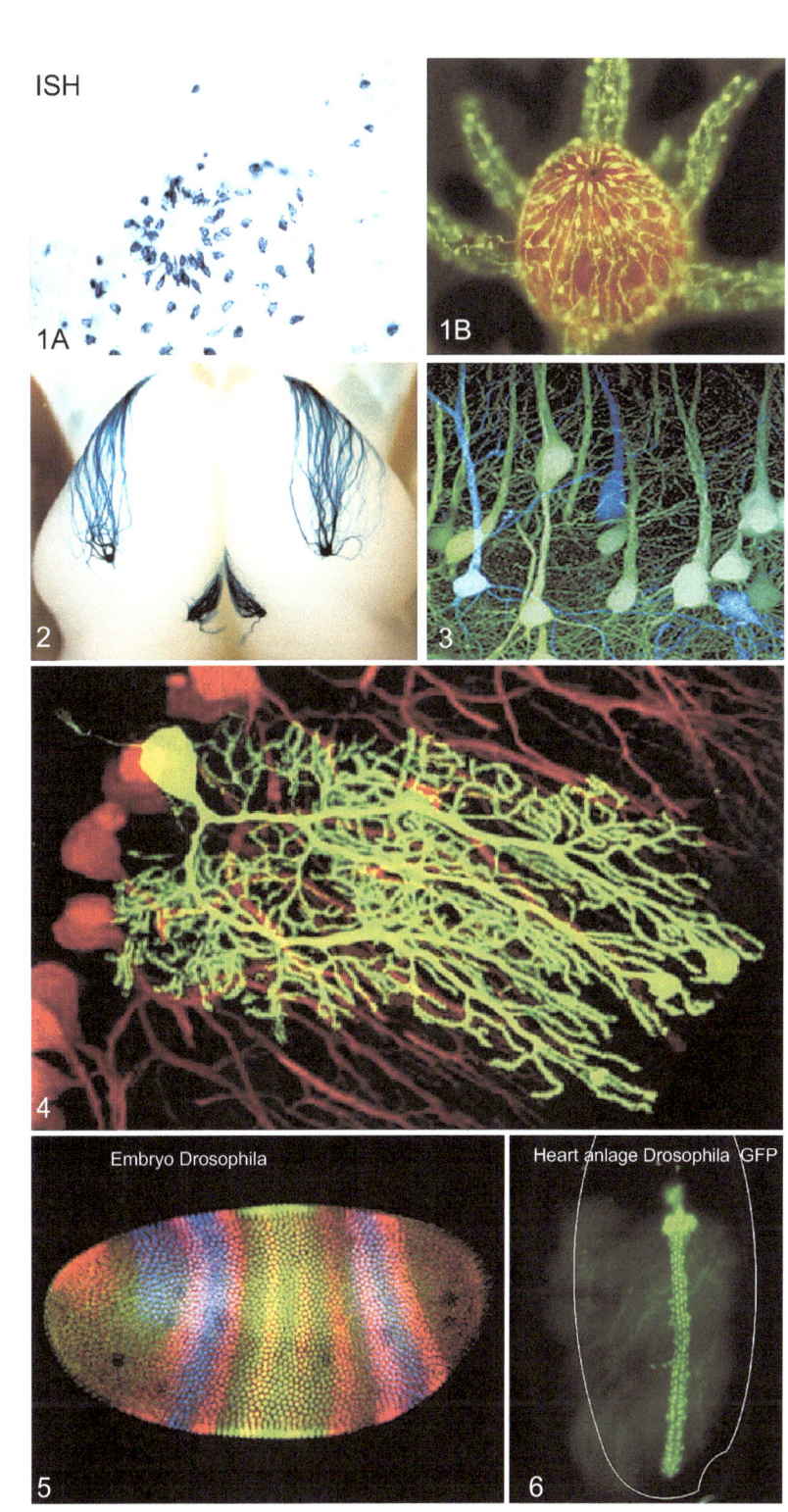

ISH

1A

1B

2

3

4

Embryo Drosophila

Heart anlage Drosophila GFP

5

6

Box 12.2 (continued)

cell-type-specific and temporally precise control of multiple cellular functions within intact cells and to control the action of hormones at the level of single cells. This will expand the technology to wide fields of the life sciences and supplement the already widely used luciferin-based technologies to visualize gene activities and aequorin-based genetic approaches used to visualize Ca^{2+} signalling in fertilized eggs (Chap. 3) and in developing neuronal systems. Transparent fish embryos and larvae are predestined to become pioneering

models in developmental biology. Moreover, proposals have been published to optically regulate the developmental path and terminal differentiation of embryonic stem cells (ESCs) and induced pluripotent stem cells (iPSCs, see Chap. 18) derived from human patients.

Demonstration of proteins without using genetic engineering

For many proteins antibodies are available enabling indirect demonstration of their presence in slices of 'fixed' embryos or tissue ▶

Fig. 12.11 Methods of molecular and cell biology frequently used in developmental biology. (**1A**) *In-situ* **hybridisation**. By means of a labelled antisense probe the presence of an mRNA for a neuropeptide in nerve cells is visualized. All those cells are visible which contain mRNA for neuropeptides of the RFamide (arginine-phenylalanine-amide) class. The object shown is a whole-mount preparation of a polyp of the hydrozoan *Hydractinia echinata*. The shot shows a detail from the mouth region. Preparation by Prof. Günter Plickert, University of Cologne. (**1B**) **Immunocytochemistry**. RFamide nerve cells in the mouth region of a polyp of *Hydractinia echinata* are visualized using antibodies that recognize RFamide neuropeptides with the C-terminal end arginine-phenylalanine-amide. The antibodies were provided by Dr. C. Grimmelikhuijzen. The antibodies on their part are bound to a secondary antibody coupled with a fluorescing dye. Preparation by the author WM of this book. (**2A,B**) **Reporter gene and targeted mutagenesis** for examining how axons of smell cells find the correct glomeruli, i.e. the first stations of data processing in the bulbus olfactorius (olfactory brain). These axons contain in their cell membrane smell receptors identical to those present in the cilia of differentiated smell cells and used there for odour detection. In this experiment the hypothesis was examined that the molecular smell receptors in the cell membrane along the axons contribute to target finding. In a sophisticated multi-step procedure first in mice the gene for such a receptor was exchanged by a mutated gene using the method of homologous recombination. The mutated gene was linked with a neomycin resistance gene framed by *LoxP* sequences. Subsequently, the neomycin cassette was removed using Cre-mediated recombination (method similar to that shown in Fig. 12.7) and exchanged for a *LacZ* construct. When the introduced *LacZ* gene is expressed the presence of the LacZ protein (=bacterial β-galactosidase) can be visualized by

means of a dye reaction. 2A Shows growing axons, 2B shows how these axons form synapses in the target area of the brain. Result of the experiment: A single exchange of one amino acid can suffice to guide the bundled axons to another target area (after Feinstein and Mombaerts 2004, Feinstein et al. 2004). The photos were kindly provided by Prof. Peter Mombaerts, Rockefeller University New York. (**3–4**) **Fluorescing tags.** (**3**) Brain of a mouse with blue and green fluorescing neurons. The mouse had inherited a gene for *GFP* (*Green Fluorescent Protein*) from its father and a gene for *CFP* (*Cyan Fluorescent Protein*) from its mother. Image copied with kind permission of Dr. Jeff W Lichtman, Washington University, School of Medicine (rights reserved). (**4**) Purkinje-neurons in the cerebellum of the mouse, arisen from the spontaneous fusion of a GFP stem cell with a nerve cell. Image kindly provided by Dr. Clas B. Johansson, Karolinska University Hospital Stockholm, with permission of Prof. Dr. Helen Blau, Stanford. For method see Johansson CB et al. (2008) Nature Cell Biol 10(5): 578–83. (**5**) Embryo of *Drosophila* at the blastoderm stage. Threefold immuno-fluorescence with antibodies to HAIRY protein in *red*, to KRÜPPEL in *green*, and to GIANT in *blue*, Laser scanning-confocal microscopy. Image kindly provided by Dr. Stephen Paddock, Lab. Cell & Mol. Biol., Wisconsin University. (**6**) Bright heart of *Drosophila* with luminescent GFP. The cells of the heart of living embryos of *Drosophila* were made fluorescent by placing the reporter *GFP* gene under the control of the *hand* promoter. The *hand* gene codes for a transcription factor of the bHLH protein family highly conserved in the animal kingdom. The promoter of the *hand* gene is switched on in the cardiomyoblasts, in the heart-associated pericardial cells, and in the cells of lymph glands. When GFP is matured the heartbeats can be observed *live* as well as the further development of the heart. The image was kindly provided by Prof. Dr. Achim Paululat and Julia Sellin, University of Osnabrück

Box 12.2 (continued)

samples by means of 'immunocytochemistry' or 'immunohistochemistry' (Fig. **12.11**). The antibodies either stem from mammals immunised with the selected protein yielding 'polyclonal' antibodies because the animal produces a collection of antibodies to different domains of the foreign protein. Or the antibodies are harvested as monoclonal antibodies from cell cultures. Production of monoclonal antibodies starts with immunising mice with the selected protein. Subsequently, the potential antibodies producing B lymphocytes are removed from the spleen and fused with lymphoma cells, that is with lymphoblasts that have acquired traits of cancer cells. Occasionally this or that fused hybridoma cell arises which combines the capability to proliferate indefinitely (immortality) of a cancer cell with the capability of a B lymphocyte to produce antibodies. From such hybridoma cells clones of cells can be grown which all produce exactly the same antibody to a distinct domain (epitope) of the protein. Such antibodies are known as monoclonal.

Whether polyclonal or monoclonal, these 'primary' antibodies are specific but invisible. The primary, mouse-derived antibodies are recognized and bound by (commercial) secondary antibodies produced by another species (e.g. goat) and directed to mouse-specific epitopes of the primary antibody. The secondary antibody either is linked with a fluorescent dye or with an enzyme that catalyses a colour or light reaction. For observing and scanning fluorescent microscopic slices multichannel confocal laser scanning microscopy is particularly useful for obtaining high-resolution optical images with depth selectivity. The procedure is known as optical sectioning. Images are acquired point-by-point and reconstructed with a computer, allowing three-dimensional reconstructions of complex objects. Using several different antibodies linked with different dyes different proteins can be analysed simultaneously (Fig. **12.9**, picture 5).

12.1   Differential Gene Expression as the Basis of Differentiation

12.1.1   Initially Cells Are "Genomically Equivalent" But They Become Programmed to Utilize Different Sets of Genes in the Course of Determination

In sexual as well as in asexual reproduction of a multicellular organism, all cells arise by mitotic divisions. Cleavage of a fertilized egg is a series of mitoses. If developing systems follow our textbooks and undergo nothing but regular mitotic divisions, the genomic information of all cells is identical as long as nothing is changed after DNA replication. In fact, initial genomic equivalence

appears to occur for the vast majority of embryonic cells. Yet, although every cell contains the entire set of genetic instructions, in each particular cell only a small fraction of the total genetic information is actually being used and expressed in a collection of proteins specific for the respective cell type. In different cell types, different sets of genes are expressed. Which parts of the genome will be used and which will not, is programmed in the process of **determination**.

The program specifying the genes to be activated does not need to be run immediately. Frequently, cells committed to a distinct developmental path undergo many rounds of cell division without revealing in their appearance their identity and destiny. The daughter cells inherit the program from their parent cell. This phenomenon is called **cell heredity**.

Frequently, the inherited program is actually executed after a considerable lapse. Only while undergoing **terminal differentiation** are the cells actually provided with a characteristic molecular inventory and acquire their characteristic shape and function.

- **Myoblasts** fuse with several other myoblasts and synthesize muscle-type molecules such as sarcomeric actin, myosin, tropomyosin and troponin; using these molecules the developing muscle fibre constructs its contractile machinery.
- **Erythroblasts** are equipped with haemoglobin, carbonic anhydrase and spectrin, and turn into erythrocytes, the red blood cells.
- **Neuroblasts** produce, while differentiating to neurons, among other molecules, α- and β-tubulin, microtubule-associated protein (MAP), neurofilament proteins, and a set of proteins required to form and operate synapses such as specific cell adhesion and recognition proteins and receptors for transmitters.
- **Retinal photoreceptors** form the discs of photoreceptive membranes inserting rhodopsin into them and attaching transducin molecules.

The number of cell type-specific proteins produced during terminal differentiation is large and steadily growing, and may exceed 1,000.

Potentially the acquisition of a cell-specific set of molecules can be controlled at many sites and levels of the cellular biochemical machinery. The equipment with active enzymes, for example, can be controlled at three levels:

- **Transcription**, control by activation of protein-coding genes, or by their silencing, that is their enduring suppression;
- **Translation** at the ribosomes – maternal mRNA, for example, stored in the cytoplasm of the egg cell must first be unpacked from the RNP particles before being translated;
- **Posttranslational modification** of the enzymes such as **phosphorylation** by protein kinases, methylation, acetylation, glycosylation or processing (splitting by proteases).

Such posttranslational modifications can be merely functional and transitory. Only modifications that are effective for long periods of time can bring about cell differentiation. As a rule, once a program is installed and terminal differentiation attained, they are maintained for life. **Transdetermination** (change of a program) and **transdifferentiation** (exchange of the particular type of differentiation or change of cell type) are infrequent.

The developmental biologist is presented with a number of intriguing questions:

1. How are the genes to be expressed activated in a coordinated manner, and how is the expression of other genes silenced?
2. How are developmental decisions 'frozen'; how can the state of determination be faithfully transmitted to the daughter cells during cell division?
3. Is differentiation accompanied by secondary qualitative and quantitative changes in the structure or accessibility of the genome?

12.1.2 Protein-Coding Genes Represent Only a Small Portion of The Genome; The Significance of Non-Protein-Coding Regulatory RNA Molecules Is Becoming Obvious Only Recently

The human genome enclosed in the nucleus of the cells contains in the haploid set of chromosomes about 22,000 "classic" protein-coding genes (in the diploid embryo 44,000 of which 22,000 have been supplied by the mother and 22,000 by the father) and further 23,000 non-coding genes (Fig. 12.12, Table 12.1). Originally the function of these non-coding genes was unknown and they were described by Francis Crick as 'junk' DNA. Now it is known that many non-coding genes perform important functions. The classical protein-coding genes amount only to 2–3 % of the 3.3 billions of base pairs. The residual non-coding part of the genome, that is 3.25 billions of base pairs, contains the following: genes for non-coding RNA including rRNA, t-RNA, Piwi-interacting RNA, and micro-RNA; cis- and trans-regulatory elements; introns; pseudogenes; telomeres; and repeat sequences.

Transposons (transposable elements TE) are DNA sequences that can change their position

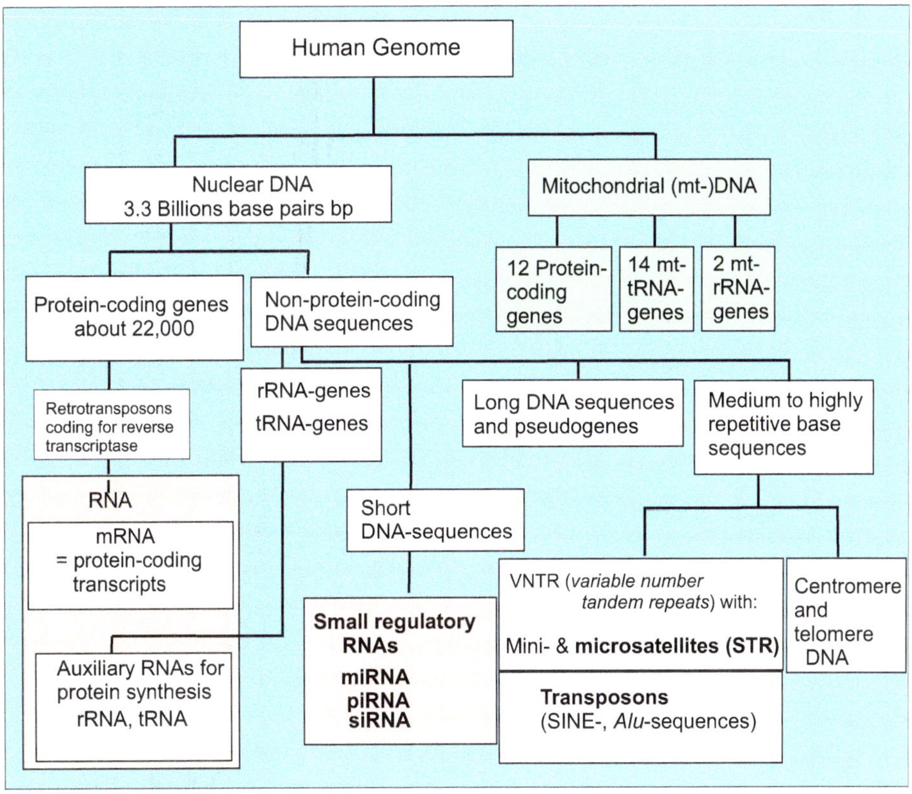

Fig. 12.12 Overviews of the composition of the human genome

within the genome, sometimes creating mutations when they insert into a gene and disrupt its message. The structure of many transposons often reveals their viral origin, for example by containing sequences coding the envelope of viruses (*gag*, *env*) or other sequences resembling typical viral sequences (*pol*, *ORF*). According to quantitative estimates transposons and related sequences encompass about 45 % of the human genome.

Non-coding DNA is faithfully copied by the replication machinery and is transmitted to the succeeding generation via the germ line. The non-coding **introns** between the coding exons of the 'true' genes were well-known examples for such seemingly useless DNA sequences as for long a time no function could be assigned to them. In the genome of other eukaryotic organisms also sequences with lacking or unknown function dominate.

There are indeed sequences that in all likelihood do not have any useful function in normal development. Examples are pseudogenes (non-functional, incomplete copies of normal genes), those many, frequently disturbing transposons, and the numerous short, repetitive sequences known as mini- and microsatellites, and the *Alu* elements, stretches of DNA about 300 base pairs long.

We accentuate here the **microsatellites**, also known as **STRs** (*Short Tandem Repeats*), that are so useful for **DNA profiling** (also called DNA testing, DNA typing, or genetic fingerprinting). Although 99.9 % of human DNA sequences are the same in every person, enough of the DNA is different to distinguish one individual from another. DNA profiling uses repetitive sequences that are highly variable short repeats and in their number and exact sequence different from individual to individual. Two or more base pairs, for

Table 12.1 Haploid genome sizes

Organism	Basepairs bp x						Protein-coding genes (estimated)	Peculiarities
	10^6	10^7	10^8	10^9	10^{10}	10^{11}		
Epstein-Barr virus	0.17						80	
Escherichia coli	4.6						4,288	
Eucaryotes								
Hydra magnipapillata			1.05				20,000	57 % Transposons 71 Genes of bacterial origin, many vertebrate-like
Caenorhabditis eleg.		9.7					~20,100	First completely sequenced genome
Drosophila melanogaster		2.27					13,379	
Sea urchin			8.14				~21,000	
Tunicata, Ascidia *Ciona intestinalis*			1.6				~16,000	
Danio rerio Zebrafish				1.2			15 761	Genome putatively 2× duplicated
Fugu rubripes Blowfish "Fugu"			3.65				~31,000	Up to now smallest vertebrate genome
Lung fish *Lepidosiren paradoxa*					7.84			Maximum in vertebrates (2010)
Mammals								
House mouse				2.7			~24,000	
Human				3.3			~25,000	97 % of DNA non-protein-coding
Phylogenetic position debated								
Placozoon *Trichoplax adhaerens*		4.8					11,514	Up to now smallest animal genome
Plants								
Arabidopsis thaliana rock cress			1				27,000	"The" model among plants
Paris japonica canopy plant						1.49	Polyploid	(2010) record among flower plants
Psilotum nudum, fern						2.5	Presumab. polyploid	Up to now absolute record (2010)
Physcomitrella patens, moss			5				35,000	Model organism

example CG or CA, are repeated up to 100 times. Such sequences may look as follows:

CG CG CG CG CG CG ....or
CA CA CA CA CA CA or.
CA AC CA AC CA AC or...
ACAAACT ACAAACT ACAAACT CCAAACT
x 10^7 in humans.

According to the triplet rule such sequences cannot encode peptides. A lack of function is also indicated because they can easily mutate, without consequence to the organism, and be it merely that at some time a single base pair gets lost in the germ line. In humans there are about 30,000 such loci distributed over all chromosomes. DNA profiling is a technique based on such sequences, and employed to assist in the identification of individuals by their respective DNA profiles in parental testing and criminal investigation. Repetitive DNA without obvious function comprises further 3–5 % of the total DNA. What is the rest of about 90 % of the DNA for?

Regulatory Small RNA Species. Since about 2006 evidence has accumulated that not just the 3 % of the human genome representing protein-coding genes, is transcribed, but 80–90 %. Countless, predominantly short copies of *non-coding*-regions of the genome are transcribed, in the present use of the term "*non-coding*" meaning '*not used for protein coding*'. The abbreviated designation "*non-coding DNA*" for these transcribed sequences should be considered outdated or misleading at least, and be replaced by more precise designations such as *non-**protein**-coding DNA* or *RNA-coding DNA*, because this DNA encodes the base sequence of short RNA molecules with specific function similar to the genes for rRNA (ribosomal RNA) and tRNA (transfer RNA).

Natural small RNA molecules with regulatory function occur in all animals; frequently they are **antisense RNA sequences**, that is complementary to the single-stranded **mRNA**. Regarding their function, several groups can be distinguished:

1. **miRNAs (micro-RNAs)**, also called **stRNA** (short temporal RNAs), and sometimes classified among the siRNAS).

   - Primarily they are single-stranded as are all RNAs. These primary and long precursors often derive from intron sequences or from transposons. Due to palindromic sections the primary transcripts, or the miRNAs excised from them, form double-stranded *hairpin loops* that are of significance for their function. The double-stranded section attaches to the enzymatic **RISC complex** which excises single-stranded segments, comprising about 20 nucleotides (see Figs. 12.6 and 12.21).

   - These short single-stranded pieces control protein synthesis predominantly in a negative way by binding to complementary sections of the 3′UTR (*UnTranslated Region*) of mRNA; attachment of the regulatory miRNA (or siRNA, respectively) causes termination of translation, or degradation of the mRNA marked in this way.

   - The miRNAs participate in the regulation of all developmentally important processes, cell proliferation, differentiation, regeneration, or apoptosis. Defective regulation can cause cancer. Because its sequence is complementary to the 3′UTR of a distinct mRNA, each miRNA indirectly but specifically turns off the expression of a distinct gene. In particular development-regulating gene products are subjected to the control by miRNAs.

   - Provided a miRNA is partially complementary to a promoter sequence on the DNA its attachment can also cause enhanced expression of a given gene. Such an upregulation can be mediated by demethylation of the promoter (see below Sect. 12.5.3).

   - Many of these miRNAs that are involved in controlling developmental events were highly conserved in evolution.

2. **piRNA (*piwi-interacting* RNAs)**. piRNAs and their associated PIWI proteins occur mainly in the germ line (see Chap. 8). Their main function appears to be protection of the germ line by silencing transposons. Many piRNAs are antisense to transposons and can suppress them by binding to them, thus preventing the transposons from inserting copies of themselves into the genome, and causing serious problems such as sterility. Recent evidence indicates that piRNAs may also exist in somatic cells and may have diverse functions including learning and memory and prevention of cancer by maintaining genome stability.

3. **siRNA (*small-interfering* RNA)**, small double-stranded RNA molecules. These are present in the genome, are mainly derived from and play a major role in silencing transposons. Similar to single-stranded miRNA single-stranded siRNA binds to complementary sequences of mRNA and induces the degradation of the partially double-stranded molecules. In the research lab this property has been exploited to make synthetic, specific siRNA molecules. These are

introduced into cells to switch off selected mRNA at the will of the experimenter and thus to terminate the action of a gene (RNAi method; see Figs. 12.6 and 12.21).

In the following sections priority is given to classic, protein-coding genes.

12.1.3 The Puffing Pattern on The Giant Chromosomes: One Sees Changing Patterns of Gene Expression but Also an Irreversible Genome Amplification

Giant chromosomes in the salivary glands of dipterans (flies, mosquito) and the **puffing activity** along these chromosomes (Fig. 12.13) were for long considered proof for **differential gene activity**. These chromosomes are still giant, and the spotty and changing blowing up is indeed the visible manifestation of gene activity. The question, however, is whether this activity is associated with differentiation and development or whether it merely reflects the momentary functional state of the respective organs.

Giant chromosomes are giant because they became polytene. First two homologous chromosomes pair as in commencing meiosis but now they remain permanently attached; the DNA strands are repeatedly replicated, about 1000 times, but the multiplied DNA strands all stay together. Polytene chromosomes remain stretched, as it is proper for interphase chromosomes, but are densely packed as they were metaphase chromosomes. When genes are to be activated first the

Fig. 12.13 Puffing of the polytene giant chromosomes in the salivary glands of *Drosophila*. Polytene chromosomes form when multiple rounds of replication produce many sister chromatids that remain paired together. Puffing is the expression of local gene activities; the pattern of puffing changes in the course of development. The puffing pattern shown refers to the 3rd larval stage when in salivary glands the chromosomes undergo many rounds of endo-reduplication, to produce large amounts of glue before pupation

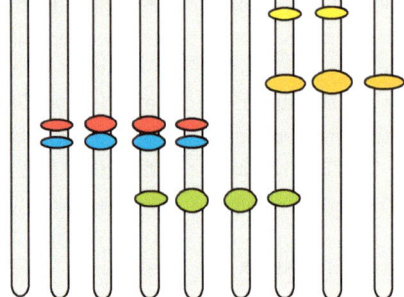

Puffing pattern

3rd Chromosome, left arm, in salivary gland of the third larval stage of *Drosophila*

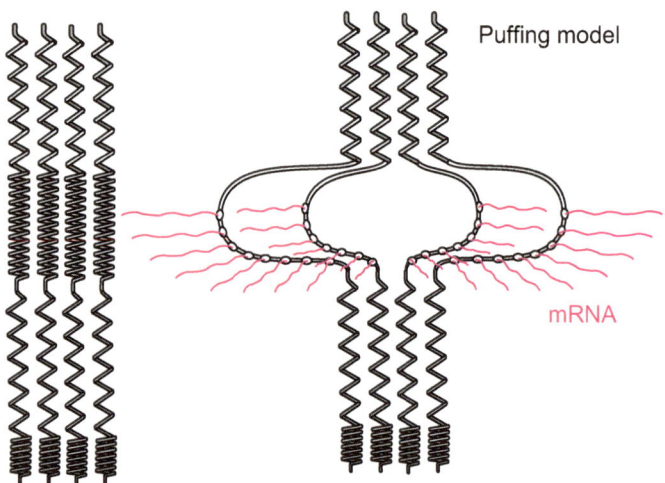

Puffing model

mRNA

DNA strands have to be untwisted. Under the microscope this unbundling is visible as **puffing**.

Polytene chromosomes are found in the salivary glands of the larvae (engaged in the production of spinning threads) but also in other organs although not of such spectacular size in all those organs. If the puffing patterns in the diverse organs are compared at different times, in the larva and the pupa, one sees changing patterns of puffing activities dependent on organ type and developmental stage. For example the pattern is different in salivary glands and Malpighian tubules.

Irrespective of whether these patterns of gene activities reflect differentiation or changing physiological states, the point is, and should be kept in mind: Giant chromosomes are polytene, that is, in the course of the development of those cells **genome amplification** took place and, thus, a quantitative change in the stock of genes. In principle, **in the course of differentiation quantitative changes in the amount of genetic material can occur**. More to that in Sect. 12.6.

in Chap. 13). The scientific result was that differentiation does not in all instances lead to loss of genetic information or to loss of possibility to retrieve and recall genetic information. On the other hand, for each particular cell type different sets of information must be called up. For differentiation particular genetic programs must be executed, as discussed in the following sections.

12.2   Genes for Specification of Body Regions and Organs: The Hox Genes

An amazing and unexpected result of developmental genetics is that some genes and gene complexes are expressed in defined body regions and compartments independently of their future cellular composition. Apparently, **the products of these genes mark positions and territories**. This applies especially to the **homeotic genes** (Figs. 12.14, 12.15, 12.16, 12.17 and 12.18) with the shortcut designation *Hom* (in *Drosophila*) or *Hox* (in other organisms).

12.1.4   Transplantation of Nuclei in *Xenopus* Were Designed with Respect to The Question of Whether Nuclei Remain Totipotent in The Course of Cell Differentiation; These Transplantations Rendered a Technique for Cloning Accessible

The question whether, and to what extent, cells uphold complete genetic potency or whether they experience irreversible alterations of their genetic makeup was the origin of those famous experiments performed in *Xenopus* in which nuclei of differentiated somatic cells were transferred into egg cells the nucleus of which had previously been removed. The question was: are such nuclei still capable of supporting a complete development, in other words, are they still **totipotent?** The result was the first vertebrates cloned using nuclei of somatic cells (more about cloning

12.2.1   Master Genes of the *Hox* Class Are Grouped Together on Chromosomes in an Order That Correlates with The Spatial and Temporal Pattern of Their Expression in The Body (Colinearity)

The existence of a particular class of genes, called **homeotic** genes, was revealed by spectacular mutants. A famous example is the *Antennapedia* mutation in *Drosophila* in which the antennae are transformed into legs (see Fig. 4.40 and Sect. 12.2.2). In the sense of the classic terminology, a homeotic gene causes, when mutated, the **transformation of a structure into another structure** appertaining to another body part. Canonical homeotic genes are not involved in either the construction of the basic body plan or in the elaboration of segments as such. This task is accomplished in *Drosophila* by the sets of maternal polarity genes and zygotic segmentation genes

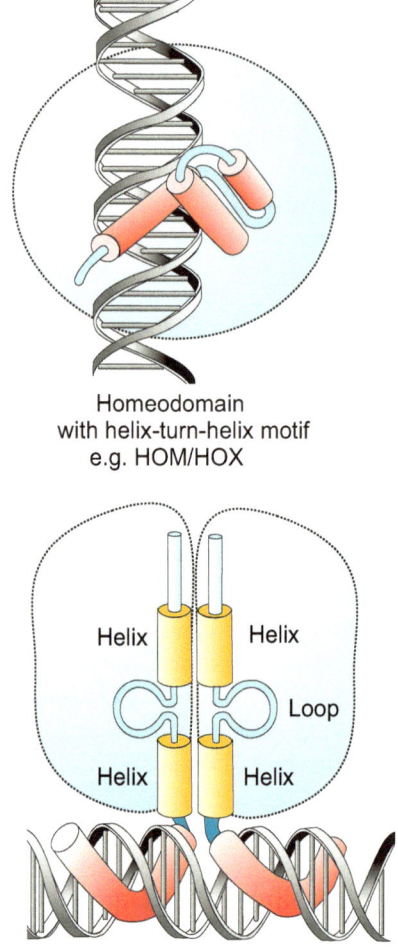

Homeodomain
with helix-turn-helix motif
e.g. HOM/HOX

Basic helix-loop-helix bHLH, dimer
e.g. Myo-D

Fig. 12.14   DNA-binding domains of two classes of transcription factors. These domains are part of larger
proteins. As examples, the HTH and bHLH motifs are
shown that are found in Hox-proteins or Myo-D protein,
respectively

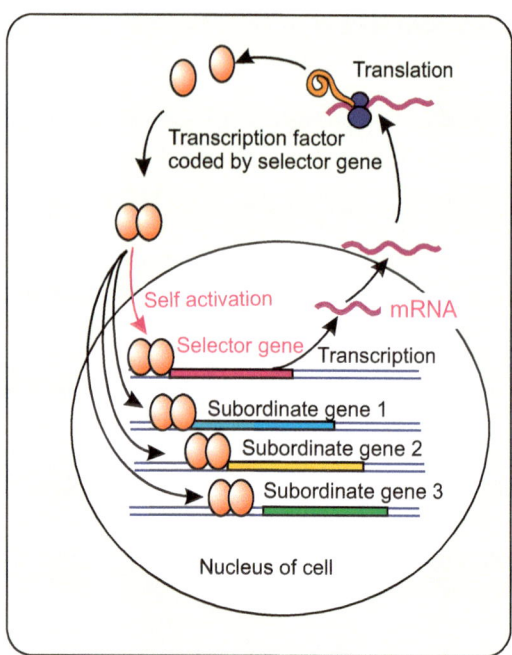

Development-controlling selector gene
(e.g. homeotic gene or *myoD*)

Fig. 12.15   Functional principle of a selector gene (master gene). The gene codes for a transcription factor that
controls downstream genes and also switches on its own
gene thus sustaining its own activity

Hom/Hox genes. (Henceforth we prefer to use
the pooling term *Hox* genes.)

It then turned out that these genes encode
transcription factors that share a DNA-binding
string of 60 amino acids termed a **homeodomain**
(Fig. 12.14) coded by a particular section of the
gene, called a *homeobox*. The homeodomain
binds DNA through a **helix-turn-helix** (**HTH**)
structure. Moreover it turned out that many other
transcription factors the mutation of which does
not cause obvious transformation of morphological traits, also contain a homeodomain and therefore are usually not classified among the
canonical homeotic genes. In humans, the
homeobox gene family contains an estimated
235 functional genes, many of which are
involved in cell differentiation, some of which
act as tumour suppressors. In this section focus is
laid on the classic homeotic genes, the genes of
the *Hom/Hox* class.

(see Fig. 4.28). Rather, the homeotic genes specify
the particular quality and attributes of a segment
or appendage – whether a segment will become
the wingless prothorax or the winged mesothorax,
for example, or whether an imaginal disc of an
appendage will acquire the characteristics of an
antenna, a foreleg, a middle leg or a hindleg. The
genes are designated *Hom* in *Drosophila* and *Hox*
in other species and collectively referred to as

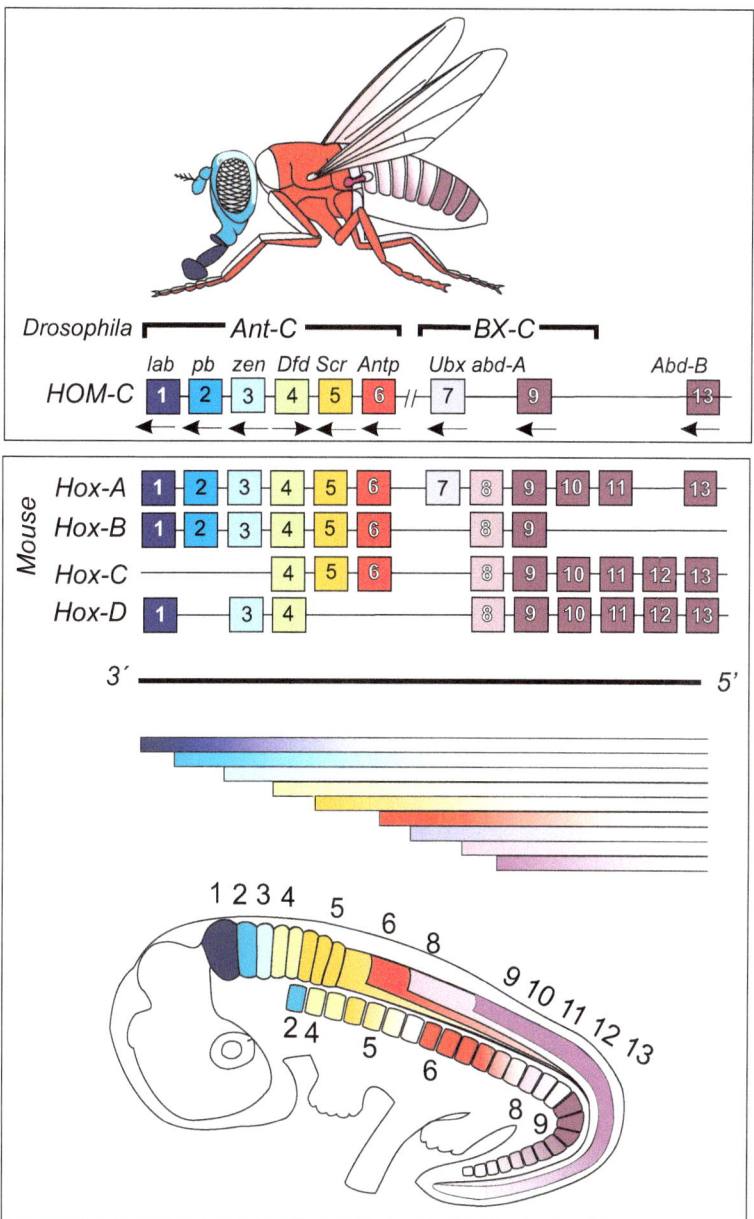

Fig. 12.16 Colinear organization and evolutionary conservation of homeotic gene organization and spatiotemporal expression patterns in *Drosophila* and the mouse. *Hom-C* comprises *Antennapedia (Ant-C)* and *bithorax (BX-C)* clusters of the homeotic genes on *Drosophila* chromosome 3. Arrows indicate the direction of transcription. The numbered members of a cluster are homologous in the sense of paralogue; in evolution their homeoboxes arose through repeated gene duplication. The corresponding *Hox* genes in the mouse are present in four clusters (*Hox-A, Hox-B, Hox-C, Hox-D*) putatively due to duplication of the entire genome in early vertebrate evolution. Matching numbers or colours in columns indicate particularly strong structural similarities across species (after McGinnis and Krumlauf (1992)). All mouse homeobox genes are successively transcribed in the same direction, starting at the 3' end. In the animal itself, the genes expressed in more anterior regions are transcribed earlier, whereas those expressed in the posterior body regions are transcribed later. In the mouse, the expression patterns in the central nervous system and in the somites are shown. Expression domains overlap and, as a rule, only the anterior boundary is sharply demarcated, while in the posterior direction the expression domains fade out. *lab = labial, pb = proboscipedia, Dfd = deformed, Scr = sex combs reduced, Antp = Antennapedia, Ubx = Ulthrabithorx, abdA = abdominal A, abdB = abdominal B* (after McGinnes, De Robertis, Gehring)

Homeotic Transformation

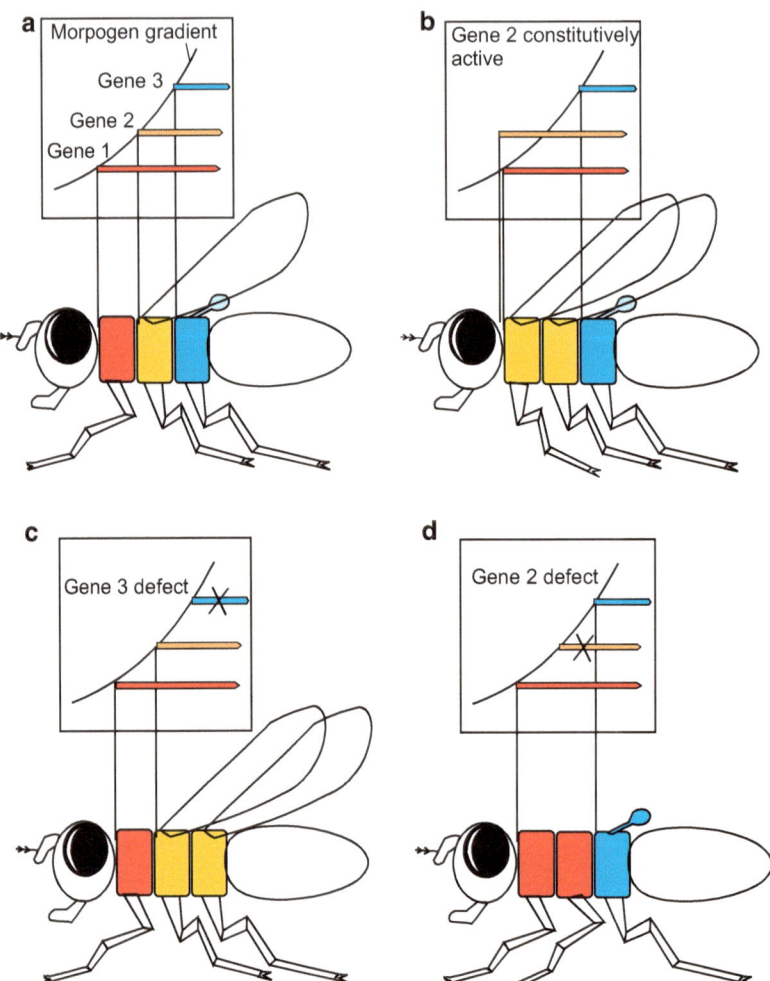

Fig. 12.17 Model tentatively explaining homeotic transformations in *Drosophila*. A basic assumption is the existence of a morphogen the concentration of which increases from anterior to posterior. With increasing concentration the (hypothetical) morphogen switches on one gene after the other of the *Hom/Hox* clusters. In the *Hom/Hox* group gene 1 is the most anteriorly located and turned on by low morphogen concentration as it exists in the anterior thorax region. The activated gene 1 causes the segment to form prothorax; the gene 2, switched on with higher concentration, causes the segment to form mesothorax, gene 3 organizes metathorax, and so forth. The rule is that a gene with a higher number dominates over the genes with lower numbers (rule of posterior dominance). If a mutated gene fails to exert dominance, the anterior gene remains in charge in the following segments until a gene with a higher number assumes command. If a gene is constitutively expressed a segment with posterior characteristics is formed too far anterior

Subsequent research has assigned a pivotal role in the evolution of metazoan phyla to the homeotic genes for such key genes are present in all 'higher' animals (arthropods, annelids, vertebrates). This view is strongly supported by an astonishing conservation not only in the nucleotide sequence within the homeoboxes of such genes but also in their arrangement within the genome. Frequently,

Hox Gene Expression in the Mouse Embryo

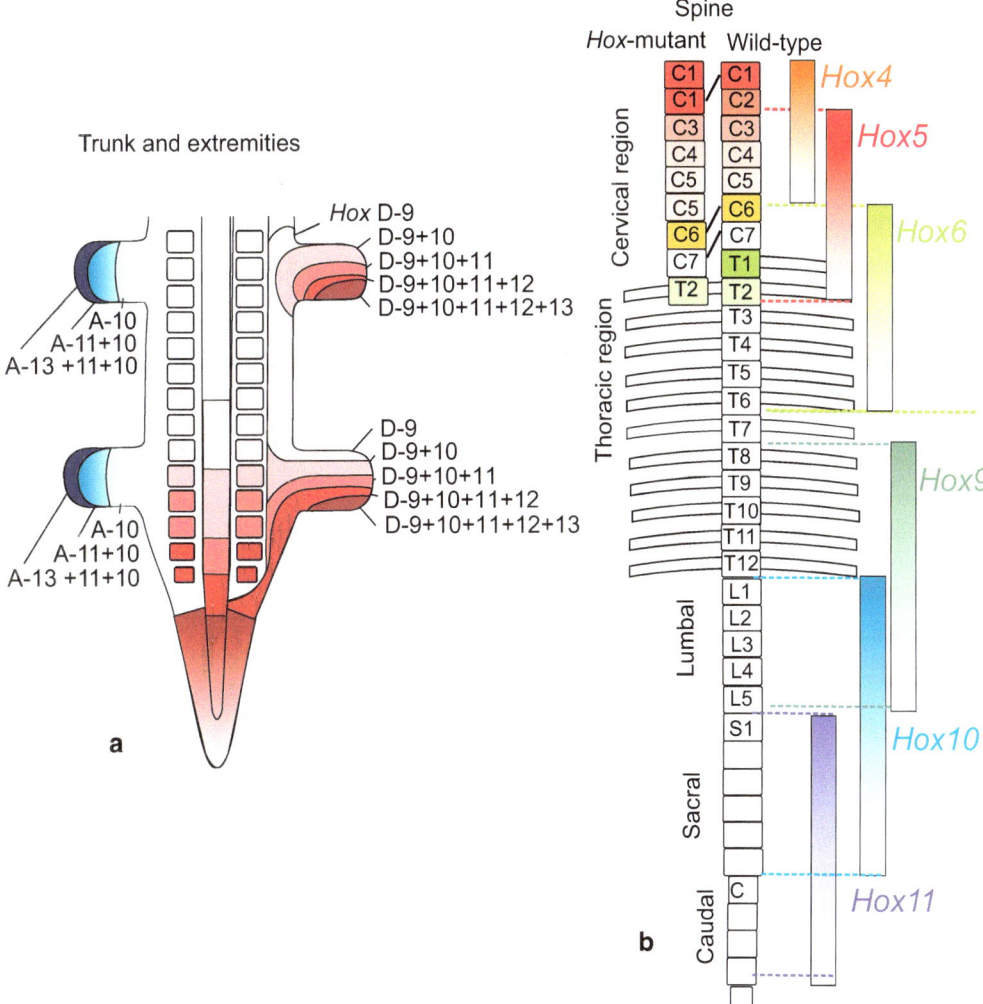

Fig. 12.18 Expression pattern of (**a**) *Hox-A* and *Hox-D* genes in the trunk and limbs and (**b**) of *Hox4* to *Hox11* in the spine of the mouse embryo. The gene with the lowest number is expressed first, the gene with the highest number is expressed last. In the limbs all expression waves of the *Hox-A* series start at the distal margin and extend more or less in the proximal direction. Likewise, the expression waves of *Hox-D* genes start in the most posterior areas and extend more or less in the anterior direction. The expression domains of genes with higher numbers are gradually less expansive than the expression domains of genes with lower numbers. In the spine the anterior boundary of the expression domains are sharply demarcated, while in the posterior direction the expression domains fade out

Hox genes are positioned along the chromosome in a sequence corresponding to the spatiotemporal pattern of their expression in the body. This applies also to vertebrates including mammals (Fig. 12.16).

The *Hox* class of genes act as **master or selector genes** that control other, downstream genes (as illustrated in Fig. 12.15). **The succession of the *Hox* genes in direction from 3′ to 5′ on the chromosome corresponds to the spatiotemporal pattern of expression along the antero-posterior axis** (This sequence concerns the **expression** in the body, not the direction of transcription along the chromosome; the individual genes are **transcribed** mostly in 5′ to 3′ direction, see arrows in Fig. 12.16). The order along the chromosome may be an accident of evolution but now represents a temporal program. The genes appear to be activated by some process that spreads along the chromosome. The distance this process propagates roughly correlates with the body region. In the most anterior head region only the first gene is expressed, and the last gene of the cluster is expressed in the most posterior body part.

One speaks of **colinearity** in the position of the genes along the chromosome and the body regions of their expression. (Exceptions are found, for example in ascidians and molluscs.) Along the main body axis a rule of **posterior dominance** is valid, meaning that in a given body part the gene situated posterior in the cluster suppresses the gene situated more anterior in the cluster (Fig. 12.17). Using the example of ANTENNAPEDIA protein it can tentatively be deduced how the sensory important antennae might have arisen in an ancient primordial arthropod that had legs on each segment (Chap. 22).

In *Drosophila* the set of homeotic master genes is pooled on the third chromosome. The complex, called *Hom-C*, is subdivided into two subclusters, called the ***Antennapedia*-complex** (*Antp-C*) and the ***bithorax*-complex** (*BX-C*). In vertebrates and other animals one speaks of ***Hox* genes**. (Frequently used but not strictly fixed nomenclature: *Drosophila* HOM, mouse *Hox*, human *Hox* or *HOX*, *Xenopus XHox*, in *Nematostella* and other cnidarians *cnHox* = *cnidarian Hox*; but according to new rules prefixes indicating a species should be avoided).

In the evolution of vertebrates a fourfold set arose by twofold duplication of the entire cluster: ***HoxA, HoxB, HoxC, HoxD*** (Fig. 12.16). The four mammalian *Hox* clusters are found on four chromosomes:

- *HoxA-1 to A-13:* Human chromosome 7, Mouse chromosome 6
- *HoxB-1 to B-13:* Human chromosome 17, Mouse chromosome 11
- *HoxC-1 to C-13:* Human chromosome 12, Mouse chromosome 15

- *HoxD-1 to D-13:* Human chromosome 2, Mouse chromosome 2

(As often observed in duplicated genes and gene complexes, not all 13 genes are present in each complex, some were deleted.)

Particularly remarkable is that along of each cluster there are genes the homeobox of which displays a particular high degree of sequence similarity to the genes at the same position in the *Antennapedia/bithorax* complex in *Drosophila*. Such sequences with high similarity are considered homologous but present science distinguishes between two kinds of homology: **Orthologous genes** (or sequences) are found in **different species**, such as *antp* in *Drosophila* and *antp* in the mouse (classical homology), whereas **paralogous genes** (or sequences) are multiplied genes found in similar edition **in one and the same species**, such as the members of the *Hom/ Hox complex* within the mouse, for example. (The terms paralogous and orthologous are further defined in the glossary.)

The high congruency of all homeoboxes suggests that the boxes once arose from a common ancestor sequence which was linked as a module with the parts of the genes outside this module by recombination (more about that in Chap. 22, Fig. 22.2). Outside the *homeobox* the genes are quite different, though genes with identical numbers and positions in the cluster in various animals also share similarity outside the box.

Congruence and compliance between the situation in vertebrates and the fly are striking:

- **The anterior boundary of the expression domains are sharply demarcated, while in the posterior direction the expression domains fade out.**
- **The anterior boundaries are tiered**; the expression domain of the next gene of a paralogous group starts a length in more posterior direction. The gene *Ultrabithorax* (*Ubx*) of the bithorax complex, for example, is in *Drosophila* expressed from the middle of the first thoracic segment T1 down to the posterior end of the abdomen, the following gene *abd-A* from the middle of segment A2, and the gene *Abd-B* from the middle of segment A4 down to the posterior end of the abdomen.

Both the *Abd-B* gene of *Drosophila* and the sequence-homologue (orthologoue) gene *Hox-B9* of the mouse are active in the posterior territory of the embryo.

In vertebrates changing expression patterns are seen spreading wave-like over large areas of the embryo:

1. **Longitudinal body axis**. All members of the *Hox* cluster are expressed with sharp anterior borders but these borders are not at exactly the same position for all corresponding genes of the four clusters. Rather, the expression patterns are different in the neural tube, in the string of somites and in the intestine (Figs. 12.16 and 12.18). It should be pointed out once more that from these embryonic structures quite different cell types will arise: nerve cells, muscle cells, cells of connective tissue, cartilage and bone, gland cells, and some more.

   Like in *Drosophila* also in the vertebrate **a mutation in the homeotic genes results in transformation of a structure into another one** belonging to a more anterior or a more posterior segment.

   For homeotic transformations in vertebrates the basic rule illustrated by the model Fig. 12.17 holds as well: A gene that in the *Hom/Hox* cluster takes a follow-up position and is expressed in more posterior territory of the body dominates over its neighbouring genes in the cluster located ahead from it. If a gene is inoperative by mutation the more anterior gene remains in charge until it is silenced by the next intact posterior gene.

   • In the fly, for example the metathorax the wings of which normally are reduced forming the halteres, can transform into a mesothorax which normotopically bears regular wings (Fig. 12.17; see also Fig. 4.40).

   • In the mouse, for example, at the upper end of the backbone where the first three vertebrae are normally formed as proatlas, atlas and axis, the atlas can be transformed into an additional protalas or an additional axis.

2. **Limb bud, antero-posterior axis** (Fig. 12.18): At the posterior border of the limb bud a centre forms in which expression of the *Hox* genes *D-9* up to *D-13* starts. This centre is virtually congruent with the signal emitter known as **ZPA** (Zone of Polarizing Activity), which emits the extracellular signalling protein SONIC HEDGEHOG. The expression zones propagate wave-like, the *D-9* zone farthest, the *D-13* zone only a short distance. This rule, however, holds only for the first phase of expression. More expression phases follow with altered patterns. Some expression zones propagate in two consecutive waves (e.g. *Hox D-11*).

   However, the proteins encoded by the *Hox* genes are transcription factors acting in the nuclei. It is therefore unlikely, that they themselves spread over areas of cells. Rather, their expression follows the spread of some kind of signalling molecules, with retinoic acid RA or SONIC HEDGEHOG as candidates.

3. **Limb bud, proximo-distal axis** (Fig. 12.18): The outgrowth of the bud is accompanied by wave-like and colinear propagation of *HoxA* gene expression in ascending order : *A-9, A-10* ... up to *A-13*; and the expression zones becoming progressively smaller.

   It might be reasonably expected *Hox* genes might turn on regional-specific gene expression. Yet, relationships between local differentiation-realising gene activities and Hox expression domains is rather indirect. This can be inferred from the observation that the genes *HoxD-1* to *HoxD-13* are required to drive development in the trunk as well as in the anterior and posterior extremities. *HoxD-13* gene products are found in the growing tail tip but also at the tip of the upper and lower limb. *Hox-A* and *Hox-C* genes are switched on in colinear order also in the intestine. Regional specificity results from regional-dependent **combinations of *Hox* genes**, the ***Hox* code**. In the succession of the *Hox* genes along the chromosomes including their cis-regulatory (– that is lying on the same

DNA strand –) sequences a **temporal expression program is encoded**.

A further, remarkable finding: In the development of the central nervous system, in the development of the axial organs such as the spine, and in the sprouting limb buds the expression patterns of the various *Hox* groups are under the control of organizing centres that emit inductive signals, such as (in temporal order):

- the Spemann organizer (upper blastopore lip in the amphibian gastrula, primitive node on the blastodisc of reptilians, birds, mammals);
- the FGF-10 transmitter in the centre of the limb field in the flank of the embryo;
- the ZPA at the posterior margin of the limb bud which emits SONIC HEDGEHOG.

The example of the limb bud demonstrates a complex interplay: *Hox* genes activate signal transmitters in the trunk the signals of which (FGFs) in turn activate *Hox* genes in the limb bud and so forth. Certainly, many more players, presumably thousands of genes, and non-appreciable numbers of signalling molecules, are involved in controlling the development of complex structures.

Historical Paradigms: The Fly with Legs on its Head and the *Antennapedia* Gene

Homeotic genes were discovered in the first place by the emergence of monster flies in inbreeding crosses of *Drosophila*. Flies appeared with two legs instead of antennae on their heads (see Fig. 4.40): **correctly shaped structures on the wrong place!** Responsible for this **homeotic transformation** were mutations in a gene that was aptly named *Antennapedia* (*Antp*). This is one of the genes of significance when the identity of body segments and body regions is specified. The expression of only one mutated gene results in – the as such correct – body structure at the wrong place.

In general terms, in wild-type *Drosophila* the *Antennapedia* gene activates downstream genes that specify and realise the structures of the mesothorax, with its legs and wings, and represses genes involved in antenna and eye development.

The function of the *Antp*-derived ANTP protein was enigmatic for a long time. Eventually the following interpretation was proposed: The head and thorax region is controlled by a **hierarchy of selector genes**. In short:

- **Wild-type:** There are selector genes that turn on 'leg program' by controlling sets of subordinate downstream genes, and there is another selector gene, *homothorax* (*hth*), that modifies the basic 'leg program' to the extended 'antenna program'. In the mesothorax functional wild-type ANTP suppresses the antenna-specifying *hth* selector gene. The basic leg program is the default option and runs in the mesothorax automatically, whereas in the head no ANTP is present to suppress *homothorax*; therefore, here the leg program is converted to antenna program.
- **Legs transform in antennae.** If *Antp* fails no functional ANTP represses *homothorax* in the mesothorax and the antenna program gets a chance. The mesothorax falsely converts its legs to antenna-like appendages.
- **Antennae transform to legs.** Normally *Antennapedia* is not expressed in the head and the *homothorax* gene switches on, besides the basic leg program, the additional antenna modifier program. In the dominant gain-of-function mutant 'Antennapedia' the *Antennapedia* gene is erroneously expressed not only in the thorax but also in the head. The ANTP protein suppresses the antenna modifier program, and the default basic leg program runs also in the head.

For a better understanding of homeotic transformations a simplified model is shown in Fig. 12.17.

The Homeotic Genes of the *Hom/Hox* Class Aid in Demarking Locations and in Specifying Peculiarities of Body Regions

We provide a definition of terms beforehand to avoid misunderstandings we repeat what has been stated above in Sect. 12.2.1: Many development-controlling genes contain a DNA sequence known as homeobox. This rather short section of the gene encodes a DNA-binding, 60 amino acid long **homeodomain** as part of a larger protein that functions as a transcription regulator (Fig. 12.14).

Not all of these genes are classified among the canonical homeotic genes in the narrow traditional sense of the word. For example many of the human homeobox containing genes are involved in cell differentiation. They do not cause, when mutated, conspicuous transformations of visible morphological traits into other traits; their mutation has more subtle effects or is lethal.

In the development of *Drosophila*, in the development of vertebrates, and, as far as current knowledge suggests, in the development of all bilateral symmetric animals classic homeotic genes specify the particular quality, the individual identity, of the originally uniform segments, whether they turn into parts of the head, of the thorax or of the abdomen. The segments are modified according to their position along the body axis. As a rule, homeotic genes are not exclusively active in one particular segment but in larger body regions. It is the **Hox-code**, the specific position-dependent **combination of Hox genes** that specifies local qualities. In other species this code may be shifted along the main body axis or more profoundly altered. Accordingly, body plans diverge (Chap. 22).

A most remarkable aspect is that all the classic homeotic genes are transitorily expressed in all cells of a given body region under their direction, whether the cells will become epidermis, muscles or nerve tissue. The same holds for the genes of the *Hox-A* and *Hox-D* families in the trunk and the extremities of vertebrates (Fig. 12.18).

In general the expression domains of several *Hox* genes mark body regions, and they are expressed in these regions in several types of future tissues. In this aspect they differ from those genes which specify cell types, such as *MyoD1* (see Sect. 12.3). Whether future cartilage, muscle or nerve cell, they express the same *Hox* genes if they inhabit the same body region marked by the same **Hox code**.

12.3 Genes for Programming Eyes and Cell Types

12.3.1 A Monster Fly with 14 Eyes Shows that Eyes in The Animal Kingdom Share More Than Morphologists Ever Dared to Think

The analysis of genes required for the development of the eye culminated in one of the most spectacular experiments of contemporary developmental biology: in the development of additional eyes on the antennae, legs, wings and halteres of a fly (Fig. 12.19).

In *Drosophila* several mutations are known leading to crippled eyes or to complete loss of the eyes. Among the genes is *eyeless (ey)*. The gene is homologous (orthologous) to the gene *small eye (Sey)* of the mouse, better known as *Pax-6* of the vertebrates in general, and to the human gene *Aniridia*. If a human embryo had inherited the mutated allele *Aniridia* twice (that is, if it is homozygous ($-/-$) with respect to this defective allele), it dies prematurely before term. Heterozygous humans ($-/+$), who have one normal and one defective allele, lack the iris.

The genes referred to are elements of a regulatory network of transcription factors that synergistically conduct eye development and are equipped with a homeodomain and, in addition, with a further DNA-binding domain called a PAX domain, derived from the *paired box* of the gene. In this network some genes take a leading role. The best known example of this *Pax* family is *Pax6*. However, *Pax* genes are not only expressed in the eye primordia but also in further parts of the brain, in the spinal cord, the head and some even in the anterior trunk. They do not denote cell types or organs but label anterior body regions.

Although these genes are interconnected forming a network the failure of one single gene can disturb eye development severely. Failure of eye development upon failure of one of those genes is not so much astonishing than is the

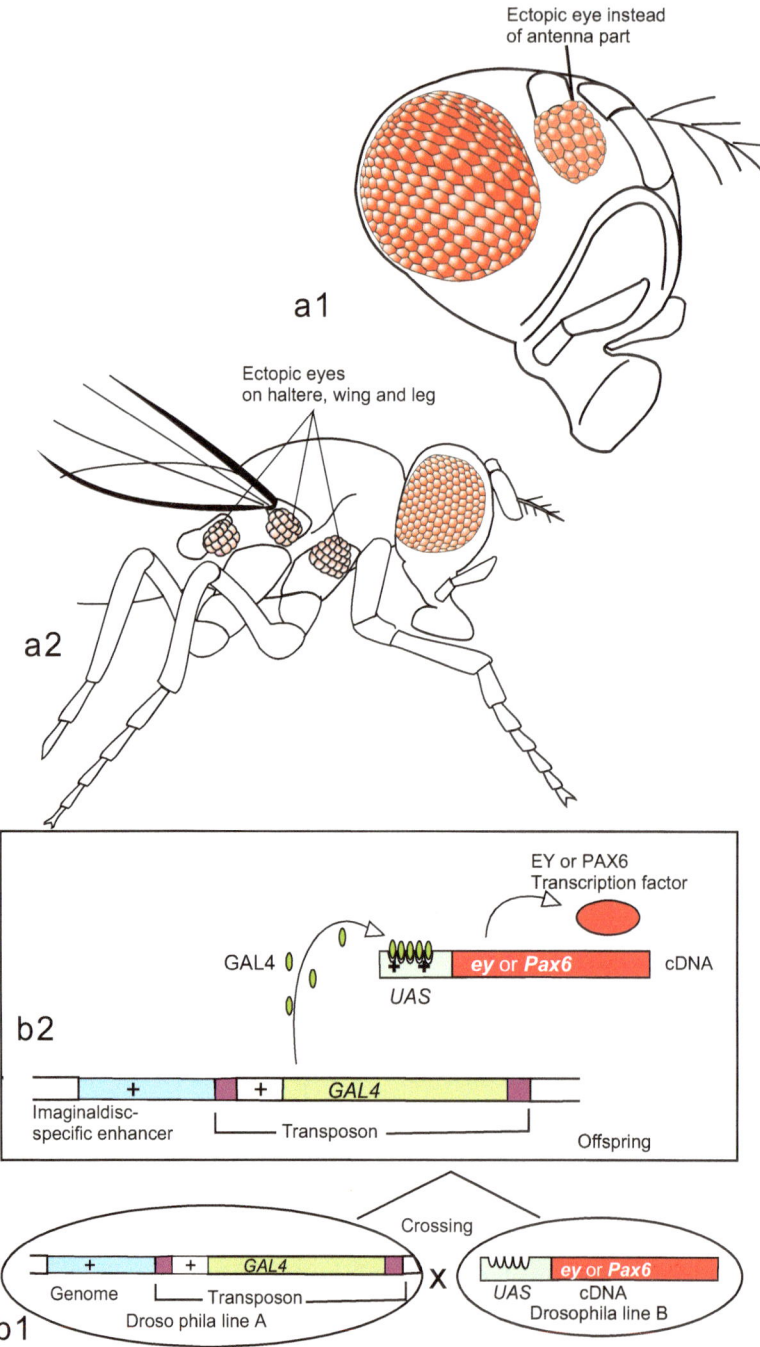

Fig. 12.19 (**a**) Ectopic eyes in *Drosophila*. A fruitfly expressing the wild-type (intact) *ey* gene in novel (ectopic) places. The ectopic expression of intact *ey* can lead to additional (small) compound eyes (after Halder et al. (1995)). (**b**) Genetic manipulation leading to supernumerary eye formation. As background for the experiment, a transposon containing the yeast gene for the transcription factor *GAL4* was randomly inserted into the *Drosophila* genome. For further crosses individuals were selected that had the *GAL4* sequence downstream of imaginal disc-specific enhancers. These could be identified because the GAL4 protein activated a *UAS* + *LacZ* reporter construct which also was inserted anywhere in the genome of those flies. When the *GAL4*

inverse situation: **Ectopic expression of only one of these genes can initiate eye development at incorrect locations**. This was shown in the following experiment:

A team guided by Walter Gehring, Basel, first produced transgenic strains of *Drosophila* the larvae of which expressed the bacterial transcription factor GAL4 in patches of their imaginal discs under the rule of imaginal disc-specific enhancers. Another strain was provided with an additional wild-type edition of the *ey* gene (in form of replicable cDNA) brought under the control of a UAS promoter; this promoter is GAL4 responsive. The respective strains were crossed. In the descendants individuals were found harbouring both the GAL4 gene and the UAS + ey construct (Fig. 12.19b). Driven by the disc-specific enhancers GAL4 was produced in patches of the imaginal discs. The expressed GAL4 factors for their part activated the UAS promoter and thus switched on the *ey* gene in those patches. Upon completion of metamorphosis, flies hatched from the pupa displaying additional compound eyes on the antennae, legs, wings, and halteres (Fig. 12.19a). The EY protein reprogrammed groups of cells to become eye precursor cells. Eyes located at the antennae grew nerve connections to the brain.

For the first time ectopic development of an organ was not evoked by transplanting inducing tissue or injecting inducing factors but by genetic manipulation. Using these tricks even in otherwise completely *eyeless* (ey−/−) mutants eye development could be induced in imaginal discs.

The most exciting result, however, was that instead of the *Drosophila* gene *ey* also the **mouse gene *Pax-6*** can be used to conjure up supernumerary eyes at incorrect places. Conversely, the ***Drosophila ey* gene can evoke eye development in vertebrates (*Xenopus*) as well**.

Astonishingly enough, the development of eyes so fundamentally different as are the compound eyes of arthropods and the camera eyes of the vertebrates, is initiated by homologous (orthologous) genes. Like other *Pax*-related genes the *ey* gene comprises a *homeobox* and a *paired box*, which provide the EY protein with a DNA-binding homeodomain and a DNA-binding PAX-domain. Apparently the *ey* and related genes control target genes that are required in the programming of both types of eyes irrespective of their different anatomical construction. In Chap. 22 we will point out, that irrespective of huge differences all animal eyes share some basic properties. But the further, detailed elaboration of the various eye types is bound to species-specific competences.

Further genetic studies revealed that the *ey* (*eless*)/*small eye*/*Aniridia* gene in turn is under the rule of genes with higher positions in the hierarchy, such as the gene ***twin of eyeless*** and the gene ***sine oculis***, a gene that even is present in hydromedusae (jellyfishes) possessing ocelli (small eyes). In flies *sine oculis* takes care that the adult fly is not only supplied with compound eyes but also with small ocelli on the forehead.

Genes controlling eye development will be in our focus again when we analyse eye development in more detail (Chap. 16, nervous system and eyes) and discuss their putative role with respect to the evolution of the eyes (Chap. 22).

Fig. 12.19 (continued) gene was activated in an imaginal disc, the GAL4 transcription factor stimulated the expression of *LacZ*; the presence of LacZ protein (bacterial β-galactosidase) is made visible by a blue stain. Flies having blue spots in any of their imaginal discs were selected and inbred yielding several *Drosophila* strains type A that expressed *GAL4* in the discs for legs, wings, halteres, or antennae. In a second *Drosophila* strain B a *UAS + ey (or Pax 6)* construct was introduced in the form of a replicable cDNA. This construct was not integrated into the host's genome but could be replicated and transmitted to offspring. Upon crossing strain type A with strain type B offspring were selected that contained both the GAL4 transgene in their genome downstream of disc-specific enhancers and also had inherited the *UAS + ey* cDNA. Now GAL4 could activate not only the *LacZ* gene but, in addition, also the *ey* or the homologous vertebrate *Pax6* gene

12.3.2 A Perfect Paradigm of Cell Type-Specific Control Genes: The *Myod/Myogenin* Family Program a Myoblast and Its Descendants Take Over the Program

How is a particular cell type programmed? One peculiar type of cell has yielded its developmental secrets more readily than any other: the **cross-striated muscle fibre**. Skeletal muscle fibres are large, multinucleated syncytial cells formed by the fusion of several myoblasts. Myoblasts are cells of mesodermal origin. They become programmed step by step: before and during gastrulation they become committed to take the mesodermal path, and in the somites they constitute the part termed myotome where they are definitively fated to become the precursors of the striated muscle fibres. This gradual programming takes place under the influence of inducing factors (Chap. 10). In the course of programming central **myogenic key genes** are switched on.

The first myogenic ('muscle generating') gene identified was the *Myoblast Determining gene number one, MyoD1*. This gene is a **master gene**, also called **selector gene**, which brings other, subordinate genes needed for terminal differentiation under its own control (Fig. 12.20). Through transfection with *MyoD1*

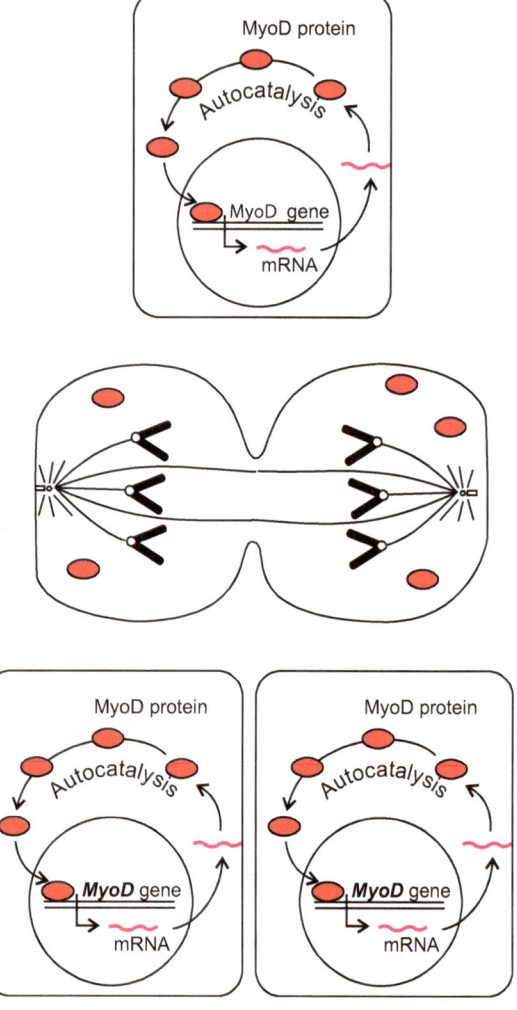

Fig. 12.20 Auto-regulatory self activation of a transcription factor depicted using *MyoD* as example. The factor activates transcription of its own gene. After cell division this positive feedback (autocatalysis) leads to resumption of MyoD synthesis, and the division-caused dilution is compensated

mRNA, fibroblasts (precursors of connective tissues and cartilage) and adipoblasts (precursors of fat cells) can be reprogrammed and prompted to become stably and heritably committed myoblasts. This is an example of a **gain-of-function** experiment.

In order to control other genes, MYO-D1 protein is provided with a **basic helix-loop-helix domain**. This domain attaches to the control region, called the E-box, of the genes to be ruled. MyoD1 protein acts as a transcription factor. It not only controls other genes, but also activates its own production. This is the cleverest aspect of the whole matter: MyoD1 physically binds to the upstream controlling region of its own gene, keeps it active, and thus amplifies and maintains its own production. This is referred to as **autocatalysis** or **positive feedback**.

During DNA replication transcription factors detach but this does not cause problems. After completion of cell division, transcription factors return into the nucleus searching for their target promoters. MyoD1 will be among them, will find the E-box upstream of its gene and increase its own production again. MyoD1 is not diluted out – both daughter cells remain myoblasts and the state of determination survives cell division. The muscle cell is stably committed to its task.

From such an important gene one expects that a loss-of-function mutation would result in offspring without muscles and incapable of life. Yet, to the surprise of researchers who conducted this experiment, targeted mutagenesis (Box 12.2) did not have this effect, even when the defective gene was homozygous due to inbreeding (**knock-out** or **null mutation**).

This result, while initially disappointing, paved the way to surprising new results and understanding. Perhaps because the gene is so important there exists not only one key gene but several similar genes that can replace one another, just as we usually have more than one key for our house and car. This phenomenon is referred to as **genetic redundancy**. In addition, various myogenic selector genes might be of significance in the programming of muscle subtypes.

At present, four myogenic selector genes are known in mammals: ***MyoD1, myf-5, MRF-4*** and ***myogenin***. All these genes belong to the same gene family – that is, they display a high degree of sequence similarity, and all four genes code for DNA-binding transcription factors containing a **bHLH (basic helix-loop-helix) domain.** When injected into fibroblasts they are each able to start the muscle differentiation program.

During embryo development, *myf-5* and *MRF-4* are expressed first, but only transiently. Subsequently, the *myoD1* and *myogenin* genes are activated. The MYOGENIN protein is exceptional: it only appears when the myoblasts fuse to form myotubes and start constructing their contractile machinery. MYOGENIN is constitutively produced, and it is indispensable: *Myogenin* null mutants are lethal. In this case there is clearly no genetic redundancy. Genes belonging to the same family cannot always replace each other, and the organism is not completely insured against all disastrous mutations.

NeuroD and the Relatedness of Genes That Program Nerve and Muscle Cells

Self-evidently, the detection of muscle-specific key genes has prompted the search for master genes launching the differentiation of other cell types. The main interest was focused on nerve cells. The tedious search has led to the identification of several genes with essential functions in nerve cell development. Such genes are generally termed **neurogenic** or **proneural genes,** and the first identified genes were given the names *NeuroD* and *neurogenin* in analogy to *Myod1* and *myogenin*. This nomenclature suggests homology, but sequence similarity is only distant though all these genes code for transcription factors possessing a bHLH motif.

In terms of function the *achaete scute-complex* (*AS-C*) of *Drosophila* which consists of four genes (*achaete*, *scute*, *lethal of scute*, and *asense*) corresponds the *NeuroD/neurogenin* genes of the vertebrates, but the relatedness is somewhat confusing as in molecular terms the *neurogenins* are more closely related to the proneural gene *atonal* of *Drosophila*. On the other hand all the mentioned muscle-specific and neurogenic/proneural genes share the bHLH motif but without revealing much congruousness in other parts of the genes. Moreover, there are cross-congruencies also in terms of function in that the *AS-C* is involved not only in specifying nerve and sensory cell development but also in muscle cell development.

Irrespective of open evolutionary and functional relationships between the genes involved, in nerve cell specification a cascade of gene activities passes off until eventually a cell is committed to a distinct task, whether it has to become a neuronal or glial cell and which subtype it has to form (more about that in Chap. 16).

The more genes are identified which are of significance in programming cell types of the nervous system, the more it becomes obvious, that several of these genes are involved not only in the development of the nervous system proper but also in the development of sensory organs and muscle cells. This applies also to the proneural genes of the *zic/odd-paired* class which code for transcription factors with zinc finger domains. In retrospect this insight is not so surprising. The common denominator appears to be that all these cell types are excitable cells capable of generating and conducting electrical signals. In the early period of metazoan evolution excitability was basic for coordinated responses to environmental cues. In the division of labour one group of excitable cells may have set priorities to receive information, another one to process and transmit information, a third group to respond by coordinated movement.

12.3.3 Developmental Genes and Transcription Control: A Résumé

Development-Governing Genes Often Are Master Genes (Selector Genes) That Control Batteries of Subordinate Effector Genes

Many essential development-controlling genes, such as the above introduced *MyoD1* or the *eyeless/Pax-6* gene, are **master genes**, also called **selector genes**. They produce proteins which for their part are also transcription factors regulating gene functions by binding to the control region (enhancer, promoter) of subordinate effector genes (Fig. 12.20).

Transcription Factors Have Specific DNA-Binding Motifs, Here Is a List

In the development of the basic body plan of *Drosophila* and the mouse, and likely in the development of all multicellular animals **homeotic genes** take crucial roles as transcription factors that are expressed and act in regionally-specific way. In spite of large variety in functional tasks transcription regulatory proteins often share similar structural motifs for binding to regulatory sequences of the DNA. In *Drosophila* more than 30 proteins have been identified which fulfil controlling functions in the construction of the basic body plan. Almost all have DNA-binding domains, many of those possess a homeodomain. At the level of the DNA the motif may be

- **a *homeobox*, in short called *Hox*;** this is a partial sequence of the gene consisting of 180 bp. The *Hox* sequence provides the protein with a characteristic DNA-binding **Helix-Turn-Helix-domain**, in short **HTH motif**, in case of the HOM/HOX proteins termed **homeodomain**. This domain is 60–70 amino acids long and includes two angled α-helix subdomains which insert into the large groove of the DNA. The homeodomains of two such transcription factors can form **mirror symmetrical dimers** (similar to the bHLH transcription factors shown in Fig. 12.14). Being specific transcription regulators HOM/HOX dimers do not bind randomly to DNA; rather the HTH motif recognizes the specific upstream regulatory sequence ahead of the genes to be controlled. Subtle differences in the homeodomain of the proteins and in the regulatory sequence of the DNA (*RE = Responsive Elements*), and cooperative assistance by cofactors, determine where exactly the transcription factor will bind and which subordinate genes it will control. Another DNA-binding domains is

- the **basic Helix-Loop-Helix bHLH** domain (please note: Helix-**Loop**-Helix here, Helix-**Turn**-Helix in case of the homeodomain). In its spatial structure the bHLH looks like the homeodomain but it is composed of different amino acids. Like the HTH domains the bHLH domains form dimers, as a rule heterodimers (Fig. 12.14). bHLH-domains are found in the cell-type specifying transcription factors **MyoD, myogenin, NeuroD, neurogenin** and the genes of the **AS-C** (see above).

Other transcription factors are equipped with

- **Pax domains** with six alpha-helical segments encoded by the *paired box* of the genes are characteristic for the PAX family of transcription factors. Many of the PAX proteins also contain a homeodomain as do the eye-specifying genes *Pax-6* and *ey(less)*. *Pax* genes are widely spread among the entire animal kingdom just like the *Hox* genes.

- **Zinc finger domains**, for instance the *Drosophila* segmentation genes *hunchback* and *Krüppel* code for proteins with zinc finger motifs of the Cys_2-His_2 type, with Cys and/or His residues folding around a zinc atom. Zinc finger has also the GAL4 factor (Fig. 12.19b)

- Also the members of the **steroid hormone receptor family** are among the gene regulatory proteins possessing **zinc fingers** (see Fig. 11.11). This family comprises the intracellular receptors for the steroid hormones, for the hormone **thyroxin (T3, T4)**, for **retinoic acid RA** and for **vitamin D3**. All these gene regulatory proteins possess different hormone-binding domains but similar zinc-enclosing fingers and similar DNA-binding domains. The DNA sequences recognized and bound by the DNA-binding domains are known as *Hormone-Sensitive Elements HSE* or *Hormone-Responsive Elements HRE*, in case of RA *Retinoic Acid Responsive Element RARE.* These

"elements" reside within the promoter sequence upstream of the protein-coding region of the respective genes.

- **T-box proteins** contain a region of conserved sequence, about 200 amino acids long, which encodes a particular sequence-specific DNA binding domain, called the **T-box domain**. These proteins are transcription factors that control developmental pathways. The prototype of this family is the mouse **Brachyury (or T) gene** product. Other T factors are involved in limb and heart development.

- **High Mobility Group (HMG)** factors differ from most other gene-regulating proteins in that they work by **bending the DNA** to bring other regulatory sites into contact with other, assisting transcription factors. Examples are the **SRY factor**, also known as Testis-Determining Factor **TDF,** the *sry* gene of which is located on the Y chromosome (Chap. 6), and **TCF** that together with the cofactor β-catenin controls WNT-dependent genes.

Textbooks of molecular biology and genetics list several more DNA-binding structures such as **leucine zipper**, **POU-domain** as found in the stem cell factor **Oct4,** LIM-domain proteins, and Winged helix proteins. The aim of this introductory book is not to present a complete listing; rather a general principle should be made clear: **Many genes that control basic developmental processes and genes that substantially contribute to the programming of cell types act by encoding transcription factors which keep in check batteries of downstream genes.**

Other development controlling genes code for signalling molecules, signal receptors or elements of the signal transducing pathways, or extracellular enzymes (Chaps. 10 and 11). Often genes for signalling molecules and for transcription factors are coupled. An example:

In the vertebrate embryo the expression domains of the genes *siamois* and *goosecoid* mark the location that is to become the Spemann organizer, the most important signal transmitter.

Both genes, *siamois* and *goosecoid*, encode transcription factors. If *siamois* or *goosecoid* mRNA is injected at other sites of early embryos a second Spemann organizer is established there. A transcription factor triggers the production of an inducing factor which in turn switches on homeobox genes, for instance members of the *Hox-D* family.

A New Chapter in Gene Regulation: Small RNAs Regulate Gene Functions at All Levels

Since the year 2000 in databases pertaining to biological and medical publications more and more articles emerge addressing the role of non-protein-coding small RNA species. They are of significance in the activation and inactivation of transcription and more so in the regulation of translation. We choose as an example the role of the **miRNAs** (microRNA), often also called endo-**siRNAs** (small inhibiting or short interfering RNAs) in translation control, because this type of control supplied the prototype for a lab method to specifically eliminate a defined mRNA molecule (**RNAi method**, Box 12.2). The natural process is depicted in Fig. 12.21, showing how the translation of a certain mRNA is specifically stopped by miRNA. The putative basis is a kind of natural protection against infection by viruses composed of double-stranded RNA. A nuclease-based natural defence mechanism shreds such double stranded molecules. In 1997 this mechanism was detected in *C. elegans* and has since been experimentally utilized to induce gene-specific 'knockdown' through introduction *antisense*-RNA duplexes, the method became popular among researchers. The method can also be applied in genetically unamenable species, and potentially even in human medicine: It yields *knockout*-like phenotypes, enabling analysis of gene function and silencing of adverse gene effects. Organisms have adapted the primal defence mechanism to control and degrade endogenous mRNAs. For that undertaking miRNAs encoded in the genome play leading roles, many miRNAs remained highly conserved in evolution and are activated in all animals investigated in this respect so far.

Development-Regulating Genes Are Connected Forming Interactive Networks: Combinatorics Create Variety

There are genes that enhance and stabilize their own activity by **positive feedback** (autocatalysis). Examples are *MyoD1* (Fig. 12.20), *myogenin, myf-5* and the segmentation gene *fushi tarazu (ftz)* of *Drosophila*. Other genes switch themselves off via negative feedback. Master genes encode transcription factors which regulate (switch on genes at other loci of the chromosome by **transactivation**, or they turn off other genes (**suppression**).

Moreover, the promoter and enhancer regions of genes to be controlled are often accessible not only to one transcription factor but to several. For instance, the hormone-responsive element of some genes can be occupied by heterodimers jointly formed by one receptor for thyroxine and one receptor for cortisol (see Fig. 11.11). The enhancer region of the *Drosophila* gene *evenskipped* possesses 14 binding sites for homeoproteins. This renders extensive interactions between homeotic genes possible. Various transcription factors that aggregate in steps along the control region of genes may synergistically enhance the activity of the regulated gene, or they may cancel each other's effect and diminish the activity of that gene.

In different body regions distinct arrays of selector genes becomes effective in a complex spatio-temporal pattern. With regard to the *Hox* gene products we have learned that in certain body regions a certain set of selector genes is effective which for their part keep various subordinate genes under control. Therewith cascades of gene activations get started (**domino effect**). The production of intermediary signalling molecules enables spatial propagation of such cascades like avalanches. This **hierarchical organization** of gene regulation during

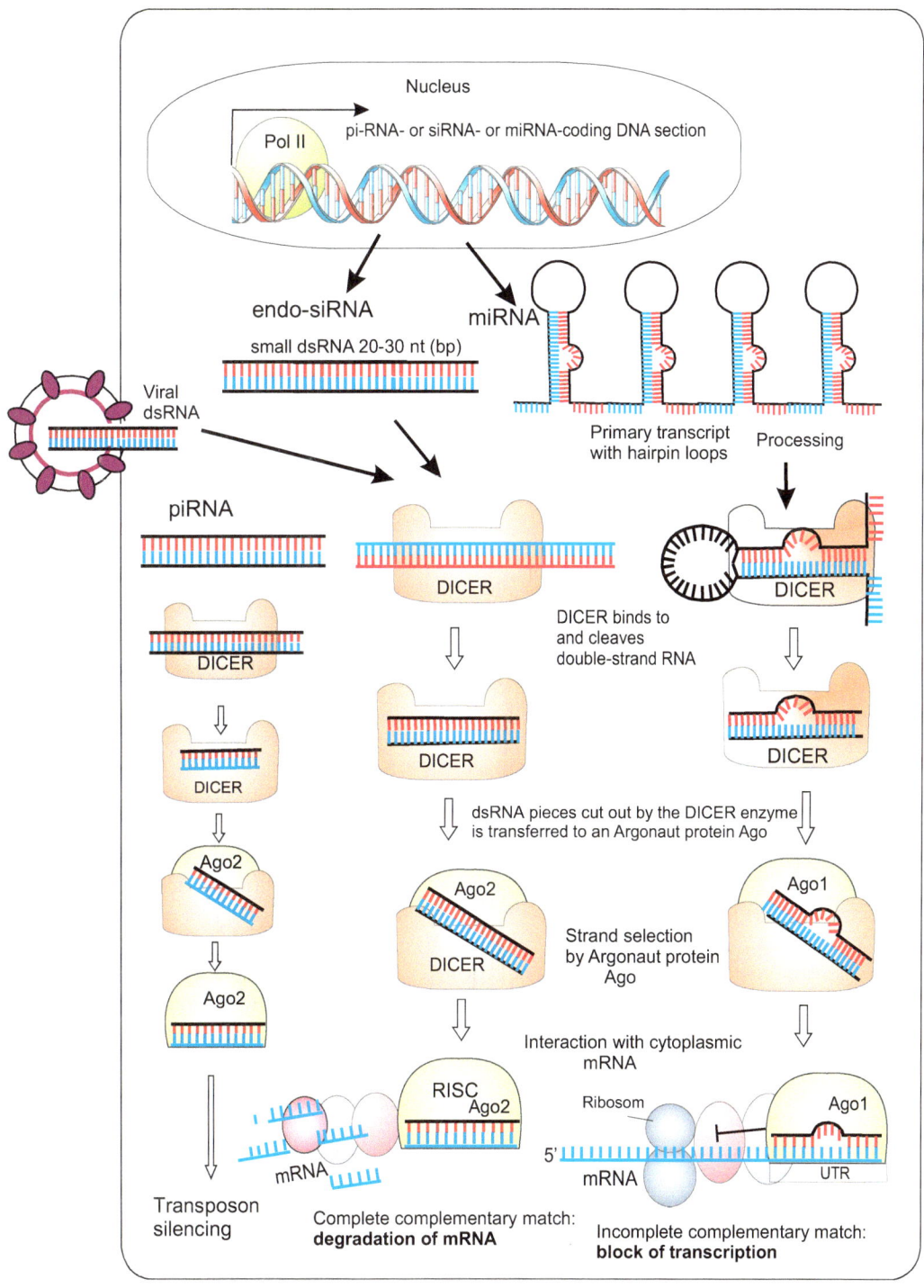

Fig. 12.21 Post-transcriptional silencing or elimination of a mRNA or of transposable elements (transposons) by means of small regulatoryRNA. These RNAs occur as piRNA (PIWI-interacting RNA), siRNA (small interfering RNA) or miRNA (microRNA). They are transcribed from non-protein-coding DNA sequences and cut in pieces by the DICER enzyme complex. miRNAs are transcribed as transcripts containing, as a rule, many *hairpin loops*. From this large transcript small pieces each containing one *hairpin* structure are

development explains why finite numbers of genes can generate a remarkable diversity of body parts, tissues, and cell types in such a precise and consistent manner. For a combinatorial analogy consider that the piano only has 88 keys. Nonetheless, innumerable pieces of music have been composed by varying the spatio-temporal activity of the keys. With a complement in the haploid state of about 10,000–20,000 genes (*C. elegans*, *Drosophila*) or 22,000 genes (human) an inexhaustible diversity of combinations is possible (though, of course, only a limited number of combinations yield a reasonable body plan).

Loss of Determination and Transdetermination: The State of Determination Can Get Lost or Experience a Sudden Change, a Transdetermination

Once programmed the state of determination is normally very stable. Nevertheless cases are known in which the program is lost and subsequently a new program is installed. Cancerous cells frequently lose, partially or completely, the ability to differentiate to maturity and instead they relapse into an embryonic state (Chap. 18). In *Drosophila* the phemomenon of **transdetermination** became known. This is covered in detail in Sect. 4.6 (with Figs. 4.40 and 4.42). Here is a summary:

The larvae have disc-shaped structures called imaginal discs from which the adult fly is assembled in a mosaic-like manner during metamorphosis. The discs stored in the larva are already committed to their respective tasks but not yet terminally differentiated. They can be propagated by dissecting them in pieces and letting the pieces regenerate: In this way clones of discs can be grown. To test their state of determination cloned discs are implanted into the body cavity of larvae about to commence metamorphosis. There they undergo metamorphosis together with their host, and after completion of metamorphosis the products are removed and examined. It will be a leg, a wing or an eye, dependent on which disc had been removed from the donor larva and was cloned. The descendants of this original disc inherited its determination program.

Occasionally it happens that the original program gets lost during the cloning procedure and another program is realized. Instead of the expected leg, a wing is harvested for example. This turn of the determination program is called **transdetermination.** This might happen because upon dissection and subsequent healing cells that are normally separated come into direct contact. The confrontation might lead to phenomena of induction such as those discussed in Chap. 10. This would mean that the cells influence each other mutually. Why mutual interference should cause installation of a new stable program is currently a matter of speculation but not of knowledge. A current hypothesis brings the phenomenon of transdetermination in connection with epigenetic modifications of the DNA, since such modifications are heritable, as discussed in the following section.

Fig. 12.21 (continued) excised. The cut out small dsRNA (double stranded RNA) pieces contain about 20 nucleotides. Exogenous sources of dsRNA can be viruses or synthetic RNAs introduced by the experimenter (see Fig. 12.6). All these short dsRNAs are transferred from the DICER to ARGONAUT (Ago) proteins. One strand of the dsRNA is selected by the Ago complex and presented to the UTR (UnTranslated Region) region of a selected mRNA (or to a transposable element). If there is complete matching between the presented piece and the UTR the Ago complex associates with further proteins, becomes the RISC complex (RNA-induced Silencing Complex) and the mRNA is degraded. If there is incomplete matching translation of the mRNA is blocked. In either case translation is terminated. After Okamura and Lai (2008), Wilson and Doudna (2013) combined, modified and supplemented

12.3.4 Epigenetics: Reversible Alterations in the Accessibility of Genes

The State of Determination Is Transmittable to Daughter Cells Over Cell Generations; One Speaks of Epigenetic Cellular Memory

Epigenetics refers to heritable changes in the phenotype of a cell or organism that are not due to a change in DNA sequence. Programming of cells to enter and follow a certain developmental path involves mechanisms which keep certain genes in a state of accessibility while other genes are decommissioned. The programming (**determination**, **commitment**) frequently takes place shortly before or in the course of DNA replication in the S-phase of the cell cycle. In other cases programming is independent of the cell cycle and proceeds stepwise over a long time period. Frequently '**blasts**' (myoblasts, neuroblasts, erythroblasts etc.) undergo several rounds of cell division until terminal differentiation commences and the program of cell typical protein synthesis is completed.

In committed but still proliferating cell lines the state of determination is transmitted from the starting cell to its daughter cells and this transmission can continue over many cell generations. The cells contain, beyond the knowledge stored in the genome, a memory of what they had learned in the course of determination. One speaks of **epigenetic memory** and **epigenetic cell heredity**. This epigenetic inheritance leads to clones of cells with identical state of determination. Classic examples are the clones of imaginal discs that emerge from the fragmentation and subsequent regeneration of larval discs in *Drosophila* (see Fig. 4.42). But also numerous cell lines kept and propagated in cell culture attest the heredity of the status of determination. From certain myoblasts billions and billions of descendants have been grown. If growth factors are withdrawn the cells fuse and become muscle fibres.

In the following sections mechanisms contributing to the cell memory are summarized.

The molecular details, however, are insufficiently known to date.

Some Genes Can Put Themselves into a State Of Permanent Activity and Repeat Self Activation Subsequent to Each Cell Division

Once more we point out the paradigm of **autocatalytic self-activation** exemplified by the muscle cell determining gene *MyoD1* (Fig. 12.20). Certain master genes (selector genes) encode proteins which immigrate into the nucleus and act as transcription factors activating not only other subordinate genes but also their own genes. Because such transcription factors are also present in the cytoplasm where they are produced on ribosomes, they end up in both daughter cells. Through positive feedback the factors alloted to the daughter cells boost their own production; thus the dilution caused by cell division is compensated. Examples of such persistent self-amplification are, besides *MyoD1* and the related myogenic genes, the segmentation gene *fushi tarazu (ftz)* and *Ultrabithorx (Ubx)* subsystem of the *HOM complex* in *Drosophila*. This system, among others, may contribute to the heritability of the determined state in growing imaginal discs.

Methylation and Heterochromatinization Lead To Shut-Down of Genes Which Outlasts Cell Divisions

Two mechanisms can lead to a permanent functional silencing of genes, **methylation** (Figs. 12.22 and 12.23) and global **heterochromatinization** (Fig. 12.24). In both cases the inactive condition can be transmitted to daughter cells through mitotic cell division.

DNA methylation. In the DNA of eukaryotes cytosine can be converted to **5-methyl-cytosine** (indicated by C*) through addition of a methyl group. Such methylation takes place only in CG nucleotide pairs.

–C-G–
–G-C–

Fig. 12.22 Chemical modification of DNA that can be copied, shown using methylation of cytosine (5-methylcytosine). After completion of the S-phase of the cell cycle, a collection of enzymes restores the original methylation pattern in so-called CpG islands whereby the conserved (i.e. not newly synthesized) DNA strand serves as template

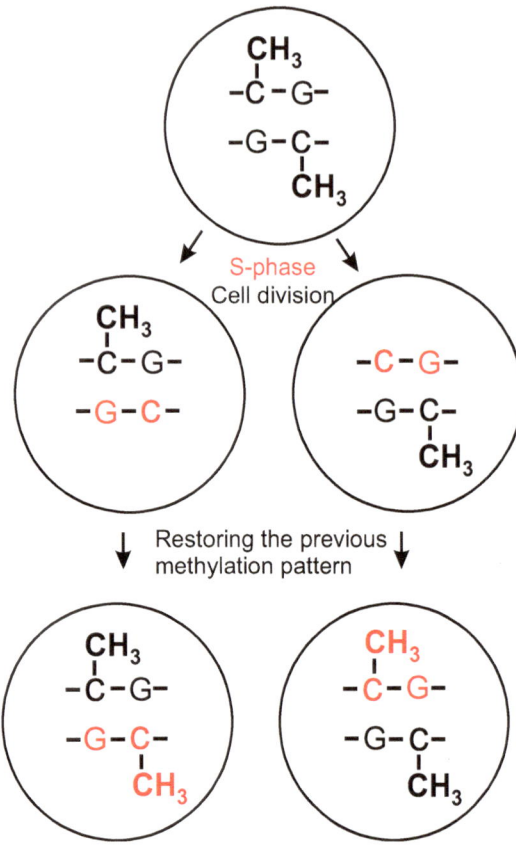

Such patterns of CG/GC nucleotide pairs are enclosed in so-called **5′CpG-3′** islands and found especially in promoter and enhancer sequences. The complementary sequence **5′CpG-3′** is mirror symmetrical (Fig. 12.22) enabling methyltransferase enzymes to copy the methylation *pattern so that both daughter cells exhibit the same pattern (Figs.* 12.22 *and* 12.23*). However, while* this mechanism makes the methylation pattern heritable, any general theory of determination cannot rely on methylation as the only mechanism. Besides DNA methylation also acetylation and methylation of **histone proteins**, and the degree of chromatin condensation are of significance (Fig. 12.23).

In mammals methylation is extensive – about 70 % of the CG sequences are methylated. The consequences of DNA methylation cannot be predicted for each single CG group or each single gene. In general, however, a high degree of methylation results in lower levels of transcription of the methylated gene.

12.3.5 There are Particular Genes the Products of Which Specifically Render Sections of the Chromosomes Accessible or Inaccessible: Once Established Such Conditions Persist Over Cell Generations

The way in which the state of determination can be specifically adjusted and made heritable was

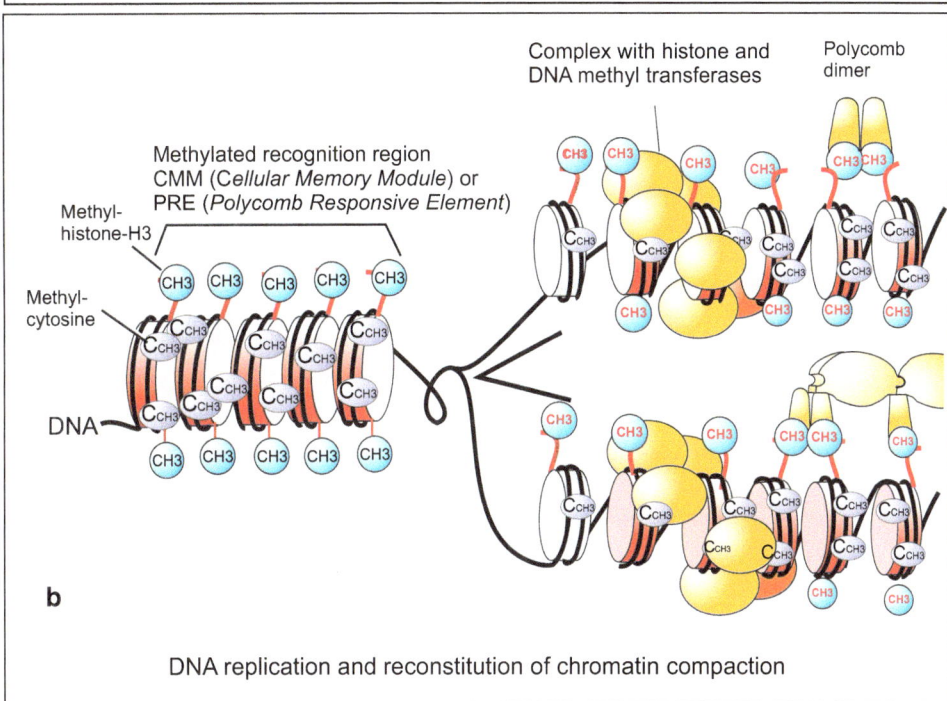

Fig. 12.23 (**a, b**) Model of gene silencing and epigenetic memory as discussed in studies on *Drosophila* development. (**a**) Certain *Hom/Hox* genes selected for silencing over cell generations have a nucleotide sequence in their promoter/enhancer region that has been called *Cellular Memory Module* CMM (Ringrose and Paro 2001). In certain phases of development a complex of proteins containing methyl-transferases binds to CMM. These transfer a methyl group to one or several lysine residues of the laterally protruding histone H3 of the nucleosomes, thus labelling the CMM. The methyl-groups are identified by polycomb proteins. Further proteins are recruited and a PRC1 complex forms that condenses the chromatin and prevents transcription. Polycomb proteins form dimers and thus contribute to the compaction of the respective chromatin region. (**b**) Upon replication of the DNA, methyl-transferases find the CMM and renew the methylation pattern in each daughter strand copying it using the pattern of the conserved strand as template. Thereby in both cells the labelled genes are silenced again (after Gibbons 2008, Morey and Helin 2010)

first revealed in *Drosophila*. "State of determination" means a status in which many gene loci along a chromosome are jointly set in a condition that allows, or prevents, gene activation.

Homeotic genes and also other clustered genes can be **permanently fixed in their state of expression**. To comply with this task specifically qualified proteins aggregate along chromosomal sections and facilitate, or impede, the accessibility of genes in those sections (Fig. 12.23). There are two **antagonistically** acting groups of genes or proteins, respectively, the **Polycomb group** and the **Trithorax group**.

• Genes of the *Polycomb group (PcG)* code for proteins which condense the package in certain chromosomal sections and thus curtail the accessibility of genes. POLYCOMB is widely used to silence genes in development. For example it suppresses expression of abdomen-specific homeotic genes in the anterior body. Defects in the *polycomb* gene result in transformation of thoracic segments into abdominal segments.

This example indicates a rule. It is the promoter region of *Hox* genes that are most frequently occupied. The POLYCOMB proteins possess a **chromo-domain**, with which they attach to DNA sequences in the promoter region that have been called *Cellular Memory Modules (CMM)*. In other words, the genes to be silenced are marked by *CMM* sequences. Starting at the *CMM* the polycomb proteins move along the promoters of genes which are to be silenced for a long time or for ever (Fig. 12.23). With antibodies to the chromo-domain more than 100 marked loci were identified along the giant chromosomes of the salivary glands (among them positions in the *Antp*- and *BX*-complexes). A reporter gene with an upstream *Cellular Memory Module* introduced in *Drosophila* eggs was also silenced and the silencing lasted for cell generations.

• Genes of the *Trithorax-group (trxG)* supply proteins which open and keep open chromosomal loci. TRITHORAX proteins are a heterogeneous collection of proteins whose main action is to maintain gene expression. Some of the proteins act by modifying histones, others proteins bind to *Cellular Memory Modules*, but cause a state that makes the genes accessible. A reporter gene with upstream *CMM* was brought into a state of permanent hyperactivity by TRXG protein.

As the products of the *Polycomb* and *Trithorax* genes sustain the state of determination also in proliferating cells, they enable an organism to grow without much change in the composition of differently programmed cell types. Loss of the cellular memory could be one of the causes of cancerous growth (see Chap. 19).

The mechanisms by which, subsequent to cell division, the epigenetic makeup of the respective chromosomal loci is renewed, and the pattern of attached proteins in one of the two chromatids copied, are unknown to date. What is the template?

In mammals *Polycomb Group* and *TrxG* gene expression are involved in a bewildering multitude of functions and important in many aspects of development. Murine null mutants are lethal. Among the functions assigned to the TrxG proteins is implication in the development of the immune system and in X-chromosome inactivation. In general POLYCOMB and TRITHORAX proteins are not only used to silence sections of a chromosome but are also involved in global silencing of chromosomes by heterochromatinization.

Extensive Epigenetic Changes: Heterochromatinization and Inactivation of One of The Two X-chromosomes in Female Mammals

In a process called **heterochromatization** large regions of chromosomes or even entire chromosomes are brought into a state of increased condensation. This conversion of euchromatin into heterochromatin blocks or impedes transcription, but permits DNA replication and duplication of the chromosomes.

Fig. 12.24 Silencing of X chromosomes in female mammals (dosage compensation). In the blastomeres of the early embryo (upper figure) one of the two X chromosomes is inactivated through heterochromatization and is visible as the condensed Barr body. Since the selection of the X to be condensed is a random event, in one blastomere it may be the 'paternal', in the other blastomere the 'maternal' X that is inactivated. After this random event, the same X chromosome is always completely converted into heterochromatin after DNA replication in the daughter cells of the blastomeres. Therefore, in some clusters of cells (cell clones) it is the 'maternal' X chromosome that is silenced; in other clusters it is the 'paternal' X. Red-black cats (tom-cats are red-black only if they show the XXY Klinefelter syndrome) display the result on their coat. The X chromosome carries genes needed for the synthesis of melanin. If in heterozygous cats one allele is active, complete black melanin is synthesized; if the other allele is active, only incomplete, red melanin can be produced

Daughter cells can inherit the pattern of heterochromatinization.

Heterochromatin contains proteins that possess a domain called the **chromatin-organizing modifier** or **chromo-domain**. With this domain the proteins attach to repetitive DNA sequences that recur at regular distances along the chromosome. By aggregation the proteins are brought closer together and the DNA becomes folded and compressed. A similar DNA-condensing protein

is the above mentioned POLYCOMB protein encoded by the *Drosophila* gene *polycomb*.

In female mammals all cells are equipped with **two X chromosomes** while males have one X and one Y chromosome. Consequently, X chromosome-resident genes are present in females twice, in males only once, including those many genes that are needed in both sexes (for instance, genes for certain blood-clotting factors). To compensate for this imbalance in the copy number of X chromosome-resident genes, the mechanism of **dosage compensation** evolved. Through heterochromatinization, one of the two X chromosomes becomes inactivated and changes its microscopic appearance. In the interphase nucleus this condensed X chromosome is observed as **sex chromatin** or the **Barr body** (Fig. 12.24).

Heterochromatinization occurs after the egg cell has already undergone some divisions. It is a random event: chance determines whether in a particular blastomere the inactivated X chromosome is the 'paternal' X chromosome provided by the sperm, or the 'maternal' X chromosome contributed by the egg cell. Once heterochromatinization has taken place for the first time, in all descendants of a blastomere the same X chromosome will undergo heterochromatinization. **Every human female, like all mammalian females, comes to be a mosaic of cell clones in which the 'paternal' X chromosome is silent, and clones in which the 'maternal' chromosome has been silenced** and converted into a Barr body.

Red-black spotted cats (the very few red-black spotted tom-cats have the aberrant XXY Klinefelter constitution!) display the result visibly on their coat. One of the two X chromosomes bears a gene enabling the synthesis of complete, highly polymerized black melanin. The other X chromosome bears an allele with which only incomplete red melanin can be synthesized (Fig. 10.5). The synthesis of melanin takes place in chromatophores that derive from neural crest cells and immigrate into the hair follicles (see Fig. 15.3).

Since the process generating heterochromatin is based on the aggregation of DNA-binding proteins, it has a tendency to spread beyond the nominal region to be silenced. Such extended spreading results in **position effect variegation**: genes close to a silenced region suffer a reduction in their accessibility to transcription. Position effect variegation suggests the participation of an autocatalytic feedback loop in heterochromatinization. Gene silencing by heterochromatinization is considered to contribute to persistent and heritable determination and differentiation.

Related to heterochromatization is an effect described in the following section.

Epigenetics Assist in Keeping Development-Controlling Genes in a State of Easy Accessibility

Current investigations focus on mammalian sperm. The chromatin of sperm cells is methylated and acetylated to a particular high degree and it is highly condensed by means of protamines so that the entire chromatin can be enclosed in the small head of a sperm. Compared to oocytes in sperm higher degrees of methylation and acetylation are achieved. Yet, enabled by peculiar patterns of methylation and acetylation, genes for transcription factors and well-known signalling molecules are kept in a state that keeps them quiet in sperm, but become quickly freed from the state of quiescence once the sperm has entered an egg cell. The egg cell contains factors which releases genes from blocking mechanisms (and this also enables cloning).

Together with fertilisation and egg activation the gamete-specific pattern of epigenetic modification is extinguished. At any rate, this applies to the mouse embryo. Demethylases and deacetylases of the egg cell erase the gamete-specific program, and are tools to confer pluripotency to the cells of the inner cell mass. Starting with the implantation of the blastocyst into the uterine wall new, gastrula-type patterns of epigenetic modifications are generated. In due time in diverse cell lineages remethylation must wait, or must be suspended. For instance, in the somites the promoter o the *MyoD1* gene

gets demethylated. Only then can the transcription factor *MyoD1* switch on the muscle-specific differentiation program. In the course of this program also the promoter of *myogenin* is demethylated. In the completed myocytes new, not yet identified in detail, patterns of methylation stabilize the state of differentiation.

This applies generally: **When upon completion of cell differentiation genes are no longer required they are epigenetically silenced.**

In the formation of germ cells designed to engender the next generation the pattern of epigenetic modifications must be readjusted in the precursor cells to the state suitable for germ cells. There is much research to be done until all these mechanisms are elucidated.

Worker Honey Bees and the Queen Differ in the Methylation Pattern of 550 Brain-Specifying Genes, Caused by the Food

The solution to an old riddle may be imminent. In the colony of honey bees worker bees are vastly superior to the queen with respect to brain performance. Worker bees navigate in the exterior world by means of their compound eyes and an internal compass of the sky, they distinguish numerous flowers, have an excellent memory and use a complex symbolic language in their dances. But they cannot lay eggs; this privilege is granted to the queen only. On the other hand, the large queen remains idle in the hive, gets fed, and works as an egg producing machine. Now the worker bees are daughters of the queen and genetically not much different, and as young larvae they had the potency to become a queen themselves. The difference is brought about by an environmental factor. When the nursing bees feed the larvae with Royal jelly only for an initial short time the larvae will later hatch as worker bee. If the nurse bees feed the larvae permanently with copious amounts of royal jelly in specially constructed queen cells, the larvae will hatch as queens.

Recently it was discovered that in the course of development the brains of the worker bees and the queens experience different patterns of methylation and acetylation. More than 550 genes are differently methylated. Royal jelly contains, among other substances, phenylbutyrate, an inhibitor of histone-deacetylase. Thus it is conceivable that the food exerts lasting influence on gene activities.

Methylation and related epigenetic modifications are important also in later life, so in memory formation in the honey bee, and in the mouse and humans as well.

Some Epigenetic Modifications Can Persist Over Generations

Paternal and maternal genetic make-up are not in all details equal and of equal rank. For humans and mammals in general the following holds: mitochondria with their genome and the nucleolus (site of ribosome formation) are brought along by the egg cell, an indispensable centrosome is supplied by the sperm. Besides these sources of inequality a further, epigenetically encoded inequality comes in addition.

In the course of mammalian germ cell development the methylation pattern is largely erased, and the DNA becomes re-methylated. As mentioned above sperm DNA is more highly methylated than the DNA of egg cells in preparation of the subsequent condensation of sperm DNA. However, since some methylation persists past fertilization, the accessibility of maternally and paternally inherited alleles to transcription can be different. One speaks of **maternal and paternal genomic imprinting** (see Chap. 8). In *Drosophila*, some artificially induced epigenetic changes survived up to two generations.

Whether by means of epigenetic mechanisms accessibility of genes can experience modifications lasting for generations and lead to permanent adjustment of the epigenetic program to environmental conditions (Lamarckism with a new livery), is subject of controversial debates.

Recent studies indicate that the environment might exert life-long influence on gene activities also in humans. Does this have an impact on subsequent generations? There is accumulating evidence that exposure of a fetus to noxious substances such as nicotine can have effects that persist into subsequent generations (Rehan et al. 2012). However, most geneticists hold onto the notion that long-term changes of genetic information that last over long chains of generations must eventually be fixed in the structure of DNA itself.

12.4 Irreversible Alterations of the Genome and of Cell-Type-Specific Genetic Programs

12.4.1 In the Development of Lymphocytes an Irreversible Somatic Recombination Takes Place

Recombination–the rearrangement of genes–is characteristic of sexual reproduction and takes place during meiosis. Surprisingly, a similar event occurs in the development of lymphocytes, when lymphoblasts combine their individual genetic programs for preparing the future production of antigen receptors or antibodies (which are liberated antigen receptors). This process, called **somatic recombination**, occurs at the 'birth place' of these cells: the bone marrow, the thymus or in lymph nodes.

Lymphoblasts are genetic gamblers. They try their luck when they combine various short DNA sequences at random to program the variable region of their receptors or antibodies, increasing the chance of creating a combination appropriate to recognize and bind a future foreign antigen. In each B-cell or T-cell a different random combination is tried (see Fig. 18.8). Subsequently, the T-lymphoblasts emigrate and settle in the thymus; the B-lymphoblasts remain in the bone marrow for some time. Wherever lymphoblasts stay, they learn to discriminate between self and non-self. A strict negative selection eliminates all those lymphoblasts whose receptors would bind molecules belonging to the self. The surviving cells are those that can bind only non-self molecules and are thus potentially capable of recognizing foreign substances. The surviving lymphocytes migrate through the body, colonize other places such as spleen and lymph nodes, and eventually serve as T-cells or antibody producing B-cells.

Somatic recombination, to date found only in lymphocytes, leads to an irreversible alteration in the genetic constitution of the cell.

Differentiation Is Often Accompanied by Quantitative Changes in the Stocks of Genes: Gene-Amplification, Genome-Amplification, Elimination of Chromosomes

1. **Selective gene amplification.** Some differentiated tissue or cell types–for example gland cells–may require large amounts of gene product. Quantitative changes in the genome of cells and cell lines may occur in the course of differentiation. The selective replication of particular genes, although a rare event, is one of several possibilities that can lead to the non-equivalence of cell genomes that were previously identical. A textbook example is the **selective amplification** of the ribosomal 18S, 5.8S and 28S rRNA genes in the nuclei of many oocytes (Chap. 8; Fig. 8.6). This amplification is manifested by the appearance of numerous nucleoli (centres of ribosome production). Selective gene amplification is also known from follicle cells in the ovaries of *Drosophila*, where genes encoding the secreted proteins of the egg envelope (chorion) are amplified.

2. **Polyteny** refers to the enlargement of chromosomes through endoreplication: the DNA is repeatedly replicated but the newly generated strands are not separated but stick together forming cables of DNA strands. During *Drosophila* development the nurse cells in the ovaries become polyploid, and the cells of the salivary glands and the Malpighian tubules become polytene.

   • **Genome amplification**. Amplification of the entire genome by polyploidy or polyteny is relatively common.

   • **Polyploidy** refers to multiplication of the chromosomes. Instead of the two sets of chromosomes observed in diploid cells, four or more sets of chromosomes are found. In mammalian development the cells of the **trophoblast**, many gland cell and liver amplify their genomes through polyploidy. These cells are metabolically highly active and polyploidy may facilitate the continuous production of large amounts of enzymes.

3. **Elimination of genetic material.** Instead of amplifying genetic material, some cells lose chromatin or entire chromosomes. A classic example is the round worm *Ascaris (Ascaris lumbricoides = Parascaris equorum)* whose early embryos taught Theodor Boveri much about the significance of chromosomes in directing development. Boveri studied a bizarre phenomenon: The haploid egg has only two chromosomes, and two more are contributed by the sperm. But even these few four chromosomes appear to be too much for ordinary somatic cells. During the cleavages the four chromosomes are allocated only to the cells of the germ line in full length and complement. In the worm's somatic cells the chromosomes are fragmented and part of the chromatin gets lost, a phenomenon known as **chromatin diminution**. It is not known whether genes that are only needed for gametogenesis are lost, or merely non-coding DNA.

Some cases are known where entire chromosomes or the entire nuclei are eliminated. In the midge *Wachtiella persicariae* many nuclei lose 38 of their original 40 chromosomes. Well known examples of cells devoid of the entire nucleus are the mammalian red blood cells (except those of camels!) and the keratinocytes of skin, feathers and hair.

Reversible alterations by epigenetics impede cloning of organisms considerably, irreversible alterations render it impossible. Techniques of cloning will be subject of the following Chap. 13.

Summary

Differentiation starts out from genetically equivalent cells (**initial genomic equivalence**) and is based on differential gene expression by which the various cell types acquire a characteristic set of cell type-specific specific proteins. In the course of differentiation though a change in the inventory of genes can occur, for example through **genome amplification**, that is through **polyploidy** or **polyteny** (by which giant chromosomes are manufactured), or through loss of genetic material (chromatin elimination).

The program for body region-specific and cell type-specific gene expression is established in the multi-step process of determination. In the implementation of the differentiation program the activity of protein-encoding genes is not only controlled through extracellular signals but also by intracellular control systems based on **micro-RNA** derived from "*non-(protein)-coding*" DNA-sequences (e.g. introns). Among the protein-coding genes **master- or selector genes** play a crucial and prominent role, the products of which act as **transcription factors** keeping batteries of subordinate downstream genes under their control.

• Among the genes specifying positions and switching on position-pertinent developmental programs are the **homeotic genes** of the **Hom/Hox** family. Mutations in these genes can lead to **homeotic transformations** by which correctly designed structures arise at incorrect locations. For example *loss-of-function* mutations of the gene **Antennapedia** bring about flies whose legs on the mesothorax are transformed into antennae, whereas other, *gain-of-function* mutations engendering the expression of functional ANTP proteins in the head, produce flies with legs instead of antennae on their head.

The entirety of the *Hom/Hox* family specifies the characteristic qualities of a body region (segment identity genes) and encodes transcription factors with a DNA-binding homeodomain. These genes are concentrated as a cluster on the chromosome (*Drosophila: Antennapedia* complex plus *Bithorax* complex). In mammals the entire cluster is present fourfold due to genome duplication in the past. The succession of the genes along the chromosomes coincides with the spatiotemporal order in which these genes are activated along the body, the first gene of the cluster being activated first and in the most anterior territory of the body, the last gene of the cluster being expressed last and in the most posterior body region (**colinearity**). A gene located at a more posterior position of the cluster suppresses the genes positioned more anteriorly in the cluster in its own expression domain (**posterior dominance**). Mutations breaking this rule cause position-dependent transformation of structures, for example wings instead of halteres on the metathorax of flies, or a second proatlas instead of an atlas in the spinal cord of the mouse.

• Example of an organ-specifying gene is the gene *ey* of *Drosophila*. The expression of this gene at wrong (ectopic) locations (legs, wings, antennae) causes there the emergence of additional eyes at those wrong places. The sequence homologue gene *Pax-6* of the mouse can in this respect replace the *Drosophila* gene *ey*; triggering eye development in *Drosophila*, however, with fly-type compound eyes. Conversely *ey* of *Drosophila* can induce camera-type eyes in the vertebrate. EY and PAX6 elicit differentiation of light-receiving structures but do not specify the particular taxon-specific eye type; this is realized by the action of taxon-specific genes.

• Examples of **cell type-specifying selector genes** are the muscle cell-specifying genes of the **MyoD/myogenin** family, and the nerve cell specifying selector gene such as *neuroD* and *neurogenin* and the *AS-c* in Drosophila.

Many development controlling genes code for **transcription factors.** With their DNA-binding domain (e.g. **homeodomain** with its **HtH** motif, or **bHLH** motif or **Pax** motif) these factors occupy the controlling region (promoter, enhancer) of downstream genes. This applies to the homeotic genes, to the *bicoid* gene for example the product of which define the head-thorax range in the offspring of a fly, and to the cell type-specifying genes and to many other genes involved in the specification and realization of the body plan. Other development directing genes code for proteins that serve as **signalling molecules**, or fulfil functions in signal reception and signal transduction.

A state of determination once established has become a part of the **cellular memory**, which in cell division is transmitted to daughter cells and can be preserved over many cell generations (cell heredity). This **cell heredity** can be based on positive feedback of transcription factors on their own gene, restarting and boosting its own expression after each cell division. Examples are the genes of the *MyoD*, *myogenin* family. Other mechanisms of cell heredity are based on **epigenetic modifications** of the chromatin brought about by enzymes conferring methyl-, or acetyl-, or glycosyl-groups on the DNA or the DNA-supporting histones.

Determination means also shut-down of genes the information of which is transiently or definitively not required. Two mechanisms are known leading to a persistent epigenetic silencing: local **methylation** and global **heterochromatization** of the chromatin. As a rule these secondary modifications of the chromatin impede accessibility of genes, whereas acetylation facilitates accessibility. In *Drosophila* but also in mammals genes are known the products of which are important elements of the **epigenetic cellular memory** and also involved in heterochromatization. Proteins encoded by the ***polycomb group*** of genes occupy a sequence termed *Cellular Memory Modules (CMM)* in the promoters of genes to be silenced; these genes, mainly genes of the *Hom/Hox* cluster, are thus marked for silencing through a complex machinery that condenses the chromatin in the respective region of the chromosome. The antagonistically acting proteins of the *Thrithorax* group also occupy *CCM* marked genes but keep the genes accessible for transcription. The epigenetic pattern of closed or opened chromatin is copied and preserved in cell division. **Persistence** and **heredity** of such epigenetic states enables ordered growth of committed cells in detail and of the organism as a whole. In female mammalian embryos one of the two X chromosomes is condensed by heterochromatization forming the **Barr body,** and is thereby almost completely silenced (**dosage compensation**). Also this type of silencing is part of the cellular memory and transmitted from founder cells to their descendants.

Gametes, in particular sperm, are highly methylated in preparation for chromatin condensation, but especially developmentally relevant genes are prepared to be quickly woken upon fertilization. The switching on of many, perhaps of all, genes requires demethylation of the promoter as has been shown using *MyoD1* and *myogenin* as examples. In terminally differentiated cells new patterns of methylation stabilize the state of differentiation. Persisting epigenetic modifications entail that maternal and paternal alleles are not always equivalent (**maternal and paternal genomic imprinting**).

In bee colonies through epigenetic modification of genes in the brain, caused by food (Royal jelly versus normal food), it is decided which daughters of the queen will become a queen and which will become worker bees. Via epigenetic modifications environment is thought to influence on the long-term accessibility and activity of genes.

Irreversible changes of the genetic makeup are multiplication of the genome such as polyteny and **polyploidy** which occurs in the mammalian trophoblast and placenta. Another type of irreversible change is **somatic recombination** as it occurs in lymphoblasts in the process of programming antibodies. While such alterations only concern somatic cells and therefore are irrelevant for subsequent generations epigenetic changes also affect germ cells and can influence gene expression in the next generation.

Application-Oriented Experiments with Early Vertebrate Embryos: Cloning, Transgenic Animals

13

In this chapter experiments are introduced that originally were designed to answer scientific questions but subsequently turned out to be useful in stockbreeding, medicine, pharmaceutics, or biotechnology. Basic research leading, for instance, to artificial teratomas, is often not of benefit for practical medicine. Nevertheless by these experiments knowledge was gained about how malformation can develop also in humans. Some of the subsequently described and discussed experiments, in particular cloning of mammals, caused some anxiety and controversies, and therefore hit the headlines of the media.

13.1 Cloning, the Production of Identical Copies

13.1.1 Cloning Is a Natural Event in the Plant World and in Many Non-vertebrates, and Plant Breeding Has Supplemented the Natural by Artificial Procedures from Time Immemorial

In present day biology the term cloning is most frequently used in connection of work in molecular biology, and refers to the, hopefully faithful, multiplication (amplification) of DNA sequences in bacterial hosts. Molecular biologists have adopted the term from classical biology. In traditional biology a **cellular** or **organismal clone is a collection of genetically identical cells or**

multicellular organisms; cloning is the production of such clones. Natural cloning, traditionally called **vegetative reproduction**, is known from many plants and invertebrates, and is based on mitotic proliferation. Reproduction through vegetatively produced tubers (for example potato), through offshoots or stolons (for example strawberries, hydrozoan colonies, coral polyps), through buds (for example the plant *Bryophyllum* or the fresh water polyp *Hydra*) are forms of natural cloning.

In plant breeding the natural procedures are supplemented by artificial methods of reproduction. Since times immemorial currants are propagated by cutting off twigs and inserting them into humid ground where the grow roots (cloning by regeneration). Fruit-bearing trees and vine stocks are propagated by grafting, that is by transplanting twigs under the bark of a rooted tree. In this way cultivated varieties have been bred which all look alike, have the same nutrient content, and taste alike, and are distributed all over the world.

The above examples show the benefits the farmer, vintner, and gardener can expect from cloning. He/she can be sure that **all descendants have the same features as has the stem organism**. Not so in sexual reproduction via seeds. As in sexual reproduction of humans it is indeterminate how offspring will turn out.

In current plant breeding many plants which cannot be multiplied by natural vegetative reproduction are propagated via protoplasts, these are cells taken of existing plants deprived of their

W.A. Mueller et al., *Development and Reproduction in Humans and Animal Model Species*,
DOI 10.1007/978-3-662-43784-1_13, © Springer-Verlag Berlin Heidelberg 2015

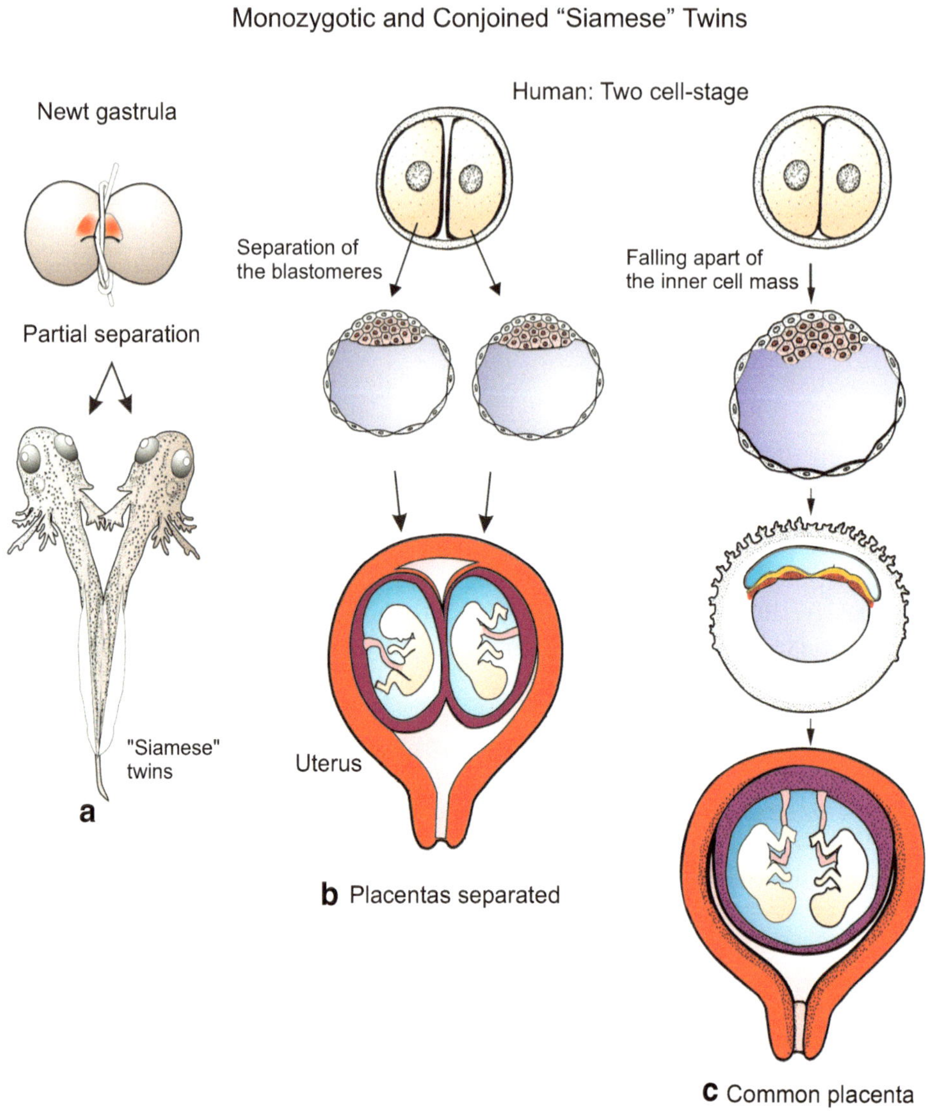

Monozygotic and Conjoined "Siamese" Twins

Fig. 13.1 Monozygotic twins in amphibians and humans. Upon partial separation of early amphibian embryos partially connected "Siamese" twins arise whereas complete separation gives origin to two complete tadpoles. In humans the first two blastomeres can fall apart, or the inner cell mass breaks into two parts. In the first case two monozygotic twins grow in separate envelopes, in the second case the twins grow in a common envelope (chorion)

cell wall and grown in culture where they form multicellular clumps, called callus. These can be induced to form vegetative embryos, without intermediary sexual processes.

In animal breeding vegetative reproduction has not yet yielded procedures that can be used in biotechnology. There is no commercial interest in cloning sponges, hydrozoans, corals or ascidians which would be easily possible. In vertebrates (except in polyembryonic armadillos)

naturally cloned animals are the result of an accident when blastomeres of early embryos fall apart and monozygotic twins are born (Fig. 13.1). Yet, the breeder would be happy if he/she could clone a successful racehorse (it is possible but extremely expensive). Currently there is much interest in cloning transgenic animals that were generated with arduous work for medical research. This will be discussed in Sect. 13.3.4.

13.1.2 A Biotechnologically Irrelevant Cloning Procedure Is Dissection of Early Embryos

Mini-clones consisting of identical monozygotic twins (actually two, four or eight individuals) arise when the first two, four or eight blastomeres fall apart or are separated experimentally. In mammals, each separated blastomere may give rise to a small but complete blastocyst which implants and develops into a fetus with its own placenta. Since mammalian embryos are supplied with nutrients multiples can grow to normal size provided space in the womb is large enough. In mammals, including humans, even the inner cell mass of the blastocyst may fall apart into two or more groups: those twins are found within a common chorion (embryonic envelope, Fig. 13.1c) and have a common placenta.

Partial division of an early embryo results in conjoined twins, in common parlance called "Siamese" twins. The emergence of such twins was elucidated by H. Spemann by partially constricting early embryos of newts (Fig. 13.1a).

13.1.3 Cloning by Transplantation of Somatic Nuclei: Pioneering Experiments in Xenopus Opened Means to Breed Many Progeny with Known Characteristics

Clones comprising many animals were first produced in the clawed toad *Xenopus* (by R.W. Briggs, T.J. King, J.B. Gurdon, Fig. 13.2). From activated egg cells the zygotic nucleus is removed with a micropipette or destroyed by UV irradiation. A substitute nucleus is removed from a suitable somatic cell type taken from a donor animal and transferred into the enucleated egg by means of a glass micropipette. Fortunately, in amphibians pricking with a micropipette suffices to activate the egg.

Originally the experiment was designed to check whether somatic cells retain totipotency. Suitable totipotent donor cells were first found in the intestine of tadpoles and later in several

tissues of adult animals. To produce numerous clones the procedure is repeated many times using the same donor animal: all offspring will be genetically identical with the donor and with each other.

13.1.4 Cloning of Mammals Meets Particular Problems; So Foster Mothers Must Cooperate

Unlike frog eggs oocytes of mammals must be surgically removed from the maternal oviduct or the ovary. Besides, the oocytes must be activated in some way, for example by fertilization with a sperm and subsequent removal of the male pronucleus, or by a well-dosed electric pulse. And mammalian eggs do not develop in water outside a mother. They must be kept in particularly prepared media up to the blastocyst stage and then transferred into the uterus of nurse females prepared to accept the egg by injection of gonadotropic hormones. The animal from which the host egg originally was removed is, as a rule, too much affected to accept and bear the child. A **surrogate mother**, also known as **foster mother** (Fig. 13.3) must step in.

13.1.5 Cloning of Mammals Using Embryonic Donor Nuclei Is Possible for Quite Some Time; Though the Dream Goal Has Not Yet Been Achieved

Using the technique developed in *Xenopus* cloning of mammals was initially only successful if the nuclei were taken from

- **Blastomeres** of two-cell through eight-cell stages, before compaction, or from
- **Embryonic stem cells** (**ES-cells**, **ESC**) that is cells removed from the inner cell mass of a blastocyst. Previous to the transfer experiment the totipotent ES cells are multiplied through several passages in culture so that many genetically identical donor cells are available and many egg cells can be provided with a foreign nucleus.

Cloning

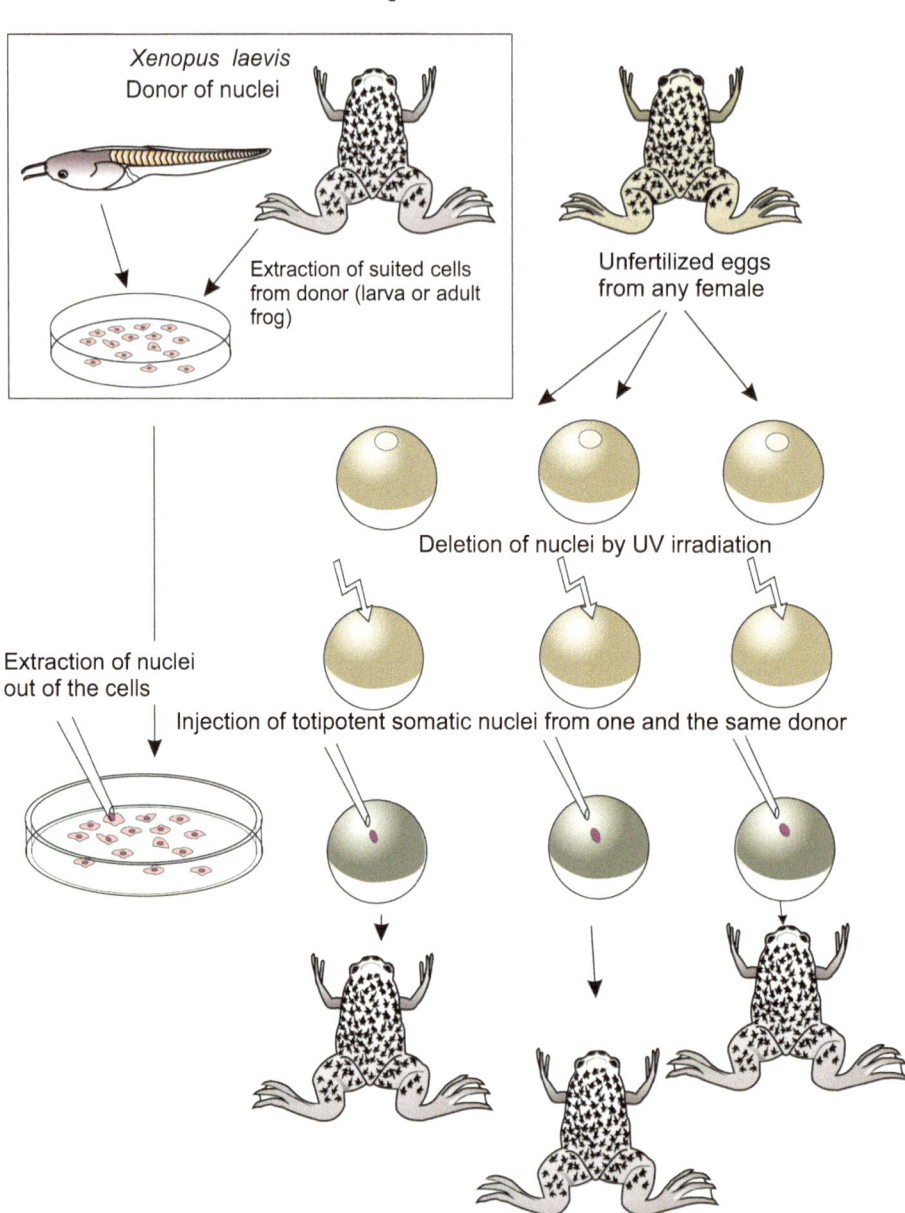

Fig. 13.2 Cloning of *Xenopus*. Totipotent nuclei taken from somatic cells of a donor are injected into oocytes the nuclei of which were previously destroyed by a UV microbeam. Suitable donor nuclei can be extracted from tadpole cells (e.g. from cells of the intestine) or even from adults (e.g. lung cells)

In a next step the difficult technique of transferring a nucleus was replaced by a simplified technique. First the nucleus of the host oocyte is removed by sucking it out with a fine glass pipette. The mammalian oocyte is surrounded by a jelly-like coat known as zone pellucida, and within this coat is inserted one cell from a donor, a blastomere or an ES-cell. Fusion of the oocyte and the donor

Fig. 13.3 Cloning of sheep. Large enucleated oocytes are fused with small somatic donor cells taken from a suitable tissue. Donor cells can be removed from a blastocyst or taken from a tissue of adult animals that contains proliferating cells, for example from milk glands. Before fusion the cells are propagated in culture and eventually locked in the G_0 phase (phase of quiescence) of the cell cycle. Fusion is induced by an electric impulse. The fusion product starts cleavage in a suitable medium. The blastocyst stage is transferred to the uterus of a ewe, where it develops to term

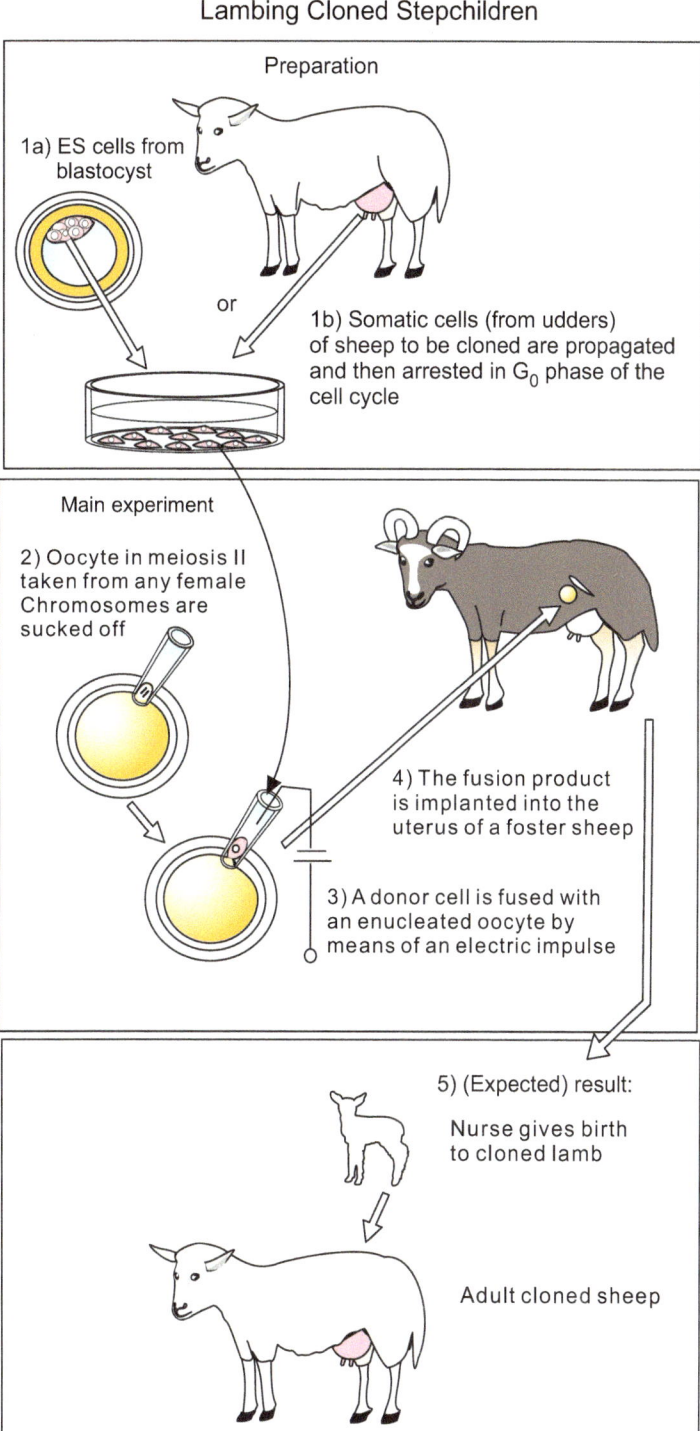

Lambing Cloned Stepchildren

Preparation

1a) ES cells from blastocyst

or

1b) Somatic cells (from udders) of sheep to be cloned are propagated and then arrested in G_0 phase of the cell cycle

Main experiment

2) Oocyte in meiosis II taken from any female Chromosomes are sucked off

4) The fusion product is implanted into the uterus of a foster sheep

3) A donor cell is fused with an enucleated oocyte by means of an electric impulse

5) (Expected) result:

Nurse gives birth to cloned lamb

Adult cloned sheep

cell is induced by administering an electric pulse. Through this electrofusion a nucleus is reintroduced into the oocyte forming a reconstituted egg. (However, the donor cell can bring along disturbing cytoplasmic components such as transcription factors, as discussed below in Sect. 13.1.6.)

After nuclear transplantation has been accomplished, the host eggs are transferred to nurse females prepared to accept the egg by injection of gonadotropic hormones. Such nurses or foster mothers have given birth to lambs, cattle, pigs and rabbits. For a long time cloning was not successful if nuclei from differentiated cells were injected. In mice, even nuclei from four-cell stage embryos were unable to support full development.

Cloning of animals using cells from embryos or fetuses is not what is desired in breeding of farm animals because the future phenotypic outcome of a genome taken from an embryonic donor genome is unknown. Therefore, the technique as described was only a preparatory step towards the main goal, cloning of adults with known characteristics.

13.1.6 Cloning with Nuclei Taken from Adults Is Possible with Restrictions: Dolly Was the First Proof

The aspiration of breeders was to clone animals (race horses for example) whose performance was known and proven. Therefore donor nuclei from adults are in demand. After many years of experimenting in 1996 eventually a first Scottish domestic sheep was cloned, widely known as Dolly. Instrumental for the success were several proceedings:

- Oocytes in the stage of final maturation (1st polar body just extruded) were selected as host cells, and proliferating cells were selected as the donors of nuclei, actually putative unipotent stem cells from the mammary milk glands which could be cultivated and multiplied before use (the meaning of unipotent is explained in Chap. 18 and in the Glossary).

- The donor nuclei were brought into a suitable state. Before use the donor cells were deprived of serum to bring them into the quiescent G_0 phase of the cell cycle. Egg cells divide rapidly, somatic cells slowly, even when kept in culture. It proved successful in

practise to arrest the donor cell in the G_0-phase of the cell cycle. In this phase the time-consuming transcription of genes comes to a standstill. Apparently from this quiescent state a quicker restart with rapid succession of nuclear replication is possible.

- With a well-dosed electric pulse not only fusion of the egg cell with the donor cell but also activation of the egg cell was achieved.

- In culture medium containing components of the uterine secretion the manipulated eggs developed up to the blastocyst stage capable of implanting itself into the uterus of a surrogate mother.

The production of Dolly showed that genes in the nucleus of a somatic partially-committed stem cell is still capable of reverting to an embryonic totipotent state, apparently rendered possible by factors present in the cytoplasm of egg cells. We will later in Chap. 18 see that even terminally differentiated cells can provide totipotent nuclei if certain pluripotency genes are reactivated.

In 1998 the first cloning of mice succeeded when potentially totipotent nuclei were found in cumulus cells. These are relicts of maternal tissue attached to the surface of freshly ovulated eggs. (In numerous previous experiments in vitro blastocysts, once implanted into a foster mother, died before term because the trophoblast did not develop a normal placenta.)

Cloning succeeded, for example, in domestic sheep (Dolly 1996), in the house mouse (1998), domestic cattle (1998), domestic goat (1998), domestic pig (2000), mouflon (2000), gaur (Indian bison) (2001), domestic rabbit (2001), domestic cat (2001), brown rat (*Rattus norvegicus*, 2002), in the mule (2003) and domestic horse (2003), in the African wild cat (2003), in the red deer (2004), ferret (2004), Asian water buffalo (2005), the Afghan hound dog named Snuppy (2005) and the grey wolf (2005). In 2007 rhesus monkeys were cloned for the first time, and in 2009 an Arabian camel.

(continued)

Currently (2012) race horses with great sporting prowess are cloned in notable numbers. In 2008 in South Korea seven cloned drug-sniffing dogs (Golden Retriever) were born who, such the hope of the customs authorities, should be as successful as was the original. The aim is also to clone endangered species such as marsupials but there is lack of donors of oocytes and of surrogate mothers.

13.1.7 There Have Been Many Disappointments Because of Hampering Epigenetics

The fact that the success rate of cloning mammals was (and remains) very low, between 0 and 4 % at the maximum – was and is unfortunate for the client and reproductive veterinarian as well. The frustrating experience, and not caused by mere technical deficiency, was that many fetuses were lost by abortion or died shortly before term. Moreover, the observation that not all offspring were exactly identical copies of the donor of the nucleus appeared to contradict classical theories. Dolly aged too early (presumably partly caused by infectious diseases). Too high birth weight, premature ageing, and physical deficiency have been diagnosed. Several causes are discussed:

- Incomplete reconstitution of the chromosomes back to the juvenile state. Chromosomes experience secondary, epigenetic modifications during differentiation such as methylation and acetylation of the DNA and the DNA-associated histones (see Figs. 12.22 and 12.23), or heterochromatinization of large parts of chromosomes. These epigenetic modifications reduce the accessibility of genes for transcription and are the cause of phenomena known as "*genomic imprinting*", "*epigenetic memory*" and "*gene silencing*" (Sect. 12.5). **In the oocyte advantageous conditions support the reconstitution of the juvenile state; after all the highly methylated and condensed chromatin of the sperm nucleus must be brought into a state allowing transcription.** For instance, the cytoplasm of the egg cell contains many demethylases. However, the resetting of the status of chromatin is often incomplete, even in the chromosomes supplied by the oocyte and the sperm. Hence maternal or paternal imprinting can be observed (Chap. 8). **In cloning by means of somatic nuclei incomplete resetting of the chromatin may hinder accessibility of important genes.** In addition in the nuclei of differentiated cells many transcription factors not yet present in the egg cell are bound to DNA.

- If cloning is started by fusion of the egg cell with a somatic cell this somatic cell may contain transcription factors acting on the genome of the zygote in a disturbing manner. We have discussed the case of the muscle-specifying factor MyoD1 in myoblasts (Sect. 12.3). Upon cell division this transcription factor immigrates into the nucleus of the daughter cells to switch on its own gene thus maintaining the state of determination. Such a factor could cause the untimely start its own expression in the egg cell.

- A contributing factor to premature ageing may be that the cloned animal could have been born with the genetic age of the adult donor. The basis for this idea is the observation that the length of the telomeres at the end of the chromosomes shrinks with progressing number of cell division, which is typically a result of the ageing process. Shortening of the telomeres does not only shorten the length of the chromosomes but also the life span according to the telomerase theory of cell senescence (Chap. 21).

The technology of cloning still has to overcome many a difficult problem until it can be considered perfect.

13.1.8 Will Cloning of Humans Be Possible and Wished-for?

Currently cloning of human beings from adult donors is performed by science-fiction novelists. However, according to present knowledge it is highly probable that human beings can actually be cloned in spite of all the discussed difficulties. Unlike sexual reproduction cloning would bring about humans with foreseeable genetic disposition. In the cloning of men, however, the egg-donating woman would contribute all the mitochondrial genes so that the child would not be a faithful copy of his father in all aspects. On the other hand cloning would enable the breeding of a purely female population ("amazons"). Aside from this hardly probable scenario, motives are given to justify cloning in certain situations on a case-by-case basis.

- If one parent is infertile, or its genome burdened with untreatable genetic defects, the second parent could bestow a nucleus alone for the desired domestic happiness. It would not be necessary to accept that a foreign, unknown sperm donor monopolizes half of the children's genome.

- To get a child without the contribution of a man might in the view of the one or another woman be sufficient reason to no longer share the ethical objections of others humans.

When women prepared to deliver oocytes and to bear children to term, urgently want to have a (cloned) child, and financial reasons do not exclude the procedure, ethical considerations and laws scarcely would prevent this in the long term.

13.1.9 Cloning Restricts Genetic Variability

Even if it is desirable in a particular case to clone an animal, a potential disadvantage should be considered. In the long term cloning diminished natural genetic diversity and flexibility. Unlike in sexual reproduction no new combinations of alleles could create a more favourable genetic makeup. To the benefit of proven quality, opportunities for better solutions are lost. Therefore, even in plant breeding sexual reproduction is not forgotten.

13.1.10 "Therapeutic Cloning" Has Not Much to Do with Cloning in the Traditional Sense

In about 2000 in the media of the world a catchphrase went around that even put parliaments in a flurry, this watchword was "therapeutic cloning". This procedure proposed by some scientists has not much to do with cloning of individuals, be it animals or humans, but has to do with the artificial production of surrogate tissue and organs from stem cells for medical purposes. Therefore, in this book this topic will be addressed in depth in Chap. 18, which is devoted to stem cells and regenerative medicine. The term "cloning" is used in this context with some validity in so far as it was proposed to remove a nucleus of the patient and insert it into an enucleated oocyte bestowed by any women, and to use the blastocyst derived from this egg cell as source of stem cells which would bear the histocompatibility markers of the patient. Such stem cells and surrogate tissues derived from them would not evoke immunological intolerance and rejection. In several countries laws forbid the use and wastage of human egg cells for that purpose.

13.2 Chimeras and Teratomas

13.2.1 Experimental Chimeras Are Intermediate Stages in Techniques Designed to Produce Transgenic Animals

A chimera is a mosaic organism composed of cells of different parental origin and, therefore, of different genetic constitution. Chimeras were produced as early as about 1920 (e.g. in H. Spemann's laboratory) by fusion of early embryos from different amphibian species. Two combined early stage embryos (two-cell stage up

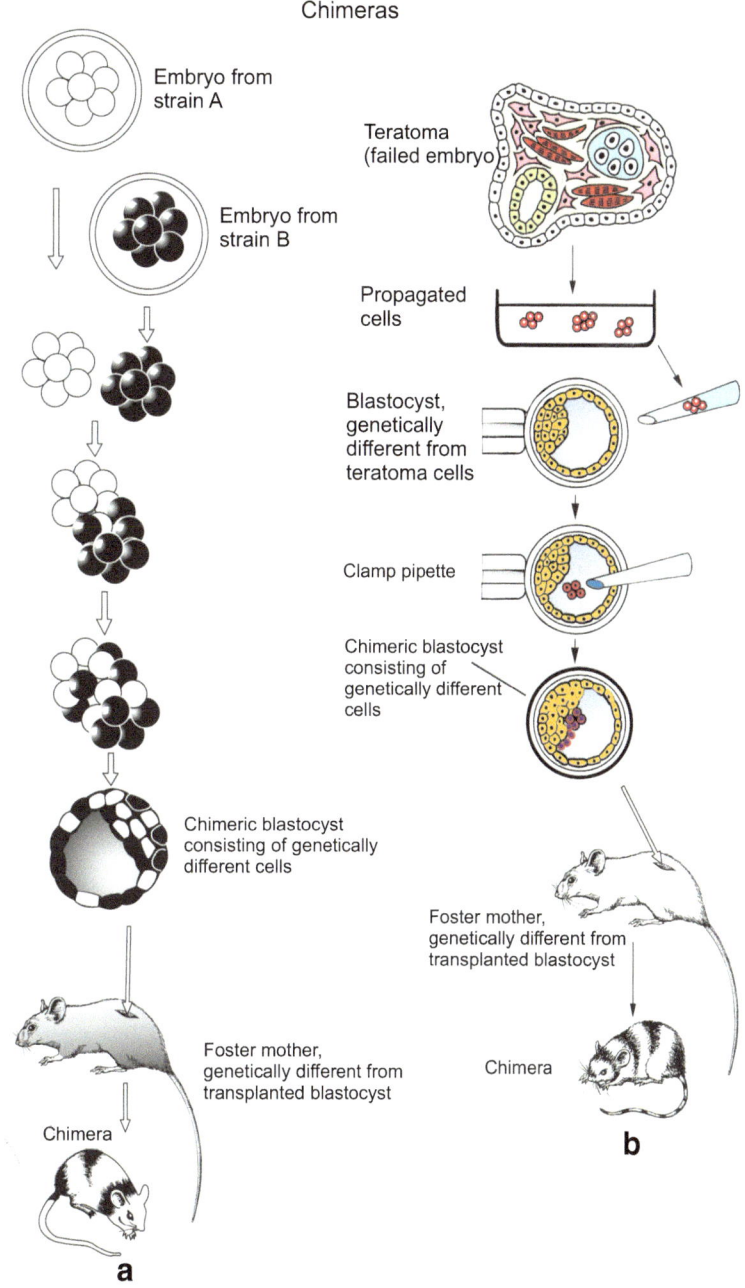

Fig. 13.4 Production of chimerical mice (**a**) through fusion of two morulae, (**b**) through introduction of totipotent cells of different origin – here from a teratoma – into a blastocyst. In the depicted cases the chimerical nature of the offspring is visualized by the striation of the coat

Chimeras

Embryo from strain A

Embryo from strain B

Teratoma (failed embryo)

Propagated cells

Blastocyst, genetically different from teratoma cells

Clamp pipette

Chimeric blastocyst consisting of genetically different cells

Chimeric blastocyst consisting of genetically different cells

Foster mother, genetically different from transplanted blastocyst

Foster mother, genetically different from transplanted blastocyst

Chimera

Chimera

a

b

to blastula) are able to reorganize themselves, forming a giant but uniform and viable blastula that develops into a chimerical individual.

Mouse experiments by Beatrice Mintz used blastocysts of two different inbred strains (albino and black) that differed by their white and black coats. The blastocysts were removed from their envelopes (*zona pellucida*) and fused. The fusion product was inserted into the uterus of a foster mouse previously prepared for pregnancy by injection of hormones. The resulting tetraparental litter displayed coats with black and white stripes (Fig. 13.4).

Even viable chimeras between goat and sheep have been produced. However, from such chimeras no novel species, strains or even genetic hybrids can be bred. If chimeras composed of A and B cells are fertile at all, their egg cells or sperm cells derive from either A or B cells. The two genomes remain completely separate. Therefore, the gametes do not contain a mixture of A and B genes.

Chimeras of a different kind, with a true mosaic of genetic information, can be generated if genetic material of a different donor is introduced into egg cells. Transgenic organisms (see below) are true genetic chimeras. Classic chimeras as depicted in Fig. 13.4b are intermediate stages in the procedure to create transgenic animals (see Fig. 13.6).

13.2.2 Teratomas Are Failed, Chaotically Disorganized Embryos That Can Develop Features of a Tumour

Teratomas are chaotically organized embryos. They derive from unfertilized cells of the germ line in the testes or ovaries. Occasionally germ line cells start embryonic development precociously. Teratomas can occur not only in ovaries of women but also in testes of men. The origin of such **germ-line derived teratomas** can be diploid spermatocytes starting cleavage instead of undergoing meiosis, or a diploid oocyte (or an oocyte having regained diploidy by rejoining with the polar body) starting cleavage without being fertilized by a sperm.

Another source of teratomas are fertilized eggs which fail to be taken up by the tube of the oviduct and instead implant anywhere in the abdominal cavity (**abdominal pregnancy**). Experimentally, in mice teratomas are produced by removing blastocysts from the uterus and implanting them into the abdominal cavity. Teratomas may develop into **malignant teratocarcinomas**, tumours which even can metastasize. Cells derived from teratocarcinomas (such as the widely used murine 3T3 or F9 cells) are frequently immortal and can easily be propagated as cell cultures.

Teratocarcinoma cells are, as a rule, still more or less intact genetically. If implanted into normal blastocysts, they may integrate into their new surroundings unobtrusively and participate in the construction of the new animal (Fig. 13.3b). Because in such experiments the implanted teratoma cells and the cells of the host blastocyst are usually different genetically, the resulting progeny will be a chimera.

13.3 Genetic Manipulation of Embryos, Knockout Mutants and Transgenic Animals

13.3.1 There Are Two Proven Methods to Introduce Altered or Foreign Genes into the Genome of an Animal

Two procedures are in use for targeted introduction of artificially altered or foreign genes by means of gene constructs (vectors) into mice (the preferred animals for such experiments):

1. **Direct injection of the construct** into the female or male pronucleus of a freshly fertilized egg cell (Fig. 13.5). Occasionally, the construct is integrated into the haploid genome. After fusion of the two pronuclei to the diploid nucleus of the zygote the fertilized egg cell will be heterozygous with respect to the introduced gene. By inbreeding homozygosity can be obtained in the F2 or F3 (compare Fig. 12.1). However, currently constructs are lacking which reliably and permanently integrate into the host's genome or which would insert into a locus determined by the researcher and not by the construct itself. This applies especially if retroviral vectors are chosen. Recently methods are being developed to cut the DNA of fertilized eggs at distinct positions using site-specific nucleases (see below).

2. **Indirect procedures via genetically engineered embryonic stem cells** (**Fig.** 13.6). The procedure is time-consuming but allows genetic engineering of almost any kind. In

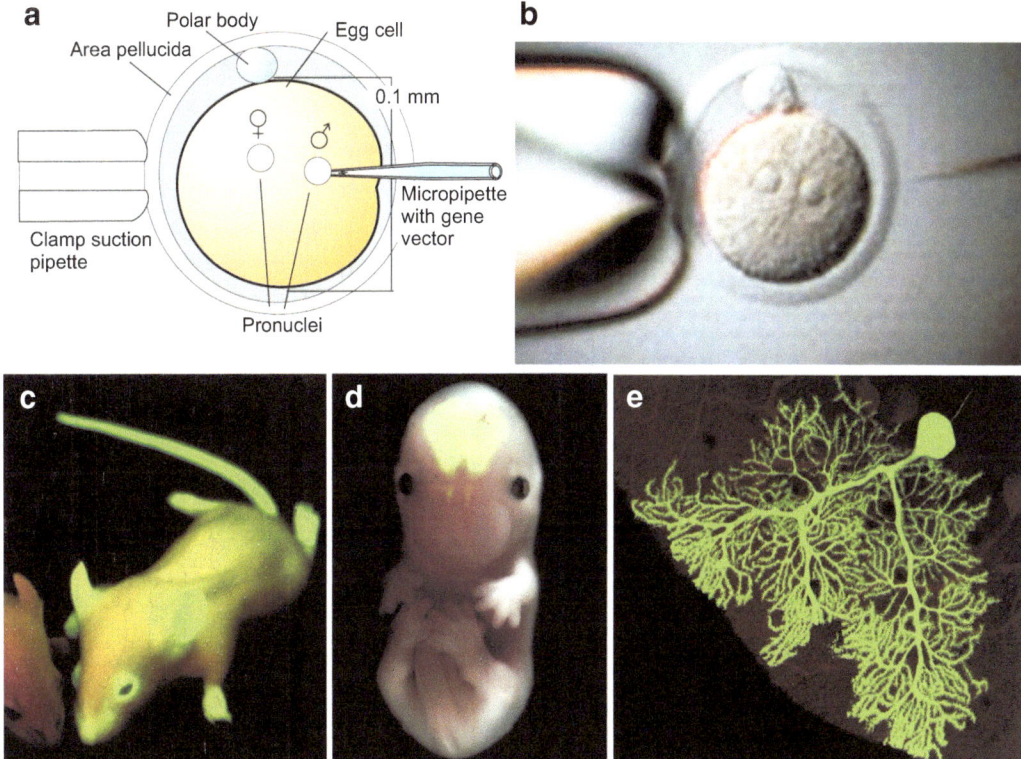

Fig. 13.5 (**a–d**). (**a, b**) Injection of a vector construct into the pronucleus of an egg cell that has previously been fertilized and freed from the cumulus cells. The egg cell is clamped by a suction pipette. (**c–e**) A GFP reporter gene marks cells in living embryos or adult mice. (**c**) GFP is linked to actin and expressed globally. Photo reproduced by the kind permission of www.ethlife.ethz.ch. (**d**) A 14 day old mouse embryo, in which neural stem and precursor cells of the brain are labelled green through expression of a GFP fusion gene. Reproduced from a research report with kind permission of Prof. Wieland Huttner and of PNAS (<©> National Academy of Sciences, USA). (**e**) Purkinje cell in the cerebellum of the mouse, arisen from the spontaneous fusion of a GFP-labelled stem cell with a nerve cell. Photo kindly provided by Dr. Clas B Johansson, Karolinska University Hospital Stockholm, and with permission of Prof. Helen Blau, Stanford. Method see Johansson CB et al., (2008) Nature Cell Biol 10(5): 578–583

mice the ability of embryonic stem cells (ESC) to integrate themselves into a host blastocyst has been exploited. The artificially mutated gene (**targeted mutagenesis**), or **transgene** (foreign gene) of interest is first inserted into a vector allowing its amplification by cloning in bacteria. Then the excised vector is introduced into ES cells either by direct injection, or by electroporation, or by means of a viral vector. The transgene is associated with a promoter that allows its controlled expression in the recipient, and it is linked with an **antibiotic resistance gene and/or a reporter gene** such as GFP (green fluorescence protein). The few ES that have successfully integrated the construct are selected by treating them with antibiotics: cells without the integrated construct die. The surviving ES cells are multiplied by culture, and then pipetted into host blastocysts. The chimerical blastocysts are implanted into the uterus of a foster mother prepared for pregnancy by hormones. When the transgenic ES cells participate in the formation of the embryo, some of their descendants may give rise to primordial germ cells and eventually to germ cells. If in mating a transgenic gamete participates in fertilization, the progeny may carry the transgene. The F1 will

Fig. 13.6 Production of knockout mutants and of transgenic mice through the procedure of homologous somatic recombination and via chimeras consisting of gene engineered donor cells and a normal blastocyst. Explained in the main text

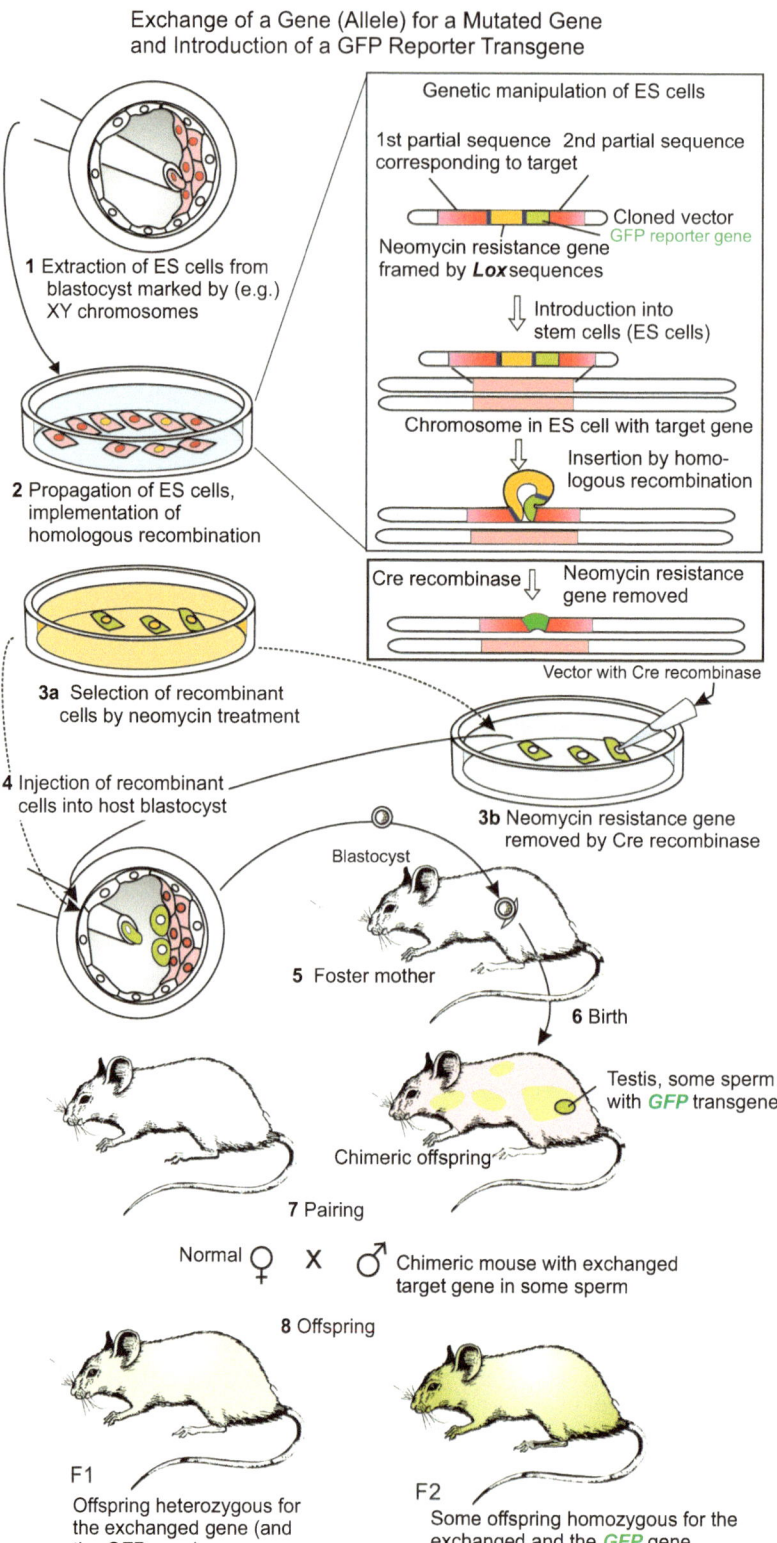

Exchange of a Gene (Allele) for a Mutated Gene and Introduction of a GFP Reporter Transgene

Genetic manipulation of ES cells

1st partial sequence 2nd partial sequence corresponding to target

Cloned vector
GFP reporter gene

Neomycin resistance gene framed by *Lox* sequences

Introduction into stem cells (ES cells)

Chromosome in ES cell with target gene

Insertion by homo-logous recombination

Cre recombinase   Neomycin resistance gene removed

Vector with Cre recombinase

1 Extraction of ES cells from blastocyst marked by (e.g.) XY chromosomes

2 Propagation of ES cells, implementation of homologous recombination

3a Selection of recombinant cells by neomycin treatment

3b Neomycin resistance gene removed by Cre recombinase

4 Injection of recombinant cells into host blastocyst

Blastocyst

5 Foster mother

6 Birth

Testis, some sperm with *GFP* transgene

Chimeric offspring

7 Pairing

Normal ♀   X   ♂ Chimeric mouse with exchanged target gene in some sperm

8 Offspring

F1
Offspring heterozygous for the exchanged gene (and the *GFP* gene)

F2
Some offspring homozygous for the exchanged and the *GFP* gene, (targeted mutants and/or transgenic)

be heterozygous, but by repeated mating of such heterozygous offspring in F3 a fraction of individuals occurs that is homozygous for the introduced allele. The foreign gene has now become part of the host strain's genome and will be passed on from generation to generation. It is even possible to introduce human genes into mice, for instance to examine the involvement of a certain mutant gene in the occurrence of a disease.

In both procedures an important step is integration of the construct into the genome of the host mouse by homologous somatic recombination explained as follows.

Homologous Recombination. When a gene is introduced into a cell of a mouse (or any other organism) that in its initial and terminal sequence is identical to a cell-resident gene, in rare cases a local recombination can take place by which the resident gene is removed and replaced by the introduced gene. This can even happen when the introduced gene is interrupted by, for example, a *neomycin resistance* and a *GFP* gene (Fig. 13.6). Such homologous recombination takes place in the course of DNA repair after a double strand break. Because this is a rare event, in the indirect procedure the constructs are not introduced into egg cells directly but into ES cells. These can be multiplied by culture with growth factors and selected by treatment with antibiotics as said above. If in addition they express GFP they can be selected under the fluorescence microscope. Homologous recombination does not work successfully in all organisms, for example not in *Drosophila*; but it succeeds in cells of the mouse capable of dividing.

Targeted Mutagenesis. The selected target gene on the chromosome is exchanged for a gene (allele) that has been artificially altered, or even rendered unusable, in the laboratory. A gene can be made unusable, for example, by inserting the *neomycin resistance* gene *neo* within the gene, thus interrupting its message. When this is done in ES cells, once more only cells that have integrated the disrupted gene by

homologous recombination will survive the neomycin treatment. The ES cells will be injected into blastocysts and the procedure continued as described above. If following inbreeding both alleles are present in unusable form (homozygous) one speaks of a **knockout mutation**, also called **null mutation** or **loss-of-function mutation**, and the condition is briefly notified –/–. Such mutations reveal whether a gene is indispensable for successful embryonic development or whether pathological symptoms do not occur until birth or until adulthood.

Cre/LoxP System for Targeted Excision of DNA Sequences. A widely-used and versatile method to excise a DNA sequence is the *Cre/loxP* method. In a wider usage *Cre-Lox* recombination is a site-specific recombinase technology used to carry out deletions, insertions and translocation of cell-resident genes or transgenes, or merely of gene promoters or enhancers. It allows the DNA modification to be targeted to a specific cell type or be triggered by a specific external stimulus. The sequences to be excised, for example, are flanked by *loxP* sequences, inserted into a vector, cloned and subsequently introduced into either egg cells or ES cells. If such a construct is integrated into the genome of a recipient by homologous recombination, and if mice are born bearing the *loxP* flanked sequence, these mice are mated with transgenic mice that can contribute a *Cre recombinase* to the genetic makeup of the progeny. In the progeny expressed Cre recombinase can excise the *loxP* flanked sequence (Fig. 13.7). If the *Cre recombinase* gene is provided with an upstream cell type-specific or organ-specific promoter the targeted gene is specifically excised in the respective cell type or organ, for example in the heart (Fig. 13.7).

Engineered Nucleases for Targeted Genome Editing

Genome editing means changing a gene sequence by adding, removing or replacing DNA bases, exploiting the cell's own DNA repair mechanisms. There are currently three mechanisms employed:

Fig. 13.7 Cre-Lox procedure for cell type- or organ-specific elimination of a selected gene. In the present case successful organ specificity is manifested by the green fluorescing heart. Explained in the main text

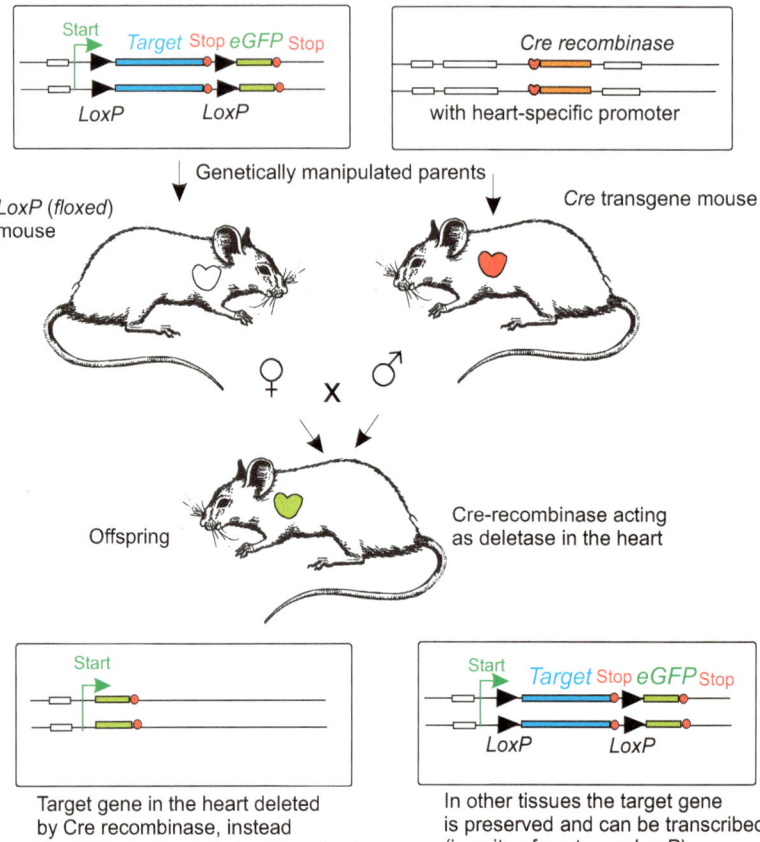

Conditional mutagenesis: here tissue-specific deletion

Target gene in the heart deleted by Cre recombinase, instead the GFP reportergene is transcribed.

In other tissues the target gene is preserved and can be transcribed (in spite of upstream *LoxP*).

1. Zinc finger nucleases (ZFNs)
2. Transcription activator effector-like nucleases (TALENs)
3. Clustered Regularly Interspaced Palindromic Repeats (CRISPR) associated nuclease (CRISPR/Cas)

All of these use an endonuclease to cut the DNA, acting like a molecular scissors. They use different mechanisms to guide the scissors to the correct site for cutting.

1. ZFNs use an array of DNA binding proteins, each of which recognises a triplet of DNA bases. Different DNA-binding proteins can be combined to design ZFNs for specific sequences (Fig. 13.8).
2. TALENs use an array of amino acids rather than proteins to bind the DNA. Each amino acid recognises and binds to a single DNA base, allowing almost unlimited targeting of sequences (Fig. 13.8).

3. CRISPR/Cas does not use protein-based DNA binding; instead it uses an RNA sequence complementary to the DNA sequence to be bound and cut (Fig. 13.9).

Following cutting of the DNA the cell tries to repair the break. This can be done by two mechanisms. The most common is non-homologous end joining. This is error prone and therefore is likely to introduce insertions, deletions or replacement of DNA bases. The result may be no effect, generation of a shortened protein or complete knockout of the protein. The second mechanism is homologous recombination, a much rarer event, and much less error-prone than non-homologous end joining. If a DNA template is injected along with the engineered nuclease then this may be inserted instead of the cell's own DNA during homologous recombination. If the template DNA contains a known human disease-causing mutation, this may be used to create an animal model of

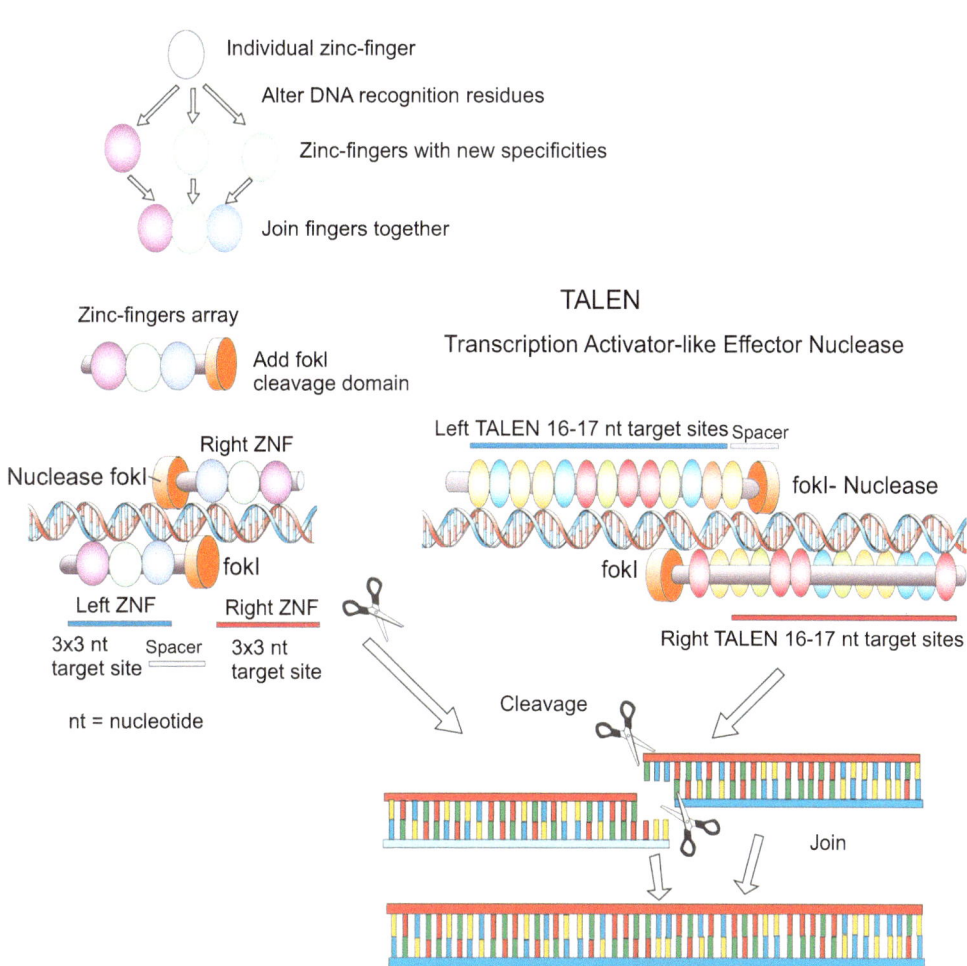

Fig. 13.8 Procedure to cleave genomic DNA at preselected sites using Zinc finger nucleases ZFN or Transcription activator effector-like nucleases TALENs. Explanation in the text. After Okamura 2008; Wilson and Doudna 2013, combined and modified

disease. In the future it may be possible to use this in gene therapy, in which case the template would be the normal form of the gene, in the hope that it would replace the mutant disease-causing form.

13.3.2 With Targeted Mutagenesis "Models" of Genetically Caused Human Diseases Are Generated

In many laboratories around the world by targeted mutagenesis "model organisms" are prepared displaying symptoms similar to genetically caused human diseases. Under therapeutic aspects such models are wanted in medical research. Other such model strains were generated by selection and inbreeding of diseased animals. Most such models are mice. They have been generated by genetic manipulation or selected by the breeder to display symptoms of, for example, diabetes, Parkinson's disease, heart diseases, kidney failure, various kinds of cancer, multiple sclerosis, and even depression.

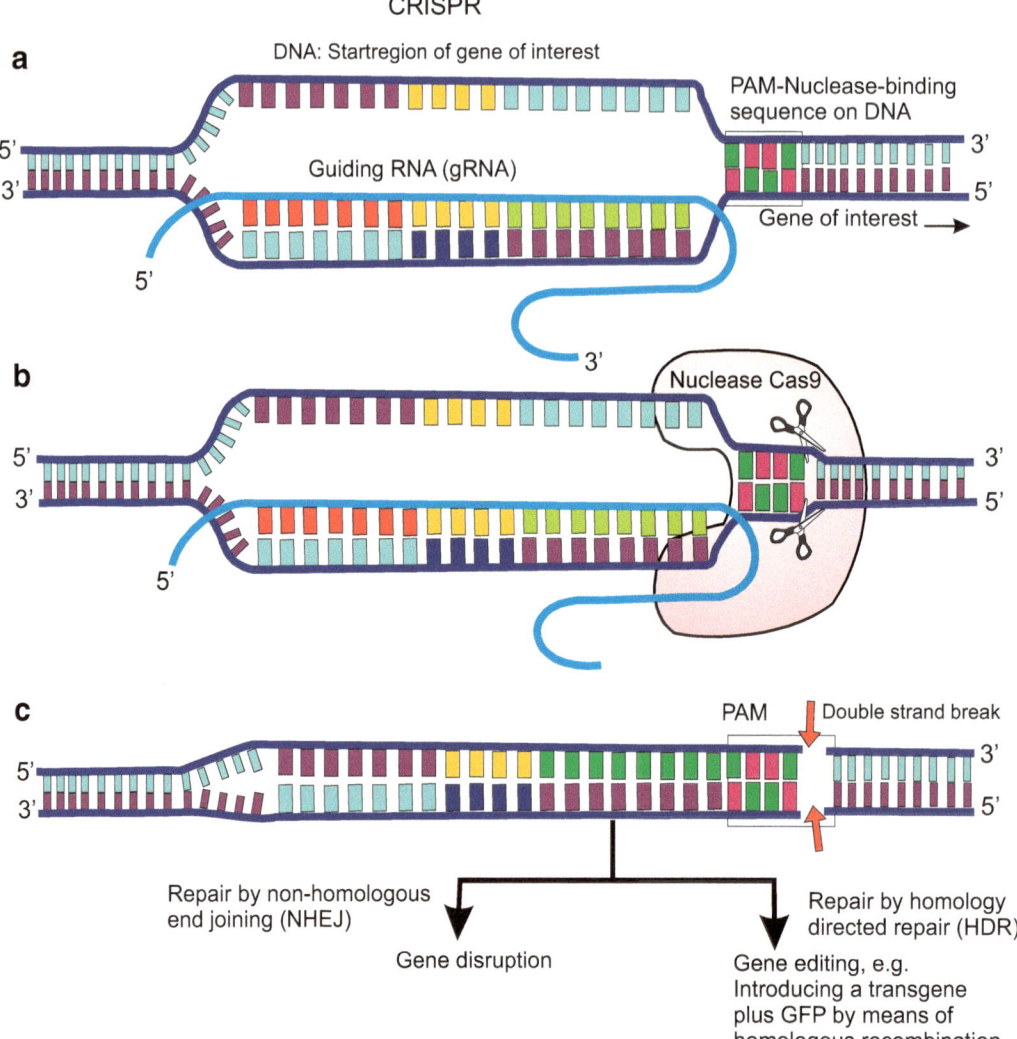

Fig. 13.9 Procedure to cleave genomic DNA at selected sites using CRISPR/Cas. The selected site is marked with a synthetic guiding RNA. Colours indicate complementarity, width of the symbols importance for the procedure. For gene editing the example of homologous recombination is shown. The strand break facilitates exchange of a genomic sequence in the host cell by a sequence constructed in the laboratory and introduced in the host cell (typically an egg cell) by injection or through a vector. The "specific change" typically is a construct with a transgene linked with a reporter gene such *GFP* or *RFP* which will be translated into the green fluorescing or red fluorescing protein

There are, however, controversial discussions whether, and to what extent, we human beings are authorized to impair the health of animals so profoundly. Replies to this question are as controversial as replies to the question as to whether animal testing can be justified or not.

13.3.3 In Transgenic Animals Also Introduction of Genes from Foreign Species, or Artificially Designed Transgenes, Succeeded

Of course, through homologous recombination instead of introducing a non-functional allele one can also introduce alleles that have been altered in the laboratory preserving functionality, being, for example, enhanced or attenuated in effectiveness. If the gene itself is not altered but its promoter, or the promoter is exchanged for a different promoter, it can be accomplished that the gene is expressed at a different time or in a different organ. If one succeeds in inserting into ES cells a gene that is not part of the genetic inventory of the species and therefore cannot be introduced by sexual mating, the prerequisite to produce transgenic animals is achieved (see following section).

13.3.4 Transgenic Animals May Bear Externally Controlled Genes and Genes Derived from Foreign Species

Organisms into which genes of foreign donors – even genes of a foreign species – have been introduced and which are now permanently carrying these genes are said to be **transgenic**. Self-evidently, foreign genes of other, non-related species, in extreme cases genes of yeasts (see Fig. 12.19) or procaryotes (tet-system, Box 12.2) cannot be introduced by sexual mating. One has to find other methods. A direct method is introduction of vectors such as plasmids, viruses or transposons into egg cells (Fig. 13.5). Another method is indirect. As explained in the previous sections, in mice and other mammals the ability of embryonic stem cells (ES) to integrate themselves into a host blastocyst has been exploited. The foreign gene (transgene) of interest is introduced into ES cells, the transgenic ES cells injected into host blastocysts, and the blastocysts implanted into the uterus of surrogate mothers. When the transgenic ES cells participate in the formation of the embryo, some of their descendants may give rise to primordial germ cells and eventually to germ cells. If in mating a transgenic gamete participates in fertilization, the progeny may carry the transgene. The F1 will be heterozygous and, as a rule, the chance-dominated allotment of the transgene not recognizable in the phenotype of the progeny. Yet it is currently quite easily

possible to verify the presence of the transgene by genetic finger printing using DNA chips (see Figs. 12.3 and 12.4), and select those individuals for inbreeding. By repeated mating of such heterozygous animals one can generate animals that are homozygous for the introduced allele. The foreign gene has now become part of the host strain's genome and will be passed on from generation to generation. It is even possible to introduce human genes into mice, for instance to examine the involvement of a certain mutant gene in the occurrence of a disease or in physical and mental performances such as memory formation.

Of particular interest are gene constructs the promoters of which are designed to be activated by external stimuli such as administering a heat shock, a hormone or the antibiotic tetracyclin. We refer to the Box 12.2.

Transgenic animals and animals generated to be representative models of diseases are not only of interest for basic research but to a growing extent also for clinical research, and for the pharmaceutical industry. Experiments with such model organisms are designed with the aim of developing novel medications and therapies. In addition, attempts are made to cause female mammals to produce human proteins with their milk, or chicks whose eggs contain antibodies designed by researchers or suppliers of biochemicals.

Certainly, such animals should be available over long periods of time and in large enough numbers. Currently it is almost impossible to *de novo* generate such model animals with equal qualities repeatedly. Here hopes are built on cloning.

Summary

Cloning means generation of genetically identical individuals. In plants and many non-vertebrates cloning is a natural event, known as **vegetative reproduction**. In vertebrates mini clones arise as **monozygotic twins or multiples** by the fragmentation of early embryos (cleavage stages); this can be reproduced in the laboratory (without much benefit). Cloning of vertebrates in large numbers and with adult genome donors was first successfully performed in the clawed frog *Xenopus*. Here the nuclei of egg cells are replaced by nuclei of somatic tissues of one and the same donor animal. All descendants are genetically identical with the donor and with each other. With similar procedures adapted to mammalian conditions also cloning of mammals (for example sheep, racing horses, mice representing human diseases) is possible. However, mitochondrial genes always derive from the oocyte-bestowing female. Lasting epigenetic modifications of the donor nuclei often curtail the rate of success and may be the reason that the cloned progeny is not always an exact copy of the donor or develops some disorders.

Chimeras are artificially produced individuals whose cells derive from different parents and therefore are equipped with different genomes. Chimeras are produced as intermediates in the procedure to generate genetically altered, for instance transgenic, mammals. For producing genetically engineered animals two procedures are used (1) Artificially designed **gene constructs are directly injected** into the ♂ or ♀ pronucleus, and occasionally integrated into the haploid genome. The zygote being diploid after the fusion of the pronuclei gives rise to progeny being heterozygous with respect to the introduced gene. By subsequent repeated inbreeding homozygous individuals are generated. (2) In the indirect procedure constructs containing, besides the gene to be introduced, marker genes for selection, are introduced into **embryonic stem cells (ES cells)**. Successfully engineered ES cells are selected and injected into blastocysts. The injected cells can participate in the development of the host animal and contribute to its germ cells. If previously foreign genes (**transgenes**) are introduced into the ES cells transgenic germ cells can occur, and these can give rise to offspring who, in all cells of their body, contain the foreign transgene as homologous partner of its own allele. Again, homozygosity is attained by repeated inbreeding; thus **transgenic animals** are produced.

Artificial or foreign genes injected into egg cells or ES cells are integrated into the host's genome by the rare event of **homologous recombination** by which a cell-resident gene is removed and replaced by the introduced gene. This is accomplished by the DNA repair machinery which heals double strand breaks, and presupposes that the introduced and the cell's own gene share extended stretches of identical sequences at their ends. Novel methods making use of **site-directed nucleases** allow site-specific exchange of genes. The exchanged gene may bear a targeted mutation and thus *knockout* **mutations** (also called *loss-off function* or *null* **mutation**) can be generated. If the introduced gene, or a promoter or any other sequence, is framed by *LoxP* sequences, the *LoxP* marked DNA sequence can be excised if a **Cre recombinase** is introduced by injection or by mating *LoxP* transgenic individuals with *Cre recombinase* transgenic individuals. From the resulting effects the normal function of a gene can be deduced. If an introduced functional gene is provided with an upstream promoter specific for another tissue or organ, the site of expression in the body changes according the responsiveness of the promoter. If an inducible promoter is used the expression can be controlled by external stimuli (see also Box 12.2). Some more newly developed methods for targeted exchange of genes such as TALENS and CRISPR are briefly introduced.

Morphogenesis: Shaping by Active Cell Movement, Differential Cell Adhesion and Cell Death

14

14.1 Active Cell Movement and Migration

14.1.1 Unlike in Plants Active Cell Movement Plays an Important Role in the Development of Animals

As a result of cell division and cell differentiation a large variety of cells with different shapes and divergent molecular constitutions and makeup appear. These cells in turn create associations of cells serving a common function. Such morphological and functional units are called a **tissue** if they are composed of one or a few cell types; they are called **an organ** if they are composed of diverse cell types and their functional role in the commission of the whole organism occupies centre stage. In cell associations, the number, size and shape of individual cells will eventually determine the shape of the whole association. In contrast to the development of plants, in animals active shaping through intracellular contractile filaments and intercellular forces of cohesion and adhesion are observed, and extensive displacement and migration of cells take place.

Active deformation is brought about by

- intracellular, in their length variable elements of the cytoskeleton (microtubules, intermediate filaments, actin filaments),
- intracellular motor proteins such as actin/myosin, kinesin, dynein and dynamin, and by

- self-regulated, variable forces of cohesion and adhesion which occur between the cells and their environment.

Furthermore in animal development extensive **migration of mobile cells** is common.

14.1.2 Displacement and Migration of Cells Enable Construction of Tissues and Organs at Distant Locations: A First Overview

Gastrulation, a most important event in animal development, consists of the displacement of cohorts of cells into the interior of the embryo where they will give rise to the inner organs. In the amphibian gastrula displacement is brought about by the combined effect of two processes, in the initial curving in of the outer cell sheet to form the blastopore and in the subsequent streaming of cell cohorts through the blastopore. When isolated by excision cells around the blastopore crawl on a suitable substrate like a caravan of migrating *Dictyostelium* amoebae. Moreover, cells can sort themselves out from their association and acquire the freedom of individual mobility. Active movement of cells into the interior is the main modus of gastrulation in fish, birds, mammals, but also in several invertebrates.

In sea urchins gastrulation begins at the vegetative pole with the immigration of cells into the cavity of the blastocoel. Descendants of the micromeres untie the bonds to their neighbours, pull out of the society of the epithelial blastoderm

Fig. 14.1 Amoeboid bottle cells in the amphibian gastrula

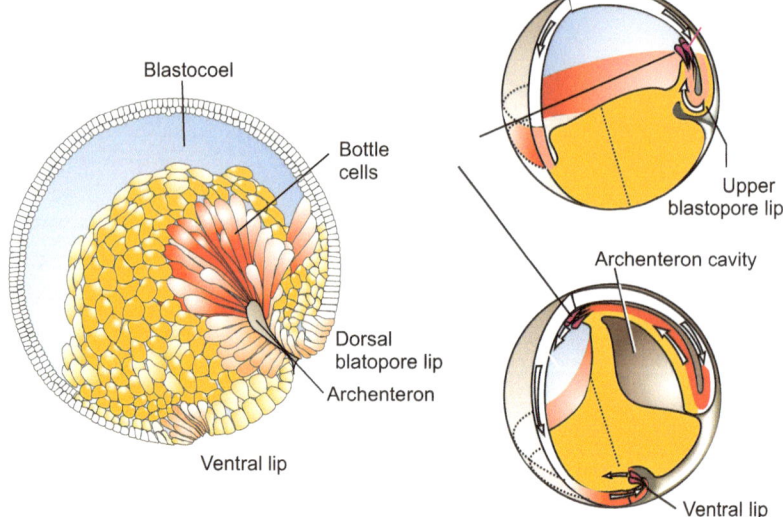

Blastocoel

Bottle cells

Dorsal blatopore lip

Archenteron

Ventral lip

Upper blastopore lip

Archenteron cavity

Ventral lip

(**dissociate**), and **immigrate** into the blastocoel. The descendants of the micromeres **reaggregate** at certain places and collectively manufacture the larval skeleton (see Fig. 4.2). In the gastrula of sea urchins and in the amphibian gastrula at the tip of the archenteron, there are cells that retain contact to epithelial sheets but extend pseudopodia and take on the function of pathfinders. In amphibians these cells are known as **bottle cells** (Fig. 14.1). Guided by these bottle cells in the course of gastrulation the roof of the archenteron slides along the overlying ectoderm (inducing it to form the central nervous system). A layer of fibronectin covering the inner surface of the ectoderm serves as lubricant.

In translucent embryos many **freely wandering cells** can be seen, such as the **deep cells** in fish embryos (see Fig. 5.22). Ameboid behaviour is also displayed by cells in the blastodisc of birds (Fig. 5.25) or mammals (Figs. 5.29, 6.8) that invade through the primitive groove into the depths and spread in the blastocoel (subgerminal cavity) to form the axial mesoderm (notochord and somites).

The **neural crest cells** (see Sect. 15.2; Fig. 15.3) of vertebrates display especially extensive migratory activity. Primordial germ cells also go on extensive tours (see Fig. 8.3).

14.2 Sliding and Sorting by Virtue of Cell Adhesion

14.2.1 Convergent Extension: Stretching and Elongation of a Cell Association on the Zipper Model

When in the amphibian gastrula the future mesoderm (marginal zone) rolls around the lip of the blastopore it stretches itself, and continues elongating while gliding along the inside of the ectoderm in an anterior direction. The analysis of each single cell movement revealed an unexpected mechanism of stretching and elongation. Imagine the following situation: Two roads converge and join forming a single major traffic pathway. In addition two roads converge from higher levels, one road entering from the left, the other from the right. Of course, at the site of convergence a traffic jam happens. How is the problem solved in such a way that everybody gets a chance to line up, without traffic lights? The solution is the zipper mode, one car from the left, the next car from the right and alternating so forth.

The multilayered cell association acts in a similar way. The cells line up coming in alternate

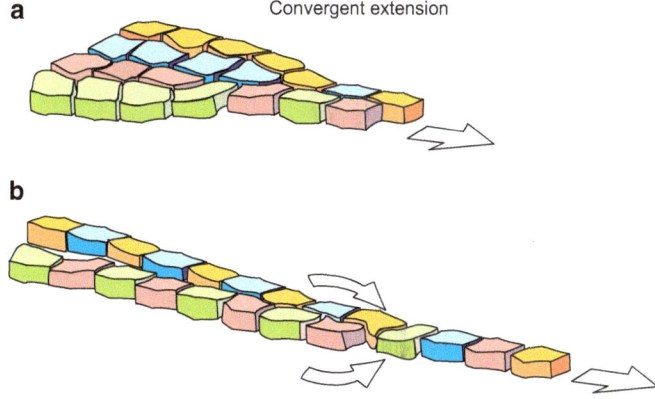

Fig. 14.2 Convergent extension and zipper procedure in the gastrulating *Xenopus* embryo

order from right or left, one by one, according to the zipper mode, so that a single layered, narrow association is formed (Fig. 14.2). In the terminology of developmental biologists the procedure is known as **convergent extension**. This mode of stretching is used in other processes as well. The archenteron of the sea urchin comprises initially 20–30 cells in its circumference, after its stretching it merely consists of 6–8 cells. In the process of intercalation changing forces of cell adhesion are of decisive significance.

14.2.2 Forces of Adhesion Can Shift the Relative Positions of Cells

A drop of oil contracts on a hydrophilic substrate to form a sphere, while it spreads on a hydrophobic (lipophilic) surface to form an extended film. A drop of water displays the opposite behaviour. Similarly, **adhesion or repulsion** of cells is influenced by physical forces such as **tension of wetting**, which is the integral result of several surface energies. In addition, **electrostatic** and **ion exchange properties** influence the strength of adhesion or repulsion. Cells can alter the degree of hydrophobic or hydrophilic interactions with neighbouring cells, and the strength of electrostatic attraction or repulsion. They can do this, for example, by inserting membrane proteins and by glycosylating or deglycosylating surface proteins. Physical forces acting at interfaces are of significance when in the course of gastrulation sheets of endoderm,

mesoderm and ectoderm slide along each other. Even if cells do not actively move, cell sheets can slide and spread over other sheets by sheer physical forces of adhesion and cohesion, thereby minimizing the strength of surface tensions. **Minimizing surface energies** also accounts for, or contributes to, the process of **sorting out** of cells of different origin in aggregates (Fig. 14.3).

During embryonic development of the vertebrates, coherent groups of cells stream to their destinations by liquid-like spreading movements. According to the "**differential adhesion hypothesis**", these movements are driven and guided by cell-adhesion-generated tissue surface tensions, operating in the same manner as surface tensions do in the mutual spreading behaviour of immiscible liquids, among which the liquid of lower surface tension is always the one that spreads over its partner. Miniature devices have been designed to measure surface tensions of cell groups. The measured values correlate with the behaviour of the cells in aggregates. Cells displaying high tensions are enveloped by cells displaying lower surface tensions. For example, limb bud mesoderm (tension 20.1 dyne/cm) is enveloped by liver (4.6 dyne/cm) which, in turn, is enveloped by neural retina (1.6 dyne/cm).

In addition cells can segregate by differential mobility. For example, in an aggregate of *Hydra* cells (Fig. 4.14) endodermal cells migrate actively and rapidly into the interior of the cell clump, while ectodermal cells hover almost immobile at the surface and eventually reorganize an outer epithelium.

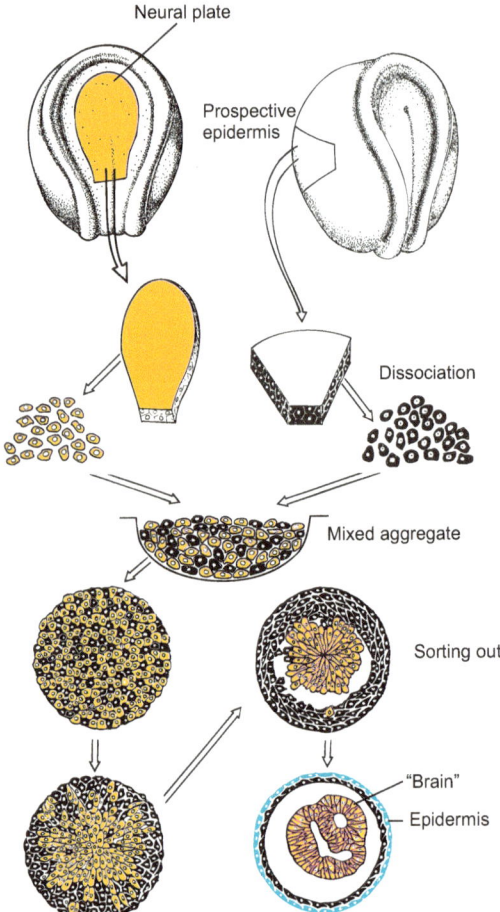

Fig. 14.3 Sorting out of cells of different origin and fate. After segregation, prospective epidermal cells form an outer epithelial layer, and the cells derived from the neural plate form hollow bodies reminiscent of neural tube or brain, respectively (after Townes and Holtfreter 1955)

In animal embryos the differential spreading of one cell population over the surface of another is a common means of embryonic **morphogenesis**.

14.3 Cell Adhesion Molecules and Cell Recognition

14.3.1 Specific Adhesion Molecules also Mediate Cell Recognition

The cell membranes of animal cells are equipped with proteins and glycoproteins. Several of them mediate the physical cohesion of cells in cell associations by forming non-covalent bounds with corresponding surface molecules of neighbouring cells. In addition, such cell adhesion molecules serve in mutual cell recognition. Specific **cell adhesion molecules (CAM)** bind to others. Binding can be **homotypic** (homophilic) between identical CAMs, or **heterotypic** (heterophilic) between different CAMs. At least five classes of CAMs have been distinguished:

1. **Cadherins** (**ca**lcium-dependent **adherins**). Cadherins enter into homotypic binding only in the presence of calcium ions. Removal of calcium in the surrounding medium breaks off the mutual adhesion of the cells, which become dissociated. The class of cadherins includes, among others, **L-CAM** (Liver Cell Adhesion Molecule) which mediates cell coherence not only in the liver. L-CAM or E-cadherin, also called **uvomorulin** in mouse embryos, is the first new cell adhesion molecule exposed on the cell surface, causing **compaction** of the mammalian 16 cell-stage embryo (see Fig. 5.28). The adhesiveness of diverse cadherins causes or facilitates sorting out of cells in cell aggregates (Fig. 14.3).

2. **CAMs of the immunoglobulin superfamily**. This superfamily includes the subfamily of the **N-CAMs** (Neural Cell Adhesion Molecules). N-CAMs (Fig. 14.4b) are found on the surface of neuronal cells including glial cells, but also appear on the surfaces of other cells, in particular during embryo development. N-CAMs are glycoproteins containing sialic acid residues. They are anchored in the cell membrane by a transmembrane domain, or coupled to membrane-resident inositol phospholipids via a glycan bridge. All N-CAMs display some sequence similarity to the **immunoglobulins** (i.e. to **antibodies**), to the **antigen receptors of B- and T-lymphocytes**, and to the molecules of the **major histocompatibility complex MHC**. As a rule, binding is homotypic: the CAMs of one cell bind to identical CAMs of adjacent cells. Calcium ions are not needed.

3. **Integrins**. These are heterodimers whose two subunits are integrated into the cell

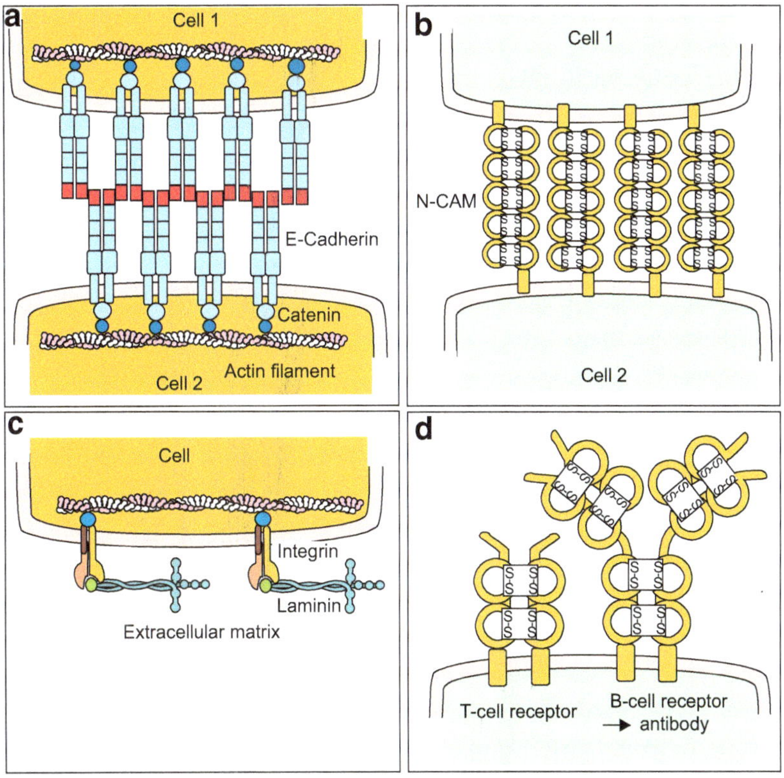

Fig. 14.4 (**a–d**) Cell adhesion molecules CAM. Various CAM's mediate adhesion and mutual recognition. (**a**) Connection of cells mediated by E-cadherins that are anchored at the actin cytoskeleton through catenin. (**b**) N-CAMs serve mutual recognition of nerve cells. (**c**) Connections between the intracellular actin cytoskeleton and the extracellular matrix ECM established via integrin-laminin interaction. (**d**) Membrane-anchored members of the immunoglobulin (Ig) superfamily. Cell recognition in the immune system is mediated by the Ig domains of the T-cell receptors and B-cell receptors

membrane. By means of integrins cells not only establish contact to other cells but also to molecules of the **extracellular matrix ECM**.

4. **Selectins** and **lectins**. Selectins are found on the surfaces of blood cells and of endothelial cells of the vascular system. Selectins are also lectins, meaning that they are capable of binding certain oligosaccharides or carbohydrate moieties of glycoproteins. Since lectins in general and selectins in particular are proteins but recognize and bind carbohydrates, they are heterotypic. Through heterotypic interaction between blood cells and the endothelial cells of blood capillaries, lymphocytes and macrophages can establish contact with the walls of blood vessels, and subsequently leave the vessels.

5. **Ephrins** and **ephrin-receptors**. lend assistance in oriented cell migration and fusion of capillaries. We will meet them when we track the growth of nerve fibres (Chap. 16) or of blood vessels (Chap. 17).

6. **Glycosyltransferases**. These are monosaccharide-transferring enzymes located on the exterior surface of the cells, and are considered to mediate reversible cell contact. For instance, a galactosyltransferase anchored in the plasma membrane of one cell can attach galactose to an acceptor molecule located on the surface of an adjacent cell. As long as no galactose is present in the surrounding medium, the enzyme-acceptor bridge mediates cell adhesion. In the presence of galactose, the sugar is transferred to the acceptor, and the binding is released. The interaction of the mouse sperm with the *zona*

pellucida of the egg envelope is thought to involve such a mechanism.

7. **Various lipids**. When research almost exclusively addressed proteins in depth this was not least because proteins can conveniently be tracked through their genes. Low molecular weight substances and lipids are more difficult to analyze. After all there are a few reports about the guiding functions of lipids in cell migration. And they influence the radiation range of signalling molecules.

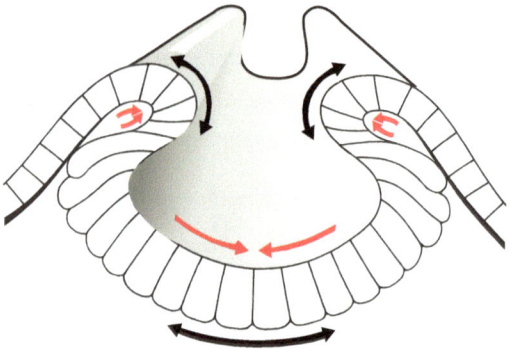

Neurulation bending moments

Fig. 14.5 Bending moments in neurulation, generated by the expansion of the cells on one side and contraction on the opposite side. In addition, sliding movements contribute to morphogenesis

14.3.2 Cell Adhesion Molecules Mediate Attachment and Detachment; They Set Limits and Mediate Signalling

Cell adhesion molecules mediate many important events in animal development. For instance, they

- mediate permanent attachment of cells to each other or to the extracellular matrix, or permit detachment;
- generate boundaries between tissues and facilitate segmentation, such as the subdivision of cell associations into repetitive units, for example somites;
- prepare the formation of cell junctions;
- act as signal molecules (as an example the NOTCH/DELTA system described in Chaps. 10 and 11);
- guide the migration of neuroblasts and the outgrowth of the nervous processes (dendrites or axons; see Chap. 16).

14.4 Formation of Curved Sheets and of Branching Tubular Structures

14.4.1 Folding and Invagination: Cells in Epithelial Associations Can Develop Coordinated Bending Forces

During invagination of the archenteron, neurulation and similar processes that create shapes from epithelial sheets, cells enlarge their surface on one side (e.g. the basal side), while on the opposite side the surface area is diminished (Figs. 14.5 and 14.6). Through this coordinated process an epithelial sheet creates bending moments leading to folding or invagination. Animal cells can be made narrower at their apical surface by the contraction of bundles of actin and myosin filaments that run beneath the upper cell surface, where the cells are strongly connected by adhesion belts.

To expand the surface area on the opposite side the cell can reorganize the cytoskeleton and make use of molecular motors. Figure 14.6 schematically illustrates such a mechanism of reshaping.

14.4.2 Three Mechanisms Enable the Formation of Tubular Structures: Folding of Epithelia, Hollowing of Massive Rods, Fusion of Vacuoles

There are many organs consisting of tubular, and often branched, structures: **tracheae of insects**, **bronchioles of the lungs**, **blood vessels**, **renal tubules**, **milk glands** and many other glands with their ducts.

1. **Invagination** or **Evagination** of an epithelial cell sheet. The above mentioned formation of the neural tube is a paradigm of one type of mechanism leading to tubular structures. As a

enncloses the lumen by establishing contact with itself.

2. A second mechanism is the **hollowing out** of a massive rod or strand. The cells of rod-like aggregates move apart opening a central channel. By forming an epithelial organization around the central channel a stable and more or less tight tube is formed. Examples are the hollow notochord and the intestine in the zebrafish. Also the majority of the blood vessels arises in this manner (see Fig. 17.3).

3. In the third mode cells arrange themselves in file, form **intracellular vacuoles** which fuse across cell borders forming a continuous intracellular channel spanning over many cells. This mechanism is used for forming fine blood capillaries (see Fig. 17.3).

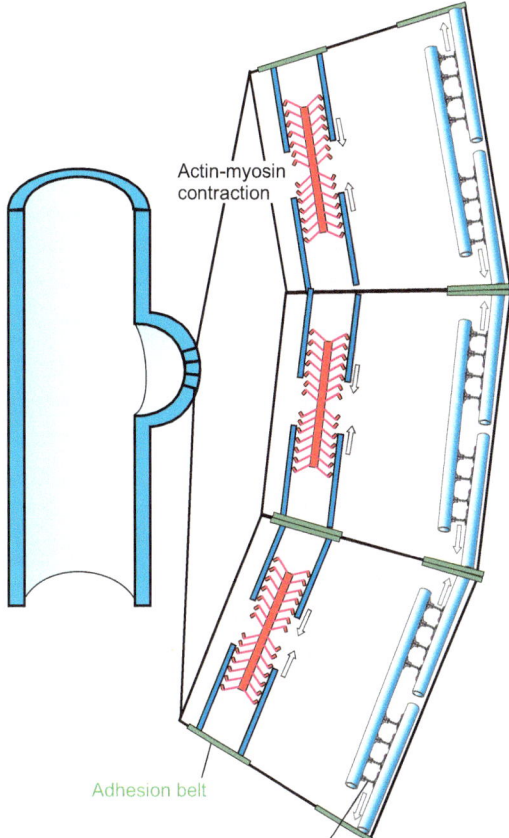

Actin-myosin contraction

Adhesion belt

Kinesin, spreading apart microtubuli

Fig. 14.6 Model for budding by evagination. At the apical side facing the lumen actin-myosin interactions effect contraction while at the basal side kinesin motors push apart microtubuli thus effecting extension

14.4.3 Branched Systems of Supply and Cleansing: Leading Terminal Cells at the Tip of the Tubes Sense Signalling Molecules and Crawl Ahead by Means of Filopodia

Many tubular structures must grow into certain target areas to ensure a supply of oxygen, nourishment, and waste disposal. The **tracheae** of insects and likewise **blood vessels** in vertebrates have to reach tissues that require oxygen and have to get rid of CO_2. And in the target area the tubes and vessels, respectively, should form many and fine branches. In an astonishing parallel one finds in both systems, in the tracheal system of insects and the blood system of vertebrates, at the tips of the tubes a **terminal cell** equipped with **filopodia**. This tip cell fulfils a **pathfinding** function, crawling ahead and sensing with their membrane-anchored receptors signalling molecules emitted by the target (Fig. 14.7, more in Chaps. 16 and 17).

The formation of lateral branches starts in that new terminal cells arise at the flanks of the tubes which crawl towards unexplored waste ground as

rule, the lateral connection of the cell sheet is strengthened by E-cadherin and reinforced by diverse junctions. At these junctions actin-myosin filaments are anchored. The cells of the sheet are polarized from the outset having an apical side facing the lumen and a basal side being supported by the basal membrane. When cells simultaneously expand one side and diminish their surface at the opposite side bending forces are generated. Example are buds of a *Hydra* which continue bending when cut from the parental body column, or the epithelial sheet which forms the channel of the nephridium like a clasping hand and

Sprouting of lateral branches

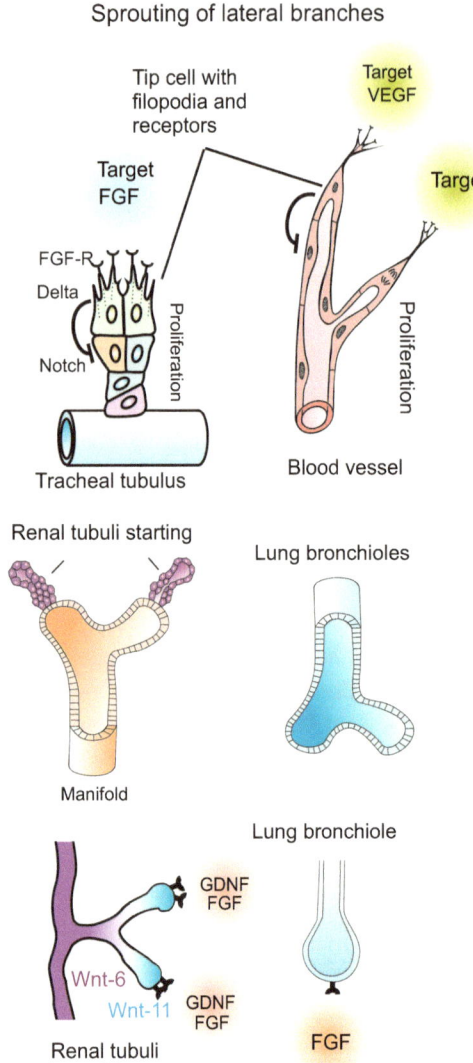

Fig. 14.7 Sprouting and branching of tracheae and blood vessels. In both cases mobile terminal cells equipped with filopodia and receptors assume leadership. Terminal cells are delimited from the stalk by NOTCH-DELTA interaction. The receptors enable orientation towards growth factor emitting targets. The elongation of the tube-like structures is brought about by proliferation behind the terminal cell and by stretching of the cells. Lateral branches arise by generating new terminal cells at the flanks of the tubules. After Affolter et al. (2009) and Miller and McCrea (2010)

pioneers. A further parallel are the filopodia of the growth cones at the tip of growing nerve fibres (Chap. 16). All these filopodia orient themselves by sensing signalling molecules such as

FGFs or other attractive signalling molecules emitted by undersupplied territories. Also the small ampullae at the end of bronchioles and growing renal tubules are fitted with receptors for growth (Fig. 14.7).

With the reception of signalling molecules systems of signal transduction are activated, the actual systems being dependent on the type of the signalling molecules. We refer to Chap. 11 and the key words such as Delta-Notch, BMP4, Sonic Hedgehog and WNT. These control the movement and elongation of the tubular structures. Growth in the sense of cell multiplication takes place behind the terminal cells.

14.5   Shaping by Cell Removal: Apoptosis, The Programmed Cell Death

14.5.1 Programmed Cell Death Is Part of Normal Development

In all multicellular organisms genetically programmed cell death, called **apoptosis**, is part of normal development. Apoptosis removes material. Nature works like a sculptor who creates a figure by removing material. In the development of the hand, for example, programmed cell death brings about the separation of the fingers (Fig. 14.8). Partial removal results in webbed hands. Moreover, apoptosis is a device for correcting failures. Even in the small round worm *Caenorhabditis elegans* (Fig. 4.19) known for the constant number of its cellular inventory, a part of the neuroblasts dies. In this animal apoptosis is initiated by the activity of two genes (*ced-3*, *ced-4*). *Loss-of-function* mutations in one of these genes keep surplus nerve cells alive.

Also in vertebrates more than 70 % of nerve cells die in a certain phase of the developing nervous system, in particular those nerve cells which are not connected with their correct target (Sect. 16.9). **We humans generate three times as many neurons as we eventually have when we are born**. Also many lymphoblasts and germ cells die. In the immune system apoptosis

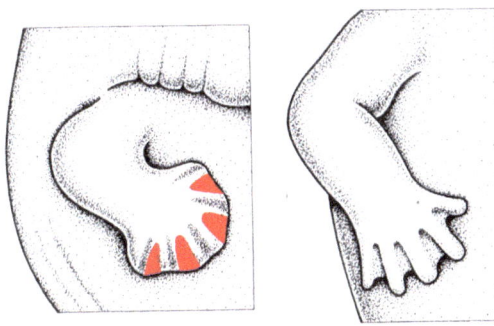

Fig. 14.8 Programmed cell death (apoptosis) in the separation of the palm in a human embryo. *Left*: 41 day old embryo; beginning cell death in coloured areas. *Right*: 51 day old embryo; separation of the fingers is still incomplete. The separation is finished by day 56

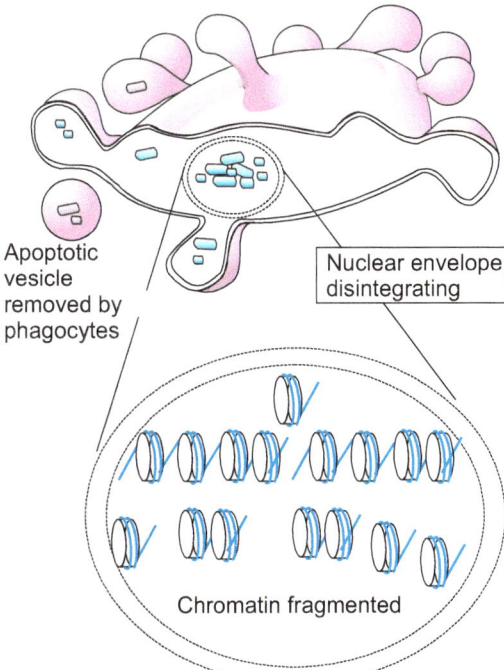

Apoptotic vesicle removed by phagocytes

Nuclear envelope disintegrating

Chromatin fragmented

Fig. 14.9 Apoptosis of a cell. In a act of self-dissolution the cell buds vesicles that are engulfed by neighbouring phagocytes (in vertebrates by macrophages). The nuclear envelope disintegrates, the chromatin condenses and is dissected by nucleases. Because nucleases first cleave between the nucleosomes, transitorily DNA fragments arise as a result which encompass one, two, three or more nucleosome windings and yield a ladder pattern after electrophoretic separation by size in polyacrylamide gels (PAGE). Residuals of chromatin are enclosed in the apoptotic vesicles and completely digested in the phagocytes

removes those lymphoblasts which could attack our own body. Deficiencies in these processes can lead to autoimmune diseases.

14.5.2 Apoptosis Is Programmed Cell Deaths Without Complications

Apoptosis is suicide with features of a well-planned deed. The suicide is preceded by active synthesis of specific proteases and other apoptosis-specific enzymes such as nucleases which cleave the DNA in pieces. The cell fractionalizes itself into vesicles, small niblets which enclose DNA fragments and are easily engulfed by neighbours, in the later life by the macrophages of the immune system (Fig. 14.9). No necrotic complications occur (necrosis = non-programmed cell death caused by damage).

Programmed cell death can be induced by external factors, or prevented by external factors. For example, the **nerve growth factor NGF** ensures the survival of neurons of the vegetative nervous system (Sect. 16.5.3).

Summary

Unlike in the development of plants, extensive **cell migration** and active **cell shape changes** contribute to development of tissues and organs in animals. Migration and changes in shape are mediated by changes in the cytoskeleton and by intracellular motor proteins (actin/myosin, dynein, kinesin). Through active deformation and displacement **bending moments** are generated causing evagination or invagination of cell sheets, and epithelia are furled forming tubes. In this way the neural tube, the primordium of the central nervous system, is formed.

In the **convergent extension** moving cell associations narrow and elongate themselves by lining up in a zipper-style, entering the major traffic pathway in alternate order from left and right.

In morphogenetic movements of all kinds **CAMs** (**Cell Adhesion Molecules**) play an important role. CAMs such as the cadherins, the CAMs of the **immunoglobulin superfamily**, and other cell membrane anchored molecules such as the **integrins**, **selectins**, **ephrins**, and **glycosyltransferases** enable selective attachment or detachment of cells, and in addition serve mutual recognition.

In mixed aggregates of previously dissociated cells, moving cells segregate and sort themselves out according to their tissue affiliation, a phenomenon called *sorting out*. By changing their makeup cells can coalesce with others. Responsible for sorting out or coalescence are, in the first instance, physical surface tensions which in turn depend on the molecular structure of the CAMs.

In many organs tubular branched structures are formed. Examples are the **tracheae of insects**, **bronchioles of the lungs**, **blood vessels**, **renal tubules**, and **milk glands**.

Tubes arise by furling of epithelia, by hollowing out of massive rods and strands, or by the fusion of intracellular vacuoles across cell borders. In growing and branching systems of supply and waste disposal such as tracheae and blood vessel **terminal cells** at the tip assume leadership. These terminal cells are fitted with filopodia and receptors to sense attracting chemical signals such as growth factors emitted by the target tissue. Elongation of the tubes occurs by cell divisions behind the terminal cell.

Apoptosis, genetically programmed cell death, is an event proceeding in characteristic way and is part of normal development. Through apoptosis cavities are formed and the fingers separated from each other. Through apoptosis also incorrectly connected nerve cells and potentially autoreactive immune cells are eliminated.

In the development of animals extensive cell migrations take place. This applies especially to vertebrate embryos. They are like cities full of tourists. In a translucent fish embryo migratory cells can be seen crawling and swarming. In avian and mammalian blastodiscs the cells of the meso'**derm'** do not form a 'skin' (Greek *derma* = skin) or a coherent germ 'layer'; instead the mesodermal cells creep around like amoebae to colonize the spaces between the ectoderm/epiblast and the endoderm/hypoblast towards their target positions and form a loose primary mesenchyme (see Fig. 5.22 fish, Fig. 5.25 bird). Only later they aggregate forming compact structures such as the somites. When the somites split up, the cells of the sclerotome and dermatome emigrate to form compact structures or soft tissues elsewhere (Fig. 15.1). Primordial germ cells, blood cells and neural crest cells travel particularly long distances. Even the germ cells and the cells of the peripheral nervous system as well as the musculature and the skeletal elements of the extremities (fingers, toes) derive from migratory precursors.

15.1 Primordial Germ Cells and Blood Cells

15.1.1 Example Primordial Germ Cells Often Migrate Long Distances to Arrive at the Gonads

By definition **primordial germ cells are the stem cells of the germ cells**. As a rule, primordial germ cells are not generated in the gonads (testis, ovary), but in another place und then migrate into the gonads.

In the colonial hydrozoan *Hydractinia*, for example, oocytes or spermatocytes originate from pluripotent **interstitial stem cells** elsewhere in the colony and immigrate into the gonads (gonophores) of sexual polyps (Figs. 4. 18 and 18.2). If a male and a female colony grow together male stem cells prowling around can enter the female colony and displace the female germ cells. This, however, occurs only rarely because genetically based histoincompatibility prevents fusion of genetically non-identical (heterogeneic) colonies and thus the invasion of **parasitic primordial germ cells**. Intruding foreign germ cell precursors would not transmit the genome of the host but their own foreign genome. Genetically fixed histoincompatibility is known also from other sessile colonial animals, for example from the colonial tunicate *Botryllus schlosserei*. A hypothesis has been proposed assuming histoincompatibility had arisen in evolution to prevent intrusion of parasitic germ cell precursors.

In *Drosophila* the stem cells of the future germ cells are the first cells to be produced at all. There is no trace of a gonad present at this early stage. The stem cells, called **pole cells**, are located at the posterior pole of the egg. These pole cells receive so called **pole plasm** or germ plasm rich with RNA-containing granules and enclosing germ line determinants. These are the products of the genes *oskar, vasa* and *nanos* or

W.A. Mueller et al., *Development and Reproduction in Humans and Animal Model Species*,
DOI 10.1007/978-3-662-43784-1_15, © Springer-Verlag Berlin Heidelberg 2015

Cells emigrating from the somites in vertebrates

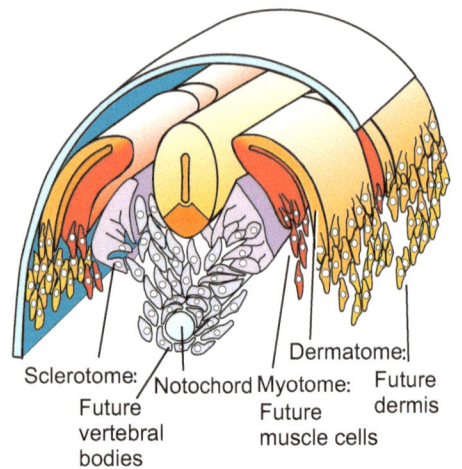

Sclerotome:
Future
vertebral
bodies

Notochord

Dermatome:

Myotome:
Future
muscle cells

Future
dermis

Fig. 15.1 Vertebrate embryo subsequent to neurulation. Fate of cells emigrated from the somites

factors dependent on these genes. During gastrulation, the pole cells are drawn into the interior of the embryo with the invaginating hindgut (Fig. 15.2a). Later, they actively leave the gut and immigrate into the developing ovarioles.

In *Xenopus* eggs a localized "**germ plasm**" has been discovered. It contains RNA-rich granules and can be made visible with certain fluorescent dyes. The granules are tethered to the precursors of primordial germ cells. Fluorescent labels have made it possible to follow the entire tour of the wandering primordial germ cells. They are first seen in the ventral archenteron; from here they migrate in an amoeba-like fashion along the mesenteries (the epithelial ligaments by which the gut is suspended in the coelomic cavity). Eventually the primordial germ cells creep into the **genital ridges**, which will give rise to the gonads.

By no means is a granular "germ plasm" found in all animal organisms as an easily identifiable luggage. The RNA-containing granules that facilitate fluorescent tracking of germ cell precursors in *Xenopus* are missing from many animal groups, including newts, birds and mammals. To trace the path of germ cell precursors in these animals, other markers are used, such as monoclonal antibodies that attach to germ-line-specific surface molecules. Or the

expression of germ line specific genes such as *vasa* (see Chap. 8) is visualized by *in-situ* hybridization and immunofluorescence (for techniques see in Box 12.2).

In the mouse embryo, primordial germ cells are first observed in the extraembryonic mesoderm; later they appear in the region of the allantois (Fig. 15.2). Subsequently, they creep along the allantois and gut toward the genital ridges.

In birds primordial germ cells are detected during neurulation along the anterior border of the blastodisc in a crescent-shaped endodermal zone (the germinal crescent). To reach the embryo, they use the blood stream for transportation over large distances. In this respect, they behave like lymphocytes and macrophages. But how do they know where to leave the blood vessels?

15.1.2 Blood Cells Emerge in the Vertebrate Embryo in Scattered Blood Islands

Blood cells originate from pluripotent stem cells, called **hematopoietic stem cells** (Chap. 18). In adult mammals, these are found in the bone marrow (in the mouse, in the spleen as well). However, the stem cells do not originate in these tissues; in the embryo, they arise anywhere in **blood islands**. In birds, blood islands are first observed outside the embryo proper in the *area opaca* where clusters of angiogenic cells form the first blood vessels. The peripheral cells of the clusters form the endothelial linings, whereas the central cells come to be blood cells. Many blood islands merge to form a capillary network.

In mammals, blood islands are detected anywhere in the extraembryonic mesoderm, especially in the mesodermal coatings of the yolk sac and in the region of the allantois. These locations coincide with places where associations of angiogenic cells form the first blood vessels (Figs. 17.2 and 17.3). Later, the stem cells of the blood cells colonize the liver, the spleen, the thymus and the bone marrow. In humans the bone marrow is the only place where new blood cells are generated lifelong. This will

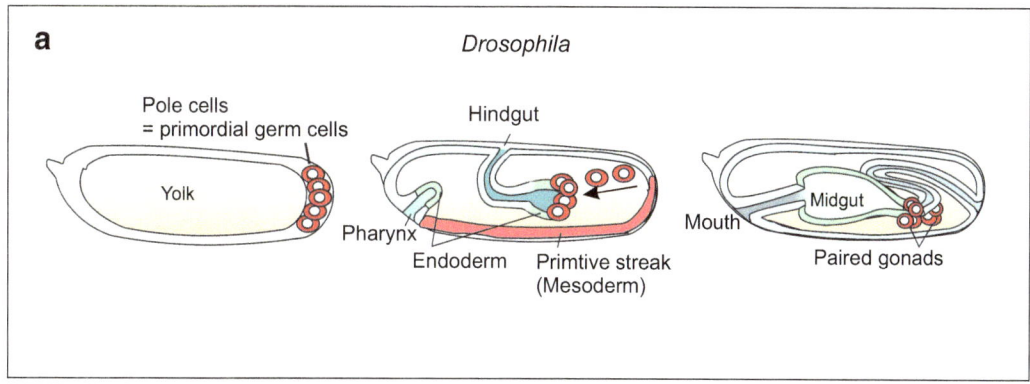

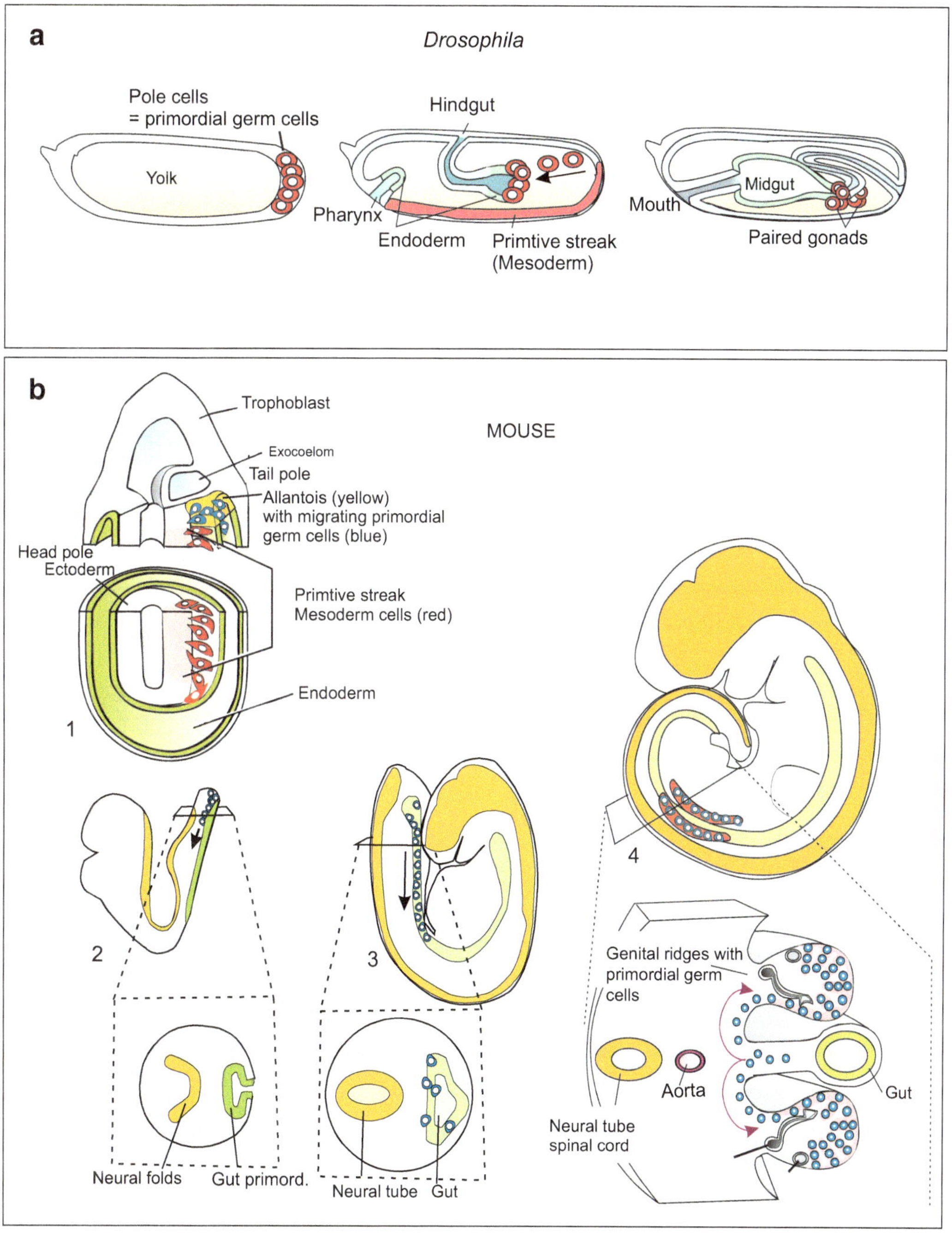

Fig. 15.2 Migrating primordial germ cells. (**a**) Embryo of *Drosophila*, (**b**) Mouse embryo. After Richardson and Lehmann (2010), and other sources

be considered in more detail in Chap. 18 (Figs. 18. 4 and 18.6).

Lymphocytes, granulocytes and macrophages (Figs. 18.5 and 18.6) are on the move life long.

They are not only present in the blood but creep through all tissues in search for foreign material to be removed and infections displaying antigenic properties, and cluster in centres of infection.

Fig. 15.3 Neural crest cells. *Above*: Singling out of neural crest cells during neurulation. *Below*: Posterior migratory route between dermamyotome and epidermis (*red*) and anterior migratory route between sclerotome and neural tube (*blue*), and cells/tissues derived from neural crest descendants

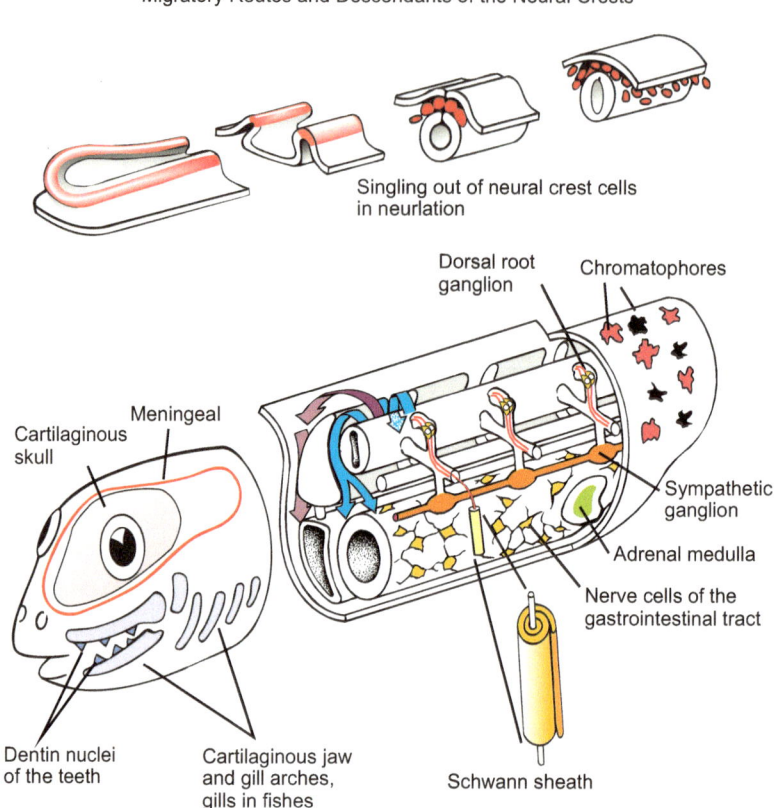

Migratory Routes and Descendants of the Neural Crests

Singling out of neural crest cells in neurulation

Dorsal root ganglion

Chromatophores

Cartilaginous skull

Meningeal

Sympathetic ganglion

Adrenal medulla

Nerve cells of the gastrointestinal tract

Dentin nuclei of the teeth

Cartilaginous jaw and gill arches, gills in fishes

Schwann sheath

15.2 Neural Crest Descendants

15.2.1 Neural Crest Cells Migrate Into Many Target Areas and Are Endowed with Manifold Developmental Potentials

With the emergence of the vertebrates a new and astonishing rich and powerful source of founder cells came into play: the neural crest. As will be illustrated in Chap. 22 (Fig. 22.10) this source enabled the vertebrates to develop an elaborated head and efficient respiratory organs. The story of neural crest cells is among the most amazing tales in all of developmental biology. When the neural plate involutes to form the neural tube during neurulation, and the neural tube detaches from the ectoderm, a series of cells is left over on either side. The cells are neither integrated into the neural tube nor taken up into the ectoderm.

Instead, they set off on travels, colonize various regions of the body and give rise to a bewildering variety of cell types, tissues and organs (Figs. 15.3, 15.4 and 15.5):

These include:

1. the **pigment cells** (chromatophores) in skin, feathers and hairs. These include the black or brown **melanophores** of our skin; they also provide our hair and the feathers of birds with their colour;

2. the **dorsal root ganglia** with bipolar afferent neurons which feed information along the dorsal root to the spinal cord coming from sensory organs of the skin, from muscle spindles or from other somatic senses; (Figs. 16.17 and 16.18). By contrast the bodies of the motor neurons which send information from the spinal cord to peripheral effectors such as muscles or glands through the ventral root are resident in the spinal cord and are not neural crest derivatives.)

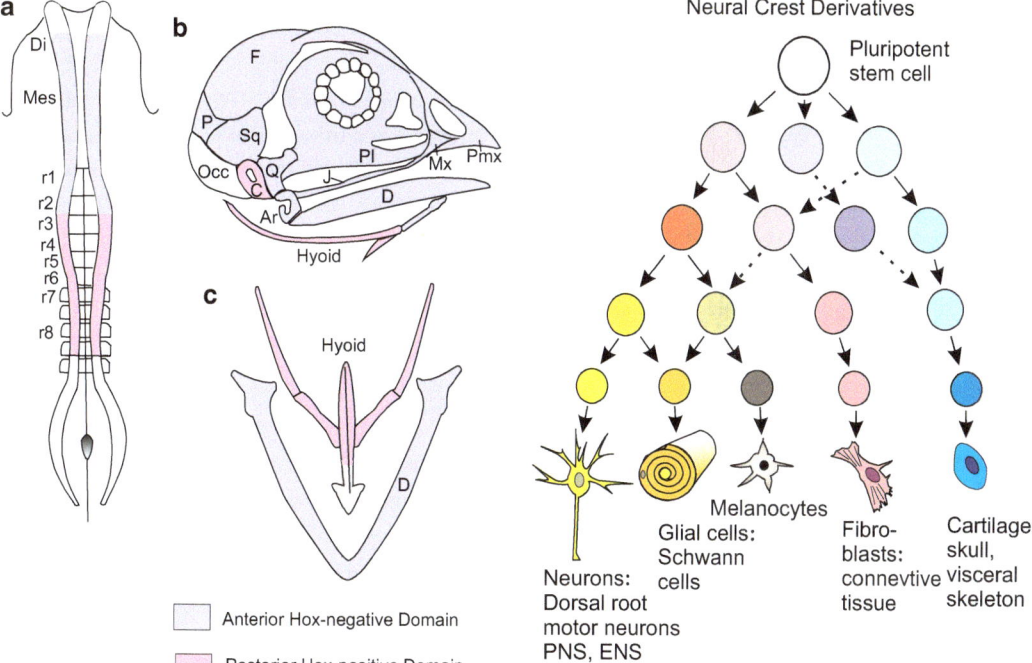

Fig. 15.4 Provenance of the cartilaginous elements of the skull and the visceral skeleton in birds, and relation to HOX-positive domains. (**a**) Neural crest viewed from above. *Di* Diencephalon, *Mes* Mesencephalon, *r1 bis r8* Rhombomeres no.1 to no. 8. (**b**) Skull: *Ar* Articulare, *C* Columella, *D* Dentale, *F* Frontale, *J* Jugale, *Mx* Maxillare, *Occ* Occipitale, *P* Parietale, *Pl* Palatinum, *Pmx* Prämaxillare, *Q* Quadratum, *Sq* Squamosum. (**c**) Visceral skeleton: Hyoid = Zungenbein; D = Dentale = lower jaw. After Le Douarin et al. (2004)

Fig. 15.5 Pedigree (*cell lineage*) of the neural crest derivatives. Cephalic neural crest cells are endowed with the potency to give rise to fibroblasts and connective tissue of the head, among other cell types. After various sources

3. the **nerve cells of the autonomic nervous system**, as far as the perikarya (cell bodies with nucleus) of these cells are located outside the central nervous system. Classified among the autonomic system are the **parasympathetic ganglia of the head** such as the ciliary ganglion, the **paravertebral sympathetic chains of ganglia**, the celiac ganglion, and the **nerve nets around the digestive tract** collectively called the **enteric nervous system ENS**.

4. The **thymus** and hormone-producing derivatives of neuroblasts, such as the producers of **adrenaline** (epinephrine) in the **adrenal medulla**, and the **neuroendocrine cells of the digestive tract** such as those producing cholecystokinin, currently viewed as part of the ENS.

5. The **peripheral glial cells** and the glial cells that accompany nervous processes, such as the **Schwann cells** which envelope and isolate long peripheral axons.

6. The **meningeal coatings** of the brain.

7. **Smooth muscle cells around blood vessels** and connective tissue in the head-neck area.

8. The respiratory **gills of fishes**, in particular the gill lamellae and the pillar cells within the lamellae (see Fig. 22.10).

9. **Skeletal of elements of the head pharynx region,** in particular **the lower jaws, the pharyngeal arches** or their derivatives, such as the mammalian sound-transmitting elements of the inner ear: **incus, malleus, stapes**, and the **laryngeal and tracheal skeleton**.

10. The **dental papilla** (producing bone-like material) of the tooth germs.

11. In the skull descendants of neural crest cells appear to contribute to some **bones of the face**. However, some authors prefer to classify the respective ectomesenchymal cells forming bones as a separate cell type.

15.2.2 The Wandering Cells Look for Targets on Preferred Migratory Routes

Accurate lineage tracing is crucial to understanding of developmental and stem cell biology, but is particularly challenging for transient and highly dispersive cell-types like the neural crest. **Xenoplastic transplantation** experiments, in which neural crest material is inserted into foreign species, have facilitated analysis of the origin, migration routes and destination of neural crest cells. Classic transplantations between the newts *Triturus torosus* and *Triturus rivularis* have been extended by transplanting neural crest cells of the **quail** (*Cortunix cortunix*) into embryos of the chick (*Gallus gallus*). Quail cells integrate themselves without difficulty into the chick, and can be easily identified by the unusually large amount of heterochromatin in their nuclei.

Tracking wandering cells is facilitated in transgenic animal strains that express **reporter genes** such as the gene for GFP, luciferase or LacZ. Embryos derived from such strains express reporter genes usually not only in the neural crest but also elsewhere. But such labelled embryos can be used as donors in transplantation studies. Labelled neural crest cells are removed from embryos and transplanted into the neural crest of normal, unlabelled embryos or albino embryos. Reporter genes permit permanent lineage-labelling in tracking the homing routes, settlement and type of terminal differentiation of the wandering cells. After labelled cells are introduced into the neural crest of unlabelled (or differently labelled) receivers migrating routes are continuously tracked with a video camera.

In their movements during embryo development, neural crest cells of the trunk follow two major routes (Fig. 15.3).

1. The **ventral route**. The majority of cells travel down the anterior portion of the somites through clefts between the myotome and the sclerotome. The cells settle and aggregate at various places. Some aggregate in clusters close to the neural tube and form the sensory **dorsal root ganglia** (s. Fig. 16.17). Others take longer pathways and assemble to form the **sympathetic paraventral chain ganglia sympathetic ganglia** (see Fig. 16.17) and the **adrenal medulla**. Other neural crest cells migrate into the muscle layers of the gut, forming the nerve net of the **enteric nervous system.**

2. The **dorsal route**. Cells travelling and spreading in the posterior region of the somites between the dermatome and the ectoderm differentiate preferentially into **pigment cells**.

Currently the **zebrafish *Danio rerio*** is about to take over the leading role in neural crest research. Not only are the animals transparent allowing to trace the routes of migrating cells in living specimens; the available genetic toolkit permits to create targeted mutations and to exchange genes of significance as described in Chap. 13. For example, using an inducible Cre/loxP system for genetic lineage tracing, the emergence of the colour stripes was, and is being, analyzed at the genetic and molecular level (see below 15.2.5 and Fig. 15.10).

15.2.3 Many Different Signallers Assume the Role of Hiking Guides

Recent research aims at finding answers to the following questions:

1. What causes cells to abandon contact to their previous neighbours and to develop organelles of amoeboid motility such as pseudopodia, filipodia and lamellipodia?

2. How do they find the way to their target place?

3. When and by which factors are the cells committed to their fate?
4. What causes the cells to stop and settle at the target place and to undergo terminal differentiation?

Since the various migrating cells have different tasks and destinations there is no single uniform answer. From experiments to stimulate and guide the migration of cells in Petri-dishes it has been inferred that neural crest cells orient themselves in at least three ways:

(a) **Contact-mediated guidance and sign posts:** Migrating cells follow paths labelled with components of the extracellular matrix (fibronectin, laminins, collagen IV, or hyaluronic acid) and by feeling out characteristic surface components (such as cell adhesion molecules) of the cells bordering the pathways. The phenomenon of orienting along threads and cables composed of extracellular matrix and along other cells has been called **contact guidance** (Figs. 15.6 and 15.9).

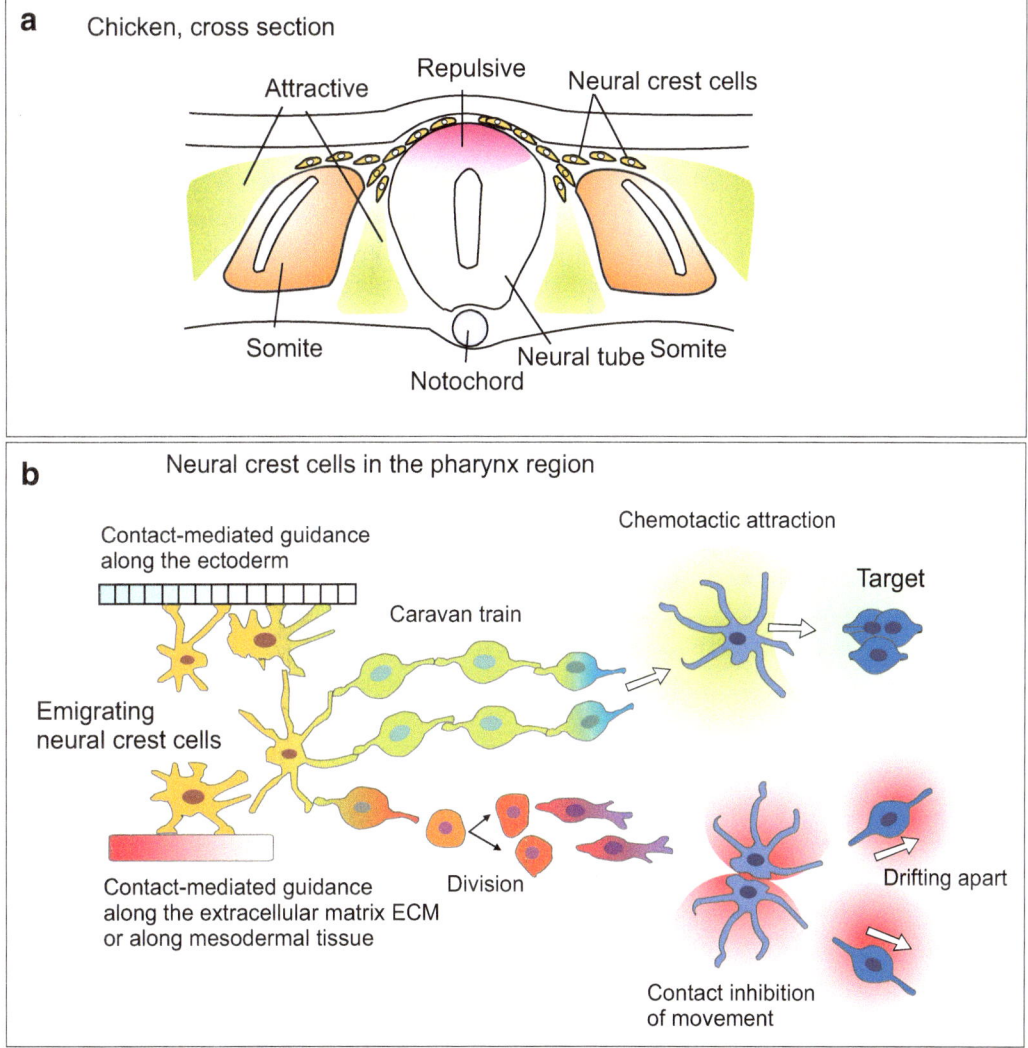

Fig. 15.6 Paths of the neural crest cells and mechanisms of pathfinding. (**a**) Migration routes exemplified by neural crest cells of the chick; the emigrating cells are guided by attractive and repulsive signals. (**b**) Emigrating neural crest cells make use of both contact-mediated and diffusible chemical signals for pathfinding. After Kulesa and Gamill (2010)

Testing chemotaxis

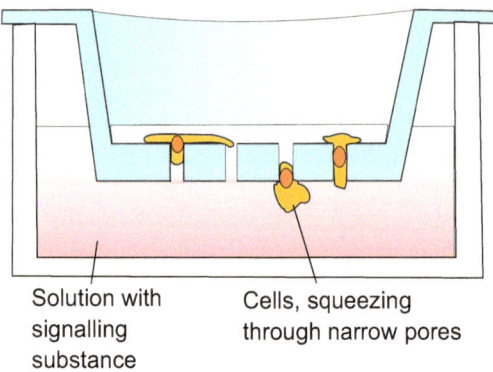

Solution with
signalling
substance

Cells, squeezing
through narrow pores

Fig. 15.7 "Transwell migration assay" with the Boyden chamber. Signalling substances diffuse from the lower chamber through narrow pores (diameter 3–5 μm) into the upper chamber. From there cells in the upper chamber try to reach the source of the signalling substances following the concentration gradient and squeeze through the pores into the lower chamber. The experimental design is used by various working groups to investigate, for example, the attractive effect of SDF-1 (Stromal-Derived Factor-1) on blood cells (macrophages, neutrophil granulocytes, and others)

(b) **By sensing long-range chemoattractants**. The existence of such long-range signals has been presumed for long, but only a few actual signal molecules have been identified so far (Fig. 15.8).

Figure 15.6 summarizes principles; Fig. 15.7 shows a simple bioassay for positive chemotactic signalling substances. If we were to name the signalling substances involved and the corresponding receptors on the surface of the migrating cells a bewildering multitude of identified or putative molecules would have to be listed. We refer to Chap. 11, and as an example to the signalling molecule WNT-11 (Fig. 15.8) because WNT-11 currently gains much attention because certain cancer cells also follow this signal. We will meet several of these signalling molecules when we track the path finding of the processes of nerve cells (see Figs. 16.18, 16.19 and 16.24), of sprouting blood capillaries (see Fig. 17.6) or of renal tubules (see Fig. 14.7). There are many overlaps attesting the conservatism of proven signalling systems.

15.2.4 Place of Origin, Migration Route and Target Determine the Fate of the Emigrants

Perhaps the most fascinating aspect of the extensive migration, besides the question of how the cells are guided to their various destinations, concerns the programming of the specialized cell types that arise from the neural crest cells.

How can these roving, unspecialized cells be induced to settle down and differentiate? Two extreme hypotheses have been put forward. One holds that neural crest cells have already been educated and trained for their future job where they were born. Those who miss the target area predetermined by their programming will undergo suicide (apoptosis); those who find their destination will be multiplied through cell division. The other view maintains that neural crest cells are initially **pluripotent** and not committed to a distinct job. Only when they cease migrating will they become committed to a distinct task.

Both of these hypotheses have evidence in their favour. For example, it is known that even **at the places of origin not all neural crest cells are identical.** Only neural crest cells of the head and neck region have the potential to form cartilage and bone. No head-specific skeletal elements are formed if the neural crest of the head region is experimentally removed or replaced by neural crest material of the trunk. On the other hand, when cells of the anterior neural crest, normally destined to form cartilage or bone, are shifted into a thoracic region, they give rise to dorsal root ganglia and sympathetic ganglia in accordance with their new position. Moreover, the careful tracking of many individually-labelled cells has shown that in the trunk region neural crest cells are indeed pluripotent. Apparently, their developmental potential is gradually narrowed toward their eventual fates while they migrate toward their target area, and is definitively determined only after they have arrived at their final destination (for examples see Chap. 16, Fig. 16.17).

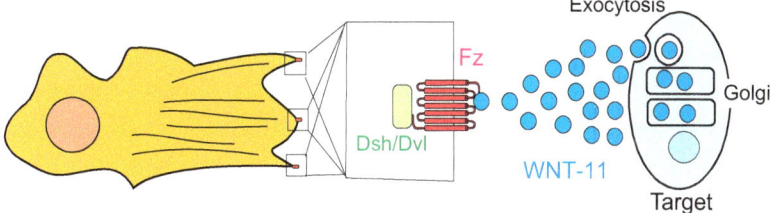

Fig. 15.8 WNT11 acting as an attractive signalling molecule in the chemotactic migration of a cancer cell. WNT 11 is liberated via vesicles and attaches to the FRIZZLED receptor on the cell membrane of the migrating cell. Signal transduction occurs as summarized in Chap. 11

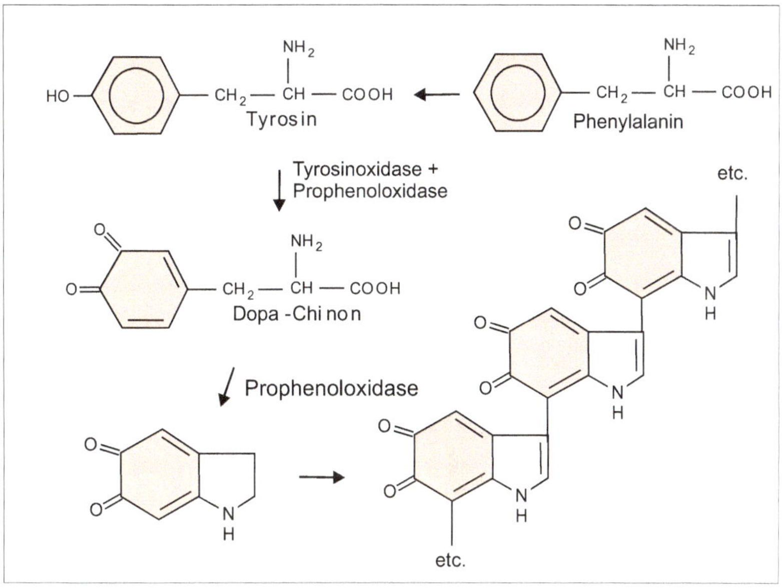

Fig. 15.9 Synthesis of melanin. The polymer threads can grow long and branch. A dense net of melanin gives a black colour, less dense nets yield a brown, red, or yellow colour

15.2.5   Colour Patterns Have a Multitude of Functions; They Are Created by Neural Crest Derivatives; in Zebrafish They Create a Striped Colour Pattern

Colour patterns are striking features of many animals; they have important functions in protection against UV irradiation, camouflage, kin recognition, shoaling and sexual selection. Colour patterns evolve rapidly and vary between closely related species.

In vertebrates colour patterns are created by derivatives of the neural crest cells. In birds and mammals they are generated by melanocytes which colonize the subdermis of the skin and the dermal papilla of hair follicles or feather primordia. By injecting melanin (Fig. 15.9) into the growing shaft of the hairs or feathers they provide the pigmentation of these structures. Different degrees of melanin polymerization yield yellow, red, brown or black pigments. Periodic activities of secretion into the elongating feathers give rise to striped or spotted patterns. In fish, amphibia and reptiles, chromatophores retain their pigment and it is their distribution in the dermis that determines the pattern.

In zebrafish neural crest cells produce three different varieties of chromatophores: black **melanophores**, yellow **xanthophores** and the

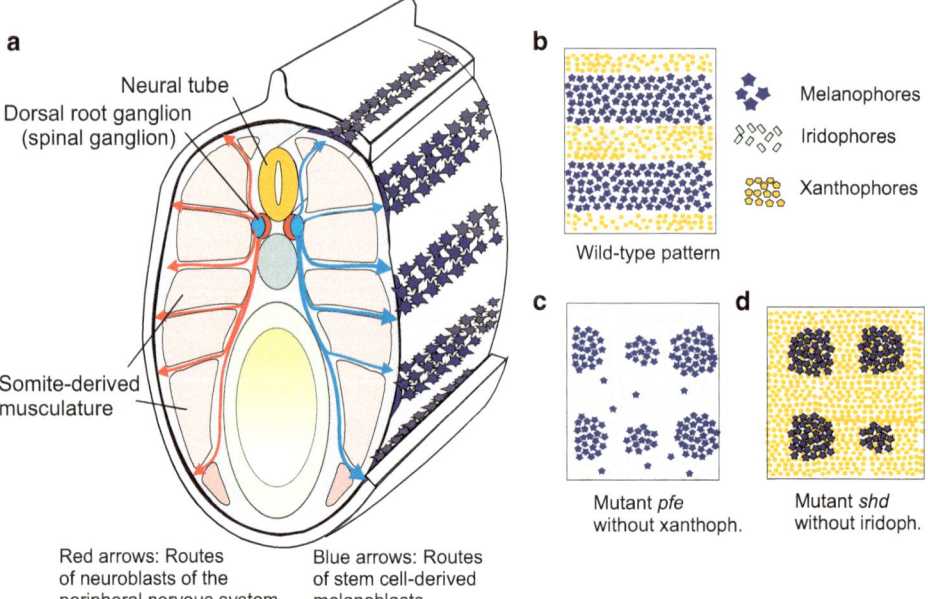

a

Neural tube
Dorsal root ganglion
(spinal ganglion)

Somite-derived
musculature

Red arrows: Routes
of neuroblasts of the
peripheral nervous system

Blue arrows: Routes
of stem cell-derived
melanoblasts

b

Melanophores
Iridophores
Xanthophores

Wild-type pattern

c

Mutant *pfe*
without xanthoph.

d

Mutant *shd*
without iridoph.

Fig. 15.10 Stripes in zebrafish. After Dooley et al. (2013) and Fronhöfer et al. (2013). In poslarval fishes melanoblasts are derived from residual neural crest stem cells which are associated with the emerging dorsal root ganglia. A complete wild-type pattern presupposes the presence of all three types of chromatophores in the skin

silvery **iridophores** (Fig. 15.10). The iridophores contain light-reflective purine platelets; these are responsible for the shiny appearance of the pattern. Melanophores are restricted to the dark stripes; a thin layer of iridophores spreads over the melanophores giving a bluish appearance to the stripes. A dense sheet of iridophores covered by xanthophores form the light interstripes.

Consistent and long-term labelling of neural crest derivatives was achieved with an inducible recombination system. Using this system GFP or other constructs were placed under the *sox10* promoter. *Sox10* is expressed in postlarval as well as in larval neural crest derivatives. Mutations in the *sox10* (also called *colourless*) gene, which encodes the HMG-box transcription factor Sox10, eliminate all three types of chromatophores. In addition, mutants defective in individual chromatophores were produced, and chimeric fishes by interchange of neural crest cells through transplantation.

Result: In zebrafish we must distinguish between larval and adult patterns. The larval pattern is derived from embryonic neural crest cells which migrate along dorsolateral and ventromedial paths. Melanoblasts migrate preferentially along the axons of motor neurons.

However, in metamorphosing juvenile animals a neural crest is no longer present. Lineage analysis and 4D in vivo imaging enabled to trace their origin of the melanophores and their future behaviour. The source of the adult melanophores are **residual neural crest stem cells associated with the dorsal root ganglia**. The ganglia provide a niche where the progenitor cells are set aside. These cells are quiescent during embryonic development but start to proliferate during juvenile development and to emigrate when metamorphosis commences. During migration they follow the spinal nerves and eventually emigrate into the subdermis of the skin where they generate the black and light stripes.

Correct and complete stripe formation presupposes the presence of all three types of chromatophores. The pattern starts to develop with a light stripe (interstripe) in the region of the horizontal myoseptum. This myoseptum serves as a morphological prepattern guiding the arriving first cohort of iridiophores. Subsequently, dark stripes appear dorsally and

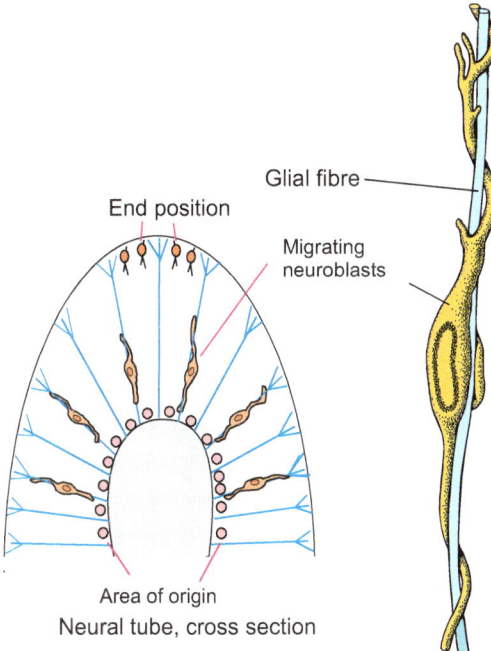

End position

Glial fibre

Migrating
neuroblasts

Area of origin
Neural tube, cross section

Fig. 15.11 Target finding mediated by contact guidance. A neuroblast in the spinal cord migrates from its birthplace close to the central canal along a glial cell process to its target area in the periphery

ventrally to this interstripe, and more interstripes and stripes are added as the fish grows. Pattern formation is the result of mutual interactions of all three types of chromatophores: Iridophores promote and sustain melanophores and attract xanthophores, whereas xanthophores repel melanophores.

15.2.6 Even in the CNS Many Cells Migrate

Surprisingly even in the central nervous system extensive cell migration takes place. In general, the early period of life of neurons and glial cells is divided in three phases.
1. Specification and switching on of cell type-specific genes at the place of origin of the founder cells;
2. Migratory phase;
3. Terminal differentiation to become a functional neuron or glial cell at the final destination.

In the CNS (Chap. 16) we will meet guiding structures provided by glial cells. In anticipation we refer to glial cells in the spinal cord which stretch themselves to serve as pathfinders for neuroblasts leading from their place of birth to their destination (Fig. 15.11). This is a striking example for **contact guidance**.

Summary

In animal embryos many cells embark on a journey. In the vertebrate embryo this applies especially to cells which originally had formed the compact **somites**, these detach and emigrate to settle at other places in order to form the **vertebral column**, the **musculature** or the dermis. Long journeys are undertaken by the **primordial germ cells** (in invertebrates as well as in vertebrates), the **stem cells of the blood cells** and the descendants of the **neural crest**.

Neural crests are cell rows along the neural plate, the precursor of the central nervous system. These cells emigrate and colonize wide parts of the embryo. They give rise to (1) the **pigment cells (chromatophores)** of the skin, feathers and hairs. In the head-neck region neural crest cells give rise to (2) the **cartilaginous skull** and **elements of the jaws and pharyngeal arches**, including the future **auditory ossicles**, (3) the **dental papilla**.

In the neck-trunk region neural crest cells give rise to (4) various sensory cells, (5) the sensory **dorsal root ganglia** and the (6) **Schwann sheets** of the peripheral nerve fibres. Furthermore they give rise to (7) the **entire peripheral nervous system** with substantial parts of the **sympathetic and parasympathetic nervous system** and the entire **enteric nervous system (ENS)** around the digestive tract and (8) the **adrenal medulla**. On their path to the target areas neural crest cells move along defined migration routes which present molecular guidelines and sign posts. In addition attractive and repulsive long-distance signals act as hiking guides. Place of origin, route and target determine the type of final cell type the emigrants eventually form.

The Nervous System and Central Sensory Organs

16

The human central nervous system is considered the most complex organ a living being has ever developed. As measured by its size it certainly is the most complex system in our field of experience. Based on weight, on the total DNA content, and on counting the number of cells in a manageable defined section and subsequent statistical projection, our brain has been estimated to encompass 100 billion (10^{11}) nerve cells and 1,000 billions (10^{12}) of glial cells and cells of the immune system (known as microglia). Each nerve cell is connected with >1,000 others. A spinal motor neuron with a relatively modest number of dendrites receives about 10,000 contacts. The dendritic tree of a Purkinje cell in the cerebellum is much larger and receives approximately 150,000 contacts. The total length of nerve fibres has been estimated to surpass 500,000 km.

To describe and understand the development of the nervous system is among the most demanding task of the contemporary biological science. In this chapter we confine ourselves to a general outline of the development of the vertebrate brain (with an eye to insects). We view the formation of the neural tube; that is the primordium of the **central nervous system (CNS)**, we track migrating cells which give rise to the **peripheral nervous system (PNS)** including the **enteric nervous system (ENS)**. We observe how neurons form their information-transmitting processes and seek synaptic contact to other cells. We pose the question how this complex order is organized.

16.1 Morphological Developmental History of the Nervous System

16.1.1 The Central Nervous System Arises From the Neural Tube, Whereas The Peripheral Nervous System Is Established by Neural Crest Cells

(a) **The central nervous system, CNS.** Recall that in vertebrate embryos, the central nervous system forms along the dorsal midline of the body axis. The morphological formation of the nervous system begins during gastrulation, when dorsal mesoderm (chordamesoderm) moves into contact with overlying ectoderm. The dorsal mesoderm emits signals which induce the overlying ectoderm to adopt a neural fate (see below). The induced neural plate forms neural folds (Fig. 16.1), the folds form the neural tube, and the neural tube forms brain and spinal cord (Figs. 16.1, 16.2, 16.3 and 16.4). The peripheral nervous system, including the neuronal network of the gut is still missing from the list. The peripheral nervous system derives from emigrated **neural crest cells** (Sect. 16.4).

(b) The peripheral nervous system, PNS comprises
- the **dorsal root ganglia**, also known as **spinal ganglia**,
- the peripheral components of **autonomic nervous system**, that is the **sympathetic**

W.A. Mueller et al., *Development and Reproduction in Humans and Animal Model Species*,
DOI 10.1007/978-3-662-43784-1_16, © Springer-Verlag Berlin Heidelberg 2015

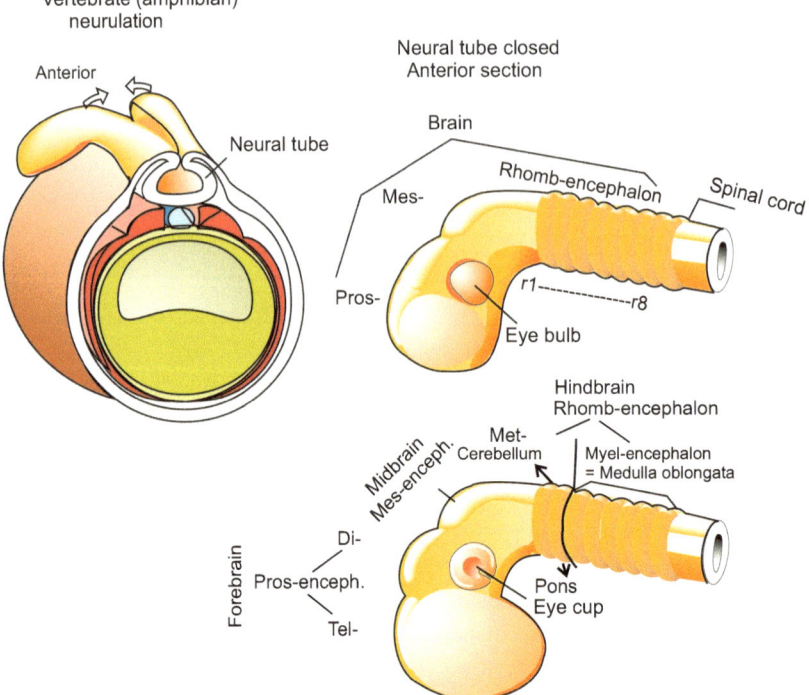

Fig. 16.1 Formation of the central nervous system exemplified by the amphibian embryo. Proceeding from anterior to posterior first the three-part brain with prosencephalon, mesencephalon and rhombencephalon is formed. Later, the three-part brain is further subdivided into five sections. The prosencephalon gives rise to the telencephalon and diencephalon, the mesencephalon remains single-section, the rhombencephalon subdivides into metencephalon and myelencephalon

system and the **parasympathetic system**, as far as the cell bodies with the cell's nucleus (called perikarya) are located outside the CNS. **Sympathetic neurons** are found in the switching stations known as **paravertebral ganglia** along the sympathetic trunk (see Figs. 16.17 and 16.18) and in visceral ganglia; **parasympathetic neurons** are found close to, or within, the target organs (Fig. 16.17);

- **the autonomous nerve net of the gastrointestinal tract.** This part of the classic autonomic (vegetative) system is currently ranked as a separate nervous system and called **ENS, enteric nervous system**, but it belongs to the PNS.

16.1.2 The Brain of the Vertebrates First Divides into Three and then into Five Segments

It seems curious that most of our brain is initially a hollow cavity. In its anterior part, the neural tube balloons into three main vesicles: **forebrain (prosencephalon), midbrain (mesencephalon), and hindbrain (rhombencephalon)**. In a second step, forebrain and hindbrain are subdivided into two parts each and we see the classic five parts of the textbook brain:

- *Prosencephalon*
  1. **Telencephalon (cerebral hemispheres, with the neocortex)**
  2. **Diencephalon**

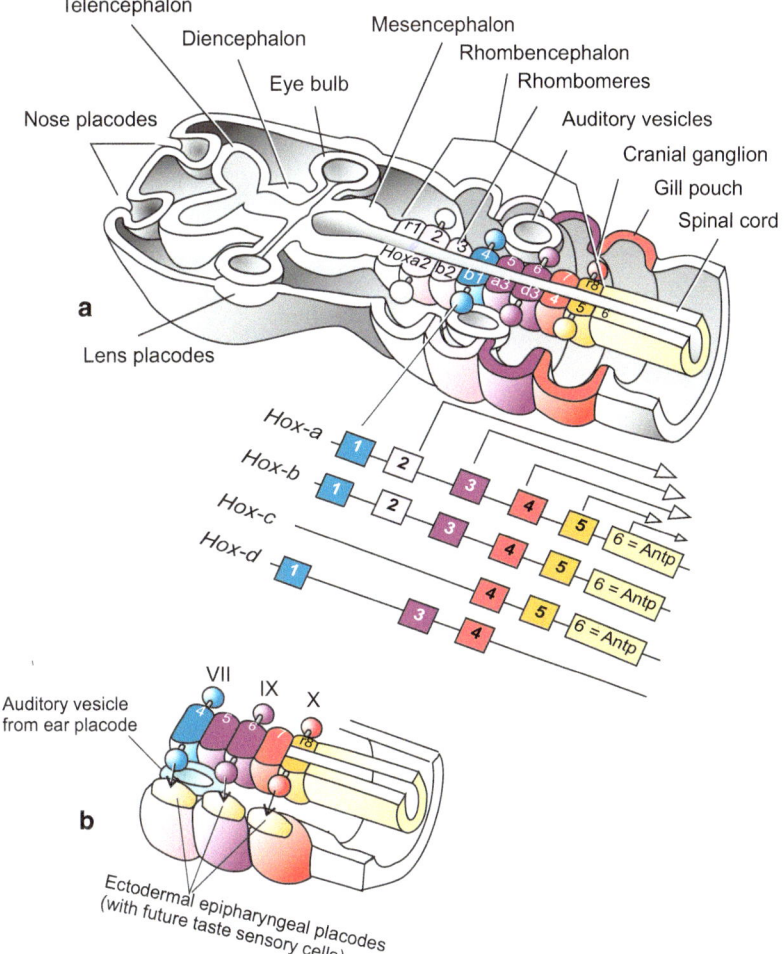

Fig. 16.2 (**a, b**) (**a**) Development of the brain and adjacent sensory organs in a generalized mammal. The posterior part of the brain, the rhombencephalon, is subdivided into repeating units called rhombomeres or neuromeres. Cranial ganglia are associated with the even-numbered rhombomeres (r2–r8). While the eye vesicle arises from the posterior forebrain (diencephalon) by evagination, the eye lens, the nasal epithelium and the inner ear arise from placodes, thickenings of the epidermis that are invaginated. In addition, the expression pattern of some *Hox* genes [after Alexander et al. (2009)] is shown. This pattern helps to identify and demarcate rhombomeres. (**b**) Excerpt, showing in addition the epipharyngeal (epibranchial) placodes, which give rise to the future taste receptor cells. They are innervated by nerve fibres (*arrows*) emanating from the cranial ganglia (= cranial nerves) VII (*nervus facialis*), IX (*n. glossopharyngeus*) and X (*n. vagus*). The otic vesicle gives rise to the inner ear and is innervated by the cranial nerve VIII (*n. vestibulooculuris*) which emanates from the myelencephalon

- *Mesencephalon*; it is not further subdivided and remains
  3. **Mesencephalon**
- *Rhombencephalon*
  4. **Metencephalon (cerebellum)**
  5. **Myelencephalon (medulla oblongata)**

In humans the following principal functions are assigned to these regions:

- **Telencephalon.** Evaluation of the information supplied by the sensory organs of smell and taste, by the inner ear with its semicircular canals and balance system, evaluation of acoustic and visual information (in lower vertebrates performed by the mesencephalon). Generation of associations, short-term, working and long-term memory, attention.

Although in humans the largest part of the telencephalon comprises the evolutionarily young **neocortex** (Fig. 16.5), in its ventral part the telencephalon also harbours old core

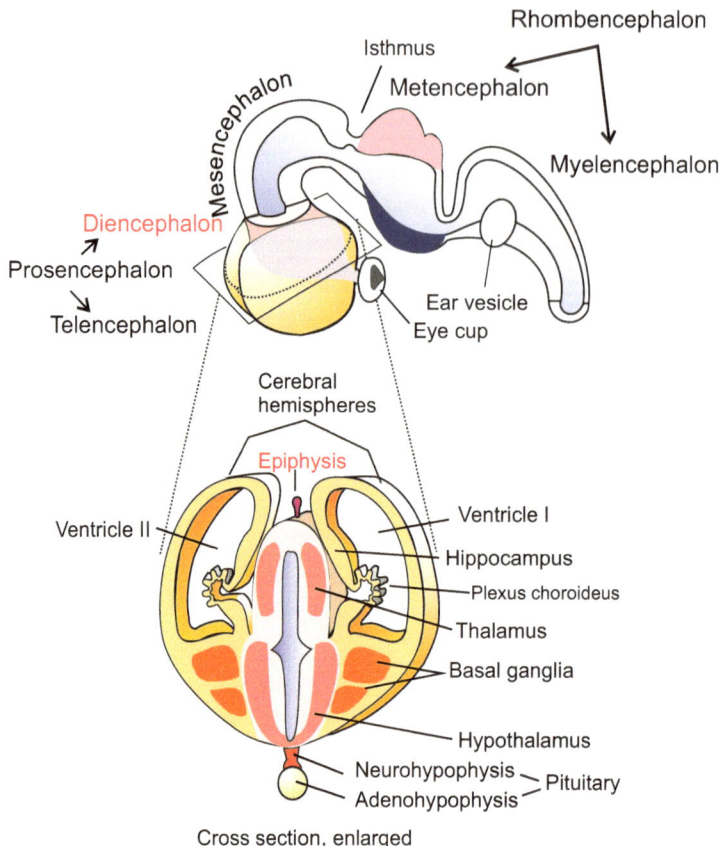

Cross section, enlarged

Human, 14 mm

Fig. 16.3 Early stage human brain. The section shown below shows the diencephalon in the centre and *left* and *right* from it the hemispheres of the paired, caudally expanding telencephalon (cerebrum). Eventually the paired hemispheres enclose the diencephalon completely

regions called "nuclei". In the traditional terminology of neuroanatomists "nuclei" are areas of "gray matter" that contain numerous densely arranged cell bodies with cell nuclei (perikarya), while the "white matter" preferentially consists of nerve fibres (nerve tracts). The phylogenetically ancient structures include the **rhinencephalon (olfactory brain**) and a series of core regions which in the traditional nomemclature are called **basal ganglia** (with striatum and pallidum) classified as part of the **limbic system**. Some often mentioned structures are shown in Fig. 16.5. Of particular significance is the paired **hippocampus** (also known as Ammon's horn), a structure of the old cortex which is displaced by the expanding neocortex to a lateral-

ventral position underneath the temporal lobe. The hippocampus has a leading role in the selection and transfer of memory contents into the long-term memory. The limbic system is responsible for the generation of emotional experiences. Its components are connected by neuronal fibres with the hypothalamus of the diencephalon.

- **Diencephalon.** The diencephalon is the origin of the eye vesicles (Figs. 16.1 and 16.2). It is subdivided into
  - **Thalamus:** Relay centre for optic and acoustic tracts; gateway for sensory input passing from the spinal cord to the cerebellar hemispheres. The mammalian metathalamus encompasses the **lateral geniculate nucleus** (*corpus geniculatum laterale*), the switching

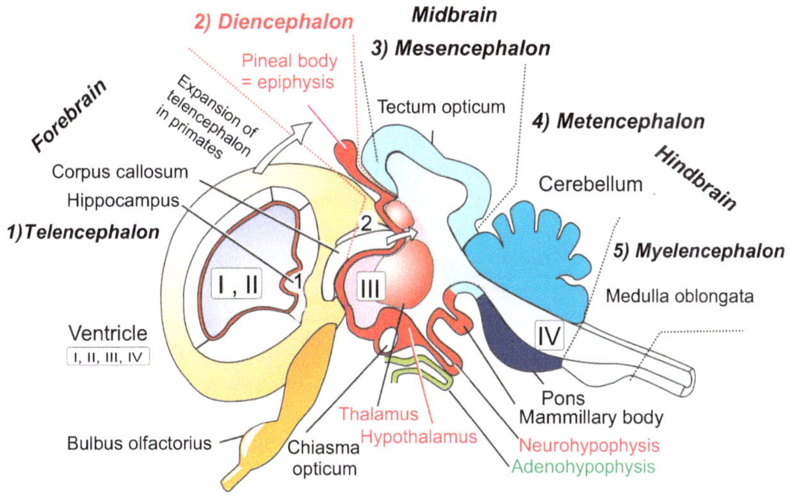

Mammalian brain   basic architecture

Fig. 16.4 Advanced mammalian brain. Median cut, but in addition one of the hemispheres is cut by a separate paramedian incision. In humans the hemispheres expand in posterior direction and eventually cover the di- and mesencephalon completely and the cerebellum partially. Correspondingly the fibre bundle connecting the two hemispheres known as corpus callosum expand posteriorly. The grey fields 1 (hippocampus) and 2 are remnants of the ancient archaeopallium. The pineal body (epiphysis) is formed as an evagination of the dorsal midbrain. The pituitary gland is composed of two different parts: The neurohypophysis (posterior lobe) arises from an evagination of the ventral diencephalon (hypothalamus); the adenohypophysis arises from an evagination of the roof of the pharynx

station for visual information, and the *corpus geniculatum mediale*, the switching station of the acoustic tract.

– **Epithalamus** and **hypothalamus**: Supreme commander of basic vegetative functions, control centre of inner homeostasis, sleep, alertness. Connects the nervous and hormonal systems. To prepare for the latter task, the

dorsal **epithalamus** forms the **pineal gland (epiphysis)** (Fig. 16.4),

the ventral **hypothalamus** extends the **infundibulum** which will form the posterior lobe of the **pituitary**, **the neurohypophysis**.

The **pituitary gland** is completed by a structure which emerges not from the brain but from the roof of the pharyngeal cavity. By a local evagination, the roof forms **Rathke's pouch**, which transforms into the anterior lobe of the pituitary gland, the **adenohypophysis**.

• **Mesencephalon.** Forms in non-mammalian vertebrates the **tectum opticum**, the main

centre of visual data processing. In mammals visual information is forwarded to the V1 region of the telencephalon via the changeover facility of the lateral geniculate nucleus (see Fig. 16.25), while the tectum opticum is merely a relay station for visual and acoustic reflexes and the acoustic tract.

• **Cerebellum.** Main location where complex automatic movements are programmed and coordinated.

• **Myelencephalon.** Reflex and control centre for vegetative functions, oscillators for breathing, control of blood circulation, reflexes for suction, swallowing, coughing.

16.1.3 Discussed Since the Days of Goethe: Is the Brain Subdivided into Segments, Like the Vertebral Column?

Since the days of Johann Wolfgang von Goethe (1749–1832) and Lorenz Oken (1779–1851) there has been continuing discussion whether or not the skull, and with it the brain, is subdivided

Enlargement of the Neocortex and Displacement of Nuclei of the Limbic System

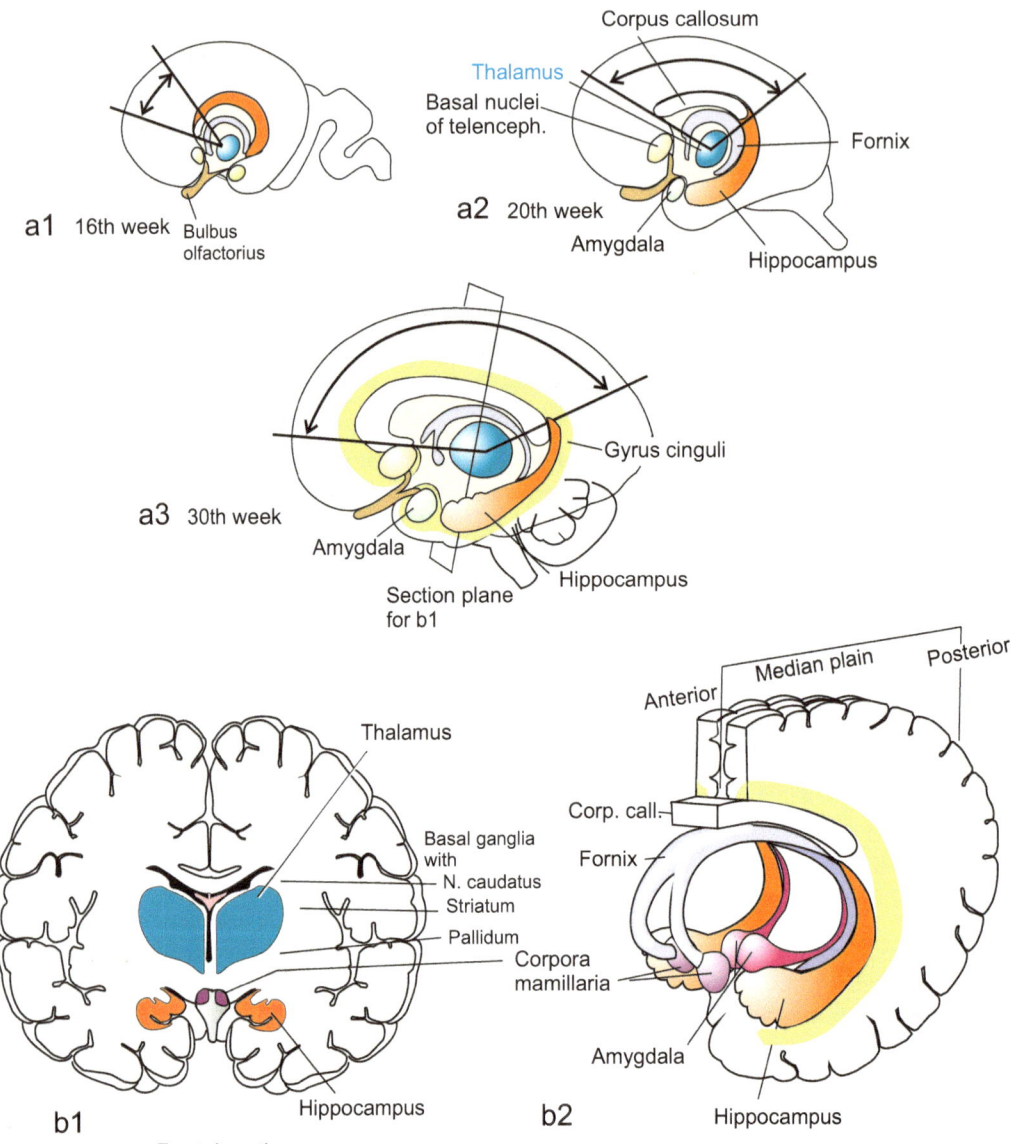

Fig. 16.5 (**a, b, a**1–**a**3) Enlargement of the neocortex in humans and limbic system. Together with the neocortex also the *corpus callosum* expands. Through insertion of new cortical areas the evolutionary ancient structures of the limbic system (*LiS*) are relatively displaced (the thalmus depicted as a structure of reference does not count among the *LiS*). Here mapped components of the limbic system are the *gyrus cinguli*, basal ganglia of the telencephalon (with septum and *nucleus accumbens*), amygdala, hippocampus, the fibre bundle of the fornix, which establishes an anatomical connection between the basal ganglia, the hippocampus and the hypothalamus, the central instance of vegetative functions. (**b**1-2) The pictures **b**1 and **b**2 refer to the corresponding structures of the adult brain and show in addition the mammillary bodies (*corpora mamillaria*), which also are attributed to the *LiS*. The functions of emotion and memory are ascribed to the limbic system. The hippocampus plays a leading role in deciding which content of the memory is transferred into the long-term memory

into segments. Not only morphological constrictions but more so the expression patterns of *Hox* genes speak for themselves: yes the hindbrain is segmented.

The hindbrain is initially patterned into seven repetitive bulges, called **rhombomeres** (Figs. 16.2 and 16.3). The first rhombomere gives rise to the **cerebellum**. Each subsequent pair of rhombomeres contains a set of motor neurons and contributes fibres to one cranial nerve root. These are connected with the **cranial ganglia** which arise from neural crest cells. The first ganglia are attached to the even-numbered rhombomeres, whereas the odd-numbered rhombomeres are connected with ganglia only later.

With respect to the anterior brain regions, prosencephalon and mesencephalon, there are no good arguments in favour of an original segmental organization as in these regions transcription factors other than *Hox*-coded ones control development, and the anatomist does not find periodically repeating structures. Based on gene expression studies some scientists proposed the forebrain to be divided into six segment-like domains called prosomeres but this proposal was not agreed by the majority of scientists. It is not currently known how the internal subdivision of the forebrain comes about.

The brain is surrounded and supplemented by the main **sensory organs** needed for long-distance orientation and for complex behavioural control (Fig. 16.2).

16.1.4 The Subdivision of the Brain Is Controlled by Local Organizing Centres

As explained in the following chapter in the arising central nervous system diverse genes are expressed that code for transcription factors highly conserved in the animal kingdom, or code for soluble, secreted factors that act directly on the neighbourhood. An important role in the antero-posterior patterning of the brain is played by the **isthmic organizer**. Isthmus is the designation of a constriction between midbrain and

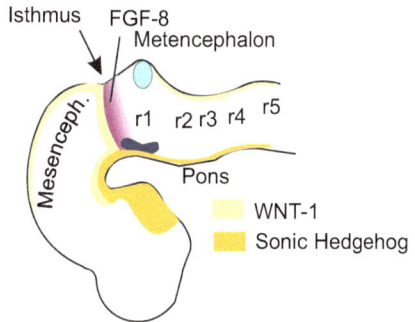

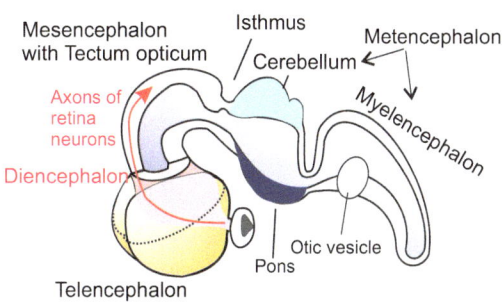

Fig. 16.6 Position of the isthmus organizer at the mid brain/hindbrain border and areas where some soluble signalling proteins are produced. Anterior and adjacent to the future isthmus a strip of cells starts expressing the signalling protein WNT-1 and a posterior strip starts expressing FGF-8 in a steep declining gradient. Both WNT-1 and FGF-8 have important functions in the establishment of the midbrain/hindbrain boundary. In addition, the boundary is determined by the activity of transcription factor genes such as *Otx2* and *Gbx2*

hindbrain (Fig. 16.6). In the tube stage of the CNS at the posterior boundary of the future midbrain, and just anterior to the future isthmus, a strip of cells begins to produce the signalling molecules **FGF-8** and **WNT-1**. These signals are later emitted by this organizing centre and control the regional specification of the brain on either side. To the anterior the centre induces **midbrain,** to the posterior it induces the **rhombomere 1 to form the cerebellum**. If a piece of the isthmic organizer is transplanted into a donor further caudally into the region of the myelencephalon a second midbrain is formed there. A similar result can be brought about by inserting a bead soaked with FGF-8. Probably more organizing centres emerge until the basic architecture of the brain is realized.

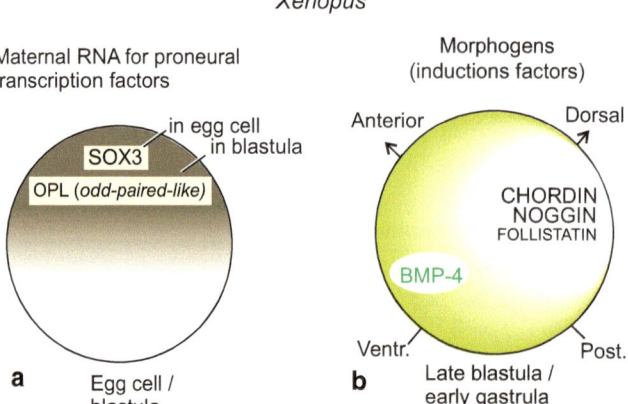

Fig. 16.7 Factors specifying the central nervous system. Factors deposited in the egg cell such as the maternal SOX3 initiate specification, supplemented by OPL subsequent to fertilization. SOX3 includes a DNA-binding domain known as HMG motif. OPL is a zinc finger protein homologous to the Odd-Paired factor of *Drosophila*. These transcription factors confer on the animal region the competence to become nerve tissue. In the ventral region of the embryo this competence is suppressed by BMP3/4, whereas the cells of the animal hemisphere are prepared to respond to neuralizing induction factors. These unveil the proneural competence. BMP3/4 is neutralised by CHORDIN and NOGGIN, which radiate from the organizer

16.2 Genetic Programming of the Nervous System

16.2.1 The Nervous System Arises From Cell Areas That Are Prepared for Their Task by Maternal Proneural Transcription Factors

The Programming of the Future Nervous System is a Multi-Step Process. Even though we place special emphasis on the development of the mammalian and human nervous system, once more first the amphibian embryos have to act as a surrogate model. It has long been established that in the late blastula when the Spemann organizer commences operation this transmitter emits signals that give the cells of the animal hemisphere the option to become nerve cells. Such signals are represented by induction factors such as NOGGIN and CHORDIN which for their part neutralize the anti-neural factor BMP3/4. This anti-neural factor is needed to suppress neural development in the ventral embryo and to program the ventral

ectoderm to become epidermis. A decades-old hypothesis implies the cells of the animal cap would have the autonomous tendency to become nerve cells but would initially be prevented from pursuing their propensity. This hypothesis was recently corroborated by molecular biological findings and experiments. Already in the egg cell maternally transcribed mRNA for proneural transcription factors is deposited, for example mRNA for the factor SOX3 which contains a DNA-binding HMG domain, and further factors providing competence for neural differentiation (Fig. 16.7).

Inherited from the mother, these factors can turn on expression of zygotic (embryo's-own), nerve cell-specifying transcription factors in the cells of the animal hemisphere. Once more we learn that in the control of developmental processes one has to distinguish between

1. extracellular factors that direct large cell associations, and
2. transcription factors located inside the cells, and eventually within the nucleus, that control gene activities.

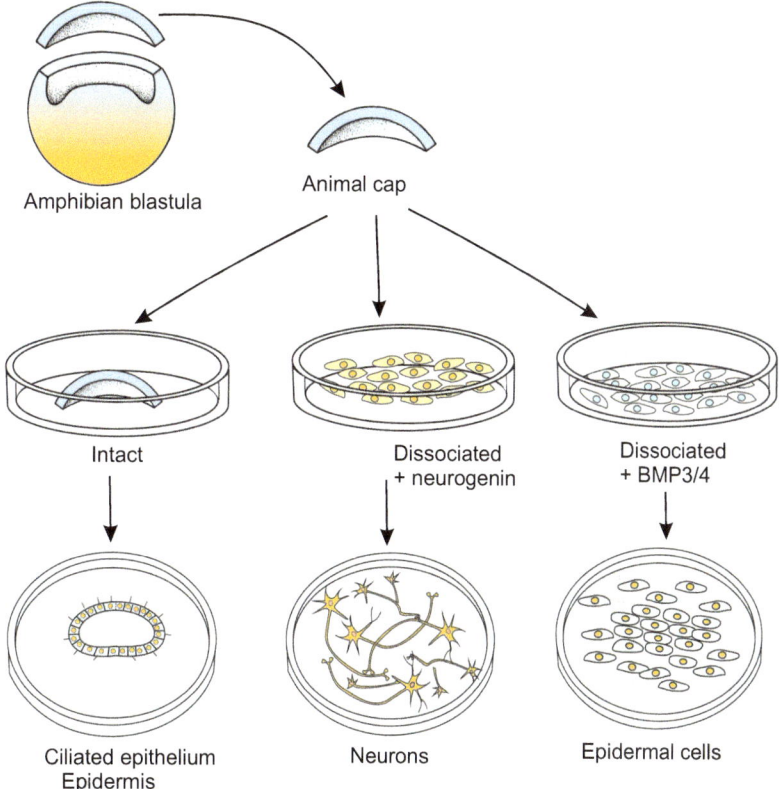

Fig. 16.8 Checking the effect of factors, for instance the anti-neuralizing action of BMP3/4, in the animal cap assay. In newts dissociation alone discloses the bias of animal cap cells to form nerve cells. The bias is suppressed by BMP3/4

Amphibian blastula

Animal cap

Intact

Dissociated + neurogenin

Dissociated + BMP3/4

Ciliated epithelium Epidermis

Neurons

Epidermal cells

16.2.2 Induction Factors Switch on Further Proneural Selector Genes

Extracellular signals involved in the specification of the vertebrate nervous system have been revealed by various methods. Originally, explants from *Xenopus* blastulae (classic animal cap assay, Fig. 16.8) were treated with solutions that putatively contained factors stimulating nerve cell differentiation. Currently, embryonic stem cells of the mouse or humans are treated with sundry factors aiming at stimulating them to differentiate into nerve cells (Chap. 18). In such assays it is checked which type of nerve cell or glial cell arises from originally "naive" cells. Without going into the experimental details we summarize:

- Besides the already mentioned anti-BMP4 factors CHORDIN and NOGGIN also factors of the **WNT** family and the **FGF** family

participate in uncovering the neuronal potency; supported by the non-protein factor **retinoic acid RA**.

- **FGF:** Subsequent to treatment with anti-BMPs *Xenopus* ectoderm expresses **FGF-4**. This factor is needed to convert cells into nerve cells.

- **WNT:** Already in hydrozoan polyps of the genera *Hydra* and *Hydractinia* the WNT-β-catenin system participates in neuronal differentiation. In vertebrates WNT molecules are at play when it comes to subdivide the central nervous system into brain and spinal cord.

- **Regionalization of the CNS** along the longitudinal body axis (Fig. 16.9). This subdivision is prepared early during gastrulation. The **WNT-8** signals emanating from a centre at the blastopore and spreading in the anterior direction favour specification of spinal cord. In the anterior region of the neurula WNT-8 molecules are bound and neutralized by the

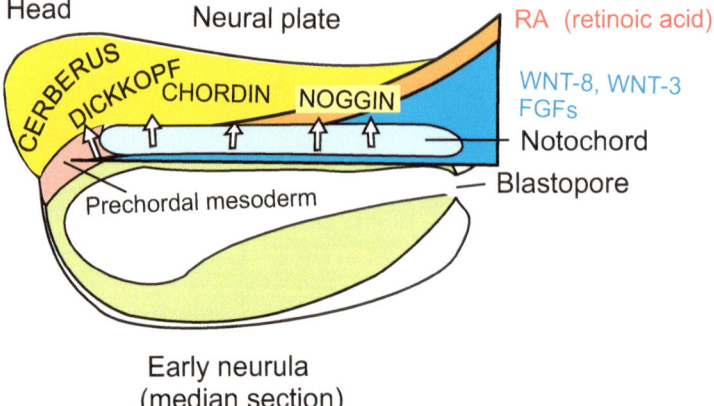

Fig. 16.9 Regionalization of the CNS in the neurula. Of particular significance are induction factors released from the notochord primordium and the prechordal mesoderm known as CERBERUS, DICKKOPF, CHORDIN, and NOGGIN. These initialize the subdivision of the central nervous system in forebrain, midbrain, hindbrain and spinal cord. In this function they are supported by further factors which are present in the form of gradients extending along the body axis from posterior to anterior such as WNT-8, WNT-3, WNT-11, FGF-4 and RA (Retinoic Acid)

factors DICKKOPF and CERBERUS and this favours brain formation. Besides permissive (permitting) factors such as CHORDIN and NOGGIN also putative positive inducing signalling molecules appear to be involved in the switching on of nerve cell-specific genes as indicated by findings made in the chick embryo.

16.2.3 Step-by-Step Further Proneural and Neurogenic Genes Coding for Transcription Factors Are Switched On

Proneural means: "in preparation of", "in favour of nerve cells"; **neurogenic** means: "nerve cell generating".

In the *Xenopus* embryo (and likewise in the fish embryo) the ontogenetic history of the nervous system starts with the maternal factor SOX3. Soon thereafter in the nuclei of the arising neural plate further **proneural transcription factors** such as **NeuroD**, **neurogenin**, Mash1, Math und NEX are found. They belong to the ATONAL group of transcription factors having a bHLH (basic-Helix-Loop-Helix) motif as DNA-binding domain. In *Drosophila* proneural genes are clustered on a chromosome forming the ***Achaete-scute-complex* (*As-C*)**. In vertebrates homologues of the *As-C*-gene cluster (such as *Asc1/lash*) are only expressed in a subclass of neurons. Probably, there exist further such fate-determining genes.

Most currently identified genes/factors are responsible for the development of the nervous system throughout the entire animal kingdom, starting with the cnidarians, the most ancient, still living animal phylum possessing a nervous system. In vertebrates all these factors emerge in changing combinations in the entire developing nervous system. They stimulate stem cells to become nerve cells, and nerve cells to differentiate into a certain subtype, including glial subtypes.

The finding that in the entire animal kingdom homologous (orthologous) genes are at play when it comes to program the nervous system is, in retrospect, not surprising. **Nerve cells share many features: they form thin processes for receiving and transmitting chemical and electrical signals; their cell membranes are equipped with ion channels of various kinds; they generate and release transmitters.** Future research has to figure out in detail where all the proneural and neurogenic genes intervene and make their contribution.

16.2.4 The Future Nervous System Is Separated From the Prospective Epidermis by the Notch/Delta Mechanism of Decision Making

At the margin of the neural plate, but also within the neural plate (as outlined below), the primary neurons must segregate from their surroundings (Fig. 16.10). This happens by means of the **Notch/Delta system**, whose function in mediating **lateral inhibition** was introduced in Fig. 10.1. In principle a future neuroblast presents the Notch receptor of its neighbours with a stronger inhibitory Delta signal than the Delta signal that the neighbours can respond with. The situation is inverse with receptors. The neuroblast with superior power to suppress neighbours becomes itself more and more insensitive to suppressing influences by its neighbours because it withdraws its own Notch receptors. With time inhibition becomes more and more unidirectional and eventually the neuroblast is definitively the winner, adjacent cells are the losers.

Freed by the initial suppression by its neighbour, the neuroblast can fully activate proneural genes such as *neurogenin* (Fig. 16.10), whereas the neighbours switch on other genes. In prospective epidermal cells of *Drosophila* the epidermis-specific *Enhancer of split* complex, **Esp-Cl,** acts as an opponent of the neurogenic *Achaete scute* complex, **As-C;** in the mouse **HES** fulfils the function of the opponent.

Later in the segregation of the future neurons from glial cells the Notch/Delta system commences activity again. The winner in this renewed duel will become the neuron, the loser will become the glial cell.

16.2.5 The Primary Neuroblasts Emerge on the Neural Plate in Six Stripes Which Express Different Homeobox Genes and Display an Intriguing Similarity to the CNS in Insects

Even though initially proneural genes are switched on in the entire neural plate this does not mean that all cells are immediately committed to become neuroblasts. When neuron-specific genes were detected and their expression pattern was visualized by in-situ hybridization unexpected patterns were found: On the neural plate **three rows each of primary neurons** on both sides of the midline became visible (Fig. 16.11). The two rows adjacent to the midline contain future **motor neurons,** the next rows contain future **interneurons** (plus glia), while the lateral rows are displaced outside the closing neural tube and become the **neurogenic cells of the neural crest**; these displaced neurogenic cells migrate from the margin of the tube to form the spinal ganglia (dorsal root ganglia) or provide the peripheral glia (see Figs. 16.17 and 16.18).

The genome of *Drosophila* encompasses, not very surprisingly, neuroblast-specific genes homologous to those of the vertebrates; they are expressed – and this was completely unexpected – in six longitudinal stripes as well (Fig. 16.12), the strongest argument in favour of a basic homology of these central nervous systems which are so widely different in their terminal structure. Even when it comes to subdivide the nervous system into brain and spinal cord, or brain and ventral cord, respectively, and to further regionalize the CNS along the longitudinal body axis of the embryo one sees homologous genes expressed in similar patterns, for example genes of *Hom/Hox B* group (Fig. 16.13).

16.3   The Growing CNS and Stem Cells

16.3.1 The Growing CNS: New, Secondary Neuroblasts Are Generated by Asymmetric Division of Stem Cells at a Rapid Speed

When in the growing human embryo the neural tube gives rise to the enlarging brain and to the extending spinal cord the existing neuroblasts must be increased in number. **In embryonic development a human brain grows at an average speed of about 250,000 new nerve cells per minute up to several billions of nerve cells.** In addition a **tenfold higher number of supporting**

Fig. 16.10 Role of the NOTCH/DELTA system and of proneural genes in alternative fate decisions along the margin of the future CNS. The decision neuroblast versus epithelial cell, and neuron versus glial cell is made in originally quite similar cells. Both produce the inhibitory DELTA signal that suppresses neuronal differentiation, and both initially had receptors of the NOTCH class to receive DELTA signal. Thanks to a slight initial advantage the future neuroblast dominates more and more by producing more and more DELTA while degrading its own NOTCH receptors and thus being protected against the DELTA presented by its neighbour. Epidermis: In the *left*, *blue* cell a presenilin protease cleaves off the intracellular domain of NOTCH, called NICD. NICD enters the nucleus and becomes part of a transcription factor complex that switches on an epidermis-specific gene cluster. In *Drosophila* this cluster is known as *Enhancer of split*, in vertebrates as *HES*. The neuroblast (*yellow*) on the other hand can switch on neurogenic genes unhampered. These are in *Drosophila* genes of the *Achaeta scute* complex, and in vertebrates genes of the *neurogenin* group

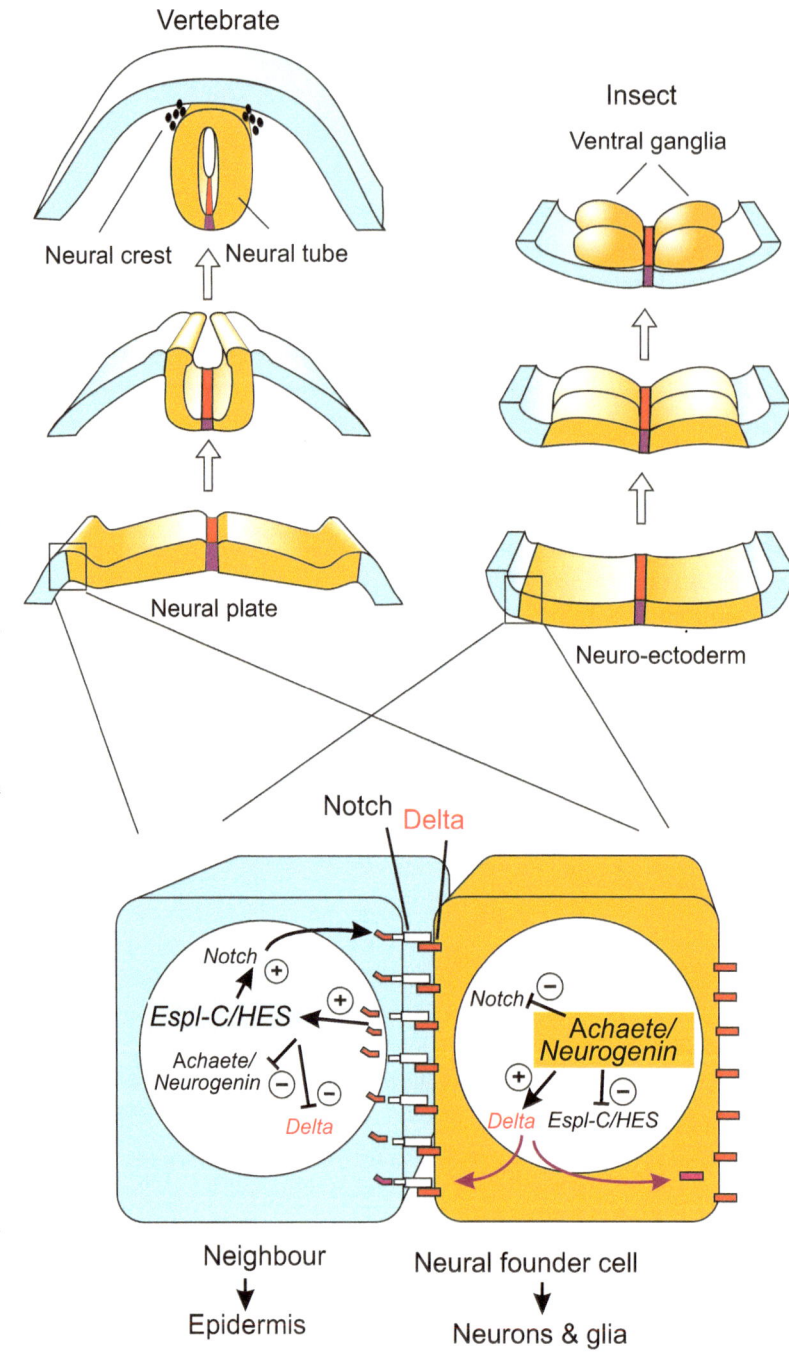

glial cells must be generated. Also subsequent to the embryonic phase and into postnatal life the brain grows in volume and mass by newly generated nerve and glial cells until by the end of puberty the maximum size of approximately 10^{11} nerve cells and 10^{12} supporting glial cells (microglia = immune cells) is reached.

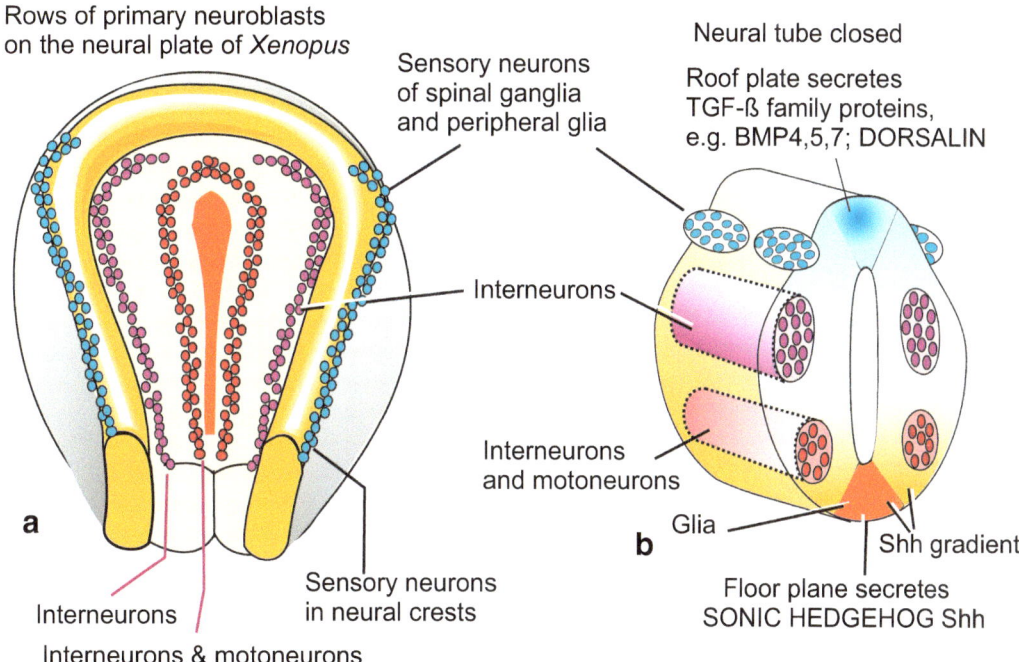

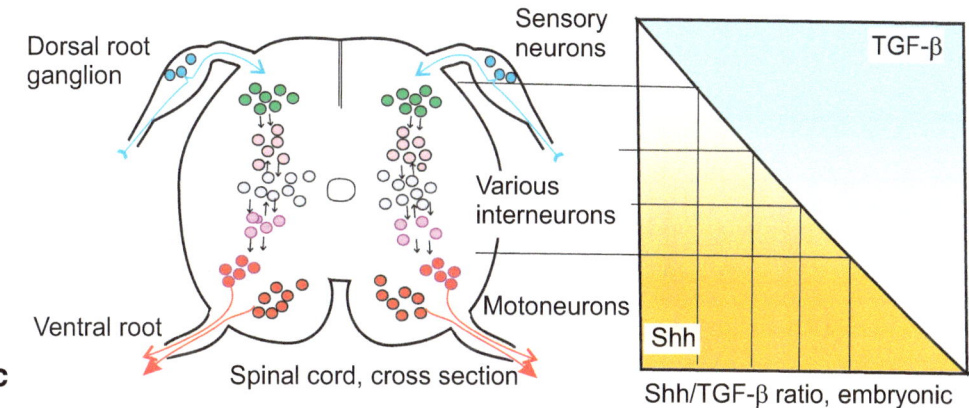

Fig. 16.11 (**a–c**) Fates of neuroblasts along the dorso-ventral axis of the developing neural tube, and factors involved in decision making. (**a**) Rows of primary neuroblasts on the neural plate of a frog neurula. (**b**) Neural tube arisen from the neural plate. (**c**) The SHH gradient and the complementary gradient of TGF-β molecules act as antagonistic morphogens. Their local ratio provides for selective and regio-specific activation of transcription factors which in turn define distinct neuronal fates

The proliferation of neuronal cells (neurons and glia cells) is brought about by an epithelium-like layer of neurogenic stem cells (progenitor cells) called **neuroepthelium or ependyma** which lines the lumen of the canal of the spinal cord and the ventricles of the brain where supply of nutrients is best. This neuroepithelium remained as a two-dimensional layer when the six stripes of primary neuroblasts were segregated out. It contains stem cells which give rise to further populations of neuronal and glial progenitors.

Fig. 16.12 Neurogenesis in insects. Also in the separation of the primary neuroblasts from the glia-type supporting cells the NOTCH/DELTA system is called into action

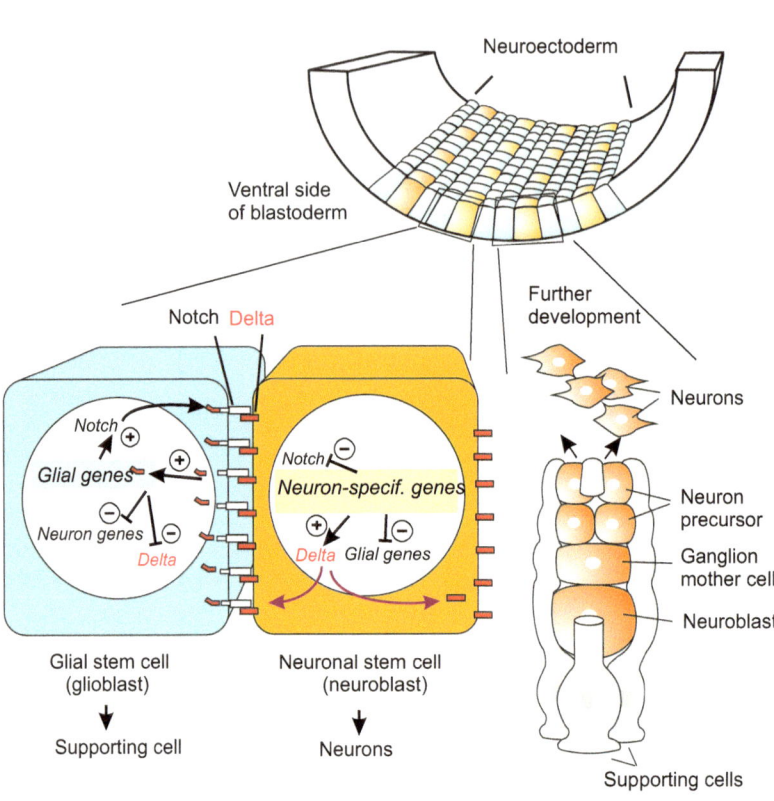

The processes of the secondary neurogenesis can, of course, not be investigated in humans. The avian embryo, and currently transgenic, transparent fishes, and with substantial restrictions the mouse embryo, must act as model and proxy. In the zebrafish *Danio* novel techniques enabled time-lapse monitoring of living transgenic individuals expressing GFP-coupled fluorescing proteins in the stem cells and their derivatives. The recordings partially corroborated, but partially corrected, findings on histological preparations of mouse embryos. The processes are strange. In the course of each mitotic cycle the nuclei in the elongated stem cells migrate from a position close to the luminal surface to a position close to the peripheral surface where they undergo the S-phase of the cell cycle. Subsequently the nuclei return to the original position where mitotic division of the cell takes place (Fig. 16.14). The mitosis can be symmetric or asymmetric.

- In **symmetric mitoses** certain cell components, for instance the protein **Numb** and the protein PAR-3 and a protein kinase aPKC, are uniformly distributed among the two daughter cells. As a result both daughter cells remain stem cells. Numb's primary function in cell differentiation is as an inhibitor of Notch signalling, and suppressing Notch signalling is essential for maintaining self-renewal potential in stem and progenitor cells.

- In the **asymmetric mitoses** these components are unequally distributed. One daughter cell retains Numb and remains stem cell. The daughter cell which inherits PAR-3 (but not Numb) becomes a **postmitotic neuroblast**. Postmitotic means that this cell will not divide anymore but will undergo terminal differentiation to become a neuron. But before this, the cell must leave its birthplace at the baseline, emigrate into peripheral layers and settle there (Figs. 16.14 and 16.15). The nucleus of the

Expression Pattern of Genes for Regional Subdivision of CNS Anlage

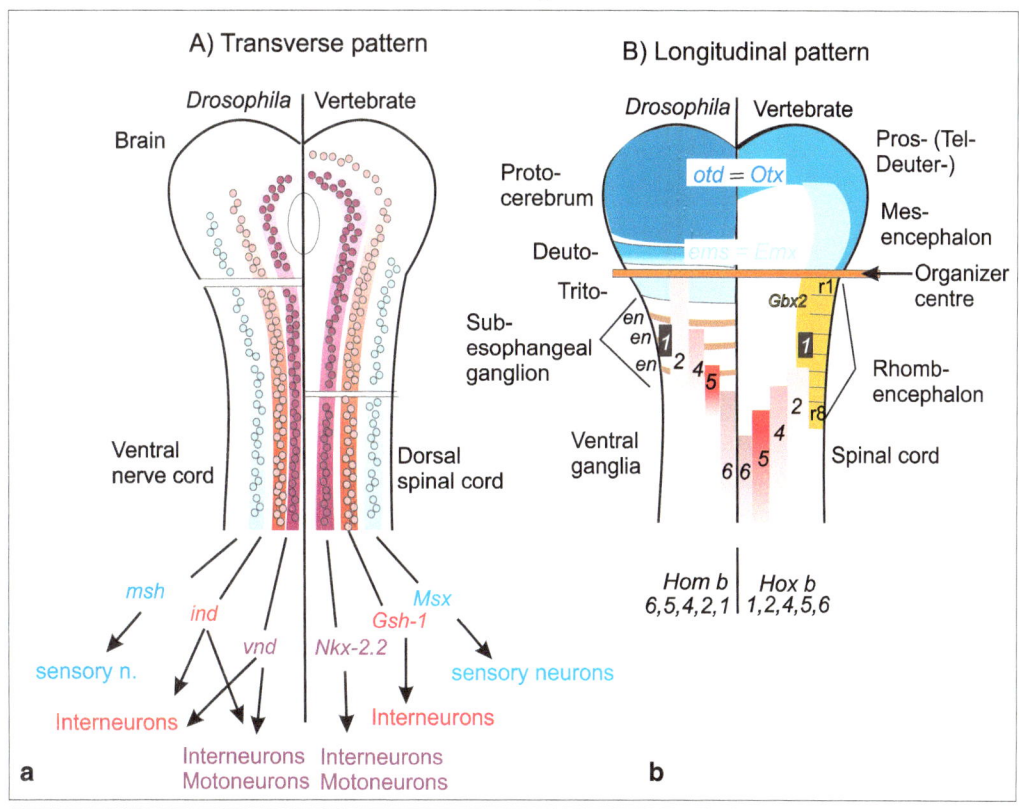

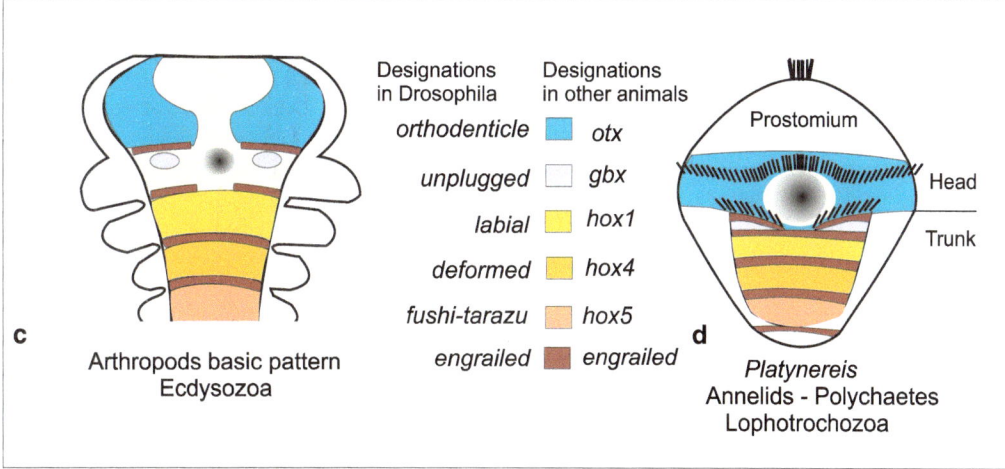

Fig. 16.13 Expression pattern of homeobox genes which contribute to the large-scale subdivision of the central nervous system CNS. (**a**, **b**) Expression pattern of homeobox genes in the anlage of the CNS in *Drosophila* and in vertebrates in comparison. (**a**) Genes programming the first rows of neuroblasts are expressed in lateral stripes. Genes marked with identical colour in *Drosophila* and vertebrates have similar nucleotide sequences and therefore are considered to be homologous. (**b**) Expression pattern of homologous *Hox* genes along the longitudinal body axis of the future CNS. In the embryo of *Drosophila* a series of genes (here *otd*, *ems*, *en*) is expressed in transverse stripes whereas the homologous genes in vertebrates are expressed in longitudinal stripes (*Otd/Otx = orthodenticle, ems/Emx = empty spiracles, en = engrailed*). The *Hox* genes (=*Hom* genes) of *Drosophila* are designated with different names on historical grounds: 1 = *lab*, 2 = *pb*, 4 = *Dfd*, 5 = *Scr*; 6 = *Antp* (see also Fig. 4.39). (**c**, **d**) Comparison of the expression and names of homeobox genes in arthropods and the polychaete *Platynereis*. (**a**) after Arendt and Nübler-Jung (1999); (**b**) after Cohen and Jürgens (1990) and Reichert (2005); (**c**, **d**) after Steinmetz et al. (2011)

Neurogenesis in Fish Spinal Cord and Brain

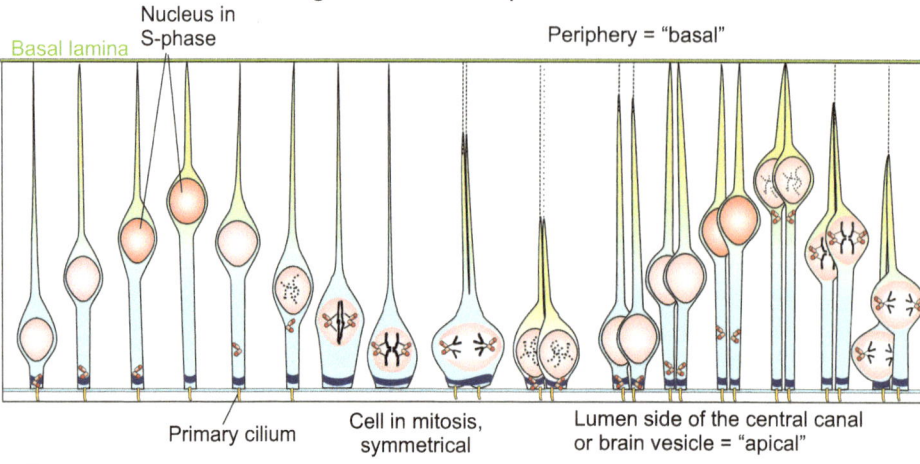

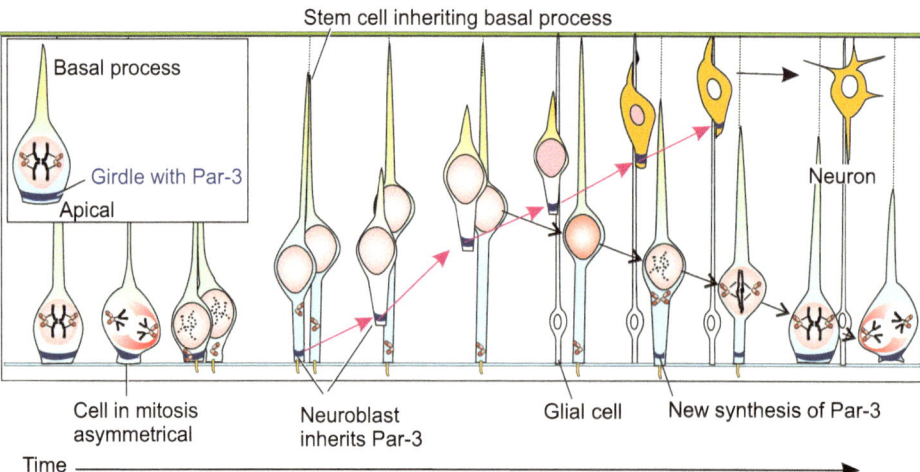

Fig. 16.14 Neurogenesis in the post-embryonic (post-natal) CNS. Enlargement is brought about by stem cells of the spinal cord located around the central channel-The mechanism has been analysed in transgenic zebrafishes by means of video recording of GFP-labelled specimens.

second cell which did not receive Par-3 but Numb, returns to the baseline and the cell re-synthesizes PAR-3, thus returning to its originals state as neurogenic stem cell.

On their way into the periphery the neuro-blasts move actively along radial glial fibres that stretch from the central canal or ventricles, respectively, to the periphery, and serve as guiding threads (Figs. 16.14, 16.15; also Fig. 15.9). In the target place they leave Ariadne's thread and integrate into the flock of already arrived neuroblasts that are about to mature to functional neurons.

16.3.2 Emigrated Neuroblasts Arrange in Laminar Levels, or Cluster in Core Areas, and Undergo Differentiation to Numerous Types of Neurons and Glial Cells

In the cortex of the telencephalon and in the cerebellum the migrating cells arrange them-selves in several layers superimposed upon each other.

When the basic architecture of the brain is established in the cortex of primates **six surface-parallel levels of neurons** can be made

Neurogenesis in Mammalian Brain

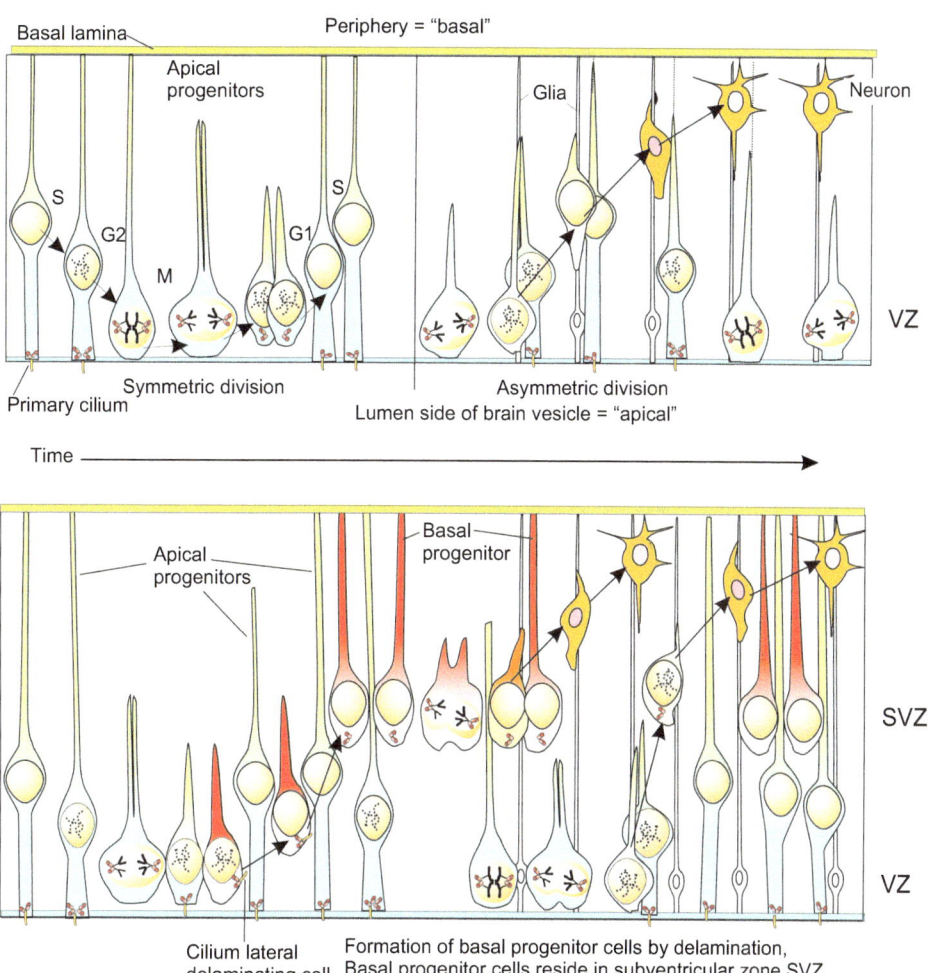

Fig. 16.14 (continued) Growth of the vertebrate brain through neurogenesis from stem cells of the ependyma, shown at the transition of the myelencephalon to the spinal cord in the fish *Danio*. In GFP transgenic fish the transparency of the body allows the tracking of stem cells labelled with GFP (Box 12.2) in vivo. By continuous video recording the migration of the cell nuclei in dividing stem cells was documented. The ependyma containing stem cells borders on the central, fluid-filled canal (in the brain on the ventricle). The side adjacent to the canal is designated "apical". Shown is the production of further stem cells by symmetric division, and of secondary neuroblasts by asymmetric division. In asymmetric division the marker Par3-GFP is allocated to the future neuroblast. The nucleus of the second daughter cell returns to the apical home position and the cell resumes synthesis of Par2-GFP. After Alexandre 2010 (and talk 2011), supplemented after Taverna and Huttner 2010 (how far the basal processes exactly stretch in basal direction is not learned from these publications)

out. The first neuroblasts born at early stages migrate to the periphery forming in the subsurface area the first layer which remains the outer layer and eventually will be the sixth layer. The second wave of neuroblasts forms the layer closest to their site of birth. When this layer is filled, the third born neuroblasts must migrate past this second one to give rise to the third layer. The fourth wave of emigrants has to penetrate two layers and will form the next, more

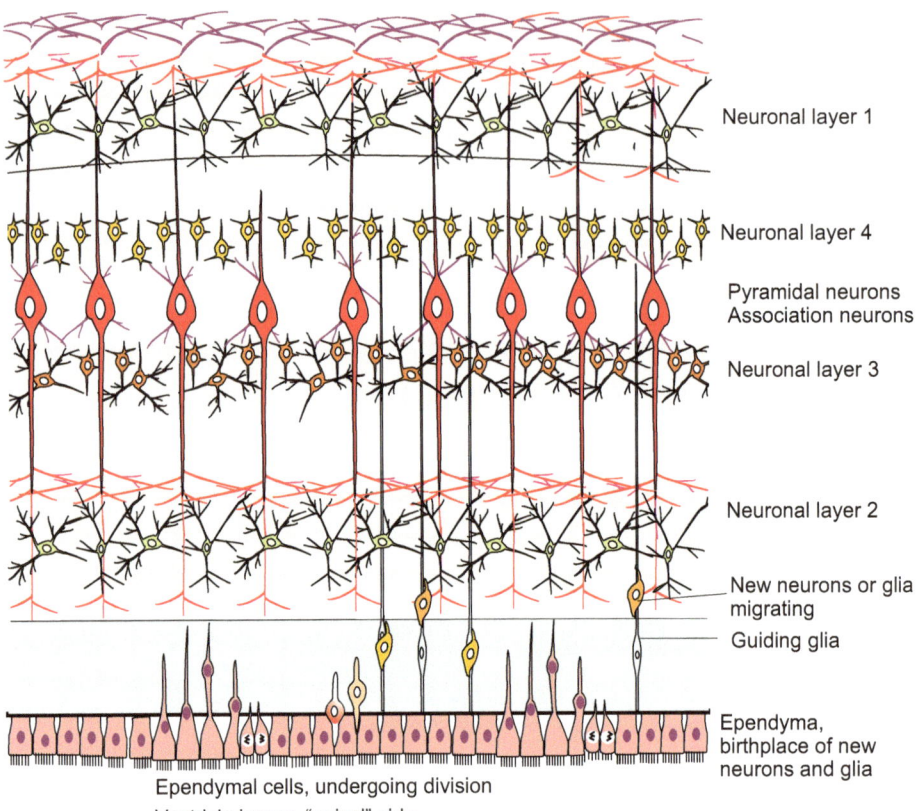

CORTEX

Surface, "basal side"

Neuronal layer 1

Neuronal layer 4

Pyramidal neurons
Association neurons

Neuronal layer 3

Neuronal layer 2

New neurons or glia
migrating

Guiding glia

Ependyma,
birthplace of new
neurons and glia

Ependymal cells, undergoing division

Ventricle lumen, "apical" side

Fig. 16.15 Organisation of the cortex in the vertebrate brain. Newly produced neuroblasts migrate along glial fibres in peripheral direction and settle down in up to six layers. The neuroblasts forming the 3rd, 4th, and 5th layer have to penetrate previously formed layers

peripheral layer underneath the outer (sixth) layer. This process continues, each new wave of emigrants penetrates the older layers and settles beneath the outer most peripheral layer until six layers are established. On speaks of *inside-out-layering*. After settling down the postmitotic neuroblasts undergo terminal differentiation.

It is currently unknown which signals stimulate neuroblasts to resume migration or to settle down at their final destination. Likewise it is an open question where the decision is taken to which subtype of neurons each individual neuroblast eventually will mature. Discussed are fate-specifying influences of the local environment (local niche).

Eventually the vast majority of neuronal cells in the six laminar layers belong to the large class of **interneurons** (e.g. the spined **stern cells,** morphologically similar to astrocytes; Fig. 16.16). The large-scale six layers are considered **functional modules**; by some researchers they are awarded the faculty of **parallel data processing**. Large **pyramidal neurons (projection neurons)**, in the cerebellum known as **Purkinje cells,** connect the six layers.

Also in vertical direction perpendicular to the surface of the cortex a modular organization can be made out in some areas of the telencephalon. Examples are the **ocular dominance columns** in the visual centre V1 (see Fig. 16.25).

Fig. 16.16 Cell types in
the CNS, a selection

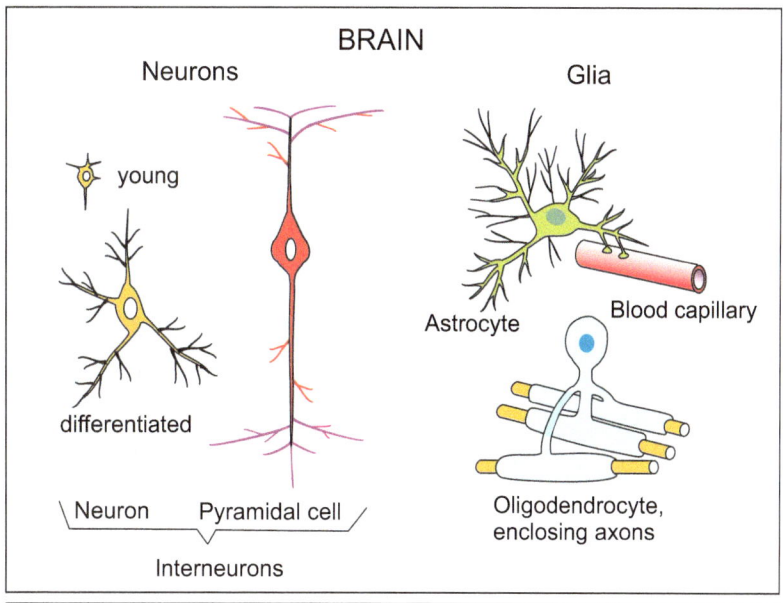

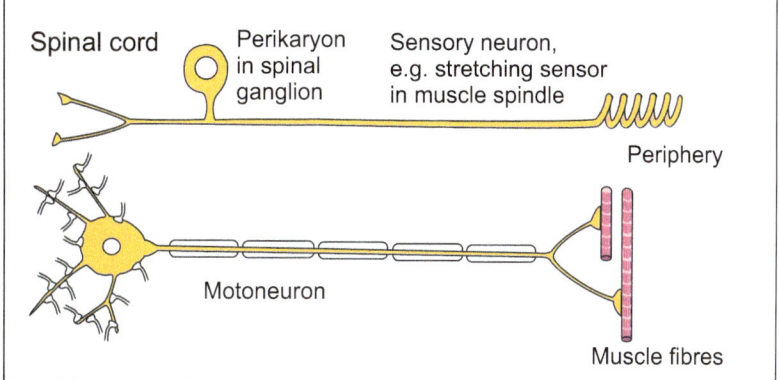

Within the brain core regions, **so called nuclei** are formed by local dense aggregations of neuroblasts. In this process cell adhesion molecules, in particular cadherins, play important roles.

16.3.3 Neurons and Glia Arise from Common Stem Cells; They Fulfil the Tasks of the Nervous System Cooperatively

Traditional textbooks often convey the impression that uptake, processing and storage of information is the sole task of the nerve cells whereas glial cells merely have assisting functions by supporting the metabolism of neurons and serving as electrical isolators. One may ask why in humans glial cells account for 90 % of the brain mass. Certainly, far-distance transmission of quantized information through series of action potentials is the specific function of nerve cells whereas pure helping functions such as providing guiding threads for migrating cells (Figs. 16.14 and 15.11), is the specific function of glial cells. Yet, their function is not exhausted with such transitory assistance. Other glial cells participate in the processing and storage of information. This applies in particular to astrocytes, less to oligodendrocytes. (The microglia, on the other hand, does not consist of neural glial cells but of immigrated cells of the immune system.) An argument for upgrading the glial cells is the finding that they derive from the same neuroepithelial stem cells as do the neurons. In the scientific literature the common stem cells usually are indistinguishably designated neuroblasts and not called neuroglioblasts. (Only when it is known that certain progenitor cells exclusively give rise to glial cells one speaks of glioblasts.) The common denomination of neuronal and glial precursors as neuroblasts underscores the close relationship of both cell types. The close relationship is more convincingly underscored by recent findings that even terminally differentiated glial cells, such as the radial glia depicted in Figs. 16.14 and 15.11, can transform into regular neurons (transdifferentiation).

When in the following there is talk of neuroblasts it is still open which specialised cells will originate from these precursors. According to site and time of their emergence one distinguishes between **primary** and **secondary neuroblasts**. The primary neuroblasts appear on the neural plate, the secondary neuroblasts are generated in the growing CNS as described Sect. 16.3.1.

16.4 The Peripheral Nervous System and Cell Migration Over Large Distances

16.4.1 The Peripheral Nervous System with the Sympathetic and Parasympathetic System and the Nervous System of the Gastrointestinal Tract Are Formed by Emigrated Neural Crest Cells Whose Fate Is Determined by Their Path and Destination

The area of origin of all cells of the nervous system is the neural plate including its margin known as neural crest. The neural crest embodies the founder cells of various cell types and tissues. Besides nerve cells and glia cells, also pigment cells of the skin, the cartilaginous cells (chondrocytes) of the head-neck region and several other cell types originate from the neural crest (Chap. 15). The place of birth and the final destination in the body can be far apart from each other. The cells have to make a journey until they can undergo terminal differentiation and commence their function.

The fate of the migrating neural crest cells is believed to be specified by
(a) the origin of the cells (lineage and place of origin), which can convey a basic predisposition,
(b) influences acting on them while they are travelling, or
(c) the conditions they encounter at the target area. Determination is accomplished by a series of alternative choices where their paths bifurcate (Fig. 16.17a).

16.4.2 The Ultimate Determination of the Type of Neurotransmitter to be Produced Is Only Made at the Destination

An experiment designed to address this issue took advantage of the fact that cervical (neck) neural crest cells normally give rise to parasympathetic, cholinergic nerve cells producing the transmitter **acetylcholine**, while thoracic (chest) cells give rise to adrenergic sympathetic ganglia producing noradrenaline. If the cells are exchanged before they leave home, the original cervical cells colonize the chest and produce **noradrenaline**. The original thoracic cells arrive in areas where parasympathetic innervation is wanted and produce acetylcholine. Such experiments suggest that the final fate and function of neural crest cells are place-dependent.

A second example for the influence of the surroundings: The cell population specified as sympathetic subdivides into two subtypes (Figs. 16.17, 16.18 and 16.19):
1. **sympathetic neurons** and
2. **adrenaline-producing cells** of the **adrenal medulla**; this second type is formed under the influence of glucocorticoid hormones released by the adrenal cortex which encase the medulla.

Fig. 16.17 Transition of the CNS to the peripheral nervous system PNS. The spinal nerves feed information coming from somatic sensory organs into the spinal cord via the sensory fibres in the dorsal roots, and transmit orders into the periphery, for examples to the muscles or the autonomic nervous system, via motor fibres through the ventral roots. The cell bodies with the nucleus (perikaryon) of the sensory fibres are located in the dorsal root ganglia, also known as spinal ganglia, whereas the cell bodies with nucleus of the motor neurons reside in the spinal cord

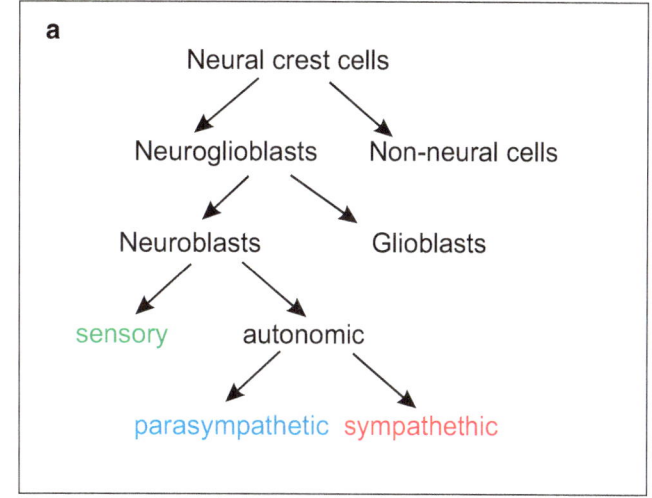

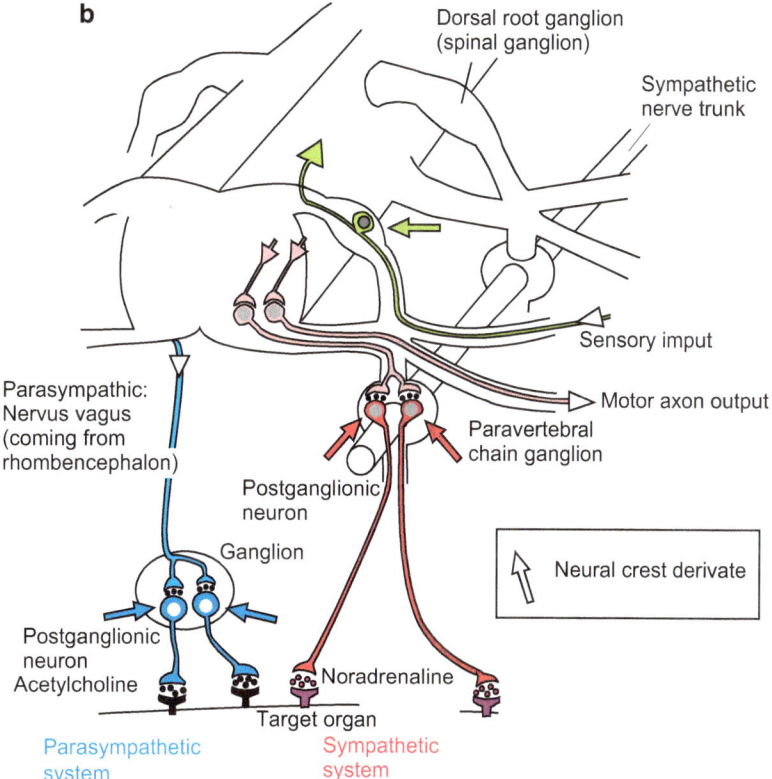

Fig. 16.18 Development of the peripheral nervous system from emigrated neural crest cells. The peripheral nervous system comprises the dorsal root ganglia (spinal ganglia), the autonomic sympathetic and parasympathetic system as far as the cell bodies lies outside the CNS, and the nerve net of the gastro-intestinal tract, called enteric nervous system ENS. Also cells of the glial class derive from the neural crest such as the Schwann cells around long axons and the adrenaline-producing cells of the adrenal medulla

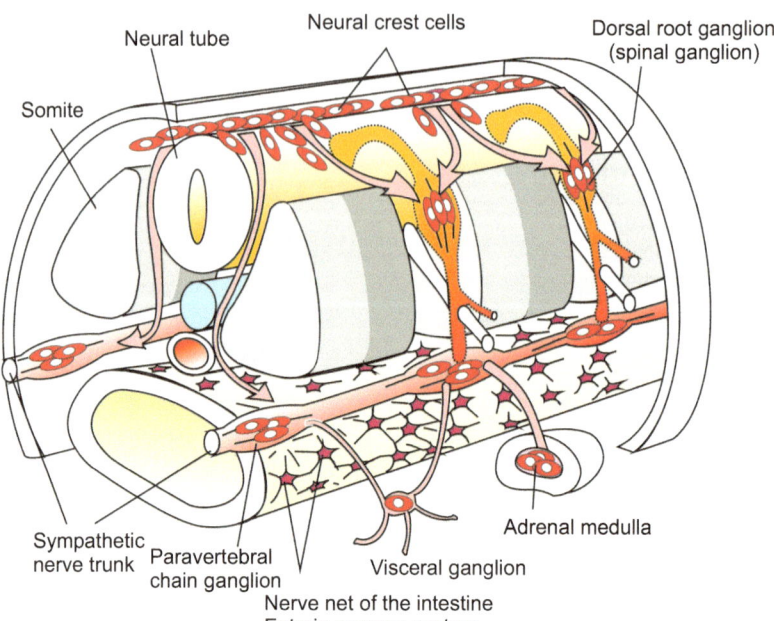

Peripheral Nervous System

16.4.3 Nerve Growth Factors Chemotactically Guide Neuroblasts and Serve as Survival Factors for Cells of the Peripheral Nervous System

The cells of the neural crest whose destiny is to establish the peripheral nervous systems are guided to their targets by numerous signposts and guides, for example by the signalling molecule WNT-11 (see Fig. 15.8). The first identified, long-range signalling molecule is **nerve growth factor NGF**. This is a polypeptide which is cleaved out of a precursor complex. The complex consists of 6 subunits; the biologically active subunit is the β-subunit; it forms a homodimer.

NGF acts on **immature dorsal root and sympathetic neuron**s. However, it does not act as a mitogenic agent stimulating cell proliferation, as initially assumed and indicated by its name. Rather, NGF guides the **direction** of growth cone movement and thus determines the direction in which a neurite (dendrite or axon) is extended. The directional cue is provided by the concentration gradient of NGF. If a micropipette

releasing NGF is placed close to a growth cone (see next section), its direction of advance can be deflected to the artificial source of NGF.

NGF binds to membrane-bound receptors on the growth cone. Bound NGF releases a cascade of signal transduction events and subsequently is internalized by endocytosis. Internalized NGF is carried in vesicles by retrograde transport into the main body of the neuron, the perikaryon. By unknown means, internalized NGF ensures the **survival** of those neurons that succeeded in arriving at the correct destination and therefore are 'fed' with NGF.

Several subtypes of NGF and NGF-like factors are known. They enable the survival of diverse subpopulations of nerve cells and are known collectively as neurotrophins. If a neuron is close to a source of NGF it is more likely to survive than if it is distant from the source. However, the term "trophic" is not very appropriate, as neurotrophins are signal molecules effective in very minute concentrations and not used for true nourishment. Simultaneously and confusingly, the known neurotrophins are neurotropic ("tropic" = guiding).

Fig. 16.19 Fate of neural crest cells. On their way they can be directed by signalling molecules of their environment towards a distinct differentiation path. By the action of FGF they become receptive to NGF (Nerve Growth Factor), which sympathetic neurons require for survival. Cells settled in the adrenal medulla receive the order to turn into producers of the stress hormone adrenaline (epinephrine) via the hormones (glucocorticoids) supplied by the adrenal cortex

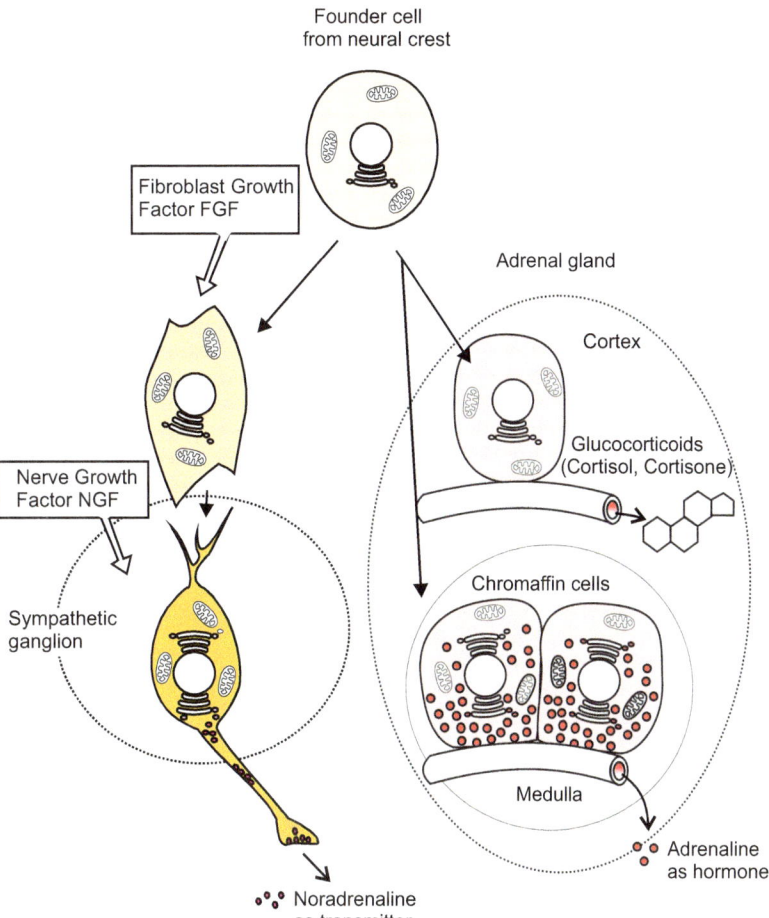

Neurotrophins are life-sustaining for nerve cells of the peripheral nervous system but other nerve cells also need help to survive; this will be explained in Sect. 16.8.2 when we take the connection of motor neurons to muscle fibres into account.

16.5   Navigation of the Nervous Processes and Interconnection of Nerve Cells

16.5.1 Patterning Nerve Connections Is a Process of Self Organization

Neuroblasts undergo terminal differentiation at their destination. One cell pole becomes the input region and forms **dendrites** for receiving information. The opposite pole is specified to become the output and forms an **axon** for transmitting information. Dendrites as well as axons must grow in the direction of their targets (such as sensory cells, other nerve cells, muscle cells or other effector cells) and have to establish synaptic contacts – a task of unimaginable complexity!

The human brain grows during embryo development at a rate of 250,000 nerve cells per minute to reach a population of a hundred billion (10^{11}) nerve cells and one trillion (10^{12}) glial cells. Every nerve cell has synaptic contacts with about 1,000–150,000 other nerve cells. It is inconceivable that the DNA encodes an exact plan for all these connections; the storage capacity of the entire genome is far too small.

The patterning of the neural connections is based on **self-organization**, the phenomenon where properties of higher order structures emerge from the interaction of lower order components. Thus brain development depends upon the interaction of neurons with other neurons, with glial cells and with supporting structures of the extracellular matrix. Genetic information enables the interacting cells to produce signalling molecules, to construct receptors for picking up signals, and to install signal transducing systems for initiating adequate responses.

16.5.2 On the Tip of Growing Axons a "Growth Cone" Serves as Antenna and Motile Pioneering Structure

Growing dendrites and axons terminate in a peculiar structure which looks like a hand with many fingers: the **growth cone** (Fig. 16.20). The fingers are thin elongated filopodia and flat lamellipodia enabling the cone to actively crawl over a substrate, to force its way through surrounding tissue and to advance towards its destination. The filopodia are equipped with membrane-associated receptors serving as molecular antennae. The filopodia are extended and retracted, bent to the left and right, and thus explore the environment. They have the function of pathfinders. Small growth cones are located at the tip of dendrites, whereas large growth cones are located at the tip of axons. The society of neurobiologists focuses on the large cones of axons.

The molecular receptors in the membrane of the growth cone pick up signals arriving from their surroundings. These signals can be **attractive** or **repulsive**: The growth cone is attracted by the signal emitter or is repelled (Figs. 16.20 and 16.21). Signal emitters are either the targets or trail signs leading the way to the target.

16.5.3 Firmly Fixed Adhesion and Recognition Molecules Mark the Route to the Targets for Pioneer Fibres; These May in Addition Emit Diffusible Signalling Substances

Substrate- and Cell-Bound Road Signs. Diffusible long-range signal molecules such as NGF cannot mediate precise long-distance guidance, if billions of neurites are to be connected with specific target cells.

Long-term observation of nerve cells growing in Petri-dishes on various substrates reveals a behaviour similar to that known from migrating neural crest cells. Growth cones at the tips of elongating neuronal processes orient themselves using cues provided by the physical and chemical qualities of the extracellular matrix covering the bottom of the dish, and by sensing the molecular surface structure of neighbouring cells. Growth cones show preference for substrates to which they can adhere firmly, and avoid areas apparently displaying a repulsive quality.

A sophisticated technique prepares fields or stripes of putative guiding biological materials on a neutral substrate (e.g. glass) by means of a silicon matrix (see Fig. 16.26). Actually, the biological materials are coatings of extracellular matrix proteins or of vesicles prepared from cell membranes derived from target areas or non-target areas. Axons are grown on such substrates and time-lapse shooting with microscopic cameras and video recorders monitor the behaviour of the growth cones. Will the axons be attracted or repelled, or does the coating have no influence on the speed and direction of the advancing cone? Analysis of the behaviour of neuronal cells *in vitro* reveals several sources of information that might also be used *in vivo* by neuronal processes to navigate in the direction of their target areas:

• Channels, folds and grooves can physically favour or restrict the path of neurite growth. One speaks of **haptotaxis**.

Fig. 16.20 Growth cone at the tip of growing axons, and examples of options for guidance

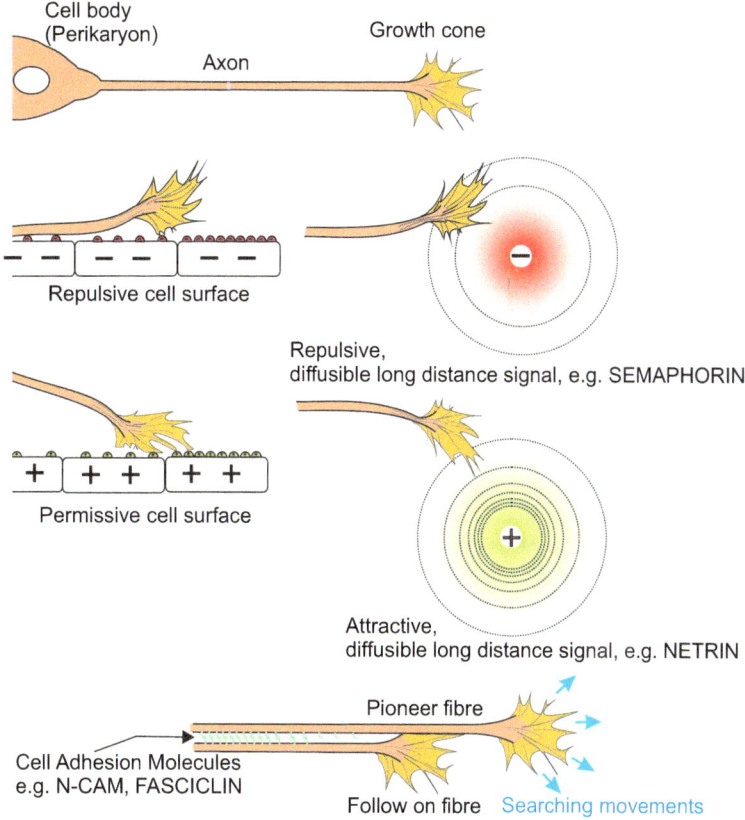

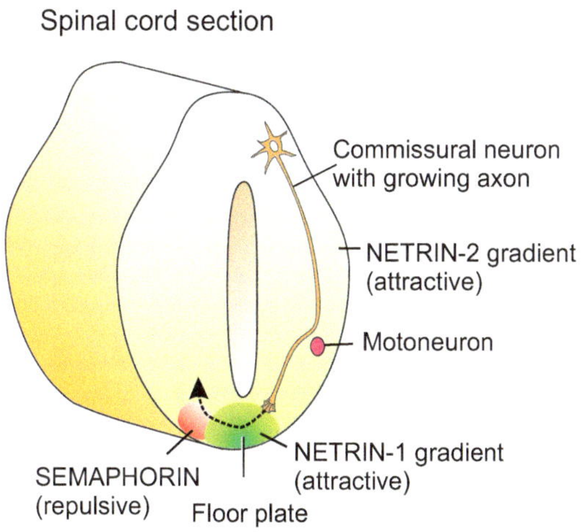

Cell body (Perikaryon)

Growth cone

Axon

Repulsive cell surface

Repulsive, diffusible long distance signal, e.g. SEMAPHORIN

Permissive cell surface

Attractive, diffusible long distance signal, e.g. NETRIN

Cell Adhesion Molecules e.g. N-CAM, FASCICLIN

Pioneer fibre

Follow on fibre   Searching movements

Spinal cord section

Commissural neuron with growing axon

NETRIN-2 gradient (attractive)

Motoneuron

NETRIN-1 gradient (attractive)

SEMAPHORIN (repulsive)   Floor plate

Fig. 16.21 Navigation of axons in the spinal cord. Gradients of netrins (attractive) and semaphorins (repulsive) direct the axons

- Of particular significance are **cell adhesion molecules (CAMs)**. Cell surface and adhesion molecules are highly diverse and suited to mediate precise **contact guidance**. The **molecular makeup of the cells bordering the migration route** provide specific information that, however, is interpreted differently by axons of different origin. When neuron A is attracted, neuron B may be repelled and caused to redirect or withdraw its growth cone.

- **ECM.** The path is coated by macromolecules that delineate the way past obstacles. Many of the roads upon which axons travel are paved with laminins, for example. In the experimental design (see Fig. 16.26) coatings prepared with **extracellular matrix ECM proteins** (laminin, fibronectin, collagens) promote or inhibit the outgrowth in a more general way. Some components of the extracellular matrix are more adhesive than others and bias axon movement. In addition, adhesive **gradients** may indicate the correct direction.

- By binding signalling molecules such as FGFs or netrins the ECM can adopt specific functions. Some signposts encountered by growing axons have a repellent label.
- **Diffusible guidance molecules.** Besides NGF mentioned above more and more long-range signalling molecules have been, or are being, identified. These are, for example, the **netrins**, **semaphorins** and the **collapsins** (their name suggests their function). For picking up signalling molecules the growth cone is equipped with corresponding receptors which frequently belong to the class of **receptor tyrosine kinases** (Chap. 11).
- **Signalling molecules positioned on the surface of cells** of the target areas and forming gradients were first found in the visual system as delineated in Sect. 16.7.2.

16.5.4 Axons of the Commissural Neurons in the Spinal Cord Are Guided by Attractive and Repulsive Long-Range Signals

One of the best investigated examples of guidance over (relatively) large distances is the targeted growth of the axons of the commissural neurons in the spinal cord. Their task is to transmit information from neurons located in the dorsal part of the spinal cord down to neurons located along the ventral midline or on the other (contra-lateral) side of the cord (Fig. 16.21). The axons creep towards an attractive **netrin-2 gradient** which is supported and continued by a steeper and higher **netrin-1 gradient**. The emitter of the netrin signals is the **floor plate** of the spinal cord (the cells of which will become glia). The emitter does not have to send netrin signals continuously to sustain the gradient. Once released the netrin molecules are bound to the extracellular matrix and thus fixed in place.

When axons are charged with the task of crossing the midline they turn avoiding netrin and are deflected in dorsal direction by **repulsive semaphorin** molecules.

16.5.5 From the Pioneer Fibres to a Cable; Cell Adhesion Molecules Serve as Glue

In many areas of the CNS millions of axonal fibres are combined forming fibre bundles. A particularly large bundle is the corpus callosum, which bridges the two hemispheres of the forebrain. Also the anatomical "nerves" which lead from the CNS into the periphery consist, as a rule, of numerous bundled conducting fibres (axons and/or dendrites) plus their accompanying glial envelopes.

In living organisms the coherent growth of bundles of nervous fibres is facilitated by **pioneer fibres** sent out to create a track that later fibres can follow. In paving the way proteases that are secreted by the growth cones may be helpful. Such enzymes remove barriers and digest channels through the extracellular matrix.

After the pioneer fibre has laid the guiding thread, the growth cones of the following fibres cling to it, or to other fibres that already adhere to it. But only the growth cones move freely; the older parts of the fibres are immobile. This situation, in which many immobile fibres attach to one another, is used to weld together all of the fibres. Cell adhesion molecules serve as glue. One of the CAMs is known as **fasciclin** (fasciclin II is identical with N-CAM). The process of bundling up is called **fasciculation**.

16.6 Sensory Organs of the Head Directly Connected with the Brain

The brain is surrounded and supplemented by sensory organs required for long-range orientation and control of nutrients.

16.6.1 The Olfactory Organ, Taste Receptors and Inner Ear Arise From Placodes

Placodes are in outline circular or elliptic areas of the ectoderm that invaginate to contact the brain. Their normotopic formation is initiated by inductive influences of the CNS. Placodes supply the material for the construction of important sensory organs:

- **The paired nasal placodes** expand and give rise to the paired **olfactory epithelia**. Later they will contain primary sensory cells with own action potentials conducting axons. These axons have to find their way to specific collection points of the rhinencephalon. More about that in Sect. 16.7.5.
- **The otic placodes** invaginate, detach from the ectoderm and form the **otic vesicles** (Fig. 16.2). Each otic vesicle undergoes complex developmental transformations to become the **vestibular apparatus** of the inner ear. It includes the **membranous labyrinth** whose **semicircular canals** mediate the sense of angular acceleration and whose **maculae sacculi** and **maculae utriculi** mediate the senses of linear acceleration, gravity and equilibrium. The third part of the labyrinth is elongated and is rolled up to become the **cochlea**. The inner ear is innervated by the cranial nerve VIII (*nervus vestibuloocularis*) which originates in the myelencephalon.
- **Epipharyngeal placodes** (synonymous **epibranchial placodes**) lie above the gill pouches (Fig. 16.2b). They will give rise to **taste buds**. These contain secondary sensory cells without their own axon. Their information is collected from dendritic nerve fibres which arise from the neighbouring cranial ganglia. Cranial ganglia are the origin of the cranial nerves VII (*nervus facialis*), IX (*n. glossopharyngeus*) und X (*n. vagus*).
- **Trigeminal chemosensory system**. This important system has only recently been detected and is currently investigated with physiological and psychological methods. So far nothing is known about its developmental origin.
- **Neural crests**. Some authors classify also the neural crests in the category of sensory placodes since in fishes and aquatic amphibians the neural crest gives rise to the

sensory lateral organs. These have the task to perceive currents, ram pressure and water sound. For that purpose they contain so called "hair cells" as does our inner ear, and it is generally assumed that in evolution an anterior part of the lateral lines gave rise to our inner ear.
- **Adenohypophysis,** albeit not being a sensory organ it may be mentioned that also the adenohypophysis arises from a placode, known as Rathke's pouch.

16.6.2 Placodes, Being Initially Undifferentiated, Are Programmed Step by Step for Their Ultimate Fate

According to comparative studies in various vertebrates including especially gene expression analyses, it turned out that all sensory placodes arise from a common ribbon-like morphogenetic field which surrounds the neural plate in a U-like fashion. In this regard the primary anlage could be seen as part of the neural crest but the future sensory cells concerned remain in the ectoderm while the classic neural crest cells emigrate. The entire U-shaped sensory field expresses transcription factors of the **SIX** and **EY** families. The initial field subdivides into partial fields. These acquire epithelial characteristics, invaginate and come in contact with the CNS. In the avian embryo in which the placodes are accessible to experimentation all placodes primarily are endowed with the competence for lens formation. Their particular toponomic specification is alloted to them by signals emanating from the CNS and further local sources in their environment. Two examples: (1) The olfactory placodes are stimulated by FGF-8 to differentiate according to their destination. (2) The otic placodes form lateral of the rhombomeres r4 to r6 under the influence of WNT-8, FGF-3 and FGF-10.

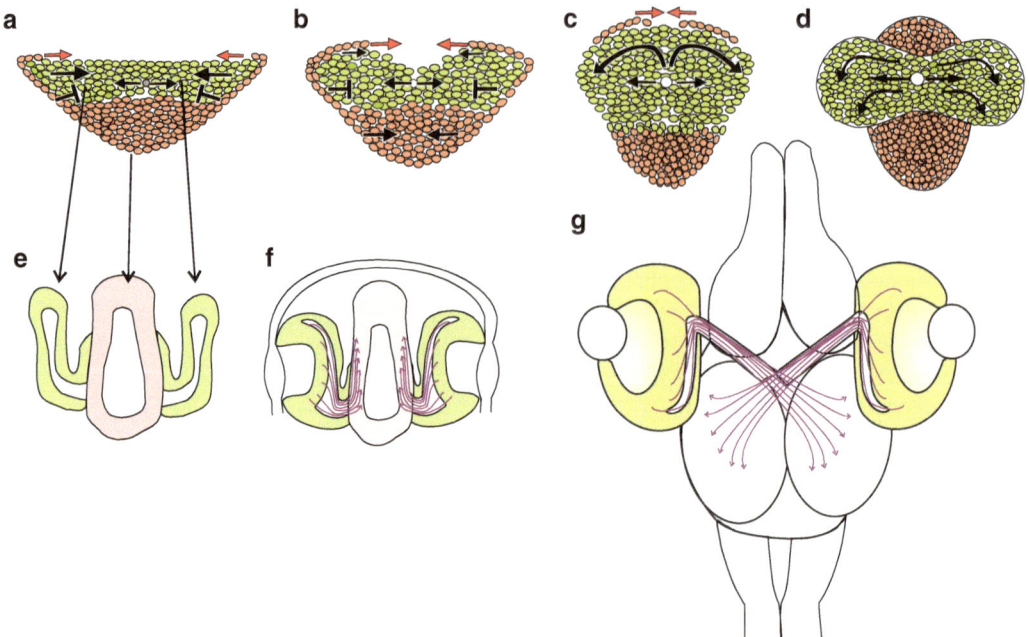

Fig. 16.22 (**a–f**) Development of the eye in the fish *Oryzias latipes*. (**a–f**) Cross sections, (**g**) Longitudinal sections. (**a–d**) Prospective retina cells are marked by green fluorescence. Labelling was performed by introducing the transgene *rx3-GFP* into the genome of the fish. The green fluorescing cells move in an area of *red* marked neural plate cells; *arrows* indicate the direction of movement [after Rembold et al. (2006)]. When the *green* marked cells have reached the outer positions they form an epithelial association and acquire the shape of an eye cup. When the cup comes in contact with the ectoderm it induces the formation of the lens. The axons sprouting from the retinal neurons form bundles forming the *nervus opticus*, extend towards the *tectum opticum* of the contralateral side, separate there from each other, and head with their growth cone for their predestined (retinotopic) targets

16.7 Connecting the Eye and Olfactory Epithelium with the Brain

16.7.1 The Brain Itself Forms the Central Part of the Eye

The eye is in its essential parts a branch of the brain. Essential parts mean the retina with its photoreceptors (cones and rods) and millions of retinal nerve cells who start data processing already within the eye. The eye as a whole arises under the organizing rule of the master genes *sine oculis, Pax6* or *ey*, respectively (Sect. 12.3) which control some thousands of downstream genes. The common origin of eyes and brain can vividly be observed with particular clarity in transgenic fishes. Transparent fishes enable monitoring and filming the behaviour of neuroblasts in living individuals from the earliest stages on up to neurogenesis in the ever growing adult eye. In preparation of such investigations in the fertilized eggs plasmid constructs with reporter genes are injected. Actually, the gene for the retina-specific transcription factor *rx3* was fused with the genes for green or red luminescent variants of the *Green Fluorescent Proteins GFP*. The expressed *rx3-GFP* gene made the neuroblasts fluoresce when subjected to UV light. The film showed: The eye anlage originates from the same area of the neural plate that gives rise to the prosencephalon. First a central unpaired field is observed. Subsequently the future retina cells begin to emigrate and push into two lateral positions. Now we see two fields which form the eye cup (Fig. 16.22).

Fig. 16.23 (**a–f**) Amphibian. Development of the eye and induction of the eye lens. Illustration (**a**) looks down upon a neurula. The retina arises from a field at the forebrain–midbrain boundary the cells of which have acquired the potency to form photoreceptors (cones and rods) and neurons (ganglionic cells). The neurons will have the task of performing a first data processing and to transmit the result to higher regions which in amphibians, fishes and birds is the *tectum opticum* of the midbrain. The axons sprouting from the retinal neurons must reach the contralateral tectum on the opposite side of the body. In its anatomy, the eye is supplemented by supporting structures that do not derive from the CNS but are supplied by surrounding tissues. Among them is the lens. The competence to form a lens is attributed to the lens field in the ectoderm. Lens formation as such is initiated by inducing signals emitted by the eye field (**a**) and subsequently from the eye vesicle and eye cup (**b–d**). The lens on its part induces in the overlaying epidermis the formation of the cornea (**e**, **f**)

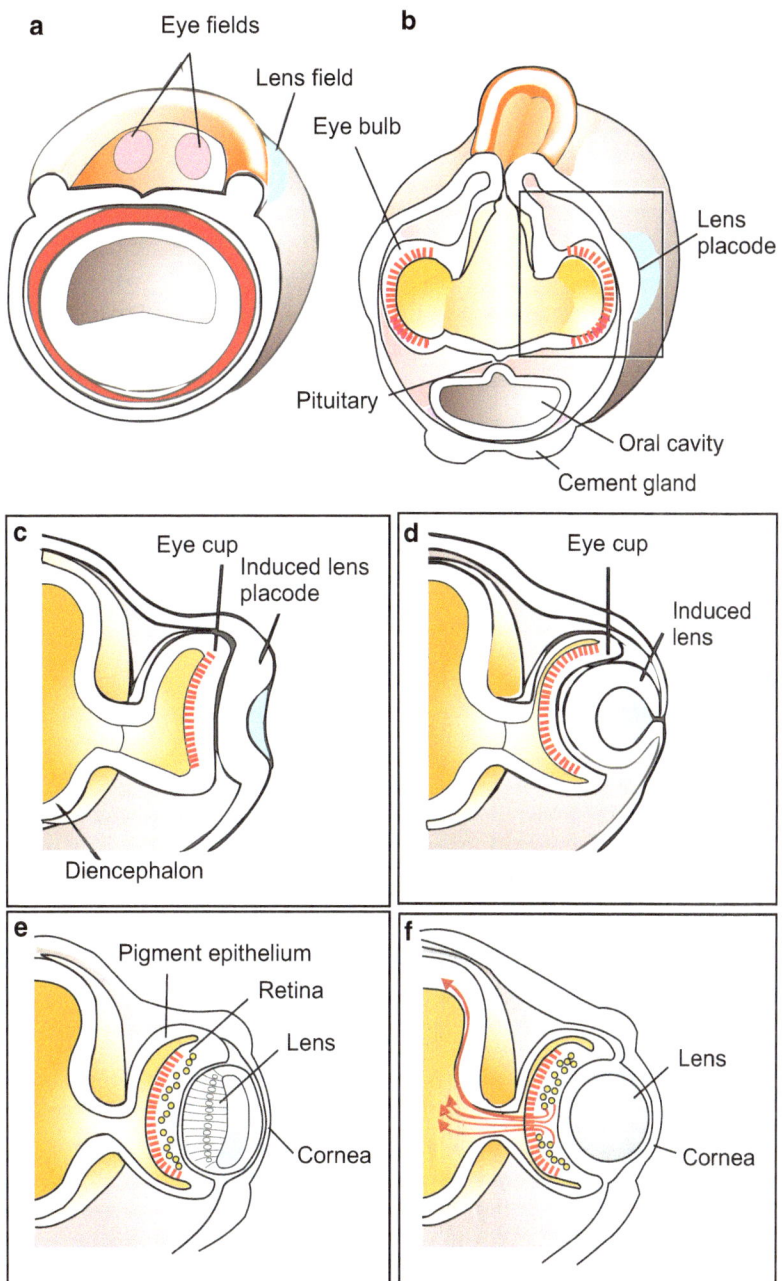

The subsequent development of the eye is shown using amphibians as an example (Fig. 16.23). Lateral **optic vesicles** emerge and balloon up from the posterior part of the prosencephalon, the later diencephalon. The bubble transforms into the double-layered **optic cup** the inner wall of which will become the retina, while the outer wall will become the pigment epithelium. The lens is supplied by the **lens placode**, a circular ectodermal thickening which forms when inducing signals of the eye field and eye vesicle reach the ectoderm. Already the area of the neural plate which later gives rise to the eye vesicle commences to emit inducing signals. The lens detaches from the ectoderm which closes over the lens. The lens in turn induces the formation of the cornea in the overlaying ectoderm.

16.7.2 Retinal Projection, the Wiring of the Eye with the Brain, Is a Great Issue of Developmental Biology

We recall: the retina of the eye is part of the brain containing, besides photoreceptors, several millions of neurons enabling a first data processing. The evaluated primary data have to be transmitted to processing stations of higher order. Therefore the bundled retinal axons leave the eye at the blind spot and grow into the brain guided by their growths cones.

The area searched for is

- the **tectum opticum** in fishes, amphibians, reptiles and birds;
- **in mammals** the axons of the retinal neurons reach only the **lateral geniculate nucleus** (*corpus geniculatum laterale*). This, however, is not the terminal station, rather it is a switching station. Another set of neurons takes over the data and forwards them to the next data processing station known as

 Primary visual area V1 in the neocortex of the occiput (back of the head). From there further processed data are forwarded to the

 Secondary visual areas (V2–V5) in the neocortex of the frontal brain.

The course of the optical tracts (and the experiential changes in the cortical input areas after birth) can be observed by an astounding method: dyes or otherwise labelled low molecular weight substances are simply injected into the eyeball. The labels are taken up by the retinal neurons and transported into the brain. Surprisingly, the labels even cross synapses in the geniculate bodies and finally arrive in the visual cortex.

If one tracks the cable of the **opticus** connecting the eye with the brain (for simplicity do this in fishes, amphibians or birds) we see the cable fraying at its end. The bundled axons detach from each other, diverging over a broad area. The axons terminate at defined neurons in the optic tectum of the midbrain. In embryonic development, axons grow out from the retina and find their way into the brain along pioneer fibres.

The pioneer fibres pilot the axons to the area of destination but not to the particular spot within this area where they will find their target cells. The task the visual system has to cope with is the following: Because of the camera-like optics of the vertebrate eye, each point on the retina receives light from a particular point in the visual field. Each point in the retina sends axons to a particular point on the tectum, so that the surface of the tectum has a one-to-one relationship to the retina and therefore also to the external visual field. The term for this correspondence is **topographic projection** or **retinotectal projection** (Fig. 16.24 and 16.25). In other words: a map on the retina corresponds to a (rotated) copy of this map on the tectum. The fibres connect point-by-point corresponding areas on both maps.

In primates including humans, axons growing to the brain first reach the optic chiasm. Here they must choose a direction in which to continue growing. The fibres coming from the nasal half of the retina must cross to the opposite side and join the contralateral optic tract, and the fibres arriving from the temporal side must continue straight and join the ipsilateral tract (Fig. 16.25). If all of these decisions are correctly made, the axons are allowed to come to rest in the lateral geniculate bodies. From here, other neurons have to find the way to the visual cortex. But the problem remains: how to find the right way and recognize the ultimate target?

If one in addition considers that the mammalian embryo in the maternal womb is hardly accessible to experimental analyses it is clear that model systems of choice are fishes and chickens though for neuropsychological investigations macaques are the most investigated model subjects.

In the visual cortex of humans and other primates, the axonal inputs coming from the nasal part of the left retina alternate with the inputs arriving from the nasal part of the right eye, and vice versa (Fig. 16.25). Thus inputs which will convey information about the same part of the seen world are brought in close juxtaposition. At birth, the various inputs must be fed into networks of neurons to evaluate the information. It is only after birth that most of the neurons which fulfil this task are generated and organized into networks. Initially, adjacent inputs project

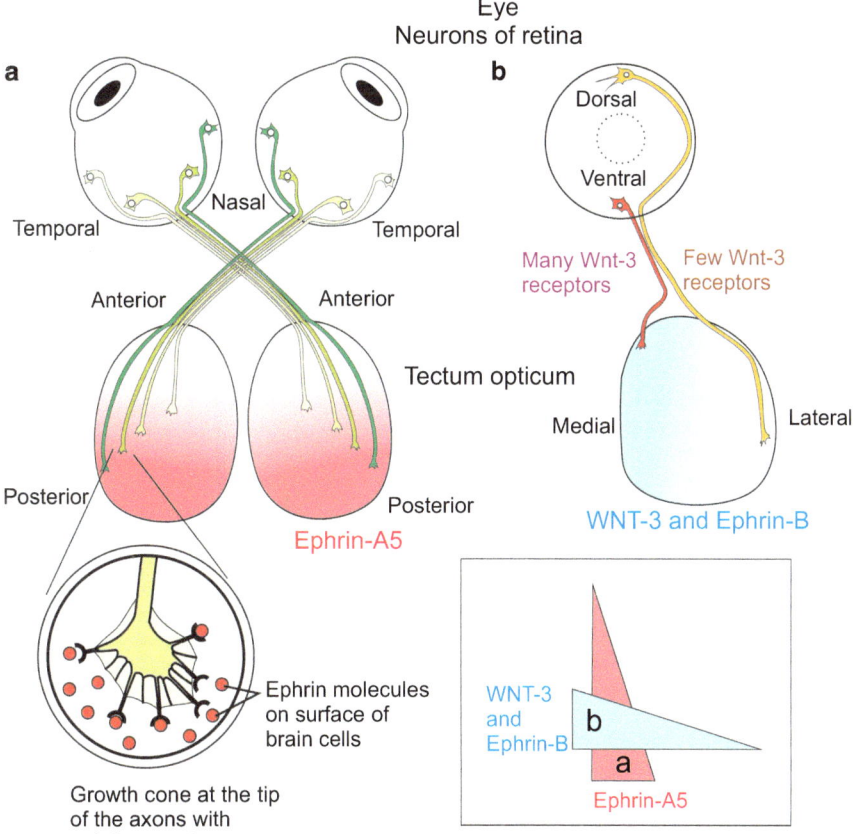

Fig. 16.24 (**a**, **b**) Retinotectal projection in the bird. Axons bundled into the nervus opticus are guided to the *tectum opticum* of the contralateral side. The growth cones find their targets by scanning the surface of the cells in the target area. The surface molecules, in particular the ephrins, form density gradients which allow an approximate orientation. As a rule, ephrins which exists in several variants call a halt to the moving growth cones. Axons of different origin are stopped by different ephrin concentrations. (**a**) Along the antero-posterior axis ephrin-A molecules represent the stopping signal. (**b**) Along the medio-lateral axis ephrin-B and Wnt-3a molecules are the signallers (Schmitt et al. 2006). Growth cones arrived at their target are withdrawn and eventually replaced by synapses with the resident brain neurons

into overlapping cortical areas. During the first few weeks after birth, the areas allocated to the left and right eye become sharpened and largely segregated from each other. **Ocular dominance columns** appear in the cortex that are alternately allocated to the left and right eye.

The diverging interconnectivity of the visual tracts in non-mammals and primates can be associated with the different orientation of the visual fields. In non-mammals (but also in steppe mammals) the eyes are located at the sides of the head, survey large territories but see two almost separated worlds. Information supplied by the left and right eye can be processed in the left and right tectum separately (combining the data enabling a unified view of the world is possible by exchanging information through the corpus callosum). By contrast, in primates the eyes are directed forwards and acquire data from largely congruent visual fields. This position facilitates estimating distances when primates brachiate. Therefore it is appropriate to compare the data supplied by the

(continued)

left and right eyes and to evaluate them together. For this reason the data arriving from the left and right eye are brought together in such a way that the data arriving from the same visual field can be analysed and assessed in the same (or in adjacent) areas of the visual centre (Fig. 16.25).

16.7.3 Signposts and an Ephrin-Based Cartesian Coordinate System Guide the Optic Fibres to Their Targets

On their way to the brain the growth cones of the optic fibres meet many signpost molecules. The literature makes reference to **netrins** and **ephrins**, for example, but also to known morphogens such as **Sonic Hedgehog** and members of the **WNT** family. We focus on the question of how the ultimate target is recognized, and on the model case of the visual system of chick embryo. The primary visual area the retinal axons must reach when they enter the avian brain is the tectum opticum of the contralateral side of the mesencephalon (Fig. 16.24a). On reaching the tectum the retinal axons must find their ultimate destination according to their origin in the retina, so that each part of the retina will be connected with a distinct corresponding part of the tectum.

The **theory of neuronal specificity** (also called the **theory of chemoaffinity** or the **theory of neurotropism**) assumes that the target cells are marked by specific surface molecules. The tectum would contain a map that can be read and interpreted by the growth cones. One also speaks of the **area-code hypothesis**. However, since there are billions of nerve cells, not every cell can have its own unique molecular identity. Instead, combinatorial cues are used to label territories, subterritories and ultimate goals. Such cues could guide the growth cones to their ultimate destination step by step and thus reduce the necessity for long searches. How are such mechanisms disclosed?

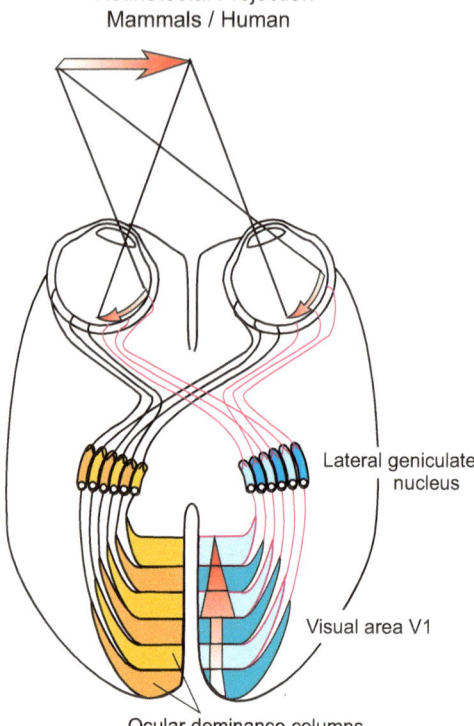

Fig. 16.25 Retinotectal projection in primates including man. The major visual pathways bring together the input from the left and right eye. All the information supplied by the left halves of both eyes is brought together in the left side of the brain, and information supplied by the right halves of the retinas is relayed to the right side of the brain. The pathways (interrupted by synapses in the lateral geniculate nuclei) terminate in the area V1 of the visual cortex, where information arrives in alternating stripes of cortical neurons, called ocular dominance columns. One column is assigned to information coming from the left eye, the next is devoted to information arriving from the corresponding point of the right eye, and so forth. In the end, each spot of the visual field is seen by both eyes, and the two corresponding streams of information arrive in neighbouring collectives of cortical neurons

Experimental Design. For detecting molecules of significance a microsystem has been developed the principle of which was introduced in Sect. 16.5.3 and now is adopted for issues of retinotectal projection (Fig. 16.26). To analyze the behaviour of retinal neurons, retinal precursor cells are placed on chips that have been covered with a laminin substrate and stripes of small 'paving stones'. The paving stones are membrane vesicles prepared from various areas of the optic

Axonal Growth on Coated Chips

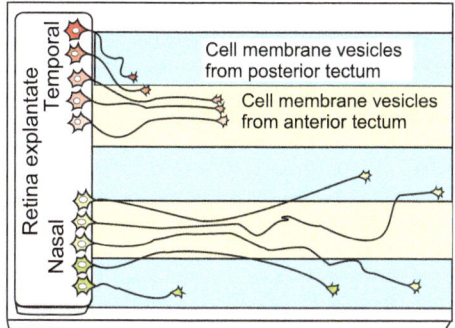

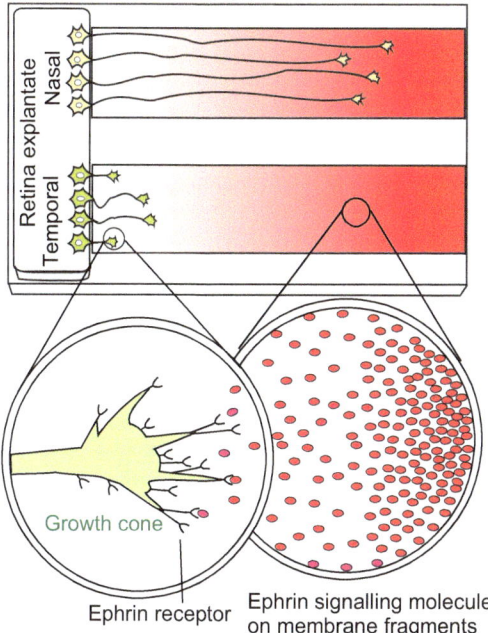

Fig. 16.26 Orientation of the growth cones of outgrowing axons on a substrate consisting of a layer of extracellular matrix (e.g. laminin) and components of the cell membranes of target cells in the brain. A chip is divided into fields and the fields covered with membrane components of different origin. These contain, among others, different amounts of ephrin molecules. The growth cones are stimulated to continue growth by attractive components but distracted by repulsive components, and having arrived at a defined density of inhibitory signalling molecules (e.g. a high ephrin concentration) the growth cones are brought to a stop

tectum (or in advanced experiments vesicles loaded with purified molecules) and; likewise, the neuronal precursor cells are taken from defined parts of the retina and the density of

receptors on their growth cone measured (e.g. by the amount of bound antibodies). It has been shown that membranes of the target cells are attractive while membranes of non-target cells can be repulsive. By assay-guided purification of molecules extracted from the membrane **ephrins** were identified as stopping signals. Eventually the following two-gradient theory was proposed.

Theoretical considerations and the just mentioned pioneering experiments have led to the notion of a relatively simple **Cartesian coordinate** system that is established by two static concentration gradients perpendicular to each other. Such a Cartesian coordinate system would allow the growth cones to recognize their current position and their target by measuring the local concentration system. However, each individual growth cone must possess prior knowledge of what combination is appropriate for it. At the target location movement must come to a standstill. To a large extent this bold theory has been proved to be essentially correct:

- **An antero-posterior gradient**. Ephrins are either members of the ephrin-A or the ephrin-B class. First we have a look at **ephrin-A5**. This variant is exposed on the surface of the tectum neuroblasts in the form of an antero-posterior gradient with highest concentrations on the posterior part of the tectum (Fig. 16.24a). The growth cone checks in which direction the density on the surface increases and moves ahead until it comes to rest at its destination. The stop signal is provided by a distinctly high ephrin A5 density. The concentration is measured by the growth cone in terms of the number of **ephA5-receptors** bound with ligand. If all receptors are occupied the movement comes to a standstill. The ingenuity of the system resides in the fact that the axons are endowed with varying numbers of receptors as a function of their point of origin in the retina. Axons arriving from a temporal position of the eye possess few receptors; these are soon occupied and the growth cone does not go very far, already in the anterior tectum a halt is called to its urge forward. By contrast, axons arriving from the nasal part of the eye are equipped with many receptors and these are not all occupied until the growth cone has advanced to the posterior area of the tectum where the ephrin-A concentration is highest. If all receptors are occupied and the growth cone has come to a dead stop synapses are established.

- **Medio-lateral gradients**. In the retina a second gradient of **ephrin-B receptors** proceeds in the layer of neurons perpendicular to the gradient of ephrin-A receptors. The axons sprouting from the retinal neurons are simultaneously equipped with **ephrin-A** and **ephrin-B** receptors as well. Axons coming from the ventral area of the retina possess more ephrin-B receptors than those coming from the dorsal retina. The ventro-dorsal gradient of ephrin-B receptors in the bundle of axons corresponds to a ventral to dorsal oriented gradient of ephrin-B signalling molecules on the surface of the tectum (Fig. 16.24b). This second gradient provides the second coordinate for localization. From the local values of the first and the second gradient the growth cone can determine its actual position – theoretically. The system for guidance, however, turned out to be more complex since parallel to the ephrin-B gradient a WNT-3 gradient runs (Fig. 16.24b). The neuroblasts of the tectum expose besides ephrin-B also the signalling molecule WNT-3 in varying concentrations, and the crawling growth cones are correspondingly equipped with varying numbers of WNT-3 receptors (i.e. receptors of the Frizzled class).

Investigations in this area of research are hampered by the fact that in the tectum several variants (isoforms) of ephrin-A and ephrin-B signalling molecules exist with overlapping expression patterns. A detailed map that would contain the distribution pattern of all ephrin types and of further signalling molecules cannot be shown at present. Besides, the responses of the axons can be ambiguous: a low concentration of ephrin molecules can be attractive, a higher concentration means "stop!" or even be repulsive. A further complication is due to the frustrating circumstance that axons can form branches which are guided to different neighbouring areas. With millions of axons and billions of possible targets it is a difficult challenge to exactly describe the targets for all axons and to determine the significance of the local molecular makeup experimentally.

The general notion is currently the following: A specific, specified combination of receptors on the growth cone specified by the source in the retina must match with a defined combination of surface molecules on the tectum. If optimal congruency is found the growth cone stops moving, is withdrawn and is replaced by a synapse.

Much of the required molecular diversity can be coded – in addition to the amino acid sequence of proteins – in the combinations of carbohydrate units of glycoproteins and glycolipids (sialoglycolipids, gangliosides). There is evidence that nature indeed uses the intricacy and diversity of such surface molecules to label paths and targets. However, positional information coded on the surface of the cells is probably not very precise. Subsequent corrections are possible (Sect. 16.8).

16.7.4 Interrupted Neurons Can Find Their Targets Again

An experiment conducted with frogs explored the issue of neuron connections and how the brain interprets action potentials arriving from the eyes. The optic nerve was cut, the eyeball rotated by an angle of, say, 180° and re-implanted. Since neurons which have only lost the distal part of their axons can regenerate, the retinal axons grow back to the optic tectum. However, they make the connections appropriate to their original, not to their new position. Thus the retinal neurons reconnect with that part of the tectum to which they had been linked before the eye was rotated. Since

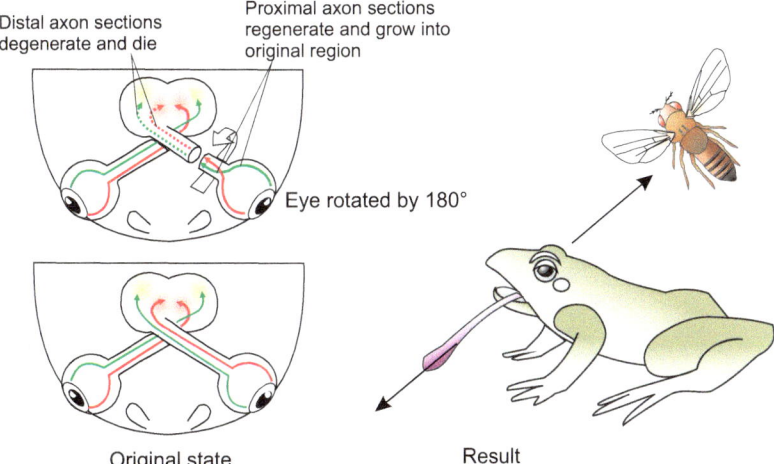

Distal axon sections
degenerate and die

Proximal axon sections
regenerate and grow into
original region

Eye rotated by 180°

Original state

Result

Fig. 16.27 Experimentally rotated eyes in the frog. Eyes are removed, rotated in their sockets and reinserted. The optic nerves regenerate back into the brain and re-establish connection to the optic tectum. If the eyes are rotated by 180°, the frog behaves as though it sees the world upside down. The frog projects its tongue downward even though the fly is above its head (after Sperry)

(a) the original connections are re-established,

(b) the brain principally evaluates visual information according to its origin in the retina, and

(c) the eye has been rotated without 'instructing' the brain of the altered situation,

the frog "sees" all things upside down. To demonstrate this result, the frog sees a fly flying above it as being instead in a position below it, and snaps downward (Fig. 16.27).

In humans a much simpler experiment can be performed: wear glasses made of prisms that show the world upside down. The high capacity of the human brain for learning by experience enables us to mentally 'rotate' the seen world back into the original position, but learning to do so takes weeks and most experimental subjects resign before having attained success.

16.7.5 Olfactory Sensory Cells Have Odorant Receptors on Their Projecting Axons. For What Purpose?

The sensory epithelium in the nose of mammals is occupied with millions of smell cells which belong to 500–1,000 different classes. Each class is specialized to receive a distinct class of odorants. The molecular receptors for picking up odorants are anchored in the membrane of the olfactory cilium. They all belong to the type of G protein-coupled seven pass transmembrane receptors (see Fig. 11.5). The receptors for different odorants are similar (paralogous homology) but not identical, and about 1,000 genes are needed to produce all of them (Nobel award 2004).

Smell sensory cells are primary sensory cells each possessing its own axon that has to find its way into the rhinencephalon, in medical terms known as *bulbus olfactorius* (olfactory bulb). All axons of a distinct class of odorant receptors terminate in the same data processing station of the bulbus (Fig. 16.28). Such a processing station is called a glomerulus (plural glomeruli; not to be confounded with the glomeruli of the kidney!). The task of specific projection arises not only in the embryonic development because smell sensory cells are short-lived. After only about two weeks a smell cell must be replaced by a new one supplied by stem cells. Also the axon of this replacement cell has to find the correct glomerulus.

Olfactory epithelium

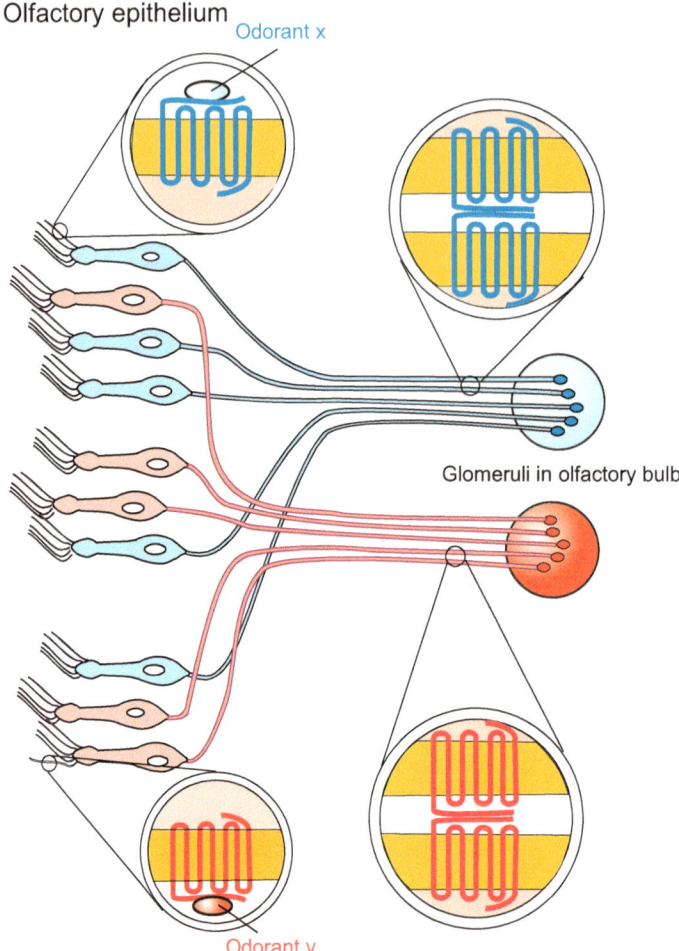

Glomeruli in olfactory bulb

Fig. 16.28 Sensory smell cells in the nasal epithelium searching their target neurons with their own growing axons. Each type of sensory cells – there are in mammals 350 to 1,000 – expresses an individual molecular receptor which is inserted into the membrane of the cilia and able to bind a distinct class of odorants. The receptor belongs to the class of seven-pass transmembrane helices. Two of 1,000 (*red* and *blue*) are chosen as examples. All sensory cells of the same type project their axons to the same processing instance (glomerulus) of the rhinencephalon. Surprisingly, the axons of the sensory cells contain odorant receptors as well, and these are the same as those exposed along the cilia. According to an experimentally founded hypothesis the molecular odorant receptors are also cell adhesion molecules (CAM). By homotypic congregation of the CAMs of neighbouring axons they serve the bringing together and coherence of matching axons. Compare colour plate Fig. 12.11(2) plus legend

Much to the surprise of all researchers involved it turned out that the molecular odorant receptors are not only installed into the membrane of the sensory cilia but also into the membrane of the axon. What is their function there? It would be hardly to believe that the target cells in the brain emit 500 to 1,000 different odorants to attract the axons to their targets. An experimentally founded hypothesis postulates that the molecular receptors acted as homotypic cell adhesion molecules enabling the axons of the same class of sensory cells to join others of their own kind (Fig. 16.28). This exactly means homotypic interaction: Two adjacent molecules of the same sort join forming dimers, and neighbouring axons coalesce. The hypothesis also proposes that in postembryonic life a distant target must not be found by self navigation.

Rather a replacement cell that expresses a distinct odorant receptor would attach to bundles of coalescent axons presenting the identical odorant receptors on their surface, and grow along those already existing axons. "*In this model, axons do not look for targets – they are the targets*" (Feinstein and Mombaerts 2004).

In the mentioned experiments transgenic mice were used which produced a distinct odorant receptor linked with a reporter protein. Useful reporter proteins are the green fluorescing GFP or the bacterial Lac-Z (see Chap. 12, Box 12.2) the result of such a labelling can be seen in the colour plate, Fig. 12.11 (part 2).

16.8   Plasticity, Corrections, Regeneration, Reserves

16.8.1   Innervation of the Musculature: Even Displaced Motor Axons Can Find the Target They Normally Have to Reach

Muscles are innervated by motor neurons of the spinal cord. The motor neurons send out axons which have to find the way to, for example, a muscle in an extremity. Following a successful pioneering axon further axons will find the way attached to it. In the extremity the fibres of the cable have to separate from each other because each individual growth cone is dedicated to a different bundle of muscles fibres. Each freed axon commences branching to innervate about 1,000 microscopic muscle fibres.

A classic experiment from the pioneering era of neurobiology shows that motor nerves can find their target tissue even when the search is made more difficult. Before axons grow out and leave the spinal cord, a piece of the neural tube is excised and re-inserted back but rotated 180°. Despite this inversion the growth cones arrived at their predestined target (Fig. 16.29). If, however, the pieces of the neural tube are transplanted to a location more distant from their home position the growth cones wander about and make connections to the wrong targets. If muscle fibres

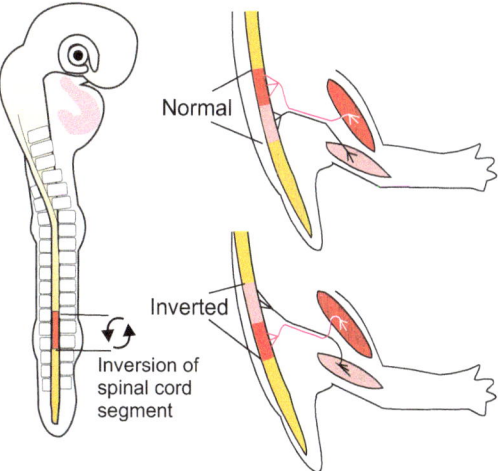

Chick: Innervation of Leg Musculature

Normal

Inverted

Inversion of spinal cord segment

Fig. 16.29 Innervation of a leg by motor fibres (axons) outgrowing from the spinal cord. If the segment responsible for the innervation of the leg musculature is rotated by 180° the fibres nevertheless find the target muscles they normally have to innervate

emit attractive signals these are either not highly specific or are not long-range signals.

16.8.2 Extra and Incorrect Neuronal Connections Are Eliminated

The phenomenon of subsequent corrections can be studied in the innervation of the skeletal muscles by the motor neurons of the spinal cord. It is common during development for too many axons to arrive at a muscle fibre and form synapses (Fig. 16.30). In the embryo, the central nervous system sends action potentials to test the muscles. Motor axons and axonal branches survive only if they are correctly connected and, therefore, frequently and regularly used to transmit action potentials.

- In the struggle for survival competition between synapses might be significant: circumstantial evidence suggests that transmitters released by synapses suppress unused synapses nearby.
  - **Stabilising and upgrading synapses**. The presynapse arising from the growth cone

Fig. 16.30 Correction of synaptic connections. Myoblasts fuse to form multi-nucleated, cross-striated muscle fibres. Initially, the fibres are innervated by several motor neurons. Incorrect connections are eliminated by apoptosis. Only those synapses survive that are correctly connected and, therefore, frequently used. These receive from their target muscle fibres survival factors. The initially ubiquitously distributed receptors for the transmitter acetylcholine accumulate below such frequently used synapses. These become neuromuscular junctions

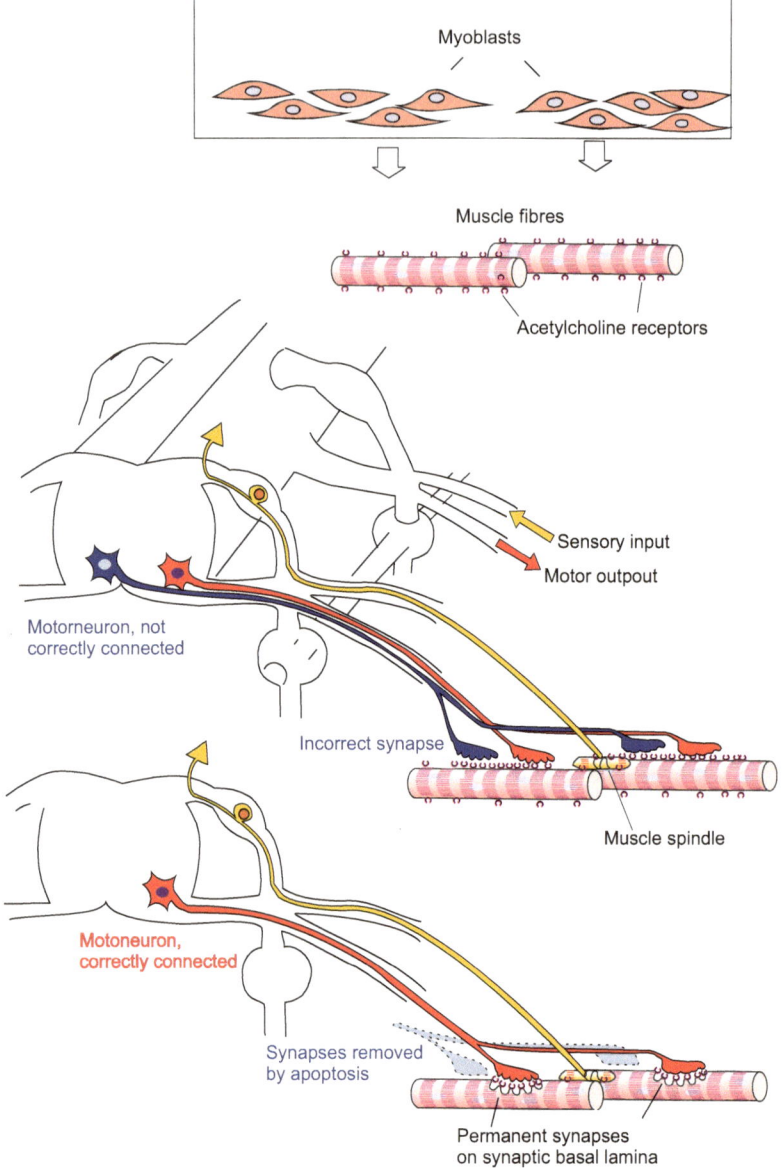

stimulates the muscle fibre to increase the production of receptors for the transmitter acetylcholine and to place them beneath the presynapse. This stimulation is thought to be mediated by a factor called BDNF (Brain-Derived Neurotrophic Factor) that is released from electrically active axon terminals and picked up by the postsynaptic membrane by means of specific receptors.

- On the other hand, muscle fibres appear to secrete **survival factors** at the synapse. By analogy to the "trophic" function of NGF (Sect. 16.4.3) and the just mentioned BNDF, survival factors secreted by active muscle fibres might be taken up by the presynaptic membrane of the motor neuron and delivered to the cell body by retrograde transport. Only correctly connected neurons survive.

Neurons which fail to establish and maintain contacts are eliminated by programmed cell death (apoptosis, see Fig. 14.9).

16.8.3 In Favourable Circumstances Regeneration of Nerve Fibres Permits Rehabilitation

As a rule, dead nerve cells and primary sensory cells (that is sensory cells with their own axon) cannot be replaced from stem cells (except the olfactory sensory cells of the nose). But nerve cells whose perikaryon (central cell body with the nucleus) is left intact can regrow dendrites and axons. (Of course, the inverse is not possible. A cut off distal axon cannot regenerate the lost mother cell with the cell nucleus).

Also in the innervation of the musculature regeneration is possible because the perikarya of the motor neurons are not resident in the musculature but in the spinal cord. Likewise the nuclei of the sensory fibres which feed information from the sensory organs of the skin and the muscle spindles (sensory organs measuring the actual length of the muscles) do not reside in the muscle itself but in the dorsal root ganglia (Fig. 16.30).

Now imagine the situation: the forearm is bruised by an accident and the nerve fibres interrupted. The distal axons of the motor neurons and of the sensory spinal ganglion cells die and the forearm and hand are paralysed for a long time. But the proximal parts of the axon can form a growth cone again. Over weeks and months axons can grow out from the intact proximal part of the arm and re-innervate the muscles of the forearm and hand and, bring back motility. Likewise, regenerating sensory fibres can bring back sensory functions. Neurosurgeons try to implant guiding structures for the growth cones. Sometimes the efforts are successful, sensing and motility return. Unfortunately, frequently barriers are formed at the wounded locations that cannot be overcome by the axons.

16.8.4 Subsequent Correction of Imprecise Connections Is a General Principle in the Self-organizing Nervous System, Even in the Brain

Growth cones do not always precisely find their target areas. Or **positional information** coded on the surface of cells is ambiguous. This is not necessarily disastrous because subsequent corrections are possible. Also in retinotectal projection numerous, incorrectly connected neuronal fibres are withdrawn or neurons die. They subject themselves to programmed cell death. Glial cells or macrophage-like microglia eliminate the residuals. The detail of how it is checked what is correctly, what incorrectly connected is still mystery of the brain.

16.8.5 Even After Birth New Links Are Established and the Pattern of Connection Is Moulded by Experience

No bird, mammal or human is born with a brain whose connections are completely finished in every detail and fixed permanently. In young animals the retinotectal projections are initially quite fuzzy and imprecise. In amphibians, fishes and birds retinal axons first branch widely and form synapses even in territories destined for neighbouring axons. The territories innervated by the axons overlap initially. Later, the territories are trimmed back by elimination of superfluous and wrong connections.

In the cortical layer of the mammalian brain most synaptic connections are established only after birth under the influence of visual input and tactile experience. Basically, the main tracts are established. When a kitten is born, axons from the retina have reached the geniculate bodies and axons from the geniculate neurons have reached the visual cortex. But the cortical architecture itself is still under construction and completed only under the influence of visual input.

At birth, the various inputs must be fed into networks of neurons to evaluate the information. It is only after birth that most of the neurons which fulfil this task are generated and organized into networks. Initially, adjacent inputs project into overlapping cortical areas. As stated above during the first few weeks after birth, the areas allocated to the left and right eye become sharpened and largely segregated from each other. **Ocular dominance columns** appear in the cortex that are alternately allocated to the left and right eye.

Dominance columns are areas of data processing and data evaluation. They are fitted with neurons which filter out distinct features of the visual world from the data supplied by the eye, for example left-right direction and angle contours. If in the course of postnatal imprinting predominantly vertical stripes are presented to the eyes of kittens the number of neurons devoted to the task of recognizing vertical lines is increased. Among the most intriguing experiments ever done in developmental neurobiology were those carried out by Hubel and Wiesel (Nobelprize 1981). When the eyelids of the right eye of newborn kittens were sewn shut and left closed for 3 months, the right eye of the kittens became functionally blind. The corresponding ocular dominance columns were reduced for the benefit of those allocated to the open left eye.

Visual experience leads to enduring changes in the visual cortex; this enables fine tuning and long-term stabilization of synapse formation in adaptation to environmental influences.

A popular current hypothesis assigns **imprinting** in birds to similar mechanisms. During the short postnatal periods when key optic or acoustic stimuli are learned for long time periods and even for life, synapses are thought to be newly established or consolidated while others are eliminated.

Nervous connections are not directly encoded by the genome, nor are all connections the result of autonomous and independent processes of self organization. **Experience can lead to the strengthening and consolidation of active connections** and to the elimination of silent synapses.

16.8.6 Long-Term Memory May Be Based on Continuing Neuronal Differentiation

In birds and mammals existing neurons send out and withdraw neurites for life; regular and frequent use strengthens and consolidates existing synapses. This permanent refinement, retuning and remodelling of synaptic connections are thought to be related to associative learning (including imprinting) and to long-term memory. In recent years novel findings have corroborated this view. Even when corporal growth has come to a standstill in some small areas of the brain new nerve cells can be generated from permanent stem cells.

16.8.7 There Are Cellular Reserves to Exploit in Favour of the Brain: Adult Stem Cells

For a long time the dogma was that beginning at the age of 20 in the brain of humans hundreds of neurons die each day, and could not be replaced. This dogma, that the adult brain would be unable to supply substitute nerve cells was some decades ago declared invalid for male song birds. A male canary bird for instance acquires its song stock anew each year, he will take this or that verse out of his program, add other verses. In the prosencephalon there are two centres (hyperstriatum ventrale pars caudale and robustus archistriatalis) which become active in acquiring and practising the new program of the year. During and before the productive phase **neuronal stem cells** in the wall of the ventricles (**subventricular zone** SVZ) supply new nerve cells. In the autumn, on the other hand, when the concert campaign comes to its end many nerve cells perish in these regions.

Currently the traditional doctrine no longer holds in mammals either. At least rodents are no longer bound to the old dogma. More and more publications report that in certain areas of the brain stem cells procure supplies for used up neurons and increase the capacity of the area concerned in case of demand.

- In mice the **subventricular zone of the telencephalon supplies replenishment for used up nerve cells in the bulbus olfactorius**. The descendants of these stem cells migrate in an endless caravan to their job site.
- At the base of the **hippocampus** of adult mammals stem cells ready to divide are found. The hippocampus is assigned a

decisive role in the transfer of experiences into the long-term memory. In mice kept in an eventful environment and stimulated to utilize and practise their learning ability new nerve cells are formed from this reservoir. Surprisingly, these stem cells look like differentiated glial cells of the astrocyte-type

Beyond that, in experiments with rodents stem cells derived from the bone marrow which normally give rise to blood cells can enter the brain via the blood circulation und give rise to neurons. This issue is addressed in Chap. 18.

Box 16.1 Genetic Networks in the Development of the Nervous System

On the anatomical and cellular level the nervous system is unimaginably complex. This holds in a similar way to the gene products used in the course of its development and their functions. Initially the development of the nervous system commences with a – seemingly – readily comprehensible event: By removing a barrier the way is opened for the covert disposition of ectodermal cells to differentiate autonomously to nerve cells. Yet, the great number of genes that commence activity is bewildering not only in the eyes of beginners.

Thus it causes bewilderment that the name of a gene often does not say anything about their function. For example, what do the designations such as *otx, sox3* or *AS-C* mean? It is also confusing that homologous genes in different organisms often bear different names. This applies in particular to genes in vertebrates as compared with corresponding genes in the genetic pioneering model *Drosophila;* for frequently the homology was only recognized after a mutant phenotype in *Drosophila*, and with it the putative gene affected by the mutation, had already got its name (which usually was chosen according to the local jargon of the laboratory), whereas independently a different phenotype in vertebrates had given reasons for choosing a different designation. Only sequencing of the genes months or years later disclosed high sequence similarity and thus homology.

Confusion is also caused by the fact that genes frequently are members of gene families whereby the different members of the family sometimes have similar, sometimes different functions. For instance, the human genome contains 22 genes for different proteins of the FGF family (27 in zebrafish) and 19 members of the WNT family which in the course of evolution partly fulfil redundant, partly different functions at different positions of the embryo, and cooperate with different partners.

In the following we try to give a function-based overview over the multitude of genes relevant for the initial development of the nervous system and to sort the proteins encoded by them to facilitate enquiry in data bases and understanding of the action spectra.

1. **Frequently mentioned diffusible, extracellular signalling molecules that act across large areas (inducers, morphogens):**
   - **BMP-4** prevents in the early embryo (blastula to neurula) neurogenesis at false places, seven more BMP isoforms are known encoded by the human genome.
   - **Chordin, Noggin**, Follistatin enable the formation of the CNS by neutralizing BMP-4.
   - Members of the **WNT family** act by modifying along the antero-posterior body axis of the CNS and participate ▶

Box 16.1 (continued)

in the formation of the spinal cord (together with and retinoic acid); in addition, WNTs participate in the programming of many, if not all, types of nerve cells and glial cells. In the human genome 19 *Wnt* genes have been identified.

- Members of the **FGF family** partially initiate and facilitate the development of the CNS, partially modify a program according to the position in the embryo. Humans possess 22 different *Fgf* genes.
- **Sonic hedgehog SHH** plus BMP fulfil important functions when location-specific genes have to be switched on or off along the dorso-ventral axis of the spinal cord. Similar functions are assumed by this pair of morphogens in the specification of endodermal organs.
- WNT, BMP, FGF and SHH are almost always involved when it comes to regionalise and subdivide differentiation potentials in sensory placodes (and in other organs as well).

2. **Genes for signal exchange between direct neighbours:**
   - The NOTCH/DELTA system (downstream genes are, among others, *HES, lunatic fringe*) is repeatedly used to forward direct neighbours onto different differentiation paths. The system effects the segregation of future neuroblasts from epidermis cells (Sect. 16.3.1) and is involved in the decision neuron versus glial cell.

3. **Gene coding for pathfinding and signpost molecules.** Factors which are released by emitter cells and often bound to the ECM such as the EPHRINS, FGFs, BMPs, SHH, WNTs or NETRIN and SEMAPHORIN as well as their receptors are found in many and various positions in

the developing nervous system. They have similar functions outside the nervous system in guiding, for instance, myoblasts or growing blood vessels (frequently nerve pathways and blood pathways are running in parallel) and direct outgrowth and differentiation also of endodermal organs such as lungs, pancreas, or liver.

4. **Genes for transcription factors (TF), which switch on (or off) neural genes in the nuclei of undifferentiated embryonic cells.** Two families of TF dominate:
   (a) **Transcription factors with a bHLH domain** (basic-Helix-Loop-Helix domain*)* **as DNA-binding structure** (see Fig. 12.14):
       - **Atonal group: Sox3, myogenin/ MyoD, atonal,** ASH, NEX, Mash1, **HES** (= Hairy and Enhancer-of-Split homologue),
       - **AS-C**-(Achaete-Scute-Complex) **group** with AS-C and ASH (human).
   (b) **Transcription factors of the Hox/paraHox class with a homeodomain as DNA-binding structure** (see Figs. 12.14 and 16.13): **Otx** (in the anterior head region and brain)**,** Emx, **PAX6** (of particular importance in the development of eyes and of further sensory organs in the head region).

5. **Surviving factors** which ensure survival of nerve cells and stabilise synapses. Factors frequently involved are **NGF** (Nerve Growth Factor*,* see Sect. 16.5.3) and **BDNF** (Brain-Derived Neurotrophic Factor).

6. **Specific human genes** required for the growth especially of the human neocortex such as micocephalin are delineated in Chap. 22 (Evo-Devo).

Summary

The development of the **vertebrate nervous systems** is comprehensively considered. In the amphibian embryo which serves as a model for the early steps of neurogenesis, maternal mRNA such as *sox3* is allocated to the cells in the surroundings of the animal pole and this maternal information confers on the animal cells a basal tendency (competence) to become nerve cells. Initially this tendency is suppressed by signals radiating from the vegetal hemisphere, in particular by BMP-4. **Inducing factors** with neuralizing power emitted by the **Spemann organizer** such as NOGGIN and CHORDIN neutralize BMP-4, so that the basic neurogenic tendency can take effect. Thereupon in the cells of the animal hemisphere genes for **proneural (neurogenic) transcription factors** such as *neurogenin, NeuroD* and many other genes (e.g. of the *Achaete scute* class AS-C are switched on. Gene expression studies disclose **surprising homologies** between the vertebrate and the insect nervous system.

In terms of morphology the nervous system of vertebrates originates from two areas of the neurula:

1. **The central nervous system CNS**. Above the inducing axial mesoderm (later forming notochord and somites) the animal cap of the balloon-shaped gastrula-stage embryo forms the **neural plate** and this forms the **neural tube** which gives rise to the **brain and spinal cord**. The brain on its part subdivides first into three parts (**Pros-, Mes-, Rhomb-encephalon**); the last section being subdivided into repeated units called rhombomeres. By subdivision of the prosencephalon into **tel- and di-encephalon** and by division of the rhombencephalon into **cerebellum and medulla oblongata** five brain sections arise.
   - **Primary neuroblasts** emerge in the vertebrate neural plate, and also in the CNS of the insects, in the form of **three stripes each on both sides of the midline** and are segregated out by means of the Notch/Delta system. The formation of the stripes in vertebrates, and in insects as well, takes place under the influence of homologous neuroblast-specific genes. The outer stripes give rise to sensory neurons of the dorsal root ganglia (spinal ganglia), the inner stripes give rise to motor neurons, the medium stripes give rise to interneurons.
   - **Secundary neurons** are produced in large numbers by **stem cells of the neuroepithelium (ependyma)** that lines the walls of the lumen of brain vesicles and the spinal cord canal. Symmetric division brings about more stem cells, asymmetric division brings about daughter cells half of which remain stem cells whereas the other half gives rise to neuroblasts. In the neocortex of the primates these neuroblasts migrate along glial fibres into the periphery and arrange themselves in six surface parallel layers proceeding from inside to outside.
2. **The peripheral nervous system PNS** arises from **neural crest cells** that initially surround the neural plate, then emigrate and colonize all parts of the body. Place of origin, wandering route and destination determine their fate. The PNS consists of the **spinal ganglia** (dorsal root ganglia), the **autonomic or vegetative nervous system** including the **sympathetic and parasympathetic system,** and the **enteric nerve system ENS** of the gastro-intestinal tract.

In the course of their terminal differentiation nerve cells sprout input and output processes: **dendrites** for receiving information, **axons** for forwarding information. At the tip of these processes **sensory and motile growth cones** fulfil pathfinding functions. The growth cone advances by means of lamellipodia and filopodia which are equipped with sensory receptors. It orients itself sensing attractive or repulsive signals of the environment. These signals can be:
- diffusible molecules emitted by target cells and acting as chemo-attractants such as

NGF (Nerve Growth Factor) and other **neurotrophins** which especially act upon the cells of the peripheral nervous system and serve, in addition, as **survival factors**.

- Further signals can be components of the **extracellular matrix ECM**, especially those which have bound specific signalling molecules such as FGFs, or netrins.
- **Surface molecules** on particular **signpost cells** and the ultimate target cells.

The target is recognized on the basis of a **molecular code** presented by the target cells. This is demonstrated using as an example the **retinotectal projection**, the interconnection of the retinal neurons of the eyes with their target areas in the brain. In the visual centres of the brain, especially in the well-studied **tectum opticum** of non-mammalian vertebrates, gradients of **ephrin-A** and **ephrin-B** surface proteins arranged perpendicular to each other, provide a **Cartesian coordinate system** for locating which is scanned by the growth cones of the optic fibres by means of **ephrin-receptors**, a type of receptor-tyrosine kinases. At points of **optimal matching** of signalling molecules on the target area and of receptors on the growth cone, the cone comes to a standstill and a **synapse** is formed.

Olfactory sensory cells insert their molecular **odorant receptors** not only into the membrane of their sensory cilium but also into the membrane of their axons where they serve as **cell adhesion molecules**. This enables bundling of the axons of sensory cells that belong to the same odorant class and enables their joint guidance to the corresponding data processing station (glomerulus) in the bulbus olfactorius (rhinencephalon).

Examples of pathfinding in the spinal cord are the **commissural axons** and the axons of the **motor neurons** which leave the spinal cord in search for the musculature they have to innervate.

Frequently, too many contacts including false contacts are established initially. **Subsequent corrections** are carried out under the influence of testing electrical impulses (e.g. action potentials sent out by the CNS). Conversely **survival factors** such as BNDF released by the target cells stabilize correct synapses. Superfluous synapses are eliminated through apoptosis of the incorrectly connected neurons.

Also after hatching or birth further synapses are established while others are withdrawn. This **fine tuning takes place under the influence of external sensory information** and can be the structural basis of postnatal **imprinting** and **long-term memory**. In certain areas of the brain such as the **hippocampus** not only are new synapses formed, but residual **stem cells** provide replenishment for used up neurons and source for novel connections. This is considered an important basis for long-term memory and adjustments of brain functions to external influences, perhaps even in advanced age of an individual.

Heart and Blood Vessels

17

17.1 From Seemingly Chaotic Origins to High Order

17.1.1 Heart and Vessels Arise in Many Separate Locations But Unite to Form One System in the End

The heart is the first organ to form and the cardiovascular system is the first functional system in the developing embryo. Both the heart and the blood vessels develop from the mesoderm at the same time so that a complete circulatory loop is ready for the start of cardiac contraction. In humans this is at about the third week of gestation. Despite divergent origins in different vertebrates, in the medium (phylotypic) and late stage of development the vascular systems faithfully repeat evolutionarily ancient structures in a highly conservative manner. Similar to the nervous system mechanisms of self organization create highly ordered structures in a reproducible manner. The anatomical construction of the circulatory system in the model animal fish and in mammals/humans is outlined in Figs. 17.1, 17.2, 17.3, and 17.4.

To avoid misunderstandings we recall basic biological knowledge, especially as the traditional terminology can be misleading. By definition arteries are vessels leading from the heart; they are usually shown in red, while veins are vessels leading to the heart, they are usually shown in blue. In terms of physiology the arterial system is the high pressure system, the venous system is the low pressure system. This is reflected in the different structure of the walls of arteries and veins (Fig. 17.4). In connection with blood the adjective "arterial" means oxygenated: arterial blood is loaded with oxygen O_2, while the adjective "venous" means deoxygenated: venous blood is unloaded from oxygen but loaded with carbon dioxide CO_2. As a rule, arteries transport arterial blood, veins venous blood, but this is merely a rule and there are important exceptions. For instance, in the mammalian embryo only the *vein* coming from the placenta carries oxygenated "*arterial*" blood, and in the adult system the lung *arteries* carry deoxygenated, "*venous*" blood, while the lung *veins* transport oxygenated, "*arterial*" blood. In the context of this chapter the colours red and blue in the Figures can have a different meaning; therefore reading the legends is important.

17.1.2 Blood Vessels and Heart Are Formed by Migratory Precursor Cells

Both the branching blood vessels and the massive heart are derived from common cardiovascular precursor cells which arise from *Brachyury*-expressing mesoderm close to the midline on both sides of the embryo. These then differentiate into either cardiac progenitors or haemangioblasts (precursors for blood vessels and blood cells). The cardiac progenitor cells

W.A. Mueller et al., *Development and Reproduction in Humans and Animal Model Species*,
DOI 10.1007/978-3-662-43784-1_17, © Springer-Verlag Berlin Heidelberg 2015

Fig. 17.1 Circulatory system in adult fishes and mammals/humans. Vessels transporting oxygenated blood are in *red*, vessels carrying deoxygenated blood are in *blue*. (In lung fishes the lung functions as an accessory respiratory organ and is supplied by blood from a side branch of the last branchial arch). From Mueller and Frings (2009)

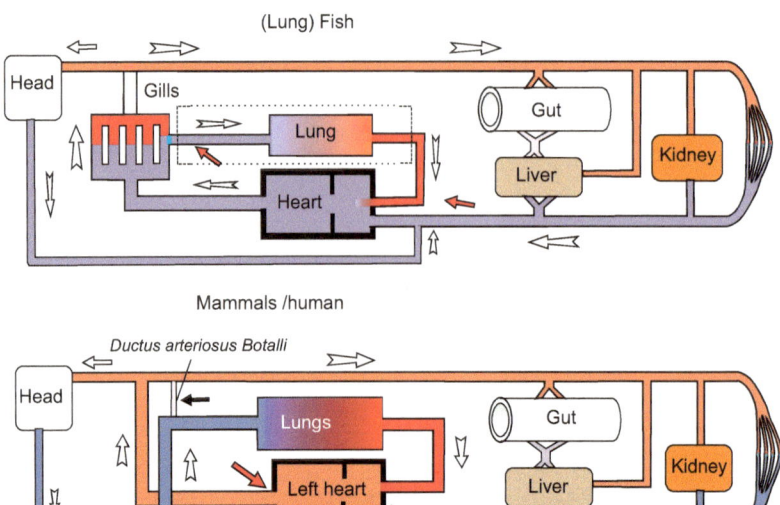

migrate anteriorly whereas the haemangioblasts migrate posteriorly (caudally). These wandering mesodermal cells have the progeny as shown in Fig. 17.5. The stem cells of the various blood cells including the cells of the immune system will be dealt with in Chap. 18 (stem cells, see Fig. 18.6).

17.2 The Heart

17.2.1 Aristotle's Jumping Point, the Heart, Arises From migratory Cardiac Progenitor Cells

Aristotle, when opening incubated eggs, observed a 'jumping point' in the blastodisc. Even without magnifying glasses he identified the jumping point as a beating heart.

The fish and avian embryos lend themselves well to studies of heart development. The events described below occur in a similar way in other vertebrates as well (Figs. 17.6 and 17.7A,B). The circulatory system is established early in development during neurulation when diffusion is no longer sufficient to supply the embryo with nutrients and oxygen.

The cardiac progenitor cells migrate anteriorly on both sides of the midline and fuse to form the primary or first heart field, also

known as the cardiac crescent. Cells from this first heart field involute to form the linear heart tube. This tube first elongates, then moves to the left side of the midline and loops to the right to form two chambers, the atrium and the ventricle. A second wave of cells derived from the same precursor population differentiates and migrates later; this is called the second heart field. Cells from the first heart field contribute mainly to the left ventricle, whereas those from the second heart field give rise to the right ventricle, the atria and the outflow tract. The embryonic heart comprises three layers (Fig. 17.6):

- **Endocardium**. The origin of the endocardium is controversial, but it is now thought that both the first heart field and endothelial lineages can give rise to endocardium. The endothelial-derived cells are found adjacent and medial to the first heart field (cardiac crescent). They migrate at the same time as the second heart field and form an inner lining on the myocardial cells.
- **Myocardium**. The power-generating myocardium of the heart, consisting of cross-striated muscle fibres, arises from precursor cells from the first and second heart fields.
- **Epicardium**. This outermost layer of the heart is formed from a transient cell cluster called the proepicardium. This cluster of cells likely arises from the second heart field under

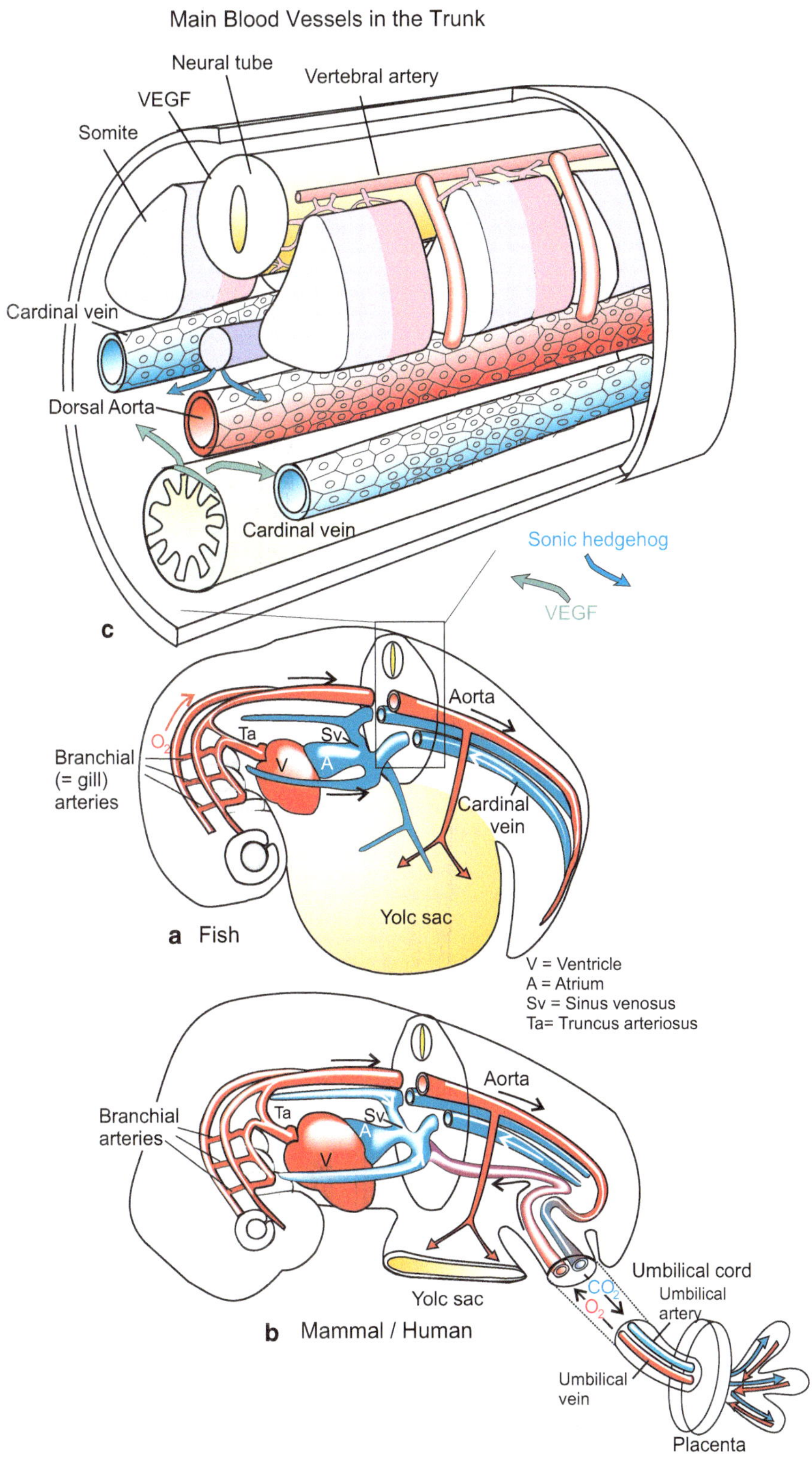

Main Blood Vessels in the Trunk

Fig. 17.2 Embryonic circulatory system in fish (**a**), and mammals/humans (**b**). Essentially, the anatomical construction is the same. However, in hatched fish larvae oxygen enters the blood in the gills and is distributed via

Fig. 17.3 Human
circulatory system after
birth. From Mueller and
Frings (2009)

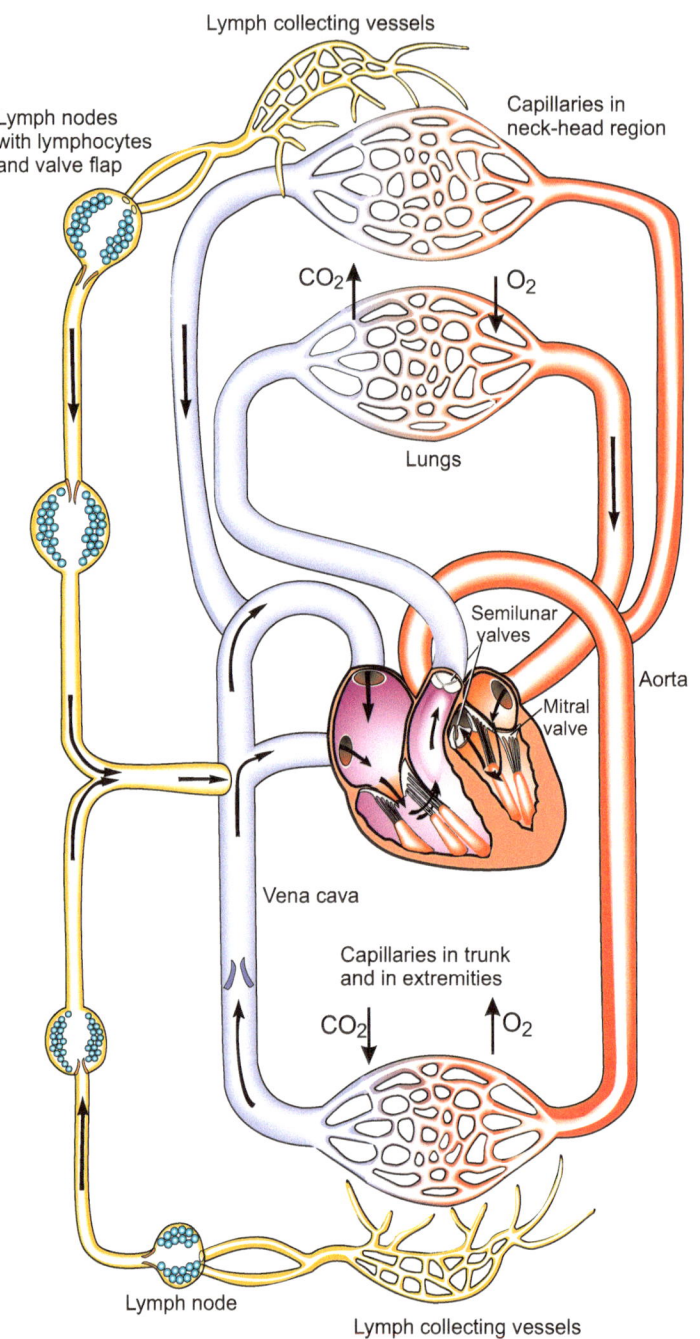

Fig. 17.2 (Continued) the dorsal aorta. In mammals oxygen enters the blood in the placental villi and is forwarded by the umbilical vein. Arteries defined as vessels leading from the heart are in red, vessels carrying blood to the heart are in blue, irrespective whether vessels transport oxygenated ("arterial") blood or deoxygenated ("venous") blood. The *upper* figure (**c**) shows a section of the trunk with the main vessels. The section is similar in larval fish and mammalian embryos (**a**, **b**, own figures, **c** after Coultas et al. 2005, modified)

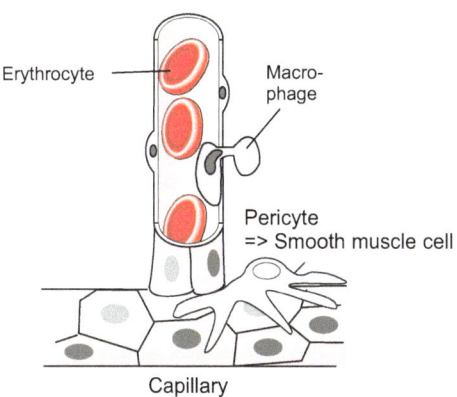

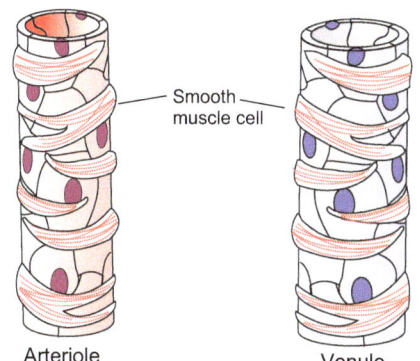

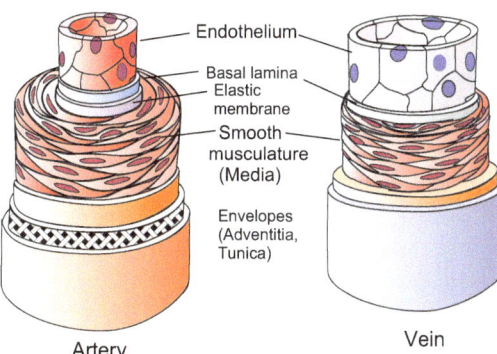

Fig. 17.4 Histological structure of blood vessels. Mueller and Frings (2009)

the control of FGF8. It is first seen at the heart looping stage near the inflow tract. The cells spread over the myocardium, forming a single epithelial layer.

17.2.2 While Beating the Bending Tube Transforms Into to a Chambered Heart

Once the heart tube has formed (initially from the first heart field) it elongates by addition of cells from the second heart field at both poles. It then loops to the right to form two chambers, the atrium and the ventricle. The atrio-ventricular valve forms and cardiac electric conduction begins. In fishes a two-chambered heart is the final stage. In mammals the septum forms and further folding results in a four-chambered heart.

The functional units in the embryonic heart are: The

- **sinus venosus and atrium** collect venous blood, forwarding it to the
- **ventricle**, the power-generating pump; it drives the blood into the
- **truncus arteriosus** which directs the expelled blood to the distributing branchial arteries.

This **simple heart**, consisting of one pump, satisfies the needs of a fish all its life. In land vertebrates the embryos also require only a single pump, since a separate circulatory system enclosing the collapsed lung is not yet necessary. But even before birth, land vertebrates – especially birds and mammals – must be prepared for a life in air.

Without missing a beat, the mammalian heart is folded and reconstructed to yield two functional pumps with two chambers each; one pump to propel the volume of blood through the body (large or systemic circulation), the other pump in anticipation of the needs after birth when the lungs must function. The process by which the folded tube is transformed into the final heart is quite complex and can be outlined here only in principle.

The **double heart** arises by in-growing of transversal partition walls. In amphibians and reptiles these walls remain incomplete, but in birds and mammals the chambers are completely separate (except the *foramen ovale*, see Sect.

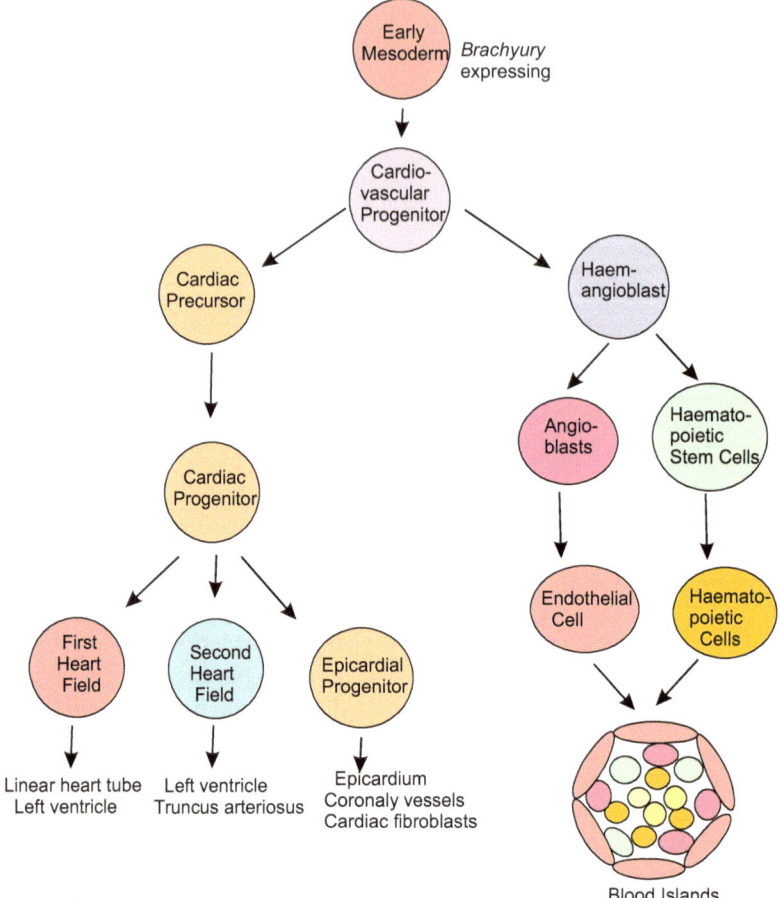

Fig. 17.5   Lineage of cells involved in the formation of the heart und circulatory system. After O'Connor et al. (2013)

Zebrafish

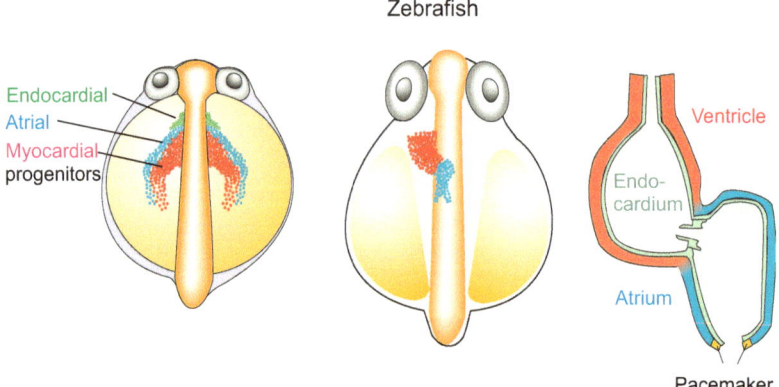

Fig. 17.6   Development of the zebrafish *Danio rerio*. Development before looping. Embryos are shown from a dorsal view. Three cohorts of migratory cells form the core of the heart. A first cohort (*green*) provides the endothelial layer, the second cohort (*red*) emigrates from the primary heart field and consist of myocardial progenitors which will form the ventricle. The third cohort (*blue*) emigrates from the second heart field and consist of myocardial cells that will form the atrium. The heart is supplemented by epicardial cells (not shown). After Staudt and Stainier (2012)

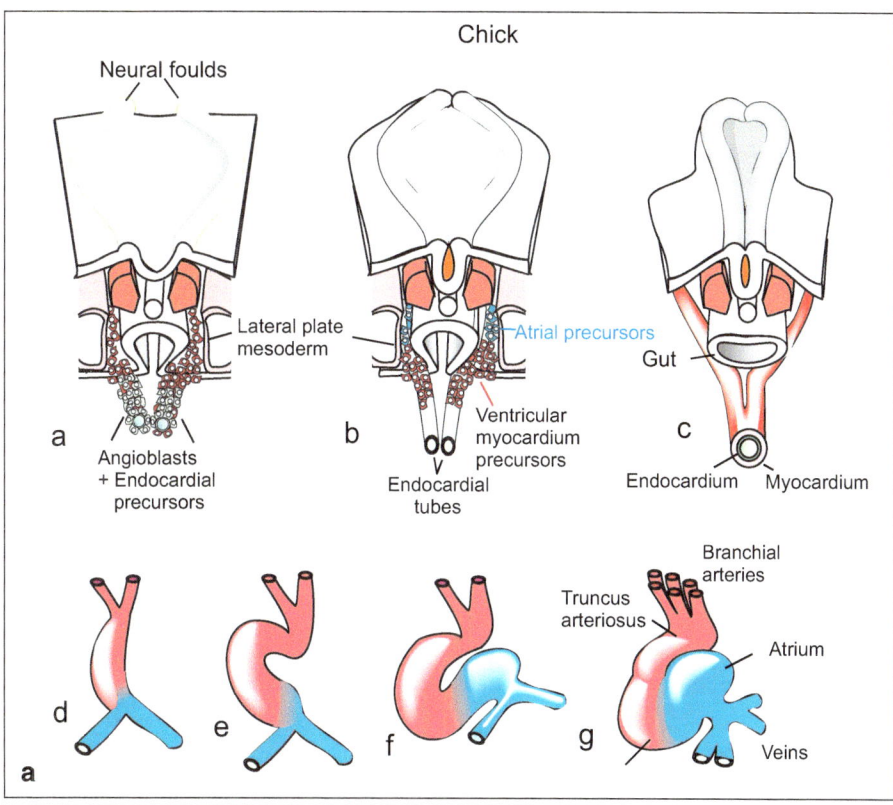

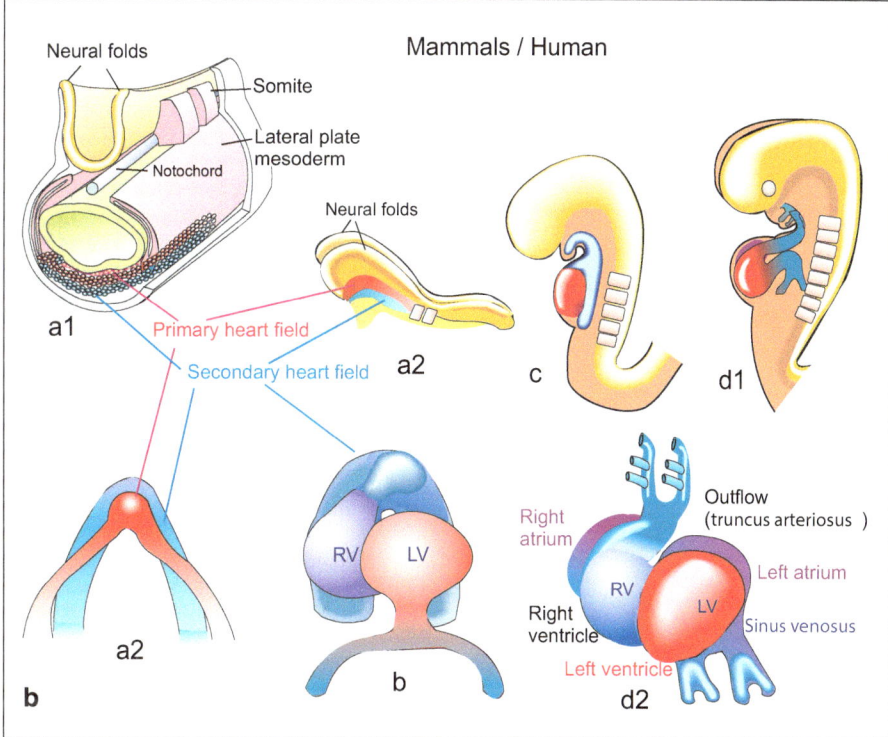

Fig. 17.7 (**A**, **B**) Development of the avian and the mammalian heart. Note that in these Figures the colours *red* and *blue* indicate the derivation of the progenitor cells from the first heart field (*red*) or from the second field (*blue*). The colours do not indicate whether the vessels will become arteries or veins by definition nor do they

17.4.2). A double pump is needed to separate oxygenated from venous blood and to overcome the high resistance in the lung capillaries. It is not sufficient, however, to separate atrium and ventricle by a partition wall. Also the venous input vessel and the arterial output vessel must be split off and the doubled vessels must be criss-crossed.

The input vessel is split off into

1. a gateway that directs the venous blood coming from the body via the vena cava into the right heart whence it can be pumped into the lungs, and
2. a second gateway that leads – after birth – oxygenated blood arriving from the lungs into the left heart whence it can be pumped into the body.

Correspondingly the exit must be split off into

1. an exit that leads from the right ventricle to the lungs, and
2. an exit that leads from the left ventricle to the body.

How these requirements are met in detail is described in special works of human embryology.

17.3 Blood Vessels: Vasculogenesis and Angiogenesis

At the same time as the heart is forming, the blood vessels and blood cells also form from mesodermal precursors. Formation of blood vessels involves vasculogenesis, angiogenesis and vascular remodelling. The heart, blood vessels and blood are all derived from a common cardiovascular precursor in the mesoderm (Fig. 17.5). Whereas the cardiac precursors

migrate anteriorly, the vascular endothelium/ blood precursors, called the haemangioblasts, migrate posteriorly and differentiate into vascular endothelial precursor cells called angioblasts which will form the blood vessels, and haematopoietic stem cells which will form the blood cells. The primary source of blood cells and capillaries are the blood islands. Amazingly, most of these islands are situated outside the embryo proper and widely dispersed in extraembryonic mesodermal layers (Fig. 17.8).

17.3.1 Blood Vessels Emerge Independently at Many Diverse Locations

With **vasculogenesis** and **angiogenesis** currently two terms are in use the present meaning of which unfortunately deviates from their literal and, primarily identical, meaning.

Current Meaning of the Terms

Angiogenesis. In the original meaning the term encompasses all processes leading to the formation of vessels (Greek: *angeion* = vessel; *genesis* = generation, emergence, genesis). In the last decades, however, a restricted meaning has got currency. In current literature the term angiogenesis is confined to processes through which **new blood vessels form as branches from pre-existing vessels**.

Vasculogenesis (Latin: *vasculum*, Plural *vascula*, = Small Vessels). In the present terminology **vasculogenesis is de novo formation of vessels from mesodermal stem cells** that give

---

Fig. 17.7 (Continued) give hints to the future state of oxygenation of the blood. (**A**) The heart of a bird. The heart primordium originates from migrating angioblasts and myocardioblasts which aggregate to form paired tubes (**a**, **b**). The first descendants of the angioblasts form the inner wall of the tubes, called endocardium, the myocardioblasts form the muscular wall, the myocardium (**c**). The paired tubes fuse forming the unpaired looped heart tube. (**d**, **e**, **f**) The heart tube subdivides into the sections sinus venosus, atrium, ventricle, and truncus

arteriosus (**g**). **B** Heart development in a generalized mammal (mouse or human). Also in mammals the primary and secondary heart fields provide the progenitors of the muscular myocardium. *Red*: Progenitors from the primary (first) heart field, blue progenitors derived from the second heart field. After several authors (e.g. Xin et al. 2013) and own figures, combined and modified

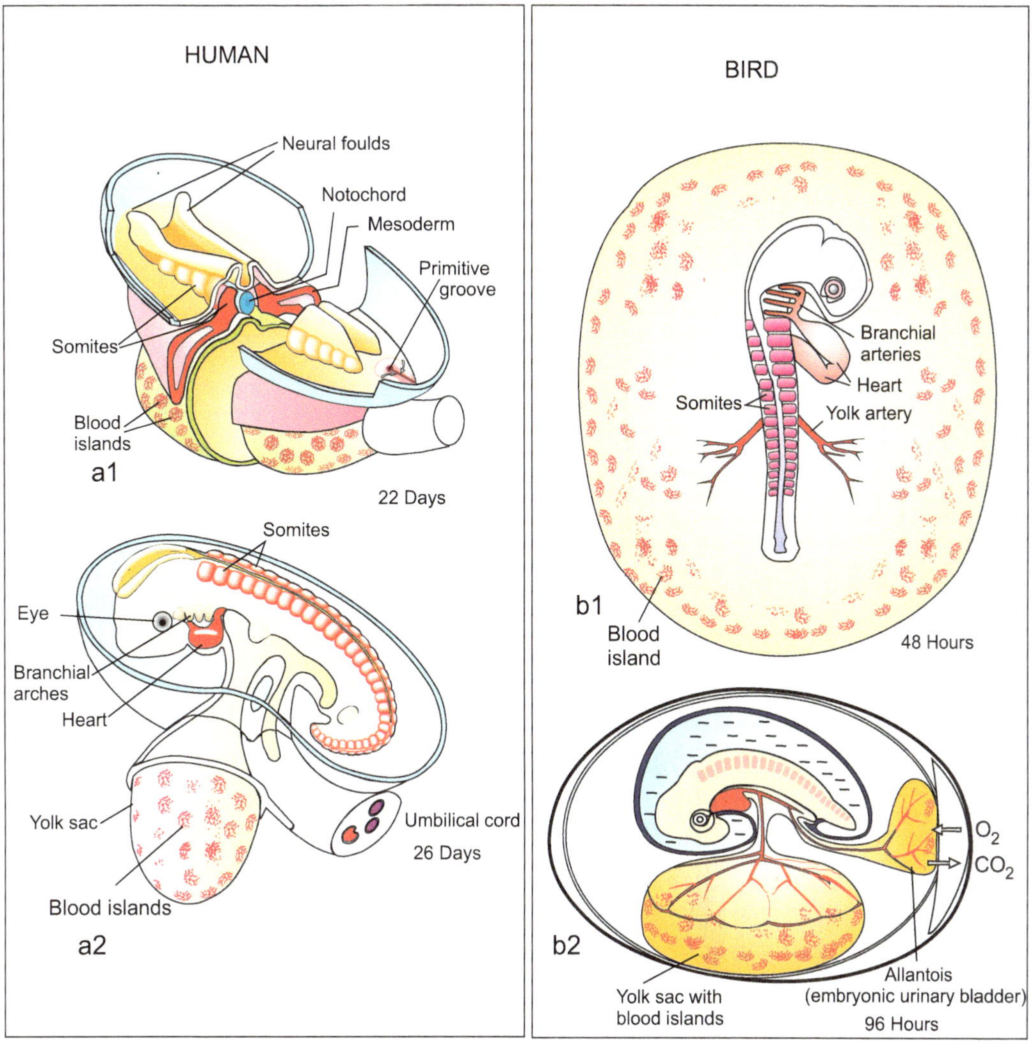

Fig. 17.8 Blood islands in comparison. Common to all vertebrates is the emergence of first blood cells outside the embryo proper in the mesodermal covering of the yolk sac (**a1**, **a2**) and, if present, of the allantois (**b1**, **b2**)

rise, among others, to endothelial cells, and are mostly found in the blood islands outside the embryo proper (Figs. 17.9 and 17.10). According to this definition the first vessels in the embryo form through vasculogenesis, that is from stem cells, while the later vessels derive from lateral branches of the primary vessels (Fig. 17.11).

Quite confusingly, the primary meaning of angiogenesis lingers on in the term "*angio*blasts" which designates those stem cells from which "*vasculo*genesis" originates (Fig. 17.9). Primary vessels are in the embryo formed by cells still designated angioblasts and not vasculoblasts.

Actual Vasculogenesis. In the process of vasculogenesis (vessel formation), angioblasts migrate to extraembryonic and embryonic sites of vascularisation and proliferate, aggregate, differentiate and contribute to the formation of capillaries in the blood islands. These contain haematopoietic cells in the centre, surrounded by angioblasts. The angioblasts will give rise to

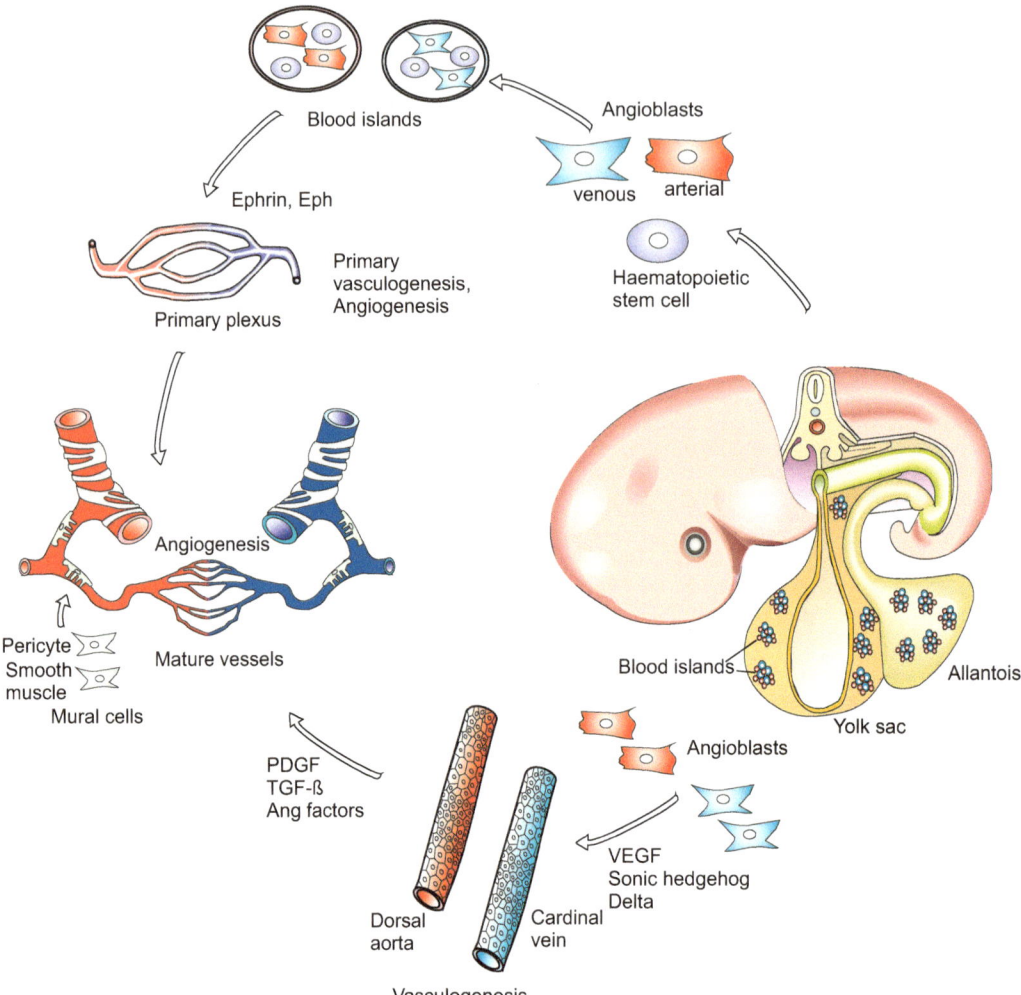

Fig. 17.9 Synopsis of events that contribute to the development of the vertebrate circulatory system. Formation of the blood vessels from endothelial progenitors. After Coultas et al. (2005), combined with own figure of a generalized mammalian embryo

endothelial cells. The capillaries of neighbouring blood islands fuse to form a larger plexus. This occurs at several embryonic and extraembryonic sites. The dorsal aorta (first main artery) and cardinal veins (first main veins) differentiate directly from angioblasts which migrate along the midline, and not via blood islands. Once formed the major arteries move towards the heart.

Actual Angiogenesis. Expansion of the network takes place from endothelial cells of existing vessels and capillaries. Endothelial cells retain the characteristics of stem cells and remain capable of dividing (Fig. 17.11).

There are astonishing parallels in the ways by which targets are sought out in the vascular and nervous systems (Figs. 17.11 and 17.12). At the tips of capillaries **moving terminal cells** are located. They are equipped with sensory filopodia and fulfil the function of pathfinders. Further terminal cells arising along the flanks of extending capillaries give rise to branches.

Vessel formation. Microscopic observations in living, transgenic-labelled fish embryos have disclosed different means of vessel formation, be it primary vessels in vasculogenesis or branches in angiogenesis. In any event first angioblasts aggregate forming strands.

Primary Vasculogenesis in Blood Islands

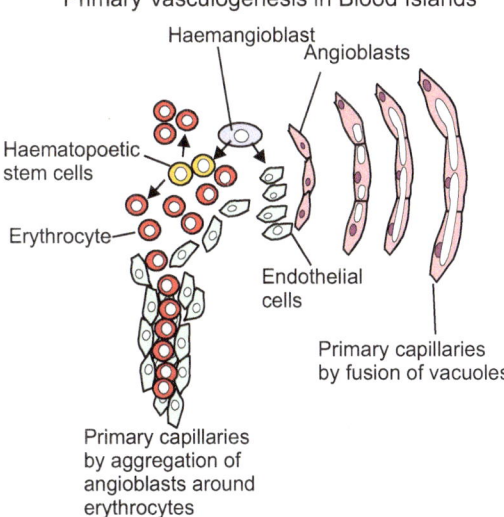

Fig. 17.10 Primary vasculogenesis. According to investigations in transparent fish embryos whose hemangioblasts were GFP-labelled the first fine blood capillaries arise in and around the blood islands from descendants of hemangioblasts, called angioblasts, which arrange themselves in file. The rows of cells turn into tubes around erythrocytes or form intracellular vacuoles the lumen of which fuse resulting elongated internal tube-like structures

In capillaries cavities form in three modes.

1. In the first mode **vacuoles** form within the cells; subsequently the vacuoles of adjacent cells fuse forming an internal canal crossing cell borders (Fig. 17.10). Viewed from their history of origins the canals within the finest capillaries are essentially intracellular structures.

2. More frequent is the second mode. Angioblasts turn into **endothelial cells** that surround rows of erythrocytes (Fig. 17.10).

3. The most frequent mode, however, is the third mode which can give rise also to larger vessels. Angioblasts aggregate forming **massive strands** (Fig. 17.11). Then the cells flatten and move towards the periphery **giving way to a central lumen**. In the periphery the cells acquire an inward-outward polarity and adopt an epithelial organization, known as **endothelium**. As the definition of an

epithelium requires, between the cells *tight junctions* form, and the peripheral surface becomes covered by a basal lamina. This happens in cooperation with **pericytes** which attach onto the capillaries. Pericytes have the ability to contract and are considered a particular type of smooth muscle cell. In larger vessels further cells associate with the pericytes adding a dense layer of **smooth muscle cells** (Figs. 17.4 and 17.9). Their future task will be to constrict or dilate the vessel to provide adequate and sufficient blood distribution. Constriction or widening of vessels in global dimensions is controlled by hormones, in local dimensions by the vegetative sympathetic and parasympathetic nervous system. Accordingly, most vessels are accompanied by nerve fibres.

17.3.2 Blood Vessels Are Stimulated to Grow and Branch by Angiogenic Factors

Guided by their sensory and motile terminal cells the capillaries extend, branch and find their way into their target areas which require blood. Arterioles find their way to venules; both find the correct pathway to the large vessels. With time, the growing and branching vessels of various origin and sizes encounter each other and fuse. But how is the overall vascular system designed and how is it all regulated so that it results in a species-specific circulatory system with a huge number of identifiable vessels?

Behind the motile terminal cell the endothelial cells divide and thus elongate capillaries. Proliferation must be adapted to the growth of the organism. Important growth stimulating and guiding factors are diffusible chemoattractants known as **angiogenic factors** (Figs. 17.9 and 17.12). Areas undersupplied with oxygen and nutrients emit such factors which stimulate proliferation of endothelial cells and the formation of lateral branches, as exemplified by tumours (Fig. 17.13).

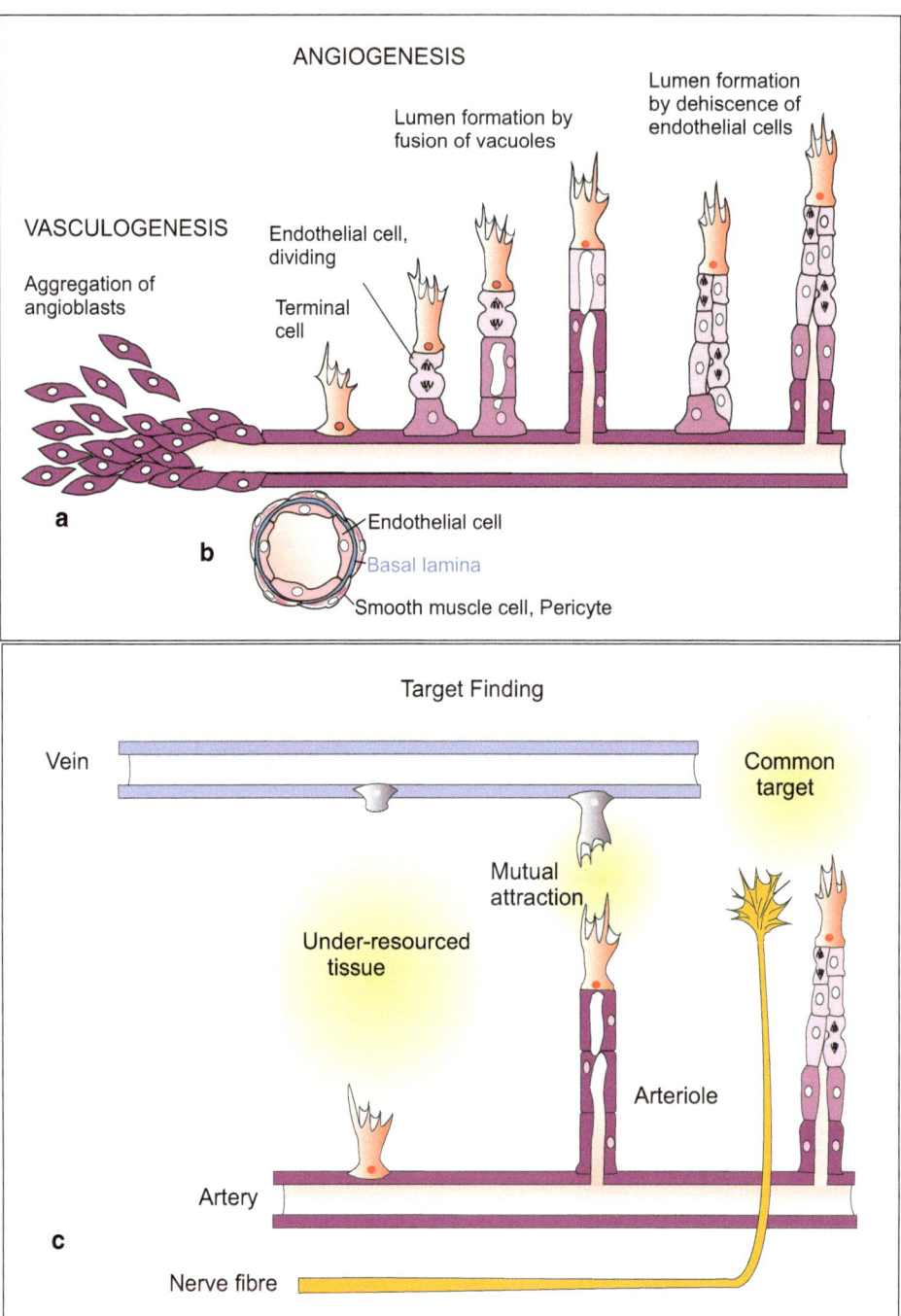

Fig. 17.11 Vasculogenesis and angiogenesis. (**a**) Vasculogenesis: Larger vessels arise and elongate from aggregating angioblasts. The initially massive aggregate surrounds itself by a basal lamina and hollows out by dehiscence of the cells into the periphery where they form a monolayered endothelium. In very thin capillaries the lumen of the vessels arises from the fusion of vacuoles. (**b**) Pericytes settle around vessels and eventually form smooth muscle cells. (**c**) In angiogenesis branches arise from capillaries and larger vessels. At the tip of growing vessels a terminal cell takes leadership and heads to a target sending out angiogenic factors. Such factors also stimulate the formation of branches. Frequently blood vessels are accompanied by nerve fibres. The growth cone of the nerve fibre and the terminal cell of the blood vessel share similar structures such as motile lamellipodia or thin filopodia, and respond to the same signals

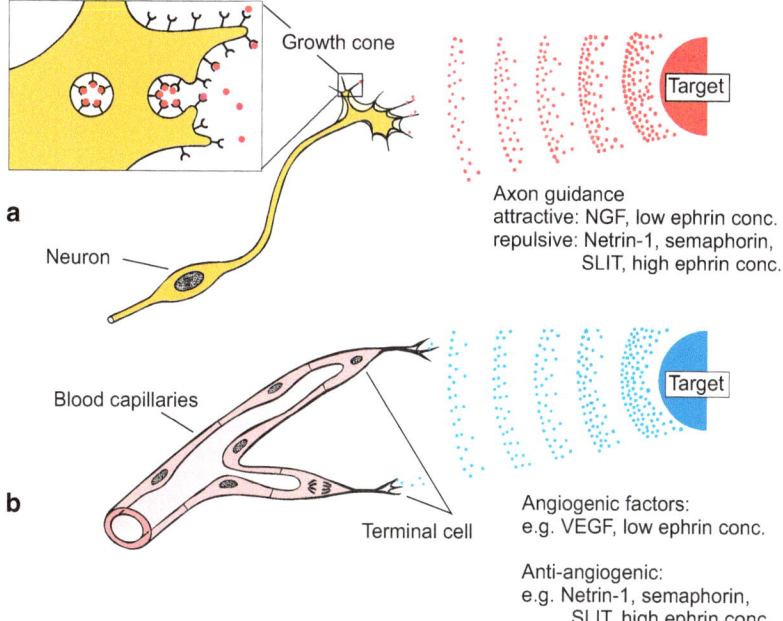

Fig. 17.12 Showing and tagging targets by chemotactic signals. (**a**) The advancing axon of a sympathetic neuron finds its NGF-emitting target by means of sensors on its growth cone. The NGF is taken up and internalized (*left insert*) and acts as survival factor. (**b**) The terminal cell of a blood capillary is also fitted with motile and sensory structures to search a target. Often neurotropic and angiotropic factors are identical. Some ephrins located at the surface of cells along the migration path have attractive properties at low concentration but deviate or stop the advancing growth cone or terminal cell at high concentration. As a rule, netrin-1, semaphorin and SLIT have repulsive effects

Many and various angiogenesis stimulating factors have been found. To present a complete list would surpass the scope of an introductory textbook. Among them are factors well-known also from other developmental processes. We mention:

– **FGF-1** and **VEGF** (Vascular Endothelial Growth Factor, existing in several isoforms),
– **HGF** (Hepatocyte Growth Factor),
– **bFGF** (basic Fibroblast Growth Factor) family,
– **TGF-α** (Transforming Growth Factor alpha),
– **Angiopoietin-1** Ang1, and several exotic factors such as
– **leptin**, known in physiology as saturation hormone; apparently it has a different function in embryonic development, and, surprisingly
– **Ephrins and NGF (Nerve Growth Factor)**, **known as factors guiding nerve fibres;** in particular ephrins procure the joining of arterioles with venules and thus assist in the formation of networks of continuous capillaries (Fig. 17.14).

Some of these factors are displayed in Figs. 17.2 and 17.9.

17.3.3 Growing Capillaries Find Their Targets Like, and Often in Parallel to, Nerve Fibres

Readers having read the Chap. 16 dealing with the development of the nervous system might experience a déjà-vu epiphany. Many factors known to direct, attract or repel nerve fibres emerge with the same or similar function when it comes to advancing blood vessels. There are astonishing parallels in path-finding. At the tips of capillaries we find the **pathfinding terminal**

Angiogenesis

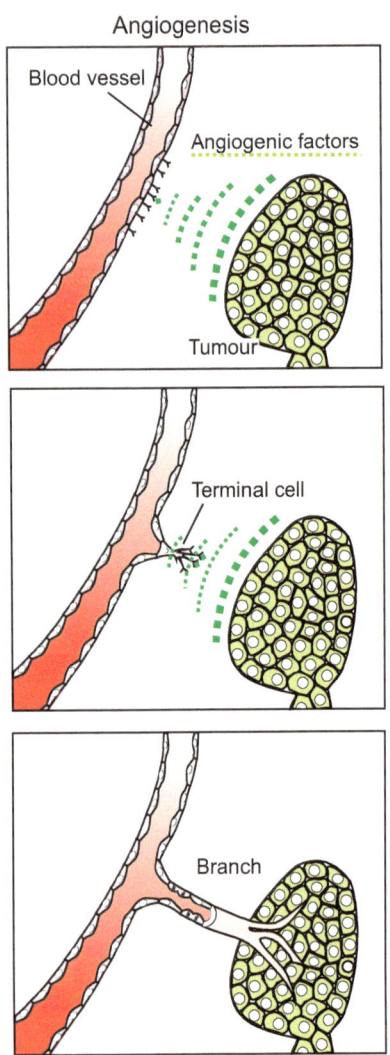

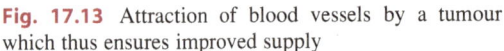

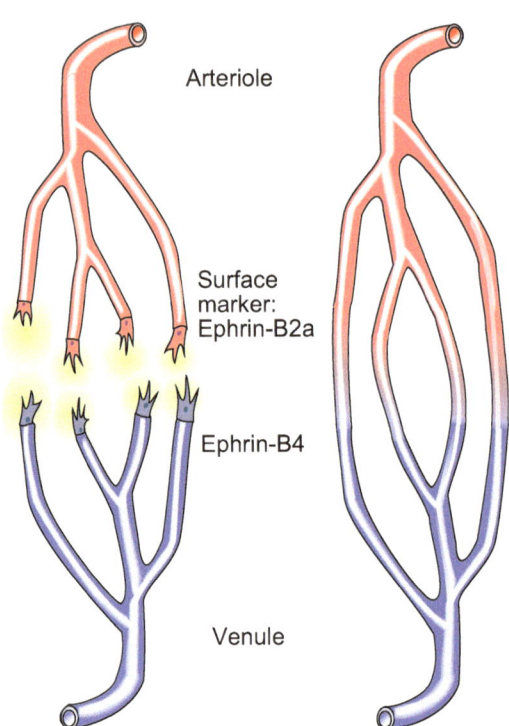

Fig. 17.14 Fusion of arterioles and venules. The terminal cells at the tips sense approaching potential fusion partners and move towards each other. With fusion the function of terminal cell expires. Certain surface molecules such as different members of the ephrin family prevent fusion of arterioles with arterioles, venules with venules. On the other hand, different ephrins promote fusion of arterioles with venules and thus connection of arteries with veins by a capillary bed

Fig. 17.13 Attraction of blood vessels by a tumour which thus ensures improved supply

cell, at the tip of nerve fibres the **pathfinding growth cone**. Both are motile and equipped with molecular receptors to receive external signals (Figs. 17.11, 17.12, and 17.14). They orient themselves sensing

- **diffusible long-range signals**, sent out by distant targets and called either angiogenic factors or neurotrophins (better neurotropins) depending on which system is considered. An example is the just mentioned Nerve Growth Factor **NGF**.

- Terminal cells like growth cones further scan **stationary short-range signals** represented by cell adhesion and cell recognition molecules such as surface-bound molecules of the **ephrin** class.

To trace and track such signals terminal cells and growth cones are equipped with receptors belonging to the category of the **tyrosine kinase receptor TRK** class (see Fig. 11.8).

On the other hand there are signals on signposts that give terminal cells and growth cones the order to avoid a certain area and to alter course. Such repulsive signals are, for example, **netrin-1** and **semaphorin**.

Actually, many target areas are colonized by blood vessels and nerve fibres in parallel.

17.3.4 Tumours Provide for Good Blood Supply by Sending Out Angiogenic Factors

The angiogenic factors emitted from targets promote the generation of branches. Thus, a wide area can be provided with blood. Among the known angiogenic factors is **VEGF** (Fig. 17.2).

Historically angiogenic factors, especially VEGF, were revealed by observing transplanted tumours. Tumours including metastases are soon vascularized by many capillaries, and thus are well supplied by oxygen and nutrients, all too well, because this promotes their growth. Tumours act like undersupplied tissues promoting their own vascularization (Fig. 17.13).

On the other hand, arduous research over decades has led to the detection of **angiogenic inhibitors**. In fact, research and development in this field has been driven largely by the desire to find better cancer treatments. Among the factors found are **angiostatin**, **vasostatin**, thrombospondin-1 (TSP-1), and angiopoietin 2. However, since proteins usually are biologically unstable and can evoke allergic responses, in current clinical applications synthetic drugs appear to be preferred. Among them is **thalidomide**, the effective component of the sleep-inducing drug contergan, a drug of ill fame because of its disastrous tertatogenic effects.

17.4 Adjustment of the Circulatory System Before and After Birth

17.4.1 The Early Embryo Has a Circulatory System Like a Fish

The circulatory system of the embryo is not constructed in such a way as to meet the demands of an independent animal. The embryonic circulatory system is adapted to the needs of the embryo, but also displays 'unnecessary' structural reminders of the evolutionary past. For instance, the **four to six branchial arches** of the embryo in terrestrial vertebrates are considered to be unneeded echoes, as they are no longer required to supply gills and must be substantially modified (Figs. 17.15 and 17.16). It is possible, of course, that the arches have retained a function. But if they have, their role – presumably in the organization of the development in the visceral head and neck region – must still be defined.

The branchial arches are in all vertebrates provided with blood by the heart initially consisting of one atrium, one main chamber (ventricle) and one output vessel called *truncus arteriosus* (Figs. 17.2 and 17.15). In terms of function, the whole construction is one single pump. In the phylotypic stage even the mammalian embryo initially has a one-chambered heart; that is one single pump. This satisfies the needs of a fish all its life but is also sufficient for the embryo as it lives in a fluid and breathes through the "gills" known as placental villi which dip into maternal blood sinuses. Here O_2 is taken up in exchange of CO_2 that is given over to the maternal blood. But even when in preparation for a life in air the heart is morphologically (almost) completely separated into two pumps the circulatory system is still a monocircular system not essentially different from the system of a fish.

17.4.2 The Necessary Adjustments After Birth Are Dramatic and Must Be Prepared

The reconstruction of the embryo's circulatory system into a double circulatory system, consisting of a systemic and a pulmonary circulation, is performed in preparation for the moment when the fetus suddenly has to switch to air breathing, but within the maternal womb the system cannot function in this way. The unborn fetus receives both nutrients and oxygen via the umbilical vein, and in exchange the embryo clears off carbon dioxide through the umbilical arteries. The umbilical vein carries oxygenated blood but leads into the right heart, where blood is pumped to the lungs after birth.

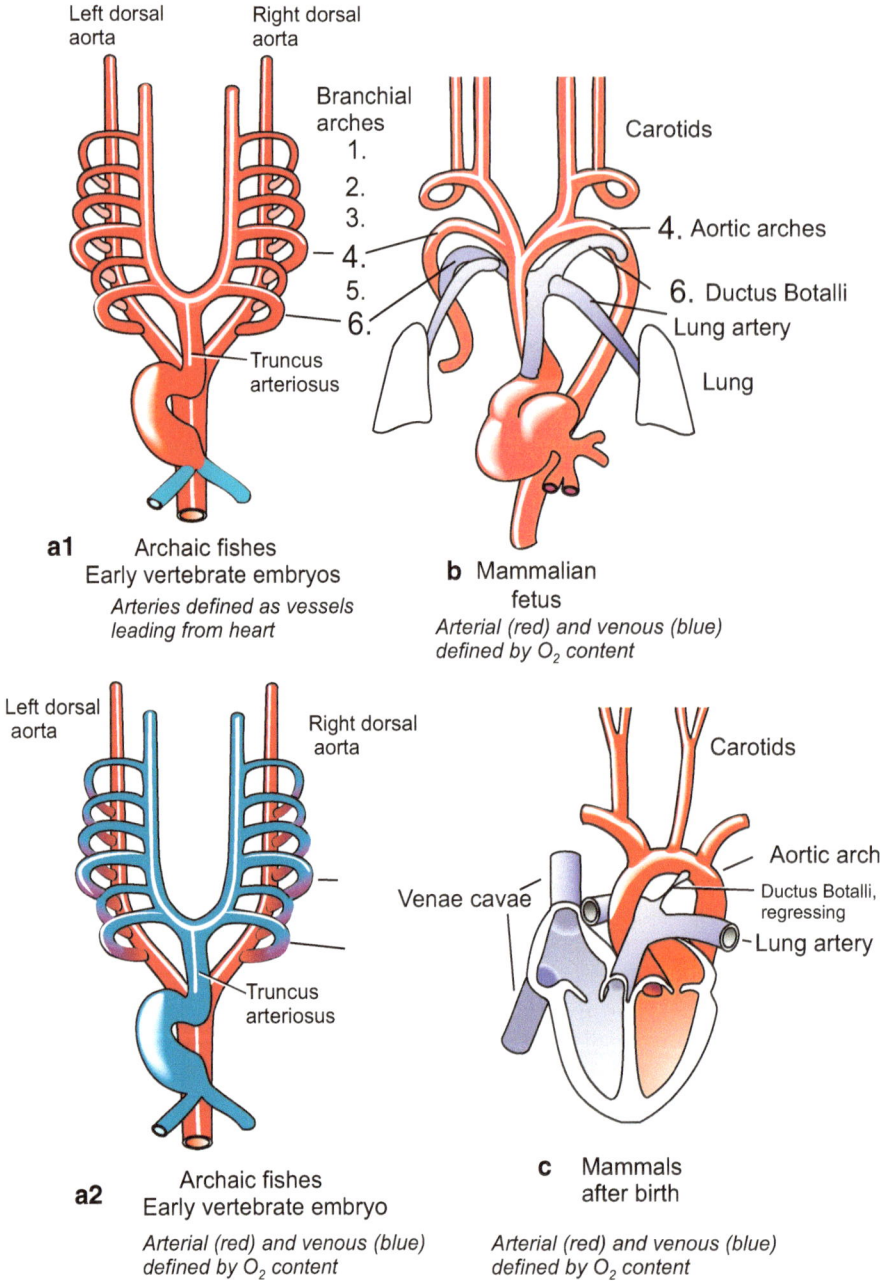

Fig. 17.15 Arterial system in the mammalian embryo and fetus. Colours of the vessels can be different, depending on the definition of arteries and veins, or whether the figure is designed to show whether a vessel transports oxygenated ("arterial") or deoxygenated ("venous") blood. **a–c:** Development of the arteries ahead of the heart, especially of the vascular aortic arches (branchial or gill arches) in the anterior body. (**a**) Branchial arteries in the embryo (sometimes less than six are present). (**b**) Conversion in the fetus before birth; the collapsed underperfused lungs are not yet connected to the main blood supply; the ductus arteriosus (also known as ductus Botalli) serves as a bypass. (**c**) After birth, the ductus Botalli is reduced

a Before Birth

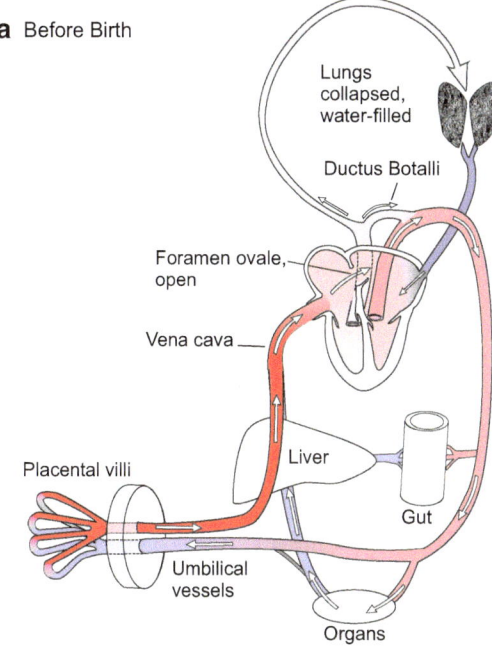

Lungs collapsed, water-filled

Ductus Botalli

Foramen ovale, open

Vena cava

Placental villi

Liver

Gut

Umbilical vessels

Organs

b After Birth

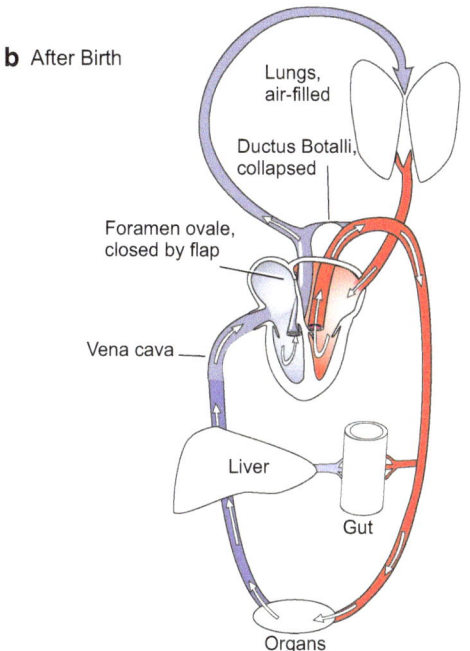

Lungs, air-filled

Ductus Botalli, collapsed

Foramen ovale, closed by flap

Vena cava

Liver

Gut

Organs

Fig. 17.16 (**a**, **b**) Switch in the blood circulation from a mono-circular to a double-circular system at birth. (**a**) In the unborn fetus, the lung circulation is practically non-functional; the wall between the two atria is perforated and the heart functions as though only one atrium and one ventricle are present just like in fish; blood is pumped to the placenta, which has the function of a gill, and flows back to the heart. The body, represented by the liver, receives only partially oxygenated blood. (**b**) After birth, pulmonary circulation is established and connected crosswise with the greater systemic circulation. The lungs can commence operating

But the lungs are still collapsed. It would not be useful, and because of hydrodynamic resistance almost impossible, to propel the entire blood supply through the non-functional, collapsed lungs. Because the embryo needs to forward the oxygen-rich blood from the umbilical vein towards the brain and body, the bloodstream must be guided from the right into the left atrium. The passage between the two atria is known as the **foramen ovale** (Fig. 17.16). This passage is a short shunt that is supplemented by the **ductus arteriosus**, also known as **ductus Botalli**, which enables bypassing the lungs.

After birth, the lungs must be inflated immediately, the blood's path to the lungs opened, and the passage between the two hearts closed. For that purpose a prepared plug consisting of fibronectin is pressed against the hole by the increasing blood pressure. In the long term mixing of venous and oxygenated blood in "blue" babies is life-endangering and must frequently be surgically prevented.

The first breath is a perilous but life-saving event.

Summary

Heart and blood vessels arise from a common cardiovascular precursor cell population expressing *Brachyury*. These differentiate into either cardiac precursor cells or haemangioblasts. The cardiac precursors migrate anteriorly on either side of the midline to form the cardiac crescent or first heart field, while the haemangioblasts migrate posteriorly. Haemangioblasts differentiate into two sub-types; angioblasts which give rise to endothelial cells, and haematopoietic stem cells which give rise to blood cells. Blood vessels arise in the early embryonic **vasculogenesis** in many widely distributed **blood islands**; in mammals these are found in the mesodermal covering of the yolk sac and allantois and in the head mesenchyme. These islands consist of haematopoietic stem cells surrounded by angioblasts. Angioblasts fuse to form the first capillaries. In **angiogenesis** the net of vessels is subsequently expanded by **branches** arising from **stem cell-like endothelial cells** along existing vessels. In the extension and branching of vessels motile and sensory **terminal cells** seek and pave the way to target tissues. Like the growth cones of nerve fibres, the terminal cells scan surface molecules such as **ephrins** presented by guidepost cells and are attracted by soluble **angiogenic factors** emitted by distant targets.

Frequently nerve fibres and blood vessels seek and find the same targets jointly. In addition, angiogenic factors stimulate targeted growth and branching of blood vessels. An important angiogenic factor is known as VEGF (Vascular Endothelial Growth Factor). It is emitted especially by tissues not sufficiently supplied with oxygen and nutrients. Tumours are a rich source of angiogenic factors.

The heart is formed when the cells of the first heart field (cardiac crescent) involute and form the linear heart tube. These mainly contribute myocardial cells that will eventually form the left ventricle. Cells of the second heart field arise from the same precursors but migrate and differentiate later. They give rise to the atria, the right ventricle and the outflow tract (truncus arteriosus). The endocardium is mainly derived from endothelial cells that migrate adjacent to the second heart field and cover the myocardial cells on the inner side of the heart. Epicardial cells arise from proepicardial cells. These adhere to and cover the myocardial cells on the outer surface. The heart tube loops to the right forming the atrium, ventricle and truncus arteriosus. In functional terms the embryonic heart is a single pump which drives blood through 4–6 branchial ("gill") arches. The circulation resembles that of fishes, but with the placental villi functioning as gills. In land vertebrates the heart is morphologically subdivided into two pumps by inserting partition walls. The doubling of the heart pump to drive the larger systemic and the shorter pulmonary circulation separately implies also a reconstruction of vessels leading to and from the heart. Yet until birth the heart still drives one, the systemic circulation as the pulmonary system begins to operate only after birth. In the unborn fetus a passage between the two atria, known as **foramen ovale**, enables forwarding oxygenated blood arriving from the placenta towards the brain and the body. Upon birth this short shunt must immediately be closed.

Stem Cells, Regeneration, Regenerative Medicine

18

18.1 Continuous Renewal of Our Organism by Stem Cells

18.1.1 At the Very Least Each Organism Has to Renew Its Stock of Macromolecules

When we think of regeneration we usually imagine the reconstruction of lost body parts. However, this is only one among several regenerative events that organisms perform. **Re-generation** means generation anew. In this general sense, regeneration occurs at all levels of life, including the level of macromolecules. Proteins suffer irreversible alterations with time (denaturation) and must be replaced by newly synthesized ones. Without renewal, slow loss of function in essential enzymes would eventually result in the death of the cell. Experimental evidence suggests that renewal of the molecular inventory and resulting **rejuvenation** takes place routinely in the course of cell divisions. **Division-mediated rejuvenation** confers potential immortality to the single-celled protist. In contrast, terminally differentiated cells, incapable of dividing, will sooner or later die. We will investigate this observation and its cause in Chap. 21. Rejuvenation by mitotic division can be supplemented or replaced by rejuvenation through sexual reproduction.

18.1.2 Stem Cells Have Two Dominant Functions: They Enable Growth and Supply Replenishment for Used Up Cells

Stem cells are undifferentiated or partially differentiated cells capable of dividing; their descendants give rise to further stem cells on the one hand, and to cells determined to mature as differentiated cells on the other hand.

> According to our inquiries the oldest scientific work in which the term **stem cells** (in German "*Stammzellen*") can be found is a comprehensive treatise of the zoologist August Weismann (1883): "The origin of sexual cells in hydromedusae" ("*Die Entstehung der Sexualzellen bei Hydromedusen*"). In this treatise not only medusae (jellyfishes) are dealt with but also species in which polyps generate germ cells such as *Hydractinia echinata* (see Figs. 4.18 and 18.12). Actually the word "*Stammzellen*" is used for the first time in the section dealing with this species and it is used to designate the (putative) precursors of germ cells. At other places of the treatise **primordial germ cells**

(continued)

("*Ur-Keimzellen*") are defined as "stem cells of germ cells". Weismann also speaks of "undifferentiated cells entailing several possibilities for differentiation" ("*indifferenzirte Zellen, die mehrere Differenzirungs-Möglichkeiten in sich bergen*", p. 279). Besides stem cells and primordial germ cells ("*Ur-Keimzellen*"), also the terms **germ line** ("*Keimbahn*") and the term **pole cells** ("*Polzellen*"), designating the primordial germ cells of *Drosophila*, were coined by Weismann. This hint to this widely unknown treatise should also be a hint that the principle of stem cells as source for germ cells and replenishment is evolutionarily very old – in fact as old as are multicellular organisms.

The issue of stem cells has led a shadowy existence in biology and medicine as well for many years, but is in the spotlight today and appeals to a broad public. In the multitude of information, with advertising appeal enclosed in therapeutic promises, the basic function of stem cells is often missed. Their primary function in the embryo to be **founders of various cell lines** only interests specialists in embryology. Only in the context of potential medical applications the term **embryonic stem cells** became familiar to a wide public. Yet, for actual therapeutic procedures and **regenerative medicine** post-embryonic, so called "**adult**" **stem cells** are of higher importance. To lay scientific foundations for discussing practical issues first two essential functions of **fetal and post-embryonic stem cells** may be outlined:

1. **Fetal and juvenile stem cells enable many organs to grow**. Thus the enormous growth of the human central nervous system, in particular of the forebrain, is based on a layer of stem cells (called ependyma) lining the canal of the spinal cord and the ventricles of the forebrain. By symmetrical division the population of stem cells is increased. By asymmetrical

division the stem cells also give rise to new neurons and glial cells (like the fish ependyma cells shown in Fig. 16.14). (Because with progressing development many embryonic stem cells lose stem cell characteristics it has been proposed to call those cells "**founder cells**" whereas the term "stem cell" should be reserved for those cells which retain stem cell characteristics life long. However world wide published literature cannot be rewritten, and so we follow current use.)

2. Stem cells known as **adult stem cells** that retain stem cell characteristics life long **enable many organs to renew and to regenerate**. This will be specified in the following sections.

The renewal of the stock of cells has in older treatises been called "physiological" regeneration in contrast to "reparative" regeneration by which lost body parts are restored.

18.1.3 Short-Lived Cells Must Be Replaced by Newly Generated Cells; Replacement Is Provided for by Stem Cells Ready to Divide

As a rule, cells, tissues and organs generated in early embryogenesis must be enlarged in the course of larval and juvenile growth. But even after obvious growth has ceased, processes of proliferation persist in most organisms, in vertebrates as well as in *Hydra*, because many cells live for only a short time and must repeatedly be replaced by new ones.

Occasionally, even fully differentiated cells such as the **hepatocytes** in our liver retain the ability to divide. Normally hepatocytes are long-lived and divide slowly. However, in response to injury or poisoning they divide rapidly to replace damaged neighbours.

Most kinds of cells lose their mitotic potential in the course of terminal differentiation. Some even lose their nucleus, such as the keratinocytes of our skin or the erythrocytes of our blood. In all such cases a population of cells must be present that remains able to divide and to terminate

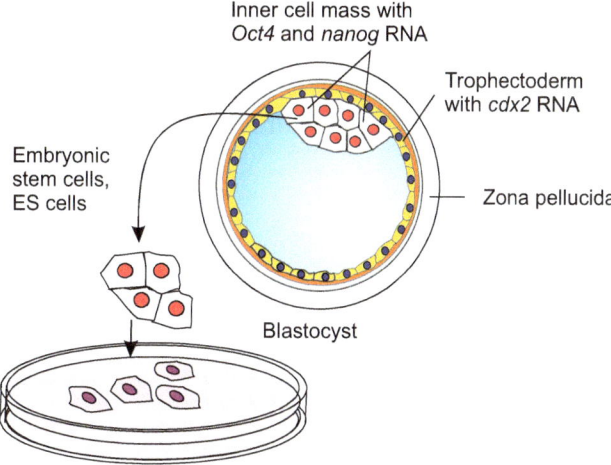

Inner cell mass with
Oct4 and *nanog* RNA

Trophectoderm
with *cdx2* RNA

Embryonic
stem cells,
ES cells

Zona pellucida

Blastocyst

Fig. 18.1   Mammalian blastocyst with inner cell mass, the source of embryonic stem cells

differentiation after being multiplied. Such cells are defined as stem cells. When stem cells retain their 'embryonic' characteristics permanently, and can even be multiplied without limit in culture, they are designated **immortal stem cells**.

Stem cells in the adult organism are characterized by the following features:

1. They serve the permanent **self-renewal** of a group of cells, of a tissue or organ.
2. They **retain the ability to divide**, but usually divide only slowly and only on demand.
3. As a rule their division is **asymmetric**. One daughter cell remains stem cell, the other undergoes differentiation. However, in several cases the proportion of 50 % stem cells, 50 % differentiating cells in the descendants is maintained only in the statistical average, while in other cases this apportionment is strictly observed. Frequently in asymmetric division the fate of the daughter cell sent onto a differentiation path is strictly determined; the cell is said to be **committed** to a task.

Dependent on whether the descendants of a stem cell can run all potential differentiation paths, or many different (but not all) paths, or only one single path, one speaks of **totipotent** (=**omnipotent**), **pluripotent**, **multipotent** or **unipotent** stem cells.

In the medical literature further criteria for distinction came into general use:

- **Totipotency (omnipotency)** (Latin: *totus* = complete, *omnis* = all, *potentia* = power,

potens = having power) is the ability of a single cell to give rise to a complete organism. In mammals this implies, that the cell in question can give rise to all cell types including germ cells and cells of the embryonic trophoblast and placenta, respectively. They have the ability to replicate in unlimited numbers without losing their total potency. In the human embryo totipotency is restricted to the cells of the cleaving embryo up to the 8-cell stage, testified by the existence of up to eight monozygotic super twins (octuplets). Cells of the germline, especially of the female germ line, are potentially totipotent until meiosis, thereafter oocytes are totipotent only if they regain a diploid set of chromosomes in fertilization (in mice one set maternally, and one set paternally imprinted).

In molecular terms totipotent cells and the cells of the germ line contain mRNA or the protein encoded by the gene *vasa* (Chap. 8) and express factors essential for maintenance totipotency such as NANOG (homeodomain protein), Oct-4 (transcription factor with POU domain), SOX-2 (transcription factor) and PIWI proteins (interacting with small piRNA).

- **Pluripotency** (Latin: *plus*, *pluris* = greater, higher, more) refers to a progenitor or founder cell that has the potential to give rise to a multitude of cell types that belong to any of the three germ layers. Whether pluripotent cells are able to give rise to germ cells is not

Fig. 18.2 Stem cells and carcinogenesis in the skin. The stem cells of the stratum germinativum produce descendants that differentiate into keratinocytes and are eventually sloughed off. In carcinomas, the differentiation is incomplete and cells derived from the stem cells continue to divide, thus forming a tumour (carcinoma). Transformed cells may dissociate from the tumour and enter the blood stream to establish colonies (metastases) in other places (after Alberts et al. 1994, redrawn and extended)

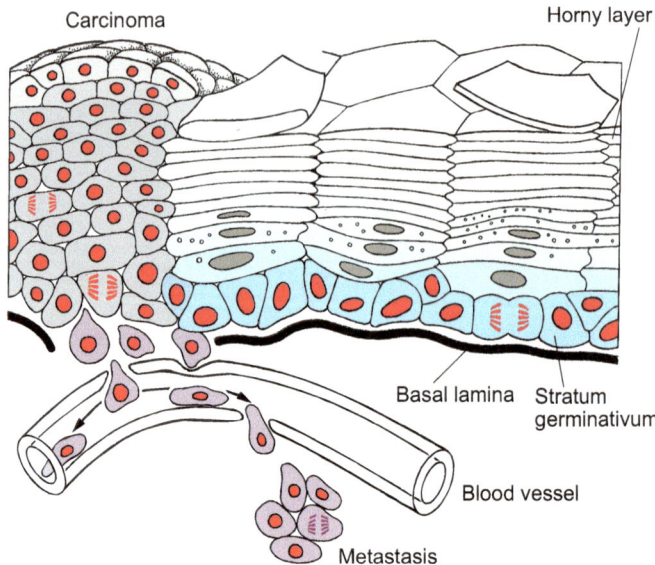

explicitly included in the definition but implicitly it is, because they derive from the inner cell mass of the blastocyst the cells of which are classified as pluripotent. In the mammalian/human embryo cells of the inner cell mass go on to create all the cells and tissues of the postnatal human body, but cells of the trophoblast, and hence of the placenta, are no longer included in the program. If removed and kept in culture cells of the inner cell mass are called **embryonic stem cells** or shortly **ES cells** or **ESC** (Fig. 18.1). Postnatal examples of pluripotent stem cells are the stem cells in the **intestinal crypts** (see Fig. 18.3).

In terms of molecular biology, pluripotent cells are featured by the activity of particular pluripotency genes. Some genes such as *Chd1* are responsible to keep the chromatin in a relaxed, accessible state. Of particular significance are the genes *Nanog, Oct-4* and *Sox-2*. In cooperation with *Sox-2, Oct-4* maintains a positive feed back loop of self activation (comparable to Fig. 12.20) and keeps the cell in the state of pluripotency. After cell division *Oct-4* continues to be active in (at least) one of the daughter cells keeping it in the pluripotent state. More about pluripotency genes in Sect. 18.5.

- **Multipotency** (Latin: *multi* = many) progenitor cells have the potential to give rise to cells from multiple, but a limited number of lineages that belong to one and the same classic germ layer. The classic paradigm of a multipotent stem cell is a **hematopoetic stem cell** of the bone marrow (see Fig. 18.6) – a blood stem cell that can develop into several types of blood cells but cannot develop into brain cells or liver cells.
- **Oligopotency** (Greek: *oligos* = few) is a comparably seldom used expression to designate stem cells that give rise to only a few cell types, for example the **lymphoid stem** cells of the hematopoetic system. Also neural progenitor cells can be assigned to this category; they are kept in the incompletely differentiate state by the gene *Sox*-2 and ready to divide.
- **Unipotency** is the restricted potency of a stem cell to give rise to only one cell type. The paradigm is the stem cells in the stratum germinativum of our skin that give rise to the horny cells (keratinocytes) of the epidermis (Fig. 18.2). Among the neural progenitor cells stem cells for a certain odorant class of the olfactory epithelium can be included in this category.

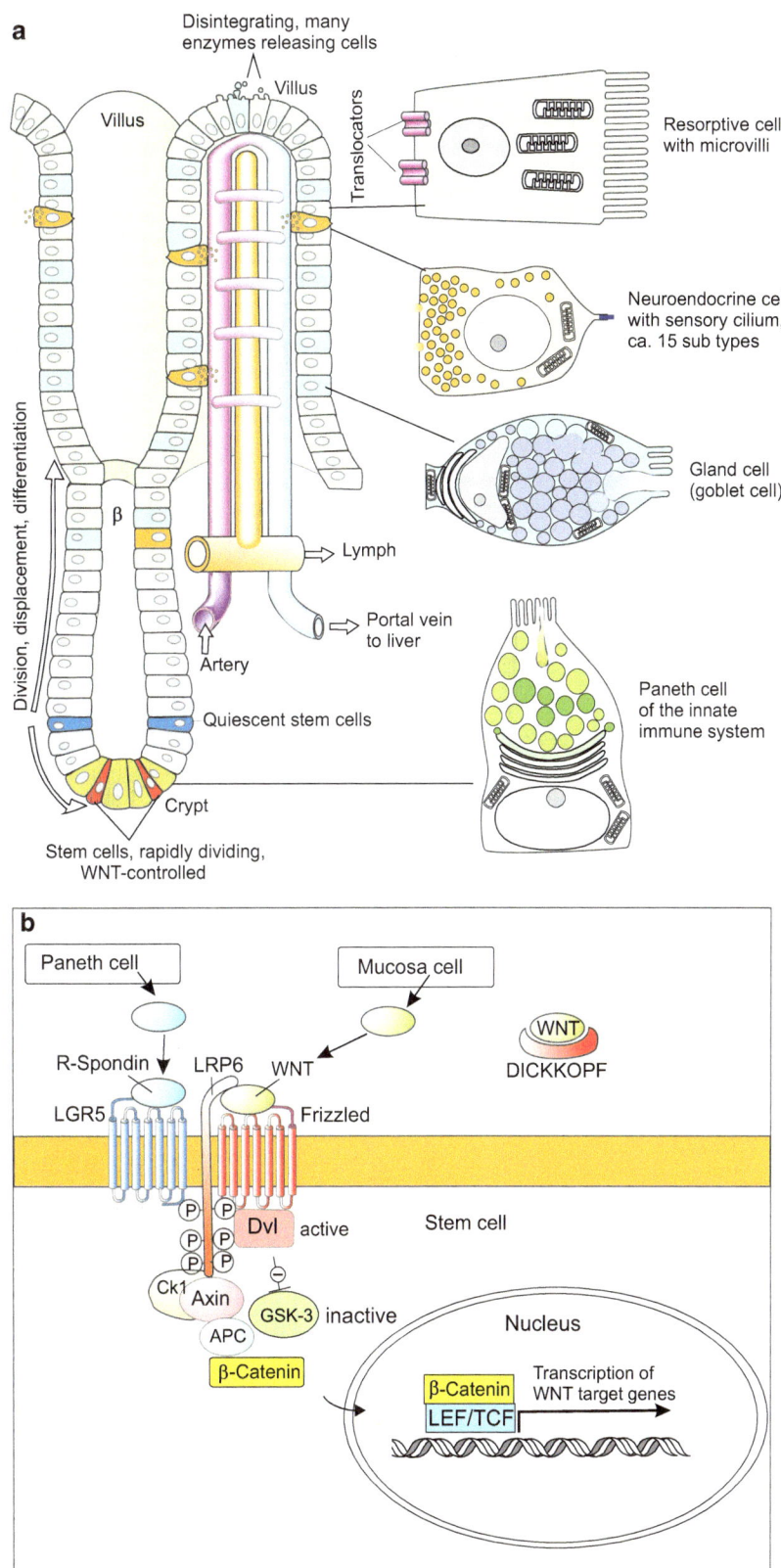

Fig. 18.3 (a) Renewal of the villi from stem cells in the small intestine. The multipotent stem cells are found not in the villi but at the bottom of adjacent crypts. The descendants of the stem cells are displaced in a distal

In practise this proposal for categorization is not strictly observed, and cannot be strictly applied to invertebrates and lower vertebrates. The definition, especially the distinction between pluripotency and multipotency, is tailor-made for placental mammals, as trophoblast and placenta are structures only found in placental mammals. In addition, linguistically the subdivision is not well founded as "pluri" and "multi" are almost synonymous ("pluri" = more, more than one, "multi" = many).

Skeletal muscles are renewed only in the event of particular need, for example when the muscle is injured. Residual myoblasts, called **satellite cells**, are left over from embryo development attached to bundles of muscle fibres. If needed, the quiescent satellite myoblasts are reactivated and multiplied. They fuse to myotubes and stop dividing. The myotubes fill their interior with bundles of actin and myosin, enlarge and take over the function of damaged muscle fibres.

18.1.4 Unipotent Stem Cells Have Only One Option: Renewal of the Skin and the Muscle as Examples

During fetal and juvenile stages skin must be expanded; throughout the whole life skin must be renewed. The skin is exposed to stress caused by mechanical strain, chemical irritation, UV irradiation, and microbial attack. Unceasing growth and renewal are accomplished through the mitotic activity of unipotent stem cells. These lie upon the basal lamina; their descendants are displaced outwards, and on their outward journey they differentiate to keratinocytes, die and finally are sloughed off after 2–4 weeks (Fig. 18.2). Stem cells are able to continue dividing in contact with laminin of the basal lamina. Stimulating growth factors such as **KGF** (Keratinocyte Growth Factor), **EGF** (Epidermal Growth Factor) and **TGFα** (Transforming Growth Factor alpha) and inhibitory signal substances, such as the stress hormone adrenaline (epinephrine), participate in the control of skin proliferation. Figure 18.2 in addition points to the danger of stem cell-derived tissues converting into tumours. This aspect is discussed below in Sect. 18.4.3 and in a more general context in Chap. 19.

18.1.5 Replenishment of Permanently Used Up Cells in the Gut and the Concept of Stem Cell Niches

Numerous protuberances in the small intestine, called **villi**, are composed of cells that move to the top of the villi, where they eventually disintegrate releasing many enzymes that supplement digestion. During their displacement, they serve the function of absorptive brush-border cells, of mucus-secreting goblet cells or of neurosecretory cells. When they arrive at the top of the villi, the cells sacrifice themselves to set free digestive enzymes; the residuals are shed into the lumen of the gut. Also the producers of the mucous layer, the goblet cells, are soon depleted and die. New cells are supplied by stem cells which lie in a protected position in the **depths of crypts** that descend into the intestinal wall in between the villi (Fig. 18.3a). As villi consist of several cell types their renewal depends on pluripotent stem cells. The term **pluripotent** is justified as the crypts contain not only derivatives of the endoderm but also neurosecretory cells that once in embryonic development derived from the neural crest cells (in traditional teachings assigned to the ectoderm).

Investigations on the crypts corroborated a concept that is currently put forward in stem

---

Fig. 18.3 (Continued) direction, enter the villi and eventually are sloughed off at the tip of the villi. Sloughed off cells release digestive enzymes. Rapidly dividing stem cells between the Paneth cells are thought to supply cells for permanent renewal of the villi while quiescent stem cells are thought to be activated upon injury (illustration by WM, based on Merlos-Suarez et al. 2011). (**b**) Model of proliferation control. Activation of WNT target genes requires the arrival of two external signalling molecules: (1) R-spondin, the ligand of the LGR5 receptor, supplied by Paneth cells, and (2) WNT, the ligand of the Frizzled receptor, supplied by mucosa cells (Illustration by WM, based on Schuijers and Clevers 2012)

cell biology, the concept of **stem cell niches** in tissues and organs composed of several different cell types. The villi are composed of a **mosaic-type pattern** of cell types including two compartments containing stem cells. Their derivatives are considered to differentiate in response to **local signals** presented by the local niche. Also the stem cells themselves must be subjected to local influences that ensure proper balance between self-renewal and differentiation. In controlling stem cell proliferation at the basis of the crypts the WNT signalling system is involved, in cooperation with the related **r-spondin** signalling system that enhances the signal strength in the deepest compartment to accelerate cell division (Fig. 18.3b).

18.2   The Hematopoetic System: Formation and Renewal of Blood Cells

In humans, each second millions of blood cells are born to replace millions of dying cells. The bloodstream carries many cell types with different functions ranging from the storage of oxygen to the combat of infections and the production of antibodies. An astonishing finding is that all these diverse cells are the offspring of a single type of stem cell. The production and differentiation of blood cells is known as **hematopoeisis** (Greek: *haimat* = blood, *poiein* = to make).

18.2.1 Hematopoietic Stem Cells Emerge in Embryonic Blood Islands; They Emigrate Into Niches of the Bone Marrow Provided by Mesenchymal Stroma Cells Where They Survive Life-long

Hematopoiesis, the production of blood cells (Figs. 18.4 and 18.5) takes place during embryonic development in scattered mesodermal **blood islands**, where the first blood vessels are also formed from angiogenic founder cells (see Fig. 17.2). Accordingly, in these islands one finds

hemangioblasts, founder cells which give rise to
- **angiogenic** progenitor cells and
- **hematopoietic (blood-forming) stem cells** (Fig. 18.6).

The first blood forming islands are found in the chick outside the blastodisc in the extraembryonic mesoderm covering the yellow yolk ball, and in humans in a corresponding position – the mesodermal layer covering the yolk sac (see Fig. 17.3). Later the stem cells migrate into the liver, the spleen and possibly the thymus. Ultimately, they populate the bone marrow. In the postnatal mammal significant numbers of hematopoietic stem cells are found only in the bone marrow. A surprisingly small number of multipotent stem cells (1 per 10,000 cells in the bone marrow) provide all of the blood cells.

A Glance at the Versatile Mesenchymal Stem Cells. Adult stem cells occupy particular **niches**. This applies also for the bone marrow. In the bone marrow mesenchymal stroma cells provide a spongy three-dimensional network with many niches where hematopoietic stem cells reside life-long (Fig. 18.5). In the present scientific literature stroma cells usually are addressed with the abbreviation **MSC**, standing for **mesenchymal stem cells**. Surprisingly, stroma cells turned out to be stem cells themselves, and to be a highly versatile type of stem cells. They can give rise to connective tissue, cartilage cells, cross-striated and smooth muscle cells, heart muscle cells and endothelial cells of blood vessels. Whether and to what extent this wide potential is actually used for tissue renewal in the adult body must still be explored. The paradigm of multipotent stem cells in the bone marrow remains the hematopoietic stem cells.

History and Medical Importance. Modern hematopoiesis research began with the search for means to protect against the disastrous effects of whole body X-irradiation. Mice whose blood forming system had been destroyed by high doses of X-rays could be rescued by injection of suspensions of bone marrow cells into the circulating blood. This was the pioneering experiment leading to procedures to cure leukaemia in

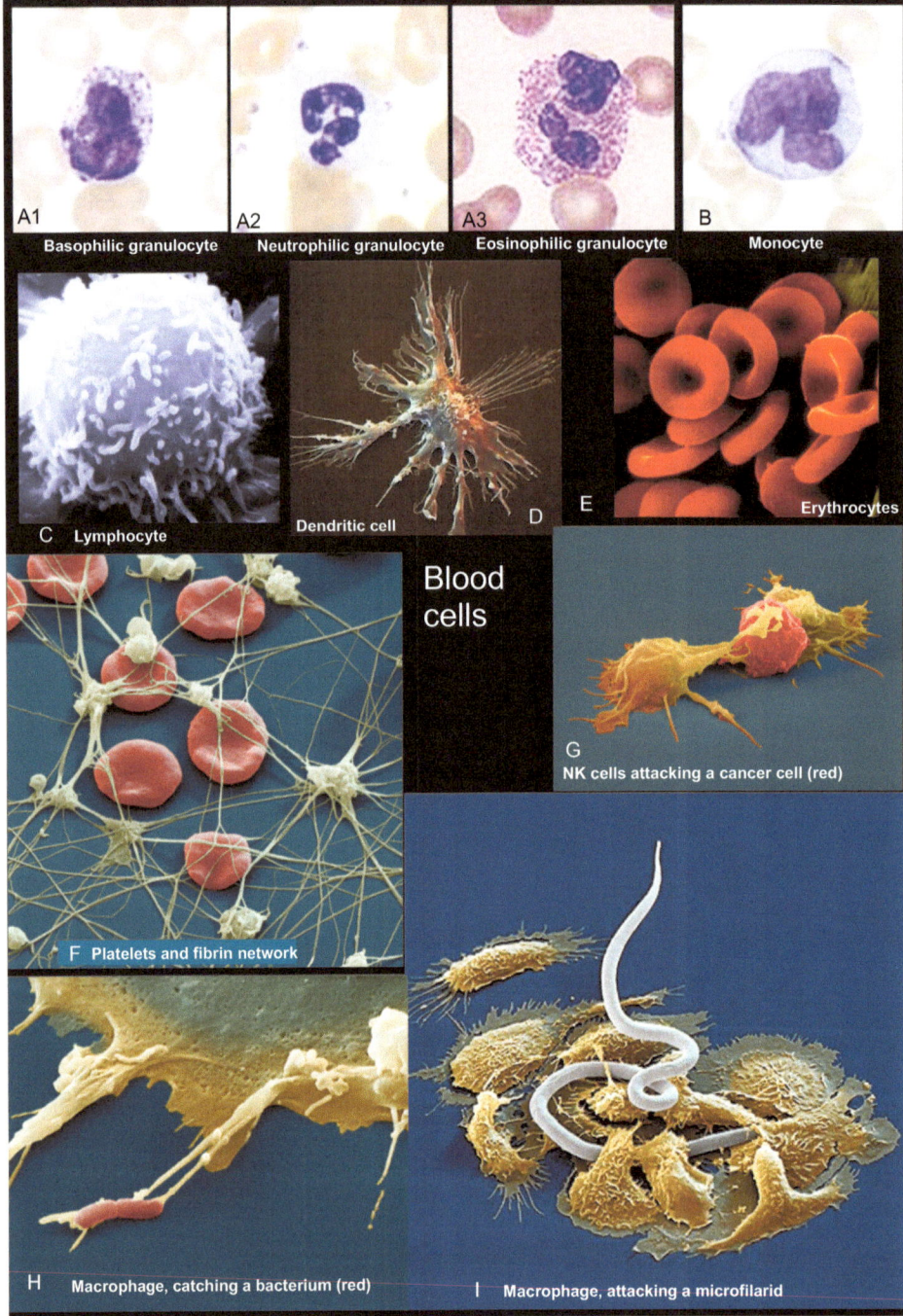

Fig. 18.4 (A–I) Cells of the human blood, illustration from Müller and Frings, Tier- und Humanphysiologie, Springer 2009. **A, B** Study collection of WM, **C** Rights reserved by Dr. Triche, National Cancer Institute; reproduced with his kind permission. **F–I** www.eyeofscience.de, purchased through Agentur-focus.de

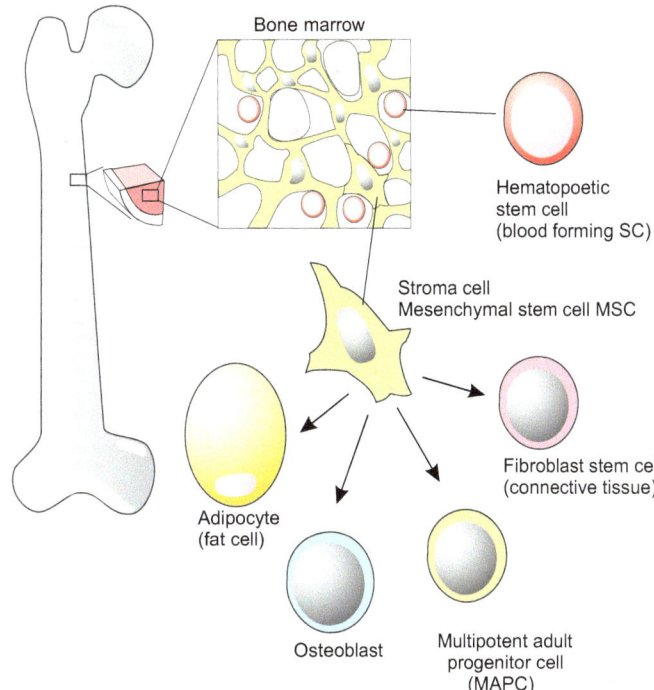

Fig. 18.5 Stem cells of the bone marrow with pluripotent stroma cells

humans. Malignant precursors are destroyed with high doses of X-rays or cytostatics and replaced by stem cells of a donor (genetic restrictions are discussed in Sect. 18.3).

However, stem cells only represent 0.01 % of the total mass of bone marrow, and much effort had to be invested to isolate and identify stem cells. Monoclonal antibodies were key tools, as they could be selected for their ability to recognize cell type or lineage-specific surface antigens.

A first indication as to whether a biopsy sample taken from the bone marrow contains stem cells, and how many, is provided by a procedure using a non-toxic dye (Hoechst 33342). Stem cells take up the dye as do other cells as well, but thereafter stem cells bleach by exporting the dye. For separation and isolation the cells removed from the bone marrow are dissociated and labelled with monoclonal antibodies raised to bind to characteristic surface components of the cells. The antibodies are coupled with fluorescent dyes, different antibodies are linked with dyes displaying different fluorescence spectra. This allows separation of the cells with a machine called FACS (Fluorescence-Activated Cell Sorting).

The surface structures recognized by the antibodies are generally designated CDx, with CD standing for "Cluster of Differentiation" and "x" standing for a number. For example progenitor cells of B-lymphocytes are characterized by the combinatorial label CD34 + CD19, T-cell progenitors are characterized by the surface structures CD34 + CD7 + CD5, whereas mature T-cells expose the surface structure CD3.

To isolate multipotent stem cells the distinguishing markers CD34 and CD90 are used. However, with these markers not only multipotent stem cells proper are

(continued)

Fig. 18.6 Pedigree of the blood cells. Haematopoesis, the formation of blood cells, starts in the bone marrow from multipotent stem cells. Natural killer cells arise from lymphoid stem cells; the intermediate stepsare still a matter of debate. Dendritic cells are considered to be phenotypes of macrophages. Other types of dendritic cells are assumed to derive from lymphoid precursors

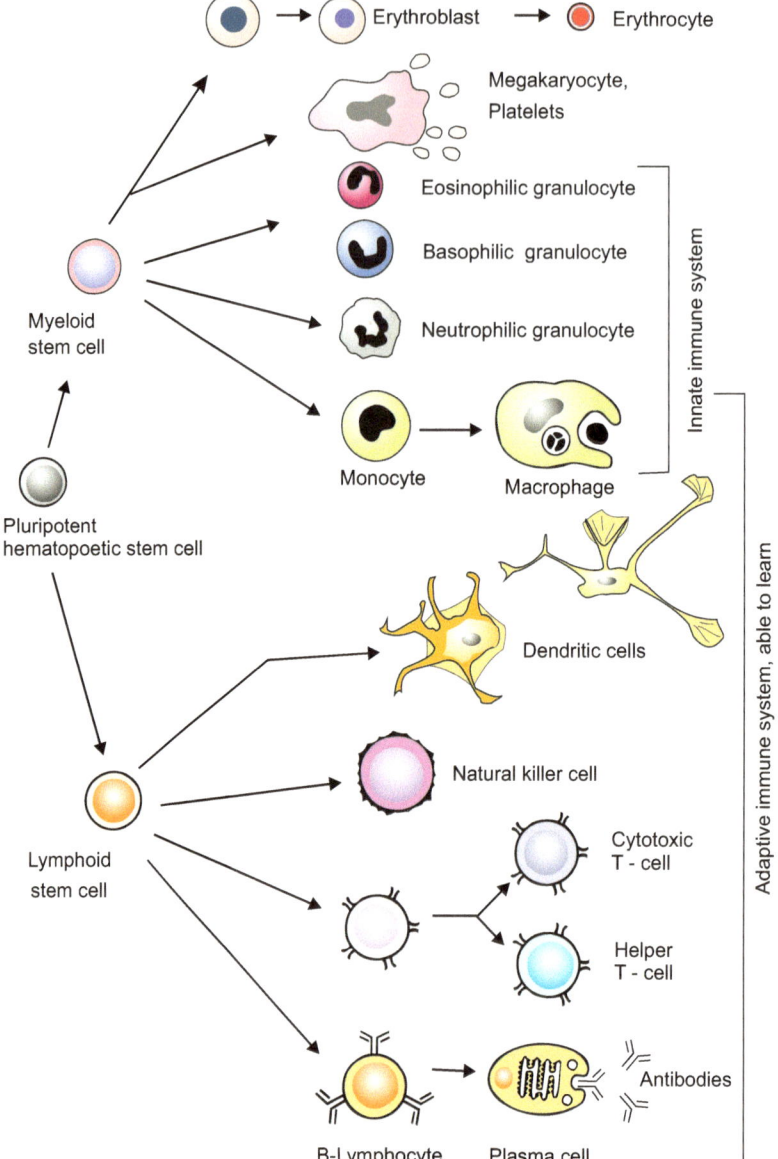

isolated but predominantly lineage restricted descendants which can undergo only a limited number of cell divisions. As a consequence the balance between the diverse blood cell types can soon be disturbed. A newly discovered marker CD49f enables to isolate the tiny fraction of the true basal multipotent stem cells.

Methods are being developed to replicate these cells in bioreactors and, if necessary, to cure genetic defects before they are reintroduced into the patient. Methods for correcting genetic defects such as homologous recombination are described in Chaps. 12 and 13.

18.2.2 The Turnover Is Huge: Up to 6 Million Erythrocytes Die Per Second and Must Be Replaced by New Ones

Without bone marrow stem cells, we would soon die because the life span of blood cells is restricted to hours, days or weeks. In humans, the half-life of

- neutrophil granulocytes is 7–14 h, of
- macrophages 5–7 days, for
- erythrocytes is 120 days. This may appear quite long, but considering the amount of blood, from this value a turnover of 2–6 million erythrocytes per second has been estimated.

Aged and dying blood cells are sorted out in the lymphatic organs, notably the spleen, and removed by macrophages through phagocytosis. The process of sorting out is called **sequestration**.

Most cells of the adaptive immune system, namely lymphocytes are also short-lived. Among the immune cells only 'memory cells' (B-memory cells and T-memory cells) are long-lived.

18.2.3 Descendants of Stem Cells Committed to Become Blood Cells Are Multiplied by Restricted Numbers of Amplification Divisions

Stem Cells, Blasts, and Terminal Differentiation. The rare multipotent stem cells of the bone marrow are endowed with the capability of dividing unceasingly. Part of their offspring remain multipotent stem cells (**self renewal**), while the remainder become committed to be progenitors (precursors) of more specialized cells with restricted developmental potential. The progenitor cells designated by the suffix -**blast** are still able to divide, but the number of rounds of divisions they can undergo is restricted. The limited reproduction of the progenitor cells serves to regulate the number of specialized blood cells on demand. One speaks of **amplification divisions**. After their last division, the cells undergo terminal differentiation.

The number of permitted cell divisions is in many precursor cell types predetermined.

18.2.4 In Mammals a Decision Tree Leads to More Than Ten Categories of Blood Cells

The pedigrees of blood cells could be traced using engineered retroviral vectors carrying a selectable marker gene (*neo*, Chap. 12). The marker virus, like other retroviruses, inserts its own DNA into the genome of the host cells and is multiplied during cell divisions together with the host genome. The virus was engineered to remove genes that would enable it to spread from cell to cell. Bone marrow cells were first removed and infected in vitro with the retroviral vector, then transferred into lethally X-irradiated mice. Offspring of the retrovirus-marked donor cells were predominantly found in the spleen, but also in the bone marrow. Clusters of blood progenitor cells in the spleen are known as **colony forming units** (**CFU**). Such units were derived – statistically – from one founder cell, but could give rise to one or several types of blood cells, depending on the position of the founder cell in the tree-like pedigree.

Blood progenitor cells can be cultured in vitro, where various factors can be added and single cells and their offspring monitored. A few cells behave like multipotent stem cells, forming colonies of various types of granulocytes, macrophages and other cell types classified below as the 'myeloid lineage'. Other isolated and seeded cells reveal more restricted potencies, generating only neutrophil granulocytes and macrophages. Still other seeded cells give rise to a single cell type such as neutrophil granulocytes or macrophages. From such studies, the whole pedigree has been reconstructed.

Starting from **one basic multipotent stem cell** (called Colony Forming Unit **CFU-M, L**) at least eleven types of blood cells are generated (Fig. 18.6):

A. **The myeloid lineage** supplies the following cell types:

1. **Erythrocytes** that carry haemoglobin and, in mammals, lose their nucleus;
2. **Megacaryocytes**. Unlike the other blood cells these "large nucleus cells" (literal translation) remain in the bone marrow, become very large and polyploid, and bud off numerous **platelets** into adjacent blood capillaries.
3. **Basophil granulocytes** which emit alarm substances such as histamine, leukotrienes and prostaglandins during inflammatory responses.
4. **Eosinophil granulocytes** which also fulfil tasks in inflammatory and allergic responses.
5. **Neutrophil granulocytes**, sometimes also called microphages;
6. **Macrophages**. Neutrophil granulocytes and macrophages together devour a large variety of infectious invaders and much waste produced in the body. **Monocytes** circulating in the blood are considered to be phenotypic modifications of the macrophages found in the interstitial spaces of tissues and in the lymphatic organs (lymph nodes, tonsils, and spleen). In the spleen the macrophages eat infectious bacteria as well as aged erythrocytes. In the liver they are known as stellate **Kupffer's cells**. In their morphology these are intermediate between macrophages and dendritic cells.
7. **Myeloid dendritic cells** are considered by several researchers to represent another phenotypic edition of macrophages or monocytes, respectively. (Another type of dendritic cells is thought to derive from the lymphoid stem cells which also give rise to B-cells and therefore is listed below under B). **Dendritic cells** share branched projections, the 'dendrites' that give them their name (but are structures distinct from the dendrites of neurons). Like macrophages dendritic cells sample the surrounding environment for pathogens such as viruses and bacteria. This is facilitated by pattern recognition

receptors (PRRs) such as the toll-like receptors (TLRs). Like macrophages dendritic cells engulf the antigenic materials and insert peptide residuals into their MHC which subsequently is exposed on their surface. The MHC + antigen peptide complex is then presented to T-cells for examination, a process called **antigen presentation**. Besides macrophages and B lymphocytes, dendritic cells are the main population of the **professional antigen-presenting cells APC**. Dendritic cells are found in large numbers in the thymus; but are met in many other places as well, especially in tissues in contact with the external environment, such as the skin (where there is a specialized dendritic cell type called **Langerhans cells**) and the inner lining of the nose, lungs, stomach and intestine.

B. **The lymphoid lineage** supplies:
8. **Dendritic cell** of putative lymphoid origin, among them **plasmacytoid** dendritic cells; they share characteristics of lymphocytes and dendritic cells.
9. **B-lymphocytes**. In the form of plasma cells, the B-cells produce and release antibodies.
10. **T-lymphocytes** are subdivided into several subtypes (T4 = T-helper cells; T5 = T-suppressor cells; T8 = cytotoxic T cells), and together constitute a system to discriminate between self and non-self and to adjust the mechanisms of defence to the current needs.
11. **Natural killer cells** presumably also derive from the lymphoid stem cells and fulfil tasks in the innate immune response attacking viruses and, under certain conditions, also cancer cells. NK cells are unique, as they are endowed with the ability to recognize stressed cells in the absence of antibodies and antigen-presenting MHC recognition molecules, allowing for a much faster immune reaction.

The various non-erythrocytes are often summarized as **leucocytes** (white blood cells).

When, as a result of disease, the blood contains too few erythrocytes physicians speak of **anaemia**. If the blood is enriched with too many but immature leucocytes physicians speak of **leukaemia**.

eventually serve as T-cells or antibody producing B-cells.

Somatic recombination, to date found only in lymphocytes, leads to an irreversible alteration in the genetic constitution of the cell.

18.2.5 In the Development of Lymphocytes a Process of Irreversible Somatic Recombination Enables Production of a High Number of Diverse Antibodies

Recombination – the rearrangement of genes – is characteristic of sexual reproduction and takes place during meiosis. Surprisingly, a similar event occurs in the development of lymphocytes, when lymphoblasts combine their individual genetic programs for preparing the future production of antigen receptors or antibodies (which are liberated antigen receptors). This process, called **somatic recombination**, occurs at the 'birth place' of these cells: the bone marrow, the thymus or in lymph nodes (Fig. 18.7).

Lymphoblasts are genetic gamblers. They try their luck when they combine various short DNA sequences at random to program the variable region of their receptors or antibodies, increasing the chance of creating a combination appropriate to recognize and bind a future foreign antigen. In each B-cell or T-cell a different random combination is tried (Fig. 18.8). Subsequently, the T-lymphoblasts emigrate and settle in the thymus; the B-lymphoblasts remain in the bone marrow for some time. Wherever lymphoblasts stay, they learn to discriminate between self and non-self. A strict negative selection eliminates all those lymphoblasts whose receptors would bind molecules belonging to the self. The surviving cells are those that can bind only non-self molecules and are thus potentially capable of recognizing foreign substances. The surviving lymphocytes migrate through the body, colonize other places such as spleen and lymph nodes, and

18.2.6 In Lymphoblasts Cell Death Serves a Process of Learning

Programmed cell death of lymphoblasts is an essential part of the learning process that enables the immune system to discriminate between self and non-self. It may seem strange, but there is good reason for nature to use a strategy based on cell death. Millions of lymphoblasts, the future B-cells and T-cells of the immune system, rearrange DNA segments at random to produce a huge variety of antigen receptors or antibodies (antibodies are released antigen receptors). This strategy will inevitably not only generate lymphocytes whose receptors fit onto foreign antigens, but also self-reactive lymphocytes with receptors that bind self-molecules. This would be fatal if such lymphocytes were not quickly eliminated because the immune system is extremely aggressive.

Most of the self-reactive lymphocytes are eliminated in humans around the time of birth when antibodies taken up with mother's milk no longer suffice. Due to this process of recombination followed by elimination, the immune system recognizes foreign substances rather reliably. How all these learning processes are accomplished in detail is subject to intensive investigations. Knowledge about these control mechanisms will be very important in understanding **autoimmune diseases**. Intolerance to self antigens in later life is thought to be due to continued production of self-reactive lymphocytes from stem cells.

Steroid hormones of the adrenal cortex, in particular cortisol and cortisone, promote the collective suicide of T-lymphocytes in the thymus. Birth and death in the immune system support the life of the entire cell community of our body.

Fig. 18.7 Locations in which cells of the immune system are prepared for their task. The lymph node and thymus represent the entire lymphatic system

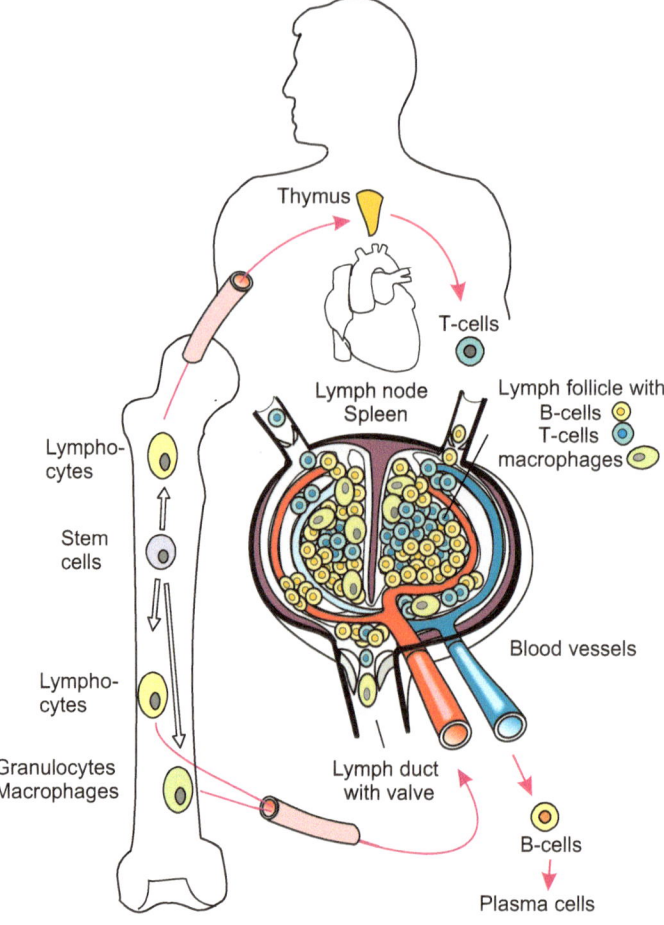

18.2.7 Blood Cell Manufacture Is Controlled by Many Cytokines and Hormones

Many soluble, extracellular polypeptides influence types and numbers of blood cells manufactured. Many of these proteins have been isolated, identified and their genes cloned. A term designating such factors in general is **cytokine**. Examples of cytokines are the

- **Stem Cell Factor (SCF)** that stimulates the stem cells of the myeloid pedigree to proliferate. Other cytokines do not manure the root of the tree but stimulate branching and growth of special branches. Such secondary cytokines are:

- **GM-CSF**, the **G**ranulocyte and **M**acrophage **C**olony **S**timulating **F**actor,
- **G-CSF**, the **G**ranulocyte **S**timulating **F**actor, and
- **M-CSF**, the **M**acrophage **S**timulating **F**actor.

Cytokines that stimulate progenitors of lymphocytes are known as **interleukins**. The name indicates that sources of such cytokines are other leucocytes. Interleukins are messengers that are exchanged between leucocytes, for instance between lymphocytes and macrophages, or between B- and T-lymphocytes.

Probably the best known cytokine is the hormone **erythropoietin** (Greek: 'making red blood cells'). Erythropoietin (known as doping substance "**epo**") is a glycoprotein that is

Fig. 18.8 Somatic recombination exemplified for the light chain (L-chain) of the B-cell receptor BCR. Without a transmembrane domain the BCR is released as an antibody. In programming the variable, antigen-binding domains of the BCR chromosomal gene segments are combined by somatic recombination according to the random principle. If by chance in a certain lymphoblast a combination is found that fits a given antigen closely this combination is retained, the lymphoblast multiplies by proliferation, and the combination is used to produce antibodies. The cell releasing mature antibodies is called a plasma cell

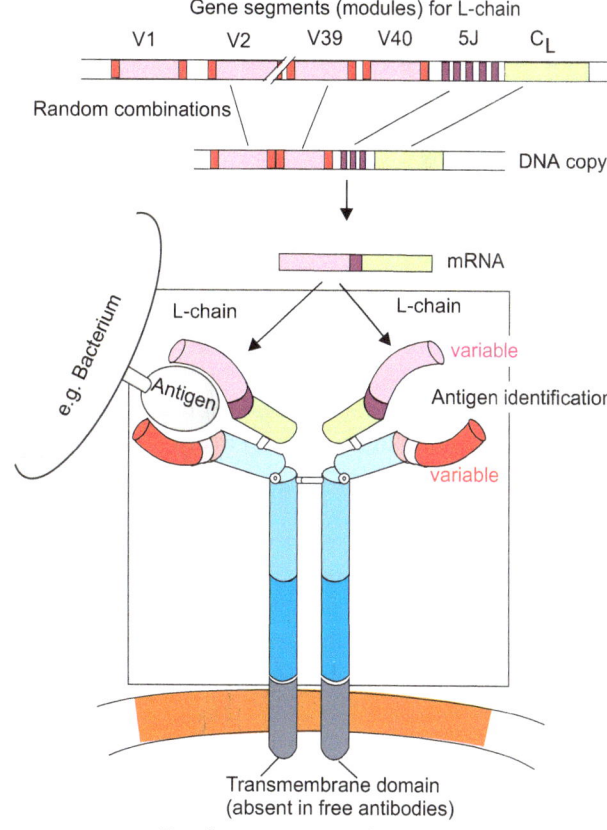

In the Nucleus of a B-Lymphoblast
(a future antibody-producing cell)

Gene segments (modules) for L-chain

V1   V2   V39   V40   5J   C_L

Random combinations

DNA copy

mRNA

e.g. Bacterium

L-chain   L-chain

Antigen   variable

Antigen identification

variable

Transmembrane domain
(absent in free antibodies)
B-cell-receptor or antibody, respectively

synthesized in the kidney under oxygen-deficient conditions and transported by the bloodstream. It stimulates the pro-erythroblasts in the bone marrow to divide more rapidly. In this way the manufacture of red blood cells is increased and the oxygen-binding capacity of the blood improved.

Such factors, of which a considerable number have been identified, are often designated hormones if they have been found in the blood, following the tradition of using the term 'hormone' to designate substances active in trace amounts and transported to their site of activity by the blood.

18.3   Regenerative Medicine: Tissue and Organ Replacement from Stem Cells

18.3.1   Stem Cells of the Adult Organism Have Higher Potentialities Than Previously Thought; If Derived from the Patient They Escape Attack by the Immune System

Infusion of foreign blood, transplantation of foreign bone marrow or foreign organs, presuppose sophisticated typing of recipient and donor. Only very few donors have markers on their cells surface similar enough to those of the patient allowing immune tolerance. In particular membrane associated proteins of the polymorphic (polygenic) **Major Histocompatibility Complexes MHC** I, in humans known as **HLA** (human Leucocyte Antigen), represent an individual specific **barcode** enabling the immune system to discriminate between self and non-self. The MHC/HLA complex consists of 16 different transmembrane proteins that in human populations occur in innumerable allelic variants and combinations. **Even between siblings the chance of tolerable matching is merely 25 %**.

Considering the complexity of the MHC/HLA system it seems unimaginable that donor stem cells can by genetic engineering be managed to express the patient's MHC/HLA instead of its own inherited complex.

Provided one could use stem cells taken from the patient's body itself such an immune intolerance is expected not to occur. Which developmental potentiality do adult stem cells actually have? Two of many examples from animal testing:

Experiment 1: Hematopoietic stem cells of the mouse are isolated and genetically marked, for example by inserting a GFP reporter gene into its genome (see Chap. 12, Box 12.2), and then injected into the blood circulation of a pinky rat baby (newborns just beginning to discriminate between self and non-self adopt foreign cells as being self cells.) Descendants of these cells light up green in the fluorescence microscope when excited with short-wave blue light.

Experiment 2: Adult hematopoietic stem cells of a male mouse are injected into female host blastocysts to enable identification of their descendants by means of the Y-chromosome. The host embryos are implanted into the uterus of foster mothers and the offspring examined at various times of their development for the presence of differentiated cells bearing a Y chromosome (also in individuals arising from chimerical blastocysts immunological rejection of host cells does not occur).

Some researchers raised hopes and arose attention by reporting that in their experiments the introduced haematopoetic stem cells not only gave rise to blood cells but also to other cell types such as cartilage cells, muscle cells, and even nerve cells if they had entered the brain. Other researchers could not confirm such a wide range of differentiation potencies.

Strange findings made in such experiments gave clues about a curious phenomenon: Occasionally stem cells of the mother can pass the placental barrier and enter the blood of the child. The child will then genetically be a **chimera**. A boy can embody some female cells. Conversely male stem cells of the fetus can enter the blood of the mother and persist in the mother's body for years. This can lead to problems associated with the Rhesus factor for the next child (see Fig. 6.13).

18.3.2 The Successful Stem Cell Therapy of Leukaemias and Lymphomas Was a Pioneering Feat

Developmental biology is going to assist medicine at an increasing rate. In 2001 media coverage directed wide public interest to ambitious plans of several (mostly American and Far Eastern) laboratories to produce spare parts from stem cells for patients with degenerative diseases or for victims of an accident. The hopes are based on experiences with stem cell therapies for years. For many years it has been possible to harvest multipotent stem cells from the bone marrow – even from adult donors – which can be replicated in cell culture and used for therapeutic purposes.

Hematopoietic stem cell transplantation is the transplantation or infusion of multipotent hematopoietic donor stem cells, usually derived from bone marrow, peripheral blood, or umbilical cord blood. This medical procedure is performed for patients with certain cancers of the blood or bone marrow, such as **multiple myeloma** or **leukaemia**.

Previous to the transfer of the stem cells, the recipient's diseased hematopoietic system is usually destroyed with X radiation (radio'therapy') or high doses of cytostatics (chemo'therapy') with the objective of destroying all cancerous cells. Unfortunately, this necessary harsh treatment inevitably destroys also healthy stem cells. Hence, no more blood would be produced and the immune system permanently paralysed. But subsequent to the chemo or radio 'therapy' the real therapy starts. Stem cells of a healthy donor tested for immunological suitability are implanted, actually infused intravenously into the bloodstream of the patient. The stem cells can repopulate the bone marrow by themselves – one speaks of **homing** – and supply new healthy blood cells.

From which source are stem cells taken? There are, or were, two procedures in use. The historically older procedure is the so called **bone marrow transplantation** in which, however, not complete bone marrow explants but suspensions enriched with stem cells were infused into the blood stream of the recipient. The procedure for enrichment is sophisticated and time-consuming (see insert above in Sect. 18.2.1), and surgical removal of bone marrow samples is associated with risk for the donor.

The procedure of therapy was much facilitated when it was shown that stem cells can be harvested from the circulating blood. To separate the tiny fraction of stem cells, drawn blood is run through a special centrifuge and the residual blood given back to the donor.

Because of possible immunological complications the physician discriminates between:

- **Autologous** transplantations. These are possible if cancer has not affected the bone marrow itself. In this case bone marrow is removed from the patient before the patient is subjected to global treatment with X rays or chemicals, and the sample is given back to the patient after completion of treatment. If, however, bone marrow is already infiltrated by cancer

cells these must be removed by elaborate and risky procedures – risky, because all cancer cells must completely be removed (this applies also to other dangerous inclusions such as HIV);

- **syngeneic** transplantations from and to monozygotic twins, truly a rare case;
- **allogeneic** donors, immunologically typed and classified as being more or less related (often taken from close relatives or from typed stem cell reserves from a stem cell bank).

18.3.3 No Longer a Utopian Dream: Replacement of Tissues Prepared from Adult Stem Cells Is Already Practiced

At present, besides autologous transplantation (infusion) of blood stem cells several more successful possibilities for therapy with stem cells are available or being developed. On a commercial basis preparations of adult stem cells are produced for

- **Artificial skin replacements**, also announced as '**tissue-engineered skin**' or '**cultured skin graft**'). Such products allow severely damaged skin, related to burns or inherited diseases and open wounds, to grow new tissue, and heal faster. Previously there was the possibility of skin grafting. Pieces of skin were removed from another area of the patient's body, where undamaged skin was available, and transferred to the area of destroyed skin. Yet, self-evidently only small pieces could be removed for grafting. Enlargement of the skin pieces by growth in culture before placing them onto the wound helped in single cases. In the further development of the procedure stem cells of the skin are isolated, augmented through replication in the laboratory, and laid on the wounded area embedded in a suitable support material. Artificial skin usually consists of synthetic epidermis and collagen-based dermis. The epidermis arises from keratinocyte stem cells, the dermis from fibroblasts seeded onto a collagen lattice.

More and more variants of the procedure are offered by commercial companies.

- **Autologous chondrocyte transplantation** aims at repairing damage in articular cartilage and replenishing destroyed cartilaginous structures such as **meniscal cartilage** (knee-cap, patella). In a first stage some milligrams of cartilage is removed arthroscopically from a less weight-bearing cartilage anywhere in the body. Stem cells are isolated enzymatically and multiplied in the laboratory by growing in a bioreactor. Growing enough cells takes about 4–6 weeks and is done on a scaffold substrate as outlined in Fig. 18.9a. The growing tissue must be subjected to mechanical forces to bring the tissue into the correct shape and orient the cell internal stress fibres according to the anticipated forces acting on them in the knee. Procedures for growing joint cartilage or intervertebral discs are in preparation.
- **Autologous heart valve tissue engineering**. At the current forefront of tissue engineering and approaching replacement of organs are successful trials to grow heart valves. They are grown in special bioreactors. As opposed to artificial valves constructed with non-living materials living valves can grow with a child.
- Some more possibilities offer pluripotent stem cells of the bone marrow which can give rise to, for example, **cardiomyocytes** under the guiding influence of biochemicals and growth factors (Fig. 18.9b). They promise some improvement of heart performance after aheart attack, and especially in congestive heart failure. Interestingly, in confluent layers of cardiomyocytes spontaneous coordinated waves of common rhythmic contractions occur.
- **Endothelial stem cells** have proved to be particularly useful. They are currently used to produce **blood vessel prostheses**. A tube-shaped matrix is prepared from biologically compatible material. Endothelial stem cells are seeded onto this, and the prostheses are cultured in bioreactors.

Several procedures are being developed or proposed but only a few have reached the stage of clinical testing. Before stem-cell therapeutics can be applied in the clinical setting much work has to be done related to safety problems and the implicit danger of cancer formation by the implanted stem cells. Nevertheless, experts propose to transplant **mesenchymal stem cells** to treat **multiple sclerosis**. We make young people aware that the efforts of bioengineers include experiments to grow a **third generation of teeth**; experiments already yielded some hopeful results in animal models.

For your own inquiries in this field we recommend the key words "*tissue engineering*", "*stem cell therapy*".

Possibilities for autologous stem cell therapy are restricted to cases in which adult stem cells are available. If no appropriate progenitor cells are found hopes may be built on embryonic stem cells harvested from one's own, stored umbilical cord, or, presumably more realistically, hopes may be set on autologous pluripotent stem cells secondarily generated from differentiated cells as outlined in Sect. 18.5.

The **umbilical cord**, previously carelessly discarded has turned out to be a rich source of stem cells of various kinds. Blood forming stem cells of the fetus arrive in the umbilical cord in the last third of gestation on their journey through the blood circulation when they change from their previous home in the spleen and liver to their postnatal destination in the bone marrow. In addition umbilical cords contain mesenchymal stroma stem cells. Yet for these cells the same applies as for ES cells: Transplanted into an allogeneic recipient they are foreign cells and subjected to attack of the host's immune system. Only when ones own umbilical cord is still available one may have a reserve source of immuno-tolerant cells. Therefore, several commercially operating companies offer to collect and keep umbilical cords frozen. However, to preserve the umbilical cord of every one of billions of

(continued)

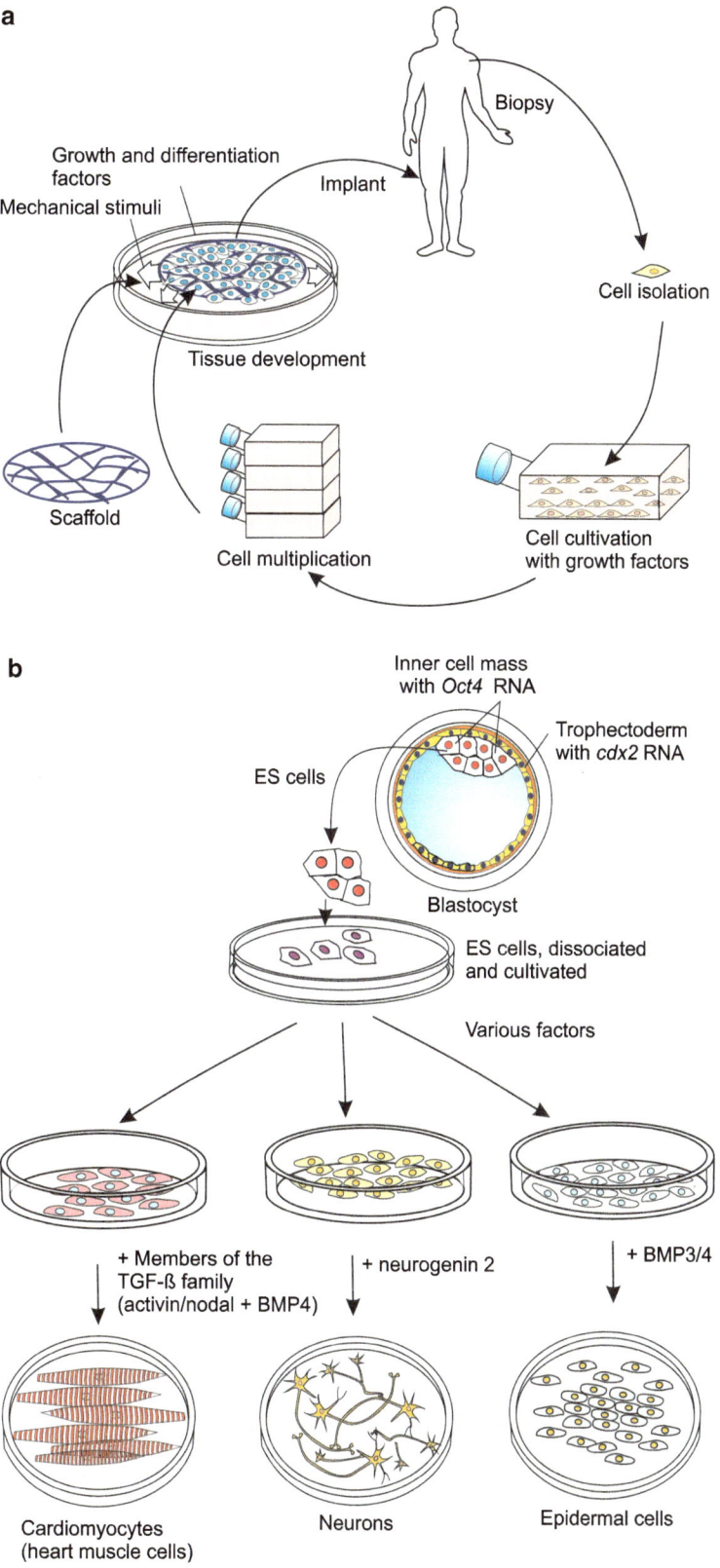

Fig. 18.9 (continued)

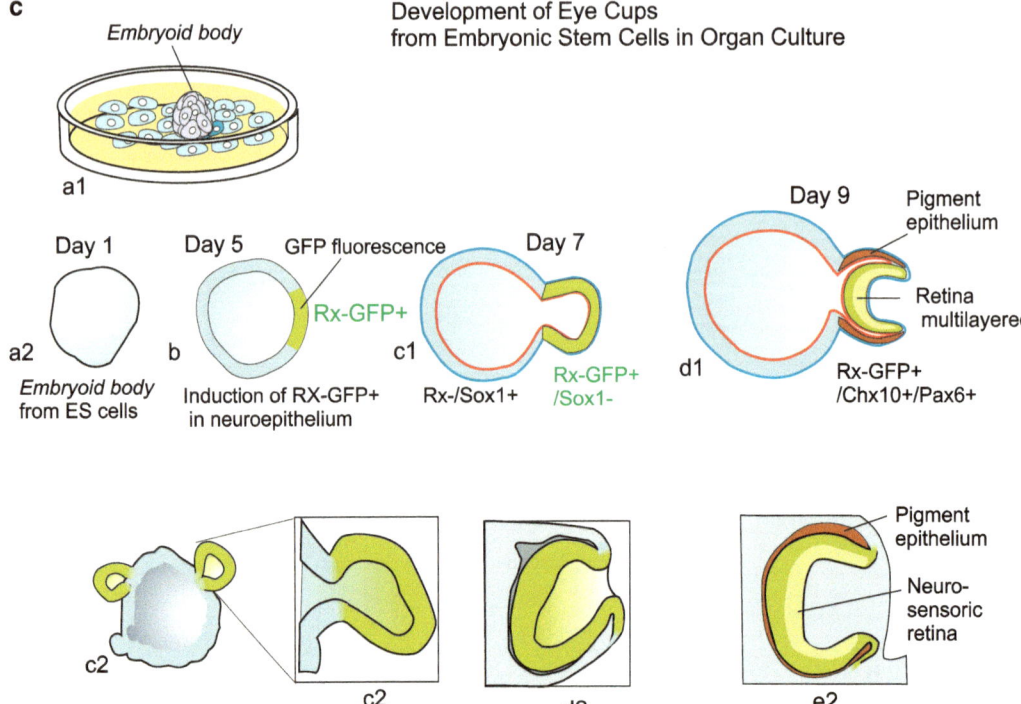

c

**Development of Eye Cups
from Embryonic Stem Cells in Organ Culture**

Embryoid body

a1

Day 1

a2

*Embryoid body
from ES cells*

Day 5

b

GFP fluorescence

Rx-GFP+

Induction of RX-GFP+
in neuroepithelium

c1

Day 7

Rx-/Sox1+

Rx-GFP+
/Sox1-

d1

Day 9

Pigment
epithelium

Retina
multilayered

Rx-GFP+
/Chx10+/Pax6+

c2

c2

d2

e2

Pigment
epithelium

Neuro-
sensoric
retina

Fig. 18.9 (**a**) Basic technique of tissue engineering, here a procedure for producing an autologous cartilage transplant, for example a kneecap. From a healthy cartilage of the patient a small piece is removed (biopsy). From this piece chondrocytes are separated and cleaned from accompanying other cells, and multiplied in culture. The chondrocytes are seeded onto a scaffold substrate that is placed into a bioreactor – for simplicity in the illustration here replaced by a Petri dish – and incubated with added growth factors. Targeted mechanical forces aid in bringing the tissue into shape and orient stress fibres within the cells. After 6 weeks the resulting kneecap can be implanted into the patient. (**b**) Factors used to direct differentiation of embryonic stem cells. (**c**) Development of an eyecup from embryonic stem cells of the mouse. Petri dish **a1** stands for a multi-gas incubator. Three-dimensional development of the cell aggregate was enabled by a three-dimensional gel prepared from extracellular matrix components. Video recordings of eye development were performed through a multiphoton fluorescence microscope. The designations Rx-GFP, Sox1, Chx10 and Pax6 stand for characteristic differentiation markers. After Eiraku et al. (2011)

human beings appears not to be a realistic future-oriented proposal. Even in the deep-frozen state living biological material ages and looses viability. And cryo-preservation in liquid nitrogen entails high costs.

In contrast, a procedure routinely performed in Great Britain and other countries does indeed matter: Stem cells harvested from umbilical cords are immunologically typed, just like organs to be transplanted, and kept for the foreseeable future in stem cell banks. In case of demand they can be awakened and multiplied in a bioreactor.

18.3.4 Organs from Stem Cells? Spectacular Eye Cups Formed in Culture Vessels Through Self-Organization, What Can They Promise?

Current research sets much hope on **embryonic stem cells (ESC, ES cells)** as these potentially give rise to any of the more than 200 different cell types of our body. A technical practice to handle such cells is to grow them in high density and let them aggregate into three-dimensional structures called **embryoid bodies**. After addition of appropriate cocktails composed of nutrients and differentiation factors such aggregates can form solid tissues of various kinds (Fig. 18.9b, c). Impressive examples:

- **Neurospheres**. Primarily a neurosphere is a free-floating aggregate of fetal cells containing neural stem cells. Since neural stem cells cannot be studied in vivo, neurospheres provide a method to investigate neural precursor cells in vitro. A neurosphere assay, or stemness assay, is used to confirm that neurospheres contain neural stem cells. When plated on a laminin substrate, and deprived of growth factors, neurogenic progenitor cells will differentiate into neurons, astrocytes, and oligodendrocytes. Depending on the source of the cells in the fetal brain they may form more complex structures even reminiscent of the adult CNS. Although currently neurospheres are the subject of basic research only, hopes are expressed that they might one day be used for cell therapy of human neurodegenerative diseases.
- 'Heart-shaped' embryoid bodies. Cardiomyocytes, either derived from ES cells or from induced pluripotent cells (Sect. 18.5) can form bodies that collectively perform rhythmic contractions. Cardiomyocytes are a mixture of spontaneously electrically active atrial, nodal, and ventricular-like myocytes that possess typical electrophysiological characteristics and exhibit expected electrophysiological and biochemical responses upon exposure to exogenous agents. Thus, these 'heart bodies' can be used in targeted drug discovery. However, they are tiny and not functional pumps.
- **Eye cups**. In 2011 Japanese researchers published a spectacular result. They succeeded in producing complete eye cups from embryonic stem cells of the mouse in the bioreactor via embryoid bodies as intermediates (Fig. 18.9c). First the aggregates were reared with a cocktail prepared from basement-membrane matrix components (matrigel) to promote the formation of rigid continuous epithelial structures. The aggregates enlarged and spontaneously formed hemispherical epithelial vesicles. In some areas the retinal epithelium formed eye fields. These fields began fluorescing green as the ES-cells were derived from transgenic mice. In these mice the gene of the eye-field marker RX had been fused with the GFP gene.

On days 8–10, the Rx–GFP positive vesicles underwent a dynamic shape change and adopted a two-walled cup-like morphology. Whereas the outer (proximal) portion differentiated into mechanically rigid pigment epithelium, the flexible inner (distal) portion progressively folded inwards to form a shape reminiscent of the embryonic optic cup. Thus optic-cup morphogenesis in this simple cell culture depended on an intrinsic **self-organizing program** involving stepwise and domain-specific regulation of local epithelial properties.

What still was lacking was a lens and cornea. Producing a lens was a success already achieved by former researchers. Whether such eyes would connect themselves with the brain remains to be seen.

With respect to possible medical applications it may be appropriate to dampen premature euphoria. It must be pointed out that these 'heart-shaped' bodies and eye cups were grown from embryonic stem cells allogeneic to any recipient, and they are very tiny embryonic structures only a few micrometers in size. How much time would it take to grow them to a transplantable size? And would the eye be able to establish contact to an adult lateral geniculate body over distances many times greater than in the embryo?

18.3.5 A Utopian Idea: "Therapeutic Cloning" Proposed to Circumvent Immunological Barriers

If medical science succeeded in producing replacement tissue from **embryonic** stem cells in sufficient quality and quantity the problem with immunological barriers still existed. As outlined above, every individual bears a unique barcode consisting of a set of MHC/HLA proteins on the surface of her/his cells, and this barcode will sooner or later also appear on the surfaces of cells derived from a blastocyst. The barcode will be different from that of any patient. And according to current knowledge it appears to be impossible to replace the complex barcode of the donor cells by that of the envisaged recipient by genetic engineering. Therefore, in the 1990s it was proposed first to clone a blastocyst with the genome of the patient as illustrated in Fig. 18.10. And this blastocyst would be used as source of stem cells. Since this procedure implies using up an embryo that could give rise to a human being the proposal met severe ethical reservations and fierce refusal. In many countries laws forbid any relevant experiments. Moreover, recent research in induced pluripotency as described in the following section, is expected to make the idea of "therapeutic cloning" obsolete.

18.4 iPSC: Reprogramming Differentiated Cells to Pluripotent Stem Cells

18.4.1 With Appropriate Protocols Current Research Increasingly Succeeds in Giving Features of Stem or Precursor Cells Back to Differentiated Cells

In 2006 and 2007 in the scientific journal Science two reports appeared presented by Japanese researchers that in professional circles instantly were perceived as a sensational discovery (in 2012 the senior author K. Takahashi was awarded the Nobel price in medicine). The articles reported that introduction of **genes for four transcription factors** into fibroblasts (in 2006 taken from adult mice, in 2007 from human skin) allowed these cells to be transformed into pluripotent stem cells (Fig. 18.11). Current science speaks of **iPSC** or **iPS, induced pluripotent stem cells**. The researchers compared differentiated cells with embryonic stem cells looking for differently expressed genes in search of genes mediating pluripotency. The genes in question were introduced into fibroblasts by means of a viral vector and the differentiation potential of the recipient cells analyzed in extensive trials. Eventually four genes were identified whose combined activity conferred pluripotency to those fibroblasts. The final successfully introduced vector comprised the DNA sequences for the transcription factors **OCT-3/4**, **SOX-2**, **c-MYC** and **KLF-4** (Krüppel-like Factor 4).

These sequences are part of the mammalian genome and therefore were already present in the fibroblasts but brought in a state of permanent silence by epigenetic modification. By contrast, the genes introduced with the viral vector were activated probably because in the fibroblasts the machinery of epigenetic silencing had already ceased to work.

However, the method of reprogramming was not very effective – only 0.0125 % of the treated fibroblasts developed features of stem cells – and the vectors were inserted into the genome at random positions.

Since then in many laboratories around the world researchers strive to improve the procedure, and success is increasing. If cells are found in which not all of these pluripotency genes are blocked less than four genes have to be introduced. On the other hand in other differentiated cell types other genes may be required such as the genes for NANOG and LIN-28. In any case the factor **OCT-4** must be active in the cell in order to regain and maintain features of pluripotent stem cells. By transfection with *oct-4* unipotent adult neurogenic precursor cells could be transformed back into cells resembling pluripotent embryonic stem cells.

Fig. 18.10 Hypothetic "therapeutic cloning" as proposed in the 1990s by some researchers. To get immune tolerant replacement cells and tissues it was proposed to fuse a pluripotent cell taken from the patient with an enucleated egg cell from any donor and to grow the product to the blastocyst stage in vitro. Embryonic stem cells removed from the blastocyst would be used as a source to raise cells or tissues which would expose the individual MHC combination characteristic of the patient on their surface and therefore would be tolerated by his/her immune system. The remnant of the blastocyst would be waste

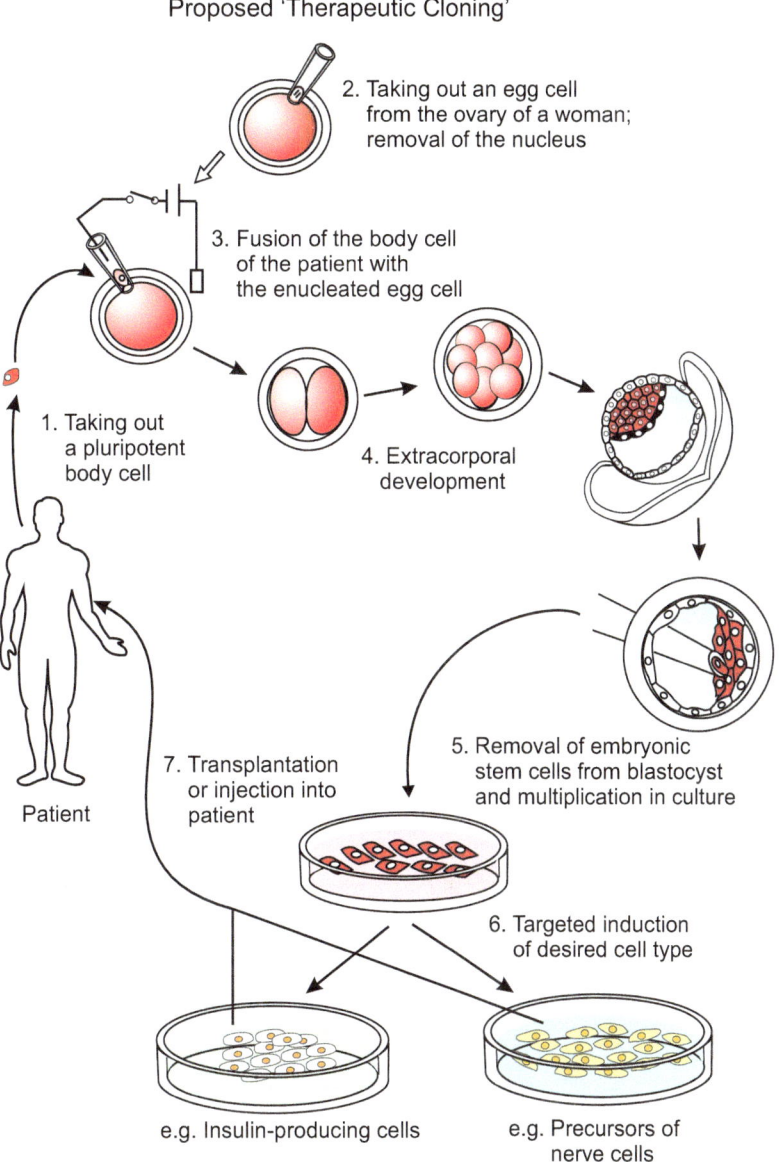

Proposed 'Therapeutic Cloning'

2. Taking out an egg cell from the ovary of a woman; removal of the nucleus

3. Fusion of the body cell of the patient with the enucleated egg cell

1. Taking out a pluripotent body cell

4. Extracorporal development

Patient

7. Transplantation or injection into patient

5. Removal of embryonic stem cells from blastocyst and multiplication in culture

6. Targeted induction of desired cell type

e.g. Insulin-producing cells

e.g. Precursors of nerve cells

To reduce the danger of vector integration at inappropriate sites of the genome vectors are being developed that insert at preferred or predefined positions using zinc-finger nuclease technology or technology based on homologous recombination combined with the *Cre-Lox* system to remove the vector sequences after the target sequence is integrated in the genome (Chap. 12, Box 12.2).

Alternatively, not DNA sequences but mRNAs for the desired factors, or ready recombinant proteins (recombinant = produced by bacteria) are introduced into the cells by established methods.

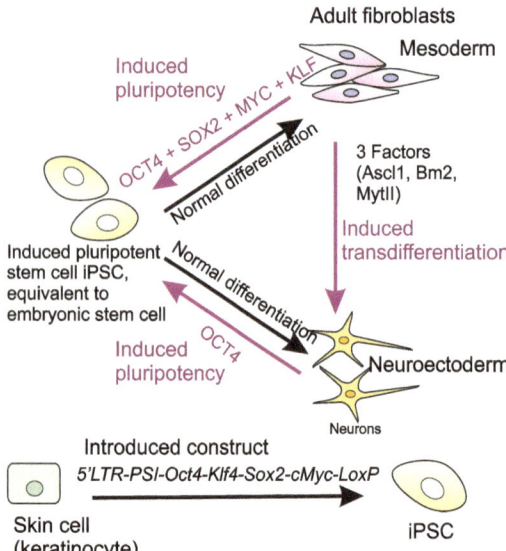

Fig. 18.11 Reprogramming differentiated cells. After Masip et al. (2010)

Viable fertile mice produced from iPS cells starting with four parents. This report may be disconcerting; nevertheless we report it here for the sake of intellectual integrity, and because the result testifies the high developmental capacity of iPS cells.

In 2009 Asian researchers first fused two 2-cell-stage embryos derived from 2 × 2 parents of a strain with white coloured coat by applying an electric pulse to produce tetraploid embryos. These developed to the blastocyst stage. In a uterus such a tetraploid blastocyst can form a trophoblast and eventually a placenta, structures which also in normal development are polyploid, but a tetraploid blastocyst does not contain a diploid inner cell mass and therefore cannot form an embryo proper. Now into these tatraploid blastocysts diploid iPS cells derived from a mouse strain with a black coloured coat were injected and the blastocysts transferred into the uterus of pseudopregnant foster mothers. These gave birth to black

coloured viable fertile mice. This is the most stringent proof that the iPS cells had attained the full developmental capacity of embryonic stem cells.

Reprogramming using transcription factors is laborious and currently rather inefficient. Therefore, a different approach tries to **abolish the epigenetic blockade of the cell's own pluripotency genes** without genetic intervention. Indeed several researchers report success in pioneering experiments using **microRNA** or chemicals that block enzymes of epigenetic chromatin modification such as **valproic acid**. The long-term goal is to enable switching on and off pluripotency genes at the will of medical therapists (see Chap. 12, Box 12.2) or to cause the adult differentiated cell to directly undergo **transdifferentiation** to the desired cell type.

Transdifferentiation is an event long known from invertebrates and amphibians (see Sect. 18.7 below) but only recently reported from human cells. Furthermore, it is not always advantageous to roll differentiated cells back to the initial state of pluripotent stem cells. Restoring the state of a committed progenitor cell capable of dividing saves labour in correct reprogramming and may reduce the risk of tumour formation.

The list of reports on successful restorations and reprogramming is growing in length and includes, for example, hepatocytes and nerve cells arising from fibroblasts (Fig. 18.11). However, much work must be done before clinical applications are reached.

Parkinson's and Other Neurodegenerative Diseases, Cure by Stem Cells?
For around a decade, scientists have been trying to regrow nerve cells lost in neurodegenerative diseases such as Parkinson's, Alzheimer's and amyotrophic lateral sclerosis (ALS) from stem cells. Researchers are already using stem cells to grow

(continued)

dopamine-producing nerve cells in the lab, those cells which are lost in patients suffering from Parkinson's disease. But can these cells be caused to immigrate into the right areas such as the substantia nigra, and to form the correct synapses?

Experiments in which dopamine neurons were created from mouse stem cells have not been successfully reproduced in humans. Moreover, up to now (2014) no experiments done with humans, not even those done with animals, fulfilled the criteria of stringent scientific standards which include double blind studies in a large numbers of stem cell treated probands versus sham treated control probands. There have also been safety concerns, with signs that dopamine neurons developed from human stem cells can be the origin of tumours. Stem cell therapies for Parkinson's disease are not yet ready for use in patients. Much work still needs to be done before clinical trials can go ahead.

Nevertheless several private clinics in industrial countries offer treatments with crude suspensions of bone marrow samples. Serious societies including the International Society for Stem Cell Research ISSCR, the International Parkinson and Movement Disorder Society, the German Parkinson Society DPG and the German Society of Neurology DGN issued repeatedly unequivocal warnings about such treatments.

For genetically caused or age-related diseases additional restrictions apply. The body's own genome can be the cause of a disease, be it that the genome includes mutations inherited or acquired in the course of one's life, or be it that the genome suffered irreversible epigenetic modifications. These defects may reside also in adult stem cells or iPS cells.

Unlike in *Hydra* in humans stem cells are not the fountain of eternal youth. Human beings still are mortal.

18.4.2 In Animal Models Some Surprising Reprogramming Has Succeeded, Such as Reverse Development of Multinucleated Muscle Fibres Back to Mononucleated Stem Cells

The hope that differentiated cells can be reprogrammed to re-attain characteristics of stem cells is corroborated by findings with animal models. In the hydrozoan *Hydractinia* targeted expression of an *Oct-4*-homologous gene in differentiated epithelial cells caused them to convert into pluripotent stem cells in amounts resembling tumours (neoplasias). More and more cells are being detected which can be reprogrammed even without genetic interventions. The decisive barrier which must be overcome is the epigenetic restriction of the developmental potentials. The barrier must be raised and complete reprogramming succeed.

The possibilities of transdifferentiation have long been underestimated, understandably enough. After all there are many cell types which suffer considerable and irreversible changes of their genetic potentialities such as endopolyploidy in placenta cells, somatic recombination in lymphoblasts, or loss of genetic material in erythrocytes and keratinocytes. On the other hand, who ever had anticipated that **multinucleated, cross-striated muscle fibres** can undergo **reembryonalization**? In the embryo cross-striated muscle fibres arise from the fusion of several myoblasts forming multinucleated myotubes. Before fusion and during differentiation, the homeodomain-containing **transcription repressor MSX-1** disappears, and instead muscle-specific selector genes such as *MyoD* and *myogenin* are activated. In the experiment the **gene *msx-1* was reactivated resulting in reversal of development**. The levels of MyoD and MYOGENIN dropped, the multinucleated myotubes disassembled, lost their contractile apparatus and converted back to mononucleated pluripotent cells capable of dividing. Through administering appropriate factors cell types arose the molecular inventory of which was characteristic of adipocytes (fat cells), chondrocytes (cartilage cells), osteocytes (bone cells) and also of myoblasts again. However, the first successful experiments along this line were done with muscle fibres of newts that were genetically manipulated in such a way that the suppressor gene *msx-1* could be switched on and thus the muscle-specific differentiation program extinguished (Chap. 12, TET system, Box 12.2). But subsequently also in mouse myotubes, the multinucleated precursors of muscle fibres, could be caused to revert development and disassemble into separate single cells, and in this case without genetic intervention but by applying extracts from regenerating newt extremities or administering low molecular purines.

Eventually (2011) it turned out that **mere injury of the muscle can suffice for multinucleated muscle fibres to revert development and break up into reembryonalized single cells (myoblasts, satellite stem cells)**.

For reprogramming adult cells also **extracts from oocytes** of the clawed frog *Xenopus* are in trial, because in oocytes resetting of the epigenetic program is operating and cloning experiments, in particular the transfer of nuclei from differentiated cells into oocytes as described in Chap. 13 and illustrated in Fig. 13.2, testify the presence of reprogramming factors in oocytes (for providing evidence for reprogramming J Gurdon was awarded the shared Nobel price in 2012, together with K Takahashi). Also extracts from other sources are being tested.

Reverse development back to a former developmental stage is a natural phenomenon observed in cnidarians. We introduce reverse development in Chap. 20 (Metamorphosis).

Perspectives: To the extent that tissue engineering succeeds in manufacturing replacement cells and tissues from adult stem cells or from induced pluripotent cells derived from the patient, and to produce replacement tissues in sufficient quality and quantity, important therapeutic goals can be attained without immunological problems and ethical conflicts. The authors of this book expect that in the time span between writing of this Chapter and its reception by the readers many reports will appear about new findings and news releases announcing imminent significant successes. On the other hand, there are severe reasons to oppose premature hopes.

18.4.3 Implanted Stem Cells Can Be Sources of Tumours

Mice can be easily subjected to experimentation, easily compared to other mammals. Embryonic stem cells (ES-cells) can be implanted in pink mice at any place, or the ES-cells can simply be injected into the blood stream of newborn recipient mice. Just like cells of the immune system ES cells can leave the blood circulation, migrate and settle down anywhere. Wherever stem cells settle if they are totipotent or pluripotent they can grow to tumour-like **teratomas (neoplasms)** made up of different types of tissue, none of which is native to the area in which it occurs. Or they may grow to malignant **teratocarcinomas**. They behave like fertilized egg cells who do not implant into the uterus but implant 'ectopically' at inappropriate locations, for example in the body cavity (abdominal pregnancy). If prevention of teratoma formation is aimed for, only carefully selected cells without extensive developmental potencies should be transferred into recipients. But also stem cells with restricted potencies can be sources of tumours; this possibility is almost inevitably associated with cells having the capability to divide. Experience teaches that the risk of cancer formation decreases with restriction of developmental potencies but not down to zero. Remarkably in many cancer cells the pluripotency gene *Oct-4* is switched on. More about results of cancer research can be read in Chap. 19.

18.5   Regeneration of Body Parts: Invertebrates and Vertebrates in Comparison

18.5.1   Some Animals Can Regenerate Whole Body Parts; This Capability Is Not Clearly Correlated with the Evolutionary Level

Replacing lost body parts is routine in many invertebrates. The regenerative capacity is most spectacular in sponges, cnidarians such as the well-known fresh water polyp *Hydra*, and turbellarians (planarians), suggesting that the faculty for regeneration reflects the evolutionarily 'primitive' position of those organisms. But this picture is too simplistic. For example, among the turbellarians *Mesostoma* is unable to regenerate. On the other hand, nematodes display a low capacity to regenerate but are not more complex in their body organization than planarians, which display excellent regenerative capabilities. Together with nematodes, the ascidians among the tunicates are considered to represent the prototype of the mosaic-type of development, with low regulative capabilities in embryogenesis. Yet adult ascidians (*Ciona intestinalis*, *Styela picta*, *Clavelina lepediformis* among others) regenerate quite well, and colonial ascidians such as *Botryllus schlosseri* are endowed with excellent capability of regeneration. Like Hydrozoans colonial ascidians possess totipotent stem cells. Various echinoderms such as sea stars, crinoids (brittle stars), and holothurians (sea cucumbers) are also endowed with good powers of regeneration. Echinoderms and ascidians belong to the group of deuterostomes as do the vertebrates which exhibit only restricted

capacities to regenerate. Thus, there is no straightforward correlation between the ability of an organism to repair and supplement body parts and its phylogenetic ranking.

The lack of the ability of most vertebrates to replace lost body parts may be correlated with their size rather with their rank in the animal kingdom. A momentum of selection for good regeneration might also have been the probability to lose a body part in an accident or upon attack by a predator.

- Among the **amphibians**, the urodeles (salamanders and newts) preserve an obvious faculty for regeneration even after metamorphosis. They are capable of regenerating amputated limbs and the lost tail which can fall prey to predators. But in addition amphibians can regenerate a removed eye lens, the spinal cord and even the heart ventricle.
- Arthropods are able to supplement incomplete legs as long as they undergo moults.
- **Autotomy**, the self-mutilation and shedding of body parts (tails in lizards, legs in arthropods) to escape an enemy, is in fact exceptional. The capacity to regenerate such structures allows such self-sacrifice.

Nature is not always perfect. Sometimes the wrong organ is formed. This phenomenon – a different organ developing than the one that was removed – is called **heteromorphosis**. For example, in shrimp an antenna may be formed in place of an amputated eye stalk. Heteromorphosis is reminiscent of the phenomenon of **homeotic transformation** (see Fig. 18.19).

Asexual reproduction is also a type of **regeneration** in the literal sense of the word. Asexual propagation is natural, organismal cloning. Self-cloning is accomplished in several ways: through fission (various turbellarians and annelids), through budding (*Hydra*) or through multicellular encysted bodies (gemmulae in sponges, statoblasts in bryozoans). One also speaks of **agametic cloning**.

18.5.2 Reconstitution: Cell Suspensions Derived from Dissociated Tissues, Organs and Animals Can Reconstitute the Original Structure by Self-organization

Reconstitution is a special case in that it is induced experimentally. The process of reconstitution documents an astonishing faculty for self-organization of multicellular associations. If embryos (e.g. sea urchin blastulae), excised organs (e.g. eye of an amphibian tailbud stage larva), or even entire animals (e.g. *Hydra*, Fig. 4.14) are dissociated carefully into single cells and reaggregates prepared from the resulting cell suspension, the original structures can be more or less completely rebuilt. This is accomplished by the cell society through mutual interactions and without help from outside. How this occurs is described for *Hydra* in Sect. 4.3.2. In Nature such a record-breaking achievement is in little demand. On the other hand, intriguing questions arise: What is the basis of such flexibility? How is it all organized and controlled?

18.5.3 Local Regeneration of Tissues and Organs Is Also Possible in Mammals Including Humans

In listings of performances based on regeneration frequently well-known phenomena are missing. To provide examples, it may be recalled

- the replacement of primary milk teeth by (hopefully) permanent dentition in humans,
- the periodic seasonal moult in birds and the seasonal change of the hair covering in most mammals in high geographic latitudes,
- the yearly shedding and replacement of antlers in deers and roebucks,
- the regrowing hairs, nails and claws,
- the healing of wounds and bone fractures,
- finally the **liver** has a markedly good capacity of regeneration in case a part has to be removed because of a tumour (The liver's capacity to regenerate was exploited by the cruel eagle of the ancient mythology who daily yanked and ate a piece of the liver of

Prometheus fettered on a rock). Also strong intoxication causing local cell death can raise the need for regeneration. In the liver hepatocytes resume division. The origin of the signals inducing resumption of growth is unknown, though one of the signalling molecules has been identified. It is the **Hepatocyte Growth Factor HGF**.

- The permanent regeneration of blood cells, of the skin, of the intestinal villi, of the odorant receptors and other tissues was issue of Sect. 18.1. However, in Chap. 21 it will be pointed out that this faculty of self-renewal has its limits.

18.6 Cellular Basis of High Regenerative Capacity

18.6.1 Basic Questions Are: Stem Cells or Transdifferentiation (Metaplasia), Epimorphosis or Morphallaxis?

The processes of regeneration are among the most enigmatic phenomena in developmental biology. Intriguing questions include:

- How does an organism recognize that a part is missing, which part is missing and how much of the part is missing? This raises the question how regeneration is initiated and the pattern controlled.
- Where is the material for the substitute taken from? Is the substitute derived from residual embryonic founder cells, from permanently dividing adult stem cells, or from differentiated cells undergoing **dedifferentiation** and redifferentiation, or undergoing short circuit **transdifferentiation** (also called **metaplasia**) changing the type of differentiation?
- If dedifferentiation takes place one asks: Do newly differentiated cells originate from their own kind, did, for example, chondrocytes of new cartilage elements arise from reembryonalized chondrocytes, muscles from reembryonalized muscle fibres? On the other hand transdifferentiation from one cell type to another type would presuppose an act of

transdetermination. Real cases of transdifferentiation are described below.

- Is the original structure rebuilt by recruitment and reorganization of non-dividing cells already present (**morphallaxis**) a process leading to a complete albeit smaller structure, or is the missing structure restored from a few cells in the wound area which proliferate and subsequently differentiate (**epimorphosis**)? In fact, actual regenerative achievements are based on a combination of transdifferentiation, cell migration and cell proliferation rather than on either pure recruitment or pure proliferation.

Long teaching tradition adduces *Hydra* as paradigm for morphallaxis since some regenerative reorganization takes place even when cell division is blocked experimentally. On the other hand regeneration of extremities in arthropods and amphibians is considered a paradigm of epimorphosis. Yet, morphallaxis and epimorphosis are conceptual extreme cases (the terms were coined by Thomas Hunt Morgan, who performed excellent regeneration studies before doing genetics with *Drosophila*). Also in epimorphosis the first cells commencing divisions can arise from reembryonalized cells. On the other hand no organism will restore a body part only through reorganzing existing cells without any cell division if cell division is not blocked experimentally.

Actual regeneration events are based to different degrees on existing stem cells or stem cells arisen from reembryonalized cells, on transdifferentiation, cell migration and cell proliferation.

18.6.2 Stem Cells Are Main Sources of Regeneration; The Principle Is Shared by the Most Advanced and the Most Primeval Organisms, and Also Differentiated Cells Can Share Properties of Multipotent Stem Cells

In the past, the high regenerative capacity of 'lower' animals has often been attributed to their hypothetical possession of 'embryonic reserve cells', sometimes collectively called neoblasts. As pluripotent stem cells, neoblasts were thought capable of giving rise to every sort of cell and thus being able to supply every lost cell type. In fact, pluripotent stem cells are found in sponges, hydrozoans, turbellarians, tunicates, and vertebrates – including humans!

Already in **sponges**, the most primeval metazoans, totipotent cells are found, known as **archaeocytes** and **choanocytes**. Remarkably, choanocytes have a differentiated appearance just like the single-celled choanoflagellates whose ancestral relatives are thought to have given rise to multicellular animals (Metazoa). Choanocytes have the capacity to divide as have the chonoflagellates, but in addition have the capacity of transforming into other cell types. If we look beyond vertebrates we come to the conclusion that **differentiation as such does not necessarily exclude stem cell properties**.

This statement is corroborated by findings in the basic eumetazoan phylum **Cnidaria**. In the orders Anthozoa (including corals), Scyphozoa (including large jelly fishes) and Hydrozoa (including small polyps and small jellyfishes) one finds self-renewing epithelial cells acting like stem cells, capable of self-renewal but also generating other cell types such as nerve cells or stinging cells through transdifferentiation. In addition, one finds amoebocytes of unknown origin and with unknown functions.

Histologically identifiable stem cells **displaying features of embryonic stem cells** are found in several hydrozoan species such as the well-known fresh water polyp *Hydra* and the marine colonial species *Hydractinia echinata* (Fig. 18.12; see also Fig. 4.18). In these two species interstitial cells, abbreviated to i-cells, are found residing in the interstitial spaces between the epithelial cells. These interstitial stem cells (and likewise the neoblasts of planarians) are small, rounded cells with large nucleus and filled with RNA, Their derivatives are nerve cells, stinging cells, certain gland cells and the gametes (Figs. 4.13 and 18.12). Remarkably, in contrast to mammals *Hydra* is able to even replace all of its nerve cells. This unique capacity is the basis for the potential immortality

Fig. 18.12 Stem cells (i-cells) of the hydrozoan *Hydractinia*, in which A. Weismann 1883 had described stem cells for the first time and named them stem cells. The population of i-cells in *Hydractinia* comprises totipotent cells. After Künzel et al. (2010)

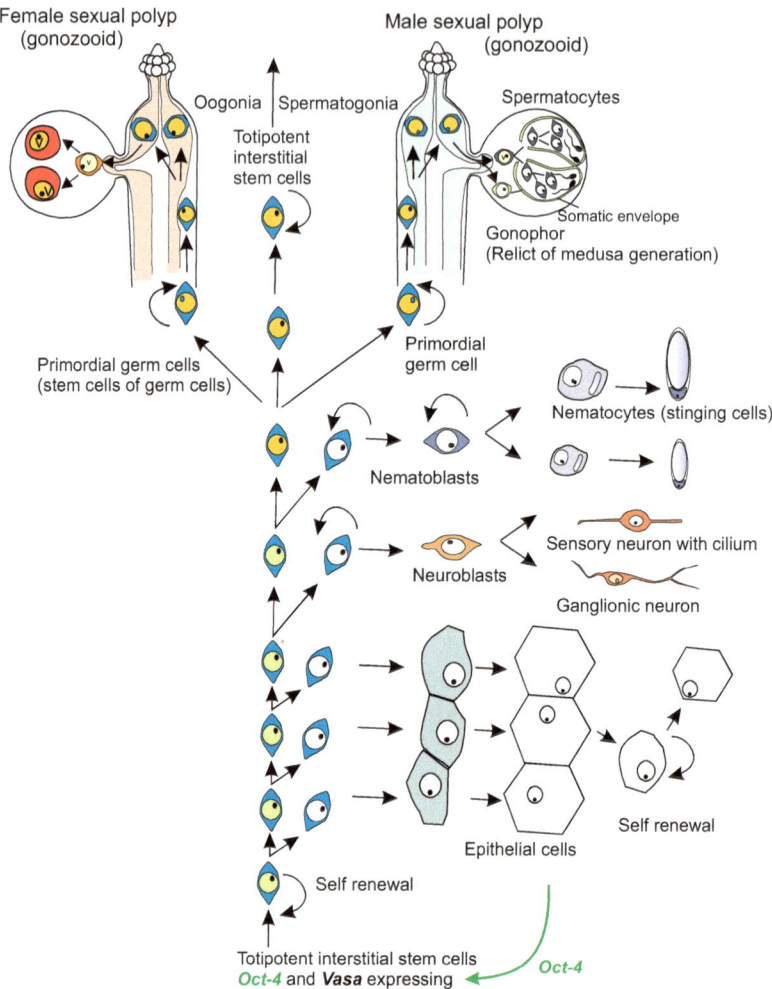

Female sexual polyp (gonozooid)

Male sexual polyp (gonozooid)

Oogonia | Spermatogonia

Spermatocytes

Totipotent interstitial stem cells

Somatic envelope

Gonophor (Relict of medusa generation)

Primordial germ cells (stem cells of germ cells)

Primordial germ cell

Nematocytes (stinging cells)

Nematoblasts

Sensory neuron with cilium

Neuroblasts

Ganglionic neuron

Self renewal

Epithelial cells

Self renewal

Totipotent interstitial stem cells
Oct-4 and ***Vasa*** expressing   *Oct-4*

of *Hydra*. Interstitial stem cells can be selectively eliminated. As a consequence, eventually nerve cells as well as stinging cells are lost. Amazingly, nerve-free hydras survive, if force-fed, and retain the ability to regenerate and to bud new, albeit nerve-free, animals!

But *Hydra* cannot produce all types of cells from multipotent i-cells. The ectodermal and endodermal epithelial cells, which form the tube-like body column, originate from their own kind. Like the choanocytes of sponges, epithelial cells *Hydra* preserve the ability to divide and to adopt different tasks. In the middle of the body column, the differentiated state does not exclude the ability to proliferate: only in the head and foot is this ability lost. On the other hand, the state of

differentiation can change. Epithelial cells are gradually displaced from the gastric region into the head or foot. While the descendants are shifted towards to oral pole of the body column (like the cells in the villi of our gut, Fig. 18.3) or towards the aboral pole they gradually change their appearance according to the region they enter and eventually convert into derivatives characteristic for their final destination such as gland cells in the foot or tentacle cells in the head region. This conversion is dependent on positional information. Certainly, this **plasticity of the state of differentiation** significantly contributes to the high power of regeneration.

An even higher differentiation potential has been shown for the migratory interstitial stem cells of *Hydractinia* (Fig. 18.12). The stem cells of this colonial species are pluripotent or even totipotent as they can give rise to all cell types including germ cells. To provide evidence, the whole population of the stem cells was eliminated using alkylating cytostatics ('chemotherapy') and replaced by stem cells from histocompatible mutant colonies. Donor and recipient were different with respect to morphology (e.g. multi-headed polyps in a mutant) and by sex (female versus male). In a few weeks the recipient colonies adopted the phenotype of the mutant donor. The recipients also underwent sex reversal according to the donor's sex.

The hypothesis that the **neoblasts of planarians** are totipotent as well, at least in their entirety, was long debated but is increasingly corroborated by recent investigations. A high regeneration potential is exhibited by the species of the genus *Dugesia*, and these species possess particularly high numbers of neoblasts. As a further owner of totipotent stem cells surprisingly a representative of the phylum Chordata was identified. It is the colonial **tunicate (ascidian)** *Botryllus schlosseri*. Isolated blood cells clustered forming blastula-like spheres and these gave rise to complete organisms within weeks.

In echinoderms stem cell features were assigned to various cell forms: amoebocytes, coelomocytes, phagocytes, granulocytes. In all these cases the attribute 'pluripotent' or 'totipotent' applies for pools of cells which mediate the power of restoration as entireties. The pools may well be heterogenous and encompass subtypes with different, restricted potentials. That even a single, defined cell type can in experimental conditions display powers of totipotency is shown in the following section.

18.6.3 Plasticity of the Differentiated State and Transdifferentiation Are Further Sources of High Regeneration Capabilities

Principal possibilities of dedifferentiation and transdifferentiation have been demonstrated in several cases. The most convincing and record-breaking case is the following:

In hydromedusae (a type of small jelly fish) of the species *Podocoryne carnea* an amazing gift for **transdifferentiation** has been documented. Cross-striated muscle cells isolated from the underside of the bell-shaped umbrella are capable of undergoing dedifferentiation. Dedifferentiated muscle cells recover the ability to divide and can give rise to a variety of different cell types, including nerve cells and germ cells (Fig. 18.13).

Entire animals have not (yet) been recovered from somatic cells (apart from the mice raised from iPS-cells mentioned above in Sect. 18.5.1). In horticultural and agricultural breeding, plants that cannot be multiplied by natural asexual reproduction or grafting can be cloned from isolated single somatic cells. Small pieces of phloem or leaves are excised and dissociated into single cells by enzymatic removing the cell wall. The resulting protoplasts are inoculated into a fluid containing nutrients and growth hormones. In an appropriate medium, the cells divide and give rise to a conglomerate of largely undifferentiated and disorganized cells, called callus. Surprisingly, within such a callus an embryo can form in the absence of true germ cells or fertilization. The embryos can be seeded onto agar, where they develop into plantlets that grow up, flower and ultimately produce seed. Such success has not yet been achieved with isolated somatic cells of animals. The transdifferentiation of striated muscle cells in hydrozoans mentioned suggests that in these organisms a method of cloning analogous to the methods used in plant breeding could be developed. However, such research is not promoted by economical interests.

Frequently, but not in all instances, transdifferentiation assumes previous DNA replication.

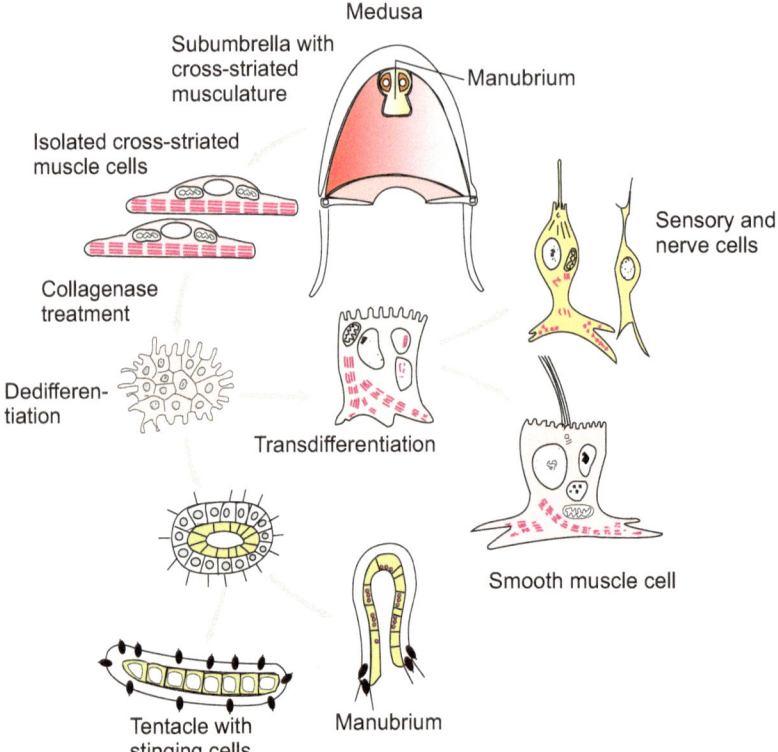

Fig. 18.13 Transdifferentiation of isolated cross-striated muscle cells in the jellyfish *Podocoryne carnea*. Muscle cells isolated from the subumbrella are destabilized by treatment with collagenase or tumour-promoting phorbol esters. Upon isolation the muscle cells lose their state of differentiation and can, after division, convert into many other cell types. Remnants of contractile filaments within the transdifferentiated cells testify their origin. The transdifferentiated cells can even form organs such as tentacles or a manubrium through self-organization. After Schmid et al. in: Regulatory Mechanisms in Developmental Processes, Elsevier 1988

DNA replication would facilitate reprogramming and the acquisition of a new state of differentiation. Transdifferentiation of jelly fish striated muscle fibres into nerve cells depends on previous cell division but not transdifferentiation into smooth muscle cells.

18.6.4 Best Known Case of Transdifferentiation in Vertebrates Is Lens Regeneration

Transdifferentiation (metaplasia), the transformation of one differentiated cell type into another is observed even in vertebrates. A classic example is **Wolffian lens regeneration** (Fig. 18.14). During embryo development of newts and salamanders the lens is formed from the epidermis after the underlying eye cup has emitted an inducing signal. If the lens is later removed, a new lens forms from the dorsal iris rather than from the epidermis. The iris is a derivative of the optic cup and thus of the brain, and consists of a type of smooth muscle cell that contains melanosomes. The transformation begins with dedifferentiation: melanosomes and muscular filaments are lost. Cell divisions are followed by differentiation into lens cells producing crystallin proteins. Apparently it is not a stimulus emanating from the wound that initiates the process of transdifferentiation but the lack of a lens (Fig. 18.14).

In the course of lens regeneration genes are activated that play a leading role in eye development, especially the genes for the transcription

Fig. 18.14 Wolffian lens regeneration in an amphibian. While in normal embryogenesis the lens is formed from the ectodermal epithelium, in regeneration it is formed from the margin of the iris through trans-differentiation. A non-denatured (living) lens can prevent lens formation even from a distance while a denatured (killed) lens does not have an inhibitory influence

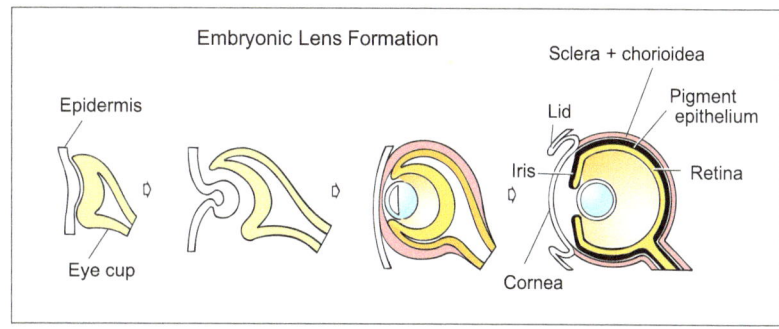

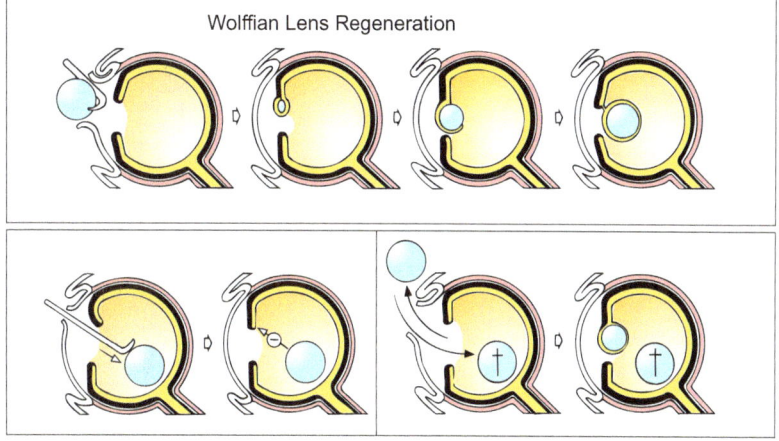

Displaced, living lens inhibits lens regeneration

Killed lens does not inhibit lens regeneration

factors Pax-6 and SOX-2, and genes coding for the signalling molecules of the TGF-ß and the FGF family.

Besides the lens further structures of the eye, namely the ciliary body and the iris, can be restored by transdifferentiation from adjacent retina cells.

Even in mammals, transdifferentiation is possible, such as transformation of lymphocytes into macrophages and of pancreatic cells into liver cells, the latter transformation being induced by artificial injury.

18.6.5 The Limits of Regeneration Are the Seeds of Death

Apart from a few organisms such as sponges and *Hydra*, multicellular animals – including of course humans – are not able to renew of all of their cell types. For instance, aged nerve cells cannot be replaced by new ones. By the time

we are 20, our nerve cells begin to decay (not in all parts of our brain as recent studies indicate. Exceptions are, as outlined above in Sect. 16.8.7, sensory nerve cells of the olfactory epithelium, photoreceptors of the eye, cells in the subventricular zone SVZ and in the hippocampus). The successive loss of nerve cells leads inevitably to death. In the brain there is no sustainable economy, no balance between birthrate and mortality. To date no efficient fountains of youth are known for human beings.

18.7 Controlling Systems

18.7.1 The Hydra and Planarians; Their Body Organization Is Supervised by Similar Controlling Systems

Hydra: Ancient Greek mythology tells the story of Hercules who was sent out to perform his heroic deeds. Once he was confronted with the frightening water snake hydra which regenerated two heads for each head that was cut off. Hydras of the present times, about 5 mm in length, are perilous only for small animals such as water fleas, and when a forceful animal attaches to its tentacles and the polyp is torn in pieces every piece regenerates the lost part faithfully. Only by tricky experiments, for example by repeatedly splitting the head lengthwise or by inference with the system of pattern control can multi-headed polyps be generated, namely by treating the animals with tumour promoters (see Fig. 10. 20a) or by stimulating the canonical WNT/β-catenin system.

A paragon of high regenerative ability was and still is *Hydra* ever since in 1735 regeneration studies with this animal rang in the era of experimental developmental biology. For example, if a ring is cut out from the middle of the body column comprising only 1/20 of the body length, the excised ring will regenerate a head at its upper end and a foot at its lower end.

> What is the benefit of high regeneration capability in the wild? It may well happen that a prey animal in his attempt to wrest itself free disrupts its captor. In particular marine cnidarian species (colonies of hydroids and corals) frequently lose their heads when slugs or fishes browse them. *Hydra* shares with its marine relatives the capacity to regenerate a head repeatedly.

The high regenerative potential of *Hydra* and of its relatives is based on (a) the plasticity of the state of cellular differentiation and the stores of pluripotent stem cells, (b) a peculiar property of the freshwater polyp: in normal growth the polyp continuously replaces the whole inventory of its cells. Therefore a system of **pattern control** must operate throughout its life. This system must supervise the replacement of old cells by young cells: the replacement must be quantitatively balanced and occur at the correct places – that is, it must be regulated by **positional information**. An important parameter in this system of control is the **gradient in positional value**:

this gradient ensures that the potential to form a head is higher in near-head regions of the body column, whereas the potential to form a foot is higher in the near-foot regions. Thus, the new head will always be made at the end of an excised segment that previously had been closer to the original head, while the foot will be made at the opposite end. The system of pattern control was introduced in more detail in Sects. 4.3 and 10.9.

Planarians (catch-all phrase for non-parasitic flatworms, representatives of the class Turbellaria) have eyes, a brain, and a rich internal organization. In spite of this structural complexity, some planarians have high powers of regeneration. This applies, for example, to members of the genera *Dugesia* (*D. gonocephala* = common flatworms found in freshwater habitats, *D. dorotocephala*, *D. tigrina*, *D. japonica*), it applies to some species of the genus *Planaria* (*P. maculata*) and the species *Macrostomum lignano* and **Schmidtea mediterranea**. The latter species is preferred model animal in molecular biological laboratories because it is diploid and encompasses strains that reproduce sexually and strains that reproduce asexually.

The listed planarians have a regenerative ability comparable to that of *Hydra*, and pattern reconstitution after cutting follows similar rules.

Parallels between hydrozoan and planarians apertain (1) possession of pluripotent or even totipotent stem cells, in planarians called **neoblast**s, and (2) principles of pattern control.

1. **Neoblasts**, are rich in ribonucleic acid (RNA) as are the interstitial stem cells of hydroids. Amputation induces stem cell mobilization to sites of injury. Similar cells reside in the trunk of whole organisms and are activated in regeneration. In the species *Microstomum lignano* regeneration is possible if 4,000 from an initial stock of 25,000 cells including 160 neoblasts are leftover. To analyze the developmental potentials of neoblasts rescue experiments were performed. Single neoblasts, isolated by flow cytometry from different body regions of one strain of the planarian *Schmidtea mediterranea*, were transplanted into lethally irradiated hosts of a different strain of the same species. Injection

of single neoblasts resulted in the formation of large descendant neoblast colonies in vivo that later gave rise to several differentiated cell types restoring the ability to regenerate. This study is the first clear evidence that neoblasts are fully pluripotent, and even totipotent, as they also give rise to the germ cells. In parallel, a flood of molecular markers, labelling either all neoblasts or subsets of them, were reported. Prominent among them were the homologues of the germ line genes *piwi*, *vasa*, **nanos**, *pumilio*, and genes for Sox and Jun-like transcription factors. Moreover, genes important for pluripotency in ESCs, including regulators as well as targets of *Oct-4*, were well conserved and upregulated in neoblasts. Considering also the presence of *Oct-4* in the interstial stem cells of hydroids it has been concluded that **molecular determinants of pluripotency are conserved throughout evolution**.

2. **Pattern control**: Planarians may be cut across or lengthwise, and each part of the body will regenerate the missing part (Fig. 18.15A). When the cut is made, a regeneration blastema is formed at the cut surface, and the missing part is developed from the blastema. The remaining part of the body is reorganized on a diminished scale. A lateral incision may cause the development of either an additional head or an additional tail. A head forms from a triangular part that faces anteriorly, the tail develops from a part facing posteriorly. Very narrow pieces cut out by two transverse cuts from the anterior body region may form a head on both the anterior and posterior cut surface: the internal gradient in positional values is not sufficiently steep to give the anterior blastema a start. Only a temporal advantage enables it to suppress competitive head formation at the posterior cut in time.

To explain pattern control in formal terms much could be said about gradients, positional values and intercalation. We refer to Sects. 4.3 and 10.9. In addition, in planarians as well in hydrozoans the overall polarity of the body is governed by the canonical WNT/β-catenin system.

Yet, in hydroid polyps the source of WNT signals is the setting of the mouth, that is in the 'head' (hypostome), whereas in planarians it is at the tail tip (Fig. 18.15B). In planarians, however, WNT does not specify polarity alone. Transport HEDGEHOG (HH) directed from anterior to posterior body specifies the position of the WNT emitter. At the posterior end HH induces expression of several paralogous variants of *wnt*. If transport of HH is interrupted in middle sections cut out of the body no WNT emitter emerges at the posterior pole and a second ectopic head is formed there instead of a tail (Fig. 18.15B). Such a second mirror-image head also is formed when the WNT system is downregulated at the posterior end by means of RNAi (see Chap. 12) against β-catenin. Upon strong blockage of β-catenin by means of RNAi individuals arise with many ectopic heads – exactly the counterparts to *Hydra*, in which not blockage but activation of the WNT/β-catenin system results in multi-headed monsters. A comparative reflection on the significance of the WNT system for orientation of body axes including other organisms is compiled in Chap. 22 (Figs. 22.6 and 22.7).

Further Differences. Hydrozoa and planarians differ in symmetry properties. Hydrozoa possess only one polar axis, planarians are bilateral symmetric, having an antero-posterior axis, a dorso-ventral axis and two mirror symmetrically oriented left-right body halves. In the control of this bilateral symmetry a parallel to vertebrates is striking. Similar to the neurula of *Xenopus* also in the planarians a Cartesian system of coordinates is realized with a roof-shaped BMP gradient perpendicular to the WNT gradient (Fig. 18.15B).

Unlike in *Hydra* in planarians at the cut surface a blastema is formed by cells that dedifferentiate at their place but are supplemented by neoblasts arriving from distant body parts. The remaining body is reorganized and reduced in size until harmonic proportions are re-established.

Fig. 18.15 **A** (a–e)
Classic regeneration
experiments with
planarians. **B**. Planarians. **a**
A WNT signalling centre is
located at the posterior end.
A flow of HEDGEHOG
(HH) aids in specifying this
position. **b** When the flow
of HH is interrupted or the
emitter of WNT shut down
by means of RNAi against
β-catenin, a second, ectopic
head emerges at the
posterior pole. **c** Activation
of the canonical WNT/β-
catenin system at the
anterior end causes this end
to form a second tail. **d**
Gradients of WNT and
BMP-4 in perpendicular
orientation specify the
bilaterally symmetric
organization of the body.
After DeRobertis 2010,
Rink et al., 2009. **e** WNT
expression in hydroids for
comparison; in *Hydra* the
position of WNT emitters
marks the head end (oral
pole)

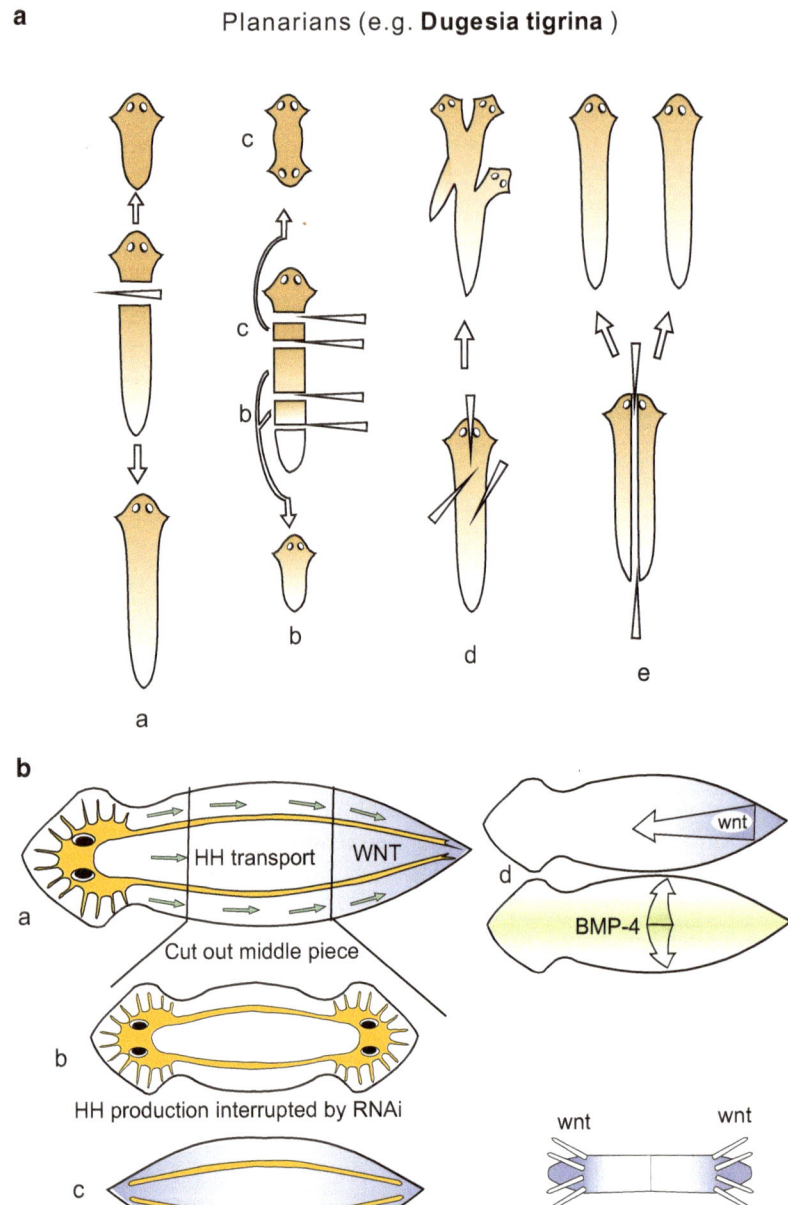

a   Planarians (e.g. **Dugesia tigrina**)

b

HH transport   WNT

Cut out middle piece

HH production interrupted by RNAi

wnt

BMP-4

wnt   wnt

e   Hydra

Activation of the WNT system at the anterior end

18.7.2 Regeneration of Limbs in Amphibians Is Dependent on the Supply of Neurotrophins

The larvae of urodeles (newts, salamanders, and axolotls) are able to completely reconstruct amputated limbs, a capacity they retain all their life. Limbs, especially those of the larvae but also of mating adults, are sometimes bitten off by various predators present in the pond. In anurans (frogs and toads), early larvae are capable of supplementing limb buds; with age, however,

Fig. 18.16 Regeneration (epimorphosis) of an amphibian leg (newt, salamander or axolotl). A blastema is formed from dedifferentiated cells. Normally the blastema generates exactly the structures removed. The capability of regeneration is dependent on the supply of neurotrophic factors by the leg nerves, more precisely, on the supply of the nerve fibres by the Schwann cell envelope

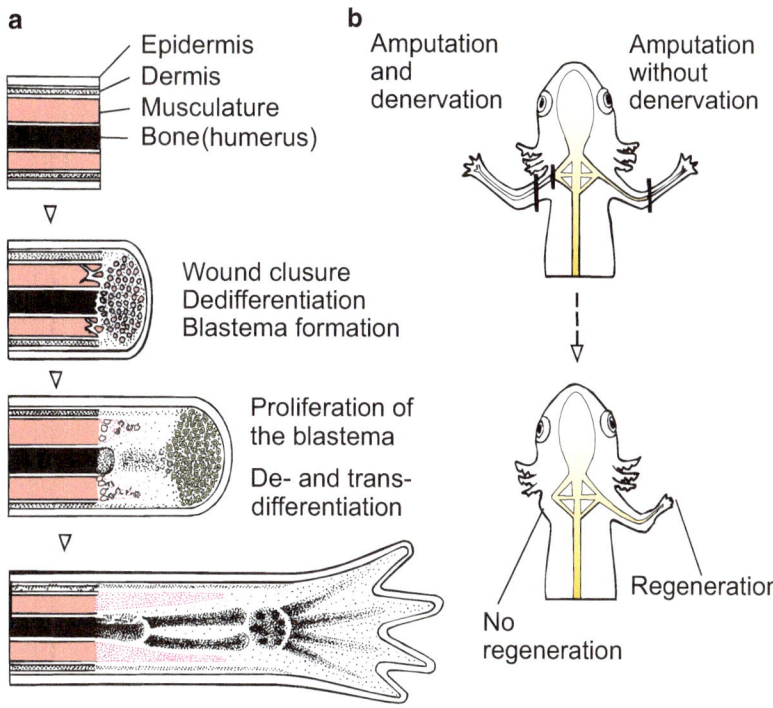

a
Epidermis
Dermis
Musculature
Bone(humerus)

Wound clusure
Dedifferentiation
Blastema formation

Proliferation of
the blastema

De- and trans-
differentiation

b
Amputation
and
denervation

Amputation
without
denervation

Regeneration

No
regeneration

the potential to regenerate diminishes and disappears completely after metamorphosis.

Regeneration of limbs takes place in several steps (Fig. 18.16).

1. In the freshly cut limb, migrating epidermal cells cover the wound and protect the remaining stump.

2. Beneath this protecting layer, some cells such as chondrocytes and osteocytes are lysed or undergo apoptosis. Most cells survive, yet disconnect from each other and undergo **dedifferentiation**. Irrespective of their origin (dermis, muscle, cartilage, connective tissue) the cells take the uncharacteristic shapes of fibroblasts or mesenchyme cells, and accumulate as a cell mass known as regeneration **blastema**.

3. In the cells of the blastema the activity of cell type-specific genes is down regulated and instead the gene of the **transcription factor msx-l** switched on. This gene remains active in the following phase of proliferation and is switched off with termination of growth.

4. The dedifferentiated blastema cells proliferate and ultimately give rise to a limb bud. In contrast to limb buds appearing in embryogenesis, limb formation from blastemas is also dependent on the presence of residual nerve cells. A limb stump lacking its nervous supply does not regenerate; it depends on supply of **neurotrophins**. These are delivered by nerves or glial cells, respectively, which accompany nerve fibres forming Schwann's sheaths. In contrast to neurotrophins described in Chap. 16, the neutrophins meant here are not survival factors *for* nerve cells but products *of* nerve or glial cells. In salamanders the cocktail enabling regeneration contains a factor called **nAG protein** (newt Anterior Gradient Protein). A denervated leg stump does not regenerate. Local injection of an expression vector with the gene for nAG enables denervated stumps to regenerate (Fig. 18.17). The growth factor nAG is picked up by the blastema cells by the receptor Prod-1 exposed on their surface. Prod-1 is a member of the

Fig. 18.17 (**a, b**) **a** Production of the neurotrophic factor nAG (newt Anterior Gradient) in the leg stump of a salamandrid (*Notophtalmus viridescens* and *Ambystoma mexicanum*). **a1** First the factor is released from Schwann's sheaths of the leg nerves. **a2** When a new epithelium covering the blastema has formed gland cells also produce nAG. (**b**) Denervated stumps (**b1**) do not regenerate (**b2**). In **b3** the (inducible) gene for nAG was introduced into the stump. Now also the denervated stump is enabled to regenerate. After Kumar et al. (2007, 2010)

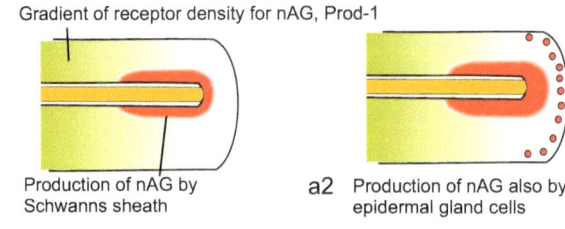

Regeneration of Foreleg in Salamanders

Gradient of receptor density for nAG, Prod-1

a1   Production of nAG by Schwanns sheath

a2   Production of nAG also by epidermal gland cells

nAG = newt Anterior Gradient protein

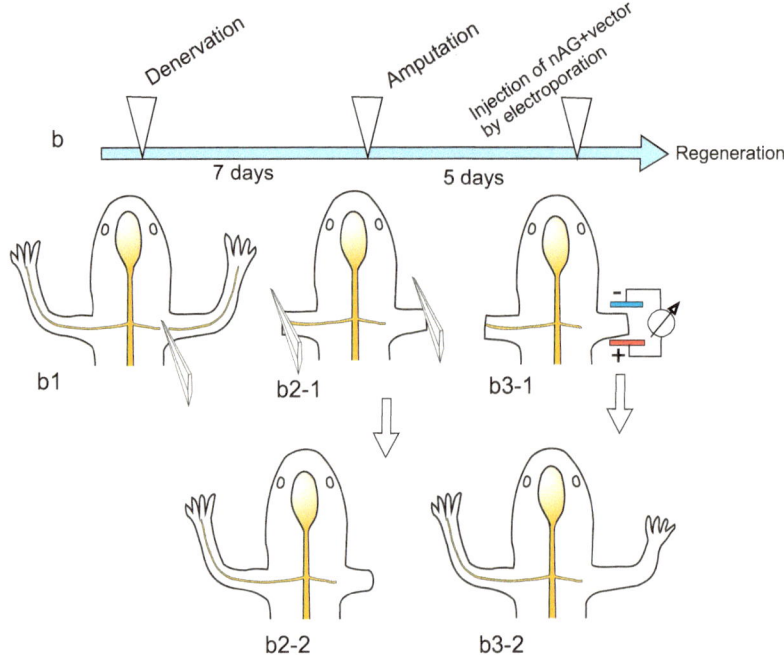

receptor family EGFR (Epidermal Growth Factor Receptor). The density of receptor molecules on the cell surfaces establishes a gradient declining from proximal to distal.

5. According to other studies FGFs also exert a supporting functions. FGF producing cells stimulate themselves mutually as they do in embryonic limb buds (see Fig. 10.11). From now on parallels to embryonic development continue. For coordinated pattern formation SONIC HEDGEHOC SHH adopts a leading role; it controls the axis from finger 4–1. On the other hand the longitudinal axis extending from the shoulders to the fingers is controlled by a **double gradient system**: a gradient of

11-cis-retinoic acid RA declining from proximal to distal is opposed by two parallel running gradients declining from distal to proximal; these are established by the signalling molecules FGF8 and WNT3a (see Chap. 10, Fig. 10.11). During regeneration externally applied retinoic acid exerts dramatic influences on patterning (see below Sect. 18.7.4).

6. During the outgrowth of the limb bud sequentially those structures are specified that normally are present distally of the cut level, and they are laid down completely and in the correct order. To address this phenomenon developmental biologists have formulated

the **rule of distal transformation**. Only structures that ought to lie distal to the cut and had been removed can be realized in the regeneration program.

Positional value is believed to be based in the expression pattern of homeobox-containing genes (*HOM/Hox* genes, Chap. 10) and coded in the equipment of cell membranes with cell adhesion molecules (CAM, Chap. 12) as outlined in the following section.

18.7.3 Completeness of a Structure, Here of a Leg, Is Controlled by Gradients of Positional Values

An intriguing and enigmatic phenomenon remains to be explained: why does the blastema form successively more distal structures, beginning at exactly the appropriate level so that nothing will be missed and nothing will be supernumerary? For example, a blastema on a stump containing a humerus normally does not form another humerus and does not form digits prematurely.

A formal explanation refers to the term **positional value**. Just as in regenerating hydras or planarians in the course of regeneration new positional values are laid down in an orderly manner so that the sequence from zero value (shoulder) to maximum (tip of digits) is complete and correct.

Positional values are correlated with the expression profiles of homeobox-containing genes *Hox-A-* and *Hox-D* (see Figs. 12.16 and 12.18) and thought to be dependent on these genes. Yet, HOX factors are transcription factors and hidden in the nuclei of the cells. Positional values must be recognized and measured by adjacent cells. If a distal fragment of a leg is grafted onto a proximal stump and the middle segment is lacking, cells with different positional values collide. Strong discrepancies are recognized and induce insertion of missing structures. This phenomenon, known as **intercalation**, is shown in Figs. 10.19 and 10.23.

Legs also can regenerate in juvenile hemimetabolous insects such as crickets. In crickets evidence was produced that **positional values are encoded in cell adhesion molecules CAMs**, in particular in the density ratio of two opposite gradients of CAMs, actually in the proximal to distal *decreasing* gradient of the cadherin **FAT** and the proximal to distal *increasing* gradient of the cadherin **DACHSOUS**. These are shown in Fig. 10.7. When the expression of these gradients is disturbed by means of RNAi (see Chap. 12) shortened legs are formed and intercalation is not perfectly accomplished.

18.7.4 Retinoic Acid Can Induce Reprogramming Revealing Hidden Regeneration Capacities

A surprising experimental result shows that this sophisticated developmental program can be reset: when the blastema of an amputated forearm is treated with **retinoic acid** (RA, vitamin A acid), the positional value of the blastema cells is set to zero. The stump ignores the existing humerus and residual radius and ulna, and forms a complete limb beginning with the shoulder and ending with digits (Fig. 18.18a).

Nucleus resident receptors for retinoic acid (**RAR**) are present and if occupied by RA, previously closed developmental paths are reopened. Treatment of the blastema at the tail end of a tadpole with RA can conceal even more extensive developmental faculties. In the course of metamorphosis the tail end may develop ectopic legs (Fig. 18.19).

In organs that cannot normally regenerate such as the adult mammalian lung, **RA induces the complete regeneration of alveoli** that have been destroyed by various noxious treatments. In the **mammalian central nervous system (CNS)**, which is another tissue that cannot regenerate, RA does not induce neurite outgrowth as it does in the embryonic CNS, because one of the retinoic acid receptors, **RARbeta2**, is not upregulated. When *RARbeta2* is transfected into

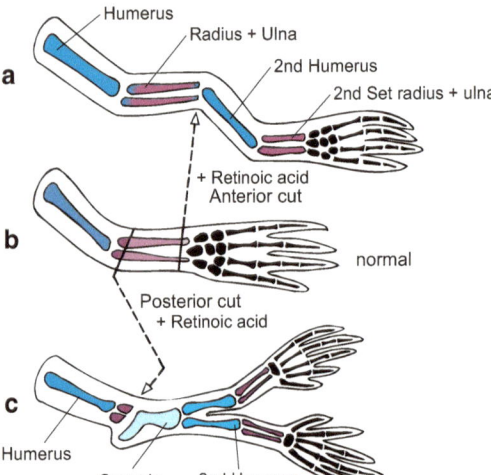

Fig. 18.18 (**a–c**) Leg regeneration in salamanders without and with treatment of the blastema with retinoic acid (RA). (**a**) After treatment with RA in the proximo-distal axis of the stump resetting of the positional value to shoulder level can occur, and humerus and radius/ulna are formed again, (**c**) occasionally even another shoulder blade is formed. In the antero-posterior axis mirror image duplication can occur. (**b**) Normal case; here the cutting levels are shown for the experiments done in **a** and **c**. Experiments by Maden and Hind 2003

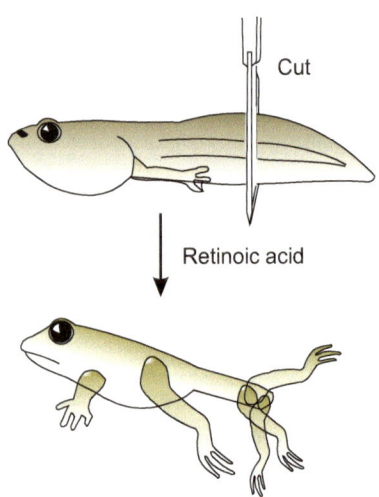

Fig. 18.19 Homeotic transformation at the tail of a regeneration blastema into leg structures after administering retinoic acid. Experiment in tadpoles of *Rana temporaria* by Maden 1993

the adult spinal cord in the dorsal root ganglia in vitro, then outgrowth of dendrites and axons is stimulated by RA. In all these cases, RA is required for the development of the organ, in the first place suggesting that the same gene pathways are likely to be used for both development and regeneration.

On the other hand lacking capability to regenerate and reorganize brain structures may enable installation of **stable programs for instinctive action** and stable **long-term memory** lasting many years.

A great and ambitious aim of current regenerative medicine is to understand and master the natural controlling systems to render regeneration of complex structures such as extremities and spinal cord possible.

Summary

Growth and renewal of tissues and organs is based on continuing division of **stem cells** which have preserved features of embryonic cells, but often also on the self-renewal capacity and plasticity of existing differentiated cells. Regenerative renewal takes place at several levels: Within the cells the stock of macromolecules is renewed. At the level of cells and tissues **stem cells enable growth and provide replenishment** for terminally differentiated cells whose life span is shorter than that of the whole body. For example, the **growth of the central nervous system** is enabled by a layer of neurogenic stem cells known as **ependyma** which lines the canal of the spinal cord and the ventricles of the brain. Asymmetrically dividing stem cells give rise in part to more stem cells, in part to neuroblasts which emigrate into the CNS and give rise to neurons. Stem cells procure replacement of our short-lived **odorant receptors**. Unipotent stem cells of our **skin** supply **keratinocytes** that eventually will be shed, while multipotent stem cells residing in **intestinal crypts** enable permanent re-growth of the **intestinal villi**. Multipotent **hematopoietic stem cells** ensure permanent replenishment of all types of blood cells. In the postnatal body all the various blood cells arise from one type of stem cell in the bone marrow in the form of a branched pedigree (summary in Fig. 18.6). The type and amount of replacement cells are controlled on demand by means of hormone-like factors, called **cytokines**, including interleukins and **erythropoetin**.

The chapter reports of increasing successes of medically oriented research to procure replacements for damaged, attrophied, or diseased tissues and organs such as **kneecaps** or **heart valves**. The power of self-organization is illustrated by 'heart-shaped' bodies and almost complete **eye cups** grown from murine embryonic stem cells. Limits for therapies based on embryonic stem cells and posed by histo-incompatibility are illustrated.

The proposal of "**therapeutic cloning**" to prevent immuno-intolerance is explained and the proposed procedure commented as obsolete because more and more **adult stem cells** are being identified, and differentiated cells are given back properties of precursor cells able to divide and thus being a potential source of immunotolerant replacement cells. It is even possible to roll the status of differentiated cells back to pluripotency. Cells can be reprogrammed to become **iPS cells (induced pluripotent stem cells)** by reactivating particular pluripotency genes, in particular *Oct4*, in other cases also *Sox2*, *cMyc* and *Klf*. Activation of the gene *msx-1* can cause multinucleate muscle fibres to disassemble and revert to single-celled stem cells. Several successful pioneering trials to rejuvenate differentiated cells without genetic intervention promise further progress. But it is also pointed out that stem cells and derivatives that retain the ability to divide often are sources of tumours or other forms of cancer.

Pluripotent and even totipotent stem cells are already found in organisms considered evolutionarily primeval such as sponges, hydroid polyps like *Hydra*, and **planarians**. Their stem cells, called **interstitial stem cells** in hydroids and **neoblasts** in planarians, share the possession of homologous pluripotency genes with human (mammalian) stem cells. Beyond being able to replace used up cells, some invertebrate species, notably *Hydra* and planarians, can replace all parts of their body from small fragments. Newts and salamanders among the amphibians can restore a few lost body parts (tail, limbs, eye lenses).

The basis of excellent capability to regenerate is not only the presence of pluripotent stem cells but in addition a high plasticity of differentiated cells, their ability to resume division, to **dedifferentiate** and to **transdifferentiate** into other cell types. For instance, in jelly fishes isolated striated muscle cells can transdifferentiate to almost all cell types including germ cells without genetic intervention. Newts can restore an eye lens from muscle cells of the iris. In the regenerating amphibian limb a protein called **nAG** (newt Anterior Gradient) supplied by leg nerves enables regeneration. Gradients of positional values, presumably encoded in gradients of cell adhesion molecules (comparable to the FAT and DACHSOUS gradients in insect limbs) appear to control correctness and completeness of restored structures. By applying **retinoic acid** RA in vertebrates **concealed potencies to regenerate** have been disclosed and, for instance, re-growth of nerve fibres in the spinal cord induced. The aim of regenerative medicine is, through intervention into the systems of control, to render regeneration possible of complex structures such as limbs or spinal cord also in humans.

The limited and insufficient faculty of our body to replace used up cells, in particular to replace all dying nerve cells, eventually causes death of the individual.

Growth Control and Cancer

19

Of all diseases, cancer is perhaps the most relevant to the study of developmental biology since it represents alterations in otherwise normal processes of growth. **Growth** is defined as increase in mass. Growth can result from more than one process. For example, enlargement of individual cells is a cause of growth. Muscle fibres grow by expanding in length and diameter. Also, the deposition of extracellular matrix material contributes to enlargement of the growing juvenile. However, most increase in mass is based on cell division followed by the enlargement of the daughter cells to the size characteristic of the cell type. In the following growth is understood as increase in cell number, not in cell size or individual cell mass.

Increase in cell number can result from the activity of certain 'oncogenes' (from the Greek *oncos* = mass and *genein* = to generate), and such an increase can be dangerous as it can result in the formation of a tumour.

19.1 Growth Control

19.1.1 Multicellular Living Beings as a Whole and the Size of all Their Organs Are Subjected to Growth Control

In multicellular organisms the individual cells are subject to social control. Unlike free-living unicellular organisms, the members of a multicellular community must reduce or cease their multiplication at the right time and place. The process of **terminal differentiation** alone is associated with a delay in the cell cycle and thus causes a slowdown of cell multiplication. Frequently, terminal differentiation results in a complete cessation of growth. In mammals, for example, mature blood cells and nerve cells completely lose the ability to divide. Termination of growth in these cells is part of their normal developmental program. In contrast, stem cells and several types of progenitor cells must continue to divide.

Apoptosis, programmed cell death, terminates the life of individual cells in many cell types. Apoptosis was introduced in Chap. 14 (Fig. 14.9) and will further be discussed in Chap. 21 (Ageing and Death) but must also be considered in theories on carcinogenesis.

Besides such endpoints programmed at the outset in the developmental progress cells remaining potentially immortal either self-regulate their proliferation rate or are controlled by their neighbours.

19.1.2 Cell Populations Self-control Their Multiplication Rate

A simple model provides a plausible indication of how a population can automatically constrict its growth itself. Imagine all cells of a still small and loosely arranged cell society produce inhibitory substances, be it trivial metabolic end products or specific negative growth factors; the

W.A. Mueller et al., *Development and Reproduction in Humans and Animal Model Species*,
DOI 10.1007/978-3-662-43784-1_19, © Springer-Verlag Berlin Heidelberg 2015

substances are released into the surrounding interstitial diffusion space having restricted capacity. Initially the density of the population would be low and so also the concentration of the inhibitory substance. With increasing cell number more and more producers are active but the diffusion space around each individual cell will not increase but shrink resulting in a steep slope upwards in concentration of the inhibitor. Eventually a threshold will be reached above which growth comes to a standstill. In addition, in large massive organs such as the liver, outflow of inhibitory substances decreases with enlargement of the organ since in relation to mass and cell number the surface area decreases.

19.1.3 The Neighbourhood Also Intervenes in Growth Control

In the scientific literature there is only rarely talk of signalling substances preventing growth probably because often clear distinctions cannot be made. Many growth factors stimulates proliferation of certain cells at low concentration but stop it at high concentration. In precursor cells capable of dividing (indicated by the suffix '-blasts'), for example in the blood forming hemangioblasts, myeloblasts and erythroblasts (Chap. 18) some of these factors promote proliferation at low concentration but induce terminal differentiation at high concentration. Hence such proteins are called **differentiation factors** or **differentiation hormones**. Moreover, in a given case it is difficult or even impossible to discriminate between specific and non-specific inhibition. Growth inhibiting signalling substances are not always soluble factors. On the contrary, in most tissues **cell adhesion molecules CAM** and components of the extracellular matrix provide such signals. This is indicated by the behaviour of dissociated cells which in cell culture resume dividing upon separation from the social society, and shown by the behaviour of cancer cells.

19.2 Cancer: Essence, Incidence, Terms

"Cancer" is a collective term comprising more than 200 different diseases the common denominator of which is excessive growth of one or more cell types. If tumours result they may throttle the supply of other tissues, disrupt blood vessels and nerve fibres, or disrupt and break organs such as the skin (see Fig. 18.2). About 25 % of the population of industrial countries will in the course of their life have to suffer complaints caused by apparent or concealed tumours (In reading statistics one has to distinguish between incidence and mortality).

The behaviour of cancer cells can also be observed in cell culture.

19.2.1 Cancer Results from Disturbed Growth and Differentiation Control

Cancer arises when basic rules of the cell-specific or social control of proliferation are violated. Excessive multiplication may occur in the following situations:

1. **Progenitor cells multiply too rapidly or frequently**. Terminal cell differentiation cannot cope with this excessive cell number and remove enough of the descendants from the pool of cells capable of dividing.
2. Even if the cell cycle is not accelerated, uncontrolled growth can arise
   - when both daughters of a dividing stem cell **retain the traits of stem cells**. Normally, on average only one of two daughter cells retains the characteristics of a stem cell, while the other becomes committed to cell differentiation;
   - when **differentiating cells do not stop dividing**. Cell divisions continue, even when the program of differentiation is largely finished. For example, **melanomas** develop from nearly mature derivatives of neural crest cells which synthesize black

melanin but nevertheless do not stop dividing;

- when differentiated cells partly **reembryonalize** regaining capability to divide;

- when cells are **not eliminated by programmed cell death** in time; this holds especially for mutated precursors of cancer cells that should subject themselves to suicide by apoptosis.

19.2.2 Glossary of Cancer Terms Reveals Which Cell Types Tend to Cancerous Growth

Transformation. The deterioration of a civilized cell into an antisocial cancer cell is called **neoplastic transformation** (Greek: *neo* = new; *plastein* = to form). Agents causing **carcinogenesis** are called **carcinogenic** (Latin/Greek: cancer-generating) or **oncogenic** (Greek: *oncos* = mass; *genein* = generating). However, the common use of the term 'oncogene' may be confusing. While the adjective *oncogenic* means "cancer generating", the noun **oncogene** refers to a gene causing cancer.

A tumour (Latin = swelling) is an association of cancerous cells, generally deriving from one transformed founder cell. A tumour may be **benign** (Latin *benignus* = mild, good-natured) or **malignant** (Latin *malignus* = malicious, ill-natured). A tumour becomes malignant when cells migrate from the primary tumour to invade other tissues and establish colonies of secondary tumours, called **metastases** (Greek: *meta* = subsequent, *stasis* = location).

Depending on its origin, a tumour may be a

- **Carcinoma**: a malignant tumour derived from epithelial tissue (Fig. 18.2)
- **Adenoma**: a benign tumour of epithelial origin and glandular appearance
- **Adenocarcinoma**: a malignant tumour of glandular appearance
- **Sarcoma**: derived from connective tissue
- **Hepatoma, hepatocarcinoma**, arisen from hepatocytes or other liver cells

- **Melanoma**: derived from melanocytes (pigment cells of the skin)
- **Neuroblastoma**: derived from neuroblasts
- **Glioma**: derived from glial cells (most brain tumours are gliomas)
- **Myoma**: derived from myoblasts including satellite cells
- **Myeloma**: derived from blood progenitor cells, in particular from myeloid stem cells
- **Lymphoma**: derived from lymphoblasts
- **Leukaemias**: derived from various, undefined blood progenitor cells but mostly from lymphoid stem cells and lymphoblasts (see Fig. 18.6) but may also include immature erythroblasts.

Strikingly, tissues and cell types **capable of self renewal** and **derived from stem cells** suffer neoplastic transformation more frequently than others. Among them, **carcinomas** are the most frequent tumours. Epithelia contain many dividing stem cells in which mutations may disturb the mechanisms of control. In addition, epithelia are exposed to carcinogenic agents – such as ultraviolet solar radiation – to a particularly high degree.

19.3 Particular Features of Cancer Cells and Tumours

19.3.1 Cancer Cells Frequently Are Immortal, Independent of Growth Factors, and Evade Apoptosis

In cell culture cancer cells display behaviour that reveals their reluctance being controlled and they ignore social rules.

1. **Immortality.** Cells of established cultures multiply as long as depletion of nutrients, available space and inhibitory metabolic products do not set limits. These cells do not age and retain the ability to divide. However, this holds not only for cancer cells. In vitro also non-transformed **stem cells** retain unlimited ability to divide, and established cell lines derive from stem cells or cells that regained

features of stem cells. Transplanted into test animals immortal cell lines do not necessarily display uncontrolled growths, thus, immortality as such is not a sufficient criterion of cancerous deterioration.

2. **Independence of growth factors.** Some cell lines produce their own growth factors (autocrine stimulation), or cells become completely independent of growth stimulating factors as are the cells of the early embryo during the phase of cleavage. Yet, if such cells did not originate from embryos but from suspicious swellings this independence can be a criterion of cancerous transformation. How such independence might be achieved is discussed in Sect. 19.4.

3. **Bypassing automatic self-destruction.** In frequently repeating divisions the mechanisms for DNA repair occasionally do not fully comply with their task. Damage to DNA accumulates, or chromosomes are unequally distributed in cell division (aneuploidy). In such cases the cells should turn on the suicide program (apoptosis) or be forced by the immune system to undergo self-destruction. Cancer cells evade such demands. In other words, cells which escape the imprinted suicide program by chance, for instance because of mutations, proliferate in spite of internal damage.

4. **Absence of contact inhibition.** Cancer cells display an unusual and selfish behaviour. Cancer cells neglect and ignore many social rules, for instance do not obey the rule of **contact inhibition**. Most cell types maintained in cell culture stop moving when they encounter other cells and stop dividing when they are completely surrounded by adjacent cells. Adhesiveness links the cells in a **confluent monolayer**. Cancer cells are different: they show persistent locomotion, crawl over their neighbours and continue to divide, producing **foci** (small cell clusters). Transformed cells are rounder in cross section, lack adhesion plaques and stress fibres, and are less covered with fibronectin. Such foci are sure signs of the presence of cancerous cells. The lack of contact inhibition is a prerequisite for the invasion of a tumour into healthy tissue and for the occurrence of metastases.

19.3.2 The Ability to Form Metastases Is Another, and Particularly Dangerous Property of Many Cancer Cells

The ability to perform amoeboid movements is another characteristic of many cancer cells. This behaviour enables them to colonize other body regions and to form **metastases**. Crossing the walls of blood vessels and invading solid tissues is facilitated by the ability of many malignant cells to produce enzymes which digest holes into the basal lamina of blood vessels and channels through connective tissues. Cancer cells share these properties with some normal cell types, notably cells of granulocytes, macrophages and lymphocytes. Thus, several types of cancer cells can spread through the body like white in cells but unlike white blood cells, cancer cells continue dividing. Cancerous **melanocytes,** for example, can not only give rise to **black skin cancer** but also resume properties of their ancestors, neural crest cells, wander about and give rise to many **metastases,** especially in the lungs.

Altogether, **cancer cells resume and reactivate many faculties characteristic of embryonic cells.** Thus, besides being a serious medical problem, cancer is a matter of developmental biology.

19.3.3 Many Tumours Can Grow Rapidly Because They Ensure Ample Provision with Blood Vessels

In vivo selfish behaviour of tumours is manifested in another feature. Solid tumours not supplied by blood capillaries cannot grow beyond a certain size, generally 1–2 mm^3. But they can promote their own growth by another trick: like embryonic tissues, they produce and emit **angiogenic factors** which stimulate blood capillaries to grow towards and into the tumour (see Fig. 17.7). Thus, nutrient supply is improved.

Much hope is set on **anti-angiogenic factors**, which effect regression of blood capillaries and thus decouple the tumour from the blood supply. In clinical applications and trials pharmaceutical, rather than endogenous, inhibitors of vascularization are currently used.

Other research projects aim at developing strategies for cure or prevention at least by investigating the various causes of cancerous transformation.

19.3.4 Currently the Existence of Particular Tumour Stem Cells Is Discussed

Over the past decade **cancer stem cells** or **tumour-initiating cells (TIC)** have been identified in many tumour types. This has led to much discussion and the development of a new dynamic model of cancer which predicts that only the cancer stem cells can initiate growth at the primary or metastatic sites. The model proposes that metastasis is started by a subpopulation of cancer stem cells called metastasis initiating cells. . In addition induced stem cells **iPSC** (Sect. 18.5) can transform into cancer stem cells. Presumably some cancer stem cells arise from partly reembryonalized differentiated cells through loss of epigenetic restrictions and reactivation of certain transcription factors such as **c-Myc** and **Sox-2**. Also very small cells resembling embryonic primordial germ cells and expressing the **pluripotency gene** *Oct-4* became suspect of being the origin of cancer. Previously there were reports that activity of *Oct-4* was characteristic of tumours based on cancerous primordial germ cells. Such findings support the notion that many tumour forming cells have preserved characteristics of stem cells, or have secondarily regained stem cell properties. Yet, such findings do not tell much about the causes of uncontrolled growth.

19.4 Causes of Carcinogenesis

19.4.1 The Majority of Cancers Is Induced by External Mutagenic Agents

Cancer is induced by (at least) four classes of agents:

1. **A plethora of chemicals**. Not only is the number of carcinogenic chemicals unimaginable, also their chemical nature is diverse. It is not only artificial products of the chemical industry that are listed in the long record of ascertained carcinogenic substances but also substances of biological origin. Known natural carcinogens include **caffeic acid**, found in a variety of plants, and **aflatoxins** produced by fungi that infest nuts and grains. Several carcinogenic agents are found when meat is fried or grilled (i.e. benzopyrenes). So far, chemists have identified only a fraction of the myriad natural compounds in food. And only a fraction of those have been adequately tested for their ability to cause or prevent cancer.

   Most identified and tested carcinogenic agents are **mutagenic**. Conversely, most mutagens can initiate cancer formation when certain genes known as proto-oncogenes are affected. DNA damage as such is not necessarily immediately dangerous. Many substances are converted only in the body by processes of detoxification into highly reactive metabolites that attach to many biomolecules including DNA, RNA and proteins. Such conversions are often due to the activity of cytochrome P450 isoenzymes in the liver whose task is to convert lipophilic substances into more hydrophilic versions that can be excreted by the kidney.

2. Among the few special cases of per se non-mutagenic agents are tiny fibres of **asbestos**

564 19 Growth Control and Cancer

and glass which cause, when inhaled, repeatedly microscopic injuries in the lung.

> The carcinogenic action of mineral fibres are explained as follows: In the permanently moving lungs the sharp and rigid fibres cause microscopic injuries again and again. Damaged cells must be exchanged by mitotically generated replacement cells. In the forced, too frequent cell divisions not all mistakes in DNA replications are corrected by the repair mechanisms which normally check the correctness of replication and correct mistakes. Mutations accumulate whether they occur spontaneously or are induced by exogenous mutagenic agents. Local injury-induced released of growth factors might cause accelerated growth of mutated cells. If proto-oncogenes were affected cancer would arise. In addition mechanically induced disturbances in the distribution of chromosomes are discussed as a potential cause of lung cancers.

3. **Ionizing radiation** (alpha-, beta-, gamma irradiation, X-rays, short-wavelength UV),
4. **Viruses**, especially DNA viruses and RNA retroviruses. Such viruses can influence their host cells so that the cells multiply more frequently (for the benefit of the virus) and do not subject themselves to programmed cell death. Many retroviruses have collected eukaryotic **oncogenes** in the course of evolution. Such oncogenes once supported cell division of the eukaryotic cells and are designated **c-oncogenes (cellular oncogenes)** in this original function. Now viruses drag them along in a version that no longer allows control by the host cell but conversely helps the virus to control the host cell. Retro- and DNA viruses insert their genes into the genome of the host cells, thus the viral genes are multiplied together with their host cells.

5. **Cancer causing genes (oncogenes),** once integrated into the host's genome, can in reproduction be handed down to offspring via the germ line, transmitted through the egg cell or the sperm. Thus oncogenes can become an integral part of the genome. Suspicion of such burdensome legacy is raised,
   - when a certain type of cancer clusters in a family and occurs more frequently than average,
   - when cancer develops early in life.

   Prime examples are:
- *rb-1 (retinoblastoma-1)*, a defective allele of a normal and vital gene. If one gets this defective allele from the father and the mother gruesome consequences will occur. Early in life a retionoblastoma, that is a tumour of the retina develops. Children, even babies, die with an outsized tumour in their eyes. If only one single defective allele is inherited cancer risk increases; as mutagenic agents from the environment affect the second allele with higher probability than if there were two normal alleles, or when the second allele is shut down by epigenetic mechanisms and no normal allele can help to prevent too frequent cell division.
- *APC (Adenomatous Polyposis Coli)* and intestinal cancer. In certain families in the colon or rectum of young adults, females and males, hundreds of small swellings develop known as 'polyps' that sooner or later deteriorate. The disease is known as **familial adenomatous polyposis**. Responsible is a defective allele of the tumour suppressor gene *APC*. The APC protein is a regulatory element of the WNT signalling cascade; this signalling system (see Fig. 11.6) is frequently turned on when descendants of stem cells are induced to undergo amplification divisions. Frequently (in about 75 % of cases) a further defective tumour suppressor gene, the gene for the **protein p53** is involved in malignant outbreak of the disease. This p53 protein should drive cancerous cells into suicide (apoptosis) as will be exposed in Sect. 19.4.4.

As a rule, tumour-inducing genes are designated with a three letter code as are many of the first genes detected in former times. For example the first identified cancer causing gene was designated *src* (enunciate sarc) with reference to its potency to induce sarcoma tumours. The normal (wildtype) proto-oncogene that does not induce tumours is denoted with the additional letter *c* (c = cellular): *c-src*.

19.4.2 Traditional Cancer Research Discriminates Two Classes of Tumour-Generating Genes: Oncogenes and Defective Tumour-Suppressor Genes

Variants of genes causing cells to elude appropriate growth and differentiation control can be transmitted to offspring via the germ line, but more frequently they are imported by viruses. Alternatively, mutagenic chemicals generate such gene variants in single cells of the organism. Remarkably, all cancer causing genes are in their original form (wild type allele) not cancer generating but are involved in the control of growth and/or differentiation. Such genes turn dangerous

- if they mutate at certain positions of the gene thus losing their complete functionality,
- if they are over expressed,
- if they are translocated onto other chromosomal locations where they evade normal control,
- if the program of epigenetic silencing fails.

Genes whose mutations or mismanagement leads to cancer are grouped into two classes:

1. **Oncogenes** in the narrow sense of the word. They derive from normal, non-cancer causing **proto-oncogenes** when a mutation creates a *gain-of-function* allele. This newly acquired function is **positive stimulation of cell multiplication** through

- **increase in rate of division** and/or through
- **blocking programmed cell death.**

Oncogenes in this sense are often **dominant**: It suffices when one of the two alleles gives rise to a modified protein that stimulates proliferation or suppresses apoptosis. Most viral tumour genes belong to this category. The gene imported by the virus may stimulate a quiescent cell to re-enter the cell cycle and to undergo mitosis. Or the altered protein prevents normal apoptosis in a cell upon termination of its individual development or upon severe damages of its DNA.

In particular RNA retroviruses such as the HIV virus causing AIDS, which insert their genome into the genome of the host cell, benefit from growth stimulation because with the genome of the host also the genome of the virus is replicated and multiplied. This positive selection may be the reason why viruses often bear oncogenes. DNA tumour viruses on the other hand often have the faculty to switch on proliferation promoting genes of the host cell.

2. **Tumour suppressor genes.** As the term suggests among the tasks of the protein coded by the normal gene is **to suppress cell multiplication** and/or **to support apoptosis.** Cancer cells and tumours arise when a mutation leads to *loss-of-function* and both alleles are defective. As a rule tumour suppressor genes are **recessive**, because one intact gene may supply just sufficient normal protein for suppressing growth. Heterozygous individuals who got a defective allele in onefold dose are afflicted with increased cancer risk but, of course, less than are individuals homozygous for the defective gene.

Tumour-suppressor genes include the above mentioned retinoblastoma-inducing *rb-1* gene, the gut cancer-inducing *APC* gene and the *p53* gene discussed in the following section.

Potential epigenetic causes. Independently of whether a growth controlling gene is classified as oncogene or mutated tumour suppressor gene recently also changes in the DNA or chromatin, respectively, are being taken into account which are not mutations in the strict sense but may have similar effects. In Chap. 12 we addressed enzymatically effected 'epigenetic' modifications of the chromatin by which the accessibility of genes is regulated. Having fulfilled their tasks genes are decommissioned. If genes involved in growth control which are epigenetically silenced in differentiated cells are released from the epigenetic blockage and resume activity, cancer may result.

19.4.3 Many Proteins Encoded by Potentially Cancer Causing Genes Are Involved in the Control of the Cell Cycle

Although in no case has the complete chain of causal events been elucidated, a general framework for explaining the origin of cancers can be outlined: cancer originates from disturbed control of growth or an insufficient change from proliferation to terminal differentiation. Defects have been found at all levels and in all members of the controlling system:

* **Growth factors**: too much or not inactivated quickly enough. Many tumour cells kept in culture constantly produce growth factors themselves such as PDGF (Platelet-Derived Growth Factor) of TGF-α (Transforming Growth Factor alpha). By producing such factors cancer cells stimulate themselves to proliferate (autocrine stimulation).
* **Disrupted signal reception**: Some oncogenes code for defective receptors that are not deactivated quickly enough after reception of a mitogenic signal, or are constitutively active even in the absence of growth factors (examples: *Erb-B2, RET*) and cannot be switched off.
* **Disrupted signal transduction**: A variety of known oncogenes code for subsequent elements of signal transduction pathways

(examples: *src, ras*). In about one quarter of human tumours a defective *ras* is found. The normal **c-RAS** protein is a member of a signalling chain which transmits proliferation-stimulating signals from membrane-associated receptors into the nucleus. Defective RAS fires ceaselessly. Another element of a signal transduction cascade is the (above mentioned) **APC** protein; it is a regulatory element of the canonical WNT signal transduction system and often involved when stem cells undergo divisions and daughter cells are directed into differentiation pathways.

* **Disrupted signal response in the nucleus**: The ultimate instance that has to decide whether DNA should be replicated and cell division prepared, or whether the cell cycle should be shut down and the cell be arrested in the G_0 phase, resides in the nucleus and employs transcription factors. Several oncogenes code for **transcription factors involved in cell cycle control** (examples of oncogenic transcription factors: are *myc, myb, jun, fos*). Thus, the proteins **JUN** and **FOS** form heterodimers known as **AP-1** protein, which prompts initiation of DNA replication. Interestingly, inverse mechanisms have been proposed for the mechanisms of the less known tumour suppressor genes: Production, reception or transduction of signals that normally slow down cell division and foster cell differentiation, are disrupted. In all these cases, the common ultimate defect is that the cell cycle starts too frequently.

19.4.4 Also Disrupted Gene Functions Which Should Regulate the Balance Between Self-renewal and Differentiation Can be Cause of Cancerous Transformations

In this context once more the versatile **Wnt/beta-catenin system** must be mentioned. It does not only fulfil tasks in establishing and patterning of body axes (compiled in Chap. 11, also in Fig. 22.6). It also participates in the control of proliferation and is linked with the machinery that regulates the cell cycle. A disrupted Wnt/β-catenin system can cause breast cancer.

One of the possible functions of **tumour suppressor genes** is to provide stop signals for termination of cell division. Accordingly, the balance between proliferation and differentiation also is disturbed when stopping signals such as differentiation factors or receptors for such signals are defective. The often addressed **TGF-β** (Transforming Growth Factor beta) can prompt stem cell descendants of the gastro-intestinal tract to leave the cell cycle and to undergo terminal differentiation. Some forms of intestinal cancer trace back to defective receptors for stopping and differentiation-inducing signals.

19.4.5 DNA Repair and Quality Control Are Further Tasks of Tumour Suppressor Proteins

In every round of DNA replication mistakes are made. DNA repair mechanisms check newly synthesized DNA strands and correct faults with astonishing efficiency. Yet, DNA double-strand breaks can not always be repaired. Damaged DNA is subject to guardian mechanisms which abort the cells. Checkpoints in the cell cycle decide whether a cell is permitted to enter a new cell cycle or is withdrawn from service and prompted to undergo apoptosis. As a rule, dangerous runaway cells subject themselves to programmed cell death and disintegrate. A pivotal role in this process has been assigned to the protein coded by the **tumour suppressor gene**

p53. The *p53* gene is frequently mutated in many types of human cancers in particular in many (about 75 %) malignant intestinal tumours. Strangely, although p53 protein is a transcription factor that can activate many genes, mice unable to express any functional p53 undergo normal embryonic and fetal development but develop cancers early in their juvenile life.

19.4.6 There Are Tumours Conceivably not Caused by Mutations: Germ Cell-derived Teratocarcinomas

It was pointed out above that not only mutations in the classical sense but also failures in epigenetic modification of the chromatin may have effects resembling those of true mutations. But even factors not directly related to control of gene activity may be the cause of uncontrolled cell behaviour. Here we point to dysplasias that can become fatal to individuals concerned, especially to women.

We recall: teratomas are failed and chaotically organized embryos. They may arise from unfertilized germ cells in the ovary or from fertilized eggs which did not implant in the uterus but at wrong places, in the Fallopian tube or in the abdominal cavity. Some of these teratomas grow into **teratocarcinomas** that behave like tumours and even may give rise to metastases. One can liberate cells from such teratocarcinomas and grow them in culture. They may behave normally in their new environment, require growth factors and do not form foci on confluent monolayer films as do cancer cells. Injected into blastocysts, teratocarcinoma cells from mice participated in normal development.

Box 19.1: Multi-stage Models of Carcinogenesis

Based on experimental findings several models have been proposed and mathematically formulated to explain carcinogenesis and as a basis of risk assessments. All data collected in animal testing, epidemiological studies and in experiments with cell cultures suggested that the formation of a malignant tumour is not a 'one-hit' event. The various models differ in the emphasis laid on relevant parameters but share some basic properties. The main steps are (1) initiation, (2) promotion, (3) progression.

1. **Initiation:** The primary paradigm is the concept of 'carcinogens as mutagens'. Models are mainly centred around **mutations,** and the main focus is on the chemical environment, radiation and viruses. As a rule, it is assumed that a normal cell must undergo **several mutations** to become a neoplastic cell. In experiments up to six mutations had to be introduced to transform healthy cells, and it was supposed that accumulation of several mutations over time is one cause for increased cancer occurrence in elderly people.

   Refined models centre on genes related to growth control and include distinction between oncogenes and tumour suppressor genes, and take epigenetic phenomena into account.

   Current modifications and expansion of the models is the **stem cell theory** of carcinogenesis. Other models lay emphasis on **genome instability** and focus on familiality.

2. **Promotion:** Several models assume that tumour growth is dependent on additional non-genetic factors such as intake with food of **tumour-promoting substances**. The paradigm of such substances are phorbol esters, heterocyclic molecules of plants in the family of Euphorbiaceae (the milkweeds) that were components of herbal teas made from the plant *Croton flavens* are significant cofactors in cancer incidence. These substances interfere with the PI-signal transducing system (see Fig. 11.5b). Of course, only a few components of food (e. g. Vitamin A) can be taken into consideration in model calculations.

3. **Progression:** Recent models on progression are based on **'natural selection'** ('Darwinian model') that is on increased proliferation rates and potential immortality of cancer cell lines.

   Several models are tuned and refined to account for the incidence and progression of distinct types of cancer such as skin or breast cancer. Advanced models further take into account time-scale of progression, age of the population and environmental risk factors.

   Advanced models of multistep processes have to consider loss of ability to terminally differentiate, loss of contact inhibition, ability to migrate to distant locations and to invade other tissues. Of course, models will increase in complexity. The main aim, however, remains to support risk assessments by computer simulations and to evaluate the importance of single factors.

Chimeric mice were born which did not develop tumours (see Fig. 13.4b).

These findings provide evidence that tumour formation is based on disturbed control of proliferation and differentiation. Mutations or failures in epigenetic programming can disturb or render mechanisms of control ineffective, but the systems of control can get mixed up also in other ways: when cells **do not sit in their correct niche but get in wrong places** where the inadequate neighbourhood cannot exert controlling functions. Even in testes primordial germ cells can surprisingly commence dividing and give rise to teratomas (testicular cancer). The issue of cancer will remain an issue of developmental biology.

Summary

In multicellular organisms cells have to undergo self-restriction in proliferation or are kept in check by their neighbours. When the control of cell amplification and cell differentiation is disturbed cancer may result, for instance when in terminal differentiation the program to switch off the cell cycle is not implemented or damaged cells do not subject themselves to programmed cell death (apoptosis).

- Like stem cells cancer cells are immortal, often independent of growth factors, evade contact inhibition by neighbours and are often motile displaying amoeboid movements. These properties enable cancer cells to wander about and establish metastases. Recently particular tumour stem cells are considered which, by contrast to normal stem cells, flee control of proliferation and proper differentiation.

- Cancer is predominantly initiated by mutagenic agents (chemicals, ionizing radiation) when these agents impair genes required for proliferation control. Cancer can also be induced by viruses which import proliferation supporting genes of eukaryotic origin (RNA retroviruses) or activate such genes in the host cells (DNA viruses). Occasionally cancer causing genes are transmitted through the germ line to offspring causing familial cancers such as retinoblastomas or colon cancer.

Cancer causing genes are subdivided into two classes:

1. **Oncogenes** which mostly are dominant and derive by mutation from normal, non-cancer-causing proto-oncogenes that are involved in the decision between proliferation and terminal differentiation. Mutated proteins encoded by oncogenes stimulate cell proliferation in various ways, for example by setting a condition that simulates a permanent supply of growth factors.

2. **Tumour suppressor genes**, which normally are involved in cell cycle control and suppress cell multiplication but are defective in cancer cells. As a rule they are recessive when inherited. Further genes can become carcinogenic when they do not fulfil their tasks properly. This applies for

3. **Genes coding for enzymes which effect DNA repair** and/or

4. **Genes required for initiation and implementation of programmed cell death;** such a program (apoptosis) should be executed when damage to DNA cannot be repaired.

Not only true mutations can be the cause of cancer. After termination of growth normally genes required for growth control are permanently switched off by **epigenetic** mechanisms. When these fail, cells may unduly resume proliferation.

Tumour formation is considered to be a multi-step process: Starting events are mutations (or mutation-equivalent alterations in the epigenetic chromatin modification) that affect genes involved in cell cycle control or regulate apoptosis of potentially dangerous cells. Secondary factors such as the supply of growth factors, improved nourishment by induced blood vessels, and intake of tumour-promoting food constituents accelerate the growth of tumours.

There are, however, also tumours which presumably are not based on genetic factors. These are teratocarcinomas that derive from teratomas, which are disorganized embryos with chaotic patterns of partially differentiated cells; such teratomas develop when blastocysts implant in the wrong place. Also inadequate environment can cause cells to escape from normal proliferation control.

Altogether, cancer cells resume and reactivate programmes that are characteristic of embryonic stem cells but without subjecting themselves to mechanisms of control. Thus, besides being a medical problem cancer is a matter of developmental biology.

Metamorphosis and Its Hormonal Control

20.1 "A Second Embryogenesis" Creates a Second Phenotype

20.1.1 Most Amphibians and Invertebrates Undergo Major Phenotypic Transformations Changing from a First to a Second Elaborated Form

Metamorphosis (Greek: transformation, change of form) refers to fundamental remodelling of the whole body, and is associated with a fundamental change in the mode of life. The new phenotype occupies a new ecological niche and colonizes a different habitat. Larvae and adults utilize different nutritional resources and often live in different environments.

Most benthic marine invertebrates produce planktonic larvae. These free-living stages enable the species to first make use of planktonic prey and subsequently to exploit a scattered or transient ecological niche suited for the adult phase. Planktonic invertebrate larvae include those of echinoderms, such as the pluteus of sea urchins (Figs. 4.1 and 4.2); the trochophore of the polychaete annelids (sandworms, tube worms; Fig. 4.22); the veliger larvae of molluscs; the actinotrocha of the phoronids (slender worm-like sedentary creatures); the nauplius of the crustaceans (lobsters, crabs, barnacles). Planktonic larvae enter the adult population through metamorphosis and settlement. The planula larvae of the Cnidaria are also free-living. They undergo metamorphosis into the sessile stage of polyps,

as do the tadpole-like larvae of the tunicates which also undergo metamorphosis into sessile phenotypes.

In **insects** and **amphibians**, the acquisition of a new phenotype in metamorphosis implies the transition to a different habitat or at least a change in the nutritional resources that are exploited. Though the amphibians were the first vertebrates to spend a significant portion of their lives in terrestrial habitats, most present-day amphibians still return to the water to reproduce. The larvae live in water and feed preferentially on algae and plants. The terrestrial adults are carnivorous. Likewise, no adult insect (imago) competes with its larval stage for food.

Regardless of the species in which it is taking place, the process of metamorphosis implies several universal events:

- destruction of specifically larval structures (abdominal legs of caterpillars, gills and tail of tadpoles),
- adaptive remodelling of tissues that persist in the adult stage (nervous system, excretory organs),
- development of structures unique to adults (wings in insects, lungs in amphibians).

In **insects** and other arthropods, **moults** occur during development and growth. Arthropods are covered with a cuticle: a hardened extracellular shell that cannot expand. The cuticle must be shed and replaced periodically to allow growth and changes in body form. At each moult, the epidermis withdraws from the old, partially

W.A. Mueller et al., *Development and Reproduction in Humans and Animal Model Species*, DOI 10.1007/978-3-662-43784-1_20, © Springer-Verlag Berlin Heidelberg 2015

dissolved cuticle and subsequently expands, forming folds and secreting a new, soft cuticle beneath the old one. When the old cuticle is sloughed off, the epidermis and the new, still elastic cuticle expand. Once the individual has hatched, it supports expansion and unfolding of epidermis and cuticle by generating hydrostatic pressure with compressed body fluid.

Metamorphosis in insects is the transformation of the larval into the imaginal (adult) phenotype. Among the various insect orders, metamorphosis takes two broadly different forms:

1. **Hemimetabolous development** (Greek: *hemi* = half; *metabole* = change).

   In its basic organization the larva already resembles the imago in hemimetabolous species and even has compound eyes, although the early larval body is very small and lacks structures such as wings and genitalia. With each moult, the size and appearance of the larva (often called a **nymph** in hemimetabolous insects) approaches the size and appearance of the final imago (Fig. 20.1e). Sometimes, only minor remodelling of the larval body is necessary (**paurometabolic development**); in other species conspicuous larval organs must be destroyed (such as gills in may-flies or the extensible labium of dragonflies, with its hook-like jaws). Frequently, the nymph enlarges the wing pads rather suddenly in the last two moults, but a pupa does not occur.

   Examples of hemimetabolous development include wingless insects (*Apterygota*), grasshoppers (*Orthoptera*), cockroaches (*Blattidae*), termites (*Isoptera*), earwigs (*Dermaptera*), may-flies (*Ephemoptera*), dragonflies (*Odonata*), bugs (*Hemiptera*), sucking lice (*Anoplura*), and cicadas (*Homoptera*).

2. **Holometabolous development** (Greek: *holos* = whole, complete). Adult structures arise in holometabolic forms from larval cell areas specified for their future task and known as **imaginal discs** (Figs. 20.2 and 4.41). From the egg, the young hatches with a worm-like segmented body (called the caterpillar, grub, or maggot). The larva may have short legs; it lacks wings and compound eyes. The successive larval instars increase in size through several moults but do not abandon their worm-like appearance until complete metamorphosis takes place in two major steps. In the first step, the last instar larva transforms into an immobile **pupa** that is encased in a protective cuticle with a form and colour characteristic of this stage. In the second step the **imago is formed within the pupa cuticle**. Beneath the cuticle, many larval organs break down and are absorbed by phagocytic cells, while adult structures arise concurrently. At the end of pupation, the finished imago hatches. Some of these dramatic events are described for *Drosophila* in more detail below.

Holometabolous development is characteristic of goldeneyes and ant lions (*Neuroptera*), caddis flies (*Trichoptera*), moths and butterflies (*Lepidoptera*, Fig. 20.1f, g), true flies and mosquitos (*Diptera*), fleas (*Siphonoptera*), beetles (*Coleoptera*), wasps, ants and bees (*Homoptera*).

It may be emphasized that all these phenotypic forms, embryonic stage, larva, pupa and imago derive from the same genome. At present it is inconceivable that we will ever be able to logically deduce alone from the sequence of base pairs in the DNA in which forms of appearance an organism can exist.

20.1.2 In Metamorphosis, Dramatic Remodelling Occurs Down to the Molecular Level

Metamorphosis entails reshaping and restructuring at every level of the organism, from the body's morphology and anatomy to its physiology and cellular machinery. Such multiple changes are initiated and synchronized by hormonal signals, which are discussed in the next chapter. But as in all hormonally-controlled events, various tissues and organs respond quite differently to changing levels of hormones, depending on their programming in previous stages of development. To exemplify the remarkable range of metamorphic events, we will compare the metamorphoses of an amphibian (frog) and an insect (the fly).

Fig. 20.1 (**a–g**) Model animals related to metamorphosis and its hormonal control. (**a**) *Xenopus laevis*, the clawed frog. (**b**) Axolotl (*Ambystoma mexicanum*); (B1) neotenic form with lifelong gills, albino; (B2) Axolotl after thyroxine-induced complete metamorphosis. (**c**) The hydrozoan *Hydractinia echinata*. Planula larva and

Fig. 20.2 Holometabolous development of a leg imaginal disc in the fly; its transformation in the course of metamorphosis. *Colour codes* indicate that distal structures are pulled out from the central area of the disc on the model of a telescope

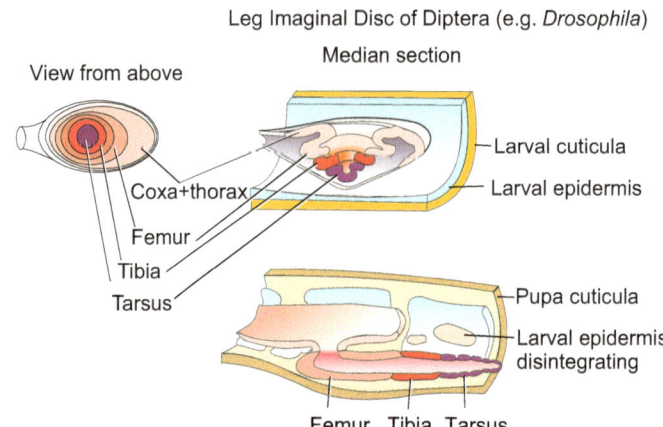

Leg Imaginal Disc of Diptera (e.g. *Drosophila*)

Median section

View from above

Coxa+thorax

Femur

Tibia

Tarsus

Larval cuticula

Larval epidermis

Pupa cuticula

Larval epidermis disintegrating

Femur Tibia Tarsus

Frog Destructive and constructive metamorphic processes proceed gradually and take weeks. There is no quiescent stage; the tadpole has to manage a smooth changeover to the adult organization while it is moving and feeding.

- Since the type of locomotion will change from smooth swimming to jumping with legs, hindlimbs emerge in prometamorphosis. During the climax of metamorphosis the tadpole extends its previously prepared but hidden forelimbs out of the branchial cavity, and gradually resorbs its tail (Fig. 20.3).
- When the **lungs** are formed, the gills vanish. The transition in the mode of breathing is accompanied (and made feasible) by **reconstruction of the circulatory system**. The aortic arches and several large body vessels are restructured. In the erythroblasts a new isoform of **haemoglobin**, that binds oxygen less avidly, is synthesized.
- To avoid desiccation, the skin is made tighter with adult keratins and interspersed with

mucus glands; the eye becomes protected by an eyelid.

- The horny teeth for tearing plants disappear; a long tongue develops and the intestinal tract is adjusted to a carnivorous diet.
- While the tadpole can readily release ammonia into the surrounding water by diffusion (**ammoniotelic excretion**), in metamorphosis the liver and kidney are retooled with new sets of enzymes enabling the frog to convert ammonia into urea (**ureotelic excretion**).
- The lateral line sensory system used by the larva to detect slow wave or current movements in water is reduced. The first branchial pocket is remodelled and becomes the **ear tube** (pharyngotympanic tube), closed by the **tympanic membrane** and incorporating the sound-transmitting **columella**.
- In the retina of the eye the visual pigments are exchanged: the fish-type **porphyropsin** (opsin + retinal A2 molecule) is replaced by **rhodopsin** (opsin + retinal A1) characteristic of terrestrial vertebrates.

Fig. 20.1 (Continued) primary polyp, founder of a new colony, arisen from the planula. Blue fluorescence marks the chitin envelope of the stolons which elongate by growth, branch like blood vessels, and bud new secondary polyps. (**d**) *Aurelia aurita*, moon jellyfish; (**d**1) asexually produced ephyra larva of *Aurelia*; (**d**2) Mature jellyfish. (**e**) *Schistocerca gregaria*, desert locust, example of a hemimetabolous insect with stepwise approach to the mature adult form (imago). In the sub-adult stage wings (*arrow*) are being formed but are not fully developed;

(**e**2) The adult imago is capable of flight and sexually mature. (**f**) *Manduca sexta*, a holometabolous butterfly. (**g**) *Araschnia levana*, the map butterfly; (**g**1) Spring form, (**g**2) Summer form, also known as *Araschnia prorsa*, (**g**3) Caterpillar, (**g**4) Pupa, which hibernates. Pictures from teaching collection of WM, (**f**) *Manduca sexta* photos kindly provided for reproduction by Prof. Richard G Vogt, University of Washington; he is owner of the copyright

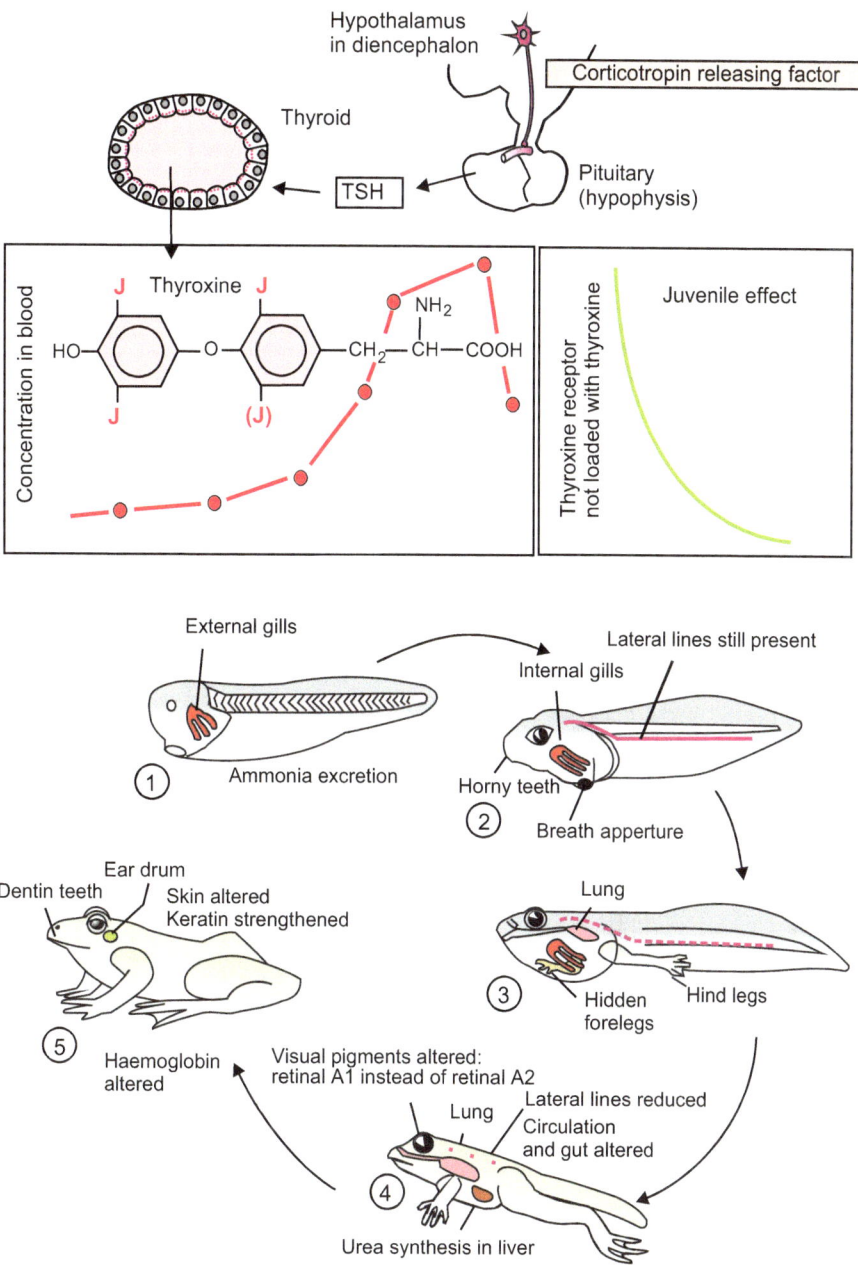

Fig. 20.3 Metamorphosis of the frog and its hormonal control (from Mueller and Frings: Tier- und Humanphysiologie, Springer Science 2009, slightly modified)

Drosophila The larva undergoes a holometabolous (complete) metamorphosis. When entering the prepupal stage, the third and last instar wanders around seeking a place in which to transform into a pupa. Having found a suitable location, it cements itself onto the substrate with a glue produced in its salivary glands. Now the larva softens its last larval cuticle and inflates into a barrel-like shape. The expanded cuticle is re-hardened and is now called the **puparium**. Beneath the protective envelope most larval tissues are destroyed. The fly is constructed like a mosaic, from residual **imaginal discs** (Figs. 20.2 and 4.41) and imaginal cells.

The exterior epidermal structures that become covered with a cuticle are manufactured from imaginal discs. In *Drosophila* most of the adult is built from 10 pairs of such discs (Fig. 4.41). Unlike caterpillars of butterflies, in dipteran larvae these discs are not seen as epidermal thickenings but lie hidden inside the body. However, the discs do not originate inside the body.

As early as the embryonic blastoderm stage small patches of 20-70 blastoderm cells are segregated out. As discussed in Chap. 10 (see Fig. 10.10), imaginal disc fields are specified at the intersection between stripes of cells secreting certain signal molecules (encoded by the genes *hedgehog, decapentaplegic,* and *wingless*). These clusters of cells remain diploid, become invaginated, are wrapped in thin epithelial sheets and remain stored inside the larval body where they grow but do not further develop until hormonally stimulated in the pupa. Within the puparium, the discs are everted and elongate. The cells in the centre of the disc telescope out to become the most distal portion of an antenna, leg or wing. Shield-like extensions at the base of the discs expand and assemble to form the capsule of the head and the thoracic segments.

The abdominal epidermis and many internal structures are generated by **imaginal cells** (also called **histoblasts**) that lie interspersed among larval cells. In its basic structure, the nervous system is taken over from the larva but becomes substantially modified.

20.2   Hormonal Control of Metamorphosis

20.2.1 Hormonal Control of Metamorphosis Shows Many Analogies Between Insects and Vertebrates

Metamorphoses are complex events and so are the systems of control. In both groups, insects and amphibians, the principle of dual control is realized. There are hormones that foster growth but curb the transformation of the body into the adult phenotype; there are also hormones promoting this transformation. Both systems are governed by neurosecretory cells in the brain. The neurohormones released by the neurosecretory cells control the liberation of development-controlling hormones from subordinate hormonal glands (Figs. 20.3, 20.4, and 20.5).

1. **Releasing the brake**. In the larvae of insects as well as in amphibian tadpoles metamorphosis is made possible by lowering the 'titre' (level or concentration) of a hormone that allows growth but inhibits the differentiation of adult structures.

   • In **insects** this is **juvenile hormone**. The hormone is a terpenoid (sesquiterpene) and as such has a structure reminiscent of many secondary metabolites in plants. The hormone is produced by a pair of glands called the *corpora allata*. As long as juvenile hormone is present in sufficient concentration, moults result in another larval instar. In the last larval instar, an axonal nerve from the brain inhibits the *corpus allatum* from releasing juvenile hormone. The lowered hormone level allows metamorphosis to proceed. Timely experimental removal of the *corpora allata* causes larvae to undergo premature metamorphosis from which small adults arise, while implantation of additional glands results in additional larval moults, delayed metamorphosis and giant adults.

     In the **adult**, production of **juvenile hormone** is resumed, but now the hormone acts as a **gonadotropic hormone** stimulating the development of the gonads.

   • In **amphibians prolactin** has long been implicated in *Xenopus* metamorphosis as an anti-metamorphic and juvenilizing hormone preventing premature metamorphosis. Prolactin is well-known to be used in other regulatory circuits. Its name points to its role in stimulating the production of milk in mammals. In other mammals prolactin acts as luteotropic hormone LTH stimulating the conversion of an ovarian follicle into a *corpus luteum* (yellow body).

     Its function as a juvenile hormone in amphibians has recently been challenged.

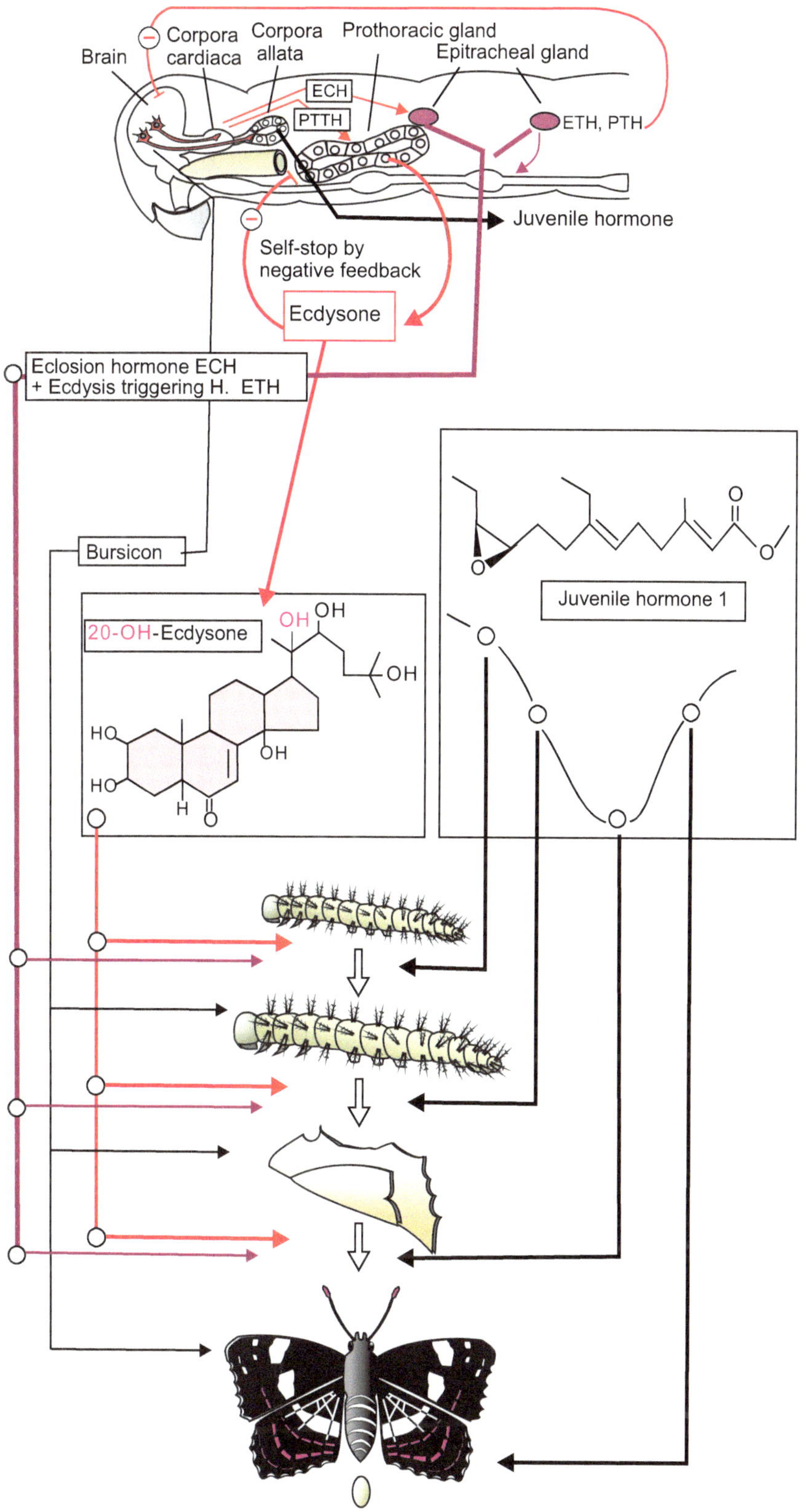

Fig. 20.4 (**a, b**) Hormonal control of metamorphosis in holometabolous insects (from Mueller and Frings: Tier- und Humanphysiologie, Springer Science 2009)

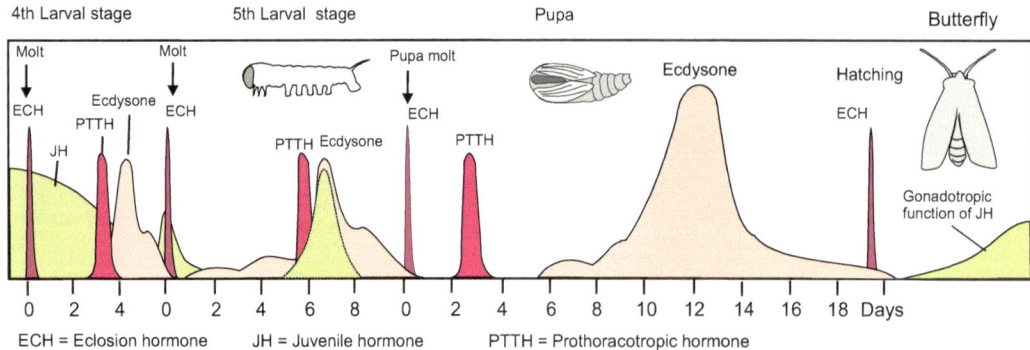

Fig. 20.5 Hormonal control of the butterfly *Manduca sexta*

During metamorphosis its level in the blood initially rises up to a climax point and only then decreases, and likewise the equipping of various tissues with prolactin receptors is initially increased. This transitory increase has been proposed to be interrelated with further functions of the hormone:

– Various tissues and organs required also in the future life on land must continue to grow in the long weeks of the metamorphosis. Prolactin is a member of the **growth hormone** family and future adult organs respond differently to those which will be out of use.

– Prolactin prepares some tissues to respond to thyroxine and supports in the initial phase of thyroxine-induced metamorphosis by inducing the **expression of proteases (collagenases)** that are needed for the degradation of larval tissues.

– Prolactin is known to be involved in **osmoregulation** in fish and presumably also in amphibians. This function may be modified in metamorphosis but remains a life-long task.

2. **Stimulating developmental progress.** The timely switching on of genes needed to redirect development toward the adult phenotype is triggered in both insects and amphibians by neurohormones delivered by the brain.

   In amphibians the chain of signalling starts with

   • a **neurohormone** produced in the **hypothalamus**. It functions as a releasing factor, stimulating the adenohypophysis (anterior pituitary gland) to liberate **TSH** (Thyroid-stimulating Hormone).

   • This peptide hormone conveys the message to the **thyroid gland**, which in turn emits the metamorphosis-inducing hormone **thyroxine T4** (T4 = tetraiodothyronine). This hormone is distributed via the blood circulation system in the entire body. In target tissues T4 is enzymatically converted into the final hormone **thyroxine T3** (T3 = triiodothyronine). T3 induces activity of genes required to accomplish metamorphosis. Administering T3 to tadpoles elicits precocious metamorphosis, resulting in small frogs. Blocking the thyroid gland by thyreostatica (drugs blocking the thyroid gland) delays metamorphosis and results in giant axolotl-like tadpoles.

In insects the chain of signals starts with the **neurohormone PTTH** (Pro-Thoracico-Tropic Hormone). This peptide is released via the corpora allata and functions as a 'glandotropic' hormone stimulating the **prothoracic gland** to liberate **ecdysone**. Ecdysone is then converted in the target tissues into the ultimately effective **ecdysterone (20-hydroxyecdysone)**. For simplicity, we will use only the term ecdysone.

The role of ecdysone in metamorphosis has been intensively investigated. Ecdysone is present prior to the beginning of metamorphosis: in the larval stage, each moult is triggered by pulses of ecdysone (Fig.20.5), but the presence of juvenile hormone prevents the larva from forming a pupa.

20.2.2 Also in the Molecular Mode of Action There Are Parallels and Even Homologies

Parallels in the molecular mechanisms by which **ecdysone/ecdysterone** in insects and **thyroxine** in vertebrates exert their action are surprising. The molecular structure of these hormones is very different – thyroxine is a derivative of the amino acid tyrosine, whereas ecdysone belongs to the family of steroid hormones. Therefore, a common mechanism of action was not expected. However, **ecdysone, thyroxine** and the chemically unrelated **retinoic acid** all bind to the same type of **homologous intracellular receptors**. Occupied by their respective ligand, the receptors become transcription factors **controlling gene activity**. The receptors contain zinc fingers which enable them to bind to DNA (see Fig. 11.11).

Considering the conserved structure of this receptor family one can justifiably speak of homology at the molecular level. The parallels go one step further:

- The receptors for thyroxine, retinoic acid and steroid hormones occupy, as a rule, the controlling region of downstream genes as dimers, often as heterodimers, and this holds also for the 20-OH-ecdysone receptors (EcR). The partner of EcR in the dimeric complex can be USP (Ultraspiracle), in parallel to RXR in vertebrates.
- It also holds for 20-OH-ecdysone as well as for thyroxine that among the target genes are the genes for their own receptors and thus the target cells are stimulated to express more receptors. By this positive feedback adaptive responsiveness increases. More parallels concern other downstream genes; many code for transcription factors.

Such far reaching parallels are not observed when juvenile hormone and prolactin are compared. Being a protein prolactin acts via membrane-anchored receptors and activates a signal transducing cascade starting with a receptor of the tyrosine kinase type (Fig. 11.9). Juvenile hormone is lipophilic and can enter, and perhaps pass through cell membranes. For juvenile hormone an anti-ecdysone mechanism is discussed (in crustaceans). Like the ecdysone receptor the JH-binding receptor (known as methylfarnesoate receptor MfR) is found within the cell. It is thought to bind Ultraspiracle and thus to prevent its association with the EcR. Without Ultraspiracle as a partner for dimerization ecdysone would not be able to exert its full power and many downstream genes needed for metamorphosis could not be activated.

Juvenile hormone is a terpenoid consisting of three isoprenoid moieties, a lipopophilic molecule unique in the animal kingdom. In plants such molecules are often found and play a role in biological pest control as discussed below in 20.2.5. Much effort has been applied to identify terpenoid receptors in insects and plants. A receptor according to classic criteria is a protein (or member of a family of homologous proteins) that binds the ligand with high specificity and affinity. Several proteins were found binding juvenile hormone but none with the expected high affinity. But is high affinity a *conditio sine qua non*? At least, among these low affinity receptors has been the transcription cofactor Ultraspiracle which is proposed to be responsible for the anti-ecdysone effect of juvenile hormone. Other low affinity binding proteins are a transcription factor with the suggestive name BROAD and protein kinase C (PKC), a lipid-activated key enzyme of the PI-PKC signalling cascade (Fig. 11.5). JH can activate PKC. Therefore the notion was proposed the multitude of binding partners would enable the multitude of effects. JH plays a role not only in the larval life but also in the adult insect. It promotes maturation of the gonads and has putative functions in the release of migratory behaviour in migrating insects, and in diapause of insects that interrupt development by a phase of rest, and in caste determination of social hymenoptera and termites.

20.2.3 Different Tissues Respond Differently to Hormonal Signals

The transcription-stimulating activity of ecdysone is visible as the pattern of puffs in the polytene chromosomes (see Chap. 12, Fig. 12.13). A close look at this pattern reveals that each stage of metamorphosis and each type of tissue is characterized by a specific combination of puffs.

Various tissues respond differently to hormonal signals. For example, when a limb bud is experimentally transplanted into the tail region of a tadpole, the bud does not grow unless stimulated by thyroxine at beginning metamorphosis. In metamorphosis both structures – tail and leg bud – are exposed to the same titre of hormones, but the tail is reduced while the leg bud persists and grows out. Reduction of larval structures is accomplished by **apoptosis.**

This different programming is reflected by the different makeup of the target tissues with variants of hormone receptors and/or transcription regulators. For example tissues responding to juvenile hormone are equipped with different variants of the BROAD transcription factor.

20.2.4 More Hormones Are Involved in Controlling Metamorphosis

When comparing amphibians and insects, differences should be noted in addition to the parallels described above, and more hormones taken into account which have different functions.

In amphibians, for example, hormones of the adrenal gland, namely **cortisol** and **aldosterone**, participate in preparing the animal for life on land and new environmental conditions. Cortisol is known to adjust the organism to long-term stress and long-term deprivation, aldosterone controls renal functions; it functions in the fine-tuning of water balance and in osmoregulation. In this function aldosterone and prolactin appear to cooperate.

In insects the shedding of the old cuticle and the acquisition of a new one are controlled not only by the classic ecdysone and juvenile hormone but also by several additional neurohormones. For example, **eclosion hormone** delivered by the brain stimulates in ca cells in scattered epitracheal cells to successively release two further hormones which induce ecdysis, the hatching from the nymph and pupal cuticula. These are the peptide hormones **PETH** (Pre-Ecdysis-Triggering Hormone) and **ETH** (Ecdysis-Triggering Hormone).

PETH and ETH trigger via the central nervous system a sequence of movements by which the armour-like cuticula is burst and can be left behind. Finally, the central nervous system supplies another neuropeptide, called **bursicon.** This hormone induces the hardening (sclerotization) of the new cuticle after the moult.

20.2.5 Phyto-ecdysone and Phyto-juvenile Hormones: Fake Hormones and Plant Defences

A remarkable story involving plant biochemistry and insect development is exemplified by the plant bug *Pyrrhocoris apterus*. When brought from Europe into an American laboratory, these bugs failed to undergo metamorphosis. Instead, they underwent additional larval moults, and finally died as giant nymphs. The systematic search for possible causes eventually took into account the paper lining the rearing dishes. It was discovered that larvae reared on European paper, including pieces of the distinguished British journal *Nature*, underwent metamorphosis. However, bugs reared on American paper, including *Nature's* competitor *Science*, refused to metamorphose. The American paper was made from balsam fir, a North American tree that produces compounds with juvenile hormone-like activity.

Numerous plants have been found to produce natural pesticides that act by interfering with the hormonal system of insects. Some members of the plant family Asteraceae produce **precocene**s, which cause the *corpora allata* of several hemimetabolous insect species to wither. As a result, the level of juvenile hormone drops too early, metamorphosis is initiated precociously, and the result is dwarf imagines unable to reproduce. The Indian neem tree, *Azadirachta indica*, and related African trees synthesize a large variety of terpenoid compounds, collectively named **azadirachtin**s, that kill a number of insect pests by interfering with the moulting process, disrupting development into adults or reducing fecundity and egg hatching. The compounds include a large complement of terpenoids including sterol-like tetra-nor-triterpenoids, pentanor-triterpenoids, and diter-penoids as well as a number of nonterpenoidal ingredients. The terpenoids are thought to act by interfering with ecdysone-mediated processes.

In addition, many of the compounds depress feeding activities in insects.

20.3   Release of Metamorphosis

20.3.1 In Insects Day Length Often Determines the Start, Course and Result of Metamorphosis, and a Phase of Quiescence Can Be Inserted

Numerous links between environmental cues, the hormonal system and development are known from insects. Often scientific consensus assumes that metamorphosis starts when the larva has accumulated enough mass. Certainly, in insects and amphibians metamorphosis is part of the developmental program and is triggered mainly by internal factors. But the period when meta-morphosis occurs and even the precise moment (think of May-flies!) must be adjusted to envi-ronmental conditions such as temperature and day length, and adjustment must be prepared in time before, for example, a sudden cold spell freezes water and animals alike. To synchronize the internal developmental program with exter-nal cues, the hormonal system is placed under the control of the nervous system. Sensory input can thus be used as trigger signals.

In many insects, a period of rest is inserted into the life cycle. In ecophysiology two forms of rest periods that facilitate survival during adverse environmental conditions such as cold or extreme heat are distinguished

- **Quiescence:** A period of rest directly induced and terminated by adverse environmental conditions (e.g. low temperature) and which can be gradual.
- **Dormancy** or **diapause:** A sustained phase of rest that is begun **prospectively** in provision for imminent raw and inhospitable times. When wintertime is coming up precautionary measures must be made because an amphibian or an insect (except honey bees) do not have heaters or body firing to prevent freezing. Non-homeothermic animals synthesize anti-freezing substances, and must do this in time. Diapause can be inserted into the life cycle at any stage: egg, pupa, or imago.

Diapause is induced by

- long-lasting adverse atmospheric conditions and/or
- shortening day length, in particular transition from long-day to short-day condition (Fig. 20.6).

Diapause is lifted,

- when subsequent to an obligatory cooling period temperature rises, and/or
- when day length surpasses a critical length, in particular when short-day condition changes to long-day condition.

Insects have provided evidence for intriguing mechanisms coupling environmental cues, the hormonal system and development. For example, the imagines of the marine chironomid midge *Clunio marinus* hatch from the pupal envelope on distinct days of the **lunar month**. These days are correlated with the spring-neap tide cycles. Ecdysone is involved in programming the time point for synchronized hatching in a local population.

In some butterfly genera (*Araschnia, Precis, Bicyclus*, and *Polygonia*) two very differently coloured forms ('morphs') appear each year (**seasonal polyphenism**). In *Araschnia levana*, the map butterfly (Fig. 20.6), butterflies with red pigmented wings hatch from diapause pupae in the spring, whereas butterflies with black and white patterned wings hatch from non-diapause pupae in the summer. The colour pattern is modified in response to ecdysone signals.

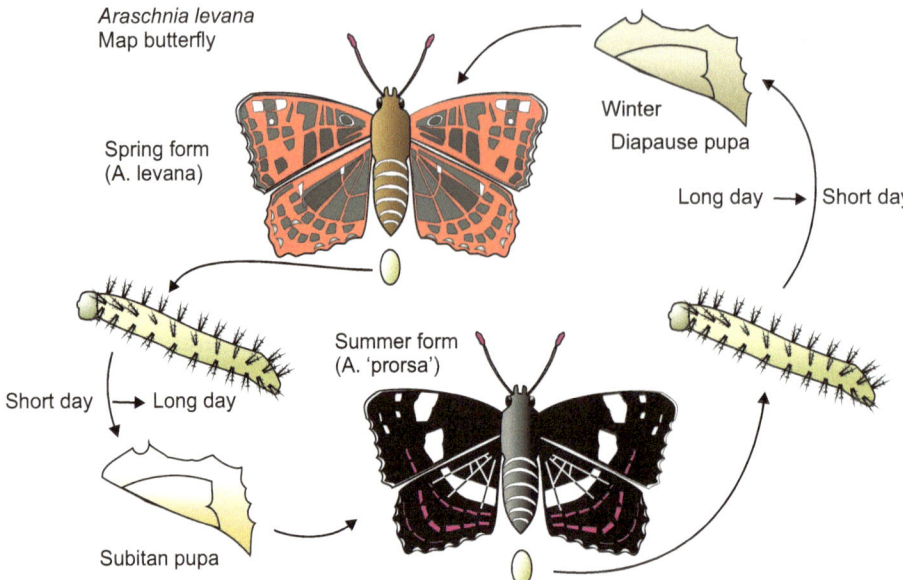

Fig. 20.6 Life cycle of the map butterfly *Araschnia levana*. This European species occurs in two phenotypes whereby increasing or decreasing day length specifies the form that will hatch out of the pupa (from Mueller and Frings: Tier- und Humanphysiologie, Springer 2009)

20.3.2 In Numerous Invertebrates External Cues Often Induce Metamorphosis; Neurosecretory Hormones Synchronize the Internal Events

Changes in the ecological roles of larvae and adults can reflect prevailing environmental conditions. The free-living larvae of sedentary marine animals must locate a suitable substrate on which to settle, since a poor choice cannot be corrected after metamorphosis. Frequently, specific chemical cues promote settling and induce metamorphosis.

An example of settlement using specific cues for recognizing and selecting a habitat is the hydrozoan *Hydractinia* (Figs. 20.7 and 4.17). The sensory inventory of the free-living larva is very limited, consisting merely of a few sensory cells fitted with a cilium. Distant cues cannot be perceived, only short-range key stimuli can be indicative of an appropriate substrate on which to settle. Surprisingly, it turned out that such stimuli derive from environmental bacteria which cover hard substrates. This principle turned out to be valid throughout marine environments. For examples it applies also to the planula larvae of corals. In addition, larvae of annelids, tentaculates, barnacles, oysters, sea urchins and tunicates respond to bacterial films with settlement and metamorphosis. The larvae of epiphytic animals may search for plant-derived chemical cues. Thus, the larvae of the hydroid *Coryne uchidai* settle on *Sargassum* algae that produce metamorphosis-inducing heterocyclic compounds.

In **amphibians** metamorphosis is considered to be internally controlled. When the tadpole has surpassed a critical size metamorphosis is thought to commence independently of environmental cues. But in Northern latitudes the water temperature must surpass 7–10°C. Rising temperature, increasing day length and sinking water level speed development considerably.

In order that the changes and adjustments associated with metamorphosis take place in time and in a coordinated manner **external signals** acting upon the sensory system are converted into **internal** signals that can reach many receivers in the body, even all cells. Such internal signals are neurotransmitters and/or **neurohormones**.

Fig. 20.7 Metamorphosis of the hydrozoan *Hydractinia echinata*. **1** The larva anchors itself by pulling along on stinging threads on a mollusc shell, and settles down. **2, 3** Metamorphosis of the planula larva to primary polyp is induced by bacteria of the genus *Alteromonas*, which colonize and settle on the surface of the shell. The presence of such bacteria is perceived through neurosecretory cells which subsequently release a hormonal signal that triggers and synchronizes the internal processes of metamorphosis (after Mueller and Leitz 2002). **4, 5** Further development of the colony through stolon formation and budding of polyps

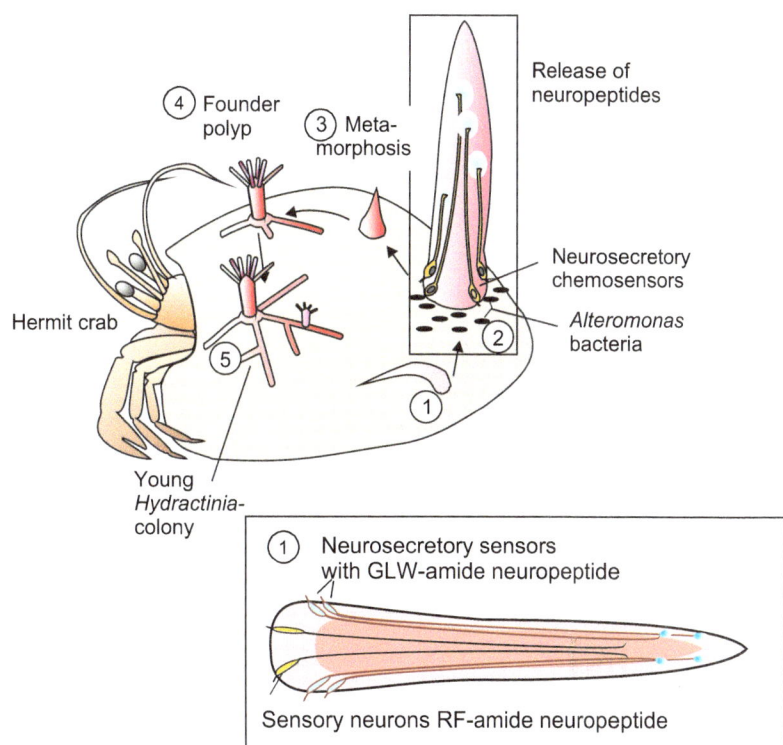

20.4   Reverse Development

The term reverse development means development back from advanced stages to a state closer to the ontogenetic origin. Such a reverse development was long considered to be impossible. Since re-transition of terminally differentiated cells back to stem cells exhibiting embryonic features is known – this was issue of Sect. 18.5 – reports on reverse development should not be dismissed as unbelievable or the phenomenon as such not considered as mere curiosity.

20.4.1 In Certain Cnidarians Reversal of Development Back to Previous Stages Is Possible and Can Even Be a Natural Phenomenon

A striking manifestation of cnidarian developmental plasticity is their unequalled potential to undergo reverse development, such as cases of medusae buds or free swimming medusae that undergo reverse development into the polyp stage (e.g. in *Podocoryne carnea*) or isolated tentacles in *Aurelia* that can convert into metamorphosis-competent planula larvae. These examples and others of reverse development are stress-induced. However, in several cnidarian species reverse development is part of the normal life cycle: *Stephanoscyphus planulophorus* is a member of the scyphozoan group Coronata. The animals lack germ cells. Instead young medusae, known as ephyrae, transform into planula larvae that metamorphose into polyps. Likewise, in the scyphozoan *Cassiopea andromeda* asexually produced buds look and behave like sexually produced planula larvae. Sexual and asexual borne larvae undergo metamorphosis into polyps, induced by the same external cues.

Summary

Metamorphosis is a kind of second embryogenesis that converts a first larval phenotype into a second adult or imaginal phenotype which colonizes another ecological niche and becomes sexually mature. In marine ecosystems the larva also serves geographical dissemination of the species.

In amphibians degradation of larval structures and emergence of new adult structures take place step by step. In insects reconstruction can either take place gradually in small steps (**hemimetabolous** or **pauro-metabolous development**) or in two dramatic phases of complete reconstruction from the larva to the pupa and from the pupa to the imago (**holometabolous development**).

As in each metamorphosis numerous structures must be reconstructed according to a fixed schedule and many physiological functions adapted to the new life, controlling signals are required that reach many receivers. Hormones are qualified for that task. In amphibians and insects a dual system of control is established:

- There is a hormone that enables cell division and thus growth but on the other hand prevents precocious transformation of the larva into the adult form. In insects this hormone is delivered by the *corpora allata* and is known as **juvenile hormone JH.** A decreasing concentration of JH enables metamorphosis. In amphibians a comparable function has been assigned to the hormone **prolactin** that is provided by the adenohypophysis, but its concentration in the blood does not constantly decrease but displays a transitory increase at beginning

metamorphosis followed by a decline, a course that suggests additional functions.

- Metamorphosis is driven forward by the steroid hormone **ecdysone** (or 20-hydroxy-ecdysone = ecdysterone, respectively) in insects and by the iodine-containing hormone **thyroxine** in amphibians. Ecdysone, delivered by the prothoracic gland, and thyroxine, delivered by the thyroid gland, use the same, homologous type of intracellular receptors to control downstream gene activities (see Fig. 11.11). It is a matter of previous programming how different tissues will response, which genes will be activated and which tissues will be degraded by apoptosis. To initiate details of metamorphic events in time several more hormones orchestrate metamorphosis such as **cortisol** and **aldosterone** in amphibians, **eclosion hormone,** ETH (Ecdysis-Triggering Hormone) and **bursicon** in insects.

The hormonal system in the amphibians as well as insects is under the control of the brain. **Neurosecretory hormones** transmit orders to hormonal glands, therefore external stimuli and environmental conditions such as day length can impinge on the start and course of metamorphosis.

Finally cases of **reverse development** back to stages closer to the origin in cnidarians are pointed out.

Immortality or Ageing and Death: What Is Nature's Aim?

21

21.1 Possibility and Impossibility of Immortality

21.1.1 Life Without Death Is Tied to Incessant Cell Divisions

Fundamental concepts linking programmed death to the evolution of multicellularity were advanced as early as 1881 by August Weismann, a zoologist and pioneer of genetic theories designed to explain development and cell differentiation. Weismann proposed that aging and decay are not inherent to life itself but are events that became integral to development only in the course of evolution of multicellular organisms. Only the multicellular organism would inevitably be doomed, through **senescence** – the process of aging.

A single-celled organism such as an amoeba increases in size when supplied with ample food, and divides into two separate cells. Does division imply the end of individual life? We may debate whether the doubling of an individual means that the original individual's life has ended. After all, a corpse is not left behind. In a sense, organisms like bacteria or protists are **potentially immortal**: their lives are ended only being devoured, destroyed by parasites and/or environmental destruction, not aging.

In protists, an amoeba for example, Weismann argued, there is no regular process (we would now say 'program') of death in order to terminate the individual life, because every individual cell is simultaneously a **generative cell** which secures the continuation of the species. By contrast, in the multicellular organism a segregation of generative and somatic cells has evolved. Somatic cells can focus on a few functions and optimize them because they do not need to meet all requirements and accomplish all functions of life, including reproduction. **Somatic cells** can be fine-tuned to perform their particular task for the benefit of the entire cell community: some cells secrete enzymes and maximize the exploitation of food; other cells optimize sensory functions to exploit the environment and to recognize dangers or opportunities.

This focus on specialized functions to the detriment of cellular reproduction was possible because generative cells took over the task of reproducing the entire organism, standing proxy for the whole community. Gametes focus on the job of reproduction and propagation without being restrained by other occupations. Although both male and female gametes are highly specialized cells, they give rise to a totipotent zygote. The zygote is unable to perform any somatic functions by itself but is able to transmit to its daughter cells the ability to adopt every specialization and to differentiate into every cell type that is provided for by the genetic program of the species. Universal potency and perfection of particular somatic functions are mutually exclusive. A single-celled organism remains bound to compromises.

W.A. Mueller et al., *Development and Reproduction in Humans and Animal Model Species*,
DOI 10.1007/978-3-662-43784-1_21, © Springer-Verlag Berlin Heidelberg 2015

Death in metazoans is the regular fate of somatic cells. But even in metazoans there are, surprisingly, exceptions to this rule. **Cells of metazoans maintained in culture** for long periods of time, even human cells, are often **immortal** just like protists. Because of their immortality such cells can be kept alive and grown for years. And the freshwater polyp *Hydra* is, as a whole, potentially immortal.

However, cells grown in culture flasks have been subjected to selection: they represent **stem cells** with the programmed ability to divide, or are cells on the way to becoming cancer cells which have recovered immortality. *Hydra* is another and peculiar case: all of its terminally differentiated cells in fact die, but every decaying cell is replaced by a substitute generated from immortal stem cells (Sect. 4.3). **Only cells that have preserved the ability to divide are immortal**. Apparently, perpetual cell divisions are a prerequisite to immortality, while loss of the capacity for cell division leads to death.

21.1.2 We Know Several Molecular, Cellular and Organismal Causes of Aging, e.g. DNA Damage

Enzymes and other proteins are not in the energetically lowest conformational state when they are biologically active. Rather, they are in a **metastable state** created during their synthesis, sometimes with the assistance of chaperones. Many influences such as thermal energy, changing ionic strength and the pH of the solution, or collision with radicals may throw a protein out of its metastable condition into a deeper energetic level: it **denatures**. The probability of spontaneous renaturation is minimal. Denatured proteins are useless and must be replaced by newly synthesized ones if the cell's metabolic or developmental programs still require their presence.

A terminally differentiated cell will have difficulties replacing all of its proteins. How could a heart muscle cell replace its contractile apparatus without interrupting its rhythmic beating? How could a nerve cell in the brain repair its numerous dendritic and axonal fibres and hundreds of synaptic connections while staying prepared to fire? And could such a nerve cell even procure all the needed building materials, crowded together with myriads of other competing nerve cells?

Limits set by the stability of genetic information. Above all, limits on the replacement of proteins comes from the condition of the DNA – the source of information needed for the synthesis of proteins.

DNA itself suffers damage by
- thermal collision with water molecules,
- ionizing and UV irradiation-induced DNA strand breaks,
- aggressive oxygen radicals changing the chemical structure of DNA bases,
- mistakes made by DNA polymerase not properly corrected,

these factors all cause mutations and irreversible loss of information. Every day a human cell loses about 5,000 purine bases (A or G) by induced depurinization, and 100 cytosine bases are converted into uracil. Cytosine spontaneously deaminates to uracil. After birth, when neuroblasts stop proliferating, the enzymes responsible for repair largely disappear from these cells.

A cell possesses multiple mechanisms of DNA repair. The mechanisms function as long as one of the two DNA strands remains intact and can serve as a template of the correct base sequence. **A comprehensive correction of defects is only possible when the entire genome is replicated**. This would explain why the only immortal cells are those that divide again and again. In mammals a correlation has been found between the capacity of the DNA repair system and the average lifespan of the species.

In addition, in unicellular organisms and cell communities alike, **natural selection eliminates faulty cells**. This may be the reason why so many cells undergo cell death in the germline: only cells with (more or less) intact genetic information survive. Thus, sexual reproduction helps to perpetuate a species. On the other hand, positive natural selection of cells with intact DNA presupposes continuing cell multiplication by mitosis. Therefore, the principle is invalid in tissues composed entirely of terminally differentiated cells, such as nervous tissue.

Limits set by the accumulation of damage at the cellular and organismic levels. Numerous age-associated irreversible changes and accumulating deficiencies have been identified. For example,

- Accumulation of **heart cells that are deficient in respiration**: the heart becomes weak.
- Deterioration of the thymus and disappearance of functional lymphocytes. As a result, the **immune system is weakened** and infectious diseases are less effectively resisted and overcome.
- Decomposition of proteoglycans and structural proteins (in particular hexa-uronates and elastic fibres) in the extracellular matrices of cartilage, dermis and blood vessels. Bond water is lost, skin and cartilage shrink, elasticity is lost, and **arteriosclerosis** (the hardening of arteries) is favoured. The resulting hypertension and mechanical stress favour **atherosclerosis**, the deposition of encrusting plaques in blood vessels.
- Renal blood flow and glomerular filtration decrease: the efficacy of the kidneys decreases by 1 % annually.
- **Nerve cells decay daily** in our brain without being replaced, most abundantly in Alzheimer's disease but to a lesser extent in every human: the performance and productivity of our brain diminishes. The ability to cope with the tasks of everyday life decreases, and degenerative disorders may eventually result in dementia.

The two last causes alone restrict the **maximum longevity** of humans to approximately 120 years. Death is to be expected.

21.2 Theories of Ageing

21.2.1 Mitochondria-based Theories, Oxygen Radicals, and Benefits of Fasting

In all eukaryotic organisms and tissues with advancing age disturbed mitochondrial functions are recorded. Over many years this was attributed to irreparable damages caused by toxic oxygen radicals. Highly reactive oxygen species ROS (Reactive Oxygen Species) are the superoxide-ion $^\bullet O_{2\ minus}$, hydrogen peroxide $^\bullet H_2 O_2^\cdot$ and the free hydroxyl radical $^\bullet OH$. These can cause oxidative damages to lipids and mitochondrial DNA. When upon abundant food intake and high ATP expenditure the mitochondrial respiratory chain operates at full stretch, inevitably more reactive oxygen species emerge and effect damage to the cells. However, normally these oxygen species are detoxified by catalases and peroxidases. Recent studies suggest that only extreme physical strain would lead to lasting oxidative damages while mild physical activity would induce mechanisms of protection and counteract age-related decrease of the energy-generating metabolism.

Recent studies provide compelling and concurring evidence that an ascetic life style with meagre fare **caloric restriction** prolongs the average life expectancy, in the round worm *Caenorhabditis elegans*, in the fly *Drosophila* and in the mouse, and it is supposed that this applies to humans as well. There are speculations about possible causes, such as the just discussed oxygen radical hypothesis, but assured facts are scarce and no theory is commonly accepted.

21.2.2 The Telomere Theory of Cell Senescence

Apart from stem cells and cancer cells, cells in culture usually show a striking reluctance to continue proliferating forever, even when supplied with abundant food. After some lively rounds of division they slow down, cease dividing, and die. This can be observed even in cell culture. In blood-forming cells (type not exactly specified) the number of divisions they undergo has been reported to be

- 50 if the cells are taken from a human fetus;
- the number may be 40 if the cells are taken from a 40-year-old human;
- it may be 30 if the cells derive from a 80-year old man.

This age-related limit of the number of potential divisions is known as **Hayflick number** or

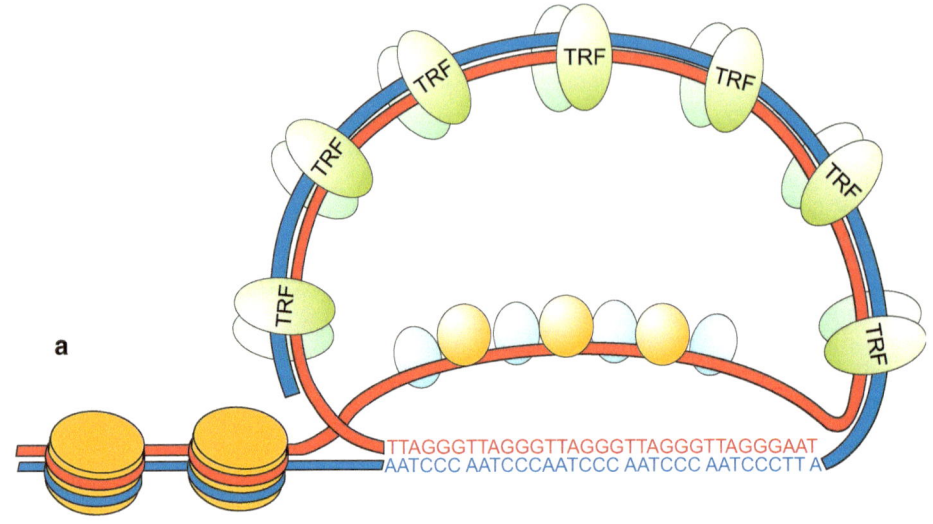

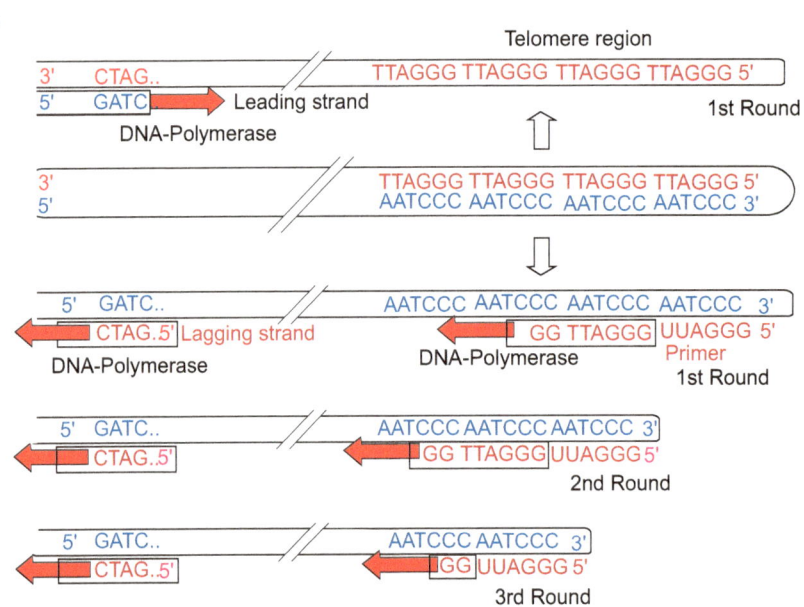

Fig. 21.1 (**a, b**) Telomere model. (**a**) Loops join both strands of DNA and prevent fraying of the chromosome ends. (**b**) Model of the telomere clock. In each round of replication one repeat of the telomere region at least gets lost in the lagging strand. Repeats are required for the attachment of the primer. The new DNA strand can be elongated by the primer only from 5′ to 3′, but not in the direction of 3′ to 5′; therefore the DNA is shortened by one telomere repeat after each round of replication

Hayflick limit. The diminishing readiness to divide is correlated with the gradual loss of nucleotides at the ends of the chromosomes.

Eukaryotic chromosomes have a particularly structured end region, called the **telomere**. Telomeres are made of repeating sequences of TTAGGG on one strand of DNA bound to AATCCC on the other strand. Thus, one section of a telomere is a "repeat" made of six base pairs. The telomeric region joins and seals the two DNA strands at the end of the chromosome (Fig. 21.1).

The telomere region may contain many such repetitive sequences, in human cells, for example, the telomere region comprises some hundred and up to 2,000 repeats. These were synthesized onto the chromosome ends by means of a telomerase in the course of embryonic development. Yet in each round of DNA replication the telomeres become shorter as illustrated and explained in Fig. 21.1b. This is due to the 'end-replication problem'; DNA polymerase cannot fill in the gap at the end of the chromosome left by the 5'-most RNA primer. In stem cells and cancer cells a special reverse transcriptase that has its own RNA template, called **telomerase**, does this job. Most somatic cells do not express telomerase, therefore the telomeres get shorter with each successive round of cell division, eventually reaching a critical length where they stop dividing. This critical length has been found to be 12.8 repeats. With continuing divisions eventually the chromosomes lose not only non-coding repeats but also coding genes. Such cells normally can divide only about 50–70 times. Once the chromosomes fall short of a critical length, the chromosomes become unstable, the cell suffers more and more damage and eventually subjects itself to **apoptosis**, that is it undergoes programmed cell death.

In somatic cells that are not immortal but able to divide the telomeres represent a counter of replication rounds that eventually stops at zero (hourglass principle).

The non-coding overhang sequences comprised in the telomere are required for the complete replication of linearized DNA as it is present in chromosomes. While the leading strand can in forward gear be replicated in one go, the lagging strand can only be replicated step by step in small pieces (Okazaki fragments). The polymerase has to jump a length forward to the next RNA starter and synthesize the intermediate DNA section 'backward' in the 5'–3'-direction. The pieces are joined and thus the DNA double strand

complemented. For each segment first an RNA primer is synthesized at the end of the lagging single DNA strand and subsequently the RNA piece becomes replaced by the corresponding DNA piece. Thus also the last segment of the 5'–3' strand is completed to a double strand segment. Yet, the very last RNA primer cannot be replaced by a corresponding DNA segment because the DNA polymerase cannot work in the 3'–5' direction. As a consequence this overhang sequence gets lost.

Telomeres do not shorten with age in tissues such as heart muscle in which cells do not continually divide. Nor do immortal stem cells and cancer cell inevitably arrive at a point where lacking telomere sequences put an end to the cell's life. Clearly death caused by telomere shortening is not an inevitable event principally inherent to life. Therefore, it is still a matter of discussion to what extent this phenomenon contributes to the termination of our life.

21.3   Death as Genetically Preprogrammed Event

21.3.1   Species-Specific Longevity Is a First Indication That Death Is a Genetically Programmed Event

Although at the molecular, cellular and organismic level many causes for irresistible aging can be found, we still do not understand why **longevity**, the maximum life span, is species-specific. Medical science and public sanitation have increased our **life expectancy**, the average number of years an individual may expect to live, but they have scarcely increased longevity. And it is still enigmatic why in some organisms (humans, for example) senescence and aging occur slowly, while in others (mayflies and salmon) aging and death occur suddenly and dramatically.

The range of longevity among metazoans is considerable, of course: *Hydra* is potentially

Table 21.1 Species-specific longevity in years

Primates		Other mammals		Non-mammalian vertebrates	
Treeshrew *Tupaja glis*	7	House mouse *Mus domesticus*	3.5	House sparrow *Passer domesticus*	13
Marmoset *Callithrix, Leontides*	15	Rat laboratory rat	5	Bird of paradise	12
Guenon *Cercopithecus*	21	Rabbit	13	Buzzard *Buteo buteo*	25
Rhesus monkey *Macaca mulatta*	29	Sheep	20	Rock dove *Columbia livia*	35
Baboon *Papio*	36	Cat	28	White stork *Ciconia ciconia*	30–35
Gibbon *Hylobates*	32	Cattle	30	Europ. Alpine newt *Triturus (Ichthyosaura) alpestris*	3
Orang-Utan *Pongo pygmaeus*	50	Dog	34	Europ. Fire salamander *Salamandra salamandra*	6
Gorilla *Gorilla*	40	Europ. brown bear *Usus arctos*	37 (47 in capt.)	Japan. giant salamander *Andrias japonicus*	55
Chimpanzee *Pan troglodytes*	45	Horse	62	European pond turtle *Emys orbicularis*	30–120
Human	95	Indian elephant	70	Greek tortoise *Testudo graeca*	120
Human, record	122	Bowhead whale (Arctic whale)	200	Galapagos giant tortoise	150

immortal, *Caenorhabditis elegans* lives for 3 weeks, and *Homo sapiens* has a life span potential of at least 120 years. Humans are the longest-lived among the primates, and among the longest-lived vertebrates together with giant tortoises and bowhead whale. Among the longest-lived invertebrate animals count clams in cold waters such as the Iceland clam (ocean quahaug, *Arctica islandica*). The oldest still living animal is deemed to be the giant sponge *Scolymastra joubini*; it is thought to be at least 10,000 years old and lives in the arctic deep sea.

Apparently, the maximum life span in mammals is correlated with body size. Small mammals whose consumption of energy and metabolic turnover are high, and whose heart beats fast, arrive at the end of their life earlier than big animals with their slower way of life. Even among a taxonomically narrow clade such as the primates life spans range over an order of magnitude and generally follow body size (see Table 21.1).

21.3.2 There Are Genes Whose Mutations Lead to Rapid Ageing and – in Fungi – Genes the Failure of Which Enables Immortal Life

Do organisms embody a genetically released suicide program?

Apoptosis, programmed cell death, affects certain cells as early as embryogenesis (Chap. 14). Later in life many cells, for example blood cells (Chap. 18), die a few days or weeks after their "birth" (last cell division). However, because blood cells and other short-lived cells are replaced by substitutes generated from stem cells, it is not so much the short-lived but the long-lived, irreplaceable cell that sets the limit to the life span.

Ever since it became possible to grow cells in culture, we learned that cell populations below a critical cell density die. Recent results have enabled a new interpretation of those observations: most **cells survive only in social communities**. Cells cut off from society commit suicide. The actual cause is the presence of **survival factors**, with which cells prevent each other from switching on their suicide program. Only in dense cultures do these factors accumulate sufficiently to exceed the threshold of critical concentration. Survival factors have been found, for instance, in cultures of oligodendrocytes, eye lens cells, and kidney cells.

However, all of the survival factors should be present in our body in sufficient quantities. Are we also under the regime of genetically fixed internal clock that eventually starts a programmed death of the entire body?

In fact, it seems to be so. This is suggested by rare autosomal recessive congenital diseases in humans leading to precocious ageing and death, known under the collective term progeria.

- In *progeria adultorum* (**Werner syndrome**) senescence commences at the age of 15; life expectancy is 47 years.
- In *progeria infantum* (**Hutchinson-Gilford syndrome**) signs of senescence occur at 1.5–3 years of age. Progeria infantum is a rare hereditary disease that affects the skin, musculature, skeleton, and vasculature. Affected individuals die at 12–18 years of age with all the attributes of old people. Persons concerned turn grey, shed hair, have wrinkled skin, and go blind because of cataracts. Significant morbidity and mortality result from accelerated atherosclerosis of the carotid and coronary arteries, leading to premature death.

The genetic defect associated with this Hutchinson-Gilford syndrome is a mutation in the intermediate filament protein of the nuclear lamina. The mutated form, called progerin, is irreversibly farensylated and does not incorporate correctly into the nuclear lamina. Expression of progerin causes nuclear blebbing, chromosome instability and cell senescence, but the molecular mechanism has not yet been identified.

The gene responsible for the WERNER syndrome codes for a **helicase**. Its function is to untwist DNA for replication and to facilitate transcription. This is a prerequisite for complete DNA replication and thus complete DNA repair,

and for synthesis of transcripts to produce new protein for rejuvenation. WERNER protein is also found in *Caenorhabditis elegans* and contributes, if mutated, to its ageing. On the other hand, life does not regularly end because of a mutation.

As said above, calorie restriction prolongs life span in these animals. More than 100 further life span-influencing genes have been identified, among them 50 which, when mutated prolong life. Nevertheless it is not clearly established whether specific age-inducing genes exist in this animal or in others.

An example of genetically determined ageing is a fungus of the clade Ascomycetidae, *Podospora anserina*. It is a filamentous fungus used in several molecular genetics laboratories to study various aspects cell biology including ageing and cell death. The wild type lives 25 days, but senescence is prevented when (only) two mutations are introduced by hybridization. The double mutant is **immortal**, yet, only in the lab, because the mutations reduce natural fitness.

21.3.3 Several Mechanisms by Which Death Might Be Genetically Programmed Are Discussed

To sum up only two hypotheses will briefly be recapitulated to discuss how genetic programs of self-destruction might work.

1. **There might exist a genetically fixed life clock** that determinates life after a species-specific life span. Protagonists of this theory point to *Podospora anserina*. In their wild-type edition those two genes the mutation of which does not shorten but prolong life, code for proteins that cause degradation of mitochondrial DNA and thus rapid ageing. The genes are switched on by a molecular clock of still unknown construction after the species-specific life span of 25 days is reached.

   According to indications provided by the **progeria WERNER-syndrome** in humans **programmed turning off of this helicase gene** might render the mechanisms of DNA repair inoperative and thus surrender the DNA to accumulation of genetic defects and ageing. A further highly remarkable function of the WERNER syndrome protein is **inactivation of the pluripotency factor Oct-4.** WERNER protein induces methylation of the *oct-4* promoter, thus the self-renewal potency of stem cells is reduced by epigenetic mechanisms. Furthermore in the WERNER as well as in the HUTCHINSON-GILFORD syndrome participation of the telomere clock is discussed.

2. **Telomerase hypothesis**. In addition to the explanation given above the question awaits an answer whether there are specific genes that induce switching off telomerase activity in somatic cells thus bringing the telomerase clock to a standstill. Knock out mutations in plants (*Arabidopsis*) indicate that ending telomerase activity does not terminate life in every species.

Perhaps the process of ageing in living beings is similar to the ageing process in our cars. It is not a single pre-programmed defect that sets limits to its life, rather a compendium of defects: rust, used up tyres, brakes and piston seals, defective shock absorbers, loss of oil, and so forth. Whoever prophecies future human immortality may first demonstrate his/her power by constructing and producing an immortal car.

21.3.4 Death Has an Important Biological Significance and Implication

In multicellular organisms death has become an integral part of the life cycle. Through selection and evolution, mechanisms of programmed cell death have emerged and have been fixed in the genome. What then is the meaning of death?

Organisms have to make room for their offspring: old humans have to clear the way for their children. But why has potentially immortal life not emerged with offspring generated sparingly (for instance through asexual budding as in *Hydra*) in order to compensate losses due to accidents and to generate colonists for new living spaces?

Hydra has persisted over millions of years, but its achievements were modest in terms of the development of complexity and conquest of novel ecological niches. In the history of life, sexual reproduction emerged as the dominant means by which eukaryotes perpetuate and evolve. Sexual reproduction makes possible the introduction of mutations into populations and their rapid spread; more importantly, it facilitates extensive recombination of allelic variants of genes. Sexual propagation enables organisms to endlessly offer new variants to natural selection. It favours adaptability in a changing world, and favours optimization of lifestyles even in a constant environment.

Death makes way for new life, new not only in the sense of fresh and **renewed** but also in the sense of **novel**. An incessant sequence of ontogenies, an incessantly repeating sequence of being born, giving birth and dying, has given rise to humans and to the inspiring diversity that the developmental patterns explored in this book have made possible.

Summary

Cells must constantly renew their stock of proteins because proteins denature with time. For that purpose correct information provided by intact DNA is needed. But DNA suffers damage continuously, for example through oxygen radicals and ionizing radiation. Damage can be repaired only in the course of DNA replication, therefore immortality is bound to incessant cell divisions.

In multicellular organisms aged and used up cells can be replaced by stem cells that are incessantly able to divide, but only *Hydra* can replace all somatic cells including nerve cells by freshly produced substitutes and thus rejuvenate its body permanently thus being potentially immortal.

Furthermore, there are putatively genes and molecular mechanisms which set limits to life after a species-specific life span. In theories on pre-programmed death the existence of a life clock is being discussed and the progressive loss of DNA sequences in the telomeres at the ends of chromosomes taken in account. **Progeria** alleles of certain genes induce ageing in humans already in adolescence (WERNER syndrome) or even in early infancy (HUTCHINSON-GILFORD syndrome).

Death as a programmed event fulfils an important biological function. Sexual reproduction permanently creates new phenotypes which might be better fitted in future. In the long term this chance must not be excluded by existing living beings.

Evolution of Development

22

22.1 A View Back to History

The theory of evolution referred to developmental biology ever since it was founded. Classic developmental biology provided many important contributions. Textbooks of classic comparative anatomy, comparative embryology and zoology provided fundamental knowledge about the action and results of evolution. In the last 20 years or so, molecular genetics and studies on expression patterns of genes open new and often unexpected insights into the mechanisms and putative paths of evolutionary changes.

In presenting classic pieces of evidence we confine ourselves to a brief recapitulation of remarks distributed in this book among various sections. Here we will focus on molecular genetic aspects and the molecular "toolkit" the content and meaning of which will be explained with a few selected examples.

22.1.1 "Ontogeny Recapitulates Phylogeny": Phylotypic Stages, Conserved Paths Versus Novelties

In Chap. 5 (Vertebrates) and Chap. 6 (Human) we pointed out that **in all vertebrates in a middle period of their embryonic development a similar basic body construction, a conserved 'Bauplan' (construction plan, body plan), can**

be recognized. Conserved characteristics are: **dorsal neural tube, notochord, somites, branchial pharynx with gill pouches, cartilaginous pharyngeal arches (gill arches), from 4 to 6 gill arteries, and a ventral heart**. In their subsequent ontogeny these basal characteristics develop quite differently in class-, family- and eventually in species-specific ways, reflecting the phylogeny.

Vertebrates of all taxonomic groups transitorily pass developmental stages reminiscent of their common ancestor. Temporarily it appears as if the tiny human embryo (1–20 mm in size) would be about to develop into a fish, subsequently into a salamander, a reptile, a monkey and finally into a human being (see Figs. 6.24 and 6.9). Such observations prompted anatomists and zoologists of the eighteenth and nineteenth century such as **Carl Ernst von Baer** and Ernst Haeckel to emphasize these astonishing similarities and to formulate ideas and hypotheses which culminated in Haeckels much debated "basic biogenetic law" – which in fact is a rule – stating ontogeny be an abbreviated form of phylogeny (see Sect. 6.6). However different the very early and the very late developmental stages may be, in the middle period a **phylotypic stage** (Figs. 5.9 and 6.8) can be recognized in which the embryos are amazingly similar to each other (Fig. 6.17). The concept of phylotypic stages has first been applied to vertebrates but phylotypic stages can also be defined in other groups. In **insects,** for example, the stage of the

W.A. Mueller et al., *Development and Reproduction in Humans and Animal Model Species*,
DOI 10.1007/978-3-662-43784-1_22, © Springer-Verlag Berlin Heidelberg 2015

extended germ band is considered to represent the phylotypic stage. In molluscs it is the **veliger** larva, in echinoderms **the gastrula** or the **pluteus larva** can be considered to be phylotypic though, or because, these pre-metamorphic stages display bilateral symmetry thus representing the ancestral primordial trait of the phylum.

Back to vertebrates: Already the fathers of developmental biology not only emphasized common traits but also pointed out that it is just the earliest phase of development in which development proceeds highly differently and in class-specific patterns, being in mammals very different compared to amphibians. For instance in human embryogenesis, very early preparations are made for developing the **placenta**, even before the embryo proper is formed. In the blastocyst the outer layer becomes an extraembryonic precursor of the placenta, called the **trophoblast** (Greek: *trophein* = to nourish) as it serves to take up nutrients provided by the uterus of the mother. Accordingly, the embryo does not arise from the entire ball-shaped embryo, called the blastula in amphibians and blastocyst in mammals; instead in the mammalian blastocyst the embryo actually forms from a small area in the interior (see Fig. 6.5). Such differences reveal adaptations to habitats as different as the pond and the womb.

22.1.2 Functional and Gene-Regulatory Constraints Restrict Possibilities of Large Evolutionary Changes

In all vertebrates a **transitory notochord** is formed prior to the vertebral column. This apparent detour has a practical function. The notochord assists in mechanically stretching the head-tail axis and it is an indispensable emitter of development-organizing signals. It orchestrates the formation and subdivision of the spinal cord (Fig. 16.1), summons the cells of the somites to migrate and to assemble around the notochord where they form the vertebral bodies (Fig. 10.6). The notochord even assists in determining the position of inner organs such as liver and pancreas. Ontogenetic inputs and presetting cannot

arbitrarily be modified; the processes are too complex and must take place in a sequence of progressive steps.

Another example of constraints: In all vertebrates capable of flight, pterosaurians, birds, and bats, wings could only be developed at the expense of the forelimbs. The gift of grace of another extra pair of extremities remained reserved for mythological figures such as dragons and winged unicorns, and the celestial angels. The range of attainable morphologies is confined.

Fossil records provide evidence that the basic traits of all recent body plans already existed in the Cambrian period about 540–480 million years ago. They thus arose from a putative common ancestor that already existed previously. Fossils in the form of imprints resembling recent medusae are known from the Precambrian (Vendium, 640 million years ago) but fossil records do not provide information on how these constructions once arose, simply because residues of soft tissue do not persist or merely persist as imprints in rock. Clues to the earliest history of life can only be deduced from the developmental pathways and the inventory of conserved genes of recent animals.

22.2 New Possibilities of Knowledge Acquisition: Evo-Devo Biology und Its Basis

22.2.1 "Evo-Devo" Biology Based on Molecular Genetics Aids in Elucidating Evolution by Comparing Gene Sequences, Expression Patterns and Regulatory Networks

The remarkable congruency of basic developmental mechanisms and gene expression patterns in organisms belonging to different taxa gave a new lease of life to an area of research that became known as "evo-devo" biology and increasingly attracts attention. Though awareness that ontogeny partly recapitulates phylogeny enabled and stimulated synergies between developmental and evolutionary biology for a long time, only the comparative analysis of gene expression patterns and regulatory networks merged these complementary disciplines on a broad level.

With entire genomes sequenced from a yearly increasing number of animals that belong to different taxa it is becoming apparent that type and complexity of a body plan does not correlate with the total number of genes. Evolution generates **duplications of key developmental genes**, change of function of the duplicated copies, and altered integration into regulatory networks. For example changes in the body plan may be caused be the acquisition of new binding sites for transcription factors (as in case of lens proteins, see below) or by an altered time span or time sequence by which signalling molecules, receptors or other proteins are produced and active.

Comparative analyses including many different species show that a molecular "**toolkit**" has existed since the early phase of history. Already at the level of the Cnidaria this toolkit contains almost the entire stock of instruments that enabled the triploblastic animals to develop an amazing diversity of body plans. (But of course, as this basal toolkit as such did not per se give rise to triploblasty in cnidarians, the kit must contain additional tools in the triploblastic clades.) According to the hypothesis of a common toolkit some genes are expected to fulfil similar tasks in morphologically different organisms. There is hope that by detecting such conserved genes and identifying their function it might be possible to define an ancestral "**zootype**" – an ancestral prototype animal expected to be similar to the common ancestor of all multicellular animals, the Metazoa.

In contrast to classic developmental biology with only few but intensely investigated model organisms evo-devo biology analyses a broad spectrum of recent taxa and species that are positioned close to branching points in the phylogenetic tree or whose position is a matter of debate. Alternatively, or in addition, the function of developmentally relevant proteins is compared across the entire animal kingdom. By comparing the expression patterns of key marker genes and the regulatory interactions of gene products, it is possible to identify traits of common ancestors at different evolutionary levels. In

parallel, in comparing gene sequences by means of computer programs (algorithms), phylogenetic relationships can be deduced, compared and matched with existing pedigrees construed on the basis of morphological or biochemical traits (Fig. 22.1).

Functionally highly conserved gene products are, as a rule, integrated in conserved molecular mechanisms of regulation and control similar structures. On the other hand, molecular biology also reveals **convergences** by showing that **the morphology and function of a structure can be attained and ensured by different ways**. Thus, in the course of evolution, different molecules may have been utilized to generate, for example, a refractive eye lens.

Before we consider the procedure and results of evo-devo research by discussing actual examples we first summarize the basic assumptions underlying the theory of evolution and discuss problems the researcher is confronted with, whether he/she is trained in molecular biology or/and in zoology.

22.2.2 Homologies Based on Morphology and Homologies Based on Molecular Genetics Are Ascertained Using Different Criteria

Evolutionary changes have their origin in changes in the genome. Relationships and changes in morphology and biochemistry should be reflected in the genome and in the spectrum of the encoded proteins. These, in turn, are the building blocks which bring about the changing morphological traits in ontogeny and the eventual lasting phenotype. Thus, the discovery of the *Hox* genes and their putative proto-*Hox* precursors in all multicellular animals (Metazoa) investigated so far pointed to a parallel molecular and morphological developmental sequence correlated with the increasing complexity of body plans (see Sect. 22.4).

Homology in Morphology. When morphologists speak of homology they mean affiliation of a structure to an evolutionary developmental

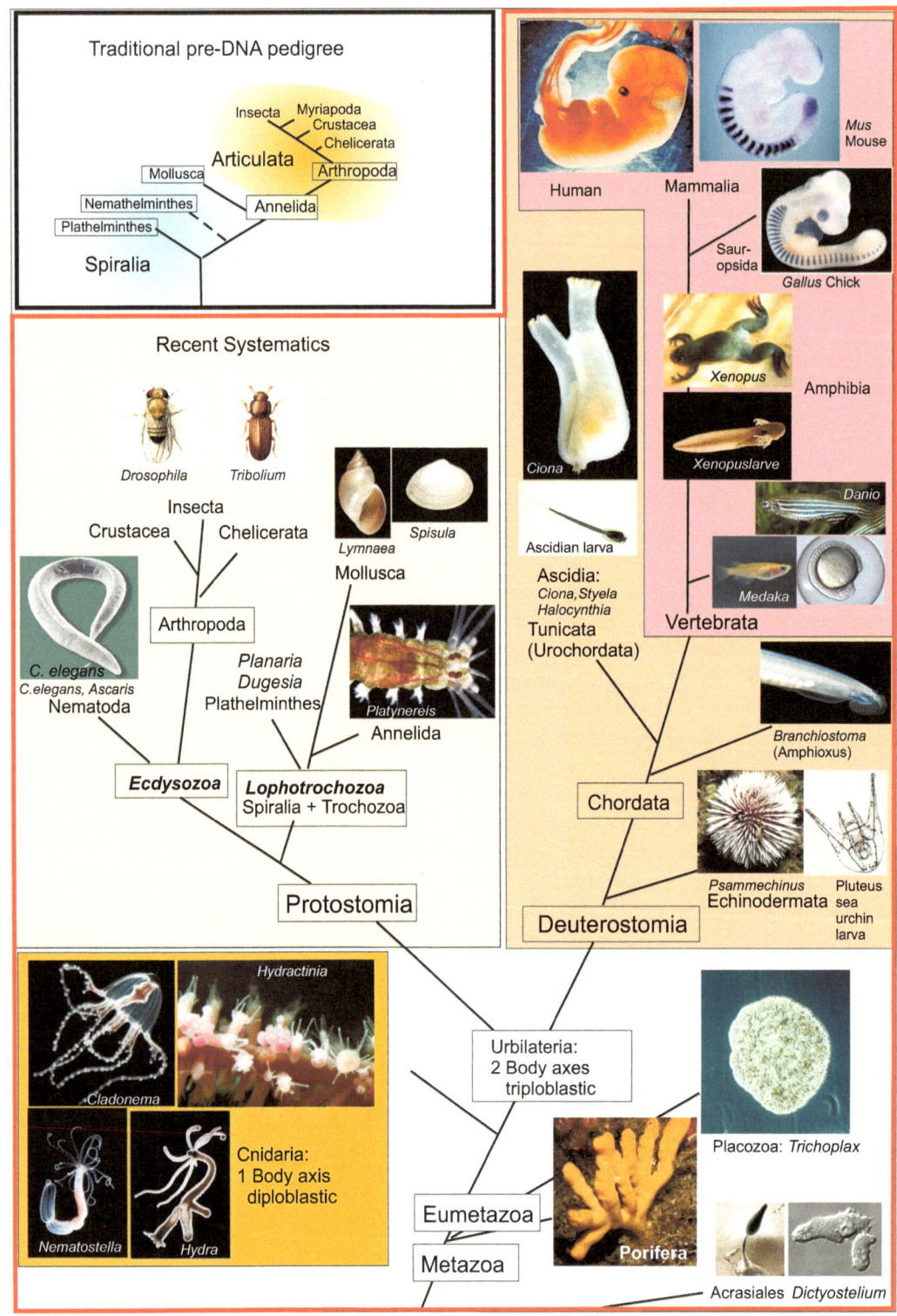

Fig. 22.1 Phylogenetic tree of the animal kingdom with the positions of model organisms mentioned in this book. The *insert* at the *upper left* shows a traditional phylogeny with Articulata as superorder comprising segmented animals including the segmented spiralian Annelida as bridge unit

sequence. Homologous anatomical structures in various organisms derive from a common original stem structure. Criteria for homology in the sense of traditional morphology/anatomy are:

1. **The criterion of location** of a structure within the fabric of neighbouring structures
2. **The criterion of the special quality** (e.g. functioning as a sensory organ or a gonad)
3. **The criterion of continuity when a structure underwent a change.** Structures altered in evolution should be connected with the ancestral structure by intermediates.
4. **Criterion of cell type.** Not explicitly listed in traditional treatises is a criterion that currently links molecular biology with morphology most closely: the **criterion of cell type.** Intuitively and self-evidently nobody would homologize a muscle with a bone or a brain. And one may subsume the criterion of cell type under the criterion of specific quality. But, for example, what is the evolutionary origin of various sensory cells? The hypothesis has been proposed that photoreceptors once derived from odorant receptors. In both cell types molecular receptors consisting of seven-pass transmembrane proteins of the same protein family are essential (Chap. 11). Retinal of photoreceptors can be considered to be a permanently attached ligand of an odorant receptor, a ligand that is converted into the active state by light.

Homology in Molecular Biology When molecular biologists and biochemists speak of homology they mean high sequence identity in genes or polypeptides, respectively. Nobody doubts that genes displaying sequence identity over many hundreds or thousands of nucleotides derive from a common primordial gene. In genes or proteins, respectively, with low sequence similarity but similar function, the three-dimensional folding structure of the protein might have been preserved; but in this case convergence must be also taken into consideration. Moreover, recombination of DNA modules by exon shuffling may cause partial homology to several different genes or proteins (see below Sect. 22.3.1) and thus make reconstruction of a phylogenetic tree a difficult task.

22.2.3 Homology of Organs Is Not Always Based on Homologous Genes and Homologous Genes Can Contribute to Building Non-Homologous Organs

Does homology at the level of genes not per se imply homology at the level of the phenotype? Unfortunately it does not! Four examples will provide proof.

Example 1 Non-homologous genes for homologous structural elements – cartilaginous versus ossified skeletal elements. When, in the course of evolution, a skeletal element is optimized or experiences a successive change in shape and function, its stock of molecular components can change, just as it does in ontogeny. A cartilage that undergoes ossification exchanges its entire chemical composition, even related molecules such as the collagen isoforms are replaced (cartilage has collagen type II whereas bone has type I).

No morphologist or embryologist would designate the bony femur and its cartilaginous ontogenetic precursor as convergent structures because in the course of life the composition of the building material changes and this restructuring uses other genes than in the original embryonic edition. Cartilage and bone, whether primarily formed with embryonic sets of genes or secondarily reformed with adult sets of genes derive in evolution from the same ancestral origin. **Homologous structures can in the course of ontogeny and phylogeny exchange their molecular components and this can reflect an evolutionary sequence,** such as the transition from Chondrichthyes (cartilaginous fishes) to the Osteichthyes (bony fishes).

Example 2 Non-homologous lens proteins for eye lenses. The lenses of the vertebrate eyes are homologous in the sense of morphology, meaning they are deemed to be derived from a common ancestral primordium. The biochemist on the other hand, who analyses the components of the lenses in different species, discovers a surprising number of different proteins which confer glassy transparency to the lens. What has

been said about the exchange of bone for cartilage applies also here: a function can be assumed
by other materials and thereby become optimized
for a special purpose.

Basically, many low molecular weight
proteins are suited for constructing a lens
because most proteins, at least in their native
state, do not absorb light in the visible range of
wavelengths and thus are transparent. A stable
protein conformation is essential. Denaturation
and precipitation of α-crystallin is the main
cause of cataract, i.e. clouding of the lens, in
humans. Furthermore, the protein concentration
along the optical axis must vary to confer
adequate refractory properties to the lens.

Lens proteins in different animals can belong
to different protein families, and not only in
animals which are commonly placed at distant
positions on the phylogenetic tree but also in
vertebrates. In the vertebrate eye changing
combinations of **up to 11 different proteins** are
found that collectively are called **crystallins**.
Yet, these crystallins are not members of a single
protein family but are quite different in their
amino acid sequence and thus encoded by non-
homologous genes. Most crystallins are sequence
related to certain enzymes or stress proteins
(chaperones). Some crystallins are active
enzymes, while others lack activity but show
homology to known enzymes. Those from birds
and reptiles are related to **lactate dehydrogenase**
(in crocodiles called δ-crystallin) and arginino-
succinate lyase, those of mammals to **alcohol
dehydrogenase** and quinone reductase.

In the convergently evolved lenses of
invertebrates other sets of proteins are used.
Cephalopods (squids) for example use glutathi-
one S-transferase and aldehyde dehydrogenase
documenting that in fact many different proteins
are suited for this purpose. In retrospect not surprisingly, the corresponding genes are expressed
in other tissues as well. Evolution recombines
proteins in new patterns in response to functional
needs. Constraints demand only that a transparent and refractory structure must arise.

**For the construction of structures classified
as homologous in terms of morphology different macromolecules can turn out to be suited,
and newly recruited macromolecules can
replace previously used macromolecules**.

**Example 3 Homologous genes for non-
homologous structures.** Macromolecules
counted by molecular biologists as homologous
are often expressed in different tissues and
organs which in the understanding of
morphologists are not homologous. The ***HOX-
D13*** gene for example is in the mouse embryo
expressed exclusively in somite No. 7, but also at
the tip of the outgrowing tail and at the tip of the
outgrowing limbs. **Homology at the level of
genes and proteins is not in each and every
case congruent with homology at the level of
tissues and organs!**

**Example 4 Homologous genes in structures
with cryptic homology**. The comparison of
expression patterns occasionally aids in
recognizing a common origin of organs of which
the homology was controversial or not expected.
A good example is the **common genetic origin of
the central nervous system in insects and
vertebrates**. Here it is the large number of homologous genes expressed in similar patterns
that provide evidence for cryptic homology
(Chap. 16, Fig. 16.13). Another example are
photoreceptors and eye types the development
of which is triggered by key genes conserved
across the entire animal kingdom, namely ***Pax6***
and/or ***sine oculis***, and the subsequent differentiation of these photoreceptors is controlled by several common downstream genes (see Fig. 12.19,
and below Sect. 22.7). If development of complex
organs in various animals belonging to different
taxa is triggered and controlled by the same set of
conserved homologous genes, sole convergence
can be excluded. Convergence is based on a series
of random genomic changes which cannot by
chance be identical at all relevant sites in the
genome.

Controversial and unexpected homology of structures can be verified by a compendium of cooperative homologous genes.

To conceive evolutionary coherence and relationship it is not possible to rely on **one single trait** or one **single gene**. In the worst case one could focus on a gene that late in evolution was imported by a retrovirus across species boundaries and now pretends genealogical relationships. In the euphoria of the early molecular genetics founding period some conclusions were drawn prematurely. Molecular genetics had to learn the same lesson as had the morphologists of the old school. The reconstruction of a phylogenetic tree on the basis of one single trait or one single gene can easily be misleading.

22.2.4 Orthologous and Paralogous Genes and Structures Must Be Evaluated Separately

In general terms, **homolog**y means: two entities have a common origin in the history of life. For example in various vertebrates, say zebrafish, grass frog, and hummingbird, the spinal column and the forelimbs (pectoral fin, foreleg, and wing) are homologous to each other. In a restricted sense one speaks of **orthologous** structures, meaning structures derived from a common ancestor and found **in different species**. Yet, within one and the same individual the somites and vertebral bodies are also homologous repetitions of the same basic theme. Likewise, in tetrapods forelimbs and hindlimbs are variations of a basic structure – one speaks of **serial homology** or of **paralogous structures**.

In parallel the molecular geneticist finds homologous genes in different species and taxa. The *Hox* gene *labial*, for instance, exists in *Drosophila* and the mouse, and likewise the gene *Antennapedia*. Both genes are members of the *Hox* cluster and arose from a common primordial gene, that was serially duplicated, and in annelids, arthropods and vertebrates is found in the anterior part of the *Hox* cluster (Figs. 22.2 and 22.3). One speaks of

- **orthologues,** when the genes in question derive from a (putative) ancestor common to two or more divergent lineages of descendants and, therefore, are found in the genomes of *different* species. One speaks of
- **paralogues,** when the genes were *multiplied within the genome* of an ancestor and are found in *one and the same individual* at different locations of its genome. The duplicates may become different with time and undergo cooption for different tasks, or turn into nonfunctional **pseudogenes** by loss-of-function mutations.

Genetic relationships between organisms can reliably only be reconstructed when orthologous genes are compared! To reconstruct a phylogenetic tree of the *Hox* cluster, for example, the *labial* gene of a given species must be compared with the *labial* gene of other species, or the *Antennapedia* genes with *Antennapedia* genes of other species, but not *labial* genes with *Antennapedia* genes. (A comparison of paralogues can be used to reconstruct the branching routes within the *Hox* cluster). Orthologous key genes with conserved functions, in particular orthologous master- or selector genes, point to common principles in the control mechanisms.

22.3 Novelties Derived from Legacies in the Genetic Program

22.3.1 Novel Genes Arise from Gene and Genome Duplication and Mutation-Induced Alteration of the Duplicates

If natural relationships and evolutionary changes are reflected in the genomes we may expect that genes exist which vary to some extent from species to species but are conserved on the whole. On the other hand, we also expect novelties, for humans are neither hummingbirds nor insects.

Gene synthesis de novo is not known up to date (at the utmost genes might arise from non-

Model of Evolution of the *Hox* and *ParaHox* Cluster
Duplication, Adaptation and Change of Function

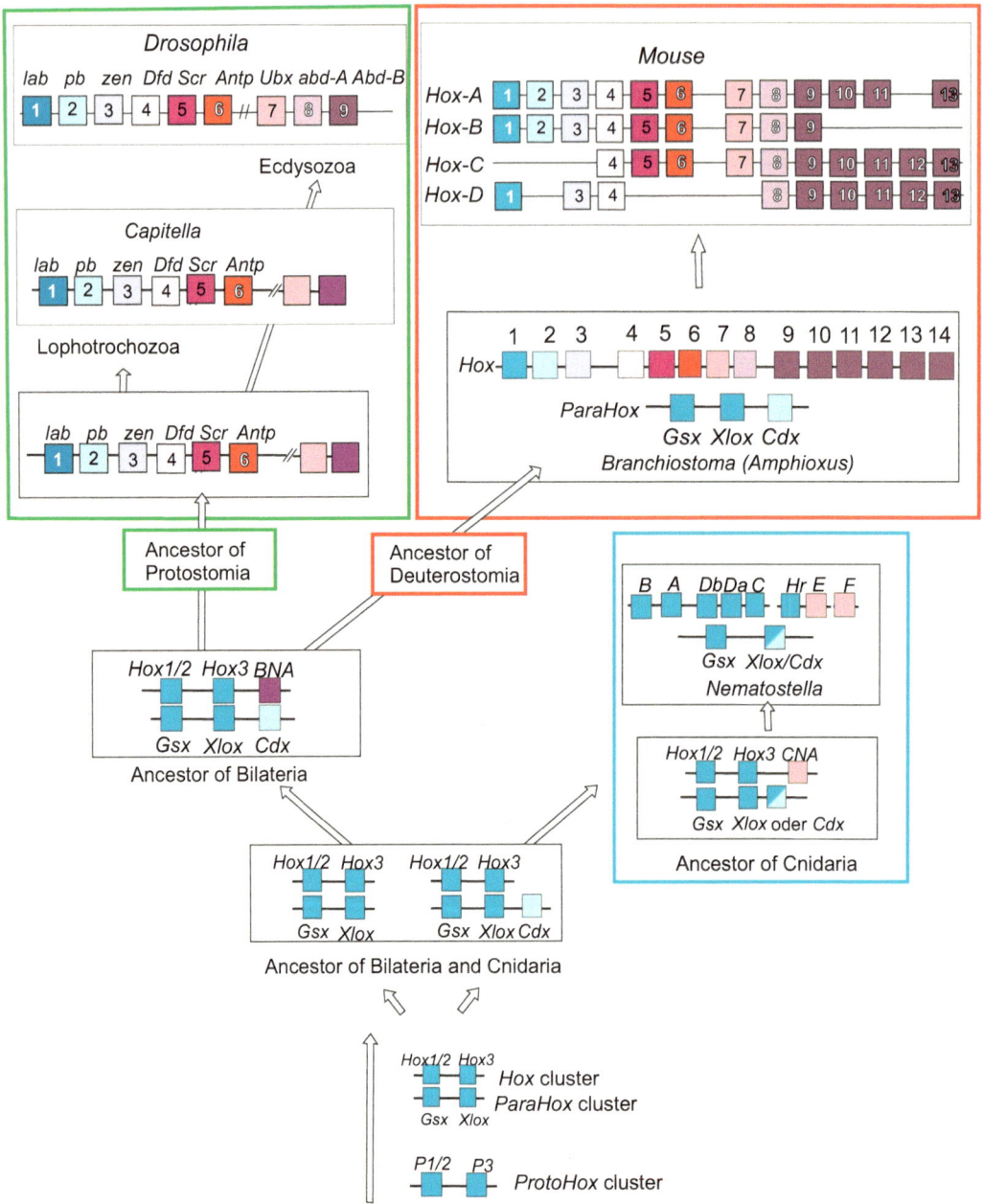

Fig. 22.2 Hypothesis concerning the evolution of *Hox* genes from simple *proto-Hox* precursors. The *proto-Hox* genes persist in extant animals. After Chourrout et al. (2006), compiled by WM

coding sequences, see framed text in Sect. 22.3.3). So how can novelties emerge?

Basic mechanisms creating something new may be comprehensible when we consider how conserved gene clusters with several paralogs and how multidomain proteins arose. Based on the crystalline examples above, we may also envision how non-homologous proteins can be

Fig. 22.3 *Hox* cluster in various organisms. Note: The almost complete congruence in number and sequence of the *Hox* genes in the annelid *Capitella* and *Drosophila* documents that already the common ancestor possessed a complement of 9–10 *Hox* genes. Despite difference in detail, their expression pattern in developmental stages of the annelid *Capitella* corresponds to insects. The general principle of colinearity is retained (after Fröbius et al. 2008a). In mammals the *Hox* cluster exists fourfold, one of several indicators that in the evolution of mammals the complete genome was duplicated twice. Compiled by WM

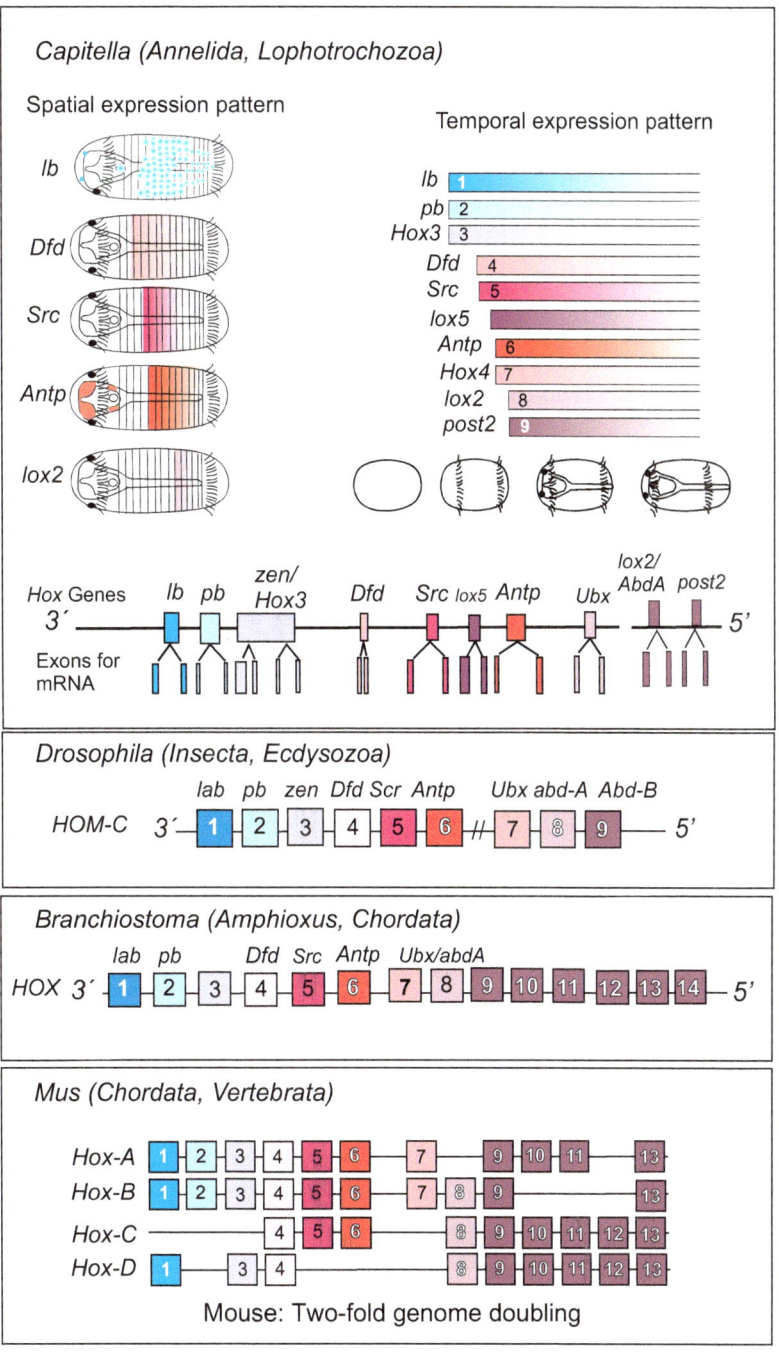

used to optimise the function of homologous structures.

- **Gene duplication**. The duplication of a primordial gene by **imprecise crossing-over**

(Fig. 22.4 or by means of transposons (see textbooks of genetics) generates subsequent to duplication new perspectives for natural selection. One copy is left to perform the

Gene Doubling by Inequal Crossingover

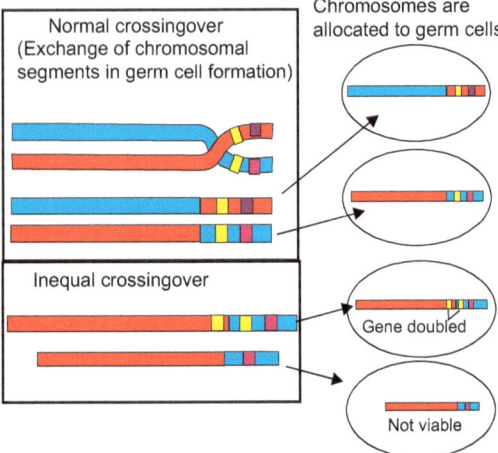

Fig. 22.4 Gene duplication by inprecise crossing over

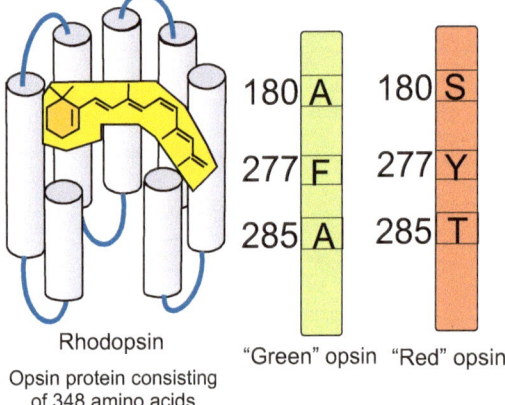

Rhodopsin

Opsin protein consisting
of 348 amino acids

"Green" opsin "Red" opsin

Fig. 22.5 The different opsins arose by duplication of one ancestral opsin gene; the variants allow analysis of wave length composition (*colour*) of light. Composed by WM

genes. In the *Hox* gene cluster all the descendants of the primordial gene still code for transcription factors but mutation-mediated changes in the DNA-binding domain enabled activation of different sets of downstream genes by different members of the *Hox* family. Presumably, the entire *Hox* cluster arose from two *proto-Hox* genes (Fig. 22.2). As said above, the members of this group are classified as paralogues if compared with other members along the cluster; they are classified as orthologues if one distinct member of this group is compared with the corresponding member in another species. (However, the introduced classification and nomenclature of the *Hox* genes was based on the comparison of the DNA-binding homeodomain, and does not exclude more extensive differences in other regions of the genes).

• **Genome duplication**. Duplication of genetic information can happen not only at the level of single genes but also at the level of chromosomal segments, chromosomes and the whole genome. In the mouse and in humans the *Hox* cluster is present four times, each cluster located on a different chromosome (see Figs. 12.16 and 22.3). Because *Branchiostoma* (Amphioxus), a representative of the Cephalochordata and as such closer to the origin of the phylum Chordata, possesses only one single cluster. The fourfold quantity in mammals indicates that during the evolution of vertebrates the entire genome was duplicated twice. Duplication of the entire genome is, for example, possible if in the first meiotic division of primordial germ cells the chromosomes are not distributed to two daughter cells, and diploid gametes result. In the very rare cases this happens in both sexes, a tetraploid organism such as *Xenopus laevis* can arise. Moreover, many polyploid organisms are known, or polyploid tissues in diploid organism such as liver cells in humans.

• **Point mutations**. It is not necessarily a huge event that will have a profound effect. Already the emergence of a new allelic

original function under the pressure of natural selection. The second copy can be varied by mutations and **assume modified or even new functions**. Examples are the different **rhodopsin alleles** (Fig. 22.5) that are expressed in the cones of our retina and enable colour discrimination.

• **Multiple duplications:** In case of the **Wnt genes** and other **multigene families**, and in case of the **Hox genes** such duplications with subsequent mutations in the duplicates happened several times leading to **paralogous**

variant can have a strong impact on developmental processes. Imagine an enzyme that degrades a morphogen. A new variant of the enzyme, working quicker or slower, can make a morphogen gradient steeper and shorter, or flatter and longer. This can lead to a shortening or an extension of body regions (compare Fig. 4.34). The implication of new gene variants, however, cannot be deduced from the primary sequence but must be elucidated through functional and biochemical investigations.

22.3.2 Novelties Arise from Recombination of Existing DNA Sequences, for Instance by *Exon-Shuffling*

Sequence analysis reveals in many proteins a **modular composition with similarity to different functional domains of other** proteins. In particular, a modular composition is evident in multimodal growth factor receptors, which combine ligand-binding and enzyme activity. A growth factor receptor of the FGFR family (see Fig. 11.8) for example possesses an extracellular ligand-binding domain, a transmembrane domain, various intracellular docking sites, and a kinase domain. The partial sequence similarity to receptors of other types and families suggests that these proteins are composed of modules. In the course of replication and recombination new genes were created in evolution by recombining segments of already existing genes. A plausible and widely accepted hypothesis proposes *exon-shuffling* as a mechanism of recombination. As a rule, genes of eukaryotes are interrupted by several interspersed introns. Exon-intron boundaries frequently correlate with those of protein domains. Through recombination of initially separate exons highly complex multimodal protein might have arisen (a model of proteins consisting of segments derived from recombined DNA segments are the antibodies, see Fig. 18.8). According to recent reports about the origin of the growth factor binding receptor tyrosine

kinases (RTK) first events of recombination took place already in the common ancestor of multicellular animals and their sister group, the choanoflagellates. Such processes warn us about using partial sequences in genes and proteins as sole indicators of homology. On the other hand comparing partial sequences found in extant organisms enable reconstruction of the evolution of modular proteins using algorithms of bioinformatics, and to correlate the result with phylogenetic trees constructed based on morphological, biochemical and genetic criteria.

22.3.3 Horizontal Gene Transfer: Many DNA Sequences Are Imported from Foreign Genomes

In the genome of all species investigated so far DNA sequences of apparently foreign origin are found. Frequently foreign (heterologous) DNA is transferred by viruses from one organism into another organism, also irrespective of borders between species or even phyla. Such sequences persist in the form of transposons or sequences partially reminiscent of transposons. Their viral origin is evident or suggested by characteristic subsequences. Even in non-functional remnant sequences partial sequences coding for viral envelope proteins (*gag*, *env*) or other typical viral sequences (*pol*, *ORF*) or characteristic palindromic flanking sequences remain traceable testifying their viral origin. According to quantitative estimations transposons and related sequences sum up to about 45 % of the human genome. They are commonly looked upon as junk DNA, but among the sequences imported into our genome sadly there also are cancer-generating **oncogenes** (Chap. 19). Beneficial for humans is, on the other hand, the likely cooption of transposons in lymphocytes. The many short sequences framed by palindromic sequences are used to program the variable, antigen-binding parts of B-cell receptors and hence of antibodies by somatic recombination (Fig. 18.8). The same applies for the variable domains of the T-cell receptors. Apparently, transposons were put into service of the adaptive immune systems during evolution of the vertebrates. In biotechnology, transposons are regularly used as vectors to produce transgenic animals.

Surprises are to be expected in evolutionary biology. In 2009 the emergence of a new gene was reported in mouse, which had been generated from a previously functionless intron in a long process that included random mutations. Researchers concerned with the less-noticed phylum of Cnidaria report numerous sequences of prokaryotic origin included in the cnidarian genome. Which mechanisms caused or allowed their uptake and integration into the host genome is unknown.

22.3.4 The Species-Specific Stock of Genes Is Also Determined by Loss of Genes

When the first genomes were sequenced, an astonishing number of conserved genes was found to be shared by the fly *Drosophila*, the round worm *Caenorhabditis elegans* and man. On the other hand in the human genome genes were detected that also were found in other mammals but not in the fly or the worm. These genes therefore were counted as mammalian-specific. Yet, many of the allegedly mammalian or vertebrate-specific genes were subsequently identified in the genomes of various members of the phylum Cnidaria (*Acropora*, *Nematostella*, *Hydra*, *Hydractinia*) as well as in the genome of the annelid *Platynereis dumerilii*. Apparently these genes were already present in the common ancestor of the Cnidaria and Bilateria but were lost in the branch leading to the Ecdysozoa including *Drosophila* and *C. elegans*.

Once more it must be emphasized that phylogenetic relationships can be reconstructed reliably only by comparing multiple genes from many different taxa. These restrictions must be taken into account in any evo-devo projects and reconstructions of phylogenetic trees.

22.4   The Common Toolkit

In this section we will show, using four selected examples, that the origin of conserved development-regulating genes dates back prior to the Cambrian period 550 million years ago when the basic body plans of all recent animals already were established. By taking the example of a few gene families it becomes evident that small changes in the genetic regulation or integration in genetic networks can have strong effects on the manifestation of molecular and morphological traits.

If one compares the genomes of all metazoa available to date, from sponges and cnidarians up to the mammals, one finds that genes coding for **transcription factors** are highly conserved among the genes relevant for developmental control. For example the eye development-triggering transcription factors PAX6 and SINE OCULIS can be traced back to the origin of the Metazoa. This will be explained in Sect. 22.7.

22.4.1 The Wnt/β-Catenin System Acts as a Conductor in Early Axis Specification Across the Metazoa

In many animal groups the WNT signalling system is concerned with the establishment of the primary body axis, or in bilaterally symmetric animals, the establishment of one of the three body axes, the longitudinal anterior-posterior axis. Signalling centre is the posterior pole (Fig. 22.6). Moreover, the system is concerned with the control of stem cells in all animals investigated.

If one wants to understand how body plans developed from the early to the complex animals it is reasonable to look at animal taxa that early in evolution diverged from the main lineage of the tree leading to bilateral symmetric animals. It is obvious that presence of a conserved gene in representatives of early side branches and also in the Bilateria lineage indicates its presence in the inventory of the last common ancestor that existed before the branches diverged. Among the taxa that branched off early in evolution are the Porifera

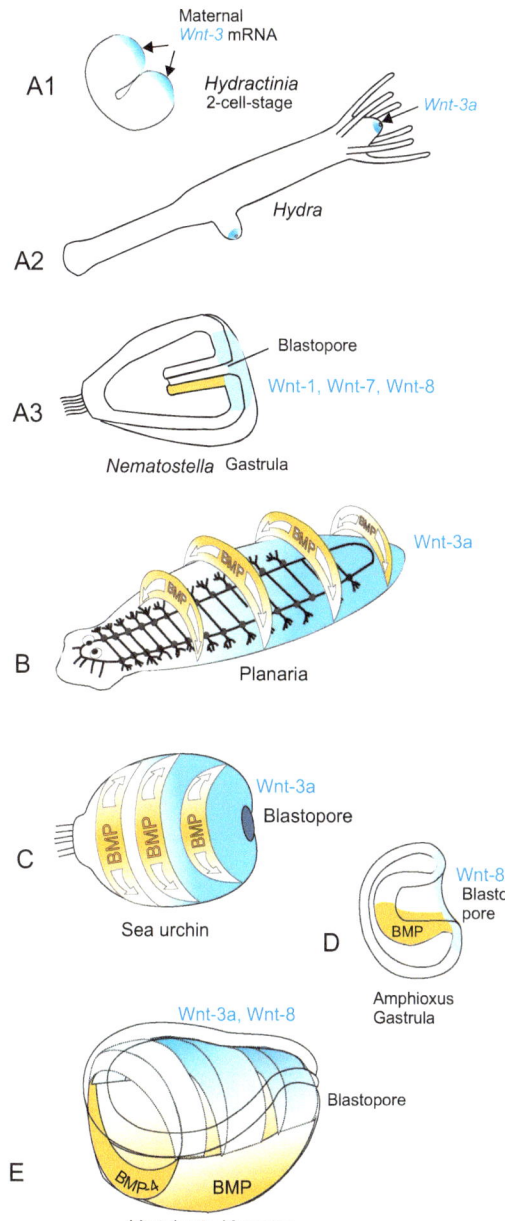

Fig. 22.6 WNT expression in the establishment of one/the body axis in various organisms. Composed by WM

(sponges), the Placozoa represented by the tiny two-layered *Trichoplax*, and the Cnidaria with *Clytia*, *Hydra*, *Hydractinia*, the coral *Acropora*, and the sea anemone *Nematostella*. The majority of multicellular animals share an indirect development with a simple motile ciliated larva. This larva has an anterior end pointing in the direction

of forward movement, and has an elongated body. Whether this longitudinal axis is called antero-posterior axis, oral-aboral axis, or head-tail axis is a matter of convention and tradition. It is this main longitudinal body axis that is established under the regime of the canonical WNT/β-catenin system (Fig. 9.5).

> In this context the question is of interest whether the oral-aboral body axis of the Cnidaria corresponds to the antero-posterior axis of higher evolved bilaterian animals, and therewith, whether the head of *Hydra*, sea anemones and of corals is equivalent to the head of annelids, insects, or vertebrates. This question is discussed in Box 22.2.

The Wnt/β-catenin signalling pathway (see Fig. 11.6), once detected as a signalling system that is required for correct outgrowth of wings in *Drosophila*, turned out to be essential for the establishment of an embryonic body axis in all animals investigated with respect to the genetic basis of axis formation, except in *Drosophila* itself. WNT signalling molecules are expressed early in embryogenesis – in part by maternal, in part by zygotic genes. In previous Chapters we referred to the examples of the hydroid *Hydractinia* (Fig. 9.5), sea urchin (Fig. 4.6) and *Xenopus* (Figs. 5.18, 5.19, and 10.4) and found a common principle: **The WNT system directs and patterns the body axis that extends from the blastopore to the opposite pole** (Fig. 22.7). The head-tail axis (a-p axis) in vertebrates can only be implemented correctly when WNT is emitted from the posterior pole at the blastopore, and on the other hand at the anterior pole anti-WNT factors neutralize the WNT signals thus enabling head formation. Working in the opposite direction, WNT and anti-WNT gradients share responsibility in the subdivision of the body into head, trunk, and tail. However, when we compare representatives of different taxa we see an important difference. While in triploblastic animals WNT is expressed with maximum intensity at the posterior pole and

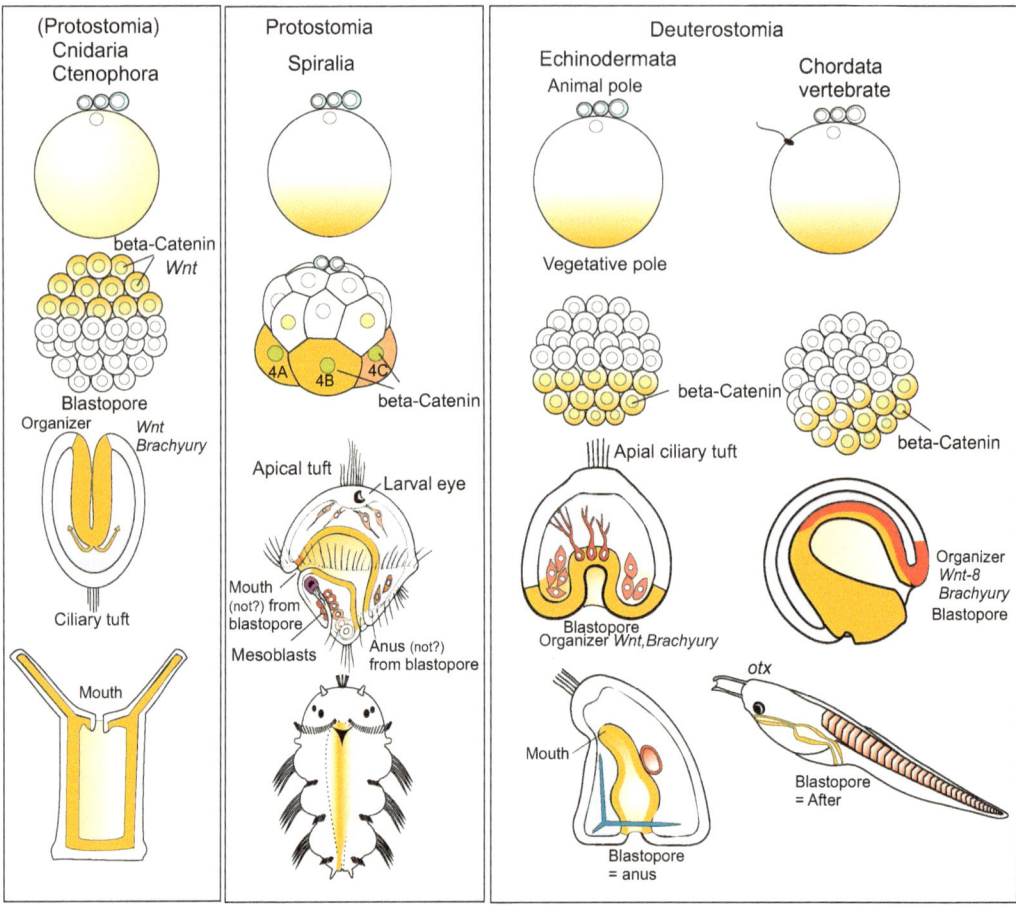

Fig. 22.7 Position of the blastopore (gastrulation) and development of the longitudinal body axis in various representatives of the animal kingdom. Note that in Cnidaria the place of gastrulation is at the pole where the polar bodies form and which is designated animal pole in other animals. Here also the head (hypostome) will arise. By contrast in the bilateral organisms gastrulation takes place at the vegetal pole or close to it. After Martindale and Hejnol (2009), and other authors, compiled, modified and extended by WM

excluded in the head region, in Cnidaria we see the obviously inverse situation: *Wnt* is expressed at the head pole. Such an inversion is not detected when we take the blastopore as point of reference and invert the polyp upside down: *wnt* expression indicates that *Hydra* is standing on its head.

Unexpectedly, already early in phylogeny an ample collection of WNT signalling molecules occurred through repeated duplication of the *wnt* gene. Already in Cnidaria 11 *wnts* are present! The successive and overlapping expression of all these WNT ligands along the body axis in *Nematostella* points to a potential function in the subdivision of the body column into functionally different zones (Fig. 22.8). The canonical WNT system apparently is a system that was acquired early in the history of life for specification and regionalization of the primary body axis. In triploblastic animals this system was supplemented or replaced (*Drosophila*) by HOX transcription factors.

Besides components of the canonical WNT signalling system also non-canonical WNT pathways are present already in Cnidaria: one of those is involved in the outgrowth of tentacles and buds in cooperation with canonical WNT signaling. As non-canonical WNT signalling systems are responsible for directed movements

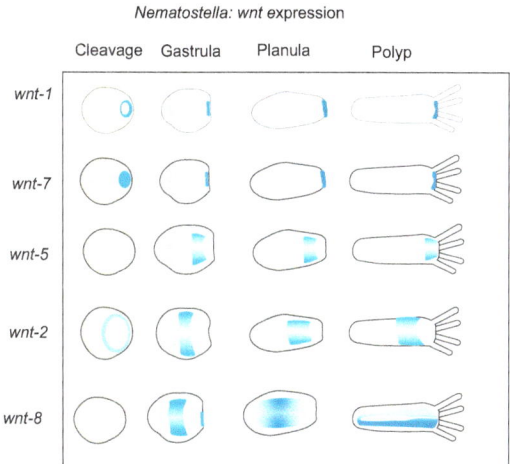

Nematostella: wnt expression

Fig. 22.8 Expression patterns of various members of the WNT family in the sea anemone *Nematostella vectensis*. After Kusserow et al. (2005)

of tissues in triploblastic animals, too, the components of this system can be also counted as instruments of the ancestral toolkit.

22.4.2 The HOM/HOX Homeotic Transcription Factors Subdivide the Body of Triploblastic Animals in Topographic Regions

In bilaterian animals the task to regionalize the main body axis is assumed by the HOX system, or at least regionalization is fine-tuned by HOX factors. In (internally) segmented animals, including the vertebrates, HOX transcription factors specify the identity of segments compliant with position along the antero-posterior axis (Chap. 12, Fig. 12.16). In eumetazoans that developed after the phylum Cnidaria split off the main lineage, paralogous genes coding for HOM/HOX transcription factors are grouped together in the genome forming one cluster. The members of the cluster are expressed in a colinear pattern, meaning that the anterior limits of the expression domains are progressively shifted in the posterior direction in an order that reflects the sequence of the genes on the chromosome. In common illustrations genes placed 'left'

(i.e. closer to the 3' end) are expressed earlier and in more anterior body regions than genes placed closer to the 'right' (i.e to the 5' end) of the *Hox* cluster (Figs. 12.16 and 22.3). Moreover, the rule of posterior dominance is valid. This implies that a newly activated posterior gene dominates over previously activated anterior genes in and posterior to the body region in which the posterior gene is activated (see Fig. 12.17). Once the *Hox* cluster was arranged in the history of the bilaterians, it was conserved under the pressure of natural selection with respect to colinear expression and function in specifying segment identities. Only in non-segmented animals such as molluscs the highly conserved arrangement of *Hox* genes within the cluster may be relieved probably due to missing constraints. Comparative studies showed that the nine members of the insect *Hox* cluster were already present in the common ancestor of annelids and arthropods (Fig. 22.3) for only a common ancestor can have handed down the cluster to both the superphyla Lophotrochozoa and Ecdysozoa. As said above, the vertebrates possess four *Hox* clusters (Fig. 22.2, see also Fig. 12.16). This indicates a twofold duplication of the cluster in ancestral vertebrates.

Below the level of the bilaterians no extensive clusters are found and comparison of the sequences of individual *Hox* genes yields ambiguous possibilities of homologization. After all there are strong indications that the series of *Hom/Hox* genes arose from a single *proto-Hox* cluster (Fig. 22.2). Repeated duplication and subsequent modifications of the multiplied members of the group enabled the development of the amazing multitude of recent body plans.

With reference to Chap. 12 (and also Fig. 22.2) it may once more be pointed out that HOM/HOX factors do not program certain cell types, tissues or organs but denote body regions to modify cell types, tissues and organs according to their position within the body. They also enabled the evolution of a large number of highly specialized body appendages from originally simple appendages as explained in the following sections.

22.5   Variations of Body Plans Within Phyla

22.5.1 Legs Become Mouthparts, and Some More Spectacular Transformations of Appendages in Arthropods

The duplication of *Hox* genes, combined with functional modifications and integration into various genetic networks, enabled the arthropods to develop a large array of appendages that arose from original simple leg-like appendages. Thus we find mouthparts that in dragonflies are optimized for biting-chewing tasks with strong jaws, in mosquitoes piercing-sucking mouthparts formed as concentric tubes with a food channel, a salivary duct and a complex stylet, in house flies we see sponging mouthparts, and in butterflies long sucking mouthparts that can hydraulically be unrolled, extended and furled. Not least also the spinnerets and the book lungs at the legless hind-body (opisthosoma) of the cobweb spiders are thought to be derived from former segmental appendages (Fig. 22.9).

A comparison of *Hox* gene expression in arthropods shows that they are one variable in specifying the quality of appendages, but *Hox* genes are not the sole determinants; eventually the quality is specified by species-specific sets of downstream genes.

The possibility of developing different tools for food intake through mere variation of a basic structure enabled the arthropods to colonize a multitude of new ecological niches. The evolution of the body appendages in the arthropods provides impressive examples for the connection of developmental mechanisms with ecological adaptability – under the term eco-devo (linking ecology and development), also an increasingly attractive field of research.

Defects in individual *Hox* genes **disclose concealed ancestral differentiation programs**. We already had addressed the *Antennapedia* mutant of *Drosophila*, in which the **antennae turn into legs,** and addressed the **four-winged fly** in which the halteres are transformed back to wings (see Fig. 4.40). This fly mutant

recapitulates the archetype of the four-winged insect that clearly can be realized in the recent four-winged dragonflies.

Concealed morphological traits of an evolutionary ancestor disclosed by mutations are known as **atavisms**, and also are known from humans (such as rows of nipples, open gill slit, tail, fur).

Back to arthropods. Several groups of researchers try to recapitulate evolutionary events by experimentally interfering with gene expression through blocking genes by RNAi or overexpressing certain genes. In the laboratory, spiders develop legs at their hind-body (opisthosoma). Will it ever be possible to re-generate the putative myriapod-like ancestor of chafers?

22.5.2 An Elaborated Head and Gills Derive from a Novel Cellular Source, the Neural Crest a Hallmark Innovation in Vertebrates

At the transition of the cephalochordates (also called protochordates, with the lancelets = *Amphioxus*) to the vertebrates, a new predatory lifestyle and increased body size coincided with the appearance of an elaborated head. Characteristic innovations of this head are a skull protecting and accommodating a centralized nervous system, a jaw for prey capture and gills as respiratory organs. Cephalochordates are suspension feeders employing a cilia-based food collecting mechanism similar to that of tunicates. Water is driven into the mouth and pharynx and out through the pharyngeal gill slits into a surrounding atrium; it exits the body through a ventral atriopore. Up to 200 gill slits are separated by cartilaginous rods, the gill bars.

While in lancelets, the feeding currents are generated by pharyngeal cilia, gill ventilation currents in aquatic vertebrates depend on muscle contractions. The neural crest became a major ontogenetic source for the 'new head' (Fig. 22.10) and was fundamental to subsequent vertebrate evolutionary history. It helped to tremendously accelerate the emergence of a predatory lifestyle as well as the utilization of

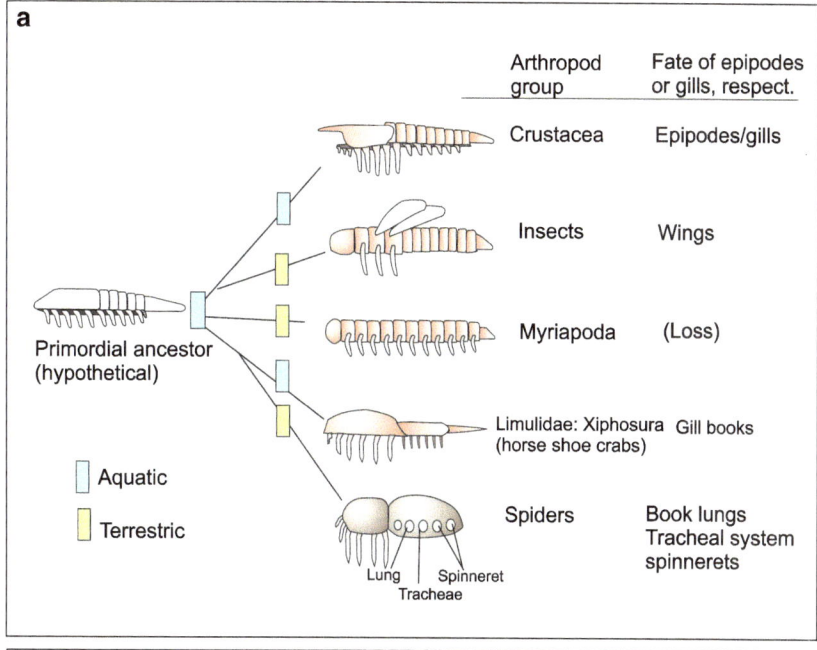

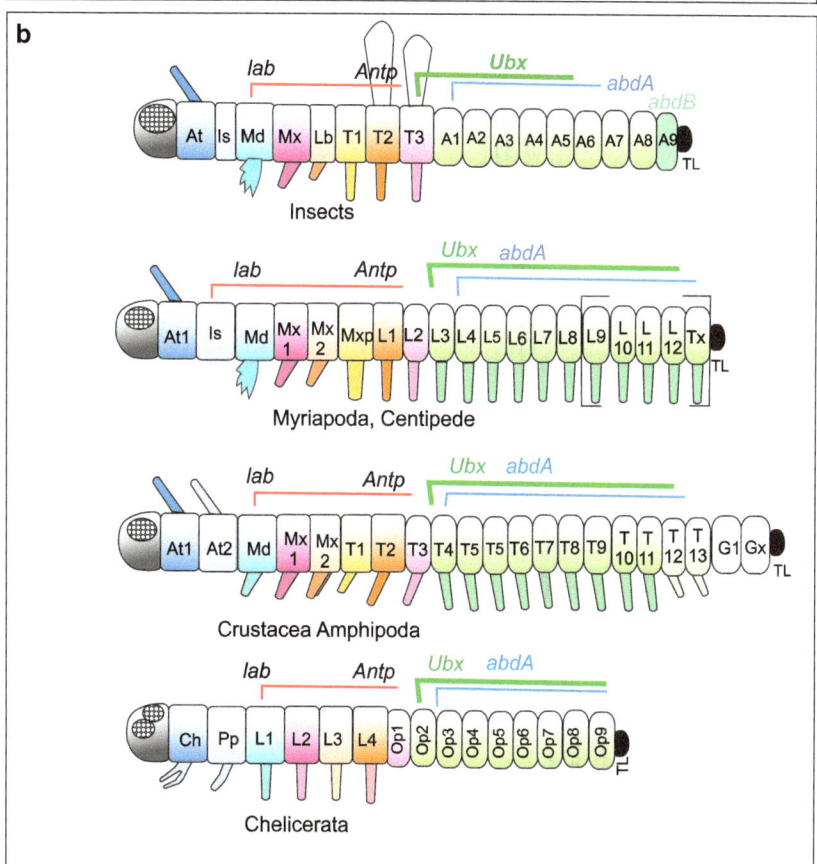

Fig. 22.9 (**a**) Model of the evolution of arthropods and their appendages. After Pechmann et al. (2010). (**b**) Correspondence of segments in various arthropods. After Hughes CL, Kaufmann TC (2002), Pavlopoulosa A (2009) and other sources. *At* antenna, *Is* intersegment, *Md* mandible, *Max* maxilla, *Mxp* maxillipeds, *lb* labial,

Fig. 22.10 Emergence of new structures in the head-pharynx region of vertebrates. Many innovations are derivatives of the neural crest cells, an innovation of the vertebrates. After Mongera et al. (2013), supplemented after Olson (2002), composed by WM

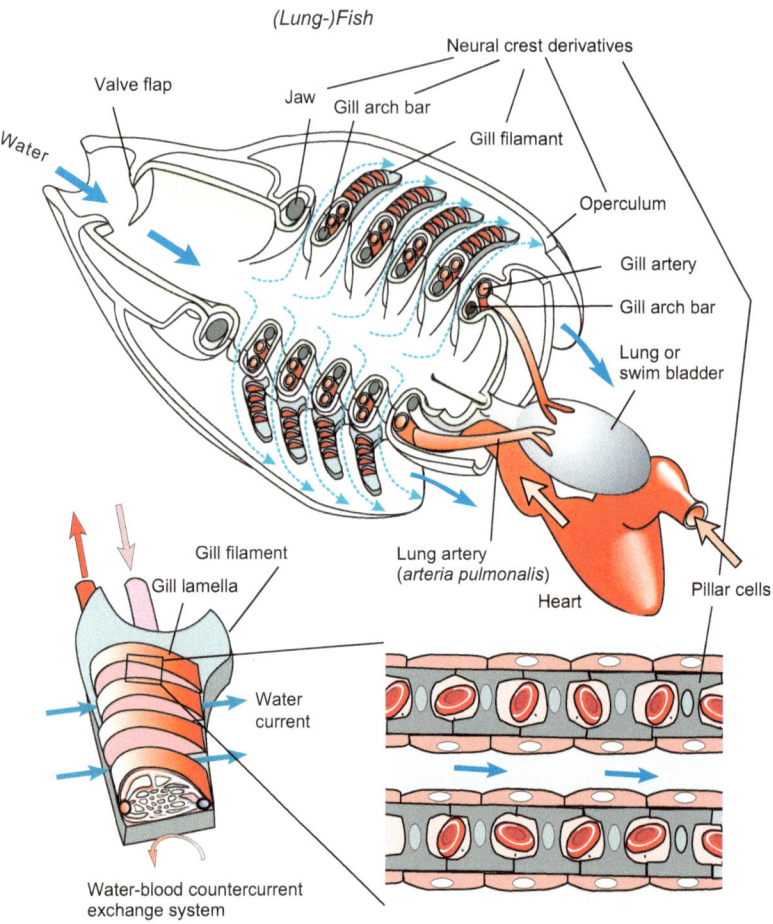

ecological niches. Key innovations that endowed vertebrates with predatory capacity were the acquisition of muscular ventilation coupled with respiratory surface expansion, the emergence of a skull vault and a dorso-ventrally articulated jaw. The neural crest supplies pillar cells which form small channels in which the gill vessels run. They also provide smooth muscle cells around the gill capillaries enabling the capillaries to expand or contract and thus to control blood flow. This switch from ciliary to muscular ventilation increased the metabolic capacities of the vertebrates, enabling an increase in body size. In addition, neural crest cells

contributed to the cranial bones of the skull, to the lower jaw and the bony gill arches, and to the otic capsules.

22.5.3 Swordtail Fishes and Platys: Examples for the Action of Sexual Selection and the Possibility to Reactivate Buried Developmental Programs

How changes in the feasibility of gene activation can lead to the formation of particular traits can be seen in fishes of the genus *Xiphophorus*. In males

Fig. 22.9 (Continued) *L* leg, *T* thorax, *A* abdominal, *G* genital, *Ch* chelicere, *Pp* palps, *L* leg, *Op* opisthosoma. *Hox* genes: *lb* labial, *Ubx* ultrabithorx, *abdA* abdominal A. *Ubx* causes reduction of legs in insects and spiders but

not in myriapods and amphipods probably because the final result depends on downstream genes which are not identical in the various taxa. Compiled by WM

Fig. 22.11 Development of the sword in *Xiphophorus helleri* and reactivation of a corresponding concealed potency in platyfish (*Xiphophorus maculatus*). Development of the sword depends on a local organizer in the tail fin and on testosterone which normally releases this potency in male swordtails but not in male platyfish. Addition of testosterone into the aquarium water induces a small sword in platyfish as well. This effect requires the activation of an FGF signalling system and of the transcription factor MsxC. In the platyfish mutant *brushtail* the internal factors are enhanced and the latent potency for sword formation manifests even without external stimulation. After Offen et al. (2008, 2009), composed by WM

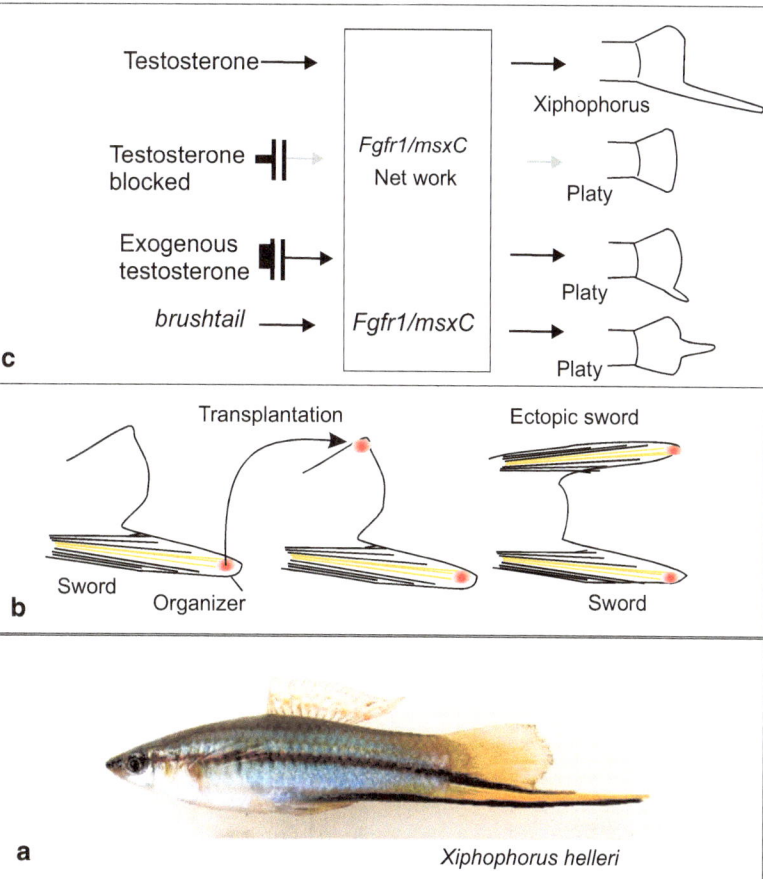

Testosterone-dependent Fin Organizer in Swordtail and Platy

of *Xiphophorus helleri*, known as swordtail, the ventral brightly coloured part of the initially symmetric tail fin elongates to the "sword". This happens under the influence of the male sexual hormone testosterone until it has reached the length known to aquarists (Fig. 22.11). Females of the species appraise this pageantry and mate preferentially with particularly gorgeous males so that the expenditure of energy finally counts.

By transplantation studies it was shown that tissue at the tip of the longest and most ventral ray acts as organizer. Transplanted into the dorsal part of the fin it stimulates dorsal rays to grow to form a second, ectopic sword. Outgrowth is dependent on androgen hormones and is possible only in males.

Remarkably this developmental path can also be reactivated by administering androgens in females and in the related platy fish which normally do not develop a sword. Exogenous testosterone induces in female *Xiphophorus helleri* and in the normally sword-less platy *Xiphophorus maculatus* the formation of a (small) sword.

The genetic network and signal relay system that underlie sword development are triggered by the androgen hormone, relayed by the growth factor FGF and lead to the activation of the transcription factor MsxC. The gene for this transcription factor is present in the genome of the swordless platy fish but lacks a testosterone-sensitive promoter. In the platy mutant *brushtail* the internal factors are enhanced and the latent potency for sword formation manifests even without external stimulation. This indicates that predisposition for sword formation was, and still is, present in both species.

Apparently, in evolution females of *X. maculatus* succumbed to the showing off trait of a long sword less than the females of *X. helleri*. The example shows that sexual selection and female preference directly influence developmental processes. It is a rewarding task for post-genomic times to return attention to links between developmental biology, physiology, sociobiology, and ecology.

22.6   Reconstruction of Large-Scale Developmental Series and Phylogenetic Trees

22.6.1   Millions of Animal Species Only Classified Among Three Main Groups (Superphyla)?

Zoological textbooks subdivide the bilaterians into 20 or more phyla – dependent on the criteria the respective authors consider weighty enough to elevate an animal group to the rank of a phylum. But how the various phyla should be arranged in the entire phylogenetic tree is still (and perhaps will remain) a matter of dispute.

An example how the evolutionary developmental biology can lead to revisions in the topology of the phylogenetic tree is the arrangement of the trees that comprise the annelids and arthropods. Both phyla were once joined and subsumed under the term **Articulata** (Fig. 22.1 insert upper left). In 1997 a new phylogenetic tree deduced from preliminary data on 18S rDNA proposed that the animal kingdom early split into only three branches, now known as **Ecdysozoa**, **Lophotrochozoa** and **Deuterostomia**. Initially, the new phylogenetic tree was not well founded and resolved. It turned out that the 18S rDNA is a problematic gene for phylogenetic reconstructions and frequently produces bizzare results. But more recent studies using additional genes, such as 12S rDNA, 28S rDNA, elongation factor-1α, RNA polymerase II, and ubiquitin, all corroborated the

18S rDNA result. Altogether the monophyletic (unitary) origin of these three groups was in the last 20 years supported by an increasing number of additional molecular, physiological and developmental markers. While the Deuterostomia were long regarded a closely related group and beyond dispute, the so called Protostomia (a disputed term per se) are now subdivided into only two comprehensive groups (Fig 22.1). These are the

- **Lophotrochozoa**, that share the formation of a **lophophor** (a circle or horseshoe-shaped band of ciliated tentacles surrounding the mouth and used for the capture of food), a **trochophore** larva and **spiral cleavage**, traits which, however, are lost or replaced in this or that subgroup;

- **Ecdysozoa**, whose characteristics are a three-layered cuticle and the occurrence of growth-related **ecdysis** = moult. The moult of the chitinous cuticle is in arthropods, and nematodes as well, triggered by the steroid hormone **ecdysone** (more precisely by 20-OH-ecdysone; see Fig. 20.4). The Ecdysozoa often lack locomotory cilia and then produce mostly amoeboid sperm. Their embryos do not undergo spiral cleavage.

Some zoologists still hold to the original view that **Arthropoda** should be classified with **Annelida** in a group called the **Articulata**, and they give good reasons. Both phyla, the arthropods and the annelids, possess a segmented body, the metameric subdivision of which arises from mesoderm bags or packages. The nervous system consists of a dorsal supra-esophageal ganglion, a ventral paired sub-esophagial ganglion and two strands of ventral nerve cord that connect segmental ganglia. Common to both the Annelida and Arthropoda also are a dorsal tubular heart and the occurrence of proto- and/or metanephridia that, however, in arthropods are highly modified or lacking. As *connecting link* between annelids and the

(continued)

Euarthropoda the Onychophora (e.g. *Peripatus*) were discussed. These have legs similar to caterpillars, have subepidermal body muscles that combine smooth and obliquely striated fibres and are arranged similarly to those of annelids. A pair of nephridia lie in each leg-bearing body segment. Yet, recent molecular data place the Onychophora into a sidelineage of the Euarthropoda.

For developmental biologist the Articulata concept was always problematic due to some inconsistencies. Thus, a spiral cleavage as it is found in the annelids is also found in flatworms and molluscs – but not in arthropods. Upon a close look mechanisms of segmentation are not identical among the various animal taxa and segmentation has been lost in some groups. Provided the dichotomy of Lophotrochozoa and Ecdysozoa persists, the common traits may be attributed to a common basal ancestor, perhaps to the hypothetical primordial bilaterian (see Box 22.3).

Molecular data such as presence of the *Hox* cluster (Figs 22.2 and 22.3) strongly suggest a common ancestor of annelids and arthropods.

22.6.2 Fundamental Novelties in the Body Plans: Two Versus Three Germ Layers, Radial Symmetry Versus Bilateral Symmetry

We will use the example of two fundamental novelties to show the stimulation and food for thought evo-devo biology can provide. The selected revolutionary reinventions concern

(a) the **switchover from diploblasty to triploblasty**, that is the introduction of the mesoderm as third germ layer,

(b) the **switchover from radial to bilateral symmetry** (or vice versa).

Both features appear to be closely coupled. Animals with advanced complexity are bilaterally symmetric and triploblastic. Bilateral symmetry means presence of a longitudinal antero-posterior axis which can be split into two halves so that one part is a mirror image of the other. Perpendicular to the antero-posterior axis, an asymmetric dorso-ventral axis separates a dorsal from a different ventral side and thereby defines the third, left-right axis. Bilateral symmetry permits streamlining, favours the formation of a central nervous system, contributes to cephalization, and promotes active movement. All bilateral animals are in addition triploblastic. The mesoderm conveys high complexity of inner organs due to its high flexibility and power of differentiation. In contrast, diploblastic animals display radial or biradial symmetry.

22.6.3 When Did the Mesoderm (Triploblasty) Arise?

In the construction of body plans the occurrence of the mesoderm marks a quantum leap forward in evolution. Conventional wisdom assumes a simple scenario: After Porifera (sponges), Parazoa (*Trichoplax*) and Cnidaria, perhaps also the Ctenophora (comb jellies), seceded from the main lineage of animals, a multitude of triploblastic phyla emerged with more and more complex body plans. They can make use of three germ layers as building material. All these triploblastic animals are also bilaterally symmetrical. (The pentamerically radial echinoderms such as sea stars and sea urchins are bilaterally symmetrical in embryogenesis up to the larval stage). In contrast, radial symmetric animals such as Cnidaria and the essentially bisymmetric (biradial) Ctenophora are said to be diploblastic; their body is built by only two germ layers, ectoderm and endoderm. There arises the question when and how bilateral symmetry and the mesoderm arose and whether these traits are connected. To discuss this question we first should have a closer look at the Cnidaria.

Radial Symmetry and Diploblasty. We recall: In their basic architecture the Cnidaria display radial symmetry. This means, one single

Asymmetric Expression of *dpp/Bmp-4*

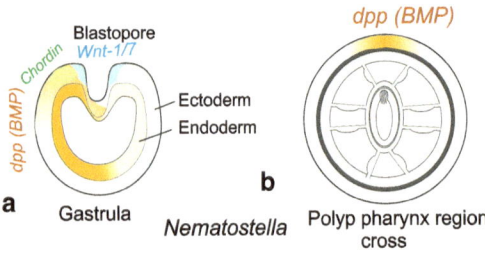

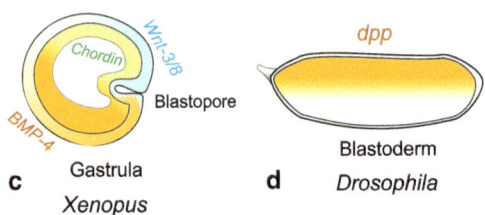

Fig. 22.12 Asymmetric expression of dpp/BMP in the sea anemone *Nematostella* (Cnidaria, Anthozoa), in *Xenopus* (vertebrates, amphibians) and *Drosophila* (arthropods, Diptera). Expression pattern in *Nematostella* after Rentzsch et al. (2006)

asymmetric body axis extends from the centre of the mouth-bearing oral surface to the centre of the opposite, aboral, end, like an axle through a wheel. All structures that are repeatedly present such as the tentacles around the mouth of polyps, the tentacles and sensory organs along the margin of the bell in jellyfishes, and inner organs such as mesenteries in corals, are arranged in circles around the oral-aboral body axis. Feeding structures and sensory receptors are predominantly located around the mouth, and in sessile forms a pedal disc for attachment to a substrate. These features establish a clear asymmetry along the oral-aboral axis. But there is no apparent dorso-ventral body axis. In the Anthozoa (sea anemones, corals), by most zoologists considered to be the most basal class, a concealed tetra-radial and partial biradial symmetry can be detected (Fig. 22.12).

As stated above Cnidaria are diploblastic, as in embryogenesis they form only two epithelial layers (ectodermis = ectoderm, and gastrodermis = endoderm). The ectoderm forms the epidermis as in all the Eumetazoa. The gastrodermis serves

for digestion and in addition forms in colonial species and in large jellyfishes a logistic network of channels, the gastrovascular system, comparable to the blood system in higher evolved animals. These two, mostly mono-layered epithelia, epidermis and gastrodermis, are separated by an originally cell-free sheet composed of extracellular matrix components.

This intermediate layer is called mesoglea in cnidarians and is in its molecular composition similar to the basal lamina in bilaterians. It is secreted by the epithelial cells of both epithelia, and occupies in large jellyfishes a large part of the body. In medusa it is colonized by immigrating cells, which form a loose mesenchyme but fail to organize in compact tissues or organs.

Though a mesoderm as an embryonic germ layer is lacking, in the endoderm of planula larvae genes are expressed that commonly are counted as mesoderm-specific. For example in the endoderm of the planula larvae of the hydrozoan *Podocoryne carnea* transcripts of *twist* homologues were found, and in the endoderm of the larvae of sea anemone *Nematostella vectensis* transcripts of *twist* and *snail*, genes that were first detected in *Drosophila* and are expressed in the mesoderm, and in the mesoderm of other triploblastic animals as well. Such observations have given rise to the suggestion that the mesoderm originated as a subdivision of the endoderm – a suggestion supported by the way the mesoderm arises in sea urchins and amphibians – namely as mesendoderm.

Mesoderm in Cnidaria? There is in the development of hydromedusae a structure that shares attributes of a stand-alone mesoderm. It is the **entocodon** (bell interior) in the medusa bud (Fig. 22.13). This tissue is inserted between the ectoderm and endoderm. Some authors consider the formation of a separate layer of smooth and cross-striated musculature from cells of the entocodon as specific evidence of mesodermal

Fig. 22.13 The entocodon of hydromedusae possesses striated and smooth muscle cells, i.e. a cell type forms that is considered mesoderm-specific in triploblastic animals. After Seipel and Schmid (2006, 2006)

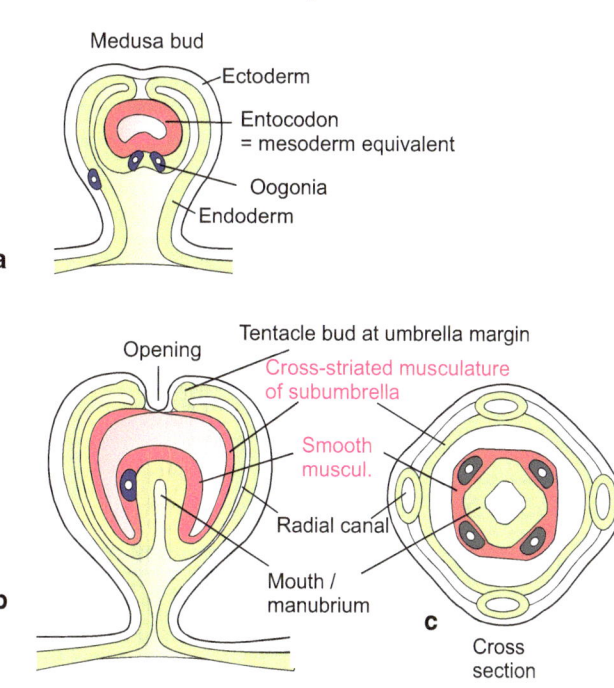

Podocoryne carnea

Medusa bud

Ectoderm

Entocodon = mesoderm equivalent

Oogonia

Endoderm

a

Opening

Tentacle bud at umbrella margin

Cross-striated musculature of subumbrella

Smooth muscul.

Radial canal

Mouth / manubrium

b

c

Cross section

features. Molecular genetics revealed that typical mesodermal genes such as *Brachyury*, *snail* (codes for a zinc finger protein) and *twist* (bHLH transcription factor) are expressed in the entocodon of the medusa of *Podocoryne carnea*, and in addition a collection of genes known to be responsible for the programming of muscle cells, such as *MyoD* and *Mef2*. In spite of these congruencies, currently the majority of authors adhere to the notion that all members of the cnidarian phylum are diploblastic. This adherence to traditional views is reasoned and explained by the fact that up to now no case is known where mesoderm is formed in embryogenesis and present in larvae. Moreover, in 2012 strong molecular evidence was given that the striated muscles of medusae do not correspond to skeletal muscle and probably arose by convergence.

Different Ways of Mesoderm Formation. By comparing the emergence of the mesoderm, for instance in the spiralians (see Fig. 4.22), in insects (Figs. 4.26, 4.27, and 4.36), in amphibians (Figs. 5.7, 5.8, and 5.10) and in mammals (Figs. 5.29 and 6.5) substantial

differences become obvious. Even in vertebrates there is more than one source of mesodermal tissues. Besides the classic marginal zone in the amphibian gastrula (Fig. 5.7) the anterior neural crest also contributes to the inventory of typical mesodermal tissue such as cartilage and muscle (Figs. 15.3 and 15.4). **In the history of biology, either intuitively or consciously, those groups of cells have been designated mesoderm which give rise to the (cross-striated) musculature** (discussed Box 22.1). Because musculature is always located in the interior of the body for obvious functional reasons, such cell groups must in some way be inserted between ectoderm and endoderm. In other words, actually mesoderm is defined in retrospect by looking at the cell types and tissues formed by the material shifted into the interior and between ecto- and entodermal layers. It would be better to speak of homologous cell types instead of homologous germ layers.

As to the question how many germ layers exist, not in textbooks but in reality, and how the nervous system is integrated into the concept of germ layers, Box 22.1 provides a discussion.

Box 22.1: The Saga of the Three Germ Layers

Inconsistencies and misleading terminology in comparative developmental biology

Textbooks all over the world adhere to traditional opinions and terms. Didactic necessities foster this continuance. Learners need a readily comprehensible scaffolding. Only when this is established and consolidated, variations can be inserted, "ifs, ands, or buts" considered, restrictions and inconsistencies named.

Such a traditional basic term is "**germ layer**". Word is all around that in diploblastic animals two germ layers existed, in tripoblastic animals **three** germ layers, **never four or more**! Subsequent to gastrulation these are

1. **Ectoderm** (ectoblast), the cellular outer layer which gives rise to the epidermis;
2. **Endoderm** (endoblast), the future epithelial layer of the gastro-intestinal tract (+liver and pancreas), and
3. **Mesoderm** (mesoblast), the cell material that moves from the exterior layer into the interior of the embryo and inserts itself between ectoderm and endoderm. Mesoderm supplies the material for many inner organs, in vertebrates for example for skeletal elements, connective tissue, blood vessels, heart, and kidneys. And in all instances, including invertebrates, the mesoderm is said to give rise to the **musculature**.

But also the nervous system has its origin in material that moves from the surface into the interior of the embryo and inserts between the ectoderm and endoderm. Nevertheless no textbook ascribes the nervous system to the mesoderm. Rather textbooks of zoology and developmental biology use to designate the nervous system as ectodermal entities if they do not desist to assign the nervous system to a defined germ layer at all.

A nervous system associated with the outer body layer, but also with the inner layer, is found in the phylum Cnidaria in form of a nerve net distributed among the interstitial spaces between the epithelial cells. And in other animals? One finds so-called secondary sensory cells without an own axon in the lateral lines of fishes and aquatic amphibians such as *Xenopus*. Primary sensory cells with own axons are found in the olfactory capsules of fish and in our olfactory epithelium. But **wherever in the world is the nervous system part of the outer skin?** *Ecto-derm* **means, literally translated, outer skin!** Now we ask, are there good reasons to assign the nervous system to the ectoderm based on the ontogenetic history of the nervous system?

In the amphibian development it is the marginal zone along the equator of the blastula that in gastrulation moves into the interior to become mesoderm, and it moves along with the future endoderm. Subsequently in neurulation the "neuroectoderm" sinks down, while forming the neural tube, into the space between the ectoderm and endoderm just as did the mesoderm. During this process of invagination neural crest cells detach and migrate in the interior to many places. There follow nose and ear placodes to sink into the interior, and finally the lens of the eye. One could designate the neural tube (+neural crest), as **second mesoderm**, in analogy to the events in sea urchin and other organisms where one speaks of a **multiphasic formation of the mesoderm**. Or one could speak of a **fourth gem layer**, but this is not done.

In *Drosophila* the precursor material of the musculature and the nervous system is displaced almost synchronously in a common process of gastrulation into the interior. Nevertheless, also in the fly the musculature is said to be mesodermal but the nervous system to be ectodermal.

Quite curious is the classic assignment when it comes to subsume the neural crest derivatives into the concept of three germ layers. In the head region neural crest cells give rise to "ectodermal" cranial nerves but ▶

Box 22.1 (continued)

also to "mesodermal" skeletal elements and musculature (Chap. 15, Figs. 15.3, 15.4 and 16.17). Among the neural crest cells of the trunk region several migrate into the ectodermal epidermis giving rise to pigment cells, other form the peripheral nervous system including the enteric nervous system in the mesodermal muscle sheet of the gastrointestinal tract or immigrate into the endodermal lining of the gut to become chemosensory neurons that control the composition of the food. They do never have contact to the ectodermal epidermis but nevertheless are in traditional terminology classified as ectodermal. Apparently, in assigning cells, tissues and cell-derived structures such as skeletal elements to germ layers, and to integrate this classification into the concept of **homology**,

intuitively the **final fate** is taken into account, not so much the exact origin. This origin is highly different in various animal phyla, for example in annelids (Fig. 4.27) compared to vertebrates. **Muscle tissue** and internal skeletal elements are always classified as **mesodermal** independently of their ontogenetic origin and the mode of their displacement into the interior, while **nervous elements** are constantly classified as ectodermal though they never stay there. For this conventionality there are **no logic arguments, it's simply fixed tradition**.

Recently, researchers laying focus on biochemistry and molecular genetics propose to speak of **homology of cell types** rather than of homology based on the traditional concept of three germ layers.

Box 22.2: Unsolved Riddles: Blastopore, Mouth and Primary Body Axes – Homologous or Not?

Among the much debated controversies in attempts to reconstruct evolutionary changes of the basic body construction is question whether the elementary body architecture, the Bauplan, of the various phyla can be considered homologous in the entire animal kingdom. Already in 1874 the zoologist Ernst Haeckel had released relevant controversies with his speculative gastraea hypothesis, a fictional gastrula-like animal thought to be origin of higher evolved animals. Now – in this digression into the speculative biology – it is about the question whether in the animal kingdom there is a consistent animal-vegetal body axis that is already specified in the structure of the egg.

Paragon of animal development and determinative for the designation of the suggestive designations of the egg poles still used today was the germ of the sea urchin (Fig. 22.12, see also Figs. 4.2–4.7). Sea urchin eggs and embryos provide unmatched advantages for the observer at the microscope. Meiosis takes

place only after spawning, the polar bodies remain at the place of origin, attached to and protected by the transparent envelope of the egg. At this pole the apical ciliary tuft, a larval sensory organ, forms and here most of the typical "animal" nerve cells will appear. At the opposite pole one sees how in the course of gastrulation the endoderm, that is the material for the "vegetative" gut, is shifted through the blastopore (primordial mouth) into the interior of the embryo. To transpose these conditions to amphibians was not difficult although in amphibians not only the animal pole but also the point of sperm entry specifies the exact position of the blastopore. Sea urchins and amphibians also turned out to be prototypes of deuterostome development, with a secondarily formed definitive mouth and the blastopore functioning as anus.

As counterpart of the deuterostomes the model of protostomes was introduced. In these the blastopore was said to become the definitive mouth, and the anus, if present, ▶

Box 22.2 (continued)

formed secondarily. Besides the (assumed) differences also conformities were seen. For example in both groups the starting material for the construction of the central nervous system is found in the animal hemisphere. By comparing the trochophore larva of annelids with the sea urchin larva, a hypothetical **amphistoma** was constructed (Fig. 22.14). In this animal both the mouth and the anus have their origin in a common slit-shaped blastopore. Prototype for this construct was the larva of the annelid *Platynereis*.

In the establishment of protostomes as a group many a improper fact was left out of consideration. For example in various protostomes sometimes the mouth, sometimes the anus is not a derivative of the blastopore but a novel opening formed at different sites. In particular, until about 1980 it was not recognized, at least not generally accepted, that in the Cnidaria and Ctenophora (formerly combined as phylum Coelenterata), which factually are protostomes, the site of gastrulation is not at the 'vegetal' pole but at the 'animal pole' (Fig. 22.12) if the position of the polar bodies is taken as reference.

This raises a question that currently is debated controversially and is left without a definitive answer even if the expression pattern of important developmental genes is taken into account: **Does the oral-aboral body axis of cnidarians correspond to the antero-posterior axis of more advanced animals?** And, if this could be approved with respect to the general longitudinal direction of the body axis then the question remains: **Is the head (hypostome) of a *hydra* homologous to the head of an annelid, insect and vertebrate, or is it located at the opposite pole?** If the various arguments are taken separately, one can arrive at different assertions. For example the cnidarian planula expresses *nanos* at the aboral pole, the embryo of *Drosophila* at the posterior pole (= aboral) and the gastrula of *Xenopus* at the vegetative pole. But transcripts of genes of Wnt signalling molecules and beta-catenin mark in cnidarians the oral pole, in deuterostomes such as sea urchins, amphioxus and vertebrates they mark the opposite, the aboral pole (Fig. 22.4) where the blastopore will become the anus. The genes *Brachyury*, *goosecoid* and *otx* are in cnidarians as well as in deuterostomes expressed in the area of the blastopore. According to our (the authors of this book) assessment the oral pole of the cnidarians is equivalent to the blastopore in other animal groups irrespective of whether the site of blastopore formation will eventually come to be the anterior or posterior pole of the finished larva (anterior pole defined with reference to the direction the animal moves).

On the other hand, in the controversial literature arguments are adduced to support attempts to homologize the mouth of the cnidarians with the definitive mouth of other metazoa (except the Chordata) but not necessarily with the blastopore. To render such a correspondence possible a displacement of the blastopore in its function as place of gastrulation from the animal to the vegetal pole was proposed and its definitive function as mouth or anus or both considered of secondary significance. For readers interested in the evolution of body constructions (Baupläne) the current discussion certainly is exciting, but fixation to a distinct doctrine would be premature.

22.6.4 What Is the Momentum of Selection in the Development of Bilateral Symmetry?

From time immemorial zoologists tried to deduce bilateral symmetry from radial symmetry, simply because in the conventional phylogenetic tree Cnidarians ranked "below" the Bilateria. In this conventional view bilateral symmetry is seen in context of a presumed transition from a sessile life style (of a polyp for instance) or a life form that drifts passively with water currents (such as jellyfishes) to a free-livings form. For the latter active unidirectional movement above or underground becomes essential to move towards a target such as a prey, or to move quickly away from a predator. Bilateral symmetry permits streamlining and promotes active movement. Active, directional movement is facilitated by positioning sensory organs at the front end for locating the target, and in further consequence by cerebralization thus favouring the formation of a central nervous system and cephalization.

Targeted movement is, to a certain degree, known from phototactic planula larvae (e.g. *Hydractinia*) and from the sea anemone *Nematostella vectensis*. During and subsequent to metamorphosis this sea anemone glides over sediments, driven by cilia, in search for a place it can settle and burrow into the sediment with its aboral end ahead. In the embryonic development of the anemone the gene *decapentaplegic* (*dpp*) is transiently expressed asymmetrically indicating an asymmetry also found during development of a dorso-ventral polarity (Fig. 22.12). This molecular finding is not so much surprising as morphologists long-since see approaches to biradial/bilateral symmetry in the internal anatomy of some Anthozoa. There is the 6-way or 8-way symmetry in the arrangement of the mesenteries and muscles, and a bisymmetry in ciliated grooves in the pharynx called siphonoglyphs. All these traits can be subsumed under the term **biradial symmetry**, as it is used to describe such traits in Ctenophora.

It must be said that many researchers consider this biradial symmetry and the mere local and transient dorso-ventral gene expressions like those shown in Fig. 22.12 not sufficient to speak of classic bilateral symmetry, as there is no clear, and never a persistent, dorso-ventral asymmetry along the entire body axis.

In this context an old controversial question may be taken up. What is the original adult form of the Cnidaria, is it the polyp or is it the medusa (jellyfish)?

- The polyp party argues: Anthozoa lack a medusa and available molecular data allow classification of this group as basal class. Medusae as produced in Scyphozoa and Hydrozoa are considered to be evolved polyps adapted to pelagic life and to distribute offspring. The striated muscle fibres are interpreted as convergence with reference to molecular data.

- The medusa party argues. In these forms we find advanced structures and organs such as striated muscles the development of which is triggered by the same key genes as in the bilaterians. Likewise, the eyes of many anthomedusae (e.g. *Cladonema radiatum*), and the highly complex eyes of the Cubomedusa *Tripedalia* (see Fig. 22.15B) are not only similar to those of the bilaterians. Their development is turned on by homologous key enzymes (*sine oculis*, see below 22.7). It is almost inconceivable that mere convergence brings about homologous key genes. In the conception of this party polyps would be incompletely developed medusae, a kind of neoteny.

Evolutionary biologists argue following the principle of parsimony: There should be as few steps as possible to reach a defined goal. According to this principle one can assume that striated muscle cells and eyes were among the stock of the common ancestor of the Cnidaria and Bilateria. Beliefs about the organization of the hypothetical Urbilateria (primordial Bilateria) are compiled in Box 22.3.

Box 22.3: How Did the Urbilateria Look Like?

Since 1996 speculations based on molecular data are discussed which, and to which extent, common attributes were shared by the **Urbilateria** (German ur = primordial), and their subsequent descendants. The term Urbilateria refers to the hypothetical last common ancestor of all animals having a bilateral symmetry. Only few of the assumed traits such as microscopic size appear to be beyond dispute.

Taking not only morphological traits but also highly conserved genes into consideration in most elaborated proposals the urbilaterians displayed the following characteristics:

- The urbilaterians possessed an antero-posterior axis subdivided into serial sections through the activity of *Hom/Hox* gene clusters (Fig. 22.9). An early and leading role of Wnt signalling (Fig. 22.6) in directing and patterning the axis is discussed.

- The gastrointestinal tract began with a mouth at the anterior pole and ended with an anus at the posterior pole. These traits, however, are disputed with respect to the Acoela, small in their appearance flatworm-like worms without a gut. Although the position of the Acoela in the tree of life is contentious, many researchers believe them to be basal among the Bilateria. And the Cnidaria and Ctenophora don't have a separate anus. In the model illustrated in Fig. 22.14 it is assumed that mouth and anus derived from a common slit-like blastopore in a fictive urbilaterian called "amphistoma".

- The dorso-ventral polarity was effected by antagonistic morphogens SHORT GASTRULATION (SOG)/CHORDIN versus DPP/BMP-4.

- Near their forward end the Urbilateria possessed a pair of pigment-spot ocelli capable of detecting light; their development was released by *Pax6/eyeless* and *sine oculis* (*six*) genes. These ocelli might have consisted of only one photoreceptor and one pigment cell like the larval ocelli of

the annelid *Platynereis* (Fig. 22.15A). Such eyes might have given rise to compound eyes of the arthropods as well as to camera-type eyes of vertebrates and advanced molluscs (Fig. 22.15). Linked to the eyes was a paired cerebral ganglion that derived from the sensory apical tuft of the larva.

- They possessed smooth and cross-striated muscle cells the development of which was triggered by genes of the *MyoD-(+myogenin)* group.

- Cerebral ganglion and muscles were connected by a subepidermal nerve net. The development of this nervous system was controlled by the proneural genes of the *neurogenin/achaete scute*-complex (Chap. 16).

- The Urbilateria had a blood circulatory system with a contractile vessel surrounded by cardiomyocytes as driving pump. For their development genes were available that still also are involved in the development of our heart. However, a circulatory system is rated as a later attribute of the bilaterians by those researches who consider the Acoela as model closest to the primordial bilaterians for the Acoela lack a circulatory system.

- Perhaps the animals already developed elongated, more or less tapered appendages, a potency that in the further evolution enabled the formation of extremities and other appendages (as shown in Fig. 22.9).

- Whether the Urbilateria already were segmented is a matter of dispute, but common features of segmentation processes (growth zones, oscillating segmentation clocks, Notch/HES as system for demarcation) point to common specialities in the mechanisms that date back to an early common ancestor.

In spite of many indications it is an uncleared issue whether there existed one common primordial type (a "zootype" as it once was called) at all or whether several ▶

Box 22.3 (continued)

metazoans might have developed a bilateral symmetric organization convergently, driven by natural selection in animals moving forwards over a ground. The debate may never come to a definitive, generally accepted theory for no representative has been, or is ever likely to be, identified in the fossil record; and the reconstructed morphology largely depends on whether the Acoela are included within the bilaterian clade. At least currently all reconstructions are non-committal speculations.

Box 22.4: Are Vertebrates and Arthropods Inverted with Respect to Ventral and Dorsal?

Already classic morphologists raised interesting hypotheses. E Geoffroy Saint-Hilaire, one of the founders of the theory of evolution, published in 1822 a treatise in which a longitudinally sectioned lobster was shown upside-down. The illustration was designed to tell the message: when the lobster is inverted, its heart is facing the earth and the central nervous system is facing the sky just like in vertebrates. Thus, the ventral side in arthropods corresponds to the dorsal side of vertebrates (Fig. 22.14). Arising hypotheses to explain animal evolution occasionally postulated an inversion of the dorso-ventral axis in certain taxa. In particular, such hypotheses tried to deduce vertebrates from annelid-like ancestors by assuming that this ancestor turned upside-down (discussed and rejected in, e.g.: Alfred Sherwood Romer: The Vertebrate Body, Saunders 1949, 1955).

This seemingly curious conception of an inversion of the dorso-ventral axis found an unexpected revival based on surprising experimental data raised by molecular developmental biology.

On the one hand, convincing arguments were provided by the expression pattern of homologous genes involved in dorso-ventral patterning of the embryo in *Drosophila* and in *Xenopus*; these genes are expressed in reciprocal patterns: Paradigms are

- the pair **decapentapegic (dpp)** of *Drosophila* and **Bmp-4** of *Xenopus*. *Dpp* as well as *Bmp-4* code for diffusible signalling molecules of the TGF-β family; but *Dpp* is expressed in the *Drosophila* embryo at the rear side along the midline, while the homologous *BMP-4* is expressed in the *Xenopus* embryo predominantly on the ventral side.

- Corresponding reciprocal patterns are found in the expression of the pair **short gastrulation (sog)** of *Drosophila* and **chordin** of *Xenopus*. The genes *sog* as well as *chordin* code for proteins (SOG and CHORDIN, respectively), which trap and bind DPP or BMP-4, respectively. SOG signalling molecules are produced and released at the ventral side of the *Drosophila* embryo, the homologous CHORDIN protein has its source in the Spemann organizer localized in the dorsocaudal region of the *Xenopus* embryo.

On the other hand, experimental evidence showing functional conservation of the molecules further supports this notion: injection of *chordin* mRNA into the *Drosophila* egg causes ventralization, that is, a molecule dorsalising the amphibian embryo (including ectopic CNS formation), promotes ventral structures (including CNS) in the fly.

The inversion hypothesis is supported by the expression pattern of further genes. For example, in Chap. 16 we learned that in the primordial anlage of the CNS three rows of primary neuroblasts are programmed lateral to the midline by three sets of homologous genes. In *Drosophila*, these genes are expressed, of course, ventrally where the ventral nerve cord will arise, in *Xenopus* their expression occurs dorsally in the region of the future brain and spinal cord (see Figs. 16.10 and 16.13).   ▶

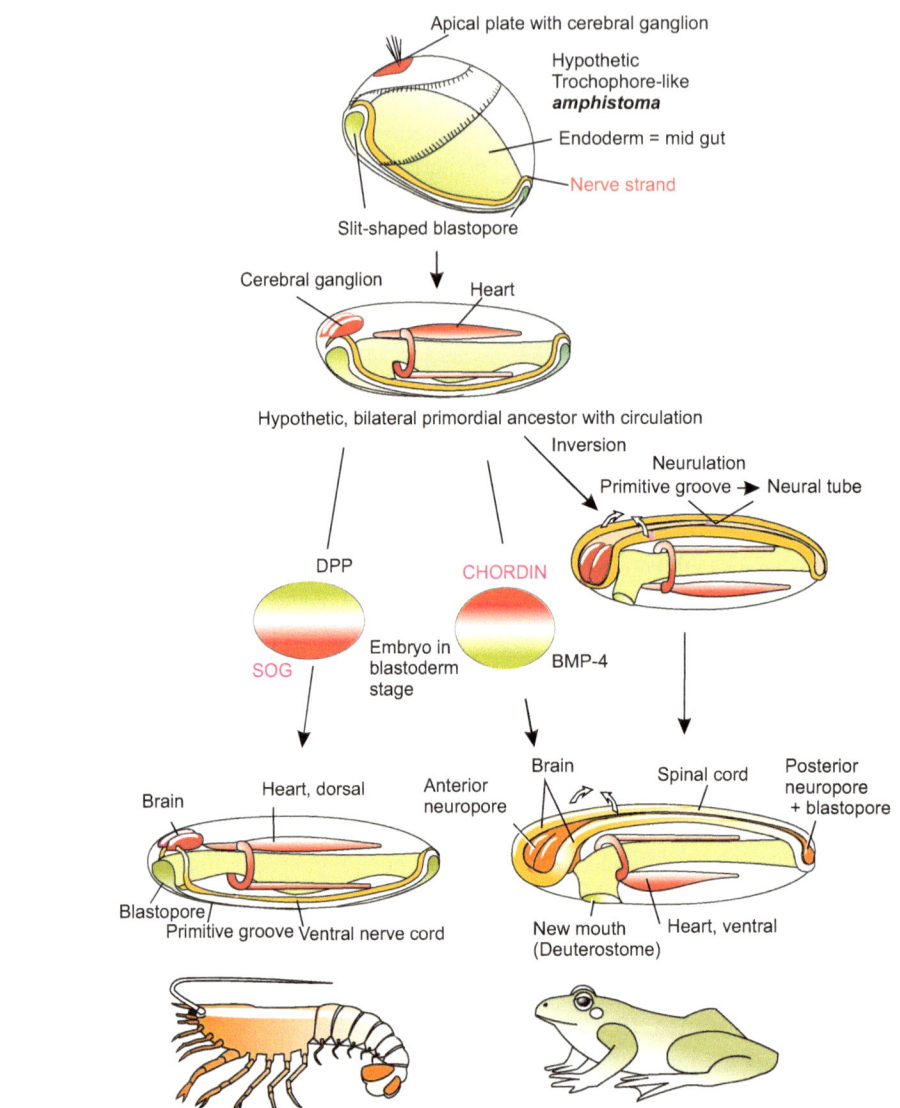

Fig. 22.14 Inversion hypothesis proposed to explain the occurrence of ventral nerve cord and dorsal heart in most invertebrates (here arthropods) but of dorsal spinal cord and ventral heart in vertebrates. After Arendt and Nübler-Jung (1994, 1997), composed by WM

Box 22.4 (continued)

However, there are traits that only with bold fantasy could be integrated into the inversion hypothesis. For example, in arthropods the cerebral ganglion is not positioned ventral but dorsal of the foregut and the optic lobes do not evaginate from the brain as do the eye vesicles in vertebrates, but are formed separately through imaginal discs. On the other hand, both the cerebral ganglion of insects and the brain of vertebrates express *orthodenticle* (in insects) =

Otx2 (in vertebrates) and *ems* (in insects) = *EMX* (in vertebrates) (see Fig. 16.13). The anterior most area of the brains of both groups express *six3* and *rx*.

One may speculate deliberately even in science, and fence with arguments, but one has to make clear that the proposed explanations and ideas are mere hypotheses. As a mere hypothesis to stimulate discussions Fig. 22.14 should be considered which illustrates the inversion hypothesis.

22.6.5 Insect Versus Vertebrate: If Only Because the Central Nervous Systems Show Homologies, the Question Arises: Are the Extremities Homologous, Too?

The highly surprising finding that the central nervous systems of insects and vertebrates show striking congruity with respect to the genes used for its development, and also the spatio-temporal patterns of their expression show extensive and unexpected matches (see Fig. 16.14), gave a boost to search after further cryptic homologies. It also prompted the question whether in the development of extremities homologous genes are involved as well, an idea that in terms of classic morphology would have been considered absurd.

Today no zoologist doubts that the forelimbs of tetrapods arose from the pectoral fins of ancestral fishes and the hindlimbs from the pelvic fin. In accordance with this notion there are several extant fish species that waddle along the bottom with their pectoral fins. And the palaeontologist shows intermediate structures between fins and limbs. The notion that likewise the presumably homologous parapodia of the annelids, the legs of butterfly caterpillars, and the legs of adult arthropods might show homology to our extremities certainly would never have penetrated anybody's mind if molecular biologists had not discovered homeobox genes such as **distalless (dll)** and the *dll*-dependent *aristaless (al)*. Genes with sequence similarity to *distalless* are expressed at the leg tips of all the just listed animals. Loss-of-function mutations lead in *Drosophila* and the mouse as well to mutilated extremities: a defective *dll* clips the tips of the extremities off. A mutation in the *aristaless*-homologous *Alx4* of the mouse causes polydactyly. Apparently, the expression of this gene is needed for correct formation of extremities as they are known today. Yet, it is a necessary but not a sufficient condition. More extensive and precise investigations soon caused disillusion, because *aristaless* is expressed also in the head of the fly, the orthologous *Alx* gene of the mouse also in the skull and in parts of the brain. To top it all, *aristaless* is present also in the legless roundworm *Caenorhabditis elegans*. The situation is similar with *distalless* and its sequence relatives. **Once more we must realize that sequence homology of one or a few genes does NOT straightforwardly imply homology of morphological structures**. What about eyes? In Chap. 12 (Fig. 12.19) the intriguing experimental result was reported, that a mouse gene homologous to the *Drosophila* eye inducer *pax6* induces formation of additional eyes in the fly.

22.7 Evolution of Eyes: Darwin's Dilemma

22.7.1 True Eyes in Headless Jellyfish?

There are medusae, small jellyfishes among the Hydrozoa and Cubozoa, that possess ocelli: light-sensitive organs that do not form images but which can detect light, and are used to determine up from down, responding to sunlight shining on the water's surface. Most ocelli are similar to those found in other invertebrate taxa (Fig. 22.15A–C) joining photoreceptors and pigment cells or they possess photoreceptors with one side pigmented. *Tripedalia* (Fig. 22.15B) among the box jellyfishes (Cubozoa), has $4 \times 4 = 16$ eyes, the related notorious sea wasp *Chironex fleckeri* has $4 \times 6 = 24$ eyes, some of which are capable of discriminating colour, and together making the medusa one of the few creatures to have a 360° view of its environment. Two times four vertically aligned, spherical camera eyes that contain cornea and retina are thought to be connected to a nervous system enabling the four aggregations of nerve cells ('brains') to process images. It is unknown how this works, as these creatures have a unique central nervous system. The eyes are positioned on stalks with heavy crystals on one end, acting like a gyroscope to orient the eyes skyward. They look upward to navigate from roots in mangrove swamps to the open lagoon and back, watching for the mangrove canopy, where they feed.

Are these eyes homologous to ours? Until recently every zoologist would have said: "No"! But let's see.

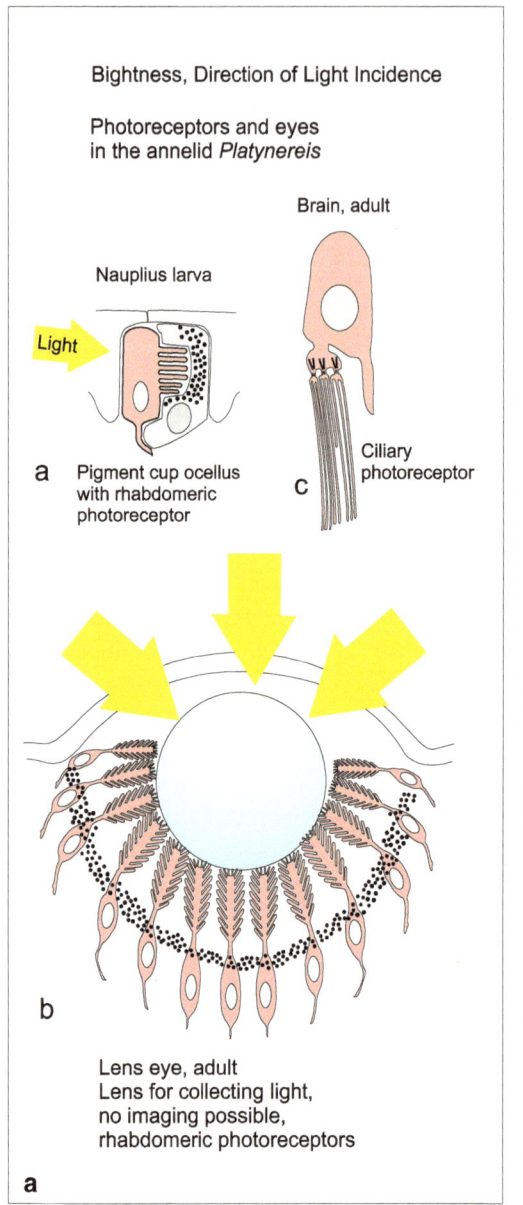

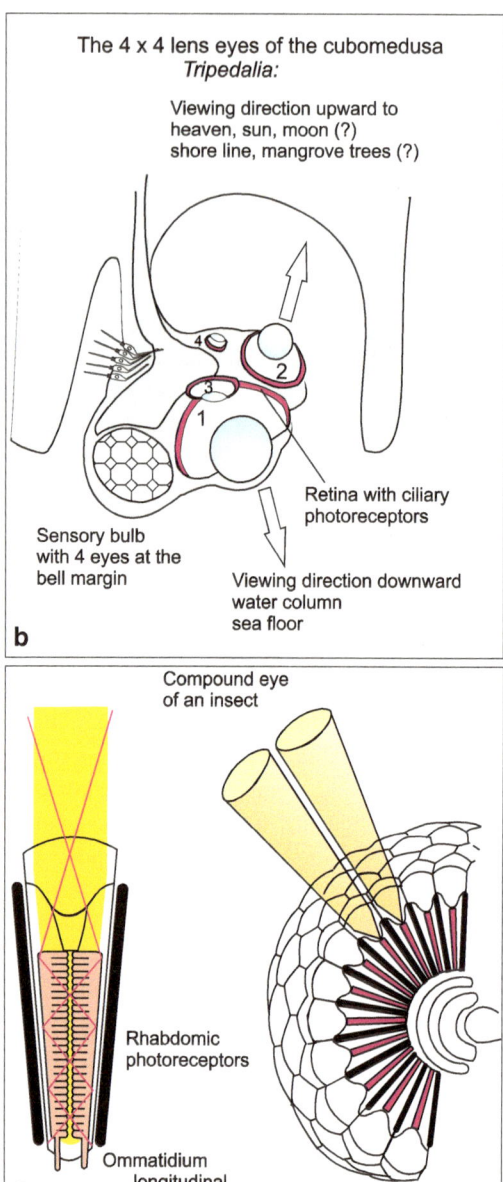

Fig. 22.15 (continued)

22.7.2 Since Darwin in Fervent Dispute: Homology of Eyes or Convergence? Perfection Through Natural Selection or Through an "Intelligent Designer"?

How can evolution in which blind chance prevails create such complex organs in highly different editions and optimize them to perfection? In all major phyla we find simple ocelli as illustrated in Fig. 22.15A, but also advanced eyes fitted with lenses, and eyes capable of projecting an image of the surrounding world onto a retina, or some other devices of imaging like the compound eyes of the arthropods.

Darwin: *"To suppose that the eye, with all its inimitable contrivances for adjusting the focus*

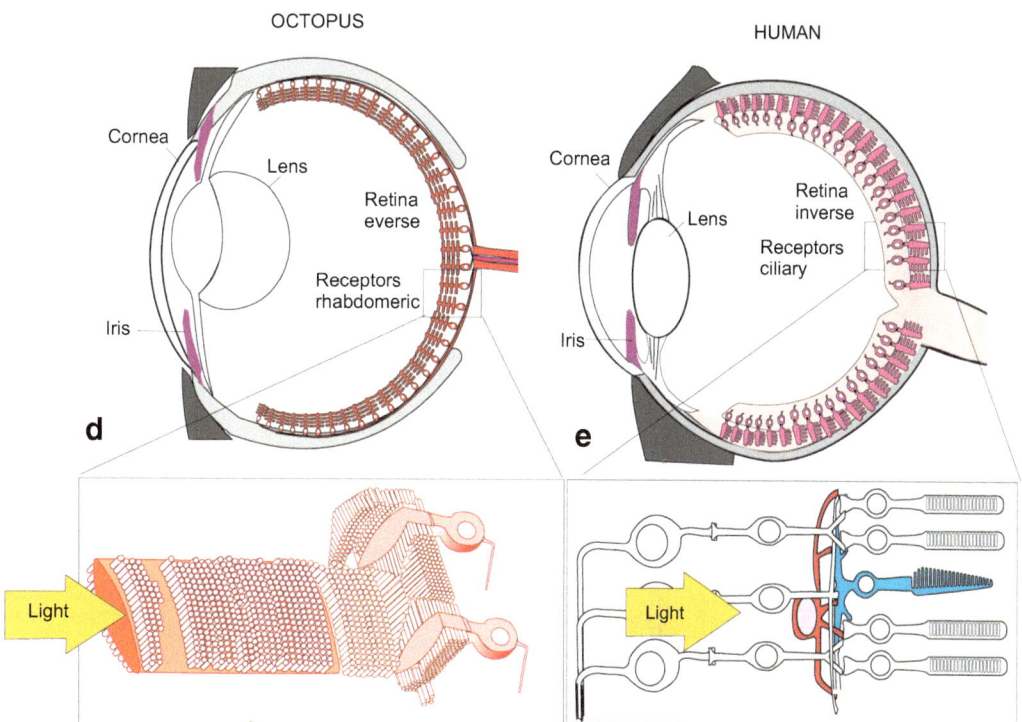

Fig. 22.15 (**A–C**) Selection of various eye types in invertebrates. (**A**) light sensory organs of the polychaete annelid *Platynereis dumerilii* (after Arendt 2003, Arendt et al. 2004). (**B**) The cubomedusa (sea wasp, box jellyfish) *Tripedalia cystophora* possesses 4 sensory bulbs (pedalia, rhopalia) at the margin of the bell with 4 lens eyes and one statocyst each. (**C**) Insect, compound eye. (**D–E**) Eyes of *Octopus* and man in comparison. All pictures from Müller and Frings, Tier- und Humanphysiologie, Springer 2009

to different distances, for admitting different amounts of light, and for the correction of spherical and chromatic aberration, could have been formed by natural selection, seems, I freely confess, absurd in the highest possible degree. Yet reason tells me, that if numerous gradations from a perfect and complex eye to one very imperfect and simple, each grade being useful to its possessor, can be shown to exist; if further, the eye does vary ever so slightly, and the variations be inherited, which is certainly the case; and if any variation or modification in the organ be ever useful to an animal under changing conditions of life, then the difficulty of believing that a perfect and complex eye could be formed by natural selection, though insuperable by our imagination, can hardly be considered real.

And later: *"Although the belief that an organ so perfect as the eye could have been formed by natural selection, is more than enough to stagger any one; yet in the case of any organ, if we know of a long series of gradations in complexity, each good for its possessor, then, under changing conditions of life, there is no logical impossibility in the acquirement of any conceivable degree of perfection through natural selection* (Darwin: On the Origin of Species, 1st edition, Chap. 6.)

Today Darwin would be highly delighted. Irrespective of high diversity in anatomy, development and function, irrespective of the many different solutions to generate aiding structures such as lenses, all organs capable of light perception are based on some fundamental homologies.

22.7.3 Common to All Eyes Is the Light-Absorbing Retinal-Opsin System: And There Is, Beyond all Diversity, a Common Set of Eye-Specific Genes

When one compares the eyes of medusae, insects and vertebrates with all other eyes in the animal kingdom at the biochemical level, one will be aware that all eyes share **rhodopsin visual pigments**. For the production of these pigments *opsin* genes are required; because rhodopsin is a compound consisting of the vitamin A-derivative **retinal** with a 7-transmembrane protein of the **opsin** family (Fig. 22.5). Moreover, for synthesis of retinal and its linkage to rhodopsin, genes are required which encode enzymes and intracellular transport systems.

As primordial origin of light-sensitive cells prokaryotes are discussed. Several representatives of Archaea, Proteobacteria and Cyanobacteria possess light-sensitive pigments fitted with rhodopsin or rhodopsin-like pigments. As endosymbionts, prokaryotes may have brought along their faculty into their eukaryotic hosts.

Moreover, all eyes share components of the signal transduction cascade that is operating after light quanta have been trapped. In fact, opsin genes and genes of signal transduction elements are among the downstream genes of the transcription factor **PAX6** and/or **SINE OCULIS** (genes *so* or *six* in vertebrates). These master genes turn on eye development in organisms as different as medusae, molluscs, annelids, insects and vertebrates. Other genes required for eye development, for example genes for lens proteins or taxon-specific subsidiary proteins, were in the course of evolution adopted by the central control system governed by PAX6 or SINE OCULIS, respectively. Thus in vertebrates the gene for a stress protein, now called α-crystalline, and the genes for several enzymes have acquired binding sites for the transcription factor PAX6, and are now used as lens proteins.

In addition, all eyes appear to share elements of signal transduction. Opsins are associated with G-proteins. Divergence begins in the cell's interior when the signal is transduced to various compartments. Nevertheless an analysis of expressed genes (more precisely of EST's = *expressed sequence tags*), revealed that in spite of the convergence of their overall architecture, the **eyes of *Octopus* and *Homo sapiens* share at least 729 homologous genes**. The hypothetical eyes of the primordial Bilateria have been estimated to use more than 1,000 genes that still are available for higher evolved animals.

The almost perfect construction of certain eyes including ours tempts some people to assume that an intelligent being must have created the eyes, natural selection could not have had such a result. Increasing knowledge about homologies at the molecular level, intermediate steps and increments in the design, but also the knowledge of imperfect solutions in detail (for example inverse arrangement of the photoreceptors in the vertebrate eye, bad optical properties of the dioptric apparatus) make conjectures about "intelligent design" anything but mandatory. On the contrary, data of anatomical, genetic and developmental research corroborate the view that known mechanisms of evolutionary changes eventually can be brought about a multitude of acceptable solutions.

22.7.4 The Light Sensitive Sensory Cells (Photoreceptors) Are Subdivided Into Two Basic Types

Comparison of eye types in the animal kingdom has a long history. Which other organ would exert more fascination than does the eye? Eyes were subjected to structural studies under the electron microscope, to functional studies with the toolkit of electrophysiology, to biochemical analyses, and eventually to the methods of molecular genetics. In these comparative studies, two basic types of photoreceptors came to light. They are encountered among many taxa and in changing combinations (Fig. 22.16). Both types share the property that rhodopsin molecules are inserted into membranes in high density, whereby two different variants of rhodopsin are inserted into two different membrane compartments. In some photoreceptors, that is the rods in our eyes, such membranes, well-

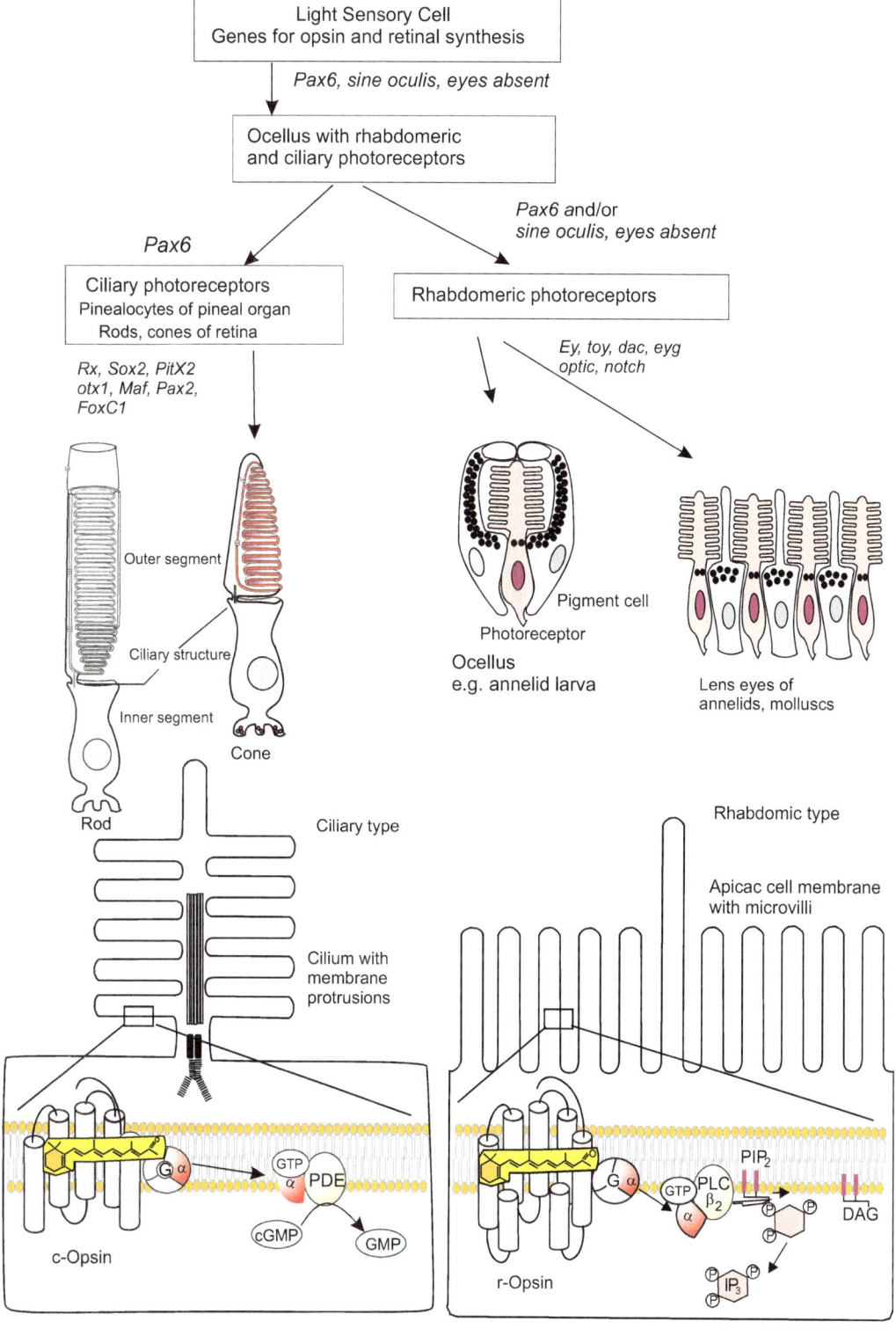

Fig. 22.16 Rhabdomeric and ciliary photoreceptor cells. After Arendt et al. (2004), Purschke et al. (2006)

stocked with rhodopsin, are secondarily folded into the cell's interior.

1. **The ciliary type**. The membrane stocked with rhodopsin is the outer membrane of a cilium. Absorption of light and first steps of signal transduction take place in a cilium, or, in the case of the rods and cones of our retina, in the outer segment of the photoreceptor that derives from a cilium. The protein component of rhodopsin is called **c-opsin**, with c standing for 'ciliary'. Signal transduction within the cell is mediated by cGMP molecules that control the opening state of ion channels. As a rule, the first electrical response is a **hyperpolarization** (enhancement) of the electrical potential (voltage) across the cell membrane.

Besides the rods and cones of the vertebrate eye, also the pinealocytes of the pineal organ (epiphysis) belong to the ciliary type. Ciliary photoreceptors are found in invertebrates, too, for example in the eyes along the margin of the mantle of the clam *Pecten* and in the brain of the annelid *Platynereis* (Fig. 22.15A).

2. **The rhabdomeric type**. It is the most frequent type in invertebrates. The rhodopsin molecules, composed of 'rhabdomeric' **r-opsin** and retinal, are inserted into the membrane of microvilli, which are protrusions of the cell membrane. In the ommatidia of insects a fringe of microvilli extends laterally along the extended retinula cells (Fig. 22.15E). This fringe is called rhabdomere and hence the designation rhabdomeric type. Physiologically rhabdomeric eyes are characterized by a response opposite to that measured in ciliary photoreceptors: Absorption of light causes **depolarization** (lowering) of the electrical membrane potential. Signal transduction from the illuminated rhodopsin down to electrochemical responses is mediated by the second messengers DAG (diacylglycerol) and IP$_3$ (inositol trisphosphate) that are formed by hydrolysis of the membrane lipid PIP$_2$ (see Fig. 11.5).

Photoreceptors of the rhabdomeric type are present in the ocelli of numerous invertebrates. In the simplest design they consist of one photoreceptor and one pigment cell allowing entrance of light only from one side. The fringe of microvilli can be facing the light (**everse type**) or be turned away and facing the pigment cell (**inverse type**). Examples of such simple ocelli are found in the trochophore larvae and juvenile stages of annelids (Fig. 22.15A) but also in the chordate *Branchiostoma* (amphioxus).

Many eyes fitted with light collecting lenses use rhabdomeric photoreceptors, for example eyes of advanced design in medusae, annelids and molluscs. The highest degree of perfection is reached in the eyes of cephalopods (Fig. 22.15D). They enable squids and octopus to perceive images, and are in their overall construction stunningly similar to the vertebrate eye, possessing cornea, lens, iris and retina. The retina, however, is equipped with everse photoreceptors of the rhabdomeric type, whereas the vertebrate eye is equipped with inverse photoreceptors of the ciliary type (Fig. 22.15D, E).

In a few stray instances both the ciliary and the rhabdomeric types are found in one and the same eye, for example in the dorsal eye of the gastropod *Onchidium*. This suggests that the eyes of the primordial bilaterians did possess both types, and subsequently evolutionary divergence alternatively favoured one or the other type. Stunningly, our eye also appears to harbour both types. While rods and cones represent the ciliary type, presumably some of the nerve cells in our retina, namely the photosensitive ganglion cells, the horizontal and amacrine cells, derive from rhabdomeric receptors. Among them are photosensitive cells containing **melanopsin** which fulfils the task to measure day length thus assisting in adjusting our inner clock to the natural light-dark cycle. Arguments used to support their derivation from rhabdomeric receptors are: 1. Melanopsin is an orthologue relative of the r-opsins; 2. All these cells permanently express *Pax6*, as do the rhabdomeric photoreceptors, and in addition ganglionic nerve cells share expression of further transcription factors with photoreceptors.

22.7.5 Eyes Display Partial Homologies and Partial Convergences

For the sake of balance, we ask: to what extent are the various eye types in the animal kingdom homologous, to what extent convergent? We see a patchwork of homologous and convergent traits.

- Homologous and common to all eyes are genes that enable cells to produce the complex compound molecule rhodopsin, to insert it into membranes and to transliterate light signals into changes of the membrane potential.
- Common to all eyes is: that these genes are placed under the control of the key transcription factors PAX6 and/or SINE OCULIS (or of homologous derivatives).
- Many eyes share the rhabdomeric type of photoreceptors, others the ciliary type. What the putative common ancestor might have looked like is a matter of speculation.
- All eyes share shielding pigment cells. Many, perhaps all, of these appear to share pigments composed of polymeric melanin. For the synthesis of this black pigment the amino acids tyrosine or phenylalanine and a set of conserved enzymes are required.
- Most of the further accessory structures with similar functions arose by convergence. In particular, lenses have different origin, be they lenses for singe ocelli, for the ommatidia of compound eyes, or lenses for camera-type eyes. Their different ontogenetic emergence, their different morphological structure in detail, and their different equipment with various lens proteins testify convergent evolutionary origins. As explained above, even the homologous lenses of the vertebrate eyes contain different proteins encoded by non-homologous genes.
- The genes needed for the production of the accessory structures that complete the eye morphologically and functionally, and genes for differing elements in the signal transduction systems within the photoreceptors, have

secondarily been placed under the control of master key transcription factors. Thus, the genes for the various, non homologous lens proteins of the vertebrate eye acquired modified promoters that now can be activated by PAX6. It is even possible to induce ectopic compound eyes in imaginal discs of *Drosophila* with PAX6 of the mouse (see Fig. 12.19).

22.8 On the Evolution of Humans

22.8.1 Quantitatively There Are Only Minor Differences in the Stock of Genes Between Chimpanzee and Humans

Comments and treatises tell that the stocks of classic, protein-encoding genes in humans and chimpanzees coincide up to 99 %, whereby no assertion can be made with regard to the range of individual variation in populations of humans and chimpanzees. After all, it is noticeable that in humans at least **136 duplications of genes** were found that are lacking in chimpanzees. Some of them are of significance in the development of the brain. Also the question of whether larger differences might exist in the huge abundance of non-protein-coding DNA sequences cannot be answered at present. Criteria are lacking to decide which differences might be of significance. After all, the first sequence identified in an arduous work of research and found to exert influence on the development of the brain, called *HAR1 (human accelerated region 1)*, codes for a microRNA and not for a protein. The sequence exhibits 18 alterations compared to that of chimpanzees. Loss-of-function mutations cause flattening of folding of the cerebral cortex. But also protein-coding genes found to be different in humans and other primates exert influence especially on the development of the telencephalon, as explained in the following.

22.8.2 Several Genes Specifically Influence the Growth of the Cerebral Hemispheres

Upright walking, and especially the enormous growth of the brain from 500 ml in *Australopithecus* (and in chimpanzees) up to1,300–1,600 ml in *Homo sapiens* are the most important anatomical novelties in the evolution of human beings. The enlargement of the brain volume is mainly due to the extraordinary enlargement of the cerebral cortex (*cortex cerebri*). Are there correlations in the stock of genes between great apes and humans – or between gene mutations and pathological failures?

The decoding of the human and chimpanzee genomes was heralded as an opportunity to truly understand how changes in DNA resulted in the evolution of our cognitive features. However, in spite of much detective work the functional consequences of such changes have proved elusive, with a few exceptions. Survey and scanning of the human genome in search for correlations between atypical genetic constitutions and joined findings of pathology and human genetics have recently (from about 2006) yielded only a few results but some of these are stunning.

From linked studies it was known that a gene designated **SRGAP2** is involved in brain development and that humans possess at least three similar copies of the gene, whereas non-human primates carry only one. The gene is involved in the formation of **dendritic spines**, that is in the multitude of synaptic structures found along dendritic processes of nerve cells in the brain (Fig. 22.17a). A multitude of spines permits entrance of a mass of information delivered by other nerve cells. Previous studies had documented the fact that humans differentiate greater numbers and higher densities of dendritic spines than other primates and rodents. The ancestral mammalian *SRGAP2A* promotes the maturation of spines and slows down the migration of neurons within the developing cerebral cortex, whereas the additional new human-specific *SRGAP2C* has opposite effects and so favours the formation of further spines. These

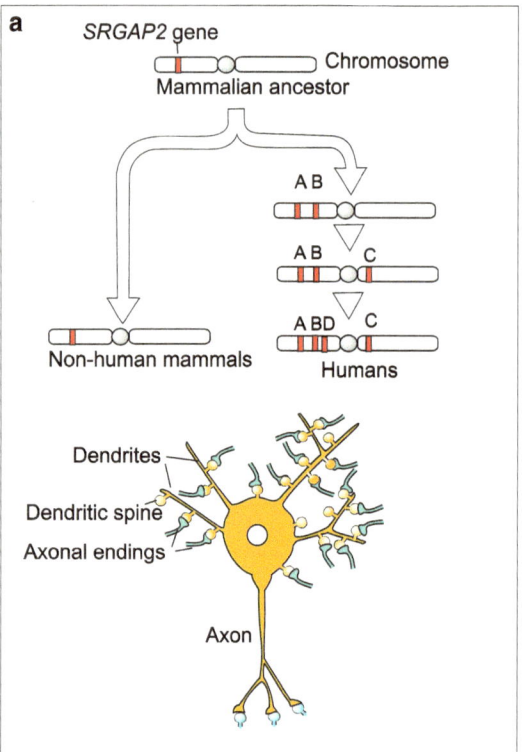

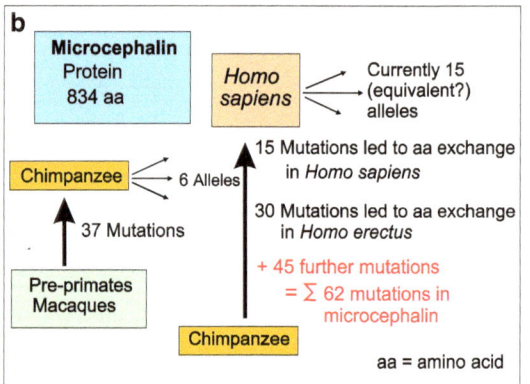

Fig. 22.17 (**a**, **b**) Effects of gene duplication on human brain development. (**a**) Two fold duplication of the gene *SRGAP2* yielded a variant *C* that leads to more dendritic spines and thus to more synapses in the human brain compared to other primates. After Geschwind et al. (2012). (**b**) Evolution of *microcephalin*, a gene essential for the enormous growth of the human forebrain. According to calculations based on mutation rates and the time period needed 37 mutations in the *microcephalin* gene appeared since the divergence of macaques and chimpanzee, and further 45 mutations in the human *microcephalin* gene since the separation of chimpanzee and human from their common ancestor. The mutations effected enhance growth of neuronal precursors. In recent humans up to now 15 allelic variants have been detected. After Evans et al. (2004, 2005)

results suggest that the emergence of *SRGAP2C* could have contributed to expansion of the cortex and increase in spine numbers in human ancestors.

With respect to the **growth of the cerebral cortex** another set of duplicated genes has even more dramatic effects. Six protein-encoding genes were found of fundamental significance for the development of the cerebral cortex, called *ASPM, CDK5RAP2, CENPJ, MCPH1, MCPH2, MCPH4*. We focus on ***MCPH1***, also called ***microcephalin*** as an example (Fig. 22.17). Loss-of-function mutations in both alleles, in the allele inherited from the father and in that inherited from the mother, leads to **microcephaly** (Greek: *micros* = small, *kephalē* = head). In particular the size of the cerebral cortex is reduced and sinks down to a volume of 650 ml, the volume characteristic for early members of *Homo erectus*. The same applies for the other genes listed above. The role of the normal form of those genes in development must still be figured out.

Understandably, it was tried to correlate various allelic variants with mental abilities and the frequency of failures in the mental world. Convincing indications were not found, but a completely unexpected finding. MCPH1 protein is found in the centrosomes of cells capable of dividing in a certain phase of the cell cycle, that is, the protein is associated with elementary cell physiological events and not with specifically human brain functions. Like most, if not all genes, this gene is utilized for various purposes, it may even have a function in the repair of DNA strand breaks. Its impact on the size of the brain is considered to derive from its role in the control of proliferation of neuronal stem cells. Prolonged expression of the intact *microcephalin* causes prolonged growth: Great events often come from little causes.

Final Remark
The evolution of developmental paths as they can be reconstructed today discloses many constraints which set limits to further restructuring and elaboration of **the arduously gained achievements. This tenacity is confronted with the necessity of all organisms to create novelties again and again, to comply with the challenges imposed by an ever changing world and to defend the achievements against increasing competition. With the creation of transgenic organisms humans begin to interfere with natural processes. What will be the impact for the future evolution?**

Summary

"Evo-Devo" (evolution of/and development) designates a new concept of research aimed at deducing the evolution of body plans from data provided by comparative morphology and developmental genetics, and aimed at deducing evolutionary relationships among the organisms. A particular goal is the reconstruction of the ancestral animal body plan and genetic tool kit of primordial forms. Evo-devo research also tries to reveal correlations between changes in morphology and changes in the genome.

Highlights in the reconstruction of ancestral and fundamental body plans based on comparative embryology were compiled in Chap. 6 (Human). Irrespective of their highly diverging initial and late stages, all vertebrates pass a common **phylotypic stage** in the middle phase of their embryonic development. In this phase all vertebrates display a common fundamental body plan. Conserved characteristics are: dorsal neural tube, notochord, somites, branchial pharynx with gill pouches, cartilaginous gill arches, 4–6 gill arteries, and ventral heart. Before and after this stage the embryos are very different corresponding to long-term adaptation to their habitat.

The question of **homology** is discussed at the levels of morphology, biochemistry and molecular genetics. It is pointed out that morphological and genomic homologies are not always coincident. The construction of morphologically homologous structures, for example, the formation of an initially cartilaginous and subsequently ossified skeletal element, requires different proteins. For the differentiation of eye lenses in vertebrates various proteins are suited, and newly recruited proteins can replace traditional ones allowing adaptation to changing challenges.

In constructing basic body plans evolution uses a set of conserved genes as a **toolkit**. Mechanisms by which **new genetic programs** can arise are **gene duplication, genome duplication, and recombination of modular gene segments**. In duplicated DNA sequences mutations in the duplicated copy can bring about modified or new functions while in the other copy the present function remains preserved.

Occasionally mutations reveal **concealed former developmental programs**, thus causing atavism: Flies develop four wings, platy fishes a sword like the closely related swordtails.

By comparing homologous genes discrimination must be made between **orthologous** and **paralogous** genes. **Orthologous genes**, like those for BMP (in *Xenopus*) and DPP (in *Drosophila*), are **found in different species** whereas **paralogous genes**, for example the members of the *Hom/Hox* clusters, are multiplied homologous genes once arisen from (repeated) gene duplication and are **found in one and the same individual**. This parallels ortholog**ous** homologous organs (e.g. forelimbs and wings in vertebrates) and paralog**ous** organs (e.g. vertebral bodies, forelimbs and hindlimbs in the same individual).

Conserved developmental stages are found in many taxa, e.g. spiral cleavage, trochophore or trochophore-like larvae in the superphylum **Lophotrochozoa** (Annelida, Mollusca, Tentaculata and several other phyla). The two

other superphyla, the **Ecdysozoa** and the **Deuterostomia,** are characterized by similar postembryonic properties. Tentatively, on the basis of molecular genetic data all three major groups are deduced from a hypothetical ancestral basis group, called **Urbilateria (primordial Bilateria)**.

A keen hypothesis assumes, **arthropods and vertebrates**, though being members of highly divergent phyla, would share a basic common body plan, save that the dorsal side of vertebrates with brain and spinal cord would correspond the ventral side of the arthropods with their ventral nerve cord. In fact, the central nervous systems of both groups display stunning homologies at the level of gene activities, and the **inversion hypothesis** is supported by the expression pattern of the signalling molecules BMP-4 at the ventro-anterior side of the *Xenopus* embryo, and the homologous DPP at the dorsal side of the *Drosophila* embryo.

Discussed is also the transition of **diploblast**s (ecto- + endoderm) to **triploblasts** (ecto- + endo- + mesoderm), the divergence of **radial symmetry** and **bilateral symmetry,** and the question whether the **main body axis** in the various phyla can be equated with one another. In most phyla**,** that is in vertebrates, in other deuterostomes (e.g. sea urchins) and in cnidarians, the orientation and patterning of the main body axis is directed by the canonical **WNT/beta-catenin system** (summarized in Fig. 22.6). It is pointed out that in Cnidaria the site of gastrulation (blastopore) is opposite to the site of gastrulation in other phyla (summarized in Fig. 22.14).

In comparing the various **eye types** existing in the animal kingdom it is conspicuous that irrespective of the often highly divergent construction, unexpected homologies occur at the level of cell types and the genes needed to differentiate the photoreceptors. All eyes share genes for the production of the common visual pigment **rhodopsin**, and genes required for signal transduction and to translate light signals into signals based on modulation of the electric membrane potential. These genes are placed under the control of the key transcription factors **PAX6** and/or **SINE OCULIS**. In their internal organization the **photoreceptors** are subdivided into two subtypes, the **rhabdomeric type**, which dominates in invertebrates, and the **ciliary type**, which includes the **rods and cones** of our retina. There are indications that the **melanopsin-containing retinal ganglionic cells** that are involved in the regulation of circadian rhythms, derive from rhabdomeric photoreceptors. Additional structures of the various eye types such as lenses developed several times independently and contain a large array of various proteins.

In the development of the **human cerebral cortex** a significant influence has been traced for a certain group of duplicated genes. They arose by duplication of genes which existed in other primates in only a single copy, and appear to promote the formation of synapse proliferation of neuronal stem cells. **Microcephalin** is presented as an example.

The developmental paths of the various organism are subjected to constraints that set limits to future transformations and elaboration of the achievements. On the other hand, organisms also are inventive in that they create new properties by modification of duplicated genes and structures and by persisting recombinations of inherited genetic components.

Glossary

Acrosomal reaction Rupture of the acrosomal vesicle releasing enzymes to enable penetration of the sperm through the egg envelope

Acrosome Vesicle located at the sperm tip filled with enzymes to dissolve the egg's envelope

AER Apical Ectodermal Ridge: edge of the avian wing bud, producer of growth factors

Aggregate Cluster of previously separated cells

Allantois Embryonic urinary bladder, in embryos of reptiles and birds also serving as respiratory organ, in humans rudimentary and integrated into the umbilical cord

Allel(es) Alternative forms (variants) of a gene once in the history of life arisen from an ancestor gene by mutation(s); various alleles occupy the same locus on chromosomes and differ, as a rule, by a few base pairs only. Animals have, as a rule, two alleles, one on the respective chromosome inherited by the mother, the other on the corresponding (homologous) chromosome inherited by the father; the allele most frequently found in a population is called wild type allele.

Allograft Transplantated tissue (graft) taken from a genetically different donor, likely provoking an immune response

AMDF Anti-Müllerian Duct Factor; synonymous: *AMH* Anti-Müllerian (duct) Hormone, or *MIS* Müllerian (duct) Inhibiting Substance. Hormonal factor causing reduction of the Müllerian duct (precursor of the oviduct) in the male sex, produced by Sertoli cells of the testis

AMH See AMDF

Amniocentesis Samplingof cells from the fluid in the amnion for screening of chromosomal anomalies

Amnion Membrane consisting of an ectodermal and mesodermal layer enclosing waters, protecting the embryo; formed by extra-embryonic parts of the early germ, the trophoblast.

Amniotes Group of vertebrate animals that, in embryogenesis, form an amnion and allantois, actually the land vertebrates (reptiles, birds, mammals).

Angiogenesis Original meaning: formation of all kinds of blood vessels; present use: formation of blood vessel branches from existing vessels. See vasculogenesis by contrast.

Animal hemisphere The upper hemisphere of the egg cell, contains, as a rule, yolk-free cytoplasm and the nucleus; hemisphere that in vertebrates and many other animals (also) gives rise to the nervous system considered to be a typically animal organ.

Animal pole The pole of an oocyte or egg cell that usually is shown as uppermost pole ('north pole') and where in meiosis the polar bodies are given off

Animalization Treatment of embryos leading to oversized animal body parts at the expense of the gastrointestinal tract.

Anlage Primordium, rudiment: embryonic, still undifferentiated precursor are or structure of complex structure or an organ

Antisense Sequence of single-stranded DNA or RNA complementary to a target sequence and therefore attaching to (hybridizing with) it

Apical Upper side, mostly of epithelial cells, the side opposite the basal lamina and often facing a free space (lumen), also top of a structure

Apoptosis Programmed cell death

W.A. Mueller et al., *Development and Reproduction in Humans and Animal Model Species*, DOI 10.1007/978-3-662-43784-1, © Springer-Verlag Berlin Heidelberg 2015

Area opaca Region of the avian blastodisc surrounding the embryo and expanding enclosing the yellow yolk to form the yolk sac

Archenteron Hollow primordium of the gastrointestinal tract in vertebrates, formed in gastrulation

Aster Configuration of microtubules which serve to separate and distribute the duplicated chromosomes in cell division

Autologous Referring to the same, or genetically identical (cloned) individual; for example autologous transplantation = transplantation within the same individual (donor and recipient identical); antonym (opposite in meaning) to heterologous

Autosomes All chromosomes except the sex chromosomes

Axial organs Embryonic organs or primordia located along the longitudinal body axis of vertebrates below the back line, that is neural tube, notochord, and the two rows of somites on both sides of the notochord

Basal (1) Being at the base of epithelia adjacent to the basal lamina, or (2) being at the base of an evolutionary branch

Bauplan German term meaning construction plan, mostly translated as body plan

bFGF Basic Fibroblast Growth Factor

bHLH Basic Helix-Loop-Helix domain (motif), a type of DNA-binding domain of transcription factors

Blastema Bud-like accumulation of dedifferentiated cells and/or stem cells forming at the wound of an amputated animal (e.g. planarians) or limb; starting material for regeneration

Blastocoel Inner, fluid-filled cavity of the early embryo, typically of the ball-shaped blastula, also called primary body cavity

Blastocyst Vesicular early mammalian embryo filled with fluid, and an inner cell mass as precursor of the embryo, and encased by an outer extraembryonic cellular wall called trophectoderm. The blastocyst is ready to implant into the wall of the uterus. Not to be confused with blastula of non-mammals (see Chapter 6)

Blastoderm Cellular outer sheet of invertebrate embryos after cleavage; precursor of the embryo

Blastodisc A flat cellular disc that becomes multi-layered during gastrulation by cells descending from the epiblast through a primitive groove forming a hyopoblast, and in the middle of which the embryo arises. While in fishes, reptiles and birds the blastodisc lies on top of the yolk-rich egg, in the embryogenesis of mammals the blastodisc is identical with the floor of the amniotic cavity.

Blastomeres First, initially relatively large cells of the embryo arising from cleavage divisions

Blastopore Primordial mouth through which cells enter the interior of the embryo during gastrulation; can give rise to the definitive mouth (protostomes) or to the anus (deuterostomes).

blastula Early ball-shaped stage of many invertebrate embryos, arisen from the cleaving egg cell, consisting of cellular wall, the blastoderm, enclosing a fluid. Found in sea urchins and amphibians, not in *Drosophila*, fishes, reptiles, birds, and not in mammals (see blastocyst).

BMP Bone Morphogenetic Protein – protein or protein-family promoting growth of bones in foetal and postnatal life, acting as a morphogen in the early embryo of mammals and many other, including (with other name such as DpP) invertebrate animals

Branchial Referring to the pharynx region (pharyngeal region) in mammals

Branchial arches See pharyneal arches

Branchial pouches See pharyngeal pouches

Branchiogenic organs Organs arisen from structures from the branchial arches or walls of the gill pouches, respectively, of the embryonic pharynx region, comprising thymus and parathyroid gland

Budding In developmental biology a type of asexual reproduction by means of a multicellular, detaching bud

Capacitation A process enabling the sperm to fuse with an egg cell

Carcinoma Cancer tissue derived from an epithelium

Caudal Being at, or close to, the tail, or direction pointing to the tail, often equivalent to posterior

Cell cycle Cyclic events in which the cell duplicates its chromosomes and divides in two. It normally consists of four phases G1, S, G2 and mitosis. DNA synthesis occurs in S phase and cell division occurs in M (mitosis) phase, giving two daughter cells.

Cell lineage Line or tree of cells, derived from a common precursor cell. The term refers to the developmental history of cells, where they come from.

Centromere Region of the chromosome where in cell division the spindle fibres attach to transport the previously duplicated chromosomes into two sister cells

Cerebral Referring to the brain, or developing brain

Chemokine Signalling molecules activating white blood cells (lymphocytes, macrophages, granulocytes)

Chimaera (chimera) Organism consisting of genetically different cells, for example consisting of cells arisen from embryos derived from parents that differ genetically, for instance from parents of different species

Chorda dorsalis See notochord

Chorion Envelope of the embryo; in insects the non-cellular egg shell, in land vertebrates the cellular, extraembryonic envelope which not only surrounds the embryo but also the amniotic cavity and which develops the placenta in placental mammals

Chromatin Chromosomal material; complex of DNA and proteins

Clade A group of organisms consisting of an ancestor and all its descendants, a single branch on the genealogical tree, also called a monophyletic group

Cladistic Referring to a method or approach to group organisms according to their putative genealogical descent, constructing dendograms, also known as phylogenetic systematics

Cleavage The first cell divisions a fertilized egg cell undergoes leading to smaller and smaller cells

Clone A population of genetically identical cells or individuals

Cloning (1) Procedure to produce multiple cells or organisms with identical genetic makeup; (2) the term cloning is also used to designate procedures to multiply sequences of DNA leading to identical copies.

CNS Central Nervous System comprising the brain and the spinal cord, or, in insects, the brain (supraintestinal = supraesophageal ganglion) and the ventral nerve cord

Coelom Body cavity lined by an epithelium (e.g. the mesodermal peritoneum)

Commitment Programming (determination) of cells to follow a distinct, predetermined developmental pathway, for example to become a nerve cell

Compaction Strong cohesion of the blastomeres in the 8 to 32 cell stage mammalian embryo (species-dependent) effected by newly produced cell adhesion molecules (uvomorulin = E-cadherin)

Compartment In developmental biology as a rule an area or space occupied by a clone of cells that is by the descendants of a single founder cell

Competence Ability to enter a distinct developmental path and to respond to inducing signals

Conceptus A mammalian embryo including the extraembryonic structures such as the placenta, derived from the fertilized egg

Cortex Outer cytoplasmic layer of a cell, especially of an egg cell, or outer layer of an organ such as adrenal gland or forebrain

Cortical reaction, cortical rotation Reaction of the egg cortex upon contact with the sperm

Cranial Referring to the skull or head, or direction pointing to the head, often equivalent to anterior

Cre-lox system Molecular tool for controlled genetic manipulation, especially for introducing a gene into the DNA of a distinct cell type and subsequent excision of the gene vector; comprises the recombinase enzyme

Cre that recognizes lox sequences previously introduced into the DNA

Crossing-over Exchange of segments between homologous chromosomes, that is between maternal and paternal chromosomes carrying the same genes but frequently in form of different alleles; occurs during meiosis

CSF Colony-Stimulating Factor(s) – growth factors stimulating multiplication of the blood-forming stem cells and their derivatives. Eponymous is their division-stimulating effect on precursor cell colonies on agar plates

Cumulus Cells surrounding the mammalian egg cell derived from the follicle

Cuticula A non-cellular, chitin-containing sheet covering the body of animals, in particular the body of the Ecdysozoa including the arthropods and nematodes

Cyclins Proteins involved in regulation of the cell cycle; their concentration rises and falls in the rhythm of cell division.

Cytoplasmic determinants Localized components of the cytoplasm of an egg cell that assign a certain fate to the descendants to which they are allocated; typically containing maternal mRNA or protein deposited in the egg at distinct locations

Decidua (deciduum) Part (endometrium) of the pregnant uterus that is shed at parturition

Dedifferentiation Backward development of differentiated cells to an embryonic state, also called reembryonalization

Dermatome Cells of the somites that emigrate to form the dermis

Dermis The layer of connective tissue below the epidermis of the skin

Determination Selection, programming and fixation of the developmental fate of a cell or a group of cells; if single cells are concerned the term commitment is recently displacing the term determination.

Deuterostomia Group of animal phyla where the blastopore gives rise to the anus while the mouth is secondarily formed at a distant anterior site; comprises the Echinodermata, Urochordata = Tunicata, and Chordata including the vertebrates

Diapause stage of developmental quiescence

DIF Differentiation-Inducing Factor in *Dictyostelium*

Differentiation (1) Divergence of developmental paths, becoming different compared to other cells or body parts, (2) acquisition of the molecular components and shape of a mature, functional cell type, also called cell maturation

Diploblastic In embryogenesis of diploblastic animals only two cellular layers occur, ectoderm and endoderm but no mesoderm, actually these are the phyla Cnidaria and Ctenophora.

Diploid A two-fold set of chromosomes, the one coming from the egg cell, the other from the sperm

Distal More distant from the main body (trunk), close to, or at, the tip (e.g. of an extremity)

Dominant-negative mutation The protein derived from the defective allele blocks the function even in the presence of a wild-type allele. Occurs frequently when receptors or transcription factors consist of dimers, the one partner of the dimer derived from the wild type allele but the other from the defective allele making the heterodimer non functional.

Dorsal The upper (back) side of an animal, or direction pointing to the back

DPP Decapentaplegic – morphogen in the *Drosophila* embryo

dsRNA Double-stranded RNA

Dysplasia Abnormal pattern of growth and differentiation

Ecdysis Moult in insects and other arthropods (and in nematodes), shedding of the cuticule

Ecdysone, Ecdysterone Steroid hormones releasing moult in insects and other arthropods

Ecdysozoa Group of animal phyla whose body is covered by a non-cellular envelope consisting of chitin and protein. Growth depends on moults.

Ectoblast Synonymous to ectoderm

Ectoderm Outer cellular layer of a gastrula, becoming the epidermis of the skin after having segregated the nervous system and some sensory cells

Ectopic Outside the normal position

EGF Epidermal Growth Factor – factor stimulating cell division of epidermal cells, in the foetal and postnatal life of stem cells in the stratum germinativum of the skin and in stem cells of the hair follicles and sweat glands

Embryoblast, inner cell mass Part of the mammalian blastocyst that gives rise to the embryo proper

Embryonic shield Nodular thickening at the posterior end of the primitive streak in fishes, homologous, and functionally equivalent, to the Spemann organizer in the amphibian embryo and to Hensen's node on the avian blastodisc

Embryonic stem cells, ES-cells Pluripotent cells capable of dividing and taken from the inner cell mass of a mammalian blastocyst

Endocrine Referring to the source of a hormonal substance

Endoderm Germ layer, as a rule in form of the archenteron of the gastrula, that eventually gives rise to the lining of the gastrointestinal tract and some associated organs such as liver and pancreas.

Endometrium Cellular lining of the uterus wall enclosing the uterine cavity and comprising uterine glands; shed in menstruation but grows to a thick layer in pregnancy, is commonly said to give rise to the placenta which, however, essentially derives from the fetus.

Endothelium Cellular (epithelial) lining of blood vessels

Enhancer Region of DNA upstream, sometimes downstream, of a gene, where regulatory proteins controlling expression bind

ENS, Enteric Nervous System Multilayered nervous system accompanying the gastrointestinal tract; comprising the Auerbach plexus and the Meissner plexus

Enterocoel Lining of the coelomic cavity originating from vesicles detaching from the archenteron

Ependyma Cellular layer lining the cerebral vesicles and the central spinal cord canal, comprising stem cells for nerve and glial cells

Epiblast Upper cell layer of the avian or mammalian blastoderm (in fish of the germ shield), equivalent to ectoderm after having segregated cells that emigrate through the primordial groove to form the mesoderm

Epiboly Expansion of the outer cell sheet (blastoderm) over the yolk mass or over yolk-rich cells which thus are shifted into the interior; is observed in gastrulation of amphibians and in early fish development.

Epidermis Outer epithelial layer covering an animal, in vertebrates outer layer of living cells of the skin

Epigenesis (1) Original meaning (term coined by Aristotle; Greek: *epi* = on, onto, upon; *genesis* = emergence, generation) formation of a shape, actually the embryo, from unformed material, bottom-up development leading to higher levels of organization

Epigenetic (1) Original meaning: bottom-up development, increase in complexity; (2) present meaning: secondary, enzymatically effected modification of DNA and chromatin proteins through methylation, acetylation or phosphorylation; these modifications influence accessibility of genes (opening or silencing).

Epiphysis (1) The pineal gland of the diencephalon, (2) terminal, growing part of long bones

Epithelium (plural epithelia) Two-dimensional sheet of tightly connected cells, facing with their apical side a lumen or the exterior space and lying with their basal side on a basement membrane. In arthropods (and other animals) epithelia may be covered with a non-cellular cuticule.

ES-cells Pluripotent Embryonic Stem cells; in mammals taken from the inner cell mass of the blastocyst, and multiplied in culture vessels

Euchromatin Decondensed parts of chromosomes containing genes accessible for transcription

Expression (of a gene) Fruition of genetic information, synthesis of a protein using the information of a gene, actually transcription + translation

Extracellular matrix ECM Padding material filling the space between cells and containing macromolecules such as laminin, fibronectin and collagens

Extraembryoic membranes/organs Membranous structures or organs such as amnion, chorion and placenta which arise from the fertilized egg but are not part of the embryo proper

FACS Fluoresence-activated cell sorting: method to separate cell types onto which antibodies linked with fluorescent dyes have been attached

Fascicle Bundles of nerve fibres attached to each other by means of cell adhesion molecules

Fate map Topographic projection of future body parts and organs onto an early stage of embryonic development, the resulting diagram does not mean that the fate of such regions is already fixed

Fertilization Fusion of a sperm with an egg cell; in strict sense fusion of the haploid sperm nucleus (male pronucleus) with the haploid egg nucleus (female pronucleus) resulting in the diploid nucleus of the zygote

Fetus, Foetus From Latin *fetus* (therefore the etymologically correct original spelling is fetus and is now the standard English spelling throughout the world in medical journals); a fetus is an embryo in the late phase of prenatal development. In humans, the fetal stage commences two months after fertilization; the embryo is typically 30 mm in length and limbs can be seen

FGF Fibroblast Growth Factor – protein family; in the early embryo FGFs act as inducing factors and/or morphogens, in fetal and postnatal life stimulating proliferation of fibroblasts and, thus, stimulating growth of connective tissue

Floor plate Ventral part of the spinal cord along the midline

Floxed A DNA site flanked by loxP sequences enabling Cre-mediated recombination

Follicle Cellular envelope of the mammalian growing egg cell (oocyte)

FSH Follicle-Stimulating Hormone – hormone stimulating maturation of the follicle in the ovary including maturation of the egg cell; in addition stimulating production of oestrogenic hormones (oestradiol, oestrone) by the follicle

Gain of function A mutation causing enhanced function or conferring additional functions to a gene

Gametes Haploid germ cells, that is egg cells or sperm, as a rule used for germ cells that are released into the exterior water

Gametogenesis Development of gametes

Ganglion Local aggregation of nerve cell bodies

Gap genes Class of genes that cause, when defective, gaps in the body architecture in *Drosophila* and other arthropods

Gap junctions Channels between adjacent cells that allow passage of low molecular weight substances

Gastrula Early embryo undergoing gastrulation, a stage in which cells are shifted into the interior of the embryo and the germ layers segregated

Gastrulation Process through which cells are shifted into the interior of the embryo to enable the formation of inner organs. A first step is the segregation of the germ layers ectoderm, endoderm, and (in most animals) mesoderm. Term derived from Greek *gaster* = stomach, as a representative inner organ

G-CSF Granulocyte Colony-Stimulating Factor

Gene expression Realisation of the information contained in the gene; actually the formation of mRNA along the gene with transfer of the information of the gene to the mRNA (transcription) plus the formation of the protein in the cytoplasm using this information (translation).

Gene Section of the DNA in the chromosomes bearing the entire DNA sequence (dictating in which sequence amino acids have to be joined) to produce a certain polypeptide (protein or peptide), or in which sequence nucleotides (bases) have to be joined to produce a certain RNA (ribosomal RNA, micro RNA)

Generative Referring to reproduction; for example: generative cells = cells serving reproduction, in contrast to somatic cells.

Genital ridges Bands of mesoderm in the embryo that will form the gonads

Genome Totality of genomic information, sum of the genes an organism possesses.

Genotype Particular genetic outfit of an individual, sum of its genes, in contrast to its phenotype, its appearance, which is the sum of its expressed features

Germ band Band-shaped blastoderm along the ventral side in the insect embryo that in gastrulation is shifted into the interior; term also used to designate the early embryo in fishes

Germ layers Cell sheets or cell packets in the embryo that become separated from each other at gastrulation and will give rise to different organs in the subsequent development. Traditionally two or three germ layers are distinguished: outer ectoderm, innermost endoderm, and mesoderm in between. Animals developing in embryogenesis only ecto- and endoderm are classified as **diplobastic** (Cnidaria, Ctenophora), those which develop all three layers are called **triploblasic** (all more evolved animals)

Germ line The cell lineage leading from the fertilized egg to the primordial germ cells and eventually to the gametes

Germ plasm Cytoplasmic determinant in the egg of several animals (nematodes, insects, amphibians) causing the cells to which it is allocated to become cells of the germ line and eventually primordial germ cells. Contains mRNA of maternal origin. (Term coined by A. Weismann with a different meaning, designating the totality of components bearing genetic information)

Germ ring The thickened margin of the fish embryo surrounding the yolk ball

Germinal vesicle Traditional designation of the large nucleus in oocytes prior to meiosis

GFP Green-Fluorescent Protein: fluorescent protein derived from jellyfish, often artificially attached to a cell's own proteins and used as reporter

GM-CSF Granulocyte/Macrophage Colony-Stimulating Factor

Gonad Testis or ovary

Gradient (1) General meaning: quantifiable property (value on the Y-axis of a diagram), which decreases or increases continuously along a length (X-axis); (2) special meaning: continuously decreasing or increasing concentration of a morphogen along a length

Graafian follicle Follicle, i.e. cells surrounding the growing oocyte, in mammals

Graft A piece of tissue transplanted to another place within the same individual (autologous graft) or into another individual (heterologous graft).

Gray (grey) crescent Region in the amphibian blastula where gastrulation commences and that in some species contains less black pigments than has the upper, black hemisphere of the embryo

Growth cone Leading, mobile structure at the tip of growing nerve processes, being most prominent at the tip of axons.

Haploid Having only a single set of chromosomes, a regular feature of gametes, in contrast to diploid where two sets of chromosomes are present.

Hemangioblast Common multipotent stem cell of endothelial stem cells and blood cells-generating hematopoietic stem cells

Hematopoiesis (hemopoiesis) Generation of blood cells from stem cells

Hemimetabolous Adjective indicating insects that undergo gradual transformation ('metamorphosis') from the larval to the adult stage; changes occur during moults.

Hemolymph Body fluid ('blood') of arthropods

Hensen's node A thickening caused by condensation of cells at the anterior end of the primitive groove (blastopore) of a gastrulating avian embryo; equivalent to the upper blastopore lip (Spemann organizer) of the amphibian gastrula.

Hermaphrodite In terms of zoology an individual or species capable of producing both sperm and egg cells; in medical terminology referring to intersexuals

Heterochromatin Condensed regions on chromosomes in which no, or only weak, transcription takes place

Heterochrony Different times in the relative beginning and duration of developmental processes in different taxa, temporal shift of developmental events in evolution

Heterologous Referring to genetically different individuals or tissue grafts; heterologous transplantation = donor and recipient are genetically different; antonym to autologous

Heterozygous Genetic constitution in diploid organisms when two alleles, the one inherited from the father, the other from the mother, differ. Designated by, e.g., a/+, with + wild-type allele and **a** the altered (mutated) allele, in contrast to homozygous (a/a or +/+)

HH HedgeHog: morphogen in many animals, named after the appearance of a *Drosophila* mutant

HLH Helix–Loop–Helix domain (motif), DNA-binding domain of a large group of transcription factors

HMG High Mobility Group, DNA-binding domain of certain transcription factors

Holoblastic cleavage Type of cleavage where the entire egg cell is divided into smaller cells (called blastomeres)

Holometabolous adjective indicating insects that undergo metamorphosis in Two dramatic steps from larva to a pupa and from the pupa to the adult.

Hom/Hox *Hom* in *Drosophila, Hox* in other animals: genes coding for transcription factors possessing a homeo domain for DNA binding. Their expression confers unique characteristics to different body regions.

Homeobox Sequence of base-pairs of a gene encoding the DNA-binding homeodomain of genes encoding an important class of transcription factors involved in the programming of body parts

Homeodomain Part of a transcription factor that mediates DNA binding through its Helix-turn-Helix motif that is encoded by the homeobox of the corresponding gene

Homeotic genes Genes conferring a particular identity to a body region, for example to become a wing-bearing mesothorax instead a wingless prothorax in the fly. Mutations in these genes can lead to the transformation of a body part into another part (homeotic transformation). Homeotic genes contain a homeobox, but not all genes containing a homeobox are classified as homeotic genes, since this classification is based on homeotic transformations in mutants

Homeotic transformation Conversion of a structure into a structure characteristic of other body parts (e.g. transformation of a wing into a leg in the fly). The transformation can be caused by a mutation, or by an epigenetic modification of a gene, or by injury, or by treatment with chemicals such as retinoic acid.

Homologous chromosomes Maternally and paternally inherited chromosomes bearing the same set of genes (though frequently in form of different alleles) and that pair with each other at the beginning of meiosis

Homologous genes Genes found in different species (orthologous genes) or being present repeatedly within the genome of a species (paralogous genes) that display high sequence similarity or sequence identity along long stretches, and are thought to be derived from a common ancestor gene in evolution

Homologous organs Organs found in different species (orthologous organs), or found in repeated editions within an individual (paralogous organs) derived from a common ancestor organ but which may have become more or less different in evolution

Homologues Characters that share descent from a common ancestor

Homology Fundamental concept in comparative biology meaning "being derived from a common ancestor"

Homozygous Genetic constitution in a diploid organisms when two alleles, the one inherited from the father, the other from the mother, are identical. Designated a/a or +/+in contrast to a/+in heterozygous constitution.

Hox **genes** Family of genes with a homeobox, specifying body regions in most animals

HtH Helix-turn-Helix – morphological structure of a DNA-binding domain (motif); various HtH-motifs are not necessarily homologous to each other

Hybridization (a) In molecular biology the joining of two single complementary DNA strands forming a double stranded molecule; (b) in breeding the crossing of individuals with different properties yielding offspring with mixed properties

Hypoblast Cell layer in the blastodisc of fishes or birds or other amniotic organisms lying below the epiblast and on top of the residual egg cell, comprising mesodermal and, putatively, also endodermal cells.

i-cells See interstitial cells

IGF Insulin-like Growth Factor, acting in the embryo and fetus

IL-8 Interleukin-8; signalling substance produced by leukocytes (white blood cells) and acting on other types of leukocytes

Imaginal discs Packets of diploid epithelial cells in the larvae of holometabolous insects that give rise to the epidermis and cuticule of the adult body at metamorphosis

Imago Adult insect hatched out from the pupa

Imprinting, genetic Epigenetic modification of the chromatin leading to different accessibility and therefore to varyingly strong expression (penetrating power) of maternal and paternal genes

In situ In place, within the body

In vitro In a culture vessel, not within a living organism

In vitro-**fertilization IVF** Artificial insemination of egg cells with sperm in a dish

In vivo Within a living organism.

Inducer Cell group, or a substance released by a cell group, that exerts an inducing influence in the sense of embryonic induction

Induction (1) Embryonic induction: initiation of a developmental pathway by an adjacent region that releases an inducing substance; (2) induction in molecular biology: activation of a gene, typically a gene coding for an enzyme

Inner cell mass Part of the mammalian blastocyst that gives rise to the embryo proper, also called embryoblast

Insemination Adding of sperm to egg cells

Intercalation Insertion of missing body parts when distant tissues are brought in juxtaposition to each other; insertion by inducing growth of those body parts that normally lie in between the apposed tissues

Interstitial cells Small cells residing in the interstitial spaces of the epithelia in Hydrozoa, contain pluripotent stem cells and early derivatives of them

Interstitium Space between cells

Invagination Inward bulging of an epithelial cell sheet giving rise to a curved, vesicular or tubular structure in the interior of the embryo. Often part of gastrulation

Involution Internalization of an epithelial cell sheet by rolling movement around an edge, typically around the blastopore lip in gastrulation

Jumping gene See transposon

Juvenile hormone Hormone in insects produced by the corpora allata that inhibits or slows down the development of a larva to the adult imago

Knockout mutation Complete deactivation or elimination of both alleles of a gene, artificially brought about by targeted mutagenesis, also called loss-of-function mutation

LacZ Gene of bacterial origin encoding β-galactosidase, often used as reporter gene

Lampbrush chromosomes Particular structure of decondensed chromosomes in transcriptionally active oocytes

Lateral inhibition Inhibition originating at a location and acting on the neighbourhood

Leucine-zipper DNA-binding section (domain) of a class of transcription factors

LH Luteinizing Hormone – hormone controlling the corpus luteum in the ovary, stimulates the production of the hormone progesterone in the second half of the ovarian cycle

Ligand General designation of external molecules, such as growth factors, hormones or odorants, that are specifically bound by molecular receptors of signal-receiving cells

Lophotrochozoa Group of animal phyla characterized by spiralian cleavage and the occurrence of larvae having ciliary bands (trochophore larvae) or a lophophore, a fan of ciliated tentacles surrounding the mouth, and a set of similar (homologous) ribosomal genes; comprises the phyla plathelminthes, annelida, mollusca and some minor phyla.

LTH Luteo-Tropic Hormone. Used interchangeably for the hormone prolactin

Marginal zone Band of cells along the equator of the amphibian blastula that in gastrulation is shifted into the interior of the embryo to give rise to the mesoderm

Marker gene Introduced gene, or manipulated endogenous gene, the expression of which can be visualized even in living cells

Master gene Gene that controls the activity of subordinate downstream genes, also called selector gene

Maternal mRNA or factor Component of the egg cell derived from genes of the mother

Maternal-effect mutation Mutation in the maternal organism that affects components of the oocyte, and hence the child, irrespective of the child's own genetic constitution.

M-CSF Macrophage Colony Stimulating Factor: factor inducing multiplication of macrophages

Meiosis The final two consecutive cell divisions in the development of gametes by which the diploid set of chromosomes of the oogonia or spermatogonia is reduced to the haploid set of the gametes; often associated with mutual exchange of chromosomal sections by crossing-over; also called maturation divisions

Meroblastic cleavage Incomplete cleavage in yolk-rich egg cells where the cleavage furrows do not continue down through the

yolk, leaving the cells continuous with the yolk..

Mesenchyme Loose assemblage of not yet fully differentiated cells, mostly of mesodermal origin

Mesoblast Another term for mesoderm

Mesoderm The middle of the three classic germ layers that gives rise to many tissues and organs such as connective tissue, muscles and blood cells, in vertebrates also to cartilage and bones

Mesonephros Second, transitory generation of embryonic kidney. In males the mesonephric duct becomes the Wolffian or spermatic (sperm-draining) duct

Metagenesis Life cycle in which an asexually propagating form alternates with a sexually reproducing form. One speaks of a metagenetic alteration of generations

Metamorphosis Transformation of an organism from a larval stage (larval phenotype) to the adult stage (imaginal or adult phenotype)

Metaplasia Conversion of a state of differentiation or tissue type and adoption of another state or tissue type, for example transformation of iris muscle cells into lens cells in Wolffian lens regeneration; also called transdifferentiation

Metastasis The emigration of daughter cancer cells (tumours) from the origin of cancer to other places in the body

Midblastula transition Stage in the late amphibian and fish blastula in which translation of maternal mRNA declines and transcription of the embryo's own (zygotic) genes commences

MIS Müllerian duct Inhibiting Substance, see AMDF

Mitosis Normal type of cell division in which both daughter cells get a whole set of chromosomes

Morphogen A term coined by A. Turing designating a then hypothetical substance involved in pattern formation by regulating the spatial order of cell differentiation according to its concentration distribution. In the meantime many morphogens are identified, often acting as a gradient causing at least two different types of differentiation according to its local concentration

Morphogenesis Generation of shape of a structure involving movement of cells or of cell sheets

Morphogenetic field Area in the embryo in which a complex structure such as an extremity or an organ arises, essentially by self-organization and involving morphogens. Initially a morphogenetic field has the faculty of regulation and, if separated into two parts, gives rise to two complete albeit smaller structures

Mosaic development, determinative development The idea that the development of an embryo is (predominantly) specified by the spatial pattern of cytoplasmic determinants in the egg cell

MPF Maturation-Promoting Factor, also called Meiosis-Promoting Factor or Mitosis-Promoting Factor; protein complex regulating the cell cycle (see Fig. 3.7) with three equivalent meanings

MSC Mesenchymal stem cells, synonymous: marrow stroma cells

Müllerian duct Embryonic precursor of the oviduct

Multipotent Term used to indicate the developmental potency of stem cells, meaning having many options of development; synonymous to pluripotent except in mammals where a proposed strict definition designates stem cells as multipotent if they give rise to cell types belonging to one and the same germ layer, while pluripotent stem cells can give rise to cell types belonging to all three germ layers

Mutagen Agent (radiation or chemical) inducing permanent alterations (mutations) of DNA

Mutant An organisms carrying a mutated gene compared to the normal wild-type gene

Mutation Permanent and heritable alteration of a gene (or other sections of DNA)

Myoblast Cell committed to become a muscle cell and having one nucleus

Myogenic Leading to muscle cells

Myotome Part of the somites that gives rise to the cross-striated musculature of the trunk, tail and extremities

Myotube Multinucleated elongated precursor of muscle fibres arisen from the fusion of several myoblasts

Nasal At, or close to, the nose, or direction pointing to the nose

Neoblasts Pluripotent stem cells of planarians

Neoplasia Abnormal proliferation of cells, synonymous to (benign) tumour

Neoplasm An abnormal mass of tissue as a result of neoplasia

Nephric duct Duct of embryonic kidneys (pronephros and mesonephros), in males persisting as Wolffian duct and becoming the spermatic duct (vasa deferens)

Neural crest cells Cells derived from the neural crest that emigrate to become various cell types in the body (see Chapter 15 and16)

Neural crest Cell rows edging the neural plate; subsequent to neurulation identical to the cell rows on both sides of the neural tube that will become migratory neural crest cells

Neural plate, neural tube The initially planar then tubular primordium of the central nervous system in vertebrates

Neuroblast Precursor of a nerve cell, still able to divide

Neurogenic Leading to nerve cells including glial cells

Neurosphere Artificial structure derived from embryonic stem cells, autonomously formed in cell culture and containing nerve cell precursors

Neurotrophins Secreted proteins that promote the development and survival of nerve cells.

Neurula Embryo undergoing neurulation

Neurulation Formation of the primordium of the central nervous system in vertebrates, includes the formation of the neural plate and its subsequent rolling up to form the neural tube

Node Condensation of cells at the anterior end of the primitive streak of the gastrulating mammalian embryo, homologous to the upper blastopore lip in amphibians and Hensen's node in the avian embryo

Notochord Embryonic rod along the longitudinal body axis, also known as **chorda dorsalis,** structure characteristic of the animal phylum Chordata, in the subphylum Vertebrata eventually replaced by the vertebral column

Null mutation Mutation causing complete loss of function

Ontogeny Development of an individual beginning with the fertilization of the egg and ending with its death

Oocyte Embryonic egg cell precursor, immature egg cell

Oogenesis Development of the egg cell

Organizer Region or cell group in a developing organism, for example in the gastrula or in regeneration of a body part, that induces and controls the development of different structures in its neighbourhood by emitting signalling substances, generally called inducers.

Organogenesis Formation of organs in embryogenesis

Orthologous genes Genes being present in different species (e.g. *Drosophila* and mouse) and displaying a high sequence similarity; considered to be homologous, that is to be derived from a common ancestor gene in a common stem organism before the divergence of the evolutionary branches. Compare paralogous genes

Orthologous organs/body parts Organs or body parts found in different species and considered to be homologous, that is to be derived from a common structure in an ancestor before their divergence in evolution; examples are pectoral fins in fishes and forelegs and wings in terrestrial vertebrates. Compare paralogous organs

Ovulation Release of an egg cell from the ovary, in vertebrates release into the oviduct

Paired domain DNA-binding section (domain) of Pax transcription factors

Paralogous genes Group of genes (gene family) being present in one and the same species and exhibiting high sequence similarity,

considered to have arisen from a founder gene by its repeated duplication; examples are the clusters of *Hox* genes in 'higher' animals

Paralogous organs Organs or organ primordia that are repeatedly present in one and the same species such as the somites, the vertebral bodies, or the skeletal elements of the forelegs and hindlegs; considered to be organized by a set of identical or similar genes.

Parthenogenesis From Greek: *parthenos* = virgin, *genesis* = emergence, generating: development of an organism from an unfertilized egg

Paternal (effect) Derived or specified by genes of the father

Pattern formation Events leading to an ordered and reproducible spatial arrangement of different cell types or of supra-cellular structures such as scales and feathers

Pax Member of a family of genes coding for transcription having a PAX motif as DNA-binding domain: e.g., *eyeless* = *Pax6*

PDI See prenatal diagnosis

P-element Transposon-like DNA sequence flanked by inverted repeats that can be found in *Drosophila*, but that is not a normal sequence of its genome; thought to be introduced in historical times by horizontal gene transmission, perhaps by virus or mites.

P-granules electrondense fibrous organelles found in the cytoplasm close to the nucleus of germ line cells, for example in the primordial germ cells of nematodes, flies and amphibians. Also called germinal granules or nuage (cloud). May undergo a phase transition to liquid-like droplets

Peripheral nervous system (PNS) Nervous system comprising the spinal ganglia (see ganglion), the enteric nervous system ENS around the gastrointestinal tract, and the sympathetic and parasympathetic nervous system if and when the central bodies with the nucleus of the nerve cells rest outside the CNS

Pharyngeal arches (=branchial arches) Structures between the gill pouches (pharyngeal pouches) comprising cartilaginous braces, arterial vessels, and cranial nerve fibres; in land vertebrates also giving rise to endocrine glands such as thymus and parathyroid

Pharyngeal pouches (=**branchial pouches**) Four to six pairs of pouches or grooves in the pharyngeal region of vertebrates, also called gill pouches and becoming the gill clefts in fishes and larval amphibians; transformed in land vertebrates with the auditory tube deriving from the first pouch

Phenotype Appearance of a cell or an organism; sum of expressed (visible) features, in contrast to genotype, the sum of the (invisible) genes underlying the phenotype

Phylogeny Evolutionary history of a phylum or of phyla

Phylotypic stage Stage in the development of an animal phylum (or a lower cladistic category) showing characteristic basic feature of the phylum. Mostly used for vertebrates where post neurula embryos show a common set of morphological characteristics, such as dorsal neural tube, notochord, somites and ventral heart

Placenta Organ of the mammalian fetus by which it establishes intimate contact with the uterus of the mother to receive oxygen and nutrients and dispose of carbon dioxide and waste products. The placenta derives from the fertilized egg but is formed by the extraembryonic cellular envelope (trophoblast, chorion) that surrounds the embryo proper. In birth it is disposed as afterbirth

Pluripotent Term used to indicate the developmental potency of stem cells, meaning having many options of development; synonymous to multipotent except in mammals where a proposed strict definition designates stem cells as pluripotent if they can give rise to cell types belonging to all three germ layers, while multipotent stem cells give rise to cell types belonging to one layer.

PNS See peripheral nervous system

Polar bodies Miniature cells extruded from the mature oocyte in the course of the two meiotic cell divisions. Are pinched off at that pole below which the diploid nucleus of the oocyte (called germinal vesicle) was located, and which is now designated animal pole; polar bodies are discarded.

Pole cells Cells at the posterior end of the *Drosophila* embryo fated to become primordial germ cells

Positional cloning Identification and cloning of an unknown gene by analysing the inheritance of a (mutated) phenotypic trait in relation to other, closely linked, known gene-related traits (linkage analysis) until its position on the chromosome is localized and the gene can be multiplied by cloning. Positional cloning involves the isolation of partially overlapping DNA segments from genomic libraries to progress along the chromosome (chromosomal walk) toward the gene in question.

Positional information Any source of information by which cells are informed about their position relative to others enabling fate specification according to their location. Information can emanate from adjacent cells or from distant emitters of signals (whereby, according to L. Wolpert, positional information is derived from the local concentration of a morphogen).

Posterior Tail end, or direction pointing to the tail

Prenatal diagnosis (PDI) Evaluation of the health of an embryo or fetus

Prenatal screening Procedures aimed at detecting chromosomal aberrations (Down's syndrome, trisomy 18), genetic or other defects such as an open neural tube

Primitive node Condensation of cells at the anterior end of the primitive groove on the avian or mammalian blastodisc, also known as Hensen's node in bird, and simply "node" in mammals

Primitive groove Groove at the ventral side in the blastoderm of the insect embryo, or groove on the avian or mammalian blastodisc; the groove is equivalent to the blastopore in other animals and is the location where in gastrulation cells move from the blastoderm or epiblast, respectively, downward into the interior to become mesoderm.

Primordial germ cells Still diploid precursor cells of germ cells, capable of dividing and migration

Probe In molecular biology a labelled antisense single stranded DNA or RNA that can be used to detect (and isolate) endogenous complementary nucleic acid by hybridization

Programmed cell death Apoptosis, natural type of cell death to prepare free spaces or to remove used up cells, involves caspase enzymes. The remnants are endocytosed by other cells, typically by macrophages.

Proliferation of cells Repeated cell divisions

Promoter Regulatory DNA sequence ahead of a gene where polymerase starts to transcribe the information of a gene to mRNA, and where also regulatory transcription factors may bind (instead or in addition to binding to an enhancer)

Pronephros First transitory embryonic kidney, anterior segment of the mesonephros

Pronucleus Plural pronuclei...designation of the haploid nucleus of the egg cell (maternal pronucleus) or the haploid nucleus supplied by the sperm (male pronucleus) prior to their fusion to the diploid nucleus of the zygote

Protostomia Group of invertebrate animal phyla where the blastopore gives rise (or has been thought to give rise) to the definitive mouth while the anus, if present, is secondarily formed near the posterior end

Proximal Close to the body, near the base (e.g., of an extremity)

RA Retinoic Acid, vitamin-A-acid; acting as morphogen in the embryo

RAR Retinoic Acid Receptor, vitamin-A-acid receptor

Rostral situated at, or pointing to, the beak/snout

SCF Stem Cell Factor, factor stimulating growth (cell division) of embryonic stem cells

Selector gene Gene that controls the activity of subordinate downstream genes, also called master gene

SHH Sonic Hedgehog – morphogen in vertebrate development

Signal transduction Processes transferring a message across the cell membrane from the exterior into the interior of a cell; is initiated upon binding of a signalling molecule (ligand) by a molecular, membrane-associated receptor

Sclerotome Region of the somite that gives rise to the vertebral bodies

Soma, somatic Referring to the body except the reproductive cells

Somites Two rows of packets of cells on both sides of the neural tube and notochord extending from the neck region to the tail tip. Somites give rise to the vertebral bodies (sclerotome part), to the cross-striated musculature of the body (myotome part) and to the connective tissue layer of the skin (dermatome part).

Sox Member of a family of genes coding for transcription factors with a SOX domain as DNA-binding part, with, e.g., *SRY*

Spemann-organizer Area of the amphibian blastula/gastrula possessing the power of inducing a second blastopore and in further consequence a second (Siamese) embryo if transplanted into the opposite side of a recipient blastula. Emitter of several inducing factors and source of morphogens. Equivalent cell groups in other vertebrates are the shield in fishes, Hensen's node in birds, and 'node' in mammals.

Sperm Term derived from the Greek word *sperma* = seed, (meaning "seed"), referring to the male reproductive cells: if motile by a flagellum also called spermatozoon.

Spermatogenesis Development of sperm in the testis

Spermatozoon (plural spermatozoa): Motile sperm equipped with a flagellum

Specification First programming pointing the way to a certain goal of differentiation but not yet implying strong fixation of the fate. The program is realized in a neutral environment but can be changed in other environments

SRY Transcription factor encoded by the *SRY* gene in vertebrates, directing sexual development in the male direction, also known as TDF

Taxon, plural **taxa** A group of organisms classified on the basis of shared characteristics (taxonomic criteria)

Taxonomy Is the biological discipline (term derived from Greek *taxis,* arrangement, and *nomia,* method) for grouping organisms on the basis of shared characteristics and giving ranks and names to those groups. The hierarchical rankings are primarily based on practical criteria but with the aim of reconstructing phylogenetic trees in the background. The groups classified are referred to as taxa (singular taxon).

TDF Testis-Determining Factor, used interchangeably for SRY, causes the initially neutral gonadal primordium to develop into a testis

Temporal Located at, or pointing to, the temples, also meaning relating to time

Teratocarcinoma Disorganized mass of embryonic cells that develops tumour-like properties. Most teratocarcinomas derive from blastocysts ectopically implanted in the oviduct or abdominal cavity instead in the uterus. In rare cases derived from an oogonium or spermatogonium in the gonads (ovarial or testis cancer).

Teratogenic Causing malformations

Teratoma Malformed embryo, mass of cells including chaotically arranged differentiated cell, often with properties of a (benigne) tumour

TF See transcription factor

TGF-ß Transforming Growth Factor beta – representative of a large protein family that comprises many morphogens; named after its effect to promote cancerous growth of some cell types in postnatal life

Therapeutic cloning (once) proposed procedure to enable curing of genetic defects or cancer in patients. The procedure involves transfer of a nucleus taken from a patient into an enucleated oocyte taken from any female, raising the oocyte to the blastodisc stage in vitro, removal of embryonic stem cells having the genome of the patient, replacing defective genes by healthy ones, and introducing the stem cells or derivatives of them into the patient.

Totipotency, totipotent Having all developmental potentials, including the option to bring about a complete individual

Transcription Manufacturing an RNA copy of a DNA sequence, typically making a mRNA copy of a gene

Transcription factor (TF), protein binding to the controlling region (promoter, enhancer) of genes, controlling their activity, i.e. the transcription of its information into an RNA copy

Transdetermination Change of the state of determination of a cell or cell group, for example conversion of a wing imaginal disc into a leg disc in *Drosophila*

Transdifferentiation Regression of a state of differentiation and adoption of another state, for example transformation of iris muscle cells into lens cells in Wolffian lens regeneration; also called metaplasia

Transformation (in developmental biology) (1) Conversion of a normal cell into a cancer cell (cancerous, neoplastic or oncogenic transformation) (2) introduction of a DNA into a cell (genetic transformation)

Transgene Foreign gene introduced into the genome of a host (cell, egg cell)

Transgenesis, transgenic animal Procedure to create transgenic animals, that is animals carrying a foreign gene

Transposon Mobile pieces of DNA that can change its location in the genome, mediated by an enzyme called transposase, often causing mutations when inserted within a gene; also called **transposable element** or jumping gene.

Triploblastic Referring to animals developing in embryogenesis three germ layers (ectoderm endoderm and mesoderm), actually congruent with animals exhibiting bilateral symmetry

Trophoblast Outer cellular wall of the mammalian blastocyst, giving rise to the chorionic villi and eventually to the placenta

TSH Thyroid-Stimulating Hormone – hormone controlling the thyroid

Upper blastopore lip Upper edge of the blastopore in amphibians, comprising the Spemann organizer

Uterus Cavity enclosed by the uterine wall; place where in mammals the embryo (fetus) grows connected to the wall initially by chorionic villi, later by the placenta

Vasculogenesis Formation of blood vessels

Vegetal (In botanics and medicine *vegetative,* literally 'plant-related'), in animal developmental biology (e.g. in the terms 'vegetal egg pole', 'vegetal hemisphere') related to the 'lower' (south) part of the embryo that usually gives rise to the gastro-intestinal tract

Vegetalization Treatment of an embryo leading to enlarged gastrointestinal tract at the expense of animal (dorsal) organs such as nervous system t

Vegetative reproduction Asexual propagation, not accomplished with a fertilized egg but, for example, through a mitotically produced bud, a kind of natural cloning

VEGF Vascular Endothelial cell-derived Growth Factor – growth factor produced by the walls (endothelia) of blood vessels

Ventral Located at, or pointing to, the belly

Visceral Related to the bowels; also used to designate the parts of the head-neck region ventral of the skull and upper vertebral column, that is the region of the throat, pharynx, larynx, trachea.

Vitelline membrane non-cellular envelope of a mature egg cell, in mammals known as zona pellucida

Vitellogenins Proteins produced by the maternal organism, in mammals by the liver, and deposited in egg cells as part of the yolk

WNT Proteins of the Wingless/WNT-family, in the embryo acting as morphogens, also involved in the regulation of supplies from stem cells

Wolffian duct Embryonic primordium of the urinary and spermatic duct in male mammalian embryos; derivate of the pronephrotic duct

Wolffian lens regeneration Regeneration of a removed lens from the iris in amphibians

Yolk sac Membranous, ball-shaped extraembryonic organ in fishes, reptiles and birds growing beneath the blastodisc and enclosing the yolk-containing residual egg; in mammals an almost yolk-free relict that eventually becomes integrated into the umbilical cord.

Yolk Material stored in the egg cell enclosed in vesicles, providing energy and molecular components for the growth of the embryo

Zinc finger DNA-binding section (domain) of certain transcription factors

Zona pellucida Envelope encasing the mammalian egg cell

ZPA zone of polarizing activity region at the posterior margin of the avian wing bud, signal emitter specifying the quality and sequence of fingers

Zygote fertilized egg cell

Zygotic genes Endogenous genes of the embryo in contrast to genes of the mother (maternal genes)

References and Further Readings

Electronic Databases (Selection)
Biosis Previews
Medline Advanced
Pubmed
Databases for Genes
Mouse: http://www.informatics.jax.org/mgihome/nomen/
Rat: http://rgd.mcw.edu/nomen/nomen.shtml
Chicken: http://www.agnc.msstate.edu/
Xenopus: http://www.xenbase.org/gene/static/gene Nomenclature.jsp
Zebrafish: https://wiki.zfin.org/display/general/ZFIN+ Zebrafish+Nomenclature+Guidelines
Drosophila: http://flybase.org/static_pages/docs/nomen clature/nomenclature3.html
C. elegans: http://wiki.wormbase.org/index.php/ Nomenclature

Monographics, Textbooks Referred To

Alberts B et al (2007) Molecular biology of the cell, 5th edn. Taylor & Francis, New York

Hinrichsen KV (1990) Human embryology. Springer, Berlin

Mueller W, Frings S (2009) Tier- und Humanphysiologie, 4th edn. Springer, Berlin

Sadler TW, von Leland J (2009) Langman's medical embryology. Lippincott Williams & Wilkins, Philadelphia, PA

Spemann H (1938a) Embryonic development and induction. Hafner, New York, Reprint 1962

Box 1 History of Developmental Biology

Aristotle: De anima (1984) Generation of animals. In Barnes J (ed) The complete works of Aristotle: the revised Oxford translation. Bollington series LXXI, Princeton University Press, Princeton, CT

von KE B (1828) Über Entwickelungsgeschichte der Thiere. Den Gebrüdern Bornträger, Königsberg

Driesch H (1892) The potency of the first two cleavage cells in echinoderm development. Experimental production of partial and double formations. In: Willer BH, Oppenheimer JM (eds) Foundations of experimental embryology. Hafner, New York, pp 38–50

Driesch H (1908) The science and philosophy of the organism. A. & C Black, London, Reprint BiblioBazaar 2009

Driesch H (1913) The problem of individuality: a course of four lectures delivered before the University of London. Macmillan, London, Reprint BiblioBazaar 2009

Fäßler PE (1996) Hans Spemann and the Freiburg school of embryology. Int J Dev Biol 40:49–57

Gilbert SF (2013) A conceptual history of modern embryology, vol 7. Springer, Berlin

Hamburger V (1988) The heritage of experimental embryology: Hans Spemann and the organizer. Oxford University Press, New York

Harvey W (1651) De generatione animalium. Clarendon Press, Oxford, English translation by R. Willis in: Encyclopedia Brittanica, 1952

McCabe BW (2012) Haeckel, his life and work. Ulan Press, London

Moritz KB, Sauer H (1996) Boveri's contributions to developmental biology – a challenge for today. Int J Dev Biol 40:27–47

Sander K et al (2012) Landmarks in developmental biology 1883–1924: historical essays from Roux's archives. Springer, Berlin

Sander K (1996) On the causation of animal morphogenesis: concepts of German-speaking authors from Theodor Schwann (1839) to Richard Goldschmidt (1927). Int J Dev Biol 40:7–20

Chapters 1–3 Reproduction General, Egg Activation, MPF Oscillator, Asymmetric Cell Division: Fertilization

Bahat A, Eisenbach M (2010) Human sperm thermotaxis is mediated by phospholipase C and inositol trisphosphate receptor Ca^{2+} channel. Biol Reprod 82 (3):606–616

Bates RC et al (2014) Activation of Src and release of intracellular calcium by phosphatidic acid during Xenopus laevis fertilization. Dev Biol 386 (1):165–180

Botchkina IL, Kirichok Y (2011) Progesterone activates the principal Ca^{2+} channel of human sperm. Nature 471(7338):387

Clark GF (2013) The role of carbohydrate recognition during human sperm-egg binding. Hum Reprod 28 (3):566–577

Goodwin CN et al (2012) The CatSper channel: a polymodal chemosensor in human sperm. EMBO J 31(7):1654–1665

Gupta SK et al (2009) Human zona pellucida glycoproteins: functional relevance during fertilization. J Reprod Immunol 83(1–2):50–55

Muro Y et al (2012) Function of the acrosomal matrix: zona pellucida 3 receptor (ZP3R/sp56) is not essential for mouse fertilization. Biol Reprod 86(1):23

Masuru O (2013) The cell biology of mammalian fertilization. Development 140(22):4471–4479

Oren-Benaroya R et al (2008) The sperm chemoattractant secreted from human cumulus cells is progesterone. Hum Reprod 23(10):2339–2345

Publicover S, Barrat C (2011) Progesterone's gateway into sperm. Nature 471:313–314

Ramos I et al (2013) Calcium pathway machinery at fertilization in echinoderms. Cell Calcium 53:16–23

Santella L et al (2012) Fertilization in echinoderms. Biochem Biophys Res Commun 425(3):588–594

Suarez SS (2008) Regulation of sperm storage and movement in the mammalian oviduct. Int J Dev Biol 52 (5–6):455–462

Vacquier VD (2012) The quest for the sea urchin egg receptor for sperm. Biochem Biophys Res Commun 425(3):583–587

Wassarman PM (2008) Fertilization: welcome to the fold. Nature 456:586–587

Wolkowicz MJ et al (2008) Equatorial segment protein (ESP) is a human alloantigen involved in sperm-egg binding and fusion. J Androl 29(3):272–282

Transmission of Mitochondrial DNA and of the Nucleolus

Shi-Ming L et al (2013) Sperm mitochondria in reproduction: good or bad and where do they go? J Genet Genomics 40(11):549–556

Ogushi S et al (2008) The maternal nucleolus is essential for early embryonic development in mammals. Science 319(5863):613–616

Ramalho-Santos J, Amaral S (2013) Mitochondria and mammalian reproduction. Mol Cell Endocrinol 379:74–84

Schwartz M (2002) Paternal inheritance of mitochondrial DNA in man. J Med Genet 39(Suppl 1):14

Sato M, Sato K (2013) Maternal inheritance of mitochondrial DNA by diverse mechanisms to eliminate paternal mitochondrial DNA. Biochim Biophys Acta 1833(8):1979–1984

Role of Centrosomes and Centrioles

Clift D, Schuh M (2013) Restarting life: fertilization and the transition from meiosis to mitosis. Nat Rev Mol Cell Biol 14(9):549–562

Coelho PA et al (2013) Spindle formation in the mouse embryo requires Plk4 in the absence of centrioles. Dev Cell 27(5):586–597

Schatten H, Sun Q-Y (2009) The role of centrosomes in mammalian fertilization and its significance for ICSI. Mol Hum Reprod 15(9):531–538

Egg Activation

Kashir J et al (2010) Oocyte activation, phospholipase C zeta and human infertility. Hum Reprod Update 16 (6):690–703

Nomikos M et al (2012) Starting a new life: sperm PLC-zeta mobilizes the Ca2+signal that induces egg activation and embryo development. Bioessays 34 (2):126–134

Uhlen P (2010) Biochemistry of calcium oscillations. Biochem Biophys Res Commun 396(1):28–32

Parthenogenesis, Reproduction General

Adams M et al (2003a) The Australian scincid lizard *Menetia greyii*: a new instance of widespread vertebrate parthenogenesis. Evolution 57:2619–2627

Brevini TA, Gandolfi F (2013) Parthenogenesis and parthenogenetic stem cells. In: Trounson A, Gosden RS, Eichenlaub-Ritter U (eds) Biology and pathology of the oocyte: role in fertility, medicine, and nuclear reprogramming, 2nd edn. Cambridge University Press, Cambridge, pp 250–260

Engelstaedter J (2008) Constraints on the evolution of asexual reproduction. Bioessays 30(11–12): 1138–1150

Kono T et al (2004) Birth of parthenogenetic mice that can develop to adulthood. Nature 428(6985):860–864

Leeb M et al (2012) Germline potential of parthenogenetic haploid mouse embryonic stem cells. Development 139(18):3301–3305

Lima J et al (2013) Embryo production by parthenogenetic activation and fertilization of in vitro matured oocytes from *Cebus paella*. Zygote 21:162–166

Miura T et al (2003) A comparison of parthenogenetic and asexual embryogenesis of the pea aphid. J Exp Zool B Mol Dev Evol 295:59–81

Cleavage and Cell Cycle Control

Abdelalim EM (2013) Molecular mechanisms controlling the cell cycle in embryonic stem cells. Stem Cell Rev 9(6):764–773

Balbach ST et al (2012) Nuclear reprogramming: kinetics of cell cycle and metabolic progression as determinants of success. PLoS One 7(4):e35322

Beckhelling C, Ford C (1998) Maturation promoting factor activation in early amphibian embryos: temporal and spatial control. Biol Cell 90:467–476

Hoermanseder E et al (2013) Modulation of cell cycle control during oocyte-to-embryo transitions. EMBO J 32(16):2191–2203

Asymmetric Cell Division, Centrosome

Almonacid M et al (2014) Actin-based spindle positioning: new insights from female gametes. J Cell Sci 127 (3):477–483

Bonifacino JS (2014) Adaptor proteins involved in polarized sorting. J Cell Biol 204:7–17

Rong L (2013) The art of choreographing asymmetric cell division.Dev. Cell 25(5):439–450

Lu Michelle S, Johnston CA (2013) Molecular pathways regulating mitotic spindle orientation in animal cells. Development 140(9):1843–1856

Noatynska A et al (2013) Coordinating cell polarity and cell cycle progression: what can we learn from flies and worms? Open Biol 3(8):130083

Pelletier L, Yamashita YM (2012) Centrosome asymmetry and inheritance during animal development. Curr Opin Cell Biol 24(4):541–546

Williams SE, Fuchs E (2013) Oriented divisions, fate decisions. Curr Opin Cell Biol 25(6):749–758

Chapter 4 Model Organisms I: Invertebrates: sea urchin, Books

Ettensohn CE et al (2004) Development of sea urchins, Ascidians, and other invertebrate deuterostomes: experimental approaches 74, Methods in cell biology. Academic, Amsterdam

Czihak G (1975) The sea urchin embryo. Springer, Berlin

Hörstadius S (1973) Experimental embryology of echinoderms. Clarendon, Oxford

Matranga V (ed) (2010) Echinodermata, Progress in molecular and subcellular biology/Marine Molecular Biotechnology. Springer, Berlin

Articles

Alford L et al (2009) Cell polarity emerges at first cleavage in sea urchin embryos. Dev Biol 330(1):12–20

Byrum CA, Wikramanayake AH (2013) Nuclearization of beta-catenin in ectodermal precursors confers organizer-like ability to induce endomesoderm and pattern a pluteus larva. Evodevo 4:31

Coffman JA et al (2004) Oral-aboral axis specification in the sea urchin embryo. II. Mitochondrial distribution and redox state contribute to establishing polarity in *Strongylocentrotus purpuratus*. Dev Biol 273:160–171

Coffman JA et al (2014a) Oral-aboral axis specification in the sea urchin embryo, IV: hypoxia radializes embryos by preventing the initial spatialization of nodal activity. Dev Biol 386(2):302–307

Croce J et al (2006) Frizzled5/8 is required in secondary mesenchyme cells to initiate archenteron invagination during sea urchin development. Development 133:547–557

Croce J et al (2010) Dynamics of Delta/Notch signaling on endomesoderm segregation in the sea urchin embryo. Development 137(1):83–91

Davidson E (2009) Network design principles from the sea urchin embryo. Curr Opin Genet Dev 19 (6):535–540

Bernadini D et al (2000) Homeobox genes and sea urchin development. Int J Dev Biol 44:637–643

Duboc V et al (2010) Nodal and BMP2/4 pattern the mesoderm and endoderm during development of the sea urchin embryo. Development 137(2):223–235

Juliano C et al (2010a) Nanos functions to maintain the fate of the small micromere lineage in the sea urchin embryo. Dev Biol 337(2):220–232

Lhomond G et al (2012) Frizzled1/2/7 signaling directs beta-catenin nuclearisation and initiates endoderm specification in macromeres during sea urchin embryogenesis. Development 139(4):816–825

Materna SC, Davidson EH (2012) A comprehensive analysis of Delta in pre-gastrular sea urchin embryos. Dev Biol 364(1):77–87

McIntyre DC et al (2013) Short-range Wnt5 signaling initiates specification of sea urchin posterior ectoderm. Development 140(24):4881–4889

Molina M et al (2013) Nodal: master and commander of the dorsal-ventral and left-right axes in the sea urchin embryo. Curr Opin Genet Dev 23(4):445–453

Pennisi E (2006) Sea urchin genome confirms kinship to humans and other vertebrates. Science 314 (5801):908

Range RC et al (2013) Integration of canonical and non-canonical Wnt signaling pathways patterns the neuroectoderm along the anterior-posterior axis of sea urchin embryos. PLoS Biol 11(1):e1001467

Ransick A, Davidson EH (1993) A complete second gut induced by transplanted micromeres in the sea urchin embryo. Science 259:1134–1138

Riaz SS, Mackey M (2014) Dynamic spatial pattern formation in the sea urchin embryo. J Math Biol 68 (3):581–608

Royo LJ et al (2012) Transphyletic conservation of developmental regulatory state in animal evolution. Proc Natl Acad Sci USA 108(34):14186–14191

Warner JF et al (2014a) Hedgehog signaling requires motile cilia in the sea rchin. Mol Biol Evol 31 (1):18–22

Zheng W et al (2012) Axial patterning interactions in the sea urchin embryo: suppression of nodal by Wnt1 signaling. Development 139:1662–1669

Yajima M, Wessel GM (2011) The DEAD-box RNA helicase Vasa functions in embryonic mitotic progression in the sea urchin. Development 138:2217–2222

Dictyostelium: Books

Kessin H et al (2010) *Dictyostelium:* evolution, cell biology, and the development of multicellularity, Developmental and Cell Biology Series. Cambridge University Press, Cambridge

Articles

Annesley SJ, Fisher PR (2009) Dictyostelium discoideum – a model for many reasons. Mol Cell Biochem 329 (1–2):73–91

Brzostowski JA et al (2013) Phosphorylation of chemoattractant receptors regulates chemotaxis, actin reorganization and signal relay. J Cell Sci 126 (20):4614–4626

Brookman JJ, Jermyn KA, Kay RR (1987) Nature and distribution of the morphogen DIF in the *Dictyostelium* slug. Development 100:119–124

Jang W, Gomer R (2008) Combining experiments and modelling to understand size regulation in *Dictyostelium discoideum*. J R Soc Interface 5(suppl 1):S49–S58

Kim L et al (2011) Combinatorial cell-specific regulation of GSK3 directs cell differentiation and polarity in *Dictyostelium*. Development 138:421–430

Konijin TM et al (1967) The acrasin activity of adenosine-3',5'-cyclic phosphate. Proc Natl Acad Sci USA 58:1152–1154

Oohata AA et al (2009) Differentiation inducing factors in *Dictyostelium discoideum*: a novel low molecular factor functions at an early stage(s) of differentiation. Dev Growth Differ 51(9):743–752

Schaap P (2011) Evolutionary crossroads in developmental biology: *Dictyostelium discoideum*. Development 138:387–396

Schaefer E et al (2013) Chemotaxis of *Dictyostelium discoideum*: collective oscillation of cellular contacts. PLoS One 8(1):e54172

Wang Y et al (2013) Rho GTPases orient directional sensing in chemotaxis. Proc Natl Acad Sci USA 110 (49):E4723–E4732

Cnidaria, Hydra: Cnidaria general

Burton P (2008) Insights from diploblasts; the evolution of mesoderm and muscle. J Exp Zool 310B(1):5–14

Gold DA, Jacobs DK (2013a) Stem cell dynamics in Cnidaria: are there unifying principles? Dev Genes Evol 223:53–66

Holstein TW, Hobmayer E, Technau U (2003a) Cnidarians: an evolutionarily conserved model system for regeneration? Dev Dyn 226:257–267

Philippe H et al (2009) Phylogenomics revives traditional views on deep animal relationships. Curr Biol 19 (8):706–712

Primus A, Freeman G (2004) The cnidarian and the canon: the role of Wnt/beta-catenin signaling in the evolution of metazoan embryos. Bioessays 26:474–478

Technau U, Steele RE (2011) Evolutionary crossroads in developmental biology: Cnidaria. Development 138:1447–1458

Watanabe H et al (2009a) Cnidarians and the evolutionary origin of the nervous system. Dev Growth Differ 51 (3):167–183

Wikramanayake AH et al (2003a) An ancient role for nuclear beta-catenin in the evolution of axial polarity and germ layer segregation. Nature 426:446–450

Yanze N et al (2001) Conservation of Hox/ParaHox-related genes in the early development of a cnidarian. Dev Biol 236:89–98

Hydra, Hydrozoa

Aufschnaiter P et al (2009) Wnt/beta-catenin and noncanonical Wnt signaling interact in tissue evagination in the simple eumetazoan Hydra. Proc Natl Acad Sci USA 106(11):4290–4295

Augustin R et al (2006a) Dickkopf related genes are components of the positional value gradient in *Hydra*. Dev Biol 296(1):62–70

Bode HR (1996) The interstitial cell lineage of hydra: a stem cell system that arose early in evolution. J Cell Sci 109:1155–1164

Bode HR (2003) Head regeneration in *Hydra*. Dev Dyn 226:225–236

Boehm A-M et al (2012) FoxO is a critical regulator of stem cell maintenance in immortal Hydra. Proc Natl Acad Sci USA 109(48):19697–19702

Böttger A, Hassel M (2012) Hydra, a model system to trace the mechanisms controlling the emergence of boundaries in eumetazoa. Int J Dev Biol 56 (6–8):583–591

Broun M, Bode HR (2002) Characterization of the head organizer in hydra. Development 129:875–884

Chapman JA et al (2010) The dynamic genome of *Hydra*. Nature 464(7288):592–596

Chera S et al (2009) Apoptotic cells provide an unexpected source of Wnt3 signaling to drive hydra head regeneration. Dev Cell 17(2):279–289

Duffy DJ, Millane R, Frank U (2012) A heat shock protein and Wnt signaling crosstalk during axial patterning and stem cell proliferation. Dev Biol 362 (2):271–281

Frank U, Leitz T, Mueller WA (2001) My favorite model organism: *Hydractinia echinata*. Bioessays 23: 963–971

Frank U, Plickert G, Mueller WA (2009a) Cnidarian interstitial cells: the dawn of stem cell research. In: Rinkevich B, Matranga V (eds) Stem cells in marine organisms. Springer, New York, pp 33–59

Gauchat D et al (2000) Evolution of *Antp*-class genes and differential expression of *Hydra Hox/paraHox* genes in anterior patterning. Proc Natl Acad Sci USA 97:4493–4498

Galliot B (2012) Hydra, a fruitful model system for 270 years. Int J Dev Biol 56(6–8):411–423

Galliot B (2013) Injury-induced asymmetric cell death as a driving force for head regeneration in Hydra. Dev Genes Evol 223:39–52

Gee L et al (2010) β-catenin plays a central role in setting up the head organizer in hydra. Dev Biol 340 (1):116–124

Gierer A et al (1972a) Regeneration of hydra from reaggregated cells. Nat New Biol 239:98–101

Glauber KM et al (2013) A small molecule screen identifies a novel compound that induces a homeotic transformation in Hydra. Development 140 (23):4788–4796

Gloria-Soria A et al (2012) Evolutionary genetics of the hydroid allodeterminant alr2. Mol Biol Evol 29 (12):3921–3932

Hassel M (1998a) Upregulation of *Hydra vulgaris* cPKC gene is tightly coupled to the differentiation of head structures. Dev Genes Evol 207:489–501

Hassel M et al (1998a) The level of expression of a protein kinase C gene may be an important component of the patterning process in *Hydra*. Dev Genes Evol 207:502–514

Hemmrich G et al (2012) Molecular signatures of the three stem cell lineages in *Hydra* and the emergence of stem cell function at the base of multicellularity. Mol Biol Evol 29(11):3267–3280

Hobmayer B et al (2000a) WNT signalling molecules act in axis formation in the diploblastic metazoan *Hydra*. Nature 407:186–189

Hobmayer B et al (2012) Stemness in Hydra – a current perspective. Int J Dev Biol 56(6–8):509–517

Juliano C et al (2014) PIWI proteins and PIWI-interacting RNAs function in Hydra somatic stem cells. Proc Natl Acad Sci USA 111(1):337–342

Kanska J, Frank U (2013) New roles for Nanos in neural cell fate determination revealed by studies in a cnidarian. J Cell Sci 126(14):3192–3203

Kuenzel T et al (2010) Migration and differentiation potential of stem cells in the cnidarian *Hydractinia* analysed in eGFP-transgenic animals and chimeras. Dev Biol 348(2010):120–129

Lengfeld T et al (2009a) Multiple Wnts are involved in *Hydra* organizer formation and regeneration. Dev Biol 330(1):186–199

Lim RS et al (2014) Analysis of Hydra PIWI proteins and piRNAs uncover early evolutionary origins of the piRNA pathway. Dev Biol 386(1):237–251

Meinhardt H (2002) The radial-symmetric hydra and the evolution of bilateral body plan: an old body became a young brain. Bioessays 24:185–191

Meinhardt H (2012a) Modeling pattern formation in hydra: a route to understanding essential steps in development. Int J Dev Biol 56(6–8):447–462

Miglietta MP, Cunningham CW (2012) Evolution of life cycle, colony morphology, and host specificity in the family Hydractiniidae (Hydrozoa, Cnidaria). Evolution 66(12):3876–3901

Millane CR et al (2011a) Induced stem cell neoplasia in a cnidarian by ectopic expression of a POU domain transcription factor. Development 138:2429–2439

Mochizuki K et al (2000) Expression and early conservation of *nanos*-related genes in *Hydra*. Dev Genes Evol 210:591–602

Mochizuki K et al (2001) Universal occurrence of *vasa*-related genes among metazoans and their germline expression. Dev Genes Evol 211:299–308

Momose T, Schmid V (2006a) Animal pole determinants define oral–aboral axis polarity and endodermal cell-fate in hydrozoan jellyfish *Podocoryne carnea*. Dev Biol 292(2):371–380

Mueller WA (1989) Diacylglycerol induced multihead formation in *Hydra*. Development 105:306–316

Mueller WA (1991) Stimulation of head-specific nerve cell formation in *Hydra* by pulses of diacylglycerol. Dev Biol 147:460–463

Mueller WA (1996a) Head formation at the basal end and mirror-image pattern duplication in *Hydra vulgaris*. Int J Dev Biol 40:1119–1131

Mueller WA (1996b) Competition-based head versus foot decision in chimeric hydras. Int J Dev Biol 40:1133–1139

Mueller WA (2002) Autoaggressive, multi-headed and other mutant phenotypes in *Hydractinia echinata* (Cnidaria: Hydrozoa). Int J Dev Biol 46:1023–1033

Mueller WA, Leitz T (2002a) Metamorphosis in the Cnidaria. Can J Zool 80:1755–1771

Mueller WA, Teo R, Frank U (2004a) Totipotency of migratory stem cells in a hydroid. Dev Biol 275:215–224

Mueller WA, Teo R, Möhrlen F (2004b) Patterning a multi-headed mutant in *Hydractinia*: enhancement of head formation and its phenotypic normalization. Int J Dev Biol 48:9–15

Mueller WA et al (2007a) Wnt signaling in hydroid development: ectopic heads and giant buds induced by GSK-3beta inhibitors. Int J Dev Biol 51:211–220

Muender S et al (2013) Notch-signalling is required for head regeneration and tentacle patterning in Hydra. Dev Biol 383(1):146–157

Nakamura Y et al (2011a) Autoregulatory and repressive inputs localize Hydra Wnt3 to the head organizer. Proc Natl Acad Sci USA 108(22):9137–9142

Nicotra ML et al (2009) A hypervariable invertebrate allodeterminant. Curr Biol 19(7):583–589

Philipp I et al (2009) Wnt/beta-catenin and noncanonical Wnt signaling interact in tissue evagination in the simple eumetazoan *Hydra*. Proc Natl Acad Sci USA 106(11):4290–4295

Plickert G et al (2006a) Wnt signaling in hydroid development: formation of the primary body axis and its subsequent patterning. Dev Biol 298:368–378

Plickert G, Frank U, Mueller WA (2012) Hydractinia, a pioneering model for stem cell biology and reprogramming somatic cells to pluripotency. Int J Dev Biol 56(6–8):519–534

Reber-Mueller S et al (2006) BMP2/4 and BMP5-8 in jellyfish development and transdifferentiation. Int J Dev Biol 50(4):377–384

Rentzsch F et al (2007) An ancient chordin-like gene in organizer formation of hydra. Proc Natl Acad Sci USA 104(9):3249–3254

Rudolf A et al (2013) The Hydra FGFR, Kringelchen, partially replaces the Drosophila Heartless FGFR. Dev Genes Evol 223(3):159–169

Schmich J, Trepel S, Leitz T (1998a) The role of GLWamides in metamorphosis of *Hydractinia echinata*. Dev Genes Evol 208:267–273

Seipel K, Schmid V (2006a) Mesodermal anatomies in cnidarian polyps and medusae. Int J Dev Biol 50 (7):589–599

Soza-Ried J et al (2010) The transcriptome of the colonial marine hydroid *Hydractinia echinata*. FEBS J 277 (1):197–209

Spring J et al (2002) Conservation of *Brachyury, Mef2*, and *snail* in the myogenic lineage of jellyfish: a connection to the mesoderm of bilateria. Dev Biol 244:372–384

Sudhop S et al (2004) Signalling by the FGF-R-like tyrosine kinase, Kringelchen, is essential for bud detachmant in *Hydra vulgaris*. Development 131:4001–4011

Technau U, Bode HR (1999) *HyBra1*, a *Brachyury* homologue, acts during head formation in *Hydra*. Development 126(5):999–1010

Technau U et al (2000) Parameters of self-organization in *Hydra* aggregates. Proc Natl Acad Sci USA 97:12127–12131

Teo R et al (2006a) An evolutionary conserved role of Wnt signaling in stem cell fate decision. Dev Biol 289:91–99

Weismann A (1883a) Die Entstehung der Sexualzellen bei Hydromedusen. Jena Fischer, Leipzig

Anthozoa: Nematostella

Artamonova II, Mushegian AR (2013) Genome sequence analysis indicates that the model eukaryote *Nematostella vectensis* harbors bacterial consorts. Appl Environ Microbiol 79(22):6868–6873

Finnerty JR et al (2004a) Origins of bilateral symmetry: *Hox* and *Dpp* expression in a sea anemone. Science 304:1335–1337

Fritz A et al (2013) Mechanisms of tentacle morphogenesis in the sea anemone Nematostella vectensis. Development 140(10):2212–2223

Fritzenwanker JH et al (2007a) Early development and axis specification in the sea anemone *Nematostella vectensis*. Dev Biol 310(2):264–279

Fritzenwanker JH, Technau U (2002) Induction of gametogenesis in the basal cnidarian *Nematostella vectensis* (Anthozoa). Dev Genes Evol 212:99–103

Genikhovich G, Technau U (2009) The starlet sea anemone *Nematostella vectensis*: an anthozoan model organism for studies in comparative genomics and functional evolutionary developmental biology. Cold Spring Harb Protoc 2009(9):pdb.emo29

Hand C, Uhliger KR (1992) The culture and asexual reproduction, and growth of the sea anemone *Nematostella vectensis*. Biol Bull 182:169–176

Yan J et al (2014) The evolutionary analysis reveals domain fusion of proteins with Frizzled-like CRD domain. Gene 533:229–239

Layden MJ et al (2012a) *Nematostella vectensis* achaete-scute homolog NvashA regulates embryonic ectodermal neurogenesis and represents an ancient component of the metazoan neural specification pathway. Development 139(5):1013–1022

Magie CR et al (2007) Gastrulation in the cnidarian *Nematostella vectensis* occurs via invagination not ingression. Dev Biol 305(2):483–497

Marlow HQ et al (2009) Anatomy and development of the nervous system of *Nematostella vectensis*, an anthozoan cnidarian. Dev Neurobiol 69(4):235–254

Marlow H et al (2013) Ectopic activation of the canonical wnt signaling pathway affects ectodermal patterning along the primary axis during larval development in the anthozoan *Nematostella vectensis*. Dev Biol 380 (2):324–334

Martindale MQ, Pang K, Finnerty JR (2004) Investigating the origins of tribloblasty: 'mesodermal' gene expression in a diploblastic animal, the sea anemone *Nematostella vectensis* (phylum Cnidaria; class Anthozoa). Development 131:2463–2474

Matus DQ et al (2007) FGF signaling in gastrulation and neural development in *Nematostella vectensis*, an anthozoan cnidarian. Dev Genes Evol 217(2):137–148

Matus DQ et al (2008) The Hedgehog gene family of the cnidarian, *Nematostella vectensis*, and implications for understanding metazoan Hedgehog pathway evolution. Dev Biol 313(2):501–518

Miller DJ (2008) Cryptic complexity captured: the *Nematostella* genome reveals its secrets. Trends Genet 24(1):1–4

Moran Y et al (2012a) Recurrent horizontal transfer of bacterial toxin genes to eukaryotes. Mol Biol Evol 29 (9):2223–2230

Putnam NH et al (2007a) Sea anemone genome reveals ancestral eumetazoan gene repertoire and genomic organization. Science 317(5834):86–94

Rentzsch F et al (2006a) Asymmetric expression of the BMP antagonists chordin and gremlin in the sea anemone *Nematostella vectensis*: implications for the evolution of axial patterning. Dev Biol 296(2):375–387

Rentzsch F et al (2008a) FGF signalling controls formation of the apical sensory organ in the cnidarian *Nematostella vectensis*. Development 135(10):1761–1769

Ryan JF et al (2007a) Pre-bilaterian origins of the *Hox* cluster and the *Hox* code: evidence from the sea anemone, *Nematostella vectensis*. PLoS One 2(1):e153

Scholz CB, Technau U (2003) The ancestral role of *Brachyury*: expression of *NemBra1* in the basal cnidarian *Nematostella vectensis* (Anthozoa). Dev Genes Evol 212:563–570

Sinigaglia C et al (2013a) The bilaterian head patterning gene *six3/6* controls aboral domain development in a cnidarian. PLoS Biol 11(2):e1001488

Steele RE, David CN, Technau U (2011) A genomic view of 500 million years of cnidarian evolution. Trends Genet 27(1):7–13

Steinmetz PR et al (2012a) Independent evolution of striated muscles in cnidarians and bilaterians. Nature 487(7406):231–234

Sullivan J (2007) Conserved and novel *Wnt* clusters in the basal eumetazoan *Nematostella vectensis*. Dev Genes Evol 217(3):235–239

Sullivan JC, Finnerty JR (2007) A surprising abundance of human disease genes in a simple "basal" animal, the starlet sea anemone (*Nematostella vectensis*). Genome 50(7):689–692

Caenorhabditis Elegans: **Books**

Hall DH, Altun ZF (2008) *C. elegans* Atlas. Cold Spring Harbor Laboratory Press, Cold Spring Harbor, NY

Hope IA (2000) *C. elegans*: a practical approach. Oxford University Press, Oxford

Strange K (2010) *C. elegans*: methods and applications, Methods in molecular biology. Humana Press, Totowa, NJ

Articles

Bischoff M, Schnabel R (2006) Global cell sorting is mediated by local cell–cell interactions in the *C. elegans* embryo. Dev Biol 294(2):432–444

Hanna M et al (2013) Worming our way ín and out of the Caenorhabditis elegans germline and developing embryo. Traffic 14(5):471–478

Hendriks G-J et al (2014) Extensive oscillatory gene expression during C. elegans larval development. Mol Cell 53(3):380–392

Modzelewska K et al (2013) Neurons refine the Caenorhabditis elegans body plan by directing axial patterning by Wnts. PLoS Biol 11(1):e1001465

Noble S et al (2008) Maternal mRNAs are regulated by diverse P body-related mRNP granules during early *Caenorhabditis elegans* development. J Cell Biol 182 (3):559–572

Oldenbroek M et al (2013) Regulation of maternal Wnt mRNA translation in C. elegans embryos. Development 140(22):4614–4623

Sawa H (2012) Control of cell polarity and asymmetric division in C. elegans. Curr Top Dev Biol 101:55–76

Schnabel R et al (2006) Global cell sorting in the *C. elegans* embryo defines a new mechanism for pattern formation. Dev Biol 294(2):418–431

Sheth U et al (2010) Perinuclear P granules are the principal sites of mRNA export in adult *C. elegans* germ cells. Development 137(8):1305–1314

Updike D, Strome S (2010) P granule assembly and function in *Caenorhabditis elegans* germ cells. J Androl 31 (1):53–60

Voronina E, Seydoux G (2010) The *C. elegans* homolog of nucleoporin Nup98 is required for the integrity and function of germline P granules. Development 137 (9):1441–1450

Walston TD, Hardin J (2006) Wnt-dependent spindle polarization in the early *C. elegans* embryo. Semin Cell Dev Biol 17(2):204–213

Spiralia, *Platynereis*

Ackermann CH et al (2005a) Clonal domains in postlarval *Platynereis dumerilii* (Annelida: Polychaeta). J Morphol 266:258–280

Arendt D (2003a) Spiralians in the limelight. Genome Biol 5(1):303

Arendt D, Wittbrodt J (2001) Reconstructing the eyes of Urbilateria. Philos Trans R Soc Lond B Biol Sci 356:1545–1563

Arendt D et al (2002a) Development of pigment-cup eyes in the polychaete *Platynereis dumerilii* and evolutionary conservation of larval eyes in Bilateria. Development 129:1143–1154

Asadulina A et al (2012) Whole-body gene expression pattern registration in Platynereis larvae. Evodevo 3:27

Backfisch B et al (2013a) Stable transgenesis in the marine annelid *Platynereis dumerilii* sheds new light on photoreceptor evolution. Proc Natl Acad Sci USA 110:193–198

Conzelmann M et al (2013) The neuropeptide complement of the marine annelid Platynereis dumerilii. BMC Genomics 14:Article No. 906

Demilly A et al (2013) Involvement of the Wnt/beta-catenin pathway in neurectoderm architecture in *Platynereis dumerilii*. Nat Commun 4:1915

Dray N et al (2011) Hedgehog signaling regulates segment formation in the annelid *Platynereis*. Science 329(5989):339–342

Ferrier DE (2012) Evolutionary crossroads in developmental biology: annelids. Development 139:2643–2653

Fischer A, Dorresteijn AC (2004) The polychaete *Platynereis dumerilii* (Annelida): a laboratory animal with spiralian cleavage, lifelong segment proliferation and a mixed benthic/pelagic life cycle. Bioessays 26 (3):314–325

Fischer AHL, Heinrich T, Arendt D (2010) The normal development of *Platynereis dumerilii*. Front Zool 7: Article No. 31

Freeman G, Lundelius JW (1982) The developmental genetics of dextrality and sinistrality in the gastropod *Lymnea peregra*. Roux's Arch Dev Biol 191:69–83

Gazave E et al (2013) Posterior elongation in the annelid *Platynereis dumerilii* involves stem cells molecularly related to primordial germ cells. Dev Biol 382:246–267

Hauenschild C, Fischer A (1996) *Platynereis dumerilii*. Mikroskopische Anatomie, Fortpflanzung, Entwicklung. Großes Zoologisches Praktikum. Fischer-Verlag, Stuttgart

Hui Jerome HL et al (2012a) Extensive chordate and annelid macrosynteny reveals ancestral homeobox gene organization. Mol Biol Evol 29:157–165

Lambert JD (2010a) Developmental patterns in spiralian embryos. Curr Biol 20(2):R72–R77

Martin-Duran JM (2012) Planarian embryology in the era of comparative developmental biology. Int J Dev Biol 56:39–48

Nielsen C (2010) Some aspects of spiralian development. Acta Zool 91(1):20–28

Pfeifer K et al (2012) Activation of genes during caudal regeneration of the polychaete annelid. Dev Genes Evol 222(3):165–179

Raible F, Arendt D (2004a) Metazoan evolution: some animals are more equal than others. Curr Biol 14: R106–R108

Randel N et al (2013a) Expression dynamics and protein localization of rhabdomeric opsins in *Platynereis* larvae. Integr Comp Biol 53:7–16

Rebscher N et al (2012) Hidden in the crowd: primordial germ cells and somatic stem cells in the mesodermal posterior growth zone of the polychaete Platynereis dumerillii are two distinct cell populations. Evodevo 3:9

Schneider SQ, Bowerman BA (2006) Asymmetry of beta-catenin localization in anterior-posterior cell divisions in the spiral-cleaving polychaete *Platynereis*. Dev Biol 296(1):343

Simakov O et al (2013a) Linking micro- and macro-evolution at the cell type level: a view from the lophotrochozoan *Platynereis dumerilii*. Brief Funct Genomics 12(5):430–439

Simakov O et al (2013b) Insights into bilaterian evolution from three spiralian genomes. Nature 493 (7433):526–531

Tessmar-Raible K, Arendt T (2003) Emerging systems: between vertebrates and arthropods, the Lophotrochozoa. Curr Opin Genet Dev 13:1–10

Tessmar-Raible K, Arendt D (2005a) New animal models for evolution and development. Genome Biol 6(1):303

Zantke J et al (2013) Circadian and circalunar clock interactions in a marine annelid. Cell Rep 5:99–113

Drosophila Melanogaster: Books

Campos-Ortega JA, Hartenstein V (2013) The embryonic development of *Drosophila melanogaster*, 2nd edn. Springer, Berlin

Dahmann C (2010) *Drosophila*: methods and protocols, vol 420, Methods in molecular biology. Humana Press, Totowa, NJ

Articles

Barckmann B, Simonelig M (2013) Control of maternal mRNA stability in germ cells and early embryos. Biochim Biophys Acta 1829(6–7):714–724

Becalska A, Gavis E (2009a) Lighting up mRNA localization in *Drosophila* oogenesis. Development 136 (15):2493–2503

Bejsovec A (2013a) Wingless/Wnt signaling in Drosophila: the pattern and the pathway. Mol Reprod Dev 80 (11):882–894

Bischoff M et al (2013a) Cytonemes are required for the establishment of a normal Hedgehog morphogen gradient in Drosophila epithelia. Nat Cell Biol 15 (11):1269

Cha B et al (2001) In vivo analysis of Drosophila *bicoid* mRNA localization reveals a novel microtubule-dependent axis specification pathway. Cell 106:35–46

Jun C et al (2013) Cytoplasmic polyadenylation is a major mRNA regulator during oogenesis and egg activation in Drosophila. Dev Biol 383(1):121–131

Deng W, Lin H (2001) Asymmetric germ cell division and oocyte determination during *Drosophila* oogenesis. Int Rev Cytol 203:93–138

Driever W, Nüsslein-Volhard C (1988a) A gradient of bicoid protein in *Drosophila* embryos. Cell 54:83–93

Driever W, Nüsslein-Volhard C (1988b) The *bicoid* protein determines position in the *Drosophila* embryo in a concentration-dependent manner. Cell 54:95–104

Driever W, Siegel V, Nüsslein-Volhard C (1990) Autonomous determination of anterior structures in the early *Drosophila* embryo by the bicoid morphogen. Development 109:811–820

Dubuis JO et al (2013a) Positional information, in bits. Proc Natl Acad Sci USA 110(41):16301–16308

Grimm O, Coppey M, Wieschaus E (2010a) Modelling the bicoid gradient. Development 137(14):2253–2264

Halder G, Callaerts P, Gehring W (1995a) Induction of ectopic eyes by targeted expression of the *eyeless* gene in *Drosophila*. Science 267(5205):1788–1792

Jaeger J, Reinitz J (2012) Drosophila blastoderm patterning. Curr Opin Genet Dev 22(6):533–541

Jones J et al (2007) Oskar controls morphology of polar granules and nuclear bodies in *Drosophila*. Development 134(2):233–236

Kimberly D et al (2007) Transdetermination: *Drosophila* imaginal disc cells exhibit stem cell-like potency. Int J Biochem Cell Biol 39(6):1105–1118

Kozlov K et al (2012a) Modeling of gap gene expression in Drosophila Kruppel mutants. PLoS Comput Biol 8 (8):e1002635

Kugler J-M, Lasko P (2009) Localization, anchoring and translational control of *oskar, gurken, bicoid* and *nanos* mRNA during *Drosophila* oogenesis. Fly 3 (1):15–28

Lasko P (2012a) mRNA localization and translational control in Drosophila oogenesis. Cold Spring Harb Perspect Biol 4(10):a012294

Layalle S et al (2011) Engrailed homeoprotein acts as a signaling molecule in the developing fly. Development 138:2315–2323

Lewis EB (1978) A gene complex controlling segmentation. Nature 276:565–570

Lindsay SA, Wasserman SA (2014) Conventional and non-conventional Drosophila Toll signaling. Dev Comp Immunol 42:16–24

Feng L et al (2013) Dynamic interpretation of maternal inputs by the Drosophila segmentation gene network. Proc Natl Acad Sci USA 110(17):6724–6729

Lopes F et al (2012) The role of bicoid cooperative binding in the patterning of sharp borders in *Drosophila melanogaster*. Dev Biol 370(2):165–172

Lucas T et al (2013) Live imaging of bicoid-dependent transcription in Drosophila embryos. Curr Biol 23 (21):2135–2139

Maeda R et al (2009) The Bithorax complex of *Drosophila*: an exceptional *Hox* cluster. Curr Top Dev Biol 88:1–33

Mani SR (2014) PIWI proteins are essential for early Drosophila embryogenesis. Dev Biol 385(2):340–349

Marques G et al (1997) Production of a DPP activity gradient in the early *Drosophila* embryo through the opposing actions of the SOG and TLD proteins. Cell 91:417–426

Papatsenko D (2009a) Stripe formation in the early fly embryo: principles, models, and networks. Bioessays 31(11):1172–1180

Perry M et al (2012) Precision of hunchback expression in the Drosophila embryo. Curr Biol 22(23):2247–2252

Pfeiffer S, Vincent JP (1999) Signalling at a distance: transport of Wingless in the embryonic epidermis of *Dosophila*. Sem Cell Dev Biol 10:303–309

Rangan P et al (2009) Temporal and spatial control of germ-plasm RNAs. Curr Biol 19(1):72–77

Rosenberg M et al (2009) Heads and tails: evolution of antero-posterior patterning in insects. Biochim Biophys Acta 1789(4):333–342

Sanghavi P et al (2013) Dynein associates with oskar mRNPs and is required for their efficient net plus-end localization in Drosophila oocytes. PLoS One 8 (11):Article No: e80605

Shvartsman S, Berezhkovskii AM (2008) Dynamics of maternal morphogen gradients in *Drosophila*. Curr Opin Genet Dev 18(4):342–347

Snee MJ, Macdonald PM (2004) Live imaging of nuage and polar granules: evidence against a precursor-product relationship and a novel role for Oskar in stabilization of polar granule components. J Cell Sci 117:2109–2120

Sokolowski TR (2012a) Mutual repression enhances the steepness and precision of gene expression boundaries. PLoS Comput Biol 8(8):Article No.: e1002654

Sorge S et al (2012) The cis-regulatory code of *Hox* function in Drosophila. EMBO J 31(15):3323–3333

Struhl G, Struhl K, MacDonald PM (1989) The gradient morphogen bicoid is a concentration-dependent transcriptional activator. Cell 57:1259–1273

Surkova S et al (2013) Quantitative dynamics and increased variability of segmentation gene expression in the Drosophila *Kruppel* and *knirps* mutants. Dev Biol 376:99–112

Ai-Guo T et al (2013a) Efficient EGFR signaling and dorsal-ventral axis patterning requires syntaxin dependent Gurken trafficking. Dev Biol 373(2):349–358

Urbach R, Technau GM (2003a) Segment polarity and DV patterning gene expression reveals segmental organization in the Drosophila brain. Development 130:3607–3620

Vanzo NF, Ephrussi A (2002) Oskar anchoring restricts pole plasm formation to the posterior of the Drosophila oocyte. Development 129:3705–3714

Tunicates, Ascidians: Books

Sawada H, Yokosawa H (2001) The biology of ascidians. Springer, Berlin

Articles

Conklin EG (1905) Mosaic development in ascidian eggs. J Exp Zool 2:145–223

Ikuta T et al (2010) Limited functions of *Hox* genes in the larval development of the ascidian *Ciona intestinalis*. Development 137(9):1505–1513

Lemaire P (2011) Evolutionary crossroads in developmental biology: the tunicates. Development 138:2143–2152

Meedel TH, Farmer SC, Lee JJ (1997) The single MyoD family gene of *Ciona intestinalis* encodes two differentially expressed proteins: implications for the evolution of chordate muscle gene regulation. Development 124:1711–1721

Nishida H (2005) Specification of embryonic axis and mosaic development in ascidians. Dev Dyn 233(4):1177–1193

Ogura Y, Sasakura Y (2013) Ascidians as excellent models for studying cellular events in the chordate body plan. Biol Bull 224(3):227–236

Prodon F et al (2006) Establishment of animal–vegetal polarity during maturation in ascidian oocytes. Dev Biol 290(2):297–311

Rinkevich Y et al (2007) Urochordate whole body regeneration inaugurates a diverse innate immune signaling profile. Dev Biol 312(1):131–146

Rinkevich B et al (2012) The candidate *Fu/HC* gene in *Botryllus schlosseri* (Urochordata) and ascidians' historecognition – an oxymoron? Dev Comp Immunol 36(4):718–727

Sardet C et al (1989) Fertilization and ooplasmic movements in the ascidian egg. Development 105:237–249

Schubert M et al (2006) Amphioxus and tunicates as evolutionary model systems. Trends Ecol Evol 21(5):269–277

Yamada A, Nishida H (1996) Distribution of cytoplasmic determinants in unfertilized eggs of the ascidian *Halocynthia roretzi*. Dev Genes Evol 206:297–304

Yasuo H, Satoh N (1998) Conservation of the developmental role of *Brachyury* in notochord formation in a urochordate, the ascidian *Halocynthioa roretzi*. Dev Biol 200:158–170

Chapter 5 Model Organisms II: Vertebrates: *Xenopus*, Amphibians, Books

Hausen P, Riebesoll M (1991) The early development of *Xenopus laevis*. Springer, Berlin

Kloc M, Kubiak JZ (2014) Xenopus development. Blackwell, Hoboken, NJ

Nieuwkoop PD, Faber J (1975) Normal table of *Xenopus laevis* (Daudin), 2nd edn. North-Holland, Amsterdam

Spemann H (1938b) Embryonic development and induction. Yale University Press, New Haven (Reprinted by Hafner, New York, 1962)

Articles

Acosta H et al (2011a) Notch destabilises maternal β-catenin and restricts dorsal-anterior development in *Xenopus*. Development 138:2567–2579

Bestman JE, Cline HT (2014) Morpholino studies in *Xenopus* brain development. Methods Mol Biol 1082:155–171

Blum M et al (2009a) *Xenopus,* an ideal model system to study vertebrate left-right asymmetry. Dev Dyn 238(6):1215–1225

Bilge B et al (2006) Vg1 is an essential signaling molecule in *Xenopus* development. Development 133:15–20

Sang-Wook C et al (2008) Wnt5a and Wnt11 interact in a maternal Dkk1-regulated fashion to activate both canonical and non-canonical signaling in *Xenopus* axis formation. Development 135(22):3719–3729

Danilchik M et al (2013a) Blastocoel-spanning filopodia in cleavage-stage *Xenopus laevis*: potential roles in morphogen distribution and detection. Dev Biol 382(1):70–81

De Robertis EM et al (2000a) The establishment of Spemann's organizer and patterning of the vertebrate embryo. Nat Rev Genet 3:171–181

De Robertis EM, Kuroda H (2004) Dorsal-ventral patterning and neural induction in *Xenopus* embryos. Annu Rev Cell Dev Biol 20(1):285–308

Dosch R et al (1997) BMP-4 acts as a morphogen in dorsoventral mesoderm patterning in *Xenopus*. Development 124:2325–2334

Dosch R, Niehrs C (2000) Requirement for anti-dorsalizing morphogenetic protein in organizer patterning. Mech Dev 90:195–203

Gerhart J et al (1989a) Cortical rotation of the *Xenopus* egg: consequences of the antero-posterior pattern of embryonic dorsal development. Development 107(Suppl):37–51

Glinka A et al (1998) Dickkopf-1 is a member of a new family of secreted proteins and functions in head induction. Nature 391:357–362

Hashiguchi M, Mullins MC (2013) Anteroposterior and dorsoventral patterning are coordinated by an identical patterning clock. Development 140(9):1970–1980

Heasman J (2006a) Patterning the early *Xenopus* embryo. Development 133(7):1205–1217

Heeg-Truesdella E, LaBonnea C (2006) Neural induction in *Xenopus* requires inhibition of Wnt-β-catenin signaling. Dev Biol 298(1):71–86

Henry JJ, Grainger RM (1990a) Early tissue interactions leading to embryonic lens formation in *Xenopus laevis*. Dev Biol 141:149–163

Gun-Hwa K et al (2013) β-Arrestin 1 mediates non-canonical Wnt pathway to regulate convergent extension movements. Biochem Biophys Res Commun 435(2):182–187

Linker C et al (2009a) Cell communication with the neural plate is required for induction of neural markers by BMP inhibition: evidence for homeogenetic induction and implications for *Xenopus* animal cap and chick explant assays. Dev Biol 327(2):478–486

Wenyan M et al (2013) Maternal Dead-End1 is required for vegetal cortical microtubule assembly during Xenopus axis specification. Development 140(11):2334–2344

Niehrs C et al (2001) Dickkopf 1 and the Spemann-Mangold head organizer. Int J Dev Biol 45:237–240

Niehrs C (2004) Regionally specific induction by the Spemann-Mangold organizer. Nat Rev Genet 5:425–434

Niehrs C (2010a) On growth and form: a Cartesian coordinate system of Wnt and BMP signaling specifies bilaterian body axes. Development 137:845–857

Nieuwkoop PD (1999) The neural induction process; its morphogenetic aspects. Int J Dev Biol 43(7):614–623

Nijjar S, Woodland HR (2013a) Localisation of RNAs into the germ plasm of vitellogenic Xenopus oocytes. PLoS One 8(4):e61847

Pere EM et al (2014a) Active signals, gradient formation and regional specificity in neural induction. Exp Cell Res 321:25–31

Piccolo S et al (1997) Dorsoventral patterning in *Xenopus:* inhibition of ventral signals by direct binding of chordin to BMP-4. Cell 86:589–598

Plouhineca J-L et al (2011) Systems control of BMP morphogen flow in vertebrate embryos. Curr Opin Genet Dev 21:1–8

Plouhineca J-L et al (2013) Chordin forms a self-organizing morphogen gradient in the extracellular space between ectoderm and mesoderm in the Xenopus embryo. Proc Natl Acad Sci USA 110(51):20372–20379

Reid CD et al (2012) Transcriptional integration of Wnt and Nodal pathways in establishment of the Spemann organizer. Dev Biol 368(2):231–241

Revinski DR et al (2010a) Delta-Notch signaling is involved in the segregation of the three germ layers in *Xenopus laevis*. Dev Biol 339(2):477–492

Schneider S, Steinbeisser H, Warga RM, Hausen P (1996a) ß-Catenin translocation into nuclei demarcates the dorsalizing centers in frog and fish embryos. Mech Dev 57:191–198

Simeonia L, Gurdon JB (2006) Interpretation of BMP signaling in early *Xenopus* development. Dev Biol 308(1):82–92

Tada M, Heisenberg C-P (2012a) Convergent extension: using collective cell migration and cell intercalation to shape embryos. Development 139(21):3897–3904

Vonica A, Gumbiner BM (2007a) The *Xenopus* Nieuwkoop center and Spemann–Mangold organizer share molecular components and a requirement for maternal Wnt activity. Dev Biol 312(1):90–102

Wallingford JB et al (2002) Convergent extension: the molecular control of polarized cell movement during embryonic development. Dev Cell 2:695–706

Wills A et al (2010a) BMP antagonists and FGF signaling contribute to different domains of the neural plate in *Xenopus*. Dev Biol 337(2):335–350

Winklbauer R, Mueller H-AJ (2011) Mesoderm layer formation in *Xenopus* and *Drosophila* gastrulation. Phys Biol 8(4):No: 045001

Ying C et al (2007) POU-V factors antagonize maternal VegT activity and beta-catenin signaling in *Xenopus* embryos. EMBO J 26:2942–2954

Danio rerio (zebrafish) and medaka: Books

Bryson-Richardson R et al (2011) Atlas of Zebrafish Development. Academic, London

Holden J et al (2013) The Zebrafish. Cambridge University Press, Cambridge

Naruse K et al (2011) Medaka: a model for organogenesis, human disease, and evolution. Springer, Berlin

Nuesslein-Volhard C, Dahm R (2002) Zebrafish: a practical approach. Oxford University Press, Oxford

Articles

Abrams EW, Mullins MC (2009a) Early zebrafish development: it's in the maternal genes. Curr Opin Genet Dev 2009:396–403

Behrndt M et al (2012) Forces driving epithelial spreading in zebrafish gastrulation. Science 338(6104):257–260

Brunet T (2013) Evolutionary conservation of early mesoderm specification by mechano-transduction in Bilateria. Nat Commun 4:2821

Castanon I, González-Gaitán M (2011) Oriented cell division in vertebrate embryogenesis. Curr Opin Cell Biol 23(6):697–704

Dohn MR et al (2013) Planar cell polarity proteins differentially regulate extracellular matrix organization and assembly during zebrafish gastrulation. Dev Biol 383:39–51

Furutani-Seiki M, Wittbrodt J (2004) Medaka and zebrafish, an evolutionary twin study. Mech Dev 121:629–637

Grandel H et al (2006) Neural stem cells and neurogenesis in the adult zebrafish brain: origin, proliferation dynamics, migration and cell fate. Dev Biol 295(1):263–277

Hamada H et al (2014a) Involvement of Delta/Notch signaling in zebrafish adult pigment stripe patterning. Development 141(2):318–324

Hisano Y et al (2014a) Genome editing using artificial site-specific nucleases in zebrafish. Dev Growth Differ 56:26–33

Holley SA, Nüsslein-Volhard C (2000) Somitogenesis in zebrafish. Curr Top Dev Biol 2000:47247–47277

Hong E, Brewster R (2006) N-cadherin is required for the polarized cell behaviors that drive neurulation in the zebrafish. Development 133:3895–3905

Kelly C et al (2000) Maternally controlled ß-catenin-mediated signaling is required for organizer formation in the zebrafish. Development 127:3899–3911

Lee MT et al (2013) Nanog, Pou5f1 and SoxB1 activate zygotic gene expression during the maternal-to-zygotic transition. Nature 503(7476):360–364

Lepage SE et al (2010) Zebrafish epiboly: mechanics and mechanisms. Int J Dev Biol 54(8–9):1213–1228

Lepage SE et al (2014) Zebrafish Dynamin is required for maintenance of enveloping layer integrity and the progression of epiboly. Dev Biol 385:52–66

Lindeman RE, Pelegri F (2010a) Vertebrate maternal-effect genes: insights into fertilization, early cleavage divisions, and germ cell determinant localization from studies in the zebrafish. Mol Reprod Dev 77 (4):299–313

Xiuli L et al (2013) Nodal promotes mir206 expression to control convergence and extension movements during zebrafish gastrulation. J Genet Genomics 40 (10):515–521

Martin ED, Grealy M (2004) Plakoglobin expression and localization in zebrafish embryo development. Biochem Soc Trans 32(5):797–798

Moriarty MA et al (2012) Loss of plakophilin 2 disrupts heart development in zebrafish. Int J Dev Biol 56 (9):711–718

Nüsslein-Volhard C (2012) The zebrafish issue of development. Development 139:4099–4103

O'Boyle S et al (2007) Identification of zygotic genes expressed at the midblastula transition in zebrafish. Biochem & Biophys Res Commun 358(2):462–468

Onichtchouk D (2012) *Pou5f1/oct4* in pluripotency control: insights from zebrafish. J Genet Dev 50(2):75–85

Prince VE et al (1998) Zebrafish *hox* genes: genomic organization and modified colinear expression patterns in the trunk. Development 125:407–420

Ramialison M et al (2012) Cis-regulatory properties of medaka synexpression groups. Development 139 (5):917–928

Rembold M et al (2006a) Transgenesis in fish: efficient selection of transgenic fish by co-injection with a fluorescent reporter construct. Nat Protoc 1 (3):1133–1139

Roberts JA et al (2014) Targeted transgene integration overcomes variability of position effects in zebrafish. Development 141(3):715–724

Schmidt R et al (2013) Neurogenesis in zebrafish – from embryo to adult. Neural Dev 8:Article No. 3

Solnica-Krezela L (2006) Gastrulation in zebrafish – all just about adhesion? Curr Opin Genet Dev 16 (4):433–441

Takeda H (2008) Draft genome of the medaka fish: a comprehensive resource for medaka developmental genetics and vertebrate evolutionary biology. Dev Growth Differ 50(suppl 1):S157–S166

Wittbrodt J, Meyer A, Schartl M (1998) More genes in fish? Bioessays 200:611–613

Wittbrodt J, Shima A, Schartl M (2002) Medaka – a model organism from far east. Nat Rev Genet 3:53–64

Wylie AD et al (2014a) Post-transcriptional regulation of wnt8a is essential to zebrafish axis development. Dev Biol 386(1):53–63

Bird, Chick: Books, Methods

Bronner-Fraser M (2011) Avian embryology. Academic, London

Freeman BM (2013) Developments of the avian embryo: a behavioural and physiological study. Springer, Berlin

Articles

Benazeraf B, Olivier P (2013) Formation and segmentation of the vertebrate body axis. Annu Rev Cell Dev Biol 29:1–26

Boettger T et al (2001a) The avian organizer. Int J Dev Biol 45(1):281–287

Chuai M et al (2012) Collective epithelial and mesenchymal cell migration during gastrulation. Curr Genomics 13(4):267–277

Kain KH (2014) The chick embryo as an expanding experimental model for cancer and cardiovascular research. Dev Dyn 243(2):216–228

Khaner O (2007) The importance of the posterior midline region for axis initiation at early stages of the avian embryo. Int J Dev Biol 51:131–137

Kochav S, Eyal-Giladi H (1971) Bilateral symmetry in chick embryo: determination by gravity. Science 171:1027–1029

Le Douarin N (2008) Developmental patterning deciphered in avian chimeras. Dev Growth Differ 50 (suppl 1):S11–S28

Le Douarin NM, Dieterlen-Lievre F (2013) How studies on the avian embryo have opened new avenues in the understanding of development: a view about the neural and hematopoietic systems. Dev Growth Differ 55:1–14

Stern C (2006) Evolution of the mechanisms that establish the embryonic axes. Curr Opin Genet Dev 16 (4):413–418

Stern CD, Canning DR (1990) Origin of cells giving rise to mesoderm and endoderm in chick embryo. Nature 343:273–275

Streit A, Stern CD (2014) Transplantation of neural tissue: quail-chick chimeras. Methods Mol Biol 1082:235–251

Theveneau E, Mayor R (2012) Neural crest delamination and migration: from epithelium-to-mesenchyme transition to collective cell migration. Dev Biol 366:34–54

Tsikolia N et al (2012) Paraxial left-sided nodal expression and the start of left-right patterning in the early chick embryo. Differentiation 84(5):380–391

Viebahn C (2001) Hensen's node. Genesis 29:96–103

Mouse: Books, Methods

Gertsenstein M, Nagy K (2013) Manipulating the mouse embryo: a laboratory manual, 4th edn. Cold Spring Harbor Laboratory, Cold Spring Harbor, NY

Kaufman MH (1992) The atlas of mouse development. Elsevier, San Diego

Kubiak JZ (2012) Mouse development: from oocyte to stem cells. Springer, Berlin

Lewandoski M (2013) Mouse molecular embryology. Humana Press, New York

Wassarman PM, Soriano PM (2010) Guide to techniques in mouse development, Part A: mice, embryos, and cells: 476. Academic, San Diego

Articles

Arkell RM et al (2013) Wnt signalling in mouse gastrulation and anterior development: new players in the pathway and signal output. Curr Opin Genet Dev 23(4):454–460

Engert S et al (2013) Wnt/beta-catenin signalling regulates Sox17 expression and is essential for organizer and endoderm formation in the mouse. Development 140(15):3128–3138

Hiramatsu R et al (2013a) External mechanical cues trigger the etablishment of the anterior-posterior axis in early mouse embryos. Dev Cell 27(2):131–144

Hudson QJ et al (2010) Genomic imprinting mechanisms in embryonic and extraembryonic mouse tissues. Heredity 105(1):45–56

Ichikawa T et al (2013) Live imaging of whole mouse embryos during gastrulation: migration analyses of epiblast and mesodermal cells. PLoS One 8(7):e64506

Lei L et al (2013) The maternal to zygotic transition in mammals. Mol Aspects Med 34:919–938

R'ada M et al (2014a) Morphogenetic movements in the neural plate and neural tube: mouse. Wiley Interdiscip Rev Dev Biol 3:59–68

Nowotschin S et al (2010) Cellular dynamics in the early mouse embryo: from axis formation to gastrulation. Curr Opin Genet Dev 20(4):420–427

Saiz N, Plusa B (2013) Early cell fate decisions in the mouse embryo. Reproduction 145(3):R65–R80

Sasaki H (2010) Mechanisms of trophectoderm fate specification in preimplantation mouse development. Dev Growth Differ 52(3):263–273

Stephenson RO et al (2012) Intercellular interactions, position, and polarity in establishing blastocyst cell lineages and embryonic axes. Cold Spring Harb Perspect Biol 4(11):a008235

Chapter 6 Human Embryology: Books

Gardner DK et al (2013) Human gametes and preimplantation embryos: assessment and diagnosis. Springer, Berlin

Reyes DE (2013) Embryo development: stages, mechanisms and clinical outcomes. Nova Science, New York

Varley H (2004) Human embryology. CBS, New York

Articles

Conaghan J (2014) Time-lapse imaging of preimplantation embryos. Semin Reprod Med 32(2):134–140

Filicori M (1999) The role of luteinizing hormone in folliculogenesis and ovulation induction. Fertil Steril 71:405–414

Ishida M, Moor G (2013) The role of imprinted genes in humans. Mol Aspects Med 34(4):826–840

Munn DH et al (1998) Prevention of allogeneic fetal rejection by tryptophan catabolism. Science 281:1131–1193

Niakan KK et al (2012) Human pre-implantation embryo development. Development 139:829–841

Rouas-Freiss N et al (1997) Direct evidence to support the role of HLA-G in protecting the fetus from maternal uterine Natural Killer cells cytolysis. Proc Natl Acad Sci USA 94:11520–11525

Schumacher A et al (2013) Human chorionic gonadotropin as a central regulator of pregnancy immune tolerance. J Immunol 190(6):2650–2658

Van de Velde H et al (2008) The four blastomeres of a 4-cell stage human embryo are able to develop individually into blastocysts with inner cell mass and trophectoderm. Hum Reprod 23(8):1742–1747

Xue L et al (2013) Global expression profiling reveals genetic programs underlying the developmental divergence between mouse and human embryogenesis. BMC Genomics 14:568

Articles

Cabej N (2012) Neural control of postphylotypic development. Epigenetic principles of evolution. Elsevier, Amsterdam, pp 147–228

Fleming A et al (2004a) A central role for the notochord in vertebral patterning. Development 131:873–880

Graham A, Richardson J (2012) Developmental and evolutionary origins of the pharyngeal apparatus. Evodevo 3:Article No. 24

Richards RJ (2009) The tragic sense of life: Ernst Haeckel and the struggle over evolutionary thought. University of Chicago Press, Chicago, IL

Richardson MK (1995) Heterochrony and the phylotypic period. Dev Biol 172:412–421

Richardson MK (2012) A phylotypic stage for all animals? Dev Cell 22(5):903–904

Svorcova J (2012) The phylotypic stage as a boundary of modular memory: non mechanistic perspective. Theory Biosci 131:31–42

Human Reproduction Box 6A-C: Books

Elder K et al (2010) In-vitro fertilization. Cambridge University Press, Cambridge

Wolf DP (2013) Human in vitro fertilization and embryo transfer. Springer, Berlin

Articles

Aiken C, Ozanne SE (2014a) Transgenerational developmental programming. Hum Reprod Update 20:63–75

Amato P et al (2013) Oocyte or embryo donation to women of advanced age: a committee opinion. Fertil Steril 100(2):337–340

Caserta D et al (2008) Impact of endocrine disruptor chemicals in gynaecology. Hum Reprod Update 14(1):59–72

Chiang T et al (2012) Meiotic origins of maternal age-related aneuploidy. Biol Reprod 86:3

Hales BF (1999) Thalidomide on the comeback trail. Will new insights into Thalidomide's teratogenic mechanism help make its return a safe one? Nat Med 5(5):489–490

Jessberger R (2012) Age-related aneuploidy through cohesion exhaustion. EMBO Rep 13(6):539–546

Jones KT (2008) Meiosis in oocytes: predisposition to aneuploidy and its increased incidence with age. Hum Reprod Update 14(2):143–158

Kalmbach KH et al (2013) Telomeres and human reproduction. Fertil Steril 99:23–29

Krausz C, Chianese C (2013) The role of aging on fecundity in the male. In: Carrel DT (ed) Paternal influences on human reproductive success. Cambridge University Press, Cambridge, pp 70–81

Chuan L et al (2014) Bisphenol A exposure at an environmentally relevant dose induces meiotic abnormalities in adult male rats. Cell Tissue Res 355(1):223–232

Manfo F et al (2014) Adverse effects of Bisphenol A on male reproductive function. Rev Environ ContamToxicol 228:57–82

Matthiesen L et al (2012) Multiple pregnancy failures: an immunological paradigm. Am J Reprod Immunol 67(4):334–340

Meldrum DR (2013) Aging gonads, glands, and gametes: immutable or partially reversible changes? Fertil Steril 99:1–43

Nelson SM et al (2013) The ageing ovary and uterus: new biological insights. Hum Reprod Update 19:67–83

Ozturk S et al (2014) Telomere length and telomerase activity during oocyte maturation and early embryo development in mammalian species. Mol Hum Reprod 20:15–30

Sartorius GA, Nieschlag E (2010) Paternal age and reproduction. Hum Reprod Update 16(1):65–79

Seidel GE Jr (2009) Sperm sexing technology-the transition to commercial application. An introduction to the symposium "update on sexing mammalian sperm". Theriogenology 71(1):1–3

Shiverick KT, Salafia C (1999) Cigarette smoking and pregnancy I: ovarian, uterine and placental effects. Placenta 20:265–272

Sickmann HM et al (2014) Prenatal ethanol exposure has sex-specific effects on hippocampal long-term potentiation. Hippocampus 24(1):54–64

Stouffs K et al (2009) Male infertility and the involvement of the X chromosome. Hum Reprod Update 15(6):623–637

Talerman A et al (1990) True hermaphrodite with bilateral ovotestes, bilateral gonadoblastomas and dysgerminomas, 46XX/46 XY karyotype and a succesful pregnany. Cancer 66:2668–2672

Tilly JL, Sinclair DA (2013) Germline energetics, aging, and female infertility. Cell Metab 17(6):838–850

van Gelder MM et al (2014) Drugs associated with teratogenic mechanisms. Part II: a literature review of the evidence on human risks. Hum Reprod 29(1):168–183

Wyns C et al (2010) Options for fertility preservation in prepubertal boys. Hum Reprod Update 16(3):312–328

Chapter 7 Sex Determination, Sexual Development: Books

Maria I (2013) Hormonal and genetic basis of sexual differentiation disorders and hot topics in endocrinology. Springer, Berlin

Articles: Sex Determination, Sex Development

Biason-Lauber A (2012) WNT4, RSPO1, and FOXL2 in sex development. Semin Reprod Med 30(5):387–395

Bowles J, Koopman P (2010) Sex determination in mammalian germ cells: extrinsic versus intrinsic factors. Reproduction 139(6):943–958

Chassot A-A et al (2008) Activation of beta-catenin signaling by *Rspo1* controls differentiation of the mammalian ovary. Hum Mol Genet 17(9):1264–1277

Cutting A, Chue J, Smith CA (2013) Just how conserved is vertebrate sex determination? Dev Dyn 242 (4):380–387

Godwin J (2010) Neuroendocrinology of sexual plasticity in teleost fishes. Front Neuroendocrinol 31 (2):203–216

Kim Y et al (2006) Fgf9 and Wnt4 act as antagonistic signals to regulate mammalian sex determination. PLoS Biol 4(6):e187

Ting J et al (2013) The SOX gene family: function and regulation in testis determination and male fertility maintenance. Mol Biol Rep 40(3):2187–2194

Koopman P et al (1991) Male development of chromosomally female mice transgenic for *Sry*. Nature 351:117–121

Koopman P (2010) The delicate balance between male and female sex determining pathways: potential for disruption of early steps in sexual development. Int J Androl 33(2):252–258

Lau Y-F C, Li Y (2009) The human and mouse sex-determining SRY genes repress the Rspo1/beta-catenin signaling. J Genet Genomics 36(4):193–202

Matsumoto Y et al (2013) Changes in gonadal gene network by exogenous ligands in temperature-dependent sex determination. J Mol Endocrinol 50 (3):389–400

Mork L et al (2014) Predetermination of sexual fate in a turtle with temperature-dependent sex determination. Dev Biol 386(1):264–271

Navara KJ (2013) Hormone-mediated adjustment of sex ratio in vertebrates. Integr Comp Biol 53(6):877–887

Parma P, Radi O (2012) Molecular mechanisms of sexual development. Sex Dev 6(1–3):7–17

Salz H, Erickson JW (2010) Sex determination in *Drosophila*: the view from the top. Fly 4(1):60–70

Schwanz LE et al (2013) Novel evolutionary pathways of sex-determining mechanisms. J Evol Biol 26 (12):2544–2557

Sekido R, Lovell-Badge R (2013) Genetic control of testis development. Sex Dev 7:21–32

Svingen T, Koopman P (2013) Building the mammalian testis: origins, differentiation, and assembly of the component cell populations. Genes Dev 27 (22):2409–2426

Tomizuka K et al (2008) *R-spondin1* plays an essential role in ovarian development through positively regulating Wnt-4 signaling. Hum Mol Genet 17 (9):1278–1291

Trukhina A et al (2013) The variety of vertebrate mechanisms of sex determination. Biomed Res Int 2013:587460

Verhulst EC (2010) Insect sex determination: it all evolves around transformer. Curr Opin Genet Dev 20 (4):376–383

Wapstra E (2010) Sex allocation and sex determination in squamate reptiles. Sex Dev 4(1–2):110–118

Warner DA, Shine R (2008) The adaptive significance of temperature-dependent sex determination in a reptile. Nature 451(7178):566–568

Wilhelm D (2007) *R-spondin1* – discovery of the long-missing, mammalian female-determining gene? Bioessays 29:314–318

Yamauchi Y et al (2014) Two Y genes can replace the entire Y chromosome for assisted reproduction in the mouse. Science 343(6166):69–72

Zhenguo Z et al (2014) Evolution of the *sex-lethal* gene in insects and origin of the sex-determination system in Drosophila. J Mol Evol 78(1):50–65

Sexual Hormones

Carreau S, Galeraud-Denis I (2010) Aromatase, oestrogens and human male reproduction. Philos Trans R Soc Lond B Biol Sci 365(1546):1571–1579

Christensen A et al (2012) Hormonal regulation of female reproduction. Horm Metab Res 44(8):587–591

Morohashi K et al (2013) Steroid hormones and the development of reproductive organs. Sex Dev 7 (1–3):61–79

Simpson ER (2002) Aromatization of androgens in woman: current concepts and findings. Fertil Steril 77(Suppl 4):6–10

Simpson ER (2003) Sources of estrogen and their importance. J Steroid Biochem Mol Biol 86(3–5):225–230

Sinchak K, Wagner EJ (2012) Estradiol signaling in the regulation of reproduction and energy balance. Front Neuroendocrinol 33(4):342–363

Stocco C (2012) Tissue physiology and pathology of aromatase. Steroids 77(1–2):27–35

Visser J et al (2006) Anti-Müllerian hormone: a new marker for ovarian function. Reproduction 131(1):1–9

Gender, Brain, Behaviour: Books

Baum M (2014) Brain development and sexual preference. Morgan & Claypool, San Rafael

Balthazart J (2012) Brain development and sexual orientation. Morgan & Claypool Life Sciences, San Rafael

Articles

Argiolas A, Melis MR (2013) Neuropeptides and central control of sexual behaviour from the past to the present. Prog Neurobiol 108:80–107

Auyeung B et al (2013) Prenatal and postnatal hormone effects on the human brain and cognition. Pflugers Arch 465(5):557–571

Bancroft J, Graham CA (2011) The varied nature of women's sexuality: unresolved issues and a theoretical approach. Horm Behav 59(5):717–729

Ai-Min B, Swaab DF (2011) Sexual differentiation of the human brain: relation to gender identity, sexual orientation and neuropsychiatric disorders. Front Neuroendocrinol 32(2):214–226

Barha CK, Galea LAM (2010) Influence of different estrogens on neuroplasticity and cognition in the hippocampus. Biochim Biophys Acta 1800 (10):1056–1067

Bava S et al (2011) Sex differences in adolescent white matter architecture. Brain Res 1375:41–48

Berenbaum SA, Beltz AM (2011) Sexual differentiation of human behavior: effects of prenatal and pubertal organizational hormones. Front Neuroendocrinol 32 (2):183–200

Borrow AP, Cameron NM (2012) The role of oxytocin in mating and pregnancy. Horm Behav 61(3):266–276

Bramen JE et al (2011) Puberty influences medial temporal lobe and cortical gray matter differently in boys than girls matched for sexual maturity. Cereb Cortex 21(3):636–646

Kun-Hsien C et al (2011) Sex-linked white matter microstructure of the social and analytic brain. Neuroimage 54(1):725–733

Chung WC, Auger AP (2013) Gender differences in neurodevelopment and epigenetics. Pflugers Arch 465(5):573–584

Faisal M et al (2014) Sexual differences of imprinted genes' expression levels. Gene 533(1):434–438

Filova B et al (2013) The effect of testosterone on the formation of brain structures. Cells Tissues Organs 197(3):169–177

Griffin GD et al (2011) Ovarian hormone action in the hypothalamic ventromedial nucleus: remodelling to regulate reproduction. J Neuroendocrinol 23 (6):465–471

Handa RJ et al (2012) Roles for oestrogen receptors in adult brain function. J Neuroendocrinol 24 (1):160–173

Henley CL et al (2011) Hormones of choice: the neuroendocrinology of partner preference in animals. Front Neuroendocrinol 32(2):146–154

Hines M (2011) Prenatal endocrine influences on sexual orientation and on sexually differentiated childhood behaviour. Front Neuroendocrinol 32(2):170–182

Jannini E et al (2010) Male homosexuality: nature or culture? J Sex Med 7(10):3245–3253

Jordan-Young RM (2012) Hormones, context, and "brain gender": a review of evidence from congenital adrenal hyperplasia. Soc Sci Med 74(11):1738–1744

Kim HJ et al (2012) Sex differences in amygdala subregions: evidence from subregional shape analysis. Neuroimage 60(4):2054–2061

Kinsley CH et al (2012) Sex steroid hormone determination of the maternal brain: effects beyond reproduction. Mini Rev Med Chem 12(11):1063–1070

Menzler K et al (2011) Men and women are different: diffusion tensor imaging reveals sexual dimorphism in the microstructure of the thalamus, corpus callosum and cingulum. Neuroimage 54(4):2557–2562

Tuck N et al (2011) The genetics of sex differences in brain and behavior. Front Neuroendocrinol 32 (2):227–246

Varlinskaya EI et al (2013) Puberty and gonadal hormones: role in adolescent-typical behavioral alterations. Horm Behav 64(2):343–349

Wade J, Arnold AP (2004) Sexual differentiation of the zebra finch song system. Ann NY Acad Sci 1016:540–559

Pregnancy, Care for the Child

Neumann ID (2008) Brain oxytocin: a key regulator of emotional and social behaviours in both females and males. J Neuroendocrinol 20(6):858–865

Box 7A Disturbed Sexual Development, Transsexuality, Homosexuality

Amudha S et al (2012) SRY (Sex determining regions in Y) basis of sex reversal in XY females. Int J Hum Genet 12(2):99–103

Ray B (2012) A possible second type of maternal-fetal immune interaction involved in both male and female homosexuality. Arch Sex Behav 41(6):1507–1511

Castronovo C et al (2014) Gene dosage as a relevant mechanism contributing to the determination of ovarian function in Turner syndrome. Hum Reprod 29 (2):368–379

Corsello SM et al (2011) Biological aspects of gender disorders. Minerva Endocrinol 36(4):325–339

Ghorbel M et al (2012) Chromosomal defects in infertile men with poor semen quality. J Assist Reprod Genet 29(5):451–456

Hare L et al (2009) Androgen receptor repeat length polymorphism associated with male-to-female transsexualism. Biol Psychiatry 65(1):93–96

Kruijver FP et al (2000) Male-to-female transsexuals have female neuron numbers in a limbic nucleus. J Clin Endocrinol Metab 85:2034–2041

Zheng L et al (2008) "Micro-deletions" of the human Y chromosome and their relationship with male infertility. J Genet Genomics 35(4):193–199

McCarthy MM, Arnold AP (2011) Reframing sexual differentiation of the brain. Nat Neurosci 14 (6):677–683

Meyer-Bahlburg H (2008) Sexual orientation in women with classical or non-classical congenital adrenal

hyperplasia as a function of degree of prenatal androgen excess. Arch Sex Behav 37(1):85–99

Nordenstrom A et al (2002) Sex-typed toy play behavior
correlates with the degree of prenatal androgen exposure assessed by CYP21 genotype in girls with congenital adrenal hyperplasia. J Clin Endocrinol
Metabol 87:5119–5124

Praveen EP et al (2008) Gender identity of children and
young adults with 5 alpha-reductase deficiency.
J Pediatr Endocrinol Metab 21(2):173–179

Rametti G et al (2011) White matter microstructure in
female to male transsexuals before cross-sex hormonal treatment. A diffusion tensor imaging study.
J Psychiatr Res 45(2):199–204

Rice WR et al (2013) Homosexuality via canalized sexual
development: a testing protocol for a new epigenetic
model. Bioessays 35(9):764–770

Tomaselli S et al (2008) Syndromic true hermaphroditism
due to an *R-spondin1* (RSPO1) homozygous mutation.
Hum Mutat 29(2):220–226

Valenzuela CY (2010) Sexual orientation, handedness,
sex ratio and fetomaternal tolerance-rejection. Biol
Res 43(3):347–356

Box 7B Hormone-Like Environmental Substances, Xenoestrogens

Besse JP, Garric J (2009) Progestagens for human use,
exposure and hazard assessment for the aquatic environment. Environ Pollut 157(12):3485–3494

Crain DA et al (2008) Female reproductive disorders: the
roles of endocrine-disrupting compounds and developmental timing. Fertil Steril 90(4):911–940

Dickerson AM et al (2011) Endocrine disruption of brain
sexual differentiation by developmental PCB exposure. Endocrinology 152(2):581–594

Frye CA et al (2012) Endocrine disrupters: a review of
some sources, effects, and mechanisms of actions on
behaviour   and   neuroendocrine   systems.   J
Neuroendocrinol 24(1):144–159

Lawson C et al (2011) Gene expression in the fetal mouse
ovary is altered by exposure to low doses of Bisphenol
A. Biol Reprod 84(1):79–86

Leet J et al (2011) A review of studies on androgen and
estrogen exposure in fish early life stages: effects on
gene and hormonal control of sexual differentiation. J
Appl Toxicol 31(5):379–398

Lilienthal H et al (2013) Sexually dimorphic behavior
after   developmental   exposure   to   characterize
endocrine-mediated effects of different non-dioxin-
like PCBs in rats. Toxicology 311:52–60

Marino M et al (2012) Susceptibility of estrogen receptor
rapid responses to xenoestrogens: physiological
outcomes. Steroids 77(10):910–917

Massart F (2010) Oestrogenic mycotoxin exposures and
precocious pubertal development. Int J Androl 33
(2):369–376

McGinnis CL, Crivello JF (2011) Elucidating the mechanism of action of tributyltin (TBT) in zebrafish. Aquat
Toxicol 103(1–2):25–31

Nohynek GJ et al (2013) Endocrine disruption: fact or
urban legend? Toxicol Lett 223(3):295–305

Rochester JR (2013) Bisphenol A and human health: a
review of the literature. Reprod Toxicol 42:132–155

Schoeters G et al (2008) Endocrine disruptors and
abnormalities of pubertal development. Basic Clin
Pharmacol Toxicol 102(2):168–175

Scott HM et al (2009) Steroidogenesis in the fetal testis
and its susceptibility to disruption by exogenous
compounds. Endocr Rev 30(7):883–925

Soeffker M, Tyler CR (2012) Endocrine disrupting
chemicals and sexual behaviors in fish – a critical
review on effects and possible consequences. Crit
Rev Toxicol 42(8):653–668

Box 7C Fate of the Y-chromosome

Bachtrog D (2013) Y-chromosome evolution: emerging
insights into processes of Y-chromosome degeneration. Nat Rev Genet 14(2):113–124

Chapter 8 Germ line, Oogenesis, Spermatogenesis, Genomic imprinting: Germ Line, Germ Plasm, piRNA, vasa, nanos, Oct4

Anderson RA et al (2007) Conserved and divergent
patterns of expression of DAZL, VASA and OCT4
in the germ cells of the human fetal ovary and testis.
BMC Dev Biol 7:Art No 136

Arkov AL, Ramos R (2010) Building RNA–protein
granules: insight from the germline. Trends Cell Biol
20(8):482–490

Beyret E, Lin H (2011) Pinpointing the expression of
piRNAs and function of the PIWI protein subfamily
during spermatogenesis in the mouse. Dev Biol 355
(2):215–226

Bortvin A (2013) PIWI-interacting RNAs (piRNAs) – a
mouse testis perspective. Biochemistry 78(6):592–602

Braat AK et al (2000) Vasa protein expression and localization in the zebrafish. Mech Dev 95:271–274

Cheng L et al (2007a) OCT4: biological functions and
clinical applications as a marker of germ cell neoplasia. J Pathol 211(1):1–9

Extavour CG, Akam M (2003) Mechanisms of germ cell
specification across metazoans: epigenesis and preformation. Development 130:5869–5884

Tingting G, Elgin SC (2013) Maternal depletion of Piwi, a
component of the RNAi system, impacts heterochromatin formation in Drosophila. PLoS Genet 9(9):
e1003780

Gill ME et al (2011) Licensing of gametogenesis, dependent on RNA binding protein DAZL, as a gateway to sexual differentiation of fetal germ cells. Proc Natl Acad Sci USA 108(18):7443–7448

Gustafson EA, Wessel GM (2010) *Vasa* genes: emerging roles in the germ line and in multipotent cells. Bioessays 32(7):626–637

Hayashi Y et al (2004) *Nanos* suppresses somatic cell fate in *Drosophila* germ line. Proc Natl Acad Sci USA 101 (28):10338–10342

Juliano CE et al (2010b) A conserved germline multipotency program. Development 137 (24):4113–4126

Kehler J (2004) Oct4 is required for primordial germ cell survival. EMBO Rep 5(11):1078–1083

Kirino Y et al (2010) Arginine methylation of Vasa protein is conserved across phyla. J Biol Chem 285 (11):8148–8154

Knaut H et al (2000) Zebrafish *vasa* RNA but not its protein is component of the germ plasm and segregates asymmetrically before germline specification. J Cell Biol 149:875–888

Kusz-Zamelczyk K et al (2013) Mutations of *NANOS1*, a human homologue of the Drosophila morphogen, are associated with a lack of germ cells in testes or severe oligo-astheno-teratozoospermia. J Med Genet 50 (3):187–193

Fangfang L et al (2012) Xenopus *Nanos1* is required to prevent endoderm gene expression and apoptosis in primordial germ cells. Development 139 (8):1476–1486

Lasko P (2013a) The DEAD-box helicase vasa: evidence for a multiplicity of functions in RNA processes and developmental biology. Biochim Biophys Acta 1829 (8):810–816

Leclere L et al (2012) Maternally localized germ plasm mRNAs and germ cell/stem cell formation in the cnidarian *Clytia*. Dev Biol 364(2):236–248

Leitch HG et al (2014) On the fate of primordial germ cells injected into early mouse embryos. Dev Biol 385 (2):155–159

Lesch BJ, Page DC (2012) Genetics of germ cell development. Nat Rev Genet 13(11):781–794

Lidke AK et al (2014) 17 beta-Estradiol induces supernumerary primordial germ cells in embryos of the polychaete *Platynereis dumerilii*. Gen Comp Endocrinol 196:52–61

Molyneaux K, Wylie C (2004a) Primordial germ cell migration. Int J Dev Biol 48:537–543

Naeemipour M et al (2013) Expression dynamics of pluripotency genes in chicken primordial germ cells before and after colonization of the genital ridges. Mol Reprod Dev 80(10):849–861

Nijjar S, Woodland HR (2013b) Protein interactions in *Xenopus* germ plasm RNP particles. PLoS One 8(11): e80077

Onohara Y et al (2010) Localization of mouse Vasa homolog protein in chromatoid body and related nuage structures of mammalian spermatogenic cells

during spermatogenesis. Histochem Cell Biol 133 (6):627–639

Shibataa N et al (1999) Expression of *vasa(vas)*-related genes in germline cells and totipotent somatic stem cells of planarians. Dev Biol 206(1):73–87

Sinsimer K et al (2013) Germ plasm anchoring is a dynamic state that requires persistent trafficking. Cell Rep 5(5):1169–1177

Soto-Suazo M, Zorn TM (2005a) Primordial germ cells migration: morphological and molecular aspects. Anim Reprod 2(3):147–160

Strome S, Lehmann R (2007) Germ versus soma decisions: lessons from flies and worms. Science 316 (5823):392–393

Tsunekawa N et al (2000) Isolation of chicken *vasa* homolog gene and tracing the origin of primordial germ cells. Development 127:2741–2750

Wang Z, Lin H (2004a) *Nanos* maintains germline stem cell self-renewal by preventing differentiation. Science 303:2016–2019

Weismann A (1892) Das Keimplasma. Eine Theorie der Vererbung. Fischer, Jena, Leipzig

Yakushev EY et al (2013) Multifunctionality of PIWI proteins in control of germline stem cell fate. Biochemistry 78(6):585–591

Oogenesis

Adhikari D, Kui L (2014) The regulation of maturation promoting factor during prophase I arrest and meiotic entry in mammalian oocytes. Mol Cell Endocrinol 382 (1):480–487

Bukovsky A (2011) Ovarian stem cell niche and follicular renewal in mammals. Anat Rec 294(8):1284–1306

De Felici M, Barrios F (2013) Seeking the origin of female germline stem cells in the mammalian ovary. Reproduction 146(4):R125–R130

Edson MA et al (2009) The mammalian ovary from genesis to revelation. Endocr Rev 30(6):624–712

Chun-Wei F et al (2014) Control of mammalian germ cell entry into meiosis. Mol Cell Endocrinol 382 (1):488–497

Lampbrush Chromosomes, Multiple Nucleoli

Gall JG (2012) Are lampbrush chromosomes unique to meiotic cells? Chromosome Res 20(8):905–909

Hill RS, MacGregor HC (1980) The development of lampbrush chromosome-type transcription in the early diplotene oocytes of *Xenopus laevis*: an electron microscope analysis. J Cell Sci 44:87–101

Kaufmann R et al (2012) Superresolution imaging of transcription units on newt lampbrush chromosomes. Chromosome Res 20(8):1009–1015

Ji-Long L, Gall JG (2012) Induction of human lampbrush chromosomes. Chromosome Res 20(8):971–978

Mais C, Scheer U (2001) Molecular architecture of the amplified nucleoli of *Xenopus* oocytes. J Cell Sci 114 (4):709–718

Penrad-Mobayed M et al (2009) Working map of the lampbrush chromosomes of *Xenopus tropicalis*: a new tool for cytogenetic analyses. Dev Dyn 238 (6):1492–1501

Penrad-Mobayed M et al (2011) The tips and tricks for preparing lampbrush chromosome spreads from *Xenopus tropicalis* oocytes. Methods 53(3):325

Trendelenburg MF et al (1996) Multiparameter microscopic analysis of nucleolar structure and ribosomal gene transcription. Histochem Cell Biol 106 (2):167–192

Spermatogenesis

Awe S, Renkawitz-Pohl R (2010) Histone H4 acetylation is essential to proceed from a histone- to a protamine-based chromatin structure in spermatid nuclei of *Drosophila melanogaster*. Syst Biol Reprod Med 56:1–18

Cheng CY, Mruk D (2010) The biology of spermatogenesis: the past, present and future Introduction. Philos Trans R Soc Lond B Biol Sci 365(1546):1459–1463

Cheng CY et al (2010) Regulation of spermatogenesis in the microenvironment of the seminiferous epithelium: new insights and advances. Mol Cell Endocrinol 315 (1–2):49–56

Chuma S et al (2009) Ultrastructural characterization of spermatogenesis and its evolutionary conservation in the germline: germinal granules in mammals. Mol Cell Endocrinol 306(1–2):17–23

Chuma S, Nakano T (2013) piRNA and spermatogenesis in mice. Philos Trans R Soc Lond B Biol Sci 368 (1609):20110338

Fraser LR (2010) The "switching on" of mammalian spermatozoa: molecular events involved in promotion and regulation of capacitation. Mol Reprod Dev 77 (3):197–208

Garcia TX, Hofmann MC (2013a) NOTCH signaling in Sertoli cells regulates gonocyte fate. Cell Cycle 12 (16):2538–2545

Griswold M, Oatley JM (2013a) Concise review: defining characteristics of mammalian spermatogenic stem cells. Stem Cells 31:8–11

Kerr GE et al (2014a) Regulated Wnt/beta-catenin signaling sustains adult spermatogenesis in mice. Biol Reprod 90(1):Article No. 3

Phillips BT et al (2010) Spermatogonial stem cell regulation and spermatogenesis. Philos Trans R Soc Lond B Biol Sci 365(1546):1663–1678

Verhoeven G et al (2010) Androgens and spermatogenesis: lessons from transgenic mouse models. Philos Trans R Soc Lond B Biol Sci 365(1546):1537–1556

Walker WH (2009) Molecular mechanisms of testosterone action in spermatogenesis. Steroids 74 (7):602–607

Yeh J et al (2011) Wnt5a is a cell-extrinsic factor that supports self-renewal of mouse spermatogonial stem cells. J Cell Sci 124(14):2357–2366

Genomic Imprinting, Reprogramming

Apostolou E, Hochedlinger K (2013) Chromatin dynamics during cellular reprogramming. Nature 502 (7472):462–471

Biliya S, Bulla L (2010) Genomic imprinting: the influence of differential methylation in the two sexes. Exp Biol Med 235(2):139–147

Dunn GA, Bale TL (2011) Maternal high-fat diet effects on third-generation female body size via the paternal Lineage. Endocrinology 152(6):2228–2236

Guibert S, Weber M (2013) Functions of DNA methylation and hydroxymethylation in mammalian development. Epigenetics and Development. Curr Top Dev Biol 10:47–83

Ohhata T, Wutz A (2013) Reactivation of the inactive X chromosome in development and reprogramming. Cell Mol Life Sci 70(14):2443–2461

Rathke C et al (2010) Distinct functions of Mst77F and protamines in nuclear shaping and chromatin condensation during *Drosophila* spermiogenesis. Eur J Cell Biol 89:326–338

Saitou M et al (2012) Epigenetic reprogramming in mouse pre-implantation development and primordial germ cells. Development 139:15–31

Seisenberger S et al (2013a) Conceptual links between DNA methylation reprogramming in the early embryo and primordial germ cells. Curr Opin Cell Biol 25 (3):281–288

Seisenberger S et al (2013b) Reprogramming DNA methylation in the mammalian life cycle: building and breaking epigenetic barriers. Philos Trans R Soc Lond B Biol Sci 368(1609):Article No. 201103303

Wasson JA et al (2013) Restoring totipotency through epigenetic reprogramming. Brief Funct Genomics 12 (2):118–128

Chapter 9 Maternal Factors, Specification of Body Axes (See Also *Drosophila, Xenopus*)

Abrams EW, Mullins MC (2009b) Early zebrafish development: it's in the maternal genes. Curr Opin Genet Dev 19(4):396–403

Barrowa JR et al (2007) Wnt3 signaling in the epiblast is required for proper orientation of the anteroposterior axis. Dev Biol 312(1):312–320

Becalska AN, Gavis E (2009b) Lighting up mRNA localization in *Drosophila* oogenesis. Development 136 (15):2493–2503

Biase FH et al (2014) Messenger RNAs in metaphase II oocytes correlate with successful embryo development to the blastocyst stage. Zygote 22(1):69–79

Jing C et al (2013) Somatic cells regulate maternal mRNA translation and developmental competence of mouse oocytes. Nat Cell Biol 15(12):1415–1423

Cuykendall TN, Houston D (2009) Vegetally localized *Xenopus* trim36 regulates cortical rotation and dorsal axis formation. Development 136 (18):3057–3065

Dietrich JE, Hiiragi T (2008) Stochastic processes during mouse blastocyst patterning. Cells Tissues Organs 188 (1–2):46–51

Farr GH et al (2000) Interaction among GSK-3, GBP, axin, and APC in *Xenopus* axis specification. J Cell Biol 148:691–702

Jiang F et al (2009) Reciprocal regulation of *Wnt* and *Gpr177/mouse Wntless* is required for embryonic axis formation. Proc Natl Acad Sci USA 106 (44):18604–18609

Fuentes R, Fernandez J (2010) Ooplasmic segregation in the zebrafish. Dev Dyn 239:2177–2189

Gerhart J et al (1989b) Cortical rotation of the *Xenopus* egg: consequences of the antero-posterior pattern of embryonic dorsal development. Development 107 (Suppl):37–51

Hiiragi T, Solter D (2004) First cleavage plane of the mouse egg is not predetermined but defined by the topology of the two apposing pronuclei. Nature 430:360–364

Hikasa H, Sokol S (2013) Wnt signaling in vertebrate axis specification. Cold Spring Harb Perspect Biol 5(1): a007955

Hiramatsu R et al (2013b) External mechanical cues trigger the establishment of the anterior-posterior axis in early mouse embryos. Dev Cell 27(2):131–144

Houston DW (2013) Regulation of cell polarity and RNA localization in vertebrate oocytes. Int Rev Cell Mol Biol 306:127–185

Kugler J-M, Lasko P (2009b) Localization, anchoring and translational control of *oskar, gurken, bicoid* and *nanos* mRNA during *Drosophila* oogenesis. Fly 3 (1):15–28

Lindeman RE, Pelegri F (2010b) Vertebrate maternal-effect genes: Insights into fertilization, early cleavage divisions, and germ cell determinant localization from studies in the zebrafish. Mol Reprod Dev 77 (4):299–313

Marikawa Y (2006) Wnt/beta-catenin signaling and body plan formation in mouse embryos. Sem Cell Dev Biol 17(2):175–184

Momose T, Schmid V (2006b) Animal pole determinants define oral–aboral axis polarity and endodermal cell-fate in hydrozoan jellyfish *Podocoryne carnea*. Dev Biol 292(2):371–380

Niehrs C (2010b) On growth and form: a Cartesian coordinate system of Wnt and BMP signaling specifies bilaterian body axes. Development 137(6):845–857

Plickert G et al (2006b) Wnt signaling in hydroid development: formation of the primary body axis in embryogenesis and its subsequent patterning. Dev Biol 298(2):368–378

Sardet C et al (2003) Maternal mRNAs of PEM and *macho 1*, the ascidian muscle determinant, associate and move with a rough endoplasmic reticulum network in the egg cortex. Development 130(23):5839–5849

Schneider S, Steinbeisser H, Warga RM, Hausen P (1996b) ß-Catenin translocation into nuclei demarcates the dorsalizing centers in frog and fish embryos. Mech Dev 57:191–198

Stern CD (2006b) Evolution of the mechanisms that establish the embryonic axes. Curr Opin Genet Dev 16(4):413–418

Stern CD, Downs KM (2012) The hypoblast (visceral endoderm): an evo-devo perspective. Development (Cambridge) 139(6):1059–1069

Takaoka K, Hamada H (2012) Cell fate decisions and axis determination in the early mouse embryo. Development 139:3–14

Wacker S, Berking S (1994) The orientation of the dorso/ventral axis of zebrafish is influenced by gravitation. Roux's Arch Dev Biol 203:281–283

Wylie AD et al (2014b) Post-transcriptional regulation of *wnt8a* is essential to zebrafish axis development. Dev Biol 386(1):53–63

Zernicka-Goetz M (1998) Fertile offspring derived from mammalian eggs lacking either animal or vegetale pole. Development 125:4803–4808

Zonies S et al (2010) Symmetry breaking and polarization of the *C. elegans* zygote by the polarity protein PAR-2. Development 137(10):1669–1677

Chapter 10 Positional Information II, Pattern Formation by Cell Interactions, Embryonic Induction (See Also References to *Hydra*, *Drosophila*, *Xenopus* and Mouse): General

Dubuis JO et al (2013b) Positional information, in bits. Proc Natl Acad Sci USA 110(41):16301–16308

Stathopoulos A, Iber D (2013) Studies of morphogens: keep calm and carry on. Development 140 (20):4119–4124

Wolpert L (1969a) Positional information and the spatial pattern of cellular differentiation. J Theor Biol 25:1–47

Wolpert L (2011a) Positional information and patterning revisited. J Theor Biol 269(1):359–365

Lateral Inhibition and Help, Notch/Delta, Ommatidium: *Sevenless*

Almudi I et al (2009) SOCS36E specifically interferes with sevenless signaling during *Drosophila* eye development. Dev Biol 326(1):212–223

Basler K, Hafen E (1989) Ubiquitous expression of *sevenless*: position-dependent specification of cell fate. Science 243:931–934

Ehebauer M et al (2006a) Notch, a universal arbiter of cell fate decisions. Science 314(5804):1414–1415

Hamada H et al (2014b) Involvement of delta/notch signaling in zebrafish adult pigment stripe patterning. Development 141(2):318–324

Tomlinson A et al (2011) Three distinct roles for Notch in *Drosophila* R7 photoreceptor specification. PLoS Biol 9(8):e1001132

Drosophila, Morphogens (See Also Chapter 4 *Drosophila*)

Alexandre C (2014) Patterning and growth control by membrane-tethered Wingless. Nature 505:180–185

Cheung D et al (2011) Scaling of the bicoid morphogen gradient by a volume-dependent production rate. Development 138(13):2741–2749

Grimm O, Coppey M, Wieschaus E (2010b) Modelling the bicoid gradient. Development 137:2253–2264

Little SC et al (2011) The formation of the bicoid morphogen gradient requires protein movement from anteriorly localized mRNA. PLoS Biol 9(3):e1000596

Piddini E, Vincent JP (2009) Interpretation of the Wingless gradient requires signal-induced self-inhibition. Cell 136:296–307

Wartlick O et al (2011a) Dynamics of Dpp signaling and proliferation control. Science 331(6021):1154–1159

Vertebrates: Spemann Organizer, Inducers, Morphogens (See also Chapter 5 *Xenopus*, Amphibians)

Agius E et al (2000a) Endodermal Nodal-related signals and mesoderm induction in *Xenopus*. Development 127:1173–1183

Bae S et al (2011) Siamois and Twin are redundant and essential in formation of the Spemann organizer. Dev Biol 352(2):367–381

Beddington RS (1994) Induction of a second neural axis by the mouse node. Development 120:613–620

Blumberg B et al (1997) An essential role for retinoid signaling in anteroposterior neural patterning. Development 124:373–379

Boekel C, Brand M (2013a) Generation and interpretation of FGF morphogen gradients in vertebrates. Curr Opin Genet Dev 23(4):415–422

Boettger T et al (2001b) The avian organizer. Int J Dev Biol 45:281–287

Danilchik M et al (2013b) Blastocoel-spanning filopodia in cleavage-stage *Xenopus laevis*: potential roles in morphogen distribution and detection. Dev Biol 382 (1):70–81

De Robertis EM et al (2000b) The establishment of Spemann's organizer and patterning of the vertebrate embryo. Nat Rev Genet 3:171–181

De Robertis EM (2006a) Spemann's organizer and self-regulation in amphibian embryos. Nat Rev Mol Cell Biol 7:296–302

Fleming A et al (2004b) A central role for the notochord in vertebral patterning. Development 131:873–880

Heasman J (2006b) Patterning the early *Xenopus* embryo. Development 133:1205–1217

Linker C et al (2009b) Cell communication with the neural plate is required for induction of neural markers by BMP inhibition: evidence for homeogenetic induction and implications for *Xenopus* animal cap and chick explant assays. Dev Biol 327(2):478–486

Luxardi G et al (2010) Distinct Xenopus Nodal ligands sequentially induce mesendoderm and control gastrulation movements in parallel to the Wnt/PCP pathway. Development 137(3):417–426

Medina A, Wendler SR, Steinbeisser H (1997) Cortical rotation is required for the correct spatial expression of *sia* and *gsc* in *Xenopus* embryos. Int J Dev Biol 41:741–745

Niehrs C (2004b) Regionally specific induction by the Spemann-Mangold organizer. Nat Rev Genet 5:425–434

Niehrs C (2010c) On growth and form: a Cartesian coordinate system of Wnt and BMP signaling specifies bilaterian body axes. Development 137(6):845–857

Reid CD, Kessler DS (2010) Transcriptional integration of the Wnt and Nodal pathways during organizer formation. Dev Biol 344(1):455

Simeonia I, Gurdon JB (2007) Interpretation of BMP signaling in early *Xenopus* development. Dev Biol 308(1):82–92

Stern CD (2005a) Neural induction: old problem, new findings, yet more questions. Development 132:2007–2021

Vonica A, Gumbiner BM (2007b) The *Xenopus* Nieuwkoop center and Spemann–Mangold organizer share molecular components and a requirement for maternal Wnt activity. Dev Biol 312(1):90–102

Dong Y, Xinhua L (2009) Shaping morphogen gradients by proteoglycans. Cold Spring Harb Perspect Biol 1 (3):a002493

Zakin L, De Robertis EM (2010a) Extracellular regulation of BMP signaling. Curr Biol 20(3):R89–R92

Lens Induction

Donner AL, Maas R (2006) Lens induction in vertebrates: Variations on a conserved theme of signaling events. Semin Cell Dev Biol 17(6):676–685

Gunhaga L (2011) The lens: a classical model of embryonic induction providing new insights into cell determination in early development. Philos Trans R Soc Lond B Biol Sci 366(1568):1193–1203

Henry JJ, Grainger RM (1990b) Early tissue interactions leading to embryonic lens formation in *Xenopus laevis*. Dev Biol 141:149–163

Planar Polarity

Gomez-Orte E et al (2013a) Multiple functions of the noncanonical Wnt pathway. Trends Genet 29(9):545–553

Goodrich LV, Strutt D (2011) Principles of planar polarity in animal development. Development 138:1877–1892

Hogan J et al (2011) Two Frizzled planar cell polarity signals in the *Drosophila* wing are differentially organized by the Fat/Dachsous pathway. PLoS Genet 7(2):e1001305

Repiso A et al (2010) Planar cell polarity: the orientation of larval denticles in *Drosophila* appears to depend on gradients of Dachsous and Fat. Development 137:3411–3415

Sharma P, McNeill H (2013) Regulation of long-range planar cell polarity by Fat-Dachsous signalling. Development 140(18):3869–3881

Wansleeben C, Meijlink F (2011) The planar cell polarity pathway in vertebrate development. Dev Dyn 240(3):616–626

Witze ES et al (2008) Wnt5a control of cell polarity and directional movement by polarized redistribution of adhesion receptors. Science 320(5874):365–369

Left-Right Asymmetry

Antic D et al (2010) Planar cell polarity enables posterior localization of Nodal cilia and left-right axis determination during mouse and *Xenopus* embryogenesis. PLoS One 5(2):e8999

Beyer T et al (2012) Serotonin signaling is required for Wnt-dependent GRP specification and leftward flow in *Xenopus*. Curr Biol 22:33–39

Blum M (2009a) *Xenopus*, an ideal model system to study vertebrate left-right asymmetry. Dev Dyn 238(6):1215–1225

Blum M (2009b) Evolution of leftward flow. Semin Cell Dev Biol 20(4):464–471

Blum M et al (2011) Serotonin signaling is required for Wnt-dependent development of the ciliated gastrocoel roof plate and leftward flow in *Xenopus*. Dev Biol 356(1):209

Gros J et al (2009) Cell movements at Hensen's node establish left/right asymmetric gene expression in the chick. Science 324(5929):941–944

Katsu K et al (2012) BMP inhibition by DAN in Hensen's node is a critical step for the establishment of left-right asymmetry in the chick embryo. Dev Biol 363:15–26

Komatsu Y, Mishina Y (2013) Establishment of left-right asymmetry in vertebrate development: the node in mouse embryos. Cell Mol Life Sci 70(24):4659–4666

Pulina MV (2011) Essential roles of fibronectin in the development of the left-right embryonic body plan. Dev Biol 354(2):208–220

Schweickert A et al (2012) Linking early determinants and cilia-driven leftward flow in left-right axis specification of *Xenopus laevis*: a theoretical approach. Differentiation 83(2):S67–S77

Song H et al (2010) Planar cell polarity breaks bilateral symmetry by controlling ciliary positioning. Nature 466(7304):378

Takao D et al (2013) Asymmetric distribution of dynamic calcium signals in the node of mouse embryo during left-right axis formation. Dev Biol 376:23–30

Yoshiba S, Hamada H (2014) Roles of cilia, fluid flow, and Ca^{2+} signaling in breaking of left-right symmetry. Trends Genet 30:10–17

Neural Tube, Sonic Hedgehog

Aviles E et al (2013) Sonic hedgehog and Wnt: antagonists in morphogenesis but collaborators in axon guidance. Front Cell Neurosci 7:86

Dessaud D et al (2010a) Dynamic assignment and maintenance of positional identity in the ventral neural tube by the morphogen Sonic Hedgehog. PLoS Biol 8:e1000382

Ribes V et al (2010) Distinct sonic hedgehog signaling dynamics specify floor plate and ventral neuronal progenitors in the vertebrate neural tube. Genes Dev 24(11):1186–1200

Kwanha Y et al (2013) Floor plate-derived sonic hedgehog regulates glial and ependymal cell fates in the developing spinal cord. Development 140(7):1594–1604

Morphogenetic Fields: Insect Extremities, Wing Patterns

Alexandre C et al (2014a) Patterning and growth control by membrane-tethered Wingless. Nature 505:180–185

Baker N (2011) Proximodistal patterning in the *Drosophila* leg: models and mutations. Genetics 187 (4):1003–1010

Bando T et al (2011) Regulation of leg size and shape: Involvement of the Dachsous-Fat signaling pathway. Dev Dyn 240(5):1028–1041

Basler K, Struhl G (1994) Compartment boundaries and the control of *Drosophila* limb pattern by Hedgehog protein. Nature 368:208–214

Bejarano F, Milan M (2009) Genetic and epigenetic mechanisms regulating *hedgehog* expression in the *Drosophila* wing. Dev Biol 327(2):508

Campbell G, Tomlinson A (1998a) The roles of the homeobox genes *aristaless* and *Distal-less* in patterning the legs and wings of *Drosophila*. Development 125:4483–4493

Held LI (2013) Rethinking butterfly eyespots. Evol Biol 40(1):158–168

Otaki Joji M (2011) Color-pattern analysis of eyespots in butterfly wings: A critical examination of morphogen gradient models. Zoolog Sci 28(6):403–413

Wartlick O et al (2011b) Dynamics of Dpp signaling and proliferation control. Science 331(6021):1154–1159

Morphogenetic Fields: Vertebrate Limbs

Andrey G et al (2013) A switch between topological domains underlies *HoxD* genes colinearity in mouse limbs. Science 340(6137):1195

Bastida MF et al (2009) A BMP-Shh negative-feedback loop restricts *Shh* expression during limb development. Development 136:3779–3789

Bénazet JD (2009) A self-regulatory system of interlinked signaling feedback loops controls mouse limb patterning. Science 323(5917):1050–1053

Cooper KL et al (2011) Initiation of proximal-distal patterning in the vertebrate limb by signals and growth. Science 332:1083–1086

Cunningham TJ et al (2013) Antagonism between retinoic acid and fibroblast growth factor signaling during limb development. Cell Rep 3(5):1503–1511

Mariani F (2010) Proximal to distal patterning during limb development and regeneration: a review of converging disciplines. Regen Med 5(3):451–462

McEwan J et al (2011) Expression of key retinoic acid modulating genes suggests active regulation during development and regeneration of the amphibian limb. Dev Dyn 240(5):1259–1270

Ohuchi H et al (1997) The mesenchymal factor FGF10 initiates and maintains the outgrowth of the chick limb bud through interaction with FGF8, an apical ectodermal factor. Development 124:2235–2244

Roselló-Díez A et al (2011) Diffusible signals, not autonomous mechanisms, determine the main proximodistal limb subdivision. Science 332:1086–1088

Roselló-Díez A et al (2014) Diffusible signals and epigenetic timing cooperate in late proximo-distal limb patterning. Development 141:1534–1543

Suzuki T (2013) How is digit identity determined during limb development? Dev Growth Differ 55(1):130–138

Tamura K et al (2011) Embryological evidence identifies wing digits in birds as digits 1, 2, and 3. Science 331 (6018):753–757

Thaller C, Eichele G (1987) Identification and spatial distribution of retinoids in the developing chick limb bud. Nature 327:625–628

Towers M et al (2012a) Gradients of signalling in the developing limb. Curr Opin Cell Biol 24(2):181–187

Wolpert L (2002) The progress zone model for specifying positional information. Int J Dev Biol 46:869–870

Pattern Control and Positional Memory in Hydra and Other Cnidarians (See Also Chapter 4)

Augustin R et al (2006b) *Dickkopf* related genes are components of the positional value gradient in hydra. Dev Biol 296(1):62–70

Broun M et al (2005) Formation of the head organizer in hydra involves the canonical Wnt pathway. Development 132(12):2907–2916

Gee L et al (2010b) beta-catenin plays a central role in setting up the head organizer in hydra. Dev Biol 340 (1):116–124

Gierer A et al (1972b) Regeneration of hydra from reaggregated cells. Nat New Biol 239:98–101

Hassel M (1998b) Upregulation of *Hydra vulgaris* cPKC gene is tightly coupled to the differentiation of head structures. Dev Genes Evol 207:489–501

Hassel M et al (1998b) The level of expression of a protein kinase C gene may be an important component of the patterning process in *Hydra*. Dev Genes Evol 207:502–514

Hobmayer B et al (2000b) Wnt signalling molecules act in axis formation in the diploblastic metazoan *Hydra*. Nature 407:186–189

Kaesbauer T (2007) The Notch signaling pathway in the cnidarian *Hydra*. Dev Biol 303(1):376–390

Lengfeld T et al (2009b) Multiple Wnts are involved in *Hydra* organizer formation and regeneration. Dev Biol 330(1):186–199

Mueller WA (1990) Ectopic head formation in *Hydra*: diacylglycerol-induced increase in positional value and assistance of the head in foot formation. Differentiation 42:131–143

Mueller WA (1995a) Competition for factors and cellular resources as a principle of pattern formation in *Hydra*. I. Increase in the potentials for head and bud formation and rescue of the regeneration-deficient mutant reg-16 by treatment with diacylglycerol and arachidonic acid. Dev Biol 167:159–174

Mueller WA (1995b) Competition for factors and cellular resources as a principle of pattern formation in *Hydra*. II. Assistance of foot formation by heads and a new model of pattern control. Dev Biol 167:175–189

Mueller WA (1996c) Head formation at the basal end and mirror-image pattern duplication in *Hydra vulgaris*. Int J Dev Biol 40:1119–1131

Mueller WA (1996d) Competition-based head versus foot decision in chimeric hydras. Int J Dev Biol 40:1133–1139

Mueller WA et al (2004c) Patterning a multi-headed mutant in *Hydractinia*: enhancement of head formation and its phenotypic normalization. Int J Dev Biol 48:9–15

Mueller WA et al (2007b) Ectopic heads and giant buds induced by GSK-3beta inhibitors. Int J Dev Biol 51:211–220

Nakamura Y et al (2011b) Autoregulatory and repressive inputs localize Hydra Wnt3 to the head organizer. Proc Natl Acad Sci USA 108(22):9137–9142

Plickert G et al (2006c) Wnt signaling in hydroid development: Formation of the primary body axis and its subsequent patterning. Dev Biol 298:368–378

Rentzsch F et al (2007b) An ancient chordin-like gene in organizer formation of Hydra. Proc Natl Acad Sci USA 104(9):3249–3254

Dequéant M-L, Pourquie O (2009) Segmental patterning of the vertebrate embryonic axis. Nat Rev Genet 9:370–382

Gibb S et al (2010) The segmentation clock mechanism moves up a notch. Trends Cell Biol 20(10):593–600

Gomez C, Pourquie O (2009) Developmental control of segment numbers in vertebrates. J Exp Zool B Mol Dev Evol 312B(6):533–544

Hamada H et al (2014c) Involvement of Delta/Notch signaling in zebrafish adult pigment stripe patterning. Development 141(2):318–324

Harima Y et al (2013) Oscillatory links of Fgf signaling and Hes7 in the segmentation clock. Curr Opin Genet Dev 23(4):484–490

Herrgen L et al (2010) Intercellular coupling regulates the period of the segmentation clock. Curr Biol 20 (14):1244–1253

Krol AJ et al (2011) Evolutionary plasticity of segmentation clock networks. Development 138:2783–2792

Oates AC et al (2012) Patterning embryos with oscillations: structure, function and dynamics of the vertebrate segmentation clock. Development 139 (4):625–639

Sick S et al (2006) WNT and DKK determine hair follicle spacing through a reaction–diffusion mechanism. Science 314(5804):1447–1450

Intercalation

Bohn H (1976) Regeneration of proximal tissues from a more distal amputation level in the insect leg (*Blaberus craniifer*, Blattaria). Dev Biol 53:285–293

Maden M (1980) Intercalary regeneration in the amphibian limb and the rule of distal transformation. J Embryol Exp Morphol 56:201–209

Mueller WA (1982) Intercalation and pattern regulation in hydroids. Differentiation 22:141–150

Walck-Shannon E, Hardin J (2014) Cell intercalation from top to bottom. Nat Rev Mol Cell Biol 15 (1):34–48

Periodic Patterns, Segmentation Clock

Aulehla A, Pourquie O (2008) Oscillating signaling pathways during embryonic development. Curr Opin Cell Biol 20:632–637

Aulehla A, Pourquie O (2010) Signaling gradients during paraxial mesoderm development. Cold Spring Harb Perspect Biol 2(2):a000869

Bajard L et al (2014) Wnt-regulated dynamics of positional information in zebrafish somitogenesis. Development 141:1381–1391

Chisholm RH et al (2011) When are cellular oscillators sufficient for sequential segmentation? J Theor Biol 279(1):150–160

Signal Transmission and Propagation, Gradients, Exosomes, Nantubes, Filopodia

Affolter M, Basler K (2011a) Cytonemes show their colors. Science 332(6027):312–313

Beckett K et al (2013a) Drosophila S2 cells secrete Wingless on exosome-like vesicles but the wingless gradient forms independently of exosomes. Traffic 14 (1):82–96

Belting M, Wittrp A (2008) Nanotubes, exosomes, and nucleic acid-binding peptides provide novel mechanisms of intercellular communication in eukaryotic cells: implications in health and disease. J Cell Biol 183(7):1187–1191

Bischoff M et al (2013b) Cytonemes are required for the establishment of a normal Hedgehog morphogen gradient in *Drosophila* epithelia. Nat Cell Biol 15 (11):1269

Danilchik M et al (2013c) Blastocoel-spanning filopodia in cleavage-stage *Xenopus laevis*: Potential roles in morphogen distribution and detection. Dev Biol 382 (1):70–81

Dobrowolski R, De Robertis EM (2012a) Endocytic control of growth factor signalling: multivesicular bodies. Nat Rev Mol Cell Biol 13:53–60

Fico A et al (2011) Fine-tuning of cell signaling by glypicans. Cell Mol Life Sci 68(6):923–929

Gradill AC, Guerrero I (2013) Cytoneme-mediated cell-to-cell signaling during development. Cell Tissue Res 352:59–66

Greco V et al (2001) Argosomes. A potential vehicle for spread of morphogens through epithelia. Cell 106:633–645

Gross JC et al (2012a) Active Wnt proteins are secreted on exosomes. Nat Cell Biol 14(10):1036–1045

Gurdon JB et al (1999) Single cells can sense their position in a morphogen gradient. Development 126:5309–5317

Honda H, Mochizuki A (2002) Formation and maintenance of distinctive cell patterns by coexpression of membrane-bound ligands and their receptors. Dev Dyn 223:180–192

Hooper C et al (2012) Wnt3a induces exosome secretion from primary cultured rat microglia. BMC Neurosci 13:144

Kornberg TB (2013) Cytonemes extend their reach. EMBO J 32(12):1658–1659

Kornberg TB, Roy S (2014) Cytonemes as specialized signaling filopodia. Development 141:729–736

Korkut C et al (2009) Trans-synaptic transmission of vesicular Wnt signals through Evi/Wntless. Cell 139 (2):393–404

Kruse K et al (2004) Dpp gradient formation by dynamin-dependent endocytosis: receptor trafficking and the diffusion model. Development 131:4843–4856

Ru-Tao L et al (2013) Exosomes: the novel vehicles for intercellular communication. Prog Biochem Biophys 40(8):719–727

Mueller P et al (2013) Morphogen transport. Development 140(8):1621–1638

Nowak M et al (2011a) Interpretation of the FGF8 morphogen gradient is regulated by endocytic trafficking. Nat Cell Biol 13(2):153

Paine-Saunders S et al (2002) Heparan proteoglycans retain Noggin at the cell surface: a potential mechanism for shaping bone morphogenetic proteins gradients. J Biol Chem 277:2089–2096

Pellon-Cardenas O et al (2011) Endocytic trafficking and Wnt/beta-catenin signaling. Curr Drug Targets 12 (8):1216–1222

Pere EM et al (2014b) Active signals, gradient formation and regional specificity in neural induction. Exp Cell Res 321:25–31

Port P, Basler K (2010a) Wnt trafficking: new insights into Wnt maturation, secretion and spreading. Traffic 11(10):1265–1271

Princivalle M, de-Agostini A (2002) Developmental roles of heparan sulfate proteoglycans: a comparative review in Drosophila, mouse and human. Int J Dev Biol 46:267–278

Rainero E, Norman JC (2011) New roles for lysosomal trafficking in morphogen gradient sensing. Sci Signal 4(171):1–4

Record M et al (2014) Exosomes as new vesicular lipid transporters involved in cell-cell communication and various pathophysiologies. Biochim Biophys 1841 (1):108–120

Roberts CT, Peter K (2013) Vesicle trafficking and RNA transfer add complexity and connectivity to cell-cell communication. Cancer Res 73(11):3200–3205

Roy S et al (2011) Specificity of Drosophila cytonemes for distinct signaling pathways. Science 332 (6027):354–358

Rustom A et al (2004) Nanotubular highways for intercellular organelle transport. Science 303:1007–1010

Sanders TA et al (2013a) Specialized filopodia direct long-range transport of SHH during vertebrate tissue patterning. Nature 497:628–632

Schilling TF et al (2012) Dynamics and precision in retinoic acid morphogen gradients. Curr Opin Genet Dev 22(6):562–569

Scholpp S, Brand M (2004) Endocytosis controls spreading and effective signaling range of Fgf8 protein. Curr Biol 14:1834–1841

Shimozono S et al (2013a) Visualization of an endogenous retinoic acid gradient across embryonic development. Nature 496(7445):363

Simons M (2009) Exosomes – vesicular carriers for intercellular communication. Curr Opin Cell Biol 21 (4):575–581

Sokolowski TR (2012b) Mutual repression enhances the steepness and precision of gene expression boundaries. PLoS Comput Biol 8(8):Article No.: e1002654

Steinberg MS, Takeichi M (1994) Experimental specification of cell sorting, tissue spreading, and specific spatial patterning by quantitative differences in cadherin expression. Proc Natl Acad Sci USA 91:206–209

Towers M et al (2012b) Gradients of signalling in the developing limb. Curr Opin Cell Biol 24(2):181–187

Willnow TE et al (2012) Endocytic receptor-mediated control of morphogen signalling. Development 139:4311–4319

Yan D, Lin X (2009) Shaping morphogen gradients by proteoglycans. Cold Spring Harb Perspect Biol 1(3): a002493

Yu Shuizi R et al (2009a) Fgf8 morphogen gradient forms by a source-sink mechanism with freely diffusing molecules. Nature 461(7263):533

Zakin L, De Robertis EM (2010b) Extracellular regulation of BMP signalling. Curr Biol 20:R89–R92

Zhu AJ, Scott MP (2004a) Incredible journey: how do developmental signals travel through tissue? Genes Dev 18(24):2985–2997

Models of Biological Pattern Formation: Books

Meinhardt H (1982) Models of biological pattern formation. Academic, London

Meinhardt H (1995) Algorithmic beauty and seashells. Springer, Berlin

Maini PK (2013) Mathematical models for biological pattern Formation. Springer, Berlin

Articles

Berking S (2003) A model of budding in hydra; pattern formation in concentric rings. J Theor Biol 222:37–52

Drocco JA, Wieschaus EF, Tank DW (2012) The synthesis-diffusion-degradation model explains Bicoid gradient formation in unfertilized eggs. Phys Biol 9(5):055004

Harvey SA, Smith JC (2009) Visualisation and quantification of morphogen gradient formation in the zebrafish. PLoS Biol 7(5):e1000101

Howard J et al (2011) Turing's next steps: the mechano-chemical basis of morphogenesis. Nat Rev Mol Cell Biol 12(6):392–398

Mani M et al (2013) Collective polarization model for gradient sensing via Dachsous-Fat intercellular signalling. Proc Natl Acad Sci USA 110(51):20420–20425

Marciniak-Czochra A (2006) Receptor-based models with hysteresis for pattern formation in hydra. Math Biosci 199(1):97–119

Meinhardt H, Gierer A (2000) Pattern formation by local self-activation and lateral inhibition. BioEssay 22:753–760

Meinhardt H (2006) Primary body axes of vertebrates: Generation of a near-cartesian coordinate system and the role of Spemann-type organizer. Dev Dyn 235 (11):2907–2919

Meinhardt H (2008) Models of biological pattern formation: from elementary steps to the organization of embryonic axes. Curr Top Dev Biol 81:1–63

Meinhardt H (2009a) Models for the generation and interpretation of gradients. Cold Spring Harb Perspect Biol 1(4):a001362

Meinhardt H (2012b) Modeling pattern formation in hydra: a route to understanding essential steps in development. Int J Dev Biol 56(6–8):447–462

Mercker M, Hartmann D, Marciniak-Czochra A (2013) A mechanochemical model for embryonic pattern formation: coupling tissue mechanics and morphogen expression. PLoS One 8(12):e82617

Papatsenko D (2009b) Stripe formation in the early fly embryo: principles, models, and networks. BioEssays 31(11):1172–1180

Sherratt JA, Mueller WA (1995) A receptor based model for pattern formation in *Hydra*. Forma 10:77–95

Turing AM (1952) The chemical basis of morphogenesis. Philos Trans R Soc Lond B Biol Sci 237:37–72

Umulis DM, Othmer HG (2013) Mechanisms of scaling in pattern formation. Development 140 (24):4830–4843

Wolpert L (1969b) Positional information and the spatial pattern of cellular differentiation. J Theor Biol 25:1–47

Wolpert L (2011b) Positional information and patterning revisited. J Theor Biol 269(1):359–365

Woolley T et al (2014) Mathematical modelling of digit specification by a sonic hedgehog gradient. Dev Dyn 243(2):290–298

Chapter 11 Signalling Molecules, Signal Transduction into the Cell: See Also Chapter 10: BMP, NODAL, TGF-β

Bond A et al (2012) The dynamic role of bone morphogenetic proteins in neural stem cell fate and maturation. Dev Neurobiol 72(7):1068–1084

Gun-Sik C et al (2013) BMP signal attenuates FGF pathway in anteroposterior neural patterning. Biochem Biophys Res Commun 434(3):509–515

Coffman JA et al (2014b) Oral-aboral axis specification in the sea urchin embryo, IV: hypoxia radializes embryos by preventing the initial spatialization of nodal activity. Dev Biol 386(2):302–307

Hassani S-N et al (2014a) Inhibition of TGF beta signaling promotes ground state pluripotency. Stem Cell Rev Rep 10:16–30

Miyazono K (2010) Bone morphogenetic protein receptors and signal transduction. J Biochem 1:35–51

Paulsen M et al (2011) Negative feedback in the bone morphogenetic protein 4 (BMP4) synexpression group governs its dynamic signaling range and canalizes development. Proc Natl Acad Sci USA 108 (25):10202–10207

Rider C (2010) Bone morphogenetic protein and growth differentiation factor cytokine families and their protein antagonists. Biochem J 429(1):1–12

Schier AF (2009) Nodal morphogens. Cold Spring Harb Perspect Biol 5:a003459

Shimmi O, Newfeld SJ (2013) New insights into extracellular and post-translational regulation of TGF-beta family signalling pathways. J Biochem 154:11–19

Watabe T (2009) Roles of TGF-beta family signaling in stem cell renewal and differentiation. Cell Res 19 (1):103–115

Zakin L, De Robertis EM (2010c) Extracellular regulation of BMP signaling. Curr Biol 20(3):R89–R92

FGF

Bertrand S et al (2014) FGF signaling emerged concomitantly with the origin of eumetazoans. Mol Biol Evol 31(2):310–318

Boekel C, Brand M (2013b) Generation and interpretation of FGF morphogen gradients in vertebrates. Curr Opin Genet Dev 23(4):415–422

Doreyand K, Amaya E (2010) FGF signalling: diverse roles during early vertebrate embryogenesis. Development 137:3731–3742

Dyer C et al (2014a) A bi-modal function of Wnt signalling directs an FGF activity gradient to spatially regulate neuronal differentiation in the midbrain. Development 141(1):63–72

Katoh M, Nakagama H (2014) FGF receptors: cancer biology and herapeutics. Med Res Rev 34(2):280–300

Mason I (2007) Initiation to end point: the multiple roles of fibroblast growth factors in neural development. Nat Rev Neurosci 8(8):583–596

Neugebauer JM, Yost HJ (2014) FGF signaling is required for brain left-right asymmetry and brain midline formation. Dev Biol 386:123–134

Nowak M et al (2011b) Interpretation of the FGF8 morphogen gradient is regulated by endocytic trafficking. Nat Cell Biol 13(2):153

Streit A et al (2000a) Initiation of neural induction by FGF signalling before gastrulation. Nature 406:74–78

Turner N (2010) Fibroblast growth factor signalling: from development to cancer. Nat Rev Cancer 10 (2):116–129

Yu SR et al (2009a) Fgf8 morphogen gradient forms by a source-sink mechanism with freely diffusing molecules. Nature 461:533–536

Hedgehog, EGF, Ephrins, and Further Protein Factors

Bischoff M et al (2013c) Cytonemes are required for the establishment of a normal Hedgehog morphogen gradient in *Drosophila* epithelia. Nat Cell Biol 15 (11):1269

Briscoe J, Therond P (2013) The mechanisms of Hedgehog signalling and its roles in development and disease. Nat Rev Mol Cell Biol 14(7):416–429

Dessaud D et al (2010b) Dynamic assignment and maintenance of positional identity in the ventral neural tube by the morphogen sonic Hedgehog. PLoS Biol 8: e1000382

Hruska M, Dalva MB (2012) Ephrin regulation of synapse formation, function and plasticity. Mol Cell Neurosci 50:35–44

Ingham PW et al (2011) Mechanisms and functions of Hedgehog signalling across the metazoa. Nat Rev Genet 12(6):393–406

Klein R (2012) Eph/ephrin signalling during development. Development 139:4105–4109

Palm W et al (2013) Secretion and signaling activities of lipoprotein-associated Hedgehog and non-sterol-modified Hedgehog in flies and mammals. PLoS Biol 11(3):e1001505

Sanders TA et al (2013b) Specialized filopodia direct long-range transport of SHH during vertebrate tissue patterning. Nature 497:628–632

Shigeki H et al (2008) Membrane-anchored growth factors, the epidermal growth factor family: Beyond receptor ligands. Cancer Sci 99(2):214–220

Suetterlin P et al (2012a) Axonal ephrinA/EphA interactions, and the emergence of order in topographic projections. Semin Cell Dev Biol 23:1–6

Sun Lai Wing K et al (2011) Netrins: versatile extracellular cues with diverse functions. Development 138:2153–69

Warner JF et al (2014b) Hedgehog signaling requires motile cilia in the sea rchin. Mol Biol Evol 31 (1):18–22

Zhu L et al (1998) Molecular cloning and characterization of *Xenopus* Insulin-like growth factor-1 receptor: its role in mediating insulin-induced *Xenopus* oocyte maturation and expression during embryogenesis. Endocrinology 139:949–955

Notch

Baron M (2012) Endocytic routes to Notch activation. Semin Cell Dev Biol 23(4):437–442

Bigas A et al (2013a) Notch and Wnt signaling in the emergence of hematopoietic stem cells. Blood Cells Mol Dis 51(4):264–270

Dunwoodie SL (2009) The role of Notch in patterning the human vertebral column. Curr Opin Genet Dev 19 (4):329–337

Ehebauer M et al (2006b) Notch, a universal arbiter of cell fate decisions. Science 314(5804):414–1415

Fortini ME (2009) Notch Signaling: the core pathway and its posttranslational regulation. Dev Cell 16 (5):633–647

Garcia T et al (2013b) NOTCH signaling in Sertoli cells regulates gonocyte fate. Cell Cycle 12(16):2538–2545

Hamada H et al (2014d) Involvement of Delta/Notch signaling in zebrafish adult pigment stripe patterning. Development 141(2):318–324

Kopan R (2009) The canonical Notch signaling pathway: unfolding the activation mechanism. Cell 137 (2):216–233

Lewis J (2009) Notch signaling, the segmentation clock, and the patterning of vertebrate somites. J Biol 8:44

Palermo R et al (2014) The molecular basis of Notch signaling regulation: a complex simplicity. Curr Mol Med 14:34–44

Revinski DR et al (2010b) Delta-Notch signaling is involved in the segregation of the three germ layers in *Xenopus laevis*. Dev Biol 339(2):477–492

Retinoids, Arachidonic Acid and Derivatives

Albalat R (2009) The retinoic acid machinery in invertebrates: Ancestral elements and vertebrate innovations. Mol Cell Endocrinol 313(1–2):23–35

Campo-Paysaa F et al (2008) Retinoic acid signaling in development: tissue-specific functions and evolutionary origins. Genesis J Genet Dev 46(11):640–656

Das BC et al (2014) Retinoic acid signaling pathways in development and diseases. Bioorg Med Chem 22 (2):673–683

Eichele G (1989) Retinoic acid induces a pattern of digits in anterior half wing buds that lack the zone of polarizing activity. Development 107:863–867

Lara-Ramirez R et al (2013) Retinoic acid signaling in spinal cord development. Int J Biochem Cell Biol 45 (7):1302–1313

Rhinn M, Dollé P (2012) Retinoic acid signalling during development. Development 139:843–858

Shimozono S et al (2013b) Visualization of an endogenous retinoic acid gradient across embryonic development. Nature 496:363–366

WNT-Cascade, Canonical, Planar

Alexandre C et al (2014b) Patterning and growth control by membrane-tethered Wingless. Nature 505:180–185

Beckett K et al (2013b) Drosophila S2 cells secrete Wingless on exosome-like vesicles but the Wingless gradient forms independently of exosomes. Traffic 14(1):82–96

Bejsovec A (2013b) Wingless/Wnt signaling in Drosophila: the pattern and the pathway. Mol Reprod Dev 80 (11):882–894

Buechling T, Boutros M (2011) WNT signaling: Signaling at and above the receptor level. In: Birchmeier C (ed) Growth factors in development, vol 97, Current topics in developmental Biology., pp 21–53

Dobrowolski R, De Robertis EM (2012b) Endocytic control of growth factor signalling: multivesicular bodies. Nat Rev Mol Cell Biol 13:53–60

Dyer C et al (2014b) A bi-modal function of Wnt signalling directs an FGF activity gradient to spatially regulate neuronal differentiation in the midbrain. Development 141(1):63–72

Gomez-Orte E et al (2013b) Multiple functions of the noncanonical Wnt pathway. Trends Genet 29 (9):545–553

Gross JC et al (2012b) Active Wnt proteins are secreted on exosomes. Nat Cell Biol 14(10):1036

Gross JC, Boutros M (2013) Secretion and extracellular space travel of Wnt proteins. Curr Opin Genet Dev 23 (4):385–390

Hirata A et al (2013a) Dose-dependent roles for canonical Wnt signalling in de novo crypt formation and cell cycle properties of the colonic epithelium. Development 140(1):66–75

Jiyuan K et al (2013) Lipid modification in Wnt structure and function. Curr Opin Lipidol 24(2):129–133

Kerr GE et al (2014b) Regulated Wnt/beta-catenin signaling sustains adult spermatogenesis in mice. Biol Reprod 90(1):3

Krausova M, Korinek V (2014a) Wnt signaling in adult intestinal stem cells and cancer. Cell Signal 26 (3):570–579

Kuehl M et al (2001) Antagonistic regulation of convergent extension movements in Xenopus by Wnt/ß-catenin and WNT/Ca^{2+} signaling. Mech Dev 106:61–67

Kuehl SJ, Kuehl M (2013a) On the role of Wnt/beta-catenin signaling in stem cells. Biochim Biophys 1830(2):2297–2306

Kusserow A et al (2005a) Unexpected complexity of the Wnt gene family in a sea anemone. Nature 433:156–160

Lawrence PA (2013) The mechanisms of planar cell polarity, growth and the Hippo pathway: Some known unknowns. Dev Biol 377(1):1–8

Wen-Hui L et al (2014) In vivo transcriptional governance of hair follicle stem cells by canonical Wnt regulators. Nat Cell Biol 16(2):179–190

Miller RK (2010) Wnt to build a tube: contributions of wnt signaling to epithelial tubulogenesis. Dev Dyn 239(1):77–93

O'Connor E, Moriarty MA, Grealy M (2013a) Beta-catenin and cardiovascular development. In: Braunfeld A, Mirsky GR (eds) Beta-catenin: structure, function and clinical significance. Nova Science Publ, Hauppauge, NY

Petersen C, Reddien PW (2009) Wnt signaling and the polarity of the primary body axis. Cell 139 (6):1056–68

Port F, Basler K (2010b) Wnt trafficking: New insights into Wnt maturation, secretion and spreading. Traffic 11(10):1265–1271

Raggioli A et al (2014) Beta-catenin is vital for the integrity of mouse embryonic stem cells. PLoS One 9(1): e86691

Rosso S, Inestrosa NC (2013) WNT signaling in neuronal maturation and synaptogenesis. Front Cell Neurosci 7:103

Saito-Diaz K et al (2013) The way Wnt works: components and mechanism. Growth Factors 31:1–31

Stewart S et al (2014) Sequential and opposing activities of Wnt and BMP coordinate zebrafish bone regeneration. Cell Rep 6(3):482–498

Swarup S, Verheyen EM (2012) Wnt/Wingless signaling in Drosophila. Cold Spring Harbor Perspect Biol 4(6): a007930

Su-Yi T et al (2014) Wnt/beta-catenin signaling in dermal condensates is required for hair follicle formation. Dev Biol 385(2):179–188

Vonica A et al (2000a) TCF is the nuclear effector of the ß-catenin signal that patterns the sea urchin animal-vegetal axis. Dev Biol 217:230–243

Wehner D et al (2014) Wnt/beta-catenin signaling defines organizing centers that orchestrate growth and differentiation of the regenerating zebrafish caudal fin. Cell Rep 6(3):467–481

Wodarz A, Nusse R (1998) Mechanisms of Wnt signaling in development. Ann Rev Cell Dev Biol 14:59–88

Wylie AD et al (2014c) Post-transcriptional regulation of wnt8a is essential to zebrafish axis development. Dev Biol 386(1):53–63

Xavier CP et al (2014a) Secreted Frizzled-related protein potentiation versus inhibition of Wnt3a/beta-catenin signalling. Cell Signal 26(1):94–101

Signal Transmission, Signal Propagation from Cell to Cell: See Chapter 10, Signal Transduction from Exterior Into Interior: Books

Gomperts BD et al (2009) Signal transduction. Academic, Amsterdam

Marks F et al (2008) Cellular signal processing: an introduction to the molecular mechanisms of signal transduction. Taylor & Francis, New York

Articles

Andersson ER (2012) The role of endocytosis in activating and regulating signal transduction. Cell Mol Life Sci 69(11):1755–1771

Bae JH, Shlessinger J (2010) Asymmetric tyrosine kinase arrangements in activation or autophosphorylation of receptor tyrosine kinases. Mol Cell 443–448

Gomez-Orte E et al (2013c) Multiple functions of the noncanonical Wnt pathway. Trends Genet 29 (9):545–553

Greenwald EC et al (2014) Scaffold state switching amplifies, accelerates, and insulates protein kinase C signalling. J Biol Chem 289(4):2353–2360

Groves JT, Kuriyan J (2010) Molecular mechanisms in signal transduction at the membrane. Nat Struct Mol Biol 17(6):659–665

Lyon AM et al (2014) Strike a pose: G alpha (q) complexes at the membrane. Trends Pharmacol Sci 35:23–30

Jung H, Sunjoo J (2013) Multitasking beta-catenin: from adhesion and transcription to RNA regulation. Animal Cell Syst 17(5):299–305

McLean S, Di Guglielmo GM (2014) TGF beta in endosomal signaling. Methods Enzymol 535:39–54

Petrova IM et al (2014) Wnt signaling through the Ror receptor in the nervous system. Mol Neurobiol 49 (1):303–315

Reyland ME (2009) Protein kinase C isoforms: Multifunctional regulators of cell life and death. Front Biosci 14:2386–99

Rosse C et al (2010) PKC and the control of localized signal. Nat Rev Mol Cell Biol 11(2):103–112

Sands WA, Palmer TM (2008) Regulating gene transcription in response to cyclic AMP elevation. Cell Signal 20(3):460–466

Sarfstein R, Werner H (2013) Minireview: nuclear insulin and insulin-like growth factor-1 Receptors: a novel paradigm in signal transduction. Endocrinology 154 (5):1672–1679

Tomas A et al (2014) EGF receptor trafficking: consequences for signaling and cancer. Trends Cell Biol 24:26–34

Wilson CW, Chuang PT (2010) Mechanism and evolution of cytosolic Hedgehog signal transduction. Development 137(13):2079–94

Xavier CP et al (2014b) Secreted Frizzled-related protein potentiation versus inhibition of Wnt3a/beta-catenin signaling. Cell Signal 26(1):94–101

Primary Cilia and Signalling

Delling M et al (2013) Primary cilia are specialized calcium signalling organelles. Nature 504 (7479):311–314

Goetz S (2010) The primary cilium: a signalling centre during vertebrate development. Nat Rev Genet 11 (5):331–344

Warner JF (2014) Hedgehog signaling requires motile cilia in the sea urchin. Mol Biol Evol 31:18–22

Chapter 12 Genetics of Development (See also Lit. *Drosophila*): Nomenclature

Wain HM (2002) Guidelines for human gene nomenclature. Genomics 79(4):464–470

Genes for Specification of Body Regions and Organs, Eye, Hox Genes

Arendt D (2003b) Evolution of eyes and photoreceptor cell types. Int J Dev Biol 47:563–571

Casaca A et al (2014) Controlling *Hox* gene expression and activity to build the vertebrate axial skeleton. Dev Dyn 24:24–36

Butts T, Holland PW, Ferrier DE (2008a) The urbilaterian super-*Hox* cluster. Trends Genet 24(6):259–262

Durston AJ et al (2012) Time space translation: a *Hox* mechanism for vertebrate a-p patterning. Curr Genomics 13(4):300–307

Gehring WJ (2012) The animal body plan, the prototypic body segment, and eye evolution. Evol Dev 14 (1):34–46

Halder G, Callaerts P, Gehring W (1995b) Induction of ectopic eyes by targeted expression of the *eyeless* gene in *Drosophila*. Science 267(5205):1788–92

Mallo M, Alonso CR (2013) The regulation of Hox gene expression during animal development. Development 140:3951–3963

Martin-Duran M et al (2012) Morphological and molecular development of the eyes during embryogenesis of the freshwater planarian *Schmidtea polychroa*. Dev Genes Evol 222:45–54

McGinnis W, Krumlauf R (1992) Homeobox genes and axial patterning. Cell 68:283–302

Montavon T, Duboule D (2013) Chromatin organization and global regulation of *Hox* gene clusters. Philos Trans R Soc London B Biol Sci 368(1620):20120367

Philippidou P, Dasen JS (2013) *Hox* Genes: choreographers in neural development, architects of circuit organization. Neuron 80:12–34

Soshnikova N (2014) *Hox* genes regulation in vertebrates. Dev Dyn 243(1):49–58

Viczian A, Solessio R (2005) Induction of functional ectopic eyes in a vertebrate. Dev Biol 283(2):670–671

Weasner BM et al (2009) Transcriptional activities of the *Pax6* gene *eyeless* regulate tissue specificity of ectopic eye formation in *Drosophila*. Dev Biol 334 (2):492–502

Genes for Programming Cell Types

D'Amico LE et al (2013) The neurogenic factor NeuroD1 is expressed in post-mitotic cells during juvenile and adult *Xenopus* neurogenesis and not in progenitor or radial glial cells. PLoS One 8(6):e66487

Yun-Ling H et al (2014) Proneural proteins Achaete and Scute associate with nuclear actin to promote formation of external sensory organs. J Cell Sci 127 (1):182–190

Jarman AP, Groves AK (2013) The role of Atonal transcription factors in the development of mechanosensitive cells. Semin Cell Dev Biol 24 (5):438–447

Kim Euiseok J et al (2011) Spatiotemporal fate map of *Neurogenin1* (*Neurog1*) lineages in the mouse central nervous system. J Comp Neurol 519(7):1355–70

Korrzh V, Strahle U (2002) Proneural, prosensory, antiglial: the many faces of neurogenins. Trends Neurosci 25:603–605

Layden MJ et al (2012b) *Nematostella vectensis Achaete-scute* homolog *NvashA* regulates embryonic ectodermal neurogenesis and represents an ancient component of the metazoan neural specification pathway. Development 139(5):1013–1022

Xin L et al (2013) Temporal patterning of Drosophila medulla neuroblasts controls neural fates. Nature 498 (7455):456

Messmer K et al (2012) Induction of neural differentiation by the transcription factor NeuroD2. Int J Dev Neurosci 30(2):105–112

Philpott A, Jones PH (2011) Hes6 Is required for the neurogenic activity of neurogenin and NeuroD. PLoS One 6(11):e27880

Nieber F et al (2013) NumbL is essential for *Xenopus* primary neurogenesis. BMC Dev Biol 13:36

Seipel K et al (2004) Developmental and evolutionary aspects of the basic helix-loop-helix transcription factors Atonal-like 1 and Achaete-scute homolog 2 in the jellyfish. Dev Biol 15:331–345

Velkey JM, O'Shea KS (2013) Expression of *neurogenin 1* in mouse embryonic stem cells directs the differentiation of neuronal precursors and identifies unique patterns of down-stream gene expression. Dev Dyn 242(3):230–253

Wood WM et al (2013) *MyoD*-expressing progenitors are essential for skeletal myogenesis and satellite cell development. Dev Biol 384(1):114–127

Epigenetics, Cell Heredity, Gene Silencing

Aiken C, Ozanne SE (2014b) Transgenerational developmental programming. Hum Reprod Update 20(1):63–75

Baumann K (2010) Epigenetics: unravelling demethylation. Nat Rev Mol Cell Biol 11:87

Benetatos L et al (2014) Polycomb group proteins and MYC: the cancer connection. Cell Mol Life Sci 71 (2):257–269

Budhavarapu V et al (2013) How is epigenetic information maintained through DNA replication? Epigenetics Chromatin 6:32

Devasthanam AS, Tomasi TB (2014) Dicer in immune cell development and function. Immunol Invest 43 (2):182–195

Di Croce L, Helin K (2013) Transcriptional regulation by Polycomb group proteins. Nat Struct Mol Biol 20 (10):1147–1155

Ehrlich M, Lacey M (2013) DNA methylation and differentiation: silencing, upregulation and modulation of gene expression. Epigenomics 5(5):553–568

Furuhashi H, Kelly WG (2010) The epigenetics of germline immortality: lessons from an elegant model system. Dev Growth Differ 52(6):527–532

Gagnon K et al (2014) RNAi factors are present and active in human cell nuclei. Cell Rep 6(1):211–221

Gruber AJ, Zavolan M (2013b) Modulation of epigenetic regulators and cell fate decisions by miRNAs. Epigenomics 5(6):671–683

Harmston N, Lenhard B (2013) Chromatin and epigenetic features of long-range gene regulation. Nucleic Acids Res 41(15):7185–7199

Leeb M, Wutz A (2013) Haploid genomes illustrate epigenetic constraints and gene dosage effects in mammals. Epigenetics Chromatin 6:41

Hong-De L et al (2013) Patterns of nucleosome and chromatin modifications and their effects on chromatin states. Prog Biochem Biophys 40(11):1088–1099

Lyko F et al (2010) The honey bee epigenomes: differential methylation of brain DNA in queens and workers. PLoS Biol 8(11):e1000506

Mondal T, Kanduri C (2013) Maintenance of epigenetic information: a noncoding RNA perspective. Chromosome Res 21(6–7):615–625

Pelechano V et al (2013) Non-coding RNA Gene regulation by antisense transcription. Nat Rev Genet 14 (12):880–893

Rehan VK et al (2012) Perinatal nicotine exposure induces asthma in second generation offspring. BMC Med 10:129

Sievers C, Paro R (2013) Polycomb group protein body-building: working out the routines. Dev Cell 26 (6):556–558

Ruby Y et al (2014) Determinants of heterochromatic siRNA biogenesis and function. Mol Cell 53 (2):262–276

Regulatory RNA, microRNA

Eulalio A et al (2007) Target-specific requirements for enhancers of decapping in miRNA-mediated gene silencing. Genes Dev 21:2558–2570

Gruber AJ, Zavolan M (2013c) Modulation of epigenetic regulators and cell fate decisions by miRNAs. Epigenomics 5(6):671–683

Gurtan AM, Sharp P (2013) The role of miRNAs in regulating gene expression networks. J Mol Biol 425 (19):3582–3600

Hale BJ et al (2014) Small RNA regulation of reproductive function. Mol Reprod Dev 81:148–159

Horabin JI (2013) Long noncoding RNAs as metazoan developmental regulators. Chromosome Res 21 (6–7):673–684

Yong H et al (2011) Biological functions of microRNAs: a review. J Physiol Biochem 67(1):129–139

Ivey KN, Srivastava D (2010) MicroRNAs as regulators of differentiation and cell fate decisions. Cell Stem Cell 36–41

Jinbo L et al (2014) MicroRNAs as novel biological targets for detection and regulation. Chem Soc Rev 43(2):506–517

Meister G (2013) Argonaute proteins: functional insights and emerging roles. Nat Rev Genet 14:447–459

Neeb Z et al (2014b) An expanding world of small RNAs. Dev Cell 28(2):111–112

Nowak JS, Michlewski G (2013) miRNAs in development and pathogenesis of the nervous system. Biochem Soc Trans 41(4):815–820

Ross R et al (2014) PIWI proteins and PIWI-interacting RNAs in the soma. Nature 505(7483):353–359

Sami U et al (2014) MicroRNAs with a role in gene regulation and in human diseases. Mol Biol Rep 41 (1):225–232

Sandoval PV (2014) Functional diversification of Dicer-like proteins and small RNAs required for genome sculpting. Dev Cell 28(2):174–188

Suh N, Blelloch R (2011) Small RNAs in early mammalian development: from gametes to gastrulation. Development 138:1653–61

Jr-Shiuan Y et al (2014) Intertwined pathways for Argonaute-mediated microRNA biogenesis in Drosophila. Nucleic Acids Res 42(3):1987–2002

Box 12B Gene Technology (See Also Lit Chapter 13)

Bedell VM et al (2012) In vivo genome editing using a high-efficiency TALEN system. Nature 491:114–118

Chappell J et al (2013) The centrality of RNA for engineering gene expression. Biotechnol J 8(12):1379–1395

Yi-Guang C et al (2014) Gene targeting in NOD mouse embryos using zinc-finger nucleases. Diabetes 63 (1):68–74

Masanori I et al (2011) Site-specific integration of transgene targeting an endogenous lox-like site in early mouse embryos. J Appl Genet 52(1):89–94

Gaj T et al (2014) Expanding the scope of site-specific recombinases for genetic and metabolic engineering. Biotechnol Bioeng 111:1–15

Gerety SS et al (2013) An inducible transgene expression system for zebrafish and chick. Development 140 (10):2235–2243

Kozlova EN, Berens C (2012) Guiding differentiation of stem cells in vivo by tetracycline-controlled expression of key transcription factors. Cell Transplant 21 (12):2537–2554

Meyer M et al (2010b) Gene targeting by homologous recombination in mouse zygotes mediated by zinc-finger nucleases. Proc Natl Acad Sci USA 34:15022–26

Moore FE et al (2012b) Improved somatic mutagenesis in zebrafish using transcription activator-like effector nucleases (TALENs). PLoS One 7(5):e37877

No D et al (1996) Ecdysone-inducible gene expression in mammalian cells and transgenic mice. Proc Natl Acad Sci USA 93(8):3346–51

Okuyama T et al (2013) Controlled Cre/loxP site-specific recombination in the developing brain in Medaka fish, Oryzias latipes. PLoS One 8(6):e66597

Pauwels K et al (2014) Engineering nucleases for gene targeting: safety and regulatory considerations. New Biotechnol 31(1):18–27

Tallafuss A et al (2012) Turning gene function ON and OFF using sense and antisense photo-morpholinos in zebrafish. Development 139:1691–1699

Wilson RC, Doudna JA (2013b) Molecular mechanisms of RNA interference. Annu Rev Biophys 42:217–239

Optogenetics

Kleinlogel S et al (2011) A gene-fusion strategy for stoichiometric and co-localized expression of light-gated membrane proteins. Nat Methods 8(12):1083

Portugues R et al (2013) Optogenetics in a transparent animal: circuit function in the larval zebrafish. Curr Opin Neurobiol 23(1):119–126

Williams S, Deisseroth K (2013) Optogenetics. Proc Natl Acad Sci USA 110(41):16287

Chapter 13 Cloning, Chimeras, Transgenic Animals (See Also Lit Chapter 12)

Belizario JE et al (2012) New routes for transgenesis of the mouse. J Appl Genet 53(3):295–315

Bouabe H et al (2013) Gene targeting in mice: a review. Methods Mol Biol 1064:315–336

Briggs R, King TJ (1952) Transplantation of living nuclei from blastula cells into enucleated frog eggs. Proc Natl Acad Sci U S A 38:455–463

Campbell KHS et al (1996) Sheep cloned by nuclear transfer from a cultured cell line. Nature 380:64–66

Capecchi MR (2005) Gene targeting in mice: Functional analysis of the mammalian genome for the twenty-first century. Nat Rev Genet 6:507–512

Chaible LM et al (2010) Genetically modified animals for use in research and biotechnology. Genet Mol Res 9 (3):1469–82

Egli D et al (2011) Reprogramming within hours following nuclear transfer into mouse but not human zygotes. Nat Commun 2:488

Gurdon JB (2013b) The cloning of a frog. Development 140:2446–2448

Gurdon JB (2013c) The egg and the nucleus: a battle for supremacy. Development 140:2449–2456

Halley-Stott RP et al (2013) Nuclear reprogramming. Development 140:2468–2471

Hoffman RM (2014) Imageable clinically relevant mouse models of metastasis. Methods Mol Biol 1070:141–170

Markoulaki S et al (2008) Somatic cell nuclear transfer and derivation of embryonic stem cells in the mouse. Methods 45(2):101–114

Mullins LJ, Wilmut I, Mullins JJ (2004) Nuclear transfer in rodents. J Physiol 554(1):4–12

Ogura A et al (2013) Recent advancements in cloning by somatic cell nuclear transfer. Philos Trans R Soc Lond B Biol Sci 368(1609):20110329

Okae H et al (2014) RNA sequencing-based identification of aberrant imprinting in cloned mice. Hum Mol Genet 23(4):992–1001

Scheikl T et al (2010) Transgenic mouse models of multiple sclerosis. Cell Mol Life Sci 67(23):4011–4034

Sosa MA et al (2010) Animal transgenesis: an overview. Brain Struct Funct 214(2–3):91–109

Zhang W et al (2014) Effects of neural stem cells on synaptic proteins and memory in a mouse model of Alzheimer's disease. J Neurosci Res 92(2):185–194

Gene Editing ZFN, TALENS

Hisano Y et al (2014b) Genome editing using artificial site-specific nucleases in zebrafish. Dev Growth Differ 56:26–33

Hockemeyer D et al (2011) Genetic engineering of human pluripotent cells using TALE nucleases. Nat Biotechnol 29(8):731–734

Joung JK, Sander JD (2013) TALENs: a widely applicable technology for targeted genome editing. Nat Rev Mol Cell Biol 14(1):49–55

Meyer M et al (2010c) Gene targeting by homologous recombination in mouse zygotes mediated by zinc-finger nucleases. Proc Natl Acad Sci U S A 34:15022–26

Moore FE et al (2012c) Improved somatic mutagenesis in zebrafish using transcription activator-like effector nucleases (TALENs). PLoS One 7(5):e37877

Wilson RC, Doudna JA (2013c) Molecular mechanisms of RNA interference. Annu Rev Biophys 42:217–239

Gene Editing CRISPR

Auer T et al (2014) Highly efficient CRISPR/Cas9-mediated knock-in in zebrafish by homology-independent DNA repair. Genome Res 24(1):142–153

Bassett A et al (2014) CRISPR/Cas9 and genome editing in Drosophila. J Genet Genomics 41:7–19

Cong L et al (2013) Multiplex genome engineering using CRISPR/Cas systems. Science 339(6121):819–23

Mali P et al (2013) RNA-guided human genome engineering via Cas9. Science 339(6121):823–6

Sampson TR, Weiss DS (2013b) Exploiting CRISPR/Cas systems for biotechnology. Bioessays 36:34–38

Sung YH et al (2014b) Highly efficient gene knockout in mice and zebrafish with RNA-guided endonucleases. Genome Res 24:125–131

Sampson TR, Weiss DS (2013c) Exploiting CRISPR/Cas systems for biotechnology. Bioessays 36:34–38

Sander JD, Joung JK (2014b) CRIRISPR-Cas systems for editing regulating and targeting genomes. Nat Biotechnol 32(4):347–355

Sung YH et al (2014c) Highly efficient gene knockout in mice and zebrafish with RNA-guided endonucleases. Genome Res 24:125–131

Wang H et al (2013b) One-step generation of mice carrying mutations in multiple genes by CRISPR/Cas-mediated genome engineering. Cell 153(4):910–8

Chapter 14 Morphogenesis by Active Movement, Cell Adhesion, Branching

Affolter M et al (2009a) Tissue remodelling through branching morphogenesis. Nat Rev Mol Cell Biol 10:831–842

Aguilar-Cuenca R et al (2014) Myosin II in mechanotransduction: master and commander of cell migration, morphogenesis, and cancer. Cell Mol Life Sci 71(3):479–492

Chung S, Andrew DJ (2008) The formation of epithelial tubes. J Cell Sci 121:3501–3504

Dong X et al (2013) An extracellular adhesion molecule complex patterns dendritic branching and morphogenesis. Cell 155(2):296–307

Hannezo E et al (2014) Theory of epithelial sheet morphology in three dimensions. Proc Natl Acad Sci USA 111(1):27–32

Hardin J, Keller R (1988) Behavior and function of bottle cells during gastrulation of *Xenopus laevis*. Development 103:211–230

Holtfreter J (1946) Structure, motility and locomotion in isolated amphibian cells. J Morphol 79:27–62

Luschnig S, Uv A (2014) Luminal matrices: An inside view on organ morphogenesis. Exp Cell Res 321 (1):64–70

Maitre J-L (2012) Adhesion functions in cell sorting by mechanically coupling the cortices of adhering cells. Science 338(6104):253–256

R'ada M et al (2014b) Morphogenetic movements in the neural plate and neural tube: mouse. Wiley Interdiscp Rev Dev Biol 3(1):59–68

Mettouchi A (2012) The role of extracellular matrix in vascular branching morphogenesis. Cell Adh Migr 6 (6):528–534

Miller RK, McCrea PD (2010) Wnt to build a tube: contributions of Wnt signaling to epithelial tubulogenesis. Dev Dyn 239(1):77–93

Ninomiya H (2012) Cadherin-dependent differential cell adhesion in Xenopus causes cell sorting in vitro but not in the embryo. J Cell Sci 125(8):1877–1883

Papusheva E, Heisenberg C-P (2010) Spatial organization of adhesion: force-dependent regulation and function in tissue morphogenesis. EMBO J 29(16):2753–2768

Sawyer JM et al (2010) Apical constriction: a cell shape change that can drive morphogenesis. Dev Biol 341 (1):5–19

Singh A et al (2012) Eph/ephrin signaling in cell-cell and cell-substrate adhesion. Front Biosci 17:473–497

Steinberg MS (2007) Differential adhesion in morphogenesis: a modern view. Curr Opin Genet Dev 17 (4):281–286

Tada M, Heisenberg CP (2012b) Convergent extension: using collective cell migration and cell intercalation to shape embryos. Development 139:3897–3904

Townes PL, Holtfreter J (1955) Directed movements and selective adhesion of embryonic amphibian cells. J Exp Zool 128:53–120

Twiss F, de Rooij J (2013) Cadherin mechanotransduction in tissue remodeling. Cell Mol Life Sci 70 (21):4101–4116

Walck-Shannon HJ (2014) Cell intercalation from top to bottom. Nat Rev Mol Cell Biol 15(1):34–48

Wolfenson H et al (2013) Dynamic regulation of the structure and functions of integrin adhesions. Dev Cell 24(5):447–458

Wu Selwin K et al (2014) Cortical F-actin stabilization generates apical-lateral patterns of junctional contractility that integrate cells into epithelia. Nat Cell Biol 16(2):167–178

Chapter 15 Cell Migration, Neural Crest and Derivatives

Chang-Joon B et al (2014) Identification of *Pax3* and *Zic1* targets in the developing neural crest. Dev Biol 386 (2):473–483

Bonaventure J et al (2013) Cellular and molecular mechanisms controlling the migration of melanocytes and melanoma cells. Pigment Cell Melanoma Res 26 (3):316–325

Dooley CM et al (2013) On the embryonic origin of adult melanophores: the role of ErbB and Kit signalling in establishing melanophore stem cells in zebrafish. Development 140:1003–1013

Frohnhöfer HG et al (2013) Iridophores and their interactions with other chromatophores are required for stripe formation in zebrafish. Development 140:2997–3007

Gammill LS, Roffers-Agarwal J (2010) Division of labor during trunk neural crest development. Dev Biol 344 (2):555–565

Hagiwara K et al (2014) Molecular and cellular features of murine craniofacial and trunk neural crest cells as stem cell-like cells. PLoS One 9(1):e84072

Kwak J et al (2013) Live image profiling of neural crest lineages in zebrafish transgenic lines. Mol Cells 35 (3):255–260

Le Douarin NM, Dupin E (2012) The neural crest in vertebrate evolution. Curr Opin Genet Dev 22 (4):381–389

Lwigale PY et al (2004) Graded potential of neural crest to form cornea, sensory neurons and cartilage along the rostrocaudal axis. Development 131:1979–1991

Mayor R, Theveneau E (2013) The neural crest. Development 140(11):2247–2251

McKeown SJ et al (2013) Expression and function of cell adhesion molecules during neural crest migration. Dev Biol 373(2):244–257

Molyneaux K, Wylie C (2004b) Primordial germ cell migration. Int J Dev Biol 48:537–543

Mongera A et al (2013) Genetic lineage labeling in zebrafish uncovers novel neural crest contributions to the head, including gill pillar cells. Development 140:916–925

Olovnikov AM (2013) Why do primordial germ cells migrate through an embryo and what does it mean for biological evolution? Biochemistry 78 (10):1190–1199

Plouhinec J-L et al (2014) *Pax3* and *Zic1* trigger the early neural crest gene regulatory network by the direct activation of multiple key neural crest specifiers. Dev Biol 386(2):461–472

Richardson BE, Lehmann R (2010) Mechanisms guiding primordial germ cell migration: strategies from different organisms. Nature Rev Mol Cell Biol 11: 37-49

Rodriguez LL, Schneider IA (2013) Directed cell migration in multi-cue environments. Integr Biol 5 (11):1306–1323

Soto-Suazo M, Zorn TM (2005b) Primordial germ cells migration: morphological and molecular aspects. Anim Reprod 2(3):147–160

Tada M, Heisenberg CP (2012c) Convergent extension: using collective cell migration and cell intercalation to shape embryos. Development 139:3897–3904

Vermillion KL et al (2014) Cytoplasmic protein methylation is essential for neural crest migration. J Cell Biol 204(1):95–109

Chapter 16 Nervous System, Olfactory Organ, Eyes (See Also Lit. Chapter 10)

Galizia G, Lledo P-M (2013) Neurosciences – from molecule to behavior: a university textbook. Springer, Berlin

Kandel E et al (2012) Principles of neural science, 4th edn. McGraw-Hill, New York

Articles: Early Neurogenesis

Arendt D, Nübler-Jung K (1999) Comparison of early nerve cord development in insects and vertebrates. Development 126:2309–2325

Cohen M et al (2010) The role of FGF-signaling in early neural specification of human embryonic stem cells. Dev Biol 340(2):450–458

Imayoshi I et al (2013) Genetic visualization of notch signaling in mammalian neurogenesis. Cell Mol Life Sci 70(12):2045–2057

Nieber F et al (2013c) NumbL is essential for Xenopus primary neurogenesis. BMC Dev Biol 13:36

Patthey C et al (2008) Early development of the central and peripheral nervous systems is coordinated by Wnt and BMP signals. PLoS One 3(2):e1625

Pere EM et al (2014c) Active signals, gradient formation and regional specificity in neural induction. Exp Cell Res 321:25–31

Sharman AC, Brand M (1998) Evolution and homology of the nervous system: cross-phylum rescues of *otd/Otx* genes. Trends Genet 14:211–213

Seo S et al (2007) Neurogenin and NeuroD direct transcriptional targets and their regulatory enhancers. EMBO J 26(24):5093–5108

Streit A et al (2000b) Initiation of neural induction by FGF signalling before gastrulation. Nature 406:74–78

Central Nervous System, Adult Neurogenic Stem Cells

Alexandre P et al (2010) Neurons derive from the more apical daughter in asymmetric divisions in the zebrafish neural tube. Nat Neurosci 13(6):673–679

Belvindrah R et al (2009) Postnatal neurogenesis: from neuroblast migration to neuronal integration. Rev Neurosci 20(5–6):331–346

Boyl PP (2001) *Otx* genes in the development and evolution of the vertebrate brain. Int J Dev Neurosci 19(4):353–363

Braun SM, Jessberger S (2014) Review: adult neurogenesis and its role in neuropsychiatric disease, brain repair and normal brain function. Neuropathol Appl Neurobiol 40:3–12

Breunig JJ et al (2011) Neural stem cells: Historical perspective and future prospects. Neuron 70(4):614–625

Centanin L, Wittbrodt J (2014a) Retinal neurogenesis. Development 141(2):241–244

Crespo-EnriquezI I et al (2012) Fgf8-Related secondary organizers exert different polarizing planar instructions along the mouse anterior neural tube. PLoS One 7(7):e39977

Del Bigio MR (2010) Ependymal cells: biology and pathology. Acta Neuropathol 119(1):55–73

Delaunay D et al (2014) Mitotic spindle asymmetry: a Wnt/PCP-regulated mechanism generating asymmetrical division in cortical precursors. Cell Rep 6(2):400–414

Deng W et al (2010) New neurons and new memories: how does adult hippocampal neurogenesis affect learning and memory? Nat Rev Neurosci 11(5):339–350

Di Giovannantonio LG et al (2014) *Otx2* cell-autonomously determines dorsal mesencephalon versus cerebellum fate independently of isthmic organizing activity. Development 141(2):377–388

Durston AJ et al (1998) Retinoic acid causes an anteroposterior transformation in the developing central nervous system. Nature 340:140–144

Dworkin S, Jane S (2013) Novel mechanisms that pattern and shape the midbrain-hindbrain boundary. Cell Mol Life Sci 70(18):3365–3374

Faigle R, Song H (2013) Signaling mechanisms regulating adult neural stem cells and neurogenesis. Biochim Biophys Acta 1830(2):2435–2448

Fietz SA (2011) Cortical progenitor expansion, self-renewal and neurogenesis a polarized perspective. Curr Opin Neurobiol 21:23–35

Guerout N et al (2014) Cell fate control in the developing central nervous system. Exp Cell Res 321(1):77–83

Hirth F (2010) On the origin and evolution of the tripartite brain. Brain Behav Evol 76(1):3–10

Hubel DH, Wiesel TN, LeVay S (1977) Plasticity of ocular dominance columns in monkey striate cortex. Philos Trans R Soc Lond B Biol Sci 278:377–409

Osorio J et al (2010) Phylotypic expression of the bHLH genes *Neurogenin2, NeuroD*, and *Mash1* in the mouse embryonic forebrain. J Comp Neurol 518(6):851–871

Philippidou P, Dasen JE (2013c) *Hox* Genes: choreographers in neural development, Architects of circuit organization. Neuron 80:12–34

Ranade SS et al (2008) Analysis of the *Otd*-dependent transcriptome supports the evolutionary conservation of CRX/OTX/OTD functions in flies and vertebrates. Dev Biol 315(2):521–534

Reichert H (2005) A tripartite organization of the urbilaterian brain: Developmental genetic evidence from Drosophila. Brain Res Bull 66:491–494

Scott C et al (2010) SOX9 induces and maintains neural stem cells. Nat Neurosci 13(10):1181–1189

Seri B et al (2001) Astrocytes give rise to new neurons in the adult mammalian hippocampus. J Neurosci 21 (18):7153–7160

Suda Y et al (2009) Evolution of *Otx* paralogue usages in early patterning of the vertebrate head. Dev Biol 325 (1):282–295

Sunmonu N et al (2011) Numerous isoforms of Fgf8 reflect its multiple roles in the developing brain. J Cell Physiol 226(7):1722–1726

Tabler JM et al (2010) PAR-1 promotes primary neurogenesis and asymmetric cell divisions via control of spindle orientation. Development 137 (15):2501–2505

Takahashi M, Osumi N (2011) Pax6 regulates boundary-cell specification in the rat hindbrain. Mech Dev 128 (5–6):289–302

Taverna E, Huttner WB (2010) Neural progenitor nuclei in motion. Neuron 67(6):906–914

Toyoda R et al (2010) FGF8 acts as a classic diffusible morphogen to pattern the neocortex. Development 137:3439–3448

Wilkinson G et al (2013) Proneural genes in neocortical development. Neuroscience 253:256–273

Wills A et al (2010b) BMP antagonists and FGF signaling contribute to different domains of the neural plate in *Xenopus*. Dev Biol 337(2):335–350

Wurst W, Bally-Cuif L (2001) Neural plate patterning: Upstream and downstream of the isthmic organizer. Nat Rev Neurosci 2:99–108

Yan B (2010) Numb - from flies to humans. Brain Dev 32 (4):293–298

Placodes – Sensory Organs, Olfactory Axons

Bailey A et al (2006) Lens specification is the ground state of all sensory placodes, from which FGF promotes olfactory identity. Dev Cell 11(4):505–517

Bhattacharyya B-FM (2008) Competence, specification and commitment to an olfactory placode fate. Development 135(24):4165–4177

Dubacq C et al (2014) Making scent of the presence and local translation of odorant receptor mRNAs in olfactory axons. Dev Neurobiol 74(3):259–268

Yun-Ling H et al (2014c) Proneural proteins *Achaete* and *Scute* associate with nuclear actin to promote formation of external sensory organs. J Cell Sci 127:182–190

Jidigam VK, Gunhaga L (2013) Development of cranial placodes: Insights from studies in chick. Dev Growth Differ 55:79–95

Maier E et al (2010) Opposing Fgf and Bmp activities regulate the specification of olfactory sensory and respiratory epithelial cell fates. Development 137(10):1601–1611

Paschaki M et al (2013) Retinoic acid regulates olfactory progenitor cell fate and differentiation. Neural Dev 8:13

Puche AC, Baker H (2007) Olfactory cell derivation and migration. J Mol Histol 38(6):513–515

Schlosser G (2006) Induction and specification of cranial placodes. Dev Biol 294(2):303–351

Sen S et al (2013a) Conserved roles of *ems/Emx* and *otd/Otx* genes in olfactory and visual system development in Drosophila and mouse. Open Biology 3:120177

Eye Development Vertebrates

Carl M et al (2002) Six3 inactivation reveals its essential role for the formation and patterning of the vertebrate eye. Development 129:4057–4063

Centanin L, Wittbrodt J (2014b) Retinal neurogenesis. Development 141:241–244

Eiraku M et al (2011a) Self-organizing optic-cup morphogenesis in three-dimensional culture. Nature 472:51–56

Gehring WJ (2011a) Chance and necessity in eye evolution. Genome Biol Evol 3:1053–1066

Gehring WJ (2014a) The evolution of vision. Interdisc Rev Dev Biol 3:1–40

Hoffmann A et al (2014a) Intrinsic lens forming potential of mouse lens epithelial versus newt iris pigment epithelial cells in three-dimensional culture. Tissue Eng Part C Methods 20(2):91–103

Huang J et al (2011) The mechanism of lens placode formation: A case of matrix-mediated morphogenesis. Dev Biol 355(1):32–42

Martinez-Morales JR, Wittbrodt J (2009) Shaping the vertebrate eye. Curr Opin Genet Dev 19(5):511–517

Matsushima D et al (2011) Combinatorial regulation of optic cup progenitor cell fate by SOX2 and PAX6. Development 138:443–454

Rembold M et al (2006b) Individual cell migration serves as the driving force for optic vesicle evagination. Science 313(5790):1130–1134

Sinn R, Wittbrodt J (2013a) An eye on eye development. Mech Dev 130(6–8):347–358

Yip HK (2014) Retinal stem cells and regeneration of vision system. Anat Rec 297(1):137–160

Growth Cone, Axonal Guidance, Retinotectal Projection, Olfactory Axons

Crispino M et al (2014) Local gene expression in nerve endings. Dev Neurobiol 74(3):279–291

Feinstein P, Mombaerts P (2004) A contextual model for axonal sorting into glomeruli in the mouse olfactory sytem. Cell 117(6):817–831

Feinstein P et al (2004a) Axon guidance of mouse olfactory sensory neurons by odorant receptors and the ß2 adrenergic receptor. Cell 117(6):833–846

Fothergill M et al (2008) Growth cone chemotaxis. Trends Neurosci 31(2):90–98

Franze K (2013) The mechanical control of nervous system development. Development 140:3069–3077

Gebhardt C et al (2012) Balancing of ephrin/Eph forward and reverse signaling as the driving force of adaptive topographic mapping. Development 139:335–345

Gumy LF (2014) New insights into mRNA trafficking in axons. Dev Neurobiol 74(3):233–244

Heckman CA, Plummer HK (2013) Filopodia as sensors. Cell Signal 25(11):2298–2311

Imai T et al (2006) Odorant receptor–derived cAMP signals direct axonal targeting. Science 314 (5799):657–661

Sheng-Jian J, Jaffrey SR (2014) Axonal transcription factors: novel regulators of growth cone-to-nucleus signaling. Dev Neurobiol 74(3):245–258

Lewis T et al (2013) Cellular and molecular mechanisms underlying axon formation, growth, and branching. J Cell Biol 202(6):837–848

Schaupp A et al (2014) The composition of EphB2 clusters determines the strength in the cellular repulsion response. J Cell Biol 204(3):409–422

Schmitt AM et al (2006) Wnt–Ryk signalling mediates medial–lateral retinotectal topographic mapping. Nature 439:31–37

Scicolone G et al (2009) Key roles of Ephs and ephrins in retinotectal topographic map formation. Brain Res Bull 79(5):227–247

Simpson HD et al (2013) A quantitative analysis of branching, growth cone turning, and directed growth in zebrafish retinotectal axon guidance. J Comp Neurol 521(6):1409–1429

Suetterlin P et al (2012b) Axonal ephrinA/EphA interactions, and the emergence of order in topographic projections. Semin Cell Dev 23:1–6

Thompson-Steckel G, Kennedy TE (2014) Maintaining and modifying connections: roles for axon guidance cues in the mature nervous system. Neuropsychopharmacology 39(1):246–247

von Philipsborn AC et al (2006) Growth cone navigation in substrate-bound ephrin gradients. Development 133:2487–2495

Yam PT, Charron F (2013) Signaling mechanisms of non-conventional axon guidance cues: the Shh, BMP and Wnt morphogens. Curr Opin Neurobiol 23 (6):965–973

Peripheral Nervous System, Regeneration

Chauvet S et al (2013a) Navigation rules for vessels and neurons: cooperative signaling between VEGF and neural guidance cues. Cell Mol Life Sci 70 (10):1685–1703

Coulombe JN, Bronner-Fraser M (1987) Cholinergic neurones aquire adrenergic neurotransmitters when transplanted into an embryo. Nature 324:569–572

Dekkers MP et al (2013) Death of developing neurons: new insights and implications for connectivity. J Cell Biol 203(3):385–393

Kang H, Lichtman JW (2013) Motor axon regeneration and muscle reinnervation in young adult and aged animals. J Neurosci 33(50):19480–19491

Levi-Montalcini R (1987) The nerve growth factor 35 years later. Science 237:1154–1161

Synapse Formation, Stabilization, Corrections

Cisterna P et al (2014) Wnt signaling in skeletal muscle dynamics: myogenesis, neuromuscular synapse and fibrosis. Mol Neurobiol 49(1):574–589

Favero M et al (2010) The timing of impulse activity shapes the process of synaptic competition at the neuromuscular junction. Neuroscience 167(2):343–353

Garcia N et al (2010) Involvement of brain-derived neurotrophic factor (BDNF) in the functional elimination of synaptic contacts at polyinnervated neuromuscular synapses during development. J Neurosci Res 88 (7):1406–1419

Jay M et al (2014) Effects of nitric oxide on neuromuscular properties of developing zebrafish embryos. PLoS One 9(1):e86930

Lai K-O, Ip NY (2009) Synapse development and plasticity: roles of ephrin/Eph receptor signaling. Curr Opin Neurobiol 19(3):275–283

Lilley B et al (2014) SAD kinases control the maturation of nerve terminals in the mammalian peripheral and central nervous systems. Proc Natl Acad Sci USA 111 (3):1138–1143

Smith IW et al (2013) Terminal Schwann cells participate in the competition underlying neuromuscular synapse elimination. J Neurosci 33(45):17724–17736

Takeuchi Y et al (2014) Large-scale somatotopic refinement via functional synapse elimination in the sensory thalamus of developing mice. J Neurosci 34(4):1258–1270

Drosophila – Nervous System

Urbach R, Technau GM (2003b) Segment polarity and DV patterning gene expression reveals segmental organization of the *Drosophila* brain. Development 130:3607–3620

Yu-Chun W et al (2014) Drosophila intermediate neural progenitors produce lineage-dependent related series of diverse neurons. Development 141(2):253–258

Chapter 17 Heart, Blood Vessels

Radwan A-I (2014) Heart fields: spatial polarity and temporal dynamics. Anat Rec 297(2):175–182

Beets K et al (2013) Robustness in angiogenesis: Notch and BMP shaping waves. Trends Genet 29(3):140–149

Zhifei C et al (2013) Tumor cell-mediated neovascularization and lymphangiogenesis contrive tumor progression and cancer metastasis. Biochim Biophys Acta 1836(2):273–286

Coultas L et al (2005) Endothelial cells and VEGF in vascular development. Nature 438:937–945

Benazzi C et al (2014) Angiogenesis in spontaneous tumors and implications for comparative tumor biology. Scientific World J 2014:919570

Bosisio D et al (2014) Angiogenic and antiangiogenic chemokines. In:Angiogenesis, Lymphangiogenesis and Clinical Implications. Chemical ImmunolAllergy 99:89–104

Brade T et al (2013) Embryonic heart progenitors and cardiogenesis. Cold Spring Harbor Perspect Med 3 (10):a013847

Chauvet S et al (2013b) Navigation rules for vessels and neurons: cooperative signaling between VEGF and neural guidance cues. Cell Mol Life Sci 70 (10):1685–1703

Daniele G et al (2014) New biological treatments for gynecological tumors: focus on angiogenesis. Expert Opin Biol Ther 14(3):337–346

Dejana E (2010) The role of Wnt signaling in physiological and pathological angiogenesis. Circ Res 107 (8):943–952

Ellertsdóttir E et al (2010) Vascular morphogenesis in the zebrafish embryo. Dev Biol 341:56–65

Fraisl P et al (2009) Regulation of angiogenesis by oxygen and metabolism. Dev Cell 16(2):167–179

Forrai A, Robb L (2003) The hemangioblast – between blood and vessels. Cell Cycle 2:86–90

Gomes F et al (2013) Tumor angiogenesis and lymphangiogenesis: tumor/endothelial crosstalk and cellular/microenvironmental signaling mechanisms. Life Sci 92(2):101–107

Harvey RP (2002) Patterning the vertebrate heart. Nat Rev Genet 3:544–556

Kamei M et al (2006) Endothelial tubes assemble from intracellular vacuoles in vivo. Nature 442:453–456

Laugwitz K-L (2008) Islet1 cardiovascular progenitors: a single source for heart lineages? Development 135:193–205

Jingwei L et al (2013) Neovascularization and hematopoietic stem cells. Cell Biochem Biophys 67 (2):235–245

Matsumoto G et al (2014) Control of angiogenesis by VEGF and endostatin-encapsulated protein microcrystals and inhibition of tumor angiogenesis. Biomaterials 35(4):1326–1333

Nakajima Y et al (2009) Heart development before beating. Anat Sci Int 84(3):67–76

Nakayama M et al (2013) Spatial regulation of VEGF receptor endocytosis in angiogenesis. Nat Cell Biol 15(3):249–260

Nikolov DB et al (2013) Eph/ephrin recognition and the role of Eph/ephrin clusters in signaling initiation. Biochim Biophys 1834(10):2160–2165

O'Connor E, Moriarty MA, Grealy M (2013b) Beta-catenin and cardiovascular development. In: Braunfeld A, Mirsky GR (eds) Beta-catenin: structure, function and clinical significance. Nova Science, Hauppauge

Paffett-Lugassy N et al (2013) Kinetic framework of spindle assembly checkpoint signalling Heart field origin of great vessel precursors relies on nkx2.5-mediated vasculogenesis. Nat Cell Biol 15:1362–1369

Reis M, Liebner S (2013) Wnt signaling in the vasculature. Exp Cell Res 319(9):1317–1323

Sheng G (2010) Primitive and definitive erythropoiesis in the yolk sac: a bird's eye view. Int J Dev Biol 54 (6–7):1033–1043

Später D et al (2013) A HCN4+ cardiomyogenic progenitor derived from the first heart field and human pluripotent stem cells. Nat Cell Biol 15:1098–1106

Staudt D, Stainier D (2012) Uncovering the molecular and cellular mechanisms of heart development using the ebrafish. Annu Rev Genet 46:397–418

Xin M et al (2013c) Mending broken hearts: cardiac development as a basis for adult heart regeneration and repair. Nat Rev Mol Cell Biol 14:520–541

Chapter 18 Stem cells, Regenerative Medicine, Regeneration: Stem cells: General, Embryonic Stem Cells, Pluripotency

Abu-Remaileh M et al (2010a) *Oct-3/4* regulates stem cell identity and cell fate decisions by modulating Wnt/beta-catenin signalling. EMBO J 29(19):3236–3248

Acampora D et al (2013) *Otx2* is an intrinsic determinant of the embryonic stem cell state and is required for transition to a stable epiblast stem cell condition. Development 140:43–55

Blake JA, Ziman MR (2014) *Pax* genes: regulators of lineage specification and progenitor cell maintenance. Development 141:737–751

Faunes F et al (2013) A membrane-associated β-catenin/Oct4 complex correlates with ground-state pluripotency in mouse embryonic stem cells. Development 140:1171–1183

Gomez-Lopez S et al (2014a) Asymmetric cell division of stem and progenitor cells during homeostasis and cancer. Cell Mol Life Sci 71(4):575–597

Greve TS et al (2013) microRNA control of mouse and human pluripotent stem cell behaviour. Annu Rev Cell Dev Biol 29:213–239

Hassani S et al (2014b) Inhibition of TGF beta signaling promotes ground state pluripotency. Stem Cell Rev Rep 10:16–30

Ishiuchi T et al (2013) Towards an understanding of the regulatory mechanisms of totipotency. Curr Opin Genet Dev 23(5):512–518

Jauch R, Kolatkar PR (2013) What makes a pluripotency reprogramming factor? Curr Mol Med 13(5):806–814

Juliano CE et al (2010c) A conserved germline multipotency program. Development 137(24):4113–4126

Kraushaar DC et al (2013) Heparan sulfate: a key regulator of embryonic stem cell fate. Biol Chem 394(6):741–751

Kuehl SJ, Kuehl M (2013b) On the role of Wnt/beta-catenin signaling in stem cells. Biochim Biophys Acta 1830(2):2297–2306

Le Bin GC et al (2014) Oct4 is required for lineage priming in the developing inner cell mass of the mouse blastocyst. Development 141:1001–1010

Oenal P et al (2012a) Gene expression of pluripotency determinants is conserved between mammalian and planarian stem cells. EMBO J 31(12):2755–2769

Saunders A et al (2013) Concise review: Pursuing self-renewal and pluripotency with the stem cell factor Nanog. Stem Cells 31(7):1227–1236

Serrano L et al (2013) Chromatin structure, pluripotency and differentiation. Exp Biol Med 238(3):259–270

Tantin D (2013) Oct transcription factors in development and stem cells: insights and mechanisms. Development 140:2857–2866

Wright JE, Ciosk R (2013) RNA-based regulation of pluripotency. Trends Genet 29(2):99–107

Guangming W et al (2013) Establishment of totipotency does not depend on Oct4A. Nat Cell Biol 15(9):1089

Epigenetic Programming, Somatic Versus Germ Cells (See also Lit Chapter 8)

Cantone I, Fisher AG (2013) Epigenetic programming and reprogramming during development. Nat Struct Mol Biol 20(3):282–289

Garcia TX, Hofmann M-C (2013b) NOTCH signaling in Sertoli cells regulates gonocyte fate. Cell Cycle 12(16):2538–2545

Griswold MD, Oatley JM (2013b) Concise review: defining characteristics of mammalian spermatogenic stem cells. Stem Cells 31:8–11

Hanna CB, Hennebold JD (2014) Ovarian germline stem cells: an unlimited source of oocytes? Fertil Steril 101:20–30

Imamura M et al (2013) Cell-intrinsic reprogramming capability: gain or loss of pluripotency in germ cells. Reprod Med Biol 12:1–14

Imamura M et al (2014a) Generation of germ cells in vitro in the era of induced pluripotent stem cells. Mol Reprod Dev 81:2–19

Kawasaki YY et al (2014) Active DNA demethylation is required for complete imprint erasure in primordial germ cells. Sci Rep 4:3658

Lasko P (2013b) The DEAD-box helicase Vasa: Evidence for a multiplicity of functions in RNA processes and developmental biology. Biochim Biophys Acta 1829(8):810–816

Marques-Mari AI et al (2009) Differentiation of germ cells and gametes from stem cells. Hum Reprod Update 15(3):379–90

Nakaki F et al (2013) Induction of mouse germ-cell fate by transcription factors in vitro. Nature 501(7466):222

Solana J (2013) Closing the circle of germline and stem cells: the Primordial Stem Cell hypothesis. EvoDevo 4:23

Ten Berge D et al (2011) Embryonic stem cells require Wnt proteins to prevent differentiation to epiblast stem cells. Nat Cell Biol 13(9):1070–1075

Valli H et al (2014) Germline stem cells: toward the regeneration of spermatogenesis. Fertil Steril 101:3–13

Wang Z, Lin H (2004b) Nanos maintains germline stem cell self-renewal by preventing differentiation. Science 303:2016–2019

Saiyong Z et al (2014) Small molecules enable OCT4-mediated direct reprogramming into expandable human neural stem cells. Cell Res 24(1):126–129

Stem Cells of the Blood, Hematopoiesis, Immune Cells

Bigas A et al (2013b) Notch and Wnt signaling in the emergence of hematopoietic stem cells. Blood Cells Mol Dis 51(4):264–270

Blau O (2014) Bone marrow stromal cells in the pathogenesis of acute myeloid leukaemia. Front Biosci 19:171–180

Jagannathan-Bogdan M, Zon LI (2013) Hematopoiesis. Development 140:2463–2467

Knapp DJ, Eaves CJ (2014) Control of the hematopoietic stem cell state. Cell Res 24:3–4

Nakajima-Takagi Y et al (2014a) Manipulation of hematopoietic stem cells for regenerative medicine. Anat Rec 297:111–120

Tavian M et al (2010) Embryonic origin of human hematopoiesis. Int J Dev Biol 54 (6–7):1061–1065

Ueno H, Weissmann L (2010) The origin and fate of yolk sac hematopoiesis: application of chimera analyses to developmental studies. Int J Dev Biol 54:1019–1031

Further Adult Stem Cells, Plasticity, Neuronal Stem Cells

Abou-Khalil R, Brack AS (2010a) Muscle stem cells and reversible quiescence. The role of sprouty. Cell Cycle 9(13):2575–2580

Barker N (2014) Adult intestinal stem cells: critical drivers of epithelial homeostasis and regeneration. Nat Rev Mol Cell Biol 15:19–33

Brand AH, Livesey FJ (2011) Neural stem cell biology in vertebrates and invertebrates: More alike than different? Neuron 70(4):719–729

Braun SM, Jessberger S (2014b) Review: Adult neurogenesis and its role in neuropsychiatric disease, brain repair and normal brain function. Neuropathol Appl Neurobiol 40:3–12

Castro-Munozledo F (2013) Review: Corneal epithelial stem cells, their niche and wound healing. Mol Vis 19:1600–1613

Charbord P (2010) Bone marrow mesenchymal stem cells: historical overview and concepts. Hum Gene Ther 21(9):1045–1056

Forostyak S et al (2013) The role of mesenchymal stromal cells in spinal cord injury, regenerative medicine and possible clinical applications. Biochimie 95 (12):2257–2270

Goossens E, Tournaye H (2013) Adult stem cells in the human testis. Semin Reprod Med 31:39–48

Hori Y (2009) Insulin-producing cells derived from stem/progenitor cells: therapeutic implications for diabetes mellitus. Med Mol Morphol 42(4):195–200

Krausova M, Korinek V (2014b) Wnt signaling in adult intestinal stem cells and cancer. Cell Signal 26 (3):570–579

Krishnakumar R, Blelloch RH (2013) Epigenetics of cellular reprogramming. Curr Opin Genet Dev 23 (5):548–555

Xiaoyan S et al (2013) Epidermal stem cells: an update on their potential in regenerative medicine. Expert Opin Biol Ther 13(6):901–910

van den Berge SA et al (2013) Resident adult neural stem cells in Parkinson's disease. The brain's own repair system? Eur J Pharmacol 719(1–3):117–127

Varela-Nallar L, Inestrosa NC (2013) Wnt signaling in the regulation of adult hippocampal neurogenesis. Front Cell Neurosci 7:100

Reprogramming, iPSC, Transdifferentiation, Medical Issues

Anokye-Danso F et al (2011) Highly efficient miRNA-mediated reprogramming of mouse and human somatic cells to pluripotency. Cell Stem Cell 8 (4):376–388

Bayart E, Cohen-Haguenauer O (2013) Technological overview of iPS induction from human adult somatic cells. Curr Gene Ther 13(2):73–92

Braun SM, Jessberger S (2014c) Review: Adult neurogenesis and its role in neuropsychiatric disease, brain repair and normal brain function. Neuropathol Appl Neurobiol 40:3–12

Eguizabal C et al (2013a) Dedifferentiation, transdifferentiation, and reprogramming: future directions in regenerative medicine. Semin Reprod Med 31:82–94

Garate Z et al (2013) New frontier in regenerative medicine: Site-specific gene correction in patient-specific induced pluripotent stem cells. Hum Gene Ther 24 (6):571–583

Guo J et al (2013) Reprogramming and transdifferentiation shift the landscape of regenerative medicine. DNA Cell Biol 32(10):565–572

Jingying H et al (2013) Cardiac stem cells and their roles in myocardial infarction. Stem Cell Rev Rep 9 (3):326–338

Huo JS et al (2013) Pivots of pluripotency: the roles of non-coding RNA in regulating embryonic and induced pluripotent stem cells. Biochim Biophys Acta 1830 (2):2385–2394

Imamura M et al (2014b) Generation of germ cells in vitro in the era of induced pluripotent stem cells. Mol Reprod Dev 81:2–19

Inagawa K, Ieda M (2013) Direct reprogramming of mouse fibroblasts into cardiac myocytes. J Cardiovasc Transl Res 6:37–45

Jiang Z et al (2014) Induced pluripotent stem cell (iPSCs) and their application in immunotherapy. Cell Mol Immunol 11:17–24

Hyun-Jung K (2011) Stem cell potential in Parkinson's disease and molecular factors for the generation of dopamine neurons. Biochim Biophys Acta 1812 (1):1–11

Kim JB et al (2009) Oct4-induced pluripotency in adult neural stem cells. Cell 136:411–419

Koudstaal S et al (2013) Concise review: heart regeneration and the role of cardiac stem cells. Stem Cells Transl Med 2(6):434–443

Kulbatski I (2010) Stem/precursor cell-based CNS therapy: the importance of circumventing immune suppression by transplanting autologous cells. Stem Cell Rev Rep 6:405–410

Masip M et al (2010) Reprogramming with defined factors: from induced pluripotency to induced transdifferentiation. Mol Hum Reprod 16 (11):856–868

Maucksch C et al (2013) Concise Review: The Involvement of SOX2 in direct reprogramming of induced neural stem/precursor cells. Stem Cells Transl Med 2 (8):579–583

Millane RC et al (2011b) Induced stem cell neoplasia in a cnidarian by ectopic expression of a POU domain transcription factor. Development 138:2429–2439

Nakajima-Takagi Y et al (2014b) Manipulation of hematopoietic stem cells for regenerative medicine. Anat Rec 297(1):111–120

Obokata H et al (2014a) Stimulus-triggered fate conversion of somatic cells into pluripotency. Nature 505:641–647

Obokata H et al (2014b) Bidirectional developmental potential in reprogrammed cells with acquired pluripotency. Nature 505:676–680

Okita K, Yamanaka S (2010) Induction of pluripotency by defined factors. Exp Cell Res 316(16):2565–2570

Park H-J (2014) Nonviral delivery for reprogramming to pluripotency and differentiation. Arch Pharm Res 37:107–119

Petit GH et al (2014) Review: The future of cell therapies and brain repair: Parkinson's disease leads the way. Neuropathol Appl Neurobiol 40:60–70

Rong Z et al (2014) An effective approach to prevent immune rejection of human ESC-derived allografts. Cell Stem Cell 14(1):121–130

Sisakhtnezhad S, Matin MM (2012) Transdifferentiation: a cell and molecular reprogramming process. Cell Tissue Res 348(3):379–396

Shu J, Deng H (2013) Lineage specifiers: New players in the induction of pluripotency. Genomics Proteomics Bioinformatics 11(5):259–263

Takahashi K et al (2007) Induction of pluripotent stem cells from adult human fibroblasts by defined factors. Cell 131:861–872

Takahashi K, Shinya Y (2013) Induced pluripotent stem cells in medicine and biology. Development 140:2457–2461

Tucker BA et al (2011) Transplantation of adult mouse iPS cell-derived photoreceptor precursors restores retinal structure and function in degenerative mice. PLoS One 6(49):e18992

Watanabe S et al (2011) MyoD gene suppression by Oct4 is required for reprogramming in myoblasts to produce induced pluripotent stem cells. Stem Cells 29 (3):505–516

Zhao X-Y (2010) Viable fertile mice generated from fully pluripotent iPS cells derived from adult somatic cells. Stem Cell Rev Rep 6:390–397

Zhu S et al (2014) Small molecules enable OCT4-mediated direct reprogramming into expandable human neural stem cells. Cell Res 24:126–129

Tissue and Organ Culture, Tissue Engineering

Biazar E (2014) Use of umbilical cord and cord blood-derived stem cells for tissue repair and regeneration. Expert Opin Biol Ther 14(3):301–310

Buikema JW et al (2013) Concise review: engineering myocardial tissue: the convergence of stem cells biology and tissue engineering technology. Stem Cells 31 (12):2587–2598

Dunn DA et al (2014) Biomimetic materials design for cardiac tissue regeneration. Wiley Interdiscipl Rev Nanomed Nanobiotechnol 6:15–39

Eiraku M et al (2011b) Self-organizing optic-cup morphogenesis in three-dimensional culture. Nature 472:51–56

Georgiadis V et al (2014) Cardiac tissue engineering: renewing the arsenal for the battle against heart disease. Integr Biol 6(2):111–126

Harris GM et al (2013) Strategies to direct angiogenesis within scaffolds for bone tissue engineering. Curr Pharm Des 19(19):3456–3465

Hoffmann A et al (2014b) Intrinsic lens forming potential of mouse lens epithelial versus newt iris pigment epithelial cells in three-dimensional culture. Tissue Eng Part C Methods 20(2):91–103

Humes HD et al (2014) The bioartificial kidney: current status and future promise. Pediatr Nephrol 29 (3):343–351

Keller L et al (2013) Whole tooth engineering and cell sources. Stem Cells Craniofacial Dev Regen 431-446

Lechguer AN et al (2011) Cell differentiation and matrix organization in engineered teeth. J Dental Res 90 (5):583–589

Song L et al (2014) Vascular tissue engineering: from in vitro to in situ. Wiley Interdiscp Rev Syst Biol Med 6:61–76

Jinling M et al (2014) Concise review: cell-based strategies in bone tissue engineering and regenerative medicine. Stem Cells Transl Med 3(1):98–107

Morin KT, Tranquillo RT (2013) In vitro models of angiogenesis and vasculogenesis in fibrin gel. Exp Cell Res 319(16):2409–2417

Mototsugu E et al (2011) Self-organizing optic-cup morphogenesis in three-dimensional culture. Nature 472:51–56

Naito Y et al (2014) Tissue engineering in the vasculature. Anat Rec 297(1):83–97

Ohashi K, Okano T (2014) Functional tissue engineering of the liver and islets. Anat Rec 297(1):73–82

Pak J et al (2014) Regenerative repair of damaged meniscus with autologous adipose tissue-derived stem cells. Biomed Res Int 2014:436029

Sasai Y et al (2012) In vitro organogenesis in three dimensions: self-organising stem cells. Development 139:4111–4121

Sato T, Clevers H (2013) Growing self-organizing mini-guts from a single intestinal stem cell: mechanism and applications. Science 340(6137):1190–1194

Serbo JV, Gerecht S (2013) Vascular tissue engineering: biodegradable scaffold platforms to promote angiogenesis. Stem Cell Res Ther 4:8

Shahrokhi S et al (2014) The use of dermal substitutes in burn surgery: acute phase. Wound Repair Regen 22:14–22

Steindorff NM et al (2014) Innovative approaches to regenerate teeth by tissue engineering. Arch Oral Biol 59(2):158–166

Viczian A et al (2009) Generation of functional eyes from pluripotent cells. PLoS Biol 7(8):e1000174

Ling W et al (2013) Regeneration of articular cartilage by adipose tissue derived mesenchymal stem cells: Perspectives from stem cell biology and molecular medicine. J Cell Physiol 228(5):938–944

Therapeutic Cloning

Lisker R (2003) Ethical and legal issues in therapeutic cloning and the study of stem cells. Arch Med Res 34 (6):607–611

Stem Cells and Cancer; s. Lit Chapter 19 (Regeneration in Animals)

Morgan TH (1901) Regeneration. MacMillan, New York

Stem Cells and Regeneration

Baguñà J (2012) The planarian neoblast: the rambling history of its origin and some current black boxes. Int J Dev Biol 56:19–37

Frank U, Plickert G, Mueller WA (2009b) Cnidarian interstitial cells: the dawn of stem cell research. In: Rinkevich B, Matranga V (eds) Stem cells in marine organisms. Springer, Berlin, pp 33–59

Geneviere A-M et al (2009) Cell dynamics in early embryogenesis and pluripotent embryonic cell lines: from sea urchin to mammals. In: Rinkevich B, Matranga V (eds) Stem cells in marine organisms. Springer, Berlin, pp 215–243

Gold DA, Jacobs DK (2013b) Stem cell dynamics in Cnidaria: are there unifying principles? Dev Genes Evol 223:53–66

Kumar A et al (2010) A comparative study of gland cells implicated in the nerve dependence of salamander limb regeneration. J Anat 217(1):16–25

Kunzel T et al (2010) Migration and differentiation potential of stem cells in the cnidarian *Hydractinia* analysed in eGFP-transgenic animals and chimeras. Dev Biol 348(2010):120–129

Millane CR et al (2011c) Induced stem cell neoplasia in a cnidarian by ectopic expression of a POU domain transcription factor. Development 138:2429–2439

Mueller WA et al (2004d) Totipotent migratory stem cells in a hydroid. Dev Biol 275:215–224

Oenal P et al (2012b) Gene expression of pluripotency determinants is conserved between mammalian and planarian stem cells. EMBO J 31(12):2755–2769

Rink JC (2013) Stem cell systems and regeneration in planaria. Dev Genes Evol 223:67–84

Rinkevich B et al (1995) Whole-body protochordate regeneration from totipotent blood cells. Proc Natl Acad Sci USA 92:7695–7699

Rinkevich Y et al (2009) Stem cells in aquatic invertebrates: common premises and emerging unique themes. In: Rinkevich B, Matranga V (eds) Stem cells in marine organisms. Springer, Berlin, pp 61–103

Solana J et al (2012) Defining the molecular profile of planarian pluripotent stem cells using a combinatorial RNA-seq, RNA interference and irradiation approach. Genome Biol 13(3):R19

Teo R et al (2006b) An evolutionary conserved role of Wnt signaling in stem cell fate decisision. Dev Biol 289:91–99

Weismann A (1883b) Die Entstehung der Sexualzellen bei Hydromedusen. Jena Fisher, Leipzig

Transdifferentiation, Lens and Eye Regeneration

Eguchi G (1988) Cellular and molecular background of Wolffian lens regeneration. Cell Differ Dev 25:147–158

Eguizabal C et al (2013b) Dedifferentiation, transdifferentiation, and reprogramming: future directions in regenerative medicine. Semin Reprod Med 31:82–94

Liua H et al (2007) Ciliary margin transdifferentiation from neural retina is controlled by canonical Wnt signaling. Dev Biol 308(1):54–67

Kodama R, Eguchi G (1995) From lens regeneration in the newt to in-vitro transdifferentiation of vertebrate pigmented epithelial cells. Semin Cell Biol 6 (3):143–149

Kondoh H et al (2004) Interplay of Pax6 and SOX2 in lens development as a pradigm of genetic switch mechanisms for cell differentiation. Int J Dev Biol 48:819–827

Maki N, Tsonis PA (2010) Changes in global histone modifications during dedifferentiation in newt lens regeneration. Mol Vis 16(204–05):1893–1897

Martinez-De Luna RI et al (2011) The retinal homeobox (*Rx*) gene is necessary for retinal regeneration. Dev Biol 353(1):10–18

McGann CJ et al (2001) Mammalian myotube dedifferentiation induced by newt regeneration extract. Proc Natl Acad Sci USA 13699–704

Panagiotis AT (2004) A newt's eye view of lens regeneration. Int J Dev Biol 48:975–980

Perez OD et al (2002) Inhibition and reversal of myogenic differentiation by purine-based microtubule assembly inhibitors. Chem Biol 9:475–483

Schmid V et al (1988) Transdifferentiation from striated muscle of medusae in vitro. Cell Differ Dev 25 (Suppl):137–146

Schmid V, Reber-Muller S (1995) Transdifferentiation of isolated striated muscle of jellyfish in vitro: the initiation process. Semin Cell Biol 6:109–116

Schubiger M et al (2010) Regeneration and transdetermination: the role of wingless and its regulation. Dev Biol 347:315–324

Sousounis K et al (2013) Transcriptome analysis of newt lens regeneration reveals distinct gradients in gene expression patterns. PLoS One 8(4):e61445

Suetsugu-Maki R et al (2012) Lens regeneration in axolotl: new evidence of developmental plasticity. BMC Biol 10:103

Thomas AG, Henry JJ (2014) Retinoic acid regulation by CYP26 in vertebrate lens regeneration. Dev Biol 386 (2):291–301

Limb Regeneration

Blum N, Begemann G (2013) The roles of endogenous retinoid signaling in organ and appendage regeneration. Cell Mol Life Sci 70(20):3907–3927

Christen B et al (2003a) Regeneration-specific expression patterns of three *XHox* genes. Dev Dyn 226:349–355

Hirata A et al (2013b) Accessory limb induction on flank region and its muscle regulation in axolotl. Dev Dyn 242(8):932–940

Kragl M et al (2009) Cells keep a memory of their tissue origin during axolotl limb regeneration. Nature 460 (7251):60

Maden M (1980b) Intercalary regeneration in the amphibian limb and the rule of distal transformation. J Embryol Exp Morphol 56:201–209

Maden M (1993a) The homeotic transformation of tails into limbs in *Rana temporaria* by retinoids. Dev Biol 159:379–391

Makanae A et al (2013) Nerve independent limb induction in axolotls. Dev Biol 381(1):213–226

Mochii M (2007) Tail regeneration in the *Xenopus* tadpole. Dev Growth Differ 49(2):5–161

Roensch K et al (2013) Progressive specification rather than intercalation of segments during limb regeneration. Science 342(6164):1375–1379

Satoh A et al (2007) Nerve-induced ectopic limb blastemas in the axolotl are equivalent to amputation-induced blastemas. Dev Biol 312(1):231–244

Tamura K (2010a) Limb blastema cell: a stem cell for morphological regeneration. Dev Growth Differ 52 (1):89–99

Wang KC (2009) Regeneration, repair and remembering identity: the three Rs of *Hox* gene expression. Trends Cell Biol 19(6):268–275

Pattern Control in Regeneration, Invertebrates (See also lit. to Chapter 4.3 *Hydra*, Chapter 10.9 (Pattern Control) and Chapter 10.10 (Intercalation)

De Robertis EM (2010) Wnt signaling in axial patterning and regeneration: lessons from planaria. Sci Signal 3 (127):Art No. pe21

Duffy DJ et al (2010a) Wnt signaling promotes oral structures and suppresses aboral ones in *Hydractinia* metamorphosis and regeneration. Development 137 (18):3057–3066

Egger B et al (2006) The regeneration capacity of the flatworm *Macrostomum lignano* on repeated regeneration, rejuvenation, and the minimal size needed for regeneration. Dev Genes Evol 216(1):565–577

Fraguas S et al (2011) EGFR signaling regulates cell proliferation, differentiation and morphogenesis during planarian regeneration and homeostasis. Dev Biol 354(1):87–101

Gaviño MA, Reddien PW (2011) A Bmp/Admp regulatory circuit controls maintenance and regeneration of dorsal-ventral polarity in planarians. Curr Biol 21 (4):294–299

Gee L et al (2010c) β-catenin plays a central role in setting up the head organizer in hydra. Dev Biol 340 (1):116–124

Gurley KA et al (2008) β-Catenin defines head versus tail identity during planarian regeneration and homeostasis. Science 319(5861):323–327

Gurley KA et al (2010) Expression of secreted Wnt pathway components reveals unexpected complexity of the planarian amputation response. Dev Biol 347(1):24–39

Holstein TW, Hobmayer E, Technau U (2003b) Cnidarians: An evolutionarily conserved model system for regeneration? Dev Dyn 226:257–267

Lengfeld T et al (2009c) Multiple Wnts are involved in *Hydra* organizer formation and regeneration. Dev Biol 330(1):186–199

Molina D et al (2011) Noggin and noggin-like genes control dorsoventral axis regeneration in planarians. Curr Biol 21(4):300–305

Mueller WA et al (2004e) Patterning a multi-headed mutant in *Hydractinia*: enhancement of head formation and its phenotypic normalization. Int J Dev Biol 48:9–15

Mueller WA et al (2007c) Wnt signaling in hydroid development: Ectopic heads and giant buds induced by GSK-3beta inhibitors. Int J Dev Biol 51:211–220

Rink JC et al (2009a) Planarian Hh signaling regulates regeneration polarity and links Hh pathway evolution to cilia. Science 326(5958):1406–1410

Salo E et al (2009) Planarian regeneration: achievements and future directions after 20 years of research. Int J Dev Biol 53(8–10):1317–1327

Tasaki J et al (2011) ERK signaling controls blastema cell differentiation during planarian regeneration. Development 138:2417–2427

Wenemoser D, Reddien PW (2010) Planarian regeneration involves distinct stem cell responses to wounds and tissue absence. Dev Biol 344:979–991

Witchley JM et al (2013) Muscle cells provide instructions for planarian regeneration. Cell Rep 4 (4):633–641

Chapter 19 Growth Control, Cancer

Anisimov SV (2010) Risks and mechanisms of oncological disease following stem cell transplantation. Stem Cell Rev Reports 6:411–424

Arend R et al (2013) The Wnt/beta-catenin pathway in ovarian cancer: A review. Gynecologic Oncology 131 (3):772–779

Barcellos-de-Souza P et al (2013) Tumor microenvironment: Bone marrow-mesenchymal stem cells as key players. Biochim Biophys Acta 1836(2):321–335

Bojang P, Ramos KS (2014) The promise and failures of epigenetic therapies for cancer treatment. Cancer Treatment Rev 40(1):153–169

Catalano V et al (2013) Tumor and its microenvironment: a synergistic interplay. Semin Cancer Biol 23 (6):522–532

Cheng L et al (2007b) OCT4: biological functions and clinical applications as a marker of germ cell neoplasia. J Pathol 211:1–9

Chin AR, Wang SE (2014) Cytokines driving breast cancer stemness. Mol Cell Endocrinol 382(1):598–602

ElShamy WM, Duhe RJ (2013) Overview: cellular plasticity, cancer stem cells and metastasis. Cancer Lett 341:2–8

Fessler E et al (2013) Cancer stem cell dynamics in tumor progression and metastasis: Is the microenvironment to blame? Cancer Lett 341:97–104

Gomez-Lopez S et al (2014b) Asymmetric cell division of stem and progenitor cells during homeostasis and cancer. Cell Mol Life Sci 71(4):575–597

Gschweng E et al (2014) Hematopoietic stem cells for cancer immunotherapy. Immunol Rev 257:237–249

Jin H et al (2014) Emerging role of microRNAs in cancer and cancer stem cells. J Cell Biochem 115(4):605–610

Ishizawa K et al (2010) Tumor-initiating cells are rare in many human tumors. Cell Stem Cell 7(3):279–282

Kampen KR et al (2013) Vascular endothelial growth factor signaling in acute myeloid leukemia. Cell Mol Life Sci 70(8):1307–1317

Kim J et al (2010) A Myc Nnetwork accounts for similarities between embryonic stem and cancer cell transcription programs. Cell 143(2):313–324

Krausova M et al (2014) Wnt signaling in adult intestinal stem cells and cancer. Cell Signal 26(3):570–579

Lee N et al (2014) Melanoma stem cells and metastasis: mimicking hematopoietic cell trafficking? Lab Invest 94:13–30

Ratajczak MZ et al (2010) Epiblast/germ line hypothesis of cancer development revisited: lesson from the presence of Oct-4+ cells in adult tissues. Stem Cell Rev Rep 6:307–316

Thieu K et al (2013) Cells of origin and tumor-initiating cells for nonmelanoma skin cancers. Cancer Lett 338:82–88

Wend P et al (2010) Wnt signaling in stem and cancer stem cells. Semin Cell Dev Biol 21(8):855–863

Ch 20 Metamorphosis: Books

Gilbert Lawrence I (2009) Insect development: morphogenesis, molting and metamorphosis. Academic, Amsterdam

Articles

Atawodi SE, Atawodi JC (2009) *Azadirachta indica* (neem): a plant of multiple biological and pharmacological activities. Phytochem Rev 8(3):601–620

Bernal J, Morte B (2013) Thyroid hormone receptor activity in the absence of ligand: physiological and developmental implications. Biochim Biophys Acta 1830(7):3893–3899

Chun-Yan L et al (2013) Upregulation of the expression of prodeath serine/threonine protein kinase for programmed cell death by steroid hormone 20-hydroxyecdysone. Apoptosis 18(2):171–187

Denver RJ (2013) Neuroendocrinology of amphibian metamorphosis. Curr Top Dev Biol 103:195–227

Dewey EM et al (2004) Identification of the gene encoding bursicon, an insect neuropeptide responsible for cuticle sclerotization and wing spreading. Curr Biol 14(13):1208–1213

Di Cara F, King-Jones K (2013) How clocks and hormones act in concert to control the timing of insect development. Curr Top Dev Biol 105:1–36

Duffy DJ et al (2010b) Wnt signaling promotes oral but suppresses aboral structures in *Hydractinia* metamorphosis and regeneration. Development 137(18):3057–3066

Erwin PM, Szmant AM (2010) Settlement induction of *Acropora palmata* planulae by a GLW-amide neuropeptide. Coral Reefs 29(4):929–939

Fuchs B et al (2014) Regulation of polyp-to-jellyfish transition in Aurelia aurita. Curr Biol 24(3):263–273

Grattan DR, Kokay IC (2008) Prolactin: a pleiotropic neuroendocrine hormone. J Neuroendocrinol 20 (6):752–763

Grimaldi A et al (2013) Mechanisms of thyroid hormone receptor action during development: lessons from amphibian studies. Biochim Biophys Acta 1830 (7):3882–3892

Hiruma K, Kaneko Y (2013) Hormonal regulation of insect metamorphosis with special reference to juvenile hormone biosynthesis. Curr Top Dev Biol 103:73–100

Holstein TW, Laudet V (2014) Life-history evolution: at the origins of metamorphosis. Curr Biol 24(4): R159–R161

Holzer G, Laudet V (2013) Thyroid hormones and postembryonic development in amniotes. Curr Top Dev Biol 103:397–425

Jindra M et al (2013) The juvenile hormone signaling pathway in insect development. Annu Rev Entomol 58:181–204

Kaneko Y et al (2011) Stage-specific regulation of juvenile hormone biosynthesis by ecdysteroid in *Bombyx mori*. Mol Cell Endocrinol 335(2):204–210

Koch PB, Bückmann D (1987) Hormonal control of seasonal morphs by the timing of ecdysteroid release in *Araschnia levana*. J Insect Physiol 33:823–829

Kulkarni S, Buchholz DR (2012) Beyond synergy: corticosterone and thyroid hormone have numerous interaction effects on gene regulation in *Xenopus tropicalis* tadpoles. Endocrinology 153(11):5309–5324

Loveall BJ, Deitcher DL (2010) The essential role of bursicon during *Drosophila* development. BMC Dev Biol 1:Article No 92

Marchal E et al (2010) Control of ecdysteroidogenesis in prothoracic glands of insects: a review. Peptides 31 (3):506–519

Mueller WA, Leitz T (2002b) Metamorphosis. In: Biology of the Cnidarians. Can J Zool 80:1755–1771

Nakajima K et al (2012) Regulation of thyroid hormone sensitivity by differential expression of the thyroid hormone receptor during *Xenopus* metamorphosis. Genes Cells 17(8):645–659

Pascual A, Aranda A (2013) Thyroid hormone receptors, cell growth and differentiation. Biochim Biophys Acta 1830(7):3908–3916

Peabody N, White BH (2013) Eclosion gates progression of the adult ecdysis sequence of Drosophila. J Exp Biol 216(23):4395–4402

Riddiford LM et al (2003) Insights into the molecular basis of the hormonal control of molting and metamorphosis from *Manduca sexta* and *Drosophila melanogaster*. Insect Biochem Mol Biol 33 (12):1327–1338

Roller L et al (2010) Ecdysis triggering hormone signaling in arthropods. Peptides 31(3):429–441

Schmich J, Trepel S, Leitz T (1998b) The role of GLWamides in metamorphosis of *Hydractinia echinata*. Dev Genes Evol 208:267–273

Shi YB (ed) (2013) Animal metamorphosis. Curr Topics Dev Biol 103: 195–227 (amphibia); 73–100 (insects)

Spindler K-D et al (2009) Ecdysteroid hormone action. Cell Mol Life Sci 66(24):3837–3850

Tata JR (2000) Autoinduction of nuclear hormone receptors during metamorphosis and its significance. Insect Biochem Mol Biol 30:645–651

Tata JR (2013) The road to nuclear receptors of thyroid hormone. Biochim Biophys Acta 1830(7):3860–3866

Tebben J et al (2011) Induction of larval metamorphosis of the coral *Acropora millepora* by tetrabromopyrrole isolated from a *Pseudoalteromonas* bacterium. PLoS One 6(4):e19082

Visser WE et al (2008) Thyroid hormone transport in and out of cells. Trends Endocrinol Metab 19(2):50–56

Mei-Xian W et al (2013) Phenol oxidase is a necessary enzyme for the silkworm molting which is regulated by molting hormone. Mol Biol Rep 40(5):3549–3555

Yamanaka N (2013) Ecdysone control of developmental transitions: lessons from Drosophila research. Annu Rev Entomol 58:497–516

Puberty as Metamorphosis

Mann DR, Plant TM (2010) The role and potential sites of action of thyroid hormone in timing the onset of puberty in male primates. Brain Res 1364:175–185

Chapter 21: Ageing and Death: Books

Carey JR, Judge DS (2000) Life spans of mammals, birds, amphibians, reptiles and fish. Odense University Press, Odense

Articles

Allsopp R (2008) Telomere-induced senescence of primary cells. In: Rudolph KL (ed) Telomeres and telomerase in ageing, disease, and cancer. Springer, Berlin, pp 23–42

Ari B (2010) DNA damage, neuronal and glial cell death and neurodegeneration. Apoptosis 15(11):1371–1381

Bratic I, Trifunovic A (2010) Mitochondrial energy metabolism and ageing. Biochim Biophys Acta 1797 (6–7):961–967

Burtner CR, Kennedy BK (2010) Progeria syndromes and ageing: what is the connection? Nat Rev Mol Cell Biol 11(8):567–578

Yi-Fan C (2011) Longevity and lifespan control in mammals: lessons from the mouse. Ageing Res Rev 9:S28–S35

Comai L, Li B (2004) The Werner syndrome protein at the crossroads of DNA repair and apoptosis. Mech Ageing Dev 125:521–528

De Meyer T et al (2011) Telomere length and cardiovascular aging: the means to the ends? Ageing Res Rev 10 (2):297–303

Fulcher N et al (2014) If the cap fits, wear it: an overview of telomeric structures over evolution. Cell Mol Life Sci 71(5):847–865

Gems D, Partridge L (2013) Genetics of longevity in model organisms: debates and paradigm shifts. Annu Rev Physiol 75:621–644

Gordon LB (2013) Progeria: translational insights from cell biology. J Cell Biol 27–31

Gordon LB (2014) Progeria: a paradigm for translational medicine. Cell 156(3):400–407

Gruber H et al (2014) Telomere-independent ageing in the longest-lived non-colonial animal, *Arctica islandica*. Exp Gerontol 51:38–45

Hulbert AJ et al (2014) Polyunsaturated fats, membrane lipids and animal longevity. J Comp Physiol B 184 (2):149–166

Ishida T et al (2014) Role of DNA damage in cardiovascular disease. Circ J 78:42–50

Ishikawa N (2011) Accelerated in vivo epidermal telomere loss in Werner syndrome. Aging 3(4):417–429

Rongrong L et al (2014) Enhanced telomere rejuvenation in pluripotent cells reprogrammed via nuclear transfer relative to induced pluripotent stem cells. Cell Stem Cell 14:27–39

Se-Jin L et al (2004) A Werner syndrome protein homolog affects C. elegans development, growth rate, life span and sensitivity to DNA damage by acting at a DNA damage checkpoint. Development 131:2565–2575

Lezzerin M, Budovskaya Y (2014) A dual role of the Wnt signaling pathway during aging in *Caenorhabditis elegans*. Aging Cell 13:8–18

Zhi L et al (2013) Werner complex deficiency in cells disrupts the nuclear pore complex and the distribution of lamin B1. Biochim Biophys Acta 1833(12):3338–3345

Guang-Hui L et al (2011) Recapitulation of premature ageing with iPSCs from Hutchinson-Gilford progeria syndrome. Nature 472(7342):221

Ljubuncic P, Reznick AZ (2009) The evolutionary theories of aging revisited – a mini-revue. Gerontology 55(2):205–216

Maures TJ et al (2014) Males shorten the life span of C. elegans hermaphrodites via secreted compounds. Science 343(6170):541–544

Mohaghegh P et al (2001) The Bloom's and Werner's syndrome proteins are DNA structure-specific helicases. Nucleic Acid Res 29:2843–2849

Monaghan P (2014) Organismal stress, telomeres and life histories. J Exp Biol 217:57–66

Neill D (2010) A proposal in relation to a genetic control of lifespan in mammals. Ageing Res Rev 9(4):437–446

O'Sullivan RJ, Karlseder J (2010) Telomeres: protecting chromosomes against genome instability. Nat Rev Mol Cell Biol 11:171–181

Olshansky SJ (2010) The law of mortality revisited: interspecies comparisons of mortality. J Comp Pathol 141 (1):S4–S9

Pawelec G, Derhovanessian E (2010) Senescence of the human immune system. J Comp Pathol 141(1): S39–S44

Chun QY et al (2013) Hutchinson-Gilford progeria syndrome and its relevance to cardiovascular diseases and normal aging. Biomed Environ Sci 26(5):382–389

Raffaello A, Rizzuto R (2011) Mitochondrial longevity pathways. Biochim Biophys Acta 1813(1):260–268

Ratajczak M et al (2011) The role of pluripotent embryonic-like stem cells residing in adult tissues in regeneration and longevity. Differentiation 81 (3):153–161

Redman LM, Ravussin E (2011) Caloric restriction in humans: impact on physiological, psychological, and behavioral outcomes. Antioxid Redox Signal 14 (2):275–287

Ristow M, Schmeisser S (2011) Extending life span by increasing oxidative stress. Free Radic Biol Med 51 (2):327–336

Sherratt MJ (2009) Tissue elasticity and the ageing elastic fibre. Age 31(4):305–325

Smith JA et al (2010) A role for the Werner syndrome protein in epigenetic inactivation of the pluripotency factor Oct4. Aging Cell 9(4):580–591

Swanson EC et al (2013) Higher-order unfolding of satellite heterochromatin is a consistent and early event in cell senescence. J Cell Biol 203(6):929–942

Sugimoto M et al (2011) Involvement of WRN helicase in immortalization and tumorigenesis by the telomeric crisis pathway (Review). Oncol Lett 2(4):609–611

Takeuchi H, Ruenger TM (2013) Longwave UV light induces the aging-associated progerin. J Invest Dermatol 133(7):1857–1862

Vendelbo MH, Nair KS (2011) Mitochondrial longevity pathways. Biochim Biophys Acta 1813(4):634–644

Wanagat J et al (2010) Mitochondrial oxidative stress and mammalian healthspan. Mech Ageing Dev 131 (7–8):527–535

Hui W et al (2014) Probing the anti-aging role of polydatin in Caenorhabditis elegans on a chip. Integr Biol 6:35–43

Lin Y et al (2013) Calorie restriction can reverse, as well as prevent, aging cardiomyopathy. Age 35(6):2177–2182

Yang Shao H et al (2011) Absence of progeria-like disease phenotypes in knock-in mice expressing a non-farnesylated version of progerin. Hum Mol Genet 20(3):436–444

Jinqiu Z et al (2011) A human iPSC model of Hutchinson Gilford Progeria reveals vascular smooth muscle and mesenchymal stem cell defects. Cell Stem Cell 8 (1):31–45

Guodong Z et al (2014) Evolution of human longevity uncoupled from caloric restriction mechanisms. PLoS One 9(1):e84117

Chapter 22 Evo-Devo: Books

Bertossa RC (2011) Evolutionary developmental biology (evo-devo) and behaviour. Royal Society, London

Smith JM, Szathmary E (2001) The major transitions in evolution. Oxford University Press, Oxford

Wilkins AS (2002) The evolution of developmental pathways. Sinauer, Sunderland, MA

Articles: General, Body Plans, Body Axes

Amiel AR et al (2013) An organizing activity is required for head patterning and cell fate specification in the polychaete annelid *Capitella teleta*: new insights into cell-cell signaling in Lophotrochozoa. Dev Biol 379:107–122

Arendt D (2003d) Spiralians in the limelight. Genome Biol 5(1):303

Arendt D (2008) The evolution of cell types in animals: emerging principles from molecular studies. Nat Rev Genet 9(11):868–882

Arendt D, Nübler-Jung K (1994) Inversion of dorsoventral axis? Nature 371:26

Arendt D, Nübler-Jung K (1997) Dorsal or ventral: similarities in fate maps and gastrulation patterns in annelids, arthropods and chordates. Mech Dev 61:7–21

von Baer KE (1828) Über Entwickelungsgeschichte der Thiere. Den Gebrüdern Bornträger, Königsberg

Bailly X et al (2013) The urbilaterian brain revisited: novel insights into old questions from new flatworm clades. Dev Genes Evol 223(3):149–157

Ball EE et al (2004) A simple plan − cnidarians and the origins of developmental mechanisms. Nat Rev Genet 5(8):567–577

Casaca A et al (2014c) Controlling *Hox* gene expression and activity to build the vertebrate axial skeleton. Dev Dyn 243:24–36

Davis MC (2013) The deep homology of the autopod: insights from *Hox* gene regulation. Integr Comp Biol 53(2):224–232

De Robertis EM (2008) Evo-devo: variations on ancestral themes. Cell 132:185–195

Denes AS et al (2007) Molecular architecture of annelid nerve cord supports common origin of nervous system centralization in bilateria. Cell 129:277–288

Dohrmann M, Woerheide G (2013) Novel scenarios of early animal evolution-Is it time to rewrite textbooks? Integr Comp Biol 53(3):503–511

Eibner C et al (2008) An organizer controls the development of the "sword", a sexually selected trait in swordtail fish. Evol Dev 10:403–412

Elinson R (2010) Molecular Haeckel. Dev Dyn 239(7):1905–1918

Finnerty JR et al (2004b) Origins of bilateral symmetry: *Hox* and *dpp* expression in a sea anemone. Science 304:1335–1337

Foret S et al (2011) Phylogenomics reveals an anomalous distribution of *USP* genes in metazoans. Mol Biol Evol 28(1):153–161

Freeman G (1981) The role of polarity in the development of the hydrozoan planula larva. Roux's Arch Dev Biol 190:168–184

Fritzenwanker JH et al (2007b) Early development and axis specification in the sea anemone *Nematostella vectensis*. Dev Biol 310:264–279

Fröbius AC et al (2008) Genomic organization and expression demonstrate spatial and temporal *Hox* gene colinearity in the Lophotrochozoan *Capitella sp*. PLoS One 3(12):e4004

Galliot B et al (2009) Origins of neurogenesis, a cnidarian view. Dev Biol 332(1):2–24

Giribet G (2003) Molecules, development and fossils in the study of metazoan evolution; Articulata versus Ecdysozoa revisited. Zoology 106:303–326

Haeckel E (1874) Die Gastraea-Theorie, die phylogenetische Classification des Thierreiches und die Homologie der Keimblätter. Z Naturwiss Jena 8:1–55

Hallgrimsson B et al (2012) The generation of variation and the developmental basis for evolutionary novelty. J Exp Zool 318B(6):501–517

Heffer A, Pick L (2013) Conservation and variation in *Hox* genes: how insect models pioneered the evo-devo field. Annu Rev Entomol 58:161–179

Hejnol A, Martindale MQ (2008) Acoel development indicates the independent evolution of the bilaterian mouth and anus. Nature 456:382–386

Hobmayer B et al (2000c) WNT signalling molecules act in axis formation in the diploblastic metazoan *Hydra*. Nature 407:186–189

Holland LZ et al (2008) The amphioxus genome illuminates vertebrate origins and cephalochordate biology. Genome Res 18(7):1100–1111

Holland LZ et al (2013) Evolution of bilaterian central nervous systems: a single origin? EvoDevo 4:Article No 27

Hughes CL, Kaufmann TC (2002) Exploring the myriapod body plan: expression patterns of the ten Hox genes in a centipede. Development 129(5):1225–1238

Hui Jerome HL et al (2012b) Extensive chordate and annelid macrosynteny reveals ancestral homeobox gene organization. Mol Biol Evol 29:157–165

Irschick DJ et al (2013) Evo-devo beyond morphology: from genes to resource use. Trends Ecol Evol 28(5):267–273

Janssen R et al (2010) Conservation, loss, and redeployment of Wnt ligands in protostomes: implications for understanding the evolution of segment formation. BMC Evol Biol 10:Article No 374

Kraus Y et al (2007) The blastoporal organiser of a sea anemone. Curr Biol 17:R874–R876

Kusserow A et al (2005b) Unexpected complexity of the Wnt gene family in a sea anemone. Nature 433:156–160

Lacalli TC (2008) Basic features of the ancestral chordate brain: a protochordate perspective. Brain Res Bull 75:319–323

Lambert JD (2010b) Spiralia. Curr Biol 20(2):R72–R77

Lyons DC et al (2012) Cleavage pattern and fate map of the mesentoblast, 4d, in the gastropod Crepidula: a hallmark of spiralian development. EvoDevo 3:Article No 21

Lee PN et al (2006) A WNT of things to come: evolution of Wnt signaling and polarity in cnidarians. Semin Cell Dev Biol 17(2):157–167

Mallo M et al (2010) *Hox* genes and regional patterning of the vertebrate body plan. Dev Biol 344(1):7–15

Martin-Duran JM et al (2012c) Planarian embryology in the era of comparative developmental biology. Int J Dev Biol 56(1–3):39–48

Morangea M (2011) Evolutionary developmental biology its roots and characteristics. Dev Biol 357(1):35–40

Martindale MQ, Hejnol A (2009) A developmental perspective: changes in the position of the blastopore during bilaterian evolution. Dev Cell 17:162–174

Matus Q et al (2006) Molecular evidence for deep evolutionary roots of bilaterality in animal development. Proc Natl Acad Sci USA 103:11195–11200

Meijlink F et al (1999) Vertebrate *aristaless*-related genes. Int J Dev Biol 43(7):651–663

Miller DJ, Ball EE, Technau U (2005) Cnidarians and ancestral genetic complexity in the animal kingdom. Trends Genet 21(10):536–539

Mueller WA et al (2007d) Wnt signaling in hydroid development: ectopic heads and giant buds induced by GSK-3beta inhibitors. Int J Dev Biol 51:211–220

Nesnidal M et al (2013) New phylogenomic data support the monophyly of Lophophorata and an ectoproct-phoronid clade and indicate that Polyzoa and Kryptrochozoa are caused by systematic bias. BMC Evol Biol 13:Article No 253

Nielsen C (2005) Larval and adult brains. Evol Dev 7:483–489

Nielsen C (2013) Life cycle evolution: was the eumetazoan ancestor a holopelagic, planktotrophic gastraea? BMC Evol Biol 13:Article No 171

O'Connell LA (2013) Evolutionary development of neural systems in vertebrates and beyond. J Neurogenet 27(3):69–85

Offen N (2008) Fgfr1 signalling in the development of a sexually selected trait in vertebrates, the sword of swordtail fish. BMC Dev Biol 8:Article No 98

Offen N et al (2009) Identification of novel genes involved in the development of the sword and gonopodium in swordtail fish. Dev Dyn 238:1674–1687

Olson K (2002) Gill circulation: regulation of perfusion distribution and metabolism of regulatory molecules. J Exp Zool 293:320–335

Onai T et al (2009) Retinoic acid and Wnt/β-catenin have complementary roles in anterior/posterior patterning embryos of the basal chordate amphioxus. Dev Biol 332:223–233

Patrick RH et al (2011) The segmental pattern of *otx*, *gbx*, and *Hox* genes in the annelid *Platynereis dumerilii*. Evol Dev 13(1):72–79

Pavlopoulosa A (2009) Probing the evolution of appendage specialization by *Hox* gene misexpression in an emerging model crustacean. Proc Natl Acad Sci USA 106(33):3897–13902

Pechmann M et al (2010) Patterning mechanisms and morphological diversity of spider appendagees and their importance for spider evolution. Arthropd Struct Dev 39:453–467

Plickert G et al (2006d) Wnt signaling in hydroid development: formation of the primary body axis and its subsequent patterning. Dev Biol 298:368–378

Raible F, Arendt D (2004b) Metazoan evolution: some animals are more equal than others. Curr Biol 14(3):R106–R108

Rentzsch F et al (2006b) Asymmetric expression of the BMP antagonists chordin and gremlin in the sea anemone *Nematostella vectensis*: implications for the evolution of axial patterning. Dev Biol 296:375–387

Schneider SQ, Bowerman B (2007) Beta-catenin asymmetries after all animal/vegetal-oriented cell divisions in *Platynereis dumerilii* embryos mediate binary cell-fate specification. Dev Cell 13(1):73–86

Schneider I, Shubin NH (2012) Making limbs from fins. Dev Cell 23(6):1121–1122

Schneider I, Shubin NH (2013) The origin of the tetrapod limb: from expeditions to enhancers. Trends Genet 29(7):419–426

Scholtz G, Edgecombe GD (2006) The evolution of arthropod heads: reconciling morphological, developmental and palaeontological evidence. Dev Genes Evol 216:395–415

Seipel K, Schmid V (2005) Evolution of striated muscle: Jellyfish and the origin of triploblasty. Dev Biol 282(1):14–26

Seipel K, Schmid V (2006b) Mesodermal anatomies in cnidarian polyps and medusae. Int J Dev Biol 50(7):589–599

Sinigaglia C et al (2013b) The bilaterian head patterning gene *six3/6* controls aboral domain development in a cnidarian. PLoS Biol 11(2):e1001488

Steinmetz PR et al (2012b) Independent evolution of striated muscles in cnidarians and bilaterians. Nature 487(7406):231–234

Stemple DL (2005) Structure and function of the notochord: an essential organ for chordate development. Development 132:2503–2512

Stern CD (2006c) Evolution of the mechanisms that establish the embryonic axes. Curr Opin Genet Dev 16(4):413–418

Tautz D (1998) Debatable homologies. Nature 395:17–18

Tessmar-Raible K, Arendt D (2005b) New animal models for evolution and development. Genome Biol 6(1):303

Toegel JP et al (2009) Loss of spineless function transforms the *Tribolium* antenna into a thoracic leg with pretarsal, tibiotarsal, and femoral identity. Dev Genes Evol 219(1):53–58

Veeman MT et al (2010) The ascidian mouth opening is derived from the anterior neuropore: reassessing the mouth/neural tube relationship in chordate evolution. Dev Biol 344(1):138–149

Vonica A et al (2000b) TCF is the nuclear effector of the ß-catenin signal that patterns the sea urchin animal-vegetal axis. Dev Biol 217:230–243

Watanabe H et al (2009b) Cnidarians and the evolutionary origin of the nervous system. Dev Growth Differ 51 (3):167–183

Wikramanayake AH et al (2003b) An ancient role for nuclear beta-catenin in the evolution of axial polarity and germ layer segregation. Nature 426:446–450

Yu JK et al (2007) Axial patterning in cephalochordates and the evolution of the organizer. Nature 445:613–617

Gene and Genome Evolution, Hox Genes

Amemiya CT et al (2008) The amphioxus *Hox* cluster: characterization, comparative genomics, and evolution. J Exp Zool 310B(5):465–477

Asrar Z et al (2013) Fourfold paralogy regions on human *HOX*-bearing chromosomes: role of ancient segmental duplications in the evolution of vertebrate genome. Mol Phylogenet Evol 66(3):737–747

Butts T et al (2008b) The urbilaterian *super-Hox* cluster. Trends Genet 24(6):259–262

Castro L et al (2011) A *Gbx* homeobox gene in amphioxus: insights into ancestry of the *ANTP* class and evolution of the midbrain/hindbrain boundary. Dev Biol 295(1):40–51

Dan C, Duda TF (2012) Extensive and continuous duplication facilitates rapid evolution and diversification of gene families. Mol Biol Evol 29(8):2019–2029

Cho S-J et al (2010) Evolutionary dynamics of the *wnt* gene family: a Lophotrochozoan perspective. Mol Biol Evol 27(7):1645–1658

Chourrout D et al (2006) Minimal *protoHox* cluster inferred from bilaterian and cnidarian *Hox* complements. Nature 442:684–687

DuBuc TQ, Ryan JF, Shinzato C, Satoh N, Martindale MQ (2012) Coral comparative genomics reveal expanded *Hox* cluster in the cnidarian–bilaterian ancestor. Integr Comp Biol 52(6):835–841

Garstang M, Ferrier DE (2013) Time is of the essence for *paraHox* homeobox gene clustering. BMC Biol 11:72

Heffer A et al (2013) Variation and constraint in *Hox* gene evolution. Proc Natl Acad Sci USA 110 (6):2211–2216

Heinen TJA et al (2009) Emergence of a new gene from an intergenic region. Curr Biol 19(8):1527–1531

Hueber SD et al (2013) Analysis of central Hox protein types across bilaterian clades: on the diversification of central Hox proteins from an Antennapedia/Hox7-like protein. Dev Biol 383(2):175–185

Hui JH, Holland PW, Ferrier DE (2008) Do cnidarians have a *paraHox* cluster? Analysis of synteny around a *Nematostella* homeobox gene cluster. Evol Dev 10 (6):725–730

Kaessmann H (2010) Origins, evolution, and phenotypic impact of new genes. Genome Res 20 (10):1313–1326

Matus Q et al (2007) Expression of *Pax* gene family members in the anthozoan cnidarian, Nematostella vectensis. Evol Dev 9:25–38

Murphy DN, McLysaght A (2012) De novo origin of protein-coding genes in murine rodents. PLoS One 7 (11):e48650

Moran Y et al (2012b) Recurrent horizontal transfer of bacterial toxin genes to eukaryotes. Mol Biol Evol 29 (9):2223–2230

Necsulea A et al (2014) The evolution of *lncRNA* repertoires and expression patterns in tetrapods. Nature. doi:10.1038/nature12943

Neme R, Tautz D (2013) Phylogenetic patterns of emergence of new genes support a model of frequent de novo evolution. BMC Genomics 14:Article No 117

Pascual-Anaya J et al (2013) Evolution of *Hox* gene clusters in deuterostomes. BMC Dev Biol 13:Article No 26

Putnam NH et al (2007b) Sea anemone genome reveals the gene repertoire and genomic organization of the eumetazoan ancestor. Science 317:86–94

Quiquand M et al (2009) More constraints on *ParaHox* than *Hox* gene families in early metazoan evolution. Dev Biol 328(2):173–187

Rebscher N et al (2009) Conserved intron positions in *FGFR* genes reflect the modular structure of *FGFR* and reveal stepwise addition of domains to an already complex ancestral *FGFR*. Dev Genes Evol 219 (9–10):455–468

Romain Derelle R et al (2007) Homeodomain proteins belong to the ancestral molecular toolkit of eukaryotes. Evol Dev 9(3):212–219

Ryan JF (2007) Pre-bilaterian origins of the *Hox* cluster and the *Hox* code: evidence from the sea nemone *Nematostella vectensis*. PLoS One 2:e153

Ryu T et al (2012) The evolution of ultraconserved elements with different phylogenetic origins. BMC Evol Biol 12:Article No 236

Schierwater B, Kamm K (2010) The early evolution of *Hox* genes: a battle of belief? Adv Exp Med Biol 689:81–90

Soshnikova N et al (2013) Duplications of *Hox* gene clusters and the emergence of vertebrates. Dev Biol 378(2):194–199

Tara A, Nolte C, Krumlauf R (2009) *Hox* genes and segmentation of the hindbrain and axial skeleton. Annu Rev Cell Dev Biol 25:431–456

Technau U et al (2005) Maintenance of ancestral complexity and non-metazoan genes in two basal cnidarians. Trends Genet 21(12):633–639

Wellik SM (2007) Hox patterning of the vertebrate axial skeleton. Dev Dyn 236(9):2454–2463

Eye Evolution

Arendt D (2003e) Evolution of eyes and photoreceptor cell types. Int J Dev Biol 47(7–8):563–571

Arendt D et al (2002b) Development of pigment-cup eyes in the polychaete *Platynereis dumerilii* and evolutionary conservation of larval eyes in Bilateria. Development 129:1143–1154

Arendt D et al (2004) Ciliary photoreceptors with a vertebrate-type opsin in an invertebrate brain. Science 306(5697):869–871

Backfisch B et al (2013b) Stable transgenesis in the marine annelid *Platynereis dumerilii* sheds new light on photoreceptor evolution. Proc Natl Acad Sci USA 110:193–198

Fernald RD, Russell D (2006) Casting a genetic light on the evolution of eyes. Science 313(5795):1914–1918

Gehring WJ (2011b) Chance and necessity in eye evolution. Genome Biol Evol 3:1053–1066

Gehring WJ (2014b) The evolution of vision. Wiley Interdisc Rev Dev Biol 3:1–40

Graziussi D et al (2012) The "Eyes absent" (eya) gene in the eye-bearing hydrozoan jellyfish *Cladonema radiatum*: conservation of the retinal determination network. J Exp Zool 318B(4):257–267

Goldsmith TH (2013) Evolutionary tinkering with visual photoreception. Vis Neurosci 30(1–2):21–37

Kozmik Z et al (2008) Assembly of the cnidarian camera-type eye from vertebrate-like components. Proc Natl Acad Sci USA 105(26):8989–8993

Kulakova M et al (2007) Hox gene expression in larval development of the polychaetes *Nereis virens* and *Platynereis dumerilii* (Annelida, Lophotrochozoa). Dev Genes Evol 217(1):39–54

Lamb TD (2011) Evolution of the eye. Sci Am 305 (1):64–69

Nilsson DE (2013) Eye evolution and its functional basis. Vis Neurosci 30(1–2):5–20

Atsushi O et al (2004) Comparative analysis of gene expression for convergent evolution of camera eye between octopus and human. Genome Res 14 (8):1555–1561

Osorio D (2011) Evolutionary biology: light sense. Nature 472:300–301

Piatigorsky J, Kozmik Z (2004) Cubozoan jellyfish: an evo/devo model for eyes and other sensory systems. Int J Dev Biol 48(8–9):719–729

Purschke G et al (2006) Photoreceptor cells and eyes in annelida. Arthropod Struct Dev 35:211–230

Randel N et al (2013b) Expression dynamics and protein localization of rhabdomeric opsins in *Platynereis* larvae. Integr Compar Biol 53:7–16

Sen S et al (2013b) Conserved roles of *ems/Emx* and *otd/Otx* genes in olfactory and visual system development in Drosophila and mouse. Open Biol 3:Article No 120177

Sinn R, Wittbrodt J (2013b) An eye on eye development. Mech Dev 130(6–8):347–358

Slingsby C et al (2013) Evolution of crystallins for a role in the vertebrate eye lens. Protein Sci 22(4):367–380

Stierwald M et al (2004) The *Sine oculis/Six* class family of homeobox genes in jellyfish with and without eyes: development and eye regeneration. Dev Biol 274:70–81

Suga H, Schmid V, Gehring WJ (2008) Evolution and functional diversity of jellyfish opsins. Curr Biol 18 (1):51–55

Evolution of Human Brain

Aboitiz F (2011) Genetic and developmental homology in amniote brains. Toward conciliating radical views of brain evolution. Brain Res Bull 84(2):125–136

Arquint C, Nigg EA (2014) STIL microcephaly mutations Interfere with APC/C-mediated degradation and cause centriole amplification. Curr Biol 24(4):351–360

Bishop DVM (2009) Genes, cognition, and communication insights from neurodevelopmental disorders. Miller MB; Kingstone A (editorss). Ann New York Acad Sci 1156:1–18

Breuss M et al (2012) Mutations in the β-tubulin gene TUBB5 cause microcephaly with structural brain abnormalities. J Cell Rep 2:1554–1562. doi:10.1016/j.celrep.2012.11.017

Charvet C et al (2011) Evo-devo and brain scaling: candidate developmental mechanisms for variation and constancy in vertebrate brain evolution. Brain Behav Evol 78(3):248–257

Evans PD et al (2004) Reconstructing the evolutionary history of microcephalin, a gene controlling human brain size. Hum Mol Genet 13(11):1139–1145

Evans PD et al (2005) Microcephalin, a gene regulating brain size, continues to evolve adaptively in humans. Science 309(5741):1717–1720

Daniel H, Geschwind DH, Konopka G (2012) Genes and human brain evolution. Nature 486:481–482

Geschwind DH, Rakic P (2013) Cortical evolution: judge the brain by its cover. Neuron 80(3):633–647

Grillner S et al (2013) The evolutionary origin of the vertebrate basal ganglia and its role in action selection. J Physiol 591(22):5425–5431

Gruber R et al (2011) MCPH1 regulates the neuroprogenitor division mode by coupling the centrosomal cycle with mitotic entry through the Chk1–Cdc25 pathway. Nat Cell Biol 13 (11):1325–1334

Holland LZ, Short S (2008) Gene duplication, co-option and recruitment during the origin of the vertebrate brain from the invertebrate chordate brain. Brain Behav Evol 72(2):91–105

Hrvoj-Mihic B et al (2013) Evolution, development, and plasticity of the human brain: from molecules to bones. Front Human Neurosci 7:707. doi:10.3389/fnhum.2013.00707

Zhongshan L et al (2013) Evolutionary and ontogenetic changes in RNA editing in human, chimpanzee, and macaque brains. RNA 19(12):1693–1702

Jackson AP et al (2002) Identification of microcephalin, a protein implicated in determining the size of the human brain. Am J Hum Genet 71(1):136–142

Matsuzawa T (2013) Evolution of the brain and social behavior in chimpanzees. Curr Opin Neurobiol 23 (3):443–449

Robertson JM (2014) Astrocytes and the evolution of the human brai. Med Hypotheses 82(2):236–239

Scharff C, Petri J (2011) Evo-devo, deep homology and *FoxP2*: implications for the evolution of speech and language. Philos Trans R Soc Lond B Biol Sci 366 (1574):2124–2140

Thornton GK, Woods CG (2009) Primary microcephaly: do all roads lead to Rome? Trends Genet 25 (11):501–510

Xianglin W et al (2009) Microcephalin regulates BRCA2 and Rad51-associated DNA double-strand break repair. Cancer Res 69(13):5531–5536

Index[1]

---

[1] Note: Page numbers in bold refer to definitions and/or
basic explanations; Page numbers with "f" point to figures

W.A. Mueller et al., *Development and Reproduction in Humans and Animal Model Species*,
DOI 10.1007/978-3-662-43784-1, © Springer-Verlag Berlin Heidelberg 2015